UNDERSTANDING FOOD

PRINCIPLES AND PREPARATION

SECOND EDITION

UNIVERSITY OF HAWAII AT MANOA

Australia • Canada • Mexico • Singapore • Spain
United Kingdom • United States

THOMSON
WADSWORTH

Publisher: Peter Marshall
Development Editor: Elizabeth Howe
Assistant Editor: Madinah Chang
Editorial Assistant: Elesha Feldman
Technology Project Manager: Travis Metz
Marketing Manager: Jennifer Somerville
Marketing Assistant: Melanie Wagner
Advertising Project Manager: Shemika Britt
Project Manager, Editorial Production: Sandra Craig
Print/Media Buyer: Doreen Suruki
Permissions Editor: Joohee Lee
Production: Fritz/Brett Associates
Text and Cover Designer: Lisa Devenish
Photo Researcher: Roberta Broyer
Copy Editor: Susan Defosset
Illustrations: Thurston Graphics
Cover Image: Getty Images
Compositor: New England Typographic Service
Printer: Phoenix Color Corp

Printed in the United States of America

1 2 3 4 5 6 7 07 06 05 04 03

For more information about our products, contact us at:
Thomson Learning Academic Resource Center
1-800-423-0563

For permission to use material from this text, contact us :
Phone: 1-800-730-2214
Fax: 1-800-730-2215
Web: http://www.thomsonrights.com

Library of Congress Control Number: 2003107401
Student Edition: ISBN 0-534-50609-7
Instructor's Edition: ISBN 0-534-50610-0

This book is printed on acid-free recycled paper.

Wadsworth/Thomson Learning
10 Davis Drive
Belmont, CA 94002-3098
USA

Asia
Thomson Learning
5 Shenton Way #01-01
UIC Building
Singapore 068808

Australia/New Zealand
Thomson Learning
102 Dodds Street
Southbank, Victoria 3006
Australia

Canada
Nelson
1120 Birchmount Road
Toronto, Ontario M1K 5G4
Canada

Europe/Middle East/Africa
Thomson Learning
High Holborn House
50/51 Bedford Row
London WC1R 4LR
United Kingdom

Latin America
Thomson Learning
Seneca, 53
Colonia Polanco
11560 Mexico D.F.
Mexico

Spain/Portugal
Paraninfo
Calle/Magallanes, 25
28015 Madrid, Spain

CONTENTS IN BRIEF

CONTENTS

PART II: FOOD SERVICE

PART IV: DESSERTS AND BEVERAGES

APPENDIXES

PREFACE

This book brings together the most current information in food science, nutrition, and food service to provide a comprehensive resource for students. Founded on research from more than 35 journals covering these disciplines, the text incorporates the very latest information, providing students with the broad foundation they need to launch a successful career in any of the many related fields of the food industry. Thoroughly updated with the latest food industry products and developments as well as standard terms, the text reflects both the basic principles and the newest trends in food.

The text is divided into four sections. Part I represents information related to the science of food, including food selection and evaluation, food chemistry, and food safety. Part II covers aspects of food service, from meal planning and equipment to the basics of food preparation, food preservation, and government regulation of foods. Parts III and IV cover all of the standard food items, including sweeteners, fats and oils, dairy products, meats, vegetables and fruits, soups and salads, grains, and breads, as well as desserts and beverages. Each chapter in these two sections follows a consistent organization, covering classification, composition, purchasing or selection, preparation, and storage of the food. Also integrated throughout the chapters is coverage of helpful culinary techniques, low-fat preparation methods, and the latest methods for food preservation and food safety. Extensive appendixes provide additional key information, including approximate food measurements, substitution of ingredients, flavorings and seasonings, and more.

Features

The unique features of this text allow for flexibility in teaching and create a dynamic learning environment for students.

- **How and Why inserts** answer the most frequently asked questions (FAQs) by students. They are used to spark natural curiosity, trigger inquisitive thought patterns, and exercise the mind's ability to answer.
- **Pictorial summaries** at the end of every chapter are a proven favorite with readers. Instead of the standard boring narrative summary, these pictorial summaries of the chapter use a combination of art and narrative text to encapsulate the key concepts in each chapter for student review.
- **Chemist's Corner** features provide information on food chemistry in boxes within the chapters for those students and instructors who wish to further explore the chemistry of food. These Chemist's Corners create a book with two chemistry levels, allowing for flexibility based on the chemistry requirements of the individual course.
- **Key terms** bolded in the text are defined in boxes on the same page to allow for quick review of the essential vocabulary in each chapter. A glossary at the end of the book assembles all of the key terms in the chapters.
- **Functions of ingredients** are highlighted in the introduction to each chapter to aid students in successful food product development and food preparation. They introduce a focus of the food industry that is often missing in previous books.
- **Problems and causes tables** in various food chapters summarize the problems that may occur when preparing specific food products and describe the possible causes, providing students with a handy reference tool for deciphering "what went wrong."
- **Numerous illustrations** injected throughout the text enhance students' understanding of the principles and techniques discussed.
- **Nutrient Content boxes** in each of the foods chapters provide an overview of the nutritional composition of the foods, reflecting the increased emphasis in the food industry on food as a means for health promotion and disease prevention.
- **A 16-page full-color insert** displays exotic varieties of fruits and vegetables, salad greens, flowers used in salads, traditional cuts of meats (including the lowest-fat meat cuts), and much more, all with detailed captions describing use and preparation tips.
- **Chapter review questions** are provided on the companion website for the book. The questions are organized by objective and short answer/essay format.

The dynamic world of food changes rapidly as new research constantly adds to its ever-expanding knowledge base. *Understanding Food: Principles and Preparation* is designed to meet the needs of this evolving and expanding discipline, and to provide students with a strong foundation in any food-related discipline that they select.

Acknowledgments

There are many individuals who assisted me in the development of this textbook. First and foremost I want to thank Peter Marshall, Publisher, without whose knowledge and experience this book would never have come to be.

I would also like to extend my thanks to the outstanding members of the Wadsworth Nutrition team: Elizabeth Howe, Developmental Editor, for her excellent skills in working with me to create a well-organized manuscript and for handling those last-minute emergencies; and Jennifer Somerville, Marketing Manager, Melanie Wagner, Marketing Assistant, and Shemika Britt, Advertising Project Manager, for getting the word out about this text.

I would also like to thank the tremendous production staff who worked miracles on this book: Susan Defosset, copy editor, who infused new life into the manuscript with her magical red pen; Elaine Brett, production wizard, in whose hands this project was masterfully brought to fruition; Lisa Devenish, who designed the book and created the memorable Pictorial Summaries; Roberta Broyer, our hard-working photo researcher; and Joohee Lee, Permissions Editor.

I would like to gratefully acknowledge Eleanor Whitney and Sharon Rolfes for contributing the Basic Chemistry Concepts appendix in this text.

A special thanks goes to the person who kindled my writing career, Nackey Loeb, Publisher of *The Union Leader*. Your early support and encouragement did far more than you will ever know.

A number of colleagues have contributed to the development of this text. Their thoughtful comments provided me with valuable guidance at all stages of the writing process. I offer them my heartfelt thanks for generously sharing their time and expertise.

They are:

Gertrude Armbruster (retired), *Cornell University*
Dorothy Addario, *College of St. Elizabeth*
Mike Artlip, *Kendall College*
Hea Ran-Ashraf, *Southern Illinois University*
Mia Barker, *Indiana University of Pennsylvania*
Nancy Berkoff, *Art Institute of Los Angeles*
Margaret Briley, *University of Texas*
Helen C. Brittin, *Texas Tech University*
Mildred M. Cody, *Georgia State University*
Carol A Costello, *University of Tennessee*
Barbara Denkins, *University of Pittsburgh*
Nikhil V. Dhurandhar, *Wayne State University*
Joannie Dobbs, *University of Hawaii/Manoa*
Linda Garrow, *University of Illinois/Urbana*
Natholyn D. Harris, *Florida State University*
Zoe Ann Holmes, *Oregon State University*
Wendy T. Hunt, *American River College*
Alvin Huang, *University of Hawaii*
Wayne Iwaoka, *University of Hawaii*
Karen Jameson, *Purdue University*
Faye Johnson, *California State University/Chico*
Nancy A. Johnson, *Michigan State University*
Mary Kelsey, *Oregon State University*
Elena Kissick, *California State University/Fresno*
Patti Landers, *University of Oklahoma*
Deirdre M. Larkin, *California State University, Northridge*
Lisa McKee, *New Mexico State University*
Marilyn Mook, *Michigan State University*
Martha N. O'Gorman, *Northern Illinois University*
Polly Popovich, *Auburn University*
Rose Tindall Postel, *East Carolina University*
Beth Reutler, *University of Illinois*
Janet M. Sass, *Northern Virginia Community College*
Sarah Short, *Syracuse University*
Darcel Swanson, *Washington State University*
Ruthann B. Swanson, *University of Georgia*
M. K. (Suzy) Weems, *Stephen F. Austin University*

Finally, I wish to express my appreciation to my students. Were it not for them, I would not have taken pen to paper. I am grateful to be a part of your journey toward greater knowledge and understanding.

Amy Christine Brown, Ph.D., R.D.

University of Hawaii at Manoa
amybrown@hawaii.edu

About the Author

© 2004 Carl Shaneff

Amy Christine Brown, Ph.D., R.D., received her Ph.D. from Virginia Polytechnic Institute and State University in 1986 in the field of Human Nutrition and Foods. She has been a college professor and a registered dietitian with the American Dietetic Association since 1986. Dr. Brown currently teaches at the University of Hawaii at Manoa in the Department of Human Nutrition, Food & Animal Sciences. Her research interests are in the area of Complementary Medicine and Medical Nutrition Therapy. Some of the studies she has conducted include "Potentially Harmful Herbal Supplements," "Kava beverage consumption and the effect on liver function tests," and "The effectiveness of kukui nut oil in treating psoriasis." Selected research journal publications include: "The Hawaii Diet: Ad libitum high carbohydrate, low fat multi-cultural diet for the reduction of chronic disease risk factors: Obesity, hypertension, hypercholesterolemia, and hyperglycemia" (*Hawaii Medical Journal*), "Lupus erythematosus and nutrition: A review" (*Journal of Renal Nutrition*), "Dietary survey of Hopi elementary school students" (*Journal of the American Dietetic Association*), "Serum cholesterol levels of nondiabetic and streptozotocin-diabetic rats" (*Artery*), "Infant feeding practices of migrant farm laborers in Northern Colorado" (*Journal of the American Dietetic Association*), "Body mass index and perceived weight status in young adults" (*Journal of Community Health*), "Dietary intake and body composition of Mike Pigg—1988 Triathlete of the Year" (*Clinical Sports Medicine*), and numerous newspaper nutrition columns over the last decade.

To Jeffery Blanton.

To the friend I met while editing a chapter, and the person who saw me through three of the five years that it took to write the first edition. Four years, forty thousand laughs, only one you.

Thank you Jeffery,

Amy Christine Brown

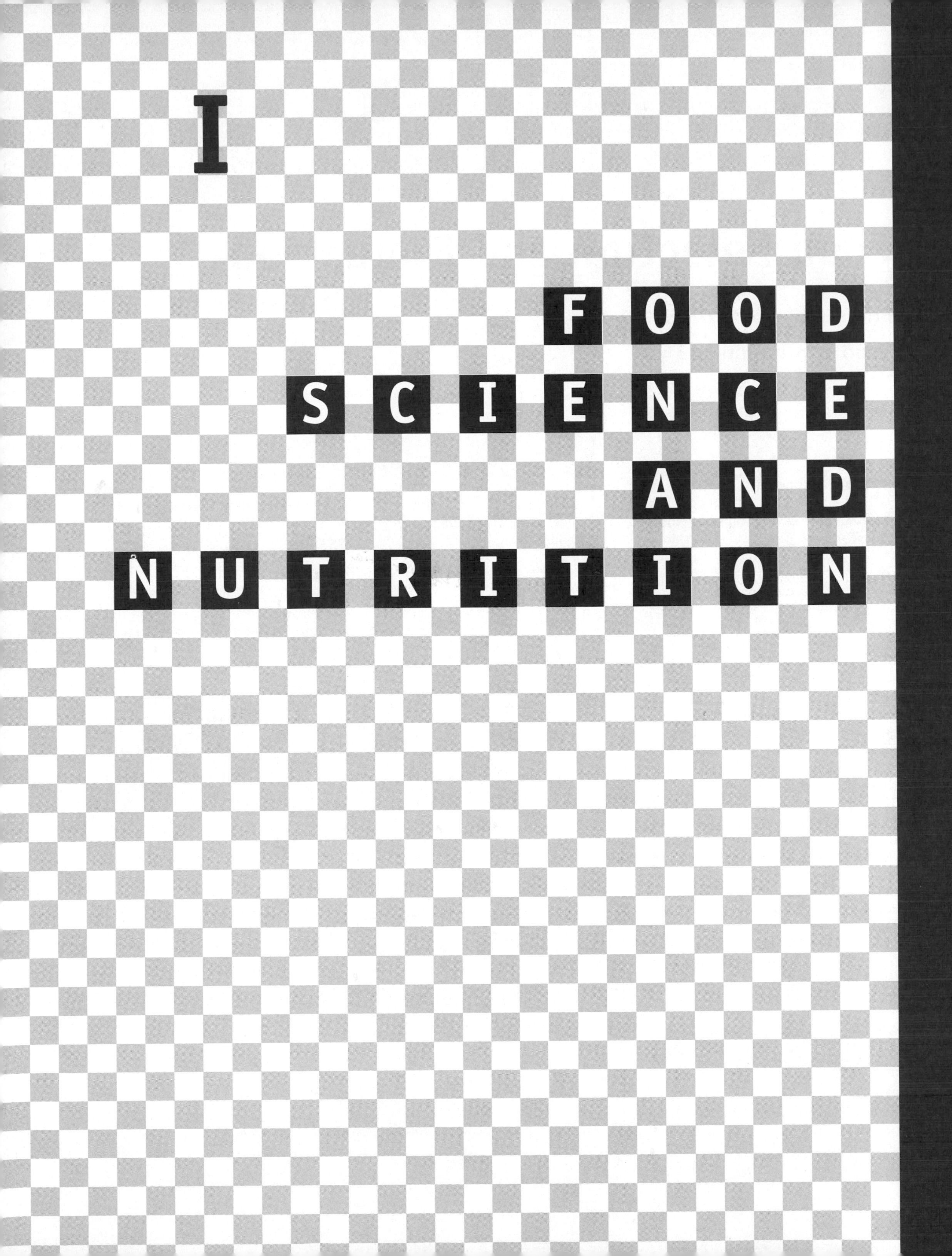

I

FOOD SCIENCE AND NUTRITION

Food Selection and Evaluation

Why do people eat what they eat? What is it about a food or beverage that causes them to choose it over others? And how does the food industry evaluate what foods consumers *will like?* The influences on food choices are many and complex, some obvious, some more subtle, but food scientists and the food industry have, with careful study, started to understand them.

Very few food choices were available several hundred years ago, with the exception of basic foods such as meats, grains, vegetables, and fruits. The number of different foods now available can actually make it more difficult, rather than easier, to plan a nutritious diet. The food industry offers thousands of foods, many of which are mixtures of the basic ones, and many of which include artificial ingredients. Food companies compete fiercely to develop ever newer and more attractive products for consumers to buy. This competition focuses on the factors that cause people to choose the way they choose. These factors of food selection, and the industry's tools of food evaluation, form the subject of this chapter.

Food Selection Criteria

People choose foods and beverages based on a number of factors. It is important how foods look and taste, but other considerations include eating for health, cultural and religious values, psychological and social needs, and budget (14). Each of these factors is now addressed in more detail.

Sensory Criteria

When people choose a particular food they evaluate it, consciously or unconsciously, primarily on how it looks, smells, tastes, feels, and even sounds (Figure 1-1). These sensory factors are more important to most consumers than nutritional considerations in making choices.

Sight. The eyes receive the first impression of foods: the shapes, colors, consistency, serving size, and the presence of any outward defects. Color can denote ripeness, the strength of dilution, and even the

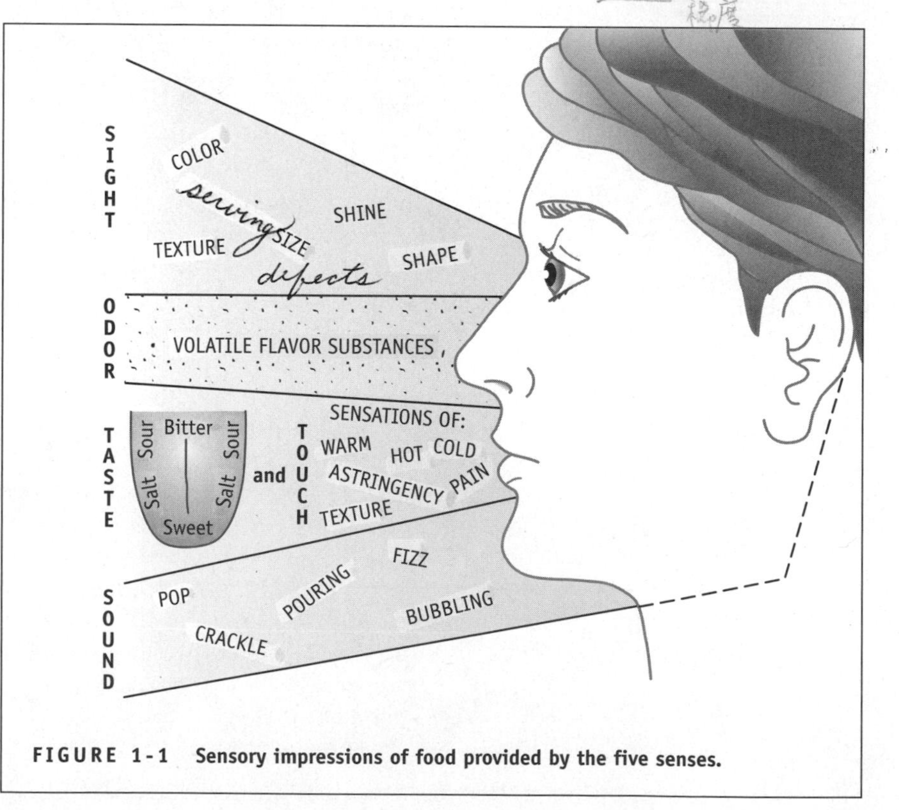

FIGURE 1-1 Sensory impressions of food provided by the five senses.

degree to which the food was heated. Black bananas, barely yellow lemonade, and scorched macaroni send visual signals that may alter a person's choices. Color can be deceiving; if the colors of two identical fruit-flavored beverages are different, people often perceive them as tasting different (69). People may judge milk's fat content by its color. For instance, if the color, but not the fat, is modified in reduced-fat (2 percent) milk, it is often judged to be higher in fat content, smoother in texture, and better in flavor than the reduced-fat milk with its original color (58).

How food is combined on a plate also contributes to or detracts from its appeal. Imagine a plate containing baked flounder, mashed potatoes, boiled cabbage, and vanilla ice cream compared to one that contains a nicely browned chicken breast, sweet potatoes, green peas, and blueberry cobbler. Most people would prefer the latter.

Odor. Smell is almost as important as appearance when people evaluate a food item for quality and desirability. Although the sense of smell is not as acute in human beings as it is in many other mammals, most people can differentiate between 2,000 to 4,000 odors, while some highly trained individuals can distinguish as many as 10,000 (5).

Classification of Odors. Since naming each of these odors separately would tax even the most fertile imagination, researchers have categorized them into major groups. One classification system recognizes six groups of odors: spicy, flowery, fruity, resinous (eucalyptus), burnt, and foul. The other widely used grouping scheme consists of four categories: fragrant (sweet), acid (sour), burnt, and caprylic (goaty) (5).

Detecting Odors. Regardless of the classifications, most odors are detected at very low concentrations. Vanillin can be smelled at 2×10^{-10} (0.0000000002) mg per liter of air (10). The ability to distinguish between various odors diminishes over the time of exposure to the smells; this perception of a continuously present smell gradually decreasing over time is called adaptation. People living near a noxious-smelling paint factory will, over time, come not to notice it, while visitors to the area may be taken aback by the odor.

We are able to detect odors when **volatile molecules** travel through the air and some of them reach the yellowish-colored **olfactory** epithelium, an area the size of a quarter located inside the upper part of the nasal cavity. This region is supplied with olfactory cells that number from 10 to 20 million in a human and about 100 million in a rabbit (10), reflecting the difference in importance of the sense of smell between people and rabbits. The exact function of these specialized cells in the sense of smell is not well understood.

Who has not experienced the feeling of bubbles tingling in the nose brought on by drinking a carbonated drink while simultaneously being made to laugh unexpectedly? This illustrates how the mouth and nose are connected and how molecules can reach the olfactory epithelium by either pathway.

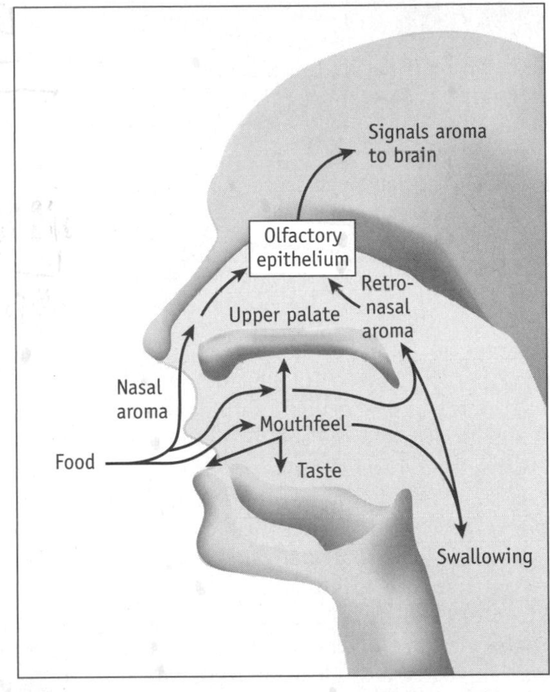

FIGURE 1-2 Detecting aroma, mouthfeel, and taste.

HOW & WHY? ?????????

Imagine the scent of chocolate chip cookies wafting through the house as they bake. *How* does this smell get carried to people? *Why* is the odor of something baking more intense than the odor of cold items like ice cream or frozen peaches? Since only volatile molecules in the form of gas carry odor, and since heat converts many substances into their volatile form, it is easier to smell hot foods than cold ones. Hot coffee is much easier to detect than cold coffee. Relatively large molecules such as proteins, starches, fats, and sugars are too heavy to be airborne, so their odors are not easily noticed. Lighter molecules capable of becoming volatile are physically detected by the olfactory epithelium by one of two pathways: (1) directly through the nose, and/or (2) during eating when they enter the mouth and flow retronasally, or toward the back of the throat and up into the nasal cavity (Figure 1-2) (56).

Taste. Taste is usually the most influential factor in people's selection of foods (17). Taste buds—so named because the arrangement of their cells is similar to the shape of a flower—are located primarily on the tongue, but are also found on the mouth palates and in the pharynx. Taste buds are not found on the flat, central surface of the tongue, but rather on the tongue's underside, sides, and tip.

Mechanism of Taste. What is actually being tasted? Many tasted substances are a combination of nonvolatile and volatile compounds. In order for a substance to be tasted, it must

KEY TERMS

Volatile molecules Molecules capable of evaporating like a gas into the air.

■ ■ ■ ■

Olfactory Relating to the sense of smell.

> **KEY TERMS**
>
> ***Gustatory*** Relating to the sense of taste.
>
> ■ ■ ■ ■
>
> ***Flavor*** The combined sense of taste, odor, and mouthfeel.

dissolve in liquid or saliva, which is 99.5 percent water. In the middle of each taste bud is a pore, similar to a little pool, where saliva collects. When food comes into the mouth, bits of it are dissolved in the saliva pools and there come into contact with the cilia, small hair-like projections from the **gustatory** cells. The gustatory cells relay a message to the brain via one of the cranial nerves (facial, vagus, and glossopharyngeal). The brain, in turn, translates the nervous electrical impulses into a sensation people recognize as "taste." As people age, the original 9,000 to 10,000 taste buds begin to diminish in number, so that people over 45 often find themselves using more salt, spices, and sugar in their food. Another important factor influencing the ability of a person to taste is the degree to which a compound can dissolve (50). The more the moisture or liquid, the more the molecules triggering flavor can dissolve and spread over the tongue, coming in contact with the taste buds (25).

The Five Taste Stimuli Different areas on the tongue are associated with the five basic types of taste: sweet, sour, bitter, salty and savory (*umami*, a Japanese word meaning "delicious") (70, 71). The fifth taste stimulus known as savory (*umami*), is found in certain amino acids. The tip of the tongue is more sensitive to sweet and sour tastes, while the sides are sensitive to salty and sour sensations, and the back to bitter taste perceptions. The time it takes to detect each of these taste stimuli varies from a split second for salt to a full second for bitter substances (10). Bitter tastes, therefore, have a tendency to linger. The chemical basis of these five categories of taste is described as follows:

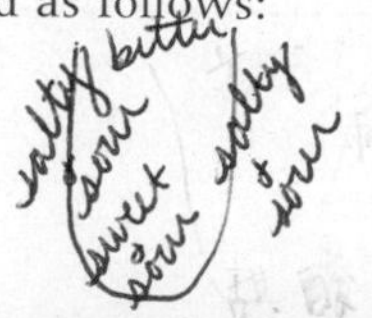

- The sweetness of sugar comes from the chemical configuration of its molecule. A long list of substances yield the sweet taste, including sugars, glycols, alcohols, and aldehydes. Little is known, however, about the sweet taste receptor and how "sweetness" actually occurs (25, 73).
- Sour taste comes from the acids found in food. It is related to the concentration of hydrogen ions (H^+), which are found in the natural acids of fruits, vinegar, and certain vegetables.
- Bitterness is imparted by compounds such as caffeine (tea, coffee), theobromine (chocolate), and phenolic compounds (grapefruit). Among the many substances that can yield bitter tastes are alkaloids that are often found in poisonous plants.
- Salty taste comes from ionized salts such as the salt ions [Na^+] in sodium chloride (NaCl) or other salts found naturally in some foods.
- Savory (*umami*) taste is imparted by certain amino acids, which is why some people can detect monosodium glutamate, which contains the amino acid glutamic acid.

Taste Interactions. Each item used in food preparation contains a number of compounds, and bringing these items together creates new tastes when all their compounds interact. Salt sprinkled on grapefruit or added to fruit pies tends to decrease tartness and enhance sweetness. Conversely, acids in subthreshold concentrations, which are present but not yet detectable, increase saltiness. Adding sugar to the point that it is not yet tasted decreases salt concentration and also makes acids less sour and coffee and tea less bitter. Some compounds, like monosodium glutamate, often used in Chinese cooking, actually improve the taste of meat and other foods by making them sweeter (46, 56).

Factors Affecting Taste. Not everyone perceives the taste of apple pie the same way. There is considerable genetic variation among individuals in sensitivity to basic tastes (14, 60). Tasting abilities may also vary within the individual, depending on a number of outside influences (50). One such factor affecting taste is the temperature of a food or beverage. As food or beverage temperatures go below 68°F (20°C) or above 86°F (30°C), it becomes harder to distinguish their tastes accurately. For example, very hot coffee tastes less bitter, while slightly melted ice cream tastes sweeter. Other factors influencing taste include the color of the food, the time of day it is eaten, and the age, gender, and degree of hunger of the taster (14).

Variety in available food choices also affects taste. This can be seen when the "taste," or appetite, for a food eaten day after day starts to diminish. Even favorite foods consumed every day for a while can lose their appeal. Some weight-reducing fad diets are based on this principle, banking on the idea that people will get tired of eating just one type of food and therefore will come to eat less. A routine of grapefruit for breakfast, grapefruit for lunch, and grapefruit for dinner quickly becomes boring and unappetizing.

Definition of Flavor. In examining the factors affecting taste, it is important to distinguish between taste and **flavor**. Taste relies on the taste buds' connection to the brain via nerve cells, which signal the sensations of sour, salt, sweet, bitter, and savory. Flavor is a broader concept with aroma providing about 75 percent of the impression of flavor (16, 30, 72). To get some idea of how the ability to smell affects flavor perception, think of having a cold with a badly stuffed up nose. Everything tastes different. The nasal congestion interferes with the function of the olfactory sense, impairing the ability to detect the aromas contributing to the perception of flavor. Some people apply this principle to their advantage by pinching their nostrils shut to lessen the bad flavor of a disagreeable medicine they must swallow.

Reduced-fat cookies, for example, taste sweeter unless they are modified to compensate (44). It is even more difficult to replace certain fats that, in addition to contributing to traditional flavor releases and mouthfeel, also have their

HOW & WHY? ?????????

Why do flavors differ in how quickly they are detected or how long they last? The amount of fat in a food or beverage determines how intense the flavor is over time. Flavor compounds dissolved in fat (fat-soluble compounds) take longer to be detected and last longer than flavor compounds dissolved in water (water-soluble compounds), which are quickly detected but disappear much more quickly (16). This explains why a reduced-fat product is much less likely to duplicate the flavor of the original food: the original fat's flavor compounds are missing, causing an imbalance of the other flavors present.

---

own distinctive flavor, as is the case for butter, olive oil, and bacon fat (46).

Flavors, regardless of the medium in which they are dissolved, do not stay at the same intensity day after day, but diminish over time. Sensory chemists and flavor technologists know that one way to keep the food products sold by manufacturers from losing their appeal is to prevent the volatile compounds responsible for flavor from deteriorating, escaping, or reacting with other substances. They look at methods in processing, storage, and cooking, all of which affect the volatile flavor compounds, to devise strategies against these occurrences. One of the major functions of protective packaging is to retain a food's flavor. Packaging guards flavor in several ways. It protects against vaporization of the volatile compounds and against physical damage that could expose food to the air and result in off odors. It keeps unpleasant odors from the outside from attaching to the food. It also prevents "flavor scalping," or the migration of flavor compounds from the packaging (sealers, solvents, etc.) to the food or vice versa (46).

Touch. The sense of touch, whether it operates inside the mouth or through the fingers, conveys to us a food's texture, consistency, astringency, and temperature.

Texture is a combination of perceptions, with the eyes giving the first clue. The second comes at the touch of fingers and eating utensils, and the third is mouthfeel, as detected by the teeth and tactile nerve cells in the mouth located on the tongue and palate. Textural or structural qualities are especially obvious in such foods as apples, popcorn, liver, crackers, potato chips, tapioca pudding, cereals, and celery, to name just a few. Textures felt in the mouth can be described as coarse (grainy, sandy, mealy), crisp, fine, dry, moist, greasy, smooth (creamy, velvety), lumpy, rough, sticky, solid, porous, bubbly, or flat. Tenderness, which is somewhat dependent on texture, is judged by how easily the food gives way to the pressure of the teeth.

Consistency is only slightly different than tenderness, and is expressed in terms of brittleness, chewiness, viscosity, thickness, thinness, and elasticity (rubbery, gummy).

Astringency, which causes puckering of the mouth, is possibly due to the drawing out of proteins naturally found in the mouth's saliva and mucous membranes (10). Foods such as cranberries, lemon juice, and vinegar have astringent qualities.

The **temperature** of a food also affects the perception of it. Most people appreciate a hot meal, but extremely hot temperatures can literally burn the taste buds, although they later regenerate. The other kind of "hot" that may be experienced with food is the kind generated by "hot" peppers (Chemist's Corner 1-1). The hotness in peppers is produced by a chemical called capsaicin (cap-SAY-iss-in). Many people enjoy the sensation of capsaicin in moderation, but it can cause real pain because it is a powerful chemical that irritates nerves in the nose and mouth. In fact, this compound is so caustic when concentrated that it is now used by many law enforcement agencies in place of the mace-like sprays.

Hearing. The sounds associated with foods can play a role in evaluating their quality. How often have you seen someone tapping a melon to determine if it is ripe or not? Sounds like sizzling, crunching, popping, bubbling, swirling, pouring, squeaking, dripping, exploding (think of an egg yolk in a microwave), and crackling can communicate a great deal about a food while it is being prepared, poured, or chewed. Most of these sounds are affected by water content, and their characteristics thus give clues to a food's freshness and/or doneness.

KEY TERMS

Consistency Describes a food's firmness or thickness.

Astringency A sensory phenomenon characterized by a dry, puckery feeling in the mouth.

Nutritional Criteria

Over the past several decades, emerging scientific evidence about health and nutrition has resulted in changing food consumption patterns in the United States (8). In one survey, six out of ten consumers reported making a major change in their diets, and another survey reported that nutrition was second only to taste in importance to shoppers (37, 79). The changing food habits are related to the increased awareness that diet can be related to some of the leading causes of death—heart disease, cancer, and diabetes—as well as to other common health conditions like osteoporosis, diverticulosis, and obesity (57, 76). Heart disease accounts for

Chemist's Corner 1-1

Hot Peppers and Body Chemistry

The warming sensation experienced by some people eating hot peppers or foods made with them is due to the body secreting catecholamines, a group of amines composed of epinephrine (adrenaline), norepinephrine (noradrenaline), and dopamine. These catecholamines activate the "fight-or-flight" response, which normally triggers increased respiration rate, a faster heart beat, slowed digestion, widened pupils, and enhanced energy metabolism (27, 61).

most of the deaths in the United States each year, and costs more than $100 billion annually in direct health care expenditures (77). Cancer is the second leading cause of death in the United States, with the amount of fat in the diet being the number one dietary factor related to cancer. Overweight is one of the biggest health problems in the United States (13), and some $33 billion dollars are spent annually by 65 million Americans on "quick fix" weight loss solutions, most of which achieve no permanent results.

In an effort to reduce dietary risk factors for some of the major health conditions affecting Americans, the U.S. government published several dietary guides. Two of these are the Dietary Guidelines and the Food Guide Pyramid.

Dietary Guidelines. The emphasis on adjusting fat and other dietary factors in the diet was reinforced by the Dietary Guidelines published every five years since 1980 by the U.S. government (Figure 1-3) (76). Dietary Guidelines for Infants are also available and take into account the differing nutrient needs at this time of life (24). The Dietary Guidelines for healthy adults encourage consumers to follow the recommendations shown in Figure 1-3. No numerical minimums or maximums are provided and no recommendations are given in terms of caffeine or artificial sweeteners. The Dietary Guidelines define healthful dietary habits for the American public, and now serve as the basis for all federal nutrition programs (4).

Food Guide Pyramid. The Food Guide Pyramid was developed in 1992 to encourage Americans to improve their diets, replacing the basic four food groups of milk, meat, vegetable/fruit, bread/cereal (20, 75). Other countries have their own versions of this type of guideline (62). The easy-to-comprehend visual concept of the Food Guide Pyramid organizes foods into five major groups. The lower-fat, complex carbohydrate foods on the bottom are emphasized, while the higher-fat, sweet foods at the tip are minimized (Figure 1-4). The 6 to 11 servings of bread suggested by the chart may seem a bit high, but the stress is on the importance of keeping carbohydrate intake high and fat intake down to somewhere between 20 and 30 percent of **calories (kcal)**. Calories (kcal) are discussed in detail in Chapter 2.

KEY TERMS

Calorie (kcal) The amount of energy required to raise 1 gram of water 1°C (measured between 14.5° and 15.5°C at normal atmospheric pressure).

■ ■ ■ ■

Antioxidant A compound that inhibits oxidation, which can cause deterioration and rancidity.

FIGURE 1-3
Dietary Guidelines for Americans.

- Aim for a healthy weight.
- Be physically active each day.
- Let the Pyramid guide your food choices.
- Choose a wide variety of grains daily, especially whole grains.
- Choose a variety of fruits and vegetables daily.
- Keep food safe to eat.
- Choose a diet that is low in saturated fat and cholesterol and moderate total fat.
- Choose beverages and foods that limit your intake of sugars.
- Choose and prepare foods with less salt.
- If you drink alcoholic beverages, do so in moderation.

*These guidelines are intended for adults and healthy children ages 2 and older. Source: USDA.

The vegetarian version of the Food Guide Pyramid is shown in Figure 1-5. This version was published in a report by the American Dietetic Association; it suggested that properly planned vegetarian diets may reduce the risk of certain chronic, degenerative diseases and conditions including heart disease, some cancers, diabetes mellitus, obesity, and high blood pressure (1). Other factors may, however, contribute to the decreased morbidity and mortality from these diseases among vegetarians. These include positive lifestyle differences such as lower rates of smoking and drinking. Any possible benefit of vegetarian diets probably comes from lower intakes of fat, saturated fat, cholesterol, and animal protein, balanced by higher levels of phytochemicals, fiber, complex carbohydrates, **antioxidants** such as vitamins C and E, carotenoids, and folate (a B vitamin) (1, 12).

Consumer Dietary Changes. As a result of these dietary guidelines and other influences, consumers have shifted their dietary concerns and intakes. People can read the Nutrition Facts on food labels to understand what they are consuming (Chapter 8). The biggest nutritional concern reported by consumers is fat, exceeding that for salt, cholesterol, sugar, and calories (kcal) (66). Overall, Americans are ingesting less red meat and whole milk, and more poultry, reduced-fat (2 percent) milk, fresh fruits, fresh vegetables, pasta, and rice. As a result, fat consumption has dropped from 40 percent of calories (kcal) to less than 33 percent since 1977, and more of that fat is coming from plant sources (64).

Despite decreased fat intakes, Americans continue to increase their calorie (kcal) consumption each year. Nevertheless, the desire of consumers to lower fat, cholesterol, and calorie (kcal) intake has led to the development of fat substitutes and many reduced-fat and low-calorie (kcal) food products (34). "Light" foods have been at the top of the list of new food product categories for the past decade (26). According to law, food companies may label products as "lower" in a given nutrient or calories (kcal), but may not link their products "with the prevention, cure, or treatment of human disease unless fitting the specific categories approved by the Food and Drug Administration" (29).

Shifting Food Focus. The desire to ensure overall good health, rather than balanced nutrition or even lower fat, is a key motivating factor in selecting foods. Foods are being viewed by more people as an integral part of the self-care movement. Despite what appears to be an increasing intake of calories, 15 percent of college students are vegetarians, 33 percent of Americans have used herbs or herb products, and about 60 per-

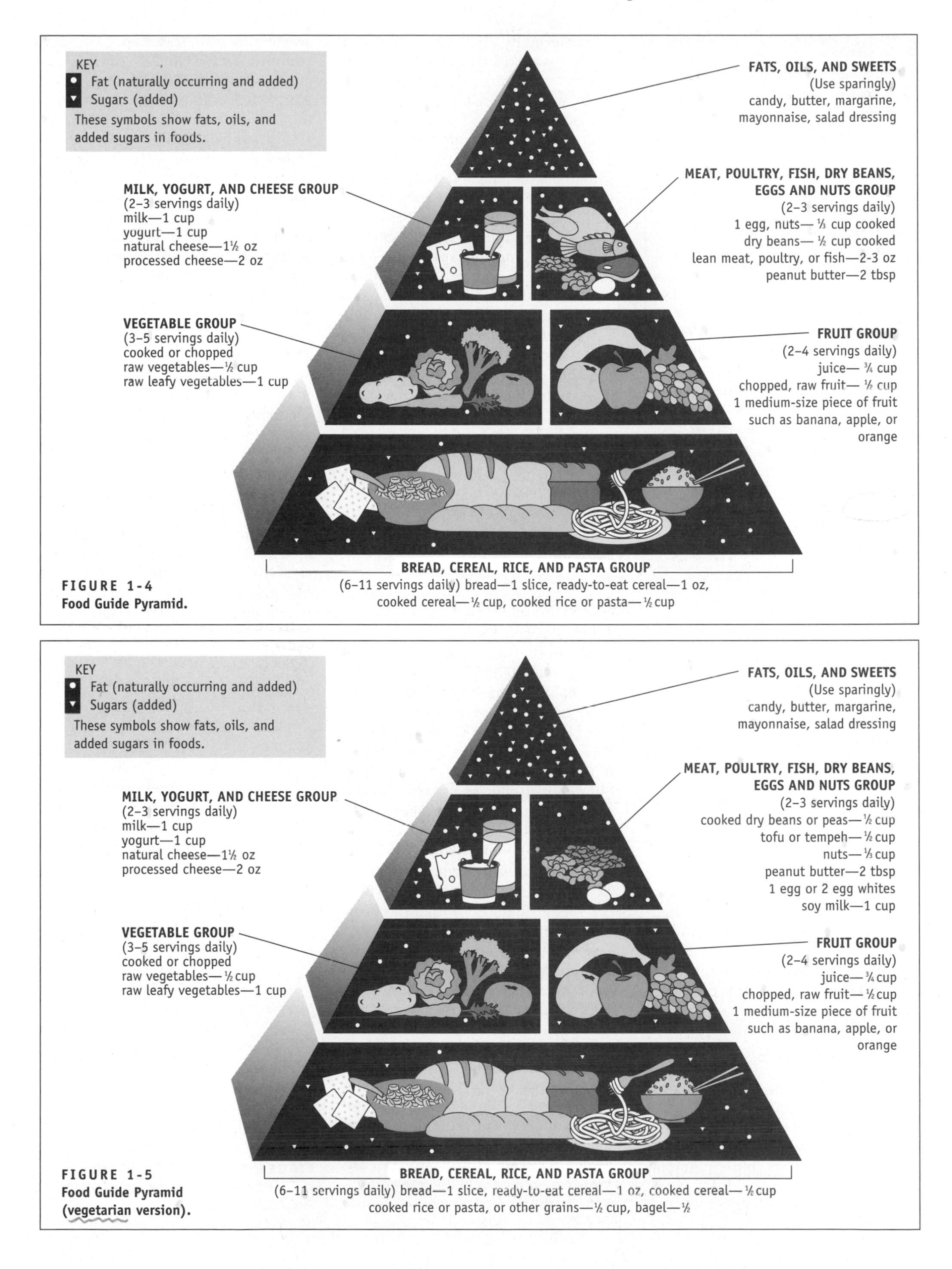

FIGURE 1-4
Food Guide Pyramid.

FIGURE 1-5
Food Guide Pyramid (vegetarian version).

KEY TERMS

Functional food A food or beverage that imparts a physiological benefit that enhances overall health, helps prevent or treat a disease or condition, or improves physical/mental performance.

Nutraceutical A bioactive compound (nutrients and non-nutrients) that has health benefits.

Monograph A summary sheet (fact sheet) describing a substance in terms of name (common and scientific), chemical constituents, functional uses (medical and common), dosage, side-effects, drug interactions, and references.

Culture The ideas, customs, skills, and art of a group of people in a given period of civilization.

cent take a multivitamin supplement (68). According to an International Food Information Council survey, 93 percent of Americans believe that some foods can have health benefits besides nutritive value and can delay the onset, or reduce the risk, of serious and chronic diseases (2). The concept that "food is medicine" is common to many cultures, and the shift from treating disease to preventing disease is a slow but growing movement.

Complementary and Alternative Medicine. Complementary and alternative medicine (CAM) is a growing trend that appears to be making permanent inroads. Terms like **functional foods** and **nutraceuticals** are becoming commonplace. Europe and Japan appear to lead the United States in their interest in how foods can benefit health beyond the intake of just carbohydrates, protein, fat, and vitamins/minerals. Japan has imported record shipments of blueberries from the United States in the past half decade. Blueberry's blue pigment, anthocyanin, is a powerful antioxidant that may benefit eyesight (36).

In Germany, the German E Commission was created in 1978 to assure product standardization and safe use of herbs and phytomedicines. Composed of a body of experts from the medical and pharmacology professions, the pharmaceutical industry, and lay persons, the E Commission studies the scientific literature for research data on herbs based on clinical trials, field studies, and case studies. It has created a collection of **monographs** representing the most accurate information available in the world on the safety and efficacy (does it work?) of herbal products. Germany defines herbal remedies in the same manner as drugs, because their physicians, and others in Europe, often prescribe herbal remedies that are paid for by government health insurance.

This trend toward food as more than something to provide calories and nutrients is not really new. About 2,500 years ago, Hippocrates said, "Let food be thy medicine, and medicine be thy food" (8).

Cultural Criteria

Culture is another factor influencing food choice. Culture influences food habits by dictating what is or is not acceptable to eat. Foods that are relished in one part of the world may be spurned in another. Grubs, which are a good protein source, are acceptable to the aborigines of Australia. Whale blubber is used in many ways in the arctic, where the extremely cold weather makes a high-fat diet essential. Dog is considered a delicacy in some Asian countries. Escargots (snails) are a favorite in France. Sashimi (raw fish) is a Japanese tradition that has been fairly well accepted in the United States. Locusts, another source of protein, are considered choice items in the Middle East. Octopus, once thought unusual, now appears on many American menus.

Ethnic Influences. Ethnic minorities comprise nearly 25 percent of the United States' population, with the four major groups being African, Hispanic (Latino), Asian/Pacific, and Native/Alaskan Americans (Figure 1-6). An increasingly diverse population in the United States, accompanied by people traveling more and communicating over longer distances, has contributed to a more worldwide community, and a food industry that continues to "go global" (59). Within the boundaries of the United States alone, many foods once considered ethnic are now commonplace: pizza, tacos, beef teriyaki, pastas, and gyros. More recently arrived ethnic foods, such as Thai, Indian, Moroccan, and Vietnamese, are constantly being added to the mix to meet the escalating demands for meals providing more variety, stronger flavors, novel visual appeal, and less fat (67).

Place of Birth. Birthplace influences the foods a person will be exposed to, and helps to shape the dietary patterns that are often followed for life. Salsa varies in flavor, texture, and color depending on whether it was prepared in Mexico, Guatemala, Puerto Rico, or Peru. Curry blends differ drastically depending on where in the world the recipe evolved. In Mexican cuisine, the same dish may taste different in different states.

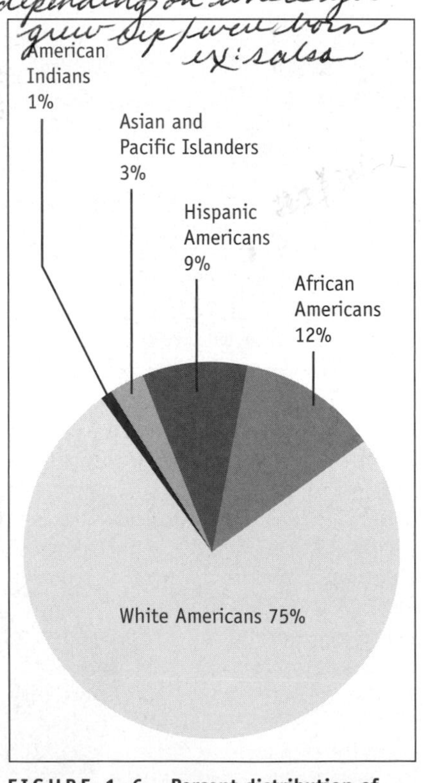

FIGURE 1-6 Percent distribution of racial/ethnic groups in the United States.
Source: U.S. Census Bureau.

Geography and Climate. Not so long ago, geography and climate were the main determinants of what foods were available to be chosen. People ate foods that were grown close to where they lived and very rarely were presented with the possibility of eating those of a more exotic nature. For example, guava fruit grown in tropical regions was not even a consideration in an area such as Greenland. Now the wide distribution of formerly "local" foods throughout the world provides many people with an incredible variety of food choices.

Cultural Influences on Manners. Culture not only influences what types of foods are chosen, but the way they are consumed and the behavior surrounding their consumption. In some parts of India, for example, only the right hand is used for eating and manipulating utensils; the left hand is reserved for restroom duties. Foods may be served on banana leaves or wrapped in corn husks. It may be eaten with chopsticks, as is the custom throughout Asia, or with spoons, forks, and knives as in Europe and the Americas. It is considered impolite in China not to provide your guest with a bountiful meal, so an unusually large number of food courses is served when guests are present. Belching heartily after a meal in Egypt is considered complimentary to the host.

Religious Criteria

Religion is another important influence on food choices. Religious beliefs affect the diets of many by declaring which foods are acceptable and unacceptable and by specifying preparation procedures. By designating certain foods for specific occasions and assigning symbolic value to some, religious principles wield further influence (19).

Over 85 percent of the American population claims to be Christian, and the bread (wafers) and wine served by many denominations during communion symbolize the body and blood of Christ. A traditional holiday meal with a turkey or ham as the main entrée is usually served at Christmas and/or Easter. Until recently, Friday was fish day for many Catholics. Some food practices of Buddhists, Hindus, Seventh-Day Adventists, Mormons, Jews, and Muslims are explored in further detail below.

Buddhism. There are over 100 million Buddhists in China and 300 million worldwide. Buddhists believe in *karuna,* which is compassion, and karma, a concept that implies that "good is rewarded with good; evil is rewarded with evil; and the rewarding of good and evil is only a matter of time" (39). Many Buddhists consider it uncompassionate to eat the flesh of another living creature, so vegetarianism is often followed; however, not all Buddhists are vegetarian. Whether a Buddhist is vegetarian depends on their personal choice, the religious sect to which they belong, and the country where they live (18).

Hinduism. Hinduism also promotes vegetarianism among some, but not all, of its followers (48). Some Hindus believe that the soul is all-important, uniting all beings, so it is against their beliefs to injure or kill an animal. Thus strict Hindus reject poultry, eggs, and the flesh of any animal. Among many Hindus, the cow is considered sacred and is not slaughtered for food. However, dairy products from cattle are acceptable and even considered spiritually pure (18). Coconut and *ghee,* or clarified butter, are also accorded sacred status, but may be consumed after a fast. Some strict Hindus do not eat garlic, onions, mushrooms, turnips, lentils, or tomatoes.

Seventh-Day Adventist Church. A vegetarian diet is recommended but not required for members of the Seventh-Day Adventist Church. About 40 percent of the members are vegetarians, the majority of them lacto-ovo-vegetarians, meaning that milk and egg products are allowed (28). Hot spices, between-meal snacks, and alcohol, tea, and coffee consumption are discouraged (7).

Church of Jesus Christ of Latter-Day Saints (Mormon Church). The Church of Jesus Christ of Latter-Day Saints supports vegetarianism and discourages the use of alcohol as well as tea, coffee, or any other caffeinated drink. A significant number of Mormons live in Utah, and several studies have shown that the death rate attributed to specific diseases for Utah residents is 40 percent below the average U.S. rate because of lower rates of heart disease and cancer. Other factors possibly affecting the death rate are the discouragement of smoking, the recommendations of regular physical activity and proper sleep, and a positive religious outlook (63). The lower fat content of some vegetarian diets and Utah's health care system also cannot be ignored as possible contributing factors.

KEY TERMS

Kosher From Hebrew, food that is "fit, right, proper" to be eaten according to Jewish dietary laws.

Judaism. The *kashruth* is the list of dietary laws adhered to by orthodox Jews. Foods are sorted into one of three groups: meat, dairy, or *pareve* (containing neither meat nor dairy) (49). Milk and meat cannot be prepared together or consumed in the same meal. In fact, separate sets of dishes and utensils are used to prepare and serve them, and a specified amount of time (1 to 6 hours) must pass between the consumption of milk and meat. Foods considered **kosher** include fruits, vegetables, grain products, and with some exceptions during Passover, tea, coffee, and dairy products from kosher animals as long as they are not eaten simultaneously with meat or fowl (65). Kosher animals are ruminants, such as cattle, sheep, and goats, that have split hooves and chew their cud. Other meats that are considered kosher are chicken, turkey, goose, and certain ducks.

Orthodox Jews are not allowed to eat non-kosher foods such as carnivorous animals, birds of prey, pork (bacon, ham), fish without scales or fins (shark, eel, and shellfish such as shrimp, lobster, and crab), sturgeon, catfish, swordfish, underwater mammals, reptiles, or egg yolk containing any blood. These foods are considered unclean or *treif.* Even the meat from allowed animals is not considered kosher unless the animals have been slaughtered under the supervision of a rabbi or other authorized individual who ensures that the blood has been properly removed. Foods that are tainted with blood, a substance considered by Jews to be synonymous with life, are forbidden (18).

Kosher foods are labeled with a logo such as those of the kosher-certifying agencies shown in Figure 1-7. People other than Jews who often purchase kosher foods include Moslems, Seventh-Day Adventists, vegetarians, individuals with allergies (shellfish) or food intolerances (milk), and anyone who perceives kosher foods as being of higher quality (49).

Food figures prominently in the celebration of the major Jewish holidays. Rosh Hashanah, the Jewish New Year, is celebrated in part with a large meal. Yom Kippur, or the Day of Atonement, requires a day of fasting preceded by a bland evening meal the night before. Passover, which is an eight-day celebration marking the exodus from Egypt, is commemorated in part by a meal whose components represent different aspects of the historic event.

Islam. Worldwide, there are over 1 billion Muslims compared to 13 million Jews (18). The Koran, the divine book of Islam, contains the food laws recommended for Muslims, some of which are similar to the food laws of Judaism. The most striking similarity is that the kosher meat consumed by Jews is permitted for Muslims because it has been slaughtered in a manner that allows the blood to be fully drained. *Halal* meat is also permitted and defined as any meat from approved animals sacrificed according to Muslim guidelines. Most meat is allowed except pork, carnivorous animals with fangs (lions, wolves, tigers, dogs, etc.), birds with sharp claws (falcons, eagles, owls, vultures, etc.), land animals without ears (frogs, snakes, etc.), shark, and products containing pork or gelatin made from the horns or hooves of cattle (11). Alcohol and products containing alcohol in any form, including vanillin and wine vinegar, are forbidden. Stimulants such as tea and coffee are also discouraged.

Ramadan is a time of the year that significantly affects diet for Moslems. Islam teaches that the ninth month of the lunar calendar is the month in which the Prophet Muhammad received the revelation of the Muslim scripture, the Koran. This month, which depends on the sighting of the new moon, is a time of religious observances that include fasting from dawn to sunset.

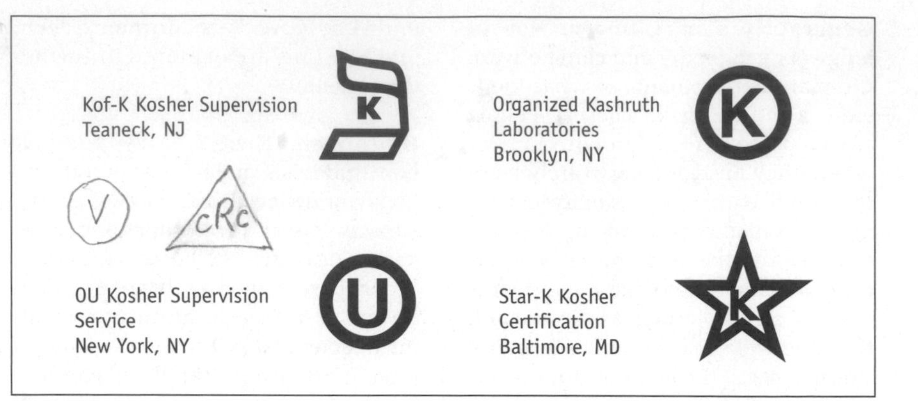

FIGURE 1-7 Examples of Kosher food symbols.

Psychological and Sociological Criteria

Social and psychological factors strongly influence food habits. For most people, the knowledge that food is readily available provides a sense of security. The aim of every food company's advertising is to develop a sense of security among consumers about its products. A soft drink held in the hand of an athlete, the cereal touted by a child's favorite cartoon character, and diet foods offered by slim, vivacious spokespeople create positive associations in people's minds for these products and assure them of their quality. Social conscience and peer pressure sometimes influence food choices. One recent trend has seen consumers moving toward more environmentally sound purchases. At a buffet, the presence of other people may influence a person's choice of food and beverages.

Psychological needs intertwine with social factors when foods are used more for a display of hospitality or status than for mere nourishment. Caviar is just fish eggs, but is esteemed by many as a delicacy. Beer tastes terrible to most people when they try it for the first time, but the social surroundings and pressures may cause it to become an acquired taste. Several studies have shown that information influences expectations, and expectations mold choices (14), so it is no surprise that consumers report that television is their predominant source of information about nutrition (42 percent), followed by magazines (39 percent), and newspapers (19 percent) (6, 54).

Psychological factors also influence people's response to two relatively recent additions to the food market: genetically modified foods and organic foods.

Genetically Modified Organisms (GMOs). Psychological and social factors are involved in the formation of public attitudes toward the **genetic engineering** of foods. **Genetically engineered foods,** commonly called **genetically modified organisms (GMOs),** are slowly gaining ground, but not everyone is knowledgeable about or accepting of the new foods (45).

History of Genetic Engineering. In the past, it took years to accomplish hybridization, or crossbreeding, by matching "the best to the best" in the plant, livestock, and fishery worlds to achieve the desired results. Cattle, corn, and even dogs were bred this way to yield desirable results. Dogs would not look the way they do without humans modify-

KEY TERMS

Genetic engineering The alteration of a gene in a bacterium, plant, or animal for the purpose of changing one or more of its characteristics.

■ ■ ■ ■

Genetically modified organisms (GMOs) Plants, animals, or microorganisms that have had their genes altered through genetic engineering using the application of recombinant deoxyribonucleic acid (rDNA) technology.

ing their genes through many years of selective breeding. Depending on the desired results, it could take decades or even centuries to develop a certain "look" and/or function in an animal or plant. Traditional ways of breeding to combine the **genes** of two species in order to obtain a specific trait were thus time consuming, cumbersome, and unpredictable (9).

Along came the age of genetic engineering, which began in the early 1970s when DNA was isolated from a bacterium, duplicated, and inserted into another bacterium. The resulting DNA, known as recombinant deoxyribonucleic acid (rDNA), allows researchers to transfer genetic material from one organism to another (40). Instead of crossbreeding for years, researchers can now identify the genes responsible for a desired trait and reorganize or insert them from the cells of one bacterium, plant, or animal into the cells of other bacteria, plants, or animals (52). The goal of this process is to produce new species or improved versions of existing ones. The U.S. Department of Agriculture envisions genetic engineering being used to increase production potential, improve resistance to pests and disease, and develop more nutritious plant and animal products (43).

Genetically Engineered Foods. What foods have been genetically engineered this way? Genetic engineering has so far resulted in benefits that increase the food's resistance to the following (40):

- Pests (less pesticide required)
- Disease (lower crop losses)
- Harsh growing conditions (drought, salty soil, climate extremes)
- Transport damage (less bruising allows more produce to make it to market)
- Spoilage (longer shelf life)

Some examples of GMO foods include ripening-delayed fruits, grains with higher protein content, potatoes that absorb less fat when fried, insect-resistant apples, and more than 50 other plant products. The first genetically engineered food to be approved by the Food and Drug Administration was Calgene's FlavrSavr tomato (Figure 1-8 on page 12). This tomato was developed so it could be left on the vine until fully ripened and flavorful, and yet still withstand the hardship of shipping without bruising (31). The Flavr Savr tomato softens at a slower rate because of genetic engineering that reduces the activity of an enzyme responsible for breaking down the cell wall during ripening (78). The previously widespread practice was to pick tomatoes while they were green to allow them to be shipped before ripening, because unripened tomatoes would be less easily damaged during transport. This is still the case with tomatoes other than the newer varieties, and it has meant that most consumers are left desiring the succulence of a vine-ripened tomato.

Other genetically engineered foods include celery without strings, squash that is resistant to a common plant virus, presweetened melons, and tomatoes resistant to both cold and hot temperatures. Genes have also been reorganized in strawberries to increase their natural sweetness. Possible genetically engineered foods of the future include cow's milk with some of the immune benefits of human milk (38), fruits containing higher amounts of vitamins A and C (43), fats and oils containing more omega-3 fatty acids (51), foods that generate proteins that could be used as oral vaccines (3), and soybeans providing a more complete source of protein (15).

Concerns About Genetic Engineering. Some consumers view genetic engineering as an invasion of nature's domain, and fear that scientists are treading on dangerous ground. Their concerns include allergies, gene contamination, and religious/cultural objections.

- *Allergies.* The concern most commonly expressed to the Food and Drug Administration by consumers was the possibility that the proteins produced by these new genes could cause allergic reactions. In one study, soy was infused with a gene from Brazil nuts, a known allergen, or allergy-causing substance (47). Some people participating in the experiment became ill, but this was a preliminary research study and the modified soy never reached the market (58). Researchers would be prudent to avoid food allergens in the process of genetically engineering foods because, even though protein food allergies affect only a small percentage of the population, they still exist and can cause problems (21).
- *Gene Contamination.* Another concern is that genetically engineered plants might "escape" the wild, take over, and change the environment. Scientists assure us, however, that such plants are no more dangerous than traditionally bred crops. The greatest fear for some is that genetic engineering will lead to researchers using this type of biotechnology to try to "improve" the human race (47).
- *Religious/Cultural.* Some people, for religious or cultural reasons, do not want certain animal genes appearing in plant foods (52). For example, if swine genes were inserted into vegetables for some purpose, those vegetables would not be considered kosher. In one instance, a group of chefs refused to use a genetically engineered tomato when they found out that its disease resistance was obtained from a mouse gene. Vegetarians may object to a fish gene being placed in a tomato to provide resistance to freezing (31).

Acceptance/Rejection of Genetically Engineered Foods. Despite the controversy over animal genes being inserted in plant foods, the line between "plant genes" and "animal genes" is already blurred. Bacteria, plants, and animals share a large number of the more than 100,000 genes found in higher organisms. Nevertheless, research surveying people's attitudes about genetic engineering repeatedly reveals that consumers are more likely to accept crop biotechnology than that conducted on animals or fish (33).

The Food and Drug Administration accepts genetically engineered foods as posing no risk to health or safety, and for this reason it does not require mandatory labeling, unless the foods contain new allergens, modified nutritional profiles, or represent a new plant (9, 58). The National Academy of Sciences has stated that genetic transfers between unrelated organisms do not pose hazards or risks different from those encountered by natural selection or crossbreeding. Currently, there is no evidence that

KEY TERMS

Gene A unit of genetic information in the chromosome.

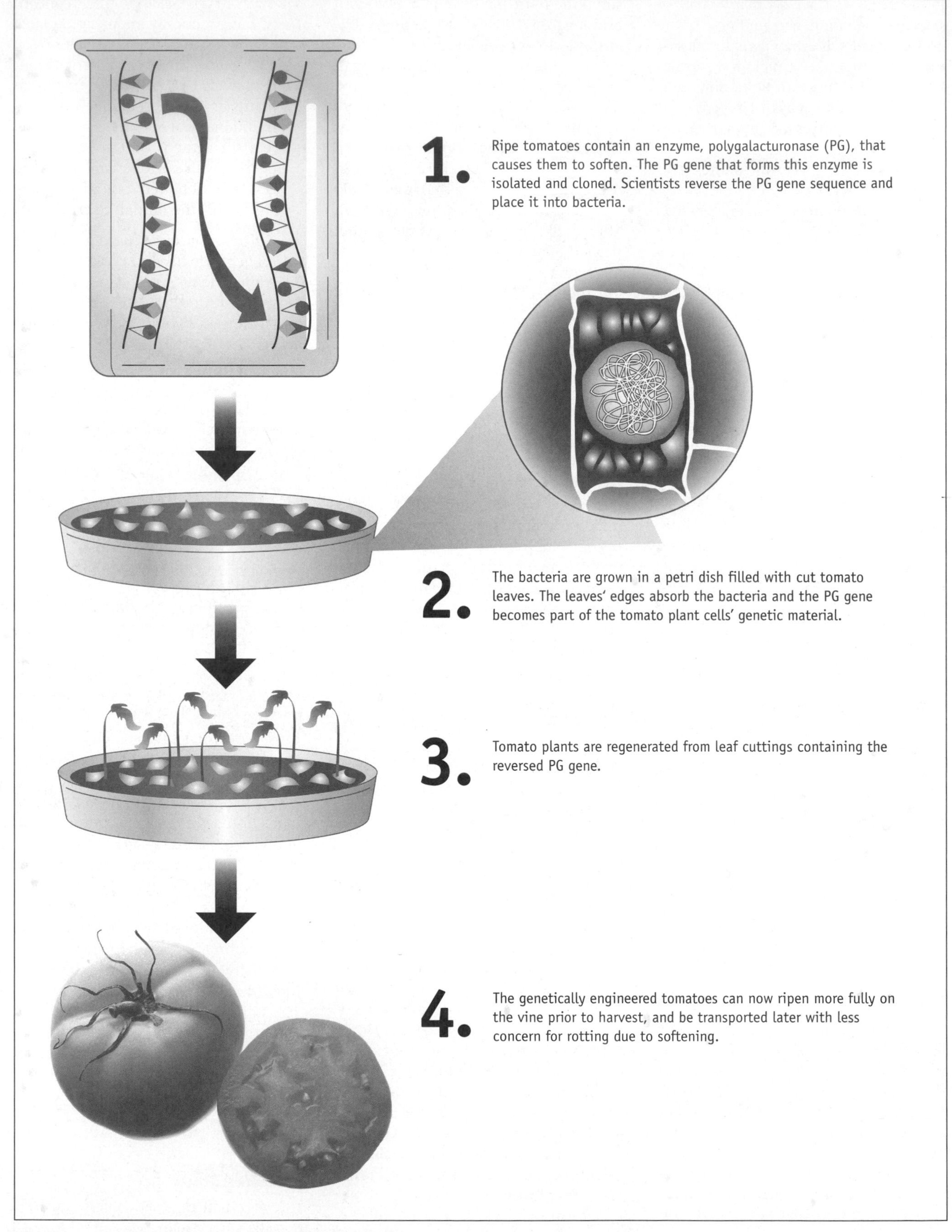

FIGURE 1-8 Genetically engineering a tomato to soften more slowly. (Adapted from FDA Consumer)

FIGURE 1-9 USDA's official organic seal.

KEY TERMS

Subjective tests Evaluations of food quality based on sensory characteristics and personal preferences as perceived by the five senses.

Objective tests Evaluations of food quality that rely on numbers generated by laboratory instruments, which are used to quantify the physical and chemical differences in foods.

transferring genes will convert a harmless organism into a hazardous one (40).

People who wish to avoid GMOs can ensure that their foods are free of this type of genetic modification by purchasing organic foods.

Organic Foods. Some people prefer to select "organic foods," a term that had no official definition until 2002 following the Organic Foods Production Act of 1990 (68). Terms commonly used in the marketplace that do not have official definitions or certification by the government include "free-range," "hormone-free," "natural," "organically produced," "pesticide free," or even "certified organic." Prior to 2002, products were often labeled "organic" by growers without any real certification, or they were certified by private agencies according to a patchwork of regulations that varied from state to state. Now, for a food to be labeled "organic" it has to fit one of the four official definitions listed by the U.S. Department of Agriculture and shown in Table 1-1. The USDA's definition of organic goes beyond just describing foods that are not sprayed with chemicals. The term "organic" now refers to food products that have been produced without most synthetic pesticides and fertilizers (including sewage sludge), crops that have not been genetically modified (no GMOs), livestock produced without antibiotics, and food products that have not been irradiated.

Organic Certification. The government agency certifying that a food is organic is the USDA, which labels such food products with the Organic seal shown in Figure 1-9. Only those food products that were organically grown or processed and certified by an accredited USDA organic certifying agent can carry the organic seal. Violators making false claims can be fined $10,000 per offense. USDA agents determine if food is organic by following the guidelines set by the USDA's Agricultural Marketing Service (AMS), published as the National Organic Program (NOP) in the Federal Register (December 21, 2000).

Budgetary Criteria

Cost is a very important limiting factor in food purchasing. In fact, food stamps obtained through the U.S. Department of Agriculture are limited by the "Thrifty Food Plan" that calculates what an average family needs to spend on food (74).

Cost helps determine the types of foods and brands that are bought and the frequency of restaurant patronage. People feeling financial strain may still eat beef, but they may choose ground beef over prime rib. "Can I afford this?" is a question that also applies to time, which can make convenience foods effectively more economical, even if the dollar price is higher. Budgeting and time management are discussed in greater detail in Chapter 4.

TABLE 1-1
The U.S. Government's Criteria for Defining Organic Food Products

Organic Term	*Definition*
100 percent organic	Cultivated and processed according to USDA standards.
Organic	95 percent or more of the ingredients are organic.
Made with organic ingredients	70 percent or more of the ingredients are organic.
Some organic ingredients	Less than 70 percent of the ingredients are organic.

Food Evaluation

The food industry uses an array of testing methods to measure the sensory factors in food selection and to evaluate food quality. These tests are conducted for research and development, product improvement, sales and marketing, quality assurance, and gauging consumer acceptance (35). Food evaluation is accomplished using both **subjective** and **objective tests** (22).

Subjective Evaluation

Subjective or sensory tests rely on the opinions of selected individuals (53, 60). There are two basic types of subjective tests: analytical and affective. Analytical tests are based on discernible differences, while affective tests are based on individual preferences (Figure 1-10). In

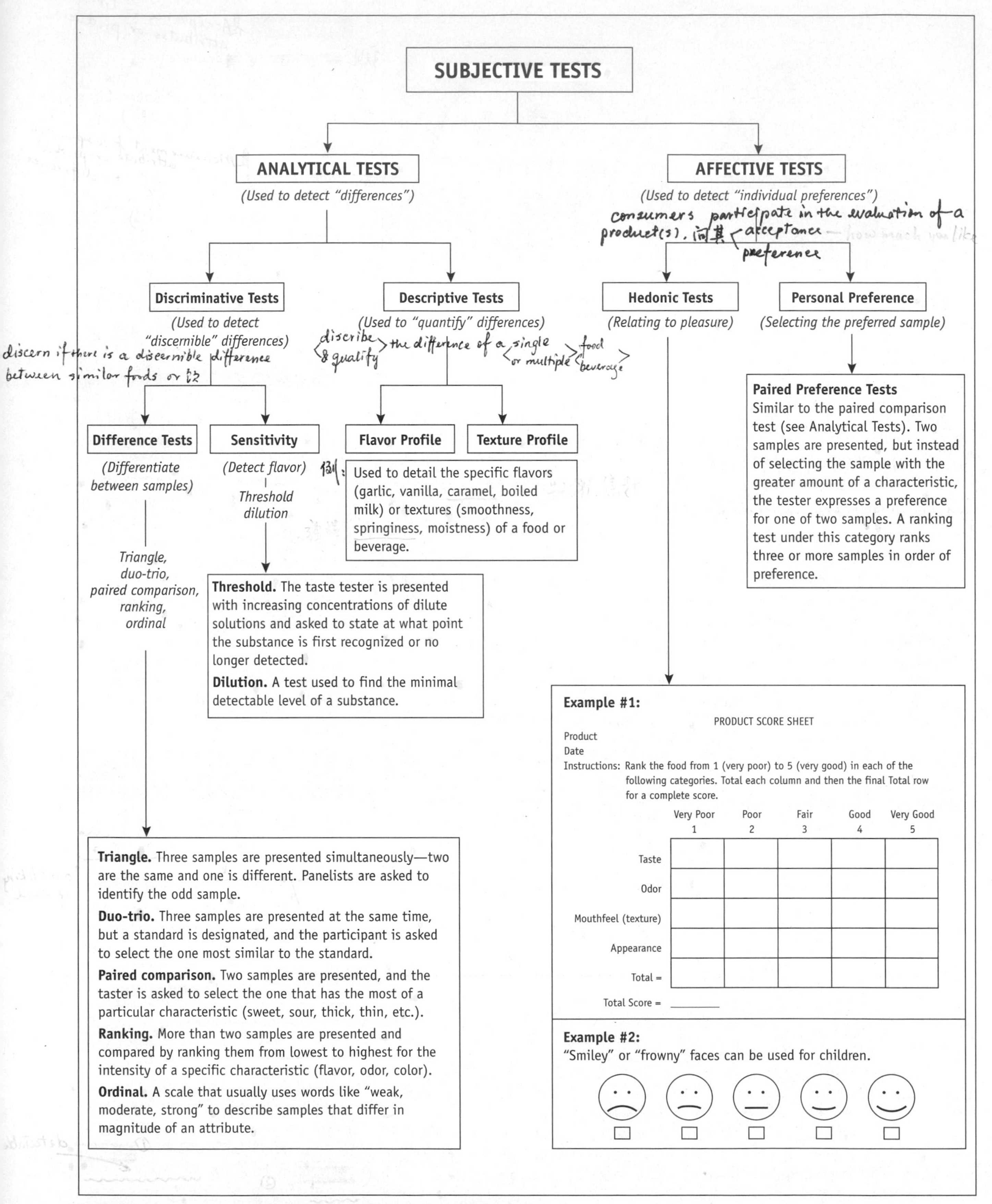

FIGURE 1-10 Summary of subjective tests for food evaluation.

both types of testing, food samples are presented to taste panel participants, who evaluate the foods according to specific standards for appearance, odor, taste, texture, and sound.

Taste Panels. The individuals on a taste panel can range from randomly selected members of the population to experts who are highly trained in tasting a particular food or beverage (Figure 1-11). Vintners and brewers rely on the latter types of skilled tasters to evaluate the proper timing for each step in the process of making wine or beer (55). The ability to detect slight differences in specific foods is a sought-after trait, prized so much that the taste buds of one gourmet ice cream taste expert are insured for $1 million. General taste panels usually consist of at least five people who meet the following criteria: They are free of colds, chew no gum immediately before testing, have not ingested any other food for at least one hour before testing, are nonsmokers, are not color blind, and have no strong likes or dislikes for the food to be tested. An equal distribution in gender is preferred, because women can usually detect sweetness better than men. Age distribution of the panel is also considered, since it may affect test results.

Sensory Computer Systems

FIGURE 1-11 **Taste test panel.**

Sample Preparation. The environment in which the taste panel evaluates foods or beverages is also carefully controlled (41). Panelists may be seated at tables, cubicles, or booths, and the food is presented in a uniform fashion. Food samples must be of the same size (enough for two bites), from the same portion of the food (middle versus outside), equally fresh, at the same temperature, and presented in containers or plates that are of the same size, shape, and color. White or clear containers are usually chosen so as not to influence panelists' perceptions of the food's color. Care is taken that the lighting in the room is uniform and that the ambient temperature is comfortable and the surroundings quiet and odor-free. Midmornings or midafternoons are considered the best times for sampling, because at these times people are not usually overly hungry or full. Samples are randomly coded and are kept to a reasonable number to avoid "taste fatigue." Room-temperature water or plain bread is made available for use between samples to prevent carryover tastes, and at least a 30-second rest period is scheduled between samples. Paper towels or napkins are provided, and since swallowing the food or beverage influences the taste of subsequent samples, small containers into which samples may be spit are provided.

Objective Evaluation

In objective evaluations, laboratory instruments instead of humans are used to measure the characteristics of foods quantitatively. The two major types of objective evaluation tests, physical and chemical, attempt to mimic the five senses, and serve as the basis of most objective food testing.

Physical tests measure certain observable aspects of food such as size, shape, weight, **volume, density**, moisture, texture, and **viscosity** (Chemist's Corner 1-2 on page 16) (22). Table 1-2 lists some of the laboratory instruments used to measure the various physical aspects of foods (Chemist's Corner 1-3). Figure 1-12 on page 17 shows an example of one such instrument.

The number of chemical tests available for use on foods is almost limitless, but Table 1-3 on page 17 lists some of these tests. Many are based on the work of the Association of Official Analytical Chemists (AOAC), which publishes a book on chemical tests, including those for determining various nutrient and non-nutrient substances in foods. Using instruments to evaluate foods provides more objective data than sensory testing, and is less costly and time consuming. Despite the benefits of objective tests, however, they cannot substitute for sensory testing by real human beings, who ultimately decide which foods and beverages they will select.

KEY TERMS

Volume A measurement of three-dimensional space that is often used to measure liquids.

■ ■ ■ ■

Density The concentration of matter measured by the amount of mass per unit volume. Objects with a higher density weigh more for their size.

■ ■ ■ ■

Viscosity The resistance of a fluid to flowing freely, caused by the friction of its molecules against a surface.

Chemist's Corner 1-2

Viscosity

Viscosity is a key term in rheology. Evaluation of certain foods is based on a branch of physics called rheology, which deals with the flow and deformation of both liquids and solids. The nature, concentration, and temperature of a liquid all affect its viscosity, which can be defined as "apparent" or "relative." Apparent viscosity is the time required for a substance like catsup to flow between two marks on the stem of a funnel. Relative viscosity compares the rate of a liquid's flow against a reference liquid (usually water).

Chemist's Corner 1-3

Analyzing Food with Chromatography

Compounds in foods can be measured using chromatography ("chrom" means color). It was first used at the turn of the century to separate plant pigments into different colored bands on a spectrum. In chromatography, a moving phase (gas or liquid) is passed over a solid, stationary phase (23). The constituents in a mixture are chemically separated when they adsorb onto the stationary phase.

Gas chromatography measures the contents of the gas produced when a food sample is evaporated. It is used to detect pesticides, cholesterol, certain fatty acids, and additives. In liquid chromatography, a liquid is created by making a solution out of the food sample. High-performance liquid chromatography (HPLC) is used to measure carbohydrates, lipids, vitamins, acids, pigments, flavor compounds, additives, and contaminants in food samples (32). Ion chromatography relies on ions being exchanged back and forth to determine sulfate, nitrate, organic acids in fruit juices, bread additives (benzoate, bromate), and sugar in various foods.

TABLE 1-2

Selected Physical Tests for Food Evaluation

Visual Evaluation	
Microscope	Used to observe microorganisms as well as starch granules, the grain in meats, the crystals of sugar and salt, the fiber in fruits and vegetables, and for any texture changes in processed foods.
Spectrophotometer	Measures color by detecting the amount and wavelength of light transmitted through a solution. Spectroscopy is based on the principle that the molecules in foods and beverages will absorb light at different wavelengths on the spectrum. The amount of absorption parallels the amount of substance found in the sample. Spectroscopy can be used to determine the amount of caffeine in coffee or the concentration of riboflavin (vitamin B_2) in milk.
Weight/Volume Measurements	
Weight	Weight is measured in pounds/ounces or milligrams/grams/kilograms.
Volume	Volume quantifies the area occupied by a mass, while density is the measure of mass (weight) in a given volume. Specific density relates a substance's density to an equal amount of water.
Texture Measurements	
Penetrometer	Simulates teeth biting into a food to measure its tenderness.
Warner-Bratzler Shear	Evaluates meat and baked product tenderness by measuring the force required to cut through a cylindrical sample.
Shortometer	Measures tenderness by determining the resistance of baked goods, such as cookies, pastries, and crackers, to breakage. Puncture testing evaluates the firmness of fruit or vegetable tissue.
Viscosity Measurements	
Line-spread test	Measures the consistency of batters and other viscous foods. Food is placed in a hollow cylinder in the middle of the spread sheet; the cylinder is then lifted, allowing the food to spread, and the spreading distance is measured in centimeters.
Viscometer (or viscosimeter)	Measures the viscosity of food such as pudding, sour cream, salad dressing, sauces, cream fillings, cake batters, and catsup.
Concentration Measurements	
Polarimeter or refractometer	Measures the concentration of various organic compounds, especially sugars, in solution by determining the angle (refractive index) of polarized light passed through the solution. Refractometers are commonly used to measure sugar concentrations in soft drinks. The Brix/acid ratio is used to measure the palatability of fruit juices that depends on the delicate balance between sweetness (sugars) and tartness (acid). This ratio is obtained by measuring the degrees Brix (determined by the use of a refractometer) divided by the total acid concentration (determined by acid titration)(42).

TABLE 1-3
Selected Chemical Tests for Food Evaluation

Test	Description
Benedict and Fehling tests	Determine the presence of sugars (reducing) such as lactose and maltose, which are more likely to be involved in a chemical reaction that turns food brown.
Chromatography	Identifies the presence of various compounds, especially those associated with flavor.
Electrophoresis	Specific proteins are characterized by passing an electrical field through a gel containing proteins and measuring the rates at which they migrate.
Enzyme tests	The peroxidase 1 test evaluates peroxidase enzyme activity in pasteurized foods: if the heat of pasteurization is adequate to destroy harmful bacteria, it should also inactivate the peroxidase enzyme. The effectiveness of briefly boiling food to destroy the enzymes responsible for vegetable deterioration can be determined by measuring the catalase enzyme activity.
Fuchsin test	Detects aldehydes in fats and oils.
Iodine value test	Measures the degree of unsaturation in fats.
Peroxide value test	Measures the extent of oxidation that occurred in a fat.
pH meter	Detects the amount of acidity or alkalinity in food mixtures or beverages.
Proximate analysis	A sequence of chemical tests to determine the macronutrient content of food.

Source: USDA

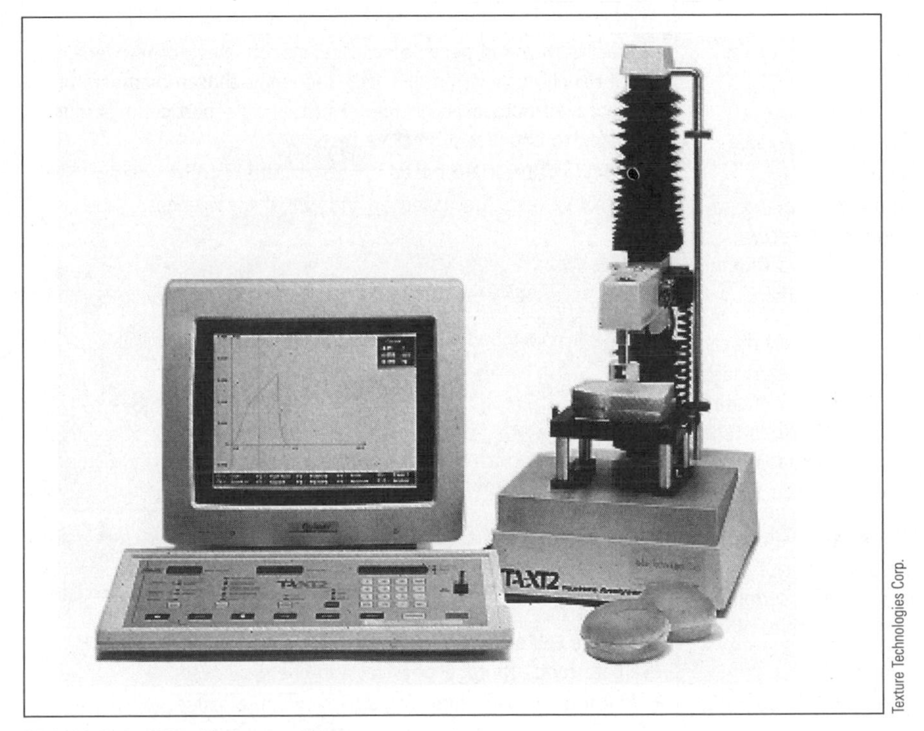

Texture Technologies Corp.

FIGURE 1-12 Texture analyzer.

PICTORIAL SUMMARY / 1: Food Selection and Evaluation

People choose foods that satisfy their senses of sight, smell, taste, touch, and hearing, their nutrient needs, cultural and religious values, psychological and social influences, and budget. As a result of today's fierce competition to develop new products, the food industry relies on evaluation methods to evaluate foods and consumer preferences.

FOOD SELECTION CRITERIA

Sensory Criteria. When most people choose a particular food, they evaluate it using the sensory reactions illustrated below rather than by considering its nutritional content.

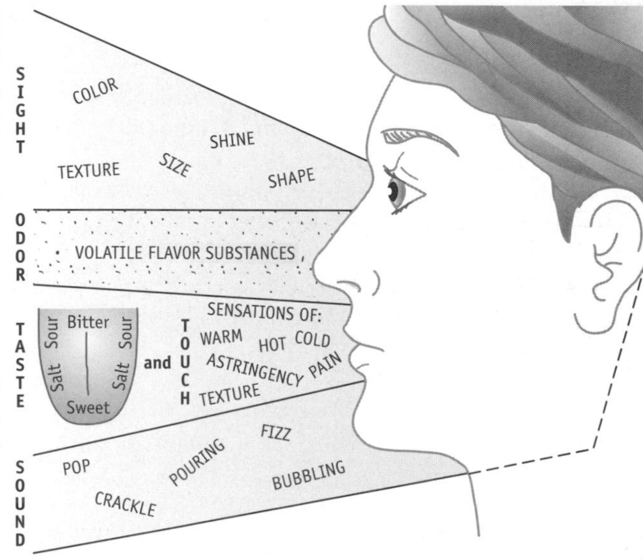

Nutritional Criteria. Over the past several decades, emerging awareness of health and nutrition has resulted in six out of ten consumers making a major change in their diets. Guidelines that reinforce an emphasis on better health through nutrition include the U. S. Government's Dietary Guidelines and the Food Guide Pyramid, developed in 1992.

Cultural and Religious Criteria. An increasingly diverse population, with greater access to travel and expanded global communication, has resulted in a huge increase in the variety of foods that are available in the United States today. Familiar taste preferences acquired in childhood as well as religious tenets affect many people's food habits throughout their lives.

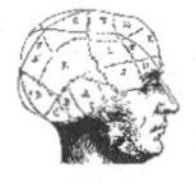

Psychological and Sociological Criteria. Advertising, social conscience, and peer pressure can all play a part in an individual's food choices. The controversies surrounding genetically engineered foods and organic foods are examples of how food products can be affected by these criteria.

Budgetary Criteria. Cost helps determine the types of food and brands that are bought and the frequency of restaurant patronage. A shortage of time for food preparation or eating out can result in greater use of convenience foods and "fast foods," even if they are often more expensive and less nutritious.

Food Guide Pyramid

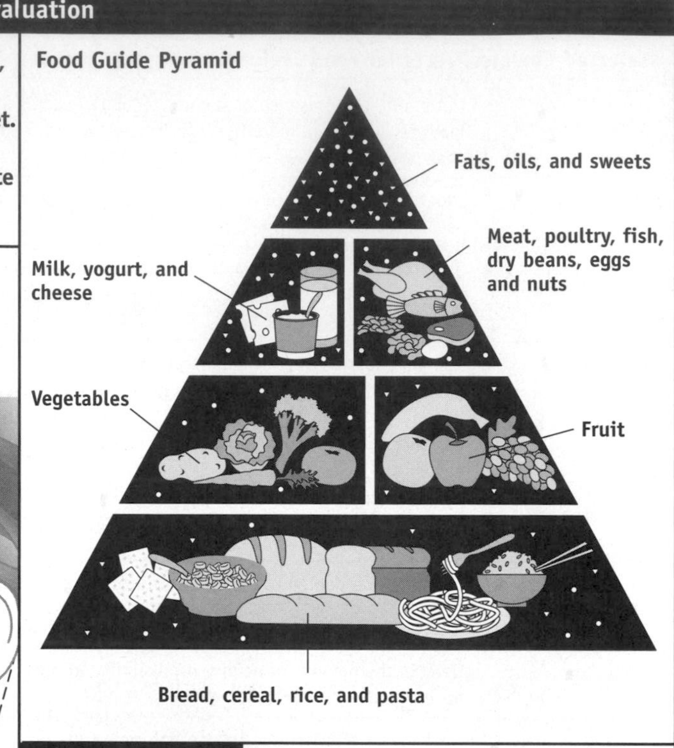

FOOD EVALUATION

Food manufacturers use both subjective and objective evaluation methods to help determine consumer acceptance of new products.

Subjective tests evaluate food quality by relying on the sensory characteristics and personal preferences of selected individuals. Taste panels, consisting of either randomly chosen members of the population or experts trained in tasting a particular product, are used to conduct subjective tests:

- *Analytical tests* are based on discernible differences.
- *Affective tests* are based on individual preferences.

Objective tests rely on laboratory methods and equipment to evaluate foods through physical and chemical tests.

- *Physical tests* measure certain observable aspects of food such as size, shape, weight, volume, density, moisture, texture, and viscosity.
- *Chemical tests* are used to determine the various nutrient and non-nutrient substances in foods.

REFERENCES

1. American Dietetic Association. Position of the American Association: Vegetarian diets. *Journal of the American Dietetic Association* 97(11):1317–1321, 1997.
2. Archer DL. The flap over functional foods. *Food Technology* 55(8):32, 2001.
3. Armtzen DJ. Edible vaccines. *Public Health Reports* 112:190–197, 1997.
4. Bialostosky K, and ST St.Jeor. The 1995 Dietary Guidelines for Americans. *Nutrition Today* 31(1):6–11, 1996.
5. Bodyfelt FW, J Tobias, and GM Trout. *The Sensory Evaluation of Dairy Products.* Van Nostrand Reinhold, 1988.
6. Borra ST, R Earl, and EH Hogan. Paucity of nutrition and food safety "news you can use" reveals opportunity for dietetics practitioners. *Journal of the American Dietetic Association* 98:190–193, 1998.
7. Bosley GC, and MG Hardinge. Seventh-Day Adventists: Dietary standards and concerns. *Food Technology* 46(10):112–113, 1992.
8. Camire ME, et al. IFT research needs report: Diet and health research needs. *Food Technology* 55(5):189–191, 2001.
9. Chapman N. Developing new foods through biotechnology. *Prepared Foods* 166(2):34, 1997.
10. Charley H. *Food Science.* Prentice-Hall, 1995.
11. Chaudry MM. Islamic food laws: Philosophical basis and practical implications. *Food Technology* 46(10):92–93, 1992.
12. Craig WJ. Phytochemicals: Guardians of our health. *Journal of the American Dietetic Association* 97(10 Suppl 2):S199–S204, 1997.
13. Dausch JG. The obesity epidemic: What's being done? *Journal of the American Dietetic Association* 102(5):638–639, 2002.
14. Deliza R, HJH MacFie, and D Hedderley. Information affects consumer assessment of sweet and bitter solutions. *Journal of Food Science* 61(5):1080–1084, 1996.
15. De Lumen BO, DC Krenz, and MJ Revilleza. Molecular strategies to improve the protein quality of legumes. *Food Technology* 51(5): 67–70, 1997.
16. De Roos KB. How lipids influence food flavor. *Food Technology* 51(5): 60–62, 1997.
17. Drewnowski A. Why do we like fat? *Journal of the American Dietetic Association* 97(7):S58–S62, 1997.
18. Eliasi JR, and JT Dwyer. Kosher and halal: Religious observances affecting dietary intakes. *Journal of the American Dietetic Association* 101(7):911–913, 2002.
19. Ensminger AH, et al. *Foods and Nutrition Encyclopedia.* CRC Press, 1994.
20. Food guide pyramid replaces the basic 4 circle. *Food Technology* 46(7):64–67, 1992.
21. Fuchs RL, and JD Astwood. Allergenicity assessment of foods derived from genetically modified plants. *Food Technology* 50(2):83–88, 1996.
22. Giese J. Measuring physical properties of foods. *Food Technology* 49(2):54–63, 1995.
23. Giese J. Instruments for food chemistry. *Food Technology* 50(2):72–77, 1996.
24. Glinsmann WH, SJ Bartholmey, and F Coletta. Dietary Guidelines for Infants: A timely reminder. *Nutrition Reviews* 54(1):50–57, 1996.
25. Godshall MA. How carbohydrates influence food flavor. *Food Technology* 51(1):63–67, 1997.
26. Gorman B. New products, new realities. *Prepared Foods New Products Annual* 160(4):14–16, 1991.
27. Hamilton EMN, and SAS Gropper. *The Biochemistry of Nutrition.* West, 1987.
28. Hardinge F, and M Hardinge. The vegetarian perspective and the food industry. *Food Technology* 46(10):114–116, 1992.
29. Harris SS. Health claims for foods in the international marketplace. *Food Technology* 46(2):92–94, 1992.
30. Hatchwell LC. Overcoming flavor challenges in low-fat frozen desserts. *Food Technology* 48(2):98–102, 1994.
31. Henkel J. Genetic engineering: Fast-forwarding to future foods. *FDA Consumer* 29(3):6–11, 1995.
32. Henshall A. Analysis of starch and other complex carbohydrates by liquid chromatography. *Cereal Foods World* 41(5):419, 1996.
33. Hoban TJ. How Japanese consumers view biotechnology. *Food Technology* 96(5):85–88, 1996.
34. Hochberg K. Health claims shift into high gear. *Prepared Foods New Products Annual* 160(4):47–53, 1991.
35. Hollingsworth P. Sensory testing and the language of the consumer. *Food Technology* 50(2):65–69, 1996.
36. Hollingsworth P. Growing nutraceuticals. *Food Technology* 55(9):22, 2001.
37. *How Are Americans Making Food Choices?* 1994 Update. The American Dietetic Association and International Food Information Council, 1994.
38. How Food Technology covered biotechnology over the years. *Food Technology* 51(4):68, 1997.
39. Huang Y, and CYW Ang. Vegetarian foods for Chinese Buddhists. *Food Technology* 46(10):105–108, 1992.
40. IFT Backgrounder. Genetically modified organisms (GMOs). *Food Technology* 54(1):42–45, 2000.
41. IFT Sensory Evaluation Division. Sensory evaluation guide for testing food and beverage products. *Food Technology* 35(11):50–59, 1981.
42. Jordan RB, RJ Seelye, and VA McGlone. A sensory-based alternative to Brix/acid ratio. *Food Technology* 55(6):36–44, 2001.

43. Katz F. Biotechnology—The new tools in food technology's tool box. *Food Technology* 50(11):63–65, 1996.
44. Katz F. The changing role of water binding. *Food Technology* 51(10):64–66, 1997.
45. Katz F. That's using the old bean. *Food Technology* 52(6):42–43, 1998.
46. Kemp SE, and GK Beauchamp. Flavor modification by sodium chloride and monosodium glutamate. *Journal of Food Science* 59(3):682–686, 1994.
47. Kendall P. Food biotechnology: Boon or threat? *Journal of Nutrition Education* 29:112–115, 1997.
48. Kilara A, and KK Iya. Food and dietary habits of the Hindu. *Food Technology* 46(10):94–102, 1992.
49. Kosher certification a marketing plus. *Food Technology* 51(4):67, 1997.
50. Leland JV. Flavor interactions: The greater whole. *Food Technology* 51(1):75–80, 1997.
51. Liu K, and EA Brown. Enhancing vegetable oil quality through plant breeding and genetic engineering. *Food Technology* 50(11):67–71, 1996.
52. McCullum C. The new biotechnology era: Issues for nutrition education and policy. *Journal of Nutrition Education* 29:116–119, 1997.
53. McEwan JA. Harmonizing sensory evaluation internationally. *Food Technology* 52(4):52–56, 1998.
54. McMahon KE. Consumer nutrition and food safety trends 1996. *Nutrition Today* 31(1):19–23, 1996.
55. Moskowitz HR. Experts versus consumers: A comparison. *Journal of Sensory Studies* 11:19–37, 1996.
56. Nagodawithana T. Flavor enhancers: Their probable mode of action. *Food Technology* 48(4):79–85, 1994.
57. Nakamura K. Progress in nutrition and development of new foods: A perspective of the food industry. *Nutrition Reviews* 50(12):488–489, 1992.
58. Nettleton JA. Warning: This food has been genetically engineered. *Food Technology* 51(3):20, 1997.
59. O'Donnell CD. Formulating products for ethnic tastes. *Prepared Foods* 166(2):36–44, 1997.
60. O'Mahony M. Sensory measurement in food science: Fitting methods to goals. *Food Technology* 49(4):72–82, 1995.
61. Onyenekwe PC, and GH Ogbadu. Radiation sterilization of red chili pepper (*Capsicum frutescens*). *Journal of Food Biochemistry* 19:121–137, 1995.
62. Painter J, JH Rah, and YK Lee. Comparison of international food guide pictorial representations. *Journal of the American Dietetic Association* 102(4):483–489, 2002.
63. Pike OA. The Church of Jesus Christ of Latter-Day Saints: Dietary practices and health. *Food Technology* 46(10):118–121, 1992.
64. Putman JJ, and JE Allshouse. *Food consumption, prices, and expenditures,* 1970–93. U.S. Department of Agriculture Statistical Bulletin no. 915. Washington, D.C.: GPO, 1994.
65. Regenstein JM, and CE Regenstein. The kosher food market in the 1990s—A legal view. *Food Technology* 46(10):122–124, 1992.
66. Schwartz NE, and ST Borra. What do consumers really think about dietary fat? *Journal of the American Dietetic Association* 97(suppl):S73–S75, 1997.
67. Sloan AE. Ethnic foods in the decade ahead. *Food Technology* 55(10):18, 2001.
68. Sloan AE. The natural and organic foods marketplace. *Food Technology* 56(1):27–37, 2002.
69. Stillman JA. Color influences flavor identification in fruit-flavored beverages. *Journal of Food Science* 58(4):810–812, 1993.
70. Szczesniak AS. Sensory texture profiling—Historical and scientific perspectives. *Food Technology* 52(8):54–57, 1998.
71. Tan T, V Schmitt, and S Isz. Electronic tongue: A new dimension in sensory analysis. *Food Technology* 55:1044–50, 2001.
72. Taylor AJ. Volatile flavor release from foods during eating. *Critical Reviews in Food Science and Nutrition* 36(8):765–784, 1996.
73. Turning sugar into sand and sweet citric acid. *Prepared Foods* 163(11):63–66, 1994.
74. U.S. Department of Agriculture. Center for Nutrition, Policy, and Promotion. *Official USDA Food Plans: Cost of food at home at four levels.* U.S. averages. Available: http://www.usda.gov/cnpp. May 1999.
75. U.S. Department of Agriculture. Human Nutrition Information Service. *USDA's Eating Right Pyramid.* Home & Garden Bulletin No. 249, Washington, D.C.: GPO, 1991.
76. U.S. Department of Agriculture and U.S. Department of Health and Human Services. *Nutrition and Your Health: Dietary Guidelines for Americans,* 5th edition. Home & Garden Bulletin No. 232, Washington, D.C.: GPO, 2002.
77. U.S. Department of Health and Human Services. *Healthy people 2000: National health promotion and disease prevention objectives.* DHHS Pub. No. (PHS)91–50213. Washington, D.C.: GPO, 1991.
78. Wilkinson JQ. Biotech plants: From lab bench to supermarket shelf. *Food Technology* 51(12): 37–42, 1997.
79. Will non-fat lead to less fat? *Food Engineering* 64(2):26, 1992.

WEBSITES

Find more information on the USDA's Dietary Guidelines:

www.usda.gov/cnpp/DietGd.pdf

Find more information on the USDA's Food Pyramid:

www.usda.gov/cnpp/pyramid2.htm

Find details about the USDA's Thrifty Food Plan:

www.usda.gov/cnpp/FoodPlans/TFP99/Index.htm.

Learn about the statistics on different ethnic groups in the United States and your state:

www.census.gov

Discover more about the National Organic Plan (NOP) from the USDA's website on the subject:

www.ams.usda.gov/nop

Find more information about food and nutrition from the USDA's Food and Nutrition Information Center (FNIC) located at the National Agricultural Library (NAL):

www.nal.usda.gov/fnic

Information about complementary and alternative medicine from the National Institutes of Health:

http://altmed.od.nih.gov

Enter keyword "Biotechnology" at USDA Biotechnology Information Center for more information on biotechnology:

www.agnic.org

Chemistry of Food Composition

"You are what you eat." When the 19th-century German philosopher Ludwig Feuerbach coined this phrase, he probably did not realize himself how true it was. Foods and people are composed of the same chemical materials, and there was a time when people served as nourishment to other animals in the food chain. All foods, including people, consist of six basic nutrient groups: water, carbohydrates, lipids, protein, vitamins, and minerals (Figure 2-1). Foods consist of varying amounts of these nutrients. For example, milk is 80 percent water, meats serve as primary sources of protein, potatoes and grains are rich in carbohydrates, and nuts are almost all fat. Actually, most foods contain a combination of the six major nutrient groups. Figure 2-2 shows the proportion of these six nutrients in humans.

Since people literally are what they eat, the main purpose of eating and drinking is to replace those nutrients used up in the body's maintenance, repair, and growth, and to obtain the calories (kcal) necessary for energy. Calories are fuel to the body, as gas is fuel to a car. Unlike cars, however, living organisms never shut down, even during sleep. Over half the calories used by the body, about 60 percent, are used solely for vital life functions such as maintaining body temperature, respiration, and heartbeat. Another 10 percent is used for digesting and absorbing the nutrients from food, and the remaining 30 percent, depending on the person, is used for physical activity.

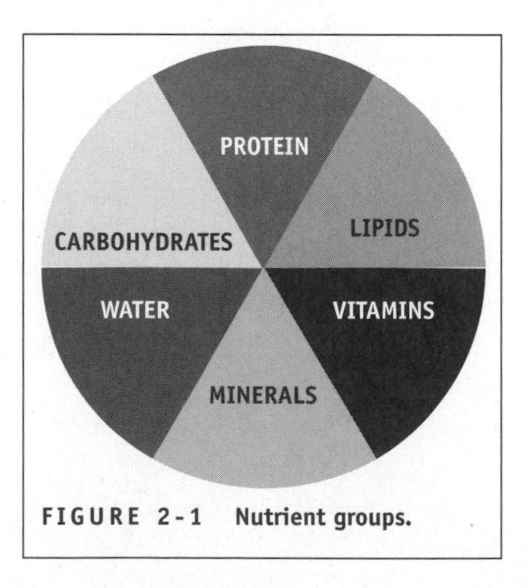

FIGURE 2-1 Nutrient groups.

Basic Food Chemistry

The body benefits from the energy and nutrients in foods at the cellular level. To comprehend how this occurs, it is necessary to know some biochemistry. While this may be a daunting term, it is simply the study of the chemistry that occurs within living organisms. Knowing something about biochemistry helps explain how nutrients from foods and beverages are assimilated in living systems.

A basic principle of biochemistry is that all living things contain six key elements (or **atoms**): carbon, hydrogen, nitrogen, oxygen, phosphorus, and sulfur (CHNOPS) (Chemist's Cor-

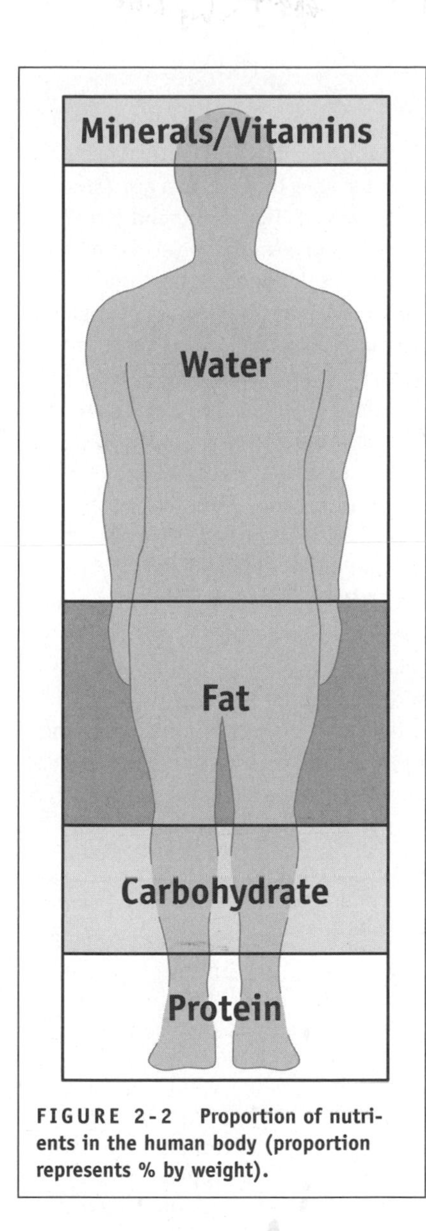

FIGURE 2-2 Proportion of nutrients in the human body (proportion represents % by weight).

ner 2-1). These are the building blocks of organic material, carbon-containing compounds that are often living material. All the elements have the capacity to join together with similar or different elements to produce **molecules** or **compounds**, which then combine to create all the substances on earth, including the focus of this book—foods and beverages.

Some organic compounds can be broken down by the body chemistry to create the energy, in the form of calories (kcal), needed to sustain life. Carbohydrates, fat, protein, and vitamins are organic molecules, but only the first three provide calories (kcal). A **gram** of carbohydrate or protein yields 4 calories (kcal), while the same amount of fat provides 9 calories (kcal). Alcohol, though not strictly a nutrient, yields 7 calories (kcal) per gram. Carbohydrates, fats, proteins, and alcohol are the only sources of calories (kcal) from the diet. No calories (kcal) are obtained from vitamins, minerals, or water. Both water and minerals are inorganic compounds, substances that do not contain carbon and cannot provide calories (kcal).

This chapter focuses on both organic and inorganic compounds by covering the six nutrient groups found in food and people: water, carbohydrates, fats, proteins, vitamins, and minerals. These **nutrients** serve as the foundation underlying all the principles in food and nutrition. They are discussed in this chapter with attention to what foods contain them, their chemical composition, and their functions in foods. Sugar (a form of carbohydrate) is discussed in greater detail in Chapter 9, and fat is covered further in Chapter 10.

Appendix H provides basic chemistry concepts to reinforce the chemistry of food found throughout this book.

Chemist's Corner 2-1

Atomic Structure

Everything physical in the universe is made from atoms, some of the smallest particles in existence. How are these smallest units of matter identified? By their number of protons and electrons. Protons are positively charged particles in the atom's nucleus, and electrons are negatively charged particles surrounding the nucleus like the rings around Saturn. The number of electrons on the outside ring of an atom dictates how many bonds that particular atom can form, and therefore, what kind of substance is formed. For example, the carbon (C) atom, the backbone of carbohydrates, fats, and proteins, usually forms four bonds. Nitrogen (N) is capable of forming three bonds, while oxygen (O) can form two, and hydrogen (H) only one (Figure 2-3). The bond holding atoms together through the sharing of electrons is called a covalent bond.

FIGURE 2-3 The number of bonds that selected atoms can form with other atoms.

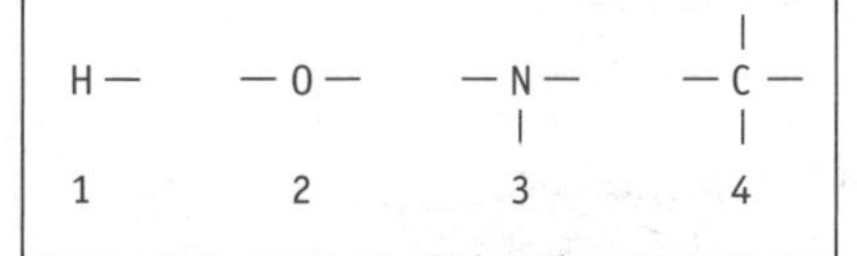

KEY TERMS

Atoms The basic building blocks of matter; individual elements found on the Periodic Table.

Molecule A unit composed of one or more types of atoms held together by chemical bonds.

Compound A substance whose molecules consist of unlike atoms.

Gram A metric unit of weight. One gram (g) is equal to the weight of 1 cubic centimeter (cc) or milliliter (ml) of water (under a specific temperature and pressure).

Nutrients Food components that nourish the body to provide growth, maintenance, and repair.

Water

Water is the simplest of all nutrients, yet it is the most important (21). Without it life could not exist. Life probably began in water billions of years ago, and it is still essential at every stage of growth and development. Water brings to each living cell the ingredients that it requires and carries away the end products of its life-sustaining reactions. The life functions of assimilating, digesting, absorbing, transporting, metabolizing, and excreting nutrients and their by-products all rely on water. The body's cells are filled with water and bathed in it. The human body averages 55 to 60 percent water, and losing as little as 10 percent of it can result in death. Water balance is maintained by drinking fluids and by eating foods, all of which naturally contain at least some water. A small portion is also obtained through metabolic processes.

Water Content in Foods

People get the water they need from foods and beverages. Although it may not always be apparent, many foods contain more water than any other nutrient. Foods range in water content from 0 to 95+ percent (Figure 2-4). Those that yield the most water are fruits and vegetables, ranging from 70 to 95 percent. Whole milk, which is over 80 percent water, and most meats, which average just under 70 percent water, are also high in water content. The foods with the least water include vegetable oils and dried foods such as grains and beans.

Free or Bound Water. The water in foods may be in either "free" or "bound" form. Free water, the largest amount of water present in foods, is easily separated from the food, while bound water is incorporated into the chemical structure of other nutrients such as carbohydrates, fats, and proteins. Examples would be the free water found in fruit and the bound water found in bread. Bound water is not easily removed and is resistant to freezing or drying. It also is not readily available to act as a medium for dissolving salts, acids, or sugars.

Composition of Water

Whether bound or free, water's chemical formula remains the same. Water is a very small molecule consisting of three atoms—one oxygen atom flanked by two hydrogen atoms (H_2O) (Chemist's Corner 2-2).

Measuring Calories. It takes heat, or its loss, to move the molecules of water through their different states, and this heat is commonly measured in the form of calories, expressed with a lowercase "c." As you learned in Chapter 1, this unit of measurement is equal to the amount of energy required to raise 1 gram of water 1° Celsius (measured between 14.5° and 15.5°C at normal atmospheric pressure). The energy values of foods are actually measured in thousands of calories, more accurately expressed as "kilocalories" and represented by the terms "kcal" or "Calories" with an uppercase "C." One kilocalorie (kcal) equals 1,000 calories.

In theory, the small "c" calorie is used by chemists, while the large "C" Calorie, or more commonly, kilocalorie (kcal), is the more accurate term for referring to the energy value of foods. In practice, however, calorie with a lowercase "c" is often used, especially with the general public, and it is assumed that those in the food field know that what is really meant is kilocalorie (kcal) or Calorie. Throughout this book, "calorie (kcal)" is used to represent the unit of measurement of the energy derived from food.

Specific Heat. It takes more energy to heat water than any other sub-

HOW & WHY? ?????????

If the atoms in H_2O do not change, how is water able to exist as a gas (steam or humidity), liquid, or solid (ice)? The distance between the molecules determines these differences, and the distances are influenced by temperature. At very low temperatures, ice forms as the water molecules line up very close together. Elevating the temperature increases the movement of the water molecules against each other, pushing them farther away from each other. When enough heat is applied, ice melts into a liquid. Continued heating transforms liquid water into a gas (steam) by giving the molecules freedom to move even farther apart (Figure 2-5). The variations of water from solid to liquid to gas are called changes in state. In spite of the obvious differences in these states, they do not involve any changes in the structure of the water molecule.

FIGURE 2-4 The percent water content of certain foods.

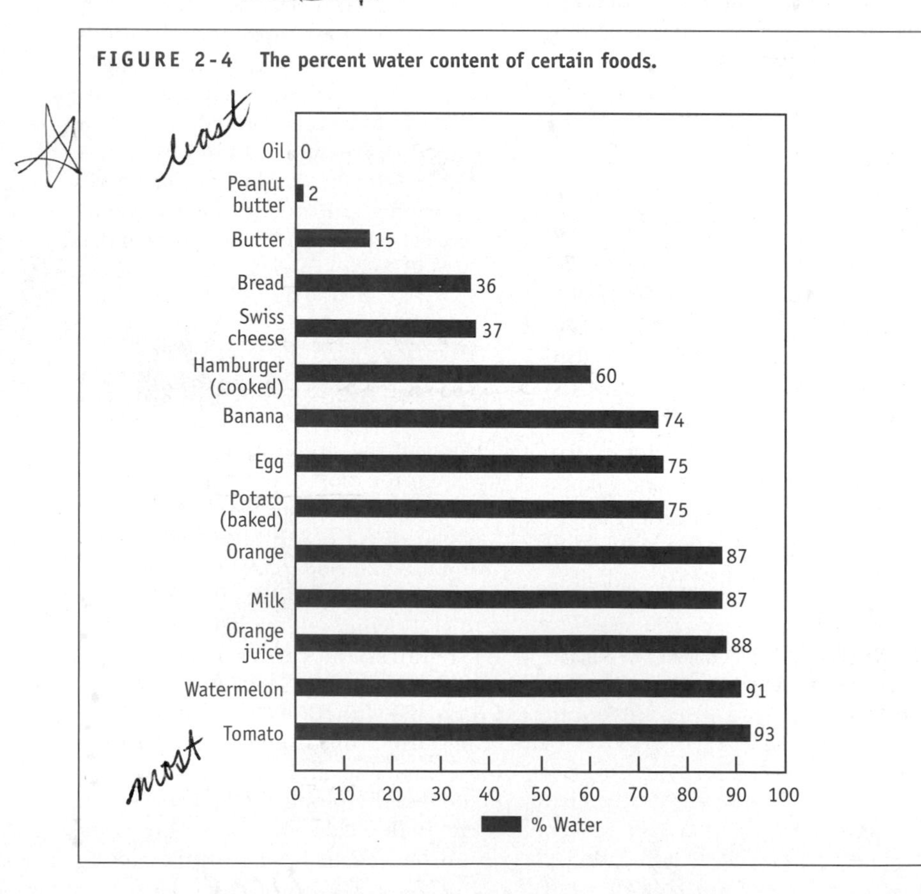

FIGURE 2-5 Molecular movement dictates whether a substance is a solid, liquid, or gas.

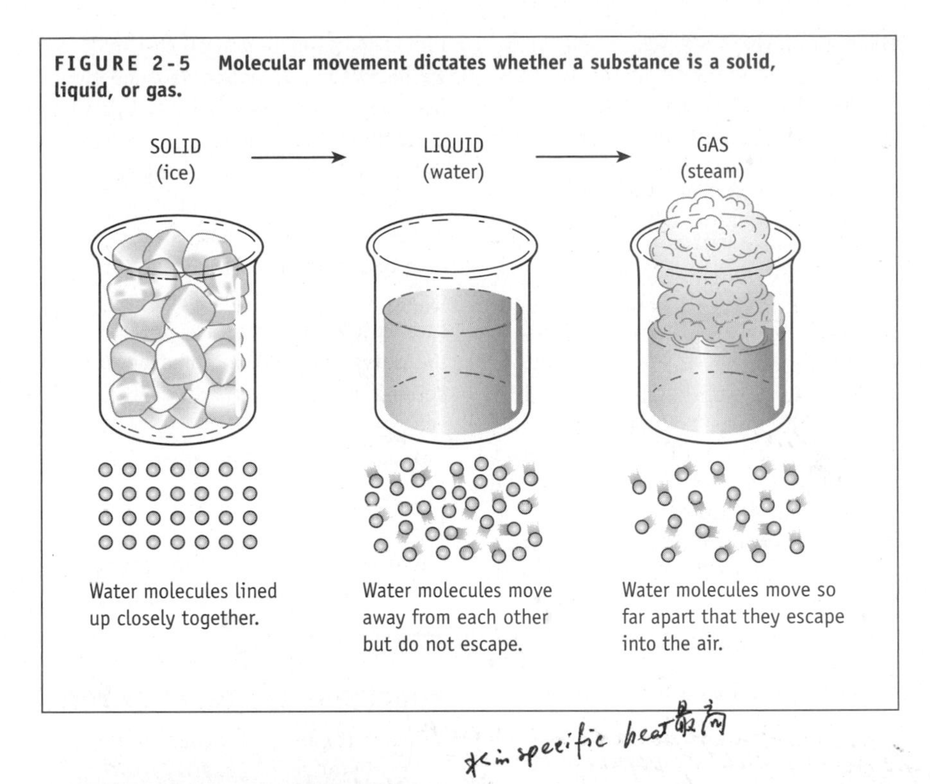

Chemist's Corner 2-2

The Chemical Structure of Water

Water has an overall neutral charge. This "neutrality" is derived from the combination of its two hydrogen (H^+) atoms, each with one positive charge, being balanced by the two negative charges of water's one oxygen (O^{-2}) atom. Overall, this gives water a neutral charge. However, it is not completely neutral in the sense that the water molecule has a negative pole and a positive pole, making it dipolar (Figure 2-6). Dipolar molecules have poles with partial charges that oppose each other and this dynamic contributes to some of water's very unique properties.

FIGURE 2-6 Two atoms of hydrogen combine with one oxygen atom, creating a dipolar molecule.

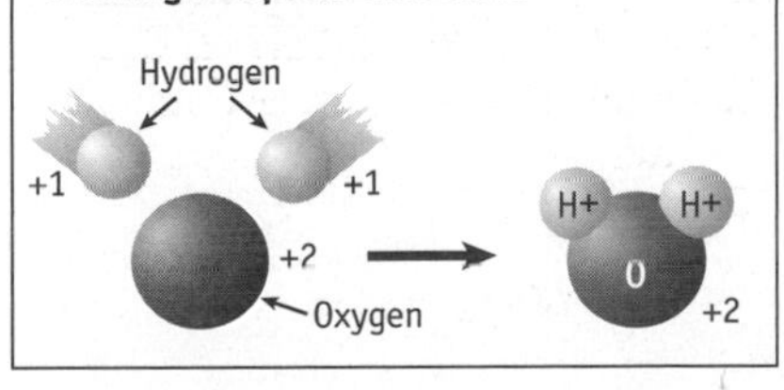

stance presently known. Water's high **specific heat** makes it unique compared to other compounds on earth. Given the same amount of heat, a metal pan or the oil in it will become burning hot, while water will become only lukewarm. This important characteristic of water enables animals, including people, with their high water content to withstand the very hot or cold temperatures sometimes found on earth. Water's specific heat of 1.00 (1 calorie will raise 1 gram of water 1°C) is used as the measure against which all other substances are compared. Similarly, water differs from other compounds in the amount of energy it takes to reach its specific freezing, melting, and boiling points.

Freezing Point. Winter in many parts of the world brings freezing temperatures. The temperature of winter can turn water into ice when its **freezing point** is reached. People living in snow country are particularly aware that when the temperature drops to freezing (32°F/0°C at normal atmospheric pressure), ice can form on the roads. The lower temperature decreases water's kinetic energy, or the energy associated with motion, which slows the movement of the water molecules until they finally set into a compact configuration. **Heat of solidification** occurs when at least 80 calories (0.08 kcal) of heat are lost per gram of water.

Unlike other substances, water expands and becomes less dense when completely frozen, which is why ice floats. The expansion of frozen water ruptures pipes and containers filled with water. It should come as no surprise, then, that it also ruptures the cells in plants and meats, diminishing the potential food's textural quality. Pure water freezes at 32°F (0°C), but adding anything else to the water changes its freezing point. The addition of **solutes** such as salt or sugar to water lowers the freezing point. Adding too much, however, slows the freezing process. Thus, frozen desserts made with large quantities of sugar take extra time to freeze.

Melting Point. If removing heat from water causes it to turn into ice, then returning the same 80 calories (0.08 kcal) of heat to a gram of ice will cause it to reach its **melting point** and turn it back into water. While the ice absorbs the 80 calories (0.08 kcal) of heat, there is no rise in tempera-

KEY TERMS

Specific heat The amount of heat required to raise the temperature of 1 gram of a substance 1°C.

■ ■ ■ ■

Freezing point The temperature at which a liquid changes to a solid.

■ ■ ■ ■

Heat of solidification The temperature at which a substance converts from a liquid to a solid state.

■ ■ ■ ■

Solute Solid, liquid, or gas compounds dissolved in another substance.

■ ■ ■ ■

Melting point The temperature at which a solid changes to a liquid.

ture. This **latent heat** does not register because it was put to work in moving the molecules of water far enough apart to change the physical structure of the solid ice into liquid water.

Boiling Point. Bubbles start to break the surface of water when it reaches 212°F (100°C) at sea level. This is its **boiling point**. The water will not get any hotter, nor will the food cook faster, no matter how much more heat is added, and this is why a slow rolling boil is often recommended. Keeping the temperature at a slow rolling boil is also more gentle on the foods and results in less evaporation.

The point at which water boils is reached when the pressure produced by steam, called vapor pressure, equals the pressure of the atmosphere pushing down on the earth. At this point the natural pressures of the atmosphere are not strong enough to push back the expanding gases of boiling water. Water requires 540 calories (0.54 kcal) of energy per gram to boil and vaporize.

This **heat of vaporization** is quite a bit higher than the 80 calories (0.08 kcal) needed to melt ice. Serious burns can result from human exposure to steam because the amount of heat required to produce it is so high.

Elevation and Boiling Point. Increasing the elevation decreases the boiling point of water. At sea level, water boils at 212°F (100°C), but drops 1°F for every 500-foot increase in altitude (an increase of 960 feet in elevation decreases water's boiling point by 1°C). Water boils in the mountains at lower temperatures than at sea level because there is less air and atmospheric pressure pushing down on the earth's surface. Steam has less resistance to overcome, and therefore occurs at lower temperatures. For example, at 7,000 feet water boils at 198°F (92°C). People at even higher elevations, such as on Mount Everest, could put a hand in a pan of boiling water and find it quite comfortable. Recipes are usually modified for elevations above 3,000 feet because the lower boiling temperature might affect ingredient actions and reactions.

Artificial changes in atmospheric pressure can be achieved by pressure cookers as well as by special equipment used only in the commercial food industry. A pressure cooker speeds up heating time by increasing atmospheric pressure to 15 pounds; thus water temperatures up to 240°F (116°C) can be achieved.

KEY TERMS

Latent heat The amount of energy in calories (kcal) per gram absorbed or emitted as a substance undergoes a change in state (liquid/solid/gas).

■ ■ ■ ■

Boiling point The temperature at which a heated liquid begins to boil and changes to a gas.

■ ■ ■ ■

Heat of vaporization The amount of heat required to convert a liquid to a gas.

Hard vs Soft Water. Most water is not pure water, but contains dissolved gases, organic materials, and mineral salts from the air and soil. The minerals in water determine whether it is hard or soft water. Hard water contains a greater concentration of calcium and magnesium compounds, while soft water has a higher sodium concentration. The temperatures at which water freezes, melts, or boils remain constant regardless of whether it is hard or soft water.

HOW & WHY? ?????????

How can you tell if water is hard or soft? Hard water leaves a ring in the bathtub, a grayish sediment on the bottom of pans, and a grayish cast in washed whites. Although permanently hard water cannot be softened by boiling or distilling, it can be converted by a water softener, which works by exchanging sodium for calcium and magnesium. Another way to determine if water is hard or soft is to call the local water department and ask how much calcium carbonate (ppm—parts per million) is in the water. The following breakdown defines whether it is hard or soft.

Water Hardness	*Calcium Carbonate (ppm)*
Soft	0–50
Medium	50–100
Hard	100–200
Very Hard	200+

Functions of Water in Food

Water is the most abundant and versatile substance on earth. Among its many uses in food preparation, its two most important functions are as a transfer medium for heat and as a universal solvent. In addition, it is important as an agent in chemical reactions, and is a factor in the perishability and preservation of foods (Table 2-1).

Heat Transfer. Water both transfers and moderates the effects of heat. A potato heated by itself in a pan will burn, but surrounding that same potato with water ensures that the heat will be evenly distributed. Water also transfers heat more efficiently, which explains why a potato heats faster in boiling water than in the oven. Because water has a higher specific heat than other substances, it

TABLE 2-1
Functional Properties of Water in Food

Heat Transfer	*Universal Solvent*	*Chemical Reactions*
Moist Heating of Foods	Solution	Ionization
Boiling	Colloidal dispersion	Changes in pH
Simmering	Suspension	Salt formation
Steaming	Emulsion	Hydrolosis
Stewing		CO_2 release
Braising		Food preservation

buffers changes in temperature. More energy is needed to increase the temperature of 1 gram of water than 1 gram of fat. For example, the specific heat of oil is 0.5; thus it heats twice as fast as water when given the same amount of heat.

Moist-Heat Cooking Methods. Almost half of the methods used to prepare foods rely on water to transfer heat, and these are known collectively as moist-heat methods. The major moist-heat methods discussed in this book include boiling, simmering, steaming, stewing, and braising. Dry-heat methods use heat in the form of radiation and include baking, grilling, broiling, and frying. Microwaving uses both dry- and moist-heat methods; microwaves are actually a form of radiation that heats the water molecules in foods, which then heat the food itself. Microwaving techniques are discussed throughout this book under moist-heat preparation methods.

Universal Solvent. The many biochemical interactions occurring in living organisms—human, animal, and plant—could not occur in the absence of a **solvent** environment. Water is considered to be the earth's universal solvent. The fluid substance, mostly water, within and around the cell is a solvent that contains many dissolved solutes.

Combining a solvent and a solute results in either a solution, a colloidal dispersion, a suspension, or an emulsion. These mixtures differ from each other based on the size or **solubility** of their solutes.

Solution. In a **solution**, the molecules of the solute are so small that they completely dissolve and will not **precipitate** from their fluid medium. They cannot be separated by filtering, but can sometimes be removed by **distillation**. If a substance is able to enter into a solution by dissolving, it is considered to be soluble.

Much of what people perceive as the taste of foods depends on the formation of solutions with solutes in foods such as sugars, salts, acids, and other flavor compounds, and their resulting enhanced ability to attach to flavor receptors. Water also forms solutions with minerals and water-soluble vitamins (B complex and C). This increases the likelihood that these minerals and vitamins may leach out of foods into cooking water, which is often discarded, causing nutrients to be lost. To the delight of tea and coffee lovers, water can also dissolve caffeine and other flavorful compounds from tea leaves and coffee beans. Higher temperatures increase the amount of solute that will dissolve in the solvent, which explains why very hot water is used for making coffee and tea.

The solubility of a substance is measured by the amount of it in grams that will dissolve in 100 ml of solvent. Raising the temperature allows more solute to dissolve in the solvent, creating a **saturated solution**. Increasing the temperature even higher sets the stage for a **supersaturated solution**, which is very unstable and must be cooled very slowly to avoid having the solute precipitate out or crystallize. Many candies, including fudge, rely on the creation of supersaturated solutions.

Colloidal Dispersions. Not all particles dissolve readily or homogeneously. Some particles, called colloids (e.g., proteins, starches, and fats), never truly dissolve in a solvent, but remain in an unstable **colloidial dispersion**. Unlike solutes in solutions, which completely dissolve, colloids do not due to their large size, but neither do they noticeably change the dispersion's freezing or boiling point. Examples of different types of dispersions include a solid in a liquid, a liquid in another liquid (salad dressing) or solid (jam, gelatin, cheese, butter), and a gas that can be incorporated into either a liquid (egg white or whipped cream foams) or a solid (marshmallow). Two types of dispersions are **suspensions** and **emulsions**.

- *Suspension.* Mixing cornstarch and water results in a suspension in which the starch grains float within the liquid. Cornstarch suspensions are often used in Chinese cooking and give Chinese food its particular shiny appearance and smooth mouthfeel.
- *Emulsion.* Another type of colloidal dispersion involves water-in-oil (w/o) or oil-in-water (o/w) emulsions. Neither water nor fats will dissolve in each other, but they may

KEY TERMS

Solvent A substance, usually a liquid, in which another substance is dissolved.

Solubility The ability of one substance to blend uniformly with another.

Solution A completely homogeneous mixture of a solute (usually a solid) dissolved in a solvent (usually a liquid).

Precipitate To separate or settle out of a solution.

Distillation A procedure in which pure liquid is obtained from a solution by boiling, condensation, and collection of the condensed liquid in a separate container.

Saturated solution A solution holding the maximum amount of dissolved solute at room temperature.

Supersaturated solution An unstable solution created when more than the maximum solute is dissolved in solution.

Colloidal dispersion A solvent containing particles that are too large to go into solution but not large enough to precipitate out.

Suspension A mixture in which particles too large to go into solution remain suspended in the solvent.

Emulsion A liquid dispersed in another liquid with which it is usually immiscible (incapable of being mixed).

become dispersed in each other, creating an emulsion. Examples of food emulsions include milk, cream, ice cream, egg yolk, mayonnaise, gravy, sauces, and salad dressings (7). These and other emulsions can be separated by freezing, high temperatures, agitation, and/or exposure to air (11). Emulsions are discussed in more detail in Chapter 10.

Colloidal dispersions, which are unstable by nature, can be purposely or accidentally destabilized, causing the dispersed particles to aggregate out into a partial or full gel, a more-or-less rigid protein structure. An example of this is seen when milk is heated; its unstable water-soluble milk proteins precipitate out and end up coating the bottom of the pot, creating a **flocculation**. Full gels such as yogurt and cheese are also made possible by the colloidal nature of milk.

Chemical Reactions. Water makes possible a vast number of chemical reactions that are important in foods. These include ionization, pH changes, salt formation, hydrolysis, and the release of carbon dioxide.

KEY TERMS

Flocculation A partial gel in which only some of the solid particles colloidally dispersed in a liquid have solidified.

■ ■ ■ ■

Ionize To separate a neutral molecule into electrically charged ions.

■ ■ ■ ■

Electrolyte An electrically charged ion in a solution.

■ ■ ■ ■

pH scale Measures the degree of acidity or alkalinity of a substance, with 1 the most acidic, 14 the most alkaline, and 7 neutral.

■ ■ ■ ■

Hydrolysis A chemical reaction in which water ("hydro") breaks ("lysis") a chemical bond in another substance, splitting it into two or more new substances.

Ionization. When particles dissolve in a solvent, the solution is either molecular or ionic in nature. Molecular solutions are those in which the dissolved particles remain "as is" in their molecular form. An example would be the dissolving of a flavored sugar mix in water to make a beverage. The sugar molecules remain unchanged in solution. Ionic solutions occur when the solute molecules **ionize** into electrically charged ions or **electrolytes**. When salt, or sodium chloride (NaCl), is dissolved in water, it ionizes into the individual ions of sodium (Na^+) and chloride (Cl^-).

This chemical reaction is written:

$$NaCl \longrightarrow Na^+ + Cl^-$$

Changes in pH—Acids and Bases. Acids are substances that donate hydrogen (H^+) ions, and bases provide hydroxyl (OH^-) ions. Another defining difference between acids and bases is that acids are proton donors, while bases are proton receptors. The **pH scale** ("pH" stands for "power" and "hydrogen") is a numerical representation of the hydrogen (H^+) ion concentration in a liquid. A solution with a pH of under 7 is considered acidic, while anything over 7 is alkaline or basic. A pH of 7 indicates that the solution is neutral, containing equal concentrations of hydrogen (H^+) ions and hydroxyl (OH^-) ions. Each number on the scale represents a tenfold change in degree of acidity (Chemist's Corner 2-3).

Water is naturally neutral, but tap water is normally adjusted to be slightly alkaline (pH 7.5 to 8.5), because acidic water causes pipe corrosion. Overly alkaline water, however, results in deposits of carbonates that may block water pipes. Many coffee connoisseurs prefer distilled water for making coffee because of its neutral nature.

Salt Formation. The universal solvent of water makes it possible to form salts, which occurs when a positive ion combines with a negative ion, as long as it is neither a hydrogen (H^+) nor hydroxyl (OH^-) ion. The primary example is sodium chloride (Na^+Cl^-), resulting from sodium (Na^+) combining with chloride (Cl^-). Salts can also be formed by combining an acid and a base, or a metal and a nonmetal.

Chemist's Corner 2-3

The Logarithmic pH Scale

The concentrations of ions in water are so small that it is awkward to speak or write about these concentrations using ordinary words. For example, water has a hydrogen ion concentration of 0.0000001 moles per liter, which is translated in terms of the negative logarithm of the hydrogen ion concentration (16):

$$pH = -\log (H^+)$$

pH is also understood in a scale of 1 to 7. The expression of 0.0000001 moles per liter using a decimal can also be written in its exponential form of 1×10^{-7} and then placed into the negative logarithm to yield a pH of 7:

$$pH = -\log(1\times10^{-7}) = \log 1/(1\times10^{-7}) = \log(1\times10^{7}) = 7$$

These single-digit numbers are much easier to fathom as long as it is understood that each number in the pH scale represents a tenfold change in the degree of acidity.

Metal salts include potassium fluoride (K^+F^-) and lithium bromide (Li^+Br^-).

Hydrolysis. Countless chemical reactions rely on **hydrolysis**. Figure 2-7 shows an example of how a water molecule is used to break a sugar into smaller molecules, and the hydrolysis of a lipid is illustrated in Figure 2-8 and Figure 2-9. Just a few of the hydrolysis applications used in the food industry include breaking down cornstarch to yield corn syrup, dividing table sugar into its smaller components to create another sugar helpful in the manufacture of some candies (see Chapter 9), and creating protein hydrosylates, smaller molecules derived from protein hydrolysis, to add to foods to improve flavor, texture, foaming abilities, and nutrient content.

Carbon Dioxide Release. Many baked products are allowed to rise before baking. One of the agents making this possi-

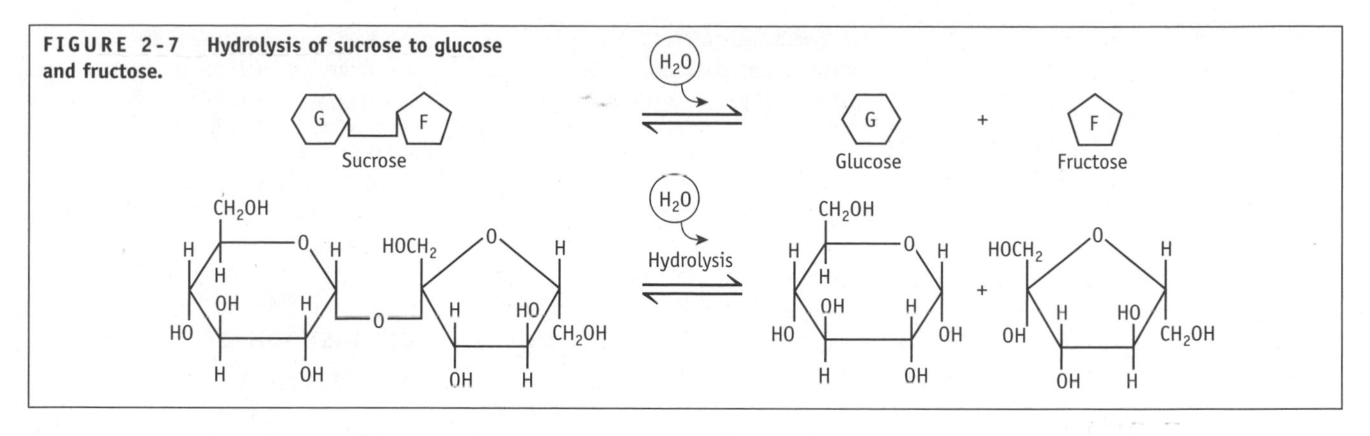

FIGURE 2-7 Hydrolysis of sucrose to glucose and fructose.

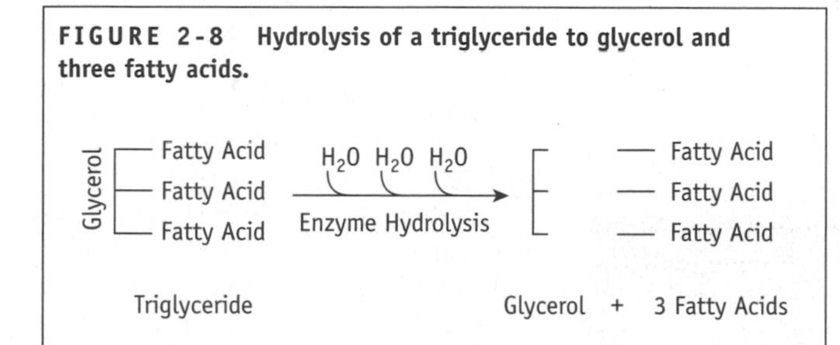

FIGURE 2-8 Hydrolysis of a triglyceride to glycerol and three fatty acids.

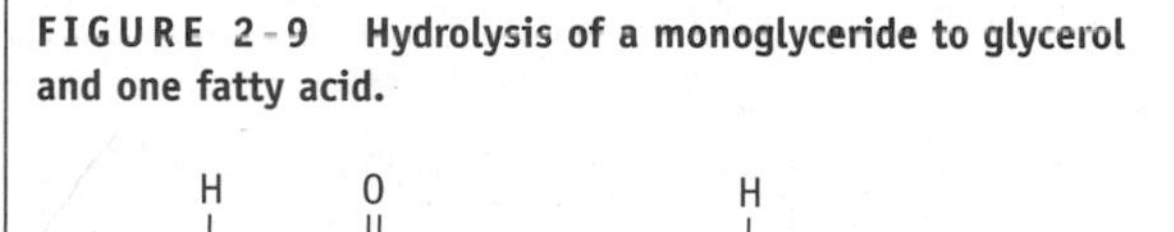
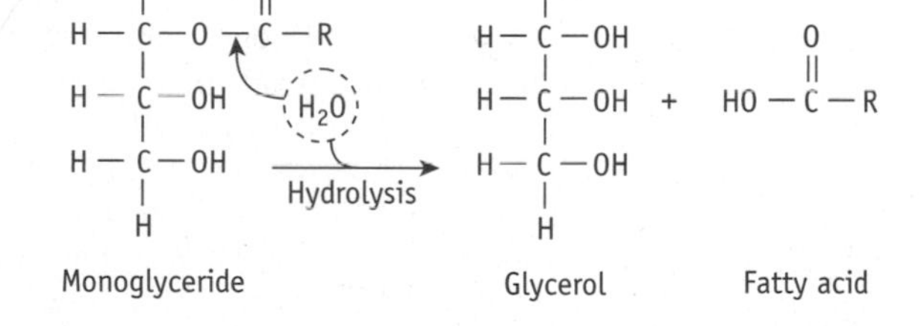

FIGURE 2-9 Hydrolysis of a monoglyceride to glycerol and one fatty acid.

ble is baking powder, which is a combination of baking soda and acid. It is only when baking powder is combined with water that the gas carbon dioxide is released, which causes baked products to rise. The chemical reaction is a two-step process:

$NaHCO_3 + HX \xrightarrow{water} NaX + H_2CO_3$
baking soda + acid ——> salt + carbonic acid
$H_2CO_3 \longrightarrow H_2O + CO_2$
carbonic acid ——> water + carbon dioxide

Food Preservation. While water is essential to the chemical reactions on which living things and many foods depend, it is also important for the life of **microorganisms** such as bacteria, molds and yeasts. The actions of these microorganisms on food cause deterioration and decay. Atmospheric humidity alone increases the likelihood of foods degenerating. For example, a relative humidity of 75 percent or more, especially if combined with warm temperatures, encourages the growth of microorganisms. Thus, removing water from fruits, vegetables, meats, and herbs was among the earliest forms of food preservation. Without water, microorganisms cannot survive, so limiting the amount of water available to them inhibits their growth. Conversely, water in a cool environment helps preserve the freshness of fruits and vegetables by preventing dehydration, hence those artificial "rain" showers we see in supermarket produce displays.

Removing dirt and other debris from fruits and vegetables by rinsing them in water or even washing them with detergent eliminates many microorganisms. Detergents lower the surface tension of water, which improves its ability to act as a cleansing agent.

Water Activity. A food's **water activity** or water availability determines its perishability. Foods high in water, such as milk, meat, vegetables, and fruits, are much more prone to microbial spoilage than drier foods such as grains, nuts, dried milk, dried beans, or dried fruits (37). Moreover, once deterioration sets in, the putrefying food itself releases water, which fuels the further growth of microorganisms. Pure water has a water activity of 1.0. Adding any solute to it decreases its water activity to below 1.00 (Chemist's Corner 2-4 on page 30) (36). Water molecules orient themselves around any added solute, making them unavailable for microbial growth. Solutes such as sugar and salt added to jams and cured meats inhibit microbial growth by lowering water activity. The food industry makes water unavailable to microorganisms by using solutes such as salt, sugars, glycerol, propylene glycol, and modified corn syrups (3).

KEY TERMS

Microorganism Plant or animal organism that can only be observed under the microscope—bacteria, mold, yeast, virus, or animal parasite.

■ ■ ■ ■

Water activity (a_w) Measures the amount of available (free) water in foods. Water activity ranges from 0 to the highest value of 1.00, which is pure water.

KEY TERMS

Osmosis the movement of a solvent through a semipermeable membrane to the side with the higher solute concentration, equalizing solute concentration on both sides of the membrane.

Osmotic pressure The pressure or pull that develops when two solutions of different solute concentration are on either side of a permeable membrane.

Osmosis and Osmotic Pressure. Salting has been used as a method of preserving foods for thousands of years because salt draws water out of foods and to itself. Of course, ancient peoples did not understand the process of **osmosis**, which causes water to be drawn to solutes; all they knew was that salting kept their food edible for long periods of time. Part of this process depends on the fact that water passes through membranes freely, but most solutes do not (Figure 2-10). The side of the membrane with more solute has more **osmotic pressure** and draws the necessary water to that side to dilute its solute concentration. Any bacteria contacting heavily salted food lose their water by the same process and die by dehydration. Beef jerky is the result of the combined processes of salting, smoking, and drying of meats. The high sugar concentration of jams and jellies acts to preserve them in the same way as salt on meats.

Chemist's Corner 2-4

Measuring Water Activity

As free water decreases, so does the water activity. Water activity is measured by dividing the vapor pressure exerted by the water in food (in solution) by the vapor pressure of pure water (P_w), which is equal to 1.00 (3).

Carbohydrates

Foods High in Carbohydrates

Carbohydrates are the sugars, starches, and fibers found in foods. Plants are the primary source of carbohydrates, with the exception of milk, which contains a sugar called lactose. The muscles from animals can also contain some carbohydrate in the form of glycogen, but much of this is converted to a substance called lactic acid during slaughter. Most carbohydrates are stored in the seeds, roots, stems, and fruit of plants. Common food sources for carbohydrates include grains such as rice, wheat, rye, barley, and corn; legumes such as beans, peas, and lentils; fruits; and some vegetables, such as carrots, potatoes, and beets. Sugar cane and sugar beets provide table sugar, while honey is derived from the nectar of flowers.

Composition of Carbohydrates

The elements making up carbohydrates are carbon (C), hydrogen (H), and oxygen (O). The word "carbohydrate" can be broken down into "carbon" (C) and "hydrate" (H_2O). This leads to the basic chemical formula of carbohydrates, which is $C_n(H_2O)_n$, where n stands for a number ranging from 2 into the thousands. Carbohydrates are found primarily in green plants, where they are synthesized through the process of photosynthesis. The chemical reaction of photosynthesis is written:

carbon dioxide + water + sun energy ——> glucose + oxygen

$$6\,CO_2 + H_20 + \text{sun energy} \longrightarrow C_6H_{12}O_6 + 6O_2$$

The carbon, hydrogen, and oxygen atoms making up carbohydrates are arranged in a basic unit called a saccharide. Carbohydrates are classified into monosaccharides, disaccharides, oligosaccharides, and polysaccharides, depending on the type and number of

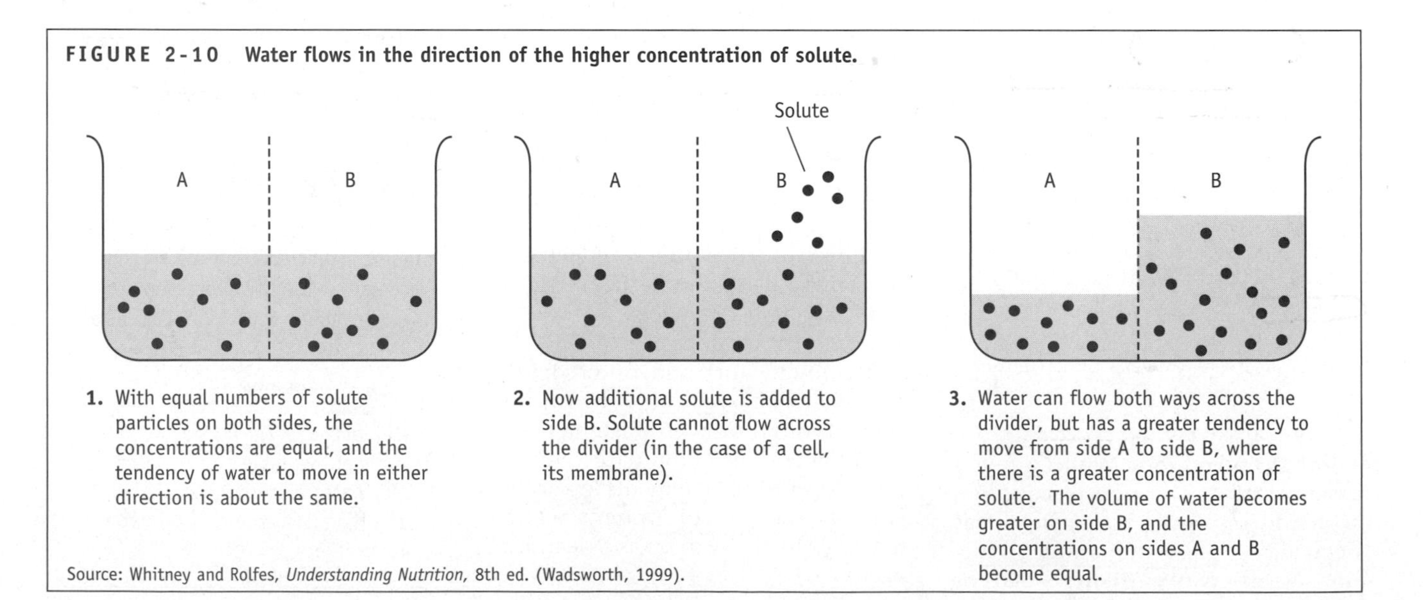

FIGURE 2-10 **Water flows in the direction of the higher concentration of solute.**

1. With equal numbers of solute particles on both sides, the concentrations are equal, and the tendency of water to move in either direction is about the same.
2. Now additional solute is added to side B. Solute cannot flow across the divider (in the case of a cell, its membrane).
3. Water can flow both ways across the divider, but has a greater tendency to move from side A to side B, where there is a greater concentration of solute. The volume of water becomes greater on side B, and the concentrations on sides A and B become equal.

Source: Whitney and Rolfes, *Understanding Nutrition,* 8th ed. (Wadsworth, 1999).

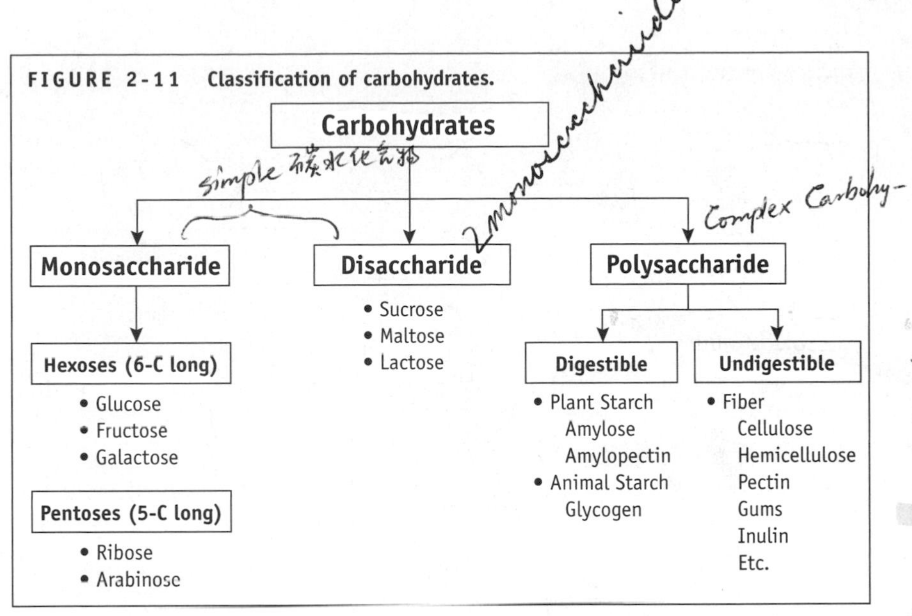

FIGURE 2-11 Classification of carbohydrates.

saccharide units they contain (Figure 2-11).

- Monosaccharides (one saccharide)
- Disaccharides (two monosaccharides linked together)
- Oligosaccharides ("few"—three to ten—monosaccharides linked together; these are not as common in foods as either mono- or disaccharides)
- Polysaccharides ("many" monosaccharides linked together in long chains; these include starch and fibers)

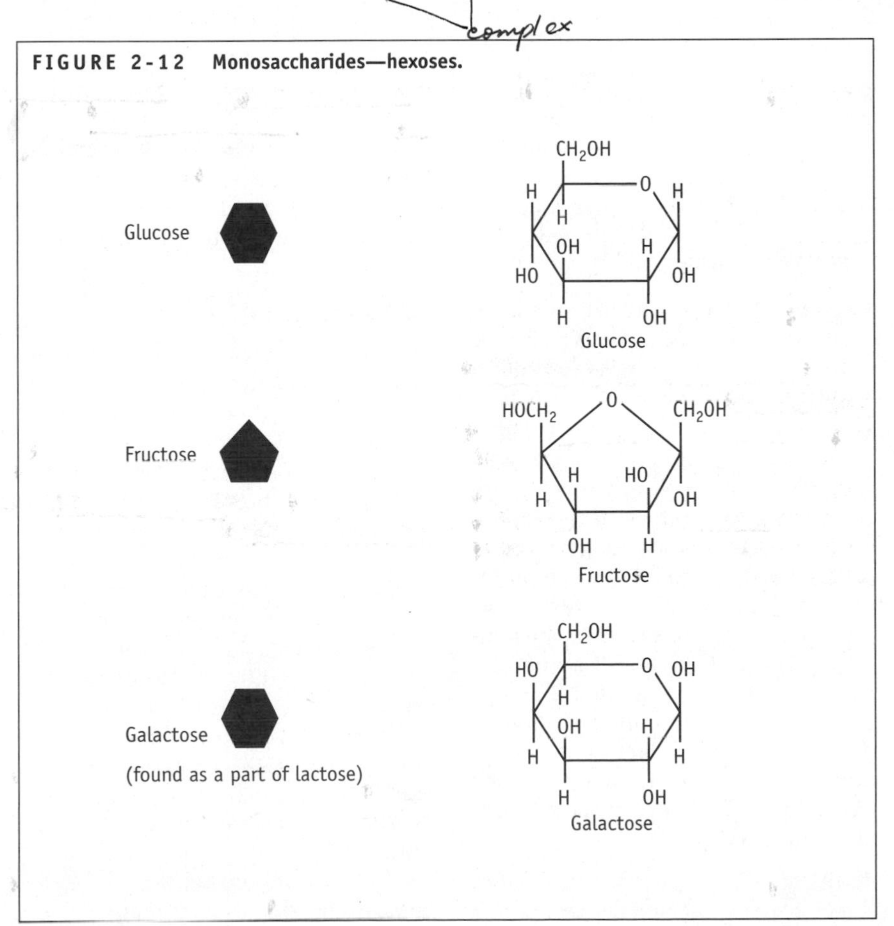

FIGURE 2-12 Monosaccharides—hexoses.

Monosaccharides

Monosaccharides are classified by the number of carbons in the saccharide unit—triose (three carbons), tetrose (four carbons), pentose (five carbons), and hexose (six carbons). The chemical names of many of the carbohydrates end in "-ose", which means "sugar" (Chemist's Corner 2-5). Pentose and hexose sugars are more common in foods, the main pentoses being ribose and arabinose, and the three most predominant hexoses being glucose, fructose, and galactose (Figure 2-12).

Ribose. Ribose is an extremely important component of nucleosides, com-

Chemist's Corner 2-5

"D and "L" Sugars

Saccharide nomenclature uses "D or L" or "alpha or beta" to describe the chemical spatial arrangement of certain saccharides. The designation D or L alludes to two series of sugars. Most natural sugars belong to the D series, in which the highest-numbered asymmetric carbon has the hydroxyl group pointed to the right (Figure 2-13). L-series sugars point to the left. The prefix "alpha" or "beta" can also be used to describe whether the hydroxyl group points to the right (alpha) or left (beta) of the saccharide.

FIGURE 2-13 The "D or L" system of nomenclature describes the "right or left" chemical configuration of a molecule.

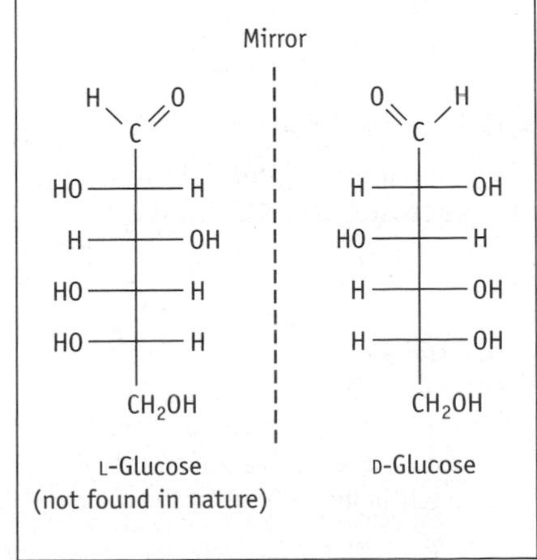

pounds that are part of the genetic material deoxyribonucleic acid (DNA), ribonucleic acid (RNA), and the energy-yielding adenosine triphosphate (ATP). Ribose also plays an important role as part of vitamin B_2 (riboflavin).

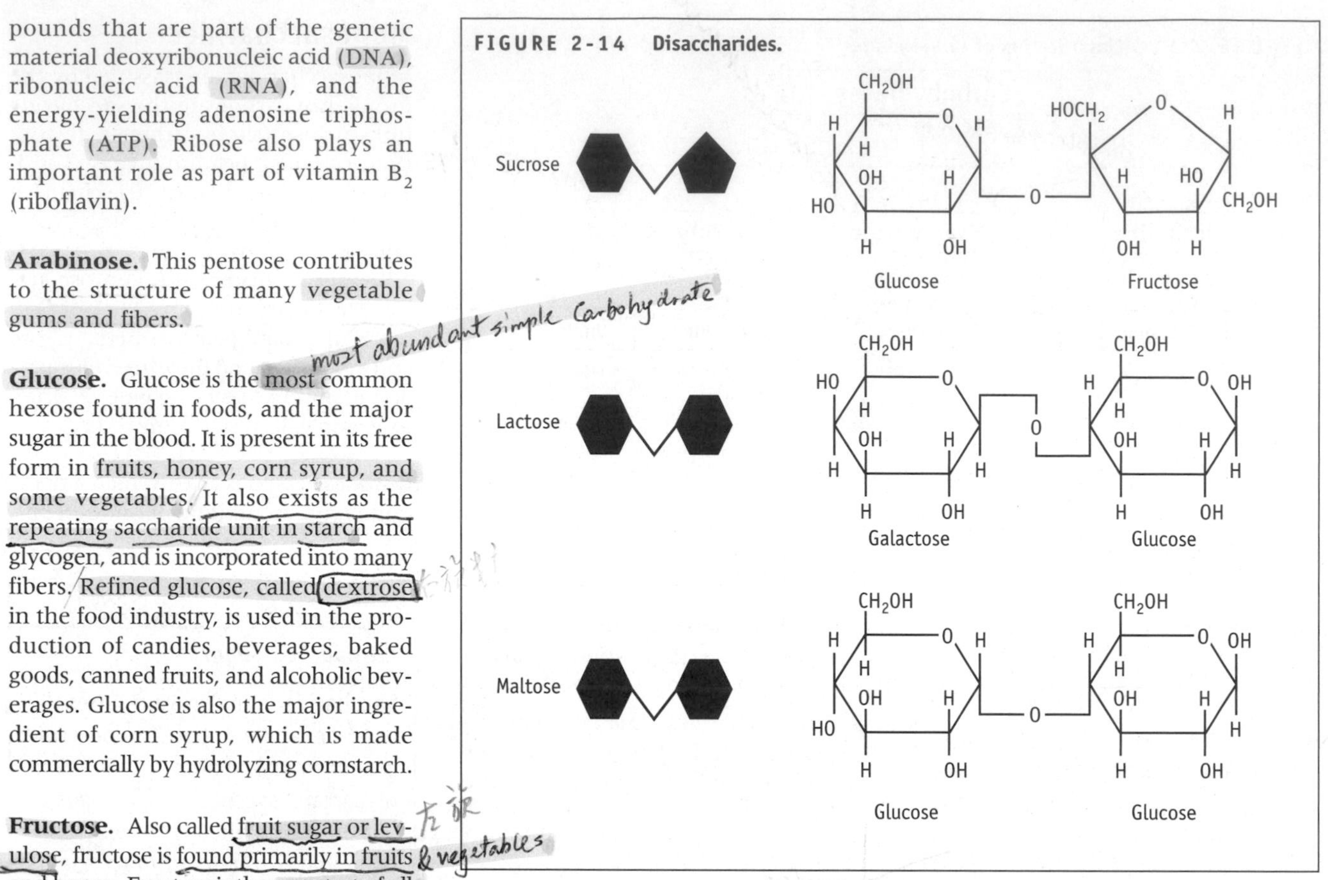

FIGURE 2-14 Disaccharides.

Arabinose. This pentose contributes to the structure of many vegetable gums and fibers.

Glucose. Glucose is the most common hexose found in foods, and the major sugar in the blood. It is present in its free form in fruits, honey, corn syrup, and some vegetables. It also exists as the repeating saccharide unit in starch and glycogen, and is incorporated into many fibers. Refined glucose, called dextrose in the food industry, is used in the production of candies, beverages, baked goods, canned fruits, and alcoholic beverages. Glucose is also the major ingredient of corn syrup, which is made commercially by hydrolyzing cornstarch.

Fructose. Also called fruit sugar or levulose, fructose is found primarily in fruits and honey. Fructose is the sweetest of all sugars, yet it is seldom used in its pure form in food preparation because it can cause excessive stickiness in candies, overbrowning in baked products, and lower freezing temperatures in ice cream. High-fructose corn syrup, however, is the preferred and predominant sweetening agent used in soft drinks.

Galactose. Seldom found free in nature, galactose is part of lactose, the sugar found in milk. A derivative of galactose, galacturonic acid, is a component of pectin, which is very important in the ripening of fruits and the gelling of jams.

Disaccharides

The three most common disaccharides are sucrose, lactose, and maltose (Figure 2-14).

KEY TERMS

Enzyme A protein that catalyzes (causes) a chemical reaction without itself being altered in the process.

Sucrose. Sucrose is table sugar, the product most people think of when they use the term "sugar." Chemically, sucrose is one glucose molecule and one fructose molecule linked together. The types of sugars derived from sucrose are explained in Chapter 9.

Lactose. A glucose molecule bound to a galactose molecule forms lactose, one of the few saccharides derived from an animal source. About 5 percent of fluid milk is lactose, or milk sugar. Some people are unable to digest lactose to its monosaccharides because they lack sufficient lactase, the **enzyme** responsible for breaking down milk sugar into glucose and galactose. The symptoms of lactase deficiency, or lactose intolerance, include gas, bloating, and abdominal pain caused by the disaccharides not being properly digested. In some cheeses, yogurt, and other fermented dairy products, the lactose is broken down by bacteria to lactic acid, which can usually be tolerated by lactase-deficient individuals.

Maltose. Two glucose molecules linked together create maltose, or malt sugar. Maltose is primarily used in the production of beer and breakfast cereals, and in some infant formulas. This saccharide is produced whenever starch breaks down; for example, in germinating seeds and in human beings during starch digestion.

Oligosaccharides

The two most common oligosaccharides are raffinose (three monosaccharides) and stachyose (four monosaccharides). These saccharides, found in dried beans, are not well digested in the human digestive tract, but intestinal bacteria do break them down, forming undesirable gas as a by-product. There are twelve classes of food-grade oligosaccharides in commercial production. These are either extracted directly from soybeans or synthesized by building up disaccharides or breaking down starch. The food industry can use oligosaccharides for bulking agents in low-calorie diet foods like confections, beverages, and yogurt, and as fat replacers in beverages (8). One benefit of oligosaccharides is that they are not cariogenic, or cavity producing, as are many of the disaccharides.

Polysaccharides

Starch, glycogen, and fiber are the polysaccharides most commonly found in foods. Polysaccharides are divided into two major groups: digestible (starch and glycogen) and indigestible (fiber).

Starch—Digestible Polysaccaride from Plant Sources. The glucose derived from photosynthesis in plants is stored as starch. As a plant matures, it not only provides energy for its own needs, but also stores energy for future use in starch granules. Microscopic starch granules are found in various foods such as rice, tapioca, wheat, and potato. A cubic inch of food may contain as many as a million starch molecules (40). Amylose and amylopectin are the two major forms of starch found in these granules. The glucose molecules in both of these starch molecules are joined together with a glycosidic bond (alpha-1, 4) that is capable of being digested by human enzymes. Amylose is a straight-chain structure of repeating glucose molecules, while amylopectin is highly branched with alpha-1, 6 bonds (every 15 to 30 glucose units) (Figure 2-15). The majority of starchy foods in their natural state

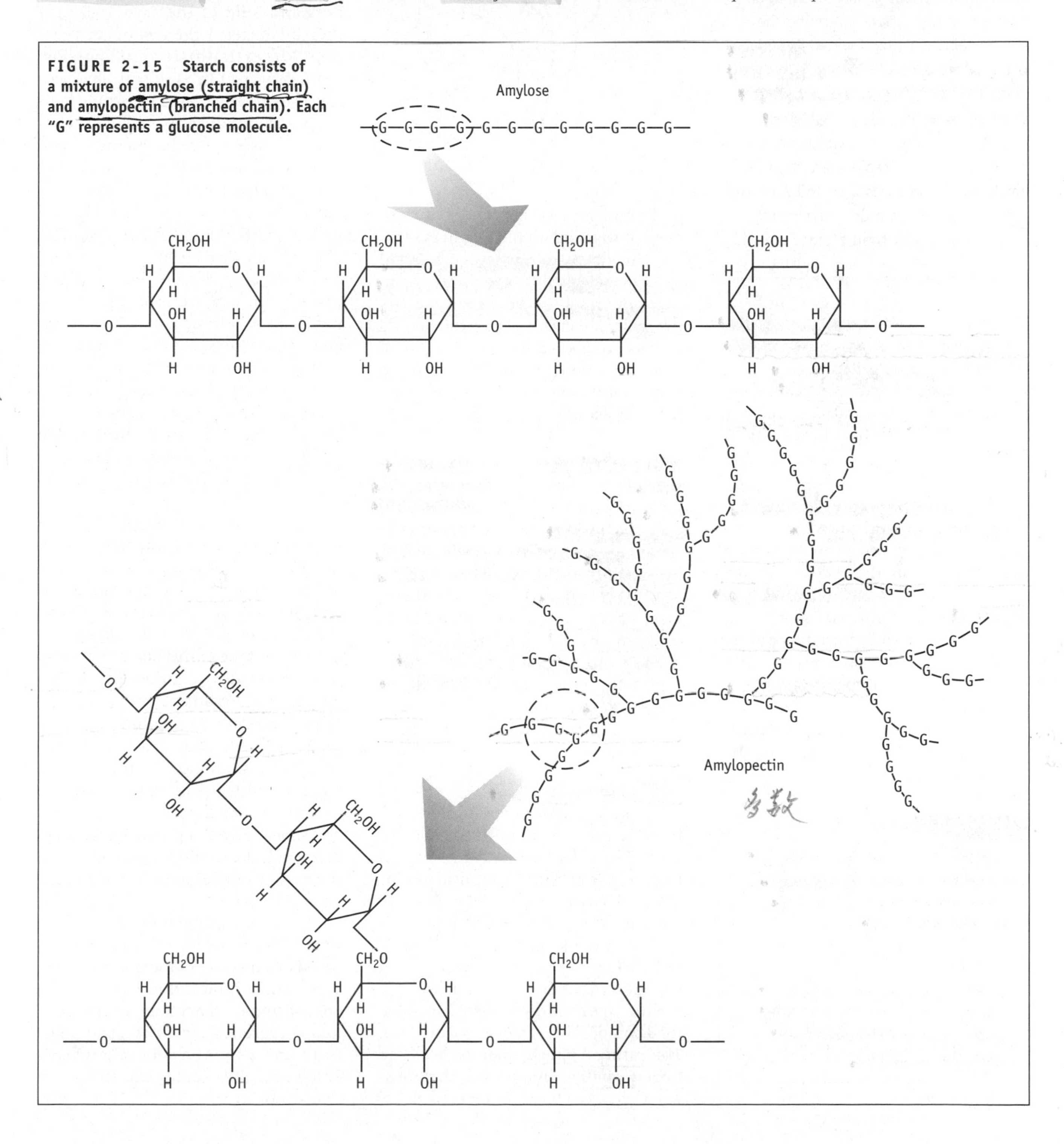

FIGURE 2-15 Starch consists of a mixture of amylose (straight chain) and amylopectin (branched chain). Each "G" represents a glucose molecule.

usually contain a mixture of about 75 percent amylopectin and 25 percent amylose. These two starches are further explained and illustrated in Chapter 20.

HOW & WHY? ?????????

Why do starches from different plant sources differ in their ability to gel? The concentrations of amylose and amylopectin in a solution determine the starch's ability to hold water. The higher the amylose content, the more likely the starch will gel (form a solid structure) when mixed with water and heated. Cornstarch is high in amylose, while potato starch and tapioca are high in amylopectin, so cornstarch will form the gels needed in custards, gravies, and other foods better than tapioca starch.

Heat, enzymes, and acid are used to break starches down into smaller, sweeter segments called dextrins. The sweeter taste of toasted bread, compared to its untoasted counterpart, comes from the dextrins formed in the toaster.

Glycogen—Digestible Polysaccharide from Animal Sources. Glycogen, or animal starch, is one of the few digestible carbohydrates found in animals. It is located only in the liver and muscles. Just as glucose is stored by plants as starch, it is stored by animal bodies in long chains of glycogen (Figure 2-16). It is a highly branched arrangement of glucose molecules, and serves as a reserve of energy. Glycogen can be quickly hydrolyzed by an animal's enzymes to release the glucose needed to maintain blood glucose levels. The glycogen in meat is converted to lactic acid during slaughtering and so is not present by the time it reaches the table. Shellfish such as scallops and oysters provide a minuscule amount of glycogen, which is why they tend to taste slightly sweet compared to other fish and plants.

FIGURE 2-16 Animals store glucose in the form of glycogen. Each circle represents a glucose molecule.

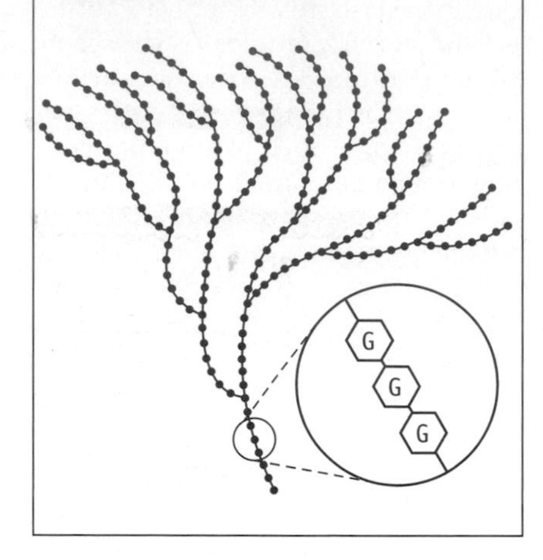

KEY TERMS

Dietary fiber The undigested portion of carbohydrates remaining in a food sample after exposure to digestive enzymes.

Diverticulosis An intestinal disorder characterized by pockets forming out from the digestive tract, especially the colon.

Fiber—Undigestible Polysaccharide. Fiber, also known as roughage or bulk, describes a group of indigestible polysaccharides. Unlike starch, the sugar units in fibers are held together by bonds the human digestive enzymes cannot break down. Most fibers, therefore, pass through the human body without providing energy. Fiber is found only in foods of plant origin, especially certain cereals, vegetables, and fruits. Plant cells rely on the fiber between their cell walls for structural strength.

Dietary Fiber vs Crude Fiber. A number of different laboratory methods are used to measure the amount of fiber in foods. The older technique consisted of treating a food with strong acid to simulate the environment of the stomach, and then treating it with a base to parallel the experience in the small intestine. The remaining weight of undigested fiber was measured as "crude fiber" and was listed in most food composition tables as "fiber" (22). This rather imprecise method has been largely replaced by a process that measures **dietary fiber**. For every 1 gram of crude fiber, there are about 2 to 3 grams of dietary fiber.

Soluble vs Insoluble Fiber. Chemists classify fibers according to how readily they dissolve in water: soluble fibers dissolve in water, while insoluble fibers do not. The insoluble fibers of foods act as a sponge in the intestine by soaking up water. This increases the softness and bulk of the stool and may thereby decrease the risk of constipation, **diverticulosis**, and possibly colon cancer (20). Scientists have also suggested that soluble fibers may benefit health by lowering high blood cholesterol levels and reducing high blood glucose in certain kinds of diabetics (41). Foods containing fiber usually have a mixture of both soluble and insoluble fiber. Foods high in soluble fiber include dried beans, peas, lentils, oats, rice bran, barley, and oranges. Insoluble fibers are found predominantly in whole wheat (wheat bran) and rye products, along with bananas.

The most common fibers are cellulose, hemicellulose, and pectic substances. A few other types of fiber include vegetable gums, inulin, and lignin, the last of which is one of the few fibers that is not a carbohydrate.

Cellulose. Cellulose is one of the most abundant compounds on earth. Every plant cell wall is composed in part of cellulose, long chains of repeating glucose molecules similar to starch. Unlike starch, however, the chains do not branch, and the bonds holding the glucose molecules together cannot be digested by human enzymes (Figure 2-17). As a result, the cellulose fiber is not absorbed, provides no calories (kcal), and simply passes through the digestive tract. The digestive systems of herbivores such as cattle, horses, goats, and sheep have the proper enzymes to digest cellulose, allowing them to use the energy from glucose found in grass and other plants.

Hemicellulose. Hemicellulose is composed of a mixture of monosaccharides. The most common monosaccharides comprising the backbone of hemicelluloses are xylose, mannose, and galactose; the common side chains are arabinose, glucuronic acid, and galac-

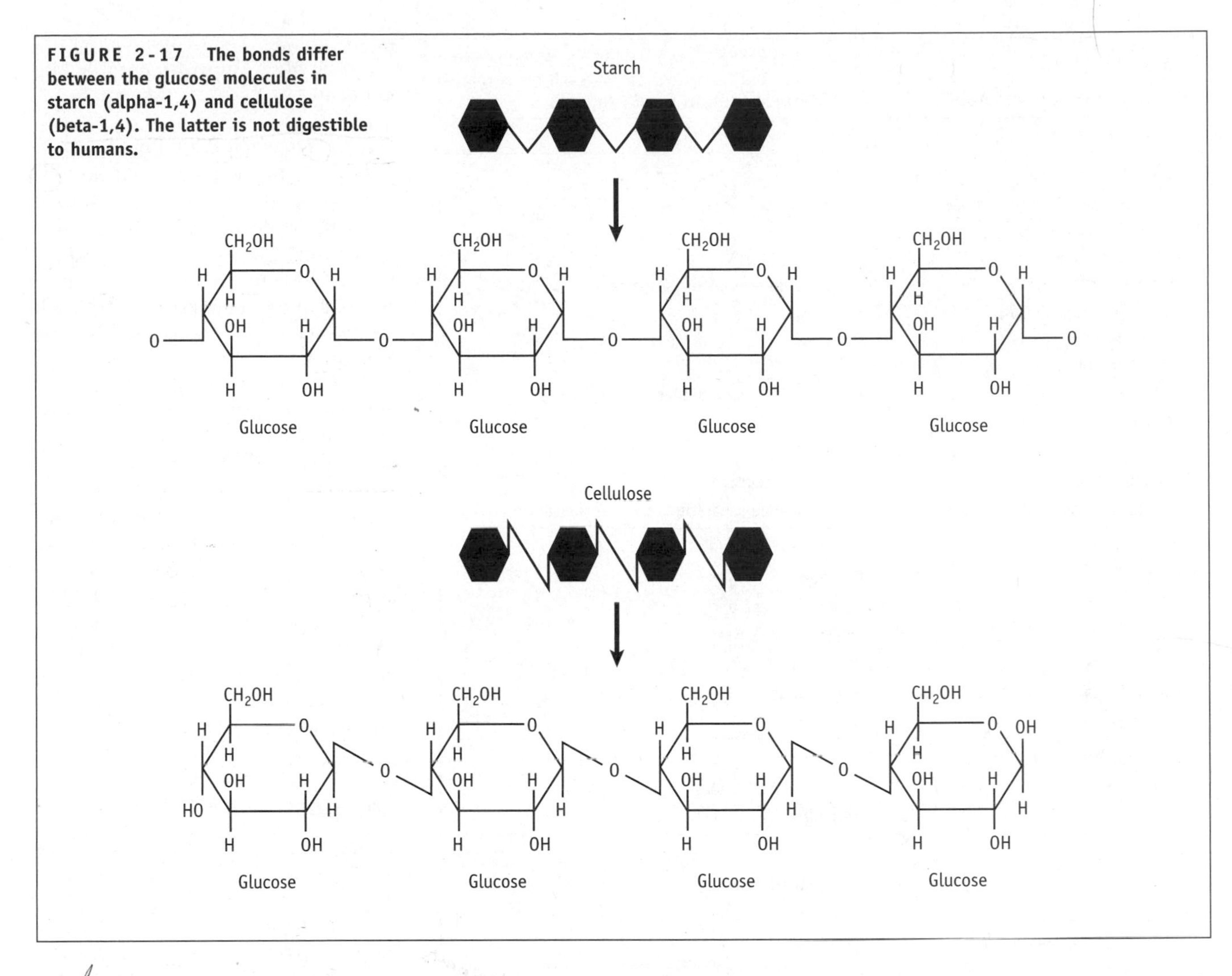

FIGURE 2-17 The bonds differ between the glucose molecules in starch (alpha-1,4) and cellulose (beta-1,4). The latter is not digestible to humans.

tose. Baking soda is sometimes added to the water in which green vegetables are boiled to maintain their color. Unfortunately, it breaks down the hemicellulose of the vegetables, causing them to become mushy.

Pectic Substances. These substances, found between and within the cell walls of fruit and vegetables, include protopectin, pectin, and pectic acid. Pectic substances act as natural cementing agents, and so are extracted from their source foods by the food industry for use in thickening jams, keeping salad dressing from separating, and controlling texture and consistency in a variety of other foods. Not all the pectic substances, however, can be used for gelling purposes, and the amounts that can be obtained vary depending on the ripeness of the fruit or vegetable. The pectin found in ripe, but not overripe, fruit is responsible for gel formation in jams. Protopectin and pectic acid are prevalent in unripe and overripe fruit respectively, and are insufficient themselves to cause gel formation.

Vegetable Gums. Vegetable gums belong to a group of polysaccharides known as hydrocolloids. They are derived from three main sources: plant gums (gum arabic, gum karaya, gum tragacanth), seeds (locust bean gum, guar gum), and seaweeds (agar, alginates, and carrageenan) (11). A bacterium (Xanthomonas campestris) serves as the source for xanthan gum. Gum fibers are composed of simple sugars, which are used by the food industry to thicken, provide viscosity, gel, stabilize, and/or emulsify certain processed foods. They impart body, texture, and mouthfeel to foods, while also making it less likely for dispersed ingredients to separate (34). The gums' "water-loving" nature combined with their ability to bind as much as 100 times their weight in water contribute a certain desirable appearance, texture, and stability to food products (9).

Vegetable gums are normally sold as a dry powder and are used extensively as stabilizers in the production of low-calorie salad dressings, confections, ice cream, puddings, and whipped cream. Gums are also used in many frozen products because they control crystal growth, yield optimum texture, and make the food more stable in the freezing and thawing process. Typical applications in the food industry of a gum, specifically carrageenan, are listed in Table 2-2). Agar gum can be used for quick-drying frostings and to reduce chipping or cracking in glazed doughnuts (38).

Inulin. Inulin consists of repeating units of fructose with an end molecule of

TABLE 2-2
Carrageenan: Typical Applications of a Natural Gum in Food Products

Application	Function
DAIRY PRODUCTS	
Whipped cream, topping, desserts	Fat and foam stabilization
Acidified cream, cottage cheese, processed cheese	Binder
Yogurt	Viscosity gelation, fruit suspension
Chocolate, eggnog, fruit, flavored milk	Suspension
Fluid skim milk	Binder
Filled milk	Emulsion stabilizer, binder
Low-calorie diet drinks	Suspension, binder
Evaporated milk (in can)	Fat stabilizer
Pudding, pie filling	Thickening gelation
Ice cream, ice milk	Prevents whey separation, controls meltdown
DESSERTS	
Pie filling	Gelation
Syrup	Binder, suspension
Imitation coffee cream	Emulsion, stabilizer
Sauces	Binder, thickener
Fruit drinks and frozen concentrates	Binder, mouthfeel
Dessert gels	Gelation
MEAT, POULTRY, AND SEAFOOD	
Meat and poultry products	Binder, fat stabilizer, fat replacer
Surimi	Binding
OTHER FOOD APPLICATIONS	
Salad dressings	Emulsion stabilizer
Fish gels	Gelation
Bakery - pastry, jam, marmalade	Viscosity
Cake glace	Controls meltdown

Source: The Carrageenan Company.

glucose. Although this fiber occurs naturally in over 30,000 plants, it is most commonly found in asparagus, Jerusalem artichoke, and garlic, but commercial processors extract it from the chicory root (18). Inulin is a soluble fiber that can be used by the food industry for giving a creamy texture to frozen dairy products such as no-fat or no-sugar ice cream, improving the textures of margarine spreads, and developing no-fat icings, fillings, and whipped toppings.

KEY TERMS

Phenolic A chemical term to describe an aromatic (circular) ring attached to one or more hydroxyl (-OH) groups.

Lignin. Lignin is the one fiber that is not a carbohydrate. Instead of saccharides, it consists of long chains of **phenolic** alcohols linked together into a large, complex molecule. As plants mature, their cell walls increase in lignin concentration, resulting in a tough, stringy texture. This partially explains why celery and carrots get tougher as they age. Boiling water does not dissolve or even soften the lignin.

Lipids or Fats

Foods High in Lipids

The fats and oils in foods belong to a group called lipids. Lipids are commonly called "fats" when discussing their content in foods; while this is not precisely accurate, this textbook will follow the generally accepted practice. Fats and oils are differentiated in two ways: (1) fats are solid at room temperature, while oils are liquid, and (2) fats are usually derived from animal sources, while oils are derived predominantly from plants. Three exceptions are coconut and palm oils, which are solid at room temperature, and fish oils, which, at the same temperature, are liquid.

The foods that are high in fats from animal sources include meats, poultry, and dairy products. Plant food sources high in fat include nuts, seeds, avocado, olives, and coconut. Most fruits, vegetables, and grains, however, contain little, if any, fat. Invisible fats are those not easily observed in foods, such as the marbling in meat. Visible fats, such as the white striations found in bacon and the outside trim on meats, are easily seen. Vegetable oils, butter, margarine, shortening, lard, and tallow are also obvious visible fats.

Composition of Lipids

Lipids, like carbohydrates, are composed of carbon, hydrogen, and oxygen atoms, but in differing proportions. One way to determine if a substance is a lipid is to test whether it will dissolve in water. Lipids will not dissolve in water, but can be dissolved in organic solvents not used in food preparation, such as benzene, chloroform, ether, and acetone. Acetic acid, which is responsible for the sour taste of vinegar, is the one lipid that will dissolve in water because its molecule is so small. Edible lipids are divided into three major groups: triglycerides (fats and oils), phospholipids, and sterols.

Triglycerides

About 95 percent of all lipids are triglycerides, which consist of three ("tri") fatty acids attached to a glycerol molecule (Figure 2-18) (Chemist's Corner 2-6). (Two fatty acids linked to the glycerol molecule form a diglyceride, while one fatty acid linked to glycerol is a monoglyceride.) The fatty acids on the glycerol can be identical (simple triglyceride) or different (mixed triglyceride).

FIGURE 2-18 A triglyceride (fat) is made by combining three fatty acids with a glycerol molecule. Water is released as a by-product.

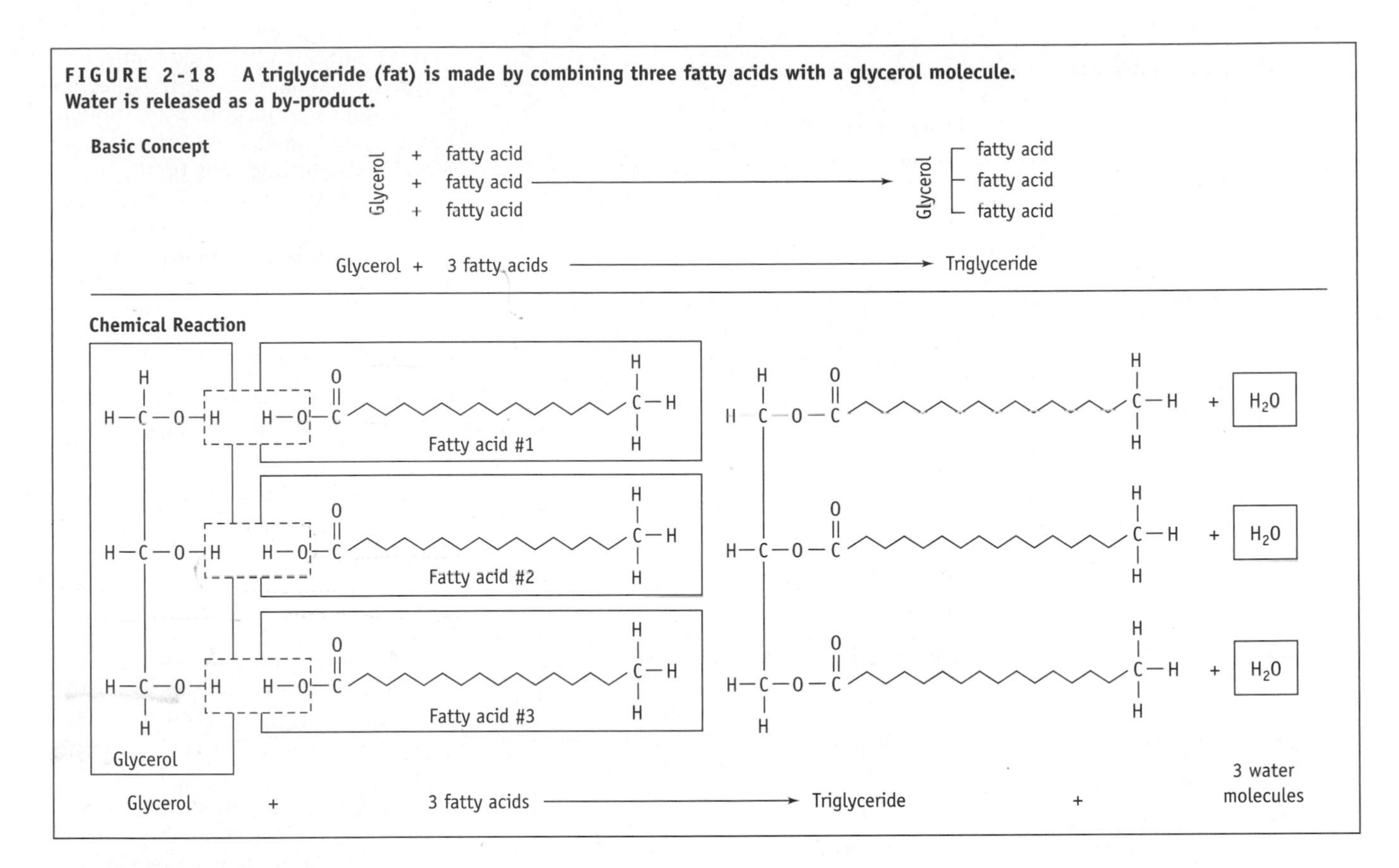

Chemist's Corner 2-6

Chemical Formation of Triglycerides

Few fatty acids occur free in foods, but rather are incorporated into triglycerides. Each fatty acid consists of an acid group (-COOH) on one end and a methyl group ($-CH_3$) on the other end. The fatty acids are attached to the glycerol molecule by a condensation reaction: the hydrogen atom (H) from the glycerol and a hydroxyl (-OH) group from a fatty acid form a molecule of water (Figure 2-18). When a fatty acid reacts like this with an alcohol such as glycerol, the resulting compound is called an ester. Since "acyl" defines the fatty acid part of an ester, what is called "triglyceride" should actually be named "triacylglycerol" (12).

Fatty Acid Structure. Fatty acids differ from one another in two major ways: (1) their length, which is determined by the number of carbon atoms, and (2) their degree of "saturation," which is determined by the number of double bonds between carbon atoms. The number of carbons can range from 2 to 22, with the number usually being even. A fatty acid is said to be saturated if there are no double bonds between carbons—every carbon on the chain is bonded to two hydrogens and therefore fully loaded. If one hydrogen from two adjacent carbons is missing, the carbons form double bonds with one another and form a point of unsaturation. A fatty acid with one double bond present is called a monounsaturated fatty acid. If there are two or more double bonds in the carbon chain of a fatty acid, the fatty acid is called polyunsaturated (Figure 2-19).

The degree of unsaturation of the fatty acids in a fat affects the temperature at which the molecule melts. Generally, the more unsaturated a fat, the more liquid it remains at room temperature. In contrast, the more saturated a fat, the firmer its consistency. Thus, vegetable oils are generally liquid at room temperature, while animal fats are solid.

Fatty Acids in Foods. Most foods contain all three types of fatty acids—saturated, monounsaturated, and polyunsaturated—but one type usually predominates (Figure 2-20). Generally speaking, most vegetable and fish oils are high in polyunsaturates, while olive and canola oils are rich in monounsaturates. The animal fats, as well as coconut and palm oils, are more saturated. Overall, foods of animal origin usually contain approximately a 50:50 **P/S ratio** of polyunsaturated and saturated fats, while for plant foods the ratio is usually 85/15. The higher the P/S ratio, the more polyunsaturated fats the food contains.

Fatty Acid Nomenclature. Each fatty acid is identified by a common name, a systematic name (Chemist's Corner 2-7 on page 39), chemical configuration (Chemist's Corner 2-8 on

KEY TERMS

P/S ratio The ratio of polyunsaturated fats to saturated fats. The higher the P/S ratio, the more polyunsaturated fats the food contains.

FIGURE 2-19 Fatty acids differ in their degree of saturation.

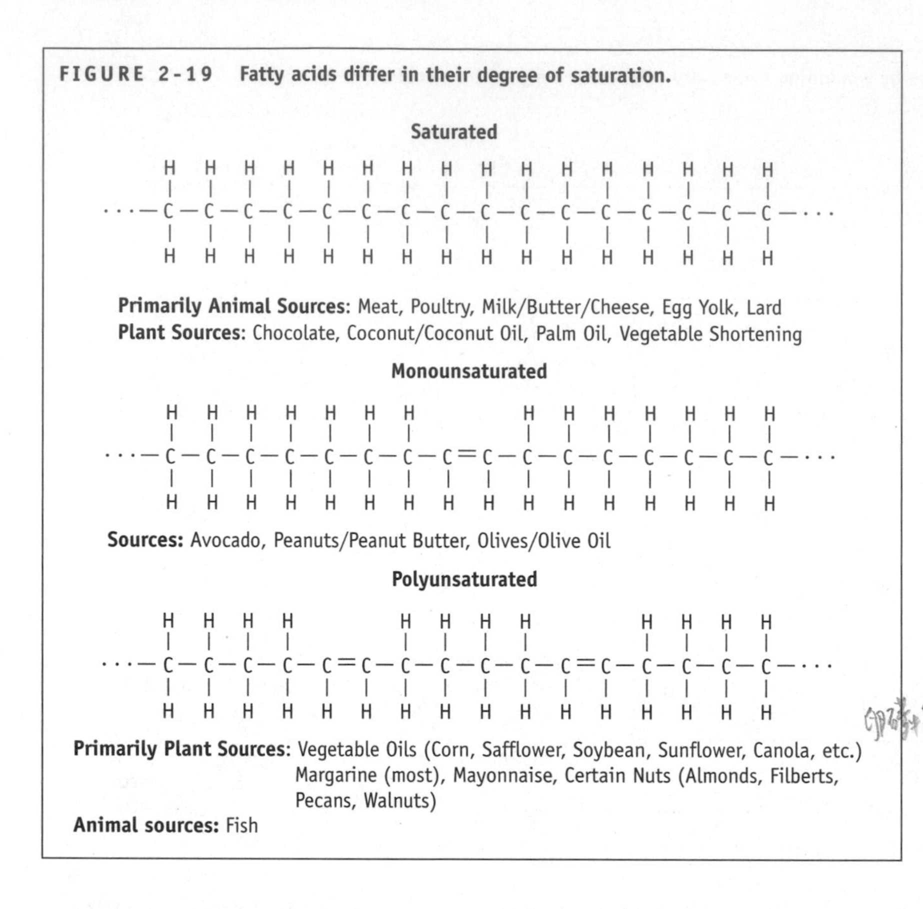

page 40), or numerical ratio (Table 2-3). Usually a fatty acid is referred to by its common name, while the systematic name is used when a more formal or correct chemical nomenclature is required. The long number of carbons is abbreviated in a type of chemical shorthand that conveys the length and saturation of fatty acids in a numerical ratio. For example, palmitic acid is a saturated fatty acid that is represented by 16:0, meaning that it is sixteen carbons long with zero double bonds. Approximately 40 fatty acids are found in nature. Some of the more common fatty acids are butyric acid, found in butter, and the two fatty acids that are **essential nutrients**—linoleic and linolenic.

Phospholipids

Phospholipids are similar to triglycerides in structure in that fatty acids are attached to the glycerol molecule. The difference is that one of the fatty acids is replaced by a compound containing phosphorus, which makes the phospholipid soluble in water, while its fatty acid components are soluble in fat (Figure 2-23 on page 41). The dual nature of phospholipids makes them ideal **emulsifiers**. The best-known phospholipid is lecithin, which is found in egg yolks. Lecithin acts as an emulsifying agent that allows **hydrophobic** and **hydrophilic** compounds to mix. Phosolipids are very important in the body as a component of cell membranes, where they assist in moving fat-soluble vitamins and hormones in and out of the cells.

Phospholipids are widely used by the food industry as emulsifiers in such products as beverages, baked goods,

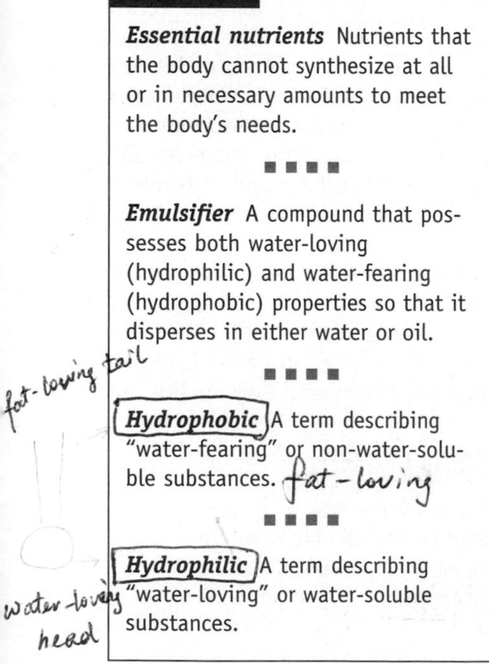

KEY TERMS

Essential nutrients Nutrients that the body cannot synthesize at all or in necessary amounts to meet the body's needs.

Emulsifier A compound that possesses both water-loving (hydrophilic) and water-fearing (hydrophobic) properties so that it disperses in either water or oil.

Hydrophobic A term describing "water-fearing" or non-water-soluble substances.

Hydrophilic A term describing "water-loving" or water-soluble substances.

FIGURE 2-20 Classification of fatty acids.

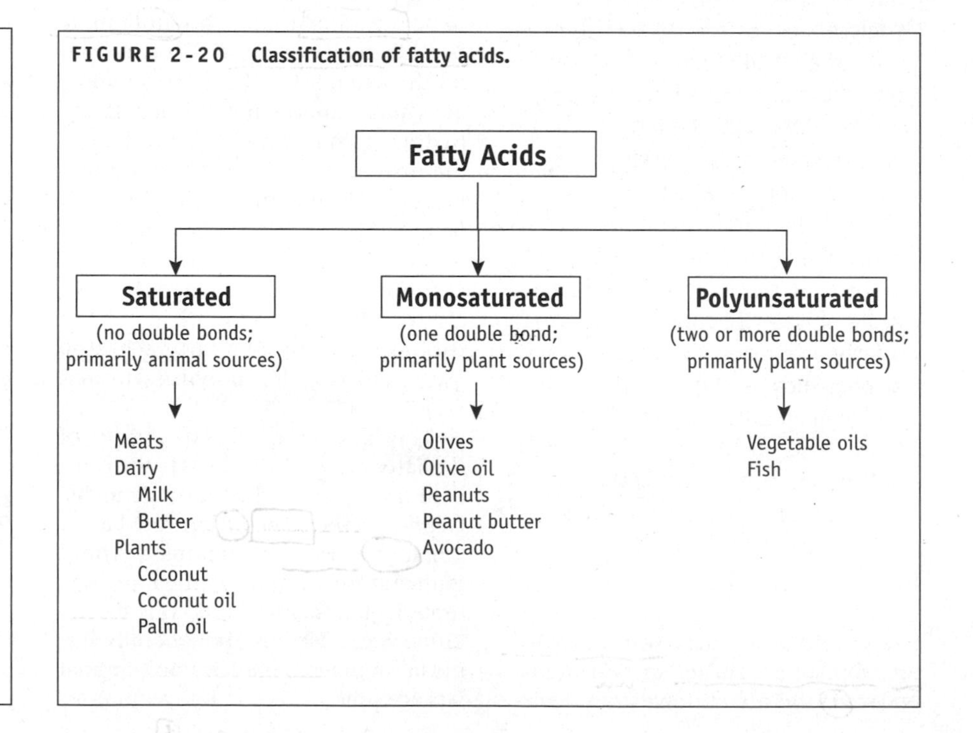

Chemist's Corner 2-7

Naming Chemical Compounds

Billions of compounds exist in nature. This vast number does not even include the compounds that are synthesized in a laboratory. At first people named compounds after people, places, and things, but that proved too cumbersome. Scientists then developed a system to name chemical substances known as "chemical nomenclature," and this is defined as the "systematic" naming of chemical compounds. In chemistry, the systematic name literally describes the chemical construction of the compound. The International Union of Pure and Applied Chemistry (IUPAC) is the official organization responsible for mandating the nomenclature of all chemical compounds. Learn more about the IUPAC through their website at http://www.iupac.org.

TABLE 2-3

Nomenclature Methods for Selected Fatty Acids

Common Name	*Systematic Name*	*Numerical Ratio #C Atoms : # Double Bonds*
Butyric	Butanoic	4:0
Linoleic	9, 12-Octadecadienoic	18:2
Linolenic	9, 12, 15-Octadecatrienoic	18:3

mayonnaise, and candy bars. Foods that naturally contain phospholipids include egg yolks, liver, soybeans, wheat germ, and peanuts.

Sterols

Sterols are large, intricate molecules consisting of interconnected rings of carbon atoms with a variety of side chains attached. Many compounds important in maintaining the human body are sterols, including cholesterol, **bile**, both sex (testosterone, estrogen) and adrenal (cortisol) hormones, and vitamin D. The sterol of most significance in foods is cholesterol (Figure 2-24 on page 41). Although both animal and plant foods contain sterols, cholesterol is found only in foods of animal origin such as meat, poultry, fish, fish roe (caviar), organ meats (liver, brains, kidneys), dairy products, and egg yolks. Plants do not contain cholesterol, but they may contain other types of sterols.

Plant Sterols. Plant sterols are found in small amounts in many fruits, vegetables, nuts, seeds, cereals, legumes, and other plant sources (38). Soybeans are not the only plants containing sterols. In 2000, the FDA allowed the use of health claims regarding foods containing these substances with regard to the role of plant sterols or plant sterol esters in reducing the risk for coronary heart disease. To qualify for a claim, the food must contain at least 0.65 gram of plant sterol esters or 1.7 grams of **plant stanol esters** per serving. It also must not contain over 13 grams of total fat per serving and per 50 grams. Spreads and salad dressings are exempted if the label refers the consumer to the product's Nutrition Facts panel (27).

Proteins

Foods High in Proteins

Proteins derive their name from the Greek word *proteos,* of "prime importance." The body can manufacture most of the necessary carbohydrates (except fiber) and lipids (except a few essential fatty acids) it needs, but when it comes to protein, the body can synthesize only about half of the compounds it requires in order to manufacture the proteins needed for the body. These substances needed for protein manufacture are called amino acids. Of the 22 amino acids, 9 are essential nutrients and thus must be obtained daily from the diet (Table 2-4).

TABLE 2-4

Essential and Nonessential Amino Acids

Classification	*Amino Acid*
Essential for all humans	Histidine
	Isoleucine
	Leucine
	Lysine
	Methionine
	Phenylalanine
	Threonine
	Tryptophan
	Valine
Nonessential	Alanine
	Arginine
	Asparagine
	Aspartic acid
	Cysteine
	Glutamic acid
	Glutamine
	Glycine
	Proline
	Serine
	Tyrosine
Related compounds sometimes classified as amino acids	Carnitine
	Cystine
	Hydroxyglutamic acid
	Hydroxylysine
	Hydroxyproline
	Norleucine
	Taurine
	Thyroxine

KEY TERMS

Bile A digestive juice made by the liver from cholesterol and stored in the gall bladder.

■ ■ ■ ■

Plant stanol esters Naturally occurring substances in plants that help block absorption of cholesterol from the digestive tract.

Chemist's Corner 2-8

"Cis," "Trans," and Omega Fatty Acids

Other notations that are frequently encountered when discussing fatty acids are cis or trans, and omega-3 or omega-6. The terms "cis" and "trans" describe the geometric shape of the fatty acid. A cis fatty acid has the hydrogens on the same side as the double bond, causing it to fold into a U-like formation. A trans fatty acid has the hydrogens on either side of the double-bond, creating a linear configuration (Figure 2-21). Most of the fatty acids in nature are in the cis or slightly V-shaped configuration, while trans fatty acids often result from hydrogenation. The difference between the omega-3 and omega-6 fatty acids is the location of the double bond between the third and fourth, or the sixth and seventh, carbon from the left of the fatty acid molecule respectively (Figure 2-22).

FIGURE 2-21 "Cis" or "trans" fatty acids are defined by their chemical configuration at the double bonds.

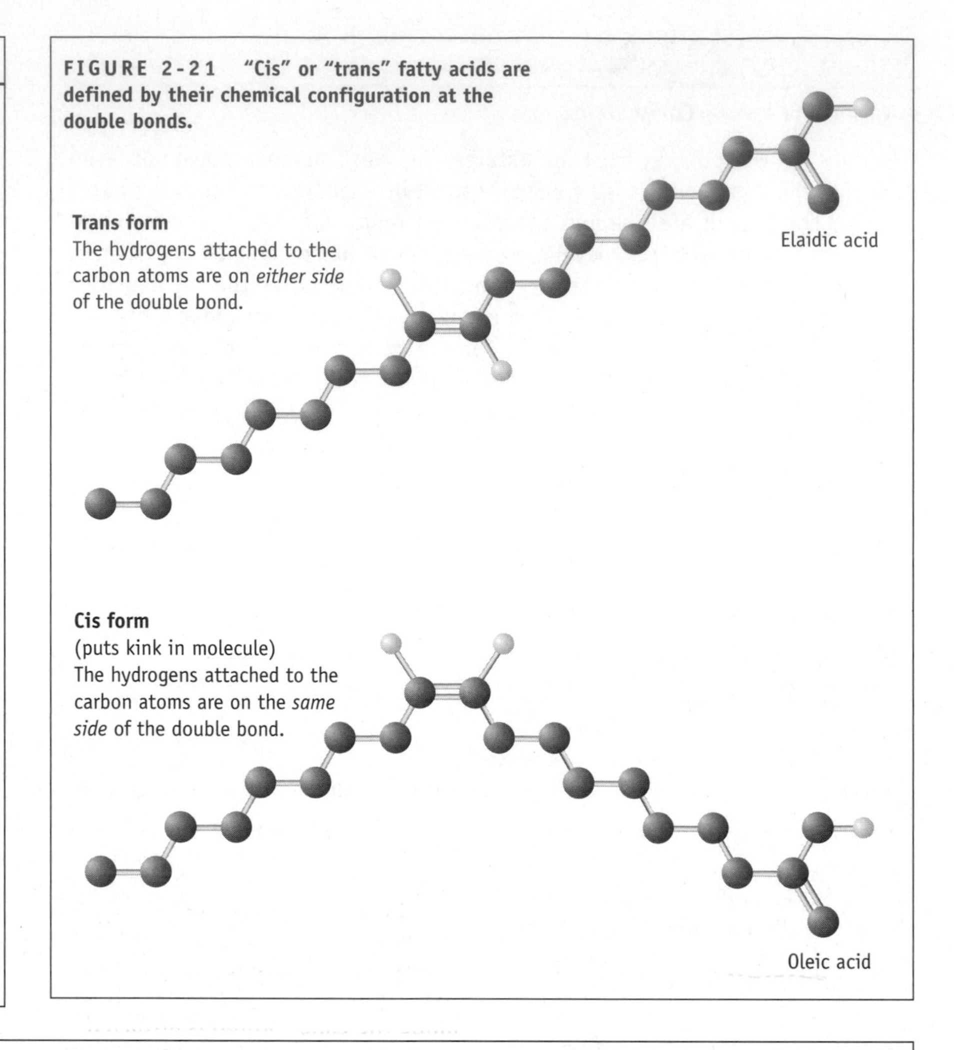

FIGURE 2-22 Omega-3 or omega-6 fatty acids are defined by the location of the first double bond.

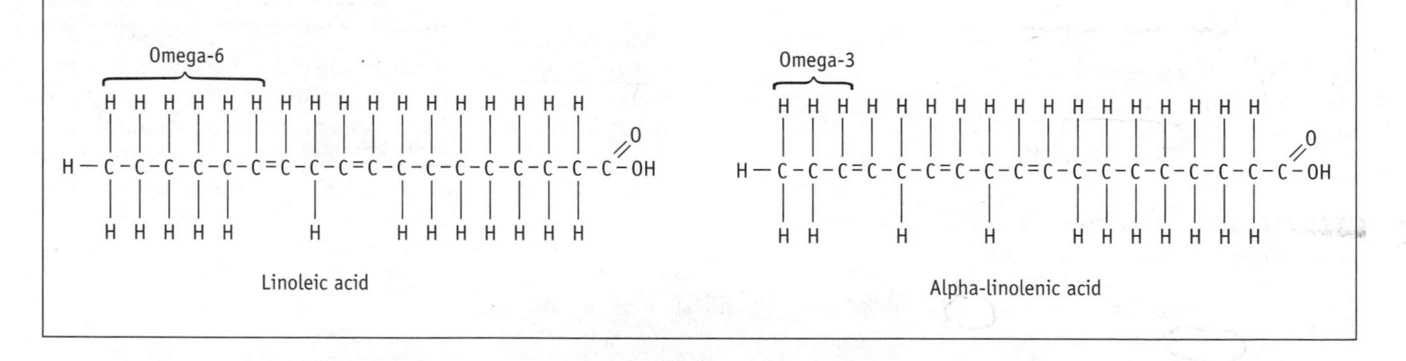

KEY TERMS

Complete protein A protein, usually from animal sources, that contains all the essential amino acids in sufficient amounts for the body's maintenance and growth.

Protein Quality. Foods vary in their protein quantity and quality. Most protein from animal sources—meat, poultry, fish and shellfish, milk (cheese, yogurt, etc.), and eggs—is **complete protein.** Gelatin is one of the few animal proteins that is not complete. Plant protein, with the exception of that from soybeans (quinoa and amaranth) and certain grains, is **incomplete protein** and will support maintenance, but not growth. In order to obtain complete protein primarily from plant sources, it is necessary to practice the strategy of **protein complementation**. The

FIGURE 2-23 Phospholipids, such as this lecithin molecule, have a phosphorous-containing compound that replaces one of the fatty acids on a triglyceride. Each compound in the lecithin molecule is derived from simpler molecules (see arrows).

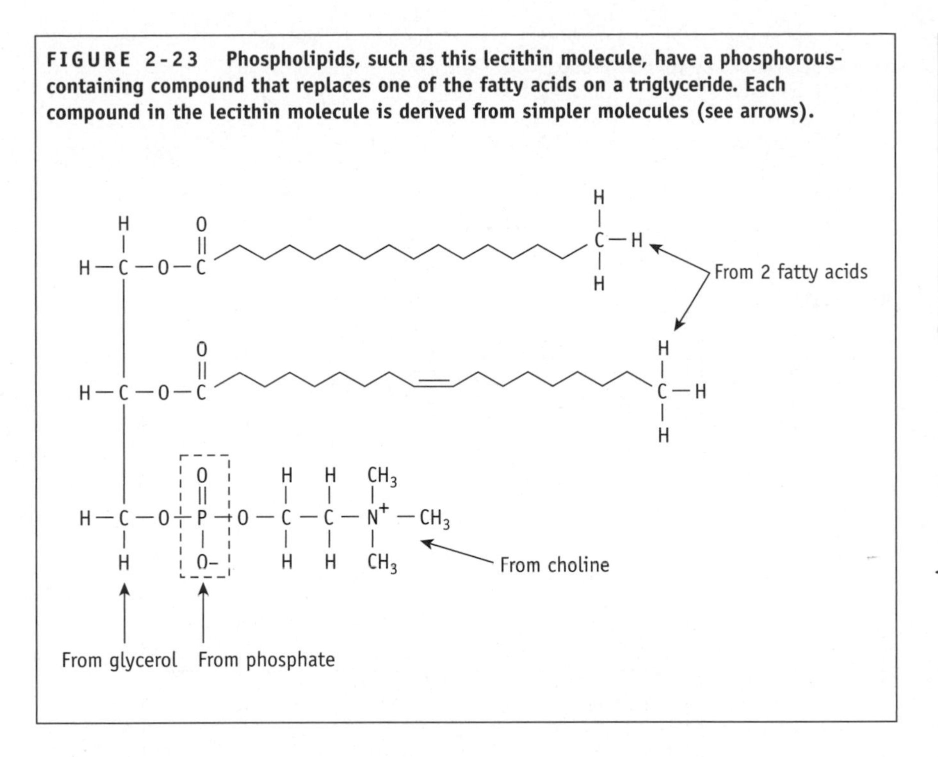

best sources of protein from plants are the legumes—beans, peas, and lentils—which are often combined with grains.

Composition of Proteins

One key way in which proteins differ from carbohydrates and lipids is that proteins contain nitrogen atoms, while carbohydrates and lipids contain only carbon, hydrogen, and oxygen atoms. These nitrogen atoms give the name "amino," meaning "nitrogen containing," to the amino acids of which protein is made. Protein molecules resemble linked chains, with the links being amino acids joined by **peptide bonds**. A protein strand does not remain in a straight chain, however. The amino acids at different points along the strand are attracted to each other, and this pull causes some segments of the strand to coil, somewhat like a metal spring. Also, each spot along the coiled strand is attracted to, or repelled from, other spots along its length. This causes the entire coil to fold this way and that, forming a globular or fibrous structure.

FIGURE 2-24 Cholesterol is a lipid found in foods of animal origin.

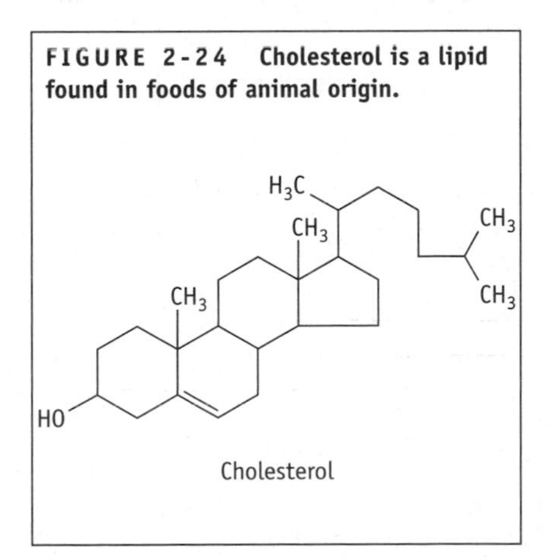

Amino Acids. Each protein has its own specific sequence of amino acids. The 22 amino acids that exist in nature are like an alphabet, forming the "letters" of the "words"—proteins—that make up the language of life itself. All amino acids have the same basic structure—a carbon with three groups attached to it: an amine group ($-NH_2$), an acid group ($-COOH$), and a hydrogen atom (H). Attached to the carbon at the fourth bond is a side chain called an R group (Figure 2-25). It is this fourth attachment, the side chain, different for each amino acid, that gives the amino acid its unique identity and chemical nature (Figure 2-26 on page 42). The simplest amino acid is glycine, with only a hydrogen for the R group. In other amino acids, the R group may consist of carbon chains or cyclic structures. Amino acids that are acidic contain more acid groups ($-COOH$) than amine groups ($-NH_2$), while alkaline amino acids contain more amine than acid groups.

FIGURE 2-25 The structure of an amino acid.

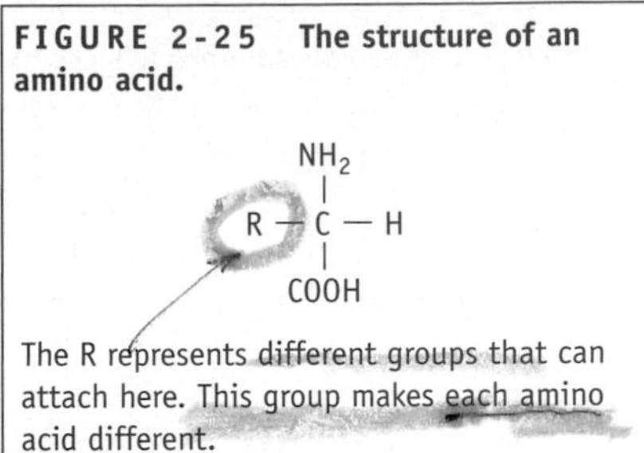

The R represents different groups that can attach here. This group makes each amino acid different.

Functions of Proteins in Food

The proteins in foods allow a number of important reactions to occur during food preparation:

- Hydration
- Denaturation/coagulation
- Enzymatic reactions
- Buffering
- Browning

Hydration. The ability of proteins to dissolve in and attract water, a process called hydration, allows them to play several important roles in foods. One of these is the capability to form a gel,

KEY TERMS

Incomplete protein A protein, usually from plant sources, that does not provide all the essential amino acids.

Protein complementation Two incomplete-protein foods, each of which supplies the amino acids missing in the other, combined to yield a complete protein profile.

Peptide bond The chemical bond between two amino acids.

FIGURE 2-26 **An amino acid's chemical nature is determined by its side group.**

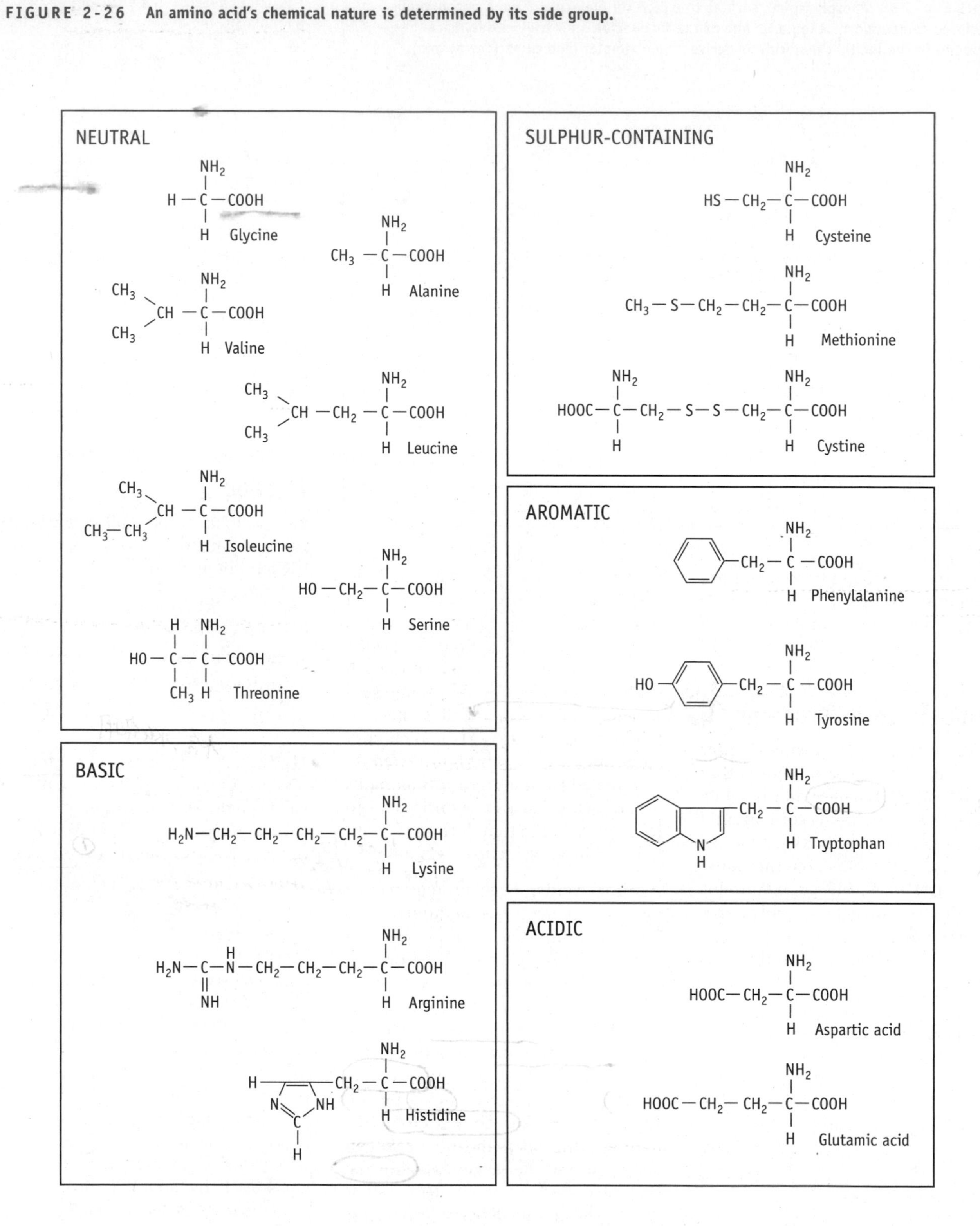

an intricate network of protein strands trapping water that results in a firm structure. Proteins from milk, meat, egg, and soy are used in a variety of gels (26, 44). The gelling ability of proteins allows them to be used as binders (6), stabilizers (42), and thickeners in a variety of foods such as preserves, confectioneries (gums, marshmallows), and desserts (ice cream, puddings, custards, pie fillings, mousses, and plain-flavored gelatins).

Another example of protein's hydrating ability in food preparation is in breadmaking. Water (or milk) is combined with yeast and the two major proteins of wheat—gliadin and glutenin—through the process of kneading to yield the protein gluten, whose elastic qualities allow it to stretch with the carbon dioxide gas produced by the yeast during fermentation. This is how bread rises, and without protein's ability to hydrate, rising would not take place.

Denaturation/Coagulation. Large protein molecules are sensitive to their surroundings. When subjected to heat, pH extremes, alcohol, and physical or chemical disturbances, proteins undergo **denaturation**. Denaturation can result in **coagulation**, which is described as a curdling or congealing of the proteins. Both denaturation and coagulation are irreversible in most proteins. Examples include the hardening of egg whites with heating, the formation of yogurt as bacteria convert lactose to lactic acid and lower the pH, and the stiffening of egg whites when they are whipped (4). Adding compounds (like sugar) to an unbeaten egg white stabilizes the denatured protein; therefore, sugar is often added near the end of whipping, just before the egg whites have reached their optimum consistency. Salt speeds the coagulation of proteins by weakening the bonds of protein structure, which is why it is frequently used by cheese makers to help produce a firm curd.

Enzymatic Reactions. Enzymes (or biocatalysts) are one of the most important proteins formed within living cells because they act as biological catalysts to speed up chemical reactions (Chemist's Corner 2-9). Thousands of enzymes reside in a single cell, each one a catalyst that facilitates a specific chemical reaction. Without enzymes, reactions would occur in a random and indiscriminate manner. The lock-and-key concept describes enzyme action (Figure 2-27). An enzyme combines with a substance, called a **substrate**, catalyzing or speeding up a reaction, which releases a product. The enzyme is freed unchanged after the reaction and is able to react with another substrate, yielding another product.

Chemist's Corner 2-9

Enzyme Classification

Most enzymes are grouped into one of six different classes according to the type of reaction they catalyze (39). Hydrolases are the most common enzymes used by the food industry; they catalyze hydrolysis reactions (31). These hydrolytic enzymes break, or cleave, a chemical bond within a molecule by adding a molecule of water. Water actually is broken apart as its two hydrogens and oxygen become part of the two new molecules formed. Examples of hydrolases include lipases that hydrolyze lipids, proteases that hydrolyze protein, and amylases that hydrolyze starch.

Another type of enzyme, oxidoreductase, catalyzes oxidation reduction reactions. This type includes dehydrogenases, which act by removing hydrogen, and oxidases, which add oxygen.

Lyases assist in breaking away a smaller molecule, such as water, from a larger substrate. Transferases, as their name implies, transfer a group from one substrate to another. Ligases catalyze the bonding of two molecules. The last type, isomerases, transfer groups within molecules to yield isomeric forms.

Enzyme Nomenclature. The names of most enzymes end in "-ase". Enzymes are usually named after the substrate they act upon or the resulting type of chemical reaction. For example, sucrase is the enzyme that acts on sucrose, and lactase breaks down lactose to glucose and galactose. This general nomenclature rule does not always apply; the enzyme papain is named after papaya, from which it is derived, and ficin gets its name from figs. These enzymes, obtained from fruits, are used in meat tenderizers to break down meat's surface proteins.

Structure of Enzymes. The overall structure of an enzyme, called the holoenzyme, contains both a protein and nonprotein portion. Most of the enzyme is protein, but the nonprotein portion, which is necessary for activity, is either a coenzyme (usually a vitamin) or a cofactor (usually a mineral).

Factors Influencing Enzyme Activity. Enzymes are readily inactivated and will only operate under mild conditions of pH and temperature. Since enzymes are primarily protein, they are subject to denaturation caused by extremes in temperature or pH or even by physical and/or chemical influences. Every enzyme has an optimum temperature and pH for its operation, but most do best in the 95° to 104°F (35° to 40°C) range and with a pH near neutral.

Enzyme Use by the Food Industry. Many foods would not be on the market if it were not for certain enzymes. Foods that can be manufactured with the aid

KEY TERMS

Denaturation The irreversible process in which the structure of a protein is disrupted, resulting in partial or complete loss of function.

Coagulation The clotting or precipitation of protein in a liquid into a semisolid compound.

Substrate A substance that is acted upon, such as by an enzyme.

FIGURE 2-27 Lactase enzyme hydrolyzing lactose to glucose and galactose.

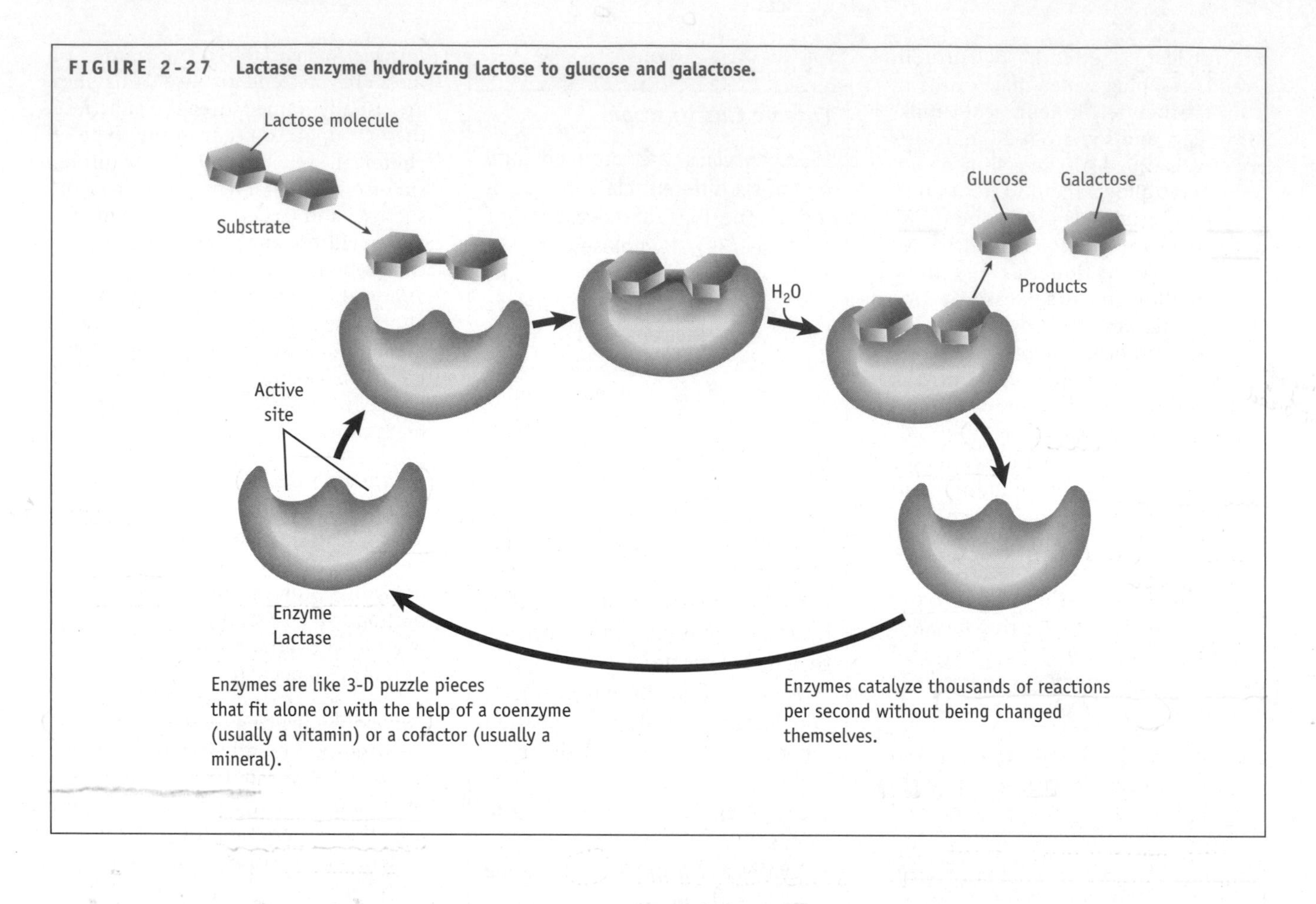

of enzymes include wines, cheeses, corn syrups, yogurt, cottage cheese, baked goods, sausages, juices, egg white replacers, the artificial sweetener Aspartame, and various Asian foods relying on molds (31, 32). Examples include:

- Rennin, also known as chymosin, aids in cheese production by converting milk to a curd. by coagulating milk
- Meats can be tenderized with the enzymes of papain, bromelain, and/or ficin.
- Phenol oxidase imparts the characteristic dark hue to tea, cocoa, coffee, and raisins.
- Glucose oxidase has been used for decades in the desugaring of eggs, flour, and potatoes, and in the preparation of salad dressings.
- Manufacturers of baked products use enzymes to retard staling, improve flour and dough quality, bleach flour, and enhance crust color.
- Enzymes can also be used in improving the filtration of beer (5).

Fruit juice processors use enzymes to increase juice yields, enhance juice clarity, improve filtration, reduce bitterness, and speed fruit dehydration. The enzymes most commonly used by fruit juice processors are pectinase, cellulase, hemicellulase, amylase, and arabinase. The bitter compounds in grapefruit juice—naringin and limonin—can be hydrolyzed with naringinase and limonase respectively (19).

Sometimes the food industry is more interested in inhibiting the action of enzymes. This is the case for lipoxygenase activity in milk, which produces off-flavors (3). The vulnerability of enzymes to high temperatures makes it easy to destroy enzymes that cause the spoilage of fruits and vegetables. Briefly submerging foods (usually vegetables) in boiling water denatures the enzymes that contribute to deterioration. Pasteurization of milk, which is intended to kill harmful bacteria, also halts enzyme activity.

Another major use of enzymes by the food industry is in quality testing of a variety of food products (Table 2-5). A test for ensuring that adequate pasteurization temperatures have been reached is to measure the activity of the phosphatase enzyme that naturally exists in milk. Lack of phosphatase activity indicates that sufficient heat was applied to destroy harmful microorganisms. Fish quality can be measured by using xanthine oxidase, which acts on hypoxanthine, a compound that increases as the fish spoils (19). A strip of absorbent paper soaked partially in xanthine oxidase can be used aboard ships, dockside, or in a food processing plant. The strip of paper is moistened in fish extracts and then observed for color intensity, which is correlated to freshness. Enzymes can also now be used to detect bacterial contamination in meat, poultry, fish, and dairy products.

TABLE 2-5

Use of Enzymes by the Food Industry to Test for Food Quality

For This Food	*Use This Enzyme*	*To Test for*
Fruits and vegetables	Peroxidase	Proper heat treatment
Milk, dairy products, ham	Alkaline phosphatase	
Eggs	ß-Acetylglucosaminidase	
Oysters	Malic enzyme	Freezing and thawing
Meat	Glutamate oxaloacetate	
Meat, eggs	Acid phosphatase	Bacterial contamination
Beans	Catalase	
Potatoes	Sucrose synthetase	Maturity
Pears	Pectinase	
Fish	Lysolecithinase	Freshness
	Xanthine oxidase	Hypoxanthine content
Flour	Amylase	Sprouting
Wheat	Peroxidase	
Coffee, wheat	Polyphenol oxidase	Color
Meat	Succinic dehydrogenase	

Buffering. Proteins have the unique ability to behave as buffers, compounds that resist extreme shifts in pH (Chemist's Corner 2–10). The buffering capacity of proteins is facilitated by their **amphoteric** nature.

Browning. Proteins play a very important role in the browning of foods through two chemical reactions: the Maillard reaction and enzymatic browning.

Maillard Reaction. The brown color produced during the heating of many different foods comes, in part, from the **Maillard reaction**. This reaction contributes to the golden crust of baked products, the browning of meats, and the dark color of roasted coffee. Temperatures most conducive to the Maillard reaction are those reaching at least 194°F (90°C), but browning can occur at lower temperatures, as seen in dried milk that has been stored too long.

Enzymatic Browning. **Enzymatic browning** is the result of an entirely different mechanism than the Maillard reaction. It requires the presence of three substances: oxygen, an enzyme (polyphenolase), and a phenolic compound (Figure 2-28 on page 46).

Another type of enzymatic browning occurs when the enzyme tyrosinase oxidizes the amino acid tyrosine to result in dark-colored melanin compounds such as that observed in browning mushrooms (19). Although the browning from either phenolase or tyrosinase is unappealing in itself, it is harmless.

HOW & WHY? ?????????

Why does an apple turn brown when you take a bite out of it and then let it sit? Enzymatic browning is responsible for the discoloration seen in certain cut fruits and vegetables. Normally, the cell structure separates the enzymes from the phenolic compounds in the fruit. When the vegetable or fruit is cut or bruised, however, the phenols and enzymes, thus exposed to oxygen, react in its presence to produce brown-colored products. Not all fruits and vegetables contain phenolic compounds, but sliced apples, pears, bananas, and eggplants turn brown rather rapidly after cutting. Potatoes turn slightly pink or gray.

Chemist's Corner 2-10

Proteins and Buffering

Proteins act as buffers to prevent extreme swings in acidity or alkalinity. This is due to their unique ability to accept or donate H^+. Specifically, amino groups on the amino acids act as bases (accept H^+ to yield $-NH^{3+}$), while the carboxyl groups act as acids (donate H^+ to yield $-COO^-$). When the amino and carboxyl groups are equally ionized (neutralized), the protein's isoelectric point is reached, defined as the point at which the protein's charges become neutral. Proteins are structurally unique because of these isoelectric points. Most proteins have isoelectric points ranging between pH 4.5 and 7.0. These different isoelectric points allow proteins to be separated by electrophoresis. This is a process in which an electrical field automatically causes the proteins to move on a plate toward each of their own neutral isoelectric points. Proteins do not all have the same neutral isoelectric points, so they stop at different points on the electrified plate and are "separated."

KEY TERMS

Amphoteric Capable of acting chemically as either acid or base.

■ ■ ■ ■

Maillard reaction The reaction between a sugar (typically reducing sugars such as glucose/dextrose, fructose, lactose, or maltose) and a protein (specifically the nitrogen in an amino acid), resulting in the formation of brown complexes.

■ ■ ■ ■

Enzymatic browning A reaction in which an enzyme acts on a phenolic compound in the presence of oxygen to produce brown-colored products.

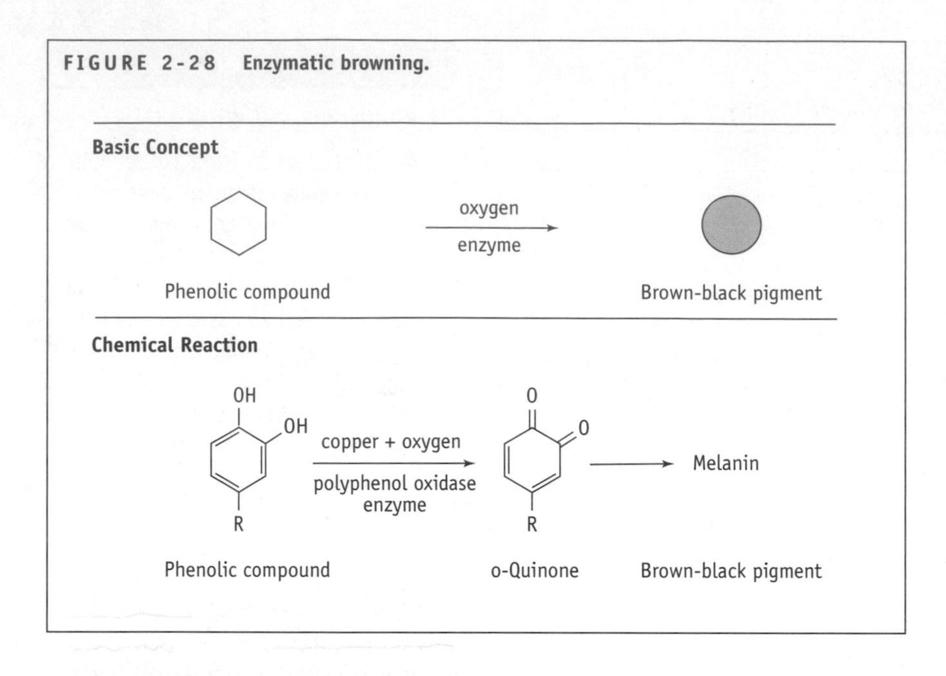

FIGURE 2-28 Enzymatic browning.

Vitamins and Minerals

Foods High in Vitamins and Minerals

Most foods contain some vitamins and minerals. Vitamins can be categorized into two major groups: fat-soluble (A, D, E, and K) or water-soluble (B complex and vitamin C). Minerals may be termed either macro or micro (Table 2-6). Meats are good sources of B vitamins, iron (Fe), and zinc (Zn). Dairy foods provide about 80 percent of the average American's daily calcium (Ca). Vitamin C (ascorbic acid) is found only in plants. All the fat-soluble vitamins (A, D, E, and K) are found in an egg yolk. Vitamin B_{12} is found only in foods of animal origin or fermented foods such as tempeh, tofu, and miso, which contain bacteria that produce vitamin B_{12} as a by-product. The two major sources of sodium (Na) in the diet are processed foods and the salt shaker.

KEY TERMS

Enriched Foods that have had certain nutrients, which were lost through processing, added back to levels established by federal standards.

Fortified Foods that have had nutrients added that were not present in the original food.

Antioxidant A compound that inhibits oxidation, which can cause deterioration and rancidity.

Free radical An unstable molecule that is extremely reactive and that can damage cells.

Composition of Vitamins and Minerals

Vitamins are organic (carbon-containing) compounds, each with a unique chemical composition. Minerals are inorganic elements and are depicted in the periodic table. Unlike vitamins, minerals cannot be destroyed by heat, light, or oxygen. Vitamins and minerals do not provide calories (kcal).

Functions of Vitamins and Minerals in Food

Enrichment and Fortification. Vitamins and minerals regulate metabolic functions. Because of the vital role these compounds play in the body's processes, many foods are now **enriched** or **fortified** with additional vitamins and minerals. During processing and preparation, foods such as wheat and rice may lose some of their vitamin or mineral content. Some of the nutrients, such as vitamin B_1 (thiamin), vitamin B_2 (riboflavin), niacin, and iron (calcium optional), may be added back (enriched) to the processed food. Fortification is intended to deliver nutrients to the general public in an effort to deter certain nutrient deficiencies (24). In 1922, salt became the first food ever fortified with the addition of iodine. Iodine deficiencies were resulting in goiter (enlarged thyroid gland) and cretinism (dwarfism, mental retardation) in children born of mothers who had not ingested sufficient iodine amounts. Other nutrients that are used to fortify foods include vitamins A and D (milk), calcium (orange juice), and/or folate, a B vitamin (cereal products) (23, 43).

The decision to fortify with a particular nutrient is a complex one. It starts with the realization that a significant number of people are not obtaining desirable levels of a specific nutrient, and the determination that the food to be fortified makes an appreciable contribution to the diet. It must be further ascertained that the fortification will not result in an essential nutrient imbalance, that the nutrient is stable under storage and capable of being absorbed from the food, and that toxicity from excessive intakes will generally not occur (33).

Antioxidants. Certain nutrients, especially vitamins A, C, and E and the mineral selenium, may also be added to foods to act as **antioxidants** (15). These compounds neutralize **free radicals** (Figure 2-29), leading to an increased shelf life (10). Foods to which antioxidants are commonly added include dry cereals, crackers, nuts, chips, and flour mixes. Consumer interest in antioxidants and health has also caused manufacturers to add additional amounts of these nutrients to other food products.

TABLE 2-6
Major Vitamins and Minerals in Foods

Vitamins		*Minerals*			
Water Soluble	**Fat Soluble**	**Macrominerals** *(minerals present in the body in relatively large amounts)*		**Microminerals** *(minerals present in the body in relatively small amounts)*	
B Complex:	Vitamin A	Calcium	(Ca)	Iron	(Fe)
Thiamin (B_1)	Vitamin D	Phosphorous	(P)	Zinc	(Zn)
Riboflavin (B_2)	Vitamin E	Potassium	(K)	Selenium	(Se)
Vitamin B_6 (pyridoxine)	Vitamin K	Sulfur	(S)	Manganese	(Mn)
Vitamin B_{12} (cobalamin)		Sodium	(Na)	Copper	(Cu)
Niacin		Chlorine	(Cl)	Iodine	(I)
Folate		Magnesium	(Mg)	Molybdenum	(Mo)
Pantothenic Acid				Chromium	(Cr)
Biotin				Fluorine	(F)
Ascorbic Acid (Vitamin C)					

Sodium. Another compound in the vitamin/mineral category that is used to preserve foods is salt, the only mineral directly consumed by people. The function of salt in foods, however, exceeds its preservation role. It provides flavor to so many processed foods that it is now the second most common food additive by weight, after sugar, in the United States. Salt can be purchased in various special forms by the food industry to add to their products: topping salt for saltine crackers, breadsticks, and snack crackers; surface-salting for pretzels, soft pretzels, and bagels; fine crystalline salt for potato chips, corn chips, and similar snacks; blending/dough salts for flour and cake mixes; light salt (potassium chloride) for reducing sodium levels; and encapsulated salt for frozen doughs (25).

FIGURE 2-29 Antioxidant action of vitamin E.

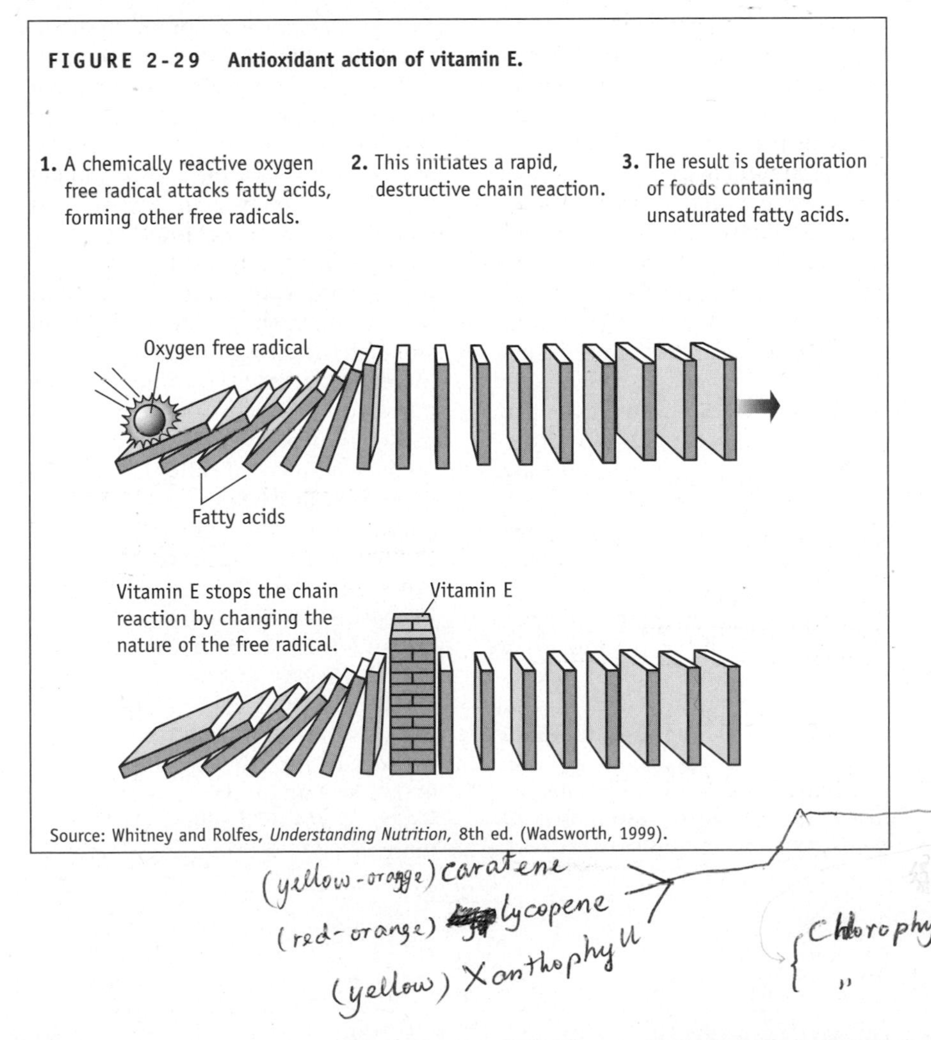

Source: Whitney and Rolfes, *Understanding Nutrition*, 8th ed. (Wadsworth, 1999).

Non-Nutritive Food Components

Foods contain some compounds that are not classified within the six basic nutrient groups. All sorts of substances can be found in food—natural, intentional, and unintentional. Only the natural compounds, whether beneficial or harmful, are discussed in this chapter. The intentional and unintentional compounds are food additives and are discussed in Chapter 8. Among the beneficial compounds naturally found in foods are those that provide color and flavor, along with certain plant compounds.

Color Compounds

Food is made more appetizing and interesting to behold by the wide spectrum of colors made possible through pigments. One way these pigments are classified by food scientists is according to the following four categories: (1) shiny (diffuse reflection), (2) glossy (specular or mirror-like reflection), (3) opaque or cloudy (diffuse transmission, or (4) translucent (specular transmission) (15). Most of the natural pigments contributing to color are found in fruits and vegetables. The colors of foods from animal products and grains are less varied and bright. The three dominant pigments found in most plants are carotenoids (orange-yellow), chlorophyll (green), and flavonoids (blue, cream, red). These pigments, their colors, and food examples are explained in more detail in Chapter 17.

Although foods of animal origin are less colorful, even meat varies in color depending on its stage of maturity (28). When first sliced with a knife, a cut of beef is purplish red from the presence of a pigment called myoglobin. As it is exposed to air, the myoglobin combines with oxygen to turn the meat a bright red color. The meat then turns grayish brown during cooking when the protein holding the pigment becomes denatured. Cured meats present an altogether different scenario as added nitrites, compounds that are used as a preservative, react with the myoglobin to cause the meat to be a red color, which converts to pink (denatured protein) when cooked. Meat pigments are covered in more detail in Chapter 14.

Milk appears white as light reflects off the colloidal dispersion of milk protein. The yellowish hue of cream comes from carotene and riboflavin (vitamin B_2). Carotene, a fat-soluble pigment, is also the substance that gives butter its yellow color.

TABLE 2-7
Phytochemicals: Potential Cancer Protectors

Phytochemical Family	*Major Food Sources*
Allyl Sulfides	Onions, garlic, leeks, chives
Carotenoids	Yellow and orange vegetables and fruits, dark green leafy vegetables
Flavinoids	Most fruits and vegetables
Indoles	Cruciferous vegetables (broccoli, cabbage, kale, cauliflower, etc.)
Isoflavones	Soybeans (tofu, soy milk)
Isothiocyanates	Cruciferous vegetables
Limonoids	Citrus fruits
Lycopene	Tomatoes, red grapefuit
Phenols	Nearly all fruits and vegetables
Phenolic acids	Tomatoes, citrus fruits, carrots, whole grains, nuts
Plant sterols	Broccoli, cabbage, cucumbers, squash, yams, tomatoes, egg plant, peppers, soy products, whole grains
Polyphenols	Green tea, grapes, wine
Saponins	Beans and legumes
Terpenes	Cherries, citrus fruit peel

Flavor Compounds

The flavors in foods are derived from both nutrient and non-nutrient compounds. These are sometimes too numerous to track as the source of a specific flavor. Among the non-nutrient compounds in foods are the organic acids that determine whether foods are acidic or basic. An acidic pH in foods not only contributes to a sour taste, but the color of fruit juices, the hue of chocolate in baked products, and the release of carbon dioxide in a flour mixture. An alkaline pH contributes a bitter taste and soapy mouthfeel to foods.

Plant Compounds

In addition to color and flavor compounds, some plants contain other non-nutritive substances that, when ingested, may have either beneficial or harmful effects.

Beneficial. Many of the possible anticarcinogens, or compounds that inhibit cancer, come from plants (1, 29). In particular, phytochemicals, a special group of substances found in plants, appear to have a protective effect against cancer (20). One class of these phytochemicals, called indoles, is found in vegetables such as cabbage, cauliflower, kale, kohlrabi, mustard greens, swiss chard, and collards. Laboratory animals given indoles and then exposed to carcinogens developed fewer tumors than animals exposed to the same carcinogens, but not given indoles. A few of the plant substances that appear to protect against cancer are listed in Table 2-7.

Harmful. There are several potentially harmful substances in plants (17, 29). The U.S. Department of Health and Human Services has gone so far as to say that natural toxins are so widespread that the only way to avoid them completely is to stop eating (see Chapter 3). Other substances, although not strictly toxins, can cause problems for certain people if ingested in excess. One such substance is caffeine (35).

Caffeine. Caffeine is a natural stimulant that belongs to a group of compounds called methylxanthines. The most widely used sources of caffeine include coffee beans, tea leaves, cocoa beans, and cola nuts. Caffeine is also found in the leaves of some plants, where it acts as a protection against insects.

Caffeine ingested at high concentrations may temporarily increase heart rate, basal metabolic rate, secretion of stomach acid, and urination. The increased secretion of stomach acids may cause problems for people with ulcers. In healthy adults, however, a moderate intake of caffeine does not appear to cause health problems. Individuals who habitually drink a lot of caffeine-containing beverages may, however, experience withdrawal headaches and irritability if they stop drinking the beverage. Another possible side effect in sensitive individuals is fibrocystic conditions in the breast, which is the painful but usually harmless occurrence of lumpiness in the breasts and under the arms (13).

Excessive caffeine intake, defined as more than five 5-ounce cups of strong, brewed coffee daily, can also cause "coffee nerves." The Diagnostic and Statistical Manual of Mental Disorders published by the American Psychiatric Association defines caffeine intoxication as exhibiting at least five of the following symptoms: nervousness, agitation, restlessness, insomnia, frequent urination, gastrointestinal disturbance, muscle twitching, rambling thought and speech, periods of exhaustion, irregular or rapid heartbeat, and psychomotor agitation (2). Infants who ingest caffeine through their mother's milk may also get the "jitters." Because infants are unable to metabolize caffeine efficiently, the compound may stay in their system up to a week, compared to about twelve hours in a healthy adult.

PICTORIAL SUMMARY / 2: Chemistry of Food Composition

Truly, we are what we eat. Food provides energy (kilocalories) and nutrients, which are needed for the maintenance, repair, and growth of cells. Understanding food chemistry is important in planning good nutrition.

BASIC FOOD CHEMISTRY

The six major nutrient groups are:

- Water
- Carbohydrates
- Proteins
- Lipids
- Vitamins
- Minerals

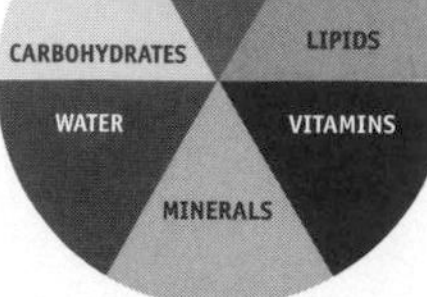

Most foods contain a combination of these six groups. The main purpose of eating and drinking is to provide calories (kcal) and to replace those nutrients used up in the body's maintenance, repair, and growth. Calories (kcal) are to the body as fuel is to a car and our bodies use calories even while we are sleeping, allocating them as follows:

- **60%** are used for vital life functions, such as maintaining body temperature, respiration and heartbeat.
- **10%** are used for digesting and absorbing food nutrients.
- **30%** (depending on the person) are used for physical activity.

Minerals
Water
Fat
Carbohydrate
Protein

Proportion of nutrients in the human body

WATER

Water (H_2O) is the simplest but most important of all nutrients. The human body contains about 55 to 60 percent water, and water concentration in foods ranges from 70 to 90 percent in fruits, vegetables, and meats to less than 15 percent in grains, dried beans, and fats. In food preparation, water acts as a heat-transferring agent and a universal solvent, plays a crucial role in preservation, and is involved in the formation of numerous solutions and colloidal dispersions, suspensions, and emulsions.

NON-NUTRITIVE FOOD COMPONENTS

Non-nutritive components in foods include color compounds (cartenoids, chlorophylls, flavonoids), flavor compounds, and beneficial or harmful plant compounds.

PROTEINS

Proteins are essential to proper growth and maintenance. They differ from carbohydrates and lipids in that they contain nitrogen. Proteins consist of amino acids linked together by peptide bonds. Plant proteins, with the exception of the protein in soybeans, lack all the essential amino acids and are therefore "incomplete."

LIPIDS OR FATS

Lipids, which are derived from both plant and animal sources, include fats and oils. Foods that are high in fat:

Animal sources

- Meats
- Poultry
- Dairy products

Plant sources

- Seeds and nuts
- Vegetable oils
- Avocados, olives, coconut

Three major lipid groups:

- Triglycerides
- Phospholipids
- Sterols

CARBOHYDRATES

Carbohydrates are the sugars, starches, and fibers found primarily in plants. The basic building block of carbohydrates is the "saccharide," which is composed of carbon, hydrogen, and oxygen. Starches consist of amylose and amylopectin. Fiber can be defined by whether or not it is crude or dietary and soluble or insoluble.

VITAMINS AND MINERALS

Both vitamins and minerals function at the cellular level. Neither contain calories (kcal) and both are found to some degree in most foods. Some foods are enriched (nutrients added that were lost in processing) or fortified (nutrients added that were not originally in the food) with vitamins and/or minerals. Salt is one of the few minerals used for a functional purpose in foods, specifically its ability to act as a preservative.

REFERENCES

1. Ahmad N, and H Muck tar. Green tea polyphones and cancer: Biologic mechanisms and practical implications. *Nutrition Reviews* 57(3):78–83, 1999.
2. American Psychiatric Association. *Diagnostic and Statistical Manual of Mental Disorders. DSM-IV.* American Psychiatric Association, 1994.
3. Ashie INA, BK Simpson, and JP Smith. Mechanisms for controlling enzymatic reactions in foods. *Critical Reviews in Food Science and Nutrition* 36(1&2):1–30, 1996.
4. Baniel A, A Fains, and Y Popineau. Foaming properties of egg albumen with a bubbling apparatus compared with whipping. *Journal of Food Science* 62(2):377–381, 1997.
5. Berne S, and CD O'Donnell. Filtration systems and enzymes: A tangled web. *Prepared Foods* 165(9): 95–96, 1996.
6. Cai R, and SD Arntfield. Thermal gelation in relation to binding of bovine serum albumin-polysaccharide systems. *Journal of Food Science* 62(6):1129–1134, 1997.
7. Cornec M, et al. Emulsion stability as affected by competitive absorption between an oil-soluble emulsifier and milk proteins at the interface. *Journal of Food Science* 63(1):39–43, 1998.
8. Crittenden RG. Functional polymers for the next millennia. *Prepared Foods* 166(6):123–124, 1997.
9. Dartey CK, and GR Sanderson. Use of gums in low-fat spreads. *Inform* 7(6):630–634, 1996.
10. Elliot JG. Application of antioxidant vitamins in foods and beverages. *Food Technology* 53(2):46–48, 1999.
11. Ensminger AH, et al. *Foods and Nutrition Encyclopedia.* CRC Press, 1994.
12. Fats and fattening: Fooling the body. *Prepared Foods* 166(7): 51–53, 1997.
13. Ferrini RL, and E Barrett-Connor. Caffeine intake and endogenous sex steroid levels in postmenopausal women. *American Journal of Epidemiology* 144(7): 642–644, 1996.
14. Giese J. Antioxidants: Tools for preventing lipid oxidation. *Food Technology* 50(11):73–78, 1996.
15. Giese J, Color measurement in foods quality parameter. *Food Technology* 54(2):62, 2000.
16. Hamilton EMN, and SAS Gropper. *The Biochemistry of Human Nutrition.* Wadsworth, 1987.
17. Horn S, et al. End-stage renal failure from mushroom poisoning with Cortinarius orellanus: Report of four cases and review of the literature. *American Journal of Kidney Diseases* 30(2):282–286, 1997.
18. Inulin is positively charged. *Prepared Foods* 166(12):85, 1997.
19. James J, and BK Simpson. Application of enzymes in food processing. *Critical Reviews in Food Science and Nutrition* 36(5): 437–463, 1996.
20. King A, and G Young. Characteristics and occurrence of phenolic phytochemicals. *Journal of the American Dietetic Association* 99(2):213–218, 1999.
21. Kleiner SM. Water: An essential but overlooked nutrient. *Journal of the American Dietetic Association* 99(2):200–206, 1999.
22. Kritchevsky D. Dietary fiber. *Annual Reviews in Nutrition* 8:301–328, 1988.
23. Labin-Godscher R, and S Edelstein. Calcium citrate: A revised look at calcium fortification. *Food Technology* 50(6):96–97, 1996.
24. Mertz W. Food fortification in the United States. *Nutrition Reviews* 55(2):44–49, 1997.
25. Niman S. Using one of the oldest food ingredients—salt. *Cereal Foods World* 41(9):728–731, 1996.
26. O'Donnell CD. Proteins and gums: The ties that bind. *Prepared Foods* 165(4):50–51, 1996.
27. Ohr LM. Nutraceuticals and functional foods. *Food Technology* 56(6):109–115, 2002.
28. Pegg RB, and F Shahidi. Unraveling the chemical identity of meat pigments. *Critical Reviews in Food Science and Nutrition* 37(6):561–589, 1997.
29. Phillips BJ, et al. A study of the toxic hazard that might be associated with the consumption of green potato tops. *Food and Chemical Toxicology* 34(5):439–448, 1996.
30. Potter JD. Reconciling the epidemiology, physiology, and molecular biology of colon cancer. *Journal of the American Medical Association* 268:1573–1577, 1992.
31. Pszczola DE. Enzymes: Making things happen. *Food Technology* 53(2):74–79, 1999.
32. Pszczola DE. From soybeans to spaghetti: The broadening use of enzymes. *Food Technology* 55(11):2001.
33. Reilly C. Too much of a good thing? The problem of trace element fortification of foods. *Trends in Food Science & Technology* 7(4):139–142, 1996.
34. Sanderson GR. Gums and their use in food systems. *Food Technology* 50(3):81–84, 1996.
35. Scientific status summary: Evaluation of caffeine safety. *Food Technology* 41(6):105–113, 1987.
36. U.S. Dept. of Health and Human Services. Food and Drug Administration. Center for Food Safety and Applied Nutrition. Factors affecting growth of microorganisms in foods. In *Foodborne Pathogenic Microorganisms and Natural Toxins Handbook* (Bad Bug Book). 1997. Available at http://www.cfsan.fda.gov.
37. VanGarde SJ, and M Woodburn. *Food Preservation and Safety: Principles and Practice.* Iowa State University Press, 1994.
38. Ward FM. Hydrocolloid systems as fat memetics in bakery products: Icings, glazes and fillings.

Cereal Foods World 42(5):386–390, 1997.

39. Whitehead IM. Challenges to biocatalysts from flavor chemistry. *Food Technology* 52(2):40–46, 1998.
40. Whitney EN, CB Cataldo, and SR Rolfes. *Understanding Normal and Clinical Nutrition.* Wadsworth/West Publishing, 1994.
41. Wursch P, and FX Pi-Sunyer. The role of viscous soluble fiber in the metabolic control of diabetes. A review with special emphasis on cereals rich in beta-glucan. *Diabetes Care* 20(11):1774–1780, 1997.
42. Xie YR, and NS Hettiarachchy. Xanthan gum effects on solubility and emulsification properties of soy protein isolate. *Journal of Food Science* 62(6):1101–1104, 1997.
43. Yetley EA, and JI Rader. Folate fortification of cereal-grain products: FDA policies and actions. *Cereal Foods World* 40(2):67–72, 1995.
44. Yuno-Ohta N, et al. Gelation properties of ovalbumin as affected by fatty acid salts. *Journal of Food Science* 61(5):906–910, 1997.

WEBSITES

Analyze your own diet for calories and nutrients at:
www.nat.uiuc.edu

The USDA Agricultural Research Service's Nutrient Data Laboratory has an online data resource for nutrient-conscious consumers. Check out how many calories and nutrients are in USDA's database of different foods:
www.nal.usda.gov/fnic/foodcomp

Find out about the composition and properties of foods at the Institute of Food Technologists (IFT) website:
www.ift.org/divisions/food_chem/

The PD Lab website was created for the food industry as a central site for food product developers:
www.pdlab.com

See the following sites for specific information on carbohydrates **(www.pdlab.com/pdcarbox.htm),** sugars **(www.pdlab.com/pdsugarsx.htm),** fats and oils **(www.pdlab.com/pdfatsx.htm),** and protein **(www.pdlab.com/pdproteinx.htm).**

3

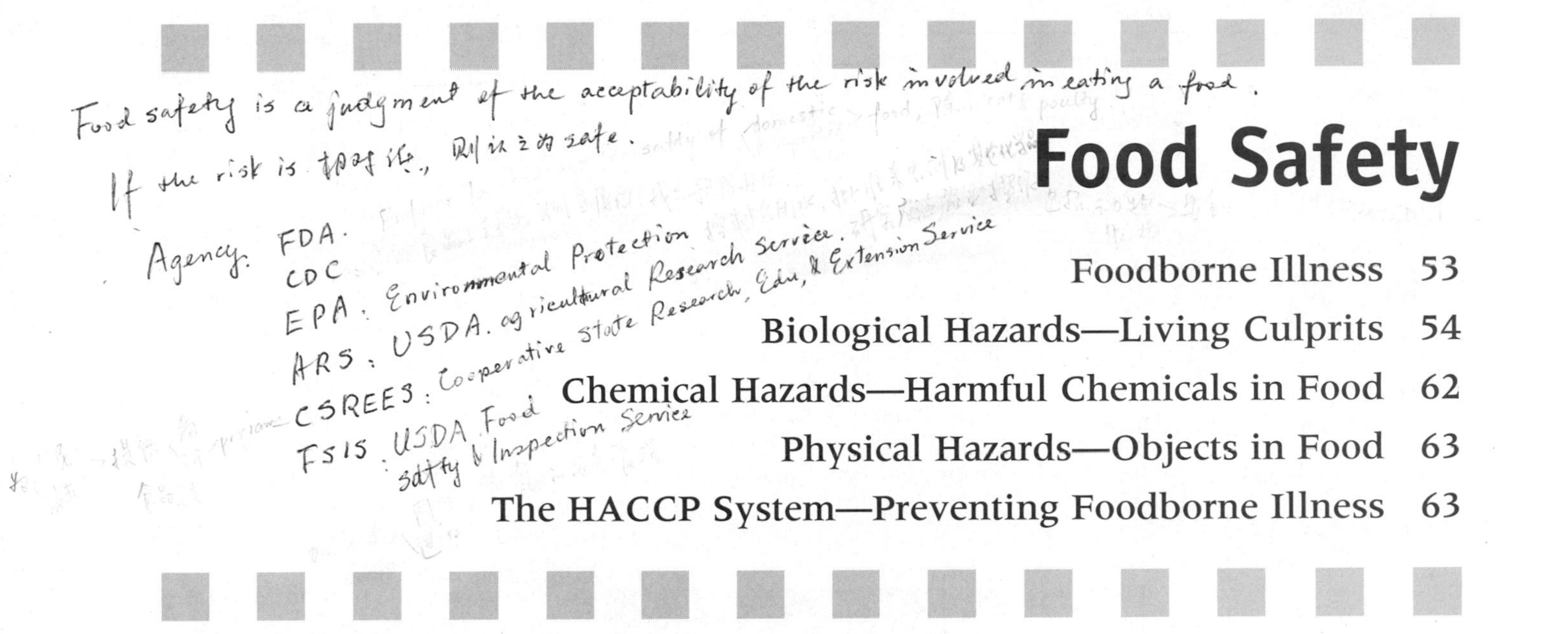

Food Safety

The United States food supply is probably the safest in the world (21). Why is this so? Federal and state regulations, along with inspections, require vigilance at all levels of food production and distribution (Chapter 8 covers government regulations in detail). Additionally, the Centers for Disease Control and Prevention (CDC) tracks down causal factors when even as few as one or two **outbreaks** of **foodborne illness** are found. Lastly, food manufacturers and distributors are motivated to avoid lawsuits due to negligence. The final result is general peace of mind for consumers.

In spite of all these preventive measures, people still get sick from food and beverages. Although it is difficult to assess the total number of people afflicted with foodborne illness each year, the General Accounting Office estimates that as many as 76 million illnesses and up to 5,000 deaths can be traced to contaminated foods (45). The CDC reports that approximately 80 percent of these foodborne illnesses originate at restaurants and other food service establishments, while most of the rest are traced to errors at home (38). There is also the perceived threat of intentional "food supply terrorism," a concern that has risen after September 11, 2001. Food biosecurity aims to keep the food supply free from planned contamination with biological, chemical, or physical hazards due to malicious and/or criminal intent.

Before developing an irrational fear of foodborne illness, consider the number of meals prepared and eaten each year. More than 273 billion meals and an inestimable number of snacks are consumed each year in the United States. Seen next to these numbers, the number of illnesses resulting from food contamination in this country is minimal. This chapter will examine the ways in which foods can cause illness, along with what can be done to prevent foodborne illness from happening.

Foodborne Illness

Many people have suffered the unpleasant experience of a foodborne illness. Symptoms of foodborne illness include inflammation of the gastrointestinal tract lining (gastroenteritis), nausea, abdominal cramps, diarrhea, and vomiting. About one-third of all reported diarrhea cases in the United States have been linked to foodborne illnesses (42). The severity of diarrhea or any of the other symptoms varies depending on the type of causative agent, the amount of it consumed, and the age and susceptibility of the immune system of the affected individual. Those most seriously affected by foodborne illness are the very young, the old, and those with immune systems compromised by diseases such as AIDS or cancer (51). Mild cases of foodborne illness usually subside with time. Dehydration resulting from diarrhea and vomiting can be treated by the consumption of electrolyte-rich liquids. Severe cases may

KEY TERMS

Outbreak Defined by the CDC as the occurrence of two or more cases of a similar illness resulting from the ingestion of a common food.

Foodborne illness An illness transmitted to humans by food.

TABLE 3-1
Types of Foodborne Hazards

Biological	*Chemical*	*Physical*
Bacteria	Plant toxins	Glass
Molds	Animal toxins	Bone
Viruses	Agricultural chemicals	Metal
Parasites	Industrial chemicals	Plastic
Prions		

result in hospitalization and even death.

What Causes Foodborne Illness?

People get sick from food that has been contaminated by one of three types of food hazards: (1) biological, (2) chemical, and (3) physical (Table 3-1). Biological hazards are living organisms or organic material that include bacteria, molds, viruses, parasites, and protein particles called prions. Some of these hazards are so small that they cannot be seen except with a microscope. Bacteria, molds, viruses, and some parasites are examples of these kinds of microorganisms ("micro" means small), and are the topics of an introductory microbiology class. Chemical hazards are chemical substances that can harm living systems. These range from agricultural and industrial contaminants (including cleaners and sanitizers) to plant and animal toxins. Physical hazards include any foreign material such as glass, bone, metal, and plastic that could potentially cause harm if ingested.

KEY TERMS

Bacteria One-celled microorganisms abundant in the air, soil, water, and/or organic matter (i.e., the bodies of plants and animals).

■■■■

Pathogenic Causing or capable of causing disease.

■■■■

Food infection An illness resulting from ingestion of food containing large numbers of living bacteria or other microorganisms.

■■■■

Food intoxication An illness resulting from ingestion of food containing a toxin.

Biological Hazards—Living Culprits

Foodborne biological hazards are organisms such as bacteria, molds, viruses, and parasites. The seriousness of these different biological hazards varies greatly as shown in Table 3-2. It is difficult to completely avoid microorganisms because they are everywhere. However, most biological hazards are inactivated or killed by adequate cooking and/or their numbers are kept to a minimum by sufficient cooling.

Bacteria—#1 Cause of Foodborne Illness

More than 90 percent of all foodborne illness is caused by **bacteria**, but only about 4 percent of identified bacteria are **pathogenic**. The remaining 96 percent are harmless, and some are even used to produce such food items as cheese, yogurt, soy sauce, butter, sour cream, buttermilk, cured meats, sourdough bread, and fermented foods such as pickles, beer, and sauerkraut. Both beneficial and pathogenic bacteria are everywhere. Pathogenic bacteria cause one of three types of foodborne illness: (1) infection, (2) intoxication or poisoning, and (3) intoxification. These are briefly summarized below and explained in greater detail in the shaded insert.

Food Infections. Approximately 80 percent of all bacterial foodborne illnesses are due to **food infections**. These types of foodborne illnesses are caused by ingesting bacteria that grow in the host's intestine, replicate, and create an infection through their colonization. Table 3-3 lists the bacteria primarily responsible for food infections.

Food Intoxication. Foodborne illnesses can also be the result of **food intoxication** or poisoning. Bacteria grow on the food and release toxins that cause illness in the person consuming the toxin-laden food or beverage. Even though certain plants and animals produce toxins, the most

TABLE 3-2
Biological Hazards Grouped According to Severity of Risk

Severe Hazards	*Moderate Hazards: Potentially Extensive Spread**	*Moderate Hazards: Limited Spread*
Clostridium botulinum	*Listeria monocytogenes*	*Bacillus cereus*
Shigella dysenteriae	*Salmonella*	*Campylobacter jejuni*
Salmonella Typhi	*Shigella*	*Clostridium perfringens*
Hepatitis A and E	Enterovirulent *Escherichia coli (EEC)*	*Staphylococcus aureus*
Brucella abortus	*Streptococcus pyogenes*	*Vibrio cholerae* (non-01)
Vibrio cholerae (01)	Rotavirus	*Vibrio parahaemolyticus*
Vibrio vulnificus	Norwalk virus group	*Yersinia enterocolitica*
		Giardia lamblia

*Although classified as moderate hazards, complications and aftereffects may be severe in certain susceptible populations.
Source: Adapted from International Commission on Microbiological Specifications for Food (ICMSF) (1986). Pierson and Corlett, eds., *HACCP Principles and Applications* (New York: Chapman & Hall, 1992).

TABLE 3-3

Bacterial Food Infections

Disease (causative agent)	*Latency Period (duration)*	*Principal Symptoms*	*Typical Foods*	*Mode of Contamination*	*Prevention of Disease*
Listeriosis (*Listeria monocytogenes*)	3–70 days	Meningoencephalitis; stillbirths; septicemia or meningitis in newborns	Raw milk, cheese, and vegetables	Soil or infected animals, directly or via manure	Pasteurization of milk; cooking
Salmonellosis (*Salmonella* species)	12–36 hr (2–7 days)	Diarrhea, abdominal pain, chills, fever, vomiting, dehydration	Raw, undercooked eggs; raw milk, meat, and poultry	Infected food-source animals; human feces	Cook eggs, meat, and poultry thoroughly; pasteurize milk; irradiate chickens
Shigellosis (*Shigella* species)	12–48 hr (4–7 days)	Diarrhea, fever, nausea; sometimes vomiting, cramps	Raw foods	Human fecal contamination, direct or via water	General sanitation; cook foods thoroughly
Streptococcus pyogenes	1–3 days (varies)	Various, including sore throat; erysipelas, scarlet fever	Raw milk, deviled eggs	Handlers with sore throats, other "strep" infections	General sanitation; pasteurize milk
Yersiniosis (*Yersinia enterocolitica*)	3–7 days (2–3 weeks)	Diarrhea, pains, mimicking appendicitis, fever, vomiting, etc.	Raw or undercooked pork and beef; tofu packed in spring water	Infected animals, especially swine; contaminated water	Cook meats thoroughly; chlorinate water

TABLE 3-4

Bacterial Food Intoxicants

Disease (causative agent)	*Latency Period (duration)*	*Principal Symptoms*	*Typical Foods*	*Mode of Contamination*	*Prevention of Disease*
Botulism (*Clostridium botulinum*)	12–36 hr (months)	Fatigue, weakness, double vision, slurred speech, respiratory failure, sometimes death	Types A&B: vegetables, fruits; meat, fish, and poultry products; condiments; Type E: fish and fish products	Types A&B: soil or dust; Type E: water and sediments	Thorough heating and rapid cooling of foods
Botulism, infant infection (*Clostridium botulinum*)	Unknown	Constipation, weakness, respiratory failure, sometimes death	Honey, soil	Ingested spores from soil or dust or honey colonize intestine	Do not feed honey to infants—will not prevent all
Clostridium perfringens	8–24 hr (12–24 hr)	Diarrhea, cramps, rarely nausea and vomiting	Cooked meat and poultry	Soil, raw foods	Thorough heating and rapid cooling of foods
Diarrheal (*Bacillus cereus*)	6–15 hr (12–24 hr)	Diarrhea, cramps, occasional vomiting	Meat products, soups, sauces, vegetables	From soil or dust	Thorough heating and rapid cooling of foods
Emetic (*Bacillus cereus*)	½–6 hr (5–24 hr)	Nausea, vomiting, sometimes diarrhea and cramps	Cooked rice and pasta	From soil or dust	Thorough heating and rapid cooling of foods
Staphyolccocal food poisoning (*Staphyloccocus aureus*)	½–8 hr (6–48 hr)	Nausea, vomiting, diarrhea, cramps	Ham, meat, poultry products, cream-filled pastries, whipped butter, cheese	Handlers with colds, sore throats or infected cuts, food slicers	Thorough heating and rapid cooling of foods

common food intoxicants originate from bacteria. Table 3-4 identifies bacteria most often associated with food intoxication.

Food Intoxification. The term "intoxification" sounds very much like "intoxication," but there is a difference. This type of foodborne illness occurs when bacteria enter the intestinal track and *then* start to produce the toxin in the intestine (Table 3-5).

TABLE 3-5

Bacterial Food Intoxifications

Disease (causative agent)	Latency Period (duration)	Principal Symptoms	Typical Foods	Mode of Contamination	Prevention of Disease
Campylobacteriosis (*Campylobacter* jejuni)	2–5 days (2–10 days)	Diarrhea, abdominal pain, fever, nausea, vomiting	Infected food-source animals	Chicken, raw milk	Cook chicken thoroughly; avoid cross-contamination; irradiate chickens; pasteurize milk
Cholera (*Vibrio cholerae*)	2–3 days hours to days	Profuse, watery stools; sometimes vomiting; dehydration; often fatal if untreated	Raw or undercooked seafood	Human feces in marine environment	Cook seafood thoroughly; general sanitation
Escherichia coli enterohemorrhagic	12–60 hr (2–9 days)	Watery, bloody diarrhea	Raw or undercooked beef, raw milk	Infected cattle	Cook beef thoroughly
Escherichia coli enteroinvasive	at least 18 hr (uncertain)	Cramps, diarrhea, fever, dysentary	Raw foods	Human fecal contamination, direct or via water	Cook foods thoroughly; general sanitation
Escherichia coli enterotoxigenic	10–72 hr (3–5 days)	Profuse watery diarrhea; sometimes cramps, vomiting	Raw foods	Human fecal contamination, direct or via water	Cook foods thoroughly; general sanitation
Vibrio parahaemolyticus	12–24 hour (4–7 days)	Diarrhea, cramps; sometimes nausea, vomiting, fever; headache	Fish and seafoods	Marine coastal environment	Cook fish and seafoods thoroughly
Vibrio vulnificus	In persons with high serum iron (1 day)	Chills, fever, prostration, often death	Raw oysters and clams	Marine coastal environment	Cook shellfish thoroughly

Bacterial Food Infections

The main bacteria that cause food infections via colonization in the intestinal tract include *Salmonella, Listeria monocytogenes, Yersinia enterocolitica,* and *Shigella* (Table 3-3).

Salmonella

Salmonella is one of the most common causes of illnesses traced to contaminated foods and water (Figure 3-1). Foods most susceptible to *Salmonella* contamination include meat, fish, poultry, eggs (especially eggnog or Caesar salad made with raw egg), and dairy products (especially custard fillings, cream, ice cream, sauces, dressings, and raw or unpasteurized milk).

Poultry is particularly vulnerable to *Salmonella* contamination (35). If birds are to be stuffed, this should be done just prior to cooking, and the stuffing removed from the cavity immediately after cooking and refrigerated as soon as possible. If reheated, it should be brought to a temperature of at least 165°F (74°C) prior to consumption. Current recommendations suggest that large birds should not be stuffed at all.

Eggs are also at risk for *Salmonella (S. enteritidis)*. Any crack or hole in an egg allows bacterial contamination to occur, so any damaged eggs in a carton should be discarded. Research suggests that *Salmonella enteritidis* can even be transmitted from infected hens to the eggs they lay (7). Consequently, some states have laws prohibiting the use of raw eggs in institutional settings. The FDA now has a regulation requiring a printed warning to consumers regarding the risk of undercooked eggs (Chemist's Corner 3.1).

Yet another source of *Salmonella* contamination is pet turtles, iguanas,

FIGURE 3-1 Salmonella.

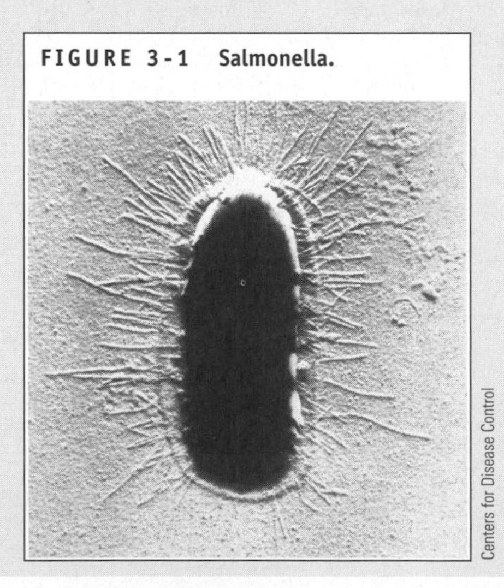

Centers for Disease Control

Chemist's Corner 3-1

Testing for Contamination

Traditional testing procedures employed by the USDA to check for the presence of contamination are tedious and time consuming. DNA analysis (genotyping) can do it much faster (29). After cutting an isolated bacterial colony's DNA with a specific enzyme, ribotyping can then be used to identify the bacterial strains by the resulting RNA fragments.

and other reptiles, so hand washing is essential after handling such pets.

Listeria Monocytogenes

Listeria monocytogenes infection can have serious consequences. The fatality rates are as high as 20 to 35 percent of those infected. The CDC recorded 400 U.S. deaths from *Listeria* infections in 1991. *Lysteria* infection may also cause pneumonia, septicemia, urethritis, meningitis, and spontaneous abortion (49). *Listeria monocytogenes* is unique for several reasons. It is a facultative bacteria (capable of growing with or without oxygen); it can survive a wide pH range (from 4.8 to 9.0); and it grows in a wide temperature range (39° to 113°F/4° to 45°C). It is one of the few bacteria that can thrive at refrigerator temperatures (54), and frozen dairy desserts have been implicated in some cases of *Listeria monocytogenes* contamination (17). Other foods associated with *Listeria* outbreaks are contaminated cabbage, pasteurized milk, luncheon meats, and Mexican-style soft cheese (6).

Yersinia Enterocolitica

Yersinia enterocolitica is destroyed by heat, but the concern is that, like *Listeria,* it too can grow in a wide temperature range (32° to 106°F/0° to 41°C). The ability of this bacteria to grow at refrigerator temperatures makes it all the more hazardous. Yersiniosis infection commonly occurs in children, resulting in gastrointestinal upset, fever, and appendicitis-like symptoms. In one outbreak, 36 children were hospitalized with apparent acute appendicitis and several underwent appendectomies before health officials determined that they had been infected with *Yersinia enterocolitica* by drinking contaminated chocolate milk (8). Yersiniosis infection can occasionally also cause septicemia (bacteria in the blood), meningitis (inflammation of the spinal cord or brain membranes), and arthritis-like symptoms.

Shigella

Poor personal hygiene by food handlers is the number-one cause of *Shigella* infection. *Shigella* is carried in the intestinal tract and transferred to the hands of food service personnel who visit the restroom and do not wash their hands (Figure 3-2).

FIGURE 3-2

Bacterial Food Intoxications

Food intoxication or poisoning occurs when a food is consumed that contains a toxin produced by bacteria such as *Staphylococcus aureus* and *Clostridium botulinum* (Table 3-4).

Staphylococcus Aureus

A major cause of foodborne illness, *Staphylococcus aureus* is ubiquitous (found everywhere) (Figure 3-3). Up to half of all healthy humans carry it, and it is a common cause of sinus infections and infected pimples and boils. It lives in the throat and nasal passages and in small cuts, so it is easily transmitted to foods through sneezing, coughing, and hand contact (52).

FIGURE 3-3 Staphylococcus aureus.

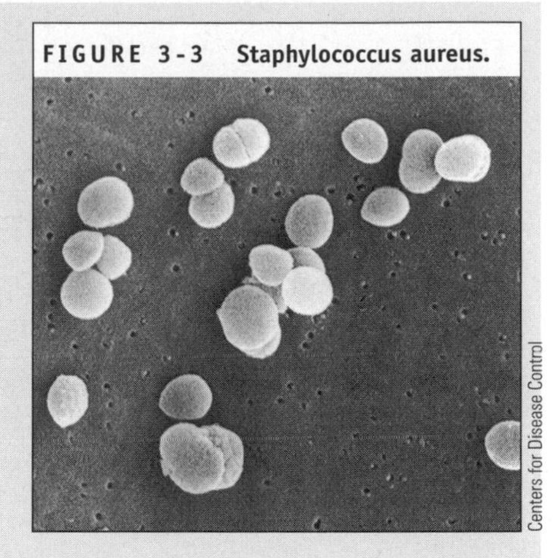

Centers for Disease Control

Clostridium Botulinum

The *Clostridium botulinum* toxin causes botulism, one of the deadliest, but fortunately rarest, forms of foodborne

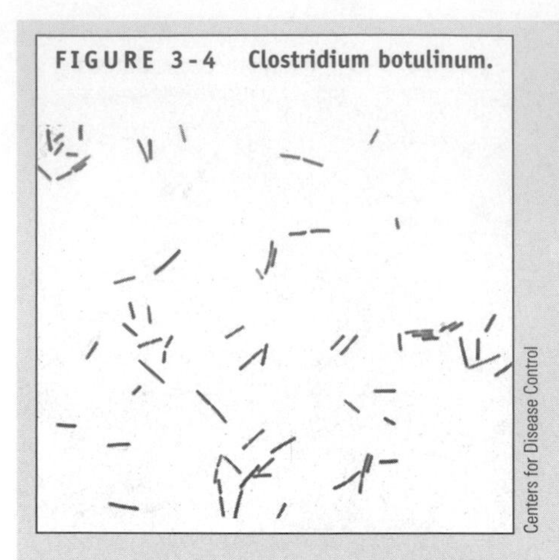

FIGURE 3-4 Clostridium botulinum.

illness (Figure 3-4). Less than a half cup of *botulinum* toxin is enough to poison every person on earth. Medical advances, including the development of an antitoxin, have contributed to reducing the death rate from botulism to less than 2 percent (57).

The most common cause of botulism is improperly home-canned food (see Chapter 7). Cans that are dented, have leaky seals, or bulge (indicating the presence of the gas produced by the bacteria) should be discarded. A foul odor or milky liquid in any can is also a sign of contamination.

Bacterial Food Intoxifications

Common examples of bacteria causing food intoxifications include *Escherichia coli, Campylobacter jejuni,* and *Vibrio* (Table 3-5).

Escherichia Coli

E. coli is dangerous, especially to children. The CDC estimates that between 7,600 and 20,400 people become ill and 120 to 360 people die each year from *E. coli* (21). The main concern for children is that certain strains of *E. coli* cause infant diarrhea, traveler's diarrhea, and bloody diarrhea. *E. coli* may also cause hemorrhagic colitis—severe abdominal cramps, vomiting, diarrhea, and a short-lived fever followed by watery, bloody diarrhea (11). A potentially deadly condition cause by *E. coli* is hemolytic uremic syndrome (HUS), which is the leading cause of acute renal (kidney) failure in children (30). Of those developing HUS, about 5 percent may die (44).

E. coli is naturally found in the intestinal tract, and only causes problems when fecal matter gets into the food or water supply (12). Most infections have been linked with undercooked meat, since contamination often occurs as a result of the meat coming into contact with the intestinal tract during the butchering of a carcass (Figure 3-5) (18). Undercooked hamburger is the most common meat source of *E. coli* contamination.

Scientists began identifying *E. coli* O157:H7 in the early 1980s, but it was

FIGURE 3-5 How *E. coli* can cause serious health problems.

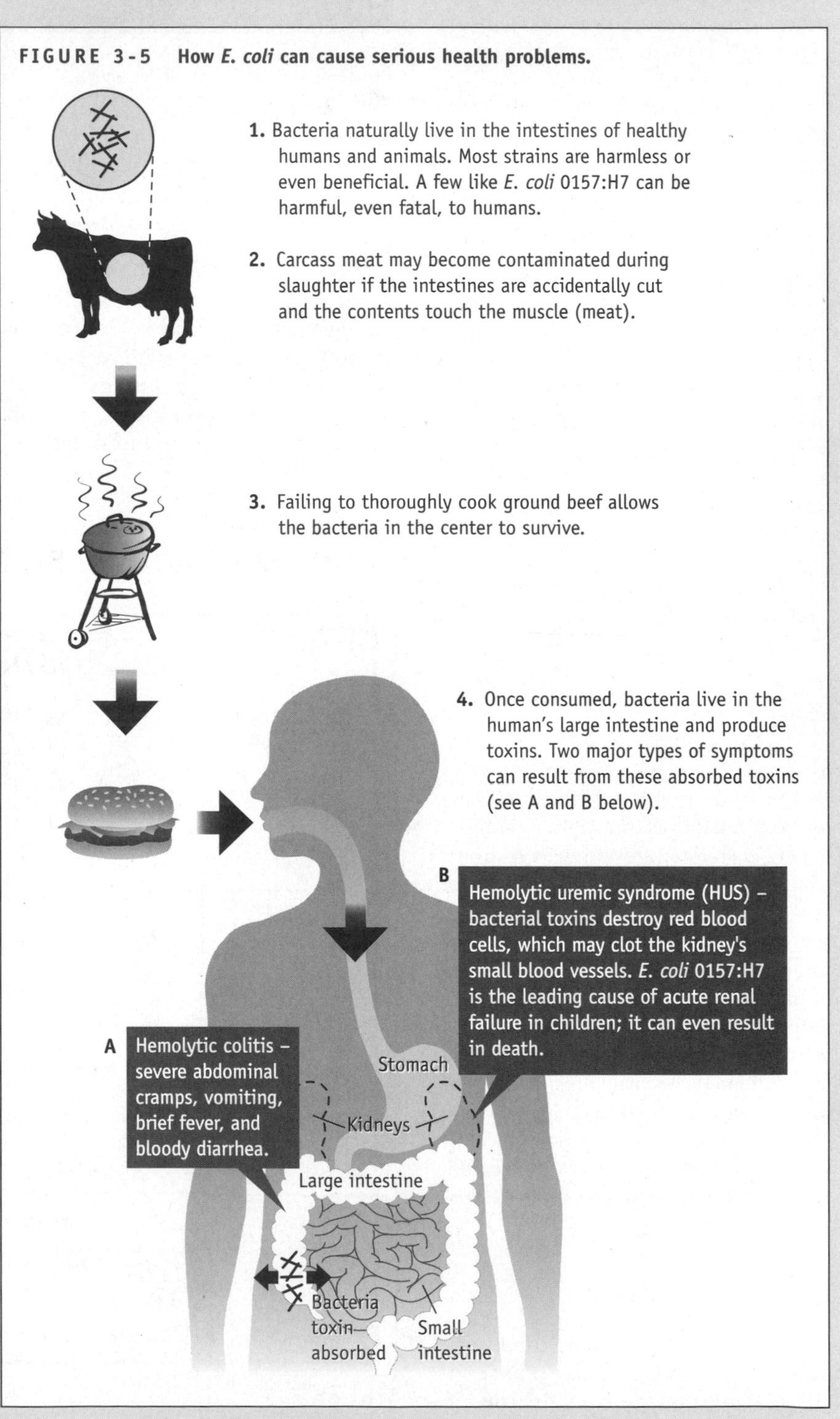

not until an outbreak in 1993 in the Pacific Northwest that national attention and preventive efforts were focused on the problem. The outbreak, which was caused by the consumption of undercooked ground beef in a fast-food restaurant, brought about a prompt response from the U.S. government. A new safety program was instituted, featuring more rapid testing of ground beef, tighter controls at slaughterhouses and processing plants, and the labeling of fresh meat products with instructions for safe handling and preparation, including increasing final cooking temperature of ground meats from 140° to 160°F/60° to 71°C (29). Health officials recommend that ground beef be thoroughly cooked so that no pink remains.

Undercooked hamburgers are not the only source of *E. coli* infection. Other foods or food-handling practices implicated in *E. coli* O157:H7 outbreaks include unpasteurized (raw) milk, unpasteurized apple juice or cider, raw sprouts (39), dry cured salami, fresh produce (especially manure fertilized), yogurt, sandwiches, and water (53).

E. coli may also be transmitted to children in child-care centers because of poor hand washing after diaper changing. Ideally, those who change the diapers should not be the same people who prepare the food (33). Other outbreaks have been reported from an improperly chlorinated swimming pool and contaminated water systems.

Campylobacter Jejuni

The number of people infected with *Campylobacter jejuni* now equals or exceeds those affected by *Salmonella* (26). *Campylobacter* species are responsible for more than 14 percent of the estimated annual food-related illness and deaths attributed to foodborne pathogens (19). Although the largest foodborne disease outbreak was traced to a municipal water supply, most other cases are linked to raw meat, undercooked poultry, unpasteurized milk, and untreated water.

Vibrio

Seafood is the major carrier of *Vibrio* infection. *V. parahaemolyticus* is the most common cause of foodborne illness in Japan (30). Cholera *(V. cholera)* is responsible for thousands of deaths each year in Asia. Poor sanitary conditions contaminate water supplies and usually account for the deaths reported in other countries (47). In the United States, very few cases are reported and they are usually associated with raw oyster consumption (34). The bacteria can also be transmitted through skin wounds during the cleaning or harvesting of shellfish, or if sea water washes over a preexisting wound. The Food and Drug Administration estimates that about 5 to 10 percent of raw shellfish on the market is contaminated.

Molds

Molds produce **mycotoxins**, which can cause food intoxication. Over 300 mycotoxins have been identified, some of which are carcinogenic (cancer causing) (46). Aflatoxin, a carcinogenic toxin made by the mold *Aspergillus flavus*, is the most potent liver carcinogen known. Foods infected with *Apergillus flavus* are most likely to be peanuts and grains.

Unlike bacteria, molds are visible, exhibiting **bloom** on affected foods. They also thrive at room temperature and need less moisture than bacteria. Foods susceptible to molds are breads, jams and jellies, and salty meats such as ham, bacon, and salami.

Black spots in the refrigerator, often called "mildew," are actually molds that can be cleaned by washing with a solution made by dissolving 1 tablespoon of baking soda in a quart of water. Musty smelling dishcloths, sponges, and mops should be thoroughly cleaned or replaced, because such odors indicate that mold has taken root.

HOW & WHY? ?????????

Why are some molds all right on foods while others are not? As a rule, foods that show signs of mold should not be eaten. The exceptions are cheeses (such as Roquefort, bleu, Brie, and Camembert), whose flavor, texture, and color depend on specific safe molds. Other foods relying on molds during processing include soy sauce, tempeh, and certain types of Italian-style salami that are coated in a thin, white mold. Cheeses such as cheddars and Swiss that become moldy can safely be trimmed 1 inch away from the mold. Soft cheeses like cottage cheese and cream cheese that have become moldy, however, should be thrown out, because the mold may penetrate through the cheese.

Viruses

Viruses are one of nature's simplest organisms. Unlike bacteria, which can exist independently, a virus needs a living cell in order to multiply. These microorganisms have been identified as causal agents in about 3 to 10 percent of foodborne illnesses (Table 3-6) (36). All foodborne viruses are transmitted via the oral-fecal route, that is, from contaminated feces to the mouth. They may be passed from person to person,

KEY TERMS

Mold A fungus (a plant that lacks chlorophyll) that produces a furry growth on organic matter.

Mycotoxin A toxin produced by a mold.

Bloom Cottony, fuzzy growth of molds.

Virus An infectious microorganism consisting of RNA or DNA that reproduces only in living cells.

TABLE 3-6

Viruses Causing Foodborne Illness

Disease (causative agent)	*Onset (duration)*	*Principal Symptoms*	*Typical Foods*	*Mode of Contamination*	*Prevention of Disease*
Hepatitis A (Hepatitis A virus)	10–50 days (2 weeks to 6 months)	Fever, weakness, nausea discomfort; often jaundice	Raw or undercooked shellfish; sandwiches, salads, etc.	Human fecal contamination, via water or direct	Cook shellfish thoroughly; general sanitation
Viral gastroenteritis (Norwalk-like viruses)	1–2 hours (1–2 days)	Nausea, vomiting, diarrhea, pains, headache, mild fever	Raw or undercooked shellfish; sandwiches salads, etc.	Human fecal contamination, via water or direct	Cook shellfish thoroughly; general sanitation
Viral gastroenteritis (rotaviruses)	1–3 days (4–6 days)	Diarrhea, especially in infants and young children	Raw or mishandled foods	Probably human fecal contamination	General sanitaiton

or through carriers such as flies, soiled diapers, water, and food. Two of the most common viruses known to cause foodborne illnesses are the hepatitis A virus and the Norwalk virus.

Hepatitis A Virus. Hepatitis A infection occurs most frequently after food is contaminated with fecal matter. (This differs from the hepatitis B virus, which is transmitted through body fluids and not through food.) A common source of hepatitis A is polluted shellfish beds and vegetable fields (14). Shellfish are a source of hepatitis A infection because they are eaten with their digestive tracts intact. Another possible source of hepatitis A contamination is child-care centers where diaper changing occurs. A vaccine is available that is 95 percent effective against the virus.

Norwalk Virus. The Norwalk virus is spread via contaminated shellfish, food handlers, and water containing raw sewage. Heating will destroy this virus, but freezing will not. Norwalk virus infection outbreaks can be large, as in the case of a Minnesota restaurant in which two salad makers contaminated the food and infected over 2,000 people (30).

KEY TERMS

Parasite An organism that lives on or within another organism at the host's expense without any useful return.

■ ■ ■ ■

Prion An infectious protein particle that does not contain DNA or RNA.

Parasites

Parasites need a host to survive. They infect people in many parts of the world, but in the United States, fewer than 500 cases of parasitic infection are reported each year. Two of the more common foodborne parasites are roundworms and protozoa (Table 3-7).

Roundworms. Roundworm infections can result from eating undercooked pork or uncooked or undercooked fish.

Trichinella spiralis. The *Trichinella spiralis* roundworm is probably the most common parasite carried in food, and is responsible for causing trichinosis. Pork products are the primary source of infection, with 1 out of every 100 swine in the United States infected. There are now relatively few cases of trichinosis each year, and most infections are thought to be contracted through the consumption of raw or improperly cooked pork, especially sausage (37). Heating pork to an internal temperature of 137°F (58°C) will kill the *T. spiralis* larvae, but the National Livestock and Meat Board recommends a final internal temperature of 160°F (71°C) as a safety margin. Microwave cooking of pork is not recommended because of uneven heating.

Herring Worms (Anisakis simplex) and Codworms (Pseudoterranova dicipiens). Japanese cooks preparing sushi (a Japanese dish of thin raw fish slices or seaweed over a cake of cooked rice) inspect fish for these tiny white worms, but because the worms are no wider than a thread, some may be missed. Therefore, not all raw fish dishes are guaranteed to be worm-free. There is no commercial method to detect all parasites. Even candling, which involves placing a fillet over a lighted translucent surface, finds only 60 to 70 percent of the worms (31).

People who consume raw or undercooked fish containing the live worms may experience a tingling throat sensation caused by the worm wriggling as it is swallowed. Other symptoms usually appear within an hour after ingestion, but may not show up for as much as two weeks. In serious cases, the worm penetrates through the stomach or intestinal wall, causing severe abdominal pain, nausea, vomiting, or diarrhea. Symptoms often continue for several days and have been misdiagnosed as appendicitis, gastric ulcer, Crohn's disease, and gastrointestinal cancer (56). After several weeks, the worm dies, or it may be coughed or vomited up by the host. It also may be removed by a physician using a fiber-optic device equipped with mechanical forceps. Despite these dramatic problems arising from worm infection from contaminated fish, only about ten cases are reported every year. The number of actual cases, however, may be significantly higher due to underreporting.

Protozoa. Protozoa are animals consisting of just one cell. They most frequently infect humans through contaminated water. Only 3 out of about 30 types of protozoa are related to food safety: *Giardia, Cryptosporidium,* and *Cyclospora*. The most common of these is *Giardia*.

TABLE 3-7
Parasites Causing Foodborne Illness

Disease (causative agent)	*Onset Principal (duration)*	*Principal Symptoms*	*Typical Foods*	*Mode of Contamination*	*Prevention of Disease*
Roundworms (Nematodes)					
Trichinosis (*Trichinella spiralis*)	8–15 days (weeks, months)	Muscle pain, swollen eyelids, fever; sometimes death	Raw or undercooked pork or meat of carniverous animals (e.g., bears)	Larvae encysted in animal's muscles	Thorough cooking of meat; freezing pork at 5°F for 30 days; irradiation
Ascariasis (*Ascaris lumbricoides*)	10 days–8 weeks (1–2 years)	Sometimes pneumonitis, bowel obstructions	Raw fruits or vegetables that grow in or near soil	Eggs in soil from human feces	Sanitary disposal of feces; cooking food
Anisakiasis (*Anisakis simplex, Pseudoterranova decipiens*)	Hours to weeks (varies)	Abdominal cramps, nausea, vomiting	Raw or undercooked marine fish, squid or octopus	Larvae occur naturally in edible parts of seafoods	Cook fish thoroughly or freeze at –4°F for 30 days
Protozoa					
Giardiasis (*Giardia lamblia*)	5–25 days (varies)	Diarrhea with greasy stools, cramps, bloat	Mishandled foods	Cysts in human and animal feces, directly or via water	General sanitation; thorough cooking
Cryptosporidiosis (*Cryptosporidium parvum*)	2–3 days (2–3 weeks)	Diarrhea; sometimes fever, nausea, and vomiting	Mishandled foods	Oocysts in human feces	General sanitation; thorough cooking
Amebic dysentary (*Entamoeba histolytica*)	2–4 weeks (varies)	Dysentary, fever, chills sometimes liver abscess	Raw or mishandled foods	Cysts in human feces	General sanitation; thorough cooking
Toxoplasmosis (*Toxoplasma gondi*)	10–23 days (varies)	Resembles mononucleosis fetal abnormality or death	Raw or undercooked meats; raw milk; mishandled foods	Cysts in pork or or mutton, rarely beef; oocysts in cat feces	Cook meat thoroughly; pasteurize milk; general sanitation
Tapeworms (Cestodes)					
Beef tapeworm (*Taenia saginata*)	10–14 weeks (20–30 years)	Worm segments in stool; sometimes digestive disturbances	Raw or undercooked beef	"Cysticerol" in beef muscle	Cook beef thoroughly or freeze below 23°F
Fish tapeworm (*Diphylliobothrium latum*)	3–5 weeks (years)	Limited; sometimes vitamin B-12 deficiency	Raw or undercooked freshwater fish	"Pierocercoids" in fish muscle	Heat fish 5 minutes at 133°F or freeze 24 hours at 0°F
Pork tapeworm (*Taenia sollium*)	8 weeks–10 years (20-30 years)	Worm segments in stool, sometimes "cysticercosis" of muscles, organs, heart, or brain	Raw or undercooked pork; any food mishandled by a *T. serum* carrier	"Cysticerol" in pork muscle; any food —human feces with *T. serum* eggs	Cook pork thoroughly or freeze below 23°F; general sanitation

Giardia lamblia. Giardia is primarily transmitted through surface streams and lakes that have been contaminated with the feces of infected livestock and other animals. This protozoan is responsible for the most common parasitic infection in the world, and is most frequently associated with the consumption of contaminated water. Another common source of infection is child-care centers. Approximately 2 percent of the population in the United States is infected (56). Infection with this organism causes recurring attacks of diarrhea and the passage of stools containing large amounts of unabsorbed fats and yellow mucus. When a *Giardia* infection is contracted, medications can be taken for the symptoms.

Prions—Mad Cow Disease

Prions are the cause of mad cow disease, or bovine spongiform encephalopathy (BSE). It is a type of transmissible spongiform encephalopathy (TSE) that riddles the brain with holes, making it look like a sponge. TSE is a group of diseases that affect the brain, resulting in symptoms that range from loss of coordination to convulsions and ultimately death. TSEs other than mad cow disease include Creutzfeldt-Jakob Disease (CJD; causes dementia), Kuru ("the laughing death" that used to be found in New Guinea due to cannibalism), new variant CJD, and scrapie (coordination loss and itching/scraping in goats and sheep).

In mad cow disease, a person ingests an infected food, allowing the prion to travel up the spinal cord to the brain. This protein material incorporates itself into the brain, causing chain reactions that create holes in the brain. The incubation period between infection to manifestation can be months, years, or decades.

Controversy exists on how prions are transmitted from food to humans. The foods most often believed to be linked with these prions have been cattle and sheep in Great Britain. Prior to the understanding of prions, the practice for some livestock growers was to kill sickly animals and feed the remains to other cattle. It is speculated that healthy cattle being fed rendered livestock would then become infected and, when slaughtered for their meat, would potentially spread this disease to the consumer.

Over 95 percent of all BSE cases have occurred in Great Britain (25). There have been no official reports of BSE in the United States. Several reasons account for this. The Department of Health and Human Services banned the use of rendered carcasses as feed for other livestock (40). There is also a strong compliance with this feed ban by the livestock industry interested in keeping their food products safe. Further measures to keep livestock safe in the United States occurred when the USDA banned imported livestock from Great Britain in 1989, and extended the ban to Europe in 1997.

New Virulent Biological Hazards

It is not unusual for microorganisms that were relatively unheard of to emerge with a new virulence (actively harmful) making them a public threat. Examples of pathogens not previously recognized as a serious cause of foodborne illness are the Norwalk virus, *Campylobacter jejuni, Listeria monocytogenes, Vibrio vulnificus, Vibrio cholera,* and *Yersinia enterocolitica* (33).

Pathogens are living organisms that rapidly evolve (3). Bacteria are constantly appearing as potential hazards to public health, so health departments and government agencies must be vigilant. Several serious outbreaks resulting from "new" microorganisms led the CDC in 1994 to implement Emerging Infections Programs (EIPs) in state health departments (23).

FIGURE 3-6 Keith Nakatano, director of the Save San Francisco Bay Association, holds up a fish caught off the wharf during a news conference where he released results of a study that found bay fish have unhealthy levels of PCBs, mercury, DDT, and other toxic chemicals in amounts up to 13 times higher than federal standards.

AP/Wide World Photos

Chemical Hazards—Harmful Chemicals in Food

Chemical hazards are any chemical substances hazardous to health. These chemicals can come from plants (herbs, fruit pits, mushrooms), animals (fish), or chemicals used in agriculture or industry that may end up in food unintentionally (Figure 3-6). Several examples of chemical toxins from seafood are explained below.

Seafood Toxins—Chemicals from Fish/Shellfish

Both fish and shellfish may harbor toxins causing foodborne illness.

Ciguatera Fish Poisoning. Ciguatera fish poisoning is the most common toxin-related food poisoning in the United States. It is caused by eating fish, usually from tropical waters, that contain a ciguatoxin that is not destroyed by heating (4). Although less than 1 percent of fish found in tropical areas with coral reefs are actually contaminated, more than one-third of Florida barracuda were found to contain ciguatera toxin, resulting in a ban on the sale of barracuda for human consumption (32).

Histamine Food Poisoning. Excessive histamine accumulation in fish (especially tuna) may result in histamine food poisoning (scombrotoxism). This is one of the most common forms of fish poisoning in the United States and occurs when the fish have not been chilled immediately after being caught. The fish become toxic when bacteria (such as *Morganella morganii*) produce histamine.

Pufferfish Poisoning. One of the most violent poisonings originating from seafood occurs when the liver, gonads, intestines, and/or skin of the pufferfish are consumed. These organs contain tetrodotoxin, which if ingested results in a mortality rate of 50 percent. Only several cases have been reported in the United States, but in Japan, where pufferfish or *fugu* is a traditional delicacy, 30 to 100 cases are reported a year (56). Most of these cases in Japan originated from people preparing the dish at home, rather than eating it at special restaurants. Licensed chefs are specially trained in these restaurants on how to remove the poisonous viscera from the pufferfish.

Red Tide. Red tide is the result of the rapid growth of a reddish marine algae, usually occurring during the summer or in tropical waters. Shellfish such as mollusks, oysters, and clams, and certain fish that consume red tide algae, become poisonous and should not be eaten until the red tide has disappeared.

Physical Hazards—Objects in Food

Physical hazards in food and beverages that can harm the consumer's health include glass (the most common), bone, metal, wood, stones, false fingernails, toothpicks, watches, jewelry, insects, staples from food boxes, and many other foreign items that have been known to find their way into the food supply.

The HACCP System—Preventing Foodborne Illness*

The acronym **HACCP** (pronounced has-sip) stands for Hazard Analysis and Critical Control Point. It is a prevention-based food safety system that includes seven principles (Figure 3-7) (42). HACCP is a systematic approach to food safety that will dramatically improve the level of food safety. The FDA recommends the implementation of a HACCP system throughout the food industry, including processing plants, food service establishments, and food corporations. A HACCP system emphasizes prevention rather than relying solely on health department inspections.

Essentially, HACCP identifies and monitors specific foodborne hazards. This hazard analysis serves as the basis for establishing **critical control points (CCPs)**. Further, critical limits are established at each CCP. Monitoring and verification steps are then included in the system, to ensure that potential risks are controlled. The hazard analysis, critical control points, critical limits, monitoring, and verification steps are documented in a HACCP plan.

FIGURE 3-7 The seven principles of the Hazard Analysis and Critical Control Point (HACCP) system

1. Assess the hazards
2. Identify the critical control points
3. Establish limits at each critical control point
4. Monitor critical control points
5. Take corrective action
6. Documentation
7. Verification

*This section is adapted from U.S. Department of Health and Human Services, Public Health Service, Food and Drug Administration, 1999 Food Code. Guidelines can be found at vm.cfsan.fda.gov/-dms/fc99-a5.html.

History of HACCP

The application of HACCP to food production was pioneered by the Pillsbury Company with the cooperation and participation of the National Aeronautic and Space Administration (NASA), Natick Laboratories of the U.S. Army, and the U.S. Air Force Space Laboratory Project Group. Application of the system in the early 1960s created food for the United States' space program that approached 100 percent assurance against contamination by bacterial and viral pathogens, toxins, and chemical or physical hazards that could cause illness or injury to astronauts. In the succeeding years, the HACCP system has been recognized worldwide as an effective system of controls. The system has undergone considerable analysis, refinement, and testing and is widely accepted in the United States and internationally.

The Seven HACCP Principles

The USDA's National Advisory Committee on Microbiological Criteria for Foods (NACMCF), established in 1988, developed the seven HACCP principles. Its members are officials from several federal agencies, which include the Food and Drug Administration (FDA), the Centers for Disease Control and Prevention (CDC), the Food Safety Inspection Service (FSIS), the Agricultural Research Service (ARS), the National Marine Fisheries Service, and the U.S. Army.

Before beginning to develop a HACCP plan for a food establishment, a team should be assembled that is familiar with the overall food operation and the specific production processes to be included in the plan. Team members should be familiar with common HACCP terms listed in Table 3-8. The following sections will review a HACCP plan and its seven principles.

KEY TERMS

HACCP Hazard Analysis and Critical Control Point System, a systematized approach to preventing foodborne illness during the production and preparation of food.

Critical control point (CCP) A point in the HACCP process that must be controlled to ensure the safety of the food.

HACCP Principle #1: Assess the Hazards

The hazard analysis process accomplishes three purposes:

1. Identify the hazards,
2. Assess the risk for these hazards.
3. Identified hazards can be used to develop preventive measures for a process or product to ensure or improve food safety.

The first step in the development of a HACCP plan for a food operation is identifying the hazards associated with the food product. Next, the risk is assessed for these hazards by determining (1) the likelihood that the hazard will occur, and (2) the severity if it does occur. Hazard analysis also involves establishment of preventive measures for control. To be effective, a HACCP plan must address each hazard in terms of prevention, elimination, or reduction to acceptable levels.

Developing Preventive Measures. After identifying the hazards, the food establishment must then consider what preventive measures, if any, can be

TABLE 3-8
HACCP Terms and Definitions

HACCP Term	*Definition*
Acceptable Level	The presence of a hazard that does not pose the likelihood of causing an unacceptable health risk.
Control Point	Any point in a specific food system at which loss of control does not lead to an unacceptable health risk.
Critical Control Point	The point at which loss of control may result in an unacceptable health risk.
Critical Limit	The maximum or minimum value to which a physical, biological, or chemical parameter must be controlled at a critical control point to minimize the risk that the identified food safety hazard may occur.
Deviation	Failure to meet a required critical limit for a critical control point.
HACCP Plan	The written document that delineates the formal procedures for following the HACCP principles developed by the National Advisory Committee on Microbiological Criteria for Foods.
Hazard	A biological, chemical, or physical property that may cause an unacceptable consumer health risk.
Monitoring	A planned sequence of observations or measurements of critical limits designed to produce an accurate record and intended to ensure that the critical limit maintains product safety. Continuous monitoring means an uninterrupted record of data.
Preventive Measure	An action to exclude, destroy, eliminate, or reduce a hazard and prevent recontamination through effective means.
Risk	An estimate of the likely occurrence of a hazard.
Sensitive Ingredient	Any ingredient historically associated with a known microbiological hazard that causes or contributes to production of a potentially hazardous food as defined in the Food Code.
Verification	Methods, procedures, and tests to determine if the HACCP system in use is in compliance with the HACCP plan.

applied for each hazard. Preventive measures are physical, chemical, or other factors that can be used to control an identified health hazard. For example, if a HACCP team were to conduct a hazard analysis for the preparation of hamburgers from frozen beef patties, pathogens on the incoming raw meat would be identified as a potential hazard. Cooking is a preventive measure that can be used to eliminate this hazard.

HACCP Principle #2: Identify the Critical Control Points (CCPs)

A critical control point is any point in a food production system where a loss of control may result in an unacceptable health risk. This can occur at any time from the point of growing and harvesting the food to consumption of it by the consumer. At any of these points, a preventative measure can be applied to eliminate, prevent, or minimize the risk of a food hazard. CCPs include cooking, chilling, specific sanitation procedures, product formulation control, prevention of cross-contamination, and certain aspects of employee and environmental hygiene.

KEY TERMS

Cross-contamination The transfer of bacteria or other microorganisms from one food to another.

CCPs must be carefully developed and documented. Different facilities preparing the same food can differ in the risk of hazards and the points, steps, or procedures, which are CCPs. This can be due to differences in each facility such as layout, equipment, selection of ingredients, or the process that is used. As a result, generic HACCP plans serve as useful guides, but it is essential that a HACCP plan be customized for each facility.

The critical control points discussed in this chapter include:

- Processing (food plants)
- Purchasing (vulnerable foods)
- Preparation (thawing, **cross-contamination**, heating, holding, serving, and cooling/reheating)
- Sanitation (cleanup, equipment, facilities, pest control, and water)
- Storage (discussed in later chapters)
- Personnel

Processing—Critical Control Point. Only about 3 percent of foodborne illnesses originate at food plants (5). The majority of outbreaks in food processing plants are usually caused by contamination of incoming foods, failure of pathogen-killing processes, or contamination of foods after sanitization.

Incoming raw foods may be contaminated in several ways. The digestive tracts of people and animals naturally contain bacteria. During the rendering of animals at the slaughterhouse, the digestive tract may be accidentally cut open or nicked, releasing bacteria that may then come in contact with meat (9). Other possible sources of contamination include any cuts, skin, feet, hair, hide, or feathers that can carry bacteria. Fruits and vegetables may be contaminated by microorganisms in the soil, or by manure used to fertilize crops.

Contamination may also occur at the plant when pathogen-killing processes, such as temperature gauges, heaters, seals, and refrigeration units, fail to work properly. And finally, foods may become contaminated *after* sanitation. Microorganisms by their nature are ubiquitous and lodge themselves in air filters, drains, equipment, floor cracks, food scraps, and even dust.

Purchasing—Critical Control Point. A quality control program in a food establishment should ensure that only foods that meet written specifications are purchased. Foods should be purchased from reputable vendors, meet temperature and humidity requirements, show no evidence of being refrozen (such as frozen fluid lining the bottom of the container or large ice crystals on the food's surface), be received in undamaged containers, and meet specifications (see Chapter 4). Suspect cans (dented, bloated, or showing signs of leakage) and foods in unmarked containers should be discarded. All foods should be in their

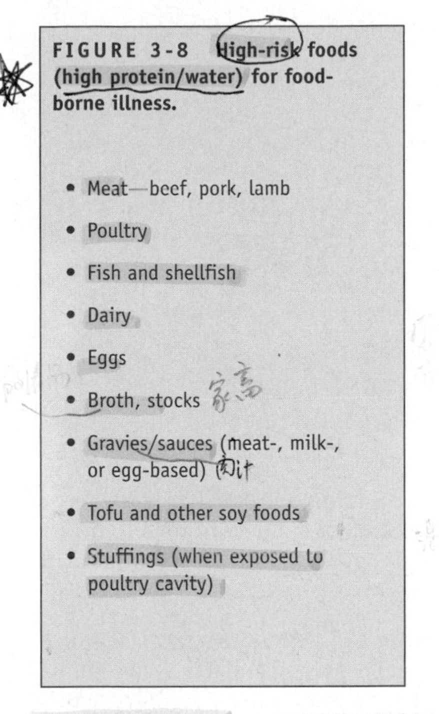

FIGURE 3-8 High-risk foods (high protein/water) for foodborne illness.

original containers or clearly labeled if they have been transferred to another receptacle.

Vulnerable Foods. In addition to the precautions that are taken when purchasing any foods, some foods should always be treated with special care. The foods that are best able to support the growth of bacteria are those containing protein and water (Figure 3-8). These foods are in the high-risk category for bacterial contamination. Proper refrigeration or freezing is a must. Vulnerable food products made with the high-risk foods and exposed to more handling include meatloaf, hamburger, salads (pasta, coleslaw, and chicken, egg, and tuna salads), Chinese and Mexican dishes, some baked goods, and cream fillings. Egg dishes likely to become contaminated include baked or soft custard, French toast, quiches, Hollandaise sauce, meringues, eggnog, and mayonnaise. Damaged eggs are good vectors for organisms that cause foodborne illnesses and should be discarded. Food establishments have the safer option of using pasteurized eggs.

In recent years, foodborne illnesses have been implicated with foods that are not high-protein/high-water. Fresh fruits and vegetables, cooked rice, sliced fruits, sautéed onions, potatoes, garlic-in-oil combinations, and apple cider have all been implicated (33). For example, illnesses from fresh-pressed apple cider, affecting hundreds of people and resulting in at least one death, have been reported over the past 20 years (50). At one time it was thought that the low acidic content of fresh apple cider (pH < 4.0) was a protective, but a strain of *E. coli* can now survive in fresh cider with a pH of 3.7. This conflicts with the FDA Food Code that has historically considered foods with a pH of less than 4.6 as generally safe (48). The FDA has given notice to the fresh juice industry regarding implementation of a HACCP system, labeling requirements, and recommended pasteurization (24).

Preparation—Critical Control Point. The various steps of food preparation—pre-preparation, heating, holding, reheating, and serving—are vulnerable to a loss of control leading to an unacceptable health risk. Two aspects of pre-preparation are now discussed—thawing and cross-contamination.

Thawing. For safe thawing, only one of the methods below should be used. Table 3-9 further illustrates safe thawing techniques related to critical control points.

- Refrigerator, on the bottom shelf to avoid contaminating other foods with any drippings. Thawing frozen meat at room temperature is considered an unsafe practice. Running cold water over meat (wrapped in protective plastic) or placing it in a bath of ice water and frequently replacing the water, are not as safe as defrosting in a refrigerator.
- Microwave oven followed by immediate cooking.
- As part of the cooking process.

Cross-Contamination. An example of a critical control point, and one of the most common causes of foodborne illness during the summer months, is the backyard barbecue. People carry raw

TABLE 3-9

Critical Control Points for Freezing and Thawing Chicken

Process	*Hazard*	*Critical Control Point*	*Standard*	*Monitoring Method*	*Action to Take if Standard not Met*
Freezer storage	Contaminating bacteria will survive	Freezer temperature	Freezer temperature ≤0°F	Measure freezer air temperature	Reset thermostat
Thawing	Insufficient thawing may lead to insufficient cooking, and pathogen survival	Thawing time-temperature	Allow 1 day in refrigerator to thaw 5 lb. chicken; check for microwave standards	Observation	Continue until thawed
	Bacteria can grow if portions are warm enough	Thawing time-temperature	Thaw in refrigerator Thaw in microwave oven Thaw under cool running water	Measure refrigerator temperature Follow appliance instructions	Lower thermostat
	Drip from thawing chicken can contaminate surfaces and other foods	Sanitation of contact surfaces, including plates, utensils, sinks, countertops	Wash, rinse, and disinfect after contact with chicken or drip	Observation	Reset water flow Rewash and sanitize

Source: American Dietetic Association

FIGURE 3-9 Three-compartment sink.

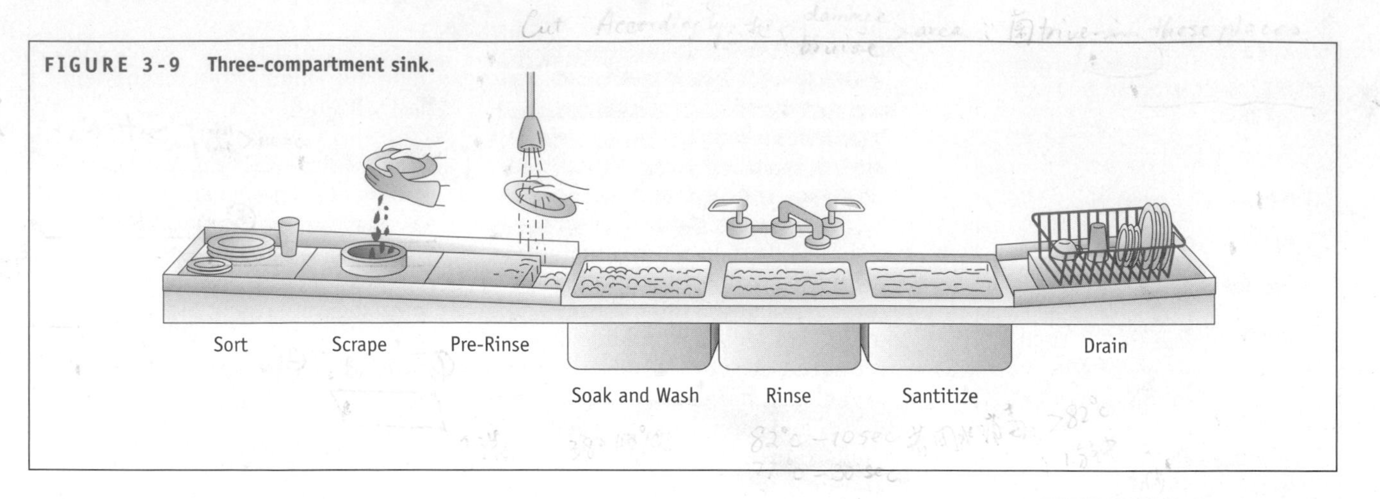

meat on a plate to the barbecue, cook the meat, and place it back on the contaminated plate. An example of commercial cross-contamination occurred in 1996, when at least 224,000 people became ill from eating ice cream contaminated with *Salmonella*. The bacteria was traced to the trucking company that had transported the pasteurized ice cream premix in trailers, which had previously carried *Salmonella*-contaminated liquid eggs (27).

To prevent cross-contamination, food should never touch contaminated surfaces unless it is going to be thoroughly cooked. Surfaces should also be regularly washed and sanitized, especially after coming into contact with raw food. Particularly susceptible surfaces include hands, utensils, tabletops, cutting boards, and slicers, as well as aprons, cleaning cloths, and sponges. Dust and soil should be washed off the tops of cans before they are opened. Raw meats should never be stored in the refrigerator above cooked or ready-to-eat foods where they may drip onto the food below.

Sanitation—Critical Control Point

Cleanup. Dishes in a food service establishment may be hand or machine washed, but whatever method is used, the process must meet certain sanitation guidelines to pass a health department food inspection. In order to kill pathogens, dishwashing temperatures should be between 140° and 160°F (60° and 71°C), and rinse temperatures at least 180°F (82°C) for 10 seconds or 170°F (77°C) for 30 seconds, in order to kill most pathogens. Despite concerns that dishwashing failures result in foodborne illness, only 5 percent of sanitation failures can be traced to faulty equipment; the remaining 95 percent are a result of human error.

Three-Compartment Sink. Manual washing in a food establishment requires a **three-compartment sink** for washing pots and pans, dishes, glasses, cutlery, and tools (Figure 3-9). The first compartment is used for soaking and washing items in water heated to 100° to 120°F (38° to 49°C). The second compartment is for rinsing, and the third area is for sanitation. The last compartment can sterilize items with either hot water or chemical sanitizers. If water is used, then temperatures must reach at least 180°F (82°C) for 1 minute. Food establishments may also sanitize with the chemical sanitizers shown in Table 3-10.

Drying. Items should always be air or heat dried. Damp cloth towels are an ideal medium for microorganism growth.

Scheduling. Schedules for cleaning should be posted and followed scrupulously to maintain a sanitary work environment. Floors, walls, windows, lights, and equipment should all be included in the frequent cleanup routine. Sanitation guidelines involving cleanup, personnel, equipment, facilities, pest control, and water could be set and routinely followed by establishing cleaning schedules to be checked off on predetermined dates.

Equipment. The National Sanitation Foundation (NSF) sets standards for equipment to be used in food service establishments. Equipment should be as free of crevices as possible, so it is best to buy equipment with rounded junctions. Wooden cutting boards are more prone to nicks and crevices than plastic or marble ones, and it once was thought that they were more prone to microbial contamination. Studies now show that once they are cleaned, bacterial levels do not differ significantly among the various types of boards (41). All equipment, utensils, contain-

KEY TERMS

Three-compartment sink A sink divided into three sections, the first for soaking and washing, the second for rinsing, and the third for sanitizing.

TABLE 3-10
Chemical Sanitizers Used in Commercial Food Establishments

Sanitizer	*How to Use*
Chlorine	200 ppm* for water and equipment
Iodine	25 ppm for handwashing and equipment
Quarternary ammonium compounds	200 ppm for walls; 500 ppm for floors
Organic acids (lactic, acetic, propionic)	130 ppm for stainless steel surfaces

*ppm = parts per million

ers, meat grinders, and slicers should be thoroughly cleaned after each use. Meat slicers should be cleaned only by trained employees, because they are the number-one cause of cuts in a food service organization.

After cleaning, utensils are best stored covered, and glasses and cups should always be stored upside down. Disposable utensils such as plastic cups, forks, knives, and plates should never be washed and reused. Freezers and refrigerators should allow at least 6 inches (10 inches is preferred) between the bottom storage shelf and the floor; this allows for adequate cleaning. They should also contain thermometers to determine that correct temperatures are being maintained.

Facilities. In order to remain sanitary, a food service establishment should be designed and maintained in ways that promote cleanliness. Floors, walls, and ceilings should have adequate ventilation, in addition to being constructed of a material that allows easy cleaning. Garbage should be discarded in covered pest-proof containers that are frequently cleaned and free of litter. Lighting fixtures must be covered to prevent dust and insects from collecting on lightbulbs and falling on the food. Food service organizations are required to have dressing rooms, restrooms, and handwashing sinks available to the employees. From the consumer standpoint, the biggest complaint many health departments receive from customers is about the poor condition of the restrooms provided for their use.

Pest Control. Even the cleanest facility can be put at risk of transmitting foodborne illness by the presence of insects, rodents, birds, turtles, or other animals. Rodents such as mice and rats can carry *Salmonella,* typhus, and the bubonic plague. Insects and cockroaches transfer microorganisms by landing, walking, and regurgitating their stomach contents on foods when feeding. Figure 3-10 shows some common pests. To discourage pests from taking up residence, all entrances should be screened or sealed, standing water should be removed, and all drains should be working properly. It also helps to cover all garbage and remove it frequently. All food should be securely stored in pest-proof containers. Even though it might be tempting to spray insects, insecticides should be applied only by a professional.

FIGURE 3-10 Common pests that may transmit foodborne illness.

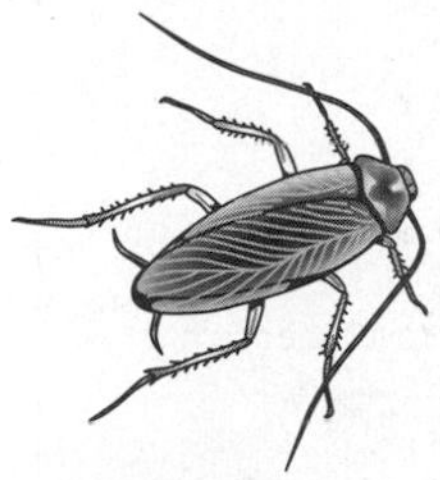

Cockroaches
Cockroaches are drawn to food crumbs and often regurgitate while eating.

Rodents
Signs of rats and mice include their droppings, urine markings, and holes in packaging.

Insects
Insects lay their eggs on decaying matter such as sewage, garbage, or rotting food.

Pantry Pests
Pests preferring pantry foods such as flour, sugar, rice, and other dry goods include beetles, mites, moths, weevils, and silverfish.

Personnel—Critical Control Point. Sanitation is largely influenced by food service personnel. In the early 20th century, a cook in New York named Mary Mallon appeared perfectly healthy, but infected about 50 people with typhoid fever. Not surprisingly, she came to be known as "Typhoid Mary." She believed that since she could not see germs, she did not have to wash her hands before cooking. As this story illustrates, a top priority for any food-serving establishment is that food workers be healthy and know how to handle food safely (Figure 3-11). Typhoid is far from the only illness that can be transferred through carelessness and poor hygiene. The common cold, mumps, measles, pneumonia, scarlet fever, tuberculosis, trench mouth, diphtheria, influenza, and whooping cough may also be spread this way. Only people who are free of colds, diarrhea, wounds, and illnesses should be involved in food handling.

Training. A Food Management Certificate obtained through a health department education class ensures that a food handler has learned safe food handling techniques and is often a job requirement for food service employees. Periodic retraining on sanitation techniques is available from local health departments, the National Restaurant Association, or the Centers for Disease Control. Training is very important in making HACCP successful in any food establishment. HACCP works best when it is integrated into each employee's normal duties rather than added as something extra. The food employee's training should provide an overview of HACCP's prevention philosophy while focusing on the specifics of the employee's normal functions.

For all employees, the fundamental training goal should be to make them proficient in the specific tasks that the

FIGURE 3-11 Personal hygiene checklist.

Self

- ❑ Stay healthy. Maintain daily sleep, well-balanced diet, and relaxation.
- ❑ Report to supervisor if you are sick.
- ❑ Stay clean. Practice daily bathing, shampoo hair regularly, use deodorant, and take care of fingernails — they should be cleaned, trimmed, and free of polish and decorations.
- ❑ Wear only clothes that are new or have been washed. Shoes should cover the foot (no sandles, open toe) and have nonskid soles.
- ❑ Wear caps or hairnets.
- ❑ Avoid items that may fall into food/beverages: hairpins, jewelry, false nails, nail polish, nail decorations, bandages on hand (cover with plastic gloves), handkerchiefs.

Food Handling

- ❑ Avoid handling food; use serving spoons, scoopers, dippers, tongs, and ladles.
- ❑ Cover all exposed food with lids, plastic, or aluminum wrap.
- ❑ Taste food with clean spoon and do not reuse.
- ❑ If gloves are used, change them between food and nonfood handling.

Kitchen

- ❑ Wash hands in the handwashing sink before starting and after breaks/meals.
- ❑ Cover all coughs/sneezes and immediately wash hands in hand sink.
- ❑ No smoking or gum chewing.
- ❑ Keep all surfaces clean.
- ❑ Use potholders for pots and dish towels for dishes.
- ❑ Keep cleaning items away from foods/beverages.

Serving

- ❑ Hold plates without touching the surface.
- ❑ Carry silverware only by the handles.
- ❑ Handle glassware without touching the rim or the inside.

HACCP plan requires them to perform. This includes the implementation of proper corrective actions when monitoring reveals violation of the critical limit. The training should also include the proper completion and maintenance of any records specified in the establishment's plan.

Reinforcement. Training reinforcement is also needed for continued motivation of the food establishment employees. Some examples might include a HACCP video training program; reminders about HACCP critical limits printed on employees' time cards or checks; and work station reminders such as pictorials on how and when to take food temperatures. Reinforcement of food safety topics should include:

- *Avoid hand-to-mouth.* Any hand-to-mouth movements should be discouraged; this includes even such simple habits as smoking, gum chewing, and eating in the food handling areas. Sampling foods, either with fingers or utensils, should not be permitted; this also transfers bacteria. *Staphylococcus* can be transferred by workers who touch their mouth, nose, a pimple, or infected cut, and then handle food. Sneezing or coughing sends millions of microorganisms into the surrounding air to settle on food.
- *Hand washing.* Hands should be washed frequently, especially before handling food and after touching raw meat or eggs, using the restroom, sneezing, or handling garbage. When food handlers touch a doorknob, handrail, telephone, counter, or any other surface that is frequently contacted by others, it necessitates another hand washing before food is touched. They also should not touch the surfaces of food-serving utensils. The forks, knives, and spoons for customer use are always placed head down in serving canisters. The same rule applies to the ice scoop handle, which should never come in contact with ice after touching an employee's hands or an unclean surface.
- *Hand-washing sink.* Food establishments have a separate sink strictly for hand washing, and it should never be used for washing foods or utensils. To ensure maximum effect from hand washing, the routine should consist of washing up to the elbow for at least 20 seconds, using a nail brush, and then drying with disposable towels or an air dryer. Cloth towels should never be used.
- *Uniforms.* Food service workers should clean their uniforms frequently, wear caps or hair nets, and avoid jewelry such as rings and bracelets, which can collect minute particles of food and dust.

Updates. Every time there is a change in a food operation, the HACCP training needs should be updated. For example, when a food establishment substitutes a frozen seafood product for a fresh one, proper thawing critical limits should be taught and then monitored for implementation.

Feedback. The HACCP plan should include a feedback loop for employees. All employees should be made a part of the continuous food safety improvement cycle because the customer's health is in their hands. Moreover, since they are on the front lines of the actual work, they are the first to see problems or potential problems. Maintaining their active awareness and involvement is critical to a successful food safety program.

HACCP Principle #3: Establish Limits at Each Critical Control Point

Critical control points are meaningless unless a standard for each is implemented and followed. This means determining a minimum standard for each critical control point that exists for the various stages of food production—purchasing, preparation, sanitation, and storage. The critical control point limits discussed in this chapter include:

- Temperature
- Time
- Water and humidity
- pH

Temperature—Critical Control Limit. The **temperature danger zone**, which includes the human body temperature of 98.6°F (37°C), allows rapid bacterial growth (Figure 3-12). To avoid the temperature danger zone, cold foods should be stored under 40°F (4°C), and hot foods should be above 140°F (60°C). Bacteria normally do not survive temperature extremes; however, some bacteria do survive below freezing temperatures (32°F/0°C).

Heating. Most foodborne outbreaks in the United States are a result of improper temperature control (10). Heating foods to a certain temperature will destroy most microorganisms. Table 3-11 shows that certain foods, such as meat and poultry, need to reach a temperature higher than 140°F (60°C) during cooking to ensure safety. Bacteria are usually destroyed by the high heat of boiling if they are subjected to it for at least 10 minutes.

TABLE 3-11
Temperatures Needed to Destroy Microorganisms in Different Foods
(Individual states are not required to adopt these recommendations, so some temperatures will vary.)

Food	*Temperature*
Poultry	180°F (82°C)
Reheated foods	165°F (74°C)
Ground meats	160°F (71°C)
Pork	160°F (71°C)
Beef	145°F (63°C)
Fish	140°F (60°C)

KEY TERMS

Temperature danger zone The temperature range of 40° to 140°F (4° to 60°C), which is ideal for bacterial growth.

Spore Encapsulated, dormant form assumed by some microorganisms that is resistant to environmental factors that would normally result in its death.

The density of the food plays a key role in heating, because heat must penetrate the entire food mass. Since microwaves only penetrate 1/2 to 2 inches into food, this limits the size and thickness of meats that can be safely cooked in a microwave oven. It is unsafe to cook turkeys in a microwave oven, even if they are unstuffed. Another concern about the safety of microwave-processed foods is the short heating time (1).

Despite precautions, some bacteria may survive environmental stresses in **spore** form. Spores are very resistant to drying and heating, and the bacteria may remain in this dormant state for long periods until their environment becomes more hospitable.

Holding. When foods are held for extended periods after cooking, they must be kept either above or below the temperature danger zone. Food handlers use thermometers to ensure that the food on the serving line meets these guidelines. To put this in perspective, room temperature is usually around 70°F (21°C), and in the kitchen it can increase up to 90°F (32°C). Critical to controlling bacterial growth is the fact that temperatures under 40°F (4°C) retard but do not kill bacteria, while those above 165°F (74°C) destroy most microorganisms.

Serving. Serving is a point of vulnerability to contamination. Good personal hygiene on the part of food

FIGURE 3-12 **Bacteria divide to reproduce, resulting in billions from just one cell in less than one day.**

Stationary phase

Billions

10 hrs.

2 hrs.

$1\frac{1}{2}$ hrs.

1 hr.

$\frac{1}{2}$ hr.

Start

Decline phase

Number of Cells

Time

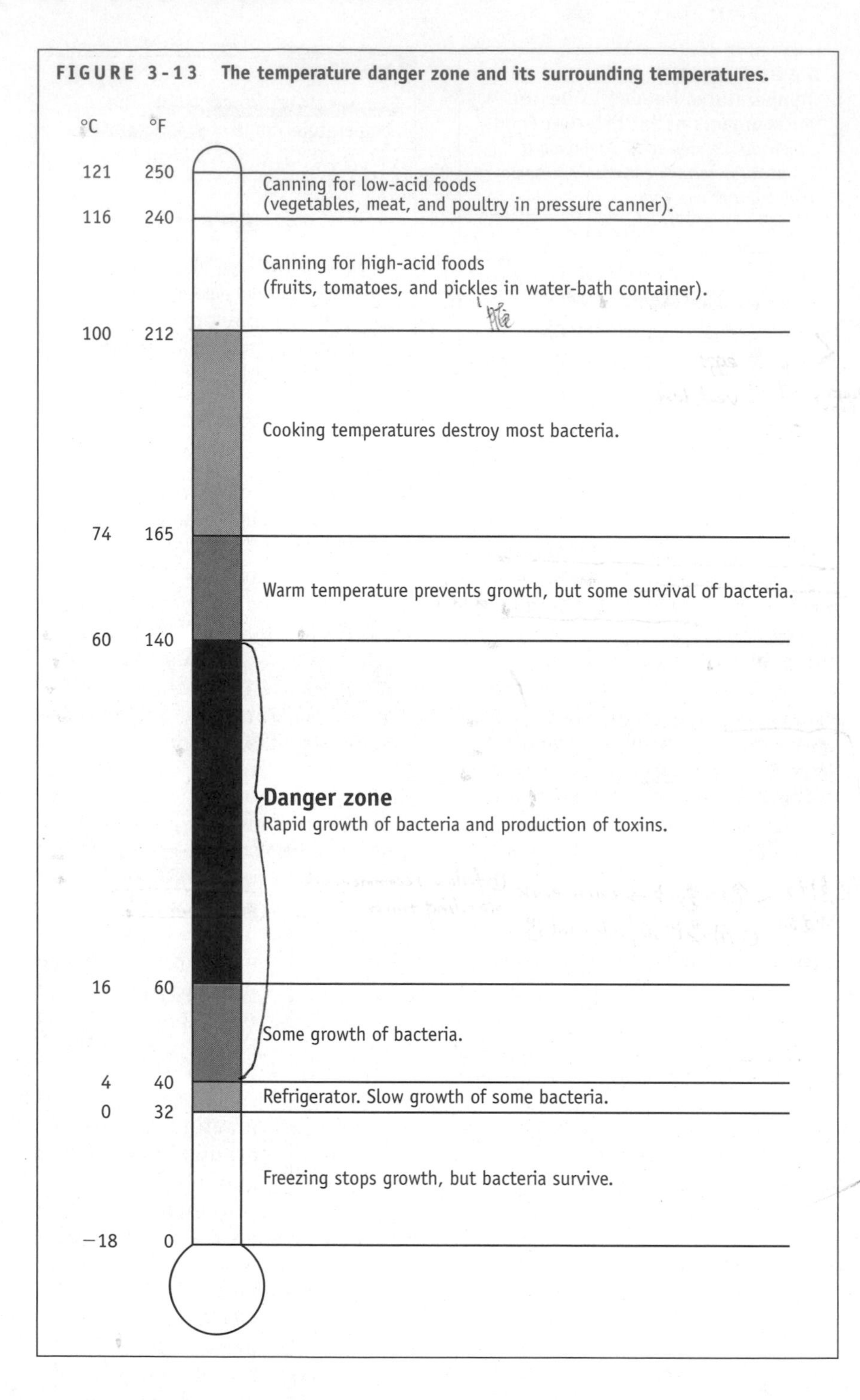

FIGURE 3-13 The temperature danger zone and its surrounding temperatures.

service employees is essential to the safety of the foods and beverages being served. Even when serving, the 140°F (60°C) and 40°F (4°C) boundaries must be observed.

Cooling/Reheating. Inappropriately cooled foods are a major cause of foodborne illnesses. Foods should be cooled to below 40°F (4°C) within four hours of removal from cooking or they pose a danger to consumers. Liquid foods should be placed in shallow pans less than 3 inches deep to cool, and thicker foods in pans less than 2 inches deep (Figure 3-14). All hot foods must be reheated to at least 165°F (74°C) within two hours before serving. In a food service establishment, untouched leftovers are sometimes discarded, because they are a potential source of microbial contamination.

Storage Temperatures. Perishable foods should be stored in the refrigerator, freezer, or dry conditions according to the following temperatures:

- Refrigerator: 40°F (4°C) or below
- Freezer: below 0°F (−18°C)
- Dry storage: 65°F (18°C)

The optimal storage temperatures of foods vary; the inside back cover describes these differences. All refrigerators and freezers, especially in a commercial food establishment, should be opened only when absolutely necessary, because frequent door-opening decreases their temperature efficiency. Refrigerator temperatures should be checked regularly to ensure that they are being maintained correctly. Studies show that 10 to 20 percent of home refrigerators are above 50°F (10°C), a temperature that cannot ensure safety (16). The refrigerator or freezer should also not be overloaded, which compromises the capacity of the unit to maintain correct temperatures. In addition, the practice of placing large amounts of hot food in a cold refrigerator will cause temperatures to fluctuate dangerously.

Time—Critical Control Limit. Microbial growth occurs exponentially; the number of bacteria can grow from harmless to staggering in a relatively short time (Figure 3-12). The Two-Hour Rule applies: any perishable food left out in danger zone temperatures for more than two hours actual time, or four hours cumulative time, should be discarded. Cumulative time includes the time from the truck to the store, the store to the freezer, the freezer to the kitchen, and the time on the counter where the food is being prepared.

Storage Times. There is a limit to how long perishable foods can be held safely in refrigerator, freezer, or dry conditions. Storage limits are provided on the inside back cover of this book, but keep in mind that recommendations for maximum storage times are not

FIGURE 3-14 Proper cooling techniques.

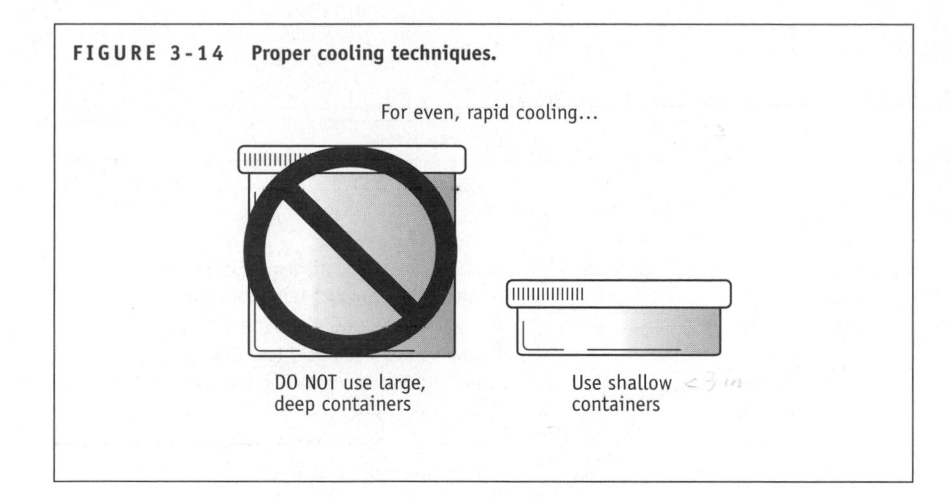

exact and vary from source to source. A general rule to follow is 3 days maximum for fresh meats and high-water-content fruits and vegetables. The "first in, first out" rule should also be followed—foods brought into the storage area at an earlier date are used before those purchased later.

Water and Humidity—Critical Control Limit. Bacteria cannot survive without water. The presence of water and the amount available to bacteria are the most important factors governing the microbial contamination of foods (13). Water must be kept safe and free of pathogenic microorganisms. Food service establishments are allowed to use water only from an approved water supply, and are required to have adequate plumbing and proper sewage and water waste disposal. They must also have mechanisms installed to prevent backflow through the pipes—an occurrence that will cause the health department to immediately close a food service establishment.

pH—Critical Control Limit. The acidity or alkalinity of a substance often determines which bacteria, if any, will grow in a food. The antibacterial action of vinegar has been known for a long time (20). Some slaughterhouses spray beef carcasses with acetic acid (vinegar) to reduce the number of microbes (55). During epidemics of cholera, a known folk remedy thought to block the transmission of the bacterium responsible for the disease is to add lime juice to the drinking water (15).

Generally, high-acid foods are less likely to engender the growth of bacteria than low-acid foods. The pH sensitivity of bacteria means that fruits and other high-acid content ($pH < 4.6$) foods are less likely to be sources of microbial contamination than foods with a lower acid content ($pH > 4.6$), such as meats and vegetables (22). Tomatoes at one time were considered to be uniformly high-acid foods, but newer varieties have been developed that are not so acidic.

HACCP Principle #4: Monitor Critical Control Points

Observations and Measurements. Monitoring is a planned sequence of observations or measurements to assess whether a CCP is under control (Table 3-12). There are three main purposes for monitoring:

1. It tracks the system's operation so that a trend toward a loss of control can be recognized and corrective action can be taken to bring the process back into control before a deviation occurs.
2. It indicates when loss of control and a deviation have actually occurred, and corrective action must be taken.
3. It provides written documentation for use in verification of the HACCP plan.

Several measures that can be taken to monitor CCP include visual observations, temperatures, times, pH, and a_w. Instrumentation used by the food establishment for measuring critical limits must be carefully calibrated for accuracy. Records of calibrations must be maintained as a part of the HACCP plan documentation.

HACCP Principle #5: Take Corrective Action

When monitoring shows that a critical limit has been exceeded, the next step is to establish the action to be taken. The purpose of a corrective action plan is to:

1. Determine any food that was produced when the deviation occurred.
2. Take corrective action to ensure that the critical control point is under control.
3. Maintain records of any corrective actions taken.

HACCP Principle #6: Documentation—Establish Record-Keeping Systems

Record keeping is essential if the HACCP system is to operate smoothly and effectively. It serves as the foundation from which proactive change and correction can occur, and may prevent small problems from escalating into large and costly ones. A written log of temperatures, times, and other critical control points at various stages of food production, maintained by individuals responsible for the various principles of the HACCP system, places accountability where it belongs. It also documents where a health problem originated and thereby protects those individuals or corporations who did meet their obligations.

Written HACCP Plan. This step requires the preparation and maintenance of a written HACCP plan by the food establishment. The plan must detail the hazards of each individual or categorical product covered by the plan. It must clearly identify the CCPs and critical limits for each CCP.

Record Keeping. The approved HACCP plan and associated records

TABLE 3-12

Monitoring Critical Control Points in the Preparation of Soups, Stocks, and Stews

Soups, Stocks, Stews	*Standards*	*Monitoring*	*Corrective Action*
Purchasing	Milk and milk products: purchase only pasteurized products.		
Cooking	Cook all hot soups and stews to recommended product temperatures.	Measure temperature.	Continue heating until temperature reaches recommended product temperatures.
Holding	Hold for service at 140°F (60°C) or higher.	Measure temperature.	Reheat one time to 165°F (73.9°C) or higher if product has been held less than 2 hours. If temperature is below 140°F (60°C) longer than 2 hours, discard. Never mix new product with old.
Cooling	Cool soups and stews to 45°F (7.2°C) or less within 4 hours.	Measure temperature.	Reheat one time to 165°F (73.9°C) or higher, if not cooled to 45°F (7.2°C) or less within 4 hours. If not cooled within 4 hours the second time, discard.
Reheating	Reheat soups and stews to 165°F (73.9°C) or higher within 2 hours.	Measure temperature.	Continue heating to 165°F (73.9°C). If temperature is not reached within 2 hours, discard.

must be on file at the food establishment. Generally, the following are examples of documents that can be included in the total HACCP system:

- Listing of the HACCP team and assigned responsibilities
- Description of the product and its intended use
- Flow diagram of food preparation indicating CCPs
- Hazards associated with each CCP and preventive measures
- Critical limits
- Monitoring system
- Corrective action plans for deviations from critical limits
- Record-keeping procedures
- Procedures for verification of the HACCP system

HACCP Principle #7: Verification

Internal Verification Procedures. Internal verification that the HACCP system is working occurs in three phases. In the first, the HACCP team verifies that critical limits at CCPs are satisfactory. The second phase requires that the HACCP plan is functioning so effectively that little end-product sampling is required. The third phase consists of documented periodic revalidations by a HACCP team on a regular basis and/or whenever significant product, preparation, or packaging changes require modification of the HACCP plan.

External Verification Procedures—Health Department Inspection. External verification occurs when state and county health departments show up unannounced to inspect the sanitary conditions of food service establishments. If there are enough violations of items on a health department inspection form, it can result in suspension or revocation of an establishment's license. A temporary suspension is served when there is an imminent hazard to health, or when there has been a failure to comply with an earlier inspection's findings. A revocation occurs with more serious or repeated violations.

Figure 3-15 shows a sample health department inspection form and the points on which inspections are graded, receiving a score from 1 to 5 for each item. The weights of the individual deficiencies are subtracted from 100 percent to obtain the score. Health department requirements and inspection forms vary from state to state and even from county to county within a state. The items most commonly checked by health inspectors involve food-holding temperatures, improper refrigeration, and improper cooling of cooked foods.

National Surveillance. The first level of surveillance consists of physicians and coroners, who are required to notify local health departments of certain disease cases. These reports are then sent to the state public health epidemiology office, where the laboratories may also receive food samples to test. Ultimately, this information goes to federal offices such as the CDC. Unfortunately, not every person who reports a foodborne illness provides a sample, so only about 38 percent of all recognized outbreaks reported to the CDC ever have their cause identified (28). Globally, the World Health Organization (WHO) has proposed the development of a food safety plan to detect global hazards (2).

FIGURE 3-15 Health department inspection form.

FOOD

		POINTS
*01	Source; sound condition, no spoilage	5
02	Original container; properly labeled	1

FOOD PROTECTION

*03	Potentially hazardous food meets temperature requirements during storage, preparation, display, service transportation	5
04	Facilities to maintain product temperature	4
05	Thermometers provided and conspicuous	1
06	Potentially hazardous food properly thawed	2
*07	Unwrapped and potentially hazardous food not re-served	4
*08	Food protection during storage, preparation, display, service, transportation	2
09	Handling of food (ice) minimized	2
10	In use, food (ice) dispensing utensils properly stored	1

PERSONNEL

*11	Personnel with infections restricted	5
*12	Hands washed and clean, good hygienic practices	5
13	Clean clothes, hair restraints	1

FOOD EQUIPMENT AND UTENSILS

14	Food (ice) contact surfaces: designed, constructed, maintained, installed, located	2
15	Non-food contact surfaces: designed, constructed, maintained, installed, located	1
16	Dishwashing facilities; designed, constructed, maintained, installed, located, operated	2
17	Accurate thermometers, chemical test kits provided, gauge cock (1/4" IPS valve)	1
18	Pre-flushed, scraped, soaked	1
19	Wash, rinse water: clean, proper temperature	2
*20	Sanitization rinse: clean, temperature, concentration, exposure time; equipment, utensils sanitized	4
21	Wiping cloths; clean, use restricted	1
22	Food-contact surfaces of equipment and utensils clean, free of abrasives, detergents	2
23	Non-food contact surfaces of equipment and utensils clean	1
24	Storage, handling of clean equipment/utensils	1
25	Single-service articles, storage, dispensing	1
26	No re-use of single service articles	2

WATER

*27	Water source, safe: hot and cold under pressure	5

SEWAGE

		POINTS
*28	Sewage and waste water disposal	4

PLUMBING

29	Installed, maintained	1
*30	Cross-connection, back siphonage, backflow	5

TOILET AND HANDWASHING FACILITIES

*31	Number, convenient, accessible, designed, installed	4
32	Toilet rooms enclosed, self-closing doors, fixtures, good repair, clean; hand cleaner, sanitary towels/tissues/hand-drying devices provided, proper waste receptacles	2

GARBAGE AND REFUSE DISPOSAL

33	Containers or receptacles, covered: adequate number insect/rodent proof, frequency, clean	2
34	Outside storage area enclosures properly constructed, clean; controlled incineration	1

INSECT, RODENT, ANIMAL CONTROL

*35	Presence of insects/rodents — outer openings protected, no birds, turtles, other animals	4

FLOORS, WALLS, AND CEILINGS

36	Floors, constructed, drained, clean, good repair, covering installation, dustless cleaning methods	1
37	Walls, ceiling, attached equipment: constructed, good repair, clean, surfaces, dustless cleaning methods	1

LIGHTING

38	Lighting provided as required, fixtures shielded	1

VENTILATION

39	Rooms and equipment — vented as required	1

DRESSING ROOMS

40	Rooms clean, lockers provided, facilities clean, located	1

OTHER OPERATIONS

*41	Toxic items properly stored, labeled, used	5
42	Premises maintained free of litter, unnecessary articles, cleaning maintenance equipment properly stored. Authorized personnel	1
43	Complete separation from living/sleeping quarters. Laundry.	1
44	Clean, soiled linen properly stored	1

*Critical items requiring immediate attention.

Score = 100 – ☐ pts. = ☐

Adapted from Texas Dept. of Health's Food Service Establishment Inspection Report.

PICTORIAL SUMMARY / 3: Food Safety

Federal and state regulations, along with regular inspections throughout the food industry, ensure that the food supply in the United States is the safest in the world.

FOODBORNE ILLNESS

Symptoms of an illness transmitted through food may include inflammation of the gastrointestinal tract lining, nausea, abdominal cramps, diarrhea, and/or vomiting.

The most common causes of foodborne illness are:

Biological
- Microorganisms
  - Bacteria
  - Molds
  - Viruses
- Animal parasites
- Prions

Chemical
- Plants
- Seafood toxins
- Agricultural/Industrial

Physical
Foreign objects in food

BIOLOGICAL HAZARDS

Biological hazards in food include bacteria, molds, viruses, parasites, and prions. The most common cause of foodborne illness is bacteria via infection or intoxication (poisoning). Food infections, responsible for about 80 percent of foodborne illnesses, occur when a person consumes a food or beverage containing large numbers of bacteria. Food intoxication and intoxification occur when a person consumes a toxin from bacteria growing on food. Other foodborne illnesses are caused by mycotoxins in molds, viruses such as the hepatitis A virus, parasites such as worms and protozoa, and protein particles called prions, which are responsible for mad cow disease.

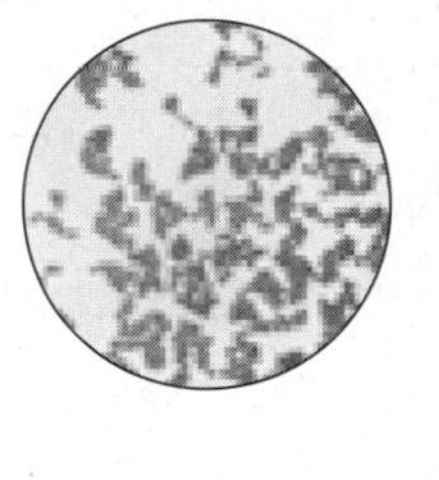

CHEMICAL HAZARDS

Chemicals that are hazardous to health can come from plants such as poisonous mushrooms, agricultural or industrial chemicals that are included in food unintentionally, or fish or shellfish that harbor dangerous toxins. Four examples of toxic seafood are ciguatera fish poisoning (the most common in the U.S.), histamine food poisoning, pufferfish poisoning, and red tide.

PHYSICAL HAZARDS

Foreign objects that inadvertently turn up in food and beverage products can threaten the consumer's health. The most common hazard is glass.

THE HACCP SYSTEM

The Hazard Analysis and Critical Control Point System (HACCP) is a proactive method of preventing foodborne illness and ensuring food safety. This system aims to prevent foodborne illness by seven methodical steps:

1. Assess the hazards.
2. Identify the critical control points—any point in the process of food production where a loss of control may result in unacceptable health risk.
3. Establish limits at each critical control point.
4. Monitor critical control points.
5. Take corrective action.
6. Documentation—establish record-keeping systems.
7. Verify that preventative and corrective measures have been taken through regular inspection. Major critical control points in the food production process:
   - Processing
   - Purchasing
   - Preparation
   - Sanitation
   - Storage

Temperatures Needed to Destroy Microorganisms in Different Foods

Food	*Temperature*
Poultry	180°F (82°C)
Reheated Foods	165°F (74°C)
Ground Meats	160°F (71°C)
Pork	160°F (71°C)
Beef	145°F (63°C)
Fish	140°F (60°C)

Dishwashing temperatures should be between 140° and 160°F (60° and 71°C), and rinse temperatures at least 180°F (82°C) for 10 seconds or 170°F (77°C) for 30 seconds to kill most pathogens.

REFERENCES

1. Adams AM, et al. Survival of Anisakis simplex in microwave processed arrowtooth flounder (Atheresthes stomias). *Journal of Food Protection* 62(4):403–409, 1999.
2. Archer DL. Global foodborne disease surveillance system faces hurdle. *Food Technology* 56:24, 2002.
3. Arthur MH. Emerging microbiological food safety systems. *Food Technology* 56(2):48–51, 2002.
4. Beadle A. Ciguatera fish poisoning. *Military Medicine* 162(5): 319–322, 1997.
5. Bean N, and P Griffin. Foodborne disease outbreaks in the United States, 1973–1987: Pathogens, vehicles, and trends. *Journal of Food Protection* 53:804–817, 1990.
6. Bernard DT, and VN Scott. *Listeria monocytogenes* in meats: New strategies are needed. *Food Technology* 53(3):124, 1999.
7. Berrang ME, et al. Eggshell membrane structure and penetration by *Salmonella typhimurium. Journal of Food Protection* 62(1):73–76, 1999.
8. Black RE, et al. Epidemic *Yersinia enterocolitica* infection due to contaminated chocolate milk. *New England Journal of Medicine* 298:76, 1978.
9. Bolton DJ, et al. Integrating HACCP and TQM reduces pork carcass contamination. *Food Technology* 53(4):40–43, 1999.
10. Bryan F. Risks of practices, procedures, and processes that lead to outbreaks of foodborne diseases. *Journal of Food Protection* 51:663–673, 1988.
11. Buchanan RL, and MP Doyle. Foodborne disease significance of *Escherichia coli* O157:H7 and other enterohemorrhagic *E. coli. Food Technology* 51(10):69–76, 1997.
12. Chapman PA, et al. A 1-year study of Escherichia coli O157 in cattle, sheep, pigs, and poultry. *Epidemiology & Infection* 119(2):245–250, 1997.
13. Chirife J, and M del Pilar Buera. Water activity, water glass dynamics, and the control of microbiological growth in foods. *Critical Reviews in Food Science and Nutrition* 36(5):465–513, 1996.
14. Cromeans TL. Understanding and preventing virus transmission via foods. *Food Technology* 54(4):20, 1997.
15. Dalsgaard A, et al. Application of lime *(Citrus aurantifolia)* juice to drinking water and food as a cholera-preventive measure. *Journal of Food Protection* 60(11): 1329–1333, 1997.
16. Daniels RW. Applying HACCP to new-generation refrigerated foods at retail and beyond. *Food Technology* 45(6):122–124, 1991.
17. Dean JP, and EA Zottola. Use of nisin in ice cream and effect on the survival of *Listeria monocytogenes. Journal of Food Protection* 59(5):476–480, 1996.
18. Delmore LRG, et al. Hot-water rinsing and trimming/washing of beef carcasses to reduce physical and microbiological contamination. *Journal of Food Science* 62(2):376–373, 1997.
19. Denton JH. Performance standards for *Campylobacter* are not warranted. 56(7):104, 2002.
20. Entani E, et al. Antibacterial action of vinegar against foodborne pathogenic bacteria including *Escherichia coli* O157:H7. *Journal of Food Protection* 61(8):953–959, 1998.
21. Espy M. Ensuring a safer and sounder food supply. *Food Technology* 48(9):91–93, 1994.
22. Food microbiology: Controlling the greater risk. *Medallion Laboratories Analytical Progress* 4(1):1–8, 1988.
23. Foodborne diseases active surveillance network, 1996. *Morbidity and Mortality Weekly Report* 46(12):258–261, 1997.
24. Fresh juice label warnings called no substitute for pasteurization. *Food Chemical News* 39(39):23, 1997.
25. Giese J. It's a mad, mad, mad, mad cow test. *Food Technology* 55(6):60, 2001.
26. Grigoriadis SG, et al. Survival of *Campylobacter jejuni* inoculated in fresh and frozen beef hamburgers stored under various temperatures and atmospheres. *Journal of Food Protection* 60(8):903–907, 1997.
27. Hennessy TW, et al. A national outbreak of *Salmonella enteritidis* infections from ice cream. *New England Journal of Medicine* 334(20):1281–1286, 1996.
28. Hill WE. The polymerase chain reaction: Applications for the detection of foodborne pathogens. *Critical Reviews in Food Science and Nutrition* 36(1&2): 123–173, 1996.
29. Hoch GJ. New rapid detection test kits speed up food safety. *Food Processing* 58(11):35–39, 1997.
30. Institute of Food Technology. New bacteria in the news: A special symposium. *Food Technology* 40(18):16–25, 1986.
31. Jenks WG, et al. Detection of parasites in fish by superconducting quantum interference device magnetometry. *Journal of Food Science* 61(5):865–869, 1996.
32. Juranovic LR, and DL Park. Foodborne toxins of marine origin: *ciguatera. Reviews of Environmental Contamination and Toxicology* 117:51–94, 1991.
33. Knabel SJ. Scientific status summary. Foodborne illness: Role of home food handling practices. *Food Technology* 49(4):119–131, 1995.
34. Lee JY, et al. Two-stage nested PCR effectiveness for direct detection of *Vibrio vulnificus* in natural samples. *Journal of Food Science* 64(1):158–162, 1999.
35. Li Y, et al. Pre-chill of chicken carcasses to reduce *Salmonella typhimurium. Journal of Food Science* 62(3):605–607, 1997.
36. Liston J. Current issues in food safety; especially seafoods.

Journal of the American Dietetic Association 89(7):911–913, 1989.

37. McAuley JB, MK Michelson, and PM Schantz. *Trichinosis* surveillance, United States, 1987–1990. *Morbidity and Mortality Weekly Report CDC Surveillance Summary* 40(3):35–42, 1991.
38. McNutt K. Common sense advice to food safety educators. *Nutrition Today* 32(3):128–133, 1997.
39. Mermelstein NH. Washington News. *Food Technology* 53(8):28, 1999.
40. Mermelstein NH. Comprehensive BSE risk study released. *Food Technology* 56(1):75, 2002.
41. Miller AJ, RC Whiting, and JL Smith. Use of risk assessment to reduce *listeriosis* incidence. *Food Technology* 51(4):100–103, 1997.
42. National Advisory Committee on Microbiological Criteria for Foods. Hazard analysis and critical control point principles and application guidelines. *Journal of Food Protection* 61(6):762–775, 1998.
43. Nightingale, S. Foodborne disease: An increasing problem. *American Family Physician* 35(3):353–354, 1987.
44. Omaye ST. Shiga-toxin-producing *Escherichia coli:* Another concern. *Food Technology* 55(5):26, 2001.
45. Peregrin T. Bioterrorism and food safety: What nutrition professionals need to know to educate the American public. *Journal of the American Dietetic Association* 102(1):14–16, 2002.
46. Pohland AE. Mycotoxins in review. *Food Additives and Contaminants* 10(1):17–28, 1993.
47. Risks misjudged in cholera epidemic. *Food Insight.* January/February 2, 1992.
48. Sado PN, et al. Identification of *Listeria monocytogenes* from unpasteurized apple juice using rapid test kits. *Journal of Food Protection* 61(9):1199–1202, 1998.
49. Schwarzkopf A. *Listeria monocytogenes:* Aspects of pathogenicity. *Pathologie Biologie* (Paris) 44(9):769–774, 1996.
50. Silk TM, ET Ryser, and CW Donnelly. Comparison of methods for determining coliform and *Escherichia coli* levels in apple cider. *Journal of Food Protection* 60(11):1302–1305, 1997.
51. Steele ML, et al. Survey of Ontario bulk tank raw milk for foodborne pathogens. *Journal of Food Protection* 60(11):1341–1346, 1997.
52. Su YC, and ACL Wong. Current perspectives on detection of staphylococcal enterotoxins. *Journal of Food Protection* 60(2):195–202, 1997.
53. Tauxe RV. Emerging foodborne diseases: An evolving public health challenge. *Emerging Infectious Diseases* 3(4):425–434, 1997.
54. Thayer DW, and G Boyd. Radiation sensitivity of *Listeria monocytogenes* on beef as affected by temperature. *Journal of Food Science* 60(2):237–240, 1995.
55. Tinney KS, et al. Reduction of microorganisms on beef surfaces with electricity and acetic acid. *Journal of Food Protection* 60(60):625–628, 1997.
56. U.S. Dept. of Health and Human Services. Food and Drug Administration. Center for Food Safety and Applied Nutrition. Foodborne pathogenic microorganisms and natural toxins handbook (Bad Bug Book). Available: http://www.cfsan.fda.gov. 1997.
57. Vangelova L. *Botulinum* toxin: A poison that can heal. *FDA Consumer* 29(10):16–19, 1995.
58. Zottola EA. Reflections on *Salmonella* and other "wee beasties" in foods. *Food Technology* 55(9):60–67, 2001.

WEBSITES

The government's food safety website:

www.foodsafety.gov

More information about the USDA's HACCP website:

www.cfsan.fda.gov/~/rd/haccp.html

Mad cow disease information:

USDA: **www.fsis.usda.gov/oa/topics/bse_thinking.htm**

US Beef Industry: **www.BSEinfo.org**

The food code established by the FDA:

www.cfsan.fda.gov/~dms/foodcolde.html

The government's food security preventative measures guidance:

www.cfsan.fda.gov/~dms/secguid.html

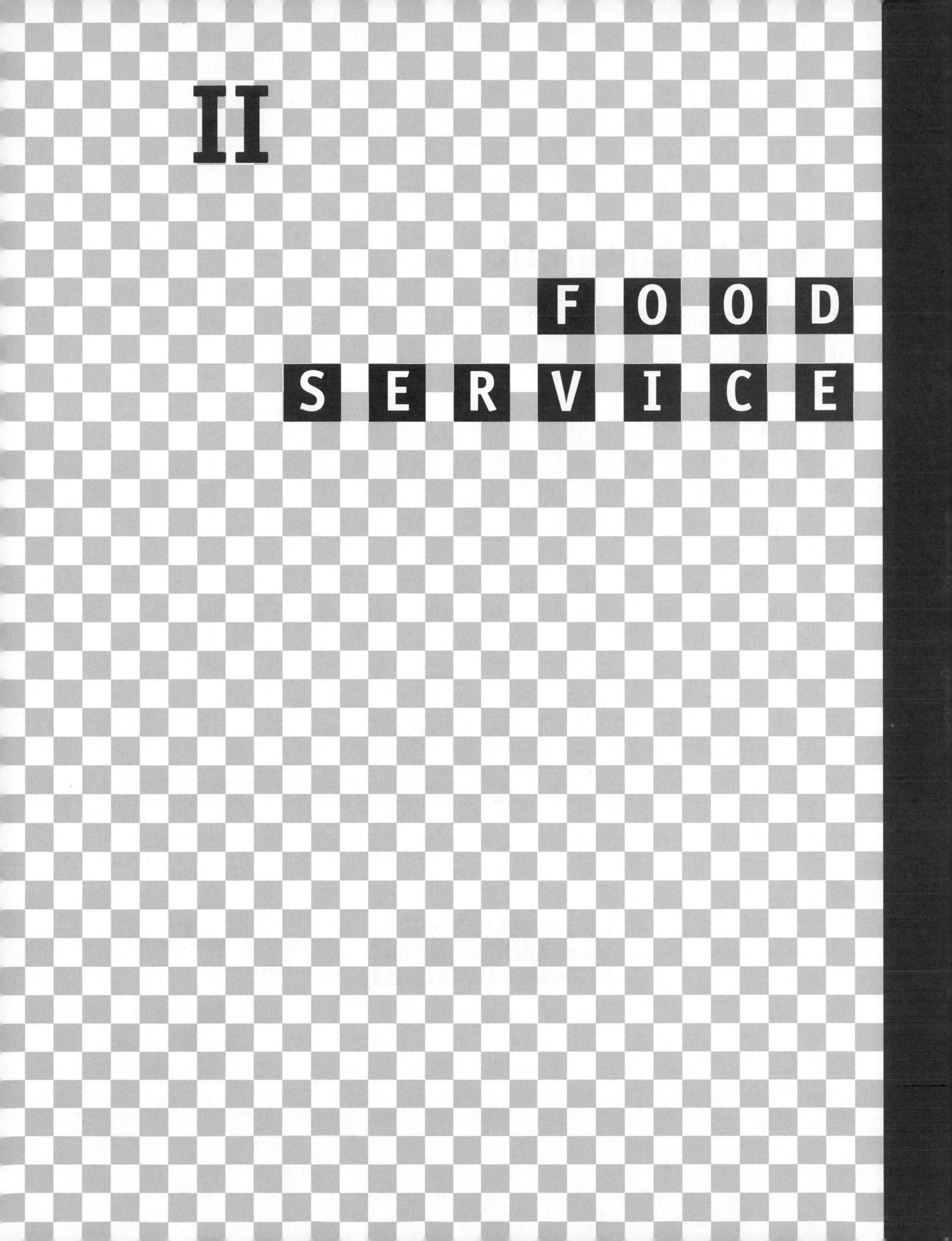

II

FOOD SERVICE

Meal Management

A successful meal is both psychologically and physiologically satisfying. Planning such meals requires a basic knowledge of food preparation, nutrition, and presentation strategies. Effective meal planning and preparation, whether for a household, an institution, or a restaurant chain, is made possible by the efficient management of money, time, and energy. These resources are used in the various steps of meal production: food procurement, storage, preparation, serving, and cleanup. All these steps require organization on the part of the individual or of the food service manager. In the case of the latter, good organization involves people working together toward the common goal of preparing and serving attractive, tasty, nutritious, and profitable meals. This chapter covers how food service establishments are organized, meal planning, purchasing, time management, types of meal service, table settings, and the art of table manners.

KEY TERMS

Job description An organized list of duties used for finding qualified applicants, training, performance appraisal, defining authority and responsibility, and determining salary.

Organizational chart A descriptive diagram showing the administrative structure of an organization.

Food Service Organization

At the core of every food service operation is an organization with a structure set up to achieve specific goals. Management determines the objectives necessary to reach those goals and then mobilizes people toward meeting them. This entails the division of work, which necessitates effective **job descriptions**. Positions are often described more fully in a job description compared to a job ad, or even how an employer describes the job to a new employee. Anyone applying for employment in such an operation should ask the employer to see the job description and check that it matches the job duties, performance evaluations, and salary offered. An **organizational chart** is also helpful to a new employee. Figure 4-1 compares the organizational charts between a hospital dietary department and a restaurant.

Commercial Food Service Organization

Large food service organizations usually follow a historical structure that was pioneered by George Auguste Escoffier (1847–1935). Escoffier, called the father of 20th-century cookery, created "stations" for particular areas of food production. Escoffier's system of dividing up large kitchens into various

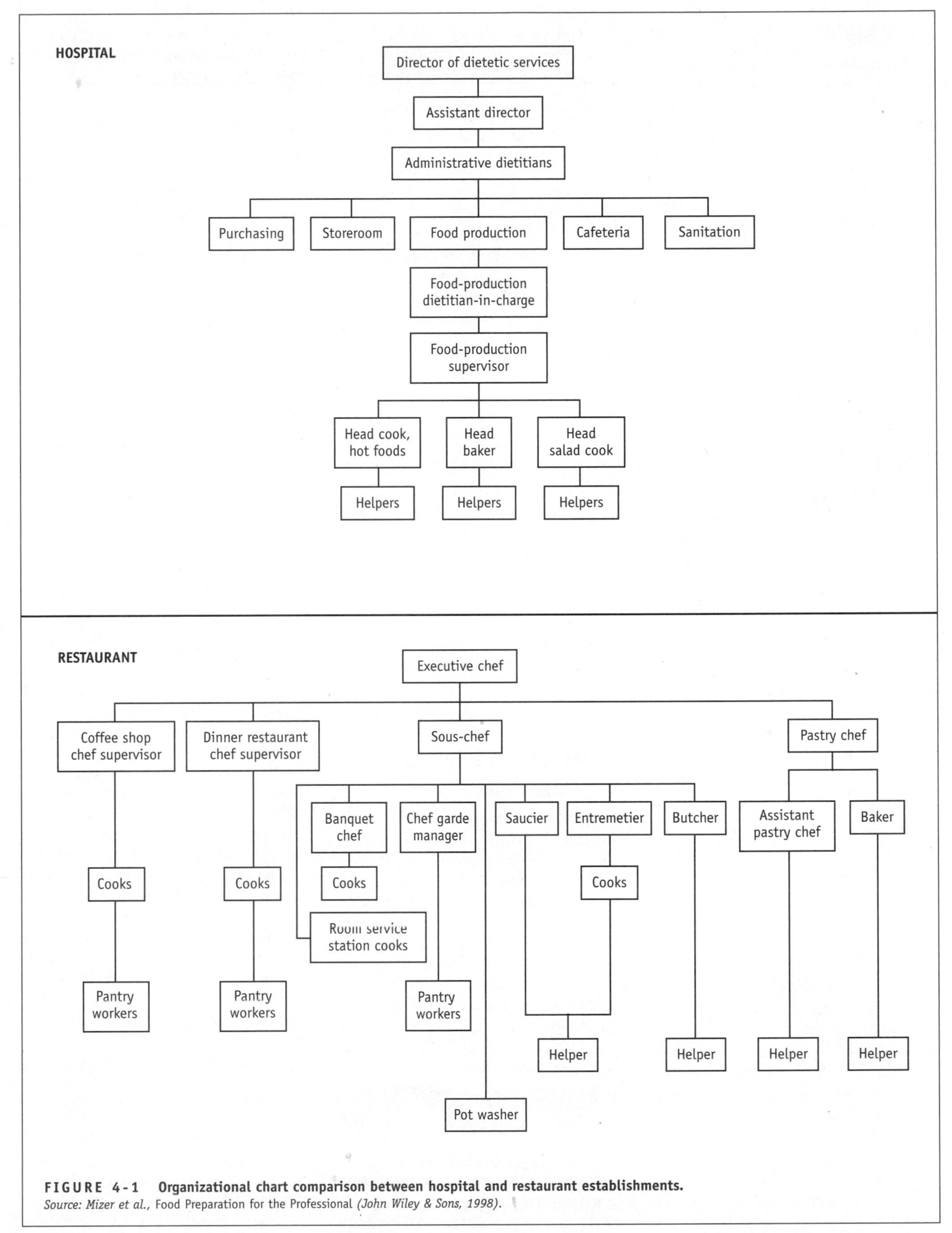

FIGURE 4-1 Organizational chart comparison between hospital and restaurant establishments.
Source: Mizer et al., Food Preparation for the Professional *(John Wiley & Sons, 1998).*

KEY TERMS

Dietitian (registered dietitian or RD) A health professional who counsels people about their medical nutrition therapy (diabetic, low cholesterol, low sodium, etc.). Registration requirements consist of completing an approved four-year college degree, exam, internship, and ongoing continuing education.

preparation areas led to the creation of jobs requiring specific skills. The kitchen team of employees under this type of food service organization is called the *brigade de cuisine* (bree-gahd-de-kwee-zeen).

Escoffier's System of Organization Via "Stations." At the head of each station in the kitchen are station chefs or heads with a particular area of expertise:

- Sauce chef/*saucier* (so-see-ay). The highest position among the stations. This chef specializes in the production of sauces, sauce-related dishes, hot hors d'oeuvres, stews, and sautéed foods.
- Fish cook/*poissonier* (pwah-so-nee-ay). Sometimes this station is covered by the sauce chef.
- Vegetable cook/*entremetier* (on-tramet-ee-ay). Prepares vegetables.
- Soup cook/*potager* (poh-ta-zhay). Prepares soups and stocks.
- Roast cook/*rotisseur* (ro-tee-sur). Responsible for meat dishes, particularly if they are roasted or braised.
- Broiler cook/*grillardin* (gree-yar-dan). Specializes in preparing grilled, broiled, or deep-fried meats, poultry, and seafood.
- Pantry chef/*chef garde manger* (guard-mon-zhay). Prepares all nondessert cold foods such as salads and cold hors d'oeuvres.
- Pastry chef/*patissier* (pa-tees-see-ay). Prepares baked goods—pastries, desserts, and breads.
- Relief, swing, or rounds cook/*tournant* (tour-non). Capable of handling any station in order to relieve one of the other chefs.

In smaller kitchens, there may be only two stations, one for hot and one for cold foods. In such kitchens the pantry chef is a major position.

Each station head has cooks and helpers or assistants that aid with that area of food production. The first step on the ladder of jobs in a large kitchen is the entry-level position of helper or assistant, which requires virtually no skills. Once the skills are acquired, however, the person may be promoted to cook. People can sign on as apprentices and receive formal training in food service by rotating through each of the kitchen stations.

Administrative Positions. In general, after about five to ten years of experience in all stations as a chef, a person can be promoted to administrative positions in the kitchen:

- Executive chef/*chef executif.* The person in charge of the entire operation, including kitchen administration, hiring, budgeting, purchasing, work scheduling, menu planning, and more.
- Production manager/*sous* (soo) chef. The second highest position in the kitchen. The sous chef is in charge of all areas of production and supervision of the staff.

Hospital Food Service Organization

Hospital food service operations differ slightly in that the person in charge is the food service director, or administrative or chief **dietitian**. Working for the chief dietitian are other dietitians, their numbers depending on the size of the hospital. Assisting the dietitians are the dietetic technicians, with at least a two-year degree, and dietary aides, entry-level employees who may start out relatively untrained.

Meal Planning

The ultimate goal of a food service organization is to plan, prepare, and serve meals. Food production begins with planning the menu. The menu dictates all other actions that will follow, such as purchasing, what equipment to use, labor, and serving.

USDA Menu Patterns

What is typically served for breakfast, lunch, and dinner? The meal pattern varies tremendously from country to country. In North America the standard fare can be divided into the selections shown in the USDA's Adult Care Meal Pattern (Figure 4-2). A similar meal pattern for infants (birth–1 year) and children (1–2 years; 3–5 years; 6–12 years) is called the USDA's Child Care Meal Pattern and is available at the USDA's website, www.fns.usda.gov/cnd/care/ProgramBasic/Meals.htm.

The USDA also distributes menu guidelines for its School Lunch Programs. These guidelines offer a choice between two meal patterns: Traditional Food-Based Menu Planning and Nutrient Standard Menu Planning ("NuMenues"). The Traditional Food-Based Menu Planning approach creates a menu of five food items from four food components: meat/meat alternate, vegetables and/or fruits, grains/breads, and milk. Portion sizes depend on the children's ages and grade groups. The trend toward healthier eating led the USDA to develop the Nutrient Standard Menu Planning or NuMenues. Instead of using food groups to create a menu, NuMenues use nutrition-analysis computer software to calculate a healthy meal. This is currently defined as one meeting the Dietary Guidelines for Americans, limiting calories (which may still be too high; e.g., 825 calories/kcal for a 7–12th grader's lunch), and providing at least one-third of the daily Recommended Dietary Allowances for protein, iron, calcium, and vitamins A and C. Percentage fat limitations depend on the grade (e.g., 7–12 graders should not have more than 30 percent fat in the course of one week).

In the same spirit of altering eating habits toward healthier goals, Figure 4-3 on page 82 suggests an Eating Right Menu in which complex carbohydrates comprise the bulk of a meal, followed by fruits, vegetables, dairy, and lastly meat.

FIGURE 4-2 USDA's Adult Care Meal Pattern.

Adult Care Meal Pattern

Breakfast for Adults
Select All Three Components for a Reimbursable Meal

1 milk	1 cup	fluid milk
1 fruit/vegetable	1/2 cup	juice,[1] fruit and/or vegetable
1 grains/bread[2]	2 slices	bread or
	2 servings	cornbread or biscuit or roll or muffin or
	1 1/2 cups	cold dry cereal or
	1 cup	hot cooked cereal or
	1 cup	pasta or noodles or grains

Lunch for Adults
Select All Four Components for a Reimbursable Meal

1 milk	1 cup	fluid milk
2 fruits/vegetables	1 cup	juice,[1] fruit and/or vegetable
1 grains/bread[2]	2 slices	bread or
	2 servings	cornbread or biscuit or roll or muffin or
	1 1/2 cups	cold dry cereal or
	1 cup	hot cooked cereal or
	1 cup	pasta or noodles or grains
1 meat/meat alternate	2 oz.	lean meat or poultry or fish[3] or
	2 oz.	alternate protein product or
	2 oz.	cheese or
	1	egg or
	1/2 cup	cooked dry beans or peas or
	4 Tbsp.	peanut or other nut or seed butter or
	1 oz.	nuts and/or seeds[4] or
	8 oz.	yogurt[5]

Dinner for Adults
Select All Three Components for a Reimbursable Meal

2 fruits/vegetables	1 cup	juice,[1] fruit and/or vegetable
1 grains/bread[2]	2 slices	bread or
	2 servings	cornbread or biscuit or roll or muffin or
	1 1/2 cups	cold dry cereal or
	1 cup	hot cooked cereal or
	1 cup	pasta or noodles or grains
1 meat/meat alternate	2 oz.	lean meat or poultry or fish[3] or
	2 oz.	alternate protein product or
	2 oz.	cheese or
	1	egg or
	1/2 cup	cooked dry beans or peas or
	4 Tbsp.	peanut or other nut or seed butter or
	1 oz.	nuts and/or seeds[4] or
	8 oz.	yogurt[5]

[1] Fruit or vegetable juice must be full-strength.

[2] Breads and grains must be made from whole-grain or enriched meal or flour. Cereal must be whole-grain or enriched or fortified.

[3] A serving consists of the edible portion of cooked lean meat or poultry or fish.

[4] Nuts and seeds may meet only one-half of the total meat/meat alternate serving and must be combined with another meat/meat alternate to fulfill the lunch requirement.

[5] Yogurt may be plain or flavored, unsweetened or sweetened.

Hospital Menu Patterns

Patients in a hospital have various dietary needs. They may be on a regular or general diet, a modified consistency diet, or a prescribed diet, depending on their current health condition (Table 4-1 on page 83). It is the responsibility of the hospital Dietary Department to ensure that all patients are being provided the appropriate diet and to counsel patients with special dietary advice if their doctor prescribed a diet as medical nutrition therapy (MNT).

Creating the Menu

Regardless of what menu pattern is followed, planning menus is the essential first step of food production. The first decision concerns what type of menu will be used: no choice, limited choice, or choice menu. The menu can reflect what is offered daily, weekly, or for several weeks (**cycle menus**).

Since dinners are usually the most prominent meal, most people plan dinner first. Meals then are built around the main entrée with the remaining items usually decided in the following order: vegetable, starch, salad, bread, soup, appetizer, and dessert (Figure 4-4 on page 84). Some provision must also be made for breakfasts, lunches, beverages, and snacks. The dietary recommendation for fruit is at least twice a day and it can be incorporated into a variety of options on the menu such as salad, dessert, snack, breakfast, and/or a beverage.

Cycle Menus. Creating several weekly menus in a row sets up a menu cycle (Figure 4-5 on page 84). This is a common practice for food service institutions, especially schools. Three-week cycles improve cost control, but four-week cycles are less monotonous, and

KEY TERMS

Cycle menu A menu that consists of two or more weeks, usually three or four, that "cycles." Cycle menus offer a combination of variety and controlled costs.

FIGURE 4-3 Eating Right Menu: selected suggestions for breakfast, lunch, and dinner.

Breakfast – Eggs (limit 4/week)
Cereal (High fiber—at least 3gm/serving)
+ milk (nonfat or lowfat)
Pancakes
Waffles
French toast
Bagel
Muffin
Scone
Cottage cheese
Yogurt
Smoothie
+ fruit (whole or juice)
+ high fiber cracker

Lunch – Sandwich (all have tomato and/or lettuce)
(bread = whole wheat, whole grain, low calorie, pita)
(meat if included not to exceed 2 oz)
Examples: Tuna/sunflower/dill/mustard
Grilled tuna/melted cheese
Turkey/cheese/cranberry sauce
Cheese/mustard
Grilled cheese/mustard
Peanut butter + jelly
Grilled cheese
Soup/salad
Examples: Pasta/rice/bean
Tuna/chicken/shrimp
Potato
Greens
Baked potato
Pasta
Rice/beans
Fruit/cheese/crackers + yogurt
+ vegetable (3 servings/day) (vitamin A & C containing at least 4x/week)
+ fruit (2 servings/day) (vitamin A & C containing at least 4x/week)

Dinner – + bread (at least 1 gm fiber/serving)

Any lunch entrée
Meat (3 oz)
Lean meat (beef, pork, lamb)
Poultry (no skin)
Fish
Pasta
Examples: Spaghetti
Macaroni
Lasagna
Rice/bean
+ vegetable (vitamin A & C containing at least 4x/week)
+ fruit (vitamin A & C containing at least 4x/week)
+ bread (at least 1 gm fiber/serving)

Source: Mizer et al., Food Preparation for the Professional *(John Wiley & Sons, 1998).*

longer ones are preferred for people who are unable to eat elsewhere, such as in nursing homes or other long-term institutions. After deciding on the number of weeks in the cycle menu, the contents of each week are then planned.

Planning a weekly menu cycle is usually done by the food service establishment management team, which attempts to balance numerous factors, such as those itemized in the checklist in Figure 4-6 on page 85.

Nutrient Value

The nutritive value of meals is another responsibility of the food service manager or director. The trend toward healthier eating has even contributed to the suggestion of a national "fat tax" aimed at fast-food restaurants delivering high-fat meals in super sizes (7). The growing national concern with increasing obesity rates and resulting health problems has the United States slowly mobilizing to address this issue.

HOW & WHY? ?????????

Why are obesity rates increasing in the United States if fat intake has decreased? Despite the successful decrease in percent fat intake from calories, Americans continue to increase their overall consumption of calories (kcals).

Portion Size. Basically, portions should be reasonably sized in order to satisfy a person's appetite without contributing to excessive calories (kcal). The percentage of calories (kcal) from carbohydrates should be at least 55 to 65 percent, with fat not exceeding 30 percent, and protein at approximately 12 percent. The way to achieve this goal is to emphasize grains, pasta, fruits, and vegetables, with carbohydrates derived primarily from complex sources and providing at least 20 to 30 grams of fiber per day. Sufficient vitamins (especially A and C) and many minerals can come from fruits and vegetables. Meat can be kept to 5 ounces per day as long as two servings of dairy foods or one of dairy foods and one of eggs are included. Food sources containing saturated fat, cholesterol, sodium, and caffeine should be minimized.

Planning Healthy Meals. Healthy meal planning is more achievable than ever because people are changing their eating patterns in a positive way. This can be seen by comparing the first food consumption records collected in 1909

TABLE 4-1
Brief Summary of Hospital Diets

Diet		*Other Common Names*	*Definition*	*Purpose*
General		Routine, House, or Selective Diet	For adults who are not on any dietary restriction	Maintain optimal nutritional status
Prescribed		Diabetic, Low Sodium, Renal Diet, etc.	Prescribed by physician as medical nutrition therapy for patient	For people with specific medical conditions (diabetes, renal disease, heart disease, etc.)
Modified Consistency	Clear Liquid	Medical Nutrition Therapy	Foods that are clear liquid Minimal fiber	Preparation for and recovery from abdominal surgery, etc.
	Full Liquid	Surgical Liquid Diet	Similar to clear except permits strained items such as milk, juice, and eggs (in drinks/custards)	Follows clear liquid diet
	Blenderized Liquid	Pureed Diet	Foods blenderized to liquid form	For those unable to tolerate solid food following oral or plastic surgery; chewing or swallowing problems; wired jaws
	Mechanically Altered	Surgical Soft, Mechanical Soft Diet	Food modified only in texture — blended, chopped, ground	Promote ease of chewing following head and neck surgery or radiation, oral or dental problems
	Soft Diet	Bland, Low-Fiber, Low Residue Diet	Low fiber, little seasoning, smooth texture, low on fried and strong-flavored foods or gas-forming vegetables	Transition between liquid and general Post-operation to prevent nausea and vomiting, gas and distension from anesthesia and gastrointestinal immobility
	High Fiber	High-Residue Diet	Increased fiber (25–35 g/day), gradual, plenty of fluid	For gastrointestinal problems—constipation, diverticulosis (if not inflamed), irritable bowel syndrome, hemorrhoids, colon cancer, heart disease, diabetes mellitus, obesity

with subsequent studies. A Food Marketing Institute study showed that 58 percent of shoppers were motivated by health reasons to make major modifications in their diets (5). U.S. Department of Agriculture surveys indicate that since 1977, food intake trends have tended toward foods that are lower in fat; fat intakes have actually dropped from 40 percent to less than 33 percent of total calories (kcal), still above the recommended 30 percent ceiling, but encouraging, nonetheless (5, 13). More recent dietary cholesterol intake values are at about 400 mg per day, which is lower than the approximately 500 mg reported in 1959, but still higher than the recommended upper limit of 300 mg. The U.S. Bureau of the Census records that, between 1973 and 1993, red meat, whole milk, and egg consumption dropped 14, 50, and 25 percent, respectively. Balancing these decreases in intake are higher consumption patterns for chicken/turkey (13 percent increase), fish (1 percent), low-fat milk (35 percent), skim milk (6 percent), vegetables (32 percent), and fruits (18 percent). During the same period, lard and butter consumption dropped 28 percent, but this was offset by a 43 percent increase in intake of vegetable oils and fats (12, 16).

Somewhat discouragingly, carbohydrates, particularly the complex ones, comprise only 49 percent of total calories (kcal), instead of the minimum recommended level of 55 to 65 percent. Fiber intake, at 12 grams per day, is far below the 20 to 30 grams suggested by the National Cancer Institute. USDA consumption surveys reveal that one-third of total carbohydrates were from sugar in the first decade of 1900, but that it increased to one-half in 1990. The primary sources of refined sugar are regular soft drinks, candy, sweet baked goods, and table sugar itself. A 69 percent rise in per capita soft drink consumption occurred in the 20 years between 1973 and 1993. During this time the consumption of fruit drinks rose 36 percent, while coffee consumption declined 20 percent (13).

From the foregoing information, it can be seen that, despite the improvement, it is imperative that more emphasis be placed on the use of fiber-containing complex carbohydrates such as breads, cereals, starches, grains, dried beans, lentils, and starchy vegetables such as peas, potatoes, and corn. One of the ways to do this is to offer vegetarian cuisine, which as restaurateurs are realizing, is no longer just a passing fad (4). In light of this lower-fat trend, numerous chefs have published books on preparing "heart-healthy" foods (9), followed by a new generation of student chefs that believes dietary guides do make a difference to the customer (14).

The United States government is also encouraging citizens to improve their diet through the USDA's Center for Nutrition Policy and Promotion. The

Center offers the Interactive Health Eating Index (IHEI), an internet site where people can analyze their own diets. This diet self-assessment tool allows users to input their daily food intake and receive an evaluation of their overall diet (11). Continually improving implementation of the Dietary Guides will benefit the nutrient intake of North Americans and reduce the dietary risk factors for degenerative diseases.

Purchasing

A third aspect of meal planning is the decision of how much of each food to buy. Budget limitations determine both the types and amounts of food to be purchased. A further consideration in food service establishments is that the **food cost** accounts for about half of all costs, with the majority of the other half incurred by labor (labor cost). These two costs are primary concerns to any food service manager who knows that the bottom line is of paramount importance (1).

Buyer

Larger food service operations usually have a buyer, a purchasing department, or a cooperative arrangement with other institutions to purchase foods according to the **forecasted** menu requirements. The buyer or purchasing department determines food needs, selects vendors, bargains for price, and negotiates contracts.

Food service purchasing may be formal or informal. Informal purchasing, or open-market buying, consists of

KEY TERMS

Food cost Often expressed as a percentage obtained by dividing the raw food cost by the menu price.

■ ■ ■ ■

Forecast A predicted amount of food that will be needed for a food service operation within a given time period.

FIGURE 4-4 Checklist for organizing a menu.

❑ **Main Entrée –** Meat (beef, pork, lamb, fish/shellfish, poultry)
Cereal (rice, wheat, oat, rye, barley, breakfast cereals)
Beans (red, kidney, pinto, lima, etc.)
Pasta (lasagna, macaroni, spaghetti, etc.)
Eggs (fried, scrambled, omelet, shirred, poached, etc.)

❑ **Vegetable –** (3 servings/day) Vitamin C and A containing at least 4x/week each

❑ **Starch**

❑ **Salad**

❑ **Bread –** At least 1 gm fiber/serving

❑ **Soup**

❑ **Appetizer**

❑ **Dessert**

❑ **Breakfasts**

❑ **Beverages –** Water, lowfat or nonfat milk, juice, soda, coffee, tea

❑ **Snacks**

* Fruit (2 servings/day) can be incorporated into breakfasts, snacks, salads, desserts, and/or beverages.

FIGURE 4-5 An example of a one-week (minus Friday) cycle menu.

Month 1. Week 1

	Breakfast	*Lunch*	*Dinner*
MONDAY	Cereals Scrambled Eggs Grapefruit Wheat Toast/Jam	Spaghetti Garlic Bread Caesar Salad Juices	Roast Beef Baked Potato Glazed Carrots Garden Salad
TUESDAY	Cereals Pancakes Apricot Cup Wheat Toast/Jam	Roast Beef Sandwiches Apple or Banana Coleslaw	Lasagna Garlic Bread Tossed Salad Fruit Pizza Dessert
WEDNESDAY	Cereals French Toast Fruit Salad Wheat Toast/Jam	Rice/Bean Combo Steamed Broccoli Whole Grain Bread Fruit Salad	Baked Fish Red Potatoes Green Beans Apple Pie (lowfat)
THURSDAY	Poached Eggs Orange Juice Scones Wheat Toast/Jam	Baked Chicken Mashed Potatoes Cranberry Salad Corn	Angel Hair Pasta Garlic Bread Broccoli/Cauliflower Peach Cobbler

Source: Ashley S. Anderson, *Catering for Large Numbers* (Reed International Books, 1995).

FIGURE 4-6 Menu cycle evaluation checklist.

- ❑ **Clientele**
  - Age
  - Religion
  - Cultural preferences
  - Regional differences
- ❑ **Cost**
- ❑ **Taste**
  - Does the entrée selection include meals that taste better than the competition?
- ❑ **Holiday meals**
- ❑ **Seasonal availability** (fruits and seafoods)
- ❑ **Nutrition guidelines**
  - Risk factors for diseases
  - Fat — less than 30% calories
  - Complex carbohydrates — at least 58% calories
  - Cholesterol — less than 300 mg/day
  - Fiber — at least 20-30 gm/day
  - Vitamins — A and C rich vegetables/fruits at least 4x/week
  - Minerals — avoid excess sodium
  - National Cancer Institute recommendations
  - Exchange list
  - Dietary guidelines
  - Dietary goals
- ❑ **Appealing menu items**
  - Flavor/color/texture/shapes (diced/strips/chopped)/temperature variation
  - Type of preparation (fried/baked/broiled/sauced/plain)
  - Are records of consumption/popularity incorporated?
  - Garnishes
- ❑ **Equipment use balanced**
- ❑ **Workload/schedules balanced** (broiling, frying, microwave, oven, etc.)
- ❑ **Cycle/day sequence**
  - Is the end of the cycle different from the beginning?
  - Are the day's options for breakfast/lunch/dinner different?
  - Is any one item repeated too frequently in the cycle?
- ❑ **Descriptive menu**
  - Steak — Sizzling Swiss Steak
  - Peas — Buttered Peas and Mushrooms
  - Potatoes — Boiled New Potatoes
  - Salad — Fresh Garden Salad
  - Brownies — Chewy Fudge Brownies

ordering food supplies from vendors on a daily, weekly, or monthly basis. Formal purchasing, or competitive-bid buying, occurs when the buyer sends vendors an invitation to quote prices on a needed food item (17). **Specifications** describe in detail the food items to be purchased and may be developed by either the buyer or the seller (8). Deadline dates are given in formal purchasing, and bids are placed in a sealed envelope that is not opened until all the qualified vendors' bids have been submitted. The lowest bid is awarded the purchasing contract.

Food Stores and Vendors/Suppliers

The cost of anything, including food, depends in part on where it is purchased. Understanding the differences among the types of retail and wholesale food supply sources allows buyers to select the ones that will give them the most for their money. The variety of food stores available to consumers includes supermarkets, warehouse stores, co-ops, farmers' markets, and convenience, specialty, and health food stores. Food service establishments rely on large food distribution centers to obtain their supplies.

KEY TERMS

Specifications Descriptive information used in food purchasing that defines the minimum and maximum levels of acceptable quality or quantity (i.e., U.S. grade, weight, size, fresh or frozen).

Supermarkets. A century or so ago, consumers went to the local grocery store and gave a storekeeper or clerk a list of what they wished to purchase. Then they waited while all the products on the list were collected for them. Although this was an accepted part of life in the community, it could be tedious and time consuming and it prevented customers from browsing and selecting items at their leisure.

In the early 1900s, large city grocery stores began allowing retailers of individual products to sell from booths inside the stores. This opened up the market for different kinds of foods, increased consumer choices, and made shopping faster and more convenient. Eventually, this arrangement developed into the modern supermarket, where the major departments include meats, produce, dairy, bakery, frozen, canned and otherwise processed foods, as well as nonfood items such as cleaning, beauty, and even car care supplies.

The easy availability of items is a major factor in consumers' selection of a supermarket. A marketing company that polled consumers nationwide, however, found that the most important consideration in this selection was the cleanliness of the store, followed by the convenience of its location, the courtesy of its clerks, its prices, and speedy checkout service. Less important were attractive displays, baggers, weekly specials, and store coupons. Those polled also indicated that they would appreciate a "medium" checkout lane for those with a quantity of groceries in between that accepted in the express and regular lanes.

Warehouse Stores. Although supermarkets are undoubtedly the most popular avenue for purchasing food, there are other options. Warehouse stores are less expensive than supermarkets because they offer the basic foods with little "glitz." Food is often found on the shelves in the original shipping containers, and shoppers may find themselves bagging their own purchases.

Co-ops. The food cooperative (co-op) is a membership arrangement that cuts out the middle, retail level by purchasing foods in bulk at wholesale prices to sell to members and, in some instances, the public. Any profits are divided among the co-op members. Some co-ops expect the members to put in several hours per week helping with the operation of the co-op, while others hire the necessary help. Co-ops have some disadvantages. They tend to offer limited choices, their management tends to be "top-heavy" and suffer from inexperience, and they are unable to offer specials that can compete with supermarkets.

Smaller Outlets. The smaller food outlets include convenience, specialty, and health food stores, and farmers' markets. Convenience stores are a mini-version of the supermarket, with easily accessible foods and fast service being the key to their success. They are the closest thing to the old-fashioned corner grocery store, but their prices are considerably higher and they carry only the fastest moving items. Specialty stores include bakeries, delicatessens, butcher shops, and ethnic food, cookie, candy, and ice cream stores. Although specialty stores are usually more expensive, they offer unique items that may not be found at the supermarket. Health or "natural" food stores offer a wide selection of foods, many of which have been produced without chemical pesticides, herbicides, fertilizers, hormones, or antibiotics. Bulk items, herbs, fermented milks, soy milk, food supplements, and "natural" cosmetics may be bought in such stores. Finally, farmers' markets and roadside stands offer fresh produce straight from the grower's hands.

Food Service Vendors. Food service distribution centers and vendors supply the food for food service establishments. These large food warehouses obtain food directly from the food companies and deliver it to various private and public food service organizations. Vendors, the purveyors or food suppliers, usually specialize in a given product or category of products, such as produce, meat, or dairy.

Keeping Food Costs Down

While the budget is certainly not the only consideration in making a purchase, it is a vital factor, and there are several methods for keeping food costs down.

Meats. The biggest expense in the food budget is meat. A money-saving and nutrition-conscious step would be to reduce daily meat intake to no more than 5 ounces per person. Realizing that a 12-ounce steak is not the most healthy or economical serving and cutting back on meat is a major move toward saving money. The less tender cuts of meat are just as nutritious, often less fatty, and less expensive than tender cuts. In addition, it is generally more economical to buy a bigger cut of meat than it is to buy the smaller cuts.

The many nonmeat substitutes available provide inexpensive protein options. They include dried beans, peas, and lentils. These legumes are high in complex carbohydrates and fiber and are the best source of plant protein. Eggs are another nutritious and inexpensive protein source.

Fish. Frozen or canned fish is often less expensive than fresh fish. Lobster, crab, and jumbo shrimp are usually more costly than other protein sources, so they are best saved for special occasions.

Dairy. The least expensive form of milk is nonfat dried milk. If the taste and texture are unacceptable, it can be mixed with fluid milk, or a teaspoon of vanilla flavoring can be added for each gallon of reconstituted nonfat dried milk. Cheeses vary widely in cost, so it is best to comparison shop among the different types. Presliced or shredded cheeses tend to be more expensive.

Bread/Grain. Creating a diet on a foundation of pasta and grains subjects the budget to an automatic belt-tightening. The more processing involved in a product, the higher the cost, so prepared goods like cakes and cookies are more expensive than those made from scratch. Ready-to-eat cereals cost quite a bit more than their uncooked counterparts. Cereals offered in mini-packages or single portions also come at a premium price. Seasoned grain and pasta mixtures cost more than plain pasta and grains with the seasonings added while cooking. A wide variety of grains is sold in bulk at very reasonable prices.

Fruits and Vegetables. Savings can be achieved in the purchase of fruits and vegetables by determining the cost of the fresh form against the processed versions. Seasonal availability, brand, grade, any other added ingredients, and a number of other factors all weigh into the equation when comparing the price between fresh, dried, canned, or frozen fruits and vegetables.

Price Comparisons

Comparing prices is accomplished by calculating the cost per serving. This is easily done when the price per unit (ounce, pound, count, etc.) is given by the supplier. It pays to check this price and not be deceived by a product's packaging, shape, or size.

Prices differ not only among brands, but among forms—fresh, dried, canned, or frozen. Convenience foods are almost guaranteed to be more costly, as shown in Figure 4-7; the last example illustrates size differences. For example, popcorn is less expensive if it is purchased for preparation on the range rather than for the microwave. It is fairly well established that fresh produce usually has a greater nutrient value and better quality than either the canned or frozen version, but the prices escalate whenever produce has been trimmed or cut, making it more expensive to purchase shredded cabbage, diced carrots, watermelon cut in portions, or strawberries and pineapple sold ready-to-be-served in plastic containers. Pound for pound, frozen vegetables and fruits are often more expensive than fresh unless they are out of season.

Reading Label Product Codes. Familiarity with the various types of dating on some packaged items helps consumers select the freshest available products (Figure 4-8). Code dates (not shown) are useful in the event of a recall because they identify the manufacturer and/or packer of the food.

FIGURE 4-7 Price comparison.

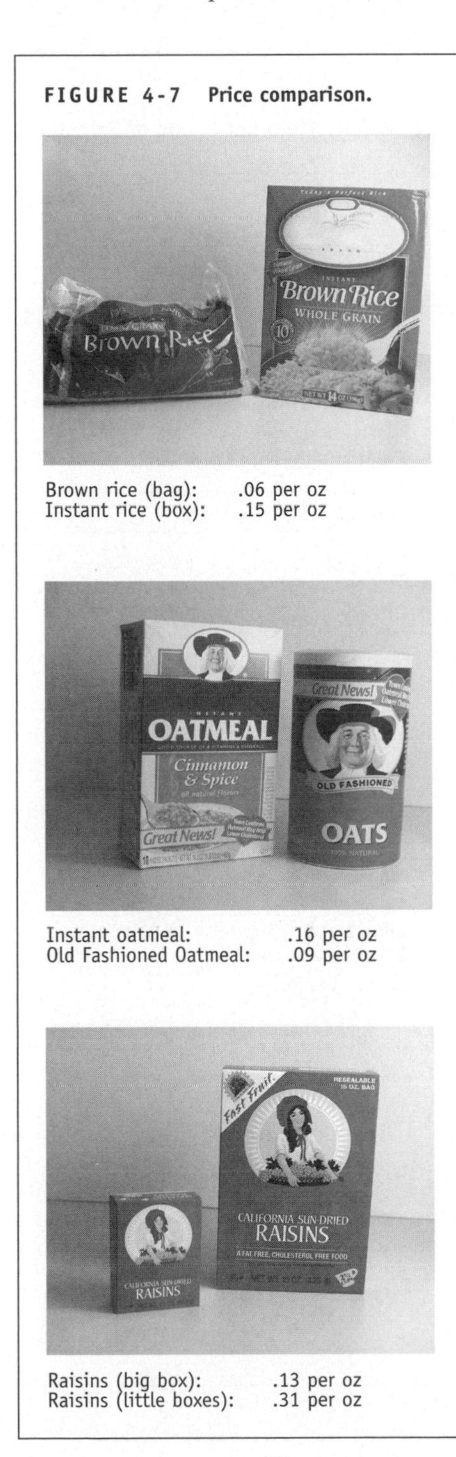

Brown rice (bag): .06 per oz
Instant rice (box): .15 per oz

Instant oatmeal: .16 per oz
Old Fashioned Oatmeal: .09 per oz

Raisins (big box): .13 per oz
Raisins (little boxes): .31 per oz

FIGURE 4-8 Dates on labels and what they mean.

Freshness or quality assurance date—The last day the product will be of optimum quality. Often preceded by "best when used by."

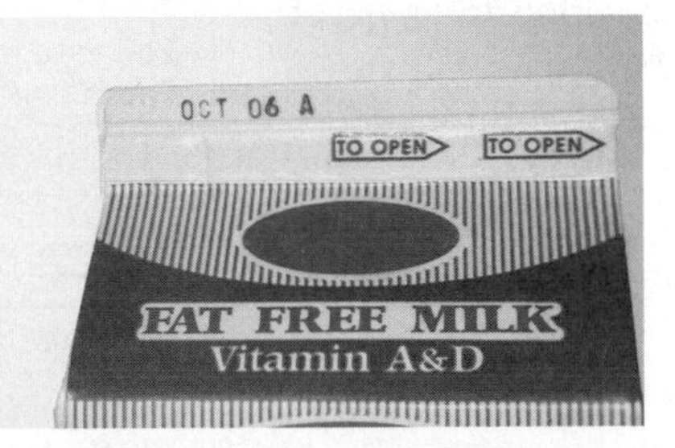

Pull date—The last day a store will sell an item, even though the food may be safe for consumption for a little while longer. Dairy and other perishable and semi-perishable items have a pull date that indicates the last day a store should sell the item. Such items are often priced very low and are a good buy if used within a short period of time.

Expiration date—The last day a food should be consumed. Certain products that will "expire," such as baking powders, yeast packages, and refrigerated doughs, need to show expiration dates to let consumers know whether or not they are still capable of making baked products rise.

Pack date—The date the food was packed at the processing plant. Canned, bottled, or frozen goods have pack dates that inform consumers how old the food is when purchased. It is often used by stores, which need to know when to rotate stock.

Reducing Waste Saves Costs

Careful purchasing can help to avoid waste. The common areas where waste occurs include the following (2):

- The overpurchasing of perishable produce and other foods. Fresh produce losses can be avoided by following the Three-Day Rule: never buy more than what can be consumed in three days.
- Losses resulting from food preparation (peeling, coring, trimming of fat, deboning).
- Losses from shrinkage during cooking.
- Losses from "plate waste," that is, food left on the plate because of too-large portion sizes or poor food quality.

"As Purchased" vs "Edible Portion." There can be a great disparity between the amount of food purchased and what ends up on the table. These different quantities are defined as **as purchased (AP)** and **edible portion (EP)**. Extra quantities of food must sometimes be purchased to make up for losses incurred during preparation, especially when buying meats, fruits, and vegetables.

Percentage Yield. In terms of waste resulting from pre-preparation, the **percentage yield** gives an estimate of how much edible food will remain after peeling and trimming (Table 4-2). The two steps to determining how much food to purchase according to percentage yield are as follows:

1. Determine the edible portion by multiplying the number of servings by the serving size.
2. Determine the amount to purchase by dividing the edible portion by the percentage yield for that particular food (Table 4-2).

For example, a meal of boneless, skinless chicken breasts for five people, each having a 3-ounce serving, calls for a total edible portion of 15 ounces (step 1). The percentage yield for chicken breast with ribs and skin is 66 percent, so 15 ounces divided by .66 results in 22.7, or approximately 23, ounces (step 2). Thus, for each guest to receive 3 ounces of edible chicken breast, a total

KEY TERMS

As purchased (AP) The total amount of food purchased prior to any preparation.

Edible portion (EP) Food in its raw state, minus that which is discarded—bones, fat, skins, and/or seeds.

Percentage yield The ratio of edible to inedible or wasted food.

TABLE 4-2

Percentage Yield: Approximate Edible Portion (EP) Yield per Pound of Selected Foods as Purchased (AP)

Food items	*% Yield* — *Pounds of EP*	*Food items*	*% Yield* — *Pounds of EP*
MEAT, POULTRY, FISH (COOKED)		*VEGETABLES (COOKED)*	
Beef, ground (no more that 30% fat)	40	Beans, green	88
Beef, chuck roast (boneless)	60	Carrots	60
Beef, round (boneless)	60*	Corn on the cob	55
Beef, stew meat	55*	Potato, baked with skin	80
Lamb, leg (boneless)	60*	Sweet potato, baked with skin	60
Lamb, stew	65*		
Pork, loin chops	40*	*VEGETABLES (RAW)*	
Pork, loin roast (boneless)	55*		
Pork, spareribs	45	Asparagus	55
Ham (boneless)	65*	Beets	45
Chicken, breast (rib and skin)	66	Broccoli	80
Chicken, drumstick (skin)	50	Cabbage	90
Turkey, whole (skin)	55	Carrots	70
Turkey, fr. rolls	65	Cauliflower	55
Fish, fr. portions (raw breaded)†	60	Celery	75
Fish, fr. sticks (raw breaded)†	60	Leeks	50
		Lettuce, head	75
FRUITS (RAW)		Mushrooms	95
		Onions	90
Apples	75	Parsley	85
Apricots	94	Peppers	82
Avocados	75	Radishes	90
Berries	95	Spinach	80
Bananas, with peel	65	Squash (summer)	90
Cantaloupe	50	Squash (winter)	75
Coconut	50	Tomato	99
Grapefruit	50	Turnips	80
Grapes	90		
Kiwi	80		
Lemons	45		
Mangoes	75		
Oranges	65		
Papayas	65		
Peaches	75		
Pears, pared	80		
Pineapple	55		
Plums	95		
Watermelon	55		

* Lean meat
† 75% fish

of 23 ounces of chicken breast with ribs and skin will need to be purchased.

HOW & WHY? ?????????

How many pounds of carrots should you buy if you need 4 ounces per serving for 50 people? First multiply 4 ounces per serving by 50 servings, for a total of 200 ounces. You know that carrots are sold by the pound, so you have to convert ounces to pounds. Divide the 200 ounces by 16 ounces per pound to get 12.5 pounds. However, 12.5 pounds of carrots would not be enough because Table 4-2 shows that the percent yield for carrots is 70%. This means that you have to divide 12.5 pounds of carrots by .70 to obtain the amount you need to buy, which is 17.86 pounds. A buyer would probably round that off to a 20-pound purchase of carrots since vendors often do not sell by the half pound or even under 5 or 10 pound increments.

Portion Control. One of the most important aspects of controlling the food budget of a food service organization is portion control. Food cost is a major expense in running a food service establishment, so it is crucial to adhere to set serving sizes. If 300 people are served 4 ounces of roast beef instead of the planned 3 ounces, this results in the consumption of almost 19 pounds of roast beef beyond what was calculated into the budget. It may also leave customer number 301 without any meat and dissatisfied. Table 4-3 shows some suggested portion sizes for various types of foods. Portion sizes may differ depending on the time of day of the meal or the meal itself. Portions are described in three ways:

1. By weight (ounce, pound) or volume (cup, pint)
2. By number (five olives, one ear of corn, two dinner rolls)
3. By size (1.8-inch cake, 2 × 2-inch brownie)

Careless measuring is ultimately reflected in the final yield of the food for the diner and the cost to the establishment. Portion guides are available to help keep food costs under control.

Time Management

Quality meals rely on timing. Food is usually best when it is prepared as soon as possible after purchase, and served immediately after preparation. Different foods are prepared at different rates, so coordination is key to an organization providing timely meal service.

Estimating Time

The preparer gains control by beginning with a realistic assessment of available time and energy. The stress of planning a meal can be minimized by logging how long it will take for each menu item to be prepared, estimating the time at which the meal will be served, and working backward to determine the time that prepara-

TABLE 4-3
Portion Guide, Common Serving Sizes*

Breakfast		
Eggs	2-4 oz (1-2 eggs)	50–125g
Meat	2–4 oz	50–125 g
Fruit	1/2 C	125 mL
Cereal	3/4 C	175 mL
Juice	1/2–3/4 C	125–175 mL
Bread	1-2 slices (1–2 oz)	30–60 g
Lunch		
Soup	4–6 oz	125–175 mL
Salad	4–8 oz	125–250 g
Salad dressing	1–2 oz	25–50 mL
Main dish	4–6 oz	125–175 g
Starch	2–3 oz	50–100 g
Vegetable	2–3 oz	50–100 g
Sauce	1–2 oz	25–50 mL
Bread	1–2 oz	30–60 g
Dessert	2–4 oz	50–125 g
Dinner		
Soup	6–8 oz	175–250 mL
Salad	4–8 oz	125–250 g
Salad dressing	1–2 oz	25–50 mL
Main dish	6–8 oz	175-250 g
Sauce	1–2 oz	25–50 mL
Starch	2–3 oz	50–75 g
Vegetable	2–3 oz	50–75 g
Bread	1–2 oz	30–60 g
Dessert	2–4 oz	50–125 g
Hors d'ouevres and Canapés†		
Lunch, with meal	2–4 per person	
Lunchtime, without meal	4–6 per person	
Dinner, with meal	4–6 per person	
Before-dinner reception	6–8 per person	
After-dinner reception	4–6 per person	

* Quantities given reflect general practice. Specific needs will vary.
† Average size is 1 oz (30 g). Total number reflects combined hot and cold items.
Source: Mizer et al., *Food Preparation for the Professional* (John Wiley & Sons, 1998).

KEY TERMS

Standardized recipe A food service recipe that is a set of instructions describing how a particular dish is prepared by a specific establishment. It ensures consistent food quality and quantity, the latter of which provides portion/cost control.

■ ■ ■ ■

Crumbing A ceremonious procedure of Russian service in which a waiter, using a napkin or silver crumber, brushes crumbs off the tablecloth into a small container resembling a tiny dust pan.

tion should be started in order to serve the meal on time. When the meal is prepared, items are usually prepared in descending order of time required.

Efficient Meal Preparation

Effective management of time can improve the efficiency of all the steps of meal preparation, which include:

- Menu planning
- Developing a purchase list
- Purchasing the food
- Storing the food
- Planning the order in which the menu items will be prepared
- Preparing the food
- Preparing the table
- Serving
- Cleaning up

The preparer can increase efficiency through menu planning and wise purchasing as described above, and through recipe consultation.

Recipes. There are four styles of recipe-writing: the descriptive, standard, action, and narrative forms (Figure 4-9). The ingredients in the descriptive method are listed in the sequence in which they are used. This method displays the ingredient, amount, and directions in three columns, which makes it easy to read. The standard recipe style lists all ingredients and amounts with the instructions in numerical order. A modification of that form is the action recipe, which gives the instruction followed by the ingredients for that step only. Probably the most tedious to decipher is the narrative form, which reads like an essay, explaining ingredients, amounts, and preparation methods in text form.

Standardized Recipes. Food service establishments rely on **standardized recipes** that have been tested and adapted for serving a large number of people (48 to 500 servings). Standardized recipes, which frequently follow the descriptive style, record ingredients, proportions, and procedures, so the number of servings can easily be increased or decreased. When standardized recipes are stored in a computer, changing the number of servings automatically changes the amount of each ingredient needed. Standardized recipes are repeatedly tested and adapted to suit a particular food service operation.

Types of Meal Service

There are six basic types of meal service. In descending order of formality they are: Russian, French, English, American, family, and buffet. Not only do the table settings differ, but so does the type of meal service and the manner in which the food and beverages are placed and removed from the table (10). No matter which type of service is employed, dessert is served only after the table has been cleared of all extraneous items, including salt and pepper shakers, all condiments, and unnecessary flatware.

Russian Service

The most formal type of meal service is Russian, also known as European, Continental, or formal service (12). The entire meal is served by well-trained waiters. Normally, the waiting staff serve and clear food items from the left with the left hand, while beverages are always served and removed from the right. The guest to the right, or the host or hostess, is served first, with the rest of the diners being served in a counterclockwise direction. Service plates, or place plates, sometimes made of or embossed with silver or gold, are part of each place setting and serve as underliners; the food is never placed directly on them.

The meal is served in courses, starting with the appetizer, and then the soup. Each of these courses has its own underliner plates, which go on top of the place plates. A fish course may follow the soup. Prior to the introduction of the main entrée, a miniature serving of chilled sorbet is provided to clear the taste buds of any lingering flavors. The place plate is then removed and the main entrée served from a platter or on plates placed before guests. Salad is served and consumed before or after the main entrée has been removed. Once the diners have finished their salads, the waiters remove all flatware, tableware, glassware, and condiments and finish clearing the table with a procedure known as **crumbing**. Filling all glasses and/or coffee or tea cups before serving dessert is a common policy, regardless of the type of meal service. After the dessert is finished and removed, a finger bowl containing cool water and, usually, a lemon slice is provided to each guest. The fingertips only are dipped into the water and dried on a napkin.

French Service

Another very formal type of meal service is French, or cart, service (13). The food is brought out on a cart *(guerdon)* to the table, where it is cooked or has its cooking completed in a small heater *(rechaud)* by the *chef de rang* (chief or experienced waiter) and *commis de rang* (assistant waiter). The French method is expensive, requires skilled personnel, and results in slower service. It tends to be reserved for elegant French restaurants or those found in Europe.

English Service

In the English service, waiters bring in the various courses, clear the table at the appropriate times, and may take servings dished out by the host and hostess to the individual guests. Frequently, when it is a family or small gathering, the host serves the meat to the guests, who pass their warmed plates to the hostess, who serves the

FIGURE 4-9 The four different styles of recipes: descriptive, standard, action, and narrative.

DESCRIPTIVE

Texas Chocolate Cake

Desserts: C-33		Oven Temperature: 350°F
Portion: 16 servings		Time: 20 minutes
Ingredients	**Amount**	**Procedure**
All-purpose flour Sugar	1 C 1 C	Sift flour and sugar together into large mixing bowl.
Water Margarine Cocoa	1 C 2 sticks 4 tbs	Melt margarine in sauce pan and add water and cocoa. Bring it to a boil while stirring constantly. Take off heat and pour into flour/sugar mixture.
Eggs Sour milk* Baking soda Cinnamon Vanilla	2 ½ C 1 tsp 1 tsp 1 tsp	In a separate bowl, beat eggs slightly and add to milk, baking soda, and flavorings. Add gradually to creamed mixture and mix with spoon until just blended smooth. Pour batter into a 13 x 9 greased baking pan. Immediately place on center rack of preheated oven. Bake until toothpick comes out clean. Cool briefly in pan on rack.

* or ½ C milk + 1 tbs lemon juice or vinegar

STANDARD

Texas Chocolate Cake

Preparation Time: 15 minutes
Cooking Time: 20 minutes
Yield: 16 servings

Ingredients:
- 1 C All-purpose flour
- 1 C Sugar
- 1 C Water
- 2 Sticks Margarine
- 4 tbs Cocoa
- 2 Eggs
- ½ C Sour milk
- or ½ C milk + 1 tbs lemon juice or vinegar
- 1 tsp Baking soda
- 1 tsp Cinnamon
- 1 tsp Vanilla

Directions:
1) Preheat oven to 350°F and grease sides and bottom of 13 x 9 pan.
2) Sift together flour and sugar into large mixing bowl.
3) Melt margarine in sauce pan and add water and cocoa. Bring it to a boil while stirring constantly. Take off heat and pour into flour/sugar mixture.
4) In a separate bowl, beat eggs slightly and add to milk, baking soda, and flavorings. Add gradually to creamed mixture with spoon until just blended smooth.
5) Pour batter into greased baking pan. Immediately place on center rack of preheated oven.
6) Bake until toothpick comes out clean. Cool briefly in pan on rack.

ACTION

Texas Chocolate Cake

Preheat oven to 350°F and grease sides and bottom of 13 x 9 pan.

Sift together flour and sugar into large mixing bowl.
1 C All-purpose flour
1 C Sugar

Melt margarine in sauce pan and add water and cocoa. Bring it to a boil while stirring constantly. Take off heat and pour into flour/sugar mixture.
2 Sticks Margarine
1 C Water
4 tbs Cocoa

In a separate bowl, blend ingredients below, add gradually to creamed mixture, and mix with spoon until just blended smooth.
2 Eggs (slightly beaten)
½ C Sour milk
or ½ C milk + 1 tbs lemon juice or vinegar
1 tsp Baking soda
1 tsp Cinnamon
1 tsp Vanilla

Pour batter into greased baking pan. Immediately place on center rack of preheated 350°F oven. Bake until toothpick comes out clean. Cool briefly in pan on rack.

NARRATIVE

Texas Chocolate Cake

Cake. Preheat oven to 350°F and grease sides/bottom of 13 x 9 pan. Sift together 1 C all-purpose flour and 1 C sugar into large mixing bowl. Melt 2 sticks of margarine in sauce pan and add 1 C water and 4 tbs cocoa. Bring to boil, stir stirring constantly, and pour liquid over the flour/sugar mixture. In a separate bowl, blend together 2 eggs (slightly beaten), ½ C sour milk (or ½ C milk + 1 tbs lemon juice or vinegar), 1 tsp baking soda (more soda results in a more cake-like cake, less produces a more brownie-like product), 1 tsp cinnamon, and 1 tsp vanilla. Add this gradually to the creamed mixture and mix with spoon until just blended smooth. Pour batter into greased baking pan. Immediately place on center rack of preheated oven and bake for 20 minutes or until toothpick comes out clean. Cool briefly in pan on rack.

Icing. Bring to boil ½ C margarine and 4 tbs cocoa. Stir constantly until mixed and remove from heat. Pour into mixing bowl and add 1 box of powdered sugar, blend with mixer on medium speed. If not blending smoothly, then add 1 tbs of milk at a time until it does. Too much milk makes runny icing, which is corrected by adding more powdered sugar. Cakes are usually cooled completely on racks before frosting however, frosting only 5–10 minutes later makes the cake more moist. Covering the sheet pan with aluminum foil further traps in moisture.

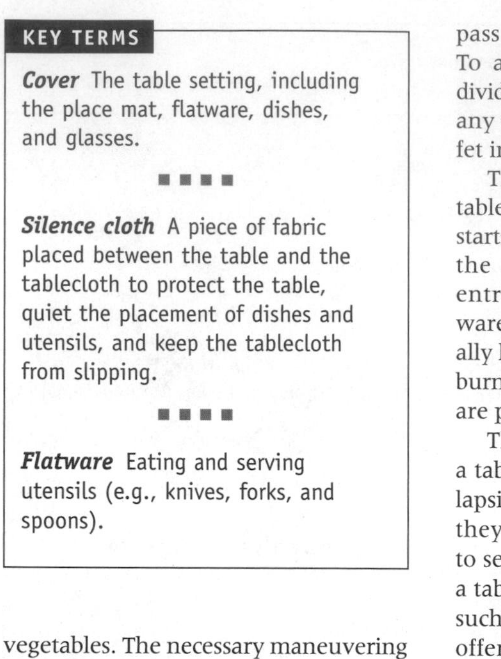
KEY TERMS

Cover The table setting, including the place mat, flatware, dishes, and glasses.

■ ■ ■ ■

Silence cloth A piece of fabric placed between the table and the tablecloth to protect the table, quiet the placement of dishes and utensils, and keep the tablecloth from slipping.

■ ■ ■ ■

Flatware Eating and serving utensils (e.g., knives, forks, and spoons).

vegetables. The necessary maneuvering of dishes between host or hostess and guests makes this type of service useful for no more than about six or eight people.

American Service

American service is that in which the meal is placed on the plates in the kitchen and then brought out to the table (12). This type of service is useful in smaller spaces, while allowing for faster service, hotter food, and fewer dishes to wash.

Family Service

Family service allows the guests to serve themselves from serving platters and bowls brought to the table and passed counterclockwise among the diners (12). The main meal-serving dishes, condiment containers, and all accompanying flatware, dinnerware, and glassware are removed after completion of the main entrée. Dessert is served at the table by the hostess or brought to the table in individual portions.

Buffet Service

The buffet service allows guests to walk to a separate buffet table from which they serve themselves. If the group of people is large, it is preferable to set the table up in such a way that guests can pass down both sides simultaneously. To avoid "the line," people can be divided alphabetically (A–O, R–Z) or by any other method and sent to the buffet in 10-minute increments.

The sequence of items on a buffet table varies considerably, but it usually starts with the plates and is followed by the vegetables, salad, bread, main entrée, condiments, beverages, flatware, and napkins. Hot items are usually kept warm in a chafing dish with a burner underneath, while chilled foods are placed on a bed of ice.

The guests may either sit down to a table, "eat off their laps," or use collapsible TV trays. Regardless of where they end up, they should never have to set glasses on the floor. Also, unless a table is available, runny menu items such as soups and sauces should not be offered. This will avoid possible spills and embarrassment. In the same vein, if paper plates are used, they should be extra sturdy and resistant to soaking through.

Table Settings

A well-prepared meal deserves to be enhanced by aesthetically pleasing table settings and surroundings. Expensive silver and china do not make up for grease marks on the utensils or an improperly set table. This section focuses on the correct presentation of the cover and linens, flatware, dinnerware, glassware, and accessories.

Cover and Linens

The table arrangement focuses on a **cover**, or table setting, for each individual. Each place setting should occupy from 20 to 24 inches to give diners elbow room. The table linen can be a tablecloth, place mats, or a combination of the two, along with a napkin. For formal service, the most common choice is a white damask cloth with a felt or other **silence cloth** placed underneath.

When a tablecloth is used, it should be centered on the table with an 8- to 12-inch overhang. Most people do not like to sit at a table where the tablecloth is any longer than lap length. Place mats, either alone or in addition to a tablecloth, are arranged so that each table setting is clearly distanced from the one(s) next to it. The napkin is placed to the left of the fork with the open edge facing the plate and its open corner at its own lower left. This makes it easy to grasp with the fingers of the right hand so it can be brought across the lap with a single motion. For formal service, the napkins are often folded attractively and placed in the center of the service plate or in the glassware. Napkins may be of linen or paper and vary in size—large for dinner, medium for lunch, and small for tea or cocktails. Paper napkins are acceptable for most meals, but linen or other cloth napkins should be used for more formal occasions.

Flatware/Dinnerware/Glassware

Flatware has assigned positions on the table setting, depending on the type of meal being served. A standard placement of flatware is shown in Figure 4-10, but most everyday and restaurant meals do not include two sets of forks, knives, and spoons. Most restaurants position their flatware according to the general rules shown in Figure 4-10, but creativity or necessity can influence the final arrangement.

Plates, saucers, bowls and other dinnerware also have generally assigned positions, as shown in Figure 4-11. Dinnerware does not have to match, but patterns should harmonize. It should also be in balance with the position of the glassware.

Accessories

Accessory items that can be distributed attractively on the table include the salt and pepper shakers, sugar bowl with spoon, cream pitcher, butter dish with butter knife, bread baskets, any decorations, and condiments removed from their containers and displayed in attractive serving dishes. Items with the potential to drip are always placed on underliners. Salt and pepper shakers are usually placed near the center of

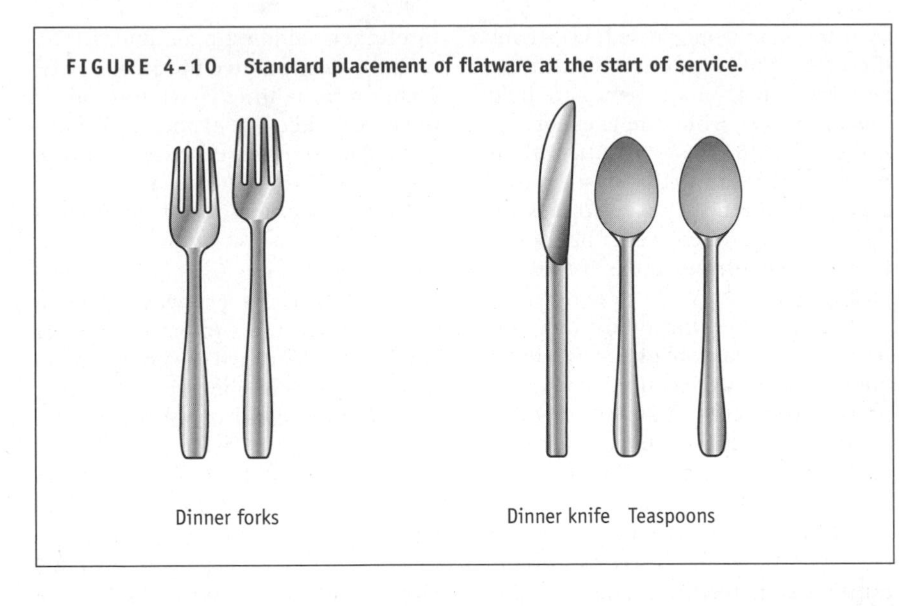

FIGURE 4-10 **Standard placement of flatware at the start of service.**

the table, with the salt placed to the right of the pepper. Individual salt and peppers are placed just above or slightly to the left on each cover. Depending on the type of service, serving dishes and serving utensils may or may not be on the table.

Centerpieces. Centerpieces and other table decorations should be given special attention, with simple elegance the rule, unless the meal has a specific theme that requires something more distinctive. Centerpieces should be in scale with the table, should not be overpowering, and should not keep guests from being able to see each other across the table. Candle flames should be kept below eye level to avoid the problem of people having to bend right or left to talk to someone across the table. Many a host or hostess failing to observe these guidelines ends up removing a centerpiece after guests are seated.

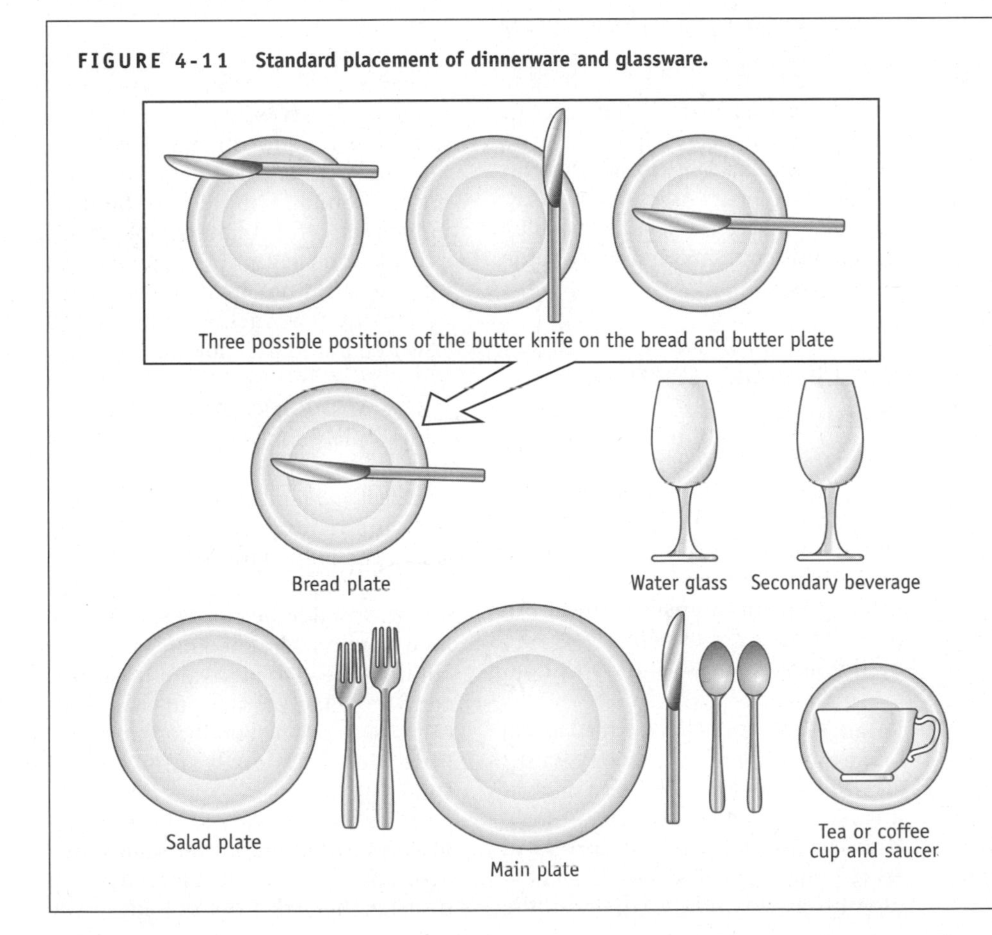

FIGURE 4-11 **Standard placement of dinnerware and glassware.**

Table Manners

Manners are important in every aspect of life, and while the rules vary from country to country (3, 7, 10, 15), the following discussion is limited to those employed at the dining table in North America. The recommendations given here are not hard and fast; when in doubt, it is always safe to follow the lead of the host or hostess.

Basic Dining Etiquette

Whether people admit it or not, they often perceive table manners as reflecting a person's upbringing. Good manners also show a general consideration of others. The best meal can be prepared and attractively presented, but if table manners, particularly those of the chef, do not meet minimum expectations, the meal may lose some of its appeal. Mastering good dining etiquette provides yet another way of enjoying a good meal, and this section provides some of the do's and don'ts of table etiquette.

Formal Invitation. The first impression of a formal dinner party comes from a written invitation. More casual dinner parties usually rely on the telephone. The "RSVP" on the invitation is French for *répondez, s'il vous plait* or "reply, if you please." It simply asks that the invitee inform the host or hostess as to whether or not attendance is planned. It is a courtesy to the host or hostess to respond to such a request, and it is considered rude not to respond. Guests should never inquire about what is being served or who else is coming for dinner, but it is acceptable to ask what kind of clothing (casual, formal, etc.) to wear. It is an accepted custom, although certainly not mandatory, that invited guests bring a small gift such as a beverage, flowers, a fruit basket, or the like.

Seating. Once at the dinner party or in a restaurant, it is polite for the lady to follow the greeter, with the gentleman following her. At the table, the gentleman helps to seat the lady on his right, and, if there are many people in attendance, he may also help to seat others to his right. If the dinner party is in a private home, the guests should not sit down until the host or hostess directs them to specific chairs and indicates that they may sit, usually by being seated him- or herself. Left-handed people are generally accommodated by seating them on the left end of the table. The napkin may be placed on the lap as soon as guests are seated, but some wait until the meal is served.

Dining Manners. Eating begins when the hostess lifts her fork. Food is chewed with the mouth closed, and talking does not begin until the food is swallowed. Any food bigger than the thumb at the second joint should be cut up. Fish bones and fruit or olive pits can be removed with the hands, but partially chewed pieces of truly unpalatable or too-tough food can be carefully stowed in the napkin as the mouth is dabbed. Any food that may end up on the face may be removed by dabbing, not wiping, the mouth corners and lips, and this is also a good practice before speaking or drinking.

If a person wishes to refuse wine or any other beverage, the correct way to communicate this to the server is to say, "No thank you," rather than to place a hand over the glass or turn it upside down.

Dishes are passed to the right, except for bread, which is passed clockwise.

If the item being passed has a handle, it should be pointed toward the recipient. Heavy food dishes are held by one person while the guest on the right serves the food with the right hand. Once eating is finished, the server should be allowed to pick up dishes, which should never be shoved to the side of the table for easier pickup.

The old rule that elbows should never be on the table has been loosened up a bit, especially if the company is relaxed and after the plates have been removed. One custom that has not changed is that the only person allowed to wear a hat in a restaurant is the chef. Hats should never be placed on the table. Also, in a restaurant it is sometimes considered impolite to visit people at another table and interrupt their meal. If the need to leave the table during dinner arises, the person excuses himself or herself and places the napkin to the left of the dinner plate or on the chair. Once finished and ready to leave, however, the diner places the napkin where the plates were before being removed.

The general rule for how long to stay after getting up from the table at a private dinner party is no less than 45 minutes, and no longer than the host or hostess appears to want company. Formal dinners should be followed up with a written thank-you note within a few days.

Flatware/Glassware Etiquette

Understanding the unique uses for each piece on the place setting is important, and what follows are the general guidelines for the salad fork, spoon, and knife and fork. The general rule is that the utensils farthest out from the plate are used first, followed by the next farthest out, and so on.

Salad Fork. The smaller fork found to the outside of the dinner fork is usually for a salad served before a meal. If it is on the inside of the dinner fork, it is to be used for dessert or for a salad served with a meal. It should be held in your dominant hand between the forefinger, middle finger, and thumb with the tines pointed up. Stabbing the salad with the tines down or holding the fork handle like a motorcycle handle, a dagger, or an exercise bar is not recommended. A fork is incorrectly held if the undersides of the fingers show.

Spoons. The correct way to use a spoon with soup is to place it on the liquid at the far rim of the bowl and dip the far edge of the spoon into the soup. The soup is carefully lifted to the mouth instead of bringing the head down to the food. The bowl may be gently tipped toward the center of the table to get the last drop or two, but it should never be picked up off the table or tipped the other way. Sipping the remaining soup by lifting the bowl to the mouth is acceptable in some areas of the world, but it is considered bad manners in Western culture. Blowing on or waving a hand wildly over the soup to cool it is both impolite and ineffective and should be forgone in favor of simply waiting for it to cool naturally.

A spoon is always placed on the saucer or plate holding the cup, soup bowl, or dessert glass. Only a spoon should be used for stirring beverages. The sugar spoon, however, is a communal item and should not be used individually. It is unacceptable to stir a drink and then put the spoon back into the sugar bowl, and it goes without saying that such a spoon should never be put in the mouth.

Knife and Fork. To cut food, the knife is placed in the dominant hand and the fork in the other. The knife handle is held in the palm of the hand with the index finger on top of the blade. The tines of the fork in the other hand are pointed down and inserted into the food to be cut. This holds the food steady while the knife is used to cut one bite-size piece at a time. It is considered bad manners to cut a piece of meat or other food into numerous little pieces, unless it is for a small child.

The next step varies. The majority of North Americans lay the knife down across the top rim of the plate, then transfer the fork from the left to the right hand for eating; while people of European or Middle Eastern cultures simply use their left hand to lift the fork to their mouths. The first style is considered American, while the latter is European or Continental.

To signal that they are finished eating, guests place both fork (possibly with tines down) and knife across the plate in one of the four positions shown in Figure 4-12. The most common is with the handles parallel at the "10 minutes past the hour" position and the tips at "5 minutes before the hour."

Glassware. After a couple of bites, many people like to take a sip of their beverage, but it should never be done with food in the mouth. A glass is held at the base and a cup by its handle. The rim should be brought to the mouth for one sip or swallow, and if another sip is desired, it should be done only after bringing the glassware back to the table. When the drink is lifted to the mouth, the head should not simultaneously tilt back for a better position.

FIGURE 4-12 Signaling meal completion by flatware positioning.

PICTORIAL SUMMARY / 4: Meal Management

Planning meals that are both psychologically and physiologically satisfying requires a basic knowledge of food preparation, nutrition, and presentation strategies. Effective meal management, whether for a household, an institution, or a restaurant, involves efficient management of resources such as money, time, and energy.

FOOD SERVICE ORGANIZATION

Food service management must set goals, determine what is needed to achieve them, and mobilize people toward these goals. Division of labor is central to good organization. *Brigade de cuisine* is a system of dividing a kitchen into stations supervised by chefs with expertise in specific areas:

- Sauce
- Fish
- Vegetable
- Soup
- Roast
- Pantry
- Pastry
- Relief

MEAL PLANNING

Planning menus and setting up menu cycles helps to control costs as well as balance nutrition. Some points to consider:

- **Nutrient recommendations:** Including correct serving sizes.
- **Individual preferences and needs:** Based on age and religious, cultural, ethnic, and regional differences.
- **Costs:** Planning nutritious, flavorful, and appealing meals within the available budget.
- **Food preparation methods:** Alternating oven-baked, boiled, and fried foods for optimum nutrition and to avoid monotony.
- **Seasonal factors:** Availability of fresh food products, method of preparation, and temperature at which food is served.

PURCHASING

Budget limitations determine both the types of food to be purchased and the amounts. Careful control of food and labor costs is critical in food service establishments and waste must be avoided. Food bills can be reduced through organized purchasing, comparison pricing, and controlling portions.

TIME MANAGEMENT

Food is usually best when it is prepared as soon as possible after purchase and served immediately after preparation.

Coordinating the preparation of different foods to be served at the same time is crucial, but equally important is an awareness of the time involved in menu planning, purchasing, preparing the table, serving, and cleaning up.

Efficiency is increased with careful menu planning and careful recipe consultation.

TYPES OF MEAL SERVICE

The six basic types of meal service differ in the manner in which the table is set and the food is served. They include, in descending order of formality,

- Russian: Most formal; entire meal is served by waiters.
- French: Food is served and or prepared from a cart brought to the table by specially trained chef/staff.
- English: Host participates in serving guests; waiters assist.
- American: Served on plates in kitchen and brought to table.
- Family: Guests serve themselves from platters brought to table and passed counterclockwise among the diners.
- Buffet: Guests serve themselves from a central buffet table.

TABLE MANNERS

The best meal can be prepared and attractively presented, but if table manners do not meet minimum expectations, the meal may lose some of its appeal. Dining etiquette provides general guidelines for handling salad forks, spoons, knife and fork, and glassware.

TABLE SETTINGS

A cover or place setting should be laid for each individual. There are specific customs for the placement of linens (tablecloth, and/or place mat, and napkin), flatware, dinnerware, glassware, and accessories such as decorations and condiments.

Three possible positions of the butter knife on the bread and butter plate.

REFERENCES

1. American Dietetic Association. Position of the American Dietetic Association: Management of health care food and nutrition services. *Journal of the American Dietetic Association* 97(12):1427–1430, 1997.
2. Ashley S, and S Anderson. *Catering for Large Numbers.* Reed International Books, 1993.
3. Claiborne C. *Elements of Etiquette.* William Morrow, 1992.
4. David C. Meatless meats. *Foodservice and Hospitality* 30(9): 25–28, 1997.
5. Harrison GG. Reducing dietary fat: Putting theory into practice—conference summary. *Journal of the American Dietetic Association* 97(7):S93–96, 1997.
6. Hollingsworth P. Burgers or biscotti? The fast-food market is changing. *Food Technology* 56(9):20, 2002.
7. Hoving, W. *Tiffany's Table Manners—for Teenagers.* Random House, 1989.
8. Katz F. How purchasing styles affect new product development. *Food Technology* 53(11):50–52, 1999.
9. Kerr T. The stealth healthy menu. *Food Management* 31(5):110–117, 1996.
10. Kinder F, NR Green, and N Harris. *Meal Management.* Macmillan, 1984.
11. Mermelstein NH. Washington News. Questionnaire on healthy eating proposed. *Food Technology* 56(9):26, 2002.
12. Payne-Palacio J, et al. *West's and Wood's Introduction to Foodservice.* Macmillan, 1994.
13. Putman JJ, and JE Allshouse. *Food consumption, prices, and expenditures,* 1970–93. U.S. Department of Agriculture. Statistical Bulletin No. 915. Washington, D.C.: GPO, 1994.
14. Reichler G, and S Dalton. Chefs' attitudes toward healthful food preparation are more positive than their food science knowledge and practices. *Journal of the American Dietetic Association* 98(2):165–169, 1998.
15. Tuckerman N. *Amy Vanderbilt's Complete Book of Etiquette.* Doubleday, 1995.
16. U.S. Bureau of the Census. *Statistical Abstract of the United States: 1993* (113th ed.). Washington, D.C.: GPO, 1993.
17. Warfel MC, and FH Waskey. *The Professional Food Buyer: Standards, Principles, and Procedures.* McCutchan, 1979.

WEBSITES

Analyze your diet. The USDA's Center for Nutrition Policy and Promotion developed an Interactive Healthy Eating Index (IHEI), an internet-based tool that people can use to analyze their own diets. Users input their daily food intakes and obtain a quick summary measuring overall diet quality:

www.usda.gov/cnpp/usda_healthy_eating_index.htm

The National Restaurant Association lists websites for the major food service organizations and associations:

www.restaurant.org/business/resources_associations.cfm

Nutrition information for planning healthy menus. The USDA's Center for Nutrition Policy and Planning:

www.usda.gov/cnpp

5

Heating and Equipment

Food preparation has come a long way since someone dropped some meat on the fire by accident, managed to retrieve it a few minutes later, and discovered that it tasted quite good that way. Of course, nobody really knows how cooking was discovered. It is safe to say, however, that cooking methods have increased steadily in sophistication since then, and knowing how to heat food properly and with the correct equipment is now essential in food preparation. This chapter provides an introduction to the principles of heating, the basic pieces of food preparation equipment, and the selection of utensils.

Heating Foods

Heat is the energy that is produced by the rapid movement of molecules. The molecules in living organisms always have some motion; heat speeds up that motion, while cold temperatures slow it down. Freezing and boiling are extremes of the range in temperatures encountered in food preparation that owe their effects to changes in this **kinetic energy** of molecules.

KEY TERMS

Kinetic energy Energy associated with motion.

Measuring Heat

Temperature Scales. The three main scales used to measure heat intensity are Fahrenheit, Celsius or Centigrade, and Kelvin (°K) (Figure 5-1). The last is used primarily in scientific research, and will not be discussed here.

Freezing Point. The freezing point of water is 32° on the Fahrenheit scale and 0° on the Celsius scale. Its boiling point at normal atmospheric pressure is 212° on the Fahrenheit scale and 100° on the Celsius scale. The boiling point changes slightly with altitude; 1°F must be subtracted for every 500-foot increase in elevation (an increase of 960 feet in elevation decreases water's boiling point by 1°C). Other compounds in the water, such as sugar or salt, influence its boiling and freezing temperatures, so all three scales pertain to pure water. Other materials have their own freezing and boiling points. Figure 5-2 summarizes various temperatures used in food preparation.

Thermometers. Thermometers are available in Fahrenheit scale (nonmetric), and Celsius or Centigrade scales (metric). Bulb thermometers work on the expansion and contraction of mercury in the bulb at the bottom of an extended glass tube marked with the specific graduated scale. Heat expands mercury; cold contracts it. Different thermometers are used for different purposes in food preparation (Figure 5-3 on page 100). A meat thermometer has a short rod for insertion into the meat and usually has an upper limit of 185°F (85°C). A candy thermometer has an upper range of 325°F (163°C), and a deep-fat frying thermometer goes up to at least 500°F (260°C).

In addition to bulb thermometers, small thermometers can hang or stand in ovens and refrigerators to check the accuracy of the equipment's thermostats. Pocket-size, instant-read thermometers can be used to check foods being held on steam tables in food service establishments. These can be carried around like a pen and give a reading in a few seconds after being inserted in the food. Care must be taken when using instant-read thermometers, because they must be sanitized between uses or contamination may result.

Pocket Thermometers. Instant-read thermometers come in two readout forms: dial (0 to 220°F/−18 to 104°C) and digital (−58 to 300°F/−50 to 149°C). The dial thermometers are not as accurate for small amounts of foods or those that are too thin or that cook quickly. They also have a tendency to malfunction when dropped or jarred. Dial thermometers are recalibrated by gripping the hex nut with pliers and twisting the dial face until it reads the correct temperature (9). Digital thermometers are slightly more expensive, but more accurate because they contain an electronic sensor near the

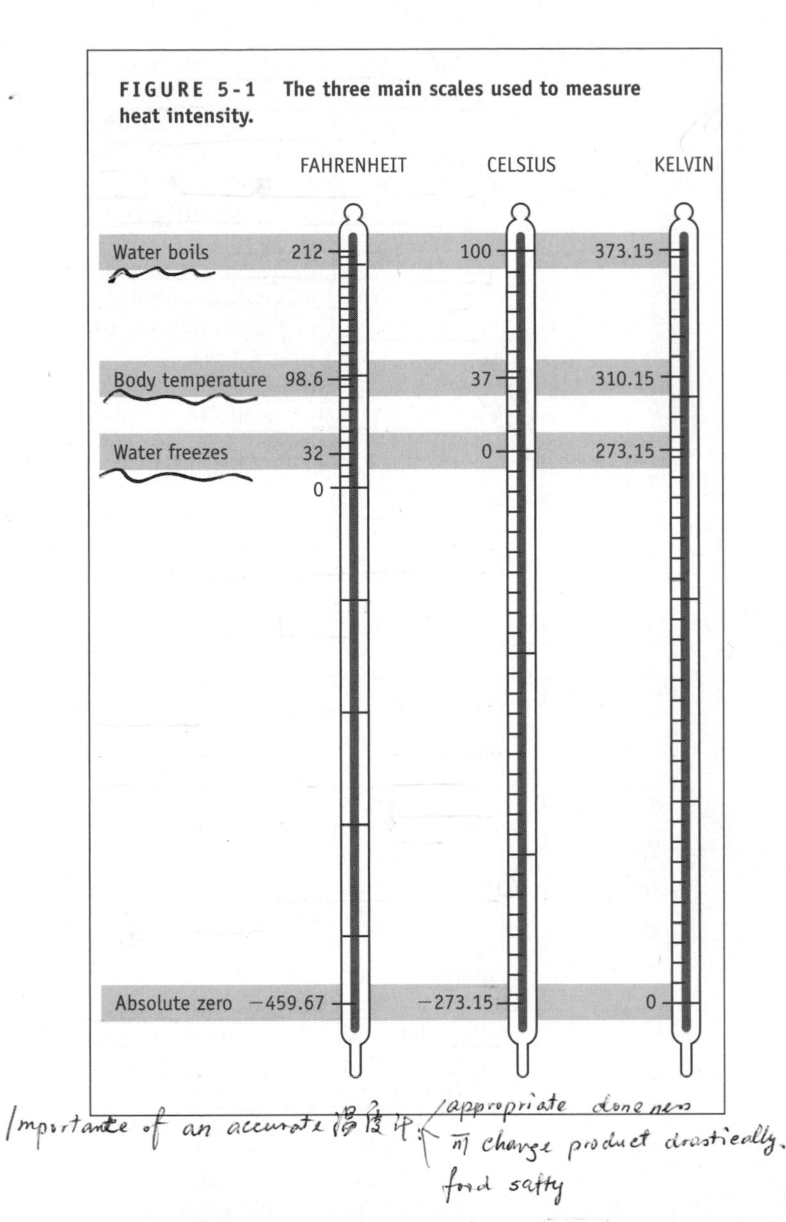

FIGURE 5-1 The three main scales used to measure heat intensity.

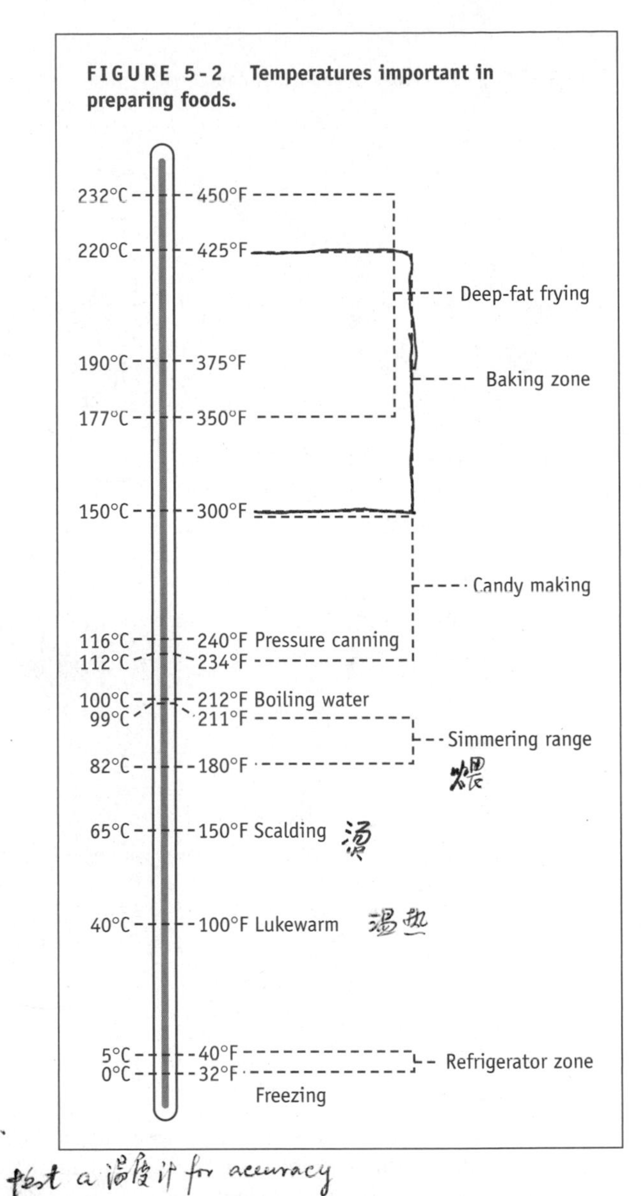

FIGURE 5-2 Temperatures important in preparing foods.

tip that is effective when inserted only 1/2 inch into food. If they drift off from calibration, they need to be replaced. Exceptions to this are battery-operated digital thermometers, or ones that are equipped with a knob for recalibration.

A few degrees difference on a thermometer can result in the success or failure of a dish, so thermometers need to be tested for accuracy. For bulb thermometers, the bulb is dipped into boiling water without touching the sides or bottom of the pan. The top of the mercury should be read at eye level, and it should reach 212°F (100°C) at sea level, with 1°F (2.2°C) subtracted for every 500 feet of increase in altitude. Digital and dial thermometers are checked by placing them 2 inches above boiling water to obtain the reading of 212°F (100°C), or in a slurry of ice water (made by blending a cup of ice in a cup of water), where they should read 32°F (0°C).

Calories. Energy can be correlated to heat and is measured in the unit of a **calorie**. For everyday use, it is more common to refer to the kilocalorie (1,000 calories), abbreviated "kcal," which is the amount of energy required to raise 1 kilogram of water 1 degree Celsius. Calories are discussed in more detail in Chapter 2.

Regardless of which term is used to quantify dietary calories (kcal), it is more accurate to speak of "energy" rather than calories (kcal), unless a specific amount is being discussed.

Kilojoule. The metric equivalent of the calorie is the joule (j) or kilojoule (kj), with 1 joule defined as the work or energy required to move 1 kilogram of mass 1 meter. One calorie is equivalent to 4.184 joules, while 1 kilocalorie equals 4.2 kilojoules.

Btu. Another measure of heat is the British thermal unit (Btu), which is the amount of energy required to raise the

KEY TERMS

Calorie (kcal) The amount of energy required to raise 1 gram of water 1°C (measured between 14.5° and 15.5°C at normal atmospheric pressure).

FIGURE 5-3 **Types of food thermometers.**

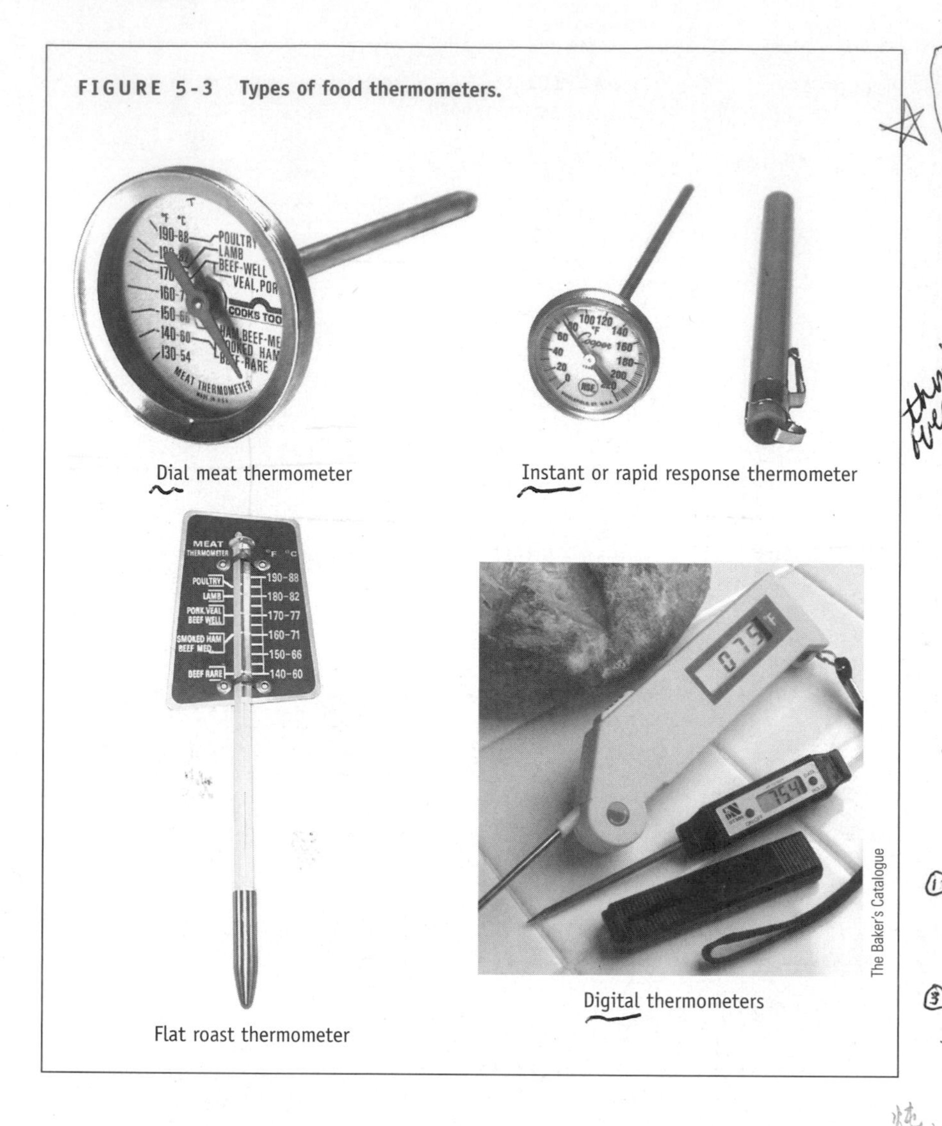

Dial meat thermometer

Instant or rapid response thermometer

Flat roast thermometer

Digital thermometers

The Baker's Catalogue

temperature of 1 pound of water 1 degree Fahrenheit. Btu is more commonly used to measure the heating capacity of fuels used in commercial or housing industries.

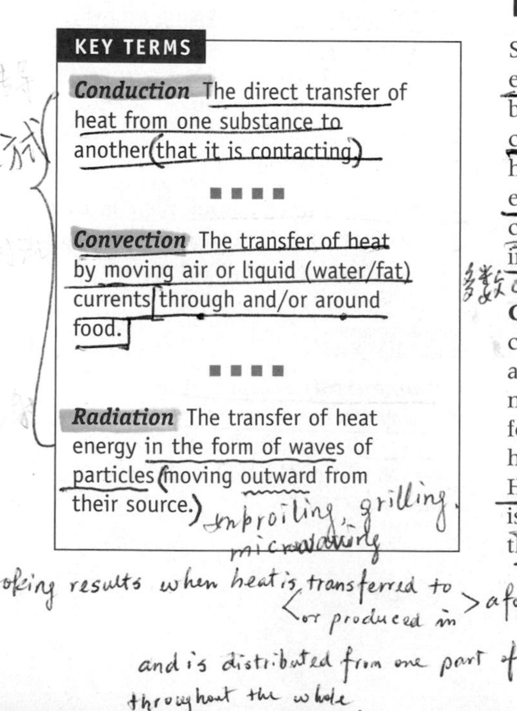

KEY TERMS

Conduction The direct transfer of heat from one substance to another that it is contacting.

Convection The transfer of heat by moving air or liquid (water/fat) currents through and/or around food.

Radiation The transfer of heat energy in the form of waves of particles moving outward from their source.

Types of Heat Transfer

Sources for heating foods are usually electricity and gas (natural or butane), but secondary sources such as wood, coal, and charcoal may also be used for heating. All of these produce heat energy that can be transferred through conduction, convection, radiation, or induction (Figure 5-4).

Conduction. Adding heat to molecules increases their kinetic energy and their ability to transfer heat to neighboring molecules. In preparing foods on the range or in the fryer, heat is transferred by **conduction**. Heat from the electric coil or gas flame is conducted to the pan or fryer and then to its contents.

The material and size of the pan greatly affect the speed and efficiency of heat transfer. Copper is an excellent heat conductor and is often used to line the bottom of stainless steel pans. Dark, dull surfaces absorb heat more readily, which shortens the baking time. Tempered glass conducts heat in such a manner that baking temperatures should be reduced by 25°F (4°C) when it is used.

Convection. **Convection** is based on the principle that heated air or liquid expands, becomes less dense, and rises to the surface. The cooler, heavier air or liquid originally on top moves to the bottom, where it is heated, thus creating continuous circular currents. A common example of the use of convection in cooking is baking in an oven. Baked goods rely on convection to allow the hot air to rise in ovens where the heating unit is at the bottom. Convection ovens, which are more common in food service institutions, have an air circulating system, while standard ovens do not. Fans move the air more quickly and evenly around the food, which speeds up baking times (15). Convection ovens do have drawbacks, however: the moving air causes foods to lose moisture, and cake batters are more prone to develop uneven tops. Injecting steam into a convection oven helps to reduce the drying and shrinking effects.

Other examples of convection cooking are simmering, steaming, and deep-fat frying. The use of water and fat to heat food relies on both conduction and convection. For example, once the heat from convection begins to heat a baked potato, conduction takes over when the heat penetrates the potato's water molecules and moves the heat to the center of the potato. Since water conducts heat more efficiently than air, it takes less time to boil than to bake a potato.

Radiant Heat. Heat is transferred by **radiation** in broiling, grilling, and microwaving. The short electromagnetic waves that are generated by microwave ovens can pass through glass, paper, and most plastic. Infrared heat lamps and ovens are other heat sources that use electromagnetic waves for heat. These are usually found in restaurants and institutional kitchens to keep foods warm and to prepare frozen foods.

FIGURE 5-4 Four types of heat transfer: conduction, convection, radiation, and induction.

FIGURE 5-5 Ranges.

Induction. Flat-surfaced ranges that have the coils buried underneath conduct heat through **induction**. The cooktop consists of a smooth, ceramic surface that allows the transfer of heat from the coiled electrical apparatus below. Since no coils are exposed on the surface, cleaning up is much easier.

Food Preparation Equipment

Primary Equipment

Ranges, ovens (conventional, convection, and microwave), and refrigerators are mandatory pieces of equipment for any kitchen. In addition, dishwashers are found in many homes and most food service establishments.

Ranges. Ranges can have open or flat top surfaces with electrical or gas burners of varying sizes underneath (Figure 5-5). Heat control and the speed with which one can raise or lower the temperature of the units vary among manufacturers and according to whether the range is fueled by gas or electricity. Personal preference and the availability and relative expense of gas and electricity usually determine which type of stove (range) is selected.

Ovens. The conventional oven usually makes up the bottom part of the range, but it can also be a separate unit placed elsewhere (Figure 5-6). It is primarily used for baking and roasting, but can also be used for braising, poaching, and simmering. The conventional oven relies on hot air for heating food, primarily by convection, but conduction and radiation can also occur. Baked foods rely on freely moving currents for the transfer of heat, so it is important to ensure that baking pans are placed on the racks in such a way as to allow the efficient flow of air currents.

Several types of ovens are available to food service establishments, some of which are shown in Figure 5-6. These include:

- Stack or deck oven. Each component of the stack has a separate thermostat.
- Convection oven. Hot air is circulated by a fan, baking contents more quickly.

KEY TERMS

Induction The transfer of heat energy to a neighboring material without contact.

FIGURE 5-6 Four types of ovens.

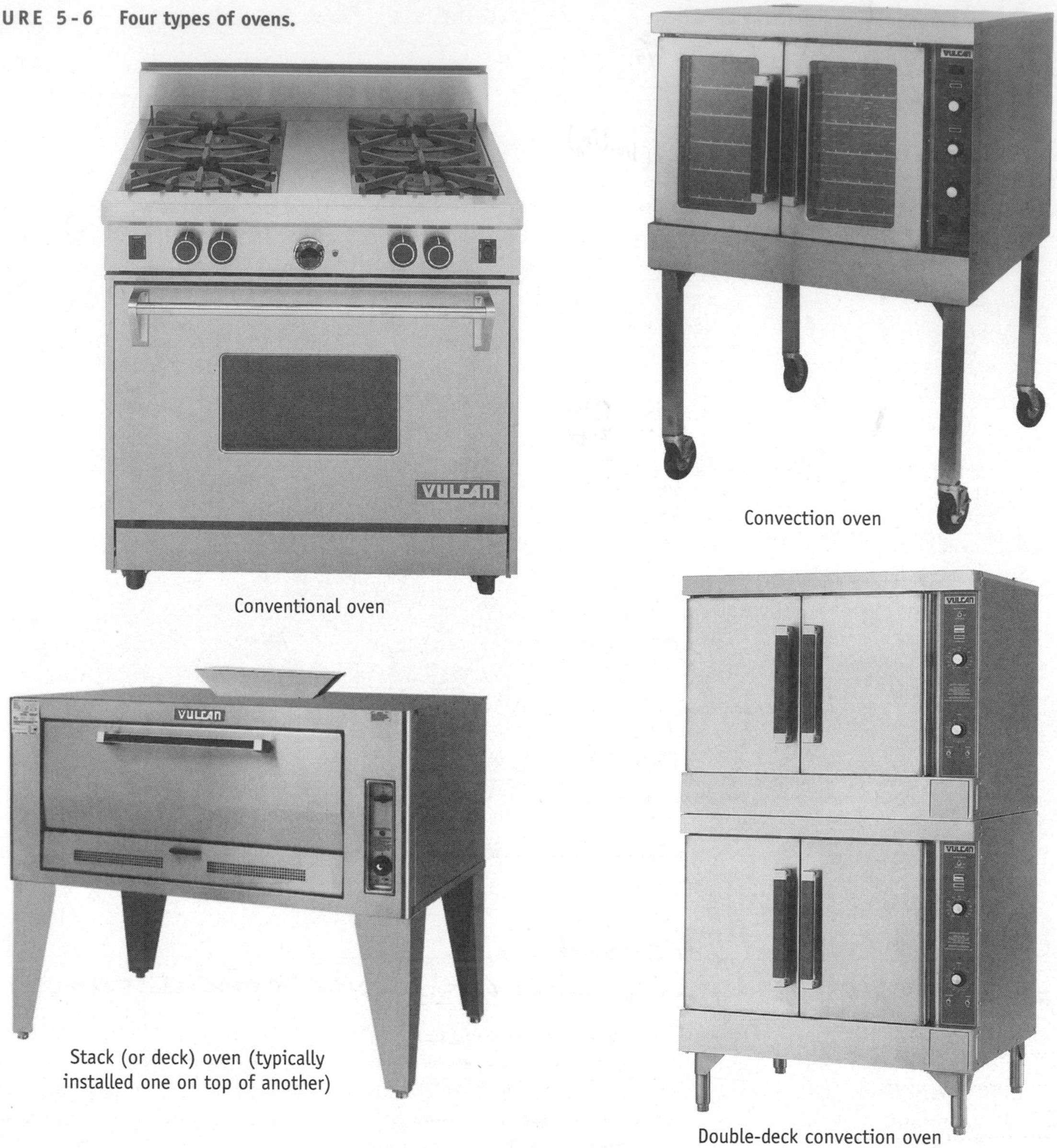

Conventional oven

Convection oven

Stack (or deck) oven (typically installed one on top of another)

Double-deck convection oven

The Vulcan-Hart Co.

- Revolving or carousel oven. Trays rotate like a ferris wheel, ensuring an even temperature.
- Impingement oven. Hot-air jets heat food more quickly (13).
- Infrared oven. Heat is generated by a very hot infrared bulb.
- Brick-lined or hearth oven.
- Pizza oven, which reaches very hot temperatures.
- Microwave oven.

Microwave Ovens. The discovery that microwaves might be useful for heating food occurred around 1945, when a Raytheon scientist noticed that his chocolate bar melted whenever he placed it beside a radar vacuum tube he was testing. In 1947, the first microwave became available for use in food service, and seven years later, microwaves were introduced into homes (10). Now, over 90 percent of all homes have a microwave oven (7, 11). These ovens contain a magnetron, a vacuum tube that generates microwaves that can be reflected, transmitted (passed through), or absorbed by the food (Figure 5-7). These waves, which are less than 5 inches in length, penetrate the food and trigger the rotation of food molecules, causing friction, which heats the food. The positive and negative charges in food molecules are attracted to the negative and positive charges found on the microwaves. Since the microwave reverses direction 2.45 billion times a second, the friction produced is quite potent at heating up foods, especially those containing water. Fat and sugar also heat rapidly because their specific heats are lower than that of water (18). Salt attracts microwaves, so microwaved soups have a tendency to boil. For this reason, it is best to add salt to microwaved foods at the end of cooking.

In general, radiating microwaves penetrate only 1.2 to 2 inches into the

Incorporates dry (radiation) / moist → heat preparation methods.

Micwaves are a form of radiation aimed at the water in [illegible]

FIGURE 5-7 Microwave oven interior.

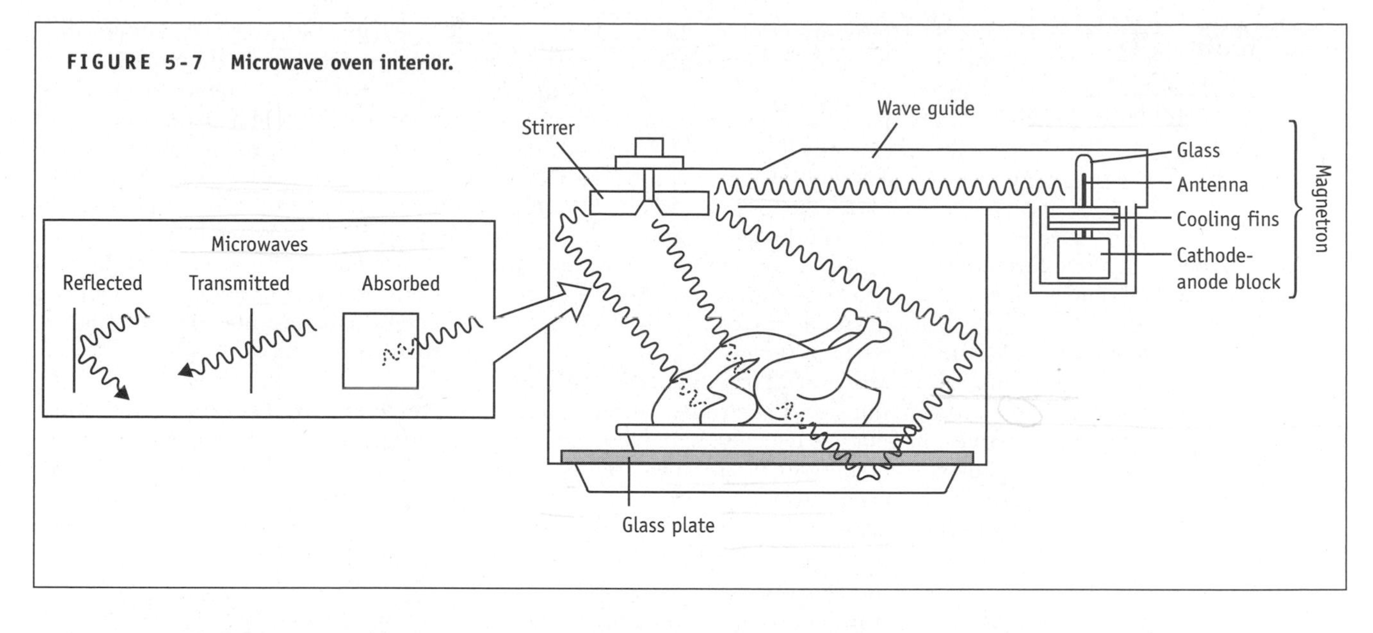

food, depending on its density; the rest of the heating occurs through conduction. The short penetration range is the reason why thick parts of the food (broccoli stalks, chicken drumsticks, etc.) should be placed on the outside rim of the dish or container where they will receive more microwaves. Molecular vibrations continue for several minutes after food has been removed from the microwave oven, so many recipes include a standing time as part of the overall preparation.

Types of Microwave Ovens. Microwave ovens come in several different styles and generate microwaves at low (10 percent), medium (50 percent), and high (90 to 100 percent) power. A 50 percent power setting means that microwaves are produced 50 percent of the time.

HOW & WHY? ?????????

Why do microwave ovens differ in the time it takes to heat the same type of food? Ovens vary in wattage. They range from about 400 to 2100 watts (usually 700), and the lower the wattage, the slower the rate of heating. Thus, "high" in one oven might be equivalent to "medium" in another. It is for this reason that microwave oven manufacturers publish their own recipes and instructions pertaining to the heating times of various foods.

The instruction booklet should give the wattage in any particular microwave oven, but if this is not available, there is another alternative: heat 2 cups of room-temperature water on high power for 3 minutes. If the water boils, the wattage is over 600 watts; if it takes longer than 3 minutes to boil, the wattage is below 600.

Microwave Containers. Foods heated by microwaves require special containers that meet specific shape and material requirements. Because the perimeter of a container heats up first, food in the corners of rectangular containers has a tendency to burn or overcook. Thus, oval or round containers are preferred. Microwaves do not penetrate metal but reflect off it, so containers made of metal or with metal trim should not be used. Containers made of glass, paper, and most plastics allow microwaves to pass through, but not all nonmetal containers are acceptable. To check whether a container is safe to use in a microwave, fill it with cool tap water and microwave it on high for 2 minutes. If the container remains cool but the water is hot, the container is microwave safe. If the container is hot to the touch, it can only withstand microwaves for short periods of time. Plastic tubs like the ones margarine, sour cream, and yogurt are sold in and styrofoam containers are not recommended because they can actually melt into the food.

Microwave Advantages. The main advantage of a microwave oven is speed. A microwave can cook a potato in 5 to 10 minutes, compared to 1 hour for a conventional oven. Most dishes can be ready within minutes rather than hours. In addition, cleanup is often easier because microwaved foods stick less, and one dish can be used for both cooking and serving. Another advantage of a microwave oven is that it is easy to clean using a damp cloth. An acceptable cleaning solution is 4 tablespoons of baking soda dissolved in 1 quart of warm water. For odor problems, a teaspoon of baking soda or a squeeze of a lemon wedge in a cup of water heated on high for 5 minutes usually eliminates unwanted smells.

Microwave ovens do not emit any heat while cooking, and burns are rare, although they occur occasionally when a burst of steam escapes as the cover is removed from the food. A research study showed that 38 percent of parents allow their children (5 to 8 years of age) to use a microwave, while only 6 percent would allow them to operate a conventional oven (11). The microwave oven is also better than other methods at retaining the nutrients and flavors of foods (8).

Microwave Disadvantages. Microwave ovens were meant to supplement, not replace, standard ovens. Several limitations of the microwave oven ensure that the conventional oven is still necessary. It is true that one potato cooks quickly, but cooking larger numbers of potatoes increases the cooking time by 5 to 10 minutes per potato. A con-

KEY TERMS

Arcing Sparks of electricity generated by microwaves bouncing off metal.

ventional oven can bake 12 potatoes in the same length of time it takes it to bake just one.

Another disadvantage of the microwave oven is that browning of meats and baked goods does not occur naturally (14). A conventional oven with its circular flow of hot air allows a flavorful, attractive-looking crust to form on meats, breads, and other baked products. In a microwave, however, steam is formed as the food cooks, producing moist heat that is not conducive to the browning and flavor development made possible with conventional ovens. A commercial browning agent or sauce can be used on microwaved meats, but these cannot duplicate the flavor of fried, grilled, barbecued, or roasted meat. Specialized browning dishes are also available for microwave ovens.

Cuts of meat that require a long, slow heating time and any foods thicker than 2 inches are best prepared by conventional methods. In general, most meats and baked foods are best heated conventionally, while vegetables, fruits, beverages, soups, and leftovers do well in the microwave oven. Another food that does not lend itself to microwave preparation is pasta, which takes almost as long to prepare in the microwave as on the range, and which develops a grainy and less viscous texture (19).

Uneven heating can also be a problem when using a microwave oven. This may be avoided by using automatically revolving turntables, by stirring, or by turning containers 90 degrees halfway through the heating period. Food will heat more evenly when covered with plastic wrap or a paper towel, which may also prevent it from splattering. Uneven heating can become a health concern when cooking meats and poultry, because there is a greater survival rate of bacteria and parasites in microwaved food compared to conventionally heated food (6). Microwaving is not ideal for the prevention of foodborne illness from bacteria and other microorganisms. In order to destroy the *Trichinella spiralis* parasite in pork, it is important to heat the meat to a final temperature of 170°F (77°C), whether in a microwave or a conventional oven. Moreover, when microwaving, the cut of pork should not be too thick, because of the short penetration range of the electromagnetic waves.

Microwave Safety Tips. One of the first things to remember when using a microwave oven is to avoid the use of metal. Metal can cause **arcing**, which can destroy the magnetron's ability to generate microwaves (in ovens built prior to 1980). Thus, aluminum foil, metal pans, dinnerware with metal trim, and twist ties should not be placed in the microwave unless the manufacturer instructs otherwise.

Some foods have a tendency to explode in microwave ovens when steam seeks to escape. Among these are foods enclosed in containers, as well as egg yolks, which have their own "container" in the form of their surrounding membrane. Containers should be vented and yolks must be pricked. Using any tightly closed container, especially glass jars, can be dangerous. There should always be at least one hole where steam can escape. Even such foods as potatoes, sausages, and squash, which have a nonporous skin, might explode if they are not punctured. Bottles of infant formula should not be heated in a microwave because the formula heats unevenly and may burn the baby.

Fire in a microwave oven is rare, but in the event one does occur, the door should be kept closed and the power disconnected by removing the electric plug from the wall socket. As a precaution against fire and other calamities, most microwave oven cords come with a grounding prong and require a three-stem socket, so extension cords should never be used nor should the grounding prong be removed. The oven itself should be placed so that the vents are not covered. It should also have two interlock switches that automatically turn off the magnetron when the door is opened. Materials that are commonly used in a microwave and that burn easily are paper, wood, and straw.

Because leaked microwaves may cause cataract formation or affect the operation of older heart pacemakers, door seals should be leak-proof, with no tears, pits, burns, or looseness at the hinges. Any heat felt around the door while the microwave is on is an indication that leakage may be a problem. Newer microwave ovens meet the government standard for leakage, which is designed to prevent excessive exposure to microwaves.

Refrigerators/Freezers. The proper refrigeration and freezing of foods is one of the most important factors in preventing bacterial infections from foods. Refrigerator temperatures should be maintained at or below 40°F (4°C), and freezers or freezer compartments at 0°F (−18°C). Household refrigerators are classified by the location of the freezer—above, below, or beside the refrigerator section (Figure 5-8). Food service establishments usually have a walk-in refrigerator and freezer, which may range in size from a small closet to a large room. Other types of freezers/refrigerators include reach-ins, roll-ins, and pass-throughs (Figure 5-9).

Most refrigerators have special compartments for meats, fruits, and vegetables. These are designed specifically for keeping the moisture content of the respective type of food at its proper level. Butter compartments hold butter separately to prevent food odors from spoiling its flavor, and at a slightly higher temperature to keep it spreadable. It is important to keep refrigerators and freezers clean of any spilled foods and raw meat drippings. As a matter of routine, all food should be removed and all parts of the refrigerator washed with water and baking soda, soap, or detergent.

Dishwashers. The four basic designs of household dishwashers are based on how they are attached to the water supply and electrical source. The most permanent design is a built-in, which, as the name implies, is integrated under the counter to match the cabinets. Connections to the hot water line, drainpipe, and electricity are permanent. A less permanent design is the portable dishwasher, which can be used as soon as the hoses are attached

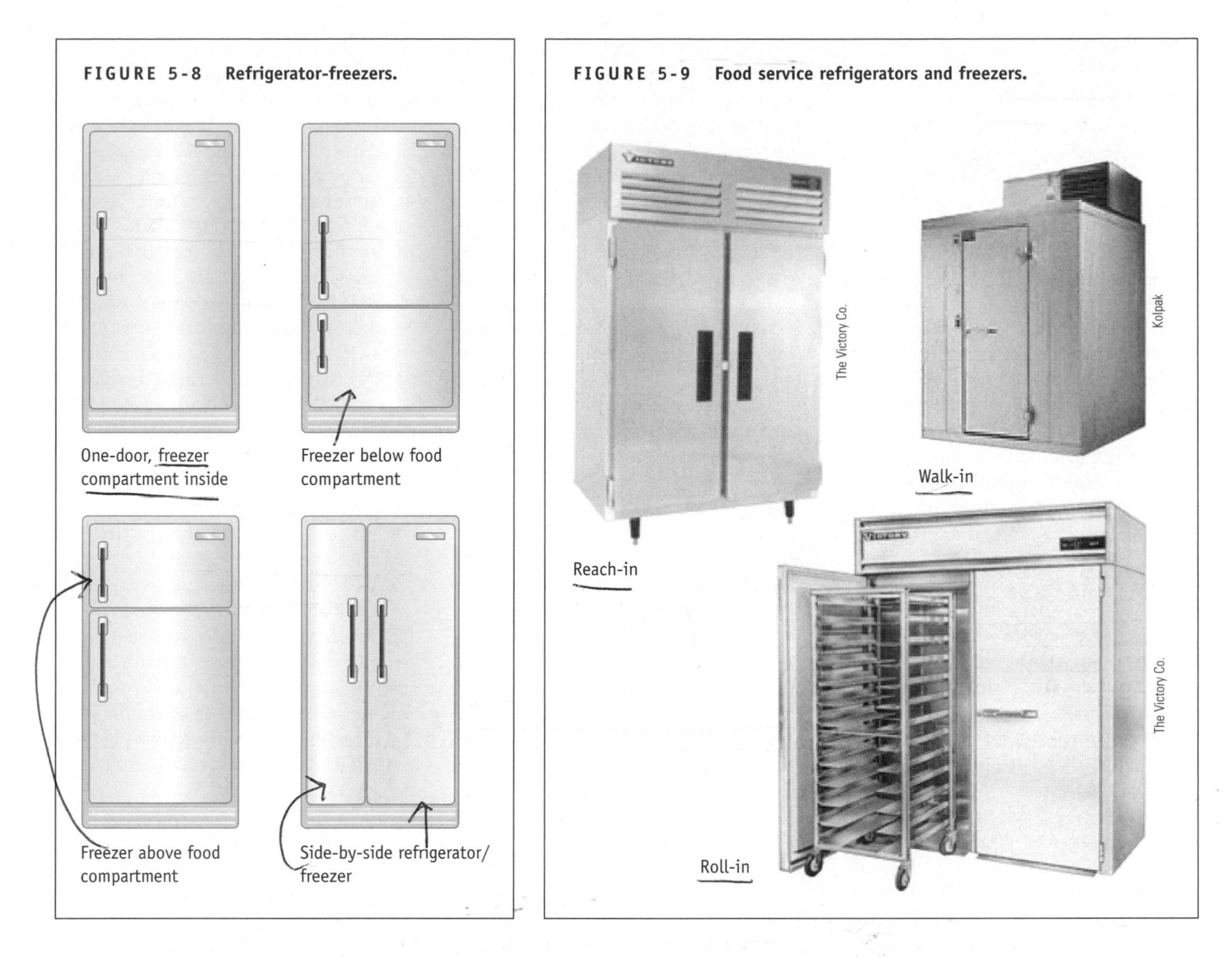

FIGURE 5-8 Refrigerator-freezers.

FIGURE 5-9 Food service refrigerators and freezers.

to the kitchen faucet; one hose drains into the sink. A convertible dishwasher can be either used as a portable or installed permanently as a built-in. Food service dishwashers are so large that they often require a separate room. In this situation, conveyor belts move the soiled dishes and utensils through the dishwasher and bring them out clean and sanitized.

Equipment Standards and Safety. The National Sanitation Foundation (NSF) seal of approval assures buyers of food service equipment that certain standards of sanitation and safety have been met in its design and production. This nonprofit organization, which is interested in the promotion of public health, has established minimum standards of construction for food service equipment (5). Information about equipment or approved manufacturers can be obtained by writing to NSF Testing Laboratory, Inc., PO Box 130140, Ann Arbor, MI 48113. Another private organization overseeing the safety of electrical equipment is the Underwriters Laboratory (UL), which ensures that an electrical appliance, cord, or plug has passed certain tests for electrical shock, fire, and other related injuries (15).

In addition to passing inspection, once equipment is in use it must be routinely cleaned and maintained on a regular basis to ensure that it is in proper working order. Perfectly designed and otherwise safe equipment can become a hazard when clinging food remains attract and support pests and microorganisms. Food service establishments are responsible for preventing foodborne illness, so one of their primary concerns is to keep the foods they serve away from such contaminants by maintaining as clean an environment as possible. A cleaning regimen executed on a regular schedule ensures that the equipment and surrounding surfaces are free from food particles and their resident population of potentially pathogenic organisms (4).

Auxiliary Equipment

In addition to the major "players" in a kitchen, other kinds of equipment may be utilized, some of which are optional or have specialized uses. These include equipment for frying, broiling, steaming, and grilling food, cutting equipment, and additional items for mixing and coffee making.

Griddles. Griddles supplement range units and have larger, flat, smooth surfaces (Figure 5-10). They are ideal for preparing eggs, hamburgers, pancakes, French toast, and hash browns. Food

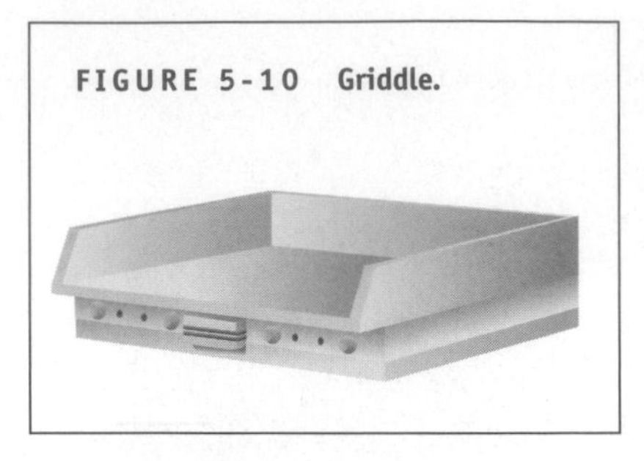
FIGURE 5-10 Griddle.

service griddles contain a drip cup to collect draining fat. Preparation is easier when grill surfaces are primed by smearing them with oil followed by a brief heating. To maintain the primed surface, griddles are never washed with soap and water, but scraped clean, wiped with a grease mop, and then polished with a soft cloth.

Tilting Skillets. Found only in large food service operations, the tilting skillet, brazier, or fry pan can be used to make anything from chili to poached eggs. The wide range of temperature settings stretches from low braising to high frying heats, thereby allowing it to be used as a fry pan, brazier, griddle, stock pot, steamer, or steam table. The entire skillet can be tilted to pour out liquid-based contents (Figure 5-11).

Broilers and Grills. The difference between broiling and grilling is the heat source: the broiler's heat is above the food while the grill's is below the food. Temperature control is achieved by moving the grid up or down. Heat for these may be provided by wood, charcoal, electricity, or gas.

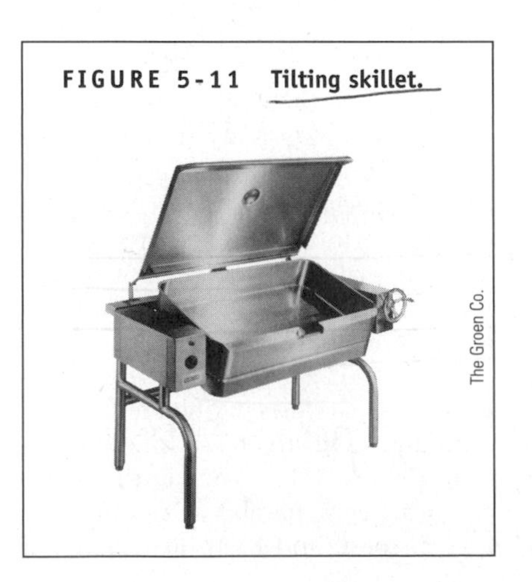
FIGURE 5-11 Tilting skillet.

The Groen Co.

Steamers. Cooked vegetables and even some fish maintain their texture, color, taste, and nutrients best when they are properly steamed or microwaved. Steam captured in an enclosed space serves to heat a variety of foods by moist heat. Most steamers in a food service establishment steam under pressure, which is measured by a gauge in pounds per square inch (psi). Pressure steamers allow food to heat to temperatures higher than boiling, which decreases cooking time. Vegetables can be cooked at pressures of 10 to 15 psi, reaching a temperature of 250°F (120°C). Lower pressures and temperatures (5 to 10 psi, 225°F/105°C) are used to cook meats, which would fall apart at the higher pressures required for vegetables. Rice, pasta, poultry, eggs, fish, and shellfish may be steamed, but the flavor of meats and poultry will usually be diminished by the process.

Two basic types of steamers are used in food service organizations: cabinet, or compartment, steamers and steam-jacketed kettles (Figure 5-12). Cabinet steamers are stacked one above the other with the door of each sealed tight with clamps. Steam-jacketed kettles are used more for fluid-type foods such as soups and stews, and range in size from 1 quart to 200 gallons. The steam is not generated inside the kettle, but is circulated between the double-layered metal plates of the kettle's outer shell. A handle is used to tilt the entire steam-jacketed kettle to pour out the food.

Safety is particularly important with steamers. They should never be run without water, and they should be periodically checked to ensure that safety valves are working. They should never be opened until the pressure has gone down, and then should always be opened away from the face.

Deep-Fat Fryers. Breaded fish and vegetables, fried chicken, and French fries are some of the foods commonly prepared in deep-fat fryers. Frying is similar to boiling, except that in frying the liquid is fat, which can reach higher temperatures than water. Food is loosely placed in a wire basket, which is then submerged in heated

FIGURE 5-12 Food service steamers.

Cabinet steamer

Steam-jacketed kettle

The Groen Co.

oil. When the food floats to the top of the oil, it can be considered cooked. The basket is then removed and set aside so that the oil can drain from the food. The fryers themselves may be small enough to be portable or so large that they are floor mounted (Figure 5-13). Most deep-fat fryers have automatic heat controls.

Woks. A wok is a large bowl-shaped pan central to Chinese cooking. It comes equipped with a metal ring to fit over a range burner, or self-contained with an electrical cord (Figure 5-14). The most time-consuming step in using a wok is cutting the foods into many small, uniform pieces. The actual cooking of the foods is a quick process, lasting approximately 5 to 10 minutes. It starts with high heat under the wok, which has been lightly coated with oil (usually sesame or peanut oil). The foods that take the longest to cook are added first. The food is stirred rapidly for a few minutes, for even cooking, and then the heat is lowered and the pan covered so the steam thus generated can complete the process.

FIGURE 5-14 Wok.

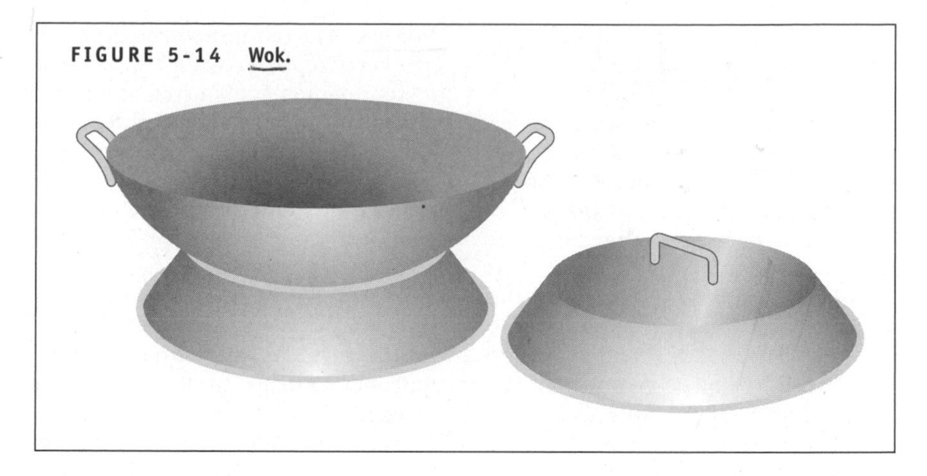

FIGURE 5-13 Deep-fat fryer.

Crockery. Crockery, or electric slow cookers, have been popular for some 40 years and are particularly good for moist-heat cooking of meat and legumes. Crockery cooking is long and slow, with controlled heat that needs little or no supervision. A meal can be started in the morning that will be ready to eat by dinnertime. Since there is some evidence that crockery may not keep food sufficiently hot for the entire duration of cooking, its use has lately been discouraged by some food experts because of the risk of foodborne illness.

Cutting Equipment. Meat slicers, food choppers, and grinders are common pieces of equipment in food service establishments (Figure 5-15).

Meat Slicer. Carelessness in the use of a meat slicer causes more food service accidents than any other kind of equipment. While great care must be taken with any piece of equipment that involves a combination of sharp blades and electric motors, the advantages of using cutting equipment—a uniform product and less time spent on food preparation—are worth the risks as long as the proper precautions are taken. The machine should always be unplugged when not in use. After the slicer is plugged in, the blade control is adjusted for the desired slicing thickness, and the blade guard positioned. The food, usually boneless meats, but possibly cheese, vegetables, fruits, and even bread, is then placed on the carriage and held there firmly with the guard before the switch is turned on. The carriage is moved back and forth by its handle in a smooth motion.

FIGURE 5-15 Cutting equipment.

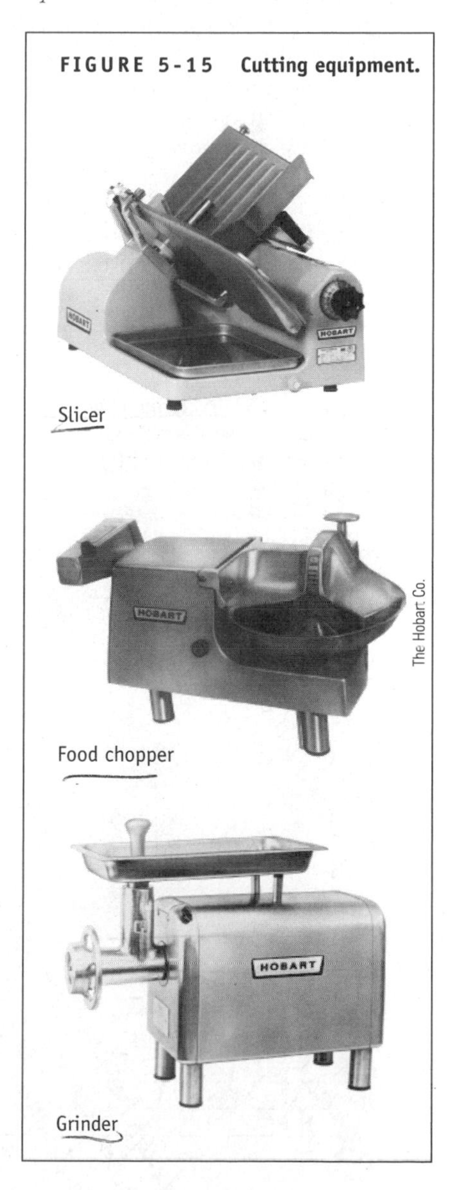

The equipment should be thoroughly sanitized after use and between different types of foods, especially with raw meats. The cord must be removed from the socket and the blade control set at zero before cleaning. Metal utensils should never be used to scrape food from the blade because they may nick the slicer. Manufacturer's instructions should be followed in removing the various parts and subjecting them to the sanitizing solution and to rinsing and drying. The blade guard should be replaced immediately to prevent any risk of cuts. The use of protective gloves through the whole cleaning process is highly recommended.

Food Chopper or Cutter. Another potentially dangerous piece of food service equipment is the food chopper or cutter. The key to preventing injuries here is to turn the machine off, allow the knife blades to come to a rest, and flip the safety catch on before removing the food with a bowl scraper. The hands should never go into the bowl. The guard can be raised to remove any remaining food. Meats with bones or gristle should not be processed with food choppers because they will damage the gears and knives.

Mixers. The old-fashioned eggbeater has been replaced with a variety of mixing equipment, including electric mixers, blenders, and food processors. Mixers are convenient for controlling the rate at which ingredients are combined, and are used to prepare whipped cream, beaten egg whites, and mashed potatoes. In the food service industry, models range in size from table-top (Figure 5-16) to floor size. Attachments vary from a paddle for general mixing, to whips for cream or eggs, to dough arms for kneading yeast dough. Additional attachments may be added, including a shredding, grating, or slicing attachment and a grinder for meats and other foods. Some home mixers have similar attachments. For safety's sake, attachments must be securely in place before the machine is turned on, and it is best to disconnect the power entirely before removing them. Spoons or hands in the bowl during mixing are not recommended, but rubber scrapers can be used occasionally to scrape down the sides of the mixing bowl.

Blenders and Food Processors. Blenders and food processors allow further refinements to mixing food (Figure 5-17). Blenders have the blades or mixing component on the bottom. They are used for everything from making milk shakes to blending the vegetables used in making gazpacho, a Spanish cold soup. Food processors are more versatile and allow cutting, chopping, grinding, slicing, and shredding of foods, and even the kneading of dough. They come with specialized blades for accomplishing all these tasks and many even come with a juicing attachment.

Mixing Bowls. No kitchen is complete without a set of mixing bowls. These can be made of glass, pottery/ceramic, or stainless steel and come in sizes ranging from 2 cups to 2 quarts—larger, of course, for food service venues.

Coffee Makers. Many homes and food service venues use automatic coffee makers daily (Figure 5-18). Food service operations serve coffee from an electric urn or automatic coffee brewer. Electric urns are connected to a hot water source and automatically shut off after the coffee is finished brewing. In a coffee brewer, the hot water running through it stops after the decanter or pot is full.

Pots and Pans

Pots and pans, many of which are identified by name (Figure 5-19 on page 110), are distinguished from one another by their size, shape, and handle. The

FIGURE 5-16 Tabletop mixer and three typical attachments: (a) wire whip—incorporates air, (b) flat beater—general mixing, and (c) dough hook—mixing heavy doughs.

FIGURE 5-17 Blenders and food processors.

FIGURE 5-18 Coffee makers and tea dispensers.

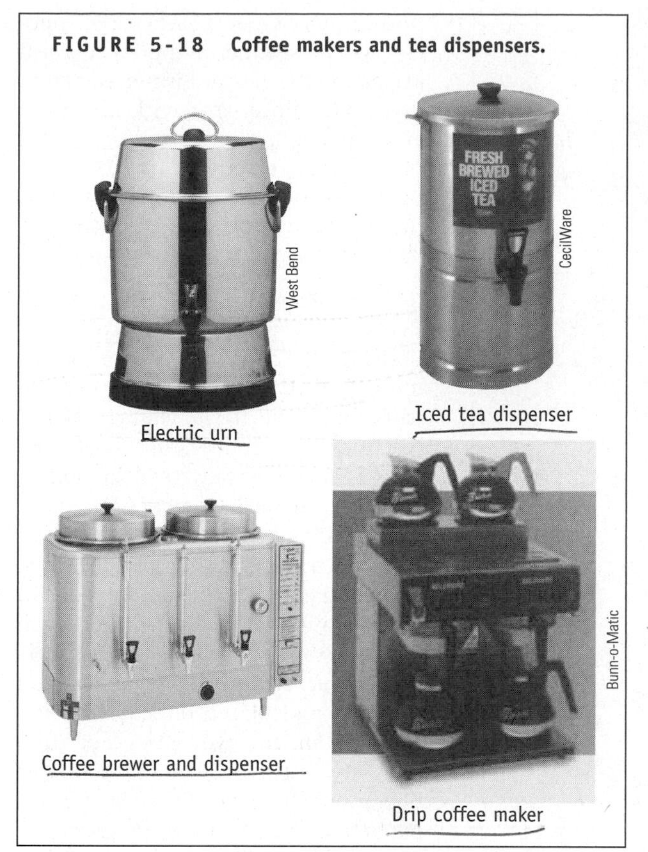

capacity of a pot, pan, or kettle is determined by the number of quarts it contains when filled with water (15). The capacity of baking pans and skillets is described in inches, as in the case of an 8- or 9-inch cake pan (16). Pots have two handles, and are used when preparing large quantities, while saucepans and frying pans have a single long handle and less capacity. A shallower sauce pot allows for easier stirring. Saucepans are usually straight-sided, while frying pans can be either straight- or slope-sided. The flattest pans are those used for baking and roasting, with the sturdiest and thickest being the roasting pan. A double boiler consists of a bottom pan in which water is heated, and a top pan containing a food item that must be kept below the boiling point. The double boiler is used for preparing certain sauces, and to keep food hot without burning.

Dessert Pans. Various pans are used to prepare desserts. Cake pans come in round, square, and oblong shapes and in a variety of sizes. There are special pans for making sponge and angel food cakes; the angel food cake pan usually has a tubular segment in the middle which separates from the sides for easy removal of the cake. Springform pans allow one to "spring" open the sides for easy removal of the cake. Cookie sheets, sometimes referred to as baking sheets, have no sides (except one or two that are raised for handling), allowing the hot air to flow evenly over the cookies. Heavy-duty sheet pans with four sides can also be used for preparing cookies and a myriad of other foods such as biscuits, bread, pizza, breadcrumbs, roasted nuts, and even some meats. Full-sheet pans are used in restaurants for bulk baking, while half-sheets (half the size of full sheets) are reserved for home use (12). Some of the half-sheets purchased at supermarkets may warp at temperatures over 300°F (149°C), but not pans made of heavy-duty aluminum or steel. The aluminum pans tend to be more popular because their lighter color reflects heat, which helps to prevent over-browning and baking (12). They also do not rust. The darker the pan, as seen in steel pans or those coated with a nonstick surface, the darker the cookies.

One advantage of the darker sheet pans is that they absorb heat, resulting in a crisper crust for pizza and the bottom crust of fruit pies. Regardless of the color, professional bakers use kitchen parchment on baking sheets to prevent sticking, to move items around with ease, and to protect against burning (12). Kitchen parchment as well as the various heavy-duty half-sheets can be purchased at restaurant or kitchen supply stores and catalogs.

Pot and Pan Materials. Pots and pans are made from a variety of materials (Figure 5-20 on page 111):

- Aluminum, copper, and stainless steel
- Nonstick coatings
- Cast iron
- Glass and glass/ceramic combinations

Best Heat Conductors. Aluminum, copper, and combinations of copper and stainless steel are the best conductors of heat (15). Aluminum accounts for over half the cookware sold in the United States, but it is very lightweight and prone to denting. Aluminum may also react chemically with many foods, particularly those high in acid, and it is not recommended for storing foods. Copper is an excellent heat conductor, but it is costly and requires special care. A further disadvantage is that excessive copper may dissolve into the food being prepared, causing nausea and vomiting. Therefore, copper pans are usually lined with stainless steel or tin. Stainless steel is known for its durability and easy cleaning; however, it is a poor conductor of heat and tends to generate hot spots, which may scorch the food. To keep this from happening, the bottoms of many stainless steel pans are coated with copper or aluminum.

Nonstick Pans. The problem of food sticking to the bottoms of pans, particularly frying pans, has been addressed by the development of nonstick pans coated with a fluorocarbon resin (Teflon, SilverStone, etc.). Nonstick pans reduce the amount of fat needed to prevent sticking, but their surfaces are easily scratched, so plastic, rubber, or wooden utensils are recommended.

Cast Iron Pans. Cast iron pots and pans are heavy and they heat slowly, rust easily, and are difficult to clean. They do, however, retain high temperatures for longer periods of time, heat evenly, and add extra iron to the diet. Acidic foods such as tomato sauces tend to absorb more iron: 5 mg of iron are absorbed for every 3 ounces of

FIGURE 5-19 Common pots and pans.

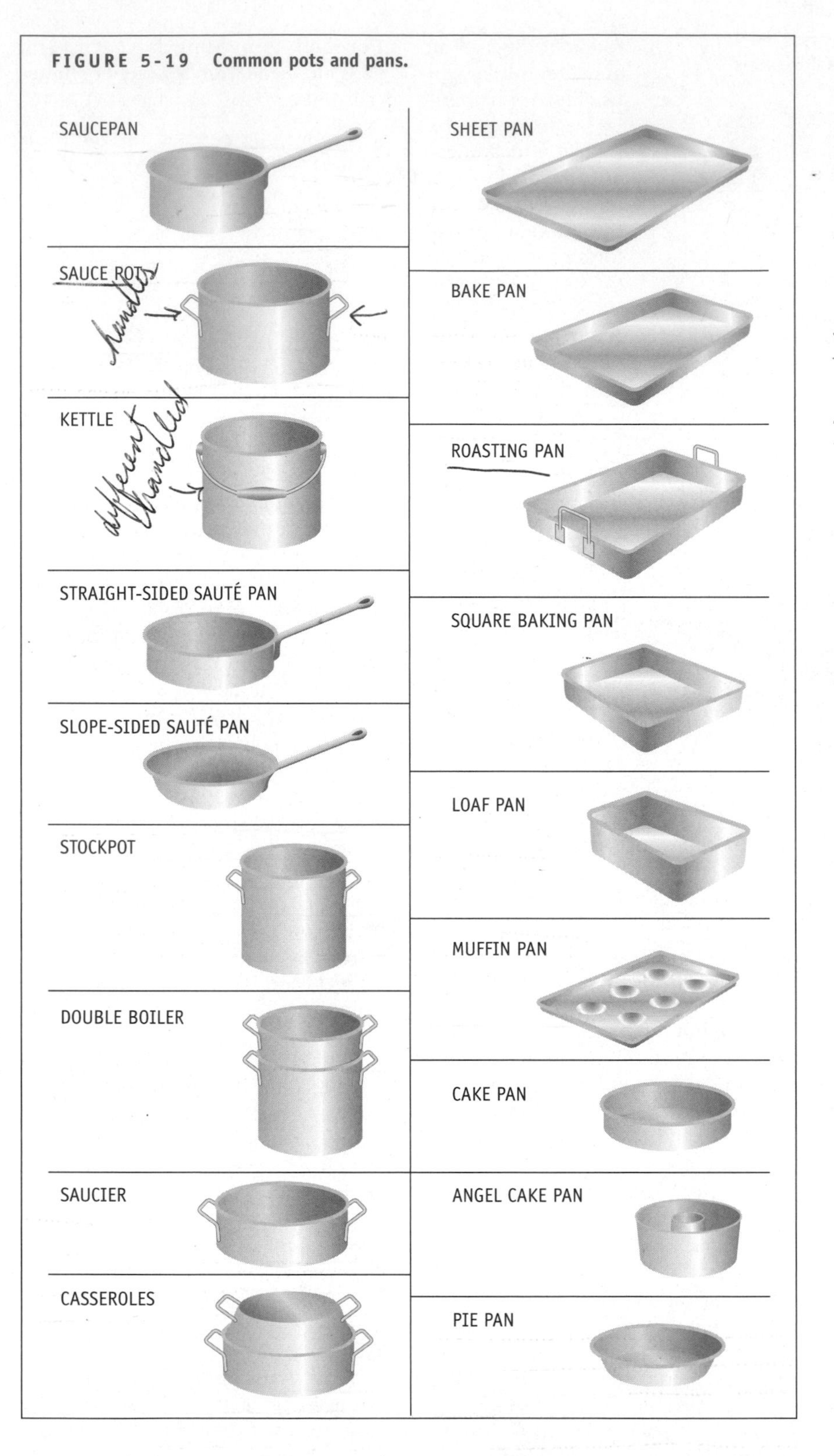

spaghetti sauce cooked in a cast iron pan.

Cast iron pots and pans may be cleaned in one of two ways. The first involves a preliminary priming or conditioning of the pan with a very thin coat of vegetable oil, after which it is heated and cooled. A primed pan is cleaned by scraping and wiping away food particles after each use. Reconditioning may be done whenever necessary. The second method is to wash the pan with soap and water, heat it to dry, and then coat it with a minute amount of oil. This second method is more likely to remove any traces of rancid fat, which can impart an off-flavor to any food subsequently prepared in the pan.

Glass Pans. Heat-proof glass, such as Pyrex, and glass/ceramic combinations, such as Corningware, break more easily than metal-based pots and pans, but have the advantage of not reacting with foods. Most casserole pans that are usually oval or oblong with low sides are made of such materials. Baking temperatures should be reduced by 25°F (4°C) when using tempered glass. The newer versions of glass/ceramic materials can be moved from the range or oven to the refrigerator or freezer, and later be taken from the cold and placed directly into the oven or microwave. Glass pots and pans are not allowed in food service operations, however, because of possible breakage and liability problems.

Pot Holders. Every kitchen must be equipped with sturdy, heat-proof pot holders. They should always be used when removing pots or pans from the stove or oven. Pot holders come mainly in the form of squares or mitts, with the latter recommended for removing roasting and baking pans from the oven.

HOW & WHY? ?????????

Why should dish towels never be used to pick up hot pans from the oven?
Dish towels were not meant to be used for this purpose. Their loose or flapping ends can catch fire easily, and, because the materials they are made of are not meant to prevent heat from penetrating, the hands may be burned and pans dropped. In addition, wet pot holders also should not be used to pick up hot pans. Water helps the heat transfer through to the hands very rapidly, and the person may instantly drop what they are holding.

FIGURE 5-20 Materials that make the pot.

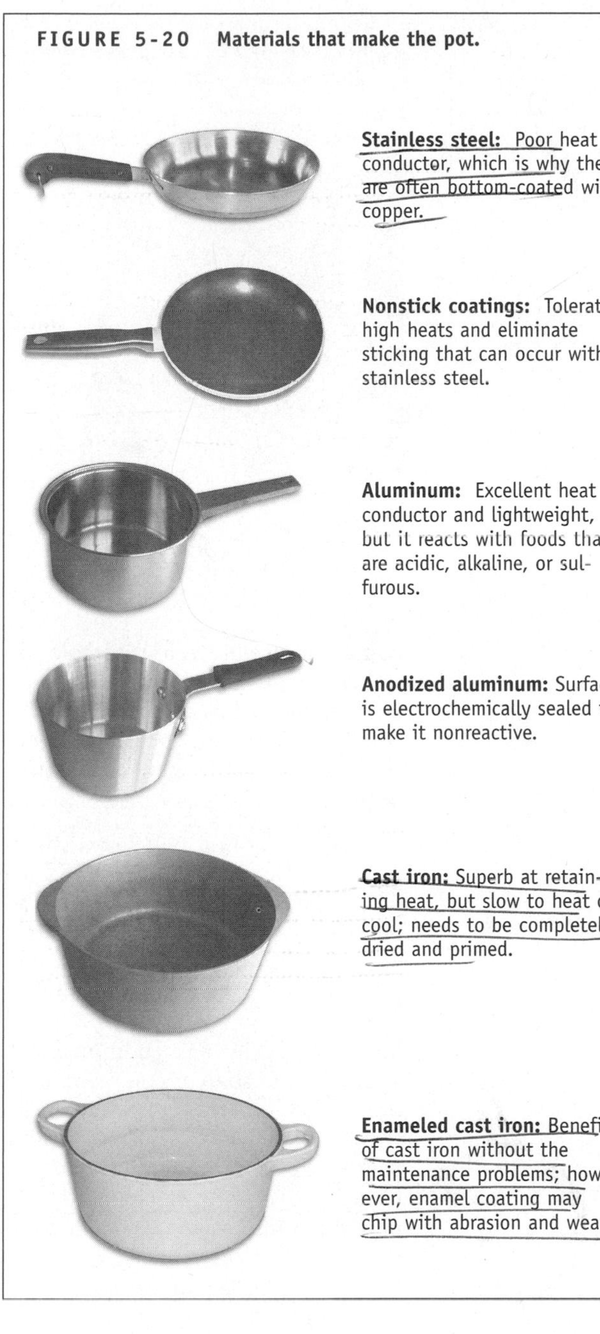

Utensils

Utensils are vital items needed for cutting, stirring, turning, measuring, and serving food. The utensils covered in this section include knives, and preparation, measuring, and serving utensils.

Knives. Knives are to the chef as brushes are to the artist. Some people consider them to be the most important tools in food preparation. Knowing the different kinds of knives, their particular tasks, and how to care for them is crucial to the preparation of foods. The food to be cut determines what type of knife should be used. The basic knife starter set consists of chef's, paring, slicing, boning, and utility knives (Figure 5-21). The first three types of knives often complete the set of many home kitchens.

Chef's Knife. The chef's, or French, knife is one of the largest and serves as an all-purpose knife for cutting meats and for mincing, dicing, and chopping a variety of foods. Chef's knives are commonly available in blade lengths of 6, 8, and 12 inches, and the side of their blades can be used to crush garlic cloves, ginger slices, and peppercorns (2).

Utility Knife. The utility knife is geared toward lighter duties such as cutting tomatoes or carving meat.

Paring Knife. The smaller, shorter, 2- to 4-inch paring knife is used for more delicate jobs that require close control, such as the trimming of vegetables, fruits, and small pieces of meat like chicken breasts.

Boning Knife. The slightly curved boning knife is handy for separating meat from bone (e.g., deboning the breast of a chicken), disjointing poultry, cutting between the joints of larger pieces of meat, and dicing raw meats.

Slicing Knife. Slicing knives are long and flexible enough to portion off thin slices of meat or poultry. Serrated slicers are useful for cutting bread or angel food cake (Figure 5-22).

Other Knives. Additional knives that are found in food service arenas include heavy cleavers for cutting through bone, a variety of butcher knives for cutting raw meats, the steak or scimitar knife used for cutting steaks from the appropriate parts of a carcass, and the oyster and clam knives used for opening these shellfish.

Purchasing Knives. Knives can range in price from a few to several hundred dollars. When selecting a knife, qualities to consider include size, weight, balance, the length of the tang, and the materials from which the blade and handle are made. While the size selected will be determined by the use for which the knife is intended, the other factors depend on more qualitative assessments.

Weight and Balance. Sometimes the "balance" or the feel of the knife in the hand is a factor in selection. Some knives are blade-heavy, others handle-heavy, and some feel evenly divided between the two. A person should select the knife that feels "right" in his or her hand (2).

Tang. Another quality that varies among knives is the length of the tang, the part of the metal blade that extends into the handle. Better-quality knives have a tang that extends the full length of the handle.

Blade. Probably the most important factor in selecting a knife is the type of steel used for the blade: carbon, stainless steel, or high-carbon stainless steel. Carbon blades are almost obsolete, because they are highly susceptible to rust and lose their edge quickly. Stainless steel, on the other hand, is rust-resistant, but difficult to sharpen and to keep sharp. This steel is actually a combination of metals, including chromium, which is added for its resistance to stains, corrosion, and heat. Despite its name, stainless steel is not stain-proof, but it does stain less

FIGURE 5-21 Common knives used in food preparation.

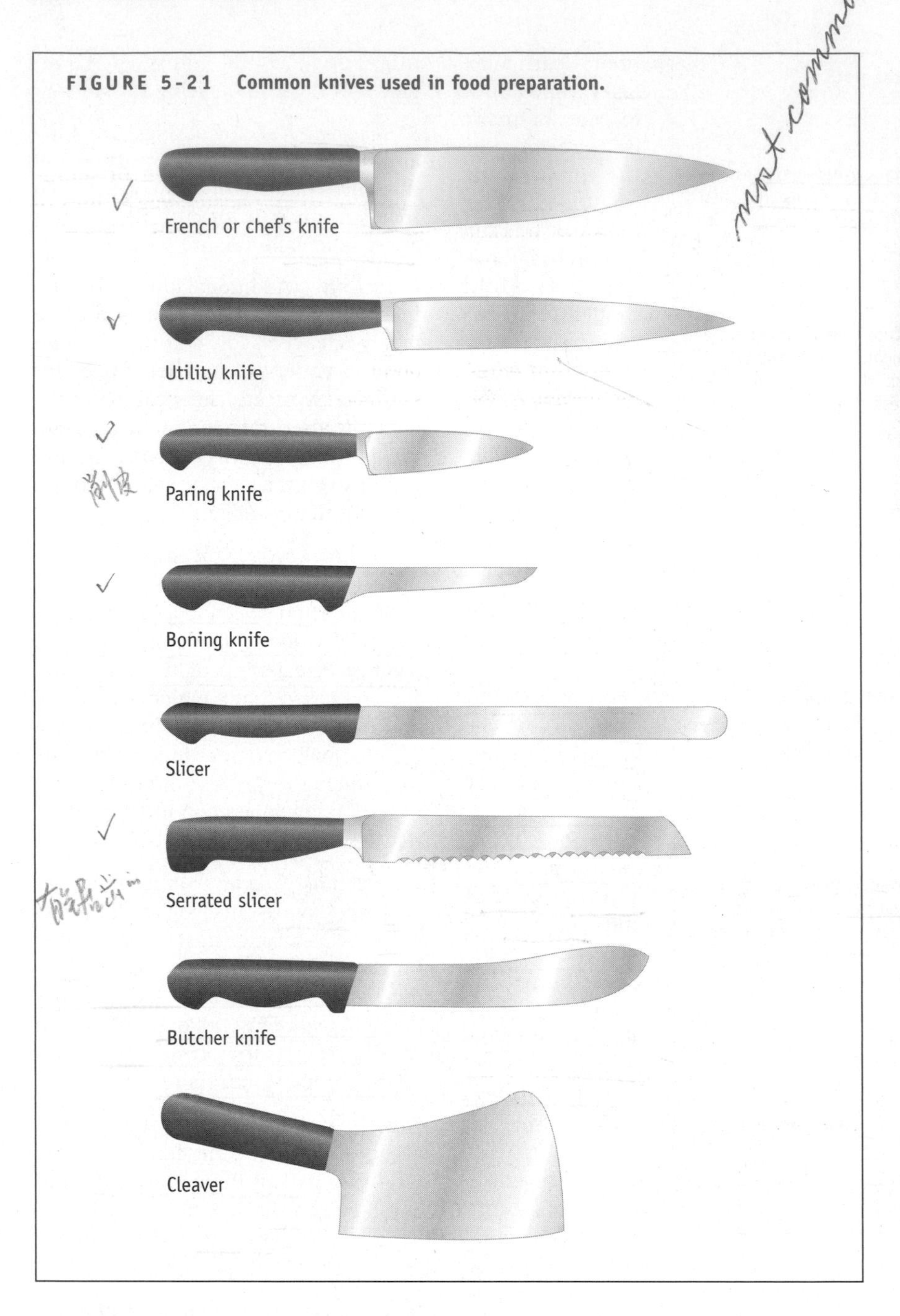

than knives not made from stainless steel when it comes into contact with food and beverages, especially salad dressing, vinegar, salt, mustard, tea, and coffee.

High-carbon stainless-steel knives are usually preferred because, in addition to not staining, they keep a sharp edge and do not rust (2). High-carbon knives are further distinguished by whether they have a stamped or a forged blade. Stamped blades are cut from a sheet of steel, ground, and polished. Because they are mass produced, they are less expensive than forged knives. Forged blades are made from a single piece of steel that has been exposed to extremely high heat, submerged in a chemical bath, and set in a die before being hand-hammered into shape. The resultant blades are more costly, but are also heavier, tougher, hold their edge longer, and require less pressure when cutting.

Handle. The knife's handle may be made of wood, carved bone, plastic, or metal. Wood is easier to hold, but water damage from frequent washing reduces the length of its life. Plastic or metal handles are more durable, but slippery. Bone-handled knives are both water- and wear-resistant, and in combination with a high-carbon stainless-steel knife, can last a lifetime or more.

FIGURE 5-22 Using a serrated slicer knife.

Care of Knives. Cutting knives should never go in a dishwasher. Strong detergents not only dull the blades, but when combined with hot water and air, can ruin wooden handles. Nicks can occur if the blade bumps against other metal utensils. Instead, knives should be washed immediately after use with soap or detergent, dried thoroughly, and then stored in such a way that their blades do not contact each other to prevent nicks. The blades may be kept separated by slipping them into a wooden knife block (blade turned upward) or a shallow knife block that fits inside a drawer, or by placing them along a magnetic strip.

Knives can be sharpened using one of the many electric and mechanical knife sharpeners on the market, but a better result can be achieved by hand. The two basic ways to hand-sharpen a knife are with a stone or a steel implement. The sharpening stone, also known as a whetstone, is used by rubbing a bit of moisture on the stone and sliding each side of the blade until the proper sharpness is acquired. A sharpening steel looks like a round sword and is held firmly in one hand while the knife, held in the other hand, is brandished against the steel (Figure 5-23). Most chefs employ both implements, for different purposes: one for getting an edge and the other for refining it.

Cutting Boards. Cutting boards are used primarily for cutting meat, poultry, vegetables, and fruits, but may also be used for kneading and rolling out dough. They may be made of wood, hard plastic, glass, or ceramic tiles; the latter three, however, are hard on knife blades (1). Cutting boards should be carefully scraped and thoroughly washed and dried every time they are used.

Preparation Utensils. Figure 5-24 shows some of the supporting utensils most commonly used in food preparation. Spoons, available in solid, slotted, or perforated versions, are used for mixing and serving. The holes in the slotted and perforated spoons allow liquids to drain. Wire whisks are used for mixing and are categorized by their shape. Straight (French) whisks are ideal for general purposes or smooth sauces. The very thin wires of balloon whisks are designed for beating the maximum amount of air into thin liquids such as egg whites and cream. Flat whisks are fashioned for creating sauces and gravies when it is important to lift up materials from the corners of a pan (17).

Spatulas come in a variety of shapes for their many purposes. Rubber spatulas or scrapers are used to scrape bowls or to fold beaten egg whites or other ingredients into each other. The straight spatula or palette knife is used for measuring ingredients, and is ideal for spreading icings onto cakes. The sandwich spreader, with a broader blade, is used, as the name implies, on sandwich fillings, butter, and jam. A pie server is an angled spatula used to lift pie, cake, or pizza wedges. Similar in design, but wider and with a larger bend, is the offset spatula, which is

FIGURE 5-23 Sharpening knives using either a stone or steel.

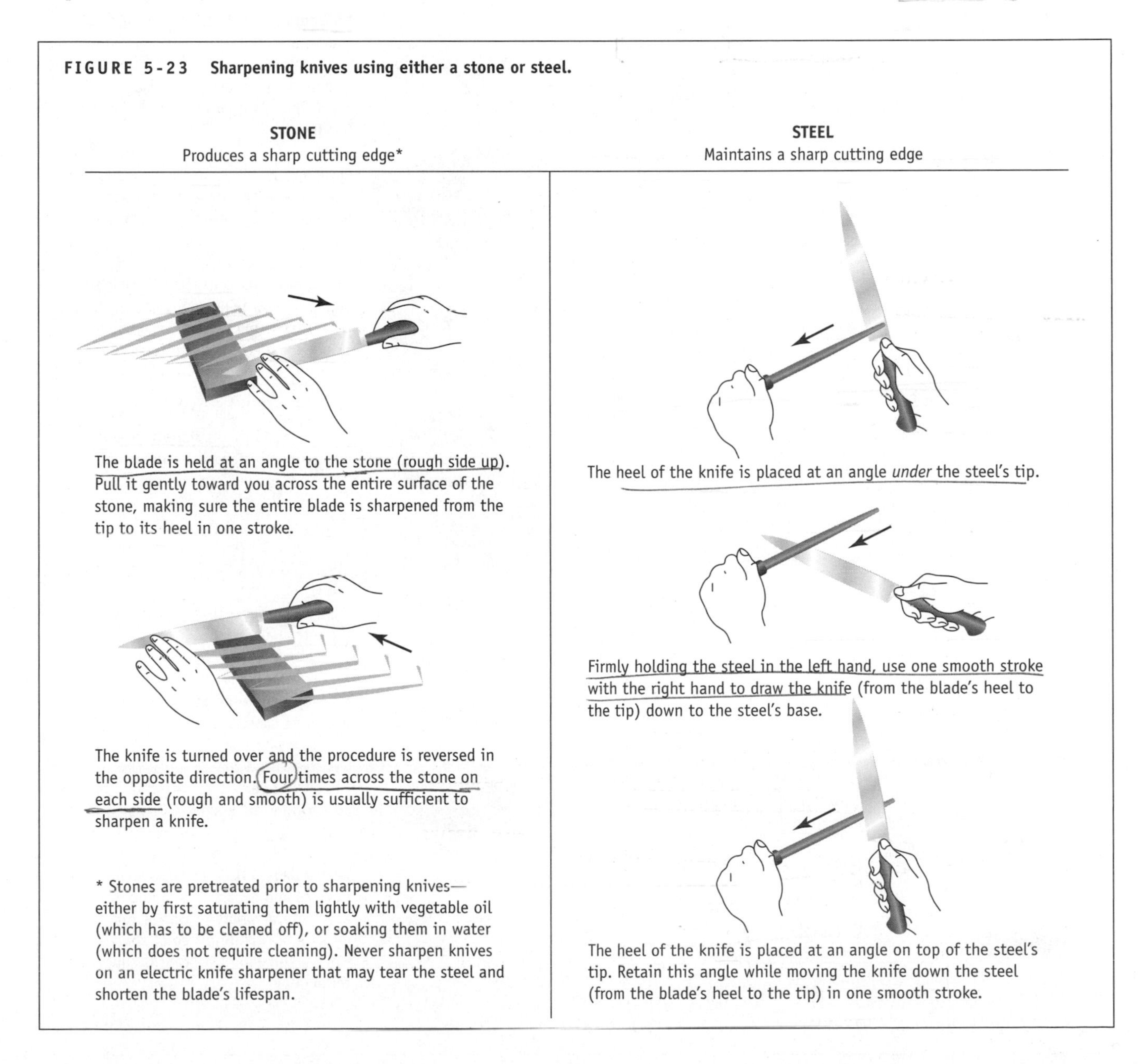

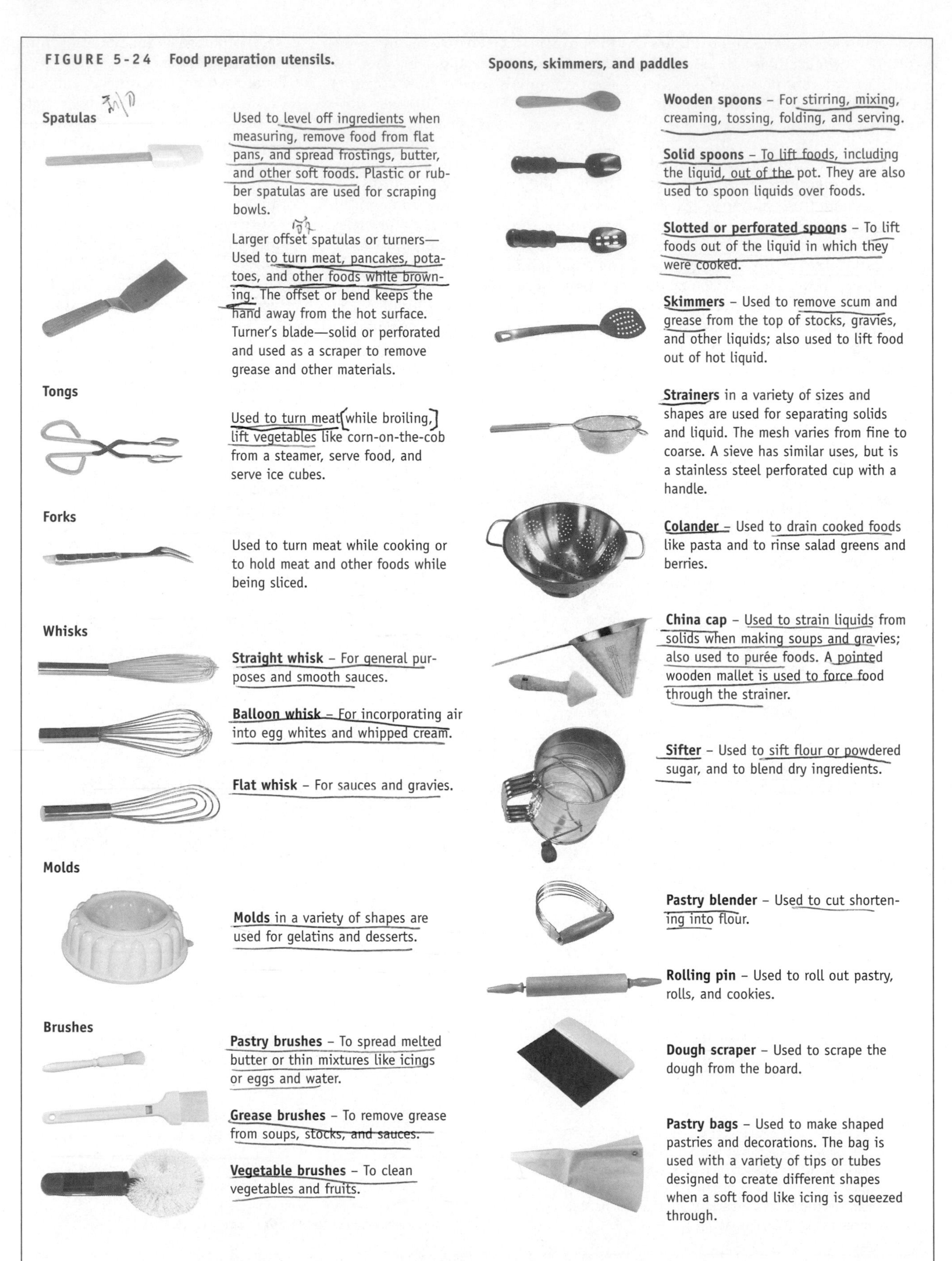

FIGURE 5-24 Food preparation utensils.

used to turn items such as hamburgers, eggs, and pancakes.

Other preparation utensils include the bench scraper, for scraping and for cutting dough; the pastry wheel, which is designed to cut pastry dough, but which can also be used to cut pizza; and the pastry brush, which is used to coat pastry with egg white or sugar glaze.

Measuring Utensils. Although different people may use the identical recipe for a cake, all of their cakes could turn out differently because of different measuring and mixing techniques. The following section explains some important measuring terms.

Mass vs Volume. Weight, commonly used to mean mass, is a much more accurate measurement than volume. As a result, many food service operations use weight rather than volume to measure recipe ingredients. Confusion between the two methods of measuring ingredients occurs because ounces can be measured either by volume, known as fluid ounces (fl), or by mass (weight), known as avoirdupois ounces (av). Water is the only substance whose fluid ounce is equal to its avoirdupois ounce. The mass of other substances will vary depending on the density, or weight per volume, of the object being measured. For example, half a cup of marshmallows weighs less than half a cup of vegetable oil.

Americans tend to measure using volume measurements such as teaspoons, tablespoons, cups, pints, quarts, and gallons. Many other countries measure ingredients by weight. Weight can be measured using a number of different available scales: spring-type scales, used principally for weighing dry ingredients like grains, beans, dried pasta, vegetables, fruits, and cheese; portion scales and balance scales for weighing ingredients; and the baker's scale, used primarily for measuring dough ingredients (Figure 5-25).

Metric vs Nonmetric. Metric measurements of volume are expressed in milliliters (ml). Metric cups come in sizes of 25, 50, 125, and 250 ml, and measuring spoons are divided into 1, 2, 5, 15, and 25 ml. A 250-ml metric cup is close to a nonmetric cup, which holds 236.59 ml. The 15 and 5 ml metric measures are almost equal to the nonmetric tablespoon and teaspoon respectively. The inside back cover of this book lists the conversions between nonmetric and metric measuring units for volume and mass.

Types of Measuring Utensils. About five different types of measuring utensils are frequently used in food preparation: liquid and dry measuring cups, measuring spoons, ladles, and scoops (Figure 5-26).

FIGURE 5-25 Various scales used for weighing ingredients.

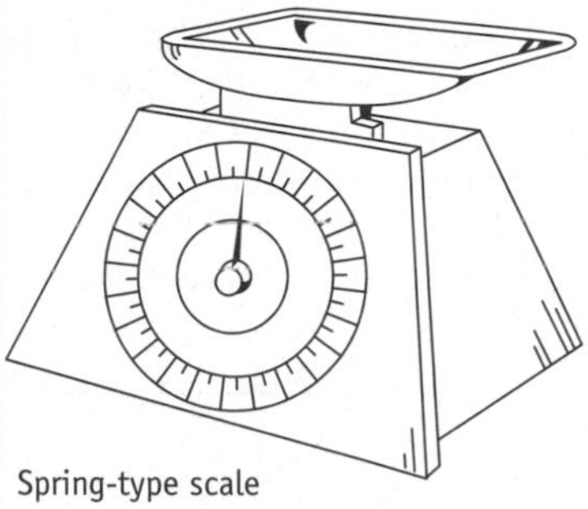

Spring-type scale

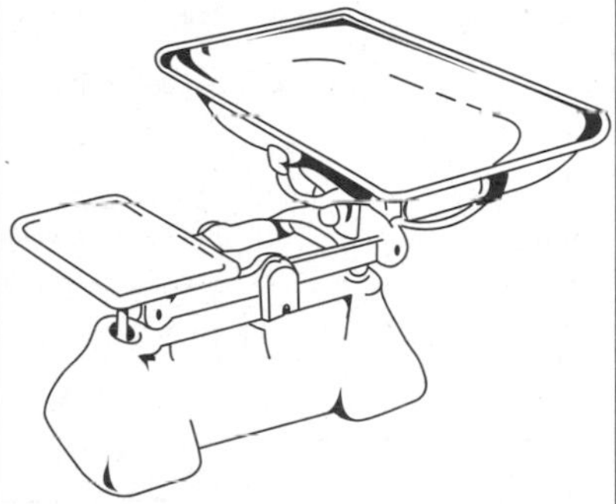

Balance scale

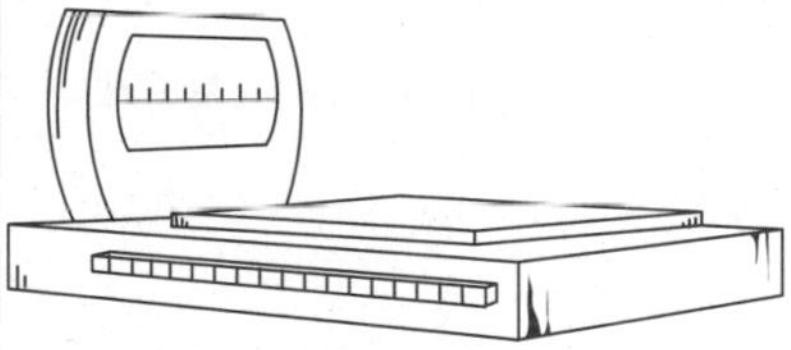

Portion scale

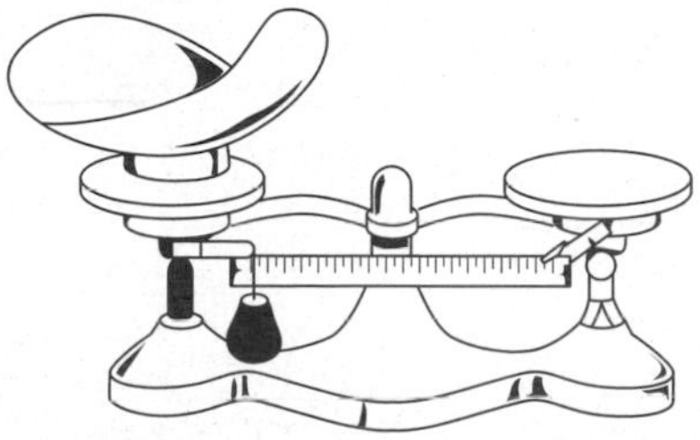

Baker's scale

Source: Texas Tech University.

FIGURE 5-26 Measuring utensils.

Liquid measuring cup

Dry measuring cups

Measuring spoons

Ladles

Dipper/Scoop

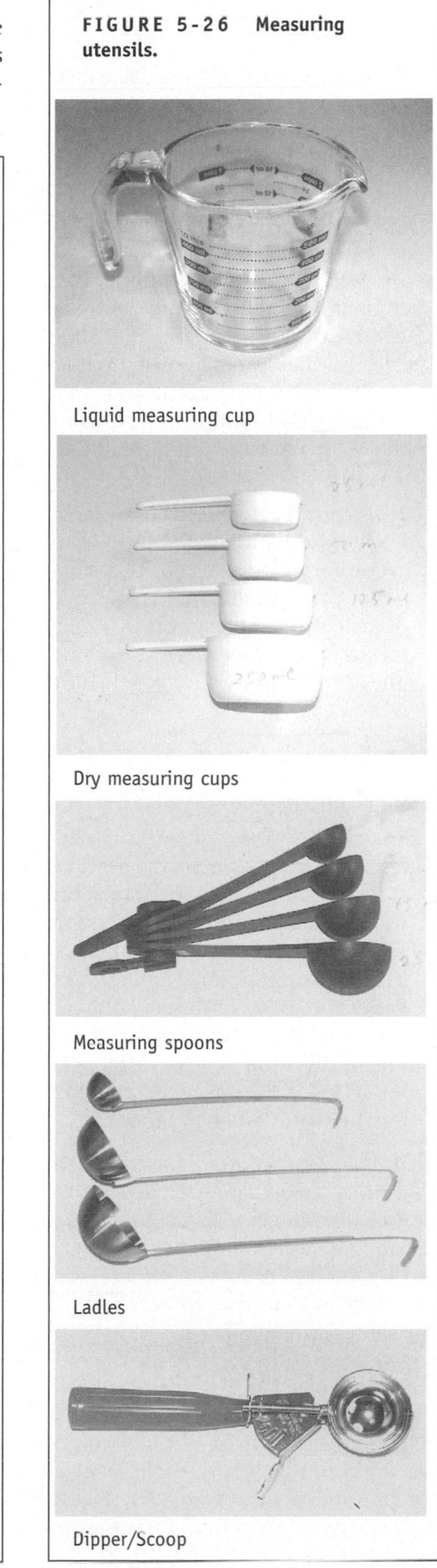

KEY TERMS

Meniscus The imaginary line read at the bottom of the concave arc at the water's surface.

- *Liquid measuring cups.* Available in 1-cup, 2-cup (1 pint), and 4-cup (1 quart) capacities. Their volumes are divided into increments of ¼, ⅓, ½, ⅔ and ¾ cup. They are usually glass, have a pouring lip, and are all-purpose.
- *Dry measuring cups.* Fractional, flat-topped (no pouring lip), single-volume cups (¼, ⅓, ½, and 1) are best because they can be leveled with a spatula for a more accurate result. Accuracy is also improved by using the 1-cup measure rather than four ¼ cups.
- *Measuring spoons.* Used to measure both liquid and dry ingredients requiring less than ¼ cup. They consist of 1 tablespoon, and 1, ½, ¼, and occasionally ⅛ teaspoon. A tablespoon is equivalent to 3 teaspoons, and 2 tablespoons equal 1 fluid ounce.
- *Ladles.* Liquids can be measured by ladles individually stamped with their capacity in ounces (Table 5-1).
- *Scoops or dippers.* The various sizes are identified by a scoop number (Table 5-2), which indicates the number of portions from a quart (e.g., a number 8 scoop yields eight ½-cup portions from 1 quart). The larger the scoop or dipper number, the smaller the serving. Used primarily by food service establishments for serving ice cream, mashed potatoes, and other soft foods.

Accuracy of Measuring Utensils. The American Association of Family and Consumer Sciences (AAFCS) has set certain tolerances for measuring the precise volume of household measuring utensils. One way to determine a cup's precise volume is to fill it with tap water and then pour it into a graduated cylinder. Both the utensil and the graduated cylinder should be on a level surface and the milliliters of water should be read at eye level at the bottom of the **meniscus** (Figure 5-27). Any measurement evaluating accuracy should be done three times and then averaged to eliminate error. The resulting number should not deviate more than 5 percent from the standard set by the AAFCS. According to these standards, a metric cup of 250 ml can deviate 5 percent, to 237.5 or 262.5 ml, and still be acceptable (3). Variations within the 5 percent range do not make any appreciable differences in ingredient proportions or in the quality of the final product.

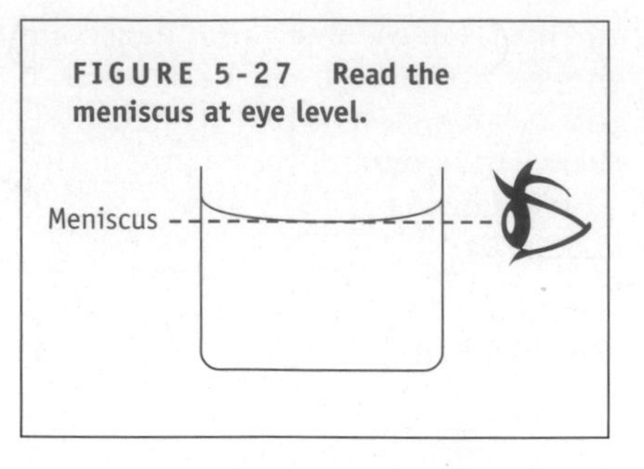

FIGURE 5-27 Read the meniscus at eye level.

Serving Utensils. At last the meal is ready to be eaten, and serving utensils enter the picture. Basic tableware includes salad forks, dinner forks, regular knives, steak knives (optional), soup spoons, and teaspoons. A more extensive "wardrobe" of tableware might include butter knives, small two-tined forks known as seafood forks, dessert spoons, luncheon knives and forks, which are slightly smaller than standard knives and forks, iced-tea spoons, and grapefruit spoons. Eating utensils, for sanitary reasons, should always be touched by the handles. Serving utensils and their uses were discussed in detail in Chapter 4.

TABLE 5-1

Ladles – Approximate Measures and Their Uses

Ladle Size	*Measure*	*Use*
1 oz	2 tbs	Sauces, salad dressings, cream
2 oz	¼ C	Gravies, sauces
3 oz	⅓ C	Cereals, casseroles, meat sauces
4 oz	½ C	Puddings, creamed vegetables
6 oz	¾ C	Stews, creamed entrées, soup
8 oz	1 C	Soup
12 oz	1½ C	
16 oz	2 C (pt)	
24 oz	3 C	
32 oz	4 C (qt)	

TABLE 5-2

Scoops – Approximate Measures and Their Uses

*Scoop or Dipper Number**	*Weight*	*Measure*	*Use*
6	6 oz	¾ C	Soups
8	4–5 oz	½ C	Luncheon entrées, potatoes
10	3–4 oz	⅜ C	Desserts, meat patties, ice cream
12	2½–3 oz	⅓ C	Vegetables, desserts, puddings
16	2–2¼ oz	¼ C	Muffins, cottage cheese, croquettes, dessert
20	1¾–2 oz	3 tbs ¾ tsp	Muffins, cupcakes, meat salads
24	1½–1¾ oz	2 tbs 2 tsp	Cream puffs, ice cream
30	1–1½ oz	2 tbs ¾ tsp	Drop cookies
40	¾ oz	1 tbs 2¼ tsp	Whipped cream, toppings, gravy
60	½ oz	1 tbs	Salad dressings, toppings
70	⅓ oz	2¾ tsp	Cream cheese, salad dressing, jelly
100	¼ oz	2 tsp	Whipped butter

*Dipper Scoop # = # Servings/Quart

PICTORIAL SUMMARY / 5: Heating and Equipment

Today's sophisticated equipment for heating and preparing food has come a long way from the smoky campfires where cooking probably first began. A thorough understanding of heating principles and a familiarity with modern food preparation tools and utensils is essential.

HEATING FOODS

Thermometers used to measure heat in cooking are available in two commonly used scales, Farenheit and Celsius or Centigrade. These scales are based on the freezing and boiling temperatures of pure water at sea level (32°F/0°C and 212°F/100°C respectively).

Basic heat sources for preparing foods are electricity, gas, wood, and coal. Heat is transferred in the following ways:

Conduction: The direct transfer of heat from one substance to another by direct contact; for example, heat from a gas flame warms the pot on the stove and then its contents.

Convection: Air or liquid expands and rises as it heats up, creating a circular current. Oven baking, simmering, steaming, and deep-fat frying are all examples of convection cooking. The use of water and fat to heat foods combines both conduction and convection.

Induction: Transferring heat energy to adjacent material without contact.

Radiation: Radiant heat in the form of particle waves moving outward is generated by broiling, grilling, and microwaving. Infrared heat lamps and ovens use electromagnetic waves to keep foods warm and prepare frozen foods.

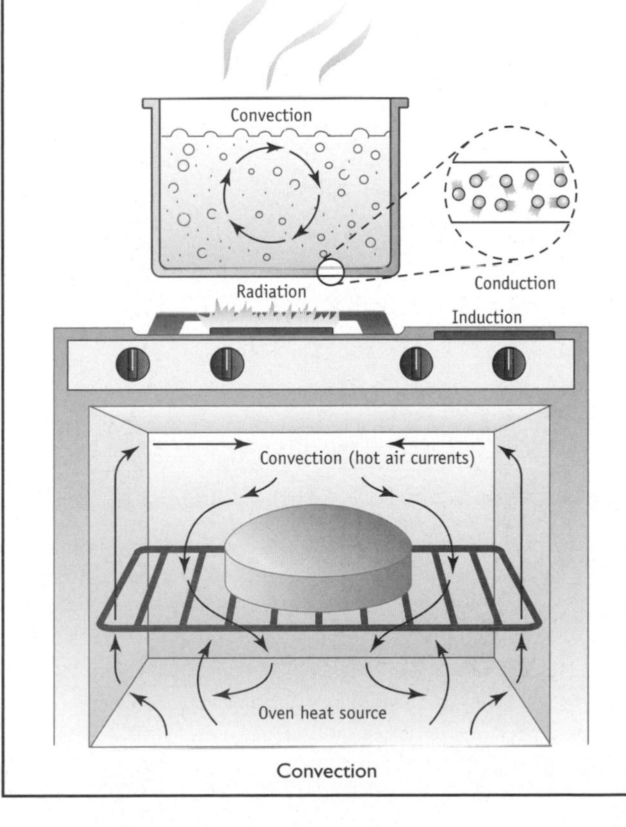

Convection

FOOD PREPARATION EQUIPMENT

Primary kitchen equipment:

- Ranges, ovens
- Refrigerators, freezers
- Dishwashers

Auxiliary equipment:

- Coffee makers, toasters
- Griddles, broilers, grills
- Steamers, woks
- Deep-fat fryers
- Food processors
- Crockpots
- Mixers, blenders
- Grinders, choppers
- Slicers

Pots and Pans | **Knives**

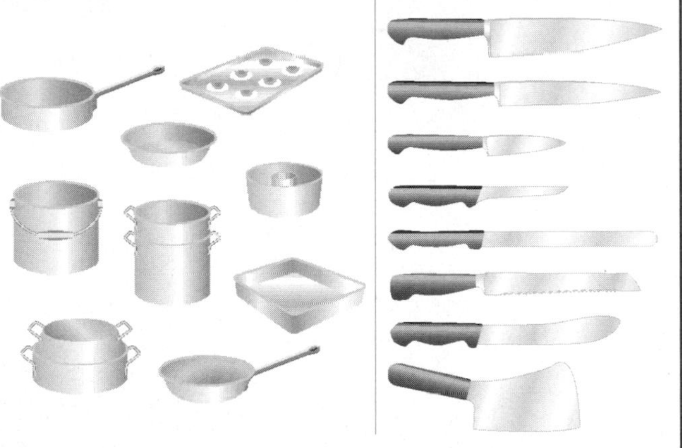

Measuring: Consistency in measuring and mixing techniques can make all the difference in a recipe's success or failure. It is important to understand the following measuring terms:

Mass vs volume: Weight, commonly used to mean mass, is a much more accurate measurement than volume and, as a result, is used by many food service operations. Confusion can occur because ounces can be measured either by volume, known as fluid ounces (fl), or by mass (weight), known as avoirdupois ounces (av). Water is the only substance with a fluid ounce equal to its avoirdupois ounce.

Metric vs nonmetric: Metric measurements of volume are expressed in milliliters. A 250-ml metric cup is close to a nonmetric cup, which holds 236.59 ml. The 15 and 5 ml metric measures are almost equal to the nonmetric tablespoon and teaspoon respectively.

REFERENCES

1. Abrisham SH, et al. Bacterial adherence and viability on cutting board surfaces. *Journal of Food Safety* 14:153–172, 1994.
2. Albert A. Choosing great knives for confident, skillful cooking. *Fine Cooking* 24:50–53, 1998.
3. American Home Economics Association. *Handbook of Food Preparation.* American Home Economics Association, 1993.
4. Ashley S, and S Anderson. *Catering for Large Numbers.* Reed International Books, 1993.
5. Birchfield JC. *Design and Layout of Foodservice Facilities.* Van Nostrand Reinhold, 1988.
6. Crespo FL, and HW Ockerman. Thermal destruction of microorganisms in meat by microwave and conventional cooking. *Journal of Food Protection* 40(7):442, 1977.
7. Giese J. Advances in microwave food processing. *Food Technology* 46(9):118–122, 1992.
8. Klein BP. Retention of nutrients in microwave-cooked foods. *Boletin: Asociacion Medica de Puerto Rico* 81(7):277–279, 1989.
9. Lydecker T. Cook to perfection with an instant-read thermometer. *Fine Cooking* 19:60–61, 1997.
10. Mermelstein NH. How Food Technology covered microwaves over the years. *Food Technology* 51(5):82, 1997.
11. Microwave meals attract children. *Journal of the American Dietetic Association* 91(6):720, 1991.
12. Middleton S. Pros pick the best baking sheets. *Fine Cooking* 26:55–57, 1998.
13. Ovadia DZ, and C Walker. Impingement in food processing. *Food Technology* 52(4):46–50, 1998.
14. Partnership yields new Maillard flavor systems for microwave foods. *Food Engineering* 65(6):36–37, 1993.
15. Pickett MS, MG Arnold, and LE Ketterer. *Household Equipment in Residential Design.* Waveland, 1986.
16. Shugart G, and M Molt. *Food for Fifty.* Macmillan, 1993.
17. Stevens M. Choosing the best whisk. *Fine Cooking* 19:72, 1997.
18. What happened to microwave foods? *Prepared Foods* 160(1):60–61, 1991.
19. Yiu SH, J Wesz, and PJ Wood. Comparison of the effects of microwave and conventional cooking on starch and beta-glucan in rolled oats. *Cereal Chemistry* 6:37235, 1.

Food Preparation Basics

"Cooking is a craft which can rise, on occasion, to an art," said Arno Schmidt, once an executive chef of the Waldorf-Astoria Hotel in New York City (18). Understanding the basics of food preparation is essential to getting a meal together, but because it is not an exact science, no matter how knowledgeable and careful the food preparer is, results vary from meal to meal. Schmidt further said it is "no wonder that seemingly similar foods taste and act differently depending upon endless factors." He cites as an example the many variables that influence the cooking of two pieces of meat. Though they both may be the same grade and size and cooked in the same way, they will not necessarily be identical when they get to the table.

Factors contributing to differences in prepared food include the type of heat used, the cooking utensils, the amount of food prepared, the fact that a pinch of freshly dried herbs can be stronger than even a larger amount of stale herbs, and that a cup of fresh leeks tastes more potent than twice the quantity of stale leeks. Add to the equation the foibles of human nature and the unique tastes and preferences of individuals, and it is easy to see how two chefs following the same recipe could come up with different products.

Food preparation most definitely approaches art at times, but until its basic techniques are learned and mastered, the results will more nearly resemble preschool finger painting than the work of the Great Masters. The purpose of this chapter is to describe the basic heating methods in food preparation, cutlery techniques, measuring and mixing techniques, the proper use of seasonings and flavorings, and the guidelines of food presentation.

Methods of Heating Foods

Heating not only destroys microorganisms that cause illness but changes the molecular structure of foods, altering their texture, taste, odor, and appearance. During food preparation, heat is transferred by either moist- or dry-heat methods. Regardless of which method is used, food should never be left unattended while it is cooking because that is the number-one cause of fire in the kitchen.

Moist-Heat Preparation

Moist-heat preparation techniques include scalding, poaching, simmering, stewing, braising, boiling, parboiling, blanching, and steaming. In these methods, liquids are used not only to heat the food, but may also contribute flavor, color, texture, and appearance to the final product. This is especially the case if broth and mixtures containing herbs, spices, and seasonings have been added. Moist-heat preparation helps to soften the fibrous protein in meats and the cellulose in plants, making them more tender. Liquids generated from heating foods can also be used as a flavorful stock to

KEY TERMS

Moist-heat preparation A method of cooking in which heat is transferred by water, any water-based liquid, or steam.

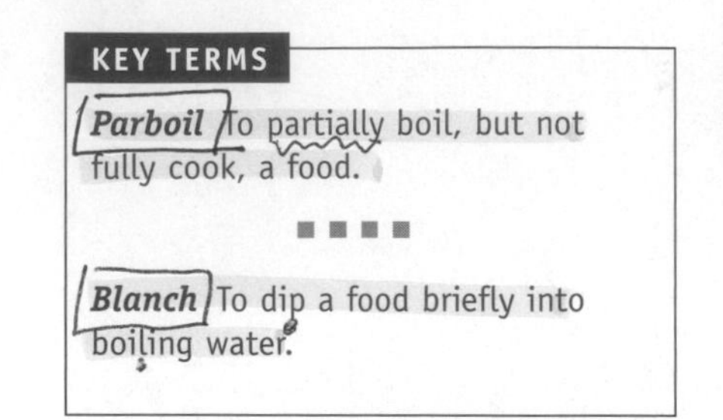

KEY TERMS

Parboil To partially boil, but not fully cook, a food.

■ ■ ■ ■

Blanch To dip a food briefly into boiling water.

make soups or sauces. One possible drawback to moist-heat methods is that color, flavor compounds, vitamins, and minerals may leach out and be lost in the liquid. However, using this liquid in a sauce or gravy retains them in the diet.

The various moist-heat preparation methods are presented below in order of increasing heat requirements, ranging from a low heat of 150°F (66°C) for scalding water to a high heat of 240°F (116°C) for pressure steaming.

Scalding. Scalding water reaches a temperature of 150°F (66°C). It is indicated by the appearance of large, but relatively still, bubbles on the bottom and sides of the pan. This process was most frequently used with milk to improve its function in recipes and to destroy bacteria. Pasteurized milk does not need to be scalded, even though many older recipes call for scalded milk. Recipes now use scalded milk to speed the combination of ingredients; in hot milk sugar dissolves more readily, butter and chocolate melt more easily, and flour mixes in more evenly without creating lumps (23).

Poaching. Water heated to a temperature of 160° to 180°F (71° to 82°C) is used for poaching in which the food is either partially or totally immersed. The water is hotter than scalding, but has not yet reached the point of actually bubbling, although small, relatively motionless bubbles appear on the bottom of the pan. Poaching is used to prepare delicate foods, like fish and eggs, which could break apart under the more vigorous action of boiling.

Simmering. Water simmers at just below the boiling point, never less than 180°F (82°C). Simmering is characterized by gently rising bubbles that barely break the surface. Many food dishes, especially rice, soups, and stews, are first brought to a boil and then simmered for the remainder of the heating time. Simmering is preferred over boiling in many cases because it is more gentle and will usually not physically damage the food, and foods will not overcook as quickly as when boiled. The lower heat of a simmer is essential when cooking tough cuts of meat that require gentle cooking in order to become tender.

Stewing. Stewing refers to simmering ingredients in a small to moderate amount of liquid, which often becomes a sauce with the food. Most stew dishes consist of chopped ingredients such as meat (often browned first) and vegetables placed in a large casserole or stock pot with some water, stock, or other liquid. The pot is covered and the food simmered for some time on the range or in a moderate oven. Stews often taste better the day after their initial preparation, because the overnight rest deepens their flavors.

Braising. Braising is similar to stewing in that food is simmered in a small amount of liquid in a covered casserole or pot. The liquid may be the food's own juices, fat, soup stock, and/or wine.

Flavors blend and intensify as foods are slowly braised on top of the range or in an oven (1).

HOW & WHY? ?????????

Why are stewing and braising called by different names if both entail simmering food in a small amount of liquid?

The primary difference between stewing and braising is that stewing generally refers to smaller pieces of meat, while braising entails larger cuts. Stews are also most often made with more liquid and served in their sauce.

In order to generate a browner color and better flavor, meats are frequently browned with a dry-heat method like sautéing before being braised. Frequently, when braising meats, the vegetables are often added during the final cooking in order to preserve some of their texture and flavor.

Boiling. In order to boil, water must reach 212°F (100°C) at sea level, a temperature at which water bubbles rapidly. The difference in the bubbles between poaching, simmering, and boiling is shown in Figure 6-1. The high temperature and agitation of boiling water are reserved for the tougher-textured vegetables and for dried pastas and beans. A common technique is to bring a liquid to a rolling boil, gradually add the food, distributing it evenly, and then bring the liquid back to a full boil before reducing the heat so that boiling becomes gentle. A lid on the pot or pan will bring the liquid to a boil more quickly by increasing the pressure. It is always recommended to reduce the heat setting once a boil has been reached, because food will not cook any faster at a higher setting than at the one required to maintain a gentle boil. Spill-overs, burns, and loss of cooking liquid from evaporation can be avoided if a gentle boil is used.

Food may also be **parboiled** in boiling water, after which it is removed and its cooking completed either at a later time or by a different heating method. Parboiling is used frequently in restaurant service when food must be prepared in advance and finished to order. Another use for boiling water is for **blanching**, which sets the color of green vegetables, loosens the skins of fruits, vegetables, and nuts for peeling, and destroys enzymes that contribute to deterioration. Foods are often blanched before being canned or frozen.

Steaming. Any food heated by direct contact with the steam generated by boiling water has been steamed. Cooked vegetables are at their best when steamed because this method helps to retain texture, color, taste, and nutrients. A common method for steaming is to place food in a rack or steamer basket above boiling water and to cover the pot or pan with a lid in order to trap the steam. An indirect technique, called *en papillote* (on pap-ee-yote), is to wrap the

FIGURE 6-1 **Bubble size and movement differ during poaching, simmering, and boiling.**

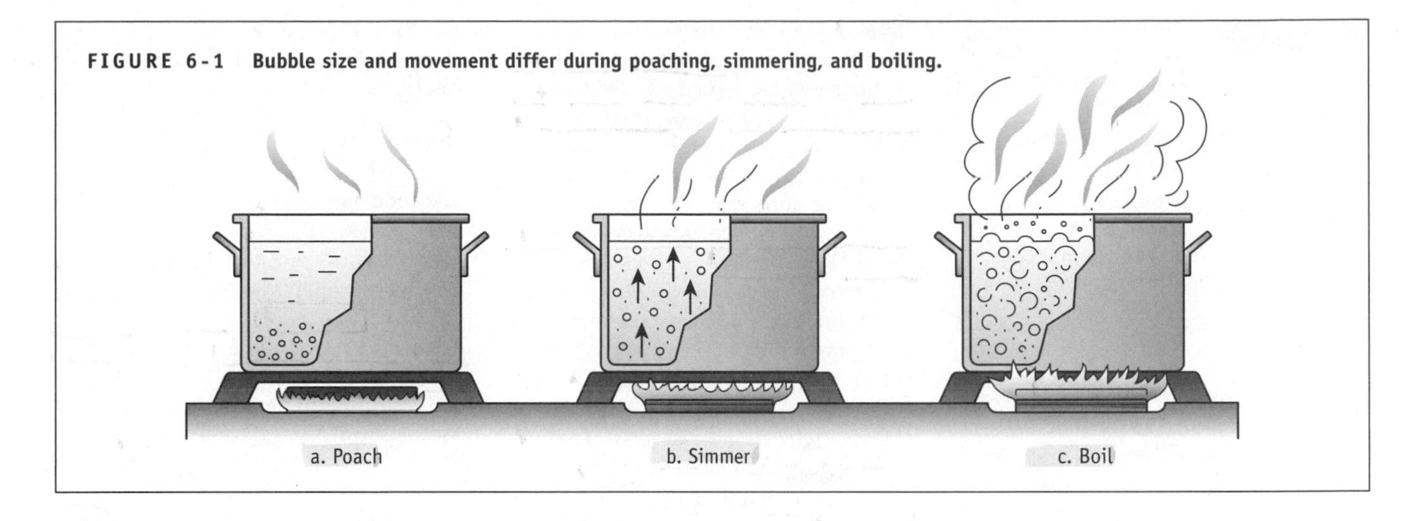

food in foil or parchment paper before it is baked or grilled. Then, in an oven or over the grill, the food cooks by the steam of its own juices, which are trapped in the packet. In a microwave oven, covering foods with plastic wrap facilitates steaming. Pressure cookers heat food by holding steam in an enclosed container under pressure. The temperature increases with increasing pounds of pressure per square inch.

Microwaving. While microwave preparation is listed under moist-heat preparation, it actually belongs in an entirely separate category because it incorporates both dry- (radiation) and moist-heat preparation methods. Microwaves are a form of radiation aimed at the water in the food or beverage. The specifics of preparing food using a microwave oven are discussed in Chapter 5 as well as in chapters on specific foods.

Dry-Heat Preparation

Examples of **dry-heat preparation** include baking, roasting, broiling, grilling, barbequing, and frying. Higher temperatures are reached in dry-heat preparation than in moist-heat methods, because water can heat only to its boiling point of 212°F (100°C), or slightly higher under pressure, while ovens can reach up to 500°F (260°C).

Baking. Baking is the heating of food by hot air in an oven. The average baking temperature is 350°F (177°C), although temperatures may range from 300° to 425°F (149° to 219°C). Baking results can be affected by rack position and the color of the pan.

Rack Position. For the best outcome, the food should be placed in the middle of the center rack (Figure 6-2). Foods placed on the uppermost rack may brown excessively on their top surface, while on the lowest rack foods are prone to burning on the bottom. It is also best to position foods using only one rack; if this is not possible, the foods should be staggered so that they are not directly over each other in order to allow hot air to flow more freely through the oven. At least 2 inches should be left between pans and between the pans and the oven walls. If these guidelines are ignored, the resulting inadequate air circulation may cause uneven browning, and food may not be thoroughly cooked.

Pan Color. In addition to rack position and placement of pans, the cooking pan material will affect the baking outcome. Shiny metal pans reflect heat and are best for cakes or cookies, where only light browning and a soft crust are desired. The darker, duller metal pans (including anodized and satin-finish) tend to absorb heat, resulting in browner, crisper crusts

FIGURE 6-2 **Oven rack positions.**

POSITION:	USED FOR:
RACK 5	Toasting bread, or for two-rack baking.
RACK 4	Most broiling and two-rack baking.
RACK 3	Most baked goods on a cookie sheet or jelly roll pan, or frozen convenience foods, or for two-rack baking.
RACK 2	Roasting large cuts of meat and large poultry, pies, soufflés, or for two-rack baking.
RACK 1	Roasting large cuts of meat and large poultry, pies, soufflés, or for two-rack baking.

5
4
3
2
1

KEY TERMS

Dry-heat preparation A method of cooking in which heat is transferred by air, radiation, fat, or metal.

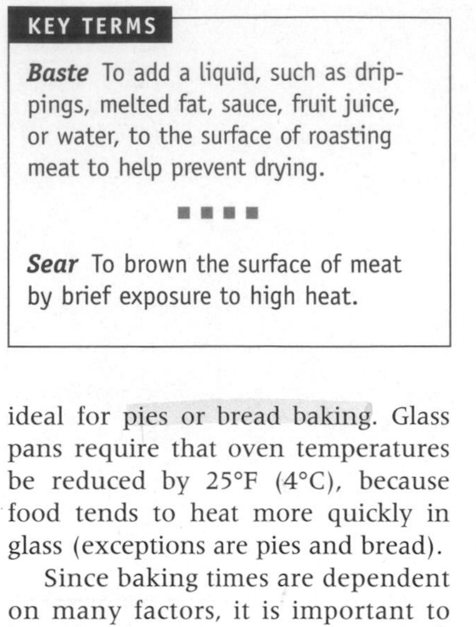
KEY TERMS

Baste To add a liquid, such as drippings, melted fat, sauce, fruit juice, or water, to the surface of roasting meat to help prevent drying.

Sear To brown the surface of meat by brief exposure to high heat.

ideal for pies or bread baking. Glass pans require that oven temperatures be reduced by 25°F (4°C), because food tends to heat more quickly in glass (exceptions are pies and bread).

Since baking times are dependent on many factors, it is important to check the food's progress at the suggested minimum baking time and then at intervals after that until the food is done. This must be done judiciously, however, because checking too soon or too frequently will allow heat and/or steam to escape from the oven, adversely affecting the baking outcome.

Roasting. Roasting is similar to baking except that the term is usually applied to meats and poultry. Roasted meats are often **basted** every 20 minutes or so to prevent the food from drying out. Some roasted meats are initially **seared** at 400° to 450°F (200° to 230°C) for about 15 minutes before reducing the heat to normal roasting temperatures. Although searing adds a desirable texture, color, and flavor to the meat's outer surface, roasts cooked at lower temperatures are juicier, shrink less, and are easier to carve than those that are seared.

"Baked" or "Roasted"? The term "roasting" can also refer to cooking on an open fire, as with roasted marshmallows and vegetables, and to cooking with a rotisserie. To make things even more confusing, "meats" such as ham, meat loaf, and fish are often referred to as "baked." Chicken may be described as either baked or roasted.

Broiling. To broil is to cook foods under an intense heat source. The high temperatures of broiling cook foods in approximately 5 to 10 minutes, so only tender meats, poultry, and fish are broiled; tougher foods require longer heating times. Temperature is controlled by moving the rack closer or farther away from the heat source. Thicker cuts are broiled farther from the heat, thinner ones closer—on the fourth or fifth rack of a home oven. Foods are often slightly oiled to prevent drying and sticking, placed under the broiler only after it has been preheated to its full heat, and then turned over only once. Food service operations often employ a "salamander," also called a cheese-melter, a low-intensity broiler used just prior to serving to melt or brown the top layer of a dish.

Grilling. Grilling is the reverse of broiling, in that food is cooked above, rather than below, an intense heat source. The grill may be a rack or a flat surface on a stove. "Grilled" can also refer to foods that are seared on a grill over direct heat (20).

Barbecuing. Barbecuing and grilling are no longer used to refer to the same heating method. Grilling over a pit used to be known as barbecuing, but now the latter term stands alone. Barbecuing now refers to foods being slow-cooked, usually covered in a zesty sauce, over a longer period of time (20). The temperature in barbecuing is regulated by adjusting the intensity of the heat source (charcoal, wood, gas, or electric), the distance between the food and the heat source, and by moving the food to different places on the grill.

Frying. Frying is heating foods in fat. Oils used in frying serve to transfer heat, act as a lubricant to prevent sticking, and contribute to flavor, browning, and a crisp outside texture (27). Although oils are "liquid," frying is a method of dry-heat preparation because pure fat contains no water. Types of frying—sautéing, stir-frying, pan-broiling, pan-frying, and deep-frying—are distinguished by the amount of fat used, ranging from a thin sheet to complete submersion. Temperatures vary among the different methods: sautéing, stir-frying, and pan-frying require only a medium or high heat—lower heat results in higher fat absorption—while deep-frying temperatures range from 350° to 450°F (177° to 232°C). Chapter 10 discusses frying with fats in greater detail.

Sautéing and Stir-Frying. These methods use the least amount of fat to heat the food. Stir-frying is predominantly used in Asian cooking; the pan is held stationary, while the food is stirred and turned over very quickly with utensils. Sautéing is done in a frying pan, a special sauté pan, or on a griddle. The foods most frequently prepared on a griddle with a little fat are eggs, pancakes, and hamburgers (with the fat derived from the meat itself).

Pan-Broiling and Pan-Frying. Pan-broiling refers to placing food, usually meat, in a very hot frying pan with no added fat and pouring off fat as it accumulates. If the fat is not poured off, pan-broiling becomes pan-frying, which uses a moderate amount of fat (up to ½-inch deep), but not enough to completely cover the food.

Deep-Frying. In deep-frying, the food is completely covered with fat. Many deep-fried foods are first coated with breading or batter to enhance moisture retention, flavor development, tenderness, browning, crispness, and overall appearance. The characteristics of the coating influence a fried food's final outcome (17, 19). A fine-crumb breading absorbs less fat, but a coarser grain produces a crisper texture. Sugar in the coating speeds up browning, but this is undesirable if the outside browns and appears done

FIGURE 6-3 Sugar in the batter darkens fried food, as seen in the fritter on the right.

while the inside is still uncooked (Figure 6-3). While the breading or batter protects the food from absorbing too much fat, it can also simultaneously protect the deep-frying oil from the deterioration that occurs when it contacts the food's natural moisture and salt content.

Because deep-frying requires high temperatures, it is best to rely on the fryer's thermostat to obtain the desired temperature. If this is not possible, another method is to place a 1-inch cube of white bread into the oil and time how long it takes to turn golden brown (Table 6-1).

FIGURE 6-4 Technique of holding a chef's knife.

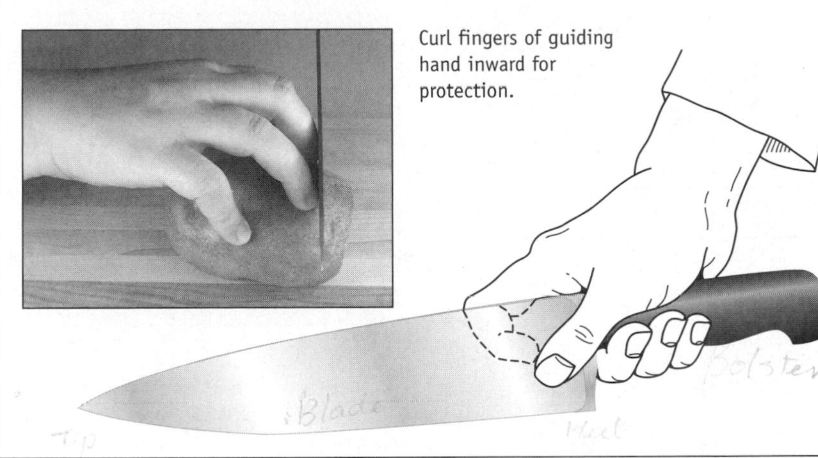

Cutlery Techniques

Another basic pillar of food preparation is the knowledge and use of cutlery. Knives and their selection were discussed in Chapter 5. The following sections cover their handling and the styles of cutting food. The techniques vary according to the type of knife selected, and this depends on the task to be performed.

Handling Knives

The most frequently used knife is the chef's or French knife. The positioning of the grip and of the food under the blade both influence the degree of control and leverage a person has over the knife (Figure 6-4). A chef's knife should be firmly held with the base of the blade between the thumb and forefinger and the other fingers wrapped around the handle. While one hand grips the knife, the other hand must hold the food and guide it toward the blade. Curling the fingers of the guiding hand under while holding the food allows the knuckles to act as a protective shield and keeps the fingertips away from the cutting edge. It is best to allow at least a half-inch barrier of food between the blade and the fingers holding the food.

Different sections of a blade are used for different tasks (Figure 6-5). Light tasks such as cutting out the stem end of a tomato can usually be accomplished with the tip of the blade, or even better, with a knife more suitable to small tasks, such as a paring knife. Heavy duties such as chopping off tough carrot ends are better accomplished by making use of the weight and thickness found at the base of the blade. Most other cutting tasks are carried out using the center of the blade.

TABLE 6-1
Deep-Fat Frying Based on Color

Temperature of Fat	*Approximate Seconds to Turn a 1-Inch Bread Cube Golden Brown*
385–395°F (196–201°C)	20
375–385°F (190–196°C)	40
365–375°F (185–190°C)	50
355–365°F (179–185°C)	60

Cutting Styles

Uniformity is the usual goal in cutting food. It allows for even heating and gives food an appetizing appearance. Cutting styles include slicing, shredding, dicing (cubing), mincing, and peeling.

- *Slice.* To move the food under the blade while keeping the point of the blade firmly on the cutting board. The base of the knife is lifted up and down with a forward and backward motion (Figure 6-6).

FIGURE 6-5 Blade position determined by the cutting task.

Tip—delicate tender work.

Center—all-purpose work.

Heel—heavy work.

FIGURE 6-6 **Slicing technique.**

Start with the blade tip down.

Press down and forward simultaneously to slice.

As soon as the blade heel touches the board, it is moved up and over (keep tip down) for the next slice.

- *Julienne.* Sliced food can be further cut up, or **julienned**, resulting in delicate sticks that are usually 1 to 3 inches long and only 1/16 to 1/8 inches thick (Figure 6-7).
- *Shred.* To cut leaf vegetables into thin strips. This may be done by first rolling the leaves into cigar-like shapes and then cutting them into shreds. Hand shredders and food processors with different sizes of shredding blades may also be used.
- *Dice.* To cut food into even-sized cubes.

KEY TERMS

Julienne To cut food lengthwise into very thin, stick-like shapes.

FIGURE 6-7 **Julienne slices.**

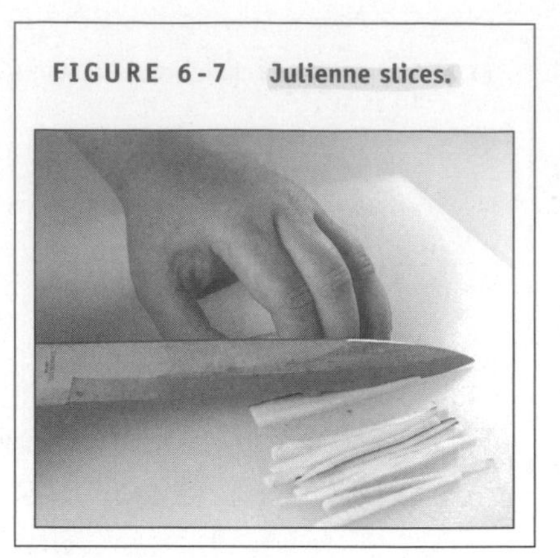

- *Mince.* To chop food into very fine pieces. This is done by placing the holding hand on the tip of the knife and rocking the base up and down in short strokes while moving it across the food several times, and then repeating as necessary. Figure 6-8 illustrates how to dice and mince an onion.
- *Peeling.* To remove the skin. The peel and rind can be cut from an orange or any thick-skinned fruit by first cutting off in a circular fashion the top of the fruit's skin, then scoring the skin through to the flesh of the fruit in four places. The skin can then be peeled in segments down from the top. Fruits can also be peeled directly with a paring knife (Figure 6-9). Avocados can be stripped of their peel by cutting the avocado from stem to stern through to the pit. Each half is cupped in the hands and twisted gently to separate the halves. The seed (nut) can be removed with the fingers or the tip of a sharp knife. At this point the avocado can be scooped out with a large serving spoon or peeled and sliced (Figure 6-10).

FIGURE 6-8 **Dicing an onion; further cuts result in mincing.**

First vertical cuts
(do not go all the way through onion)

Second vertical cuts
(go entirely through onion)

FIGURE 6-9 **Peeling with a paring knife.**

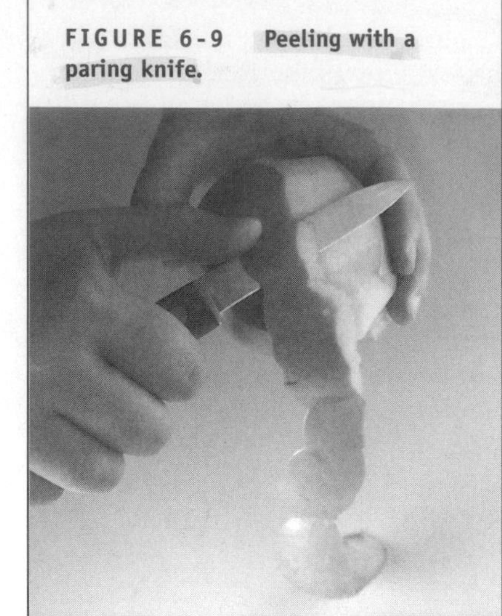

Measuring Ingredients

Correct measuring is essential to basic food preparation. The three major steps in measuring are:

1. Approximating the amount required for a specific measurement (e.g., 4 ounces of cheese yields 1 cup shredded)
2. Selecting the right measuring utensil
3. Accurate measuring technique

Whether an ingredient is liquid or dry determines the kind of measuring utensil that will be used. A graduated measuring cup with a lip for pouring is best for measuring liquid ingredients. Sets of flat-topped measuring cups are reserved for measuring dry ingredients. All dry ingredients are best measured by first stirring them to eliminate any packing or lumps. Amounts less than 1/4 cup should be measured with measuring spoons. Sifting flour with dry ingredients such as baking soda or salt is an efficient way to blend and distribute the ingredients evenly.

Whether using liquid or dry measuring cups, it is important to be able to use measuring utensils interchangeably,

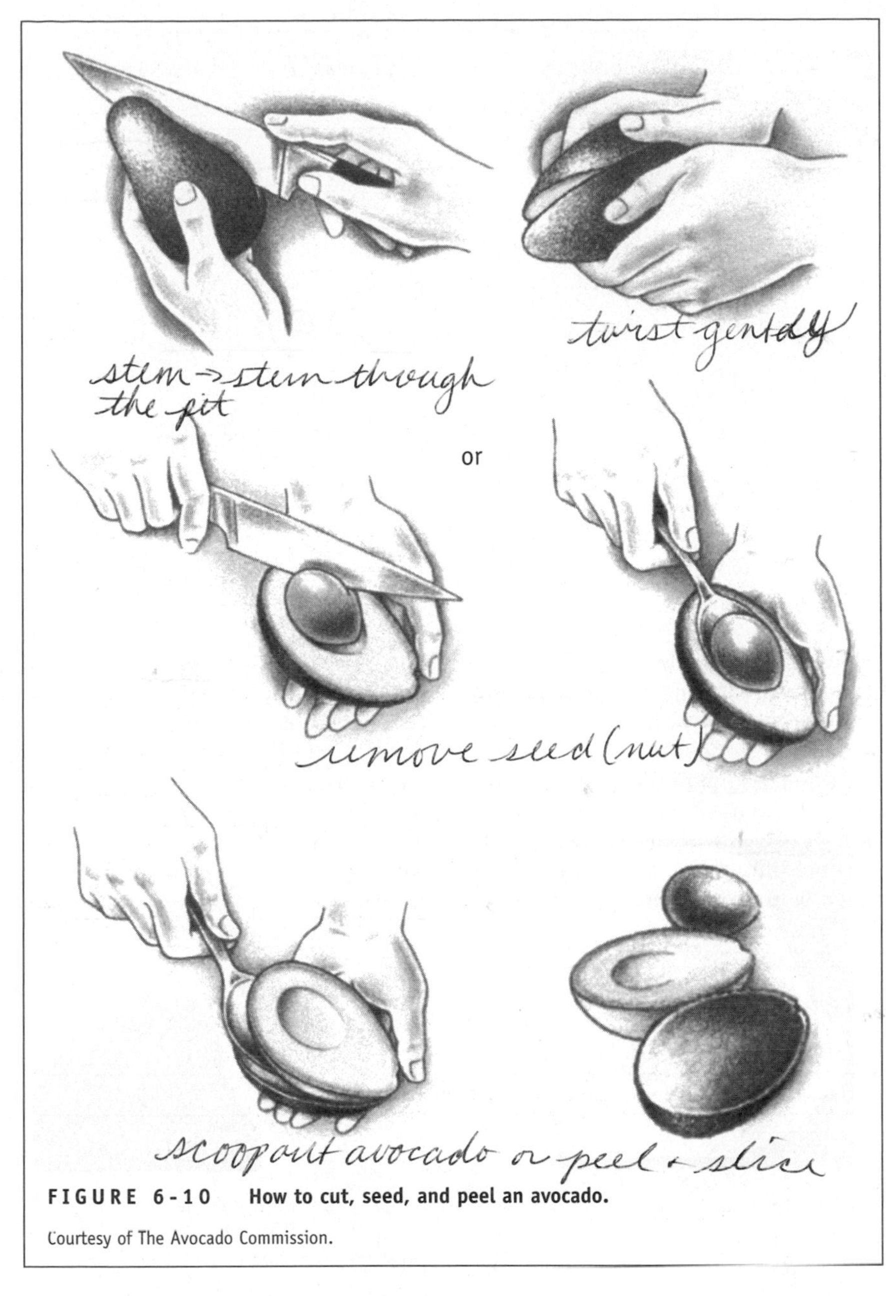

FIGURE 6-10 **How to cut, seed, and peel an avocado.**

Courtesy of The Avocado Commission.

and this is easy if a few basic equivalents are remembered (see Table 6-2, and inside back cover).

1 teaspoon	=	about 5 grams*
1 tablespoon	=	3 teaspoons
2 tablespoons	=	1 fluid ounce or 28.35 grams
¼ cup	=	2 fluid ounces
½ cup	=	4 fluid ounces
1 cup	=	8 fluid ounces, 16 tablespoons, or 48 teaspoons
1 pint	=	2 cups, or 16 fluid ounces (1 pound)

*When weighing water.

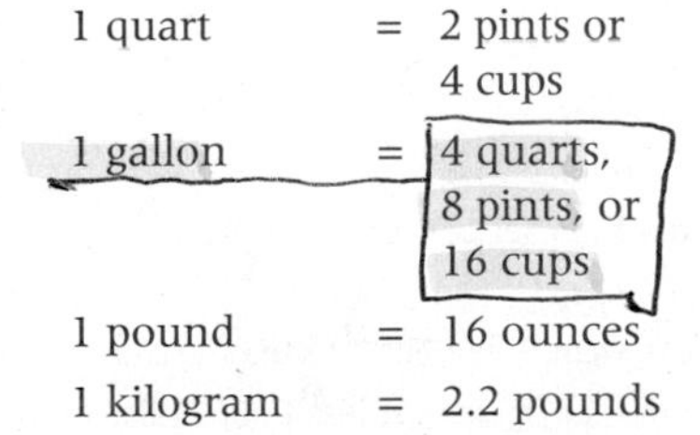

1 quart	=	2 pints or 4 cups
1 gallon	=	4 quarts, 8 pints, or 16 cups
1 pound	=	16 ounces
1 kilogram	=	2.2 pounds

Knowing the general units used in measuring allows for the next step required for accuracy—using the largest measuring device possible. For example, 3 teaspoons of sugar should be measured using 1 tablespoon; ¾ cup should be measured using ½ cup plus a ¼ cup. Accuracy is also achieved by using the guide provided in Table 6-3 for rounding off weights and measures. For even better accuracy, scales may be used to measure ingredients.

Specific volume-measuring techniques for liquids, eggs, fat, sugar, and flour are discussed below.

Liquid

Only transparent graduated measuring cups with pouring lips should be used to measure liquids. The cup should be on a flat surface and all reading done at eye level in order to accurately read the line at the bottom of the meniscus. The exception is milk, which is read at the top meniscus. Viscous liquids (such as honey, oil, syrup, and molasses) have a tendency to stick to the sides as they are poured, so the amount that was measured is diminished by the amount that stuck to the sides. Should this happen, a rubber scraper can be used to remove the remaining contents.

Eggs

Eggs range in size from "pee wee" to "jumbo," but most standard recipes are based on "large" size eggs, if not specified. When half an egg or less is called for, it can be measured by beating a whole egg into a homogeneous liquid, which can then be divided in half or smaller increments. When measuring eggs, it is helpful to remember the following volume equivalents:

- One large egg = 2 ounces
- Four large eggs = 7 ounces (just under 1 cup)
- Eight to ten egg whites, or twelve to fourteen yolks = 1 cup

Fat

Manufacturers of butter and margarine have made it easy to measure their products. Both usually come in 1-pound packages that contain four ¼-pound sticks, with each stick equivalent to ½ cup. Thus, 1 pound of butter is equivalent to approximately 2 cups. The same weight of vegetable shortening, on the other hand, is equivalent to 2½ cups by volume. The wrappings of the ¼-pound sticks are usually further marked into eight 1-tablespoon segments.

Different methods are used to measure liquid and solid fats. Liquid fats such as oil and melted butter are mea-

TABLE 6-2
Basic Measuring Equivalents

Teaspoon (tsp)	*Tablespoon (T)*	*Ounce (oz)*	*Cup (C)*	*Pint*	*Quart*	*Gallon*
⅛ tsp (dash)						
¼ tsp						
1 tsp						
3 tsp =	1 T					
	2 T =	1 oz				
		2 oz =	¼ C			
		4 oz =	½ C			
		8 oz =	1 C			
		16 oz =	2 C =	1 pint		
			4 C =	2 pints =	1 qt	
			16 C =	8 pints =	4 qt =	1 gallon

sured in glass measuring cups. Solid fats such as lard, shortening, butter, and margarine should be removed from the refrigerator and allowed to become **plastic** at room temperature. Once pliable and soft, they can be pressed into a fractional metal measuring cup with a rubber scraper. The fat should be pressed down firmly to remove any air bubbles and the top of the cup leveled with the straight edge of a spatula. As with liquids, amounts under ¼ cup should be measured with measuring spoons.

Solid fats may also be measured by using the water-displacement method. For example, if ½ cup of fat is required, a 1-cup liquid measuring cup is filled with cold water to ½ cup. The fat is added and pressed below the water line until the water line reaches the 1-cup measuring line. The colder the water, the easier the cleanup, because cold fat is less likely to stick to the sides of the cup. Some water may cling to the fat and should be shaken free or patted away lightly with a paper towel.

Sugar

The amount of sugar needed depends on its type—granulated white sugar, brown sugar, or confectioners' sugar (powdered or icing). Measuring methods differ among these sugars, because 1 pound of each yields 2, 2¼, and 4½ (sifted) cups respectively.

White granulated sugar is usually poured into fractional measuring cups and leveled with a spatula. If it becomes lumpy, it can be mashed and sifted before measuring. Brown sugar has a tendency to pack down and become hard because it contains 2 percent moisture, which has a tendency to dry out. Lumping can be prevented by placing the brown sugar in an airtight container and storing it in the refrigerator or freezer. Hardened brown sugar can be softened by placing it in a microwave oven for a few seconds, or

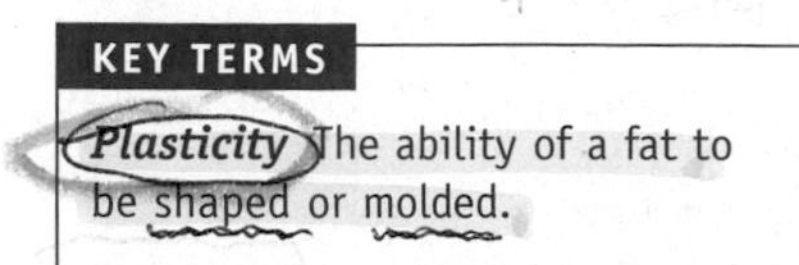
KEY TERMS

Plasticity The ability of a fat to be shaped or molded.

TABLE 6-3
Guide to Rounding Off Weights and Measures

If the total amount of an ingredient is:	*Round it to:*
WEIGHTS	
Less than 2 oz	Measure unless weight is ¼-, ½-, or ¾-oz amounts
2–10 oz	Closest ¼ oz or convert to measure
More than 10 oz but less than 2 lb 8 oz	Closest ½ oz
2 lb 8 oz–5 lb	Closest full ounce
More than 5 lb	Closest ¼ lb
MEASURES	
Less than 1 T	Closest ⅛ tsp
More than 1 T but less than 3 T	Closest ¼ tsp
3 T–½ cup	Closest ½ tsp or convert to weight
More than ½ cup but less than ¾ cup	Closest full tsp or convert to weight
More than ¾ cup but less than 2 cups	Closest full tsp or convert to weight
2 cups–2 qt	Nearest ¼ cup
More than 2 qt but less than 4 qt	Nearest ½ cup
1–2 gal	Nearest full cup or ¼ qt
More than 2 gal but less than 10 gal*	Nearest full quart
More than 10 gal but less than 20 gal*	Closest ½ gal
More than 20 gal*	Closest full gallon

* For baked goods or products in which accurate ratios are critical, always round to the nearest full cup or ¼ qt.

in a conventional oven set at about 200°F (93°C) for a few minutes. Brown sugar is best measured by pressing it firmly into a fractional metal measuring cup and leveling it. The packing should be firm enough that the brown sugar retains the shape of the measuring cup when it is turned out. Lump-free or free-flowing brown sugar, which weighs 25 percent less than regular brown sugar, is measured in the same manner as granulated white sugar.

Measuring Confectioners' Sugar. Confectioners' sugar must be sifted before measuring to break up any existing lumps. The light, airy nature of confectioners' sugar causes it to have a greater volume than the same amount of granulated sugar, which is why 1¾ cups of confectioners' sugar is equal to the weight of 1 cup of granulated sugar. After sifting, confectioners' sugar is measured the same way as granulated sugar.

Flour

White flour is one of the more difficult ingredients to measure accurately by volume, because its tiny particles not only vary in shape and size, but also have a tendency to pack. In addition, the various white flours differ in density, ranging from 88 grams per cup in soy flour to approximately 132 grams per cup in wheat flour. This influences the number of cups obtained from various flours of the same weight (Table 6-4). Although there is no standard weight for a cup of flour, 1 pound of all-purpose flour averages 4 cups. Professional bakers and chefs avoid the discrepancy in volume measurement by always weighing the flour.

White flour should be sifted before being lightly spooned into a fractional measuring cup and leveled with a spatula. The cup should never be tapped or shaken down, because doing so can pack the flour particles tightly, which may result in too much flour being used. To avoid sifting and still get consistent baking results with regular white flour, one technique is to remove 2 tablespoons from each cup of unsifted flour (12).

Not all flours are sifted prior to being used. Whole-grain and graham flours and meal should not be sifted, because sifting will remove the bran particles. These flours should simply be lightly stirred before being scooped into a fractional measuring cup. Presifted or instantized flours (discussed in Chapter 22) have already been processed into uniform particles and should not be sifted. Instantized flour should not be used in baked products.

Other Ingredients and Substitutions

Some foods to be measured do not fall into the basic categories described above. The methods for measuring foods such as cheese, nuts, chocolate, and garlic depend on their form—whole, cubed, shredded, minced, etc. Foods cut into pieces tend to occupy a greater volume. For example, 1 pound of cheese is equivalent to approximately 2 cups, but 1 pound of grated cheese is equivalent to approximately 4 cups. The following basic recipe amounts are helpful:

- Three apples usually equal about 1 pound, and six apples are needed for the average apple pie.
- A medium orange or lemon yields up to a ½ cup of juice.
- A medium orange yields approximately 1 to 2 tablespoons of grated rind (zest) while a lemon yields only ½ to 1 tablespoon zest. Appendix B gives the volume-to-weight equivalents of different types of foods.
- Another helpful reference is the number of cups found in common can sizes (inside back cover).

Often a basic ingredient turns up missing in the preparation of a dish. This can put a halt to food preparation, but in some situations, a substitution may save the day. Appendix C lists some substitutions that can be made for standard ingredients.

Mixing Techniques

"Mixing" is a general term that includes beating, blending, binding, creaming, whipping, and folding. In mixing, two or more ingredients are evenly dispersed in one another until they become one product. This, in general, is what the other processes accomplish also, but there are distinctions.

- *Beat.* The ingredients are moved vigorously in a back-and-forth, up-and-down, and around-and-around motion until they are smooth.
- *Blend.* Ingredients are mixed so thoroughly that they become one.
- *Bind.* Occurs when ingredients adhere to each other, as when breading is bound to fish.
- *Cream.* To beat fat and sugar together until they take on a light, airy texture.
- *Whip.* Very vigorous mixing, usually with a beater of some type, that incorporates air into such foods as whipping cream and egg whites.
- *Fold.* One ingredient is gently incorporated into another by hand with a large spoon or spatula.

TABLE 6-4
One Pound of Flour Varies in Volume (# cups) and Weight (grams) Depending on the Flour

Flour (1 lb)	*Volume (Approximate Cups)*	*Weight (Per Cup)*	
		gm	oz
All purpose (sifted)	4	115	4.2
Cake (sifted)	4½	96	3.4
Rice (sifted)	3½	126	4.4
Rye (sifted - light)	5	88	3.2
Rye (sifted - dark)	3½	128	4.6
Soy (low-fat)	5½	84	3.0
Whole wheat (stirred)	3⅓	132	4.8

There are many methods for combining the ingredients of cakes and other baked products, but the most commonly used are the conventional (creaming), conventional sponge, single-stage (quick-mix), pastry-blend, biscuit, and muffin methods. Methods for mixing yeast bread doughs are discussed in Chapter 24.

Conventional (Creaming) Method

The conventional method, also known as the creaming or cake method, is the most time consuming, and is the method most frequently used for mixing cake ingredients. It produces a fine-grained, velvety texture. The three basic steps are:

1. Creaming
2. Egg incorporation
3. Alternate addition of the dry and moist ingredients

The fat and sugar are creamed together by working the fat until it is light and foamy and then gradually adding small portions of sugar until all of it is well blended. A well-creamed combination of fat and sugar incorporates air while suspending sugar crystals and air bubbles in the fat. As the fat melts during baking, it creates air cells which migrate toward the liquid, resulting in a very fine-grained texture (15).

The eggs or egg yolks are then added one at a time to the creamed fat and sugar. An alternative method is to whip the egg whites separately and fold them into the cake batter after all the other ingredients have been mixed.

Finally, flour, baking powder or soda, and salt are sifted together with other dry ingredients such as cocoa in order to distribute the leavening agent evenly. The sifted dry ingredients, divided into three or five portions, are then added alternately with a liquid (usually milk) into the fat, sugar, and egg base. After one portion of dry ingredients has been incorporated, a portion of liquid is added and stirred or beaten until well blended. The process begins and ends with the dry ingredients.

Stirring Too Little or Too Much. As with any type of mixing method, too much or too little stirring can cause problems. Over-stirring a cake batter creates such a viscous mass that the cake may not be able to rise during baking, and the texture will tend to be fine but compact or lower in volume, full of tunnels, and have a peaked instead of a rounded top. Too little stirring can also result in a low-volume cake from an uneven distribution of the baking powder or soda or an incorporation of air into the foam. The texture of an under-stirred cake tends to contain large pores, have a crumbly grain, and brown excessively.

Conventional Sponge Method

The conventional sponge method, also known as the conventional meringue method, is identical to the creaming method except that a portion of the sugar is mixed in with the beaten egg or egg white, and the egg foam is folded into the batter in the end. The conventional sponge method is preferred for foam or sponge cakes because it contributes volume, and for baked goods made with soft fats whose creamed foam breaks and releases much of its incorporated air when egg yolks are added. In either case, the air in the foam that is folded in during the last stage increases volume.

Single-Stage Method

In the single-stage method, also known as the quick-mix, one-bowl, or dump method, all the dry and liquid ingredients are mixed together at once. Packaged mixes for cakes, biscuits, and other baked goods rely on the single-stage method. Only baked products containing higher proportions of sugar, liquid, and possibly an emulsifier in the shortening can be mixed by this method. Starting with the dry ingredients in a bowl, the fat (usually vegetable oil), part of the milk, and the flavoring are added and stirred for a specified number of strokes or amount of time (if an electric mixer is being used). The eggs and remaining liquid are then added, and the batter is mixed again for a specified period of time. The sequence and mixing of ingredients is important, because creaming is not a part of the process. To attain a uniform blend, the bottom and sides of the bowl should be scraped frequently. Quick-mix batters are more fluid than conventional batters.

Pastry-Blend Method

Fat is first cut into flour with a pastry blender, or with two knives crisscrossed against each other in a scissor-like fashion, to form a mealy fat-flour mixture. Half the milk and all of the sugar, baking powder, and salt are then blended into the fat-flour mixture. Lastly, eggs and more milk may then be blended into the mixture.

Biscuit Method

This method is similar to the pastry method except that all the dry ingredients—flour, salt, and leavening—are first combined. The fat is then cut into the flour mixture until it resembles coarse cornmeal. Liquid is added last. The dough is mixed just until moistened and not more or the biscuits will be tough.

Muffin Method

This is a simple, two-stage mixing method. The dry and moist ingredients are mixed separately and then combined and blended until the dry ingredients just become moist. Over-mixing will result in a tough baked product riddled with tunnels.

Seasonings and Flavorings

The most nutritious and beautifully presented meal in the world cannot be enjoyed unless it tastes good. Enhancing the flavor of foods is an art that is critical to the acceptability of foods, and a restaurant can succeed or fail depending on how that art is practiced. The most common reason for consumers to reject food is unacceptable flavor (16). **Seasonings** and **flavorings** help food taste its best.

They are rarely, however, capable of redeeming foods that are not of good quality to start with or of rejuvenating foods that have lost their quality during preparation. No amount of cinnamon will raise the flavor of an apple pie made from frozen apple slices to the level of one made from fresh, juicy apples.

Types of Seasonings and Flavorings

The number and variety of seasonings and flavorings available from all over the world would be nearly impossible to catalog, so this chapter focuses on the basics: salt, pepper, herbs and spices, flavor enhancers, oil extracts, marinades, batters, and condiments.

Salt. The value of salt was esteemed so highly in ancient times that the word "salary" is derived from "salt." Salt, or sodium chloride ($NaCl$ = 40 percent Na, 60 percent Cl), is the second most frequent food additive by weight. (Sugar, which is fully discussed in Chapter 9, is first.) Salt was originally introduced into foods as a preservative; salting, or curing, meat and fish was the only way to preserve food prior to refrigerators, freezers, or canning. The function of salt in foods expanded to those seen in Table 6-5. Salt in its most common form is a crystalline seasoning that may or may not be iodized and combined with an anticaking material.

Types of Salt. A variety of salts may be purchased including sea salt, rock salt, kosher salt, and a number of flavored salts, the most common being garlic, onion, and celery (Figure 6-11) (22). There are also some expensive and rare sea salts known as *fleur de sel* and *sel gris*, used only in the finest restaurants.

Adding Salt in Food Preparation. Regardless of the type, salt should be added in small increments because of the potential to overwhelm the taste buds when too much is added. The preparer should also be aware that any liquid such as a sauce or soup that will be reduced should be only lightly salted, because the salt will become even more concentrated as the volume of the liquid diminishes. Although removing excess salt is almost impossible, salty soup may be partially neutralized by adding a touch of sugar or by dropping in a raw, peeled potato to absorb some of the salt.

Processed Foods. Foods that are canned, frozen, cured, or pickled provide more than 75 percent of all the sodium ingested (13). Because high sodium intake has become a health concern (see the "Nutrient Content" box), food companies have many new processed food products that are now lower in sodium. To make it easier to look for these lower-sodium products, food labels carry the following terms describing sodium/salt content:

Very low salt	= 35 mg or less per serving
Low salt (sodium)	= 140 mg or less per serving
Reduced salt	= at least a 75 percent reduction compared to original food
Unsalted	= no salt added
Salt-free	= 5 mg or less

Sources of Salt in the Diet. Sources of sodium to watch out for when preparing food are seasoning salts (garlic, parsley, onion, celery), meat tenderizers, meat flavorings (barbecue sauce, smoke-flavored products), salad dressings, molasses, party spreads, dips, condiments (mayonnaise, mustard, ketchup, tartar sauce, chili sauce, soy sauce, relish, horseradish, Worcestershire sauce, and steak sauces), MSG (monosodium glutamate), and bouillon cubes.

Salt Substitutes. Finally, the salt added at the table should also be reduced. Salt substitutes are an option for some people. Many salt substitutes, however, contain potassium, which may also be inappropriate for people with kidney, heart, or liver problems. Calcium chloride is another salt substitute. "Low sodium" salt, which contains half the sodium of regular table

KEY TERMS

Seasoning Any compound that enhances the flavor already found naturally in a food.

Flavoring Substance that adds a new flavor to food.

TABLE 6-5
Functions of Salt in Foods

Function	*Description*
Flavor Enhancer	Salt's best known function is to flavor foods. Breads are less bland, cheeses are not as bitter, and processed meats owe a lot of their flavors to salt.
Preservative	Salt cures and has been used for thousands of years to preserve foods. Refrigerators and freezers were not always around, and drying out the moisture in foods with salt prevented bacteria from being able to live on the food.
Binder	Food manufacturers use salt to help form a gel on sausage and other smoked meat products.
Texture Enhancer	Salt contributes to the texture of ham, processed meats, bread, and certain cheeses.
Color Aid	The color of processed meats such as ham, bacon, hot dogs, and sausage is partially due to salt.
Control Agent	Bacteria and yeast are sensitive to salt concentrations and it is used to control their growth during fermentation in such foods as bread, cheese, pickles, sauerkraut, and sausage.

Source: Salt Institute.

FIGURE 6-11 Different types of salt used in food preparation.

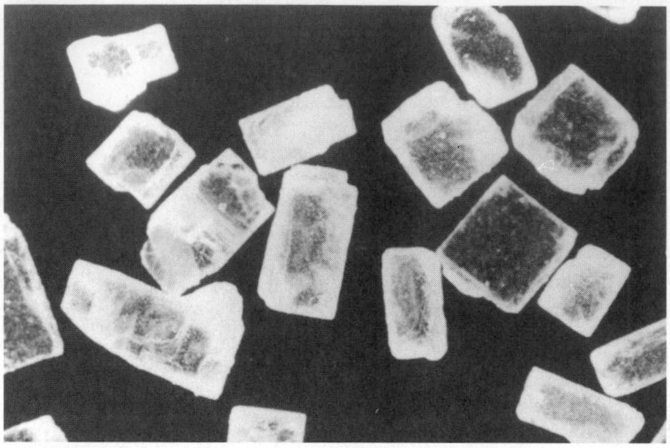

Morton Salt

Salt Varieties

Sea salt —	Obtained from evaporated sea water. Sea salt is more costly than other salts, yet often preferred because it has a pure, mineral-like taste.
Rock salt —	Derived from ancient sea beds that have long dried up and are underground.
Table salt —	Refined rock salt that is often fortified with iodine and contains additives to prevent caking.
Kosher salt —	Rock salt with no additives. Preferred by professional chefs because of its large, flaky crystals that are picked up easily with fingers (use one third more if substituting for table salt and vice versa).
Flavored salts —	Garlic, onion, and celery salt mixtures.

salt, is not considered a salt substitute. "Lite" salt products sold to replace the salt shaker should be avoided, because they also contain some sodium. Ultimately, the craving for salt is an acquired taste, so gradually cutting back on salt will eventually decrease craving.

KEY TERMS

Herb A plant leaf valued for its flavor or scent.

■ ■ ■ ■

Spice A seasoning or flavoring added to food that is derived from the fruit, flowers, bark, seeds, or roots of a plant.

Pepper. Pepper is just behind salt in popularity as a seasoning. Pepper is added most frequently to meats, soups, sauces, and salads. Ground black or white pepper comes from the berries of a tropical climbing shrub. The color of pepper depends on the berry's ripeness. Black pepper is from the dried, unripe berry, while white pepper is from the ripe berry from which the dark outer skin has been removed. Green peppercorns, a less common variety, are from under-ripe berries that are preserved in brine or freeze dried.

Peppercorns belong to an entirely different genus than the *Capsicum* family of chili peppers, which are classified as vegetables. Many varieties of *Capsicum* peppers are dried and used in chili powder, cayenne pepper, and paprika.

Herbs and Spices

Herbs. **Herbs** were described by Charlemagne as "a friend of physicians and the praise of cooks" (7). The Food and Drug Administration groups culinary herbs and spices together and considers them both to be spices. Regardless of how they are defined, herbs are well known for their seasoning capabilities in food preparation (Appendix D). The best-known seasoning herbs include basil, sage, thyme, oregano, bay leaves, cilantro, dill, marjoram, mint, parsley, tarragon, rosemary, and savory. For the best in flavor and texture, fresh herbs are generally preferred over dried.

Spices. **Spices** are distinguished from herbs, which are derived from leaves, by the other parts of the plant from which they are derived. Some examples include:

- Allspice (from a fruit)
- Saffron (flower)
- Cinnamon (bark)
- Anise, caraway, celery, cumin, fennel, mustard, poppy, and sesame (seeds)
- Ginger and turmeric (roots)

Although garlic, onions, and shallots can serve as a spice, they are officially recognized as vegetables. Appendix D also lists some of the more common spices.

History records a time when spices were greater in value than gold. In fact, they have been called "vegetable gold" and were once used as currency. A Goth leader once demanded 3,000 pounds of pepper as a partial ransom for calling off his siege of Rome (9). The search for these flavoring ingredients resulted in the carving of trade routes between countries, the founding of wealthy empires, and the exploration of far-off lands. Their value now rests in their unique ability to add a flavorful difference to dishes. The various world cuisines owe their distinctiveness to the unique combinations of spices in foods. Thai food relies heavily on hot peppers, while Central American dishes are distinguished by their use of chili peppers or powder. Mexican meals often incorporate

NUTRIENT CONTENT

Sodium is the portion of salt that has raised concerns with its possible connection to high blood pressure, or hypertension. Not everyone is susceptible to having high blood pressure caused by a high-sodium diet, but about 15 to 25 percent of Americans are genetically prone to developing the condition. The average North American diet is high in sodium, which automatically puts this subgroup at risk.

To safeguard this 15 to 25 percent of sodium-sensitive individuals, the Surgeon General recommends that all Americans lower their sodium intake. The concern is that high blood pressure, regardless of its cause, is a known risk factor for heart disease, kidney disease, and strokes (4, 10).

How much should sodium intake be reduced? Most Americans ingest 2 to 3 teaspoons (4,000 to 5,000 mg) of sodium per day. That amount can be lowered to 1,100 to 3,300 mg (½ to 1½ teaspoons) by taking three steps: (1) cutting back on high-sodium sources such as processed foods, (2) not adding salt during food preparation, and (3) removing the salt shaker from the table.

cumin, coriander, paprika, pepper, and cilantro. Indian dishes are enhanced with curry mixtures, which are combinations of spices whose exact ingredients and proportions can be closely guarded family secrets.

Purchasing Herbs and Spices. Herbs and spices can be purchased in whole, crushed, or ground form. Whole herbs retain their freshness longer than crushed, which in turn keep longer than ground. Whole seeds and leaves provide a visual and textural appeal, although the flavor release may be slow and unevenly distributed. Ground spices provide a quick infusion of flavor that is more uniform, but their aromas are easily lost when exposed to oxygen (oxidized) during storage. The natural antioxidant properties of certain herbs are also lost when exposed to oxygen (3).

Storing Herbs and Spices. According to the American Spice Trade Association, dried spices and herbs should be kept below 60°F (16°C) for optimum potency and replaced every twelve months. Herbs and spices deteriorate rapidly when exposed to air, light, and heat. They keep best in airtight, opaque containers stored in cool, dry places. Green herbs such as chives and parsley are light sensitive and will fade if exposed to light (2). Testing the freshness of a particular spice or herb is done by crushing it in the palm of the hand and then sniffing it to detect its intensity. The full-bodied aroma of fresh herbs becomes weak and barely detectable over time. If an herb or spice is to be used only occasionally, it is best to buy it in small quantities.

Flavor Enhancers. MSG (monosodium glutamate) is a compound that does not fit into any particular category. It influences flavor without contributing any flavor of its own. Hundreds of years ago in Asia, people found that food cooked in a seaweed-based soup stock had a unique flavor. In 1909, this compound was isolated from seaweed by a Japanese scientist and called *umami,* meaning "delicious" (see Chapter 1 defining *umami* as the fifth taste) (6, 14). Its scientific name, monosodium glutamate, comes from glutamic acid, an amino acid found in seaweed. It is now widely used in processed foods, including canned/dried soups, spaghetti sauces, sausages, and frozen meat dishes. It has been implicated in "Chinese Restaurant Syndrome," in which MSG-sensitive people experience nausea, diarrhea, dizziness, grogginess, sleepiness, warmth, headache, chest pain, and arthritis-like symptoms from consuming MSG (26).

Oil Extracts. Oil extracts can be used as food flavorings. These essential oils are obtained from natural sources such as flowers (orange), fruits (oranges, lemons), leaves (peppermint), roots (garlic), bark (cinnamon), buds (clove), and nuts (almonds, vanilla beans). The flavor in essential oils is so concentrated that only a small amount is required for flavoring purposes. Oil extracts are primarily used to flavor puddings, candy, ice cream, cakes, and cookies.

Vanilla beans from the cured pod of a tropical orchid provide the purest, most intense vanilla flavor. The small black specks in vanilla sauces and ice cream are the seeds of the pod. The Food and Drug Administration defines "pure vanilla extract" as at least 35 percent alcohol by volume, while those of lesser content are labeled "pure vanilla flavor." Vanilla/vanillin blends or imitation versions should be avoided, because they contribute an artificial flavor to foods.

Extracts are made by steam-distilling the oils from various plant sources and blending them with ethyl alcohol, which can evaporate. For that reason they should be stored in a cool, dark place and stored for no more than a year to retain maximum flavor.

Marinades. Marinades are seasoned liquids that flavor and tenderize foods, usually meats, poultry, and fish. A vinaigrette is a marinade used for vegetables served cold. The basic marinade consists of one or more of the following ingredients: oil, acid (lemon juice, vinegar, wine), and flavorings (herbs, spices). The food is completely submerged in the marinade and refrigerated from a few minutes to several days. The food should be turned occasionally in order to evenly distribute the marinade. Meat, fish, and poultry marinades should be discarded after use and never served raw with the cooked food.

Breading and Batters. Breading and batters enhance the flavor and moisture retention of many foods. Most foods coated in this manner are deep-fried, pan-fried, or sautéed to give them a browned, crisp outer texture.

Breadings. The flours most frequently used for breading are either wheat- or corn-based. Coating the food lightly in flour, called dredging or *à la meuniere* (ala moon-yare), results in a light, golden crust. Crumb coatings differ in that they are applied in three steps (Figure 6-12). First the food is dredged lightly in flour to seal in moisture and provide a base for the next step. The flour-coated food is then dipped quickly in an egg wash consisting of beaten eggs plus a tablespoon of water or milk. (Substituting oil for the water or milk results in a richer, more tender coating.) The proteins in the eggs or milk act as binding agents to "glue" the breading to the surface of the food (11). Finally, the sticky-coated food is placed in a bowl of crumbs for the final coating. Seasoned bread crumbs, cracker crumbs, cornmeal, or cereal (cornflakes) may be used to coat foods. Smaller, more delicate foods such as mushrooms require finer-grained breadings. Seasonings or flavorings such as salt, pepper, rosemary, thyme, sage, or others can be added at any of the three steps of breading, although mixing them into the egg wash ensures they are evenly distributed (21). Sugar can also be added, but be aware that it results in a browner product (5).

Batters. Another way to coat foods is through the use of batters, which are wet flour mixtures containing water, starch, and seasonings into which foods are dipped prior to being fried. Commercial batters are available that require simply adding water. There is no one recipe for a batter, and formulas can be extremely flexible (24). The addition of eggs to the batter will produce a darker coating due to the yolk content. Commercial batters often have added ingredients such as gums for viscosity and starches to increase adhesion by the swelling of their granules. Shortening or oils contribute to overall flavor and mouthfeel (11). Figure 6-13 shows the basic differences between using a breading or batter process to coat foods.

Condiments. Condiments are seasonings or prepared relishes used in cooking or at the table. Some of the most common condiments are mustard, catsup, mayonnaise, relish, tartar sauce, salsa, barbecue sauce, chili sauce, soy sauce, horseradish, Worcestershire sauce, chutney, and steak sauce.

Adding Seasonings and Flavorings to Food

How Much to Add? If tested recipes are available, they should be followed. If there is no recipe, start by adding ¼ teaspoon of spice (or ⅛ teaspoon for chili, cayenne, or garlic powder) for every pound of meat or pint of liquid (soup, sauce). Flavor-test and add more seasonings as desired. It is always easier to add than to subtract, and because it is important not to overpower other ingredients in a dish, it pays to be cautious. There really is no easy set rule or formula for adding seasoning and flavoring to foods. The freshness of herbs and spices will influence how much should be added, and evaporation of liquid during heating will concentrate what is already present. Successfully prepared foods have well-balanced flavors that are complementary.

HOW & WHY? ?????????

How much dried herbs should one use in place of fresh herbs? When substituting dried herbs for fresh, the general rule is to use about one-third as much as fresh herbs, because the flavor of dried herbs that have not become stale is generally more intense.

When to Add? Seasonings should be added to prepared foods early enough in the cooking to release their flavor, but not so soon that their flavor is lost. Most seasonings (especially ground) are added near the end of the heating period, while a few (whole or lightly crushed) need more time to release their flavors and aromas to blend with the other ingredients. Foods tend to better retain the flavor of seasonings and flavorings if their surfaces are partially cooked and therefore permeable to what

FIGURE 6-12 Breading—application of a crumb coating.

Roll in flour and shake off excess.

Dip floured piece in egg wash.

Dip in bowl of crumbs and toss more crumbs on top.

FIGURE 6-13 **The difference between using breadings and batters.**

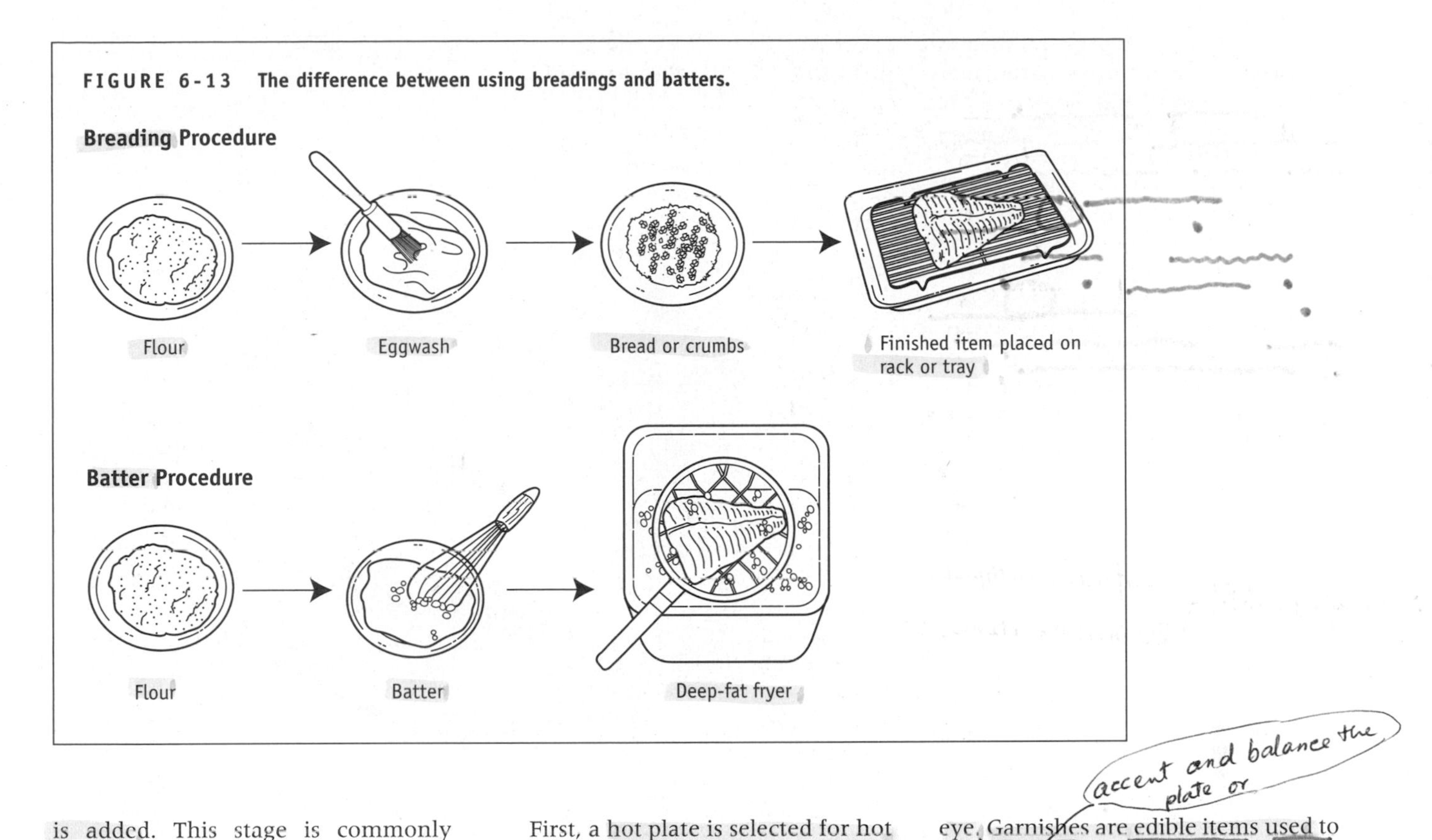

is added. This stage is commonly referred to by professional chefs as **sweating** (25). Delaying the addition of seasonings and flavorings is particularly true for salts, which tend to shrink meats if they are added too soon. Flavor retention is influenced by the length of the heating and the final temperature attained (8). Experience may well be the best teacher.

Food Presentation

Plate Presentation

The highest quality, best-prepared food is shortchanged if the plate presentation has not achieved or surpassed the same level of quality. An artistic layout of food items on the plate plays a very important role in winning over and satisfying the customer, whose first impression is based largely on sight. When plating food, the top consideration is coordination of colors, shapes, sizes, textures, and flavors. Following are some guidelines to help in achieving this coordinated balance.

First, a hot plate is selected for hot foods, while a cold plate is reserved for cold foods. The size of the plate should be sufficient so that food is not crowded, but not so large that the food looks meager in comparison. Items are placed on the plate to achieve balance. The main food item, often the meat, is set in front of the guest with the best part forward, with any fat or bone facing the back. The plate should not have to be turned in order for the main entrée to be consumed. Accompanying items are plated around the main entrée, and garnishes may be added to contribute to balance (Figure 6-14 on page 134). Space should be kept between each item on the plate, with the border of the plate serving as the frame. The border should never be covered with food; any food that does spill over onto the edges should be wiped clean. The exception is when the plate rim is dusted with chopped herbs, spices, or other decorative touches.

Garnishes

Garnishing adds color and design to a plate, making it more attractive to the eye. Garnishes are edible items used to decorate food and should generally reflect the flavors of the dish being served. For example, a rosemary sprig would be appropriate for a rosemary-scented meat sauce. Other possible garnishes, depending on what is being served, may include:

- Leaves, such as parsley sprigs, or mint leaves in iced tea
- Fruit, such as pineapple sticks, kiwifruit slices, olives, or lemon, lime, or orange wedges
- Vegetables, such as cucumbers, tomatoes, green peppers, radishes, or onions
- Pickled items, such as olives, pickles, or pimentos
- Nuts, croutons, crackers
- Hard-boiled egg slices or halves

Only fresh, high-quality foods should be used for making garnishes.

KEY TERMS

Sweat The stage of cooking in which food, especially vegetables, becomes soft and translucent.

Appendix E illustrates several garnish preparation techniques. Garnishes should be used to add balance; if the items on a plate are already harmonized, a garnish is not necessary. Plate garnishes are best when they are colorful, contrasting but not clashing, and compatible with the food being served in terms of flavor, size, and shape. Garnishes should not crowd the dish, and an odd number tends to be more visually appealing. For example, three slices of apple on a plate look better than two or four slices. To prevent any possible injuries, unfrilled toothpicks and other hard inedible items should be avoided.

FIGURE 6-14 Plate presentation.

PhotoDisc

PICTORIAL SUMMARY / 6: Food Preparation Basics

Mastering the basics of food preparation is essential to getting a meal together, but food preparation is not an exact science. Understanding and adjusting for the many variables at play in preparing even the simplest recipe can elevate food preparation from a craft to an art form.

METHODS OF HEATING FOODS

Moist-heat preparation: Heat is transferred by water, water-based liquid, or steam.

Dry-heat preparation: Heat is transferred by air, radiation, fat, or metal.

Microwaving: Usually listed as moist-heat, microwaving actually incorporates both dry- (radiation) and moist-heat methods.

Moist Heat		**Dry Heat**	
Scalding	Poaching	Baking	Roasting
Simmering	Stewing	Broiling	Grilling
Braising	Boiling	Frying	Sautéing
Parboiling	Blanching	Stir-Frying	Pan-Broiling/Frying
Steaming		Deep-Frying	

Poach Simmer Boil

CUTLERY TECHNIQUES

Knowing knives and how to use them is essential to basic food preparation.

Holding a chef's knife.

It is important to know how to hold the knife, the different sections of the blade assigned to various tasks, and the differences between slicing, shredding, dicing, mincing, chopping, and peeling.

MEASURING INGREDIENTS

The three major steps in correct measuring:

- Acquiring the amount required for a specific measurement (e.g., 4 ounces of cheese yields 1 cup shredded)
- Selecting the right measuring utensil
  - *-Wet ingredients*: transparent, graduated cup with pour spout
  - *-Dry ingredients:* flat-topped measuring cups for leveling
- Accurate measuring technique

Know your substitutions: Sometimes knowing what item can replace a missing ingredient can save the day!

MIXING TECHNIQUES

Mixing is a general term describing beating, blending, binding, whipping, and folding. The ingredients for baked goods can be mixed in a number of different ways, but the most common methods are the conventional (creaming), conventional sponge, single-stage (quick-mix), pastry-blend, biscuit, and muffin.

SEASONINGS AND FLAVORINGS

Seasoning: Any compound that *enhances the flavor* already found naturally in a food.

Flavoring: An addition that *adds a new flavor* to a food.

The major seasonings/flavorings are:

- Salt and pepper
- Herbs and spices
- Oil extracts
- Flavor enhancers
- Marinades
- Breading and batters
- Condiments

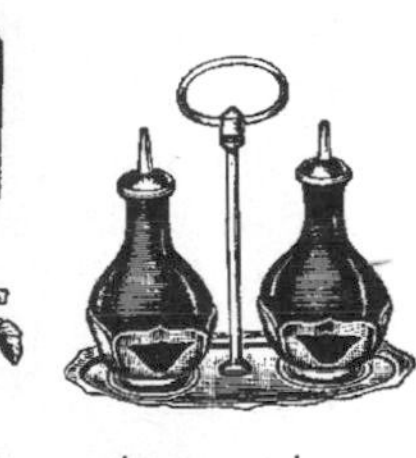

The freshness of herbs and spices will influence how much should be added, and evaporation of the liquid during heating will concentrate the effect of the flavoring/seasoning added.

It is always easier to add than to subtract! Flavor-test and add more seasoning as desired.

Most seasonings are added near the end of cooking time.

FOOD PRESENTATION

A customer's first impression is based largely on sight, and an artful arrangement of food on a plate contributes a great deal to the dining experience.

Garnishes, fresh, if needed.

Correctly sized plates; hot plates for hot food, cold plates for cold.

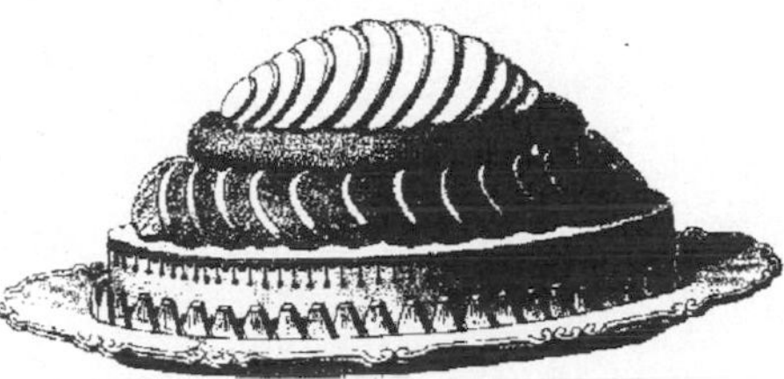

The highest-quality, best-prepared food is shortchanged if the presentation on the plate has not achieved or surpassed the same level of quality.

REFERENCES

1. Adams SJ. Slow cooking enhances winter vegetables. *Fine Cooking* 18:40–43, 1997.
2. ASTA. *The Foodservice and Industrial Spice Manual.* American Spice Trade Association, 1990.
3. Brookman P. Antioxidants and consumer acceptance. *Food Technology New Zealand* 26(10):24–28, 1991.
4. Cappiccio FP, et al. Double-blind randomized trial of modest salt restriction in older people. *Lancet* 350(9081):850–854, 1997.
5. Corriher SO. Taking the fear out of frying. *Fine Cooking* 16:78–79, 1996.
6. Corriher SO. MSG enhances flavors naturally. *Fine Cooking* 20:76, 1997.
7. Farrell KT. *Spices, Condiments, and Seasonings.* Avi, 1985.
8. Flavor loss. *Food Engineering* 62(6):38, 1990.
9. Giese J. Spices and seasoning blends: A taste for all seasons. *Food Technology* 48(4):88–98, 1994.
10. Kannel WB. Blood pressure as a cardiovascular risk factor: Prevention and treatment. *Journal of the American Medical Association* 275(20):1571–1576, 1996.
11. Loewe R. Role of ingredients in batter systems. *Cereal Foods World* 38(9):673–677, 1993.
12. Mathews RH, and R Batcher. Sifted versus unsifted flour. *Journal of Home Economics* 55:123, 1963.
13. Mattes RD. Discretionary salt use. *American Journal of Clinical Nutrition* 51:519, 1990.
14. Nagodawithana T. Flavor enhancers: Their probable mode of action. *Food Technology* 48(4):79–85, 1994.
15. Ponte JG. Sugar in baking foods. In *Sugar: A User's Guide to Sucrose,* eds. NL Pennington and CW Baker. Van Nostrand Reinhold, 1990.
16. Reineccius G. Off-flavors in foods. *Critical Reviews in Food Science and Nutrition* 29(6):381–402, 1991.
17. Saguy IS, and EJ Pinthus. Oil uptake during deep-frying: Factors and mechanism. *Food Technology* 49(4):142–145, 1995.
18. Scott ML. *Mastering Microwave Cooking.* Bantam, 1976.
19. Singh RP. Heat and mass transfer in foods during deep-fat frying. *Food Technology* 49(4):134–137, 1995.
20. Sloan AE. Grilling and slow cooking are gaining. *Food Technology* 53(6):28, 1999.
21. Stevens M. Coating food for a golden crisp crust. *Fine Cooking* 20:74, 1997.
22. Stevens M. What's the difference in salt? *Fine Cooking* 24:10, 1998.
23. Stevens M. Why scald milk? *Fine Cooking* 27:12, 1998.
24. Suderman DR. Selecting flavorings and seasonings for batter and breading systems. *Cereal Foods World* 38(9):689–694, 1993.
25. "Sweating" vegetables coaxes out flavor. *Fine Cooking* 25:76, 1998.
26. Tarasoff L, and MF Kelly. Monosodium L-glutamate: A doubleblind study and review. *Food and Chemical Toxicology* 31(12):1019–1033, 1993.
27. Varela G, and B Ruiz-Roso. Some effects of deep-frying on dietary fat intake. *Nutrition Reviews* 50(9):256–262, 1992.

WEBSITES

Culinary Institute of America, a leading chef school in the United States: **www.ciachef.edu** and **www.ciaprochef.edu**

Food preparation encyclopedia: **www.allrecipes.com/encyc/default.asp**

Food Preservation

The World Health Organization estimates that about 20 percent of all food is lost to food spoilage. Spoilage contributes to the average North American family discarding about one-fourth of the food they purchase (54). Some of the biological, chemical, and physical changes that lead to spoilage are preventable. Preservation methods that make this possible are discussed after a brief introduction on food spoilage.

Food Spoilage

Spoilage is characterized by a decline in eating quality, resulting in food that is less acceptable to the consumer in appearance, taste, texture, or odor (2). The foods we eat are all derived from living matter, making them subject to the natural process of decomposition. Food not only decomposes, but is lost or spoiled by being consumed by creatures other than humans—rats, mice, flies, and microorganisms. Food spoilage is obvious and detectable. In contrast, food contamination, which can also occur with food spoilage, is often undetectable (see Chapter 3 on Food Safety).

Not all foods spoil at the same rate. Foods are classified as perishable, semi-perishable, or nonperishable, depending on their susceptibility to spoilage. The foods most perishable are those with large concentrations of protein and/or water, which accelerate the microbial and chemical processes of decomposition. For example, fish, seafood, meat, eggs, and dairy products, which all have a high protein and water content, are very perishable. Watery fruits and vegetables such as tomatoes, peaches, berries, and leafy vegetables are also highly perishable. Semi-perishable foods contain less water and include those, such as potatoes, carrots, beets, turnips, onions, and apples, that are edible for several months if stored under the proper conditions. Processed nuts, cereals, dried tea leaves, pastas, and dried beans and peas are nonperishable because they contain very little water and will keep for many months with little loss of quality.

The biological, chemical, and physical changes in food that contribute to spoilage are now briefly discussed.

KEY TERMS

Yeast A fungus (a plant that lacks chlorophyll) that is able to ferment sugars and that is used for producing food products such as bread and alcohol.

Fermentation The conversion of carbohydrates to carbon dioxide and alcohol by yeast or bacteria.

Biological Changes

The prime biological factors involved in food spoilage are microorganisms such as bacteria, **yeasts**, and molds. Just like people, these tiny organisms need food to survive, so food is a natural target. The most common foods spoiled by bacteria include meat, eggs, milk, and opened canned goods. Foodborne illnesses caused by bacteria were discussed in Chapter 3.

Yeasts prefer high-sugar foods such as fruits, vegetables, and fruit preserves, and can cause unwanted **fermentation** of fruits and fruit juices in the presence of the proper amounts of oxygen, moisture, and acidity (pH). The naturally occurring yeasts found in the air normally pose no threat. The moisture in foods, however, can encourage their growth to a point that becomes unacceptable. Any method that keeps the moisture content low will be successful in their control.

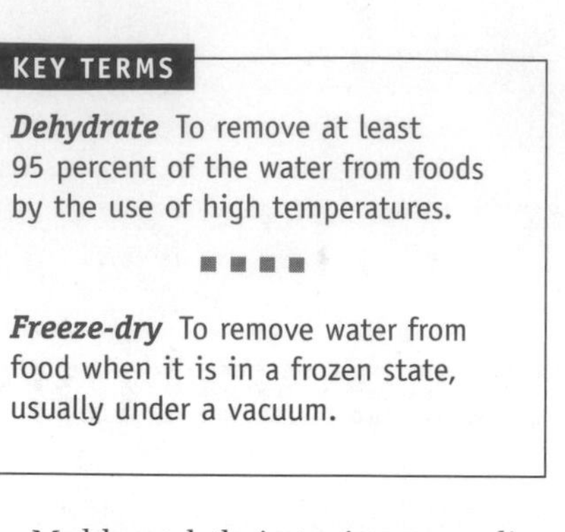
KEY TERMS

Dehydrate To remove at least 95 percent of the water from foods by the use of high temperatures.

Freeze-dry To remove water from food when it is in a frozen state, usually under a vacuum.

Molds and their toxins were discussed in Chapter 3. These microorganisms, like yeasts, prefer high-sugar foods, but are particularly drawn to cheese and bread. The appearance of their bloom on foods indicates that spoilage has begun. Molds are easily spread through the air, are very resistant to drying, and can be difficult to control by the means used for bacteria and yeasts. Commercial food enterprises employ vacuum pumps to remove oxygen from containers, because molds cannot grow in its absence.

The many weapons in the arsenal used to fight spoilage resulting from microbial action include boiling, refrigeration, drying, and curing with high concentrations of sugar or salt. These methods, which destroy or inhibit the growth of microorganisms, are discussed later in this chapter under Food Preservation Methods.

Chemical Changes

Chemical reactions or changes also contribute to food deterioration. Enzymes play a significant role in catalyzing these reactions and can be categorized by the substance on which they act (substrate) or their mode of action. Proteases, also called proteolytic enzymes, split proteins into smaller compounds (Chemist's Corner 7-1). Fish has many more active proteases than meat, which is one of the reasons fish deteriorates so quickly. Lobsters are also prone to proteolytic breakdown, which occurs the minute they expire. Unless lobsters are kept alive to the very last second, the proteases cause the lower abdomen to partially liquefy, with the result that the tail meat becomes crumbly when cooked (45).

Lipids are broken down by enzymes called lipases, which degrade the triglycerides of fat into glycerol and fatty acids. Further degradation leads to rancidity, or off-odors and tastes.

Enzymes that decompose carbohydrates are carbohydrases, each named after the particular sugar on which it acts. For example, sucrase breaks down sucrose into glucose and fructose. Another group of enzymes serves to oxidize compounds. Some of the more common oxidases include ascorbic acid oxidase, peroxidase, tyrosinase, and polyphenolase. The latter two enzymes are involved in enzymatic browning, which leads to unappetizing brown discoloration in some fruits and vegetables. Hydrolysis may also contribute to the deterioration of foods. See Chapter 2 for a more detailed discussion of these chemical reactions.

Physical Changes

Physical changes, unlike chemical changes, do not result in the formation of new compounds. Common physical changes occurring in foods as they spoil are evaporation, drip loss, and separation. Water evaporates out of improperly stored foods, creating an unattractive, dried-out appearance and possible undesirable flavor changes. Water can also be lost (syneresis) out of foods such as gelatins, yogurt, and sour cream as they age. Separation of water and oil occurs in such foods as nonhomogenized milk, mayonnaise, salad dressings, and high-moisture cheeses when they are stored too long or are frozen and later thawed. Separation is the reason sandwiches spread with mayonnaise or made with high-moisture cheeses do not freeze well.

Chemist's Corner 7-1

Proteinases

Proteases can be further classified into proteinases. These enzymes split proteins into smaller compounds, such as peptones and proteoses, which are then broken down to polypeptides and peptides. The latter two are then split by peptidases into amino acids. The various protein enzymes break protein down into a liquified mass because the amino acids are water soluble (45).

Food Preservation Methods

For over 5,000 years, humans have been preserving foods by drying, salting, and fermentation. Ironically, the demands of war have triggered the monumental developments in food preservation techniques. Napoleon's need for a safe and portable food supply for his armies in the late 1700s and early 1800s led to the discovery of canning. World War II led to the development of **dehydrated** foods such as instant potatoes and eggs. The American Red Cross provided irradiated milk in the food packages given to prisoners of war (42). The Vietnam War spurred the refinement of the process of **freeze-drying**, which allowed for the development of complete, lightweight foods that could be carried into the field easily and transformed into ready-to-eat meals by adding water.

Because of newer preservation techniques and advances in refrigeration and transportation, people now enjoy a wide variety of foods, including out-of-season and exotic foods from all over the globe.

Drying

Drying is the food preservation process that consists of removing the food's water, which effectively inhibits the growth of microorganisms. Bacteria and molds need approximately a 15 percent moisture level to survive, while yeast needs at least a 20 percent moisture content. Once dried, the food can be eaten as is, or, unlike microoganisms, be rehydrated (have water added), which changes its size, color, flavor, and texture.

Sun Drying. Many early cultures subsisted throughout the year on naturally dry foods such as nuts, grains, and dried legumes. The discovery that fruits, vegetables, and meats could be dried in the sun was a natural extension of this practice (54). The sun provided the heat for evaporation for many centuries, and continues to do so in various countries around the world. The drying process relies on some form of heat to evaporate the water, and the hotter the environment surrounding the food, the faster the rate of evaporation. Ancient Egyptian hieroglyphics show fish being dried in the sun, and tribes along the Nile buried fruits and meats just under the surface of the hot desert sands to dry before storing them in earthenware containers. More than 3,000 years ago, the Incas were sun-drying foods available in abundance in summer to provide nourishment throughout their harsh winters. The Bible records that grapes and figs were dried. Marco Polo observed that the Mongols consumed a sun-dried milk product to fuel them in their military conquests. The first food exported from the American colonies to England was sun-dried fish.

Commercial Drying. Since sun drying takes a long time and exposes foods to the weather and to the action of insects, most foods are now dried by various commercial processes, although raisins are still sun dried (Figure 7-1). The most important types of commercial drying are conventional (heat), vacuum (pulls the water out), osmotic (water drawn out by osmosis), and freeze-drying (ice crystals vaporize).

Conventional Drying. Conventional drying uses heat to evaporate the water. In one method, the food is spread on a slatted floor or on shelves within kilns or drying rooms. A blower then passes hot air from a heater over and through the food. In tunnel drying, food is placed on trays or "cars," which are moved through a tunnel of carefully controlled hot air. Liquids can be dried by either spray drying or drum drying. In the former, a fine spray of the liquid is dried very quickly in mid-air. Spray drying is used to produce such foods as nonfat dried milk and some types of instant coffee. Drum drying occurs when liquid is poured over the very hot surface of a drum dryer, an apparatus resembling a large barrel. The dried food can then be peeled off like tissue paper, ground into flakes, and packaged. Some mashed-potato flakes and quick-cooking hot cereals are dried in this way.

Vacuum Drying. Vacuum drying dehydrates foods to very low moisture levels (1 to 3 percent) through the use of a vacuum. Milk, tomato paste, orange juice, and coffee are often concentrated by vacuum drying. The food is placed in a chamber, and the surrounding pressure is reduced below atmospheric pressure, which lowers water's boiling point. The water is then more easily boiled off at this lower boiling temperature. This drying method preserves flavor and color while increasing its shelf life considerably, but it has two drawbacks: (1) it is expensive, and (2) it requires that the dehydrated food be stored in airtight containers to prevent rehydration by drawing moisture from the air.

Osmotic Drying. Osmotic drying is not often used commercially. It relies on the reuse of a strong syrup with a high sugar concentration that osmotically draws water from the object being dried. Ocean Spray's Craisins are cranberries dried through osmotic drying (10).

Freeze-Drying. Freeze-drying consists of first freezing the food and then placing it in a vacuum, where the ice **sublimates** to a vapor (Figure 7-2). This process of sublimination is the most effective method for drying foods, because it does not subject the food to high heat, which alters a food's flavor, color, and structure. The heat used is low enough to prevent melting of the ice, while warm enough to assist in evaporation of the water. As a result, freeze-dried products yield the highest quality, store indefinitely, and can be reconstituted easily. The process, however, is more costly than conventional drying.

Pretreatments. Some fruits are pretreated prior to drying. Plums have their skins "checked" before being dried into prunes by dipping them in lye or very hot water. This process cracks the skin, thereby improving skin texture and shortening the drying time by exposing more surface area to drying. Another pretreatment consists of blanching certain vegetables, such as potatoes or carrots, prior to drying

KEY TERMS

Sublimation The process in which a solid changes directly to a vapor without passing through the liquid phase.

FIGURE 7-1 Grapes drying in the sun to yield raisins.

SunMaid Raisins

FIGURE 7-2 Typical freeze-drying process.

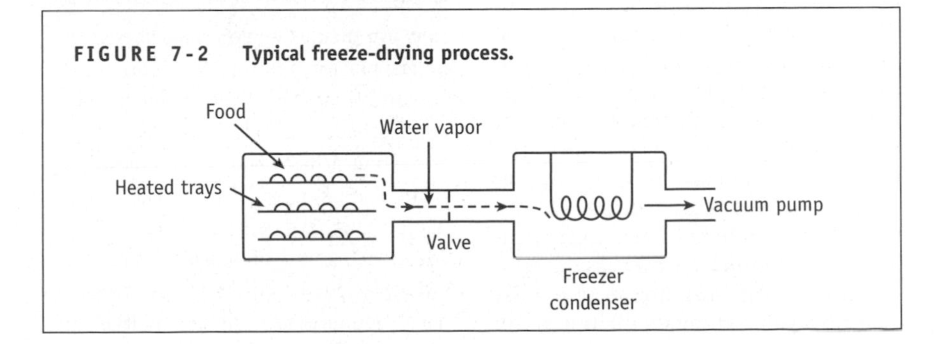

KEY TERMS

Cure To preserve food through the use of salt and drying. Sugar, spices, or nitrates may also be added.

■ ■ ■ ■

Edible coating Thin layer of edible material such as natural wax, oil, petroleum-based wax, etc. that serves as a barrier to gas and moisture.

to prevent enzymatic browning. The heat of this brief boiling disrupts the enzymes responsible for browning and inhibits the discoloration. Some fruits, such as apricots and peaches, are dipped in a sulfite solution or exposed to sulfur dioxide gas.

HOW & WHY? ?????????

Why is sulfur dioxide often listed as an ingredient in dried fruits? To preserve their natural color and prevent spoilage, some foods, such as apricots and peaches, are dipped prior to drying in a sulfite solution or exposed to sulfur dioxide gas. Sulfur protects against enzymatic browning and the loss of vitamins A and C. However, it destroys thiamin (vitamin B_1) and in certain sensitive individuals, it may cause headaches or allergic reactions, and even coma and death.

Curing

High concentrations of salt bind to the water in food, making it unavailable to microorganisms. The earliest recorded use of salt as a preservative dates back to 3000 B.C., when salt was used to **cure** fish. One of the earliest methods of preservation was to rub the surface of meats and fish with salt. Native Americans used salt to preserve some of their foods; the Hopi people traveled long distances to the salt mines within the Grand Canyon to obtain salt for this purpose.

Corned beef is a cured meat. The word "corn" refers to the Latin word for "grain," which in this case means "grains" of salt (46). Today, the most commonly cured meats include ham, sausage, hot dogs, bacon, and bologna. These and other cured foods, however, should be treated as fresh foods, because they do not contain enough salt to be stored on the shelf.

Salt is used for many other food purposes in addition to curing (see Table 6-5 in Chapter 6).

Smoking Cured Meats. Cured meats sometimes undergo the optional treatment of smoking for added flavor. Meats are placed in "smokers" where they are exposed to the smoke of burning wood. The type of wood selected (sawdust, mesquite, hickory, oak, and various combinations of these woods) determines the resulting flavor (54). There are some health concerns with smoked foods, however, as they have been linked to cancer in laboratory animals (46).

Fermentation

For thousands of years, fermentation has been used both for the production and preservation of various foods (Figure 7-3). In the third century B.C., laborers building the Great Wall of China were fed fermented vegetables as part of their rations. Throughout Asia, vegetables are still commonly fermented. In North America, foods most often preserved by fermentation are cucumbers, olives, and cabbage.

Carbohydrates are required for the fermentation process. Vegetables are not usually considered sugary foods, but they contain enough carbohydrates and natural bacteria for fermentation to occur. Although only the plants' natural carbohydrates and microorganisms are needed for fermentation, salt or vinegar (acid) may be added to regulate bacterial growth.

HOW & WHY? ?????????

How are pickles made? Cucumbers can be manufactured into sweet, sour, dill, kosher dill, and other pickles by one of two methods: the longer process of fermentation, which yields brined pickles (with acid produced from the bacteria); and pickling (see below), where acid is added in the form of vinegar, resulting in quick pickles. During the process of fermentation, a 10 percent salt solution serves as the liquid in which the cucumbers are submerged and allowed to ferment for several weeks. In this fermentation period, bacteria normally found on vegetables break down the sugar in the cucumbers. The salt penetrates the cucumbers, and the brine concentration is increased to 15 percent, except in the processing of dill pickles. Once fermented, pickles (whole or sliced) are placed in warm water, packed in glass jars, covered with a combination of vinegar, sugar, spices, and garlic, and pasteurized. It is this canning process, rather than fermentation, that preserves the pickles.

Pickling

Pickling uses vinegar to preserve foods, because the acidity of the vinegar keeps many microorganisms in check. In the Middle East, vinegar was used as early as 1000 B.C. to preserve such foods as fruits, onions, and walnuts. The food was simply covered with vinegar, boiled, and sealed in a container. It was then allowed to stand for at least three weeks to give the vinegar time enough to penetrate all parts of the food.

Most people associate "pickles" with pickles made from cucumbers, but pickled foods include beets, cauliflower pieces, green tomatoes, green beans, chiles, bell peppers, and sliced Jerusalem artichokes. These foods are preserved in vinegar and salt, with spices often added to enhance their flavor. Pickled foods set aside for long-term storage are canned for safety reasons.

Edible Coatings on Foods

A unique food preservation method is to surround the food with an **edible coating**. The purpose of edible coatings is fourfold (12, 27):

1. To increase shelf life by acting as a barrier to moisture, oxygen, carbon dioxide, volatile aromas, and other compounds whose loss would lead to deterioration.
2. To impart improved handling characteristics, such as the ability to bend more easily without breaking.
3. To improve appearance through increased gloss and color.

FIGURE 7-3 Selected food products produced by fermentation.

Milk →	Cheese	
	Yogurt	
	Buttermilk	
	Sour Cream	
Meats →	Sausages (Salami, Bologna, Cervelat)	
Grains →	Yeast Breads	
	Beer and Saké ("Japanese Wine")	
	Whiskey	
Vegetables:	Cucumbers →	Pickles
	Olives →	Green Olives
	Cabbage →	Sauerkraut
	Chinese Cabbage →	Kimchi
	Soybeans →	Miso, Soy Sauce
Fruit:	Grapes →	Wine

4. To serve as a vehicle for added ingredients such as flavors, antioxidants, antimicrobials, etc.

Composition of Edible Coatings. Edible coatings can be produced from carbohydrate or protein materials. The most common edible coatings are lipid-based (beeswax, candelilla wax, carnuba wax, rice bran wax), oils (paraffin oil, mineral oil, vegetable oils), and petroleum-based waxes (paraffin, polyethylene wax) (3, 22).

Commonly Coated Foods. Edible coatings are commonly used for vegetables and fruits such as cucumbers, tomatoes, peppers, eggplants, pumpkins, summer squash, apples, bananas, guavas, mangoes, papaya, melons, nectarines, and citrus fruits. They are also used to coat candy, cheese, nuts, dried fruit (prevents stickiness, especially in raisins and dates), eggs in their shell (as a moisture and bacterial barrier), and processed meats (especially sausages), poultry, and fish (27).

Canning

Canning is a two-step process. First the food is prepared by being packed into containers, which are then sealed. Then the containers are "canned," or heated to ensure that all microorganisms are destroyed.

As previously mentioned, during the Napoleonic wars, Napoleon was having difficulty feeding his troops. He offered a prize to the person who could discover a new method of preserving food. The winner in the late 1790s was Nicolas Appert who invented the canning process. Food was placed in glass jars or canisters (cans), boiled, and then sealed shut. The jars were then boiled a final time, creating a vacuum. Many people believed that lack of oxygen in the cans preserved the foods. Almost a hundred years later, Louis Pasteur discovered that the real reason canning was successful was because the high boiling temperatures destroyed harmful bacteria. The heat processing also destroyed the enzymes responsible for the deterioration of foods, thereby protecting canned food from both harmful microorganisms and natural spoilage (55).

Preparing Food for Canning. Fruits, vegetables, and meats are the most frequently canned foods. All are prepared prior to canning by either "hot" or "cold" packing. Hot-pack canned foods are heated to at least 170°F (77°C) in syrup, juice, or water prior to being poured into **sterilized** jars. This initial heating drives out much of the air, allows the food to be packed tightly in the jar, gives a translucent appearance to fruit, and reduces the heating time. In the cold-pack method, food is placed directly into the sterilized jars. The jars are then filled with boiling liquid. In both cases, some space is left at the top of the jar (the amount of space depending on the product and canning method). Air bubbles are removed by gently sliding a rubber spatula or knife between the jar's side and the food, and leaving a small amount of air, known as the headspace, at the top of the jar (54). This headspace is required for pulling a vacuum and ensuring a seal. Jars with nicks or cracks should not be used for canning because they are not guaranteed to provide a reliable seal.

KEY TERMS

Sterilization The elimination of all microorganisms through extended boiling/heating to temperatures much higher than boiling or through the use of certain chemicals.

Two Methods of Canning. The two major techniques for canning foods are boiling-water processing and pressure canning. The method chosen depends primarily on the pH of the food. A potentially deadly bacterium, *Clostridium botulinum,* cannot survive in an acidic environment, so boiling temperatures are sufficient to process most fruits and tomatoes whose pH falls below 4.6. For varieties of tomatoes that are less acidic, lemon juice is sometimes added to the tomatoes to avoid the possibility of food poisoning. For foods above pH 4.6, the USDA states, "To prevent the risk of botulism, low-acid and tomato foods (not canned according to USDA recommendations) should be boiled 10 minutes at altitudes below 1,000 feet. For altitudes at and above 1,000 feet, add 1 additional minute per each 1,000 feet." Another way to heat foods with pHs above 4.6 (meats and vegetables) is to use a special piece of equipment known as a pressure canner that allows higher temperatures to be reached.

The majority of *C. botulinum* food poisoning cases in the United States are traced back to errors in home canning. In an effort to discourage home canning, the U.S. Department of Agriculture no longer provides information booklets showing how to can foods at home.

What follows is a general description of the home-canning process according to the two methods. After the foods are prepared according to one of these two packing methods, they are ready to be canned.

Boiling-Water Processing. This method of canning consists of boiling filled glass jars in a water-bath canner large enough to cover them with up to 2 inches of water. The containers should be commercially sterile or boiled in water for at least ten minutes before being filled with food and placed in the boiling water bath. Continuous boiling throughout the process is necessary.

Pressure Canning. In order to prevent the risk of botulism food poisoning, foods like meats, poultry, fish, and vegetables with a pH higher than 4.6 must be canned under pressure. Pressure canners make it possible to reach temperatures higher than boiling (212°F/100°C) (Figure 7-4). To kill *C. botulinum* spores, home canning must achieve a minimum temperature of 240° (116°C), which means using at least 10 pounds of pressure per square inch. Processing time is also determined by the heat penetration of the food. The two biggest mistakes in canning are: (1) failure to check the accuracy of the pressure gauge on the pressure canner, and (2) processing at incorrect temperatures for incorrect amounts of time.

Quality/Nutrient Retention. Canning will preserve foods longer than most other methods of food preservation, with the exception of drying. Although canned foods will keep approximately one year in a cool, dry place, the heat treatment used in canning may adversely affect the food's texture, color, and flavor (49). It may also cause the loss of some heat-sensitive nutrients. Minerals are not affected by canning, nor is the nutrient effectiveness of carbohydrates, proteins, and fats, but vitamins are susceptible to loss. Thiamin (B_1) and vitamin C may experience 50 to 80 percent losses during canning. Most other vitamins experience losses ranging from 10 to 30 percent (54).

FIGURE 7-4 Pressure canner.

Presto®

Safety of Cans. One of the ways to ensure the safety of a home-canned product is to boil it for ten minutes prior to consumption. Any can that is bulging or leaking or whose contents are bubbling or have a bad odor should be discarded. Bulges may be caused by multiplying bacteria releasing gases that push out the metal. A leaky or broken seam means the food inside is no longer safe. While odor indicates that microbiological contamination has occurred, a contaminated food may also be odorless, so the lack of odor does not guarantee that the food is safe to eat. Questionable food should never be tasted to determine whether or not it is spoiled, but should be discarded immediately. The deadly toxin produced by *C. botulinum* can even be absorbed through the skin or mucous membranes.

Cold Preservation

Refrigeration

Refrigeration slows down the biological, chemical, and physical reactions that shorten the shelf life of food. About one-half of the foods consumed in the United States are refrigerated or frozen, compared to about one-third canned (24). Prior to the 1900s, refrigeration consisted of ice boxes filled with ice cut from lakes. In the 1850s, beer brewers began using mechanical refrigeration. Later, in the 1880s, meat packers started using refrigerated railroad cars to ship carcasses rather than live animals. However, home refrigeration was not widespread in the United States until the 1940s.

Refrigerating Produce. Not all foods have to be refrigerated, but it definitely helps certain fruits and vegetables high in water content. Vegetables and fruits often continue to ripen and exchange gases with the environment after harvesting, even though they are cut off from their roots or leaves. When the fruit or vegetable's stored nutrient supply is finally depleted, the cells slowly die, and spoilage occurs as enzymes are released that break down, soften, and brown the tissue. Placing produce in the crisper section of the refrigerator limits the amount of oxygen available to the foods, which slows their metabolism and prolongs their life and quality.

Plastic bags used to store vegetables or fruits should have holes in them in order to let in a small amount of oxygen to prevent tissue death, but not enough to speed up deterioration. Trapping these gases by using a closed bag speeds ripening and deterioration.

Refrigeration Temperatures and Times. For safety purposes, refrigerators should be kept between just above freezing to no more than 40°F (4°C). A study reported that one-fifth of household refrigerators surveyed were set at or above 50°F (10°C), while the majority were set at about 45°F (7°C) (28).

All perishable foods should be refrigerated as soon as possible, preferably during transport, to prevent bacteria from multiplying. Refrigerator storage times vary according to the food (see back inside book cover). Hot foods can be cooled down by placing them in shallow containers and setting the containers in cold water. Hot foods may also be quickly cooled by distributing the heat with frequent stirring during the cooling process (54).

Freezing

Serendipity played a part in the discovery that foods could be frozen commercially. Clarence Birdseye was ice fishing when he noticed that a fish he pulled out of the water froze in mid-air. He experimented and found that fish frozen immediately after being caught tasted fresh when thawed and prepared weeks later. In 1930, Birdseye started the frozen food industry (45).

In the past, people had used nature's snow and ice to preserve foods in the winter of colder climates. In fact, Sir Francis Bacon is reputed to have died of pneumonia contracted while he was trying to preserve chickens by stuffing their cavities with snow (54).

It is now known that freezing foods at 0°F (−18°C) or below is the least damaging to the food's original flavor, nutrient content, and texture compared to most other preservation methods. Freezing makes water unavailable to microorganisms. In addition, chemical and physical reactions leading to deterioration are slowed by freezing. Some oxygen is still present, however, allowing these reactions to continue, with the result that frozen foods have a shorter shelf life than canned foods.

What Foods Can Be Frozen? The food's composition determines whether, and for how long, it can be stored in the freezer. Certain fruits, vegetables, and liquid dairy products do not freeze well. The plant tissues of unblanched fruits and vegetables are irreversibly damaged during freezing. Dairy products may have their original distribution of fat and water permanently altered to the point of becoming unacceptable in quality. The higher a food's fat content, the shorter its life span in the freezer, because fat can become **rancid** even when frozen. Conversely, foods containing very little water, such as dried fruits and beans, grains, nuts, and coffee, can have their useful life span doubled if they are frozen (Chemist's Corner 7-2).

> **Chemist's Corner 7-2**
>
> ***Glass Transition Temperature***
>
> The shelf life of frozen products is extended if their water movement is limited. Deterioration results when water migrates during storage, resulting in large ice crystal formation. One way to avoid this is to raise the food's glass transition temperature (Tg), the temperature at which the food is said to be in a "glassy" state. This glassy stage is reached when the food's pure water freezes and leaves behind a very viscous concentration of solutes that trap the remaining water. The trapped water cannot crystallize. Food components that significantly raise Tg are higher molecular compounds such as protein, polydextrose, and gums. Tg is not increased by low molecular weight compounds such as sucrose, sorbitol, lactilol, and certain maltodextrins (39).

How Long Can Food Be Frozen? Specific maximum freezer storage times for various foods are listed in the book's back inside cover, but in general, foods can be kept frozen for 2 to 12 months. The "first in, first out" rule should be followed: foods should be stored so that those most recently bought or frozen are the farthest back in the freezer, thereby moving food into a more easily accessible position as the freezer time increases.

Four Problems with Frozen Food. Despite its effectiveness as a food preservation method, freezing has several potential disadvantages. These include freezer burn, cell rupturing, fluid loss, and recrystallization.

Freezer Burn. **Freezer burn** occurs when air spaces are left in a package, or the package is torn, or moisture-proof paper is not used. Once the food is thawed, the texture of freezer-burned food is spongy, often ruining its quality to the point that it may be discarded (47).

The best way to prevent freezer burn is to wrap foods properly in airtight, vapor-resistant material, tape the package tightly, date it, and use the food before the optimum storage time has passed. An effective method for meats is to triple-wrap them, first with plastic wrap, then aluminum foil, and finally freezer paper secured with freezer tape. Masking tape should not be used, because the adhesive is not made to withstand freezing temperatures. The practice of keeping an up-to-date inventory on the freezer door showing the dates each food was frozen avoids the costly waste of freezer-burned food. The vacuum packaging of commercially frozen foods has almost eliminated the problem of freezer burn.

Cell Rupturing. Water expands when it freezes, and the resulting ice crystals pierce the food's cell walls, rupturing them and causing the food to take on an inferior texture. For this reason, some foods with a very high water content, such as lettuce, milk, tomatoes, and cottage cheese, should not be frozen. One way to minimize this problem is with rapid freezing, which results in smaller ice crystals, less cell rupturing, and a higher quality (4). Rapid freezing is accomplished commercially by the use of liquid nitrogen (–280°F/ –173°C), which freezes foods in a few minutes rather than the six or more hours needed in a home freezer. Several other commercial methods of freezing that overcome the cell-rupturing problem include the following (21, 54):

- *Air-blast freezing.* Frigid air is blown on foods as they pass on a belt.
- *Plate or contact freezing.* Foods are placed between two plates of metal while being cooled by refrigerants.
- *Immersion freezing.* Foods are submerged in a low-temperature brine.
- *Cryogenic freezing.* Incorporates very low temperatures (–140°/ –60° C) with liquid nitrogen, liquid carbon dioxide, or their vapors.

Fluid Loss. Most frozen meats lose fluid when thawed. This is known as "drip," and although it is red, it is not blood, but is actually water being lost from the cells. If too much fluid is lost, the meat's texture will be dryer when cooked than the fresh product would have been.

Recrystallization. Another problem with freezing is recrystallization, which occurs with longer storage times and temperature fluctuations caused by opening and closing the freezer door. Numerous small ice crystals melt, then combine on refreezing to form larger crystals, thus affecting the texture and quality of foods such as ice cream. During a lengthy power failure, the freezer door should be kept shut, and dry ice

> **KEY TERMS**
>
> ***Rancid*** The breakdown of the polyunsaturated fatty acids in fats that results in disagreeable odors and flavors.
>
> ***Freezer burn*** White or grayish patches on frozen food caused by water evaporating into the package's air spaces.

KEY TERMS

Pasteurization A food preservation process that heats liquids to 160°F (71°C) for 15 seconds, or 143°F (62°C) for 30 minutes, in order to kill bacteria, yeasts, and molds.

Ohmic heating A food preservation process in which an electrical current is passed through food, generating enough heat to destroy microorganisms.

Irradiation A food preservation process in which foods are treated with low doses of gamma rays, x-rays, or electrons.

(solidified carbon dioxide) should be placed in the freezer and/or refrigerator as soon as possible. The addition of 25 pounds of dry ice will keep a 10-cubic-foot freezer below freezing for two to three days. A much smaller amount is required for the refrigerator and refrigerator freezer. Once the power returns, any thawed foods can be refrozen as long as the temperature has stayed below 40°F (4°C).

Heat Preservation

Heat in its various forms can be used as a preservation method, because many of the microorganisms responsible for food spoilage or foodborne illnesses are susceptible to heat. Heating methods include boiling, pasteurization, and ohmic heating.

Boiling

The simplest heat preservation method, which has been used for centuries, is boiling. Even when canning evolved, it incorporated thoroughly boiling the can's contents before sealing. Ten minutes of boiling renders most foods free from microorganisms, but to remain that way, they must not be allowed to touch any unsterile object or be exposed to the air. Boiling temperatures can be exceeded by using pressure canners, autoclaves, and other instruments.

Pasteurization

Another method of preserving liquids and extending their shelf life through heat is **pasteurization**, which destroys non-spore-forming pathogenic microorganisms. In the process, many organisms that cause spoilage are destroyed as well. Milk is the most common beverage treated by pasteurization (discussed in more detail in Chapter 11), but fruit juices and other beverages can also be pasteurized. The USDA has approved the use of steam pasteurization for meat, in which whole beef carcasses are sprayed with a blast of steam, killing microorgansims on the surface (9). Steam-pasteurized meat is not considered kosher (see Chapter 1) because there is no rabbinical supervision, and the necessary temperatures required for kosher acceptance are not reached, as such temperatures would cook the edges of the carcass meat.

High-Temperature Pasteurization. Using temperatures higher than those used for conventional pasteurization are described as high-temperature, short-time (HTST), and ultrahigh temperature (UHT) (see Chapter 11 on milk). HTST temperatures and times start at 161°F (72°C) for 15 seconds, while UHT temperatures are above 280°F (138°C). Shorter times mean that there is minimal degradation of the products (36).

FIGURE 7-5 **Ohmic heating.**

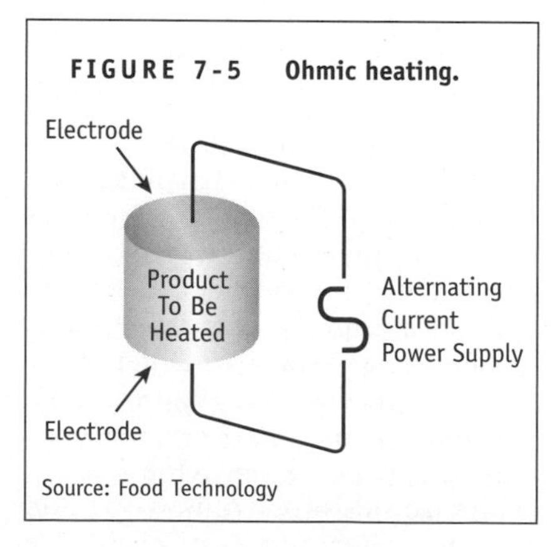

Source: Food Technology

Ohmic Heating

One of the latest heating techniques is **ohmic heating**, also known as resistance heating, joule heating, or electroheating (Figure 7-5) (46). In order to be a candidate for ohmic heating, the material must be in a suspended state such as a sauce (58). The major benefit of ohmic heating is that it minimizes overprocessing and results in higher-quality food products (26).

Other Preservation Methods

Irradiation (Cold Pasteurization)

Although **irradiation**, also known as cold pasteurization, has been used for a variety of applications, it is a relatively recent food preservation method. The Army Medical Department started to study the safety of irradiated foods in 1955, and the process was adopted in the early 1970s for foods consumed by astronauts (11). Irradiation of foods has expanded rapidly since it was first approved by the Food and Drug Administration in 1963 for the control of insects in wheat flour. Other foods now allowed to be irradiated include spices and seasonings, pork, fruits, vegetables, poultry, and red meat. In 2002, the concern with school-related foodborne outbreaks led the USDA to approve the use of irradiated meat for the school lunch program (17, 18, 40). Listed below are several examples of how irradation is used:

- Sterilization of medical equipment (instruments, surgical gloves, alcohol wipes, sutures, etc.)
- Sterilization of consumer products (adhesive bandages, contact lens cleaning solutions, cosmetics, etc.)
- Foods for immune-compromised hospital patients (e.g., AIDS, cancer, or transplant patients)
- Some foods for astronauts, who cannot risk foodborne illness
- Spices and seasonings used in products such as sausage and certain baked goods

The Irradiation Process. Food is passed through an irradiator, an enclosed chamber, where it is exposed to an ionizing energy source. A number of ionizing energy sources may be used (gamma rays, x-rays, or electrons), but gamma rays are the most commonly used (Chemist's Corner 7-3). Irradiation works by breaking down the chemical bonds within the DNA and other molecules in the cells. The treated food does not become radioactive itself, any more than does a briefcase passing through an airport security check. The energy of the radiation used is not high enough to induce changes in atomic nuclei that result in radioactivity (33).

Effects of Irradiation on Foods. Treating foods with gamma radiation renders them less susceptible to deterioration and destroys many, but not all, microorganisms. For example, while it destroys bacteria, yeasts, and molds, the dosage used on foods for pasteurization is too low to eradicate *C. botulinum* spores, and it may not eliminate smaller viruses (24). Irradiation reduces spoilage, lengthens shelf life, sterilizes or kills insects, and can protect the public from many bacterial foodborne diseases (Figure 7-6) (23). Meat can keep three to four times longer in the refrigerator when irradiated (1). Research has demonstrated that irradiated fresh fruit retained desirable quality for up to sixteen days, compared to nonirradiated fruit, which spoiled within six days. Not all produce can be irradiated successfully, however. Many fruits and vegetables become mushy and discolored, and actually spoil faster with irradiation (57). In fact, only about ten different fruits experience an extended shelf life with irradiation.

FIGURE 7-6 Irradiation inhibits mold formation in strawberries refrigerated for 17 days.

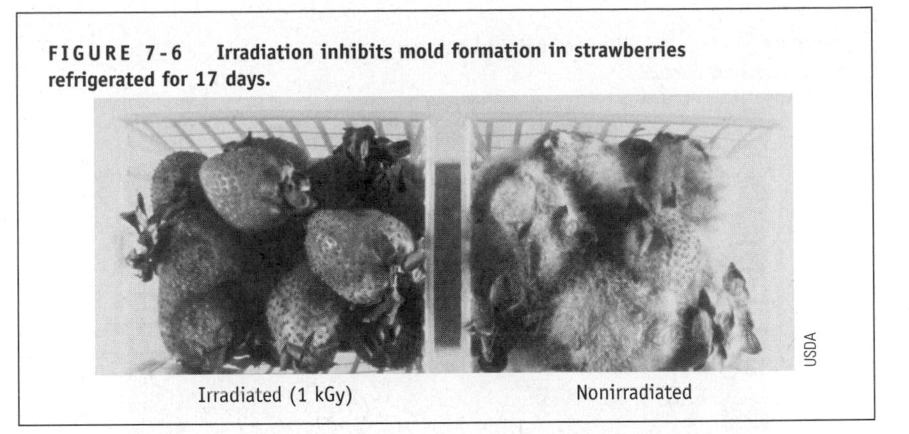

Chemist's Corner 7-3

Irradiation Dosages

Gamma rays used in irradiation are emitted from radioactive cobalt (^{60}Co), and possibly cesium (^{137}Cs). These gamma rays are expressed in kilograys (kGy). Kilograys describe the amount of radiation energy absorbed. For example, one kilogray equals 1,000 grays, and 1 gray equals 1 joule per kilogram. Irradiation dosages from low (under 1 kGy) to medium (1 to 10 kGy) kill insects, larvae, and pathogenic bacteria in wheat and wheat flour, inhibit sprouting in potatoes, and slow ripening and spoilage in fruits. Higher dosages (10 to 50 kGy) are enough to sterilize foods for use by astronauts or immune-compromised hospital patients (40).

Irradiation Pros and Cons. Despite official approval for its use by many organizations (FDA, Department of Agriculture, World Health Organization, United States Army, American Medical Association, American Dietetic Association, Institute of Food Technologists, and others) (51), food irradiation has aroused considerable controversy and is the subject of extensive studies (30, 42, 53). Many companies are reluctant to use irradiation because of consumer fears or lack of information (32). Consumer considerations include possible nutrient loss and the environmental hazards arising from the use of radioactive materials in irradiation facilities. There is also the question of free radicals produced by irradiation; however, the FDA states that these are generated by other heating methods. Proponents counter that irradiation eliminates the need for some chemical fumigants and preservatives, reduces spoilage losses, and increases food safety by destroying harmful microorganisms, especially *E.coli* and *salmonella* (30).

Consumer acceptance of irradiation may actually be higher than is indicated by media coverage. In fact, consumer concern about food irradiation has decreased (31). Survey participants of another study declared more concern for bacteria, pesticide residues, food additives, animal drug residues, and growth hormones (35). The major consideration for consumers is whether irradiated food should be clearly labeled so that they can make the choice for themselves (Figure 7-7) (50).

Pulsed Light

Another type of preservation technique, which uses the visible spectrum of radiation and is currently undergoing approval by the FDA, is called pulsed light (PureBright). This method works by exposing the food to intense, very brief flashes of light, which disrupt the cell membranes of bacterial cells, but not that of the surrounding food

FIGURE 7-7 The green radura symbol identifies irradiated food.

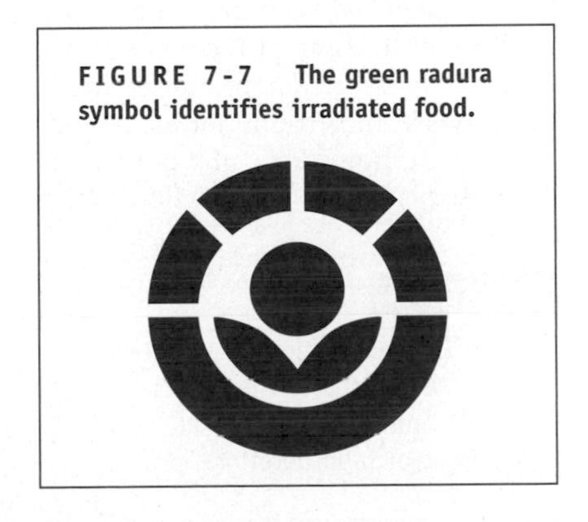

FIGURE 7-8 The effect of pulsed light* on *Staphylococcus aureus.*

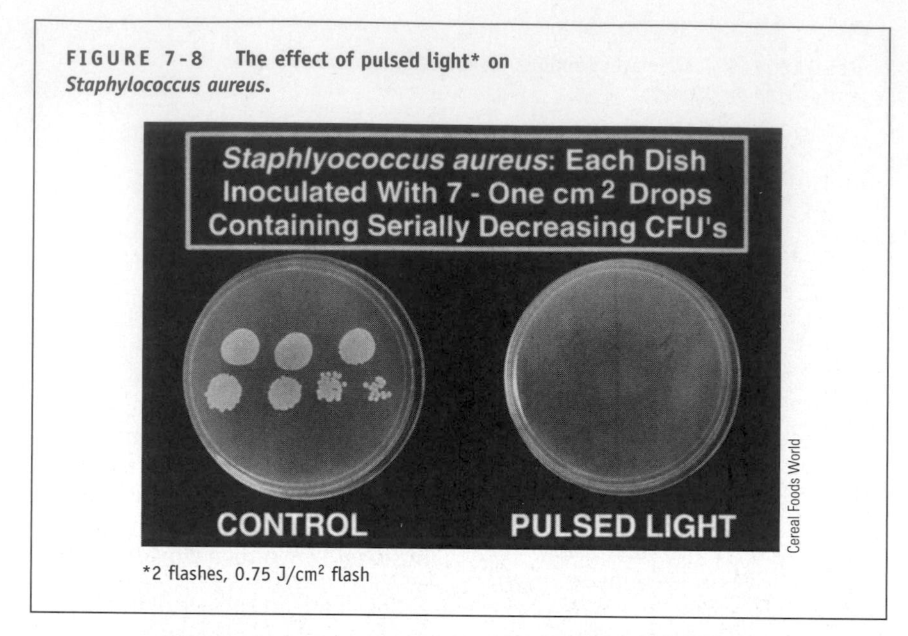

*2 flashes, 0.75 J/cm² flash

(Figure 7-8) (14). Pulsed light also kills fungi, spores, viruses, protozoa, and cysts (13). The intensity of the light, which lasts only a second, is 20,000 times brighter than sunlight, but there is no thermal effect, so quality and nutrient content are retained. A possible problem of this preservation method is that folds or fissures in the food may protect microorganims from being exposed to the pulsed light (20).

High-Pressure Processing

High-pressure processing of foods inactivates foodborne microorganisms at low temperatures without the use of chemical preservatives (52). Although discovered in 1899, treating foods with high pressure is a relatively new method of preservation, and one still under development. High-pressure processing or **pascalization** is named after Blaise Pascal, a 17th-century French scientist who described how contained fluids are affected by pressure. The pressure that pascalized foods are exposed to is tremendous. A normal car tire holds about 30 pounds of pressure per square inch, whereas the pressure applied to foods in pascalization is at least 50,000 pounds per square inch, applied for 15 minutes.

KEY TERMS

Pascalization A food preservation process utilizing ultrahigh pressures to inhibit the chemical processes of food deterioration.

HOW & WHY? ?????????

How does pascalization work?

Hydrostatic pressure deactivates various microorganisms and numerous enzymes responsible for food deterioration (40). The high pressure of pascalization kills many bacteria, yeasts, and molds (8). However, bacterial spores remain resistant and must be treated with acid to block their ability to germinate. Thus, naturally acidic foods such as yogurt, tomato products, and sliced fruit are best suited for pascalization. The process also denatures proteins and disrupts noncovalent bonds without interfering with the food's overall structure.

Foods Preserved with Pascalization. In Japan, the technique has been used since 1990 on some juices, jams, and jellies. Other foods now include fish, meat products, salad dressing, rice cakes, and yogurt (48). The first commercial food product produced in the United States that was vacuum packaged in plastic pouches and exposed to high pressures was guacamole. The texture, color, and taste of the guacamole was unaffected, while shelf life increased from 3 to 30 days (37). Other commercial products being treated with pascalization are some meats and seafoods (38).

Ozonation

The fresh, clean, invigorating smell in the air after a thunderstorm is due to ozone formed by the action of lightning (Figure 7-9). Ozone (an oxidizing agent) is commercially produced by exposing oxygen to an electrical current. The oxygen molecules (O_2) are broken apart, creating numerous oxygen fragments that reunite with other oxygen molecules to yield ozone (O_3).

It has been known for decades that ozone is an effective disinfectant and sanitizer for many food products. The FDA has allowed ozone to be used in the treatment of bottled drinking water, in meat-aging coolers where frozen beef carcasses are stored (19), and, since 1997, on food. Japan, the United Kingdom, Germany, France, the Netherlands, and Scandinavia allow ozone to be used in the storage of meat, fruit, cheese, and other products. Israel uses ozonation to reduce the decay of table grapes. In Europe, especially Germany and France, ozone is the principle means of sanitizing the public water supply, rather than chlorine, which is used in the United States (19). Advocates claim that ozone is safer and more effective than chlorine (43). Ozone is 1.5 times stronger than chlorine, is capable of killing a wider range of bacteria, and is free of chemical residues (56).

Aseptic and Modified-Atmosphere Packaging

Aseptic and modified-atmosphere packaging (MAP) has been used for many years on food products. Foods commonly packaged in this manner include processed meats and cheese, lunch kits, prepared poultry, ground turkey, pastas, sauces, snack foods, juices, and liquid dietary food (Figure 7-10).

Aseptically Packaged Food. Food that is aseptically packaged is sterilized, packed, and sealed in a sterilized

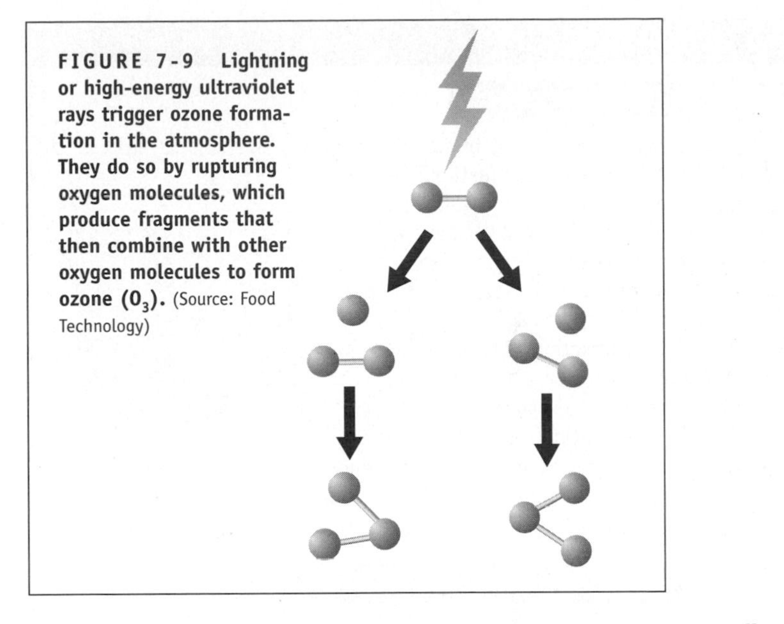

FIGURE 7-9 Lightning or high-energy ultraviolet rays trigger ozone formation in the atmosphere. They do so by rupturing oxygen molecules, which produce fragments that then combine with other oxygen molecules to form ozone (O_3). (Source: Food Technology)

container under sterile conditions (34). This technique, which has been used since the 1940s, allows a food or beverage product to be stored on the shelf at room temperatures. A classic example of an aseptically packaged product is the "juice box." The advantages of aseptic packaging is its reclosability (certain products), handling convenience made possible by the light, flexible containers, and cost savings to the manufacturer (41).

FIGURE 7-10 Aseptic and modified-atmosphere packaging.

Modified-Atomosphere Packaging (MAP). Aseptically treated foods are sometimes combined with modified-atmosphere packaging, which consists of changing the air composition around the food to prolong its shelf life; specifically, reducing oxygen and increasing carbon dioxide (7, 29). A form of modified-atmosphere packaging is conducted at the consumer level when shelf life is increased by removing as much air as possible from an opened bag of food before putting it away. The shelf life of products packaged with modified-atmosphere packaging is considerably longer than foods packaged without this technology.

Modified-atmosphere packaging is commonly used for fruits and vegetables that are "ready-to-use"—fresh, peeled, sliced, shredded, or grated produce sold within a week of preparation (16). Cured cheese is commonly packaged in modified-atmosphere packaging where the carbon dioxide has been flushed out (5). Most meat and poultry is packaged using modified-atmosphere packaging, and applications for bakery products continue to grow (6, 15). In Denmark, aseptic and modified-atmosphere packaging of fresh meat is a common practice. Instead of being cut in the store, meat is cut centrally under aseptic conditions by maintaining freezing temperatures, packaged in an atmosphere high in carbon dioxide and oxygen (oxygen maintains the red color of meat), and **hermetically sealed**. Meat packaged this way has a ten-day shelf life versus the usual two to three days (25).

Many aseptically packaged and MAP foods are combined with refrigeration, as seen with refrigerated juices and chilled puddings and pastas. Cooked rice packaged in this manner and combined with refrigeration is a popular product in Japan.

KEY TERMS

Hermetically sealed Foods that have been packaged airtight by a commercial sealing process.

PICTORIAL SUMMARY / 7: Food Preservation

The World Health Organization estimates that about 20 percent of the world's food supply is lost to spoilage. The average North American family discards about one-fourth of the food they purchase because of spoilage. Understanding the biological, chemical, and physical changes that contribute to food spoilage can help prevent food waste.

FOOD SPOILAGE

Foods are organic in nature and are subject to spoilage. Food deterioration decreases both the quality and safety of food due to a combination of changes:

Biological: Microorganisms are responsible for some of the biological reactions in foods that contribute to their deterioration.

Chemical: Chemical reactions that contribute to spoilage often involve enzymes.

Physical: Physical reactions include evaporation, water drip, and separation.

Some of these changes are preventable, some can be alleviated, and others will occur to some degree despite current preservation methods.

FOOD PRESERVATION METHODS

Keeping spoilage to a minimum is the objective of both food manufacturers and consumers. Food preservation methods include:

- Drying
- Curing
- Fermentation
- Pickling
- Canning
- Edible coatings
- Aseptic and modified-atmosphere packaging
- Heat preservation
- Cold preservation
- Irradiation
- Pulsed light
- Ozone
- Pascalization

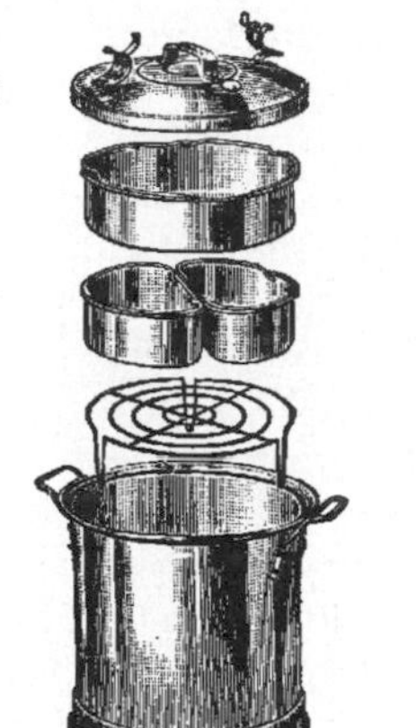

COLD PRESERVATION

One-half of the foods consumed in the United States are refrigerated or frozen. Refrigeration slows down the biological, chemical, and physical reactions that shorten the shelf life of food. For safety purposes, refrigerators should be kept between just above freezing to no more than 40°F (4°C), and foods should be frozen at 0°F (-18°C) or below, for the least damage to the food's original flavor, nutrient content, and texture.

HEAT PRESERVATION

Heat in its various forms can be used as a preservation method, because many of the microorganisms responsible for food spoilage or foodborne illnesses are susceptible to heat. Heating methods include boiling, pasteurization, and ohmic heating. Ohmic heating (also known as resistance heating, joule heating, or electroheating) is a recent development that sends an alternating electrical current through food to kill microorganisms, minimizing overprocessing and resulting in a higher-quality food product.

OTHER PRESERVATION METHODS

Other preservation methods:

Irradiation: Food is passed through an enclosed chamber, where it is exposed to an ionizing energy source, such as gamma rays, x-rays, or electrons, which renders them less susceptible to deterioration and destroys many, but not all, microorganisms. This method is controversial and under study.

Pulsed light: Currently undergoing approval by the FDA, this method works by exposing the food to intense, very brief flashes of light, which disrupts the cell membranes of bacterial cells but not the surrounding food.

Pascalization: A relatively new method, pascalization subjects food to very high pressure, a process that kills many bacteria, yeasts, and molds, although bacterial spores remain resistant. Acidic foods are best suited for this method since the acid will prevent the spores from germinating.

Aseptic and modified-atmosphere packaging: Aseptically packaged food is sanitized, packed under sanitized conditions, and stored on the shelves at room temperature. Modified-atmosphere packaging changes the air composition around the food, decreasing oxygen and increasing carbon dioxide.

REFERENCES

1. Alur MD, SP Chawla, and PM Nair. Microbiology method to identify irradiated meat. *Journal of Food Science* 57(3):593–595, 1992.
2. Ashie INA, and BK Simpson. Effects of hydrostatic pressure on alpha-macroglobulin and selected proteases. *Journal of Food Biochemistry* 18:377–391, 1995.
3. Baldwin EA, et al. Use of lipids in coatings for food products. *Food Technology* 51(6):56–62, 1997.
4. Berne S. New techniques yield a tastier freeze. *Prepared Foods* 163(7):109–115, 1994.
5. Brody AL. Say "Cheese"—and package it, please! *Food Technology* 55(7):76, 2001.
6. Brody AL. Case-ready fresh red meat: Is it here or not? *Food Technology* 56(1):77, 2002.
7. Brody AL. Integrating aseptic and modified-atmosphere packaging to fulfill a vision of tomorrow. *Food Technology* 50(4):56–66, 1996.
8. Cano MP, A Hernandez, and B De Ancos. High pressure and temperature effects on enzyme inactivation in strawberry and orange products. *Journal of Food Science* 62(1):85–88, 1997.
9. Castillo A, et al. Decontamination of beef carcass surface tissue by steam vacuuming alone and combined with hot water and lactic acid sprays. *Journal of Food Protection* 62(2):146–151, 1999.
10. Clark JP. Drying still actively researched. *Food Technology* 56(9):76, 2002.
11. Cliver DO, et al. Position of the American Dietetic Association: Food irradiation. *Journal of the American Dietetic Association* 96(1):69–72, 1996.
12. Debeaufort F, JA Quezada-Gallo, and A Voilley. Edible films and coatings: Tomorrow's packages: A review. *Critical Reviews in Food Science* 38(4):299–313, 1998.
13. Dunn J, et al. Pulsed white light food processing. *Cereal Foods World* 42(7):510–515, 1997.
14. Dunn J, T Ott, and W Clark. Pulsed-light treatment of food and packaging. *Food Technology* 49(9):95–98, 1995.
15. Ferrante MA. Modified atmosphere packaging: Putting the pieces together. *Food Engineering* 70(1):79–82, 1998.
16. Finn MJ, and ME Upton. Survival of pathogens on modified-atmosphere-packaged shredded carrot and cabbage. *Journal of Food Protection* 60(11):1347–1350, 1997.
17. Frenzen PD, et al. Consumer acceptance of irradiated meat and poultry in the United States. *Journal of Food Protection* 64:2020–2026, 2001.
18. Giese J. Identifying bioengineered and irradiated foods. *Food Technology* 51(11):88, 93, 1997.
19. Graham DM. Use of ozone for food processing. *Food Technology* 51(6):72–75, 1997.
20. Hoover DG. Minimally processed fruits and vegetables: Reducing microbial load by nonthermal physical treatments. *Food Technology* 51(6):66–71, 1997.
21. Hung YC, and NK Kim. Fundamental aspects of freeze-cracking. *Food Technology* 50(12):59–61, 1996.
22. Jangchud A, and MS Chinnan. Peanut protein film as affected by drying temperature and pH of film forming solution. *Journal of Food Science* 64(1):153–157, 1999.
23. Johnson J, and M Marcotte. Irradiation control of insect pests of dried fruits and walnuts. *Food Technology* 53(6):46–51, 1999.
24. Katz F. Battling bacteria: New ways for meat, poultry processors to attack contamination. *Food Technology* 57(6):83–84, 1996.
25. Katz F. Is it time for changes in meat packaging and handling? *Food Technology* 51(6):99–100, 1997.
26. Kim HJ, et al. Validation of ohmic heating for quality enhancement of food products. *Food Technology* 50(5):253–262, 1996.
27. Krotchta JM, and C De Mulder-Johnston. Scientific status summary. Edible and biodegradable polymer films: Challenges and opportunities. *Food Technology* 51(2):61–72, 1997.
28. Labuza TP, and AE Sloan. *Food for Thought*. Avi Publishing, 1977.
29. Lakakul R, RM Beaudry, and RJ Hernandez. Modeling respiration of apple slices in modified-atmosphere packages. *Journal of Food Science* 64(1):105–110, 1999.
30. Loaharanu P. Cost/benefit aspects of food irradiation: A summary of the report and recommendations of the working group of the FAO/IAEA/WHO International Symposium. *Food Technology* 48(1):104–108, 1994.
31. Lusk JL, JA Fox, and CL McIlvain. Consumer acceptance of irradiated meat. *Food Technology* 53(3):56–59, 1999.
32. Marcotte M. Irradiated strawberries enter the U.S. market. *Food Technology* 46(5):80, 1992.
33. McCrae R, RK Robinson, and MJ Sadler. *Encyclopaedia of Food Science, Food Technology, and Nutrition*. Academic Press, 1993.
34. Mermelstein NH. Aseptic bulk storage and transportation. *Food Technology* 54(4):107–110, 2000.
35. Mermelstein NH. Washington News. Irradiation of shell eggs approved. *Food Technology* 54(8):30, 2000.
36. Mermelstein NH. High-temperature, short-time processing. *Food Technology* 55(6):65–70, 2001.
37. Mermelstein NH. High-pressure processing reaches the U.S. market. *Food Technology* 51(6):95–96, 1997.
38. Myer RS, et al. High-pressure sterilization of foods. *Food Technology* 54(11):67–72, 2000.
39. O'Donnell CD. Stabilizing the big freeze. *Prepared Foods* 165(1): 45–46, 1996.
40. Olsen DG. Irradiation of food. *Food Technology* 52(1):56–62, 1998.

41. Palaniappan S, and CE Sizer. Aseptic process validated for foods containing particulates. *Food Technology* 51(8):60–68, 1997.
42. Pszczola DE. How Food Technology covered irradiation over the years. *Food Technology* 51(2):49, 1997.
43. Richardson SD, et al. Chemical byproducts of chlorine and alternative disinfectants. *Food Technology* 52(4):58–61, 1998.
44. Robeck MR. Product liability issues related to food irradiation. *Food Technology* 50(2):78–82, 1996.
45. Ronsivalli LJ, and ER Vieira. *Elementary Food Science.* Van Nostrand Reinhold, 1992.
46. Sastry SK, and Q Li. Modeling the ohmic heating of foods. *Food Technology* 50(5):246–249, 1996.
47. Simatos D, and G Blond. DSC studies and stability of frozen foods. *Advances in Experimental Medicine and Biology* 302: 139–155, 1991.
48. Sizer CE, VM Balasubramaniam, and E Ting. Validating high-pressure processes for low-acid foods. *Food Technology* 56(2): 36–42, 2002.
49. Stanley DW, et al. Low temperature blanching effects on chemistry, firmness, and structure of canned green beans and carrots. *Journal of Food Science* 60(2): 327–333, 1995.
50. Stevenson MH. Identification of irradiated foods. *Food Technology* 48(5):141–144, 1994.
51. Thayer DW, et al. Radiation pasteurization of food. *Council for Agricultural Science and Technology* 7:1–10, 1996.
52. Ting E, VM Balasubramaniam, and E Raghubeer. Determining thermal effects in high-pressure processing. *Food Technology* 56(2):31–42, 2002.
53. U.S. Department of Health and Human Services. Food and Drug Administration. Irradiation in the production, processing, and handling of food; Final rule (21 CFR Part 179). *Federal Register* 62(232):64107–64121, 1997.
54. VanGarde SJ, and M Woodburn. *Food Preservation and Safety: Principles and Practice.* Iowa State University Press, 1994.
55. Van Loey A, et al. Potential *Bacillus subtilis* alpha-amylase-based timetemperature integrators to evaluate pasteurization processes. *Journal of Food Protection* 59(3):261–267, 1996.
56. Xu L. Use of ozone to improve the safety of fresh fruits and vegetables. *Food Technology* 53(10):58–62, 1999.
57. Yu L, CA Reitmeier, and MH Love. Strawberry texture and pectin content as affected by electron beam irradiation. *Journal of Food Science* 61(4):844–846, 1996.
58. Zoltai P, and P Swearingen. Product development considerations for ohmic processing. *Food Technology* 50(5):263–268, 1996.

WEBSITES

United States Government website on general descriptions of food processing techniques of food and beverage manufacturers:

www.agnic.org/cc/d_q100.html

The California Food Processing Industry:

www.commerce.ca.gov/california/economy/profiles/food.html

Food Processing Machinery and Supplies Association:

www.fpmsa.org/

Prepared Foods Magazine and Food Processing Magazine—trade publications for food manufacturers:

www.preparedfoods.com and **www.foodprocessing.com**

Virtual link to the food processing industry:

www.foodonline.com

Government Food Regulations

The use of food laws and regulations dates back to ancient Greece and Rome, when it became illegal to dilute wine with water (15). During the Middle Ages, trade guilds policed themselves to ensure that teas and spices were pure, thus protecting against anyone taking unfair trade advantage with an inferior, cheaper product. In 1630, a colonist of the Massachusetts Bay Colony was whipped for selling "a water of no worth" as a cure for scurvy (36).

In the United States, the public's demand for laws to protect consumers against contaminated foods helps make this country's food supply the safest in the world (28). The government has responded at all levels—federal, state, and local—with laws and regulations to protect the foods and beverages consumed by the public (14). Companies that do not comply with the rules can be shut down until they do, or they may be put out of business. Most companies are motivated to comply voluntarily with regulations in order to maintain consumer confidence in their products (36). International food standards have been developed by the **Codex Alimentarius Commission** of the Food and Agriculture Organization and the World Health Organization of the United Nations (7).

This chapter provides an overview of the United States government food regulations and the various agencies involved—the Food and Drug Administration, the United States Department of Agriculture, the Environmental Protection Agency, Centers for Disease Control and Prevention, and other regulatory agencies.

KEY TERMS

Codex Alimentarius Commission
The international organization that develops international food standards, codes of practice, and other guidelines to protect consumers' health.

Federal Food Laws

Food and Drug Act (1906)

The federal Food and Drug Act of 1906, sometimes referred to as the Pure Food Law, was the beginning of the government's involvement in food regulation. This followed a series of nationally publicized blunders by manufacturers, including reports following the 1898 Spanish-American War that many soldiers had become ill after being fed spoiled canned meat and contaminated flour. Around the same time, it was found that some dairies were using formaldehyde as a milk preservative. One notorious incident during the early 1900s involved store owners who substituted poisonous mushrooms for edible varieties.

Upton Sinclair's novel, *The Jungle*, published in 1905 and very widely read, exposed the filthy conditions in Chicago meat-packing plants. Animals that had died of disease were sometimes processed into meat products, and it was not uncommon to find foods contaminated with filth, including the hairs, urine, and feces of rats and mice, insects, maggots, larvae, parasitic worms, and excrement from humans and animals. Sausage meat often contained ground rats, other vermin, and even human fingers.

Sinclair's book added to the public groundswell of demand for the government to take action. The passage of the 1906 Federal Food and Drug Act followed, but the list of injurious foods and manufacturing processes continued to

KEY TERMS

Food additive A substance added intentionally or unintentionally to food that becomes part of the food and affects its character.

■ ■ ■ ■

Food Code An FDA publication updated every two years that shows food service organizations how to prevent foodborne illness while preparing food.

grow, resulting eventually in the need for modifications to the original bill.

Food, Drug, and Cosmetic Act (1938)

The food laws were rewritten, and in 1938 the Food, Drug, and Cosmetic Act was passed. This was brought on, in part, by the deaths of 100 people who ingested a poisonous "Elixir of Sulfanilamide" (15). The federal government entrusted several agencies to oversee consumer food safety, but the result was a fragmented and overlapping system of regulations (18).

Numerous Government Agencies

Numerous laws were introduced, passed, and enforced over ensuing years (Figure 8-1). However, there was no systematic application from a single government agency (13). The Safe Food Act of 1997 proposed the creation of a single independent agency to oversee food safety (6). As of this date, however, four federal agencies are primarily responsible for ensuring the safety of the food supply: the Food and Drug Administration (FDA), the U.S. Department of Agriculture (USDA), the Environmental Protection Agency (EPA), and the Centers for Disease Control and Prevention (CDC). Other agencies involved in monitoring food safety include the Department of Commerce, the Federal Trade Commission, the Public Health Service, the Consumer Product Safety Commission, and the Department of the Treasury (13). These latter agencies have jurisdiction over food that travels across state lines.

FIGURE 8-1 Timeline of selected Congressional acts and amendments legislating the U.S. government's role in regulating the food supply.

1906	Pure Food and Drug Act (FDA)
1906	Federal Meat Inspection Act (USDA)
1914	Federal Trade Commision Act (FTC)
1935	Federal Alcohol Administrative Act (U.S. Dept. of the Treasury)
1938	Federal Food, Drug, and Cosmetic Act (FDA)
1944	Public Health Services Act (USDA)
1946	Agricultural Marketing Act (USDA)
1947	Federal Insecticide, Fungicide, and Rodenticide Act (EPA)
1954	Pesticide Residue Amendment to 1938 (EPA)
1956	Fish and Wildlife Act (U.S. Dept. of Commerce)
1957	Poultry Products Inspection Act (USDA)
1958	Food Additives Amendment to 1938
1960	Color Additives Amendment to 1938
1966	Fair Packaging and Labeling Act (FDA)
1967	Wholesome Meat Act (USDA)
1970	Egg Products Inspection Act (USDA)
1972	Federal Environmental Pesticide Control Act (EPA)
1974	Safe Drinking Water Act (EPA)
1977	Saccharin Study and Labeling Act
1990	Nutrition Labeling and Education Act (FDA)
1996	Food Quality Protection Act
1996	Pathogen Reduction Hazard Analysis and Critical Control Points (HACCP) Systems Final Rule
1997	Pathogen Reduction Act
1997	Safe Food Act
2002	Country of Origin Labeling (or COOL) Requirements
2002	Bioterrorism Preparedness Act

Food and Drug Administration

The Food and Drug Administration, a branch of the Department of Health and Human Services, is the oldest federal consumer protection agency in the United States. Its job is to enforce the Food, Drug, and Cosmetic Act of 1938, which ensures that the food, drugs, and cosmetics purchased by consumers are safe, wholesome, and produced under sanitary conditions. FDA has jurisdiction over the production of all foods involved in interstate marketing, with the exception of the meat, poultry, and egg industries, which are regulated by the USDA. The FDA is responsible for inspecting facilities and manufacturing processes, setting standards, labeling, and regulating **food additives**. It also conducts research and educates the public about food and nutrition.

The Code of Federal Regulations

The specific regulations enforced by the FDA are listed in the Code of Federal Regulations (CFR), Title 21. New regulations and proposals are published throughout the year in the Federal Register, updated April 1 of every year in an annual edition. A publication specific to food is the FDA's **Food Code.** The Food Code provisions are compatible with the Hazard Analysis and Critical Control Point (HACCP) system (see Chapter 3), and are used by local, state, and federal regulators to update their own food safety rules.

Research/Education

In 1883, Dr. Harvey W. Wiley, a chemistry professor at Purdue University, left that institution to become Chief of the Bureau of Chemistry, which later became the FDA. At the time, the use of chemical preservatives in food and beverages was virtually uncontrolled (15). In an attempt to generate support for the introduction and implementation of food safety laws, Dr. Wiley recruited scientists to study the safety of foods in the marketplace and took the findings to the public. Research continues to be an important function of the FDA, and the results of this research are found in two major publications: (1) *FDA Consumer*, and (2) *FDA Drug Bulletin*. The information summarized in these journals often covers products that come before the FDA for approval.

FDA Inspections

The Food, Drug, and Cosmetic Act prohibits the sale of adulterated products—any foods that are defective, unsafe, contaminated with filth, or produced under unsanitary conditions. Good Manufacturing Practice Regulations set the requirements for quality controls such as sanitation inspection, and these regulations are under constant scrutiny and subject to change (15).

Inspection is the tool for determining whether or not food manufacturers meet these standards. The FDA, based in Rockville, Maryland, has about 1,700 inspectors to cover over 90,000 FDA-regulated businesses (34). These inspectors are located in regional, district, and local offices spread throughout 157 cities across the country. Although they cannot visit every site, they do inspect more than 20,000 facilities a year to determine whether products are being prepared, processed, and packaged under sanitary conditions.

FDA Enforcement of Its Laws. Prior to the 1906 passage of the Pure Food Law, judges had no authority to stop rampant food adulteration. For example, a product called "fruit jam," consisting of water, sugar, grass seed, and artificial color, could be sold. This kind of unethical practice not only cheated the consumer, but also gave the offending manufacturer a higher profit margin than a legitimate competitor. The FDA now has the legal power to enforce the sanitation standards set by law. If a shipment is found to be unfit for consumers, it is either detained in order to determine whether it should be exported or destroyed, or conditionally released back to the manufacturer until it is brought into compliance with the law. Another tool of the FDA is **product recall** (11). Once a product is recalled, the manufacturer has three alternatives:

- It can allow the FDA to dispose of the food product.
- It can contest the government's charges in court.
- It can request permission of the court to bring the product into compliance under the law.

Over 3,000 products are withdrawn from the marketplace every year, either by recall or court-ordered seizure. The two most common reasons for the FDA recalling foods or cosmetics are mislabeling and microbial contamination (40).

Allowable Contaminants. The FDA, acknowledging that all filth cannot be eliminated, has set allowable amounts of contaminants in foods, some of which are shown in Table 8-1. This policy is based on a 1956 decision that insect and worm parts fall under the legal principle of *de minimis non curat lex*, "the law does not concern itself with trifles" (36). It is simply impossible to guarantee 100 percent compliance with the law or the absolute wholesomeness of foods. For example, it is extremely difficult, if not impossible, to ensure that all flour is free of insect parts or that spinach does not contain a single aphid. Since eliminating all contaminants is an unattainable goal, setting acceptable levels is an economic and practical necessity.

KEY TERMS

Product recall Civil court action to seize or confiscate a product that is defective, unsafe, filthy, or produced under unsanitary conditions.

Standards of Identity Requirements for the type and amount of ingredients a food should contain in order to be labeled as that food.

Standards of Minimum Quality Minimum quality requirements for tenderness, color, and freedom from defects in canned fruits and vegetables.

Standards of Fill Requirements for the amount of raw product that must be put into a container before liquid (brine or syrup) is added.

FDA Standards

The FDA has established three standards to ensure the content and quality of various foods: **Standards of Identity**, **Standards of Minimum Quality**, and **Standards of Fill**. Prior

TABLE 8-1

What the FDA Allows in Selected Foods

Don't read the information in the table if you are about to eat lunch—you may not like what may be in your food. The list below represents what the FDA considers an acceptable level of contaminants in the foods listed.*

Food	*Acceptable Levels of Filth*
Chocolate	Fewer than one rodent hair/100 g
Coffee beans	Less than 10% insect-infested
Fig paste	Fewer than 13 insect heads/100 g
Fish (fresh frozen)	Less than 5% "definite odor of decomposition"
Mushrooms (canned)	Fewer than 20 maggots of any size/100 g (drained)
Peanut butter	Fewer than 30 insect fragments/100 g
Popcorn	Less than one rodent pellet/sample or one rodent hair/2 samples
Spinach	Fewer than 50 aphids, thrips, or mites/100 g
Tomato paste	Fewer than 30 fly eggs/100 g

*Adapted from data provided by the FDA at http://www.cfsan.fda.gov/~dms/dalbook.html.

to the establishment of these standards, unscrupulous manufacturers were at liberty to replace expensive ingredients with water, or to sell cans that were only half full of food.

Standards of Identity. These standards were established to prevent economic deception for common food products purchased by consumers. Minimum amounts of specific ingredients have been established for a large number of foods, including catsup, cheese, ice cream, frankfurters, bread, mayonnaise, salad dressings, soups, and jams (Figure 8-2) (31). Standards of Identity were established to ensure quality and consistency, and to curb unfair business practices. When a food does not adhere to the agreed-upon standard, it cannot be labeled by its common name. For example, raisin bread cannot be designated as such unless the raisins equal 50 percent of the flour's weight. Fruit jam cannot be so named unless it consists of at least 45 percent fruit (31). Beef stew must have at least 25 percent beef. Even with a Standard of Identity designation, ingredients must still be listed on the food's label.

FIGURE 8-2 FDA Standards of Identity.

Food manufacturers have to list the ingredients in standardized foods, which include the following product categories.

- Milk and cream
- Cheeses and related products
- Frozen desserts
- Bakery products
- Cereal flours and related products
- Macaroni and noodle products
- Canned fruits
- Canned fruit juices
- Fruit butters, jellies, and jams or preserves
- Fruit pies
- Canned vegetables
- Vegetable juices
- Frozen vegetables
- Eggs and egg products
- Fish and shellfish
- Cacao products (cocoa, chocolate)
- Tree nut and peanut products
- Margarine
- Sweeteners and table syrups
- Food dressings and flavorings (mayonnaise, salad dressing, vanilla flavoring)

Standards of Minimum Quality. These standards limit such things as the "string" in string beans, the hardness of peas, and the "soupiness" in cream-style corn. The Standards of Minimum Quality are mandatory and are not to be confused with the voluntary grades of A, B, C (or Fancy, Choice, Standard) set by the USDA for canned fruits and vegetables. Canned foods that do not meet the Standards of Minimum Quality are labeled "Below standard in quality; good food—not high grade."

Standards of Fill. Until a Standard of Fill regulation was put into effect, manufacturers were not held legally responsible for filling a can to the top with its ingredients. Cans were often only partially packed with solid contents while the remainder of the volume was filled with fluid. Consumers are now safe from this type of manufacturing fraud because the Standards of Fill require manufacturers to have a minimum weight for solid food after the liquid is drained from the can. It is further mandated that cans must be filled to their maximum capacity (usually 90 percent) with solid ingredients. Some foods, such as nuts, applesauce, and crushed pineapple, have a minimum fill in terms of total food in the container.

Food Labeling

The FDA is responsible for ensuring that consumers are informed about a packaged product's ingredients and nutrient content. The exceptions are meat and poultry products, which are regulated by the USDA (22).

In the 1960s consumers began to be concerned about processed foods and started demanding to know exactly what was in the foods they were ingesting. As a response to this public concern, the government developed new labeling regulations. A legal requirement was also instituted, through the Fair Packaging and Labeling Act, to include the following information on the food label:

- The ingredients of packaged (canned, bottled, boxed, and wrapped) foods listed in descending order by weight. Food additives, colors, and chemical preservatives are also required to be listed on the label.
- The name and form (crushed, sliced, whole) of the product.
- The net amount of the food or beverage by weight, measure, or count. In addition to "net weight," two other types of weight measurement include "drained weight" and "solid content" (Figure 8-3).
- The name and address of the manufacturer, packer, or distributor.
- The nutrient content depicted as "Nutrition Facts" (Figure 8-4).

Nutrition Facts Label. In 1994, the nutrient label on foods changed to "Nutrition Facts" when nutrient labeling became mandatory (Figure 8-5 on page 156). Food labels provide consumers with pertinent information in making reasonable food choices.

Serving Size and Number. The first items on the food label are serving size and the number of servings per container.

FIGURE 8-3 "Drained weight" and "solid content" as weight measurements.

Drained weight. Net drained wt 9.25 oz.

Solid content. Wt of pineapple 14 oz.

FIGURE 8-4 Mandatory nutrient labels assist consumers in making healthy choices.

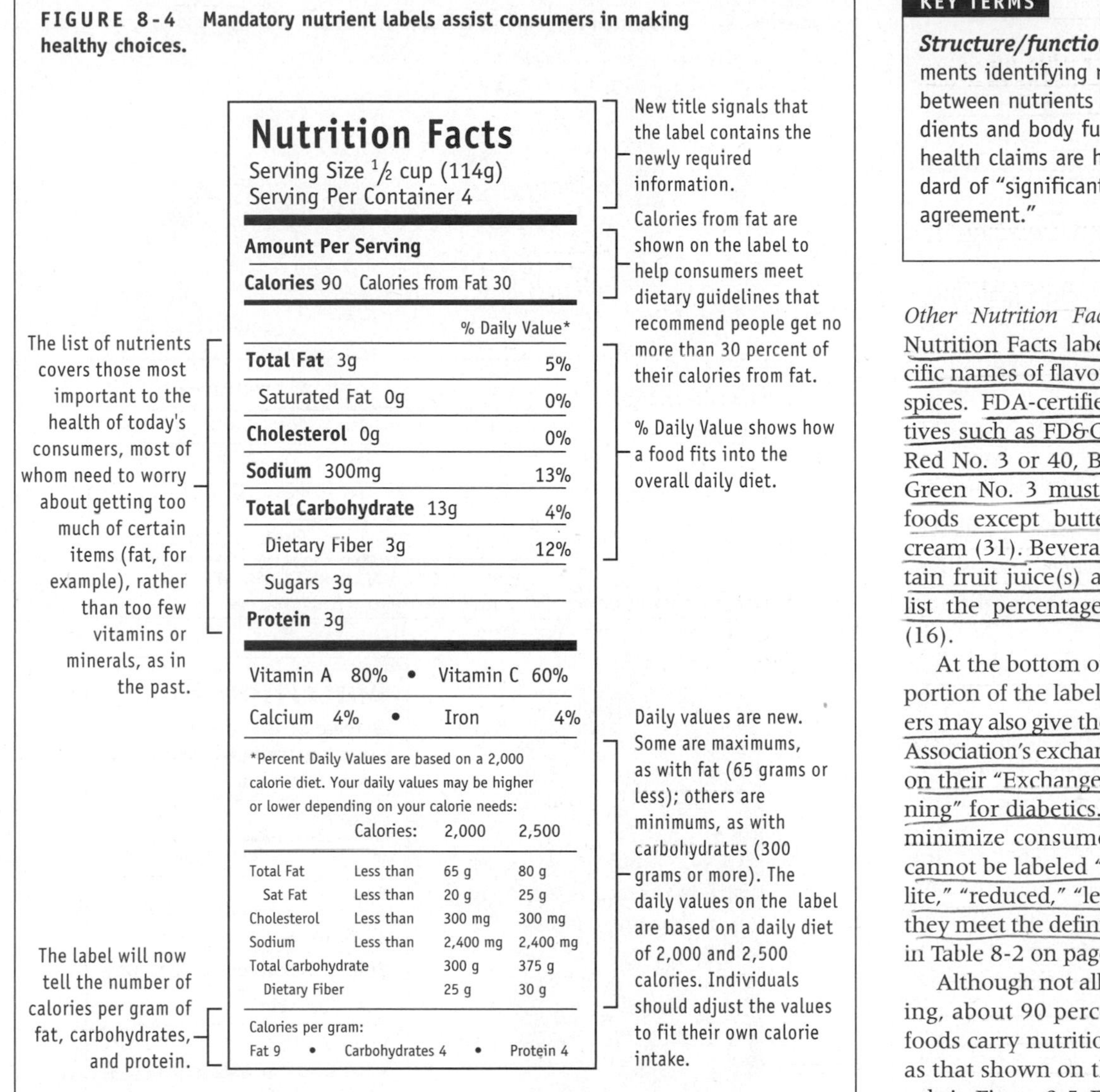

KEY TERMS

Structure/function claims Statements identifying relationships between nutrients or dietary ingredients and body function. These health claims are held to the standard of "significant scientific agreement."

These should be checked carefully because serving sizes were previously decided by manufacturers, who sometimes divided foods high in calories (kcal) or fat into an unreasonable number of serving sizes to disguise their true contents. For example, a candy bar containing 220 calories (kcal), when divided by a manufacturer into 2½ servings, would contain only 90 calories (kcal) per serving, looking more appealing to consumers eating the whole bar. New regulations require that a serving size be defined as the "amount of food customarily eaten at one time." Nevertheless, in some instances, they may still be smaller (31).

Calories (kcals) and Nutrients. The calories (kcals) are listed below the serving information on the Nutrition Facts label. Included are the calories (kcal) from fat, actual grams of fat (total and saturated), cholesterol (milligrams), sodium, total carbohydrate (dietary fiber and sugar), and protein. Also listed is the "% Daily Value" for these nutrients as well as vitamins A and C, calcium (Ca), and iron (Fe). Daily Values are nutrient standards derived from the Daily Reference Values (DRV) and Reference Daily Intakes (RDI). Daily Reference Values refer to fat, saturated fat, cholesterol, carbohydrates, fiber, sodium, and potassium. Reference Daily Intakes cover other nutrients (protein, vitamins/minerals). Daily Values are based on a daily diet of 2,000 or 2,500 calories (kcal), and are mandatory for 10 food components while optional for 22 others (21).

Other Nutrition Facts Information. The Nutrition Facts label also includes specific names of flavorings, colorings, and spices. FDA-certified food color additives such as FD&C Yellow No. 5 or 6, Red No. 3 or 40, Blue No. 1 or 2, and Green No. 3 must be declared on all foods except butter, cheese, and ice cream (31). Beverages claiming to contain fruit juice(s) are now required to list the percentage of real fruit juice (16).

At the bottom of the Nutrition Facts portion of the label, some manufacturers may also give the American Diabetes Association's exchange calculation based on their "Exchange Lists for Meal Planning" for diabetics. Finally, in order to minimize consumer confusion, foods cannot be labeled "free," "low," "light/lite," "reduced," "less," or "high" unless they meet the definition standards listed in Table 8-2 on page 157 (21).

Although not all foods require labeling, about 90 percent of all processed foods carry nutrition information such as that shown on the cans of tuna and cola in Figure 8-5. Foods not required to carry a nutrition label include raw foods (fresh fruits, vegetables, and fish), game, restaurant food, deli and bakery items, foods containing insignificant amounts of nutrients (coffee, tea, spices), bulk food, infant formula, medical foods, donated foods, and foods in packages weighing less than half an ounce, such as gum or breath mints (16).

Health Claims on Food Labels. Prior to the new food labeling, claims that linked a nutrient or food to a disease, in either a positive or a negative way, were not permitted. Now, under the new regulations, the FDA permits **structure/function claims** about the relationship between the following (37):

- Heart disease and saturated fat (negative)

- Heart disease and cholesterol (negative)
- Heart disease and fiber-containing fruits, vegetables, and grains (positive)
- Heart disease and soluble fiber from whole oats and psyllium seed husk (positive)
- Heart disease and soy protein (positive)
- Heart disease and whole-grain foods (positive)
- Heart disease and plant stanol/sterol esters (positive)
- Cancer and dietary fat (negative)
- Cancer and fruits, vegetables, and fiber (positive)
- Hypertension and sodium (negative)
- Osteoporosis and calcium (positive)
- Neural tube defects and folic acid (positive)
- Dental caries and dietary sugar alcohol (positive)

Recently, other health claims were reviewed and tentatively found not to have sufficient basis to be authorized by the FDA for use in food labeling (21). Some of these included the link between antioxidant vitamins and cancer, and zinc and the immune systems of elderly people (33).

Strict requirements now govern how and when the FDA-approved health claims may be used. The food must contain enough of the nutrient to contribute at least 10 percent of the Daily Value, and must not contain any nutrient or food substance that increases the risk of a disease or health condition. For example, whole milk cannot bear the calcium/osteoporosis claim because of its high fat content, excess fat having been linked to both heart disease and cancer. Low-fat and nonfat milk, on the other hand, do qualify for the calcium/osteoporosis claim (8).

Dietary Supplements. Under the 1994 Dietary Supplement Health and Education Act (DSHEA, pronounced "de-shay"), dietary supplements are treated as foods (43). Health claims can be made for dietary supplements as long as they are not disease claims that are only allowable for **drugs**. The FDA allows qualified health claims. An example of one such health claim is relating omega-3 fatty acids and certain B-vitamins to heart disease (37).

KEY TERMS

Drug A product able to treat, prevent, cure, mitigate, or diagnose a disease or disease symptom.

FIGURE 8-5 Nutrition Facts labels for tuna and cola.

TUNA

Nutrition Facts
Serv. Size 1/3 cup (56g)
Servings about 3
Calories 80
Fat Cal. 10

Amount/serving	% DV*	Amount/serving	% DV*
Total Fat 1g	2%	**Total Carb.** 0g	0%
Sat. Fat 0g	0%	Fiber 0g	0%
Cholest. 10g	3%	Sugars 0g	
Sodium 200mg	8%	**Protein** 17g	

Vitamin A 0% • Vitamin C 0%
Calcium 0% • Iron 6%

*Percent Daily Values (DV) are based on a 2,000 calorie diet.

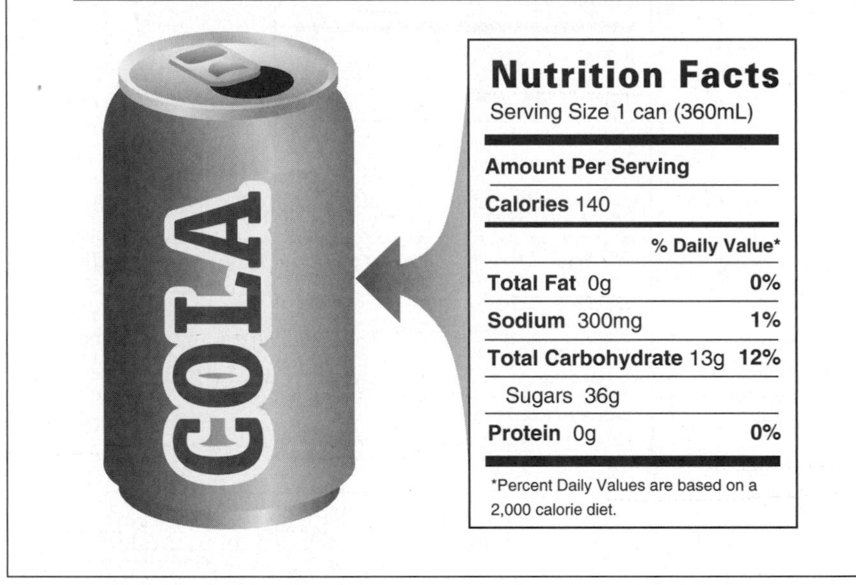

Food Additives

The FDA is responsible for regulating more than 3,000 food additives that are used in the food industry. Appendix F lists some of the more common additives and their functions in foods. The three most common additives by weight are salt, sugar, and corn syrup. Others that are commonly used include citric acid (from oranges and lemons), baking soda, vegetable colors, mustard, and pepper. These food additives account by weight for 98 percent of all food additives listed on labels (17).

Purposes of Food Additives. The purposes of food additives, according to FDA regulations, are to meet one or more of the four objectives listed below:

- Improve the appeal of foods by improving their flavor, smell, texture, or color

TABLE 8-2

New Food Labeling Descriptors

Description	*Amount/Serving*
Free	No nutrient or in very small amounts – calories <5, fat <1/2 g, saturated fat <1/2 g, cholesterol <2 mg, sugar <1/2g, and sodium <5 mg
Low	< 40 calories < 3 g fat < 1 g saturated fat < 20 mg cholesterol < 140 mg sodium (very low sodium <35 mg)
Extra Lean	< 5 g fat < 2 g saturated fat < 95 mg cholesterol
Lean	< 10 g fat < 4 g saturated fat < 95 mg cholesterol
Light/Lite	1/3 fewer calories Half the fat Can also describe color or texture
Reduced or Less	25% less than regular product
More	10% or more of Daily Value compared to reference food
High	20% or more of Daily Value for the nutrient
Good Source	10–19% or more of Daily Value for the nutrient
More Than	10% or more of Daily Value than comparison food
Healthy	< 3 g fat < 1 g saturated fat < 480 mg sodium < 60 mg cholesterol

KEY TERMS

GRAS list A list of compounds that are exempt from the "food additive" definition because they are "generally recognized as safe" based on "a reasonable certainty of no harm from a product under the intended conditions of use."

- Extend storage life
- Maximize performance
- Protect nutrient value

Improve Appeal. Food appeal can be improved by the addition of color, flavoring, and texture agents. Colors added to foods may be synthetic or natural. Synthetically certified colors include FD&C colors Blue No. 1 and 2, Green No. 3, Yellow No. 5 and 6, and Red No. 3 and 40. Natural colors, or synthetic replicas of natural colors that include those obtained from vegetable, animal, or mineral sources, are exempt from certification (25). Flavoring agents can also be synthetic, such as saccharin, or natural, such as fruit extracts, juice concentrates, processed fruits, fruit purées, spice resins, and MSG (monosodium glutamate) (26). Food additives are also used to add body and texture to foods. For example, thickeners generate a smooth texture in ice cream by preventing ice crystals from forming.

Extend Storage Life. By reducing the rancidity of fats, food storage life can be extended with additives such as butylated hydroxyanisole (BHA) and butylated hydroxytoluene (BHT), which slow or prevent food deterioration.

Maximize Performance. Emulsifiers, stabilizers, and other additives maximize the performance of foods. Emulsifiers make it possible to distribute tiny particles of one liquid into another, thereby preventing immiscible liquids from separating. For example, emulsifiers prevent the oil from separating out of peanut butter. Stabilizers and thickeners give milk shakes body and a smoother feel in the mouth. If the need exists to alter the pH of a food, certain compounds can be added to achieve the necessary acidity or alkalinity. Some additives retain moisture, while anticaking agents prevent moisture from lumping up powdered sugar or other finely ground powders.

Protect Nutrient Value. Food additives that protect nutrient value include vitamins and minerals that are added to enrich or fortify foods (23). Enriched foods have the vitamins thiamin (B_1), riboflavin (B_2), folate, and niacin, and the minerals iron and sometimes calcium, added back to levels established by federal standards for breads and cereals. Most table salt is fortified with iodine to help prevent goiter. Milk is fortified with vitamin D to help prevent rickets. Many fruit drinks are fortified with vitamin C, which tends to be missing in the diets of people who do not consume sufficient amounts of fruits and vegetables.

Safety of Food Additives. "Safety" is defined as "reasonable certainty . . . that the substance is not harmful under the intended conditions of use" (36). Two amendments to the Food, Drug, and Cosmetic Act of 1938—the Food Additives Amendment (1958) and the Color Additives Amendment (1960)—state that the FDA must approve the safety of all food additives. Prior to the 1958 Food Additives Amendment, the burden was on the FDA to prove that an additive was harmful before a court order could be issued banning its use. After the amendment was passed, the tables were turned and it became the manufacturer's responsibility to convince the FDA that the additive was safe through extensive testing on laboratory animals.

Two groups of substances were exempt from FDA approval: substances that had been used "safely" and were sanctioned before 1958, and substances termed Generally Recognized as Safe (GRAS)—the **GRAS list** (44).

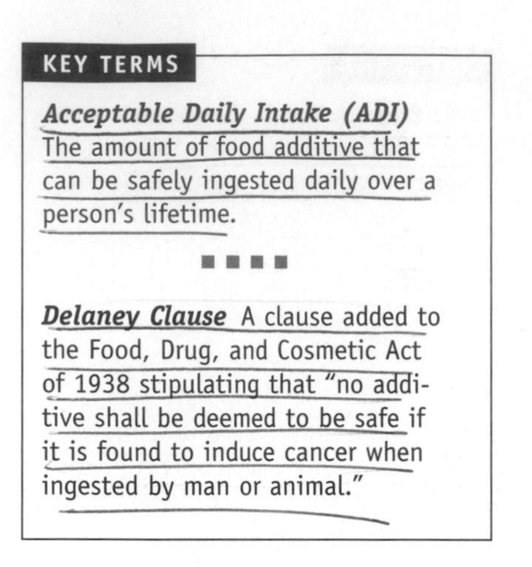

KEY TERMS

Acceptable Daily Intake (ADI) The amount of food additive that can be safely ingested daily over a person's lifetime.

■ ■ ■ ■

Delaney Clause A clause added to the Food, Drug, and Cosmetic Act of 1938 stipulating that "no additive shall be deemed to be safe if it is found to induce cancer when ingested by man or animal."

GRAS List. To avoid the time and expense of testing the safety of every substance, the FDA appointed a group of scientific researchers to compile a list of substances considered safe. Items on this GRAS list are permitted to be used until further testing proves otherwise (30). Because the compounds on the GRAS list are not officially considered food additives by definition, they are thereby exempt from the legal requirement of proving their safety (32). GRAS compounds are continually reevaluated, and as new methods for testing are developed, some substances are removed from the GRAS list.

HOW & WHY? ?????????

How do foods obtain FDA approval? Additives are never given permanent approval, but are subject to review to determine whether the approval should continue, be modified, or be withdrawn. The process of approving a new additive may take years and cost millions of dollars. An additive is first tested on laboratory animals to determine whether it affects life span, cancer rates, incidence of birth defects, allergies, or other health problems. A hundredfold margin of safety is applied; that is, if 100 mg of the substance is the minimum level that causes no harmful effect in laboratory animals, then manufacturers may use no more than 1 mg in foods given to humans (27). This safety margin sets the **Acceptable Daily Intake (ADI)** for human consumption (42). Only food additives on the GRAS list, plus those more recently approved by the FDA, are allowed in the U.S. food supply (15).

The Delaney Clause—Does the Additive Cause Cancer? The **Delaney Clause** (1958) tightened the regulations on food additives, but has subsequently been reinterpreted and modified as the techniques for identifying carcinogenic compounds have improved (33). The Department of Health and Human Services has estimated that food additives are linked to only 1 percent of all cancers. The FDA has banned all food additives that were found to be carcinogenic in laboratory animals, with the exception of saccharin and a few food colorings. Other additives whose safety has been questioned include nitrites, pickling additives, and BHA and BHT.

Saccharin. High doses of saccharin have been linked to bladder cancer in rats, but the Office of Technology and Assessment concluded that saccharin is one of the weakest food carcinogens detected (29). Nevertheless, by law, a warning label disclosing its possible carcinogenic effects is required on all saccharin products (21).

Food Colorings. Approximately 30 colors are approved for use in foods, half of which are synthetic and classified as "certified" (12). Several of the food colors have tested positive as carcinogens (29). The FDA banned the use of certain dyes in foods, but there have been temporary lifts on the bans, allowing foods containing these colors to remain on the market (5).

Nitrites. Nitrite and salts of nitrate are used as a preservative in approximately 7 percent of foods, particularly processed meats such as ham, hot dogs, bacon, sausage, bologna, salami, and cold cuts (1). They are added to these foods to prevent botulism, to give meat a reddish hue, and to create a distinctive flavor. Despite the benefits of nitrites, when these compounds are heated and combined with acidic stomach juices they may form nitrosamine compounds, which are carcinogenic. Nitrites, however, are also formed in the body and are found naturally in such foods as cabbage, cauliflower, carrots, celery, lettuce, radishes, beets, and spinach.

Pickling Salt and Vinegar. In China and Japan, it is a common practice to preserve foods by pickling in salt and vinegar or other acid. Researchers speculate that the high incidence of stomach cancer in both countries is directly linked to the high intake of pickled vegetables and fish (2).

BHA and BHT. Butylated hydroxyanisole (BHA) and butylated hydroxytoluene (BHT) act as antioxidants to help prolong storage life. They are often found on the ingredient lists of cereals, potato chips, chewing gum, and baked products (9). Scientific evidence shows that the compounds both stimulate and inhibit cancer in animals. However, the amount used in food is extremely small and does not reach the cancer-producing levels used with rats.

Pesticides—Unintentional Additives. Pesticides are among the more than 12,000 substances that unintentionally become a part of the food chain (29). Until recently, pesticides were classified as food additives, but the Food Quality Protection Act of 1996 modified the Delaney Clause by redefining "food additive" to exclude pesticide residues found on foods. The FDA regulates the use of various pesticides, and constantly monitors pesticide residue levels to ensure that the risk of food contamination remains low (10). Nevertheless, accidental poisonings do occur, and 80 percent of American shoppers perceive pesticide residues as a major health concern (41).

Allowable levels for pesticides vary from country to country. In 1987, a review of the FDA's pesticide program by the U.S. General Accounting Office and the Congressional Committee on Energy and Commerce reported that there was inadequate monitoring of pesticide residues on imported foods (40). As a result, the Pesticide Monitoring Improvements Act of 1988 was passed, which mandates pesticide monitoring, foreign pesticide usage information, and pesticide analytical methods. An FDA study in 1991 revealed that only about 3 percent of the foods sampled contained unacceptable levels of pesticides (38). However, according to the General Accounting Office, the FDA's pesticide tests do not screen for several pesticides that most seriously affect health (39).

Food Additives and Cancer. Cancer death rates overall, with the exception of lung cancer, have remained relatively the same or slightly decreased for over 50 years. If food additives influenced cancer death rates, an increase in the general population dying of cancer would most likely have been documented. It has been estimated that if there is a risk, food additives would be responsible for less than 1 percent of all cancers.

Cancer death rates are much more influenced by cigarette smoking. In fact, if smoking-related deaths were eliminated, it would reduce the death rate by cancer in the United States by one-fourth. In addition, almost half the deaths on the roadways involve alcohol. From a public health perspective, it may be wiser to focus less on the dangers of food additives and more on the very real problem and immeasurable cost of people dying by cigarettes and those driving under the influence of excess alcohol. The real question is, how can we increase the number of people making healthy lifestyle choices?

U.S. Department of Agriculture

USDA
United States Department of Agriculture

The U.S. Department of Agriculture (USDA) was initially established to:

- Increase the income of farmers by developing methods for improving productivity and by generating new markets for farm products.
- Reduce hunger and malnutrition.
- Inspect and **grade** farm products.

The USDA and its responsibilities have expanded tremendously. This chapter, however, focuses on the USDA's inspection and grading services. Meat, poultry, and eggs are regulated by the USDA's Food Safety and Inspection Service (FSIS). Food products containing relatively small amounts of meat (3 percent) and poultry (2 percent) fall under the responsibility of the FDA (36). The USDA also grades dairy products, grains, fruits, and vegetables.

USDA Inspections

Beginning in 1906, Congress has passed numerous acts of legislation that led to the inspection of meats, poultry products, and eggs, thereby ensuring that both the processes and the products were sanitary. For products that cross state lines, inspection at the federal level is mandatory. State inspection regulations for intrastate commerce must equal or exceed USDA standards. Products that pass the USDA inspection are then given the blue inspection stamp (Figure 8-6).

Legislation affecting USDA inspections includes the following:

- Federal Meat Inspection Act (1906). The USDA is responsible for inspecting meat to ensure it is safe, wholesome, and accurately labeled.
- Agricultural Marketing Act (1946). The USDA's Agricultural Marketing Service administers the inspection and grading of raw and processed foods such as cereal, dairy, fresh fruits and vegetables, poultry, eggs, fish, and shellfish.

KEY TERMS

Grade The voluntary process in which foods are evaluated for yield (a 1 to 5 grading for meats only) and quality (Prime, Choice, AA, A, Fancy, etc.).

- Fish and Wildlife Act (1956). Governs the inspection of fish and shellfish.
- Poultry Products Inspection Act (1957). Requires inspection of all poultry.
- Wholesome Meat Act (1967). Meats that do not travel interstate must meet state inspections equal to federal guidelines. This act also includes provisions for inspection of foreign processing plants.
- Egg Products Inspection Act (1970). Mandatory inspection of plants producing egg products.
- Pathogen Reduction Hazard Analysis and Critical Control Points (HACCP) Systems Final Rule (1996). Meat and poultry plants must implement mandatory HACCP programs.
- Pathogen Reduction Act (1997). A zero tolerance is set for *E. coli* in foods.

The latest laws governing meat and poultry safety radically changed the way inspections had been conducted for the last 90 years (20). Inspectors used to rely on sight, touch, and smell to detect spoiled meat, and on testing procedures for microorganisms that took hours or even days (Figure 8-7). According to the USDA's Food Safety and Inspection Service (FSIS), the HACCP Systems Final Rule requires that meat and poultry plants incorpo-

FIGURE 8-6 USDA inspection stamps.

FIGURE 8-7 USDA inspectors look for spoiled meat. The Federal Meat Inspection Act of 1906 required that all interstate beef be inspected. However, some believe that the old "sniff and poke" methods need to include more routine and rapid tests of detecting pathogenic bacteria.

Vito Palmisano/©Tony Stone Images

rate Sanitation Standard Operating Procedures (SSOPs) to ensure that their plants and equipment are clean (3). They are also required to develop and implement HACCP programs to identify critical points where microbial contamination could occur, install controls to prevent or reduce those hazards, and document the process (19). Scientific tests and microbiological tests that quickly detect *E. coli, Salmonella,* and other bacteria are encouraged. Each meat and poultry manufacturing plant must also adopt HACCP programs to eliminate hazards at every point in the production process. Similar HACCP programs have been recommended for the egg and fruit/vegetable juice industries (24).

USDA Grading

Only food that has passed inspection may be graded for quality. Unlike inspection, however, USDA grading is voluntary. Grading is available for meat, poultry, eggs, dairy products, fresh fruits and vegetables, and some fish, shellfish, and cereal. There are two types of grading, quantitative and qualitative.

Quantity Grades or "Yields." Quantitative grading is reserved for meats and describes the "yield," or the ratio of the lean or muscle tissue to fat, bone, and refuse on the animal's carcass. The yield grade ranges from 1 to 5, with 1 the highest yield and lowest waste (Figure 8-8). This type of grade is not commonly seen at the consumer level.

Quality Grades. Consumers are more familiar with quality grades, which are based on a food's appearance, texture, flavor, and other factors, depending on the particular food. Figure 8-9 shows the various quality grades and their characteristics for different types of foods (4). Figure 8-10 shows how different grades of fruit may be used for various food dishes. For example, high-quality U.S. Grade A strawberries would be used to top strawberry shortcake, while lower-quality U.S. Grade B strawberries would be reserved for strawberry pie, and even lower U.S. Grade C strawberries for jam.

FIGURE 8-8 USDA Grade Yield stamp.

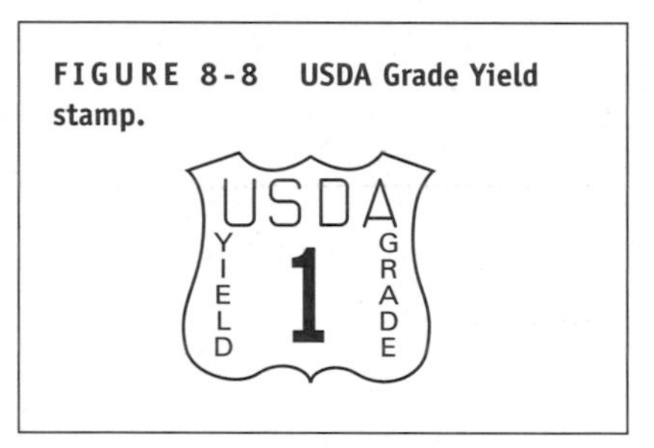

FIGURE 8-9 USDA quality grade stamps for specific food types.

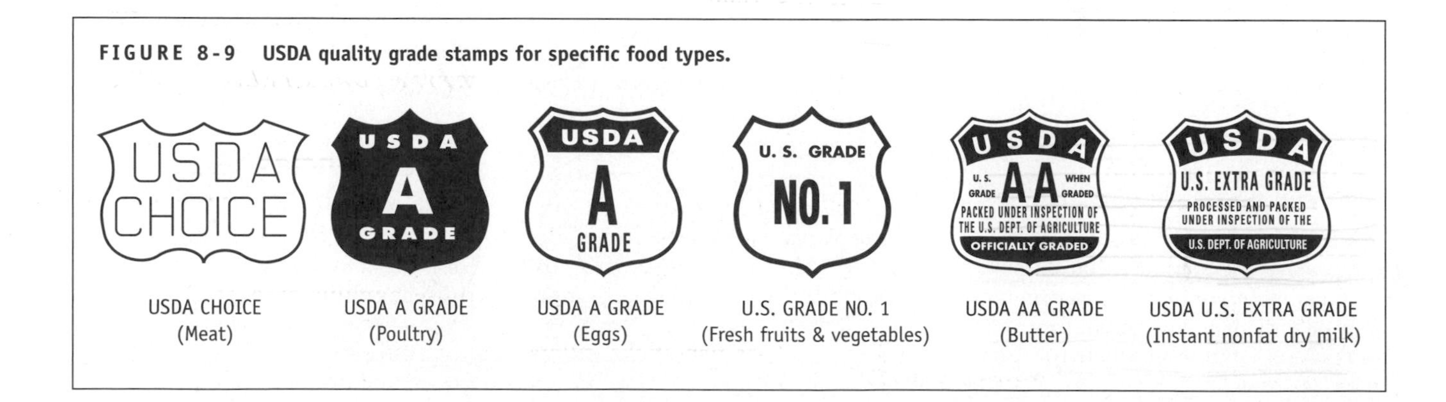

USDA CHOICE (Meat) | USDA A GRADE (Poultry) | USDA A GRADE (Eggs) | U.S. GRADE NO. 1 (Fresh fruits & vegetables) | USDA AA GRADE (Butter) | USDA U.S. EXTRA GRADE (Instant nonfat dry milk)

FIGURE 8-10 The use of different fruit grades in food preparation.

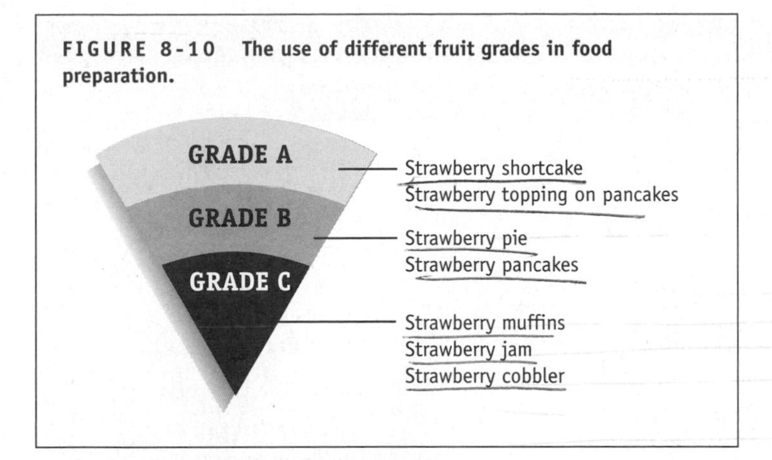

Environmental Protection Agency

EPA United States Environmental Protection Agency

The application of the Pesticide Residue Amendment (1954) to the Food, Drug, and Cosmetic Act falls under the jurisdiction of the Environmental Protection Agency (EPA). The EPA determines the safety of new pesticides and sets tolerance levels for pesticide residues in foods. It is illegal for foods to bear traces of pesticides in excess of the safe tolerance levels set by the EPA. Fresh fruits, vegetables, and grains, especially those from foreign countries, are continually inspected by the EPA for pesticide residues. The major law empowering the EPA with the authority to regulate the use of certain pesticides is the federal Insecticide, Fungicide, and Rodenticide Act. This legislation is an exception to the Delaney Clause, allowing certain pesticides that are known carcinogens to be used on raw agricultural commodities as long as the benefits from their use outweigh the risks from their presence (42).

Centers for Disease Control and Prevention

The Centers for Disease Control and Prevention (CDC) has many functions related to preventing and controlling foodborne diseases. This federal agency dates back to World War II, when the Public Health Service established the Office of Malaria Control to prevent malaria in army camps and defense installations. Today, one of the CDC's responsibilities is to track outbreaks of foodborne illnesses across the country, determine their cause, and prevent their recurrence. State health departments report their data on such outbreaks and on infectious diseases to the CDC, which uses this information to conduct national surveillance and to keep records on a national scale (35).

Other Regulatory Agencies

U.S. Department of Commerce

It may seem odd, but the U.S. Department of Commerce (USDC) is responsible for the inspection of fish and fish products. Specifically, the National Marine Fisheries Service within the USDC oversees these voluntary inspections and gradings, which are all paid for by the fish-processing plants. Fresh fish is not graded because deterioration begins the moment fish are caught. Grading is available, however, for frozen fish products such as fish sticks, fillets, steaks, and breaded/precooked fish portions. Shellfish such as oysters, clams, mussels, and whole scallops may only be sold to the consumer if they have been harvested from certified waters. The Interstate Certificate Shellfish Shipper List, published monthly by the FDA, describes these certified waters.

Federal Trade Commission

The major purpose of the Federal Trade Commission (FTC) is to maintain free and fair competition in the economy. One method of achieving this goal is to protect consumers from false or misleading advertising. The FTC enforces the sections of the Food, Drug, and Cosmetic Act that prohibit the sale of products that are misrepresented by advertising or by misleading words, designs, or pictures on the label (15). "Cease and desist" orders are issued by the FTC against a company or individual it believes to be engaging in unlawful trade practices.

Department of the Treasury

Alcoholic beverages fall under the jurisdiction of two agencies within the Department of the Treasury, the Internal Revenue Service and the Bureau of Alcohol, Tobacco, and Firearms. The IRS enforces federal laws regulating alcoholic beverages, because alcohol is subject to federal tax. The Bureau of Alcohol, Tobacco, and Firearms is also responsible in part for enforcing the laws that regulate the production, distribution, and labeling of alcoholic beverages. Wines with less than 7 percent alcohol are regulated by the FDA.

State Agencies

Products made and sold exclusively within a state often have to meet state food safety regulations similar to those administered at the federal level. Each state has slightly different laws unique to its specific locality, needs, and problems.

PICTORIAL SUMMARY / 8: Government Regulation of Food

Ever since it became illegal to dilute wine with water in ancient Greece and Rome, food laws and regulations have been instituted to protect the consumer. Thanks to public demand for legislation against contaminated foods, our country's food supply is among the safest in the world.

FEDERAL FOOD LAWS

From the Food and Drug Act of 1906 (the Pure Food Law) to the Bioterrorism Preparedness Act of 2002, the U.S. government has been involved in regulating the country's food supply. One of the most important of these laws is the Food, Drug, and Cosmetic Act, which became law in 1938.

FOOD AND DRUG ADMINISTRATION

The Food and Drug Administration (FDA) is a branch of the Department of Health and Human Services.

The FDA:

- Enforces the Food, Drug, and Cosmetic Act of 1938, which ensures that those products are safe, wholesome, and produced under sanitary conditions.
- Regulates the production of all food involved in interstate marketing with the exception of the USDA-regulated meat, poultry, and egg industries. This includes the following:
  - Inspecting facilities and manufacturing processes
  - Setting standards
  - Labeling
  - Regulating food additives
- Conducts research and educates the public about food and nutrition

ENVIRONMENTAL PROTECTION AGENCY

The Environmental Protection Agency (EPA) determines the safety of new pesticides and sets tolerance levels for pesticide residues in foods. Fresh fruits, vegetables, and grains, especially those from foreign countries, are continually inspected by the EPA for pesticide residues.

CENTERS FOR DISEASE CONTROL AND PREVENTION

The Centers for Disease Control and Prevention (CDC) has many functions related to preventing and controlling diseases, and one of them is to track outbreaks of foodborne illnesses across the country, determine their cause, and prevent their recurrence. State health departments report their data on such outbreaks to the CDC, which uses this information to conduct surveillance and to maintain records on a national scale.

U.S. DEPARTMENT OF AGRICULTURE

USDA

The U.S. Department of Agriculture was first established to help farmers improve productivity and find new markets, reduce hunger, and inspect and grade farm products. Since then, its role has expanded tremendously.

The USDA inspects and grades:

- Meat
- Poultry
- Eggs
- Dairy products
- Grains
- Fruits and vegetables

USDA CHOICE

USDA Choice (Meat)

USDA A Grade (Poultry)

USDA A Grade (Eggs)

U. S. GRADE NO. 1

U.S. Grade No. 1 (Fresh fruits and vegetables)

USDA U.S. GRADE AA WHEN GRADED PACKED UNDER INSPECTION OF THE U.S. DEPT. OF AGRICULTURE OFFICIALLY GRADED

USDA AA Grade (Butter)

USDA U.S. Extra Grade (Instant nonfat dry milk)

OTHER REGULATORY AGENCIES

- **U.S. Department of Commerce (USDC):** The National Marine Fisheries Service within the USDC oversees the voluntary inspections and grading of fish and fish products, which are paid for by fish-processing plants.
- **Federal Trade Commission (FTC):** The FTC maintains free and fair competition in the economy, and one method of achieving this goal is to protect consumers from false or misleading advertising.
- **Department of the Treasury:** The Bureau of Alcohol, Tobacco, and Firearms within the Department of the Treasury is responsible for enforcing the laws that regulate the production, distribution, and labeling of alcoholic beverages.
- **State Agencies:** Products made within a state often have to meet safety regulations unique to the state as well as federal regulations.

REFERENCES

1. Archer DL. Nitrite and the impact of advisory groups. *Food Technology* 55(3):26, 2001.
2. Baily GS, and DE Williams. Potential mechanisms for food-related carcinogens and anticarcinogens. *Food Technology* 47(2):105–118, 1993.
3. Berne S. Simplifying sanitation. *Prepared Foods* 166(3):80–84, 1997.
4. *Code of Federal Regulations.* 7 (Part 52):[Section 1]84–89. Washington, D.C.: GPO, 1995.
5. Collins TFX, et al. Teratogenic potential of FD&C red No. 3 when given in drinking water. *Food Chemistry and Toxicology* 31(3):161–167, 1993.
6. Durbin RJ. The Safe Food Act of 1997. *Food Technology* 52(1):112, 1998.
7. Ebert AG. IFT's Codex Alimentarius activities. *Food Technology* 56(1):22, 2002.
8. Farley D. Look for "legit" health claims on foods. FDA Consumer/Supplement. *Focus on Food Labeling*:21–28, May 1993.
9. Giese J. Antioxidants: Tools for preventing lipid oxidation. *Food Technology* 50(11):73–78, 1996.
10. Giese J. Pesticide residue analysis in foods. *Food Technology* 56(9):93, 2002.
11. Giese J. Testing for adulterated foods. *Food Technology* 56(2):66–68, 2002.
12. Hallagan JB, DC Allen, and JF Borzelleca. The safety and regulatory status of food, drug, and cosmetics colour additives exempt from certification. *Food Chemistry and Toxicology* 33(6):515–528, 1995.
13. Hutt PB. A guide to the FDA modernization Act of 1997. *Food Technology* 53(5):54–60, 1998.
14. Institute of Medicine/National Research Council. *Ensuring safe food: From production to consumption.* Commission to ensure safe food from production to consumption. Institute of Medicine & National Research Council. National Academy Press, 1998.
15. Janssen WF. *The U.S. Food and Drug Law: How It Came, How It Works.* DHHS pub. (FDA) 79–1054. Washington, D.C.: GPO, 1979.
16. Kurtzweil P. Good reading for good eating. FDA Consumer/Supplement. *Focus on Food Labeling*:7–13, May 1993.
17. Lehmann P. More than you ever thought you would know about food additives. *FDA Consumer* 13(3):10–16, 1979.
18. Looney JW, PG Crandall, and AK Poole. The matrix of food safety regulations. *Food Technology* 55:60–76, 2001.
19. McCue N. Getting a handle on HACCP. *Prepared Foods* 166(6):109–116, 1997.
20. McNutt K. Common sense advice to food safety educators. *Nutrition Today* 32(3):128, 1997.
21. Mermelstein NH. A new era in food labeling. *Food Technology* 47(2):81–96, 1993.
22. Mermelstein NH. Nutrition labeling. Regulatory update. *Food Technology* 48(7):62–71, 1994.
23. Mertz W. Food fortification in the United States. *Nutrition Reviews* 55(2):44–49, 1997.
24. Morris CE. HACCP update. *Food Engineering* 69(7/8):51–56, 1997.
25. O'Donnell CD. Colorful experiences. *Prepared Foods* 166(7):32–34, 1997.
26. Ohr LM. Distinctions blur between natural and artificial flavors. *Prepared Foods* 166(11):53–56, 1997.
27. Potter ME. Risk assessment terms and definitions. *Journal of Food Protection* Supp:6–9, 1996.
28. Rodricks JV. Safety assessment of new food ingredients. *Food Technology* 50(3):114–117, 1996.
29. Sandler RS. Diet and cancer: Food additives, coffee, and alcohol. *Nutrition and Cancer* 4(4):273–279, 1983.
30. Scheuplein RJ. Perspectives on toxicological risk; An example: Foodborne carcinogenic risk. *Critical Reviews in Food Science and Nutrition* 32(2):105–121, 1992.
31. Segal M. Ingredient labeling: What's in a food? *FDA Consumer Focus on Food Labeling* May suppl: 21–28, 1993.
32. Smith RL, et al. GRAS flavoring substances. *Food Technology* 55(12):34–55, 2001.
33. Stauffer JE. Food quality protection act. *Cereal Foods World* 43(6):444–445, 1998.
34. Taylor MR. Reforming food safety: A model for the future. *Food Technology* 56(5):190–194, 2002.
35. Tauxe RV. Emerging foodborne diseases: An evolving public health challenge. *Emerging Infectious Diseases* 3(4):425–434, 1997.
36. Thonney PF, and CA Bisogni. Scientific status summary. Government regulation of food safety: Interaction of scientific and societal forces. *Food Technology* 46(1):73–80, 1992.
37. Turner RE. Cholesterol claims at issue. *Food Technology* 55(11):26, 2001.
38. U.S. Department of Health and Human Services. Food and Drug Administration. FDA's pesticide residue program: Residues in foods—1990. *Journal of the Association of Official Analytical Chemists* 74:121Z–141A, 1990.
39. U.S. General Accounting Office. *Pesticides: Need to Enhance FDA's Ability to Protect the Public from Illegal Residues.* GAO/RCED–87–7. Washington, D.C.: GPO, 1986.
40. Venugopal R, et al. Recalls of foods and cosmetics by the U.S. Food and Drug Administration. *Food Protection* 59:876–880, 1996.
41. Wessel JR, and NJ Yess. Pesticide residues in foods imported into the United States. Reviews of *Environmental Contamination and Toxicology* 120:83–104, 1991.
42. Winter CK, and EJ Francis. Scientific status summary: Assessing, managing, and communicating chemical food risks. *Food Technology* 51(5):85–92, 1997.

43. Yetley EA. Implementing DSHEA—Progress made, but much remains to be done. *Food Technology* 53(7):24, 1999.
44. Whitehead IM. Challenge to biocatalysts from flavor chemistry. *Food Technology* 52(2):40–42, 1998.

WEBSITES

Department of Health and Human Services (DHHS)
www.hhs.gov

• Food and Drug Administration (FDA)
www.fda.gov

• Center for Food Safety and Nutrition (CFSAN)
www.cfsn.fda.gov

• Centers for Disease Control and Prevention (CDC)
www.cdc.gov

• National Institutes of Health (NIH)
www.nih.gov

Department of Agriculture (USDA)
www.usda.gov

• Marketing and Regulatory Programs Agriculture Marketing Service (AMS)
www.ams.usda.gov

Animal and Plant Health Inspection Service (APHIS)
www.aphis.usda.gov

Grain Inspection, Packers and Stockyards Administration (GIPSA)
www.usda.gov/gipsa

• Research, Education and Economics
www.reeusda.gov/ree

Agricultural Research Service (ARS)
www.ars.usda.gov

Cooperative State Research, Education, and Extension Service
www.reeusda.gov

Economic Research Service
www.ers.usda.gov

National Agricultural Statistics Services
www.usda.gov/nass

• Undersecretary for Food Safety, Food Safety and Inspection Service (FSIS)
www.fsis.usda.gov

Environmental Protection Agency (EPA)
www.epa.gov

• Office of Prevention, Pesticides, and Toxic Substances
www.epa.gov/opptsfrs

Office of Pesticide Programs
www.epa.gov/pesticides

Department of Commerce
www.doc.gov

• National Oceanic and Atmospheric Administration
www.noaa.gov

• NOAA Fisheries Service (NOAA FS)
www.noaa.gov/fisheries

Federal Trade Commission (FTC)
www.ftc.gov

Department of the Treasury
www.ustreas.gov

• Bureau of Alcohol, Tobacco, and Firearms
www.atf.treas.gov

• United States Customs Service
www.customs.treas.gov

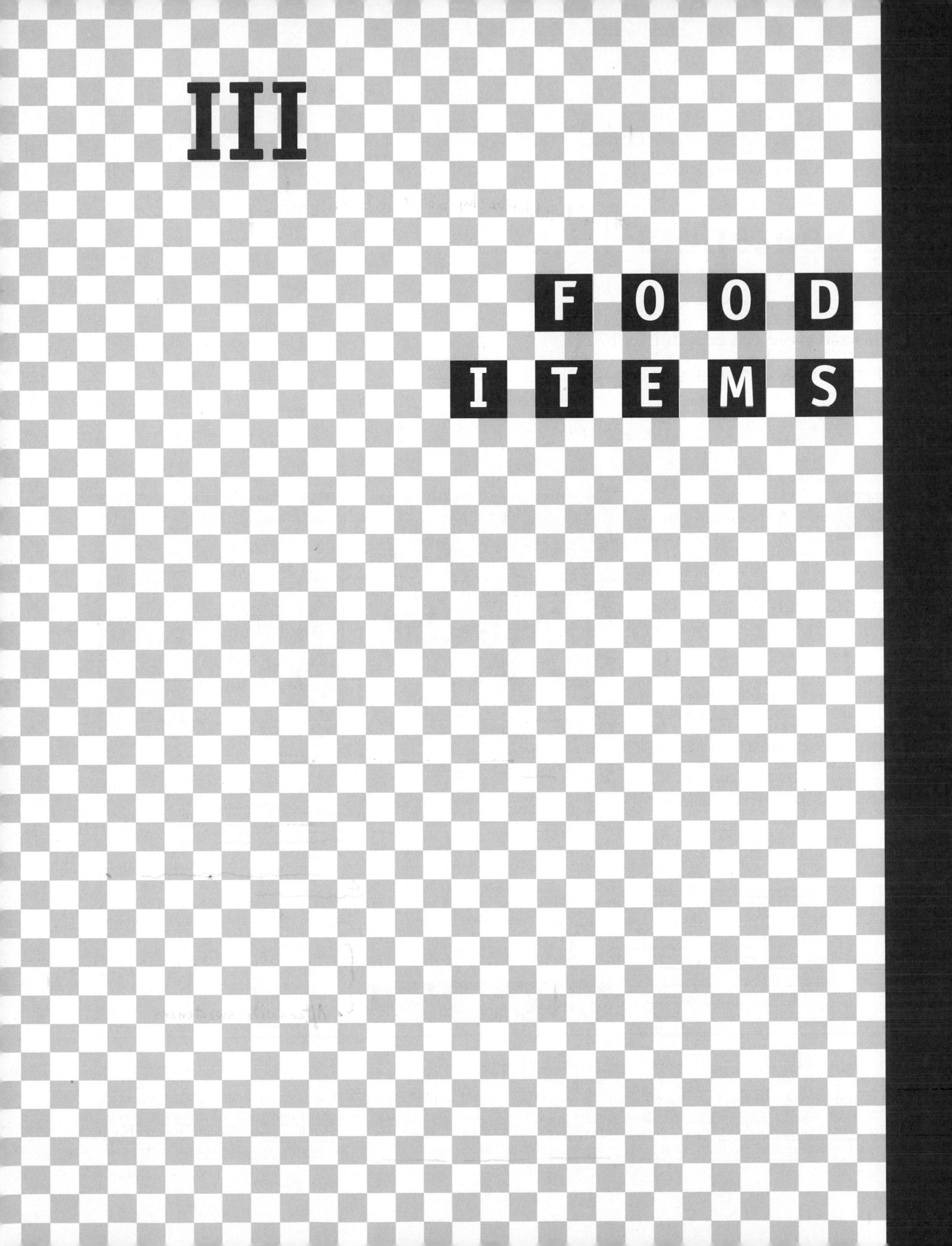

III

FOOD ITEMS

9

Sweeteners

Just as bears are drawn to honey, people seek numerous methods to extract sweeteners from the natural plant world. Sugar cane and sugar beets deliver the most sugar to satisfy the human tastebuds (Figure 9-1). The history of people's interest in these plants dates back to Alexander the Great, who was enticed by the "honey-bearing rods" (sugar cane) he encountered near the Indus River. Christopher Columbus brought sugar cane to the West Indies, where it has been a major cash crop ever since. Jesuit missionaries followed by introducing sugar cane into Louisiana in 1751, and in the mid-1800s sugar beets were planted in the United States.

While sugar is the most widely used sweetener in food preparation, other types of sweeteners may be used, including syrups, sugar alcohols, and alternative sweeteners. And sweetness is only one of the many ways in which sugar functions in foods. In baked goods, sugar produces a finer texture, enhances flavor, generates browning of the crust, promotes fermentation of yeast breads, and extends shelf life by virtue of its ability to retain moisture. Sugar also gives body to soft drinks; so much so that sugar-free drinks need additional ingredients to replace the viscosity provided by sugar so they will not seem flat (26). Sugar further offsets the acidic, bitter, or salty tastes of certain foods, such as tomato sauces, chocolate, and sodium-processed meats (ham, bacon, etc.).

FIGURE 9-1 Sugar cane and sugar beets are the primary sources of sucrose (table sugar).

Sugar's many important functions are discussed later in this chapter, following a look at the different types of sweeteners (natural and alternative) and their nutrient contributions.

Natural Sweeteners

As people searched for new ways to sweeten food, they discovered that plants produce abundant natural sugars through the process of photosynthesis (Figure 9-2). The plants providing most of the sweeteners are sugar cane, sugar beets, maple trees, and corn. Lactose in milk is the only sweetener of animal origin, and it is not very sweet.

Sweeteners can be placed into three major groups based on their different chemical structures, which influence their functions in foods and beverages:

- Sugars
- Syrups
- Sugar alcohols

Sugars

"Sugar" is what usually comes to mind when the word "sweetener" is used. Once extracted from its plant source, sugar becomes a refined carbohydrate, providing a pleasing flavor and packing 4 calories (kcal) per gram. Sugar's abil-

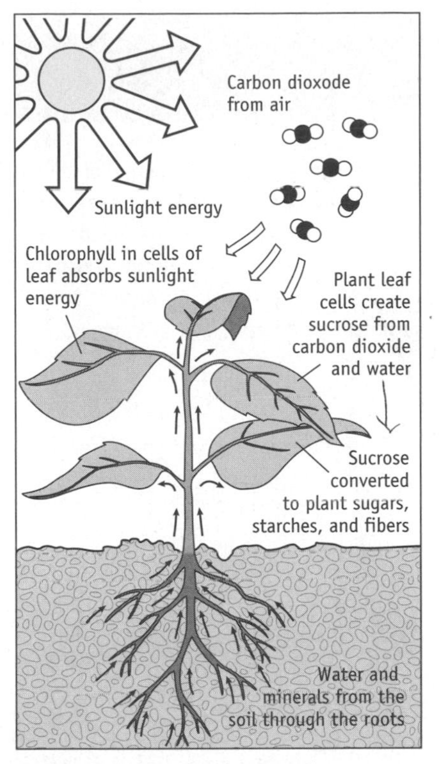

FIGURE 9-2 The process of photosynthesis.

ity to sweeten makes it the number-one food additive, by weight, in the United States. The manufacture of candy contributes to this high use of sugar, as candies continue to experience increasing sales with each passing year (36). Sugar is also found in less obvious foods, such as most commercially produced catsup (29 percent sugar), and fruit canned in heavy syrup (18 percent sugar). Many other foods that contain relatively large amounts of sugar, such as nondairy creamers (up to 65 percent sugar), do not taste particularly sweet.

The many kinds of sugars differ in their individual characteristics and functions in food. The sugars discussed now include sucrose, glucose, fructose, lactose, and maltose.

Sucrose. Sucrose, or table sugar, is derived from either sugar cane or sugar beets. Sugar cane has been used as a source of sugar for centuries. It first grew in India, where it is still a major crop. Sugar beets are a more recent addition, a German scientist having first extracted sugar from this plant in 1794. Today, sugar cane provides about 60 percent of the sugar consumed in the United States; white sugar beets, a relative of the red beet commonly eaten as a vegetable, provide the rest. The sugar obtained from sugar cane and beets is extracted through a number of different refining and purifying steps after harvesting, including crushing, extraction, evaporation, and separation. Once processed, there is no difference between the sugar these two sources.

Obtaining Table Sugar From Plants. Harvested sugar cane is washed and then machine-shredded. Heavy steel rollers crush and squeeze the sweet juice from the cane. Water is sprayed on the cut and pulverized sugar cane, allowing more juice to seep out and be collected. The process of extracting sucrose from sugar beets differs in that they are washed, sliced, and then soaked in vats of hot water to remove the sugar. The extracted juices of both cane and beets are then heated and concentrated in evaporation tanks to create a thick syrup known as molasses. Vacuum equipment lowers the boiling point of the syrup so that it may be concentrated without burning. Large sugar crystals form as the solution becomes saturated. The crystals are then separated from the molasses liquid by a centrifuge. Centrifuges spin solutions at very high speeds to separate particles and/or liquids according to density.

Types of Sucrose. Once refined, the sugar is further processed to produce the commercial products known as raw, turbinado, white, powdered, fruit, Baker's Special, sanding, liquid, and brown sugars (Table 9-1).

TABLE 9-1
Different Forms of Available Sucrose

Raw sugar	Made from sugar cane, not beets. Contains natural contaminants such as soil, insect parts, yeast, molds, waxes, and lint, and is banned by the FDA for sale to the public.
Turbinado sugar	Raw sugar that has been centrifuged and purified with steam. Sometimes labeled "raw sugar," although it is not truly raw; its color is light amber.
White sugar	Made by further refining raw sugar by repeatedly washing and filtering until the rinse liquid is a clear, colorless syrup. The syrup is then boiled until it crystallizes. Crystals are separated by size into "fine," or table sugar, and "superfine" or "ultrafine," which are used by the food industry for cake baking, dry mixes, candy coatings, and mixed drinks.
Powdered sugar	Made by pulverizing white granulated sugar. Also called confectioners' or icing sugar, it is frequently combined with an anti-caking substance such as cornstarch, silica gel, or tricalcium phosphate to keep the powder soft and pourable.
Fruit sugar	Very finely granulated sucrose. Its uniform crystal size allows it to remain evenly disbursed in a mix. Used in dry mixes such as gelatins, puddings, and drink bases.
Baker's Special	Even more finely granulated than fruit sugar. Used primarily by the baking industry in cookies, cakes, and doughnuts.
Sanding sugar	Large-granule sugar often used to decorate the tops of baked goods because it does not melt during baking and it sparkles attractively.
Liquid sugar	A solution containing a highly purified sugar that is used in canned foods, beverages, confections, baked goods, frozen foods, and ice cream.
Brown sugar	Made by adding molasses syrup to white sugar. The amount and type of molasses determine the grade, with higher-grade brown sugars having a darker color and stronger flavor. These are used when strong flavors are desired, as for baked beans, minced meat, plum pudding, and gingerbread cookies.

...Y? ?????????

...n sugar have a tendency ... become hard? This ...e even though the ... film over the sugar keeps brown sugar soft and pliable at first, it hardens as soon as the water from the molasses evaporates. Airtight containers, preferably sealed plastic bags, are best for storing brown sugar. Hard sugar can be softened by draping a moist paper towel over the sugar within its container for about 12 hours (sugar draws the moisture to itself). Brown sugar can also be softened by briefly heating it in an oven at 250° to 300°F (121° to 149°C) or in a microwave. Alternatively, a piece of bread may be placed in the sugar container for several hours so the brown sugar absorbs moisture from the bread. Storing brown sugar in a plastic bag in the refrigerator is another way to increase humidity and delay dehydration.

NUTRIENT CONTENT

White granulated sugar is 99.9 percent pure carbohydrate, so table sugar or any food made with it is a source of refined carbohydrate. Density varies quite a bit between sweeteners, but in general, 1 teaspoon of white granulated sugar yields 18 calories (kcal), while most other sugars contribute 15 to 22 calories (kcal). Confectioners' sugar has a much lower yield, only 8 calories (kcal), because it does not pack as densely as granulated sugars. The nutrient content of various sugars is summarized in the table below.

Nutrient Content of Various Types of Sugars

Sugar (1 tsp)	*Calories (kcal)*	*Calcium (mg)*	*Iron (mg)*	*Potassium (mg)*
White sugar				
Granulated (4.2 gm)*	16	0	0	0.1
Powdered (2.5 gm)	10	0	0	0.1
Fructose (3.5 gm)	14	0	0	0
Brown sugar (3.5 gm)	13	3.0	0.1	12.1
Honey (7 gm)	21	0.4	0	3.6
Corn syrup (6.8 gm)	19	1.2	0	3.0
Maple syrup (3 gm)	8	2.0	0	6.1
Molasses				
Light (6.8 gm)	17	11.2	0.3	62.4
Blackstrap (6.8 gm)	16	58.5	1.2	169.0

*Grams noted due to different gram-wts/tsp

One health problem currently associated with excess sugar consumption is dental caries (cavities). Another problem, which is debated by the Sugar Association, is the relationship of sugar to obesity (3). Obesity has reached epidemic proportions in the United States, but the emphasis has primarily been on fat, rather than also focusing on sugar and total calories. People going on weight-reducing diets often lose weight by cutting back on sweets. The Dietary Guidelines recommendation is to choose a diet moderate in sugars, while the World Health Organization has advised that sugar intake should be less than 10 percent of calories (kcal) (44). In the United States, other than table sugar, regular soft drinks are the main dietary source of simple sugars (14). Children tend to consume more added sugar (13 to 14 percent of calories (kcal)) than adults (9 to 11 percent of calories (kcal)) (5).

One way to reduce refined sugar intake is to reduce the use of table sugar and consumption of processed foods made with sugar. Food and beverage labels can list sugar as sucrose, glucose (dextrose), lactose, maltose, brown sugar, honey, molasses, and corn syrup. The words "no sugar added" on a label do not necessarily mean there is no sugar in the product; it may be present naturally.

Glucose. Glucose, also known as dextrose, is the basic building block of most carbohydrates and is the major sugar found in the blood. Chief plant sources of this monosaccharide are fruits, vegetables, honey, and corn syrup. Glucose can be obtained from starch by commercially treating it with heat and acids or with enzymes. Food companies often use glucose, which is half as sweet as sucrose, in candies, beverages, baked goods, canned fruit, and fermented beverages. The baking industry relies on glucose to enhance crust color, texture, and crumb; as a component in dry mixes for baking; and to temper the sweetness of sucrose (34).

Fructose. Fructose, also called levulose, or fruit sugar, is found naturally in fruits and honey, and is the sweetest of all granulated sugars. It is not related to the fine-textured form of fruit sugar used in dry mixes. This sugar is rarely used in food preparation, because it causes excessive stickiness in candies, overbrowning in baked products, and lower freezing temperatures in ice cream. It is used primarily in the manufacture of pharmaceutical products.

Fructose has been widely touted as a low-calorie sweetener, but this is not entirely accurate. While less of it may be used because it is sweeter than sucrose, it still provides the same 4 calories (kcal) per gram found in any other carbohydrate. Diabetics should be aware that while the blood sugar may increase more slowly with fructose than with glucose, it will still be raised.

Lactose. The sweetness of milk comes from its lactose content. Lactose, a disaccharide, is the least sweet of all sugars and is extracted from whey (a by-product of cheese production) for commercial use in baked products, where it aids in browning. It cannot be fermented, however, so it is rarely used in preparation of yeast bread products or alcoholic beverages that depend on fermentation. Lactose is also used in pharmaceutical products as a filler in pills because its gritty, hard crystals flow freely and are easily compacted for tablet pressing.

Maltose. Maltose, also called malt sugar, lends certain milk shakes and candies their characteristic malt taste. It is used primarily as a flavoring and coloring agent in the manufacture of beer. During the malting process, barley and other grains are treated to convert the grain's starch to maltose (see Chapter 29).

Syrups

Syrups are sugary solutions that vary widely in viscosity, carbohydrate concentration, flavor, and price (Chemist's Corner 9-1). The more common include corn syrup, high-fructose corn syrup, honey, molasses, maple syrup, and invert sugar. The most frequently purchased syrups in the supermarket are maple-flavored syrups, which are primarily a blend of corn syrup and cane sugar syrup, and honey (38).

Corn Syrup. A by-product of cornstarch production, corn syrups are viscous liquids containing 75 percent sugar and 25 percent water. Corn syrups, rather than granulated sugar, are often used in soft drinks and processed foods to reduce manufacturing cost. Dried corn syrups or corn syrup solids are used in dry mixes for beverages, sauces, and instant breakfast drinks.

How Corn Syrup Is Made. Corn syrup is made commercially by adding a weak acid solution to cornstarch, then boiling, filtering, and evaporating the mixture until the right sugar concentration is achieved. Adding different enzymes to cornstarch can convert the starch into glucose, maltose, or small polysaccharides known as dextrins (Figure 9-3). Beta-amylase enzymes yield more maltose, while glucoamylase enzymes result in a higher-glucose syrup, which is sweeter and less viscous. Manufacturers can modify the characteristics of a corn syrup to their specific requirements by controlling the type of saccharides formed.

Dextrose Equivalents. The sugar that comes from corn is most often called dextrose, rather than glucose. Thus, corn syrups are rated according to their dextrose content, which is measured in **dextrose equivalents (DE)** (Chemist's Corner 9-2). A complete conversion of starch to glucose has a DE value of 100. **High-conversion corn syrups** are less viscous, have greater sweetening power, are able to promote fermentation, and contribute to browning. The less viscous nature of high DE syrups is due to smaller molecules such as glucose, fructose, and maltose. In comparison, lower-conversion syrups containing a higher concentration of larger molecules are more viscous, cannot be fermented, and are less sweet.

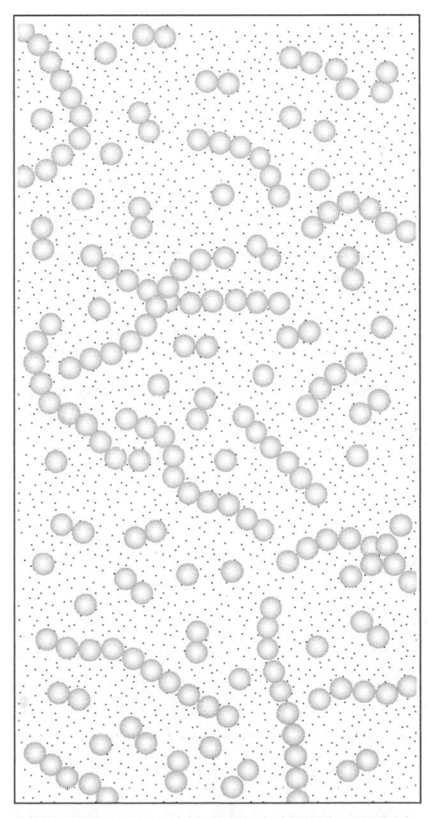

FIGURE 9-3 Corn syrup, obtained from hydrolysis of cornstarch, is a solution of sugars dissolved in water. Smaller sugars taste sweeter, while longer ones contribute to viscosity.

KEY TERMS

Dextrose equivalent (DE) A measurement of dextrose concentration. A DE of 50 means the syrup contains 50 percent dextrose.

High-conversion corn syrups Corn syrups with a dextrose equivalent over 58.

Chemist's Corner 9-1

Measuring Sugar Solutions

Various instruments are used to measure the sugar solids in a syrup or liquid. The concentration of sugar solutions can be measured using a hydrometer, refractometer, or flowmeter. The hydrometer relies on the Archimedes principle by comparing the weight of a plummet in juice with its weight in water. Refractometers measure how much light is bent by the sugars in the liquid; the more concentrated the liquid, the more light is refracted. A flowmeter judges density by how quickly a liquid flows. The concentration measured by these instruments is converted into a degrees brix value, which is the food industry's standard of identifying the sugar concentration in syrups or liquids (11).

Chemist's Corner 9-2

Dextrose Equivalent

The dextrose equivalent (DE) value reflects the proportion of reducing sugars to the total number of glucose molecules.

$$DE = \frac{\text{Number of reducing sugars}}{\text{Total number of glucose molecules}} \times 100$$

DE	*Conversion*
28–37	Low
38–47	Regular
48–57	Intermediate
58–67	High
> 68	Extra high
80–100	Crude corn sugars

High-Fructose Corn Syrup. High-fructose corn syrups (HFCSs) are intensely sweet and are frequently used in food products: baby foods, bakery products, canned fruits, carbonated beverages, confections, dry bakery mixes, fountain syrups/toppings, frozen fruits, fruit juice drinks, frozen desserts, jams, jellies, preserves, meat products, pickles, condiments, and table syrups. The beverage industry alone uses 90 percent of HFCS (9). HFCS is made by treating cornstarch with a glucose isomerase enzyme and converting it to a syrup that is approximately 40 percent fructose and 50 percent glucose. The greater sweetening power of HFCS means that less is needed, so the cost of using HFCS is well below that of sugar. Its clarity and colorlessness also contribute to its industrial popularity.

Honey. The world initially depended primarily on honey as a sweetener. Bees collect the thin, watery nectar of flowers and, during the flight back to their hive, convert it through enzymatic action into fructose and glucose molecules. The bees deposit the nectar in honeycombs, where most of the water evaporates to create a thick, sweetened syrup, which is further flavored with enzymes added by the bees. It takes 2 million flowers to produce enough nectar to make 1 pound of honey, and the average worker bee makes only $\frac{1}{12}$ of a teaspoon of honey in its entire lifetime.

HOW & WHY? ?????????

Why are there different flavors and colors of honey? The flavor and color of the honey depend on the type of flower visited by the bees. There are over 300 varieties of honey, the most popular being alfalfa and clover. Honeys are blended by the bees as they collect nectar from an assortment of flowers. Normally, honey is a golden amber, but the darker the color, the stronger the flavor. For example, Australian eucalyptus honey has a reddish-brown color and a strong, tangy flavor. Acacia honey, in contrast, is almost clear and has a very delicate flavor and aroma.

Sugars in Honey. Honey typically consists of sugars other than sucrose: fructose (40 percent), glucose (35 percent), sucrose (2 percent), and traces of other carbohydrates. To protect the consumer from honey that has been extended by the addition of sucrose, the FDA limits commercial honey to no more than 8 percent sucrose. Over 180 substances, including beeswax, minerals, and water (18 percent), are found in honey. *Clostridium botulinum* spores are also often present in honey and pose a hazard to children under 1 year of age. Honey is therefore not recommended for infants, whose systems are not yet able to handle the spores as do those of older children and adults.

Removing Honey From the Comb. Honey was originally sold in the comb, but now it is generally extracted by processors by cutting the comb on one side and releasing the honey in a centrifuge. Honey can also be collected by crushing the combs and straining out the thick fluid. The extracted honey is heated to 140°F (60°C) for 30 minutes to destroy most microorganisms, then filtered and packaged in airtight containers. Some small producers sell unfiltered honey as well as comb honey.

Whipped, Creamed, Granular, and Infused Honeys. Prior to packaging, honey may be processed into various forms. Whipped or creamed honey has had some of the fructose removed, resulting in a thicker consistency. Dried, granular honey is used in baked products, confections, and dry mixes. Infused honey has been heated with a flavor or herb to give it a unique taste. Infused honey can be made by adding 1 cup of honey to a saucepan with a flavoring (mint, ginger, lime, rosemary, cinnamon stick, orange, etc.). Heat it to boiling and immediately reduce to a simmer for 5 minutes, then let stand for 10 minutes (24 hours for cinnamon stick). Strain it while still warm. Cooled infused honeys are then good in hot or cold teas, or over cooked carrots, French toast, seasoned pork, glazed chicken, cornbread, and numerous other foods.

Storing Honey. Regardless of the form of honey, its naturally high sugar content prevents the growth of bacteria; therefore, honey can remain shelved for years without spoiling. Stored for long periods of time, however, it can harden as its sugar precipitates into crystals. If this occurs, it can be softened by warming the jar in hot water for an hour, or setting the opened jar in the microwave on low defrost.

Substituting Honey for Sugar in Recipes. Honey can be substituted for table sugar in recipes if a few guidelines are followed (8):

- In baked products, no more than half the granulated sugar should be replaced with honey.
- Use 1 part honey for every $1\frac{1}{4}$ parts sugar.
- Reduce the liquid in the recipe by $\frac{1}{4}$ cup because honey is largely water.
- Add $\frac{1}{2}$ teaspoon baking soda for every cup of honey to reduce the acidity and weight of honey.

Honey has a more pronounced flavor than sugar, and this will affect the final flavor of the product. It also has a tendency to increase the browning of baked products (6). Adding $\frac{1}{8}$ of a teaspoon of baking soda allows even browning; reducing oven temperatures by 25° helps prevent over-browning. The stickiness of measuring honey can be minimized by coating the inside of a measuring cup with water or a very thin layer of vegetable oil before measuring.

Molasses. Molasses is the thick, yellow to dark brown liquid by-product of the juice of sugar cane or beets. The liquid is repeatedly boiled, but for the end product to be called molasses, it must contain no more than 75 percent water and 5 percent mineral ash. Most of the sugar in molasses is sucrose, which renders the product darker with each boiling. The syrup's ultimate color determines its grade. Blackstrap molasses, the most concentrated in syrup and minerals, is the darkest in color, most bitter, and is used primarily for industrial purposes and cattle feed, although it is available for home consumption. Most commercial grades of molasses are actually blends of different types of molasses.

Foods Made with Molasses. Molasses is used both in food preparation and in the making of rum. Its main use is in baking, where it enhances the flavor of breads, cakes, and cookies. A few other foods that incorporate molasses are baked beans, glazes for hams and sweet potatoes, cookies, and candies such as toffees and caramels. Fermenting molasses yields rum, an alcoholic beverage that is distilled and generally aged for five to seven years. Quicker aging periods of one to four years are used for rapidly fermented light rums.

Maple Syrup. Native Americans were the first to collect the sap from maple trees and boil it into a smooth, tasty syrup. Long ago, sap was harvested by drilling holes into a maple tree, inserting a spout, and catching the fluid in a bucket positioned under the spout. Newer methods eliminate the buckets, instead utilizing a network of plastic pipelines attached to the trees. The pipeline carries the sap, a clear, almost tasteless, watery liquid, directly to the sugar house, where it is boiled down. Sap is collected in the late winter and early spring during the few weeks when the days are relatively warm, but the nights are still cold. Vermont, New Hampshire, northern New York, and parts of Canada, where the dramatic rise and fall in spring temperatures trigger the flowing of the sap, are ideal for maple syrup harvesting.

Maple Syrup Colors. The flavor and color of maple syrup develop during the boiling of the initially colorless sap. Government standards specify that maple syrup must contain at least 65.5 percent sugar among its other ingredients, such as acids and salts. Maple syrup is graded and sold by color and ranges from light amber, or Fancy, to the darkest color, known as Commercial. The darker the color, the more pronounced the flavor. The lightest-colored syrups have the most delicate flavors.

"Real" vs Blended Maple Syrups. Since it takes about 40 gallons of sap to produce 1 gallon of maple syrup, most "maple syrup" sold today is blended with corn syrup and/or cane sugar syrup. Many companies add artificial maple flavorings to foods, but real maple syrup has a unique flavor and smoothness not duplicated by substitutes. Pure or blended maple syrup is commonly poured over pancakes, waffles, and French toast or added as an ingredient in maple butter, cream, and candy.

Maple Sugar. Maple sugar is a product of maple syrup. It is made by further boiling the syrup until most of the water evaporates and the sugar **crystallizes** out of the syrup. About 8 pounds of maple sugar are produced from 1 gallon of maple syrup.

Invert Sugar. **Invert sugar** is available only in clear, liquid form and is sweeter than granulated sugar. This type of sugar resists crystallization and is commonly used by professional confectioners who need a sugar that yields a smooth, melt-in-the-mouth texture.

How Invert Sugar Is Made. Invert sugar is made commercially by dissolving sucrose in water, heating the solution, and adding an acid such as cream of tartar, or an invertase enzyme such as sucrase, which hydrolyzes the sucrose into two equal portions of glucose and fructose. This process is called inversion. The use of cream of tartar or sucrase inhibits crystallization. In addition, the acidity of cream of tartar (tartaric acid) has the added benefit of preventing the natural decomposition of monosaccharides into bitter, brown-colored substances, which occurs when they are exposed to hard water or any other alkaline source. The amount of cream of tartar added depends upon the percentage of invert sugar concentration desired.

Invert Sugar in Foods. The confectionary industry uses invert sugar to develop the soft, fluid center of certain chocolates. See Chapter 27 for more information on how invert sugar is utilized in preparing confectioneries.

KEY TERMS

Crystallization The precipitation of crystals from a solution into a solid, geometric network.

Invert sugar An equal mixture of glucose and fructose, created by hydrolyzing sucrose.

Sugar Alcohols

Sugar alcohols are not carbohydrates, but the alcohol counterparts of specific carbohydrates. They are found naturally in fruits and vegetables or are synthesized by hydrogenating certain sugars (Chemist's Corner 9-3). They include sorbitol, mannitol, xylitol,

Chemist's Corner 9-3

Sugar Alcohols

Sugar alcohols are polyols, which is short for polyhydric alcohols (32). Below are several sugars and their alcohol counterparts. These sugar alcohol compounds differ from their monosaccharides by a slight arrangement of their atoms; the carbohydrate's hydroxyl group (-OH) is replaced with an aldehyde or ketone group (C=0).

Sugar	*Alcohol Counterpart*	*Calories (kcal)/Gram*
Sucrose	Sorbitol	2.6
Mannose	Mannitol	1.6
Maltose	Maltitol	2.1
Xylose	Xylitol	2.4

Polyols occur naturally in certain foods, but for commercial use they must be prepared in a laboratory. Mannose or glucose is hydrogenated (hydrogen is added) to yield mannitol; glucose is hydrogenated to generate sorbitol; and xylitol from xylose is derived from xylans, gummy polysaccharides found in plant parts such as oat hulls, corncobs, and birch wood. Another kind of sugar alcohol are the hydrogenated starch hydrolysates, which are a mixture of hydrogenated oligo- and polysaccharides, maltitol, and sorbitol (22).

KEY TERMS

Humectant A substance that attracts water to itself. If added to food, it increases the water-holding capacity of the food and helps to prevent it from drying out by lowering the water activity.

maltitol, isomalt, lactitol, and erythritol. Although primarily sold to food manufacturers as ingredients, isomalt is available to professional chefs (22).

Sugar Alcohols in Foods. Sugar alcohols' ability to contribute sweetness, combined with their tendency to be slowly absorbed, make them useful ingredients in various dietetic foods. In addition, the cooling sensation experienced when they dissolve in the mouth makes them useful for such products as sugarless gums, dietetic candies, sugar-free cough drops, throat lozenges, breath mints, and tablet coatings. Sugar alcohols do provide some calories (kcal), although fewer than sucrose. Sorbitol, mannitol, and xylitol supply 1.6 to 3.0 calories (kcal) per gram compared to the 4 grams provided by sucrose (20). The low caloric (kcal) content and cooling sensation of sorbitol, mannitol, and xylitol make them attractive to food manufacturers as sweetening agents, especially in dietetic candies.

Sugar alcohols have other advantages besides their mouth-cooling property and lower calorie (kcal) content. They are cariostatic, or cavity preventing, because they cannot be digested by the bacteria responsible for dental caries (cavities) (32). Sorbitol, the most widely used sugar alcohol, has the added quality of acting as a **humectant** and is frequently used in marshmallows and shredded coconut to maintain moistness.

Problems with Sugar Alcohols. One drawback of sugar alcohols in dietetic foods is that they are more slowly absorbed from the small intestine than other sugars, which can lead to diarrhea, abdominal pain, and gas (30). For this reason, consumption of food products containing over 30 grams of sorbitol is not recommended, and only limited quantities of xylitol are allowed in special dietary foods (12).

Alternative Sweeteners

Only five alternative sweeteners are currently approved by the FDA for use in the United States: saccharin, aspartame, acesulfame-K, sucralose, and neotame (16). These compounds are also known as intense sweeteners, defined as those that are substantially sweeter than sucrose (by weight). Their intensity of sweetness ranges from 30 to several thousand times that of sucrose. Alternative sweeteners are nonnutritive substances, providing minimal to zero calories (kcal). Although aspartame provides the same 4 calories (kcal) per gram as sucrose, so little of the sweetener is used that its caloric contribution is negligible. Table 9-2 summarizes the properties of the five approved alternative sweeteners. One other alternative sweetener, alitame, has been developed and is awaiting approval by the FDA (12). In addition, cyclamates, which had been approved but were later banned in the United States as a potential carcinogen, are trying for a comeback.

Despite the controversy over the safety of alternative sweeteners, they continue to be in demand by diabetics, people watching their weight, and individuals trying to prevent tooth decay (13). The food industry attempts to satisfy the market by providing a wide variety of foods containing one or more of the FDA-approved alternative sweeteners. In descending order, the most common foods sold to consumers that contain alternative sweeteners are diet soft drinks, tabletop sweeteners, pudding, gelatin, yogurt, frozen desserts, powdered drinks, cakes, cookies, jams, jellies, and candy (4).

One drawback of alternative sweeteners is that they do not provide the important functional characteristics of sugar: bulking, binding, texturing, and fermenting. However, certain compounds can be added to foods to compensate for the lost characteristic of bulking. These include cellulose, maltodextrin (also used for its binding property), the sugar alcohols, and polydextrose. Polydextrose provides a texture similar to sugar, with only 1 calorie (kcal) per gram, and is currently approved for use in frozen dairy desserts, baked goods and mixes, confections and frostings, hard and soft candy, chewing gum, gelatins, puddings and fillings, and salad dressings (12). The Milky Way Lite candy bar is made with polydextrose, and as a result, it contains 25 percent fewer overall calories (kcal) and 50 percent fewer calories (kcal) from fat compared to the original Milky Way bar (35).

Saccharin

Saccharin was discovered as a sweetener in 1879 by Constantin Fahlberg (2). The researcher noticed that his dinner roll tasted strangely sweet and traced it back to a saccharin substance he had accidentally spilled on his hands while working in his university research lab (25). Saccharin is now available as acid saccharin, sodium saccharin, and calcium saccharin. It is 500 times sweeter than sucrose, contains no calories (kcal), and can be used in a variety of products, including baked or processed foods. Saccharin's major drawback, at least for some people, is its bitter aftertaste, which can be masked only partially by blending it with other sweeteners (12).

Concerns About Saccharin. The controversy over saccharin's safety peaked when researchers in a Canadian study reported an increased incidence of bladder cancer in rats fed very high amounts of saccharin (5 to 7.5 percent of the diet)—the human equivalent of drinking at least 800 diet sodas a day (17). Responding to that study, the FDA proposed a ban on saccharin in 1977. When letters of protest poured into Congress, a congressional moratorium was placed on the FDA ban, along with the requirement that all saccharin-containing products carry a public health warning (23). That moratorium was extended several times before the FDA officially withdrew its proposed ban. In 2000, President Clinton signed legislation relieving manufacturers of the requirement to include a warning label on products containing saccharin (18).

The Office of Technology Assessment, a research arm of Congress that attempts to review scientific matters objectively, concluded that saccharin is

TABLE 9-2
Approved Alternative Sweeteners

Sweetener	*Chemical Structure*	*Sweetness (sucrose = 1)*	*Taste Characteristics*	*Uses*	*ADI* (mg/kg of body weight)*
Saccharin	O, NH, SO_2	200–700	Slow onset, persistant aftertaste, bitter at high concentrations	Used in soft drinks, assorted foods, and tabletop sweeteners	2.5
Aspartame (Nutrasweet)	CH_2, CH—C(=O)OCH_3, CH, C=O, CH—CH_2, NH_2, C, OH O	180	Clean, similar to sucrose, no bitter aftertaste	Approved for use in tabletop sweeteners, dry beverage mixes, chewing gum, beverages, confections, fruit spreads, toppings, and fillings	50
Acesulfame-K (Sunette)	CH_3, O=, O, N—SO_2, K	130–200	Rapid onset, persistant, side-tastes at high concentrations	Approved for use in tabletop sweeteners, dry beverage mixes, and chewing gum	15
Sucralose (Splenda)	CH_2OH, Cl, O, OH, OH, O, $ClCH_2$, O, OH, CH_2Cl, OH	600	Can withstand high temperatures without losing flavor	Approved for use in soft drinks, baked goods, chewing gums, and tabletop sweeteners	15
Neotame	CH_3—C(CH_3)(CH_3)—CH_2—CH_2—N(H)—, COOH, O, H N, $COOCH_3$	8000	Clean, sweet, sugar-like taste; enhances flavors of other ingredients	Under review	—

*Acceptable Daily Intake (ADI) values for aspartame and acesulfame-K are FDA. Values for saccharin and sucralose are United Nations Joint FAO/WHO Expert Committee on Food Additives (JECFA).

a potential cause of cancer in humans, although it is among the weakest carcinogens ever detected. The FDA established an acceptable daily intake (ADI) of 2.5 milligrams per kilogram of body weight, equivalent to 147 milligrams for a 130-pound adult, or 205 milligrams for a 180-pound adult. A packet of Sweet-n-Low® contains 30 milligrams of saccharin; a diet soft drink averages 125 milligrams.

Aspartame

Like saccharin, aspartame was discovered by accident. In 1965, James Schlatter was doing research on ulcer drugs when he licked his finger to pick up a piece of weighing paper and noticed that the finger tasted sweet. He realized that the sweetness came from an earlier spill in the laboratory (21). What Schlatter discovered was a substance that is 180 times sweeter than sucrose.

Aspartame derives its sweetness from the synthetic combination of two amino acids, aspartic acid and phenylalanine. The amino acid content of aspartame contains 4 calories (kcal) per gram, but the number of calories (kcal) is insignificant because so little is needed to produce intense sweetness. FDA-approved in 1981, aspartame is now sold as NutraSweet®, Spoonful®, and Equal®. Spoonful combines aspartame and maltodextrin, a nonsweet bulking agent derived from cornstarch, which provides 4 calories (kcal) per gram. Equal is a blend of dextrose, maltodextrin, and aspartame. In the United States, the acceptable daily intake (ADI) for NutraSweet has been set at 50 milligrams per kilogram (23 milligrams per pound) of body weight. Since there are approximately 125 milligrams of aspartame in a diet drink, the daily limit for a 130-pound (59-kilogram) adult is 24 diet sodas.

Aspartame Side Effects. As with saccharin, several research studies have questioned the safety of aspartame, and there does appear to be a small subgroup in the population that is sensitive to one or more of its breakdown products (aspartic acid, phenylalanine, and methanol) (37, 39, 42). Common complaints among

KEY TERMS

NOEL The No-Observed-Effect Level is the level or dose at which an additive is fed to laboratory animals without any negative side effects.

this subpopulation include headaches (43), dizziness, mood changes, and nausea. In addition, research suggests a possible increased risk of brain tumors in rats (29). Although these side effects are controversial, there is no question that aspartame should not be consumed by individuals with phenylketonuria (PKU), a rare genetic disease afflicting one out of every 15,000 infants. Those with this condition lack the enzyme needed to metabolize phenylalanine. Anyone afflicted with PKU must avoid all food sources of phenylalanine, including aspartame, milk, and meat. For that reason, food products containing aspartame as an additive must carry the following warning: "Phenylketonurics: Contains Phenylalanine."

Aspartame does not have the bitter aftertaste of saccharin, but it did have a drawback. With exposure to heat or acids its sweetness is lost and therefore it could not be used in its pure form in baked goods. However, that changed when researchers were able to encapsulate the alternative sweetener in a hydrogenated fat coating that melts at the end of baking.

Acesulfame-K

Acesulfame-K was discovered in 1967 and, like the other artificial sweeteners, was stumbled upon by accident. Sold as Sunette® and Sweet One, acesulfame-K was FDA-approved in 1988 for use in tabletop sweeteners, dry beverage mixes, and chewing gum (40). It is 130 times sweeter than sucrose and is stable to heating and cooling, but it has a bitter aftertaste like that of saccharin. An ADI of 9 mg per kilogram of body weight has been established, but the FDA nevertheless recommends no more than 15 milligrams, which is equivalent to 20 diet sodas or 10 sweetener packets. The difference in recommendations results from the use of two different **NOELs** to determine safety (12).

Sucralose

Sucralose was discovered in 1976 and approved by the FDA in 1998 (19). A foreign student (Shashikant Phadnis) discovered this alternative sweetener when working in a laboratory in London at King's College where he misunderstood "testing" for "tasting." Sucralose is a modified form of sucrose, but is 600 times sweeter. Its stability at high temperatures makes it suitable for use in baked products. It is marketed in the United States and Canada as Splenda® and is used in over 100 products, such as carbonated soft drinks, cakes, muffins, juices, and gums, and as a tabletop sweetener (7).

Neotame

In 2002, the FDA approved the newest alternative sweetener on the scene, neotame (24). It is 8,000 times sweeter than sugar and has no calories. Like aspartame, neotame is made from the amino acids aspartic acid and phenylalanine. However, neotame is not metabolized to phenylalanine, so no warning label is required for people with phenylketonuria (PKU) (31). This tabletop sweetener has the potential to replace both sugar and high-fructose corn syrup (28). Only about 6 milligrams of neotame is needed to sweeten a 12-ounce soda.

Pending Alternative Sweeteners

Cyclamates were discovered in 1937 by a university graduate student, Michael Sevda. They are 30 times sweeter than sucrose and stable to heat. In 1970, cyclamates were banned for use in the United States because studies indicated they might be carcinogenic. They are, however, permitted for use in low-calorie foods in more than 50 countries, and in Canada for tabletop sweeteners and pharmaceuticals (27). In 1980, the FDA rejected a petition for cyclamate approval, but is now reviewing a second one.

Another alternative sweetener pending approval before the FDA is alitame, a peptide formed from the two amino acids aspartic acid and alanine (27).

Other Sweeteners

Around the world, the search continues for a sweetening substance without the caloric (kcal) content of sugar. Several sweeteners from a variety of sources are being investigated or are awaiting approval. The chemical structure of alternative sweeteners varies tremendously and includes peptides, amino acids, carbohydrates, inorganic salts, and synthetic compounds.

- *Glycyrrhizin.* An extract from the licorice root, it is 50 to 100 times sweeter than sucrose and is used only in confections.
- *Dihydrochalcones.* Obtained from citrus peel, these compounds are several hundred times sweeter than sucrose, with a slow taste onset and a lingering aftertaste.
- *L-sugars.* The chemical mirror image of natural sugars, these are not metabolized by body enzymes and are noncarcinogenic.
- *Stevioside.* Derived from the leaves of the South African plant Stevia rebaudiana, it is 300 times sweeter than sucrose.
- *Thaumatin.* An extract of the fruit of a West African plant, Thaumatococcus danielli, it is one of the sweetest substances known—1,600 times sweeter than sucrose. It has a licorice-like taste when used in high concentrations, is very stable to heat and acid, and is used in chewing gums (41, 45).
- *Tagatose.* This naturally occurring substance found in dairy products carries only 1.5 calories (kcals). It offers the bulk of sucrose and is almost as sweet. Unlike other sweeteners, it is considered under the GRAS list and may be used in the U.S. food supply (27).
- *Trehalose.* Also listed under the GRAS list is trehalose, which is naturally found in honey, mushrooms, and lobster but is commercially made from starch. It provides as many calories as sucrose, but is half as sweet. It benefits frozen foods by protecting their cell structure so their texture is better preserved (27).

Functions of Sugars in Foods

Sugars contribute many more functions in foods than merely providing sweetness (Table 9-3). Even when it comes to sweetness, however, various sugars differ in their sweetening ability due to their unique chemical arrangements. These structural differences also influence how each dissolves, crystallizes, browns, melts, absorbs water, contributes to texture, ferments, and preserves food.

Sweetness

Sugars are not equal in their ability to sweeten bland foods or minimize sour and bitter tastes. Table 9-4 shows this by comparing the relative sweetness of sugars, sugar alcohols, and alternative sweeteners to sucrose, which is scored as 1. Even the type of sweetness differs among sweeteners, as observed when comparing the tastes of table sugar (sucrose), honey, molasses, and corn syrup. Temperature also influences sweetness; cold foods and drinks usually taste sweeter than their hot counterparts. Other variables are the pH, other food ingredients, and the taster's sensitivity to sweetness (10). The dominant determinant of sweetness, however, is the type of sugar used and its concentration.

Solubility

Syrups owe their existence to sugar's ability to dissolve in water. Solubility is determined by measuring how many grams of sugar will dissolve in 100 milliliters of water. The degree to which a given sugar dissolves in water varies from sugar to sugar. Fructose is the most soluble, followed by sucrose, glucose (dextrose), maltose, and finally lactose (1). The solubility of a sweetener influences the perceived mouthfeel and texture of a food or beverage. For example, the least soluble sugar, lactose, is responsible for the gritty texture of ice cream that has partially thawed and been refrozen. Increasing the temperature of a sugar solution allows more sugar to dissolve, resulting in a supersaturated solution when it cools (Figure 9-4). In turn, increasing the sugar concentration raises the boiling point of water (Table 9-5). (See Chapter 2 for more discussion of saturation and solutions.)

Crystallization

Crystallization is a vital process in the manufacture of candy. The development or inhibition of crystal formation determines the finished product's quality (15). (See Chapter 27 for more on candy.) The goal in preparing noncrystalline candies is to prevent the sugar in solution from precipitating out in the form of crystals, causing an undesirable grainy texture in noncrystalline candies. The formation of one crystal can start a domino effect, triggering the

TABLE 9-3
Functions of Sugars in Foods

Functions in Foods:	*Baked Goods (Yeast)*	*Beverages*	*Cakes/ Cookies*	*Cereals*	*Confections*	*Dairy*	*Frozen*	*Jams/ Jellies*	*Processed Foods*
Aeration	X		X						
Boiling Point (Increase)					X				
Browning (Maillard Reaction)	X		X	X					
Bulking/Bodying Agent		X	X		X	X	X	X	
Caramelization	X		X		X				
Crystallization			X		X				
Fermentation	X								X
Flavor Enhancer		X	X	X	X	X	X	X	X
Foam Formation			X		X		X		
Freezing Point (Decrease)							X		
Glaze Formation	X		X		X				
Moisture Retention	X		X		X				
Preservative					X	X		X	X
Shelf Life (Extender)			X	X					X
Solubility		X			X			X	X
Sweetener	X	X	X	X	X	X	X	X	X
Tenderizer	X	X	X	X	X		X	X	

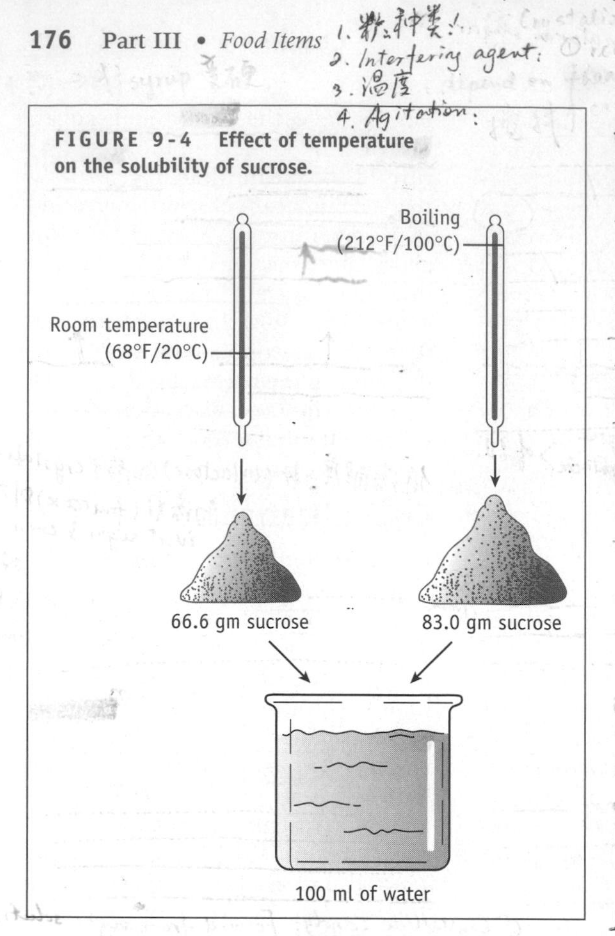

FIGURE 9-4 Effect of temperature on the solubility of sucrose.

entire mixture to crystallize. Small foreign particles, changes in temperature, or nicks or cracks on the container's surface may also serve as the starting point for crystal formation.

Preventing Crystallization. To prevent crystallization when heating sugar solutions, the sides of the pan can be kept cleared of any particles with a brush dipped in water, or the pan can be covered to generate steam, which will have the same effect. Once the condensed steam has done its job, the lid should be removed to allow water to evaporate. Heating is then continued with no further agitation such as stirring, because any agitation during the cooling process may result in crystal formation. Another way to control crystallization is by being selective in the type of sugar used. Sugars with low solubility, such as lactose, have a greater tendency to crystallize, while fructose, with its high solubility, does not. Invert sugar and corn syrup also resist crystallization, which is why they are often used in confectionary production.

KEY TERMS

Reducing sugars Sugars such as glucose, fructose, maltose, and others that have a reactive aldehyde or ketone group. Sucrose is not a reducing sugar.

■ ■ ■ ■

Caramelization A process in which dry sugar, or sugar solution with most of its water evaporated, is heated until it melts into a clear, viscous liquid and, as heating continues, turns into a smooth, brown mixture.

Browning Reactions

Browning reactions involving sugars (**reducing sugars**) and proteins (usually from milk) are due to the Maillard reaction (see Chapter 2). Food products relying on browning from the Maillard reaction include microwaved baked products that have incorporated fructose or dextrose sugars. These reducing sugars should not be added to powdered beverage mixes, because they may cause browning during storage. Instead, sucrose is added to such products.

Caramelization. Caramelization results from heating sugars. Sucrose heated in a dry pan will start to melt into a clear, viscous mass when heated to about 320°F (160°C). If heating continues to 338°F (170°C), the melted sugar mass will become smooth and glossy and start to caramelize. Sugars differ in the temperatures at which they melt. Fructose, for example, caramelizes at a slightly lower temperature, about 230°F (110°C).

Caramelized sugars are less sweet but more flavorful than the original sugar and may even be slightly bitter. The darker the caramel, the less sweet it is. The food industry relies on caramelized sugars to give a distinct flavor and color to food products such as puddings (flan), frostings, ice cream toppings, and dessert sauces. Candies made using the principles of

TABLE 9-4
The Relative Sweetness of Sweeteners Compared to Sucrose

Sweetener	*Sweetness (Sucrose = 1)*
Less Sweet than Sucrose	
Lactose	0.4
Maltose	0.5
Sorbitol	0.6
Galactose	0.6
Glucose	0.7
Mannitol	0.7
Xylose	0.7
Invert Sugar	0.7
Xylitol	0.8
Sweeter than Sucrose	
Invert Sugar Syrup	1.6
Fructose	1.7
Cyclamate	30
Glycyrrihizin	50
Acesulfame-K	130
Aspartame	180
Stevioside	300
Saccharin	500
Sucralose	600
Thaumatin	1600
Alitame	2000
Talin	2500
Monelin	3000
Neotame	8000

TABLE 9-5
Adding Sucrose to a Solution Increases Its Boiling Temperature*

% Sucrose	*% Water*	*Boiling Point °F*	*°C*
0	100	212	100
40	60	215	101
60	40	217	103
80	20	234	112
90	10	253	123
99.6	0.4	340	170

*The boiling point corresponding to each sugar concentration differs for different sugars.
Source: Food Technology.

caramelization) include, as the name suggests, caramels, as well as peanut brittle.

Moisture Absorption (Hygroscopicity)

The **hygroscopic** nature of sugars is responsible for their influence on a food's moistness and texture. The degree to which sugars draw moisture from the air differs depending on the sugar, with fructose having the most moisture absorption capability. Foods made with high-fructose sugars like honey and molasses are noted for retaining their moisture. In fact, cookies and other baked goods made with honey stay moist even to the point of losing their desired texture. These and other baked products, when made with sucrose, will retain freshness longer. The moisture-absorbing property of sugar makes it imperative that baking mixes be stored in airtight containers; otherwise, moisture drawn to the mixture will lower the quality of the baked product. Unfortunately, sugar itself gets lumpy if it absorbs too much moisture from the air, so additives are incorporated into commercial baking mixes to prevent lumping or caking of ingredients.

> **KEY TERMS**
>
> ***Hygroscopic*** The ability to attract and retain moisture.

Texture

The texture of many processed or prepared foods relies on sweeteners, especially sucrose. Without sugar, soft drinks feel flat in the mouth, so bulking agents are often added. Sometimes other carbohydrates such as inulin (a polysaccharide) are added to increase viscosity and add a creamy, fat-like consistency to a liquid. Inulin occurs naturally in plants and is used primarily as a texturizer to give body to beverages, improve the texture of low-fat ice creams, make creamier sauces, and help aerate non-fat icings (33).

Fermentation

Many alcoholic beverages and quite a few other foods around the world rely on the ability of carbohydrates to be fermented. Fermentation plays a role in producing beers, wines, cheeses, yogurts, and certain breads. The following conditions are desirable for fermentation:

- The presence of a yeast, mold, or bacteria culture. Even natural yeasts in the air can cause foods to ferment, and in some cases spoil, by fermenting their available sugars, a circumstance that probably led to the discovery of yeast-bread baking.
- A food source, usually a carbohydrate, for the microorganisms. Yeast bread rises because of the carbon dioxide gases (produced by the yeasts feeding off the carbohydrates in flour). Any sugar except lactose can be fermented to carbon dioxide and alcohol by yeast organisms.
- The correct temperature to help the microorganism to grow.
- Conducive salt and acidity concentrations.

Preservation

High concentrations of sugar can act as a preservative by inhibiting the growth of microorganisms. Sugar was used to preserve jams, jellies, and other fruit spreads long before either canning or freezing methods were developed. The osmotic pressure created by the high concentration of sugar dehydrates the bacteria or yeast cells to the point of inactivation or death.

PICTORIAL SUMMARY / 9: Sweeteners

From earliest times, the taste of sweetness has attracted people and has led them to search for new ways to extract it from the world around them. Today sugar is the number-one food additive, by weight, in the United States.

NATURAL SWEETENERS

As people searched for new ways to sweeten foods, they discovered that the only sweetener of animal origin was lactose, found in milk. Plants, however produce abundant natural sugars through photosynthesis. Once extracted from its source, sugar becomes a refined carbohydrate, with 4 calories (kcal) per gram. Other sweeteners besides sugar are used in food preparation.

Sweeteners can be grouped into four categories:

1. GRANULATED SUGARS

- **Sucrose** (table sugar) is derived from either sugar cane or sugar beets. Once processed, the resulting sugars are the same, but are then refined into these commercial products:

White sugar **Powdered sugar** **Liquid sugar**
Raw sugar **Fruit sugar** **Brown sugar**
Turbinado sugar **Baker's Special**
Sanding sugar

- **Glucose** (dextrose) is the basis of most carbohydrates and is the major sugar in our blood. Sources are fruits, vegetables, honey, and corn syrup. It is one-half as sweet as sucrose.
- **Fructose** is found naturally in fruits and honey, and is the sweetest of all the granulated sugars.
- **Lactose**, found in milk, is the least sweet of all sugars.
- **Maltose,** also called malt sugar, provides the characteristic "malt" flavor to milkshakes, but is mostly used to make beer.

2. SYRUPS

(A by-product of cornstarch)

- Corn syrup
- High-fructose corn syrup
- Honey
- Molasses
- Maple syrup
- Invert sugar

3. SUGAR ALCOHOLS

(Found naturally in fruits and vegetables, or synthesized from certain sugars)

- Sorbitol
- Mannitol
- Xylitol
- Maltitol
- Isomalt
- Lactitol
- Erythritol

ALTERNATIVE SWEETENERS

(Only five approved by the FDA)

- Saccharin
- Aspartame
- Acesulfame-K
- Sucralose
- Neotame

FUNCTIONS OF SUGARS IN FOODS

Sugars have many more functions in foods than merely providing sweetness. Even when it comes to sweetness, however, various sugars differ in their sweetening ability. The following unique chemical arrangements of the various sugars influence how they are used in food preparation:

Sweetness: The type of sugar and its concentration determines sweetness. Sucrose is the sweetest.

Solubility: Solubility is determined by measuring how many grams of sugar will dissolve in 100 ml of water. Fructose is the most soluble.

Crystallization: Sugars with low crystallization, such as lactose, have a greater tendency to crystallize, while fructose, with its high solubility, does not invert. Invert sugar and corn syrup also resist crystallization, which is why they are often used in confectionary production,

Browning reactions: Two major types of browning involving sugars are the Maillard reaction, dependent on protein (amino acids) and sugar (reducing); and caramelization, dependent on dry heat.

Moisture absorption (hygroscopicity): The hygroscopic nature of sugars influences the moistness and texture of food to which they are added. Fructose has the best ability to absorb moisture from the air and impart it to food.

Texture: Many foods rely on sucrose for body and texture.

Fermentation: Sugars (except lactose) serve as fuel for yeast during fermentation, especially in certain baked products and alcoholic beverages.

Preservation: High concentrations of sugar act as a preservative.

In baked goods, sugar produces a finer texture, enhances flavor, generates browning of the crust, promotes fermentation of yeast breads, and extends shelf life by virtue of its ability to retain moisture. Sugar gives body to soft drinks and helps offset the bitter, acidic, or salty taste of certain foods, such as tomato sauces, chocolate, and sodium-processed meats (ham, bacon, etc.).

REFERENCES

1. Alexander RJ. Sweeteners: Nutritive. *American Association of Cereal Chemists,* 1998.
2. Bakal A. A satisfyingly sweet overview. *Prepared Foods* 166(3):47–49, 1997.
3. Baker CW. Sugar Association response to "Sugar and sugars: Myths and realities." *Journal of the American Dietetic Association* 102(6):776–777, 2002.
4. Best D, and L Nelson. Low-calorie foods and sweeteners. *Prepared Foods* 162(7):47–57, 1993.
5. Black RM. Sucrose in health and nutrition: Facts and myths. *Food Technology* 47(1):130–133, 1993.
6. Cardetti M. Functional role of honey in savory snacks. *Cereal Foods World* 42(9):746–748, 1997.
7. Chapello WJ. The use of sucralose in baked goods and mixes. *Cereal Foods World* 167(5):133–135, 1998.
8. Charlton J. Discover honey's many flavors. *Fine Cooking* 22:74, 1997.
9. Coulston AM, and RK Johnson. Sugar and sugars: Myths and realities. *Journal of the American Dietetic Association* 102(3):351–353, 2002.
10. Deliza R, HJH MacFie, and D Hedderley. Information affects consumer assessment of sweet and bitter solutions. *Journal of Food Science* 61:1080–1084, 1996.
11. Demetrakakes P. Sweet success: The latest options for on-line measurement of sugar solids. *Food Processing* 57(9):83–86, 1996.
12. Giese JH. Alternative sweeteners and bulking agents: An overview of their properties, function, and regulatory status. *Food Technology* 47(1):114–126, 1993.
13. Grenby THE. Intense sweeteners for the food industry: An overview. *Trends in Food Science and Technology* 2(1):2–6, 1991.
14. Guthrie JF, and JF Morton. Food sources of added sweeteners in the diets of Americans. *Journal of the American Dietetic Association* 100:43–48, 2000.
15. Hartel RW, and AV Shastry. Sugar crystallization in food products. *Critical Reviews in Food Science and Nutrition* 1(1):49–112, 1991.
16. Hollingsworth P. Sugar replacers expand product horizons. *Food Technology* 56(7):24–27, 2002.
17. Kalkhoff RK. Symposium on sweeteners. Policy statement: Saccharin. *Diabetes Care* 1(4):209–210, 1978.
18. Karottki SL. Saccharin warning removed. *Food Technology* 55(2):10, 2001.
19. LaBell F. Sucralose sweetener approved. *Prepared Foods* 167(5):135, 1998
20. Masalin K. Caries-risk-reducing effects of xylitol-containing chewing gum and tablets in confectionery workers in Finland. *Community Dental Health* 9(1):3–10, 1992.
21. Mazur RH. Discovery of aspartame. In *Aspartame: Physiology and Biochemistry,* eds. LD Stegink and LJ Filer. Marcel Dekker, 1984.
22. McNutt K, and A Sentko. Sugar replacers: A growing group of sweeteners in the United States. *Nutrition Today* 31(6):255–261, 1996.
23. Mermelstein NH. Washington news. Advisory panel says that saccharin should remain listed as carcinogen. *Food Technology* 51(12):24, 1997.
24. Mermelstein NH. Neotame is approved as nonnutritive sweetener. *Food Technology* 56(8):22, 2002.
25. Mitchell ML, and RL Pearson. In *Alternative Sweeteners,* eds. LO Nabors and RC Gelardi. Marcel Dekker, 1991.
26. Myer S, and WE Riha. Optimizing sweetener blends for low-calorie beverages. *Food Technology* 56(7):42–45, 2002.
27. Nabors LO. Sweet choices: Sugar replacements for foods and beverages. *Food Technology* 56(7):28–45, 2002.
28. Ohr LM. A sampling of sweeteners. *Prepared Foods* 167(3):57–63, 1998.
29. Olney JW, et al. Increasing brain tumor rates: Is there a link to aspartame? *Journal of Neuropathology and Experimental Neurology* 55(11):1115–1123, 1996.
30. Payne ML, WJ Craig, and AC Williams. Sorbitol is a possible risk factor for diarrhea in young children. *Journal of the American Dietetic Association* 97(5):532–534, 1997.
31. Prakash I, et al. Neotame: The next-generation sweetener. *Food Technology* 56(7):36–40, 2002.
32. Pszczola DE. Feelin' good ingredients that add texture. *Food Technology* 51(11):82–87, 1997.
33. Pszczola DE. Products and technologies: Ingredients. Sweet beginnings to a new year. *Food Technology* 53(1):70–76, 1999.
34. Shallenberger, RS. Sweetness theory and its application in the food industry. *Food Technology* 52(7):72–74, 1998.
35. Sky's the limit for Milky Way. *Food Engineering* 64(4):33, 1992.
36. Sloan AE. Consumer/product trends. How sweet it is! *Food Technology* 51(3):26, 1997.
37. Smith RJ. Aspartame approved despite risks. *Science* 213:986, 1981.
38. Spreads and syrups. *Progressive Grocer* 70(7):78, 1991.
39. Stegink LD. Aspartame: Review of the safety issues. *Food Technology* 41(1):119–121, 1987.
40. Sweetener. *Food Engineering* 62(2):24, 1990.
41. Thaumatin: The sweetest substance known to man has a wide range of food applications. *Food Technology* 50:74–75, 1996.
42. Tollefson L, and RJ Barnard. An analysis of FDA passive surveillance reports of seizures associated with consumption of aspartame. *Journal of the American Dietetic Association* 92(5):598–601, 1992.

43. Van den Eeden SK, et al. Aspartame ingestion and headaches: A randomized crossover trial. *Neurology* 44(10):1787–1793, 1994.

44. World Health Organization. *Diet, Nutrition, and the Prevention of Chronic Diseases.* Report of a WHO study group. Technical Report Series 797. World Health Organization, 1990.

45. Zemanek EC, and BP Wasserman. Issues and advances in the use of transgenic organisms for the production of thaumatin, the intensely sweet protein from *Thaumatococcus danielli. Critical Reviews in Food Science and Nutrition* 35(5):455–466, 1995.

WEBSITES

USDA's general information page on sugar:

www.ers.usda.gov/briefing/sugar/background.htm

The Sugar Association, Inc.:
www.sugar.org

The Lobbying Group for Alternative Sweeteners:
www.caloriecontrol.org

Information on high-fructose corn syrup:
www.sbu.ac.uk/biology/enztech/hfcs.html

10

Fats and Oils

The most abundant sources of fats and oils (referred to in this chapter as fats) in the diet are those of animal origin such as meats, poultry, and dairy products. Plants also contribute to fat in the diet, and those with the highest levels include nuts, seeds, avocados, olives, and coconut. So many foods contain fats and oils that it is sometimes difficult to find foods with no fat. Add to this the diverse array of food products that incorporate fat as an ingredient and there are even more foods added to the list of fat sources—breads, dairy foods, confections, cakes, icings, cookies, pastries, and numerous processed foods, including potato and tortilla chips.

Fats further find their way into the diet when they are extracted from high-fat foods during food preparation and processing. Frying is a common way to heat foods by partially or completely submerging the food in fat depending on the type of frying method used. Food manufacturers add fats to their food products to ease the production, handling, and storage stability of their various food products (4). The preparation of various processed food products is made possible in part because fats have unique shortening powers, melting points, plasticities, and solubilities. In addition, fats are added directly to foods, as with butter or margarine on breads or vegetables, mayonnaise in salads (potato, tuna, and chicken), and salad dressings on greens.

These and other functions of fat in foods, the different types of fat, fat replacers, the use of fat in food preparation, and storage requirements are topics discussed in this chapter. The chemical composition and classification of fats were covered in Chapter 2.

Functions of Fats in Food

Nothing compares to the unique properties that fats impart to foods (Table 10-1). As a result, it is difficult to substitute any other ingredient for the functional characteristics of fats. The unique chemical configuration of the fat molecule (specifically the length and saturation of the three fatty acids

TABLE 10-1
Functions of Fats in Foods

Function	*Examples*
Heat transfer	Sautéing, pan frying, deep-fat frying
Shortening power	Biscuits, pastries, cakes, cookies
Emulsions	Mayonnaise, salad dressings, sauces, gravies, puddings, cream soups
Varying melting points	Candies
Plasticity	Confections, icings, pastries, other baked goods
Solubility	Fats do not dissolve in water, yielding unique food flavors/textures and foods such as salad dressings
Flavor/mouthfeel	Flavor (butter, bacon, fried foods), lubricity, thickness, cooling
Textures	Creaminess, flakiness, tenderness, elasticity, cutability, viscosity
Appearance	Sheen, oiliness, color
Satiety	Fats contribute to "feeling full"
Nutrients	9 calories (kcal)/g

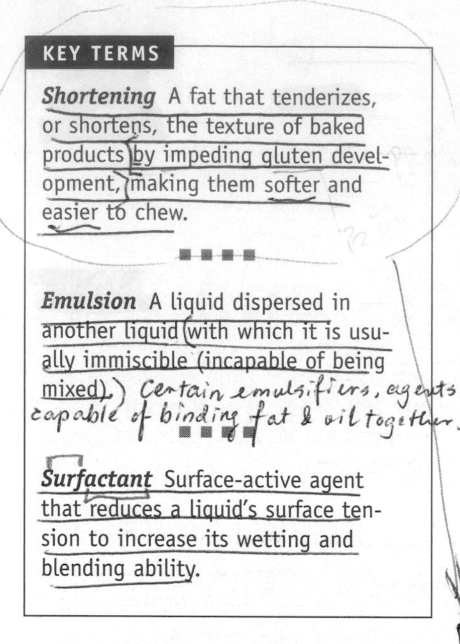

KEY TERMS

Shortening A fat that tenderizes, or shortens, the texture of baked products by impeding gluten development, making them softer and easier to chew.

Emulsion A liquid dispersed in another liquid with which it is usually immiscible (incapable of being mixed).

Surfactant Surface-active agent that reduces a liquid's surface tension to increase its wetting and blending ability.

on the glycerol molecule of a triglyceride) contributes to the numerous uses of fats in foods. It contributes to the functions of fat in heat transfer, shortening power, and emulsions and influences the fat's melting point, plasticity, solubility, flavor, texture, appearance, satiety, and nutrient content that are now discussed in greater detail.

Heat Transfer

A major function of fats is their ability to act as a medium for heat transfer. Numerous meals use fat to transfer heat to foods without burning them—butter in the frying pan, oil in the deep-fat fryer, and peanut oil in the wok. The amount of fat used can range from the minimal quantities used in sautéing, to the moderate levels used in pan frying, to enough to completely submerge a food as in deep-fat frying.

HOW & WHY? ?????????

How is food heated in deep-fat frying? In deep-fat frying, food is quickly cooked in several stages involving moisture transfer, fat transfer, crust formation, and interior cooking (32). As soon as the food is submerged in fat, the water on the food's surface vaporizes into the surrounding oil, which draws the moisture within the food toward the surface (Figure 10-1). A layer of steam forms around the food, protecting it from the high temperatures of frying and preventing it from becoming saturated with oil, although some oil is transferred into the food through the pores from which the water escaped. In the next stage the crust browns, in part because of the Maillard reaction, and becomes somewhat larger and more porous from the water being driven out of the food by the frying heat. Most of any oil that has been absorbed remains in the crust and outer layer of the fried food. Finally, the inner core cooks through heat penetration rather than by direct contact with the heated fat.

Shortening Power

The **shortening** power of certain fats makes them essential in the preparation of pastries, pie crusts, biscuits, and cakes. The more highly saturated fats tend to have greater shortening power. Mixed into a flour mixture, fat separates the flour's starch and protein and, when heated, melts into the dough. This creates air spaces that give the finished product its characteristic delicate texture. A fine grain is created from certain cake and cookie batters with the use of shortenings that gently encase the numerous air bubbles, serving as a starting point for the air to expand and increase overall volume. Baked goods become more tender, up to a point, as fat concentration increases. The exact role of fat in the formation of various bread products and their lower-fat alternatives is discussed in Chapters 23, 24, and 25.

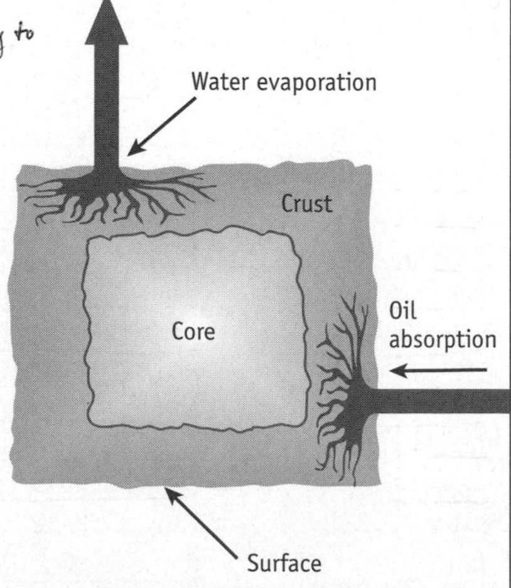

FIGURE 10-1 In fried food, oil is absorbed and water leaves as steam, contributing to a crisp, moist surface. *(Reprinted with permission from Food & Nutrition Encyclopedia)*

Emulsions

All foods contain some liquid, and if fats or oils are present, then the combination is some type of **emulsion** (29) (Chapter 2). There are two types of emulsions:

1. Oil-in-water (Figure 10-2), in which oil droplets are dispersed throughout the water
2. Water-in-oil, in which water droplets are dispersed throughout the oil

Most food emulsions are of the first type—oil-in-water. Examples of such natural emulsions include milk, cream, and egg yolks. Emulsions can be quite viscous and thick or more liquid and less stable. Examples of prepared foods emulsions (oil-in-water) include mayonnaise, salad dressings, cheese sauces, gravies, puddings, and cream soups. The less common water-in-oil emulsion, in which the smaller amount of water is dispersed in the fat, is found in foods such as butter and margarine.

Emulsifiers. There are three parts to an emulsion:

- The dispersed or discontinuous phase, usually oil
- The dispersion or continuous phase, most likely water-based
- An emulsifier, which is a stabilizing compound that helps keep one phase dispersed in the other

The two phases are kept apart by surface tension, and the boundary between them is called the interface. The emulsifier migrates to this interface and acts as a **surfactant**, lowering the surface tension between the dispersed and continuous phases so the two

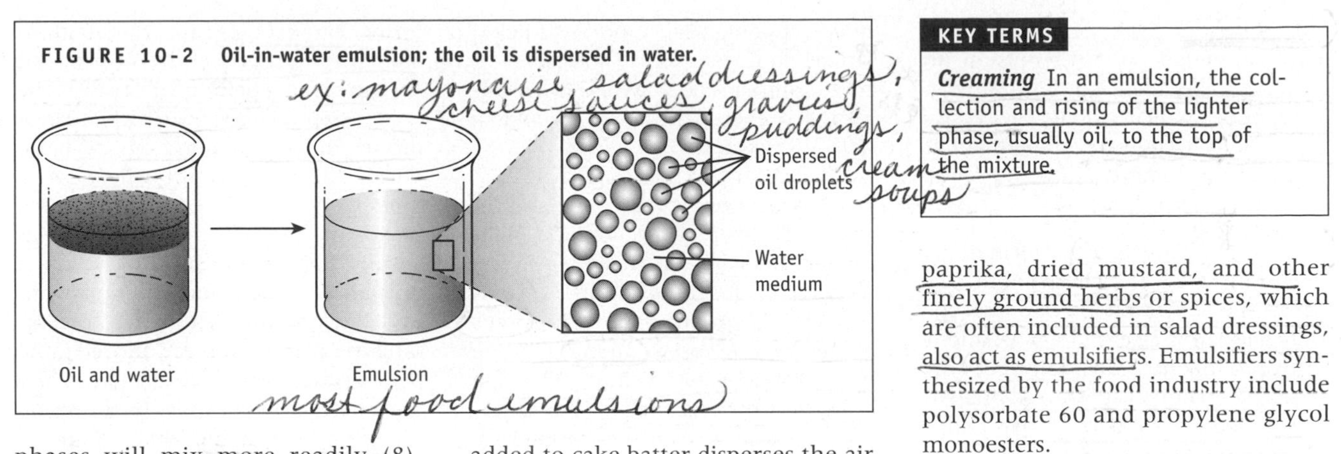

FIGURE 10-2 Oil-in-water emulsion; the oil is dispersed in water.

> **KEY TERMS**
>
> ***Creaming*** In an emulsion, the collection and rising of the lighter phase, usually oil, to the top of the mixture.

phases will mix more readily (8). Emulsifying agents act as a bridge between oil and water by being a two-part molecule, one portion being hydrophilic ("water-loving") while the other is hydrophobic ("water-fearing") (Figure 10-3). It has been suggested that this balance of water-loving and water-fearing (or lipid-loving) portions allows emulsifiers to act like a "zipper" in drawing the water and oil phases together (5).

Emulsifiers work not only with liquids, but also with gaseous phases. Figure 10-4 shows how an emulsifier added to cake batter disperses the air bubbles, resulting in a cake with a finer crumb.

Mono- and diglycerides are the most frequently used emulsifiers in the food industry (Chemist's Corner 10-1). They are added to foods for their ability to increase or improve emulsion stability, dough strength, volume, texture, and tolerance of ingredients to processing (6). Other emulsifiers are phospholipids (lecithin from egg yolks), milk proteins, soy proteins, gelatin, gluten, vegetable gums such as carageenan, and starches. Ground paprika, dried mustard, and other finely ground herbs or spices, which are often included in salad dressings, also act as emulsifiers. Emulsifiers synthesized by the food industry include polysorbate 60 and propylene glycol monoesters.

Stability of Emulsions. Emulsions can be temporary, semipermanent, or permanent and differ in their degree of viscosity and stability. Stability is defined by the degree to which the liquids stay in emulsion regardless of gravity, agitation, long storage times, extreme temperatures, surface drying, or added salt.

Temporary Emulsions. Temporary emulsions are the least viscous and stable, separating on standing, a process called **creaming**. Such temporary emulsions include oil and vinegar salad dressings, in which the oil rises to the top of the vinegar. These emulsions must be shaken each time they are used in order to reform the emulsion.

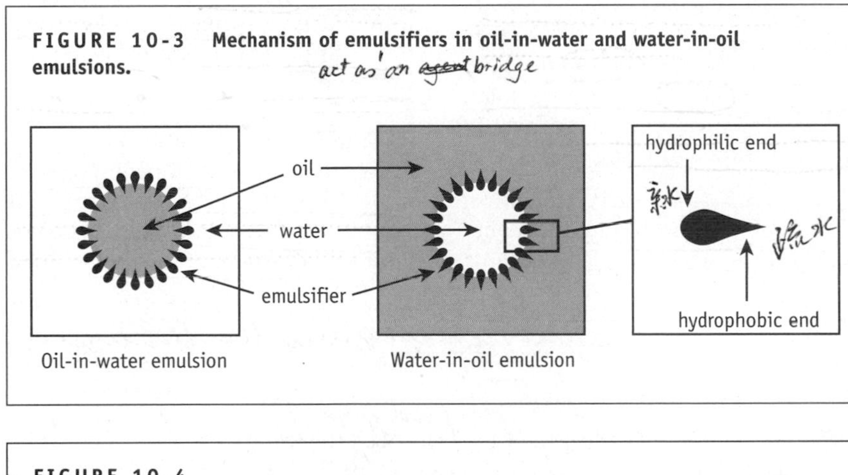

FIGURE 10-3 Mechanism of emulsifiers in oil-in-water and water-in-oil emulsions.

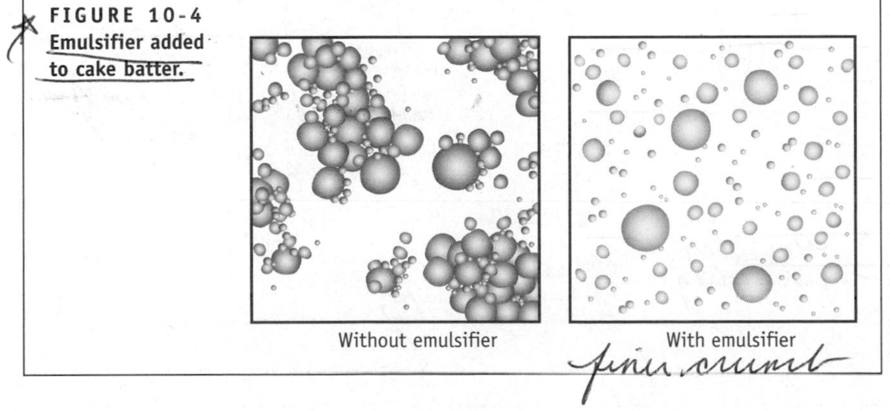

FIGURE 10-4 Emulsifier added to cake batter.

> **Chemist's Corner 10-1**
>
> ***Mono- and Diglyceride Emulsions***
>
> Mono- and diglyceride emulsifiers work because their glycerol molecules and hydroxyl groups are drawn toward the aqueous phase, while the one (monoglyceride) or two (diglyceride) fatty acids are drawn toward the lipid phases. This same principle is observed in a micelle, an aggregate of surfactant molecules arranged in a "ball fashion" so that the hydrophobic portions point toward the middle, and the hydrophilic heads are oriented toward the outside or surface of the "ball" (9).

Semipermanent Emulsions. Stabilizers may be added to an emulsion to decrease the tendency of the emulsion to separate, which creates a viscosity similar to soft yogurt. Semipermanent emulsions include commercial French and Italian salad dressings.

Permanent Emulsions. Permanent emulsions are very viscous and stable, to the point that they do not separate. Mayonnaise is a permanent emulsion in which the major ingredient is vegetable oil (dispersed phase). The added egg yolk in mayonnaise contains lecithin, which acts as an emulsifier to keep the oil dispersed in the liquid, usually vinegar or lemon (continuous phase). The high oil content of mayonnaise contributes about 100 calories (kcal) and 11 grams of fat in every tablespoon. Lower-fat alternatives are available, however.

Melting Point

Not all fats melt at the same temperature. Fatty acids are single molecules, each with a distinct melting point. Because triglycerides contain different fatty acids, food fats have a range of melting temperatures. Ultimately, a fat's melting point is determined by the following four characteristics of the fatty acid:

- Degree of saturation
- Length
- Cis-trans configuration
- Crystalline structure

Degree of Saturation. Most plant oils contain more polyunsaturated than saturated fatty acids, which causes them to be liquid at room temperature. Animal fats tend to have more saturated fatty acids, causing them to be solid at room temperature.

Length of the Fatty Acids. The length of the fatty acids can alter these general rules, as even saturated fats with shorter carbon chains can have lower melting points than those with longer ones. Butyric acid and stearic acid are saturated fatty acids found in butter. However, butyric acid has only four carbons and thus melts sooner than stearic acid, which is eighteen carbons long. Coconut oil is a saturated oil containing short fatty acids, which causes it to remain solid at room temperature. It will quickly liquefy, however, if the bottle is held in a person's warm hand.

Cis-Trans Configuration. Another structural difference that affects melting point is whether the fatty acid has more cis or trans double bonds (Chapter 2). A fatty acid with a trans configuration has a higher melting point than an identical fatty acid with a cis form at the double bond. For example, oleic acid, an eighteen-carbon fatty acid with one double bond in the cis form, has a melting point of 57°F (14°C), while the same fatty acid in the trans form is called eladidic acid and has a melting point of 111°F (44°C). Hydrogenation, a commercial process that adds hydrogen to the double bonds of the unsaturated fatty acids, changes the cis form to a trans form.

Crystalline Structure. The fourth influence on the melting point of fats is its crystalline structure—the arrangement of the fatty acids on the triglyceride molecule. How they are packed, or crystallized, in the solid phase of the fat determines when the fat will melt. This principle is important to chocolate manufacturers. The larger the fat's crystals, the higher the melting point, which allows chocolate to be held in the hand without melting.

Types of Crystals Affect Food Quality. Most fats exhibit polymorphism, the ability to exist in more than one crystalline form. Fat crystals are classified as alpha (α), beta prime (β′), or beta (β). The melting point of fats rises as the crystal sizes increase from alpha to beta prime, and eventually to beta (Figure 10-5). The rate of cooling dictates the type of crystals formed. Rapid cooling results in unstable alpha crystals with a waxy, transparent consistency. Alpha crystals are extremely fine and very unstable, melting readily and recrystallizing into the larger, more stable beta prime form. These beta prime crystals can be obtained by agitating the fat during cooling, which should be conducted at an intermediate rate. Beta prime crystals are best for food preparation, because they yield fine-textured baked goods and smooth-surfaced hydrogenated vegetable shortenings (26). Extremely slow cooling or long storage times form the most stable, or beta, crystals, which have an opaque look. Beta crystals, however, produce a sandy, brittle texture (13).

Plasticity

The plasticity of fat is its ability to hold its shape but still be molded or shaped under light pressure. Plasticity determines a fat's spreadability. It is an important characteristic in the preparation of confections, icings, pastries, and other baked products. Although most fats look solid at room temperature, they are actually composed of liquid oil with a network of solid fat crystals holding it in place. This combination allows the fat to be molded

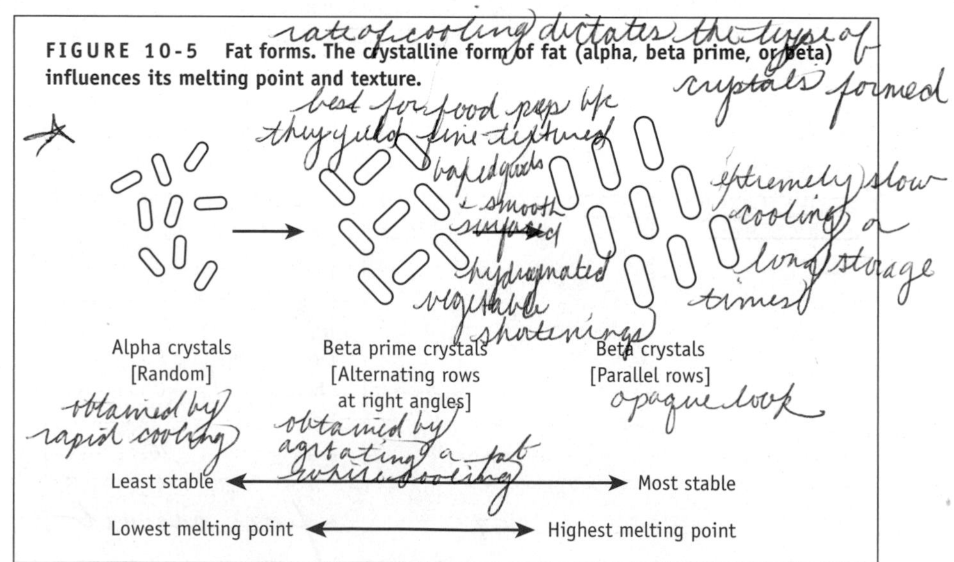

FIGURE 10-5 Fat forms. The crystalline form of fat (alpha, beta prime, or beta) influences its melting point and texture.

NUTRIENT CONTENT

Excess Dietary Fat and Health. Pure fats contain no carbohydrates, proteins, vitamins, or minerals—just fat. They also contain very little water. In contrast, foods such as butter contain about 16 percent water and small amounts of protein and other nutrients. Diet margarines can contain up to 50 percent water, which causes them to spatter when placed in a hot pan.

The many important functions and attributes of fat have led to its incorporation into many foods, perhaps more so than needed. It is generally agreed by researchers that the average American diet contains too much fat. The health concern comes from fat's probable relationship to cancer (especially breast, colon, or prostate), gallbladder disease, obesity, and high blood cholesterol, the latter two being risk factors for heart disease (12, 20). Another major issue is that many American adults are overweight, which usually reflects an imbalance between energy intake and energy expenditure.

Recommended Fat Intake. The American Heart Association, the National Academy of Science's "Diet and Health" report, and the National Institute of Health's National Cholesterol Education Program recommend that no more than 30 percent of total calories (kcal) should be derived from fat (42).

A Food Marketing Institute annual trends survey found that almost three-fourths of those who reduced fat in their diets did it primarily "to improve their family's health," while only one-fourth did so for "weight control" (37). Unfortunately, "America got fatter on fat-free" because the focus on fat has not always been accompanied by a clear message that calories (kcal), even low-fat ones, still count (36).

Major Sources of Dietary Fat. Still, in order to reach the "no more than 30 percent of calories (kcal) from fat" goal, it is important to know where fat is found in the diet. North Americans typically get most of their fat from meats; from dairy products such as whole milk, cheese, cream, ice cream, whipped cream, and butter; and from commercial fats and oils. One tablespoon of oil averages 14 grams of fat and 126 calories (kcal), while the same amount of mayonnaise, butter, and stick margarine contains about 11 fat grams and 100 calories (kcal). In addition, many processed foods made with these ingredients, such as cakes, cookies, pies, and snacks, are also high in fat. A few plant foods are rich dietary sources of fat; these include vegetable oils, nuts, seeds, avocados, coconuts, and olives.

Some fats are clearly visible: poultry skin, steak fat, vegetable oil, butter, margarine, shortening, lard, and tallow, for instance. Fat becomes invisible, however, when it acts as an ingredient in candies, cakes, sauces, snack foods, and other food products. Among all these sources of fat, the most commonly used fat-modified food products were reported to be milk, chips or snack foods, salad dressings, mayonnaise, and sauces (20).

Avoiding a Fat Deficiency. Completely eliminating all sources of dietary fat is not recommended. A total lack of fat can cause a deficiency of essential fatty acids (linoleic acid and linolenic acid), resulting in failure-to-thrive (stunted growth) in children and eczema (red, itchy, scaly skin) in both children and adults. Dietary fat should never be restricted in healthy children under two years of age (2), and at least 3 to 5 grams of essential fatty acids (especially linoleic acid)—equivalent to 15 to 25 grams of total fat—are recommended daily for adults. Keeping a certain amount of fat in the diet is also important because it serves as a carrier for the fat-soluble vitamins: A, D, E, and K (38). Another positive function of fat is seen in the case of the **omega-3 fatty acids**, as discussed in Chapter 16. These may benefit health by helping to prevent heart disease, hypertension, and thrombosis (coagulation of the blood in the heart or veins), and acting against autoimmune diseases such as lupus and arthritis (14, 16). It should be noted, however, that excess omega-3 fatty acids ingested in supplements have been reported to result in bruising under the skin (44).

KEY TERMS

Omega-3 fatty acids The polyunsaturated fatty acids eicosapentaenoic (EPA) and docosahexaenoic acid (DHA).

into various shapes. Chilled butter has very little plasticity as compared to hydrogenated vegetable oil, or shortening. The more unsaturated a fat is, the more plastic it will be. Temperature also influences plasticity. For example, hard fats such as butter become soft and more spreadable when warmed.

Solubility

Fats are generally insoluble in water. That is why oil floats above the vinegar in a salad dressing. Fats are actually defined as fats because they do not dissolve in water, but will dissolve (become soluble) in organic compounds such as benzene, chloroform, and ether.

Flavor

The flavor developed in certain foods by fats is very difficult to duplicate. For example, fats give butter, bacon, and olive oil their own distinctive tastes. Fats not only contribute their own flavor to foods, but also absorb fat-soluble flavor compounds from other foods. Sautéing garlic, onions, and herbs in oil releases their flavorful and aromatic compounds while also lending them a smooth, rich mouthfeel. One of the most obvious contributions to flavor in foods from fat is found in fried foods such as breaded poultry or fish, French fries, potato chips, and doughnuts.

Texture

Fats also contribute texture. Consider how fat gives textures to flaky pastries, smooth chocolates, half-melted ice cream, whipped cream topping, and crispy fried foods. The texture of baked products would not be the same without fat's positive influence on tenderness, volume, structure, and freshness (see Chapter 22).

The higher the fat content in ice cream, the smoother and creamier the mouthfeel. The tenderizing effect of fats on foods makes them easier to chew and causes them to feel more moist in the mouth (40). The lubricating action of fat acts to moisten certain foods such as crackers and chips in which saliva would not be enough. These dry foods are processed in the mouth much more easily if they are coated with an oil or served with a high-fat dip or spread.

Appearance

Foods are made more appealing by pigments located in a food's natural fats. Milk would be chalky white or bluish if not for its natural fat-based pigments giving it a more appealing color. The soft yellow hue of butter was so important to consumers that attempts were made to duplicate it in margarine. Fat also coats food with a sheen of delicate oil that improves the appeal of chicken, pastries, chocolate, and many other foods.

Satiety or Feeling Full

Fats induce a sense of fullness, or satiety. Foods and beverages containing fat help to delay the onset of hunger pangs signaling the next meal by two methods: (1) fats take longer to digest than carbohydrates and proteins, and (2) fats delay the emptying of the stomach contents, which makes a person feel full longer.

Types of Fats

The desirability of fat's presence in foods and its multiple roles in food preparation have led to many different types of fats being obtained from both animal and plant sources through the years. At first, people probably used the fat rendered from animal carcasses. Butter was probably not far behind once milk from domesticated animals became available. As the population grew, a more easily obtained and abundant source of fat was needed, so the oils in plants began to be extracted. When war triggered the demand for a butter replacement, margarine was introduced for the first time in the 1860s. The abundance of available fats and their incorporation into foods have not benefited the health of North Americans and Europeans, so fat replacers have recently become increasingly available in the market. The different types of fats—butter, margarine, shortenings, oils, lard, and cocoa butter—and fat replacers are discussed below.

Butter

Butter is made from the cream of milk. It takes 10 cups (2½ quarts) of milk to generate one stick (¼ pound) of butter. Butter contains about 80 percent milk fat, no more than 16 percent water, and 4 percent milk solids. Salt and coloring additives such as extract of annatto seed or carotene may or may not be added.

HOW & WHY? ?????????

How is butter made? Dramatic changes occur when the cream of milk is converted to butter, because milk is an oil-in-water emulsion that reverses in butter to a water-in-oil emulsion. For this to occur, the membranes (phospholipid) around the fat globules have to be mechanically broken down to release the milk fat so it can lump together (Figure 10-6). Methods of doing this include stirring cooled cream or, in commercial operations, centrifuging the cream to expel the water. The fatty portion of the cream separates out as a soft, yellowish solid; these are granules of butter the size of corn kernels. Liquid drained from the process is collected and called buttermilk, a tangy tasting, opaque, reduced-fat milk by-product. The butter granules are washed and then churned at slower speed until they are mixed into a smooth, homogenous paste. Any remaining water is drained, and salt is sometimes added at this point for flavor and to act as a preservative.

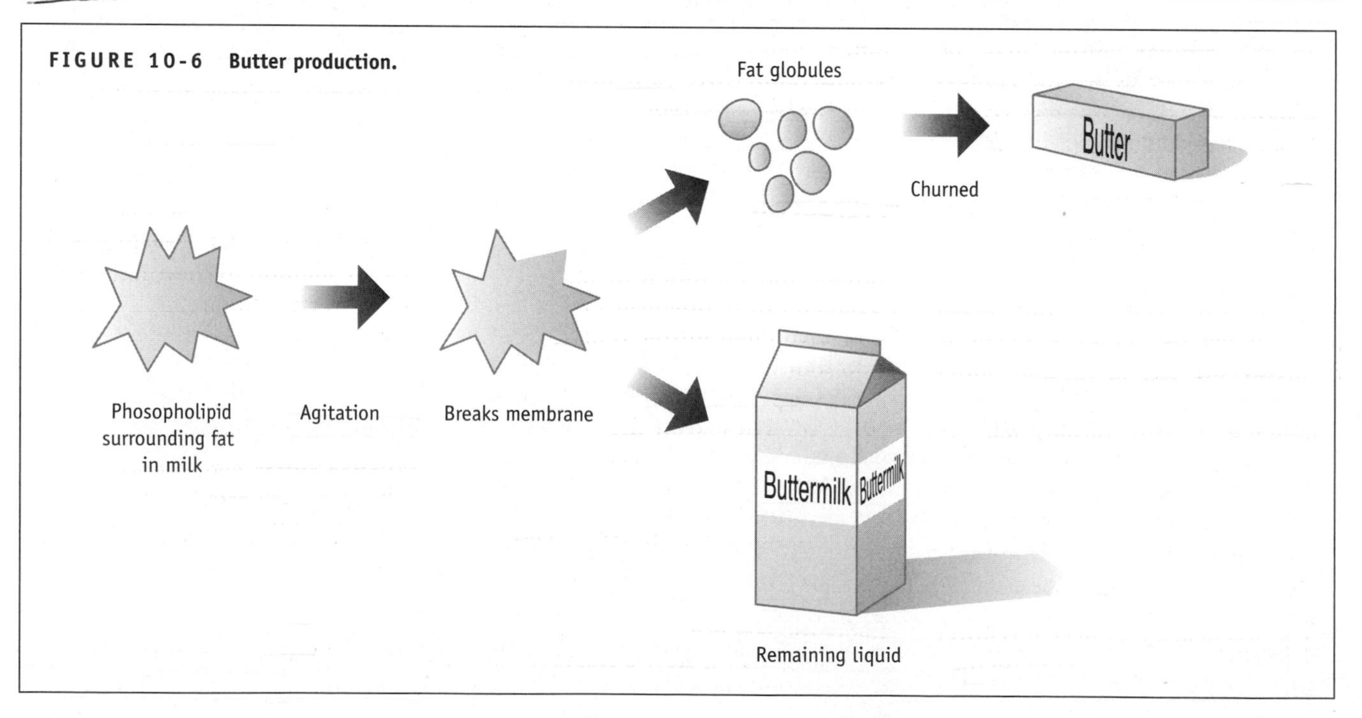

FIGURE 10-6 Butter production.

In commercial dairies, the process of making butter is begun with a cream that is concentrated up to 80 percent fat, and then further concentrated to 98 percent. It is first pasteurized to destroy pathogenic bacteria, cooled, and then recombined with the milk solids and water.

Once it is formed, commercial butter is divided into blocks that are individually wrapped, usually in waxed paper, to prevent odor absorption from other foods. Butter bought at the market has usually been cut into quarter-pound segments and rewrapped, and is sold in wrapped or boxed half-pound or 1-pound packages. Grading is voluntary on the part of the processor and is based on the butter's texture, flavor, color, and salt content (Figure 10-7).

FIGURE 10-7 **USDA grades for butter.**

Types of Butter. Butter can be purchased in a number of forms. Taste options are sweet cream butter, with or without added salt, and cultured butter, which has been acidified by lactic bacteria. A popular variant in texture is whipped butter, which is lighter in weight than regular butter and easier to spread because it has been aerated with air or nitrogen gas. Its lower density results in butter with half the calories (kcal)—50 rather than 100 calories (kcal) per tablespoon—of regular butter on a solid volume, unmelted basis. More of the whipped than the regular butter may be needed, however, to get the equivalent flavor and color. Another type of butter, called compound, composed, or flavored butter, is a softened butter mixed with one or more flavors, such as garlic, lemon, honey, wine, herbs, or nuts.

Powdered Butter. One of the newer forms, powdered butter, is made by removing the fats in natural butter and chemically treating, drying, and combining it with various ingredients—whey solids, maltodextrins, guar gum, corn syrup solids, and color additives. Powdered butter can be reconstituted with hot water and used in food service establishments to add flavor to vegetables, sauces, soups, and baked products. It is lower in calories (kcal) and cholesterol than natural butter.

Clarified Butter. Regular butter, or whole butter, must be specially treated before it can be used in certain types of food preparation involving high heats, because its milk solids burn very easily. Butter so treated, called **clarified butter** (*ghee* in Indian cuisine), is almost 100 percent fat. It is easily obtained by melting butter over low heat, allowing it to cool until the heavier, cloudy-looking milk solids have settled out, and gently pouring off the clear liquid portion. The **smoke point** of clarified butter is much higher than regular butter.

Brown or Black Butter. Sometimes regular butter is preferred over clarified butter, because the tendency of its milk solids to burn can also have a positive effect in food preparation. It allows the nuttier-flavored brown butter, or Beurre Noisette, and black butter, Beurre Noir, to be made by heating butter over low heat until it turns the desired color.

Margarine

During the Napoleonic Wars, the short supply and rationing of butter led Napoleon III to organize a contest to find a suitable butter replacement. And so it was that in 1869, a French pharmacist and chemist, Hippolyte Mege Mouries, won the contest by developing oleomargarine. During World War II, when a law prevented the coloring of food products, margarine was introduced to United States consumers in a form unappetizingly lard-like and flat white in color. Eventually the law was repealed, and yellow margarine is now a staple in the North American market.

Composition of Margarine. Standard stick margarine must contain at least 80 percent fat, about 16 percent water, and 4 percent milk solids, which is very similar to butter's general composition. Contrary to popular belief, regular margarine contains as many calories (kcal) as butter, though the fat sources differ and lower-fat versions are available. Margarine may be made from soybean, corn, safflower, canola, or other partially hydrogenated vegetable oils. In addition, margarines usually contain the following:

- Cultured skim milk
- Emulsifiers such as lecithin
- Mono- and diglycerides
- Preservatives such as sodium benzoate, potassium sorbate, calcium disodium EDTA, isopropyl citrate, and citric acid
- Vitamins A and D
- Flavorings, usually diacetyl
- Food colorings, usually annatto and/or carotene

Diacetyl is added to margarine for flavoring because it is largely responsible, in addition to short fatty acids, for butter's characteristic flavor (Figure 10-8).

KEY TERMS

Clarified butter Butter whose milk solids and water have been removed and thus will not burn.

Smoke point The temperature at which fat or oil begins to smoke.

FIGURE 10-8 Diacetyl and other short-chain fatty acids contribute to a "buttery" flavor in butter.

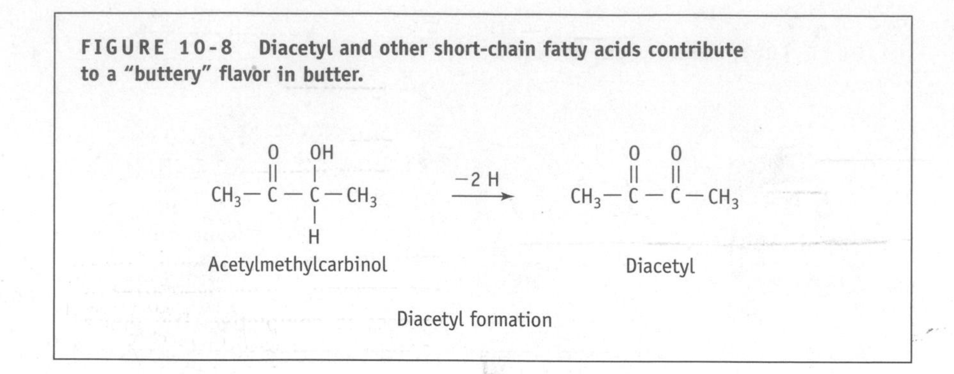

Types of Margarine. Available types of margarines include whipped margarine; light blends of margarine and butter containing about 60 percent fat; diet or reduced-calorie (kcal) margarine, which has a higher water content; and imitation margarines, identified as "vegetable oil spreads." Imitation margarines average half the fat of regular margarines.

Shortenings

Shortenings are plant oils that have been hydrogenated to make them more solid and pliable. Soybean oil is the major source of hydrogenated shortening and serves as a common frying oil. In the manufacture of shortenings, the soybean oil is hydrogenated until it reaches a solid consistency and then whipped or pumped with air to improve plasticity and give it a white color. Many shortenings are also **superglycerinated**, making them ideal for baking applications where a solid fat is needed, especially for flaky pastries and cakes containing more sugar than flour (39).

Oils

Vegetable oils are derived primarily from soybeans, rapeseed (canola oil), sunflower seed, corn, cottonseed, and safflower seed. Fruit oil sources include the avocado, coconut, palm kernel, palm, and olive. Oils differ dramatically in their taste, color, and texture, depending on their source and method of extraction.

KEY TERMS

Superglycerinated A shortening that has had mono- and diglycerides added for increased plasticity.

Extracting and Refining Oils. Oils are obtained from plant sources through extraction and refining.

Extraction. Extraction of oils is done by either mechanically pressing the seeds against a press, called cold pressing, or chemically removing the oil from the seeds with solvents. Specialty, full-flavored oils are generally cold-pressed and are often consumed unrefined, while most commercially produced oils are extracted by the use of heat and solvents.

Refining. When the goal is a neutral, clean-flavored oil, the oil is purified or refined after extraction to remove impurities such as water, resins, gums, color compounds, soil, and free fatty acids. If these compounds are not removed, they adversely affect the oil's flavor, color, clarity, smoke point, and shelf life. For example, free fatty acids detract from the oil's flavor and reduce the temperature to which an oil can be heated without smoking.

Refining, which results in oil that is 99.5 percent pure, consists of five steps: degumming, neutralizing, washing and drying, bleaching, and deodorizing.

1. *Degumming.* Certain impurities in oil form gums when combined with water. These are removed by adding hot water to the oil and spinning it at high speeds to separate the oil from the gums.
2. *Neutralizing.* Free fatty acids are removed by adding an alkaline medium to convert the fatty acid to an insoluble soap, which settles to the bottom of the neutralizing tank. Newer methods use a centrifuge to separate the major layers according to specific gravity.
3. *Washing/Drying.* Traces of soap created by the neutralizing process are removed by washing the oil with water. The water is drained, and the oil is dried under a vacuum.
4. *Bleaching.* Colored matter in the oil is removed by adding absorbent materials, such as fuller's earth or activated carbon. The absorbed colored matter is then filtered out.
5. *Deodorizing.* Volatile compounds—aldehydes, free fatty acids, hydrocarbons, ketones, and peroxides—which contribute off-odors are removed by passing steam through the heated oil.

Types of Oils. Many different types of oils are available for food preparation purposes, and the type of oil used depends on the desired outcome. The first factor to consider when selecting an oil is its flavor or lack thereof (24). The bland, mild flavor and heat stability of soybean, corn, and safflower oils make them ideal for frying. Cottonseed oil, however, is the leading choice in food service operations for frying potato chips and for producing baked goods and snacks, because of its low risk of developing and imparting off-flavors and its relative low price. Canola oil (so named because it was developed in *Can*ada) is the light, clear oil of rapeseed; it has a bland flavor and high monounsaturated fatty acid content. Rapeseeds originally contained high levels of erucic acid and glucosinolates, which, in large amounts, were found to cause cancer in laboratory animals. New genetic varieties, however, contain minimal amounts of these substances and the FDA has allowed the sale of canola oil.

Many refined oils are without any distinguishing characteristics, while unrefined, cold-pressed oils, such as peanut and olive oils, have the full flavor of the plants from which they were pressed. When choosing peanut or olive oils, it is important to be aware of a wide variance in quality and character. Because their flavors are distinctive, these oils must be used carefully in foods. Peanut and sesame seed oils are more costly than many others, but their unique flavors make them the oils most commonly used in Chinese stir-fry dishes. Refined peanut oil is less expensive and is very heat stable, making it ideal for high-heat sautéing and frying. Peanut oil's flavor is preferred by some snack food manufacturers for their products (13).

Olive Oil. Olive oil, which is considered a specialty oil, is more expensive than most other vegetable oils. Despite its higher price, olive oil consumption has increased among health-conscious consumers because of its high monounsaturated fatty acid content (78 percent) (45). Unrefined olive oils are also popular in Italian dishes and salad dressings for their full flavor.

The FDA has not established grades for olive oil, but following Italian law, olive oils are classified according to acidity: the lower the acid content, the better the grade. Extra Virgin is the best (less than 1 percent acid content), followed by Virgin (between 1 and 3.3 percent). Extra Virgin and Virgin olive oils are produced from cold pressing. The result is a high-quality oil with a strong olive flavor and a greenish tint from the presence of chlorophyll pigments.

Oil labeled only "olive oil" is a blend of refined olive oil and Virgin olive oil, resulting in a lower acidity but also less intense flavor and color than either Extra Virgin or Virgin olive oil. "Light" or "extra-light" olive oil is a refined oil made to be lighter in color than corn or safflower oil, and almost as mild. In the United States, olive oils labeled "olive oil" or "pure olive oil" account for 70 percent of all olive oils sold (19). Price does not always reflect quality; olive oils are best judged by whether or not they have a clear, deep color (usually, but not always, green) and a distinct olive aroma and flavor. Some of the compounds contributing to the flavor and aroma of Extra Virgin olive oil are volatile and are lost when heated, so it is best used in cold salad dressings and as a final flavoring to a dish. Milder olive oils are preferred for sautéing.

Another oil made from olives, Olive-Pomace, may not be called "olive oil." It is produced less expensively by extracting the oil from olives through both cold pressing and the use of solvents. The resulting oil is refined and then blended with Virgin olive oil to improve its taste, odor, and color.

Tropical Oils. Longer shelf lives are obtained in food products using tropical oils such as coconut, palm, and palm kernel oil. Their higher saturated fat content made them popular in the past with the food industry for use in cereals, candy, baked items, chocolate coatings for ice cream bars, pressurized whipped toppings, and dog and cat food. Since saturated fats contain no double bonds, they do not break down as easily. Therefore they do not become rancid as fast as unsaturated fats when subjected to oxygen, heat, and light. Tropical oils received negative publicity, however, when the consumption of saturated fats was linked to an increased risk of heart disease (19). As a result, many food manufacturers and even some fast-food enterprises have switched from tropical oils to vegetable oils for frying. Tropical oils are still used for some confectionaries, such as chocolate coatings for ice cream bars, because they become firm but melt quickly in the heat of the mouth.

Winterized Oils. Some vegetable oils, when stored in the refrigerator, do not stay completely liquid. The cooler temperatures may result in cloudiness from the crystallization of certain fatty acids that have a higher melting point than their neighboring fatty acids. This cloudiness may be eliminated by **winterizing** the oil. Commercial salad dressings and "salad oils" are usually made with winterized oil (17). Unwinterized vegetable oils that have crystallized in the refrigerator are perfectly edible and will revert to their clear character if allowed to come to room temperature.

Hydrogenation. **Hydrogenation** makes fats and oils more solid, allows them to be heated to higher temperatures before smoking, and increases their shelf life or that of the foods coated with them (Chemist's Corner 10-2). Through this process, vegetable oils may be converted to spreadable hydrogenated shortenings or margarines. Too much hydrogenation, however, will cause the product to become brittle and hard. In addition to affecting plasticity, hydrogenation contributes to making pie crusts flaky and puddings creamy.

Chemist's Corner 10-2

Hydrogenation

The process of hydrogenation is facilitated with the aid of a metal catalyst (nickel, copper, platinum, or palladium) and the presence of pressure and heat (Figure 10-9). The catalysts are removed after the process is completed. The degree of hydrogenation, or the number of hydrogen atoms added, determines the firmness of the final product.

FIGURE 10-9 Hydrogenation: Hydrogen atoms can be added to double bonds in unsaturated fatty aids in the presence of a catalyst, pressure, and heat.

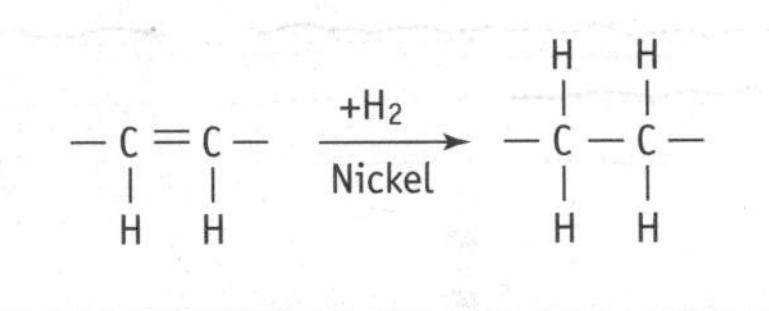

Side Effects of Hydrogenation. One of the side effects of hydrogenation is that more trans than cis configurations are created at the double bonds. The benefits of a higher concentration of trans fatty acids are a rise in the fat's melting point, increased solidity, and lengthened shelf life. The long-term health effects of trans fatty acid intakes have yet to be determined, however (7). Among other health risks, trans fatty acids have been reported to increase the risk for heart disease (30). The health concern led the FDA to propose that trans fatty acid content be listed

KEY TERMS

Winterizing A commercial process that removes the fatty acids having a tendency to crystallize and make vegetable oils appear cloudy.

Hydrogenation A commercial process in which hydrogen atoms are added to the double bonds in monounsaturated or polyunsaturated fatty acids to make them more saturated.

KEY TERMS

Interesterification A commercial process that rearranges fatty acids on the glycerol molecule in order to produce fat with a smoother consistency.

on the Nutrition Facts label of foods and beverages (10).

Lard/Tallow/Suet

Lard, which is the fat from swine, was the major shortening in use in the early 1900s. Tallow is also an animal fat, but it is derived from beef cattle or sheep. Suet is the solid fat found around the kidneys and loin of beef and sheep. These animal sources of fat are primarily saturated fat. They cannot be used for their shortening power in food preparation without first being rendered (melted down); for commercial use, the rendered fat is then deodorized. Antioxidants are often added to lard to increase shelf life. Lard produces poor textures in cakes and icings, so it is used primarily in pastry pie crusts, commercial frying, and regional cooking.

Interesterification. Lard was largely replaced by the shortenings that appeared on the market in the 1950s, but lard usage increased again in the 1960s when the process of **interesterification** was introduced. Certain fats, such as cocoa butter substitutes and lard, have unacceptable textures until they are modified by interesterification. Lard is naturally too grainy and soft at room temperatures, but becomes extremely hard when refrigerated. Interesterification creates a smoother-textured lard with a slightly higher melting point, which allows the lard to retain its shape at room temperature. Another food application dependent on interesterification is emulsifiers, which are incorporated into numerous processed foods to improve the functionality of a fat.

Cocoa Butter

Cocoa butter originates from the seeds of the *Theobroma cacao* tree. Its melting point is just below body temperature, making it perfect for "melt in the mouth" chocolate confections and candies. Cocoa butter is also used in the manufacture of nonfood products such as soaps and cosmetics.

Fat Replacers

One of the biggest challenges facing the food industry is reducing the fat content in foods (28). In the early 1990s, fat and calorie reduction was the leading research and development priority for 76 percent of food processors (3). Problems occur, however, when fat content is reduced in foods. Fat reduction affects important functional properties such as flavor, appearance, texture, mouthfeel, handling, preparation, and storage stability. In addition, food engineers become exasperated when consumers demand low-fat foods but simultaneously value taste above nutrition when it comes to selecting food products (11). Nevertheless, approximately 10 to 15 percent of all new food products are labeled "reduced in fat" (22).

The number of fat replacers and their use continue to increase as the food industry tries to meet the consumer demand for better-tasting, low-fat food items. Dietary fat consistently ranks at the top of the list of consumer nutrition concerns. Processed foods that currently incorporate fat replacers include dairy products such as cheese, sour cream, butter, and margarine; meat products such as sausages and hamburgers; frozen desserts, including ice cream and yogurt; baked goods such as cakes, biscuits, and muffins; and frostings, sauces, and gravies (47).

Types of Fat Replacers

There is no official classification of fat replacers or any standard method of naming them. "Fat substitute" was often used interchangeably with "fat replacer"; however, "replacer" is a more general term describing any ingredient used to replace fat, which can include substitutes, mimetics (imitators), and extenders. What do each of these words mean? Substitutes physically resemble fats, are often lipid-based, and usually replace the fat in foods on a one-to-one basis to duplicate the functional properties of fat (25). Fat mimetics are water-soluble, often protein or carbohydrate based, and imitate the mouthfeel of fat, which makes them useful in improving the texture of low-fat foods, especially cheeses (1). They do not, however, replace fat by weight, as do the fat-soluble substitutes and extenders (43).

What Are Fat Replacers Made From? Fat replacers are made from a variety of ingredients—synthetic fats, microparticulated proteins, starch, fiber (cellulose, gums, etc.), and even dried fruit purée—and research continues on others. Synthetic fat replacers approved by the FDA are Simplesse® and Olestra. Other synthetic fats awaiting approval are esterified propoxylated glycerols (EPG) and trialkoxytricarballate (TATCA).

Fat replacers, regardless of their name, are commonly grouped according to whether their chemical structure is carbohydrate, protein, or lipid based (Table 10-2). Some of the fat replacers are now discussed based on this classification.

Carbohydrate-Based Fat Replacers

Most of the fat replacers used by the food industry are based on carbohydrates. Fibers, gums, pectin, cellulose, and starches bind with water, swell, and impart some of the texture, mouthfeel, and opacity of fat (17). Carbohydrate-based fat replacers contain more than half the calories of fat, and may be higher in sugar.

Fruits are used as fat replacers to some degree because they naturally contain fibers, pectins, and sugars. The fiber and pectin of dried fruits such as plums, figs, and raisins provide texture, while their sugars contribute bulk and provide hygroscopic properties, which draw in water and improve the moistness of a product (21).

Protein-Based Fat Replacers

Like carbohydrate-based fat replacers, protein-based fat replacers contain fewer calories than fat. Milk (whey) or egg proteins usually serve as the source of protein-based fat replacers, but isolated soy protein has been used in foods, par-

TABLE 10-2
Fat Replacers Classified as Carbohydrate-, Protein-, or Fat-Based

Class	*Trade Names*	*Composition*	*Functional Properties*
Carbohydrate-Based			
Cellulose	Avicel® cellulose gel, Methocel™, Solka-Floc®, Just Fiber	Cellulose ground to microparticles	Water retention, texturizer, thickener, mouthfeel, stabilizer
Dextrins	Amylum, N-Oil®	Sources include tapioca	Gelling, thickening, stabilizing, texturizing
Fiber • Grain	Opta™, Oat Fiber, Snowite, Ultracel™, Z-Trim	Sources include oat, soybean, pea, and rice hulls or corn or wheat bran	Moisturizer, mouthfeel
• Fruit	Prune paste, dried plum paste, Lighter Bake, WonderSlim, fruit powder	Sources include fruits (prunes and plums)	Gelling, thickening, stabilizing, texturizing
Gums (xanthan, guar, locust bean carrageenan, alginates)	Kelcogel®, Keltrol®, Slendid™	Hydrophilic colloids or hydrocolloids	Water retention, texturizer, thickener, mouthfeel, stabilizer
Maltodextrins	CrystaLean®, Lycadex®, Maltrin®, Paselli® D-LITE, Paselli EXCEL, Paselli SA2, Star-Dri®	Sources include corn, potato, wheat, tapioca	Gelling, thickening, stabilizing, texturizing
Polydextrose	Litesse®, Sta-Lite™	Water-soluble polymer of dextrose containing minor amounts of sorbitol and citric acid	Moisture retention, bulking agent, texturizer
Starch (modified food starch)	Amalean®I & II, Fairnex™VA15 & VA20, Instant Stellar™, N-Lite, OptaGrade®, Perfectamyl™AC, AX-1, & AX-2, PURE-GEL, STA-SLIM™	Sources include potato, corn, oat, rice, wheat, tapioca starches	Gelling, thickening, stabilizing, texturizing
Protein-Based			
Microparticulated protein	Simplesse	Whey, milk, or egg protein	Mouthfeel
Modified whey protein concentrate	Dairy-Lo®	Whey protein	Mouthfeel
Other	K-Blazer, ULTRA-BAKE®, ULTRA-FREEZE™, Lita®	Egg white, milk, and corn protein	Mouthfeel
Fat-Based			
Emulsifiers	Dur-Lo®, EC™-25	Vegetable oil mono- and diglycerides	Mouthfeel
Salatrim	Benefat	Short- and long-chain acid triglyceride molecules	Mouthfeel
Lipid analogs	Olean	Sucrose and edible fats and oils	Mouthfeel
Fat extender	Veri-Lo	Oil-in-water emulsion	Mouthfeel

Adapted from Fat Replacers: Food Ingredients for Healthy Eating, Calorie Control Council, http://www.caloriecontrol.org/fatreprint.html, 2002; and the American Dietetic Association's Position Statement on Fat Replacers, *Journal of the American Dietetic Association* 98(4):463–468, 1998.

ticularly ground meat products, for many years. The USDA allows isolated soy protein at certain percentages to be added to ground meat, poultry products, cooked sausages, and cured pork (17). Protein-based fat replacers are often used in meats and other food products that have to be refrigerated or frozen.

Simplesse. An example of a protein-based fat replacer is Simplesse, which has only one-seventh the number of calories (kcal) found in an equal amount of fat. Simplesse is made from the whey of milk or from egg white proteins that have been reduced to tiny particles through a process called microparticulation (Figure 10-10). The droplets roll around on the tongue, imparting a smooth, creamy consistency that closely resembles the creamy mouthfeel of fat (27). The FDA approved Simplesse as a "generally recognized as safe" (GRAS) substance for use in frozen desserts such as ice cream, cheese foods such as cream cheese and cheese spreads, and other products (41).

Problems with Simplesse. Unfortunately, salad dressings made with Simplesse are more acidic, and frying or baking with it is out of the question because the protein droplets of Simplesse break down when heated (35). As a result, Simplesse is limited to use in frozen and uncooked food products. Also, consumers who are allergic to milk or egg protein are cautioned to avoid Simplesse.

Lipid-Based Fat Replacers

Chemically modifying the molecular structure of fats can result in fat-based fat replacers that have fewer calories (kcal) than fat. Often the chemical changes involve inhibiting absorption or shortening the length of the fat's fatty acid. Short- and medium-chain fatty acids provide fewer calories (kcal) than larger ones. This is why butyric acid (four carbon atoms) yields fewer calories (kcal) than palmitic acid (sixteen carbons) (16).

Salatrim. This principle is applied in the manufacture of the lipid-based fat replacer known as Salatrim (Benefat™). Salatrim provides 5 calories (kcal) per gram instead of fat's usual 9 calories (kcal) per gram (23). Salatrim is an acronym for short and long-chain acyl triglyceride molecules, which describes how it achieves these lower calorie (kcal) values: it consists of short-chain fatty acids combined with one long-chain fatty acid that is only partially absorbed.

Fat Extenders. One way to reduce calories (kcal) in higher-fat food products such as sauces, mayonnaise, and salad dressings is to use a fat extender. Calories (kcal) are reduced by diluting the fat with an "extender" like water in the form of an oil-in-water emulsion. As a result, fat extenders can reduce the amount of fat in mayonnaise and salad dressings by as much as 70 percent (34).

Olestra. Another fat-based fat replacer on the market is Olestra. This unique compound was discovered by accident by Fred Hugh Mattson, a research chemist for Procter & Gamble. Company researchers were trying to locate an easily digested fat for premature infants, but instead found a substance not broken down by the body at all (46). Previously known as sucrose polyester, Olestra gained FDA approval in 1996 and is marketed under the brand name Olean®. It is made from sugar and vegetable oil in a process in which the three-carbon glycerol molecule in the oil is replaced by a large sucrose with six to eight fatty acids attached. This molecule is so large that it moves through the digestive tract before enzymes have time to digest the fatty acids (13).

FIGURE 10-10 Microparticulation: The process of reducing the original size of a substance to numerous tiny particles.

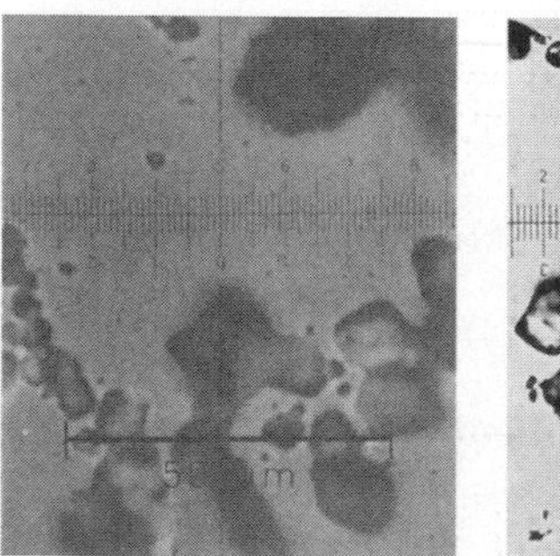

Powdered sugar (10–30 microns)

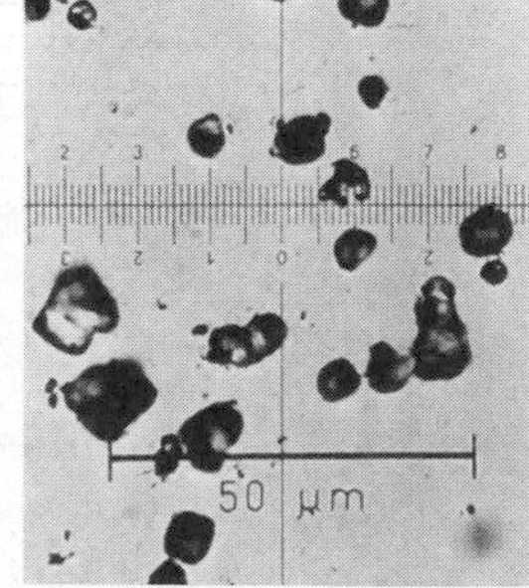

Nonfat chocolate (5–10 microns)

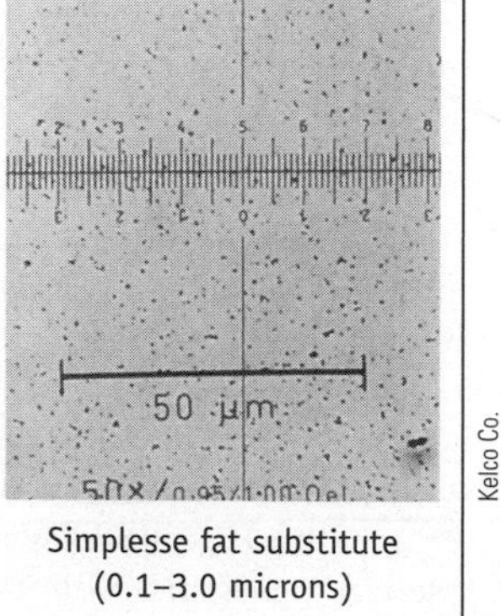

Simplesse fat substitute (0.1–3.0 microns)

Kelco Co.

Olestra is stable during heating and can even withstand the high temperatures of frying. Food companies use it in snack foods (crackers, potato and tortilla chips), fried and baked goods, and dairy products.

Problems with Olestra. A drawback of Olestra is that it reduces the absorption of vitamins A and E (27). For this reason, the FDA requires that fat-soluble vitamins be added to products made with Olestra. Side effects of too much Olestra consumption may be diarrhea, cramps, and gas.

Food Preparation with Fats

Very few fat replacers can replace the function of heat transfer used in food preparation. Fats allow the transfer of heat during frying—sautéing, stir-frying, pan-frying, and deep-fat frying (see Chapter 6). This section focuses on how to take care of the fat used in frying, and describes food preparation techniques that reduce the amount of fat transferred to the food.

Frying Care

Can Any Fat Be Used for Frying? Not every fat is suited for the very high temperatures of deep-fat frying, which average 350°F to 450°F (177° to 232°C). The higher temperatures of frying allow foods to be heated more quickly than if they were boiled. The fats commonly used for frying must be 100 percent fat. This includes vegetable oils (except for olive or sesame oil), and hydrogenated shortenings (without additives such as emulsifiers). The vegetable oils most frequently used include cottonseed, corn, canola, peanut, and safflower.

Many vegetable oils are also chosen for frying because they have little flavor of their own and will not overpower the flavor of even lightly seasoned or bland foods. On the other hand, some foods call for butter as a sautéing fat to enhance flavor, but the heat must be carefully controlled because the water and milk solids in butter cause it to spatter and burn more easily. Margarine is not recommended for frying because, in addition to containing water, it has a low smoke point. The water will splatter, and foods fried in fats with low smoke points develop unpleasant flavors.

Smoke Point. Select a fat with a high smoke point that is at least above 420°F (216°C). This temperature is much higher than the boiling point of water (212°F/100°C) and even higher than frying temperatures that range from 350° to 450°F (177° to 232°C). Failing to select a fat with a smoke point at or below the frying temperature will cause it to overheat and decompose into glycerol and its individual fatty acids. The glycerol is further broken down (hydrolyzed) to a steel-blue smoke called acrolein. Acrolein's sharp, offensive odor warns people of its presence. This smoke is extremely irritating and even harmful to the mucous membranes of the mouth and nasal passages.

Table 10-3 lists the smoke points of various fats. Selecting fats with smoke points above 420°F (216°C) for commercial frying automatically excludes olive oil, lard, and vegetable shortenings. Hydrogenated shortenings with added mono- and diglycerides are not recommended for frying, because the fatty acids are easily removed from the glycerol molecule, which is then free to form acrolein.

TABLE 10-3

Smoke Points of Selected Frying Fats and Oils

Fat/Oil	*Smoke Point*
Vegetable shortenings + emulsifier	356–370°F (180–188°C)
Lard	361–401°F (183–205°C)
Vegetable oils	441–450°F (227°–232°C)
Most olive, virgin oils	391°F (199°C)
Corn oil	440°F (227°C)
Soybean oil	492°F (256°C)

Flash Point and Fire Point. A more serious problem than smoking is when an oil is heated to its **flash point**, or to about 600°F (316°C). Increasing the heat to 700°F (371°C) will reach the fat's **fire point**. If this occurs, water should not be used to put out the fire. Fire extinguishers with a "C" designation should be kept on hand for a fire started by fat use. If an extinguisher is not available, it may be possible to smother the fat-fueled fire with a pan lid or large amounts of baking soda.

Controlling the Temperature of Frying Fats. It is difficult to detect overheating visually, because fat does not boil. Compounding this problem, heated oil is always hotter than it appears. These two problems can cause overheating, which contributes to the rapid deterioration of fat through **polymerization**. Any egg yolks used in the coating batter also contribute to the darkening effect on the fat. Further, the increased viscosity of overheated fat results in higher fat absorption rates in the fried foods, making them greasy (31). One way to control the temperature of cooking fats and prevent excess absorption is to use thermostatically controlled deep-fat fryers, but it is recommended that these thermostats be checked for accuracy routinely.

Avoid Too Low Temperatures. While it is important not to overheat frying fats, it is equally important not to let temperatures drop too low, since this may lead to excessive fat absorption, resulting in soggy, greasy fried food. Temperatures quickly drop when large quantities of frozen food are added to hot oil. To combat this problem and help stabilize the temperature, the food should be added in batches so the oil will be given sufficient time to reheat to the correct temperature. It is also important that the food pieces in a batch be the same size, so they finish cooking at the same time.

Perfect Browning of Fried Foods. If temperatures are correctly controlled, the result will be a food that has a crisp, golden crust surrounding a tender, perfectly cooked center. The key in deep-fat frying is to ensure that the food's inside is sufficiently cooked without overdoing its outside. Fried foods cook on the principle that frying temperatures convert the food's water to steam, which then escapes, keeping the food cool and preventing it from burning and/or absorbing fat. Eventually, however, the amount of steam decreases, allowing the outside to brown. Foods left too long in the fryer after all the steam has escaped will have burned crusts and excess fat absorption.

High-moisture foods such as French fries need to be cooked at lower temperatures or the outside will turn crispy before the inside has had a chance to cook. Steam trapped by the hard crust will cause the food to become limp as it cools. Conversely, low-moisture foods need higher temperatures so they will cook quickly, leaving oil no time to enter the food. Other determinants in temperature selection to obtain the best crust color are the amount of food, the length of time it is submerged in oil, the temperature of the food, the oil quality, and the food's shape and size, porosity, and type of coating (33).

Optimal Frying Temperatures. Optimal frying temperature is 375°F (191°C), with higher temperatures (375° to 390°F/191° to 199°C) required for smaller pieces of food, and lower temperatures (350° to 365°F/177° to 185°C) for larger pieces of food (32).

Recommended Equipment. Frying temperature is not the only factor in influencing the fried food product's quality. It is important to use stainless-steel equipment; iron, and especially copper or copper alloys such as brass,

KEY TERMS

Flash point The temperature at which tiny wisps of fire streak to the surface of a heated substance (such as oil).

Fire point The temperature at which a heated substance (such as oil) bursts into flames and burns for at least 5 seconds.

Polymerization A process in which free fatty acids link together, especially when overheated, resulting in a gummy, dark residue and an oil that is more viscous and prone to foaming.

may increase rancidity. Hoods or exhaust systems above the fryer should be cleaned frequently so that accumulated particles do not drip back down into the fat. Deep, narrow containers are recommended for deep-fat frying, because shallow, wide pans increase the surface area, lowering the smoke points by the greater exposure to air. The fryer should be filled no more than one-half to three-fourths full of oil. As fat is absorbed by the foods, it should be replaced with fresh fat. However, fresh fat should never be added to fat that is rancid, foaming, or dark, because it will not overcome these defects and will deteriorate very quickly.

Optimal Frying Conditions. The fats in a fryer go through stages that influence the quality of the fried product. At the "new" and "break-in" stages, foods absorb too little oil; just the right amount is absorbed at the "fresh" and "optimum" levels that follow; and then too much soaks in at the "degraded" and "runaway" phases (39). Many professional chefs claim that foods fry best in oil that has been used at least once. Desirable browned crusts occur when oils pick up proteins and carbohydrates from the fried foods. Eventually, however, the browning becomes too dark and the fat must be replaced. Also, as the fat deteriorates, the surface tension of the frying oil decreases, making foods more likely to soak up the fat. Repeated use of a frying fat will also lower its smoke point, because each heating hydrolyzes some of the triglycerides into smaller molecules.

Avoid Water. Foods should be as free of surface moisture as possible before being submerged in the heated fat. Water causes spattering of hot oil, which can cause burns; it requires more energy to maintain temperatures; it may result in longer frying times; and it causes the fat to break down chemically, reducing its frying life.

Avoid Food Particles. Inevitably, particles of food or breading break off or fall through the basket and build up in a deep-fat fryer over time. These food particles should be filtered out daily (or every eight hours of use) or they will darken the oil's color, lower its smoke point, and reduce its keeping time. On the other hand, excessive filtering introduces oxygen into the oil, resulting in rancidity, gum development, and foaming, the latter observed as a persistent layer of bubbles on the surface.

Cool the Frying Fat. A frying fat should theoretically stay fresh for several months if it is cooled immediately after use and stored in an air-tight container in a dark, cool place. Refrigeration will also increase its shelf life. Large commercial fryers contain too much fat to be cooled completely and then efficiently reheated, so they are turned down to approximately 225°F (107°C). Decreasing the temperature during down time prevents the fat's breakdown and extends its usefulness.

When to Discard the Used Frying Fat. There is no easy method for determining when oil that has been used repeatedly should be discarded. The first indication that an oil needs to be replaced is that its color and the color of the fried food starts to darken. This transformation takes place soon before the flavor and odor of the oil start to deteriorate. An experienced person can tell by looking at it if the oil needs changing, but food service establishments may purchase a commercial kit that allows anyone to determine an oil's freshness by checking its color against the kit's standard. A further indication that the oil is too old is that the food fried in it is greasier than normal because of increased oil absorption. Other factors to consider in the decision to discard oil include the type of oil used, the type of foods being fried, the number of times the oil is used, the presence of many particles, excessive foaming or smoking, and the quality of the foods cooked in the oil (32). Guidelines for preserving frying oils are summarized in Figure 10-11, while Table 10-4 provides a list of problems that may arise and what causes them.

FIGURE 10-11 **Preserving frying oils.**

Equipment
Stainless steel
Deep, narrow containers for deep-fat frying
Fill no more than one-half to three-fourths full
Clean commercial fryers at least once a week
Keep hood/exhaust system above fryer clean

Heating
Most frying occurs between 350 – 450°F (177 – 232°C)
Avoid high temperatures (undercooked inside, overly crispy/brown outside)
Avoid low temperatures (soggy fried food)
Heat no longer than necessary
Allow sufficient time for oil to reheat between batches
Avoid large quantities of frozen foods
Do not overheat to flashpoint, about 600°F (316°C)
Check thermostat for accuracy

During Frying
Use oils with smoke points above 420°F (216°C)
Avoid exposure to oxygen/air, light, salt (season after frying), certain metals (iron, copper, nickel)
Only completely dry food should be submerged
Filter particles of batter/flour and/or food
Limit egg yolks used in the batter or flour (darkening effect on the fat)
Monitor freshness of frying oils by their color against a standard
Add fresh oil daily to commercial fryers in order to replace entire amount every three to five days

Storage
Store unopened containers in cool, dry place
Store opened containers airtight in refrigerator
Commercial fryers: reduce temperatures to about 225°F (107°C)
Discard dark, gummy, or repeatedly used oil

TABLE 10-4
Problems in Deep-Fat Frying Oils and Their Causes

Problem	*Possible Cause*
Fat darkens excessively	Overheating of fat Defective equipment Inadequate filtering Inadequate cleaning of equipment Use of inferior fat Foreign material entering the fat
Fat smokes excessively	Overheating of fat Defective equipment Inadequate filtering Use of inferior fat Poor ventilation Inadequate cleaning of equipment
Life apparently gone from fat: fat won't hold the heat; fat won't brown the food	Too low a frying temperature (including faulty thermostat) Lack of proper kettle recovery Not cooking long enough Excessive foam development Overloading kettle Improper preparation of food
Fat foams excessively and prematurely	Overheating of fat or failure to reduce heat when not in use Insufficient filtering Failure to clean and rinse equipment adequately Brass or copper being used in kettle Use of inferior fat Salt or food particles getting into the fat
Food greasy; too much absorption	Frying at too low a temperature Slow temperature recovery due to poor equipment Overloading kettle Frying fat has broken down Improper preparation of food Keeping food in fat after it is done, or insufficient draining after removal
Fat has "objectionable" odor or flavor	Excessive crumb or foreign material in kettle Use of poor-quality food Detergent film due to insufficient rinsing Use of inferior or broken-down fat Excessive fat absorption Holding cooked food too long

Source: Frosty Acres

Lower-Fat Preparation Techniques

Reducing the consumption of dietary fat may be accomplished by following the dietary guidelines recommending a meal pattern that is lower in fat, especially the saturated type; relying on lower-fat or nonfat cooking methods; and reducing the fat in recipes. The lower-fat food consumption pattern discussed in Chapter 4 is the most significant step a person can take to lower dietary fat.

1. Reduce overall fat intake
2. Modify types of fat used in food preparation
3. select cooking method that 不需油 or 少需油
4. Modify recipes — substrate / reduce > fat ingredients.

Fats Preferred for Health. Once overall fat intake is reduced following these guidelines, the next step is to modify the types of fat that are ingested. Monounsaturated fats are preferred over polyunsaturated, which in turn are recommended over saturated fats. For cooking fats, Table 10-5 shows that, compared to other oils and fats, canola oil contains one of the highest levels of monounsaturated fatty acids (58 percent). In the same category, olive, avocado, almond, and apricot oils tend to impart more flavor but are more expensive. Safflower oil scores highest in the category of polyunsaturated oils. Saturated fats—coconut, palm, and palm kernel oils and butter—should be avoided according to certain dietary guides. Butter is often chosen, however, for its unique flavor or by those concerned about the trans fatty acids found in margarines and other partially hydrogenated fats. Although butter and margarine contain approximately the same number of calories (kcal) and grams of fat, the fat in butter is primarily saturated, while that from margarine is more unsaturated. Lard, the saturated fat from swine, is best replaced by vegetable shortening, but even the latter is partially saturated.

Reducing Fat by Healthy Methods. Lowering dietary fat may also be achieved by selecting a cooking method that does not rely on fat. All of the moist- and dry-heat cooking methods, with the exception of frying, lend themselves to fat-free preparation of foods. Even frying, specifically sautéing and stir-frying, is acceptable if the right type of fat is chosen and a minimal amount is used. Pan-frying and deep-fat frying are the only two methods for which it is essentially impossible to lower the amount of fat used.

Modifying Recipes to Reduce Fat. Another way to reduce fat in food preparation is to focus on the recipes. The following foods are the main contributors to fat in recipes: meats, dairy products (including whole milk, cheese, cream, ice cream, whipped cream, and butter), commercial fats and oils, avocado, coconut, olives, nuts, and seeds. Processed foods such as cakes, cookies, pies, snacks, and others that are made with these ingredients are also high in fat. Many recipes could simply have their fat content reduced or another ingredient substituted without affecting overall quality (Table 10-6 on page 197). Sometimes the fat can be removed altogether. Following the dietary guides will automatically eliminate recipes that are too high in fat that cannot be adequately modified. A good rule of thumb is that any meal exceeding 20 grams of fat is

TABLE 10-5
Comparison of Dietary Fats*

DIETARY FAT	Cholesterol mg/tbs	% Monounsaturated Fat	% Polyunsaturated Fat	% Saturated Fat
Highest in monounsaturated				
Olive oil	0	77	9	14
Canola oil	0	58	36	6
Peanut oil	0	49	38	13
Highest in polyunsaturated				
Safflower oil	0	12	79	9
Sunflower oil	0	20	69	11
Corn oil	0	25	62	13
Soybean oil	0	24	61	15
Cottonseed oil	0	19	54	27
Hydrogenated				
Margarine	0	48	34	18
Vegetable shortening (Crisco)	0	43	31	26
Highest in saturated				
Coconut oil	0	6	2	92
Butter (fat)	33	30	4	66
Palm oil	0	39	10	51
Lard	12	47	12	41

*Fatty acid content normalized to 100 percent.

probably too high in dietary fat for people consuming three meals a day. Other ways to reduce the amount or modify the type of fat in the diet include:

- Fruit preserves and honey can replace butter on breads.
- Mustard, ketchup, or low-fat salad dressing or mayonnaise may substitute for regular mayonnaise in sandwiches or salads.
- Purées of fruits such as plums, dates, apples, and figs may replace some, but not all, of the fat in recipes for baked products.
- Crumb crusts can replace standard pie crusts.
- Double-crust pies can be converted to one-crust pies, automatically cutting fat by close to 50 percent.
- A nonfat condiment such as salsa, relish, or chutney can replace some of the butter or sour cream toppings on baked potatoes.

Storage of Fats

Storage of fat depends on its type. Fats such as butter and margarine are best stored in the refrigerator. Butter will keep for months in the freezer, but margarines do not freeze as well because their emulsions may separate. Shortenings and most oils are usually stored at room temperature and should be kept tightly covered in a dark spot on the cupboard shelf; however, they will keep longer if refrigerated. Olive oil has a shorter shelf life than most vegetable oils and should be refrigerated fairly soon after opening.

HOW & WHY? ?????????

Why does refrigerated oil look cloudy?
The cloudiness that occurs in some refrigerated oils is caused by the solidification of fatty acids. The cloudiness disappears once they reach room temperature again.

Rancidity

Rancidity is the chemical deterioration of fats, which occurs when the triglyceride molecule and/or the fatty acids attached to the glycerol molecule are broken down into smaller units that yield off-flavors and odors. The longer a fat is stored, the greater the possibility of its becoming rancid. Fats and oils used in cooking tend to become rancid because they are exposed to oxygen, heat, and light. For this reason, they should be checked frequently for rancidity. Using a rancid fat or oil to make cakes, cookies, or other baked goods will adversely affect their flavor. Rancid fat will also ruin the flavor of sautéed or fried foods and cause problems during heating because of its lower smoke point.

Types of Rancidity. There are two basic types of rancidity: hydrolytic rancidity, which occurs when *water* breaks larger compounds into smaller ones; and oxidative rancidity, in which the double bond of an unsaturated fatty acid reacts chemically with *oxygen* to result in two or more shorter molecules.

Hydrolytic Rancidity. Fats become rancid when exposed to water, usually the water found frozen on food to be fried. The addition of water hydrolyzes the bonds in the triglyceride, causing it to break down into smaller compounds. Catalyzing this reaction are lipase enzymes and heat. This hydrolytic rancidity has implications for deep-fat frying. Placing cold, wet food in a heated frying oil introduces water, making the oil prone to hydrolytic rancidity. Conversely, fats that have not been heated are more prone to hydrolytic rancidity because the lipase enzymes have not yet been destroyed by heat. Room temperature is ideal for the lipase enzyme, which is why butter left out at room temperature quickly decomposes. This is why butter is often refrigerated or frozen. Water is also found in butter, which is why it has a tendency to go rancid. Butter's volatile short-chain fatty acids, such as butyric and caproic acids, create a rancid odor and off-flavor when released into the air. The long-chain fatty acids are also freed, but they are not volatile and therefore do not contribute to the odor of rancid butter.

TABLE 10-6
Replacing Fatty Ingredients in Recipes

Higher-Fat Item		*Replaced by Lower-Fat Item*	
1 oz (1 sq)	Baking chocolate	3 tbs	Powdered cocoa + 1 tbs margarine
1 C	Butter	1 C	Margarine (lowers saturated fat)
1 oz	Cheese	1 oz	Lower-fat cheese
1 C	Cream (heavy)	1 C	Evaporated skim milk
		⅔ C	Nonfat milk + ⅓ C vegetable oil
1 C	Cream cheese	1 C	Reduced-fat cottage cheese + 4 tbs margarine + salt to taste + milk for blending
1	Egg (large)	1	Egg white + 1 tsp vegetable oil
		2	Egg whites
		¼ C	Egg substitute
1 C	Fat	⅓ C	Applesauce + ⅔ C fat
1 C	Milk (whole)	1 C	Nonfat or reduced fat milk
1 tbs	Salad dressing	1 tbs	Low-calorie salad dressing
½ C	Shortening	⅓ C	Vegetable oil
1 C	Sour cream	1 C	Plain yogurt
		1 C	Reduced-fat cottage cheese (blended)

KEY TERMS

Flavor reversion The breakdown (oxidation) of an essential fatty acid, linolenic acid, found in certain vegetable oils, leading to an undesirable flavor change prior to the start of actual rancidity.

Oxidative Rancidity. Fats can also become rancid when they are exposed to the oxygen in air. The higher the degree of unsaturation, the more likely it is that the fat will be subject to oxidative rancidity. This is why saturated and hydrogenated fats used to be popular with some food manufacturers and food service establishments.

Unlike hydrolytic rancidity, the rancidity due to oxygen occurs in a series of steps (Chemist's Corner 10-3). The initiation period is slow and is triggered by light, high temperatures, table salt, food particles in the frying oil, and certain metals such as iron, copper, and nickel. This initial stage is followed by a quicker, irreversible, and self-perpetuating chain reaction. Oxygen atoms attach to the carbons next to the double bond of the fatty acid, creating very reactive and unstable molecules called free radicals. These free radicals contribute to the further breakdown of fats into smaller compounds, resulting in unpleasant odors and off-flavors. Once this process starts, it is difficult to stop, because the free radicals generated by the reaction create more free radicals, which in turn keep producing free radicals until all the double bonds have been used in the process. Antioxidants, found naturally in the fat or commercially added, inhibit oxidative rancidity and extend shelf life.

Flavor Reversion. Food manufacturers must also deal with **flavor reversion**, a type of characteristic flavor change that occurs even before actual rancidity begins. The odor and flavor of oils, particularly those having high linolenic acid levels, can be altered by light and heat, which convert the fatty acids to volatile compounds, causing off-odors. Only a small amount of oxygen needs to be present to oxidize linolenic acid. The odors and flavors produced by flavor reversion depend on the type of oil. Soybean oil initially becomes "beany" and then "fishy," but the two most com-

Chemist's Corner 10-3

Oxidative Rancidity

The three stages of oxidative rancidity, also known as oxidation, involve initiation, propagation, and termination (Figure 10-12) (18).

- **Initiation.** Initiation occurs when the loosely held hydrogen atom at the double bond of an unsaturated fatty acid is lost, forming a free radical (R•), which is very reactive.
- **Propagation.** In the propagation stage, an oxygen combines with the free radical, forming a compound (peroxide-free radicals) that can remove another hydrogen near a double bond and yield another free radical. This reaction can be repeated several thousand times until most double bonds on the fatty acids have been removed. Further breakdown of the hydroperoxide molecules into smaller units such as acids, alcohols, aldehydes, and ketones results in rancid off-odors and flavors.
- **Termination.** The chain reaction is terminated when all the free radicals have reacted with other free radicals or antioxidants and/or there are no more hydrogens at the unsaturated fatty acids' double bonds to react with oxygen.

FIGURE 10-12 **Stages of oxidative rancidity.**

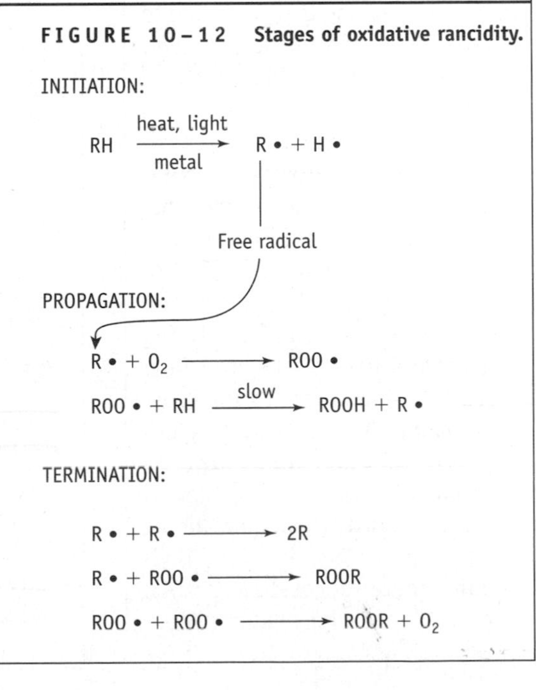

monly used oils, cottonseed and corn, are very resistant to flavor reversion.

Preventing Rancidity. Rancid products have reduced shelf lives and must be discarded. In the past, cereal manufacturers incorporated predominantly saturated fatty acids such as coconut and palm oils into their products to reduce the risk of rancidity. More recently, public concern over saturated fat and its relationship to blood cholesterol levels contributed to increasing use of unsaturated oils and new ways of deterring rancidity.

Avoid Oxygen and Heat. One method of inhibiting rancidity consists of packing such items as potato chips, tortilla chips, and other foods high in unsaturated fatty acids in vacuum packs or nitrogen to prevent contact with oxygen.

To prevent rancidity of the oils and fats themselves, there are several protective measures that can be taken. Vegetable oil bottles should be recapped immediately after use to minimize exposure to oxygen. Storing a bottle of oil on the shelf near the range, where heat is constantly being generated, is not recommended. They are best kept in cool, dry places away from air, light, high temperatures, and exposure to metals such as iron and copper. In warmer climates, they fare better in the refrigerator.

Chemist's Corner 10-4

Antioxidants in Action

Antioxidants prevent oxidation of fats either by being oxidized themselves, by donating their hydrogen to a fatty acid as a reducing agent, or by sequestering metals such as a chelating agent. Reducing agents that transfer hydrogen include vitamin C, vitamin E (tocopherols), beta-carotene, flavonoids, erythorbic acid, ascorbyl palmitate, and sulfites.

Chelating agents "attach" to metal ions such as copper or iron and prevent them from catalyzing oxidation. The word "chelate," from "chela" or claw, refers to the claw-like manner in which the agent binds to the metal through coordinate covalent bonds. Examples of chelating agents, also known as metal sequesterers, include ethylenediaminetetraacetate (EDTA), citric acid, and phosphates (18).

Antioxidants Antioxidants, natural and commercial, are added to foods containing large amounts of unsaturated fats in order to prevent rancidity (Chemist's Corner 10-4). The USDA's Code of Federal Regulations defines antioxidants as substances used to preserve food by retarding deterioration, rancidity, or discoloration due to oxidation. Foods to which antioxidants are commonly added include dry cereals, crackers, nuts, chips, and flour mixes.

Naturally occurring antioxidants include vitamins E and C, lecithin, flavonoids, and gum guaiac. Many vegetable oils naturally contain vitamin E. Commercial antioxidants permitted by the FDA include butylated hydroxyanisole (BHA), butylated hydroxytoluene (BHT), propyl gallate, and tertiary butyl hydroquinone (TBHQ) (Figure 10-13) (15).

FIGURE 10-13 **Commonly used commercial antioxidants.**

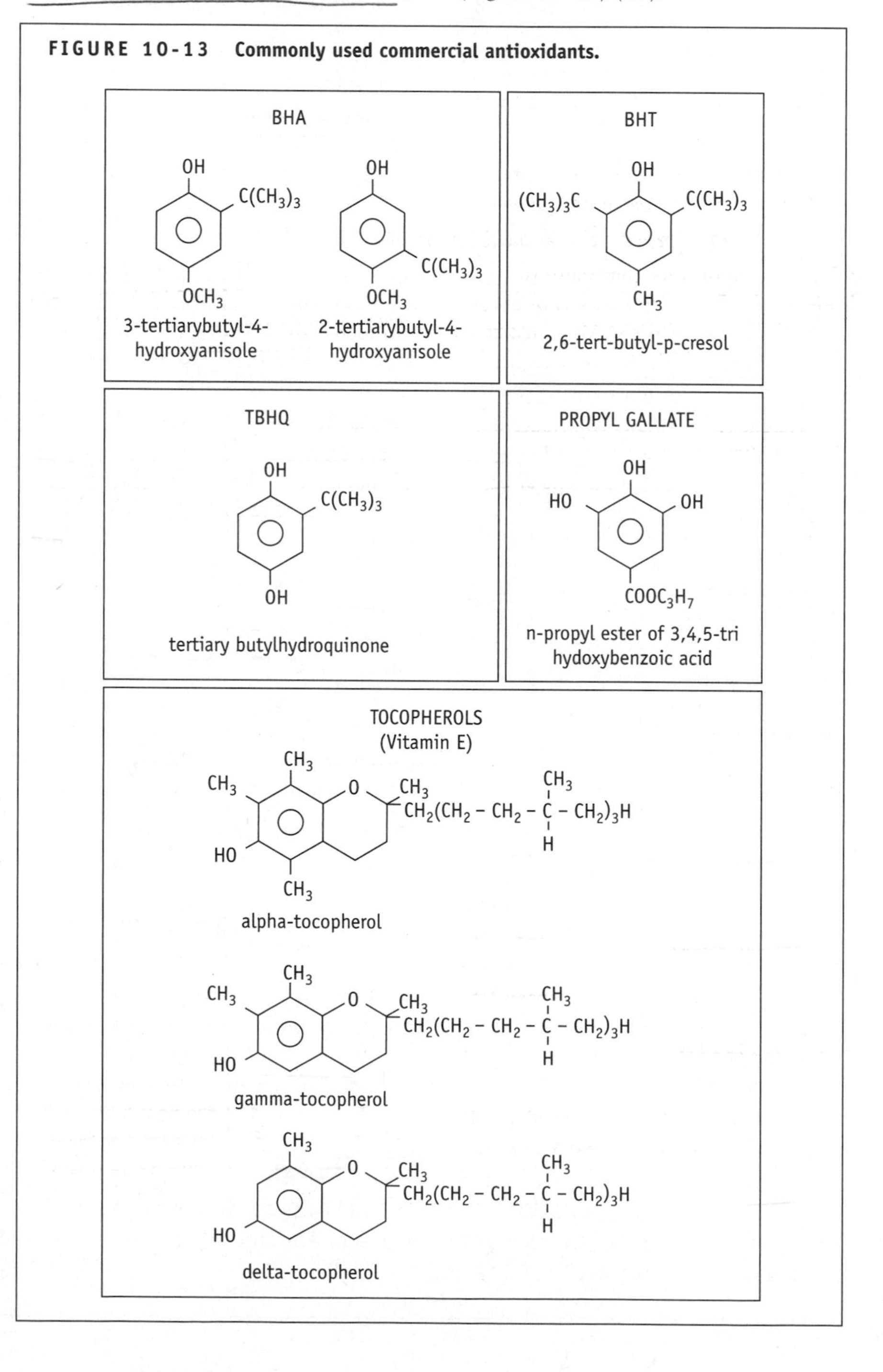

PICTORIAL SUMMARY / 10: Fats and Oils

Fats and oils are important components in well-prepared and good-tasting food. Their unique properties not only contribute to taste, texture and nutrition, but also greatly influence food preparation.

FUNCTIONS OF FATS IN FOOD

Properties of fat that affect food preparation:

Heat transfer: Fats act as a medium to transfer heat and prevent food from burning.

Shortening power: Fats are essential in the preparation of pastries, pie crusts, biscuits, and cakes. The more highly saturated the fat, the greater the shortening power.

Emulsions: All foods contain some liquid, and if fats or oils are present, the combination is an emulsion, such as mayonnaise.

Melting point: Fats have a range of melting points. Most plant oils are liquid at room temperature, while animal fats are solid.

Plasticity: The plasticity of fat is its ability to hold its shape but still be molded and spread; this influences its use in icings, etc.

Solubility: Fats are generally insoluble in water.

Flavor/Satiety: Fats contribute their own flavor as well as release the flavors and aromas of other foods when cooked with them. Fats also provide a creamy texture to foods, and because they take longer to digest than carbohydrates and proteins, they induce a sense of fullness or satiety.

Nutrition: Fats and oils contribute essential fatty acids and calories (kcal) to the diet. According to the current dietary goals, no more than 30 percent of caloric (kcal) intake in adults should be derived from fat; however, children under two years of age should not be restricted in their fat intake.

TYPES OF FATS

Fats are derived from both plant and animal sources.

Margarines are "vegetable oil spreads."

Shortenings are hydrogenated oils.

Vegetable oils are derived from plants, primarily the seeds of soybeans, corn, cottonseed, rapeseed, sunflower, and safflower.

Cocoa butter, which is used in the manufacture of chocolate candies, is made from the seeds of the cacao tree.

Butter is made by churning the cream from milk.

Lard is derived from the fat of swine, and **tallow** from beef or sheep fat.

FAT REPLACERS

Fat replacers are increasingly used in the food industry, and include synthetic fats (Olestra), proteins, starch, fiber, and even dried fruit purée.

FOOD PREPARATION WITH FATS

In food preparation, fats and oils are primarily used for frying (sautéeing, pan-frying, deep-fat frying) and as shortening agents in pastries and other baked goods.

Vegetable oils with high smoke points are used for deep-fat frying, and their care consists of several steps that must be taken to ensure their preservation. Frying temperature that is too high or too low can lead to excessive fat absorption.

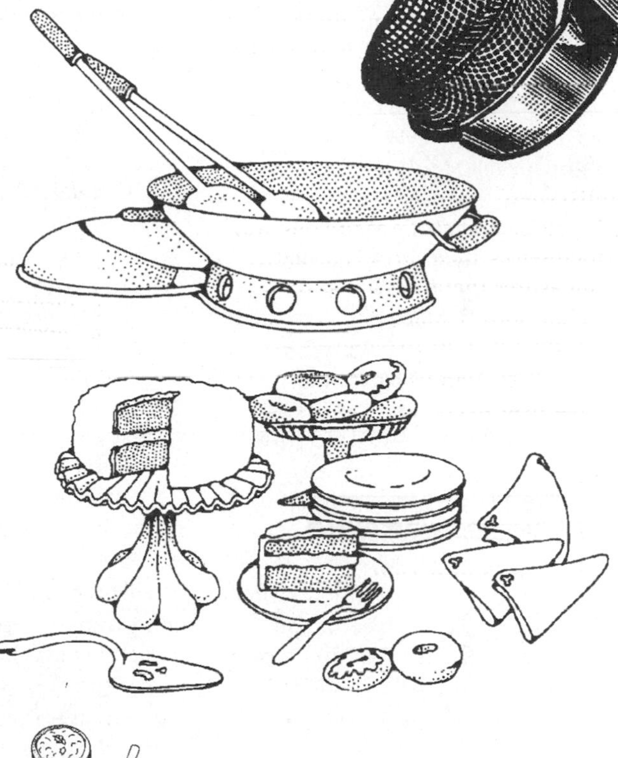

Fat content can be reduced during food preparation by carefully regulating the amount of fat and by making substitutions.

STORAGE OF FATS

Fats are best stored in the refrigerator, and butter will keep for months in the freezer. Oils, except olive oil and cold-pressed oils such as walnut oil, can be stored tightly covered for long periods of time at room temperature. The longer a fat is stored, the greater the possibility of its becoming rancid (oxidative or hydrolytic) and producing undesirable off-odors and flavors.

REFERENCES

1. Akoh CC. Scientific status summary: Fat replacers. *Food Technology* 52(3):47–52, 1998.
2. American Dietetic Association. Position of the American Dietetic Association: Vegetarian diets. *Journal of the American Dietetic Association* 97(11):1317–1321, 1997.
3. Best D. Technology fights the fat factor. *Food Development* 160(2): 48–49, 1991.
4. Best D. The challenges of fat substitution. *Prepared Foods* 160(6):72–76, 1992.
5. Boyle E. Monoglycerides in food systems: Current and future uses. *Food Technology* 51(8):52–59, 1997.
6. Boyle E, and JB German. Monoglycerides in membrane systems. *Critical Reviews in Food Science and Nutrition* 36(8):785–805, 1996.
7. Clement IP, and JR Marshall. Trans fatty acids and cancer. *Nutrition Reviews* 54(5):138–145, 1996.
8. Colbert LB. Lecithins tailored to your emulsification needs. *Cereal Foods World* 43(9):686–688, 1998.
9. Coupland JN, et al. Solubilization kinetics of triacyl glycerol and hydrocarbon emulsion droplets in a micellar solution. *Journal of Food Science* 61(6):1114–1117, 1996.
10. Dausch JG. Trans fatty acids: A regulatory update. *Journal of the American Dietetic Association 102*(1):18–20, 2002.
11. Drewnowski A. Why do we like fat? *Journal of the American Dietetic Association* 97(7):S58–S62, 1997.
12. Dunford NT. Health benefits and processing of lipid-based nutritionals. *Food Technology 55*(11):38–44, 2001.
13. Dziezak JD. Fats, oils, and fat substitutes. *Food Technology* 43(7):66–74, 1989.
14. Fat substitutes: Finding method in the madness. *Prepared Foods* 162(13):21–24, 1992.
15. Fats and fattening: Fooling the body. *Prepared Foods* 166(7):51–52, 1997.
16. Garcia DJ. Omega-3 long-chain PUFA nutraceuticals. *Food Technology* 52(6):44–49, 1998.
17. Giese J. Fats, oils, and fat replacers: Fats and oils play vital functional and sensory roles in food products. *Food Technology* 50(4):78–84, 1996.
18. Giese J. Antioxidants: Tools for preventing lipid oxidation. *Food Technology* 50(11):73–78, 1996.
19. Good chemistry: How form shapes function in food oils. *Journal of the American Dietetic Association* 91(7):777–778, 1991.
20. Harrison GG. Reducing dietary fat: Putting theory into practice. Conference summary. *Journal of the American Dietetic Association* 97(7):S93–S96, 1997.
21. Katz F. Functional fruit: Fruits as fat replacers providing texture, water holding, mouthfeel. *Food Processing* 57(7):74–75, 1997.
22. Klis JB. Continuing trend: Reducing fat and calories. *Food Technology* 47(1):152, 1993.
23. Kosmark R. Salatrim: Properties and applications. *Food Technology* 50(4):98–101, 1996.
24. LaBell F. Cottonseed oil for frying and baking. *Prepared Foods* 166(2):75, 1997.
25. Mattes RD. Position of the American Dietetic Association: Fat replacers. *Journal of the American Dietetic Association* 98(4):463–468, 1998.
26. McWilliams M. *Foods: Experimental Perspectives.* Prentice Hall, 1996.
27. Mela DJ. Nutritional implications of fat substitutes. *Journal of the American Dietetic Association* 92(4):472–476, 1992.
28. Morris CE. Balancing act: Engineering flavors for low-fat foods. *Food Engineering* 64(8):77–78, 1992.
29. McClements DJ, and K Demetriades. An integrated approach to the development of reduced-fat food emulsions. *Critical Reviews in Food Science and Nutrition* 38(6):511–536, 1998.
30. Nelson GJ. Dietary fat, trans fatty acids, and risk of coronary heart disease. *Nutrition Reviews* 56(8):250–252, 1998.
31. Orthoefer FT, S Gurkin, and K Liu. Dynamics of frying. In *Deep Frying: Chemistry, Nutrition and Practical Applications,* eds. EG Perkins and MD Erickson. AOCS Press, 1996.
32. Paul S, and GS Mittal. Regulating the use of degraded oil/fat in deepfat/oil food frying. *Critical Reviews in Food Science and Nutrition* 37(7):635–662, 1997.
33. Pinthus EJ, P Weinberg, and IS Saguy. Oil uptake in deep-fat frying as affected by porosity. *Journal of Food Science* 60(4):767–769, 1995.
34. Pszczola DE. Functional ingredients enhance value of low-fat foods and microwaved foods. *Food Technology* 46(4):116, 1992.
35. Reducing acid taste in low-fat dressings. *Food Engineering* 64(5):46, 1992.
36. Schwartz NE, and ST Borra. What do consumers really think about dietary fat? *Journal of the American Dietetic Association* 97(Suppl):S73–S75, 1997.
37. Sloan AE. Consumer project trends: Fats and oils slip and slide. *Food Technology* 51(1):30, 1997.
38. St. Angelo AJ. Lipid oxidation in foods. *Critical Reviews in Food Science and Nutrition* 36(3):175–224, 1996.
39. Stauffer CE. *Fats and Oils: Practical Guides for the Food Industry.* Eagen, 1996.
40. Stauffer CE. Fats and oils in bakery products. *Cereal Foods World* 43(3):121–126, 1998.

41. Stern JS, and MG Hermann-Zaidins. Fat replacements: A new strategy for dietary change. *Journal of the American Dietetic Association* 92(1):91–93, 1992.
42. Sullivan LW. Steps toward a fat-free future. *Journal of the American Dietetic Association* 97(7):S52–S53, 1997.
43. Turn of the lites. *Food Engineering* 42(9):41, 1991.
44. Uauy-Dagach R, and A Valenzuela. Marine oils: The health benefits of n–3 fatty acids. *Nutrition Reviews* 54(11):S102–S108, 1996.
45. Visioli F, and G Claudio. The effect of minor constituents of olive oil on cardiovascular disease: New findings. *Nutrition Reviews* 56(5):142–147, 1998.
46. Williams J, and FH Mattson. Discovered fat substitute known as olestra. *San Diego Union Tribune,* June 1, 1997.
47. Yackel WC, and C Cox. Application of starch-based fat replacers. *Food Technology* 46(6):146, 1992.

WEBSITES

American Oil Chemist's Society:
www.aocs.org

Lower the fat in your diet:
www.nalusda.gov/fnic/dga/dga95/lowfat.html

Calorie Control Council's (lobby group) glossary of fat replacers:
www.caloriecontrol.org/frgloss.html

Website for Olean fat replacer:
www.olean.com

11

Milk

People have been using milk as a food source for thousands of years. Records from ancient Babylon, Egypt, and India show evidence of cattle being raised for their milk (Figure 11-1). Milk is a unique beverage that provides complete protein, many of the B vitamins, vitamins A and D, and calcium. In fact, approximately 80 percent of the calcium ingested by Americans is derived from dairy products. Because a lack of dietary calcium causes poor bone development in children and is a risk factor for osteoporosis (porous bones) in later life, milk is a vital source of nutrition for millions of people. Although milk is rich in many nutrients, it is low in vitamins C and E, iron, complex carbohydrates, and fiber.

This chapter focuses on cow's fluid milk—its composition and variations, the purchasing of milk products, its use in food preparation, and its safe storage. Butter, cheese, and frozen dairy products are covered in Chapters 10, 12, and 28, respectively. While milk from other animals, such as goats, sheep, and camels, is a common part of the diet in some parts of the world, in this book, unless otherwise indicated, the word "milk" refers only to cow's milk.

Corbis/Bettman Archive

FIGURE 11-1 Dairy scene from ancient Egypt.

Composition of Milk

Nutrients

The basic composition of milk regardless of the source remains the same. Milk is primarily water—87.3 percent. The remaining 13 percent by weight consists of carbohydrate (4.8 percent), protein (3.4 percent), fat (3.7 percent), and minerals (0.8 percent). The high concentration of water gives milk a near-neutral pH of 6.6. Among domesticated cattle, the breed, stage of lactation, type of feed ingested, and season of the year all tend to slightly influence milk's content.

Carbohydrate. Lactose, or milk sugar, is the primary carbohydrate found in milk—12 grams per 8-ounce cup. When bacteria in milk metabolize lactose, lactic acid is produced. The flavor of cheeses and fermented milk products such as yogurt and sour cream is, in part, derived from the lactic acid. Lactose tends to be less soluble than sucrose, which may cause it to crystallize into lumps in nonfat dried milk and to produce a sandy texture in ice cream.

Lactose Intolerance. Some people suffer from lactose intolerance caused by a deficiency of the lactase enzyme, which is required to digest lactose (26). For people with this problem, fermented milk products are usually more easily digested.

Protein. The protein in milk is a complete protein; that is, it contains all the essential amino acids in adequate quantities necessary to support growth and the maintenance of life. A cup of milk contains approximately 8 grams of protein. Two servings of milk or milk products a day provide almost half the protein recommended for a healthy adult woman, and one-third that for a man.

Casein and Whey. The two predominant types of protein found in milk are **casein** and **whey** (Chemist's Corner 11-1) (5). Casein accounts for almost 80 percent of the protein in milk, while whey protein constitutes about 18 percent. Whey proteins consist primarily of lactalbumin and lactoglobulin (22). The nutritious whey protein can be isolated by putting the whey through an ultrafiltration process. These whey protein concentrates are used extensively by the food industry as emulsifiers and as foaming and gelling agents (25, 28). Adding milk proteins to other foods generally improves their texture, mouthfeel, moisture retention, and flavor (20, 38).

Chemist's Corner 11-1

Casein and Whey Proteins

Casein is actually a composite of four proteins—alpha-, beta-, kappa-, and gamma-caseins (Table 11-1). Structurally, caseins are large, amphoteric (capable of reacting as either an acid or base, depending on the pH), random coils. This differs from the shape of whey proteins, which are compact, globular, and helical (11). The large particles of casein are often referred to as phosphoproteins because, in addition to calcium, they contain phosphorus. This is only at a certain pH, because below a pH of 4.6, the casein is completely free of salts (10).

TABLE 11-1

Milk Proteins—Approximate Percentage of the Major Proteins Found in Milk

Protein	*% Total Protein*
Caseins	*79*
α_{s1}-Casein	43
β-Casein	20
κ-Casein	12
γ-Casein	4
Whey Proteins	*18*
β-Lactoglobulin	9
α-Lactalbumin	5
Immunoglobulins	2
Serum albumin	2

KEY TERMS

Casein The primary protein (80 percent) found in milk; it can be precipitated (solidified out of solution) with acid or certain enzymes.

Whey The liquid portion of milk, consisting primarily of 93 percent water, lactose, and whey proteins (primarily lactalbumin and lactoglobulin). It is the watery component removed from the curd in cheese manufacture.

Fat. The fat in milk, called milk fat or butterfat, plays a major role in the flavor, mouthfeel, and stability of milk products. The creaminess of milk chocolate, for example, is due to the milk fat, which softens the characteristic brittleness of cocoa butter (13). Milk fat consists of triglycerides surrounded by phospholipid-protein membranes (lipoproteins), which allow them to be dispersed in the fluid portion of milk, which is primarily water. Milk fat contains substantial amounts of short-chain fatty acids—butyric, caprylic, caproic, and capric acids. The fatty acids in milk fat are approximately 66 percent saturated, 30 percent monounsaturated, and 4 percent polyunsaturated.

Fat and Calorie (kcal) Content of Milks. An 8-ounce cup of fluid milk ranges from 86 to 150 calories (kcals) and 0 to 8 grams of fat. The fat and caloric (kcal) content of various milk products is listed in Table 11-2. Buttermilk, despite its name, contains only about 2 grams of fat per cup, and fewer than half the calories (kcal) of whole milk. Removing the fat from whole milk to make butter resulted in naming the remaining fluid "buttermilk." Other types of milk products vary greatly in their fat content per cup, from condensed milk, with about 27 grams, to fat-free (nonfat) milk, with less than half a gram.

Cholesterol. Like other animal products, milk contains cholesterol—an average of 33 mg in a cup of whole milk, 18 mg in reduced-fat (2 percent) milk, and 4 mg in fat-free (nonfat) milk. The fat and cholesterol content of milk and other dairy foods such as cheese, butter, and ice cream drives some consumers to seek lower-fat alternatives to dairy products. Fat content from milk products can also be reduced by choosing more of the lower-fat dairy items shown in Table 11-3 (page 205). In fact, low-fat dairy products accounted for nearly 40 percent of new food products introduced to the market in the early 1990s (40).

Vitamins. Milk contains vitamins A and D, riboflavin (B_2), and tryptophan, an amino acid important in the formation of the B vitamin niacin. It is low in vitamins C and E. Milk exposed to ultraviolet light loses riboflavin, so it is packaged in cardboard or opaque plastic containers to prevent the degradation of this vitamin by light.

A&D Fortification. Many milks are fortified with vitamins A and D. Vitamin D is found naturally in very few foods and was initially added to milk, a staple food, to reduce the incidence of rickets, a bone-softening condition in children that was at one time endemic

TABLE 11-2
Calorie (kcal) and Fat Content of Selected Milk Products

Milk Product	*Nutrients/Cup* *Calories*	*Fat(g)*
Fluid Milk		
Fat-Free (Nonfat) (Skim)	86	0
Low Fat (1%)	102	3
Reduced Fat (2%)	121	5
Whole	150	8
Flavored Fluid Milk		
Chocolate		
Low Fat (1%)	158	3
Reduced Fat (2%)	179	5
Whole	209	9
Eggnog		
Reduced Fat (2%)	189	8
Whole	342	19
Canned Milk		
Whole Evaporated	338	19
Fat-Free (Nonfat) Evaporated	199	1
Sweetened Condensed	982	27
Sweetened Condensed (Fat-Free)	632	0
Cultured Milk		
Buttermilk	99	2
Yogurt (Plain)		
Fat-Free (Nonfat)	137	0
Reduced Fat (2%)	155	4
Whole	150	8
Yogurt (Fruit Flavored)		
Fat-Free (Nonfat)	100	0
Reduced Fat (2%)	231	3
Whole	250	6
Sour Cream (1 tbs)	31	3
Cream		
Half-and-Half	315	28
1 tbs	20	2
Light Whipping	698	74
1 tbs	44	5
Heavy Whipping	821	88
1 tbs	51	6
Cream Substitute (1 tbs)	20	2

in North America. Before the fortification of milk was widely practiced, many children grew up with severely bowed legs and other effects of vitamin D deficiency.

Because vitamins A and D are fat-soluble, they are found in the milk fat of whole milk. For this reason, whole milk is not required to be fortified with either vitamin, although many milk manufacturers add both. In reduced-fat (2 percent fat) and fat-free (nonfat) milks, however, the vitamin A has been diminished, so these milks are required to be fortified with that vitamin. Fortification with vitamin D in reduced-fat (2 percent) and fat-free (nonfat) milks is optional, but 98 percent of milk processors add it anyway. Vitamin A fortification is also required in dried whole milk and evaporated skim milk, while fortification with vitamin D is required in evaporated whole and fat-free (nonfat) milk.

Minerals. The major mineral in milk is calcium, with 1 cup of milk containing, on average, 300 mg of the nutrient. Two servings of milk a day provide a substantial portion of the 1000 mg RDI for adults. Milk can also provide calcium in the form of other foods such as yogurt, pudding, ice cream, custards, hot chocolate, and cheese.

Other primary minerals found in milk and milk products include phosphorous, potassium, magnesium, sodium chloride, and sulfur. Although milk is rich in many minerals, it is low in iron.

Color Compounds

Factors that contribute to the color of milk are fat, colloidally dispersed casein and calcium complexes, and water-soluble riboflavin (B_2). These compounds, by interfering with light transmission, contribute to milk's opaque, ivory color. The amount of carotene (a pigment found in some plants) in the cow's feed influences the color of its milk. Carotenoid pigments dissolved in the milk fat provide the yellowish tinge of butter and cream (see Chapter 10).

HOW & WHY? ?????????

Why does nonfat milk have a bluish hue? Removing any of the fat eliminates a proportional amount of carotenoid pigments and solids, resulting in the color changing from a yellowish white to the bluish hue seen in fat-free (nonfat) milk.

Purchasing Milk

Grades

Milk is graded according to its bacterial count. The highest grade, Grade A, has the lowest count. The law requires that all Grade A milk and milk products crossing state lines must be pasteurized. Although Grade A is the most common grade of milk sold, Grade B is also available. In addition, different grades exist for fat-free (nonfat) dry milk: U.S. Extra and U.S. Standard. Grading is voluntary and is paid for by the dairy industry. The USDA is responsible for grading, while the U.S. Public Health Service recommends and enforces specific procedures for pasteurization (Grade A Milk Ordinance), laboratory tests, and sanitation at dairy farms and processing plants.

TABLE 11-3
Dairy Products to "Choose More" or "Choose Less" to Reduce Dietary Fat

Choose More	*Choose Less*
Milk	
Fat-free (skim)	Whole milk
Low fat (1%)	Evaporated
Reduced fat (2%)	Condensed
Fat-free dried milk	
Buttermilk	
Fat-free evaporated	
Fat-free or reduced-fat chocolate milk	
Yogurt	
Reduced fat (2%)	
Low-fat yogurt	Whole-milk yogurt
Fat-free yogurt	Custard-style yogurt
Cream	
Light cream cheese	Whipping cream
Light sour cream	Half-and-half
Mocha	Sour cream
Poly Perx (creamer/Mitchell Foods)	Sweet cream
Poly Rich (creamer/Rich Products)	Cream cheese spreads
	Nondairy creamers
	Cream soups
	Creamy dressings
Cheeses	
Lower-fat cheese (see Chapter 12)	Cheese over 6 g of fat/ounce
Frozen Desserts	
Sherbet	Ice cream
Ice milk	Frozen whole-milk yogurt
Frozen reduced-fat (2%) yogurt	

Source: National Dairy Council

KEY TERMS

Ultrapasteurization A process in which a milk product is heated at or above 280°F (138°C) for at least two seconds.

■ ■ ■ ■

Ultrahigh-temperature (UHT) milk Milk that has been pasteurized using very high temperatures, is aseptically sealed, and is capable of being stored unrefrigerated for up to three months.

Pasteurization

Milk is an excellent growth medium for microorganisms such as bacteria, yeast, and molds. In the early 1900s, it was frequently the vehicle for carrying such serious foodborne illnesses as typhoid, diphtheria, scarlet fever, and tuberculosis. Pasteurization, named after Louis Pasteur (1822–1895), was originally used to treat wine and beer, but soon came into use to treat milk as well, when it was found that heating milk for a short time to below its boiling point killed microorganisms. Pasteurization destroys 100 percent of pathogenic bacteria, yeasts, and molds and 95 to 99 percent of other, nonpathogenic bacteria. The process of pasteurization also inactivates many of the enzymes that cause the off-flavors of rancidity. Almost all milk sold commercially in North America is first pasteurized. In some states, where allowed by law, there is a small niche for unpasteurized, or raw, milk.

To ensure that sufficient pasteurization has occurred, milk processors measure the activity of a specific enzyme found in milk, alkaline phosphatase. If this enzyme is no longer active, then the milk is safe for consumption. Pasteurization temperatures and times vary, but the ones most commonly used by milk processors are the first two listed in Table 11-4. Even though pasteurized milk is no longer pathogenic, it will still spoil because the 1 to 5 percent nonpathogenic bacteria remaining convert lactose to lactic acid.

Ultrapasteurization. A process called **ultrapasteurization** uses higher temperatures than regular pasteurization temperatures to extend the shelf life of refrigerated milk products. If this same treatment is combined with sterile packaging techniques, it is called **ultrahigh-temperature (UHT)** processing. UHT processing destroys even more bacteria than standard pasteurization and increases the milk's shelf life. This milk is then packaged aseptically in sterile containers and sealed so that it can be stored unrefrigerated for up to

TABLE 11-4
Pasteurization Temperatures

*Temperature** *°F*	*°C*	*Time*	*Type of Pasteurization*	*Refrigeration Required*
145°	63°	30 minutes	Low-Temperature Longer-Time (LTLT)	Yes
161°	71.5°	15 seconds	High-Temperature Short-Time (HTST)	Yes
212°	100°	0.01 second	Higher-Heat Shorter-Time (HHST)	Yes
280°	138°	2 seconds or more	Ultrapasteurization	Yes, but product has longer shelf life
280–302°	138°–150°	2–6 seconds	Ultrahigh-Temperature (UHT)	Not until opened

*If the dairy ingredient has a fat content of 10 percent or more, or if it contains added sweeteners, the specified temperature shall be increased by 37°F/3°C.

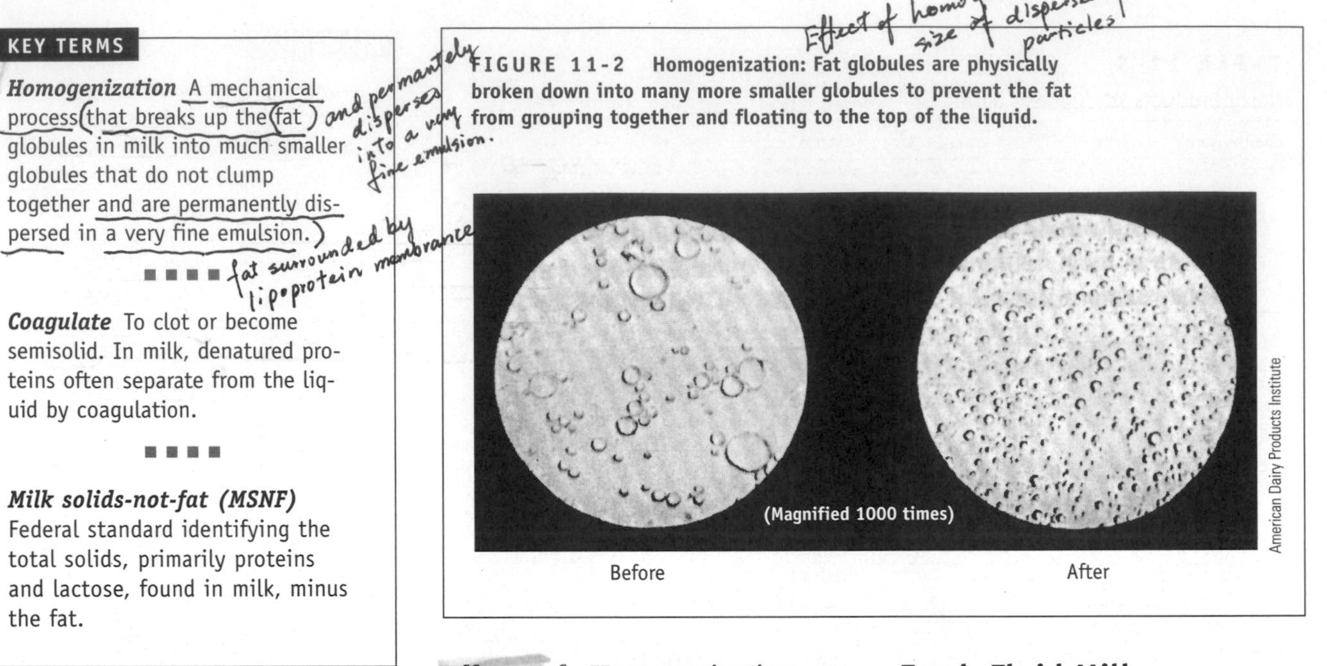

KEY TERMS

Homogenization A mechanical process that breaks up the fat globules in milk into much smaller globules that do not clump together and are permanently dispersed in a very fine emulsion.

Coagulate To clot or become semisolid. In milk, denatured proteins often separate from the liquid by coagulation.

Milk solids-not-fat (MSNF) Federal standard identifying the total solids, primarily proteins and lactose, found in milk, minus the fat.

FIGURE 11-2 Homogenization: Fat globules are physically broken down into many more smaller globules to prevent the fat from grouping together and floating to the top of the liquid.

three months (2). Once the aseptic seal is broken, the milk must be refrigerated. Originally, this preparation method was used on less frequently purchased milk products such as whipping cream, half-and-half, and eggnog, but it is now used on a wider variety of products.

Homogenization

Fat is less dense than water, causing it to float to the top of milk. This results in the thick layer of yellowish cream that rises to the top of unprocessed milk. **Homogenization** prevents this separation of water and fat.

HOW & WHY? ?????????

How is milk homogenized? The mechanical process of homogenization pumps the milk under a high pressure of 2,000 to 2,500 pounds per square inch through fine holes, which breaks up the fat globules. This decreases the fat globule size to less than 2 microns (Figure 11-2). The now very small droplets of milk fat are surrounded by a lipoprotein membrane, which prevents them from joining together and separating out. The liquid and fat components of the milk are now, in effect, homogenized.

Effect of Homogenization on Milk. Most milk in the United States is homogenized. This purely mechanical process has no effect on nutrient content; however, sensory changes do occur, resulting in a creamier texture, whiter color, and blander flavor. Homogenized milk also **coagulates** more easily, making puddings, white sauces, and cocoa more viscous. Its increased surface tension gives it a greater foaming capacity. Homogenized milk is also more prone to rancidity caused by oxygen being added to the double bonds of the unsaturated fatty acids. Pasteurizing milk before homogenization inhibits rancidity because the lipase enzymes responsible for breaking down fat are inactivated.

Types of Milk

About half the milk produced in the United States is sold as fluid milk and cream. Much of the rest comes to market as butter, cheese, and ice cream. The available market forms of milk include fluid milk—whole, reduced fat (2 percent), low fat (1 percent), fat-free (nonfat), UHT, chocolate, canned, and many others—dry milk, cream, and cultured milk products such as yogurt and buttermilk.

Fresh Fluid Milks

Whole Milk. To be classified as whole, milk must contain 3.25 percent milk fat and at least 8.25 percent **milk solids-not-fat (MSNF)** (Table 11-5). The milk is usually fortified with vitamins A and D, but this is optional (30).

Reduced-Fat and Low-Fat Milk. These milks have had some of their fat removed so that milk fat levels are decreased to 2.0 and 1.0 percent respectively and are so noted on the carton. A minimum of 8.25 percent MSNF is necessary, but if it exceeds 10 percent, then the milk must be labeled "protein fortified" or "fortified with protein." The addition of milk solids improves the consistency, taste, and nutritive content of reduced- and low-fat milks. Vitamin A fortification is required, while the addition of vitamin D is optional. New milk labeling requirements effective January 1998 require that 2 percent milk, which was previously called "low fat," now be called "reduced fat" milk. The low-fat designation is now used to describe 1 percent milk.

Consumer interest in lower-fat products has resulted in a drastic downward trend in the consumption of whole milk (14). Between 1970 and 1990, reduced- and lower-fat milk sales increased by 300 percent, while sales of whole milk

TABLE 11-5
Standards of Identity for Milk Products

		Pasteurization [a] Ultrapasteurization [b] UHT Processing [c]	Homogenization	Fat % (min./range)	MSNF % min.	Vitamin D	Vitamin A	
FLUID	Milk	M [a, b, c]	Opt.	3.25	8.25	Opt.	Opt.	
	Reduced-Fat Milk	M [a, b, c]	Opt.	0.5-2.0	8.25	Opt.	M	
	Fat-Free Milk	M [a, b, c]	Opt.	<0.5	8.25	Opt.	M	
CREAM	Half-and-Half	M [a, b, c]	Opt.	10.5-18.0				
	Light Cream [Coffee Cream or Table Cream]	M [a, b, c]	Opt.	18-30				
	Light Whipping Cream	M [a, b, c]	Opt.	30-36				
	Heavy Whipping Cream	M [a, b, c]	Opt.	36				
CAN	Evaporated Milk		M	7.5	25.0	M	Opt.	60% H_2O removed
	Condensed Milk	M [a]	Opt.	7.5	25.5	Opt.		No vitamins added
	Sweetened Condensed Milk	M	Opt.	8.0	28			15% sugar added
DRY	Fat-Free Dry Milk	M	Opt.	Max 1.5				
	Fat-Free Dry Milk Fortified with Vitamins A and D	M	Opt.	Max 1.5		M	M	
CULTURED	Sour Cream	M	Opt.	18				
	Acidified Sour Cream	M	Opt.	18				
	Sour Half-and-Half	M	Opt.	10.5-18.0				
	Acidified Sour Half-and-Half	M	Opt.	10.5-18.0				
	Yogurt	M [a, c]	Opt.	3.25	8.25	Opt.	Opt.	
	Reduced-Fat Yogurt	M [a, c]	Opt.	0.5-2.0	8.25	Opt.	Opt.	
	Fat-Free Yogurt	M [a, c]	Opt.	<0.5	8.25	Opt.	Opt.	

M = Mandatory
Opt. = Optional
a Pasteurization is mandatory but declaration of term *pasteurized* is optional.
b "Ultra-pasteurization" is an optional process—declaration of term *ultra-pasteurized* is mandatory on principal display panel, if applicable.
c UHT declaration is on label although not included in standards at this time.

Source: National Dairy Council

	% milk fat	kcal/cup
Whole milk	3.25	150
Reduced fat	2.0	122
Low fat	1	102
Fat-free/non fat	≤ 0.5	80

dropped by 50 percent (29). Overall, children in the United States are drinking less milk because they are drinking more of other beverages such as soft drinks and fruit juices (21).

Fat-Free or Nonfat Milk. "Nonfat" is synonymous with "fat-free." Removing as much fat as technologically possible results in fat-free (nonfat) milk. The term "fat-free" replaced the "skim" milk designation. Fat-free (nonfat) milk should contain no more than 0.5 percent milk fat and a minimum of 8.25 percent MSNF. Vitamin A fortification is required, while vitamin D is optional. Consumption of fat-free (nonfat) milk is less than that of reduced-fat (2 percent) milk. In fact, between 1988 and 1991, sales of reduced-fat (2 percent) milk increased 8 pounds per person, while sales of fat-free (nonfat) milk increased only 1 pound per person (39).

Ultrahigh-Temperature Milk (UHT). Unopened, aseptically sealed packages of UHT milk can be stored on shelves without refrigeration for up to three months. UHT milk has a "cooked" flavor at first, which tends to disappear with storage time. After one year, UHT milk develops off-flavors described as "sweet," "flat," "musty," "rancid," and "chalky" (16). Chilling UHT milk before serving improves its taste. Once opened, UHT milk must be refrigerated and handled with the same care used for fresh milk. Although it is slightly more expensive than fresh milk and not widely distributed, UHT milk is ideal for boating, camping, hiking, and other situations where refrigeration is not always available.

Chocolate Milk. Pasteurized milk containing 1.5 percent liquid chocolate, or 1 percent cocoa and 5 percent sugar, can be called chocolate milk. Chocolate milk is actually a suspension in which the continuous phase consists

KEY TERMS

Imitation milk A product defined by the FDA as having the appearance, taste, and function of its original counterpart but as being nutritionally inferior.

of milk fat and cocoa butter, while the dispersed phase includes the cocoa particles, fat-free (nonfat) milk solids, and sugar (12). Fortification with vitamins A and D is optional for chocolate whole milk, but fat-free (nonfat) chocolate milk must be fortified. Whole chocolate milk is high in calories (kcal), because the additional sugar and cocoa contribute approximately 58 calories (kcal) per cup to its already higher calorie (kcal) count, for a total of 209 calories (kcal) per cup. Chocolate milk made with reduced-fat (2 percent) milk contains 179 calories (kcal) and 5 grams of fat per cup, while the calories (kcal) and grams of fat in chocolate low-fat (1 percent) milk drop to 158 and 2.5 respectively.

Eggnog. Packaged, commercially produced eggnog is manufactured to replicate a traditional rich holiday beverage made with eggs, cream or milk, nutmeg, and often added spirits. It is sold predominantly during the Thanksgiving and Christmas holidays, and is defined as a pasteurized and homogenized mixture of milk, cream, milk solids, eggs, stabilizers, and spice. It contains 6 to 8 percent milk fat, about double that of whole milk, 9 percent MSNF, and 1 percent egg yolk solids. Other ingredients such as sugar, color and flavor additives, rum extract, nutmeg, and vanilla may be added. Although eggnog products vary widely and are available in low-fat versions, an average cup provides 342 calories (kcal) and 19 grams of fat, more than double that found in whole milk.

Carbonated Milk. Carbonated milk is a new product that has been called the "soft drink that is not junk food" (1, 6). It is simply milk that has had carbonated bubbles added to make it fizz like a soda. Whether or not milk is carbonated, it has a higher nutrient density than any soft drink. Figure 11-3 compares nutrients per calorie (kcal) in fat-free (nonfat) milk to those in a cola.

Imitation Milk. Imitation milk looks like milk but usually has little or no dairy content. Ingredients include

FIGURE 11-3 Selected nutrient content of fat-free milk vs soft drink.

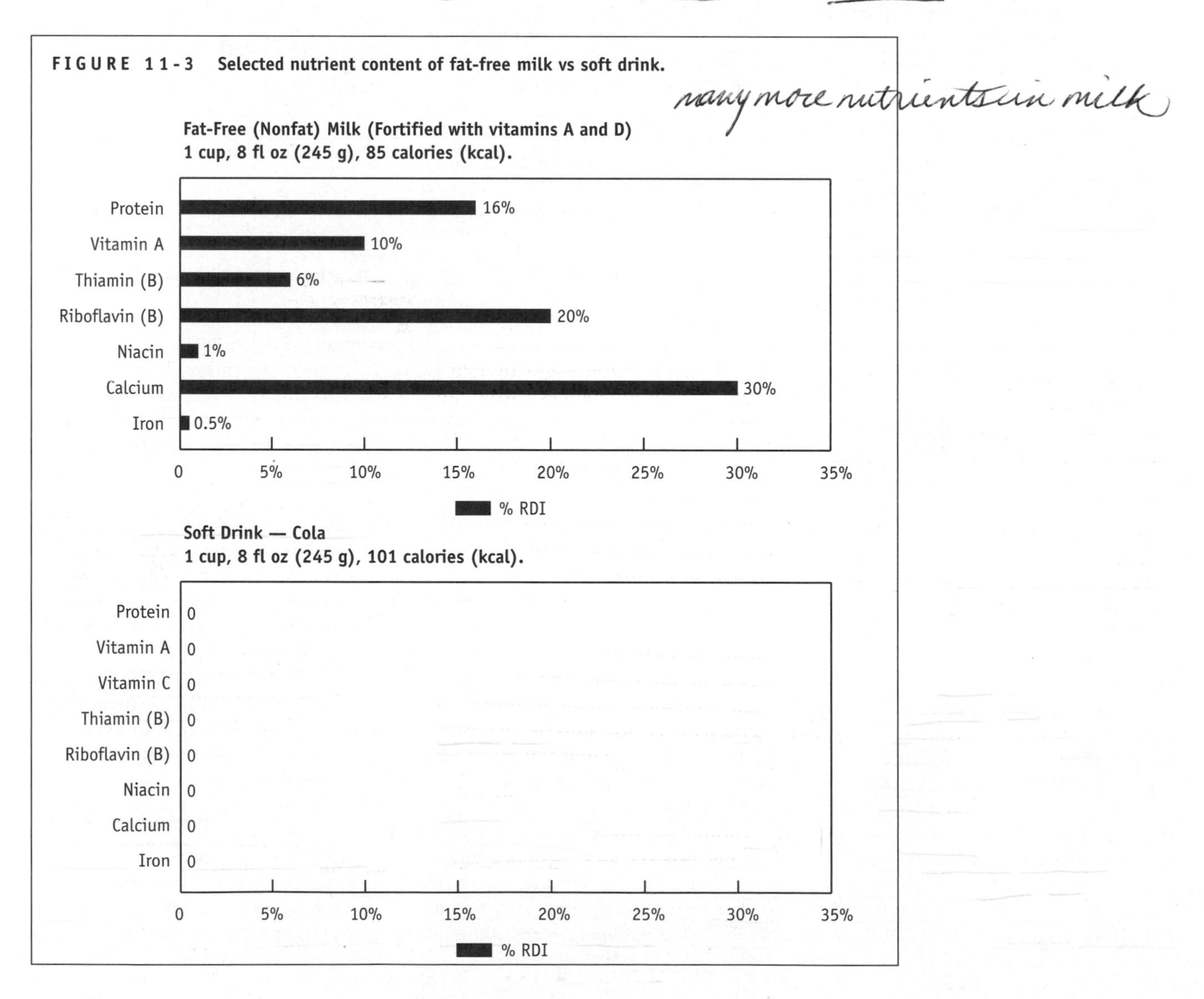

water, corn syrup solids or sugar to replace the lactose, vegetable oils to replace the milk fat, protein from the sodium caseinate in soybeans, whey or milk solids-not-fat to substitute for the protein, and some stabilizers, emulsifiers, and flavoring agents.

Nutritionally, imitation milks are lactose-free and so are useful for people with lactase deficiency, but sometimes sodium caseinate and whey are added, which makes these products inappropriate for people with milk protein allergies. The vegetable fat, like milk fat, may be high in saturated fat from coconut or palm oils used in the manufacture of these products. The calcium and protein content is about half that found in regular milk. Although the nutrient content of milk is superior to imitation milk, the latter is less expensive. To emphasize their products' superiority over imitation milk, some dairy processors place a "REAL" seal shaped as a drop of milk on their products (Figure 11-4).

Filled Milk. Filled milk is made by replacing all or part of the milk fat with a vegetable fat. Cholesterol levels drop to zero in filled milk, but if the fat substitute is a saturated vegetable oil such as coconut or palm oil, there is a higher ratio of saturated to polyunsaturated fats. Both imitation and filled milks are regulated at the state rather than at the federal level.

Reduced-Lactose Milk. Any pasteurized milk treated with the enzyme lactase will have most of its lactose converted to its two monosaccharides, glucose and galactose. This doubling of the sugar molecules results in a slightly sweeter flavor. The resulting milk is more easily digested by people who have some degree of lactose intolerance.

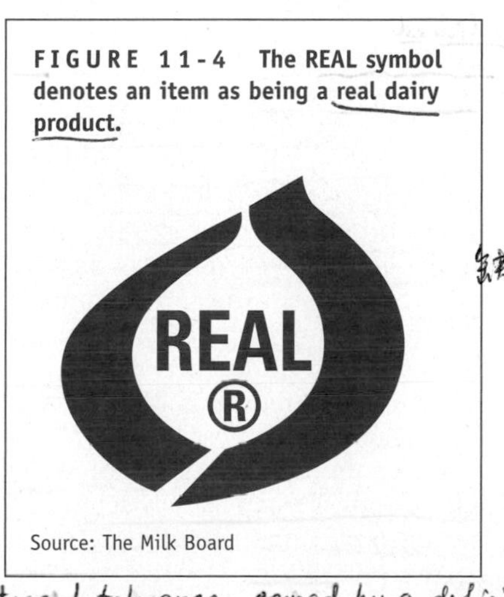

FIGURE 11-4 The REAL symbol denotes an item as being a real dairy product.

Source: The Milk Board

Low-Sodium Milk. A cup of milk contains 120 mg of sodium, so people on sodium-restricted diets must watch their milk intake. Low-sodium milks containing only 6 mg per cup are available. These are produced by an ion exchange method that removes all but 5 percent of the original sodium.

Goat, Sheep, and Other Animal Milks. The history of the use of milk from various animals dates so far back that no one knows when the practice actually started. Cows provide almost all of the milk consumed in North America, but goats rank a close second in supplying milk to other regions such as Norway, Switzerland, the Mediterranean area, Latin America, and parts of Asia and Africa. Goat's milk is low in folate and vitamins D, C, and B_{12}. Some people in Spain, the Netherlands, Italy, and the Balkans also obtain milk from sheep. Camels provide milk for some people in the Middle East and Central Asia, while reindeer are used for milk in the Arctic region. The llama in South America sometimes serves as a source of milk, as does the water buffalo in parts of the Philippines, Asia, and India.

Soy Milk. A milk-like product that is not from any mammal is soy milk, which is made from soybeans that have been soaked, ground, and strained. Soy milk has been known for centuries in China and Japan. Now it is consumed as a milk-like liquid, is used to make tofu (the cheese of soy milk), and is incorporated into some infant formulas. As useful as it is, soy milk is lacking in certain nutrients. Only soy milks fortified with methionine (an essential amino acid), calcium, and vitamin B_{12} should be substituted for cow's milk for infants and growing children. On the other hand, it is also lacking in the carbohydrate lactose, which makes it ideal for people with lactose intolerance.

Rice Milk. Another milk suitable for people with lactose intolerance is rice milk. It contains no lactose because it is made primarily from brown rice, filtered water, and a small amount of brown rice sweetener. Even though rice milk is made from rice, not all rice milks are gluten-free, which would not be acceptable to people with celiac disease, an immune condition making them allergic to gluten. Although rice contains no gluten, some rice milks are made using barley enzymes to convert the carbohydrates in brown rice to naturally occurring sugars. This is why rice milk is somewhat sweeter than soy milk. In addition, the flavor of either rice or soy milks can be enhanced by adding vanilla flavoring, making these milks even more acceptable to consumers than their plain counterparts.

Canned Fluid Milks

Whole Milk. Some ultrahigh-temperature milk is canned for export. Requirements regarding its content are similar to those for whole fluid milk sold in a carton.

Evaporated Milk. Evaporated milk is produced by evaporating at least 60 percent of the water found in whole milk. By definition, evaporated milk contains at least 7.5 percent milk fat, 25.5 percent milk solids-not-fat by weight, and is fortified with vitamin D. Evaporated milk provides 338 calories (kcal) and 19 grams of fat per cup. Stabilizers are often added to prevent separation of the fat during storage. Fat separation may also be prevented by turning the cans over every few weeks.

The evaporation process consists of initially exposing the milk for 10 to 20 minutes to a temperature of 203°F (95°C). This stabilizes the casein so that it will not coagulate during sterilization. In the next step of evaporation, the milk is heated to 121° to 131°F (50° to 55°C) at reduced atmospheric pressure, which allows the milk to boil at beneath the normal boiling temperature of 212.3°F (100°C). It is then homogenized, canned, and sterilized. The high temperatures of canning and the metal of the can may impart cooked and off-flavors to evaporated milk products.

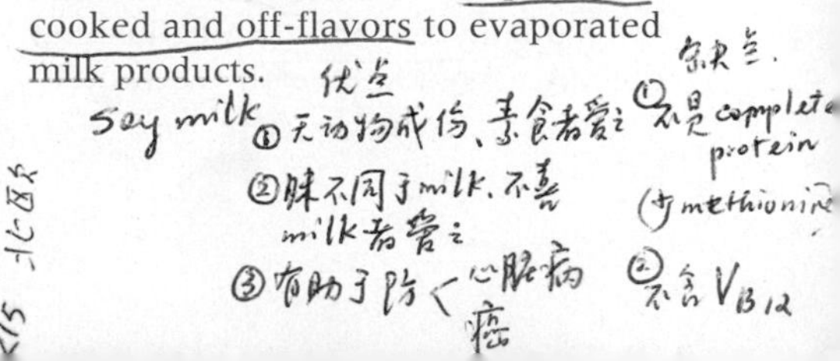

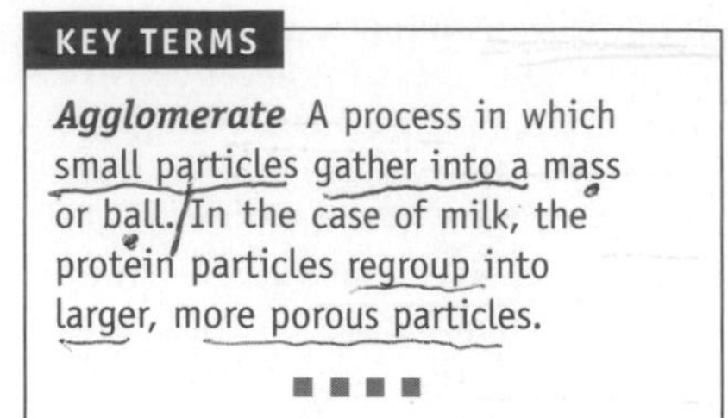

KEY TERMS

Agglomerate A process in which small particles gather into a mass or ball. In the case of milk, the protein particles regroup into larger, more porous particles.

■ ■ ■ ■

Curd The coagulated or thickened part of milk.

Newer evaporation techniques produce evaporated milk by exposing it to ultrahigh temperatures for longer periods of time, placing it in sterilized cans, and aseptically sealing the cans. Compared to evaporated milk produced by the old method, UHT-evaporated milk is less viscous, less white, and has a different flavor.

Evaporated Fat-Free (Nonfat) Milk. Fat-free (nonfat) evaporated milk is produced from fat-free (nonfat) milk. By definition, it must contain less than 0.5 percent fat, at least 20 percent total solids, and vitamins A and D. The lower fat content drops the calories (kcal) to 199 per cup and the fat to 0.5 grams. To reconstitute either regular or fat-free (nonfat) evaporated milk, combine equal volumes of the canned milk and water.

Sweetened Condensed Milk. Sweetened condensed milk contains a high quantity of added sugar, which makes it ideal for preparation of desserts, especially pies and cheesecake. After whole milk has been evaporated by 50 percent by weight, 15 percent sugar in the form of dextrose or corn syrup is added. Sweetened condensed milk is defined as containing at least 28 percent total milk solids and about 8 percent milk fat. The extra sugar and highly concentrated nature of the nutrients in sweetened condensed milk make it very high in calories (982 kcal/cup) and fat (27 grams/cup). Unlike evaporated milk, sweetened condensed milk does not have to be sterilized, because its 40 to 45 percent by weight sugar concentration prevents microbial spoilage. The sugar content, either added or in the form of lactose, also contributes to the Maillard reaction. During heating the sugar combines with the protein in the milk to give it a light-brown color. This attribute makes sweetened condensed milk ideal for creating caramel-flavored or colored desserts such as pumpkin pie.

Dry Milk

Nonfat Dried Milk. Many milk products can be dried, but by far the most common product is nonfat dried milk (NFDM). NFDM can be made from whole, reduced-fat (2 percent), low-fat (1 percent), or fat-free (nonfat) milks, as well as buttermilk, but nonfat is usually selected because it contains the least amount of fat.

HOW & WHY? ?????????

How is milk dried? Milk is dried by either spray, foam-spray, or roller drying processes. In spray drying, concentrated milk is sprayed into hot air, while foam-spray drying sends a jet of hot air into the concentrated milk. Roller drying consists of moving pasteurized fluid or condensed milk through two steam-heated rollers.

After the fat-free (nonfat) milk is dried, vitamins A and D may be added, although they are not required. Dried milks that originally contained some fat are required to be fortified with vitamin A and identified as "fortified with vitamins A and D." Overall, nonfat dried milk is nutritionally similar to fat-free (nonfat) milk, except that the levels of vitamins may be reduced by 20 percent. The absence of fat also gives it a long shelf life; nonfat dried milk can keep for about a year in a cool, dry storage area. The calcium and protein content of meals or beverages can be increased easily by adding nonfat dried milk to puddings, shakes, soups, casseroles, milk, and many other food combinations. Although it is inferior in taste to fluid milks, reconstituted nonfat dried milk is ideal for making batters and doughs for baked goods.

Instant Milk. "Instant milk" is different from regular nonfat dried milk. It is manufactured by exposing nonfat dried milk to steam, after which it is redried. The double drying helps instant milk achieve the ability to **agglomerate** and instantly dissolve in cold water. Generally, about ⅓ cup of instant nonfat dried milk will reconstitute into 1 cup of fluid milk, but manufacturers' instructions will vary. Allowing the reconstituted milk to refrigerate for several hours, and adding a teaspoon of vanilla to a half gallon of reconstituted chilled nonfat dried milk, make it taste more like regular milk. Also, more milk solids can be added by commercial food manufacturers to make it more flavorful and nutritious.

Cultured Milk Products

Cultured or fermented milk products have been used for centuries. Some cultured milk products commonly consumed in North America are buttermilk, yogurt, acidophilus milk, kefir, and sour cream. The one factor that all cultured milk products have in common is that they have had bacterial cultures added to them in order to ferment their lactose to lactic acid. The increased acid concentration causes the casein to precipitate out, which changes the milk to a more **curd**-like consistency. The type of bacterial culture inoculated into the milk largely determines the flavor of the resultant product. Bacteria also influence the quality of fermented dairy products by the amount and type of acids produced. Some protein is also broken down to provide nitrogen for bacterial growth, and this makes the curd softer and more digestible.

Buttermilk. Buttermilk contains little or no butterfat. Sweet natural buttermilk originally was the liquid left over after fresh cream had been chilled and churned to produce butter. Natural buttermilk is often dried and used in baked products and ice cream, because the phospholipids obtained from the fat-droplet membranes, which are broken down during churning, makes it an excellent emulsifier. Now, most commercially available buttermilk is cultured by adding *Streptococcus lactis* bacteria to pasteurized nonfat, reduced-fat (2 percent), or low-fat (1 percent) milk.

Flavor may be enhanced by adding other bacteria, butterfat granules or flakes, natural sweeteners, citric acid (up to 0.15 percent), salt, and artificial flavors or colors. Fortification with vitamins A and D is optional. Cultured buttermilk must contain less than 0.5 percent milk fat, at least 8.25 percent milk solids-not-fat, and an acidic pH of about 4.6. It is not mandatory that this pH be created by lactic-acid-generating bacteria, but when these bacteria are not used, the milk should be labeled "acidified buttermilk." A fat content between 0.5 and 2.0 percent changes the name to "cultured low-fat milk" or "acidified low-fat milk." A milk fat content over 3.25 percent is denoted by "cultured milk" or "acidified milk." Buttermilk has a longer shelf life than milk because the higher acid content inhibits the growth of spoilage-causing bacteria, and its lower fat content makes it less likely to go rancid. Salt may be added to further inhibit bacterial growth.

Yogurt. People in the Middle East have been eating yogurt for thousands of years. This smooth, semisolid fermented dairy product can be made from whole, reduced-fat (2 percent), or fat-free (nonfat) milks. Firm yogurt is known as the "set" style, while a more runny, semiliquid consistency is characteristic of "stirred" yogurt. Yogurt drinks are also gaining in popularity (37).

Yogurt is produced by mixing two types of bacteria, *Lactobacillus bulgaricus* and *Streptococcus thermophilous*, and adding them to pasteurized, homogenized milk to which some MSNF are also usually added. The whole mixture, including the bacteria, is held at a warm temperature (108° to 115°F/42° to 46°C) to allow fermentation to develop the desired consistency, flavor, and acidity. Although the milk used to make yogurt is usually slightly higher in lactose due to the addition of milk solids-not-fat, fermentation decreases the amount of lactose to 4 percent (34). During fermentation, the bacteria convert lactose to lactic acid, increasing the acidity (Chemist's Corner 11-2). In addition, folate (a B vitamin) levels increase as a natural by-product of bacterial growth.

"Active Culture" Yogurts. Once the yogurt has reached the desired consistency, the fermentation and accompanying changes caused by the bacteria are discontinued. To accomplish this, bacterial growth can be inhibited by either chilling or heating the yogurt. If the yogurt is chilled, the culture remains alive, while the heated yogurt's cultures are destroyed. Yogurts containing live bacteria are labeled "with active yogurt cultures," "living yogurt cultures," or "contains active cultures." Only yogurt containing viable cultures is recommended for people with lactase deficiency or those taking antibiotics (23, 24). Consuming fermented milk products containing live cultures reportedly helps restore the normal intestinal bacteria eliminated by antibiotics, and has been associated with the treatment of diarrhea for centuries (32). Intestinal bacteria are beneficial because they help to produce some B vitamins and vitamin K.

Other Ingredients Added to Yogurt. After fermentation, several ingredients, including gelatin and nonfat dried milk, may be added to yogurt to create a firmer texture, reduce the perception of acidity, or add color. The addition of sweeteners such as sugar, honey, fruit, fruit extracts, flavorings, and alternative sweeteners has increased the popularity of yogurt—over 85 percent of yogurt currently consumed in the United States is flavored. Yogurt with fruits blended throughout is called "Swiss-" or "French-style" yogurt, while yogurt with fruit on the bottom is known as "sundae-style."

Official Yogurt Definition. To be called yogurt, the milk product must contain at least 8.25 percent MSNF and 0.5 percent acid. Fat content requirements for whole, reduced-fat (2 percent), and fat-free (nonfat) yogurt are 3.25 percent, 0.5 to 3.0 percent, and less than 0.5 percent respectively. There are no federal standards for frozen yogurt, an increasingly popular dessert choice for those watching their dietary fat and calorie (kcal) intake.

Calorie (Kcal) Content of Yogurts. The sweetening and other flavoring in yogurt increases its caloric (kcal) content. Plain whole yogurt at 150 calories (kcal) per cup jumps to 250 calories (kcal) when it is sweetened.

KEY TERMS

Probiotics Live microbial food ingredients (i.e., bacteria) that have a beneficial effect on human health.

■ ■ ■ ■

Prebiotics Nondigestible food ingredients (generally fibers such as fructooligosaccharides (FOS) and inulin) that support the growth of probiotics.

Probiotics in Yogurts. Yogurt is the most common vehicle for **probiotics** (e.g., *Lactobacillus* and *Bifidobacterium*) and **prebiotics** (27). The suggestion that probiotics have a positive impact on health by improving the intestine's microbial balance goes back to the Nobel Prize-winning Russian scientist Elie Metchnikoff (1845-1916) who suggested that the long, healthy lives of Bulgarian peasants were due to their consumption of fermented milk products (33). Since then, probiotics have been suggested to benefit health by relieving diarrhea, inflammation of the stomach and intestines, food allergies, and certain cancers; and even by boosting immunity. Probiotics are theorized to exert their beneficial effects in several ways. They are known to create a healthy microbial balance in the intestines, aid digestion by producing helpful enzymes, prevent the attachment of harmful bacteria either directly as a barrier or indirectly through mucin (a mucoprotein) production, and stimulate immune function (33).

Europe and Japan have long recognized the possible benefits of probiotics and include them in a wide range of food products—yogurts, dairy-based

Chemist's Corner 11-2

Flavor Compounds in Yogurt

The tartness of yogurt is partially derived from the bacterial conversion of lactose to lactic acid and acetaldehyde (15). A buttery flavor is also generated as bacteria break down the natural citric acid in milk to produce diacetyl.

beverages, breakfast drinks, health snacks or bars, luncheon meats, teas, puddings, and even candy (36). The Japanese probiotic drink called Yakult is consumed by 24 million people daily (18). Probiotics are also sold in pill form.

Because of their newness on the market and the lack of scientific studies, potential problems exist with probiotics. One question is the ability of the microorganisms to stay alive, first through food processing and then through digestion (35). Another issue is dosage. What is correct amount without delivering too much? It has been suggested that 1 to 10 billion live cells be consumed to ensure delivery of at least 100 million live cells per dose in the gastrointestinal tract (33).

The real question lost in the excitement of more new probiotic products entering the market is: What is the effect of these extremely high microbial dosages on the natural intestinal tract environment? Is there a possibility that the artificial inflation of "good" bacteria, or unusual combinations of bacteria not seen before, might actually be "bad"?

Acidophilus Milk. Acidophilus milk is a cultured milk created with the assistance of *Lactobacillus acidophilus*. These bacteria break down lactose to glucose and galactose, resulting in twice as many sugar molecules. The resultingly somewhat sweeter milk, usually packaged in cartons, is made by inoculating pasteurized milk with *L. acidophilus* culture and letting it incubate at 99°F (37°C) until a slight curd forms. A slightly acidic, sour taste also results, but this can be eliminated by mixing the bacteria directly into cold milk.

Kefir. Kefir is a fermented milk product and probiotic that originated in Russia. It is sometimes referred to as "the champagne of milk" because of its bubbly, fizzling nature. Kefir is made by adding bacteria, *Lactobacillus caucasius*, and yeast, *Saccharomyces kefir* and *Torula kefir*, to milk (19). The milk is initially heated to 185°F (85°C) for half an hour and then cooled to 73°F (23°C), which allows the milk to ferment to a soft, foamy curd. The strong, tangy, sour taste comes from the formation of lactic acid. Kefir contains about 1 percent alcohol and a little carbon dioxide due to fermentation, and provides 250 calories (kcal) and 4.5 grams of fat per cup.

Sour Cream. Cream can be soured by *Streptococcus lactis* bacteria or some other acidifying agent. Light cream or half-and-half is fermented at 72°F (22°C) until the acidity from lactic acid reaches 0.5 percent. A thicker sour cream is produced if MSNF, vegetable gums (carrageenan), or certain enzymes are added. To be labeled "sour cream," a minimum of 18 percent milk fat is required, although when sweeteners are added, the minimum milk fat content can be lowered to 14.4 percent. Lower-fat sour creams with half the fat content of standard sour cream, and sour cream substitutes, are also available.

Creams and Substitutes

Cream. Cream is a collection of fat droplets that float to the top of non-homogenized whole milk. The heavier and thicker the cream, the higher the fat content. Cooling the cream firms its fat globules and makes it even thicker. Creams vary in their milk fat content, ranging from a low of 18 percent to a high of 36 percent.

Cream manufacturers are not required to list on the carton the percentage of fat in the cream. The fat content of light or coffee cream is 18 to 30 percent, of light whipping cream is 30 to 36 percent, and of heavy cream or heavy whipping cream is not less than 36 percent. Combining cream with pasteurized or ultrapasteurized milk yields half-and-half, which is not true cream and contains only 10.5 to 18 percent fat.

Some whipping creams are marketed containing added sugars and stabilizers to improve their taste and texture. Whipping cream is also sold in pressurized canisters, which provide the taste and texture of whipped cream in a convenient-to-use form.

Cream Substitutes. Some of the whipped toppings in pressurized cans and tubs, as well as coffee creamers, dry mixes, imitation sour cream, and snack dips, are made from nondairy ingredients. These products came into being as low-cost substitutes that would last longer on the shelf. Nondairy coffee creamers can last over a year at room temperature. They often contain saturated fats, but many lighter substitutes have half the calories (kcal) and one-third the fat of cream, and contain no cholesterol. One fat replacer, Supercreme, can substitute for up to 100 percent of the fat in reduced-fat milk (2 percent), sweet cream, sour cream, and butter. This product is made from milk protein that has been transformed into small, spherical globules; it has natural cream flavor and can be used in muffins, dressings, and low-fat soup (8). The nondairy whipped toppings in pressurized cans are made from water, vegetable oil, corn syrup solids, sodium caseinate or soy protein, emulsifiers, vegetable gums, coffee whiteners, and artificial flavors and colors.

Milk Products in Food Preparation

Flavor Changes

The bland, slightly sweet flavor of milk comes from its lactose, salts, sulfur compounds, and short-chain fatty acids. The percentage of fat determines the mouthfeel and body of a particular milk. Exposure to heat or sunlight, oxidation, the use of copper equipment or utensils, and the feed ingested by the source animal are just some of the other factors that can influence the flavor of milk (7). For example, off-flavor develops when the amino acid methionine reacts with the sunlight-sensitive riboflavin (vitamin B_2). The "cooked" flavor of heated milk develops in part because heating denatures whey proteins to release volatile sulfur compounds, which contribute to off-flavors (10). Dairy cattle allowed to feed off wild onions, ragweed, French weed, beets, potatoes, cabbage, or turnips produce off-flavored milk (31).

Coagulation and Precipitation

Some milk proteins coagulate or precipitate to form a solid clot, or curd, under certain conditions. These conditions include the application of heat and the addition of acid, enzymes, polyphenolic compounds, and salts.

Heat. When milk is heated to near the boiling point, the whey proteins lactalbumin and lactoglobulin become insoluble, mesh with the milk's calcium phosphate, and precipitate, forming a film on the bottom and sides of the pan. This film can scorch easily. Scorching can be prevented by constant stirring, slow temperature increases, or use of a double boiler.

HOW & WHY? ?????????

Why does a skin form on the surface of heated milk? This is caused by the evaporation of water, which is accompanied by an increased concentration of casein, fat, and mineral salts. This thin skin also scorches easily; in addition, it can trap steam that is trying to escape and cause the milk to boil over. Several steps can be taken to avoid this problem, including using a lid, continual stirring during heating, floating a small pat of butter on top of the milk, or, in the case of hot chocolate, adding whipped cream or marshmallows.

Chemist's Corner 11-3

Milk Coagulates as pH Drops

The calcium in milk may be in one of two forms: either combined with the casein protein, or as free calcium ions (Ca^{++}). At milk's near-neutral pH of 6.6, the casein combines with available calcium content, creating calcium caseinate. The casein complexes form a physical configuration known as a micelle, which is colloidally dispersed in the liquid portion of the milk (Figure 11-5) (3).

Coagulation of these casein proteins occurs when the negative charges, normally existing on the casein proteins and causing them to repel each other in the fluid milk, are neutralized by the hydrogen ions (H^+) from acid. When the pH drops to 4.6, casein becomes very insoluble and precipitates readily into a curd. This is milk's isoelectric point—the negative and positive charges on the molecules (i.e., casein) balance each other, and the overall charge is neutral. Milk products coagulated with acid are lower in calcium than those coagulated with an enzyme, because the acid releases calcium from the casein molecules, causing it to be lost in the whey:

FIGURE 11-5 **The role of milk proteins and calcium in coagulation.**

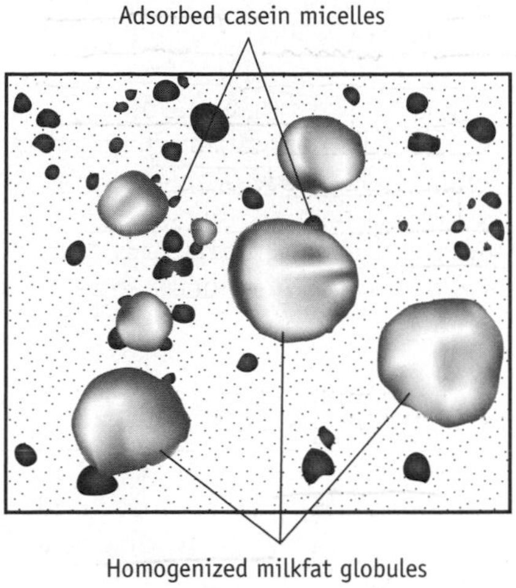

Adsorbed casein micelles

Homogenized milkfat globules

$$\text{calcium phosphocaseinate} \xrightarrow{\text{acid}} \underset{\text{(gel)}}{\text{neutral casein}} + \underset{\text{(lost in whey)}}{Ca^{++}}$$

Casein will not coagulate with heat unless it is boiled for long periods of time. Canned evaporated milk, however, which contains higher concentrations of casein, may coagulate during the high heats of sterilization. This is prevented by warming the milk prior to sterilization.

Acid. Adding acid to milk causes the casein in the milk to coagulate (Chemist's Corner 11-3). Casein precipitates when the normal 6.6 pH of fresh milk drops below 4.6. Whey proteins do not coagulate. Sources of acids include those from foods such as lemon and lime juices, tomato products, and certain fruits; or from bacteria-produced acids in cultured milk products. Some foods must be carefully prepared because of the coagulating effect of acids on milk products. For example, extra caution is required when combining milk or cream with lemon-flavored tea, tomato soup, and coffee (which is acidic). The key to preventing the milk from coagulating is to add the acid to the milk base instead of the other way around. Avoiding high temperatures after milk has been mixed with acid also helps to prevent coagulation.

Enzymes. Milk also coagulates and forms curds when it is combined with certain enzymes originating from animal, plant, or microbial sources. Enzymes used to coagulate milk include pepsin from the stomach of swine, proteases from fungal sources, and certain enzymes from fruits. The enzyme most commonly used to coagulate milk is **rennin**, which is used in the production of cheese and ice cream.

One of the major differences between coagulation caused by enzymes and that initiated with acid is that rennin-coagulated clots are rich in calcium and have a tough, rubbery texture, unlike those created by acid, which are less elastic and more fragile in consistency. Cottage cheese, which is normally coagulated by acid, contains less calcium per ounce (19 mg) than cheddar cheese (204 mg), which is usually made with rennin.

Polyphenolic Compounds. Some fruits, vegetables, teas, and coffees contain polyphenolic compounds, which, when combined with milk,

KEY TERMS

Rennin An enzyme obtained from the inner lining of a calf's stomach and sold commercially as rennet.

result in the precipitation of proteins. They also contribute to the curdling of cream or milk that is sometimes present when making scalloped potatoes or tomato or asparagus soup.

Salts. When used in combination with milk, the salts in cured ham, some canned vegetables, and some seasonings can cause milk to curdle. Here again, to prevent curdling, salt or salted foods should be added to the milk base instead of the other way around, and high temperatures must be avoided after the combination has been made.

Whipped Milk Products

Liquid milk products such as cream, evaporated milk, and reconstituted nonfat dried milk can be made into a foam by whipping air into the liquid. During whipping, the protein in these milk products is mechanically stretched into thin layers that trap air bubbles, fat particles, and liquid (Figure 11-6).

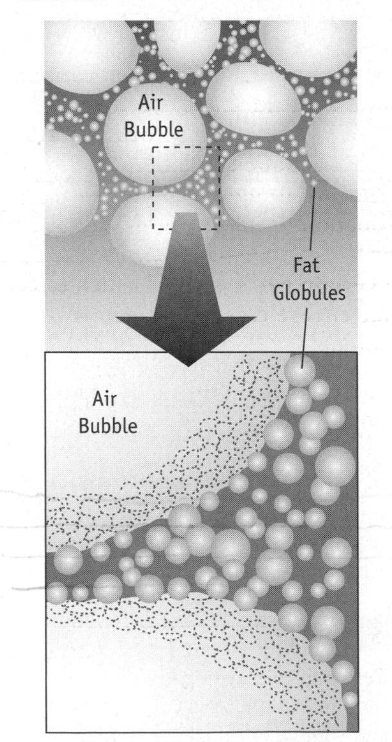

FIGURE 11-6 Whipped foam structure: The numerous fat globules in whipped cream surround the air bubbles, giving solid reinforcement to the foam.

Whipped Cream. Cream expands two to three times its volume when whipped. The stability of milk foams, especially whipped cream, is dependent on several factors: the fat content, the temperature of the cream and the equipment used, the age of the cream, the sugar content, and the length of whipping time.

Fat Content. The higher the fat content, the more stable the whipped cream, because solid fat particles provide rigidity to the foam. Heavy whipping cream beats more easily than lower-fat whipping creams, but becomes lumpy and buttery with overbeating. An advantage of using heavy whipping cream (36 to 40 percent butterfat) in one particular baking application is that its lower moisture content will prevent pastry crust from becoming soggy when it is filled with cream fillings.

Most whipping creams are sold unhomogenized to allow for easier aggregation of the fat globules. When the cream is homogenized, much of its protein surrounds the now smaller and more numerous fat globules instead of being available to envelop the air bubbles that are essential for foam formation. Vegetable gums and gelatin are sometimes added to improve the foaming ability of the commercial creams.

Temperature. Cooling cream increases its viscosity or firmness and its tendency to clump. For best results, refrigerate the cream, bowl, and beaters at 45°F (7°C) or less for at least two hours before whipping. Cream allowed to warm to room temperature or even to above 50°F (10°C) has more widely dispersed fat globules, which reduces the cream's ability to be whipped and creates a softer texture. The optimum temperatures for whipping various milk products are shown in Figure 11-7.

Most cream is pasteurized, but the heating process denatures an enzyme that helps fat globules to cluster, so nonpasteurized creams whip more readily and have a smoother texture (4). Although ultrapasteurized whipping cream takes even longer to beat to a peak-holding consistency, other ingredients can be added to improve its whipping ability and dramatically extend its shelf life.

Age. The older the cream, the greater its viscosity and ability to foam. Whipping cream must be at least one

FIGURE 11-7 Optimum temperatures for whipping.

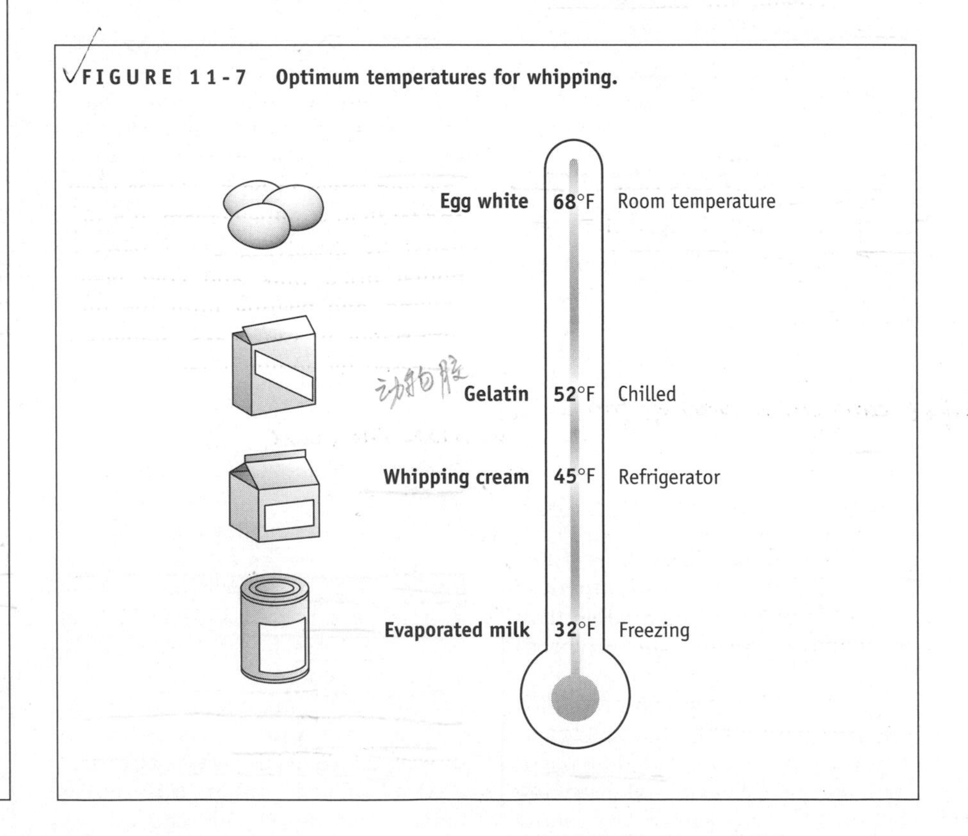

day old in order to allow it to incorporate the air necessary for the optimum increase in volume.

Sugar. Sugar increases the stability of whipped cream, but it should be added gradually, toward the end of the whipping period. Added earlier, it increases the whipping time and reduces overall volume and rigidity by delaying the clumping of fat. Sugar has the benefit, however, of lessening the likelihood of overbeating the cream. For the best stability, powdered sugar instead of granulated sugar should be used, because it dissolves more readily in the cold cream, and the cornstarch in the powdered sugar acts as a stabilizer.

Whipping Time. Physical agitation of the cream is necessary, because it disrupts the phospholipid membranes surrounding the fat globules, preventing them from aggregating (17). Overbeating, however, is a common mistake. Overbeating for even a few seconds over the peak point turns whipped cream into butter and whey. One way to salvage this problem is to add some more whipping cream and whisk it by hand gently into the overbeaten cream. To make whipped cream with an electric beater, it is best to beat on medium high and then slow to a lower speed as soon as the cream starts to thicken. To check for sufficient whipping, the beating is stopped and the beaters are lifted to see if the cream is falling into glossy, large globs with soft peaks. The formation of stiff, yet moist, peaks signals the completion of the whipping process. The cream should be underbeaten slightly if ingredients such as sugar are to be whipped into the cream.

Whipped Evaporated Milk. The high concentration of milk solids in evaporated milk makes it possible to whip it to three times its volume (Figure 11-8), but the flavor, texture, and stability are less acceptable than for whipped cream. The flavor of evaporated milk has a tendency to overpower other flavors; thus it is best used with highly flavored foods. Evaporated milk foams are less stable than whipped cream foams, partly because of the former's lower viscosity and lower fat content. This can be compensated for to some degree by chilling the can of evaporated milk in the refrigerator for 12 hours or in the freezer until ice crystals form. Adding 1½ tablespoons of sugar per chilled cup of evaporated milk can further stabilize the protein and resulting foam. Lemon juice can also be used to stabilize the foam. The main advantage of whipped evaporated milk over whipped cream is its lower cost.

FIGURE 11-8 **Average volume yields for whipped milk products.**

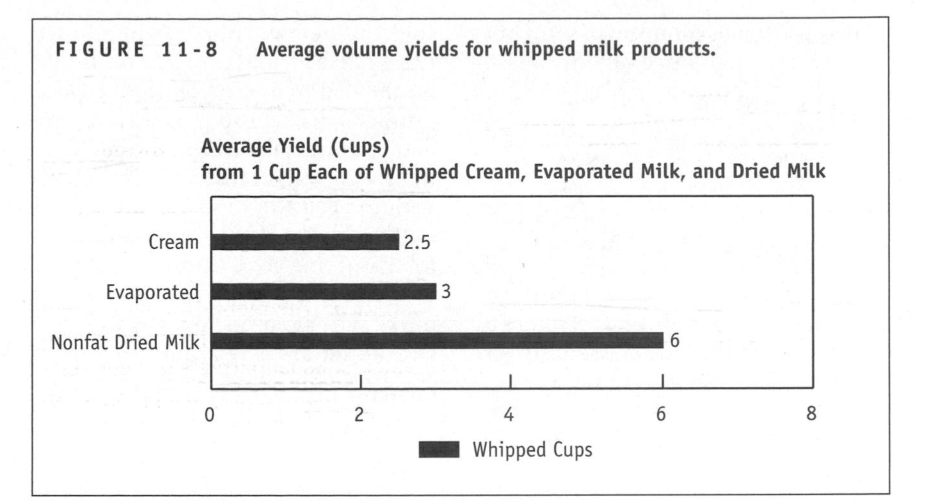

Nonfat Dry Milk (NFDM). Prepared nonfat dried milk powder can actually be whipped into a foam. This whipped milk product is very unstable, but it is much less expensive and lower in both calories (kcal) and fat than whipped cream. It is prepared by dissolving equal parts of nonfat dried milk and cold water, chilling, and beating until the mixture stands in soft peaks. Stability is increased by adding 1 tablespoon of lemon juice or 2 to 4 tablespoons of sugar during beating, which continues until the peaks bend over slightly on top.

Storage of Milk Products

Refrigerated

All fluid milk except unopened, aseptic packs of ultrahigh-temperature pasteurized milk and certain canned milk products should be stored in the refrigerator. They should be removed only long enough to take what is to be used and then quickly returned to the refrigerator. Containers should be closed or covered to avoid rancidity, microbial contamination, and the absorption of odors from other foods in the refrigerator. Both oxidative and hydrolytic rancidity are a potential problem in milk and milk products because of the substantial amounts of short-chain fatty acids. One should never drink milk directly from the container, because bacteria in the mouth can wash back into the product. Pouring unused milk products back into the original container is also not recommended, because microbial contamination could have occurred from exposure to the air or other sources. Proper opaque containers will reduce exposure to light, which can trigger oxidation, resulting in off-flavors and the loss of riboflavin (B_2). The shelf life of certain dairy products such as cottage cheese can be doubled with the addition of carbon dioxide, which disrupts microbial functions (9).

The following guidelines should be followed when storing milk products in the refrigerator:

- *Milk.* No more than three weeks.
- *Yogurt.* Best consumed within the first ten days, but can last up to three to six weeks. If it separates, simply stir the liquid back into the curd before serving.
- *Buttermilk.* Best when used within three to four days after purchase,

because it will continue to sour, but it can last up to three or four weeks.

- *Sour cream*. Unopened, up to one month, but is best when used within a few days.

Dry Storage

Nonfat dry milk, ultrapasteurized milk, evaporated milk, and sweetened condensed milk are stored at or slightly below room temperature (72°F/22°C). Nonfat dry milk should not be exposed to moisture, because humidity will cause it to turn lumpy and become stale. Keeping containers tightly closed minimizes any contact with oxygen from the air. Nonfat dry milk stored in this manner will keep for about one year. Unopened cans of evaporated and sweetened condensed milks will keep up to a year in dry, ventilated areas, but double that if refrigerated. Both should be turned over every few weeks to prevent the solids from settling, thickening, and producing clots. Ultrapasteurized milk can be stored unopened at room temperature for up to three months. Once opened, all these milks must be treated like fresh milks and refrigerated. The "sell by" or "pull" date is the last day the item should be sold by the store, and this should always be checked before purchase.

PICTORIAL SUMMARY / 11: Milk

About half the milk produced in the United States is sold as fluid milk and cream. Much of the rest comes to market as butter, cheese, and ice cream.

COMPOSITION OF MILK

Water
Carbohydrates
Proteins
Fat
Vitamins and minerals

Almost 90 percent of milk is water.

Lactose, or milk sugar, is the primary carbohydrate found in milk.

The predominant proteins in milk are of high quality and consist of casein and whey proteins.

The fat in milk is known as milk fat or butterfat.

Milk is an excellent source of vitamins A and D (depending on fortification), riboflavin (B_2), and tryptophan, important in the formation of the B vitamin niacin. About 80 percent of the calcium ingested by Americans is derived from the dairy food group.

PURCHASING MILK

USDA Grade A, the highest grade for milk, is based on the lowest bacterial count. Although grading of milk is voluntary, pasteurization, the process of heating milk to kill harmful bacteria, is required for all milk. Homogenization prevents the separation of water and fat in milk.

MILK PRODUCTS IN FOOD PREPARATION

Heat, the source animal's feed, oxidation, the use of copper equipment or utensils, and exposure to sunlight can all influence milk flavor.

The percentage of fat determines the mouthfeel and body of a particular milk.

Milk proteins coagulate to form a solid clot when exposed to heat, acid, enzymes, polyphenolic compounds, and salt.

Scorching of milk can be prevented by constant stirring, slow increases in temperatures, and use of a double boiler.

Liquid milk products such as cream, evaporated milk, and reconstituted nonfat dry milk can be whipped into a foam.

STORAGE OF MILK PRODUCTS

All fluid milk except unopened ultrahigh-temperature pasteurized milk and certain canned and dry milk products should be stored in the refrigerator. Containers should be closed or covered to avoid rancidity, microbial contamination, and the absorption of odors from other refrigerated foods. Oxidation caused by exposure to air can result in an off-flavor and losses of riboflavin (B_2). Nonfat dry milk, ultrapasteurized milk, evaporated milk, and sweetened condensed milk should be stored at or slightly below room temperature (72°F/22°C). Once opened, all these milks must be treated as fresh milks and refrigerated.

TYPES OF MILK

About half the milk produced in the United States is sold as fluid milk and cream.

Types of milk include:

- Whole
- Reduced fat (2 percent)
- Low fat (1 percent)
- Fat-free (nonfat)
- UHT
- Chocolate
- Carbonated
- Imitation
- Filled
- Reduced lactose
- Low-sodium
- Canned
  - –Evaporated
  - –Sweetened condensed

Cultured milk products

- Yogurt
- Acidophilus milk
- Buttermilk
- Sour cream

Cheese

Ice cream

Butter

Eggnog

Cream

REFERENCES

1. Adam S. Carbonated milk: Soft drink that's not junk food. *Agricultural Research* 36(3):14, 1988.
2. Amanter GF. Aseptic packaging of dairy products. *Food Technology* 37(4):138, 1983.
3. Aynie S, et al. Interactions between lipids and milk proteins in emulsion. *Journal of Food Science* 57(4):883, 1992.
4. Babcock CJ. *The Whipping Quality of Cream.* U.S. Dept. of Agriculture Bulletin No. 1075. Washington, D.C.: GPO, 1922.
5. Chandan R. *Dairy-Based Ingredients.* Eagen Press, 1997.
6. Chang MK, and H Zhang. Carbonated milk: Proteins. *Journal of Food Science* 57(4):880–882, 1992.
7. Christensen TC, and G Holmer. Lipid oxidation determination in butter and dairy spreads by HPLC. *Journal of Food Science* 61(3):486–489, 1996.
8. Dairy ingredient replaces fat, adds flavor. *Food Engineering* 64(1):26, 1992.
9. Dairy Management, Inc. Increasing milk's competiveness. *Dairy Research Focus* 2(4):1–2, 1996.
10. deMan JM. *Principles of Food Chemistry.* Van Nostrand Reinhold, 1999.
11. Euston SE, et al. Oil-in-water emulsions stabilized by sodium caseinate or whey protein isolate as influenced by glycerol monostearate. *Journal of Food Science* 61(5):916–920, 1996.
12. Full NA, et al. Physical and sensory properties of milk chocolate formulated with anhydrous milk fat fractions. *Journal of Food Science* 61(5):1068–1072, 1996.
13. German JB, and CJ Dillard. Fractional milk fat composition, structure, and functional properties. *Food Technology* 52(2):33–38, 1998.
14. Goldberg JP. Nutrition and health communication: The message and the media over half a century. *Nutrition Reviews* 50(3):71–77, 1992.
15. Hamann WT, and EH Marth. Survival of Streptococcus thermophilus and Lactobacillus bulgaricus in commercial and experimental yogurts. *Journal of Food Protection* 47:781, 1984.
16. Hansen AP. Effect of ultrahigh-temperature processing and storage on dairy food flavor. *Food Technology* 41(9):112, 1987.
17. Hinrichs J, and HG Kessler. Fat content of milk and cream and effects on fat globule stability. *Journal of Food Science* 62(5):992–995, 1997.
18. Hollingsworth P. Yogurt reinvents itself. *Food Technology* 55(3): 43–49, 2002.
19. Honer C. Now kefir. *Dairy Field* 176(9):91, 1993.
20. Hoogenkamp HW. *Milk Protein.* De Melkinclustrie Veghel BV, 1989.
21. Johnson RK, C Frary, and MG Want. The nutritional consequences of flavored-milk consumption by school-aged children and adolescents in the United States. *Journal of the American Dietetic Association* 102(6):853–856, 2002.
22. Kadharmestan C, B Byung-Kee, and Z Czuchajowska. Thermal behavior of whey protein concentrate treated by heat and high hydrostatic pressure and its functionality in wheat dough. *Cereal Chemistry* 75(6):785–791, 1998.
23. Labell F. Yogurt cultures offer health benefits: Biotechnology to transform the yogurt of the future. *Food Processing* 50:130–138, 1989.
24. Laye I, D Karleskind, and CV Morr. Chemical, microbial, and sensory properties of plain nonfat yogurt. *Journal of Food Science* 58(5):991–995,1000, 1993.
25. Matte JI, and JM Krochta. Whey protein coating effect on the oxygen uptake of dry roasted peanuts. *Journal of Food Science* 61(6):1202–1206, 1996.
26. Ming-Fen L, and SD Krasinski. Human adult-onset lactase decline: An update. *Nutrition Reviews* 56(1):1–8, 1998.
27. Ohr LM. Improving the gut feeling. *Food Technology* 56(10):167–70, 2002.
28. Otte J, et al. Protease-induced aggregation and gelation of whey proteins. *Journal of Food Science* 61(5):911–915, 1996.
29. Putman JJ. *Food Consumption, Prices, and Expenditures, 1967–1988.* U.S. Dept. of Agriculture Statistical Bulletin No. 804. Commodity Economics Division, Economic Research Service, 1990.
30. Renken SA, and JJ Warthesen. Vitamin D stability in milk. *Journal of Food Science* 58(3):552–556, 1993.
31. Ronsivalli LJ, and ER Vieira. *Elementary Food Science.* Van Nostrand Reinhold, 1992.
32. Salminen S. Probiotics: Scientific support for use. *Food Technology* 53(11):66–77, 1999.
33. Sanders ME. Probiotics: A publication of the Institute of Food Technologists' expert panel on food safety and nutrition. *Food Technology* 53(11):67–77, 1999.
34. Savaiano DA, and MD Levitt. Nutritional and therapeutic aspects of fermented dairy products. *Contemporary Nutrition* 9(6):1–2, 1984.
35. Siuta-Cruce P, Goulet J. Improving probiotic survival rates. *Food Technology* 55(10):36–42, 2001.
36. Sloan AE. The top ten functional food trends. *Food Technology* 54(4):33–62, 2000.
37. Sloan AE. Got milk? Get cultured. *Food Technology* 56(2):16, 2002.
38. Smith DM, and AJ Rose. Gel properties of whey protein concentrates as influenced by ionized calcium. *Journal of Food Science* 59(5):1115–1118, 1994.
39. State of the food industry: Dairy products. *Food Engineering* 63(6):89, 1991.
40. The 1992 supermarket sales manual. *Progressive Grocer* 71(7):80, 1992.

WEBSITES

The Dairy Industry's National Dairy Council website:
www.nationaldairycouncil.com

An industry association for consumers and their dairy products:
www.doitwithdairy.com

Solutions over the web for dairy industry people:
www.dairy.com

12

Cheese

Cheese is a preserved food made from the curd, or solid portion, of milk. Adding certain enzymes and/or acid to any type of milk causes the casein proteins and fat to coagulate and separate from the liquid portion, or whey (see Chapter 11). No one really knows when humans first started to consume cheese, but legend links its discovery to an anonymous shepherd who decided to carry milk in a bag made from a sheep's stomach. The sun warmed the bag, activating the natural rennin enzyme in the stomach lining and turning the liquid milk into a semi-solid clump. To the shepherd's surprise, the resulting mass was quite palatable. It probably did not take long after that for people to realize how useful this natural process would be in providing an edible food that could be transported.

Making cheese involves removing moisture from the curd to varying degrees after the whey is drained. The curd can then be treated in a variety of ways to produce over 2,000 varieties of cheese, some of which are shown in Figure 12-1. The whey also contains dissolved materials such as proteins, which can be processed to produce cheeses and other foods.

In the United States, the Food and Drug Administration states that cheese must be "a product made from curd obtained from the whole, partly fat-free/nonfat, or fat-free/nonfat milk of cows, or from milk of other animals, with or without added cream, by coagulating with rennin, lactic acid, or other suitable enzyme or acid, and with or without further treatment of the separated curd by heat or pressure, or by means of ripening ferments, special molds, or seasoning." This definition serves as the foundation for many different cheese varieties, but fifteen varieties now account for most of the cheese consumed today. Appendix G lists the commonly used cheese varieties along with their shape, flavor, basic ingredient, ripening (aging) period, and mode of serving.

This chapter discusses the various ways to classify cheese, along with its nutrient composition, production, purchase, use, and storage.

Classification of Cheeses

There are many ways to classify cheeses. They can be defined by their microbial characteristics, appearance, mode of packaging, even their place of origin. The most common ways of classifying cheeses, however, are by the processing method, the milk source (cow, sheep, goat, etc.), or the moisture content. The last method is the one discussed in this text.

During the production process, moisture content in cheeses decreases, especially during aging. As cheeses age, they become drier and harder. Cheeses classified according to their moisture content are described as fresh, soft, semi-hard, hard, and very hard (27).

- *Fresh.* Fresh cheeses, also called "country cheeses," are soft, whitish in color, and mild tasting. They are highly perishable, because their moisture content is over 80 percent, and they are not aged. They include cottage cheese, cream cheese, ricotta, farmer's, pot, and feta cheeses.
- *Soft.* Soft cheeses, such as Brie, Camembert, and many Hispanic cheeses, are aged for just a short time. Water content ranges from 50 to 75 percent.
- *Semi-hard.* Semi-hard cheeses contain 40 to 50 percent moisture. The best-known examples are Roquefort, blue, Muenster, brick, Gouda, Edam, Port du Salut, Gorgonzola, and Stilton.
- *Hard.* The moisture content of hard cheeses ranges from 30 to 40 percent. Cheddar and Swiss are examples of hard cheeses.

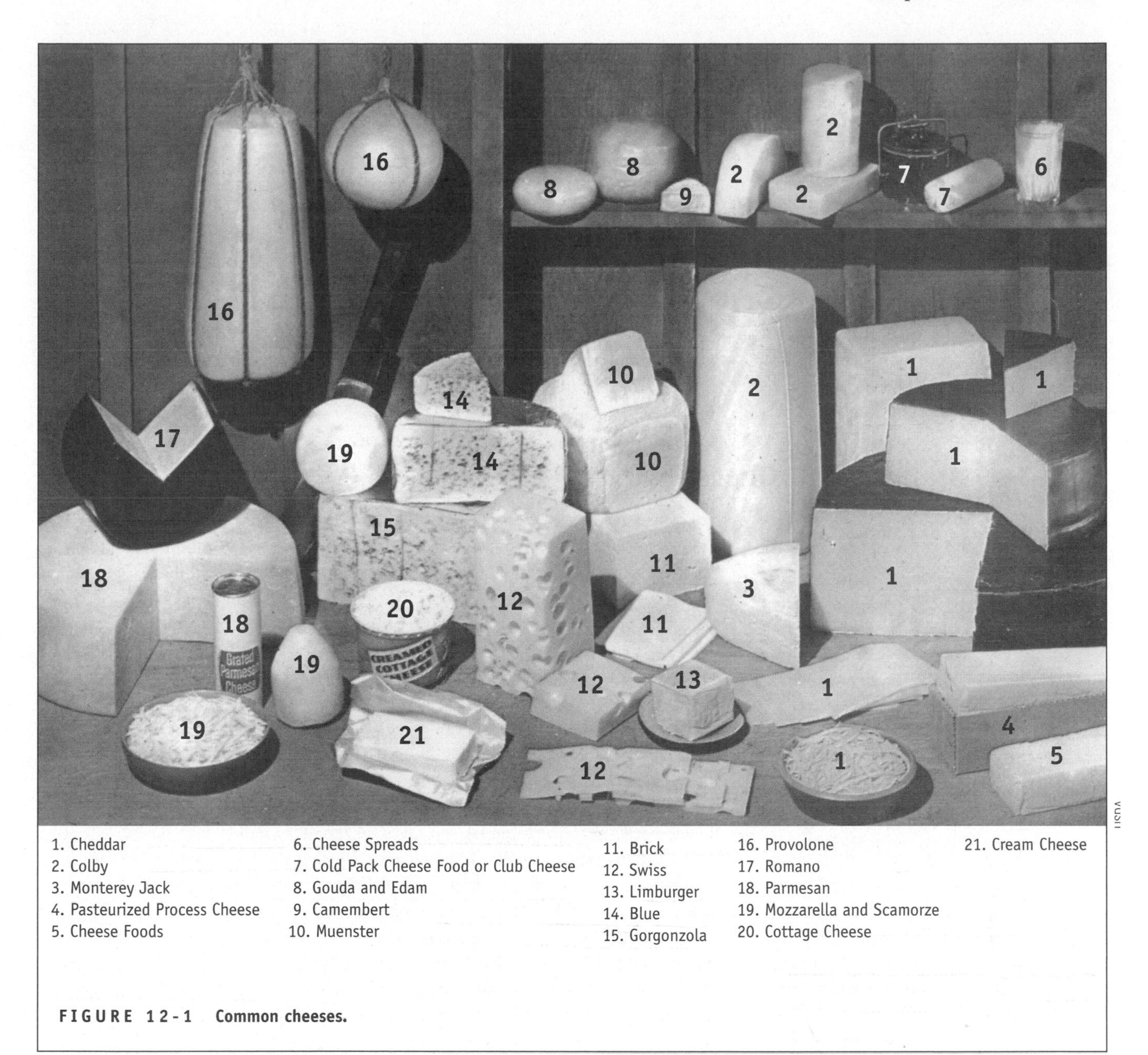

FIGURE 12-1 Common cheeses.

- *Very hard* Parmesan and Romano are classified among the hardest cheeses. Very hard cheeses will not slice easily, but are easily grated or crumbled. They are aged the longest and have a water content of approximately 30 percent.

Cheese Production

No two cheese varieties are produced by exactly the same method, but the basic steps described below are common to them all. They include milk selection, coagulation, curd treatment (cutting, heating, salting, knitting, and pressing), curing, and ripening. The yield from 10 pounds of milk is approximately 1 pound of cheese and 9 pounds of whey (24). On a smaller scale, that is equivalent to 1 cup (8 ounces) of milk providing about 1 ounce (specifically, 0.8 ounce) of cheese.

Milk Selection

The first step in making cheese, and the one that has the greatest influence on its classification, is choosing the appropriate milk. Any mammal's milk can be made into cheese, but in the United States, pasteurized cow's milk is the most common source. In Europe and the Middle East, a significant amount of cheese is made from either sheep's or goat's milk. Other animals whose milk may be used for making cheese include the camel (Iran and Afghanistan), reindeer (Lapland), horse (Mongolia), water buffalo (Philippines , India, and Italy), and the yak and zebu (China and Tibet).

The amount of fat found in cheese is determined by the type of milk from which it is made. Whole, reduced-fat (2 percent), or fat-free (nonfat) milk, buttermilk, cream, whey, milk solids-not-fat (MSNF), or any combination of these can serve as the basis for cheese-making. Homogenized milk is usually selected for making soft cheese because homogenization makes the casein proteins coagulate more easily, and the increased surface area of the broken-up fat results in a moister product. Homogenized milk is not used to make hard cheeses, because the same trait that was desirable for softer cheeses creates a brittle texture during the long aging process (18).

Currently, the FDA requires pasteurization for all fresh or soft-ripened cheeses, but allows the use of raw milk (unpasteurized) to make hard cheeses such as cheddar if they are aged for at least 60 days (23).

Coagulation

Cheese-making starts with the coagulation of the casein protein in milk. The type of method used determines many of the characteristics of the resulting cheese. The two main methods to aid coagulation are the action of enzymes or acid (Figure 12-2).

Enzyme Coagulation. The enzyme most commonly used to coagulate milk in cheese-making is rennin, obtained from milk-fed calves, specifically their fourth stomach. Rennin, also called chymosin, is sold commercially as rennet. Other sources of available rennin include cows, pigs, plant sources, and genetically engineered bacteria (11) (Chemist's Corner 12-1). Alternatively, enzymes may be derived from bacterial starters (Streptococci, Lactobacilli), from certain molds, or from any of a number of other microorganisms. The different enzymes, bacteria, molds, and/or yeasts added during coagulation influence the flavor, texture, and color developed by the cheese during ripening (10,30,31).

The milk is usually heated in large vats at temperatures from 72° to 95°F (22° to 35°C) to provide an optimal environment for enzymes and bacterial activity, which contributes to the formation of curd. Calcium chloride may be added to speed up coagulation and strengthen the curd's consistency. Coagulation with enzymes occurs in less than an hour, and creates a tough, rubbery curd. As the curd forms, the whey separates, but most of the milk's calcium remains in the curd.

NUTRIENT CONTENT

Cheese is one of the most nutrient-dense foods. The primary ingredients by weight are water, protein, and fat, and these nutrients vary in concentration according to the cheese. A pound of cheese is equivalent in protein and fat to approximately 1 gallon of milk. Cheeses with higher moisture levels are less concentrated in their nutrients than those containing less moisture. On average, 1 ounce of cheese (a 1¼-inch cube or a packaged slice of pasteurized process cheese) provides about 100 calories (kcal), most of which are derived from fat.

Fat. Fat is responsible for much of the satiety value, flavor, and texture of cheese. It is the fat content of milk, which is rather bland in taste, that is acted upon during ripening to result in a cheese with a very distinct flavor (14). The fat in cheeses averages 9 grams per ounce (primarily saturated fat), which is equivalent to two pats of butter. Cholesterol content averages 25 mg per ounce, but ranges from 0 mg in pot cheese to 40 mg per ounce in gouda cheese. Dietary fat from cheese can be reduced by using less cheese or by using lower-fat options.

Lower-Fat Cheeses. The American Heart Association and the National Cholesterol Education Program define a low-fat cheese as one containing no more than 6 grams of fat per ounce (1). In addition, cholesterol should not exceed 20 mg, and sodium should be below 500 mg per serving (8). Only a few cheeses qualify as very low in fat, at less than 2 grams per ounce. These are gammelost, a semi-soft blue cheese; sapsago, a pungently flavored, hard-textured, high-sodium cheese; and baker's, pot, or hoop cheese (21). The two most common fat-free (nonfat) milk cheeses available are ricotta and mozzarella. Replacing some of the fat in cheese with fat replacers can reduce the fat content to 3 grams per ounce. Another fat-reducing option is yogurt cheese, which contains only 1 gram of fat per ounce and is made by draining yogurt of its whey. Yogurt cheese is an excellent substitute for cheeses normally used to make cheesecake or other high-fat foods, because it lowers the fat and takes on the other ingredients' flavors.

Fat is important to texture and flavor, so reducing the fat in cheese can result in a product with a rubbery texture and either no flavor or off-flavors (9, 16). One reason that low-fat cheeses taste different is that their increased moisture content, resulting from decreased fat content, is more conducive to the growth of natural bacteria, which can produce compounds yielding flavors that are brothy, meaty, and bitter (19). Fat also masks certain off-flavors such as the bitter compounds produced from protein breakdown, which are detected when fat is reduced.

Many reduced-fat cheeses behave differently when heated, which sometimes makes it difficult to substitute them in recipes calling for full-fat cheeses. Fat also contributes to the smooth, lubricated texture of full-fat cheese by breaking up the protein matrix. Removing this fat leaves more protein and increases its influence on texture (19).

Protein. Since cheese is of animal origin, its protein is complete and of high quality: 1 ounce of cheese contains approximately 7 grams of protein, equivalent to that in 1 cup (8 ounces) of milk or 1 ounce of meat (7 grams). Long before protein was recognized as an essential nutrient by food and nutrition scientists, cheese was being used as a meat substitute by monks in monasteries.

Carbohydrate. Cheese contains very little carbohydrate because most of the lactose, the primary carbohydrate in milk, is drained off with the whey, with any remaining lactose converted to lactic acid. With the lactose largely gone, some people with lactase deficiency (lactose intolerance) can consume cheese with no ill effects. Ripened cheeses such as Swiss and cheddar are often suitable for lac-

NUTRIENT CONTENT

tase-deficient individuals (28), but processed cheeses are usually not, because milk or whey is frequently added to them.

Vitamins. Many of the milk's vitamins are lost in cheese production. When whole milk is used to make cheese, the fat-soluble vitamins A and D remain in the curd, but many of the milk's water-soluble vitamins are drained off in the whey. Only about one-fourth of the riboflavin (vitamin B_2) and one-sixth of the thiamin (vitamin B_1) remain in cheddar cheese, while much of the milk's original niacin, vitamin B_6, vitamin B_{12}, biotin, pantothenic acid, and folate drain off with the whey (22).

Minerals. Cheese has a high concentration of minerals, specifically calcium, phosphorus, and zinc. There are approximately 200 mg of calcium in 1 ounce of cheese, compared to 300 mg in 1 cup of milk. Calcium content will vary depending on whether the cheese was coagulated with enzymes or acid: enzyme-coagulated cottage cheese contains twice the calcium as acid-coagulated cheese. Cheese often has a high concentration of sodium, because salt is frequently added during manufacturing to aid in the process of ripening and to contribute to flavor. The amount of salt added to cheese varies widely, and low-sodium cheeses are available. Processed cheese products tend to be especially high in sodium, averaging 400 mg per ounce.

Acid Coagulation. There are two methods by which acid may be used to coagulate milk proteins and thus form cheese. The first method is to simply add acid directly to the milk. The second and more complex method is to inoculate the milk with cultures of bacteria that convert lactose (milk sugar) to lactic acid, which makes the milk medium more acidic (4). Bacterial cultures have been used for centuries to produce fermented foods, and are carefully selected for their characteristic influences on a cheese's flavor and texture (5). It takes from four to sixteen hours to coagulate milk with acid-forming bacteria. About one-fourth to one-half of the calcium in milk is lost in the whey during this process. The curd produced by acid has a soft and spongy texture. This texture is influenced by pH, becoming more solid and compact as the acidity increases (Figure 12-3). Acid-coagulated cheese is usually not aged because its high acidity inhibits the bacterial and mold growth that characterizes the aging process.

FIGURE 12-2 Cheese is usually produced by coagulating (clotting) the milk with either an enzyme or acid.

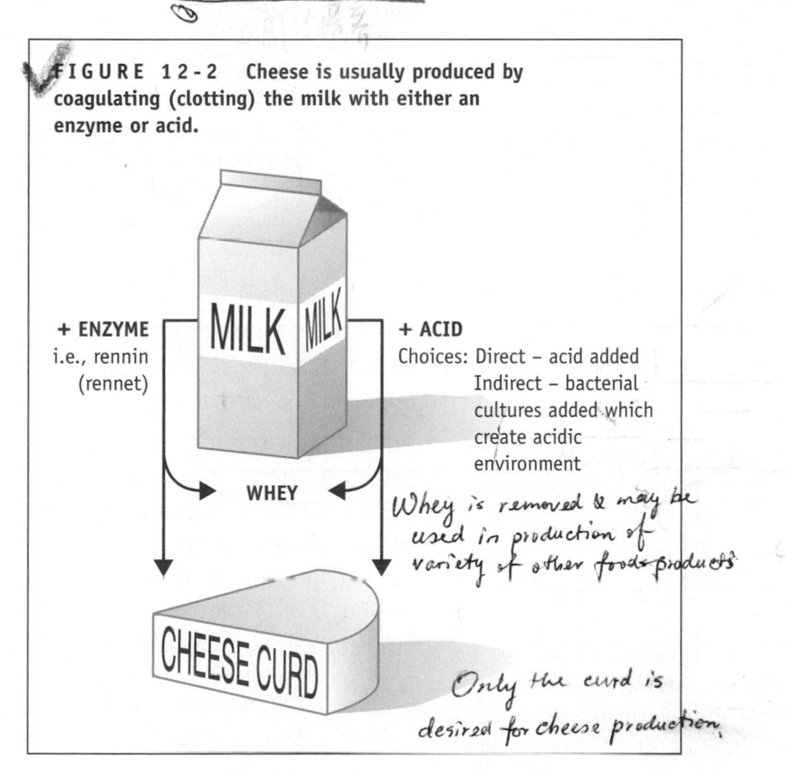

Curd Treatment

The curd may be treated to remove more whey by cutting, heating, and salting. Further optional treatment includes knitting and/or pressing. Although a few chemical tests can be made to assess the progress of the curd through each of these treatments, it is often the experienced judgment of a

Chemist's Corner 12-1

Rennin's Role

Enzymes act on milk and cause it to clot or coagulate. The most common enzyme for this job is rennin, which coagulates milk by hydrolyzing casein. The rate of this reaction depends on the pH and ionic strength or salt/NaCl concentration of the medium. Commercial rennins in the United States include calf chymosin, bovine pepsin, porcine pepsin, microbial proteases (derived from bacteria such as *Cryophonectria parasitica, Mucor miehei* var. *lindt,* and *Mucor pusillus*), and chymosins generated from the fermentation of certain genetically engineered strains of bacteria (*Asperigillus niger* var., *Awamori, Escherichia coli,* and *Kluyveromyces lactis*) (11). All of these milk-clotting enzymes that cleave the polypeptide molecule are classified as aspartic proteases, because their active sites consist of aspartic acid residues.

FIGURE 12-3 Cheese texture.

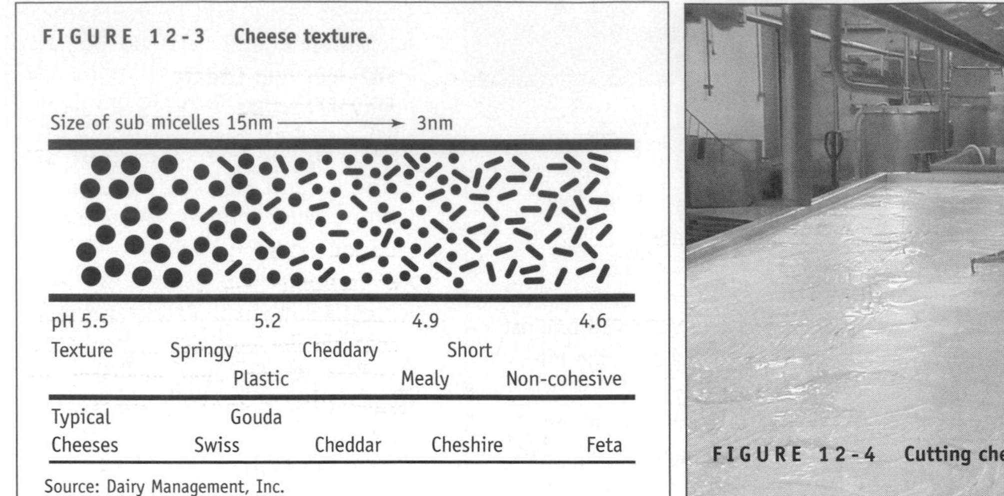

Source: Dairy Management, Inc.

FIGURE 12-4 Cutting cheese curd.

Karoun Dairies

cheese maker that determines when it is time for the next step (15).

- *Cutting.* Slicing the curd increases its surface area (Figure 12-4). Sometimes the curd is placed on strainers to remove even more whey.
- *Heating.* Encourages the evaporation of whey. This allows lactic acid to build up to create a firmer, more elastic texture. Heat also destroys certain undesirable microorganisms. After heating, the curds are washed with cold water to produce softer, higher-moisture cheeses.
- *Salting.* Salt controls the growth of bacteria and further dehydrates the curd. It also contributes to the flavor, texture, and appearance of cheese.
- *Knitting.* Some cheeses are "knitted"; that is, the curd is united or melted into a solid mass through the use of heat.
- *Pressing.* Pressing is another way to create a solid mass out of the curd, and the last step before ripening. Curds are physically pressed into compact masses by placing them in boxes or other containers under pressure.

KEY TERMS

Curing To expose cheese to controlled temperature and humidity during aging.

■ ■ ■ ■

Ripening The chemical and physical changes that occur during the curing period.

Curing and Ripening

Cheese becomes stronger in flavor as it ages.

HOW & WHY? ?????????

Why are cheddar cheeses labeled "mild," "medium," or "sharp"? Like wine, the finished character of these cheeses is determined not only by the original ingredients, but by the maturation process. The flavor of cheddar cheese can range from mild to sharp depending on the duration of aging. Although the time cheddar is held for aging varies by the cheese company, the general guideline is that mild cheddar is aged for at least 60 days, medium for 3 to 6 months, sharp for a minimum of 9 months, and extra sharp for at least 15 months.

The aging process whereby cheese is converted from a bland, tough, rubbery, fresh curd into a unique cheese with its own mature flavor, aroma, and texture is called, often interchangeably, **curing** or **ripening** (26). Depending on the variety, cheeses are subjected to different temperatures (36° to 75°F/2° to 24°C) and humidities (higher for mold-ripened cheeses such as Roquefort and blue). Certain cheeses are treated in such a way as to develop a rind, which is simply the dried surface of the cheese. Ripening times range from four weeks to two years or longer.

Cheese flavor originates from a combination of over 300 different volatile and nonvolatile compounds that develop during curing and ripening (11). It is believed that some of these compounds originate from the milk, the activity of milk enzymes, and the starter bacteria. The skillful adjustment of curing techniques, along with the maintenance of the proper environment of temperature and humidity, creates the desired flavors, textures, and aromas of the multitude of cheeses available in our markets.

During ripening, a number of elements may be manipulated to affect the final product. Added salt will draw out some of the remaining whey and inhibit bacterial growth, thereby slowing the ripening process. Bacteria and molds contribute to the development of flavors, aromas, and textures (Chemist's Corners 12-2 and 12-3). The mold *Penicillium roqueforti* added to homogenized whole milk converts free

Chemist's Corner 12-2

Bacteria and Cheese Flavor

Lactic acid bacteria are believed to impart flavor to cheese by reducing the oxidation-reduction potential, by protein hydrolysis, and by producing dicarbonyls. Certain dicarbonyls such as diacetyl, methylglyoxal, and glyoxal are thought to interact with amino acids via the Maillard reaction to produce flavor compounds (29).

FIGURE 12-5 The eyes in Swiss cheese are generated by carbon dioxide produced by the fermentation of lactate (from lactose) to proprionate and acetate.

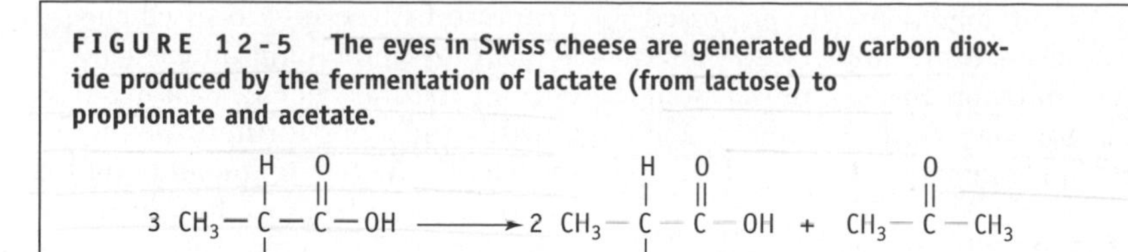

fatty acids to smaller compounds, which impart the characteristic tangy flavor to blue cheese. The holes (eyes) in Swiss cheese are produced by gas-forming microorganisms that are active during the early part of ripening when the curd is pliable (Figure 12-5). Other processes that influence flavor during ripening include the hydrolysis of proteins to peptides (smaller protein molecules) and amino acids, the conversion of lactose to lactic acid, and the breakdown of fatty acids into shorter, volatile fatty acids (12).

Cheeses may exhibit different textures due to processing techniques during production, two of which are inoculation and kneading. The blue-veined cheeses have been inoculated with mold spores, whose growth within the cheese creates the blue veins. Mozzarella and provolone are ropy in texture due to kneading, which is the pulling and stretching of the curd after it has been knitted.

Whey and Whey Products

Whey separated from its water content is low in fat and rich in nutrients. It contains the water-soluble whey proteins and most of the lactose, water-soluble vitamins, and minerals of the milk. Whey is highly perishable when fresh, so it is most often processed quickly into whey cheeses, dry whey, and modified whey products. In the past, the nutrient-rich whey was principally used to feed animals, but now the food industry is using pasteurized whey in a variety of products. Whey is found in baked goods, beverages,

Chemist's Corner 12-3

Enzymes and Cheese Flavor/Texture

One of the major biochemical processes responsible for cheese flavor and texture development during ripening is the breakdown of proteins via proteolysis. The breakdown of the casein structure into smaller peptides also reduces cheese rigidity and increases its flexibility (17). Proteolytic enzymes in cheese are derived from the milk, starter bacteria, coagulant, and other microorganisms (12). The enzymes break down casein to amino acids, which can then be further broken down to various substances, giving each cheese its unique flavor (Figure 12-6). Proteolysis produces beneficial flavors in cheese, but also some bitter compounds (13). These various factors result in the wide variety of cheeses with different flavors.

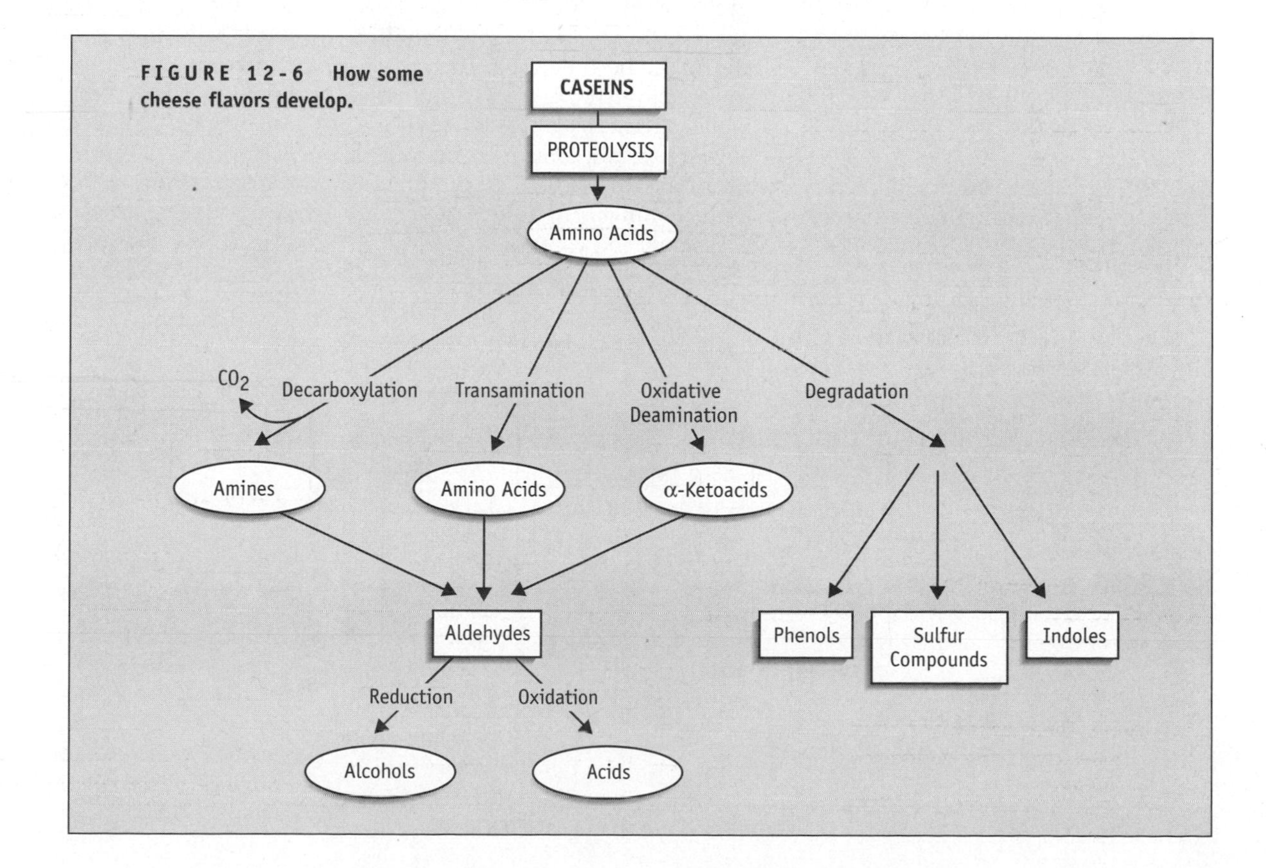

FIGURE 12-6 How some cheese flavors develop.

KEY TERMS

Processed cheese A cheese made from blending one or more varieties of cheese, with or without heat, and mixing it with other ingredients.

sauces, salad dressings, cheese products, canned fruits and vegetables, confections (caramels and other candies), dry mixes, frozen foods (fruits, vegetables, desserts), jams, jellies, meat, pasta, and milk products.

Whey incorporated into foods is available in two types, sweet and acidic whey (6). Sweet whey results from coagulating milk with rennin, while acid whey originates from acid-coagulated milk. Besides having greater acidity, mineral concentrations are also higher in acid wheys, because the acid releases the calcium from the casein molecule, causing it to be dispersed in the whey portion.

Whey Cheeses. Scandinavian cheeses such as Primost, Mysost, and Gjetost are very hard whey cheeses. They are made by evaporating the water until the whey is extremely concentrated. These cheeses tend to be sweet and lightly brown, a result of their lactose caramelizing in the process of water removal. Another type of whey cheese, ricotta, is produced by coagulating the whey with acid and high heat.

Dry Whey. A large portion of the dry whey produced in the United States is still fed to livestock. The remainder is used as an ingredient in processed foods such as confections, soups, beverages, imitation cheeses, dessert toppings, nondairy coffee creamers, and sport supplements.

Modified Whey Products. Condensed whey contains no more than 10 percent of its original water and is often used as an ingredient in processed-cheese food. Another product, sweet-type condensed whey, is used as an ingredient for certain candies. "Whey protein concentrate" is whey concentrated to a minimum of 25 percent protein. Removing some lactose results in a partially delactosed whey. Also available is partially and/or totally demineralized whey, which has had some or all of its minerals removed.

Processed Cheeses

Processed cheeses are all made from blended cheeses, but they differ based on the ingredients and manufacturing methods (Figure 12-7). Approximately one-third of the cheese produced in the United States is used for the various kinds of pasteurized processed cheeses. These include processed cheese, cold-pack cheese, process-cheese food, process-cheese spread, and imitation cheese (22). Processed cheese was patented in 1916 by James L. Kraft, who founded Kraft Foods (24). These cheeses appeal to many consumers because of their uniform taste and creamy, melted texture, longer shelf life, convenient packaging, and lower cost.

Processed Cheese. Processed cheese is made by combining different varieties of natural cheese. Heating pasteurizes and stops further ripening of the cheese. Other ingredients added to produce a stable, homogenous emulsion include emulsifying salts such as sodium citrate or sodium phosphate. During the emulsifying process, powdered milk, whey, cream or butter, and water may also be added. The moisture content of the processed cheese must not exceed 40 percent, and the fat content is similar to the natural cheese from which it is derived. When the mixture is partially cooled, it is formed into blocks, cut into slices, wedges, or other shapes, and packaged. American cheese made from blended cheddar cheese is a popular form of processed cheese.

Cold-Pack Cheese. Cold-pack or club cheeses were developed in 1918 by Hubert Fassbender, who found he could blend cheddar cheese, milk by-products, and spices to create a spreadable cheese for use on crackers and sandwiches. Since cold-pack cheeses are blended without being heated, they must be made from pasteurized milk products. In cold-pack cheese food, the original cheese may be combined with milk (whole, reduced-fat (2 percent), fat-free (nonfat), or buttermilk), MSNF, cream, or whey. It also may be sold in smoked form.

Process-Cheese Food. Process-cheese food must be at least 51 percent natural cheese by weight, which

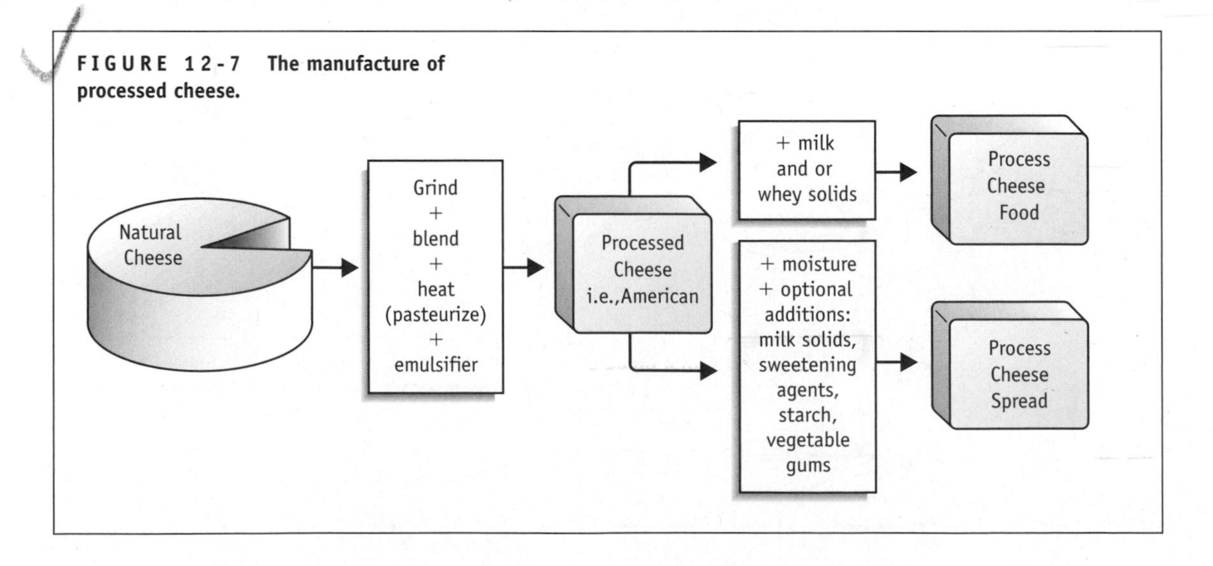

FIGURE 12-7 The manufacture of processed cheese.

is less than either processed cheese or cold-pack cheese. The remaining ingredients may include milk, cream, oil, or whey (24). The resulting products, including Cheez Whiz, Velveeta, and Kraft Singles, have a milder flavor and softer texture, and tend to melt more quickly.

Process-Cheese Spread. Process-cheese spread is a softer, more spreadable product than process-cheese food, but, like cheese food, at least 51 percent of its weight must be from natural cheese (3). Ingredients such as sugar, dextrose, maltose, and corn syrup may also be added. The higher spreadability of process-cheese spread, obtained by adding more liquid and an emulsifier and reducing the amount of milk fat, makes it ideal for use in sandwiches and on crackers.

Imitation Cheese. Cheese analogues or imitation cheeses are cheese-like products in which the milk fat in natural cheese has been replaced with vegetable oil. These analogues are less expensive than natural cheese and are manufactured using a process similar to that used to make processed cheese. Milk proteins such as calcium caseinate are mixed with a small amount of vegetable fat, water, salt, emulsifiers, and lactic acid before being heated to pasteurization temperatures for several minutes. The liquid is then poured into molds or formed into slices. The texture, flavor, and melting properties of imitation cheeses are similar to processed cheese. Nutritionally, these analogues are lower in cholesterol and sodium, but equivalent in fat, although it is less saturated than that from natural cheese (25).

Tofu. Tofu is the cheese made from soy milk. The protein in soy milk is precipitated out through the use of coagulants such as calcium sulfate dihydrate (increases volume), calcium chloride (increases firmness and calcium content), or magnesium chloride (20). After the soy milk coagulates and the liquid is removed, the curds are pressed, and these blocks of fresh tofu are sold in plastic containers filled with water, in aseptic packages with no water, or in bulk by selling several blocks together in buckets of fresh water. When refrigerated, tofu is relatively stable. Once opened, draining off the water prior to and during refrigeration and replacing it with fresh water increases its shelf life.

Tofu has a bland flavor, and its texture varies in consistency and firmness. It is available in smoked, dried, seasoned, and sweetened varieties. Tofu can be used in stir-fry dishes, sandwiches, soups, casseroles, and egg or meat dishes.

Purchasing Cheese

Grading

Not all cheeses are graded according to the USDA-defined U.S. Grades AA, A, B, and C (Figure 12-8), but those that are graded are evaluated based on their variety, flavor, texture, finish, color, and appearance. Exceptions to these criteria are Colby cheese, in which color is not considered, and Swiss cheese, which is graded additionally for its salt level and eyes (holes) (7). Table 12-1 lists the Federal standards for the maximum moisture and minimum milk fat content of various cheeses. U.S. grades have not been established for processed cheese products. A Quality Approved inspection shield on the label means only that the cheese meets minimum quality standards and has been produced in a plant meeting USDA sanitary standards. An imitation cheese is defined as one that looks and tastes like the one it is intended to replace but is nutritionally inferior. Substitute cheeses resemble the traditional product and meet the nutritional equivalency comparisons (22).

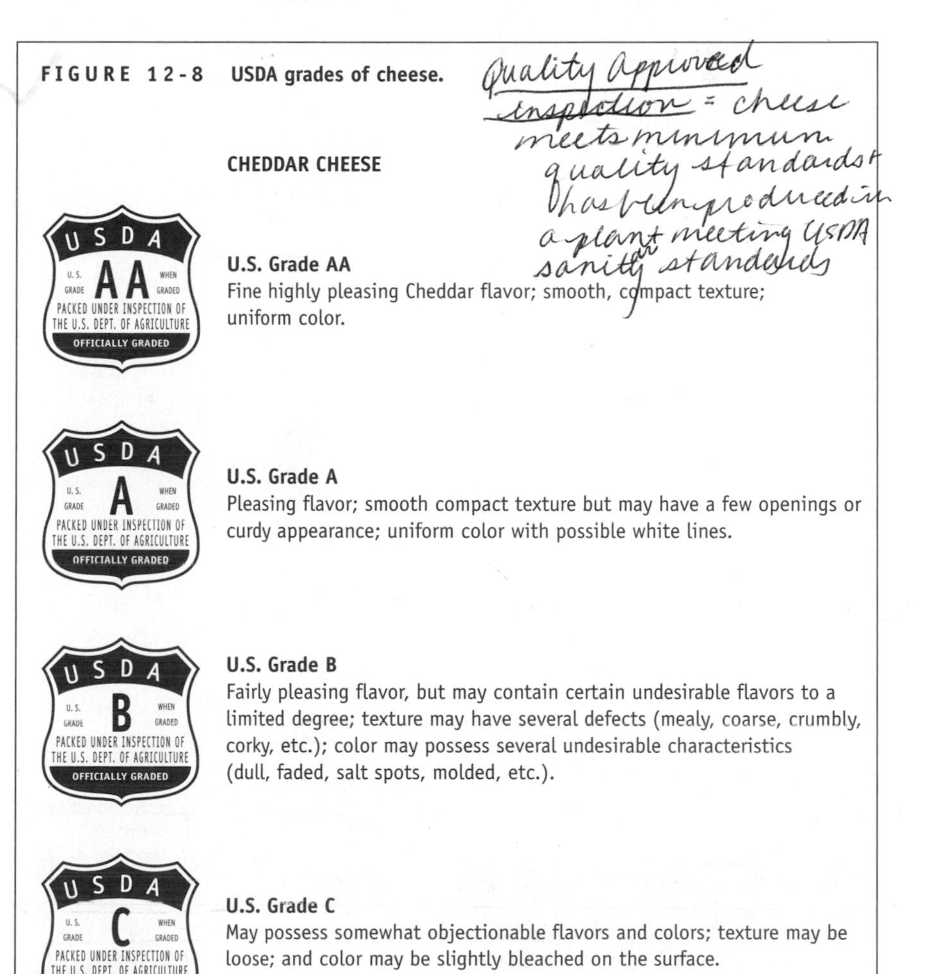

FIGURE 12-8 USDA grades of cheese.

TABLE 12-1

Federal Standards for Maximum Moisture and Minimum Milk Fat in Cheese and Cheese Products

Cheese	Maximum Moisture %	Minimum Milk Fat %
Soft, unripened		
Cottage (dry curd)	80	—
Cottage	80	4
Cottage (lowfat)	82.5	0.5
Cream	55	33
Neufchatel	65	20–33
Soft, ripened		
Brie	—	27.7 ± 1.8
Camembert	—	24.3 ± 0.6
Limburger	50	25
Semisoft, ripened		
Asiago (fresh)	45	27.5
Blue	46	27
Brick	44	28
Gorgonzola	42	29
Monterey	44	28
Muenster	46	27
Roquefort	45	27.5
Hard, ripened		
Cheddar	39	30.5
Colby	40	30
Edam	45	22
Gouda	45	25.3
Gruyere	39	27.5
Samsoe	41	26.5
Swiss	41	25.4
Very hard, ripened		
Asiago (old)	32	28.6
Parmesan	32	21.8
Romano	34	25.1
Pasta filata (plastic)		
Caciocavallo siciliano	40	25.2
Mozzarella	52–60	18
(low moisture, part skim)	45–52	14
Provolone	45	24.8
Skim milk or lowfat		
Gammelost	52	—
Sapsago	38	—
Pasteurized process		
Cheese	43	27
Pimiento cheese	41	29
Cheese food	44	23
Cheese spread	44–60	20
Cold-pack		
Cold-pack cheese	42	27
Cold-pack cheese food	44	13

Food Preparation with Cheese

Cheese is most often used as an ingredient to add flavor, color, and texture in a variety of ways—on pizza, as a taco topping, and in cheese soufflés, sandwiches, casseroles, quiches, and sauces. Mozzarella, Parmesan, and ricotta cheeses are often found in Italian dishes, feta in Greek dishes, Monterey Jack in Tex-Mex dishes, and cheddar in North American food dishes. Numerous dishes incorporate various cheeses; the most popular cheeses purchased in the United States for these purposes, in descending order, are cheddar, processed, mozarella, cream cheese, ricotta, Swiss, provolone, Muenster, Parmesan, Neufchatel, and blue.

The two most important principles when preparing foods with cheese are to select the best cheese and to keep temperatures low and heating times short.

Selecting a Cheese

The chemical composition of a cheese determines its functional properties and dictates how it will be used in food preparation. Some of these functional properties include shredability, meltability, oiling off, blistering, browning, and strechability (7).

Shredability. Not all cheeses shred uniformly, which may be important to food service operations interested in cost control. On average, 4 ounces of shredded cheese are equivalent to approximately 1 cup in volume. It may be possible to use less when using aged cheeses because of their stronger flavor. Grating or chopping cheese increases the surface area and thereby increases the ease and speed with which cheese melts.

Meltability. The higher the fat and moisture content of a cheese, the greater its meltability (7). Aged or ripened cheeses (such as cheddar and Swiss) tend to melt and blend more easily when heated than the less-ripened cheeses. Processed cheeses, which contain added water and emulsifiers, can be heated without the fat separating, blend more smoothly, and melt more easily than natural cheeses. Problems may result when using lower-fat cheeses, because they separate more easily when exposed to high heat, and their higher protein content makes them toughen as they are heated. For these reasons, they are not always good candidates for use in cooking.

Oiling Off. One drawback to using higher-fat cheeses is their greater tendency to "oil off," which occurs when some free fat is released and glistens on the surface.

HOW & WHY? ?????????

Why is mozzarella cheese used on pizza? Mozzarella cheese does not tend to oil off during cooking. Since many people find a shiny sheen of oil on pizza unappetizing, this makes mozzarella the optimum cheese for pizza production.

Blistering. Blistering is another unsightly side effect, with the number and size of the blisters depending on the cheese's age: large blisters tend to form when using excessively aged cheese, while numerous, small blisters may be a sign that the cheese has not been aged very long.

Browning. The browning of cheeses during heating, a result of the Maillard reaction, is desirable, but only up to a certain point. Too much browning occurs if there is an excess of sugars, amino acids, or lactose in the cheese.

Stretchability. The functionality of a cheese also differs based on how well it stretches. The stretchability of a cheese depends on its concentration of calcium phosphate and its protein network structure (16). A tough, grainy texture results from the presence of too much calcium, while texture turns excessively soft when undergoing too much protein breakdown during aging (7).

Temperatures

Cooking temperatures for cheeses should be kept low and heating times short. High heat or prolonged cooking toughens cheese proteins and causes the fat to separate out, creating an oily, stringy, and inferior product. When using a microwave, it is best to use lower power settings—between 30 and 70 percent—for melting cheese. Cheeses used in sauces and other dishes should be added during the last stages of preparation to prevent separation. Adding a pinch of dry mustard to cheese sauces helps to bring out their flavor. One way to soften cream cheese without too much heat is to enclose it in an airtight zip-top plastic bag and briefly submerge it in hot water.

Temperature is also important when serving cheese. Most cheeses (semihard and hard) reach their full flavor when taken out of the refrigerator and allowed to reach room temperature before serving. To prevent cheese from drying out, it is best to either cut cheese that has reached room temperature just before serving or to let people cut their own. Cream cheese, cottage cheese, and other unripened cheeses should, however, always be served chilled.

Storage of Cheese

Cheese must be stored properly to prevent deterioration. Most cheeses should be refrigerated; some can even be frozen. Processed cheese products can be stored in a cool, preferably dark, cupboard until ready for use, though refrigeration more effectively retains desirable qualities.

Dry Storage

Many process-cheese spreads, as well as dry, grated Parmesan cheese sold in cardboard containers, have long shelf storage times. Process-cheese spreads may be safely stored on the shelf in jars at room temperature (70°F/5°C) for up to four months, but the quality is better retained if the product is refrigerated. Packaged dry, grated Parmesan cheese, one of the hardest cheeses, and other very hard cheeses can be kept at room temperature for up to one year. The exception is fresh Parmesan cheese, which has such a high moisture content that it is sold in the refrigerator section.

Refrigeration

Most cheeses are best refrigerated in their original wrappers. Once opened, the cheese should be rewrapped as tightly as possible in its original wrapping or in aluminum foil, plastic wrap, or a sealable bag. This will prevent drying and absorption of odors from other foods. Because their odor may be picked up by other foods, strong-smelling cheeses should be double-wrapped.

Properly wrapped cheeses are also protected from the development of molds and their possible mycotoxins (2). Commercial efforts to reduce molds include coating the cheeses with a wax or resinous material and wrapping them in packaging film (30). In Europe, the cheese rind is sometimes coated with olive oil to protect it from bacterial contamination, and certain spices applied to the surface of some cheeses for flavor have been reported to have an antifungal effect (32).

Maximum storage time varies, because no two cheeses are alike. Ripened cheeses can be stored longer than unripened, softer cheeses. Fresh cheeses such as ricotta, cream cheese, and cottage cheese should be used within a week of their sell-by date. Processed cheeses can be stored up to four months because they have been heat treated and contain mold inhibitors. Once opened, processed cheeses should be refrigerated. Unopened Parmesan cheese stores the longest and can be kept up to a year, although the cheese may lose some of its flavor and further dry out. After being opened, Parmesan must be refrigerated. Dried-out cheeses can be salvaged by grating and storing them in the freezer for later use as toppings or in casseroles, sauces, or soups.

HOW & WHY? ?????????

Why should moldy cheeses be discarded? Cheese that is kept too long can become moldy and/or dry. Most molds that develop on cheese are harmless, but since there are some molds that may produce toxins, all mold should be removed. The FDA Model Codes for Food Service recommends cutting 1 inch beyond the moldy area. Although mold-ripened cheeses such as Roquefort and blue have had special molds purposely added to them, they may develop a different-looking mold on the outside edges, and this should be removed. When soft, unripened cheeses develop molds, they should be discarded because of possible permeation throughout the cheese.

Frozen

The water content of a cheese determines whether it can be successfully frozen. Most hard natural cheeses with a low water content can be frozen for up to two months, and processed cheeses for up to five months, but freezing is not recommended for soft cheeses having a high water content (22). The most suitable cheeses for freezing include brick, cheddar, Edam, Gouda, Gruyere, Parmesan, provolone, and Swiss. Freezing will change the texture and flavor to some degree, but the cheese should still be acceptable in quality (24). For best results, cheese should be frozen quickly, and this is best accomplished if it is in ½-pound pieces not more than 1 inch thick. Larger chunks will freeze more slowly, possibly resulting in crumbly cheese. It is best to freeze cheese in its original wrapper, but the next-best option is to use foil or plastic wrap designated for freezing. It should be wrapped tightly, with excess air being expelled, or else it can dry out.

Just as it is important to freeze quickly, so is it crucial that thawing be gradual. As a result, thawing is best done in the refrigerator over a period of a few days, after which the cheese should be used as soon as possible. Freezing certain cheeses may cause them to develop a dry and crumbly texture, but they may still serve as a shredded or cubed ingredient in a dish even if they would be undesirable in a sandwich or on crackers.

PICTORIAL SUMMARY / 12: CHEESE

Cheese is a preserved food made from the curd, or solid portion, of milk. One of the most nutrient-dense foods, cheese is most often used as an ingredient to add flavor, color, and texture to prepared foods.

CLASSIFICATION OF CHEESES

Cheeses are most commonly classified by their moisture content (fresh, soft, semi-hard, hard, and very hard), processing method (process cheeses, whey products, and texture), and milk source (cow, sheep, goat, etc.). Cheese is primarily composed of water, fat, and protein. One ounce of cheese averages:

- **7 grams of protein.** High-quality protein equivalent to that found in 1 cup of milk or 1 ounce of meat.
- **100 calories** (kcal).
- **9 grams of fat.** A low-fat cheese is defined by the American Heart Association as one containing no more than 6 grams of fat per ounce.
- **0 carbohydrates.**

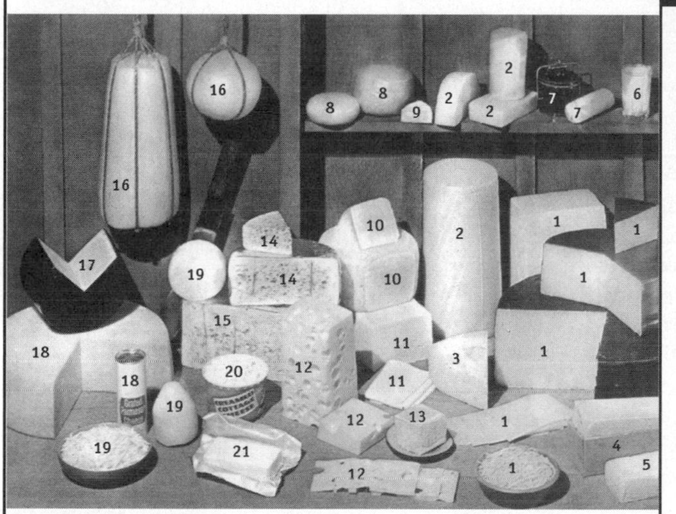

Common cheeses

1. Cheddar
2. Colby
3. Monterey Jack
4. Pasteurized Process Cheese
5. Cheese Foods
6. Cheese Spreads
7. Cold Pack Cheese Food or Club Cheese
8. Gouda and Edam
9. Camembert
10. Muenster
11. Brick
12. Swiss
13. Limburger
14. Blue
15. Gorgonzola
16. Provolone
17. Romano
18. Parmesan
19. Mozzarella and Scamorze
20. Cottage Cheese

CHEESE PRODUCTION

The basic steps of cheese manufacture include milk selection, coagulation (enzyme and/or acid), curd treatment (cutting, heating, salting, knitting, and pressing), and curing and ripening. Other products from cheese manufacture are whey products and processed cheeses. Tofu is the "cheese" made from soy milk.

PURCHASING CHEESE

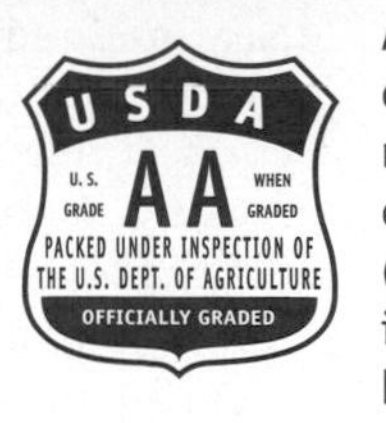

A USDA Quality Approved inspection shield on the label means that the cheese meets minimum standards of flavor, texture, finish, color, and appearance. Of four USDA grades, (AA, A, B, and C), Grade AA is the best. It is not mandatory, however, and U.S. grades have not been established for processed cheese products. The most popular cheese in the United States is cheddar.

FOOD PREPARATION WITH CHEESE

Cheese can be consumed "as is" or as an ingredient in casseroles, pizza, soufflés, soups, salads, omelets, eggs, and other dishes. Selecting a cheese for preparing a food depends on its functional factors of shredability, meltability, oiling off, blistering, browning, and stretchability. Ripened cheeses heat better than soft cheeses. Temperatures should be kept low, and heating time of cheeses should be short.

STORAGE OF CHEESE

Most cheeses are best refrigerated in their original wrappers. Freezing is not recommended for soft, high-water-content cheeses, but hard cheeses can be frozen for up to two months. Processed cheese spreads packaged in jars are commonly stored on a cupboard shelf at room temperature for up to four months, but once opened, they must be refrigerated.

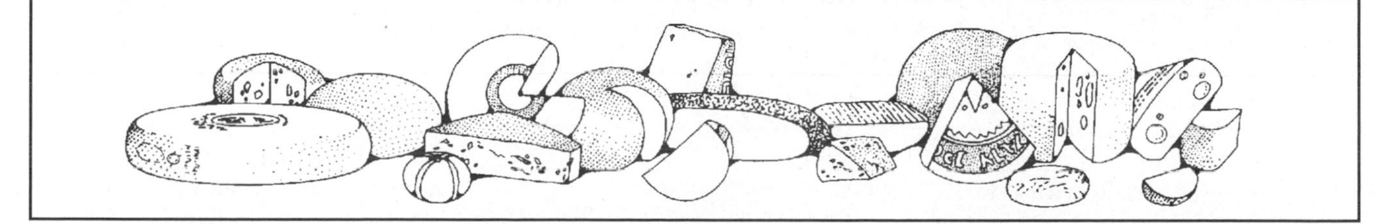

REFERENCES

1. American Heart Association. *The Heart and Stroke Guide.* American Heart Association, 1999.
2. Barrios MU, et al. Aflatoxin-producing strains of *Aspergillus flavus* isolated from cheese. *Journal of Food Protection* 60(2):192–194, 1997.
3. Blaesing D. Cheese spreads it on thick. *Prepared Foods* 166(5):70–71, 1997.
4. Branger EB, et al. Sensory characteristics of cottage cheese whey and grapefruit juice blends and changes during processing. *Journal of Food Science* 64(1):180–184, 1999.
5. Breidt F, and HP Fleming. Using lactic acid bacteria to improve the safety of minimally processed fruits and vegetables. *Food Technology* 51(9):44–48, 1997.
6. Burrington KJ. Food product developement. Winning wheys. *Prepared Foods* 167(7):83–89, 1998.
7. Chandan R. *Dairy-Based Ingredients.* Eagen Press, 1997.
8. Cowart VS, and P Gunby. Heart Association to endorse certain foods. *Journal of the American Medical Association* 260:1192, 1988.
9. Drake MA, et al. Sensory evaluation of reduced-fat cheeses. *Journal of Food Science* 60(5):898–901, 1995.
10. Farkye NY, and CF Landkammer. Milk plasmin activity influence on cheddar cheese quality during ripening. *Journal of Food Science* 57(3):622, 1992.
11. Farkye NY. Contribution of milk-clotting enzymes and plasmin to cheese ripening. In *Chemistry of Structure-Function Relationships in Cheese,* eds. EL Malin and MH Tunick. Plenum, 1995.
12. Fox PF, TK Singh, and PLH McSweeney. Biogenesis of flavour compounds in cheese. In *Chemistry of Structure-Function Relationships in Cheese,* eds. EL Malin and MH Tunick. Plenum, 1995.
13. Habibi-Najafi MB, and BH Lee. Bitterness in cheese: A review. *Critical Reviews in Food Science and Nutrition* 36(5):397–411, 1996.
14. Holsinger VH. Nutritional aspects of reduced-fat cheese. In *Chemistry of Structure-Function Relationships in Cheese,* eds. EL Malin and MH Tunick. Plenum, 1995.
15. Lee SJ, IJ Jeon, and LH Harbers. Near-infrared reflectance spectroscopy for rapid analysis of curds during cheddar cheese making. *Journal of Food Science* 62(1):53–56, 1997.
16. Ma L, et al. Rheology of full-fat and low-fat cheddar cheeses as related to type of fat mimetic. *Journal of Food Science* 62(4):748–752, 1997.
17. Malin EL, and EM Brown. Influence of casein peptide conformations on textural properties of cheese. In *Chemistry of Structure-Function Relationships in Cheese,* eds. EL Malin and MH Tunick. Plenum, 1995.
18. Malin EL, et al. Inhibition of proteolysis in mozzarella cheese prepared from homogenized milk. In *Chemistry of Structure-Function Relationships in Cheese,* eds. EL Malin and MH Tunick. Plenum, 1995.
19. Misty VV. Improving the sensory characteristics of reduced-fat cheese. In *Chemistry of Structure-Function Relationships in Cheese,* eds. EL Malin and MH Tunick. Plenum, 1995.
20. Murphy PA, et al. Soybean protein composition and tofu quality. *Food Technology* 51(3):86–88, 1997.
21. Nonfat dairy products perform in cheesecake, pizza topping. *Food Engineering* 64(9):22, 1992.
22. Olson NF. Newer knowledge of cheese and other cheese products. *National Dairy Council,* 1992.
23. Pszczola DE. Say cheese with new ingredient developments. *Food Technology* 55(12):55–64, 2001.
24. Pearl AM, C Cuttle, and BB Deskins. *Completely Cheese: The Cheeselover's Companion.* Jonathan David, 1978.
25. Potter NN, and JH Hotchkiss. *Food Science.* Chapman & Hall, 1995.
26. Rosenberg M, et al. Viscoelastic property changes in cheddar cheese during ripening. *Journal of Food Science* 60(3):640–644, 1995.
27. Ruiz LP. Technology for healthier dairy foods. *Food Engineering* 61(12):57, 1989.
28. Scrimshaw NS, and EB Murray. The acceptability of milk and milk products in populations with a high prevalence of lactose intolerance. *American Journal of Clinical Nutrition* 48(Supp):1083–1147, 1988.
29. Steele JL. Contribution of lactic acid bacteria to cheese ripening. In *Chemistry of Structure-Function Relationships in Cheese,* eds. EL Malin and MH Tunick. Plenum, 1995.
30. Tarlov R. Discover the pleasures of a cheese course. *Fine Cooking* 25:56–59, 1998.
31. Trepanier G, et al. Accelerated maturation of cheddar cheese: Microbiology of cheeses supplemented with *Lactobacillus casei* subsp. *casei L2A. Journal of Food Science* 57(2):345–349, 1992.
32. Wendorff WL, and C Wee. Effect of smoke and spice oils on growth of molds on oil-coated cheeses. *Journal of Food Protection* 60(2):153–156, 1997.

WEBSITES

A classic website for cheese information that includes a cheese library:
www.cheesenet.com

This website lists cheeses by names, countries, and milk. It also lists vegetarian cheeses, cheeses in alphabetical order, cheese facts, and cheese recipes.
www.cheese.com

The American Cheese Society provides cheese standards, educational resources, and product promotion:
www.cheesesociety.org

13

Eggs

Bird eggs, long honored as symbols of fertility and life, have been part of our diet for thousands of years. All bird eggs are edible and highly nutritious, and neatly packaged in their own shells. An egg contains everything needed to sustain the life of a new chick.

The egg's nutritious contents makes it susceptible to predation by other animals seeking food. Eggs are not always available or easy to find, so Americans solved this problem by breeding almost as many laying hens as there are people in the United States. Each of these laying hens produces about one egg per day.

Eggs are one of the most versatile prepared foods. In their most basic form, they can be cooked and eaten on their own—fried, scrambled, poached, and boiled. The unique physical and chemical properties of eggs also make them an invaluable ingredient in many prepared dishes. Just a few examples of how eggs are used in food preparation are listed below:

- Giving a foam structure to cakes and meringues
- Thickening custards and puddings
- Adding color to lemon meringue pie and eggnog
- Emulsifying mayonnaise and hollandaise sauce
- Leavening soufflés and popovers
- Binding ingredients in meatloafs and casseroles
- Coating foods prior to breading
- Glazing pastries and breads
- Clarifying liquids for soups

A solid understanding of these various roles that eggs play in food preparation requires a general knowledge of eggs. While all bird eggs are edible, this chapter is limited to those that are commonly consumed: chicken eggs—their composition, purchase, functions in foods, preparation, and storage.

Composition of Eggs

Structure

The egg has five major components: the yolk, albumen (egg white), shell membranes, air cell, and shell (Figure 13-1). Each of these plays an important role in the egg's unique attributes that make it invaluable in food preparation.

Yolk. The sunny yellow yolk situated in the center of the egg constitutes about a third (30 percent) of the egg's

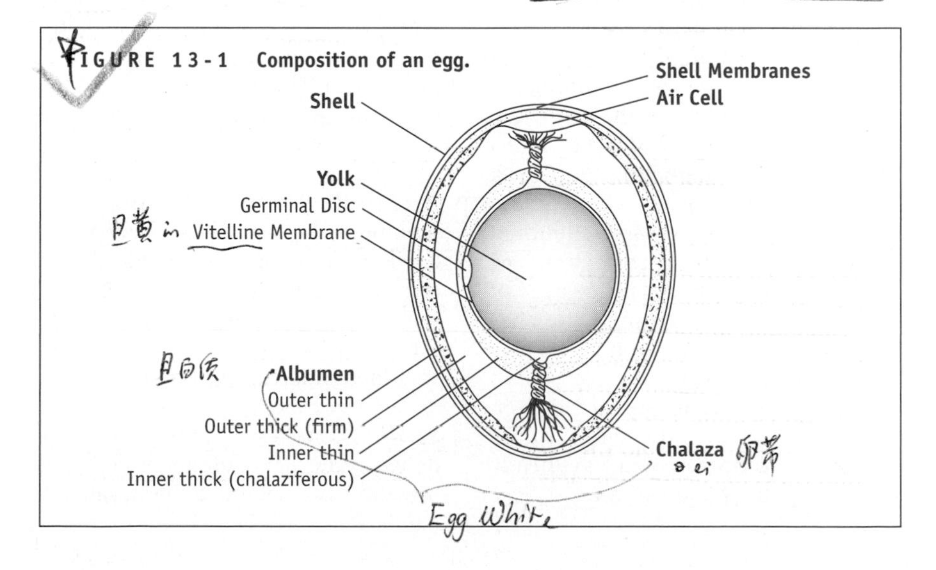

FIGURE 13-1 Composition of an egg.

KEY TERMS

Chalaza (pl. chalazae) The ropy, twisted strands of albumen that anchor the yolk to the center of the thick egg white.

Vitelline membrane The membrane surrounding the egg yolk and attached to the chalazae.

Cuticle (bloom) A waxy coating on an eggshell that seals the pores from bacterial contamination and moisture loss.

weight. Dense in nutrients, the yolk serves to nourish the chick. A white, pinhead-sized germinal disc sits on the surface of the yolk. This appears darker if the egg has been fertilized, but has no effect on the taste, functional properties, or nutritional value of the egg. Most eggs are screened to ensure they are not fertilized, but even if they do pass inspection, fertilized eggs do not develop once refrigerated. The color of the yolk depends on the hen's diet. Colors ranging from pale yellow to deep red are caused by pigments in the chicken feed, such as beta-carotene. Natural yellow-orange substances such as marigold petals are sometimes added to the chickens' feed to enhance the color of their egg yolks. Artificial color additives are not permitted.

Chemist's Corner 13-1

Egg Proteins

Eggs, and especially the egg white, are composed of dozens of different proteins. Each of these proteins has its own characteristics and functions (Table 13-1). More than half of all the protein in eggs is ovalbumin. It plays an important role in egg preparation because it gels well and denatures easily when heated (6, 19). The next two proteins of any significance are ovotransferrin (conalbumin) and ovomucoid. Ovotransferrin helps to protect against bacterial contamination by forming a complex with iron, thereby making it unavailable to bacteria, which need iron for growth. Ovomucoid proteins, which are resistant to denaturation, inhibit the activity of trypsin, an enzyme that breaks down specific proteins. Lysozyme, an enzyme, fights certain bacteria by lysing (breaking chemical bonds by the addition of water) the cell wall of the bacterium. Ovomucin helps protect against microbial damage by reacting with viruses. Avidin in uncooked eggs binds to the B vitamin biotin and prevents it from being absorbed by the small intestine, which is why excessive raw egg white consumption is discouraged. However, once cooked, the avidin denatures and no longer inhibits the absorption of biotin. About 3 grams of protein are also found in the egg yolk in the form of lipoproteins—lipovitellin and lipovitellinin—which act as emulsifiers in food preparation.

TABLE 13-1
Protein Content of Egg White

Protein	*Amount (%)*	*Properties*
Ovalbumin	54	Denatures easily
Conalbumin	13	Antimicrobial, complexes iron
Ovomucoid	11	Inhibits enzyme (trypsin)
Unidentified	8	Mainly globulins
Lysozyme	3.5	Antimicrobial
Ovomucin	1.5	Viscous, reacts with viruses
Flavoprotein	0.8	Binds riboflavin
Proteinase inhibitor	0.1	Inhibits enzyme (bacterial proteinase)
Avidin	0.05	Binds the B vitamin Biotin (raw egg white only)

Albumen. The albumen, or egg white, accounts for almost three-fifths (58 percent) of an egg's weight and is made up largely of water and protein (Chemist's Corner 13-1). Although it appears to be one mass, the egg white is actually constructed of layers differing in viscosity, alternating from thick to thin. Differences in viscosity are determined by the type and amount of proteins in various parts of the egg white. Around the yolk is a layer of thick protein called albumin. The **chalazae** (ka-lay-zee) at the top and bottom of the egg anchor the egg yolk in the thick egg white surrounding it. They also secure the yolk to its **vitelline membrane** so it stays neatly centered in the middle of the egg.

Shell Membranes. Between the egg white and the shell are two membranes, an inner and outer shell membrane. These press up against the shell and protect the egg against bacterial invasion.

Air Cell. Between the two shell membranes at the larger end of the egg is a pocket of air known as the air cell. As a freshly laid egg cools, its contents contract, causing the inner shell membrane to separate from the outer shell membrane, forming the air cell.

Shell. Nature's way of protecting the delicate internal contents of an egg is to surround it with a hard calcium carbonate shell (12 percent of an egg's weight). Egg shells are not solid but contain thousands of small pores, allowing an exchange of gases between the inner egg and the surrounding air. Shell color indicates the breed of the hen, but has no bearing on the nutrient content or taste of the egg. Brown eggs tend to cost more because they usually come from larger hens, which require more food and produce fewer eggs.

The shell is protected by the **cuticle** or bloom. Commercially sold eggs are washed, which removes this protective cuticle. Producers compensate for this loss by applying a thin coat of edible oil on the shell (11).

NUTRIENT CONTENT

Eggs are an excellent source of many nutrients—protein, fat, vitamins, minerals, and water—although they contain very little carbohydrate or fiber. An average large egg provides about 75 calories (kcal), primarily from fat and protein.

Protein. One large egg contains approximately 7 grams of complete protein—4 grams from the white and 3 from the yolk. As a life-sustaining protein, the quality of protein in eggs is so high that it has become the standard by which researchers rate all other foods.

Fat. One egg yolk contains approximately 5 grams of fat, predominantly in the form of triglycerides, phospholipids, and cholesterol. The fat is approximately 47 percent monounsaturated (2 grams), 37 percent saturated (slightly under 2 grams), and 16 percent polyunsaturated (slightly under 1 gram).

Cholesterol. Egg yolks are high in cholesterol, averaging 213 mg in a large egg (8). Dietary cholesterol has been reported by some researchers to increase blood cholesterol. As a result, the American Heart Association recommends no more than 100 mg of dietary cholesterol be consumed for every thousand calories (kcal), which generally limits egg consumption to no more than four egg yolks per week. Despite the dietary focus on lowering fat and cholesterol, eggs are not always easily replaced in recipes, and consumer surveys show that the amount of fat often determines whether or not a food is acceptable (3, 10).

Commercial food processors have also tried various methods to reduce the amount of cholesterol found in foods. One California poultry farmer developed a special chicken feed, using sea kelp, to reduce cholesterol content in large Grade AA eggs to only 125 mg each as compared with 213 mg in an average egg, but it also resulted in high iodine levels (7).

Vitamins and Minerals. Eggs are rich in certain vitamins and minerals. They are one of the few foods containing all the fat-soluble vitamins—A, D, E, and K. An egg yolk provides at least 10 percent of the RDI (Reference Daily Intake) for each of these fat-soluble vitamins. Egg yolks are one of the few foods that naturally contain vitamin D. Certain water-soluble B vitamins, found primarily in the white, are also supplied by the egg in relatively high RDI percentages: 25 percent of vitamin B_{12}, 15 percent each of riboflavin (vitamin B_2) and biotin, 12 percent of folate, and 11 percent of pantothenic acid.

In addition, eggs are a source of several minerals, especially selenium, iodine, zinc, iron, and copper. Unfortunately, the iron in egg yolks is not very available because it binds to phosvitin, an egg protein that inhibits absorption. When eggs are overcooked, the iron in the yolk interacts with the small amount of sulfur found in the egg white to produce ferrous sulfide, which causes a characteristic strong off-odor.

KEY TERMS

Candling A method of determining egg quality based on observing eggs against a light.

Purchasing Eggs

Inspection

The Egg Products Inspection Act of 1970 requires that egg processing plants be inspected and that their eggs and egg products be wholesome, unadulterated, and truthfully labeled. This law is enforced by the USDA Poultry Division and applies to all eggs, whether imported or shipped intra- or interstate.

Eggs Failing USDA Inspection. Eggs that do not pass inspection, called restricted eggs, are not allowed to be sold whole to the consumer. Examples of restricted eggs are ones with cracked shells, called "checks"; "leakers" have cracked shells as well as broken membranes; "dirties" have at least one-fourth of their shell covered with dirt or stain; "inedibles" have greenish egg whites or are fertilized, rotten, moldy, or bloody. Less than 1 percent of eggs have small blood spots, also known as "meat spots." Although these are harmless, the eggs containing them are removed when detected by electronic blood detectors used during grading. Inspection is intended to ensure that eggs with defects such as these are not sold to the consumer.

Grading

Once eggs pass inspection, a producer can pay the USDA to have them graded for quality. The best-quality eggs are graded USDA Grade AA, followed by USDA Grade A. USDA Grade B, the lowest grade, is available to food service establishments and not sold directly to consumers. Grade AA and A eggs are the grades sold at supermarkets. Their firm, high albumens and yolks make them suitable for frying, coddling, poaching, and hard cooking. Grade B eggs have thinner whites and somewhat flattened yolks, making them better for scrambling or baking, or as ingredients in other food products.

Grading Methods. Grades are determined by graders who incorporate three methods to judge the quality of eggs: **candling**, measuring Haugh (pronounced "how") units, and evaluating appearance, specifically that of the shell, white, yolk, and air cell. Commercial egg grades are assigned based on both interior and exterior quality. Interior quality is primarily determined by candling. The exterior quality is determined by the cleanliness of the shell, the shape of the egg, and the presence of shell irregularities such as pimples (calcium deposits) and weak shells.

Candling. As the name implies, the original method of candling involved holding an egg up to the light of a candle to view its contents. Although it is still a good way to determine an egg's inner quality and to detect defects, eggs are now mechanically rotated over lights, many at a time, by rollers (Figure 13-2).

Candling is based on the principle that eggs start to deteriorate the minute they are laid, and these changes can be vaguely seen through an egg held against a light. The whites become more thin and transparent as carbon dioxide departs through the shell; and the yolk membrane stretches as the yolk absorbs water from the white, which eventually will cause the yolk to break easily when the shell is cracked.

HOW & WHY? ?????????

How does candling reveal if an egg is fresh or aged? The yolk in a fresh, high-quality egg is suspended tightly by the chalazae, seen in candling only as a slight shadow. The yolks in older eggs, on the other hand, are surrounded by thinning, clearer egg whites and deteriorating chalazae. These older yolks lie closer to the shell because they are no longer suspended well by the chalazae; they are looser in consistency, and cast a darker shadow. The egg's air cell, too, becomes wider or moves as the egg ages. Grade AA eggs must have an air cell depth smaller than $\frac{1}{8}$ inch; Grade A eggs are limited to $\frac{3}{16}$ inch; and Grade B eggs have no limit to the size of their air cell.

Haugh Units. The freshness of an egg can be detected by cracking it open onto a flat surface and looking at the height of its thick albumen. Fresh egg whites sit up tall and firm, while older ones tend to spread out. Professional graders cannot evaluate every egg for freshness, so an egg is randomly selected and measured using a special instrument called a micrometer (Figure 13-3). The Haugh unit, a numerical value reflecting an egg's freshness, is obtained by mathematically combining the thick albumen height with the egg's weight, and then using a formula or table to convert this number into a Haugh unit. As Haugh units decrease, so does egg quality, and this is reflected in grading: Grade AA is given to eggs with a Haugh unit of 72 or higher, Grade A for a measure of 60 to 71, and Grade B for a measure of 31 to 59.

Appearance. Grading can also be based on the appearance of eggs broken on a flat surface. Graders evaluate the quality of an egg by observing the thickness of the albumen, the prominence of the chalazae, the roundness and firmness of the yolk, and the shape, cleanliness, and texture of the shell. Figure 13-4 summarizes the characteristics of USDA Grades AA, A, and B eggs. The vast majority of eggs are graded on interior quality by candling. Grading of broken-out eggs is reserved for research purposes or random spot checks of candled eggs.

Sizing

Sizing is not related to grading in any way. Eggs are sold in cartons by various sizes determined by a minimum weight for a dozen eggs in their shell (Figure 13-5). Unless otherwise designated, recipes are based on the use of large eggs. The contents of one large egg are equivalent to approximately $\frac{1}{4}$ cup. Table 13-2 on page 238 shows that four to six eggs will fill 1 cup, depending on the size of the egg. Table 13-3 shows how eggs of other sizes are used when large eggs are not available.

Egg Substitutes

Consumer demand for products lower in cholesterol has created a market for liquid egg substitutes, made by either omitting egg yolks, replacing egg yolks with vegetable oils, or removing some of the cholesterol in egg yolks (17). One drawback to egg substitutes is that they may be higher in sodium than regular eggs.

USDA

FIGURE 13-2 Candling: Eggs are automatically rolled against a light background during mass scanning, allowing checkers to remove those that have defects or that are cracked, soiled, or damaged.

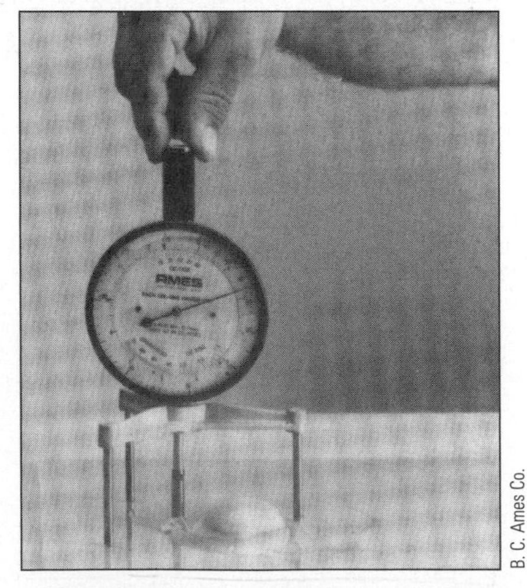

B. C. Ames Co.

FIGURE 13-3 Measuring Haugh units to determine egg quality.

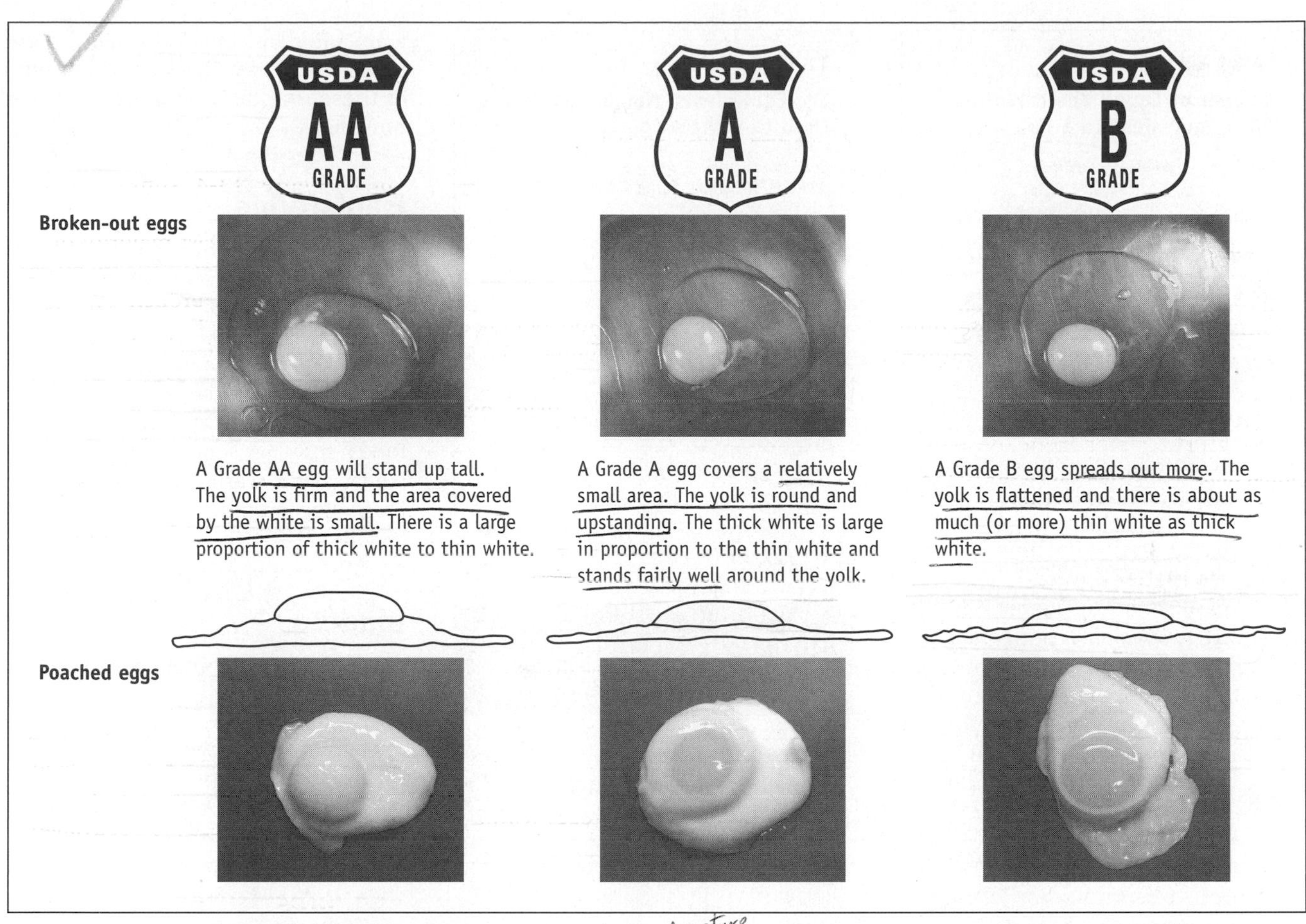

FIGURE 13-4 USDA grades for eggs.

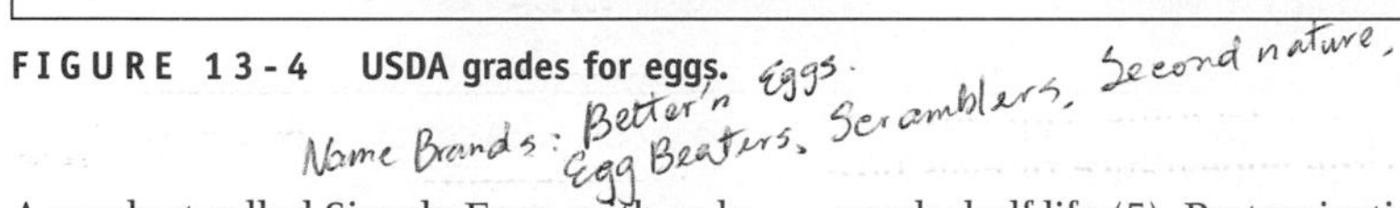

A product called Simply Eggs, with only 45 mg of cholesterol per egg equivalent, has a sodium content of 120 mg, compared with 70 mg of sodium in a fresh egg. On the positive side, egg substitutes are ultrapasteurized, packaged, and then refrigerated, preparing them for a ten-week shelf life (5). Pasteurization makes them ideal for use in uncooked foods like eggnog, mayonnaise, and ice cream when raw eggs would pose a food safety risk. The ingredients and cholesterol contents of several brands of egg substitutes are listed in Table 13-4.

Value-Added Eggs

There is a new type of egg on the market that takes into consideration the health of the consumer and that of the laying hen. "Value-added" eggs have special attributes due to their

FIGURE 13-5 Egg sizes determined by minimum weight (including shells) per dozen.

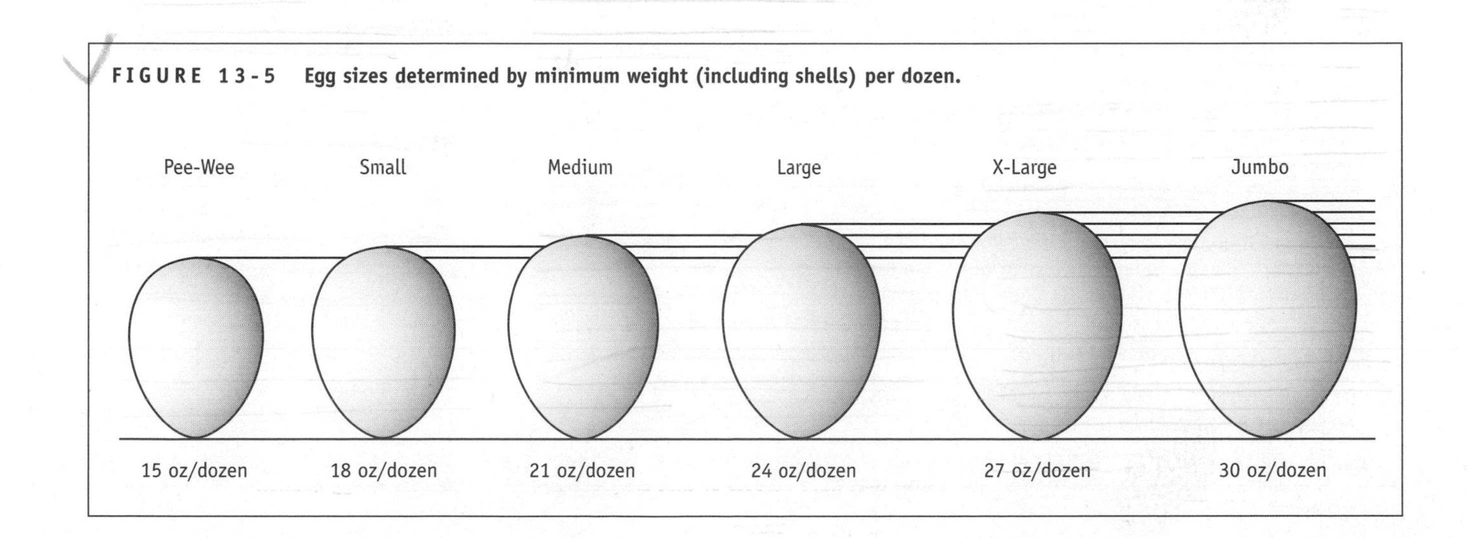

TABLE 13-2
Number of Eggs (Whole/Whites/Yolk) Equivalent to 1 Cup

Egg Size	*Whole*	*Whites*	*Yolks*
Jumbo	4	5	11
X-Large	4	6	12
Large	5	7	14
Medium	5	8	16
Small	6	9	18

nutrient content or conditions under which the hens are raised. For instance, lower-cholesterol eggs are now on the shelf along with those that have higher omega-3 fatty acid levels or 300% more vitamin E levels, achieved by special feeding practices. In addition, value-added eggs include those that have been laid by hens that are free-roaming, cage-free, or fed natural grains without animal products—a plus for animal-friendly egg eaters (15).

TABLE 13-3
Egg Equivalents: Number of Eggs Used to Replace Eggs of Other Sizes

Equivalent Number of Eggs				
Jumbo	*X-Large*	*Large*	*Medium*	*Small*
1	1	1	1	1
2	2	2	2	3
2	3	3	3	4
3	4	4	5	5
4	4	5	6	7
5	5	6	7	8
9	10	12	13	15
18	21	24	27	28
37	44	50	56	62

Functions of Eggs in Foods

Eggs are often combined with other ingredients. Their unique ability to flavor, color, emulsify or thicken, bind, foam, interfere, and clarify makes them nearly indispensable in cooking. Some of these unique qualities of eggs are now discussed.

Emulsifying

Lecithin found in the egg yolks is a natural emulsifying agent: one end of the molecule attracts water, while the other end is drawn to fat (1). Eggs thus help keep fat and water or other liquid compounds from separating, so they are often used to thicken and stabilize foods such as salad dressings, hollandaise and bearnaise sauces, mayonnaise, ice cream, cream puffs, and certain cakes.

Binding

The high protein content of eggs makes them excellent binders. Fish, chicken, vegetables, and other foods are often dipped in beaten egg and then rolled in breading, batter, flour, or cereal. During cooking, heat coagulates the eggs'

TABLE 13-4
Egg Substitutes Compared to Standard Egg

	Kcal	*Fat (gm)*	*Chol* (mg)*	*Protein (gm)*	*Sodium (mg)*
Better'n Eggs (55 gms) 98% egg whites, corn oil, water, natural flavors, sodium hexametaphosphate, calcium sulfate, guar gum, xantham gum, nisan, citric acid, hydrolyzed wheat gluten, and beta carotene.	20	0	0	5	100
Egg Beaters (61 gms) 99% egg white, less than 1%: vegetable gums (xanthan gum, guar gum), color (includes beta carotene), salt, onion powder, natural flavor, spices.	30	0	0	6	115
Scramblers (57 gms) Egg whites, nonfat milk, calcium caseinate, modified corn starch, natural and artificial flavors, corn oil, salt, mono- and diglycerides, beta carotene, vitamins and minerals.	35	0	0	6	95
Second Nature (60 gms) 99% egg whites, less than 1% of the following: magnesium chloride, sodium stearoyl lactylate, xanthan gum, annatto (color), vitamins & minerals (calcium, iron, vitamin A).	30	0	0	7	110
Simply Eggs (50 gms) Reduced cholesterol liquid whole eggs, salt, citric acid, calcium, vitamin A, thiamin.	70	5	45	6	120
Standard one large egg	75	5	215	6	65

*Cholesterol

protein, which then acts as an adhesive, binding the other ingredients to the surfaces of the cooked material. Egg also binds together mixtures such as meat loaf, meatballs, lasagna, and manicotti. When the mixture cooks, the egg proteins firm and stabilize. Too much egg, however, can cause foods to become overly firm and dry.

Foaming

The capacity of egg whites to be beaten into a foam that increases to six or eight times its original volume is invaluable in food preparation. Egg-white foams are used to aerate and leaven a number of food products, such as puffy omelets, soufflés, angel food cake, sponge cake, and meringues.

HOW & WHY? ?????????

Why do beaten egg whites make a stable foam? Vigorous beating or whisking of egg whites breaks the links between protein molecules, causing the protein molecule coil to unwind or become denatured. A foam structure is created when the unfolded proteins rearrange to construct films around the air cells (2). When the airy foam is heated, its air cells further expand, after which the egg proteins coagulate, solidifying the egg protein to create a firm, stable structure, higher in height than the same food made without egg-white foam.

The best eggs to use for an egg-white foam are fresh eggs because they have thick egg whites, which contribute to a stable foam. Older eggs have thinner whites, which beat to a larger volume but are less stable and may collapse during heating. Consistency, regardless of an egg's age, is achieved in some food service establishments by the use of dried egg whites.

The formation and stability of egg white foams also depend upon the beating technique, the temperature, type of bowl, the careful separation of yolks and whites, and whether or not sugar, fluid, salt, or acid have been added. The skillful control of all these factors yields the best possible egg-white foam.

Beating Technique. Egg whites are best beaten with an electric mixer, but a wire whisk or a double-bladed rotary egg beater can also be used. Whichever device is used, the key is to whip the egg whites into very fine, delicate bubbles. Figure 13-6 shows the consistency of the foam at the various stages of foam formation—foamy, soft, stiff, and dry.

At first, beating speed should be moderately slow and even. The foamy stage consists of very large air bubbles, but as the foam thickens, smaller, finer bubbles begin to appear. When the egg whites are half whipped and beginning to hold their shape, the beating speed should be increased to medium or fast.

Testing for Doneness. Testing for doneness consists of stopping, lifting the beaters out of the foam, and seeing how the egg-white peaks form. Initially, the egg-white foam forms soft, shiny peaks that droop over without holding their shape, and the whipped foam slides around in the bowl. Beating should continue until the peaks are stiff and shiny but not dry. Whipping is complete at the stiff peak stage when the peak tips fall over slightly but keep their shape and the whipped foam sticks to the side of the bowl. Perfectly beaten egg whites will pass the inverted bowl test: they will cling to the inside of a bowl turned upside-down.

Avoiding Overwhipping. Excessive beating of egg whites occurs when the peaks stand tall, dry, and are no longer shiny. In addition, the protein films surrounding the air cells rupture, creating bubbles that are too large and unstable. It is important to beat egg whites to the correct stage, because both over- and underbeaten foams will eventually collapse and separate into liquid pools at the bottom of the bowl.

Temperature. The bowl, beater, and eggs should be at room temperature. The decreased surface tension of room-temperature egg whites allows them to whip more easily and to a larger volume than cold eggs.

(a) Foamy

(b) Soft peaks

(c) Stiff peaks

(d) Dry peaks

FIGURE 13-6 Stages of foam formation.

KEY TERMS

Clarify To make or become clear or pure.

However, the stability of the foam deteriorates with continued exposure to room temperature. In addition, leaving eggs out for more than 30 minutes risks *Salmonella* growth. Although not highly recommended, eggs can be briefly submerged in a bowl of warm water before whipping.

Bowl. Deep bowls with rounded bottoms sloping up into the sides are best, because they allow the egg whites to be picked up by the beater. Plastic bowls should be avoided, because their porous surface may harbor a thin film of grease from previous usage, which could interfere with foam formation (11). Even the smallest amount of fat will interfere significantly with an egg white's ability to foam.

HOW & WHY? ?????????

Why were copper bowls once commonly used to whip egg-white foams? It was long a common chef's practice to use copper bowls made specifically for beating egg whites. A unique reaction between the copper and the egg whites occurs in which trace amounts of copper from the bowl combine with an egg-white protein, conalbumin, allowing the air bubbles in the foam to expand to a larger size as they are beaten. Although copper bowls did improve the whipping properties of eggs, they are no longer recommended because of toxicity risks associated with excess copper.

Separation of Eggs. Although fat contamination can come from plastic bowls or utensils, or from traces of cream, oil, butter, or other foods, the most common source is the egg yolk itself. Careful separation of the egg yolk and the white is imperative. Eggs can be separated using an egg separator, a small cup centered in a round frame. The frame catches the yolk while slots around the frame let the white slip through to a container underneath (Figure 13-7). Another method is to break the egg into a clean funnel, which allows the white, but not the yolk, to run through. Egg yolk traces are sometimes removed by using the egg's shell, which has an affinity for picking up traces of yolk remaining in the egg white. This method of removing it is not recommended because egg shells are a possible source of microbial contamination. For this reason also, eggs should not be separated by passing the yolk back and forth between the two shell halves.

Sugar. Sugar stabilizes egg-white foam, but it also inhibits the mechanical coagulation of proteins necessary for foam formation. Therefore, it is best to add sugar near the end of the whipping time or volume may be compromised. One teaspoon of granulated or superfine sugar per egg may be added only after soft white peaks have formed. Egg white sweetened with sugar is less likely to be overbeaten, and has a very fine texture with a smooth, satiny surface.

Fluid. Adding fluid to egg whites increases the foam volume up to 40 percent, but decreases its stability.

Salt. Salt decreases the stability and volume of egg-white foam, and for that reason is rarely added to egg whites.

Acid. Normally, whole eggs are relatively neutral in pH (7.0 to 7.6), but egg whites by themselves tend to be alkaline (about 8.4 pH). In fact, alkalinity increases with age as eggs lose carbon dioxide, which increases the pH of the whites (up to 9.2) (11). Egg whites whip more easily into a stable foam when their pH is lowered slightly. Adding acid in the form of lemon juice or cream of tartar (the salts of tartaric acid) decreases the pH of egg whites, causing the egg proteins to become unstable and more likely to denature and whip into a foam. Adding too much acid, however, results in delayed foam formation and a much less stable foam. Acid can be added in the form of cream of tartar—1 teaspoon per cup of egg white or ⅛ teaspoon per egg white—or by adding ⅛ teaspoon lemon juice or white distilled vinegar per egg white.

FIGURE 13-7 Egg separator.

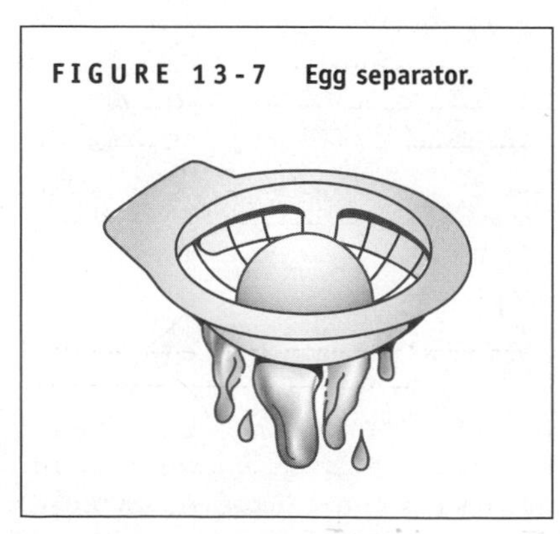

Interfering

Eggs are often used in the preparation of frozen desserts such as ice cream, because they interfere with the formation of ice crystals. Similarly, in some candies, eggs are used to block the formation of large sugar crystals to create a smoother, more velvety texture.

Clarifying

Egg whites are often used to **clarify** liquids. This is done by dissolving egg proteins, especially egg whites (albumen), in cold liquid, which is then heated. This causes the proteins to solidify, to attract other particles that may be clouding the liquid, and to rise with them to the surface for removal. This food preparation technique is used to make clear soups (see Chapter 19).

Preparation of Eggs

Eggs are extremely versatile and can be prepared alone or in combination with other foods (Figure 13-8). Countless recipes that include eggs can be cooked by either dry- or moist-heat methods. Before these are discussed, a brief overview of the changes that can occur in prepared eggs is addressed.

Changes in Prepared Eggs

Effects of Temperature and Time. The key to cooking eggs is to keep the temperature low and/or the cooking time short. Heating eggs at high tem-

FIGURE 13-8 Food preparation with eggs.

Eggs Prepared Alone:	
Fried	
Baked	
Simmered	
Coddled	
Poached	
Omelet	
Scrambled	
Foods Prepared with Eggs:	
Eggs benedict	Waffles
Meringue	Egg soup
Deviled eggs	Cream puffs
Pudding	Mayonnaise
Pancakes	Quiche
Egg salad	Fritatas
Mousse	Popovers
Eggnog	Cake
Soufflé	French toast
Crepe	Egg sandwich
Egg foo young	Cream pies
Custard	Salad dressing

KEY TERMS

Prime (season) To seal the pores of a pan's metal surface with a layer of heated-on oil.

peratures and/or for long periods of time diminishes the eggs' texture, flavor, and color. Overheated proteins become tough and rubbery and shrink from dehydration, which is why overcooked scrambled eggs look curdled and feel dry and rubbery.

Coagulation Temperatures. Egg whites and yolks coagulate at different temperatures. Egg whites first start to coagulate at about 140°F (60°C) and become completely coagulated at 149° to 158°F (65° to 70°C). Slightly warmer temperatures of about 144° to 158°F (62° to 70°C) are needed for the egg yolks to start coagulating. This difference allows eggs to be cooked so their whites are firm but the yolks remain soft. An egg may be cooked at 142°F (61°C) for an hour and still have a soft yolk. Also, beaten eggs coagulate at a slightly higher temperature (about 156°F/69°C).

Effects of Added Ingredients. Adding other ingredients to eggs changes their coagulation temperature. For example, incorporating milk into whole eggs in a custard dish increases the coagulation temperature to about 175°F (79°C). Sugar also increases coagulation temperature, while the addition of salt and/or acid lowers it. Eggs can curdle if too much of an acid ingredient, such as tomato or vinegar, is added.

Color Changes. Undesirable color changes may occur during egg preparation. Sometimes when eggs are overcooked or heated at too high a temperature, the sulfur in the egg white may combine with the iron in the yolk, the cooking water, or other iron sources to form ferrous sulfide, a green-colored compound with a strong odor and flavor. To eliminate the problem of "green yolk," it is best to use stainless steel equipment and low cooking temperatures, to avoid overcooking, to cool hard-cooked eggs quickly in cold water, and to serve them immediately. Another change, which is more difficult to prevent in heated eggs, is the slight browning that results from the Maillard reaction, in which egg proteins react with the few carbohydrates that exist in an egg.

Dry-Heat Preparation

Dry-heat preparation of eggs primarily involves frying and baking. Egg dishes that are commonly fried are fried eggs, scrambled eggs, and omelets. Baked egg dishes include shirred eggs, meringues (both soft and hard), and soufflés. These dry-heat methods are now further discussed.

Frying. A frying pan, a sauté pan (omelet pan), or even a griddle can be used to fry eggs. Cast iron pans work best for eggs if they are **primed** or seasoned. To accomplish this, a clean frying pan is rubbed with a thin layer of vegetable oil, set on moderate heat, which is then briefly increased to high. Then it is removed from the heat and allowed to cool. Washing the frying pan with soap, or cooking anything but eggs in it, removes the primed surface. Nonstick pans do not need to be primed or seasoned. Frying is used to prepare fried and scrambled eggs and omelets.

Fried Eggs. For each fried egg, about 1 teaspoon or less of butter, margarine, or oil is added to a hot pan. Clarified butter can also be used; it will not burn like regular butter. To cut down on fat, a bit of fat may be spread on the pan's surface with a paper towel or waxed paper, or a vegetable spray may be applied to its surface before heating. Too little fat causes sticking, but excessive fat will result in greasy eggs. The fat should be hot enough to prevent the eggs from running, but not so hot that it toughens the egg proteins. The temperature is just right when a drop of water dropped into a hot pan sizzles instead of either rolling around or instantly vaporizing into the air.

Yolks are less likely to break open when the eggs are cracked if the eggs are allowed to warm very briefly in a bowl of hot water. Broken yolks can also be avoided by using fresh eggs and/or by first breaking the eggs into a bowl or other container rather than dropping them directly from the shell into a frying pan or griddle. Then, once the pan and the fat have been heated to the right stage, the eggs should be slid from the bowl, no more than two at a time, onto the pan or griddle. The heat should be lowered immediately to medium-high.

Coagulation is then allowed to occur according to the following "cook-to order" stages:

- *Sunny side up.* The egg is cooked until the white is set and the yolk is still soft. The egg is not flipped. Sunny side up eggs may not be sufficiently cooked to eliminate bacteria, and thus are not allowed by some state health departments to be served to the public. Covering the pan with a lid during cooking gives the yolk a rather opaque appearance, but eliminates any risk of an undercooked egg.
- *Over easy.* The eggs are flipped over when the whites are 75 percent set. Cooking continues until the whites are completely cooked but the yolks are still soft.
- *Over medium.* The same as over easy, except that the yolks are partially set.
- *Over hard.* The same as over easy, except that the yolks are completely set.

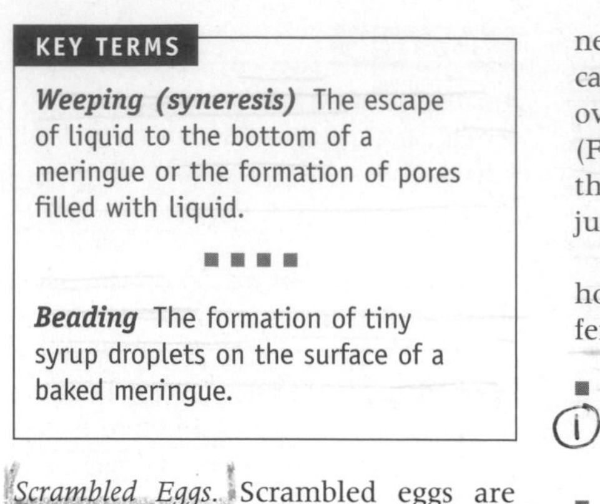
KEY TERMS

Weeping (syneresis) The escape of liquid to the bottom of a meringue or the formation of pores filled with liquid.

■ ■ ■ ■

Beading The formation of tiny syrup droplets on the surface of a baked meringue.

Scrambled Eggs. Scrambled eggs are beaten while raw until well blended and may be seasoned with salt and pepper or other seasonings. Liquid in the form of milk, cream, or water may be added to impart more body and/or flavor and a soft, creamy texture. The added liquid, a tablespoon or less for each egg, creates steam during cooking, which lifts the eggs and makes them lighter and fluffier. Too much liquid makes the eggs watery and forms small, tough, curdlike masses.

The beaten egg mixture is poured onto a heated surface, the heat is reduced, and the eggs are gently stirred as soon as they begin to coagulate. Too much stirring will break the egg into too many smaller pieces, so it is better to lift the cooked egg repeatedly with a spatula so the undercooked portions may slide underneath rather than literally to stir them. Scrambled eggs are finished cooking when they are set, yet still soft and moist. Like most egg dishes, they are best when served immediately.

Omelets. When eggs are beaten, cooked, and rolled into a cigar shape or folded into a flat half circle, the resulting dish is called an omelet. Both plain (French or American-style) and puffy (fluffy) omelets can be prepared with or without fillings. Omelet preparation is considered so important by chefs that it is not unusual for a job applicant to be asked to chop an onion and make an omelet as part of the interview process.

Plain omelets consist of whole eggs, beaten, seasoned as desired, and poured into a prepared pan heated to medium-high. Once the mixture is in the pan, the heat is lowered to medium, and the mixture is not stirred. Uncooked portions are allowed to cook by lifting just the edges of the omelet with a spatula so the runny mixture flows underneath. When the top is firm, the omelet can be folded in half, rolled and folded over itself, or rolled and slid onto a dish (Figure 13-9). If fillings are to be added, they are placed on top of the omelet just before it is folded.

There are many opinions about how French and American omelets differ, but here are four basic differences:

- French omelets are never allowed to brown, while American omelets may have some color.
- French omelets never have texture lines, while an American omelet may have a few.
- Folding a French omelet occurs in "3s" (left over center, right over center) or complete rolling (like a rug), while American omelets are simply folded in half.
- The center of French omelets is "soft," while American omelet centers are fully cooked.

The fluffiness of puffy omelets is achieved by separating the yolks from the whites and whipping each portion separately. Seasonings and liquid, if added, are incorporated into the whipped yolks. The egg whites are whipped until stiff but not dry and then gently folded into the yolks. This mixture is poured into a preheated omelet pan or suitable frying pan with sloping sides, and the heat is reduced to medium. When the omelet is browned on the bottom, it is placed in a 350°F (177°C) oven for five to ten minutes to allow additional rising and further coagulation of the surface proteins. The omelet is finished cooking when the top springs back from a gentle touch of the finger.

Baking. Baking eggs and their ingredients leads to a number of different egg dishes: shirred eggs, meringues, and soufflés.

Shirred Eggs. Whole eggs that are baked and served in individual dishes are called shirred eggs. The egg is cracked, gently placed into a cup from which it can be rolled into a container coated with butter or margarine, and then baked in an oven at 350°F (177°C) until cooked to order.

Meringue. A meringue is an egg-white foam used in dessert dishes as a pie topping, a cake layer, or as frosting. It may also serve as a dessert on its own or be combined in other ways with dessert ingredients. Meringues are made by whipping egg white into foam and adding sugar, the amount of which determines whether the meringue is soft or hard.

Soft meringues are made with about 2 tablespoons of granulated (preferably superfine) sugar per egg white and are often used as pie toppings (e.g., lemon meringue pie). The sugar is gradually added to the egg whites—three will cover an average pie—and whipped to the soft peak stage. The meringue is then spread immediately over the still-warm filling. A warm filling is necessary so the egg-white proteins can coagulate and bind to it. The whole pie with the meringue is then baked in the oven at between 325°F (163°C) and 350°F (177°C) for about fifteen minutes. Too low a temperature dries the meringue; too high a temperature shrinks it.

Some problems that can occur when preparing soft meringues are shrinking, **weeping**, and **beading**.

- *Shrinking.* To prevent the meringue from shrinking back and leaving an unsightly gap around the outside edges of the pie, it should be spread to slightly overlap the entire perimeter of the crust.
- *Weeping.* Also known as syneresis, weeping may be caused by underbeating the eggs, which leaves unbeaten whites on the bottom of the beating bowl; or by undercoagulation, created, for example, by placing meringue on a cold pie filling. A meringue can be protected from weeping by adding a teaspoon of cornstarch to the sugar before beating it into the egg whites.
- *Beading.* Undissolved sugar is the main cause of beading, but overcooking (overcoagulation) also contributes to this phenomenon. Beading can be avoided by using shorter cooking times and increasing the temperature up to 425°F (218°C).

Hard meringues are usually baked as cookies, but they can be formed into different shapes and used as decora-

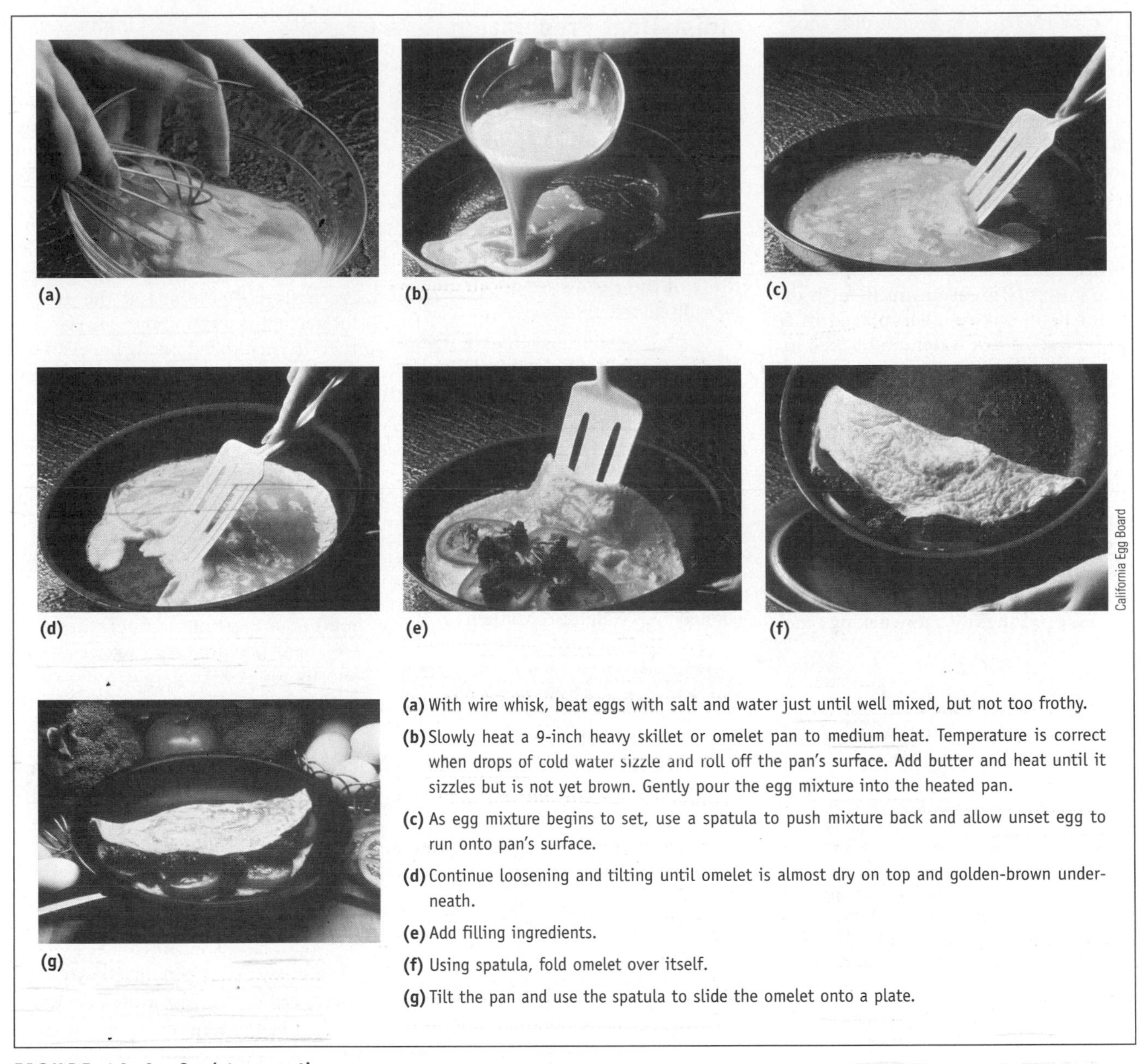

(a) With wire whisk, beat eggs with salt and water just until well mixed, but not too frothy.

(b) Slowly heat a 9-inch heavy skillet or omelet pan to medium heat. Temperature is correct when drops of cold water sizzle and roll off the pan's surface. Add butter and heat until it sizzles but is not yet brown. Gently pour the egg mixture into the heated pan.

(c) As egg mixture begins to set, use a spatula to push mixture back and allow unset egg to run onto pan's surface.

(d) Continue loosening and tilting until omelet is almost dry on top and golden-brown underneath.

(e) Add filling ingredients.

(f) Using spatula, fold omelet over itself.

(g) Tilt the pan and use the spatula to slide the omelet onto a plate.

California Egg Board

FIGURE 13-9 Omelet preparation.

tions on puddings or other desserts. They are prepared with twice the amount of sugar used in soft meringues, about 4 tablespoons (1/4 cup) per egg white. Confectioner's sugar is preferred over granulated sugar for use in hard meringues, because it is more evenly distributed through the beaten egg whites and lacks a gritty texture. Egg whites are beaten to the stiff stage, the sugar is beaten in, and the resultant meringue is shaped, placed on a parchment-covered baking sheet, and baked at the low temperature of 225°F (107°C) for about an hour or longer, depending on the size of the individual portions. When the meringue is delicately browned and the end product firm, the oven is turned off, the door left open, and the meringue left in the cooling oven for at least five minutes. Once the meringue is removed from the oven, the remainder of the cooling period should occur in a warm place free of drafts.

Soufflés. A soufflé is actually a modified omelet. The main ingredients of a soufflé are a thick base generally made from a **white sauce** or pastry cream, an egg white foam, and flavoring ingredients. Initially, the egg yolks and whites are separated. A thick white sauce or pastry cream is prepared and combined with the egg yolks. Stiffly beaten egg whites are folded into the thick egg yolk mixture

KEY TERMS

White sauce A mixture of flour, milk, and usually fat.

(Figure 13-10). For a main dish soufflé, flavoring ingredients such as diced or grated cheese, cooked meat, cooked seafood, and/or vegetables and seasonings are added to this mixture. Dessert soufflés will include sweet ingredients like sugar, chocolate, or fruit, but the process is the same.

Whichever the type of soufflé, the entire combination is gently poured into a lightly greased soufflé dish or other deep baking dish, placed in a larger pan of hot water, and baked in a moderate (350°F/177°C) oven for 50 to 60 minutes or until delicately browned and firm to the touch. Small, individual soufflés will take less time. The oven door should not be opened during baking until time to check for doneness, because it creates a draft that can cause the soufflé to fall. Doneness is determined by gently shaking the oven rack. If the center jiggles, even slightly, more baking time is required.

FIGURE 13-10 Folding egg whites: When combining beaten egg whites with other heavier mixtures, it is best to pour the heavier mixture onto the beaten egg whites. Then gradually combine the ingredients with a downward stroke into the bowl, across, up, and over the mixture using a spoon or rubber spatula. Come up through the center of the mixture about every three strokes and rotate the bowl during folding. Fold just until there are no streaks remaining in the mixture. Avoid stirring, which will force air out of the egg whites.

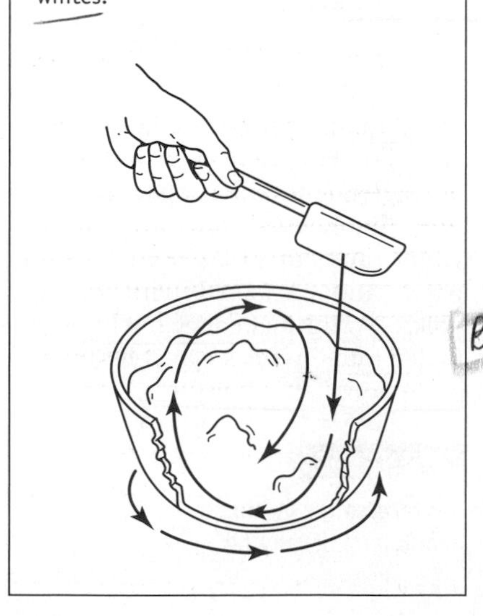

Moist-Heat Preparation

Eggs can be prepared by moist heat using a variety of methods. Most common among these are "boiled" eggs, coddled eggs prepared in a cup, poached eggs, a variety of custards, and eggs that are prepared using the microwave. In all cases, eggs are cooked at simmering temperatures (see Chapter 6). Each of these methods and some of the egg dishes produced are now discussed in more detail.

Hard or Soft "Boiled." Although the term "hard-boiled eggs" is commonly used, eggs should actually be simmered and never boiled, because they will become tough and rubbery if so treated. The high heat of boiling also transforms the iron in the egg yolk into ferrous sulfide, causing the greenish-black color and unpleasant flavor found in the yolk of overly hard-cooked eggs (Chemist's Corner 13-2).

HOW & WHY? ?????????

Why do boiled eggs sometimes crack?

Eggs crack because the pressure created by fast-heating water pops the shell. Although simmering hard-cooked eggs makes them prone to cracking, this can be avoided by first warming eggs to room temperature in a bowl of warm water. Another way to reduce the chances of cracking is to push a sterilized pin or needle through the large end of the shell where the air cell is located. To reduce cracking still further, a spoon or other utensil may be used to place the eggs gently in the water.

There are two methods for hard-cooking eggs: hot start and cold start. Each has advantages and disadvantages; each produces acceptable products.

Hot-Start Method. In the hot-start method, the water is heated to boiling and then the eggs are completely immersed in the boiling water. The heat is immediately reduced to simmer, and the eggs are cooked for 3 to 15 minutes, depending on the desired doneness:

- Soft — 3 to 4 minutes
- Medium — 5 to 7 minutes
- Hard — 12 to 15 minutes

The cooked eggs are drained and then rinsed under cold running water to stop further cooking from residual heat. The extreme temperature change from hot to cold also helps loosen the egg's membrane from the shell, making it easier to peel. To further ease peeling, the first crack should be made at the air cell located at the larger end of the egg, and then the egg rolled gently between the hands to break the shell all over. Peeling under cold running water also makes the job easier. Fresher eggs are harder to peel because the air cell is smaller and the membrane is tight against the cell wall. Although the larger air cell and higher pH of older eggs makes them easier to peel, they also tend to break more easily during heating.

The benefits of using the hot-start method are greater temperature control, eggs that are easier to peel, and a shorter total cooking time. A drawback is that lowering the eggs into boiling water may cause them to crack.

Cold-Start Method. In the cold-start method, the eggs are placed in a saucepan with enough cold water to

Chemist's Corner 13-2

Yolk Color in Boiled Eggs

The green "ring" around the yolk is from the sulfur. Egg whites contain sulfur in the form of hydrogen sulfide gas. Under the high pressure of boiling, this gas moves from the egg white to the yolk. There it combines with the iron in the egg yolk to form ferrous sulfide. The result is a greenish ring around the yolk. Under reduced pressure, however, the gas moves away from the yolk toward the shell. The green discoloration in hard-cooked eggs can thus be avoided by dipping them in cold water immediately after boiling; the rapid temperature drop lowers the pressure within the egg, and the hydrogen sulfide gas does not enter the yolk.

cover them by at least an inch. The water is brought to a boil, immediately lowered to a simmer, and the eggs are then cooked to order:

- Soft 1 minute
- Medium 3 to 5 minutes
- Hard 10 minutes

Another way to prepare hard-cooked eggs from a cold start is to remove the pan from the heat as soon as the water boils, cover it tightly, and let it stand for 20 minutes. Cold-start eggs are less likely to crack during cooking.

The advantages to the cold-start method are that less attention to the process is required, the eggs are easier to add to the water, and they are less likely to break. On the other hand, starting eggs out in cold water may cause the egg white by the shell's surface to be more rubbery, and there is a greater chance of a greenish tint forming on the egg white.

Once cooked, eggs can be cut into slices or wedges using the equipment shown in Figure 13-11. Dipping the knife in hot water before slicing keeps the hard-cooked eggs from falling apart. To tell a hard-cooked egg from a raw one, spin the egg on its side. A smoothly spinning egg is hard cooked, while one that wobbles out of balance is not.

Coddling. Coddled eggs are prepared by breaking an egg into a small cup, called a coddler, made of porcelain or heat-proof glass with a screw-on top, and submerging the whole coddler in simmering water until the egg is cooked. The coddler should be buttered or greased before adding the raw egg. Cream or other flavorings such as ham or bacon are sometimes added before cooking. Once done, the egg is eaten directly out of the coddler.

Poaching. Eggs are poached by being cracked and simmered in enough water to cover the egg by at least twice its depth. Fresh USDA Grade AA eggs are best to use for poaching, because the whites are firmer and less likely to spread out in the water and create "streamers," floating strands of partially cooked egg whites. Salt (½ teaspoon per cup) and/or vinegar (1 teaspoon per cup) may be added to the water to speed coagulation and help to maintain a compact, oval shape of the egg. On the other hand, salt or vinegar will give the cooked egg a shinier, tougher, and, perhaps more shriveled surface than one cooked in plain water. Poached eggs are cooked for three to five minutes, removed with a slotted spoon, drained, trimmed of any streamers, and served immediately. The well-poached egg should have a firm yolk and compact white. Poached eggs are commonly used for eggs Benedict, consisting of an English muffin layered first with a slice of ham or Canadian bacon, followed by a poached egg, and topped with a dollop of Hollandaise sauce.

Custards. Custards are mixtures of milk and/or cream, sweeteners (sugar, honey), flavorings (vanilla, nutmeg, etc.), and eggs or egg yolks. Custards are thickened by the coagulation of egg proteins during cooking. These egg proteins denature when heated and recombine to form a network that sets or coagulates, at the right temperature, to form the solid gel of a custard. All custard dishes are very susceptible to microbial contamination and should be covered and refrigerated as soon as possible after preparation.

Custards are distinguished by whether they are sweet or savory, and by their preparation method: stirred or baked.

Sweet and Savory Custards. Sweet custards are served as desserts in the form of puddings or as fillings for eclairs and pies. Savory (nonsweet) custards are used for dishes such as quiches. A popular quiche made with bacon and Swiss cheese is known as quiche Lorraine.

Stirred Custard (Soft Custard or Custard Sauce). The ingredients of this custard are stirred while being heated on the range over low heat or in a double boiler. The mixture retains a smooth, creamy, fluid consistency. Stirred custard is often eaten as a pudding; however, it may provide the base for many frozen desserts, be served as a sauce for cake, fruit, and other desserts, or be used to replace eggnog. The repeated stirring prevents the formation of a gel, so the custard mixture thickens instead of gels.

Baked Custard. Although discussed under the topic of moist-heat preparation, baked custards are actually an example of dry-heat preparation. Both types of custards begin with the same ingredients, but are simply heated differently. Baked custard mixes are poured into ungreased custard cups that are placed in the oven, usually in a water bath (*bain marie*), where they sit undisturbed and gel during baking. A water bath is made by filling a large, low-sided pan with 1 inch of hot water, into which are placed the cups containing the custard mix. The layer of water insulates the cups and prevents the outside of the

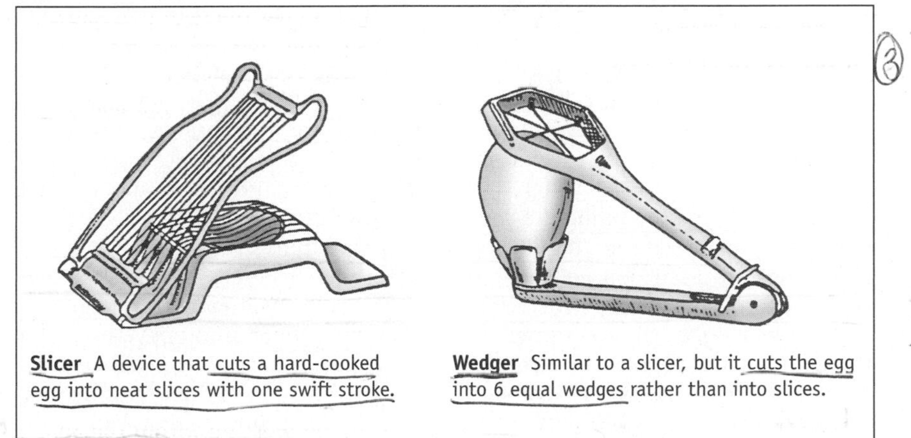

Slicer A device that cuts a hard-cooked egg into neat slices with one swift stroke.

Wedger Similar to a slicer, but it cuts the egg into 6 equal wedges rather than into slices.

FIGURE 13-11 **Egg-cutting tools.**

custard from cooking to completion before the inside has had a chance to coagulate.

The internal temperature of custards should never be allowed to rise above the point of coagulation of the egg-liquid mixtures (185°F/85°C). Overheating causes the egg proteins to shrink, allowing liquid to be released from the egg and producing a product with a curdled, weepy, porous appearance. Another problem with baked custards is that they tend to have a runny texture, which makes them unsuitable for making solid pie fillings. As a result, custards to be used as pie fillings are often thickened with starch in the form of cornstarch or flour.

Custards should be baked at 350°F (177°C) until a knife inserted in the middle of the cup comes out clean—about 23 to 25 minutes for custard cups and 35 to 40 minutes for a casserole-size dish. Just before complete doneness is reached, the custard is immediately removed from the oven and placed on a rack. Some additional cooking will inevitably occur during cooling, but can be minimized by using a cooling rack. Should the custard be overcooked, the cups can be set in ice water to stop further coagulation. Undercooking should likewise be avoided, because it will prevent the custard from setting properly.

Microwaving. Eggs cook extremely rapidly in a microwave oven, so special caution should be taken to avoid overcooking. Manufacturer's instructions should be followed for microwave egg cooking. Whole eggs with intact shells should never be microwaved, because steam expanding within the shell can cause them to burst. The same principle applies to whole eggs out of the shell, because the vitelline membrane around the egg yolk traps steam and will burst if not punctured with a toothpick or the tip of a knife prior to going into the microwave.

Fried. A browning dish is required to fry an egg in a microwave, and should be preheated on full power for two minutes, plus one additional minute for each egg being fried. About ½ teaspoon or less of fat per egg is melted in the dish before adding the cracked eggs from a bowl into the dish. The yolk membrane is punctured, and the dish is covered with plastic wrap, then microwaved on high for 45 seconds per egg or until the desired doneness is reached.

Shirred. Shirred eggs are cooked in individual containers and are ideal for cooking in a microwave oven. The egg is placed in a custard dish, the yolk is punctured, the dish covered, and the egg is heated on medium for 45 to 60 seconds. It should be rotated a quarter turn at the half-minute mark.

Scrambled. Before scrambling, 1 teaspoon of butter is melted in a 2-cup glass measure by setting the microwave very briefly on high. The beaten eggs are placed in the measuring cup and microwaved on high for 20 seconds. The egg mixture is then stirred, and the heating and stirring process is repeated one or two more times. Microwaving is completed when the eggs are just past the runny stage. They should be allowed to stand one or two minutes if a firmer set is desired.

Poached. To poach an egg, 1/4 cup of water, with a dash of vinegar and salt, is heated to a boil in a custard dish or 1-cup glass measure. The egg is dropped into the hot water, and the yolk is pierced with a toothpick. The dish is partially covered with plastic wrap and then heated at 50 percent power for about a minute, plus or minus fifteen seconds. Allowing the cup to stand two to three minutes and gently shaking back and forth helps to set the egg whites.

Omelet. Omelets can be prepared in the microwave by using a browning dish or 9-inch pie plate. Enough butter is added to slightly coat the bottom of the dish. It is melted on high and then spread evenly by tilting the container. The combined eggs, liquid, and seasonings are then poured into the container, covered with plastic wrap, and cooked on medium for two to three and a half minutes, or until the omelet is almost set. After removing the omelet from the oven, any fillers are added, and the omelet is folded over with a spatula. Puffy omelets are prepared in the same manner. An omelet cooked in a microwave will not brown unless a browning dish is used.

Quiche. A quiche dish or pie shell is filled with cooked vegetable and/or meat ingredients. Cream (preheated, unlike the cream in conventionally cooked quiches) is added to the beaten eggs, and the egg mixture is then poured over the vegetables and baked according to the manufacturer's guidelines.

Storage of Eggs

Eggs begin to deteriorate as soon as they are laid and lose quality very rapidly at room temperature. In fact, an egg will age more in one day at room temperature than in one week in the refrigerator. To ensure the freshness of whole or liquid eggs, they may be refrigerated, frozen, or dried.

Refrigerator

Whole Eggs. An eggshell is not airtight. There are as many as 17,000 tiny pores over the shell's surface, so keeping eggs in the carton and refrigerated helps retain their freshness. Several signs distinguish fresh eggs from those that have aged. Changes in proteins over time cause egg whites to thin. Fresh eggs also have more prominent, viscous chalazae on either side of the yolk than older eggs. In the process of aging, the vitelline membrane weakens and the yolk migrates or breaks. The size of an egg's air cell provides another indication of its age (Figure 13-12). The air cell gap between the membranes increases in size as the egg ages because moisture and carbon dioxide escape through the porous shell.

Proper refrigeration of eggs helps to delay these changes and protects them from microbial growth, thus helping to maintain their quality. Many home refrigerators have built-in egg containers, but eggs retain their moisture better and keep longer if stored in the carton. It also helps prevent flavors and odors from being absorbed through their porous shells. Eggs should sit in the carton with their large ends up to

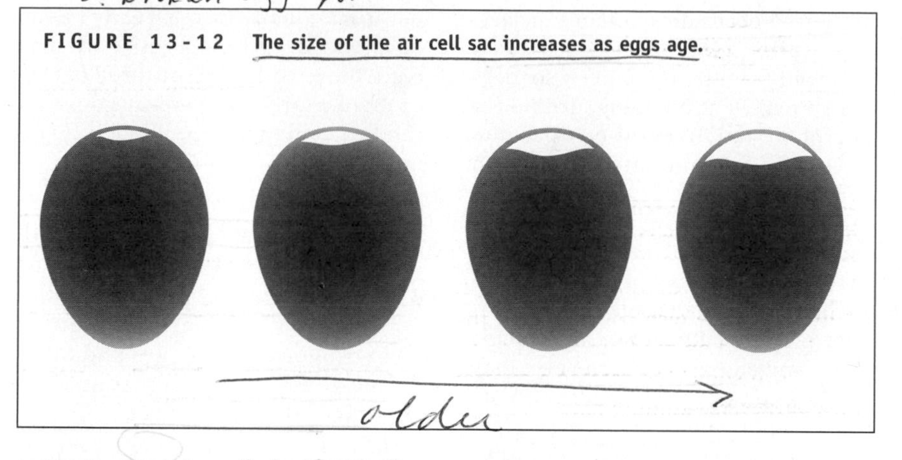

FIGURE 13-12 The size of the air cell sac increases as eggs age.

prevent the air cell from moving toward the yolk. Washing eggs is not recommended, because this will remove the oil coating applied by the processor to prevent microbial growth and moisture loss.

Shelf Life of Refrigerated Eggs. Refrigerated whole eggs should stay fresh for about a month. Separated egg yolks may be stored under water in the refrigerator, but should be used within two days. Egg whites kept tightly covered in a glass container will last up to four days.

Storage Eggs. Restaurants, food service institutions, and other food manufacturers must be especially careful about storing eggs, because they purchase such large quantities. **Storage eggs**, used by commercial food service establishments, are usually used within a month, but can be stored for up to six months. They are not available at the retail level. The coating of oil or plastic prevents microbial invasion and any loss of moisture or carbon dioxide.

Pasteurized Eggs. The USDA regulations require that all liquid, frozen, or dried eggs be pasteurized or otherwise treated to protect against *Salmonella*. Commercial outlets frequently use refrigerated liquid eggs that have been pasteurized. Typical processed food products that may incorporate pasteurized liquid egg whites include baked goods, candies, and chilled or frozen desserts. After being pasteurized, the advantages of liquid eggs over whole eggs or even frozen egg blends are convenience, consistent quality, microbial safety, and costs savings in terms of space, labor, and freezing (16). New to the scene are pasteurized eggs in their shells that are being marketed to the consumer (15).

Frozen

Freezing a whole egg is not possible because it will crack under the expanding liquids. Food manufacturers solve this dilemma by breaking the eggs open at the processing plants where the contents are frozen whole (whites and yolk mixed together) or separated as whites or yolks. Prior to being frozen, the liquid whole eggs are usually pasteurized (140° to 143°F/52° to 55°C for three and a half minutes). Egg whites by themselves denature if pasteurized, so prior to this process they are often combined with a small amount of lactic acid and aluminum sulfate (Chemist's Corner 13-3). There are several drawbacks to using frozen pasteurized eggs: they are costly to freeze and keep frozen, they must be thawed, they are cumbersome to portion, and they have lowered functional quality (9).

> **Chemist's Corner 13-3**
>
> ***Egg Pasteurization***
>
> Aluminum sulfate is sometimes added to egg whites prior to pasteurization to stabilize conalbumin, an egg protein that is labile (unstable) to heat at pH 7.0. Most of the other egg proteins are stable to heat at this pH.

> **KEY TERMS**
>
> ***Storage eggs*** Eggs that are treated with a light coat of oil or plastic and stored in high humidity at low refrigerator temperatures very close to the egg's freezing point (29° to 32°F/ −1.5° to 0°C).

Fortunately, separated egg whites are not adversely affected by freezing. Some commercially frozen egg whites have added stabilizers and whipping aids to improve their ability to form large, stable foams. For separated yolks, sugar, corn syrup, or salt is added (2 to 10 percent) to prevent them from becoming viscous and rubbery when thawed. Salt is used in frozen eggs only if they will be incorporated into sweet foods that will partially mask the salty taste. When freezing eggs at home, 1 tablespoon of sugar (corn syrup) or ½ teaspoon of salt is added for every cup of blended eggs. Raw egg whites can be frozen with no special measures taken.

Dried

Drying eggs is a simple process. Whole eggs or separated yolks are spray-dried to create a fine powder, which is mixed with anti-caking substances to prevent clumping. Egg whites are dried in different ways to form granule, flake, or milled textures. Dried eggs sometimes brown due to the Maillard reaction, but this can be prevented by removing glucose from the eggs before drying with the aid of an enzyme (glucose oxidase) or by yeast fermentation. Once dried, eggs can be stored in the refrigerator for up to a year, but they must be kept in tightly closed containers to prevent the clumping that can result from moisture accumulation.

Dried eggs, used extensively by food manufacturers, are particularly advantageous when storage and refrigeration space is limited. The major disadvantage of using dried eggs is that they lose many of the functional and sensory qualities of eggs, and are highly susceptible to bacterial contamination. Therefore, they should be used only when the end product will be thoroughly heated.

Rehydrating Dried Eggs. Dried eggs are used in food preparation by adding them to water or by sifting them with dry ingredients. One egg can be reconstituted by sprinkling 2 tablespoons plus 1½ teaspoons of sifted egg powder over an equal amount of lukewarm water and beating until smooth. Combining ½ cup each of sifted egg powder and water produces the equivalent of three eggs. The mixtures should be used within five minutes. Table 13-5 lists the amount of frozen, refrigerated, or dried eggs needed to substitute for regular large eggs.

Safety Tips

The chances of an egg being internally contaminated are relatively low, less than one in 10,000 commercial eggs (13). It is more common for contamination to occur during handling and preparation after the egg has been removed from its shell. Even so, eggs are an excellent breeding ground for microbial activity, and can become internally contaminated through a hen with a *Salmonella enteritidis* infection in her ovary or oviduct (4), or from absorbing bacteria through the pores. The latter can occur if the eggs are boiled and then cooled in the presence of infected water or an infected food handler. Externally, the eggs may also be exposed to *Salmonella enteritidis* by fecal contamination during egg laying. The Centers for Disease Control implicated eggs as the source for 73 percent of *Salmonella enteritidis* outbreaks (18), and there is an increasing possibility that *Listeria monocytogenes,* which can grow at refrigerator temperatures and has already been observed on whole eggs, may also contribute to future outbreaks (14, 15). There are many precautions that can be taken to prevent foodborne illness from eggs:

- Use an egg separator rather than passing the yolk back and forth between the two shell halves.
- Always store eggs in the refrigerator, never at room temperature. Meringue-covered pies and other egg-containing foods should be refrigerated until served.
- Raw eggs should never be consumed; this is especially the case for the very young, elderly, or immune compromised (people with conditions such as AIDS, cancer, etc).
- FDA recommends only pasteurized eggs for food items in which eggs are only lightly cooked or not at all, such as Caesar salad, uncooked hollandaise or béarnaise sauce, and homemade mayonnaise, eggnog, ice cream, etc. (12).
- Do not add raw egg to already scrambled eggs, a practice sometimes used in food service operations to increase the moisture content of dried scrambled eggs.
- Cook eggs until no visible liquid egg remains, especially when preparing French toast, scrambled eggs, poached eggs, and omelets.
- A knife inserted into baked egg dishes such as quiches, baked custards, and most casseroles should come out clean.
- Scrambled eggs should be held on cafeteria and buffet lines at appropriate temperatures.
- Be extra cautious when preparing lightly cooked egg dishes such as mousse, meringue, and other similar dishes, because they may not be sufficiently cooked to eliminate possible bacteria.
- All egg dishes should be heated to 145°F (63°C).

TABLE 13-5
Substituting Frozen, Refrigerated, or Dried Eggs for Regular Large Eggs

Frozen or Refrigerated Liquid Eggs	*Weights*	*Volume*
Whole Large Eggs:		
1	1¾ oz	3 tbs
10	1 lb 1¾ oz	2 C
12	1 lb 5½ oz	2½ C
25	2 lbs 13 oz	1 qt 1¼ C
50	5 lbs 8 oz	2 qts 2½ C
Yolks:		
10	7¼ oz	¾ C
12	8½ oz	¾ C 2 tbs
22	1 lb	2 C less 2 tbs
Whites:		
10	11½ oz	1¼ C 2 tbs
12	14 oz	1½ C 2 tbs
14	1 lb	2 C less 2 tbs
Dried Eggs	***Weight/Volume***	***Water (to reconstitute dried eggs)***
Whole Large Eggs:		
6	3 oz (1 C)	1 C
12	6 oz (2 C)	2 C
24	12 oz (1 qt)	1 qt
50	1 lb 9 oz (2 qt and ⅓ C)	2 qt and ⅓ C
100	3 lb 2 oz (2 qt and ⅔ C)	1 gal and ⅔ C
150	4 lb 11 oz (6 qt and 1 C)	6 qt and 1 C
200	6 lb 4 oz (2 gal and 1⅓ C)	2 gal and 1⅓ C

Source: American Egg Board

PICTORIAL SUMMARY / 13: Eggs

As a life-sustaining protein, the quality of protein in eggs is so high that it has become the standard for protein by which researchers rate all other foods. The versatility of eggs, whether prepared alone or in combination with other foods, makes them nearly indispensable in cooking.

COMPOSITION OF EGGS

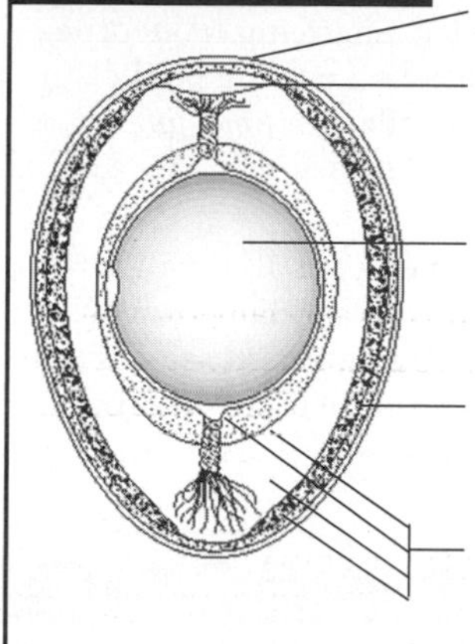

Shell: 12% of the egg's weight, porous calcium carbonate.

Air cell: located between the two shell membranes at larger end of the egg; becomes larger as egg ages.

Egg yolk: 30% of the egg's weight, dense in nutrients—one yolk provides at least 10% of the RDI for a male 25-50 years of age.

Shell membranes: inner and outer membranes protect the egg against bacterial invasion.

Albumen (egg white): 58% of the egg's weight, four layers, composed largely of water and protein.

Nutrient Content: Eggs are one of the very few foods containing all the fat-soluble vitamins and large amounts of certain water-soluble vitamins (A, E, D, and K). Eggs contain all the essential amino acids, making them a good source of complete proteins. They contain very little carbohydrate or fiber.

A large egg contains:

- Calories: 75 kcal
- Protein: 7 grams
- Fat: 5 grams
- Cholesterol: 213 mg
- Vitamins: A, D, E, K, and B vitamins
- Minerals: selenium, iodine/zinc, iron

PURCHASING EGGS

Inspection of processing plants producing egg products is mandatory. Candling, Haugh units, and appearance are used to determine quality grading. Size, unrelated to grading, is determined by the weight of a dozen eggs.

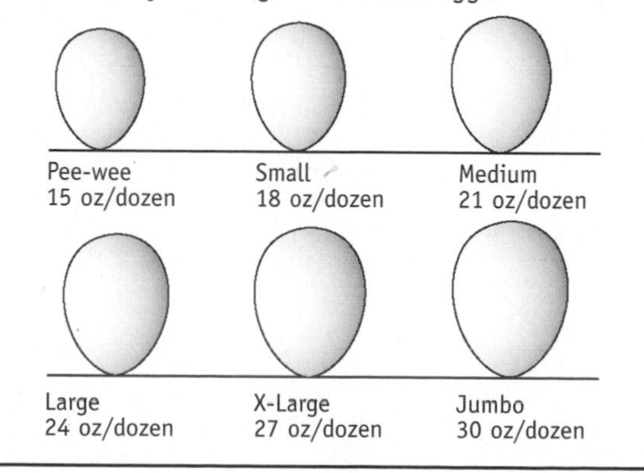

FUNCTIONS OF EGGS IN FOODS

When combined with other ingredients, eggs have a unique ability to flavor, color, emulsify or thicken, bind, foam, interfere, and clarify.

- **Emulsifying:** Lecithin, found in egg yolks, keeps liquid compounds (like fat and water) from separating, thus thickening and stabilizing foods such as sauces and salad dressings.
- **Binding:** The high protein content in beaten eggs can act as an adhesive when cooked. Examples are egg use with breaded, fried foods and the addition of eggs to bind meatloaf.
- **Foaming:** Egg whites beaten into a foam increase their original volume six to eight times, and aerate and leaven food products such as soufflés and meringues. The best egg-white foam is made from fresh eggs at room temperature.
- **Interfering:** Because they interfere with the formation of crystals (ice, sugar), eggs are used to create a smooth, velvety texture in ice cream and candy.
- **Clarifying:** Egg whites are used to make clear soups.

PREPARATION OF EGGS

To preserve the egg's texture, flavor, and color, it is best to keep cooking temperature low and the heating time short. Eggs can be prepared in a great variety of ways, using either dry-heat or moist-heat methods.

Dry-heat

- Fried, scrambled, omelets
- Baked (shirred, meringues, soufflés)

Moist-heat

- Simmering
- Coddling
- Poaching
- Custards (stirred, baked)
- Microwaving

STORAGE OF EGGS

Eggs begin to detiorate as soon as they are laid, but can be preserved by refrigeration, freezing, and drying. Refrigerated eggs retain more moisture and keep longer when they are left in the carton.

The chances of an egg being internally contaminated are relatively low, but external bacterial contamination is possible and certain precautions can be taken to reduce this risk.

REFERENCES

1. Anton M, and G Gandemer. Composition, solubility, and emulsifying properties of granules and plasma of egg yolk. *Journal of Food Science* 62(3):484–487, 1997.
2. Baniel A, A Fains, and Y Popineau. Foaming properties of egg albumen with a bubbling apparatus compared with whipping. *Journal of Food Science* 62(2):377–381, 1997.
3. Bringe NA, and J Cheng. Low-fat, low-cholesterol egg yolk in food applications. *Food Technology* 49(5):94–106, 1995.
4. Chen J, RC Clarke, and MW Griffiths. Use of luminescent strains of *Salmonella enteritidis* to monitor contamination and survival in eggs. *Journal of Food Protection* 59(9):915–921, 1996.
5. Cyclodextrins remove cholesterol from eggs. *Food Engineering* 64(8):42, 1992.
6. deMan JM. *Principles of Food Chemistry.* Van Nostrand Reinhold, 1999.
7. FDA orders egg company to halt sales. *Food Engineering* 63(3):36, 1991.
8. Gebhardt SE, and RH Matthews. *Nutritive Values of Foods.* Home and Garden Bulletin No. 72, U.S. Dept. of Agriculture, Washington, D.C.: GPO, 1994.
9. Giese J. Ultrapasteurized liquid whole eggs earn 1994 IFT Food Technology Industrial Achievement Award. *Food Technology* 48(9):94–96, 1994.
10. Light A, H Heymann, and DL Holt. Hedonic responses to dairy products: Effects of fat levels, label information, and risk perception. *Food Technology* 46(7):54–57, 1992.
11. McGee H. *On Food and Cooking: The Science and Lore of the Kitchen.* Macmillan, 1997.
12. Mermelstein NH. Pasteurization of shell eggs. *Food Technology* 55(12):72–73, 79, 2001.
13. Muriana PM, H Hou, and RK Singh. A flow-injection system for studying heat inactivation of *Listeria monocytogenes* and *Salmonella enteritidis* in liquid whole egg. *Journal of Food Protection* 59(2):121–126, 1996.
14. Palumbo MS, et al. Thermal resistance of *Listeria monocytogenes* and *Salmonella* ssp. in liquid egg white. *Journal of Food Protection* 59(11):1182–1186, 1996.
15. Pszczola DE. Emerging ingredients: Still full of surprises. *Food Technology* 54(7):72–83, 2000.
16. Schuman JD, and BW Sheldon. Thermal resistance of *Salmonella* spp. and *Listeria monocytogenes* in liquid egg yolk and egg white. *Journal of Food Protection* 60(6):634–638, 1997.
17. Smith DM, et al. Cholesterol reduction in liquid egg yolk using betacyclodextrin. *Journal of Food Science* 60(4):691–694, 1995.
18. Tauxe RV. Emerging foodborne diseases: An evolving public health challenge. *Emerging Infectious Diseases* 3(4):425–434, 1997.
19. Yuno-Ohta N, et al. Gelation properties of ovalbumin as affected by fatty acid salts. *Journal of Food Science* 61(5):906–910, 1996.

WEBSITES

The American Egg Board's Incredible Edible Egg website:
www.aeb.org

The egg industry's website for egg education:
www.enc-online.org

U.S. Poultry and Egg Association's website focusing on research, education, and product promotion:
www.poultryegg.org

14

Meat

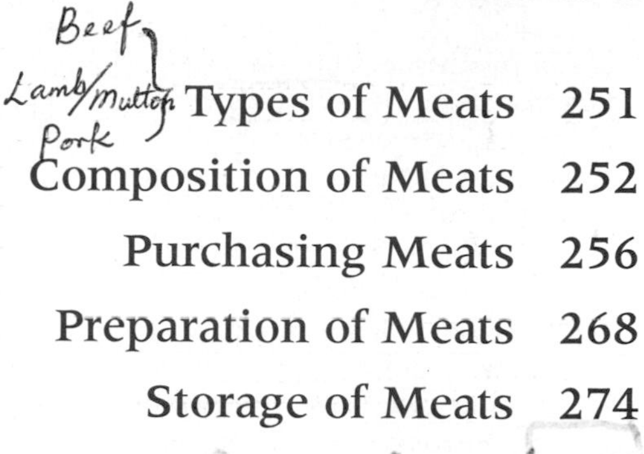

In North America and Europe, meat from herbivores such as beef cattle, sheep, and swine serves as an important source of complete protein. Meat from other animals, such as goat, rabbit, deer, elk, moose, horse, possum, and squirrel, is less commonly eaten. Significant sources of meat in other countries include the camel in the Middle East, the llama in Peru, the kangaroo in Australia, and the dog in some parts of the Far East. This chapter's content is confined to the meat from cattle, sheep, and swine. Meat is generally defined as the muscles of animals, but in a broader sense it also covers the organs and glands obtained from the animal. Although the term "meat" includes the flesh of poultry and fish, these are each covered separately in Chapters 15 and 16.

Meat is usually the most expensive item on a menu, but expense will vary based on the variety, cut, and tenderness of the portion. Some cuts are naturally tender, while others are tough, so preparation methods must vary accordingly. Tender cuts lend themselves to dry-heat methods such as roasting, broiling, grilling, and frying, while tougher cuts are better for long, slow, moist processes such as braising, stewing, or steaming.

Types of Meats

Beef

The ancestor of beef cattle was a type of wild ox domesticated in ancient Greece and Turkey during the Stone Age (around 10,000 B.C.). Since that time, hundreds of breeding lines have been specially developed to provide cattle that serve as abundant sources of good quality beef. Cattle are classified according to age and gender.

- *Steers.* Most of the consumed beef is supplied by steers, male cattle that are castrated while young so that they will gain weight quickly.
- *Bulls.* Consumers often do not see the tougher meat from bulls, older uncastrated males that provide "stag meat," usually used in processed meats and pet foods.
- *Heifers and cows.* Heifers, females that have not borne a calf, are also used for meat. The meat from cows, female cattle that have borne calves, is less desirable than that from steers or heifers.
- *Calves.* Calves three to eight months old are too old for veal and too young for beef. If they go to market between eight and twelve months, their meat is referred to as "baby beef."

Veal. Veal comes from calves of beef cattle, either male or female, between the ages of three weeks and three months. These very young animals are fed a milk-based diet or formula and have their movements greatly restricted, resulting in meat with an exceptionally milky flavor, pale color, and tender texture. Some retailers have stopped selling veal, however, because of possible consumer objections over what is perceived as the inhumane treatment of these animals. There is also a specially marketed "free-range" veal that is somewhat less tender than traditionally fed veal.

Lamb and Mutton

Lamb and mutton are the meat of sheep. The primary difference between the two is the age of the animal from which they come: in general, lamb comes from sheep less than fourteen months old, and mutton from those over fourteen months. Further confirmation of whether one is dealing with lamb or mutton may be found in

KEY TERMS

ATP Adenosine triphosphate is a universal energy compound in cells obtained from the metabolism of carbohydrate, fat, or protein. The energy of ATP, which is located in high-energy phosphate bonds, fuels chemical work at the cellular level.

Connective tissue A protein structure that surrounds living cells, giving them structure and adhesiveness within themselves and to adjacent tissues.

where the lower leg of a carcass will snap. Lamb breaks off above the joint, while mutton will break in the joint. Mutton is also darker and tougher than lamb and has a stronger flavor, which grows even stronger as the animal matures.

Pork

Most pork is derived from young swine of either gender slaughtered at between seven and twelve months of age. Technically, pigs are less than four months old, while hogs are older than four months. In recent times, pork has been bred to be leaner and more tender. This has resulted over the last 30 years in a 50 percent increase in the amount of lean meat yielded per animal. About one-third of all pork is sold fresh, while the rest is cured and provided to consumers as ham, sausage, luncheon meats, and bacon.

Composition of Meats

Structure of Meat

Meats are composed of a combination of water, muscle, connective tissue, adipose (fatty) tissue, and often bone. The proportions of these elements vary according to the animal and the part of its anatomy (represented by the cut of meat.)

Muscle Tissue. Most of the protein in animals is found in their muscles, which serve as the main sources of dietary meat. The characteristics of muscles are an important consideration in deciding how the resulting meat should be prepared.

Muscles are made up of a collection of individual muscle cells, called muscle fibers, that are each surrounded by an outer membrane called the sarcolemma (Figure 14-1). Each muscle fiber is further filled with cell fluid (sarcoplasm) in which are about 2,000 smaller muscle fibrils serving as the contractile components of the muscle fiber. If the muscle fibrils are small, the result is finer muscle bundles, which give the meat a very delicate, velvety consistency.

Muscle Contraction and Relaxation. Muscle fibrils play an important role in muscle contraction and relaxation. The muscle fibril is separated into segments called sarcomeres, which are bordered by dark bands called Z lines. The sarcomeres contain two proteins, actin (thin) and myosin (thick), that are alternately aligned. It is thought that muscle contraction occurs when the sarcomeres shorten as the thick and thin filaments "slide" past each other, forming another protein called actinomyosin (Figure 14-2) The energy for muscle contraction is provided by adenosine triphosphate **(ATP)**.

Connective Tissue. Connective tissue is a part of ligaments and tendons, and it also acts as the "glue" that holds muscle cells together. It is composed

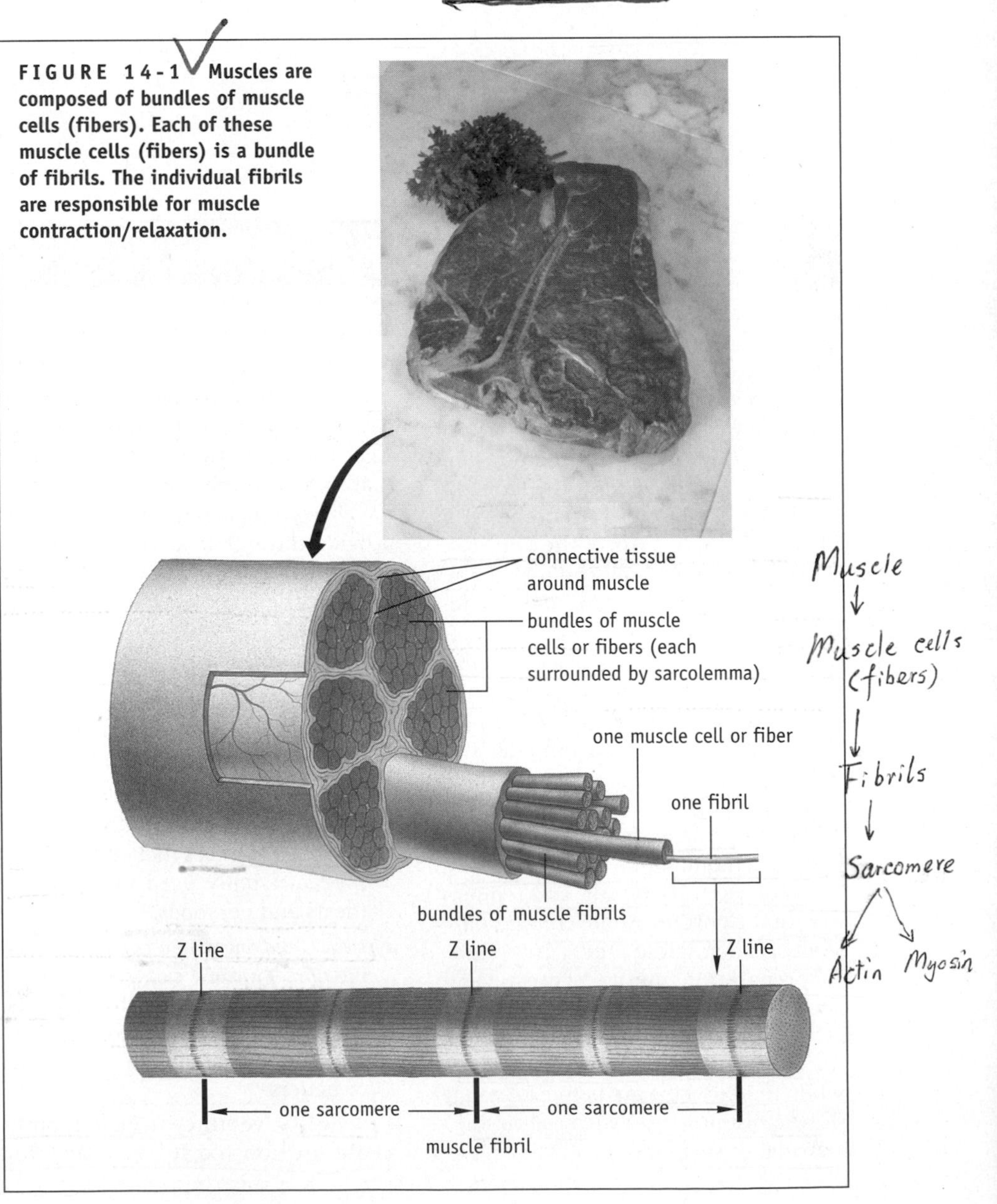

FIGURE 14-1 Muscles are composed of bundles of muscle cells (fibers). Each of these muscle cells (fibers) is a bundle of fibrils. The individual fibrils are responsible for muscle contraction/relaxation.

Muscle ↓ Muscle cells (fibers) ↓ Fibrils ↓ Sarcomere ↓ Actin Myosin

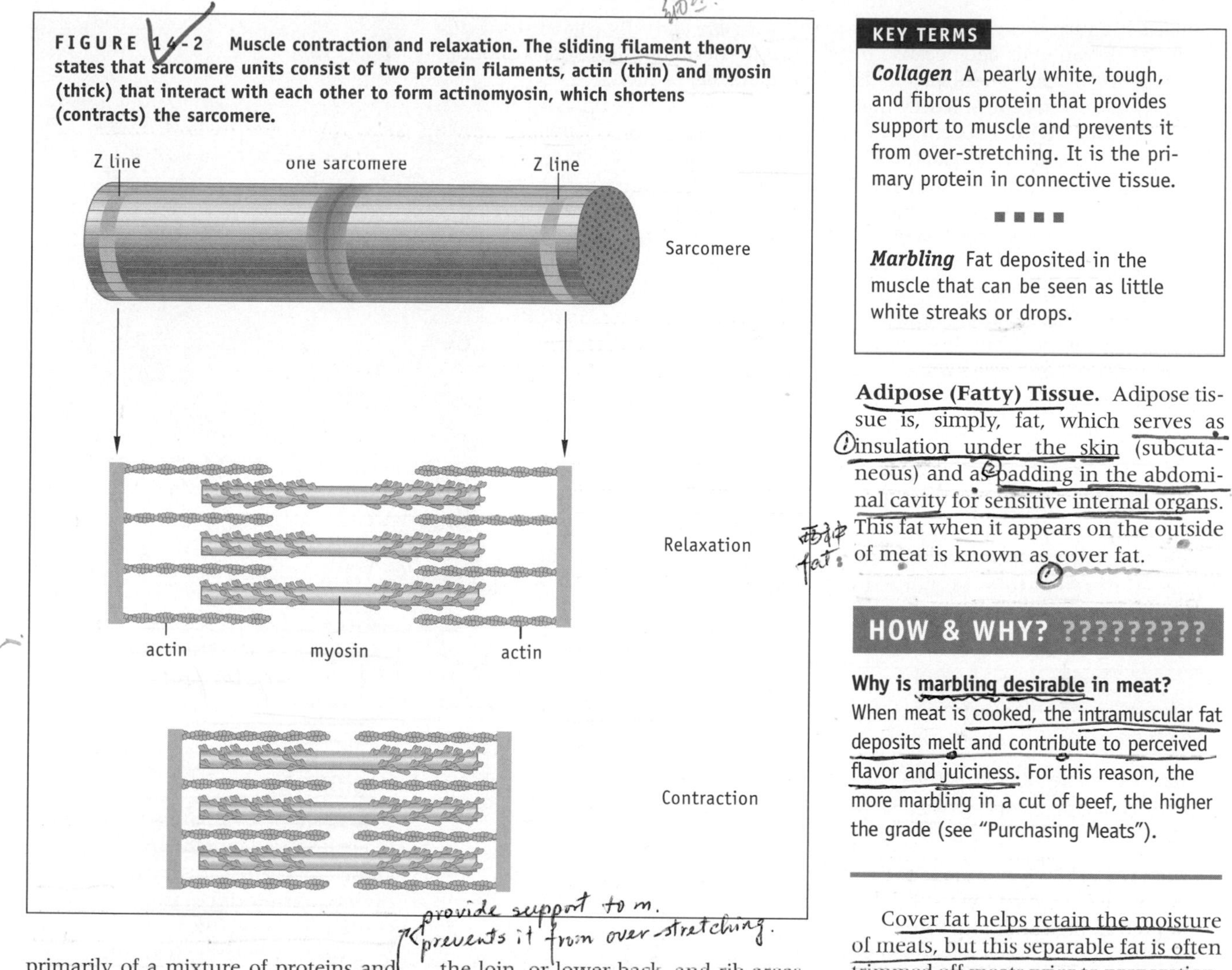

FIGURE 14-2 Muscle contraction and relaxation. The sliding filament theory states that sarcomere units consist of two protein filaments, actin (thin) and myosin (thick) that interact with each other to form actinomyosin, which shortens (contracts) the sarcomere.

primarily of a mixture of proteins and mucopolysaccharides (a type of polysaccharide). The most abundant protein in connective tissue is **collagen** (Chemist's Corner 14-1). It is tough and fibrous, but converts to a gel when exposed to moist heat. The other two main types of connective tissue proteins are elastin and reticulin. Elastin, as the name implies, has elastic qualities, and reticulin consists of very small fibers of connective tissue that form a delicate interlace around muscle cells.

Effect of Collagen on Tenderness. The type and amount of connective tissue found in a meat cut determines its tenderness or toughness and the best type of cooking method. Cuts high in connective tissue are naturally tough and need to be properly prepared in order to become more tender. Muscles used for movement, such as those found in the neck, shoulders, legs, and flank, contain more collagen and tend to be tougher than muscles from the loin, or lower back, and rib areas, which get less exercise.

Effect of Age on Tenderness. Collagen concentration also increases as animals age, which is why meat from older animals is tougher. These usually less expensive, tougher cuts require slow, moist heating at low temperatures to convert, or hydrolyze, the tough connective tissue to softer gelatin. On the other hand, the tougher cuts have more flavor than the more tender ones.

Effect of Elastin on Tenderness. The other two components of connective tissue have less effect on meats when they are cooked. Elastin, which is yellowish, rubbery, and often referred to as "silver skin," does not soften with heating, so it should be removed before preparation if possible. There is very little elastin in meats, except in cuts from the neck and shoulder, so it is less likely to affect tenderness.

KEY TERMS

Collagen A pearly white, tough, and fibrous protein that provides support to muscle and prevents it from over-stretching. It is the primary protein in connective tissue.

■ ■ ■ ■

Marbling Fat deposited in the muscle that can be seen as little white streaks or drops.

Adipose (Fatty) Tissue. Adipose tissue is, simply, fat, which serves as insulation under the skin (subcutaneous) and as padding in the abdominal cavity for sensitive internal organs. This fat when it appears on the outside of meat is known as cover fat.

HOW & WHY? ?????????

Why is marbling desirable in meat?

When meat is cooked, the intramuscular fat deposits melt and contribute to perceived flavor and juiciness. For this reason, the more marbling in a cut of beef, the higher the grade (see "Purchasing Meats").

Cover fat helps retain the moisture of meats, but this separable fat is often trimmed off meats prior to preparation. Fat found within muscles is called intramuscular fat or **marbling**. Fat content varies widely among meats and is dependent on the source animal's genetics, age, diet, and exercise, and on the cut of the meat.

Well-marbled beef fetches a higher price, so many cattle ranchers, in an attempt to improve marbling, feed cattle richer grain during the last weeks or months before slaughter. Paradoxically, however, with the consumer trend in recent years away from fatty meats, some ranchers are raising lower-fat beef to meet consumer demands. Some livestock are being bred to average 25 to 30 percent fat (18). Similarly, a recent technique in swine livestock management is the use of a growth hormone, somatotropin, which results in a leaner animal (29).

Fat Color and Texture. The animal's age, diet, and species affect the color and texture of fat. It is white in younger animals, and turns progressively more yellow as the animals age because of the presence of carotenoid pigments in the feed. Feeding-lot practices that provide swine with fats that are primarily saturated will yield pork fat that is more saturated and hard (Figure 14-4). Conversely, including more polyunsaturated fatty acids in the animal's diet will make its fat softer. The species and breed of the animal also influence the softness of fat; beef fat, for example, is very different from the hard, more brittle and dense fat observed in lamb.

Bone. Bones are used as landmarks for identifying the various meat cuts from a carcass (Figure 14-5). When buying meat, keep in mind that bone weighs more than meat and that the higher the proportion of bone to meat, the less the meat yield and the more the cost of the edible portion.

Marrow. Marrow is the soft, fatty material in the center of most large bones. The marrow found within the bone will generally be of two different types: (1) yellow marrow, found in the long bones, and (2) red marrow—red because it is supplied with many blood vessels—in the spongy center of other bones. Marrow is a valued food in many cultures and can provide much of the flavor in soups. (See Chapter 19 for more on how bones are used in soups.)

©Wolfgang Kaehler/CORBIS

FIGURE 14-4 A polyunsaturated diet will yield pork that is higher in polyunsaturated fat.

Chemist's Corner 14-1

Collagen

Collagen's molecular structure consists of three collagen strands twisted together (Figure 14-3). These strands, rich in proline, hydroxyproline, and glycine, are held together by hydrogen and covalent bonds. Older meat is less tender for two reasons: the collagen content of meat increases with an animal's age, and more covalent crosslinks are formed between the collagen strands (31).

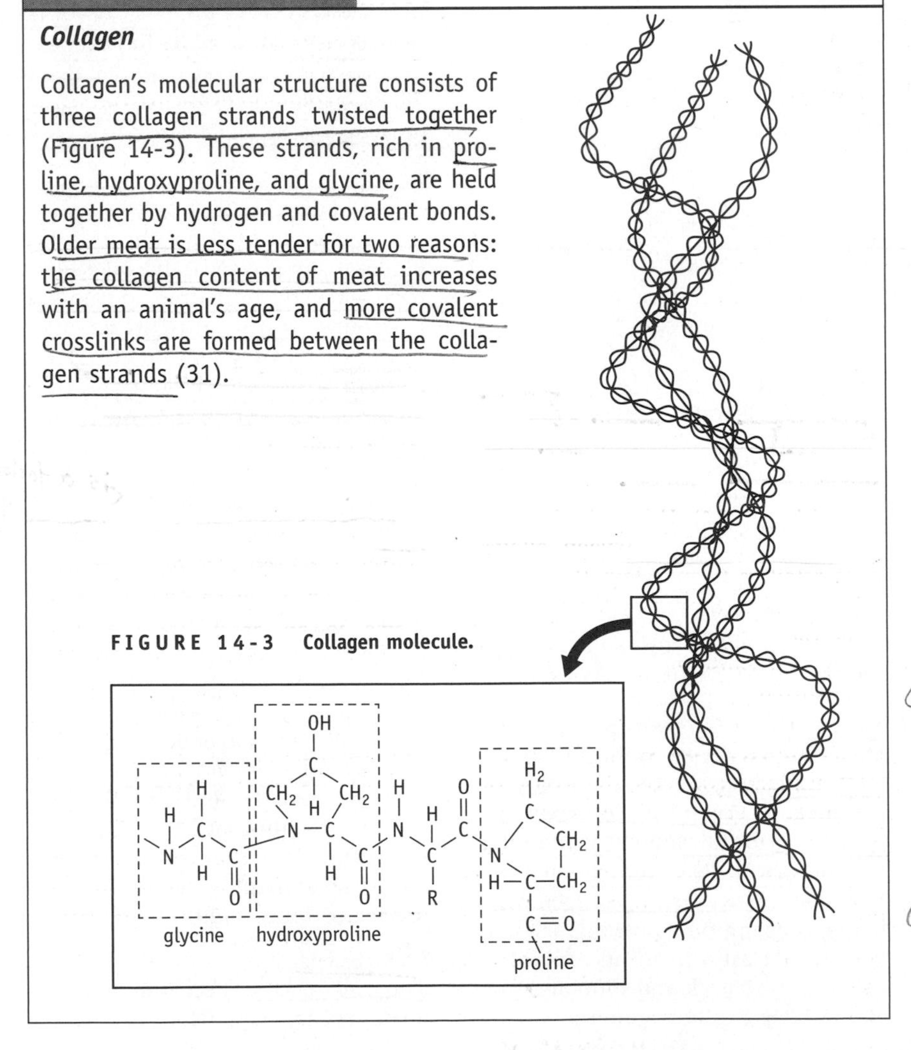

FIGURE 14-3 Collagen molecule.

Pigments

Many people evaluate a meat's color when deciding whether or not to purchase a particular meat cut. The color of meat is derived from pigment-containing proteins, chiefly myoglobin, and to a lesser extent, hemoglobin. The so-called "red meats"—beef, pork, sheep, and lamb—have more of these pigments than poultry or fish. Myoglobin receives oxygen from the blood and stores it in the muscles, while hemoglobin transports oxygen throughout the body and is found primarily in the bloodstream.

The higher the concentration of myoglobin in raw meat, the more intense its bright red color. A number of factors influence the concentration of myoglobin. Heavily exercised muscle has a higher demand for oxygen, so it is higher in myoglobin and therefore redder than the less exercised muscles. The red color of meat also increases as the animal ages, which is why beef is redder than veal, and mutton is darker than the pink hue of lamb. Meat color also varies from species to species. Beef is darker than lamb, which, in turn, is darker than pork, a meat that is on the pink side with no visible red.

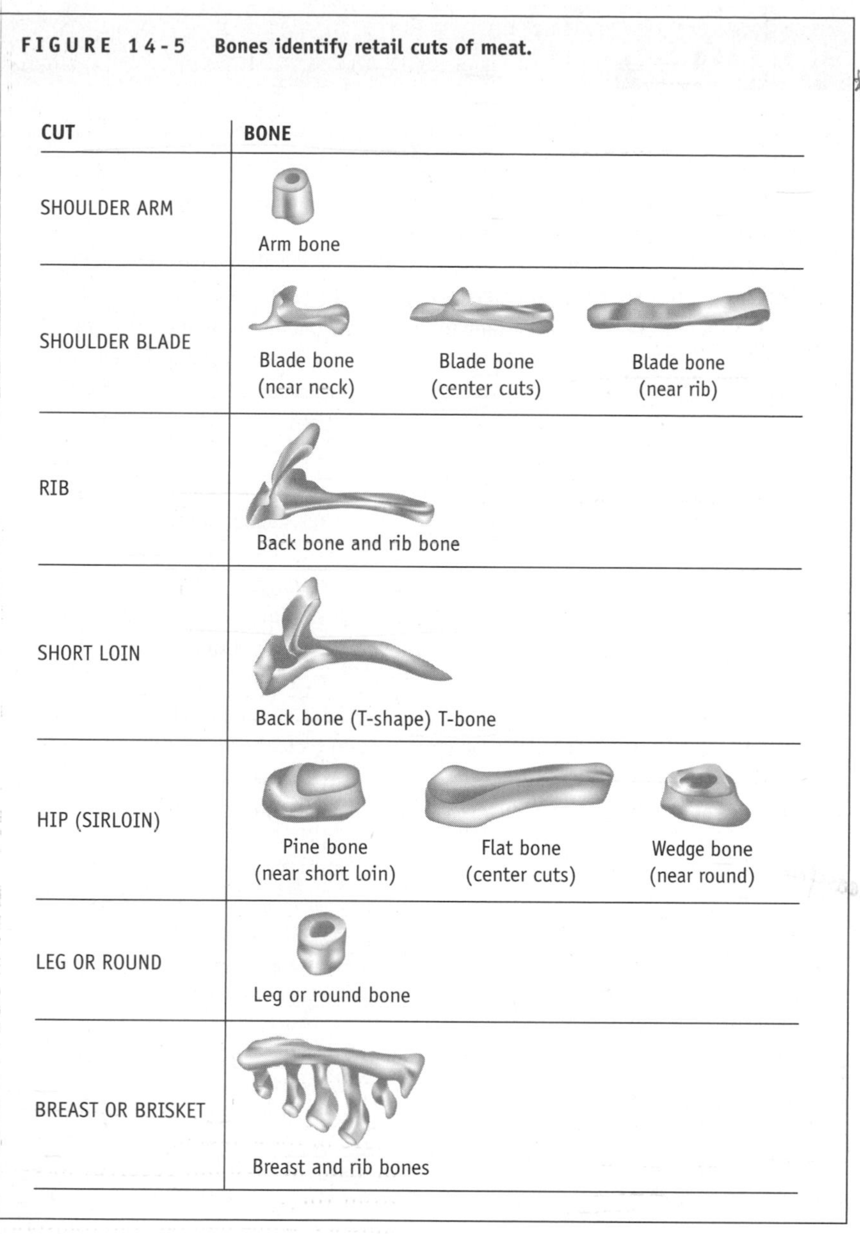

FIGURE 14-5 **Bones identify retail cuts of meat.**

Effect of Oxygen on Color. Exposure of meat to oxygen changes the color of myoglobin, and therefore the meat. After slaughter, meat undergoes several changes in color over time that are due to modifications in the molecular structure of myoglobin and/or hemoglobin (Chemist's Corners 14-2 on this page and 14-3 on page 257). Myoglobin within the meat is purplish red, but once cut and exposed to oxygen, it becomes bright red—a color indicating freshness and desired by consumers. After awhile, meats left in storage may be exposed to bacteria, less oxygen, and/or kept under fluorescent or incandescent lights, all of which turn the meat brownish-red (34). Using plastic wrap that is permeable to oxygen allows meat retailers to maintain the bright red color for a longer period of time, while vacuum wrap, which eliminates the oxygen, causes the meat to appear purplish-red.

Effect of Heat on Color. Cooking meat initially converts the color of raw meat to bright red, but then the denaturing of the pigment-containing proteins yields the classic color of well-done meat—grayish brown. Storing cooked meat too long causes the denatured protein to further break down, causing the meat to turn yellow, green, or faded.

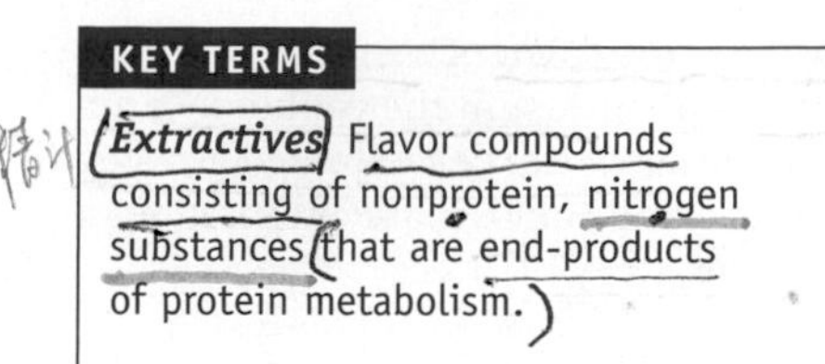
KEY TERMS

Extractives Flavor compounds consisting of nonprotein, nitrogen substances that are end-products of protein metabolism.

Nitrites in Processed Meats. The food industry uses several methods to keep meat products from turning brown. One such method is the addition of nitrites to processed meats (Chemist's Corner 14-4 on page 258). Nitrites are responsible for keeping packaged bologna and ham permanently pink, while simultaneously reducing the risk of *Clostridium botulinum* growth (34). To maintain this color, ham slices are often stored upside down with the label on the back of the Styrofoam board because grocery store lights promote undesirable color changes.

Extractives

Meat derives some of its flavor from nitrogen compounds called **extractives**. The most common extractives are creatine and creatinine, but urea, uric acid, and other compounds also contribute to the flavor of meat. The meat from older animals contains more connective tissue and extractives, and therefore yields more flavor than that from younger livestock. Extractives are water soluble, so

Chemist's Corner 14-2

Meat pigments

Each pigment-containing compound in meat consists of two parts: a protein (globin), and a nonprotein pigment (heme). The heme is an atom of iron surrounded by four connecting pyrrole rings. The difference between myoglobin and hemoglobin is that the simpler myoglobin molecule consists of one protein polymer strand and one heme (molecular weight = about 17,000), while the larger hemoglobin molecule is made of four protein polymer strands and four hemes (molecular weight = about 68,000) (31).

some of the flavor of boiled or simmered beef may be lost in the cooking water, but the flavor can be recaptured by using the cooking liquids in the preparation of soup or gravy.

Purchasing Meats

To ensure that consumers are purchasing meat that is safe, federal laws require the inspection of animal carcasses. In addition to this mandatory inspection for safety, meat may also be assigned yield grades and the later quality grades to assist consumers in selection. Meat processors submit to the grading system voluntarily.

Inspection

The Federal Meat Inspection Act of 1906 made inspection mandatory for all meat crossing state lines or entering the United States through foreign commerce. Inspections are the responsibility of the USDA Food Safety and Inspection Service. This inspection is a guarantee of only wholesomeness and does not ensure the quality or tenderness of the meat itself. A meat inspection is conducted by licensed veterinarians or by specially trained, supervised inspectors.

They examine live animals prior to slaughter, as well as animal carcasses, observe the meat at various processing stages, monitor temperatures and additives, review packaging materials and labels, determine employee and facility hygiene, and check imported meat. Meat that passes this federal inspection is marked with an inspection stamp (Figure 14-6) to distinguish it from meats that are diseased or slaughtered in unsanitary conditions. The exception is inspection for *Trichinella spiralis,* since

FIGURE 14-6 USDA meat inspection stamp. The number is assigned to the meat processing plant. Consumers rarely see the stamp, which is placed on larger wholesale cuts.

NUTRIENT CONTENT

Meat consists of water, protein, and fat, with a few minerals and some B vitamins. It contains trace amounts, if any, of carbohydrate (liver is the highest source), no fiber, and no vitamin C. Meat is about 75 percent water, and most of this water is found in the muscle cells.

Protein. High-quality protein is the second major constituent of meat after water, accounting for about 20 percent of its weight (34). Meat contains 7 grams of protein per ounce. Current recommendations on meat intake suggest 4 to 6 ounces of meat a day, or 28 to 42 grams. The Reference Daily Intake of protein for adults is 50 grams per day, so meat is an excellent source of this nutrient. The remaining protein is usually met by consuming foods from other food groups.

Fat. Fat content can vary widely, according to the grade of meat and its cut. Several cuts of beef are lower in fat than an equal amount of some poultry choices, yet consumers often select poultry over beef, thinking it is lower in fat (40). See Color Figure 6 for the cuts of meat that are lowest in fat.

The general rule of thumb is that beef cuts from the loin or round, and veal and lamb cuts from the loin or leg, are the leanest choices. Examples of lean beef cuts include sirloin, tenderloin, top loin, top round, and eye round. Lower-fat meats, including some types of wild game, are becoming more popular with consumers. The fat content of wild game compared to beef is shown in Table 14-1. Most processed meats, such as hot dogs and bologna, are not a good choice for consumers looking to lower fat consumption, because they average 30 to 50 percent fat.

TABLE 14-1
Fat Content of Wild Game Compared to Beef*

Species	*Fat (grams)*	*Calories*
Beef, T-bone, USDA choice	10	214
Antelope	4	165
Bison (buffalo)	2	143
Deer (venison)	3	158
Duck (skinless)	11	201
Elk	2	146
Moose	1	134
Pheasant, breast without skin	3	133
Rabbit	8	197
Rabbit (wild)	4	173
Squirrel	5	173

*All values shown are for a 100 gram (about 3 oz) cooked portion, with visible fat removed unless noted.

Carbohydrate. Meat contains very little carbohydrate. Glycogen, found in liver and muscle tissue, is present when the animal is alive, but the glucose that makes up the glycogen is broken down to lactic acid during and after slaughter.

Vitamins. Meat is an excellent source of certain B vitamins—thiamin (B_1), riboflavin (B_2), pyridoxine (B_6), vitamin B_{12}, niacin, and some folate. Niacin is

NUTRIENT CONTENT

obtained from tryptophan, an amino acid plentiful in meats and milk. Lean pork is an excellent source of thiamin (B_1). Fat-soluble vitamins, especially vitamin A, are found in liver. Vitamin loss during meat preparation depends on the temperature, the time exposed to the heat source, and the cooking method. Thiamin (B_1) is especially sensitive to heat, so levels of this vitamin are somewhat reduced in canned meats, which have undergone high-heat processing (41). Water-soluble B vitamins can be leached from meat into cooking liquid, but can be recaptured by making gravy or soup from that liquid.

Minerals. Meat is an excellent source of iron, zinc, copper, phosphorus, and a few other trace minerals. Liver is especially rich in iron and vitamin B12. Minerals are stable when heated, and although they can be lost in cooking water, retention of most minerals in cooked meat ranges from 80 to 100 percent.

Chemist's Corner 14-3

Color Changes in Meat

The molecular changes in the pigment-containing proteins determine the color of meat from slaughter to consumption. Color changes are dependent on the oxidation or reduction of the iron in the heme. Initially, the internal color of meat is purplish-red, because slaughter depletes oxygen concentrations in the meat. As soon as meat is cut from the carcass and exposed to the oxygen in the air, a bright red compound known as oxymyoglobin forms (Figure 14-7). Over time, the bright red oxymyoglobin is further oxidized to metmyoglobin which is a brownish-red color (42). Older meat cuts look browner because myoglobin or oxymyoglobin is converted to metmyoglobin as the iron in the pigment is oxidized from its ferrous ($^{+}2$) to ferric ($^{+}3$) state. This usually occurs during storage when the meat continues to be exposed too long to bacteria, oxygen, or light (fluorescent and incandescent). The brownish-red color resulting from metmyoglobin is undesirable to retailers (34).

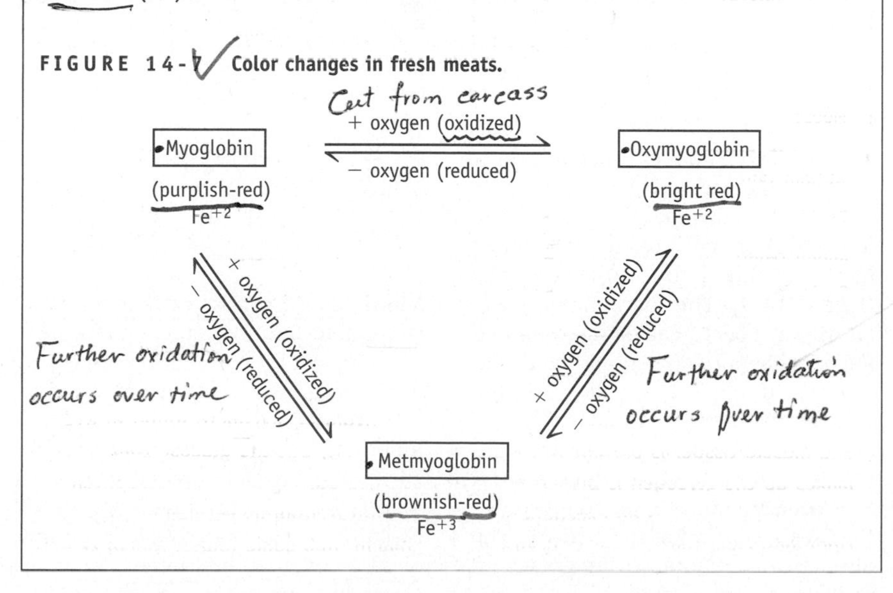

FIGURE 14-7 Color changes in fresh meats.

> **KEY TERMS**
>
> ***Quality grades*** The USDA standards for beef, veal, lamb, and mutton.

visual inspection may miss the small parasite (see Chapter 3 for information on *Trichinella spiralis*).

Other laws passed since 1906 further protect the meat supply. The USDA can oversee only those meats that are transported between states, so the Wholesale Meat Act of 1967 was passed to require that meat sold within the states meets requirements equal to the federal standards. Most recent was the implementation in 1997 of new USDA inspection regulations incorporating hazard analysis and critical control points (HACCP) within meat and poultry slaughterhouses, along with mandatory testing for *E. coli* (14).

Grading

The grading of meat is not under government mandate or control, but is a strictly voluntary procedure that the meat packer or distributor may have done under contract with the USDA (27). For purposes of grading, a cut is made between the twelfth and thirteenth rib in order to expose the rib muscle.

Quality. The **quality grades** for the different types of meat are shown in Table 14-2. Factors considered in grading are color, grain, surface texture, and fat distribution. Unfortunately, this system is not used uniformly by retailers. Instead of Prime, Choice, Select, and Standard, retailers frequently designate the quality of their meat with a descriptive word or phrase such as "5 Star," "Blue Ribbon," or "Supreme." This is purely a marketing strategy and leaves it up to the consumer to determine the validity, or lack thereof, of the designation. While a large percentage of meat sold is graded, the term "no roll" is used to indicate ungraded meat.

Any judgment of quality must be somewhat subjective, but several identifiable factors separate a poor cut of meat from one that is excellent. Top cuts of meat have the optimum color for their type, and fine-grained, smooth surfaces that are velvety, silky, or satiny to the touch. They contain fat that is evenly

Chemist's Corner 14-4

Nitrites in Processed Meats

Nitrite, a conjugate base of a weak acid named nitrous acid, provides color to processed meats by combining with the myoglobin pigment to produce nitrosylmyoglobin. This resulting compound denatures during the cooking phase of the curing process to form a pink-colored compound called nitrosyl hemochrome (Figure 14-8). One of the major problems of storing nitrite-cured meats is that continued exposure to oxygen and light oxidizes the iron from the ferrous ($^{+}2$) to ferric form ($^{+}3$), which results in a brown discoloration. Additional oxidation of the pigment-containing protein's porphyrin ring, instead of the iron, results in a very undesirable yellow or greenish color, making the meat unappealing to consumers (34).

FIGURE 14-8 Color changes in cooked meats.

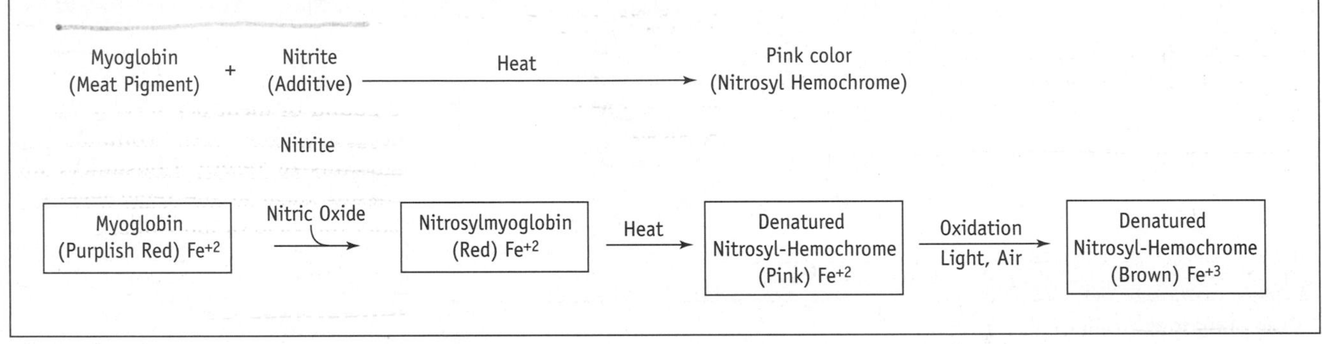

TABLE 14-2
USDA Quality Grades for Beef, Veal, Lamb, and Mutton From Highest to Lowest*

Beef	*Veal*	*Lamb*	*Mutton*
Prime	Prime	Prime	Choice
Choice	Choice	Choice	Select
Select	Select	Select	Commercial
Standard	Standard	Commercial	Utility
Commercial	Commercial	Utility	Cull
Utility	Utility	Cull	
Cutter	Cull		
Canner			

*There are no quality grades for pork.

distributed, white or creamy-white rather than yellow in color, and firm instead of brittle or runny. These factors contribute to tenderness, which is never directly measured in grading, although it remains a top concern among both retailers and restaurateurs (32).

Influence of Fat Content on Grading. Fat, especially in the form of marbling, melts during heating, thereby increasing the flavor and perceived tenderness of the meat. USDA quality grades of beef reflect this marbling. Prime, the top USDA grade, contains the most marbling and is the most expensive (Figure 14-9). The marbling and any fat trim of a beef steak being examined for possible purchase should be cream colored. If the fat is yellowish, the meat may be from an older animal and may be tough. However, when retailers trim the fat to ⅛ inch around the meat, it makes it difficult to judge the fat's actual color and texture. The differences in USDA quality grades for beef are described below.

KEY TERMS

Yield grade The amount of lean meat on the carcass in proportion to fat, bone, and other inedible parts.

Beef USDA Quality Grades. There are eight USDA quality grades for beef, with the top three—Prime, Choice, and Select—being of most concern to consumers (Table 14-3). Prime accounts for less than one-fifth of the beef marketed. It is usually sold to restaurants, because the price is not competitive enough for the average supermarket consumer. Choice and Select are the grades most commonly purchased by consumers in the supermarket. Select cuts contain 5 to 20 percent less fat than Choice cuts, and 40 percent less fat than Prime. Standard and Commercial USDA grades are not seen at the retail level, because they are usually from older, more mature, and therefore less tender cattle. USDA grades identified as Utility, Cutter, and Canner are usually used in processed foods such as canned meats, sausages, and pet foods.

Yield. **Yield grade** is the other main factor determining the grade of a meat cut (Figure 14-10). The carcasses of beef, lamb, and mutton are rated at yields of 1 to 5, with 1 providing the highest yield and 5 the lowest (Table 14-4). Pork is yield-graded from 1 to 4. Veal is not yield-graded because it contains so little fat. While 4 ounces of raw meat with little or no bone generally constitutes one serving, ¾ to

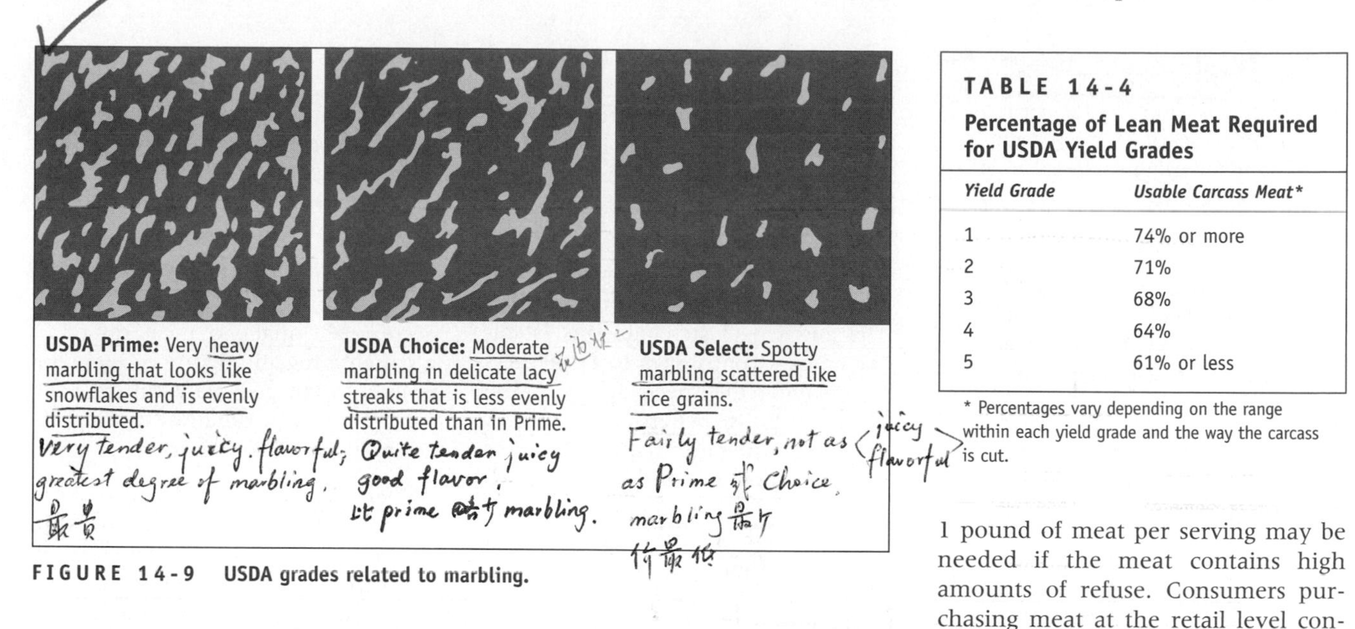

FIGURE 14-9 USDA grades related to marbling.

TABLE 14-3
Top Three USDA Quality Grades for Beef

USDA Grade	*What the Grade Means*
USDA PRIME	Very tender, juicy; flavorful; the greatest degree of marbling. The most expensive of the grades, Prime is sold to finer restaurants and some meat stores.
USDA CHOICE	Quite tender and juicy, good flavor; slightly less marbling than Prime. The grade most frequently found in retail stores.
USDA SELECT	Fairly tender; not as juicy and flavorful as Prime and Choice; has least marbling of the three, and is generally lower in price.

TABLE 14-4
Percentage of Lean Meat Required for USDA Yield Grades

Yield Grade	*Usable Carcass Meat**
1	74% or more
2	71%
3	68%
4	64%
5	61% or less

* Percentages vary depending on the range within each yield grade and the way the carcass is cut.

1 pound of meat per serving may be needed if the meat contains high amounts of refuse. Consumers purchasing meat at the retail level consider the yield of individual cuts.

Tenderness of Meats

Tender meat generally is preferred by consumers, but just because meat is given a top quality grade does not guarantee its tenderness. The only real test is how easily the meat gives way to the teeth. Extreme variations of tenderness exist in beef, even within different areas of a single meat cut, but overall, natural meat tenderness is due to factors such as the cut, age, and fat content. Meats can also be treated to make them more tender by adding enzymes, salts, and acids, or by subjecting them to mechanical or electrical treatments. Preparation temperatures and times also have an influence on tenderness.

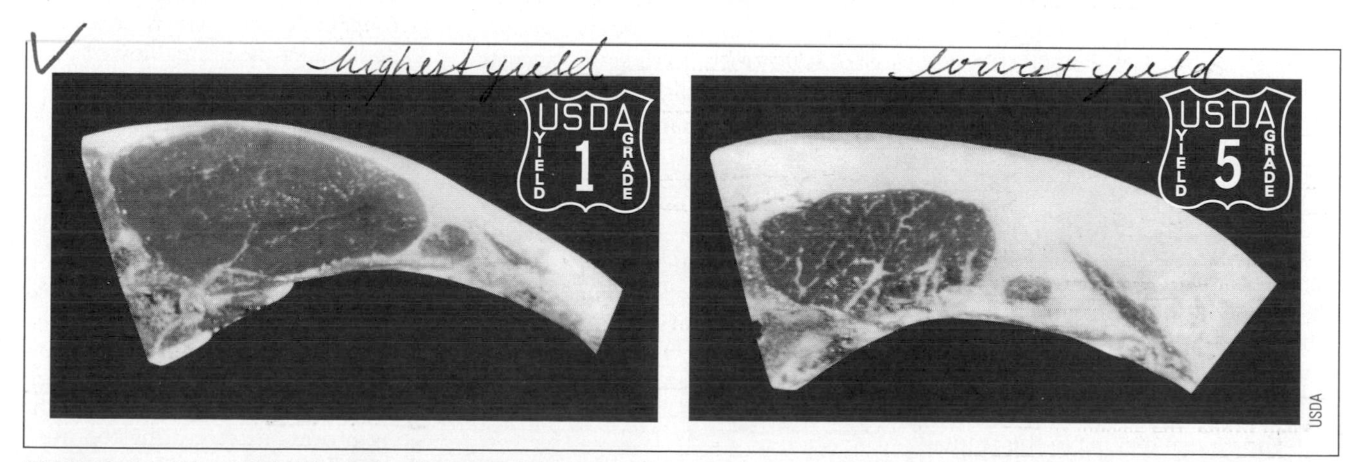

FIGURE 14-10 USDA yield grades. The amount of lean muscle is compared in a ratio to the non-meat portion—fat, bone, and inedible material.

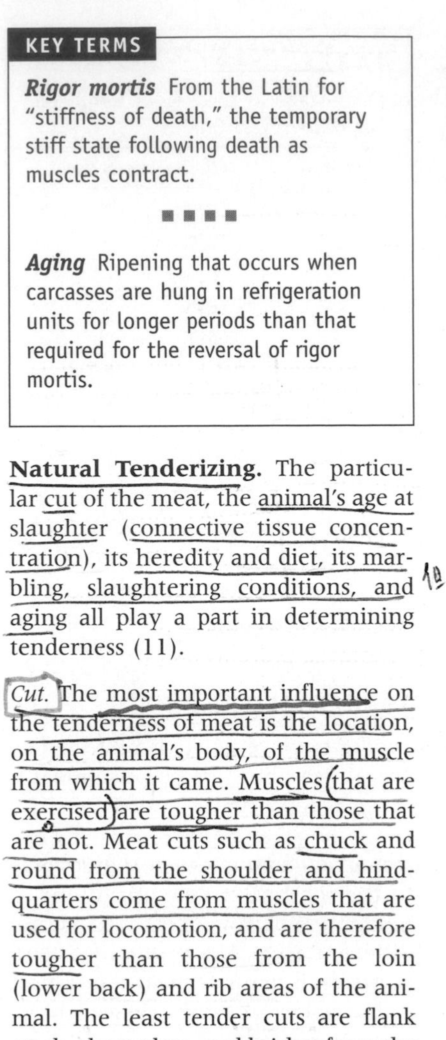

KEY TERMS

Rigor mortis From the Latin for "stiffness of death," the temporary stiff state following death as muscles contract.

Aging Ripening that occurs when carcasses are hung in refrigeration units for longer periods than that required for the reversal of rigor mortis.

Natural Tenderizing. The particular cut of the meat, the animal's age at slaughter (connective tissue concentration), its heredity and diet, its marbling, slaughtering conditions, and aging all play a part in determining tenderness (11).

Cut. The most important influence on the tenderness of meat is the location, on the animal's body, of the muscle from which it came. Muscles that are exercised are tougher than those that are not. Meat cuts such as chuck and round from the shoulder and hindquarters come from muscles that are used for locomotion, and are therefore tougher than those from the loin (lower back) and rib areas of the animal. The least tender cuts are flank steak, short plate, and brisket from the legs and underside of the animal. The most tender cuts of the carcass, such as sirloin, tenderloin, and rib eye, are found in the loin and rib areas.

Animal's Age. An animal's age at the time of slaughter contributes to tenderness, and top USDA grades usually come from relatively young animals. As muscles age, the diameter of the muscle fibers increases and more connective tissue develops, resulting in toughening of the meat (22).

Heredity. Cuts of meat will vary in tenderness because of genetic factors. For example, beef from Black Angus cattle, which are bred to be heavily muscled and marbled, will be very different from meat obtained from dairy cattle or from one of the other, larger breeds of cattle.

Diet. The type of diet fed to the animal directly influences its fat accumulation, which is one of the factors affecting the tenderness and flavor of its meat. Ranchers have long known that grain-fed cattle yield ground beef that is more tender and better flavored than that from cattle fed hay or left to feed on the range (30, 39).

Marbling. Fattening animals before slaughter is thought to increase tenderness by increasing marbling and the development of subcutaneous fat. The amount of subcutaneous fat on the carcass contributes to tenderness by delaying the speed at which the carcass chills when refrigerated. When choosing meat cuts, consumers seem to prefer lean-looking meats over more marbled ones, but usually reverse their choices in a taste test after those same cuts are prepared.

Slaughtering Conditions. Both the conditions preceding slaughter and the handling of the carcass immediately afterward affect the tenderness of meat. If the animal is under stress from fear, fasting, temperature extremes, or exercising, its muscle cells may be deprived of oxygen. When this occurs, the cells switch to anaerobic energy sources such as glycogen, which is converted to lactic acid, causing the pH to fall. A poor-quality meat will result if the glycogen has already been converted to lactic acid prior to slaughter.

Rigor Mortis. Within 6 to 24 hours after slaughter, the muscles of livestock enter the state of **rigor mortis**. This condition reverses naturally one or two days after slaughter.

HOW & WHY?

Why does a carcass stiffen? Rigor mortis is caused by a cascade of events that take place at the cellular level. Death interrupts the blood flow and prevents oxygen from reaching the cells. Changes then occur within the cells of the muscles, causing them to contract and stiffen. The rigidity of the muscles in rigor mortis occurs because the crosslinks between the actin and myosin filaments overlap and cause the sarcomeres to shorten. The automatic contraction of fibrils in the muscle cells causes the characteristic muscle stiffness.

During rigor mortis, the oxygen-deprived cells switch to glycogen as an energy source, converting it to lactic acid, the build-up of which causes the pH to fall from approximately 7.0 to 5.8. The perception of a meat's juiciness or dryness depends on the binding of water to muscle proteins, and this is influenced by pH. Water-holding capacity is best in meats with a pH of 5.8 (20).

Problems With Improperly Handled Meat. The following problems, which decrease the quality of meat, can occur if slaughtering and rigor mortis are not properly managed: dark-cutting beef, pale, soft, and exudative (PSE) pork, thaw rigor, and cold shortening.

- *Dark-cutting beef.* If glycogen stores are depleted before death because the animal is exercised or stressed, insufficient lactic acid will be produced during rigor mortis. The resulting higher pH (above 5.8) of the meat will result in a deep-purple brown meat known as dark-cutting beef, which has a sticky texture that is unacceptable to consumers (31).
- *PSE pork.* Pale, soft, and exudative (PSE) pork results if the pH drops too low. A low pH—under 5.1, or even up to 5.4—can cause the pork to become extremely pale, mushy, slimy, flavorless, and full of excess drip (7, 13).
- *Thaw rigor.* Freezing meat before it undergoes rigor mortis can cause thaw rigor, a phenomenon in which the meat shrinks violently by almost 50 percent when thawed.
- *Cold shortening.* The same thing occurs, though to a lesser degree, when meat has been chilled too rapidly before rigor mortis, resulting in cold shortening. In both cases, the meat will be tougher. Neither thaw rigor nor cold shortening meat are allowed to be sold at the consumer level.

Also, meat that is cooked while in a state of rigor mortis, called green meat or cooked rigor, will be tough. If it is prepared before stiffening begins, however, it can be quite tender.

Aging. **Aging** meats improves their juiciness, tenderness, flavor, color, and their ability to brown during heating. This treatment pertains primarily to beef. Hanging aids in the aging process by stretching the muscles (Chemist's Corner 14-5). The animal's species, size, age, and activity before slaughter influence how long

m. fibers begin to break down.

rigor mortis lasts. Beef takes about ten days to age, which is about the same amount of time it takes for meat to be transported, packaged, and sold to the consumer. Top-quality beef is often aged longer, up to six weeks. Mutton is sometimes aged, but pork and veal come from such young animals that aging is not required to increase tenderness. The fat in pork tends to go rancid quickly, and veal's lack of protective fat covering causes it to dry out too quickly—further reasons these meats are not routinely aged.

Meats are aged in one of several ways. The time required for aging depends on the method used.

- *Dry aging.* Carcasses are hung in refrigeration units at 34° to 38°F (1° to 3°C) with low (70 to 75 percent) or high (85 to 90 percent) humidity for 1½ to 6 weeks. Specialty steak houses and fine restaurants usually purchase dry-aged meat.
- *Fast aging.* Warmer temperatures of 70°F (21°C) with a high humidity of 85 to 90 percent lower the aging time to two days, but additional aging will occur during the ten or so days it takes the meat to reach the consumer. Ultraviolet lights are used to inhibit microbial growth. Most retail meat is fast-aged.
- *Vacuum-packed aging.* Less weight loss and spoilage occur in meats that are vacuum-packed (cryovac) aged. During this process, meat carcasses are divided into smaller cuts and vacuum-packed in moisture- and vapor-proof plastic bags, and then aged under refrigeration.

Artificial Tenderizing. External treatments can be applied to meats to increase their tenderness. These include the use of enzymes, salts, acids, and mechanical methods such as grinding or pounding.

Enzymes. One of the reasons that contracted muscles begin to "relax" toward the end of rigor mortis is that proteolytic enzymes work internally to break down the proteins within the muscle fibrils (20, 24). A more even distribution of enzymes may be achieved by injecting a tenderizing solution of papain, or some other proteolytic enzyme, into the bloodstream of animals ten minutes before slaughter. This optional treatment sends enzymes travelling to all the muscles through the circulatory system, but they are not activated until meat from the animal is exposed to heat during preparation. This process not only increases tenderness, but shortens the time of rigor mortis and aging (35).

Commercial meat tenderizers containing enzymes are available for consumers to use, but they are effective only on fairly thin cuts of meat because they penetrate to a depth of only ½ to 2 millimeters. They are ineffective on larger cuts such as roasts. Tenderizers are sold as a salt or liquid mixture and differ in the proteolytic enzymes they contain: papain from papayas, bromelin from pineapples, ficin from figs, trypsin from the pancreases of animals, and rhyozyme P-11 from fungi.

> **HOW & WHY? ?????????**
>
> **How Do Meat Tenderizers Work?** Meat tenderizers contain enzymes that break down muscle proteins. They are sprinkled on meat, which is then pierced with a fork to drive the enzymes below the surface, where they hydrolyze muscle cell proteins and connective tissue when activated by the heat of preparation.

The enzymes are not active at room temperature. The optimal activity temperature (highest rate of activity) for papain, the most common tenderizing enzyme, is about 131° to 170°F (55° to 76°C), which is reached only during heating. Exceeding 185°F (85°C) denatures the enzyme, thus inhibiting its activity. Uniform distribution is hard to achieve with the use of commercial tenderizers, and any attempt to get more of the enzyme to penetrate by adding excessive amounts of it can cause the meat to have an unappetizing, mealy, mushy texture.

Salts. Tenderness can also be increased by the addition of salts in the form of potassium, calcium, or magnesium chlorides. These salts retain moisture and break down the component that surrounds the muscle fibers, resulting in the release of proteins. Polyphosphates are sometimes added to the salts to improve the meat's juiciness by increased water retention ability (12), and if added to processed meats, they also increase firmness, emulsion stability, and antimicrobial activity (16, 28). However, along with increased water retention capacity is an increase in sodium concentration.

Acids. Meats can be made more tender by applying marinades containing acids or alcohol, which break down the outside surface of the meat (33). The various acids found in marinades include vinegar, wine, and lemon, tomato, or other fruit juices. Not only do marinades tenderize the meat, but they increase flavor and also contribute to color. The maximum benefit of a marinade can be obtained by increasing the surface area of the meat. This may be done by cutting the meat into small pieces, such as teriyaki strips or kabob cubes. Marinades penetrate only the surface of the meat and are therefore not effective at tenderizing large cuts of meat or poultry. Generally, the acid in a marinade is responsible for tenderizing, although some marinades rely on added enzymes from certain tropical fruits such as papayas and pineapples. The meat is then allowed to soak in the marinade, in the refrigerator, from half an hour to overnight, or for several days for sauerbraten.

> **Chemist's Corner 14-5**
>
> ***Rigor Mortis***
>
> The lack of blood flow after slaughter creates an anaerobic condition. This causes the muscles to rely on the breakdown of glycogen (glycolysis) to glucose. Muscles stay relaxed in the presence of sufficient adenosine triphosphate (ATP), but once it is used up through glycolysis, the lack of ATP causes the actin to bind irreversibly with myosin. The muscles then contract into a state of rigor mortis. The passing of rigor occurs when the muscles gradually extend again. This is facilitated by the proteases that hydrolyze proteins and disrupt the Z bands. As a result, the actin and myosin release from each other, causing the muscles to relax (20).

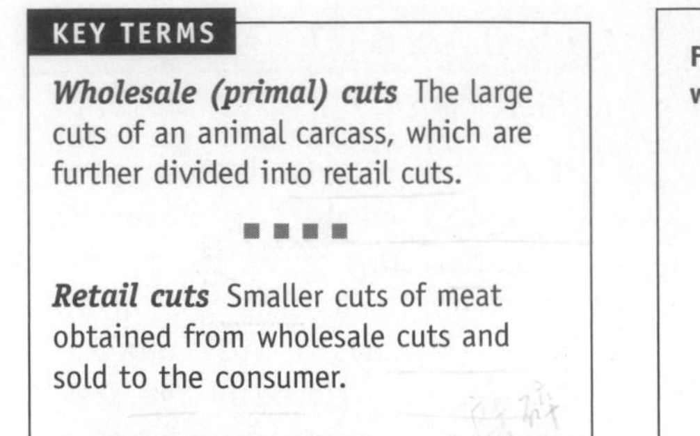

KEY TERMS

Wholesale (primal) cuts The large cuts of an animal carcass, which are further divided into retail cuts.

■ ■ ■ ■

Retail cuts Smaller cuts of meat obtained from wholesale cuts and sold to the consumer.

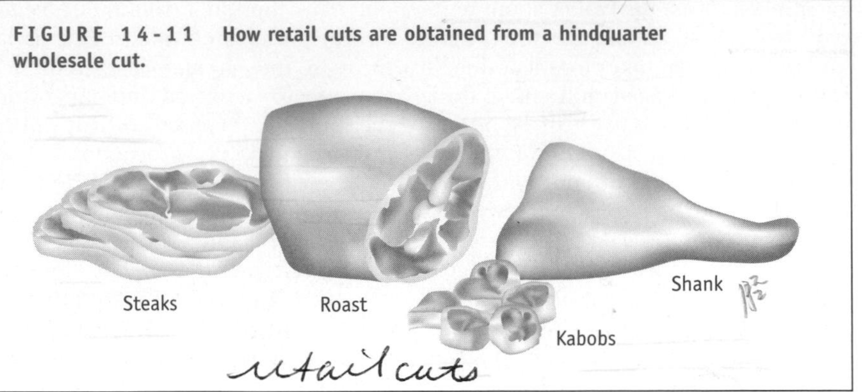

FIGURE 14-11 How retail cuts are obtained from a hindquarter wholesale cut.

Mechanical Tenderization. Meat can be tenderized mechanically by a number of methods, including grinding, cubing, needling, and pounding. These actions physically break the muscle cells and connective tissue, making the meat easier to chew. Grinding and cubing meat simply increases the surface-area-to-volume ratio, causing the teeth to have less work to do. Needling uses a special piece of equipment to send numerous needle-like blades into the meat, separating the tissues. Because of the equipment required to do this, it is usually not done at the consumer level. Another method of mechanical tenderization, which is more easily done in the home, is simply pounding the meat with a special hammer that breaks apart its surface tissue.

Electrical Stimulation. The meat of beef cattle and sheep, but not swine, becomes more tender when a current of electricity is passed through the carcass after slaughter and before the onset of rigor mortis. Electrical stimulation speeds up rigor mortis by accelerating glycogen breakdown and enzyme activity, which disrupts protein structure, making the meat more tender (17). In this way, the meat can be immediately cut up without any loss of quality.

Cuts of Meat

There are two major types of meat cuts, wholesale and retail. Prior to reaching the supermarket, a carcass is divided into about seven **wholesale** or **primal cuts**. Although the carcasses of each species are sectioned slightly differently, the basic wholesale cuts are similar to each other and are identified by the major muscles and by bone "landmarks" (see Figures 1 to 5 in the color insert). These wholesale cuts are then divided into the **retail cuts** purchased by consumers. Figure 14-11 shows how common retail cuts are obtained from the wholesale cut of a hindquarter.

Terminology of Retail Cuts. The use of the standard system of naming retail cuts is not mandatory, so consumers often face confusion at the market. The same cut of meat may be called by different names, depending on the retailer or the part of the country in which it is sold. Beef chuck cross-rib pot roast is variously known as Boston cut, bread and butter cut, cross-rib roast, English cut roast, and thick rib roast. A system of standardized names and specifications for over 300 cuts of beef, veal, pork, and lamb has been established and is known as IMPS (Institutional Meat Purchases Specifications). "The Meat Buyers Guide" containing these IMPS is published by the National Association of Meat Purveyors (NAMPS), and it serves as a key reference for those responsible for the preconsumer purchase and for the sale of meat. It is adhered to by most retail meat markets. Under this system, meat labels include the species (beef, veal, pork, or lamb), primal cut, and retail cut (Figure 14-12). Hence, rib-eye steak would be labeled "Beef, rib, rib-eye steak."

Beef Retail Cuts. Rib, short loin, and sirloin wholesale cuts lie along the back of the animal and are usually the most tender and expensive cuts of beef (Figure 14-13). Rib roasts are the most tender roasts, and tenderloin the tenderest steak. Filet mignon is the small end of the tenderloin, but some retailers incorrectly, perhaps deliberately, label any cut from the tenderloin as filet mignon. Although less tender, chuck and round wholesale cuts provide many popular retail cuts. The least tender wholesale cuts are flank, short plate, briskets, and foreshank (Table 14-5).

Ground Beef. About 44 percent of all fresh beef is sold in the form of ground beef and used extensively in fast-food restaurants, schools, military programs, and homes (3). The terms "ground beef" and "hamburger" are often used interchangeably, but there is a difference. The USDA classifies ground beef as beef that has been ground. Hamburger is ground beef that is often combined with ground fat; seasonings may be also added. Neither ground beef nor hamburger may exceed 30 percent fat by weight. Regular ground beef con-

FIGURE 14-12 Meat labeling.

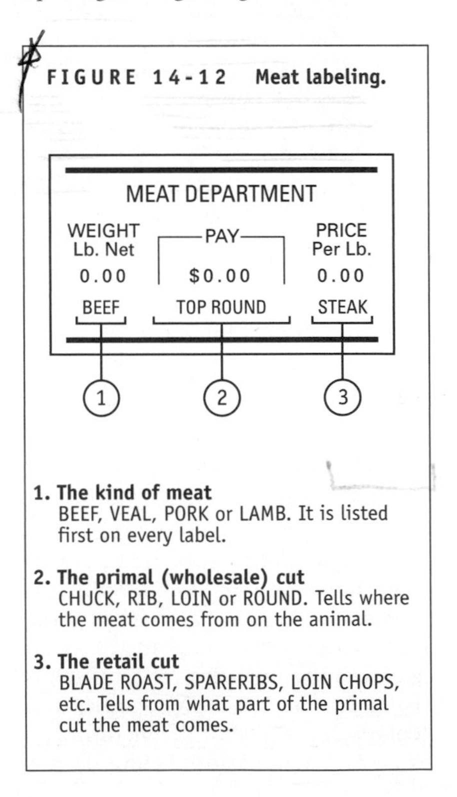

1. **The kind of meat**
   BEEF, VEAL, PORK or LAMB. It is listed first on every label.
2. **The primal (wholesale) cut**
   CHUCK, RIB, LOIN or ROUND. Tells where the meat comes from on the animal.
3. **The retail cut**
   BLADE ROAST, SPARERIBS, LOIN CHOPS, etc. Tells from what part of the primal cut the meat comes.

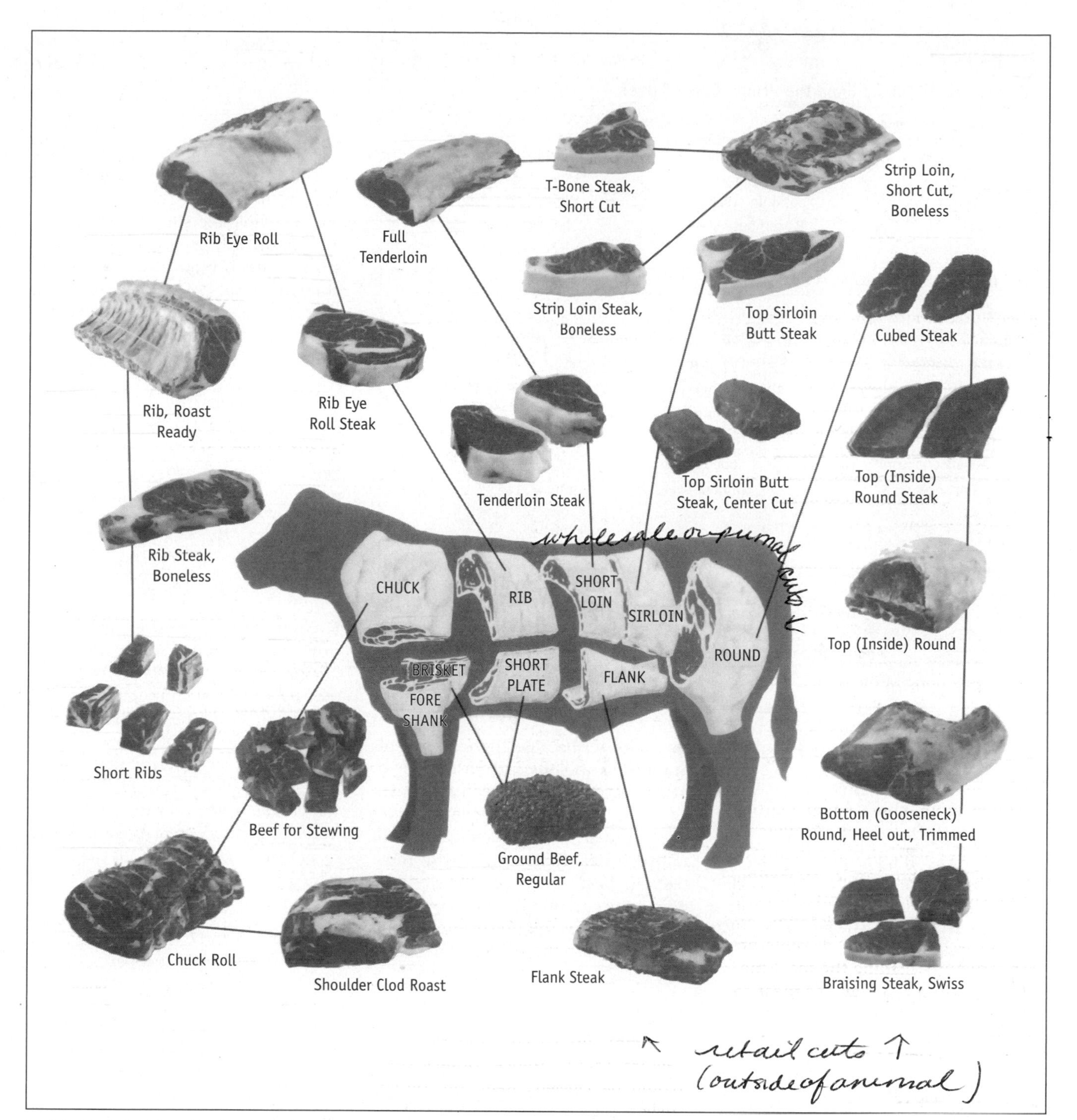

FIGURE 14-13 Wholesale and retail cuts of beef.

tains 30 percent fat, lean ground beef about 23 percent, and extra lean ground beef does not exceed 15 percent fat. Draining the fat off hamburger or ground beef during and after cooking lowers the fat content appreciably. Consumer preference studies have shown that ground beef containing 15 to 20 percent fat is preferred. Reducing the fat content below 20 percent decreases the flavor, tenderness, and juiciness of the product (4).

The fat in ground beef can be reduced by adding extenders such as nonfat dry milk solids, texturized vegetable protein (TVP), plant starches, soy proteins, oat bran or fiber, modified food starches, maltodextrins (starches), and vegetable gums (carrageenan) (13). Many of these extenders enhance the flavor as well as lower the fat content.

Veal Retail Cuts. As will also be seen with pork and lamb, the retail cuts of veal are fewer in number than those of beef because the carcasses are smaller. The hind legs of these animals are suitable for roasts, but veal

TABLE 14-5
Retail Cuts Obtained From the Primal Cuts of Beef

Some of the more tender retail cuts

Rib	Short Loin	Sirloin
Rib eye (Spencer) (Delmonico)	Tenderloin	Sirloin steak
	Top loin steak	Sirloin tip roast
Rib roast	Porterhouse steak	Tenderloin steak
Rib steak	T-bone steak	

Less tender but still popular retail cuts

Chuck	Round
Chuck roast	Top round steak or roast
Cross-rib roast	Eye of round steak or roast
Boneless chuck eye roast	Bottom round or roast
Blade roast or steak	Rump roast
Arm pot roast or steak	Heel of round
Boneless shoulder pot roast or steak	Cubed steak
Chuck short ribs	Ground beef
Stew meat	
Ground chuck	

The least tender cuts

Flank	Short Plate	Brisket	Foreshank
Flank steak	Skirt steak rolls	Brisket	Cross cut shank
	Short ribs	Corned beef	Stew meat
	Stew meat	Stew meat	

roasts are usually tender regardless of their wholesale cut origin.

Pork Retail Cuts. Pork is usually tender, regardless of the cut, because it comes from animals under one year of age. When compared to beef, veal, or lamb wholesale cuts, the wholesale loin and spare rib cuts of pork are much longer because there is no separation of the rib and sirloin as in other carcasses (Figure 14-14). In addition, modern breeders have developed an even longer swine with fourteen ribs (as compared with thirteen in beef and lamb). These two wholesale cuts provide the majority of fresh pork retail cuts:

- *Loin:* Pork loin chop or roast, Canadian-style bacon, pork loin tenderloin
- *Spare rib:* Spare ribs, bacon, salt pork

Lamb Retail Cuts. Lambs are smaller than either cattle or swine, so the leg wholesale cuts are usually cut into roasts, with leg of lamb being the most common (Figure 14-15). A rack of lamb consists of seven or eight rib chops; the backbone is usually removed to make carving easier. A fancier cut is crown roast of lamb, which consists of two rib sections or racks attached to the backbone. Formed into a circle or crown, it can be stuffed and is often decorated just before serving by covering the bone tips with paper frills, making a very handsome main dish for any table. Lamb chops are frequently cut from the loin, rack (rib), or shoulder. Loin chops are the most tender.

Kosher Meats. Kosher meats are from animals (cattle, sheep, and goats, but not swine) designated as "clean" that have been slaughtered according to Jewish religious practices dating back more than 3,000 years. The animal must be slaughtered in the presence of a rabbi or other approved individual; with a single stroke of a knife; be completely bled; and have all its arteries and veins removed. Blood, which is synonymous with life in the Jewish tradition, must not be consumed. The hindquarter is rarely used

FIGURE 14-14 Wholesale cuts of pork.

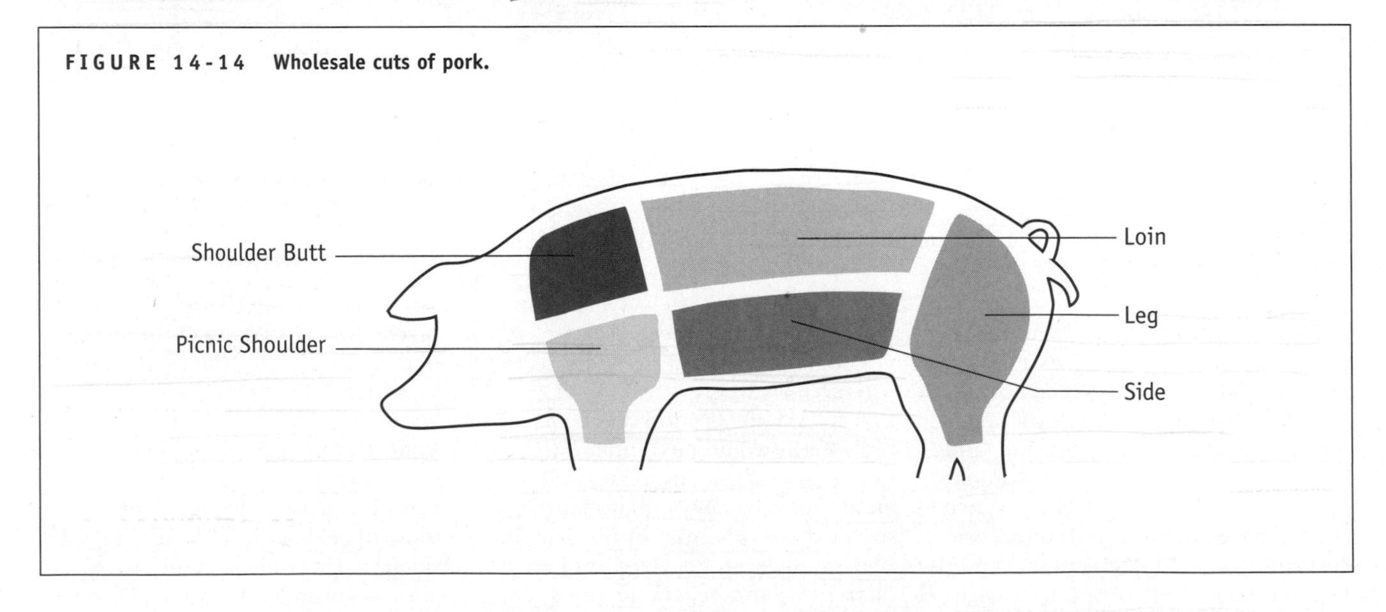

for kosher meats because it is so difficult to remove the blood vessels in this area.

Variety Meats. Variety meats, also known as organ, offal, sundry, or specialty meats, can be divided into two categories: organ meats and muscle meats (Figure 14-16). Organ meats such as liver, kidneys, and brains from young animals are generally very soft, extremely tender, and require only very short heating times. Sweetbreads can be obtained only from calves or

KEY TERMS

Variety meats The liver, sweetbreads (thymus), brain, kidneys, heart, tongue, tripe (stomach lining), and oxtail (tail of cattle).

FIGURE 14-15 Wholesale cuts of lamb.

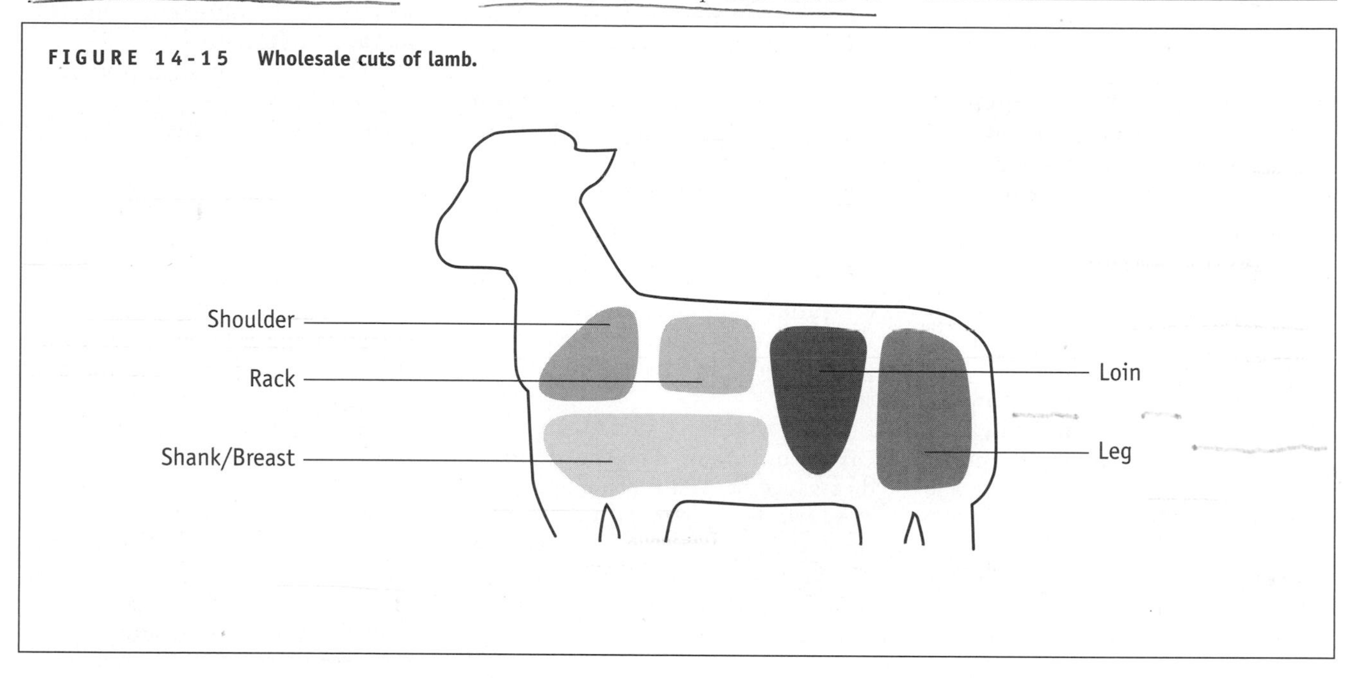

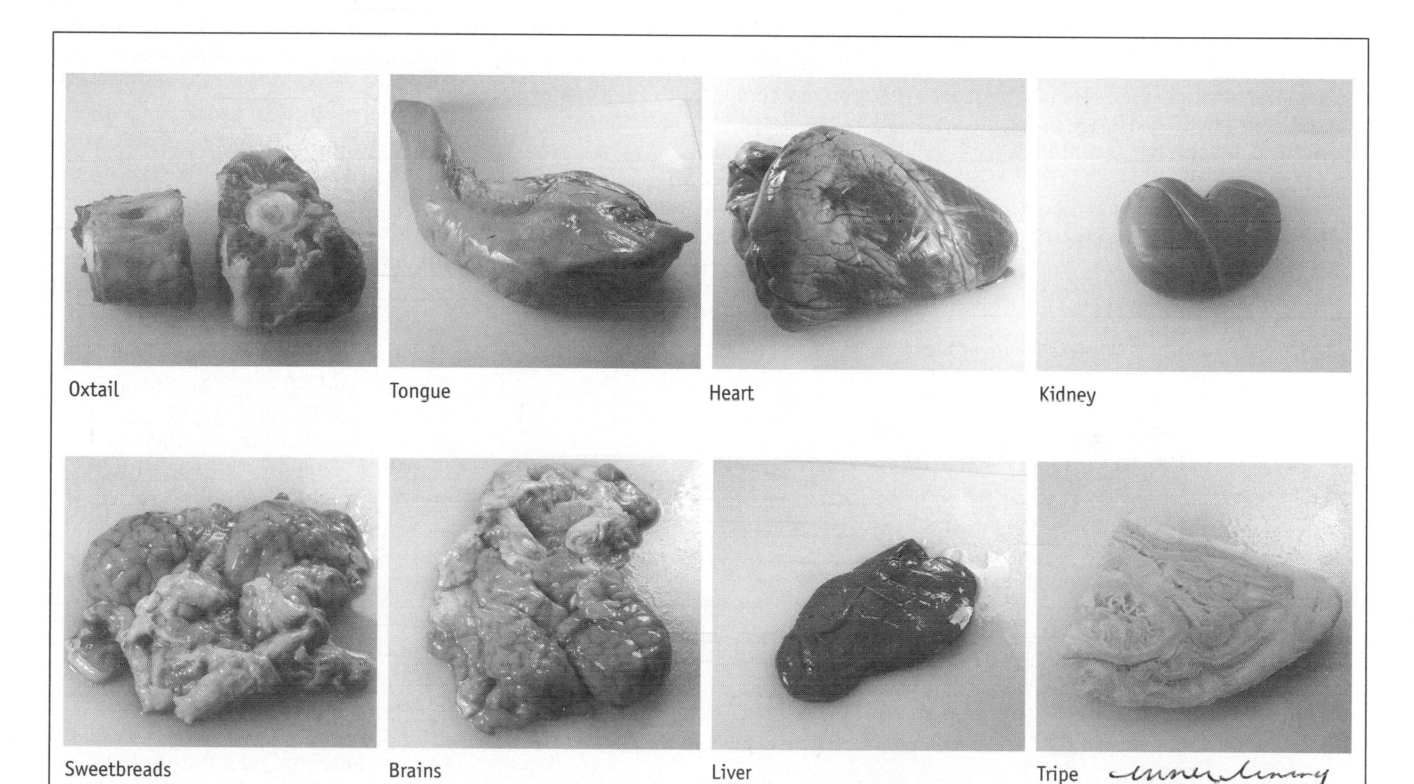

FIGURE 14-16 Variety meats.

young beef, because the thymus gland disappears as the animal matures.

The meat of heavily exercised muscles such as the tongue and heart is quite tough and requires long, slow cooking. Tripe, the inner lining of the stomach, can be smooth or honeycombed. Smooth tripe originates from the first stomach, and honeycombed tripe, which is more popular, comes from the second stomach. Both types are extremely tough and strong in flavor. Similarly to the tongue and heart, they require long, slow cooking.

Processed Meats

About one-third of all meat is processed, meaning it has been changed from its original fresh cut (Table 14-6). Ham and sausage are the most popular processed meat products. Other examples of processed meats include salami, bologna, bratwurst, and pastrami.

Processing Methods. Before the advent of refrigeration, meat was preserved by such processing methods as curing, smoking, canning, and drying.

Curing. Commonly cured meat products include ham, bacon, sausages, frankfurters, corned beef, and luncheon meats. Meat once was cured by saturating it with salt. Corned beef, a cured beef brisket, was so named because in the 16th century the word "corn" was used synonymously with "grain," so meat rubbed with coarse grains of salt was called "corned" (15). Now the same result is accomplished with a mixture of salt, sodium or potassium nitrate, sugar, and seasonings (36). Nevertheless, salt remains one of the major flavoring agents of cured meat. The different proportions and combinations of ingredients used for curing contribute to the varying flavors of cured meats, which often garner additional flavor from being smoked.

There are several ways to cure meat. Dry curing consists of mixing the ingredients together and rubbing them into the surface of the meat, from where they work their way into the center. Another method involves immersing the meat in a brine or pickling solution. The most common commercial curing technique is one in which the curing solution is mechanically pumped or injected into the meat with a machine lined with needles. Meats injected with curing solutions increase in weight. If they are not shrunk back to their original weight through heating and/or smoking, and if they contain up to 10 percent added moisture, they must be labeled "Water Added."

Although the original purpose of salting foods was to keep them from spoiling, now that refrigeration is widely available, meats are no longer cured solely for preservation. The high sodium content of many cured meat products now serves several purposes: to provide flavor, to improve texture by facilitating the binding of proteins, and to increase the proteins' water-binding capacity, which reduces fluid loss within the packages. Nevertheless, lower-sodium processed meats are becoming increasingly available on the market.

Nitrites are sometimes used in curing meat to inhibit the growth of *Clostridium botulinum*, enhance flavor, and give the product a characteristic pink color. The safety of foods containing nitrites became an issue after the discovery that carcinogenic nitrosamines can form when nitrites combine with secondary amines in the stomach acid (9) (see Chapter 8). This concern resulted in the lowering of nitrite levels used in processing, but not in their elimination, because of their role in preventing botulism poisoning. Antioxidants such as ascorbic acid (vitamin C) or vitamin E are now often added to cured meats to help reduce nitrite reactions.

Smoking. Most cured meats are also smoked and cooked. Smoke imparts flavor, aroma, and color to foods. Meats are placed in smokers, where they are exposed to the smoke of burning wood. In smoke houses, the intensity of the smoke, the humidity, and the temperature are all carefully regulated, and the type of sawdust or wood used to produce the smoke determines the product's resulting flavor. Sawdust is the most economical and is often used by commercial processors, but other woods available for smoking include mesquite, hickory, oak, apple, and various combinations. In the late 1800s, a technique was developed to distill the smoke

TABLE 14-6

Examples of Processed Meats: The meats are grouped according to their major meat ingredient. Differences within each group are based on added ingredients and processing techniques.

Beef	*Pork/Ham*	*Beef and Pork*	*Veal and Pork*	*Liver*
Beef bologna	Blood sausage	Club bologna	Bockwurst	Braunschweiger
Beef salami	Bratwurst	Cervelat	Bratwurst	Liverwurst (pork)
Pastrami	Capacolla	Frankfurters*	Veal loaf	
	Chorizo	Honey loaf	Wiesswurst	
	Frizzies	Hot dog*		
	Ham	Knackwurst		
	Ham bologna	Luncheon meat		
	Linguica	Mettwurst		
	Lola/Lolita	Mortadella		
	Luncheon meat	Olive loaf		
	Lyons	Peppered loaf		
	New England–style sausage	Pimento loaf		
		Salami		
	Old-fashioned loaf	Smokies		
	Pork sausage	Weiner*		
	Proscuitto	Vienna sausage*		
	Salsiccia			
	Scrapple			
	Thuringer			

* Terms used interchangeably.

from burning wood to create "liquid smoke," which could be spread on cured meats to achieve the same flavor as the smoke house method. Today the use of liquid smoke is more common, and it saves time and minimizes air pollution. Although the additional flavor provided by "smoked" meats is preferred by some consumers, there is some concern about its posing a possible cancer risk regardless of the type of smoking used.

Canning. Canned meats are processed through either pasteurization or sterilization. Pasteurized canned meats require refrigeration and are labeled "Perishable-Keep Refrigerated," while those that are sterilized do not need refrigeration as long as the can remains sealed.

Drying. Drying is not widely used for meats, but it has some applications for them. Certain types of sausage, including pepperoni, salami, and cervelat, are dried. They are cooked, sometimes smoked, and dried under specific conditions of humidity and temperature. Beef jerky, usually dried to a water activity of 0.7 to 0.85, is convenient, ready to eat, and requires no refrigeration (37).

Types of Processed Meat. There are three types of meats that are commonly processed: ham, bacon, and sausage. In addition, lower-fat processed meats are becoming popular with consumers.

Ham. Ham is cured pork, and according to USDA standards, only meat from the hind leg of a hog can be labeled ham. Several types of cooked ham products are available for purchase:

- *Canned ham*. Boneless, fully cooked ham that can be served cold or heated. Most are cooked only to pasteurization temperatures, so they must be refrigerated. Sterilized hams are usually available only in cans of under 3 pounds. Gelatin is often added in dry form to absorb the natural juices of the ham as it cooks.
- *Water-added ham*. Contains no more than 10 percent by weight of water added. The added moisture contributes to a moist, juicy, and tender texture.
- *Imitation ham*. Ham that retains more than 10 percent moisture after curing.
- *Country ham*. Ham cured by the dry salt method and usually hickory smoked to develop a distinctive flavor.
- *Picnic ham*. Cured pork that comes from the front leg instead of back leg of the hog, and therefore cannot be labeled simply "ham." This cut is less tender and higher in fat than regular ham.

Bacon. Bacon is cured and smoked meat from the side of a hog. It should be balanced in its proportion of fat to lean. When cooked, bacon with too much lean will be less tender, while bacon with too high a proportion of fat will shrink too much.

Sausage. Sausage is meat that has been finely chopped or ground and blended with various ingredients, seasonings, and spices. It is then stuffed into casings or skins. Traditionally, the casings were made of the intestines of pigs or sheep, but now they are often manufactured from cellulose or collagen. Beef and pork, or a combination of the two, are the usual main ingredients. Other meats and meat combinations may be used, including veal, chicken, turkey, lamb, duck, rabbit, venison, and liver from any of several animals. Other ingredients that may be added include eggs, cream, oatmeal, bread crumbs, potato flour, tripe, wine, and beer. Pork and/or beef fat is often added to boost the moisture content and enhance the texture. There are three major classifications of sausage:

- *Uncooked*. Made from ground, uncooked meat. Fresh pork sausage, bratwurst, and bockwurst are examples. New combinations of chicken, turkey, apple, and other lower-fat alternatives are available.
- *Cooked*. Made from cured meat, which may be slightly smoked before being stuffed into the casings. Examples include hot dogs, bologna, and knockwurst.
- *Dry/semidry*. Made of cured meat that has been dried. Examples are pepperoni, salami, thuringer, and cervelat. Dried, cured sausage undergoes a ripening period in which the texture changes from a soft, pliable mass into a hard, sliceable, distinctly flavored sausage. The unique flavors of dry, cured sausage result from the enzymatic breakdown of proteins, carbohydrates, and lipids to smaller compounds that exhibit intense aromas (21).

Lower-Fat Processed Meats. Many processed meats contain 30 to 50 percent fat. Consumers have challenged processed meat product manufacturers by demanding foods that are lower in fat and cholesterol. Oscar Mayer's Leanest Cuts are 95 percent fat free, and Hillshire Farm has introduced Lite Smoked Sausage, Lite Polish Kielbasa, and Deli Select Thin Sliced Lunch Meats.

Lower-fat processed meat products are produced by using leaner cuts of meat, adding more water, and/or including ingredients such as fiber, gums, modified starches, and whey protein concentrate (5). Water can be substituted for fat in processed meats as long as the total amount of fat and water does not exceed 40 percent, with a maximum fat content of 30 percent (2). Less fatty ingredients, including the new fat replacers, may also take the place of more fatty ones.

The federal government used to define certain processed products by a minimum amount of fat, but these regulations have been changed in light of dietary recommendations. For example, cooked frankfurters had been required to contain about 30 percent fat, but a 1998 change in regulations lowered it to 20 percent. Sausages used to average 43 percent fat, but sausages are now available that do not exceed 15 percent fat (6).

Mechanically Deboned Meat. The traces of meat left on the bones after butchering can be collected and sold as mechanically deboned meat. This is accomplished by grinding the remaining meat and bones together, and removing the bone by putting the mixture through a sieve. The resulting meat contains ground bone, bone marrow, and soft tissue and is most commonly used in further processed meat products. The presence of the bone increases the calcium and trace mineral content of the meat. Processed meat products containing

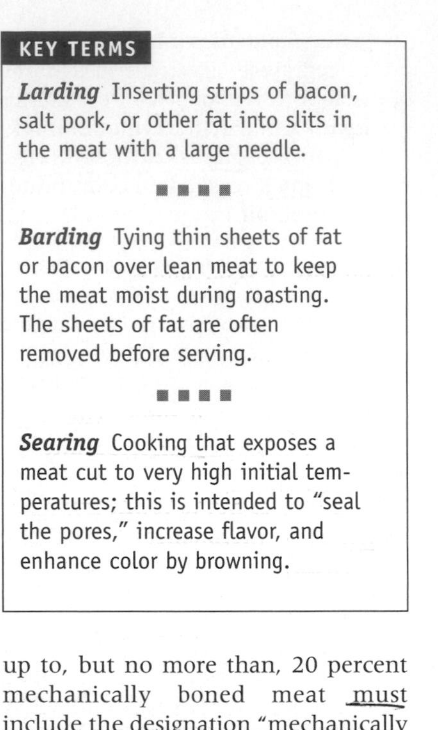
KEY TERMS

Larding Inserting strips of bacon, salt pork, or other fat into slits in the meat with a large needle.

■ ■ ■ ■

Barding Tying thin sheets of fat or bacon over lean meat to keep the meat moist during roasting. The sheets of fat are often removed before serving.

■ ■ ■ ■

Searing Cooking that exposes a meat cut to very high initial temperatures; this is intended to "seal the pores," increase flavor, and enhance color by browning.

up to, but no more than, 20 percent mechanically boned meat must include the designation "mechanically separated meat" on the food product's ingredient list.

Restructured Meat. Restructured or fabricated meat is made from meat trimmings and/or lower-grade carcasses. It is similar to real meat in texture, flavor, and appearance, but is less expensive. The meat trimmings are broken down to particle size by flaking, shredding, grinding, or chopping, and are then bound together into uniform shapes and sizes. Some natural binding between the meat's proteins occurs, but binding is further accomplished by adding nonmeat ingredients such as egg albumen, gelatin, textured soy protein, and wheat or milk proteins (38). The uniformity in shape and weight of the types of products that is made possible with restructured meat makes it ideal for the fast-food industry and food service establishments (8).

Preparation of Meats

Whether meat is prepared by dry-heat methods or by any of the various moist-heat methods, it should first be wiped with a paper towel to remove any surface moisture. Leaving water on the meat or washing it will result in a faded color and the loss of some water-soluble nutrients and flavor compounds. After it is wiped, the meat can be trimmed of any visible fat or connective tissue to reduce calories (kcal) and increase tenderness. If it is a tougher cut, it can be tenderized according to the techniques discussed earlier.

When preparing frozen meats, for best results they should be thoroughly thawed in the refrigerator or microwave before cooking. Cuts prepared from the frozen state take longer to heat and are less energy- and cost-efficient. A frozen roast may take up to three times longer to prepare than a thawed roast. Frozen cuts are more difficult to heat evenly, and the center may remain frozen even though the outside looks perfectly done.

Changes During Heating

Tenderness and Juiciness. Cooking meats at the correct temperature for the right amount of time will maximize their tenderness, juiciness, and flavor. Although heat makes meat more palatable, exposing it to high temperatures for too long will toughen, shrink, and harden meat because it shortens muscle fibers, denatures proteins, and causes it to dehydrate (1). Even with proper cooking, it is not unusual for a 4-ounce piece of meat to be cooked to 3 ounces. During heating, the collagen molecule begins to denature at 102°F (39°C), and collapses at 149°F (65°C), resulting in a considerable loss of volume and length in the meat. Another contributing factor to meat shrinkage is the freeing of some water as the meat's other proteins denature and lose their water-binding capacity. Tenderness starts to decrease as temperatures reach 104°F (40°C). Longer cooking at lower temperatures makes meat, especially the tougher cuts, more tender, because it breaks down the collagen to gelatin.

As has been mentioned, any fat in the meat melts as it is cooked, which increases tenderness, juiciness, and flavor. When meat is very lean, it may be desirable to add fat to it. This may be done by two older techniques known as **larding** and **barding**.

Searing and Blanching. It was once thought that **searing** or blanching would help to keep the juices inside a piece of meat as it cooked. It is now known that roasts heated at low temperatures for the entire cooking time retain more juices than those that are seared. Searing still remains a valuable technique for increasing the flavor and color of meat, however, because it caramelizes the outside, sealing in the flavor. Meat is blanched by boiling it very briefly, but this method is no longer recommended, because water-soluble compounds such as vitamins, minerals, and flavor substances may be lost. In the end, proponents argue that neither blanching nor searing makes any difference in moisture loss in meats exposed to prolonged heating (19).

Flavor Changes. Natural compounds in meat yield that characteristic "meat" flavor, but other factors contribute to flavor as well, including protein coagulation, melting and breakdown of fats, organic acids, and nitrogen-containing compounds. The trace amount of carbohydrates in meat contributes to the special flavor of browned meat surfaces as these sugars react with proteins in the Maillard reaction, producing the desirable brown color. Storing meat for over two days in the refrigerator or heating leftover meat can result in an unfavorable warmed-over flavor (WOF), which is best avoided by reheating the meat in a microwave oven (Chemist's Corner 14-6) (26).

Flavor Enhancements. The flavor of baked or broiled meat can be enhanced by basting and seasoning. If the seasoning includes salt, however, some professional chefs recommend

Chemist's Corner 14-6

Warmed-Over Meat Flavor

The warmed-over flavor in reheated meat is thought to be due to the oxidation of the meat's unsaturated fatty acids, resulting in various off-flavor substances (e.g., hexanal) (26). Warmed-over flavor is just one example of lipid oxidation thought to be the major cause of quality deterioration in meats (10).

collagen gelatinizes to a point, ↑tenderness
but once it collapse, 肉失水, 变tough.

adding it only after the meat has been slightly browned, because salt draws out juices and retards browning. Meat is basted by brushing the meat drippings or fat-based marinade over its surface to help it retain moisture and flavor. Self-basting can be achieved by barding. Seasoning prior to heating may improve flavor if the seasoning becomes part of the crust. Marinating meat is a flavorful way to preseason it, while prepared sauces may be served with the meat (Figure 14-17). Condiments can also be used to add flavor to meats. Those frequently served with meat include steak sauces, ketchup, seasoned butters, salsas, and chutneys and fruit sauces such as mint sauce for lamb cuts.

Determining Doneness

A number of changes occur in meat during cooking, and a multitude of factors affect the cooking times of meats: the effects of carryover cooking; differences in the type, size, and cut of meat; the presence of bones, which conduct heat faster than flesh, or of fat, which acts as an insulator; the actual oven temperature; the temperature of the meat before heating; and variations in the degree of doneness preferred by the preparer. Various methods are used to determine doneness and sometimes more than one method is used. Those discussed below include internal temperature, time/weight charts, color changes, and touch.

Internal Temperature. Using a meat thermometer is the most accurate method of determining doneness. There are several different styles of meat thermometers on the market, with some being inserted into meats before heating and others, such as instant-read thermometers, that can be inserted at any time. The thermometer should be inserted into the thickest portion of the meat and in such a way as not to touch any fat or bone. Meat thermometers should be thoroughly sanitized after each use. Table 14-7 gives the internal cooking temperatures indicating doneness for various meats. The final internal

FIGURE 14-17

Sauces for beef.

Au Jus—natural beef juices

Béarnaise—thick sauce of egg yolks, white wine, tarragon vinegar, herbs

Béchamel—seasoned white sauce

Bercy Butter—shallots cooked in white wine mixed with creamed butter and parsley

Beurre Noir—clarified butter with vinegar or lemon juice

Bordelaise—brown sauce with red wine, shallots or green onions, herbs and lemon juice

Brown (Sauce Espagnole)—flavorful beef sauce used as baste for others

Chasseur—brown sauce with mushrooms, tomato sauce, tarragon

Chili Salsa—chopped tomato, onion, green chili pepper

Choron—béarnaise sauce and tomato

Colbert—béarnaise sauce and meat glaze

Hollandaise—thick sauce of egg yolks, melted butter and lemon juice

Madeira—brown sauce and Madeira wine

Maître d'Hôtel Sauce—béchamel sauce with butter, lemon juice, parsley, and tarragon

Marchand de Vin—red wine, parsley, green onions, and lemon juice

Meunière—browned butter with lemon juice and parsley

Mornay—creamy cheese sauce

Périgueux—wine sauce with diced truffles

Robert—brown sauce with mustard, onion, tomato, and pickle

TABLE 14-7

Internal Temperatures Recommended for Cooked Meat

Meat	*Description*	*Color*	*Internal Temperature* °F	°C
Beef	Rare	Rose red in center; pinkish toward outer portion, shading into a dark gray; brown crust; juice bright red	140	60
	Medium	Light pink; brown edge and crust; juice light pink	160	70
	Well-done	Brownish gray in center; dark crust	170	77
Veal	Well-done	Firm, not crumbly; juice clear, light pink	165	74
Lamb	Rare	Rose-red in center; pinkish toward outer portion; brown crust; juice bright red	140	60
	Medium	Light pink; juice light pink	160	70
	Well-done	Center brownish gray; texture firm but not crumbly; juice clear	170	77
Pork				
Ham				
Fully cooked or canned	Heated	Pink	130–140	55–60
Cook before eating	Medium	Pink	140	60
Smoked loin	Medium	Pink	160	70
Fresh rib, loin, picnic shoulder	Well-done	Center grayish white	170	77

Source: USDA

KEY TERMS

Carry-over cooking The phenomenon in which food continues to cook after it has been removed from the heat source as the heat is distributed more evenly from the outer to the inner portion of the food.

temperatures according to the USDA for beef are as follows:

- *Rare:* 136–140°F (58–60°C)
- *Medium:* 160–167°F (71–75°C)
- *Well done:* 172–180°F (78–82°C)

Aside from pork, which must reach 160°F (71°C), most other meats are expected to reach an internal temperature of at least 140°F (60°C). In January 1993, following an outbreak of *E. coli,* health departments increased the required preparation temperature for hamburgers served by eating establishments from 140°F (60°C) to 160°F (71°C).

When measuring internal temperature, it is important to adjust for **carryover cooking**. This can result in an average temperature increase of 10° to 15°F (6° to 8°C) for average-sized roasts. Very large roasts can have as much as a 25°F (14°C) increase in temperature, while small cuts may rise only 5°F (3°C) in temperature. To adjust for this carryover cooking, most roasts should be removed from the oven when the internal temperature is 10° to 15°F (6° to 8°C) below the final desired degree of doneness. Meat cooked at low temperature such as 200° to 250°F (93° to 121°C) will experience only minimal carryover cooking. Depending on their size, roasts should be allowed to stand for 15 to 30 minutes in order to distribute the heat and juices.

Time/Weight Charts. Time/weight charts, such as the one shown in Table 14-8, are useful in estimating roughly how long it will take to cook a piece of meat, but are unreliable if used alone because of the many factors that can affect doneness. Instead, a combination of criteria is used to determine the doneness of meats.

These criteria include time/weight charts along with color changes, internal temperature, and touch.

TABLE 14-8
Time/Weight Chart for Roasting Beef

Cut	*Approximate Weight (pounds)*	*Oven Temperature (degrees F)*	*Approximate Cooking Time (minutes per pound)* Rare	Medium	Well
Rib roast	4 to 6	300 to 325	26 to 32	34 to 38	40 to 42
	6 to 8	300 to 325	23 to 25	27 to 30	32 to 35
Rib Eye roast	4 to 6	350	18 to 20	20 to 22	22 to 24
Boneless Rump roast	4 to 6	300 to 325	—	25 to 27	28 to 30
Round Tip roast	3½ to 4	300 to 325	30 to 35	35 to 38	38 to 40
	6 to 8	300 to 325	22 to 25	25 to 30	30 to 35
Top Round roast	4 to 6	300 to 325	20 to 25	25 to 28	28 to 30
Tenderloin roast					
Whole	4 to 6	425		45 to 60 (total)	
Half	2 to 3	425		35 to 45 (total)	

Color Changes. Meat pigments change color as the meat is cooked. Doneness can be determined by observing the following colors in red meats:

- *Rare.* Strong red interior. Rare meat does not reach a final internal temperature considered microbiologically safe.
- *Medium.* Rosy pink interior and not quite as juicy as a rare piece of meat.
- *Well done.* Brown interior. No traces of red or pink left. Moist, but no longer juicy.

Veal and pork are known as "white meats," in part because they change from a pinkish to a whiter color as they are heated to the well-done stage. According to the USDA, pork should be heated at least to an end-point temperature of 160°F (71°C). Color may not be a good indicator for doneness in meat from older swine, which is often grayish-brown rather than pink (23). It is not recommended that color be used to judge the doneness of hamburger, either, because of the risk of *E. coli* 0157:H7 contamination.

Touch. Doneness can be determined by the firmness of the meat. Some meat cuts such as steaks and chops can be judged for doneness based on their color and firmness. Pressing lightly on the center of the lean tissue can help to determine whether the meat is rare, medium, or well done (Figure 14-18). This technique takes a fair amount of experience to master and is most often used by professional chefs who frequently prepare steaks.

Dry-Heat Preparation

Tender cuts are usually prepared by one of the dry-heating methods: roasting (baking), broiling, grilling, panbroiling, and frying.

Roasting. Roasting is the heating of moderate-to-large tender cuts of meat in the dry hot air of an oven. A roast will usually be at least 2½ inches thick and provide more than three servings. The meat is placed, with any fat side up, on a rack in an open pan. The rack prevents the meat from sitting in its own juices, which would cause the meat to simmer rather than roast. If a rack is not available, one can be made by lining up carrots and celery stalks lengthwise across the bottom of the pan. Figure 14-19 shows examples of cuts suitable for roasting.

Roasting Temperatures. Temperatures from 300° to 350°F (149° to 177°C) are recommended for roasting and should produce an evenly cooked, easy to carve, juicy, tender, flavorful roast with a greater yield than higher temperatures would have produced. Higher temperatures of 350° to 500°F (177° to 260°C) are recommended to produce roasts with deeply seared crusts in less time, but the higher oven temperatures cause greater shrinkage. In general, it usually takes 18 to 30 minutes of roasting time for every pound of meat. As previously mentioned, roasts should be removed from the oven

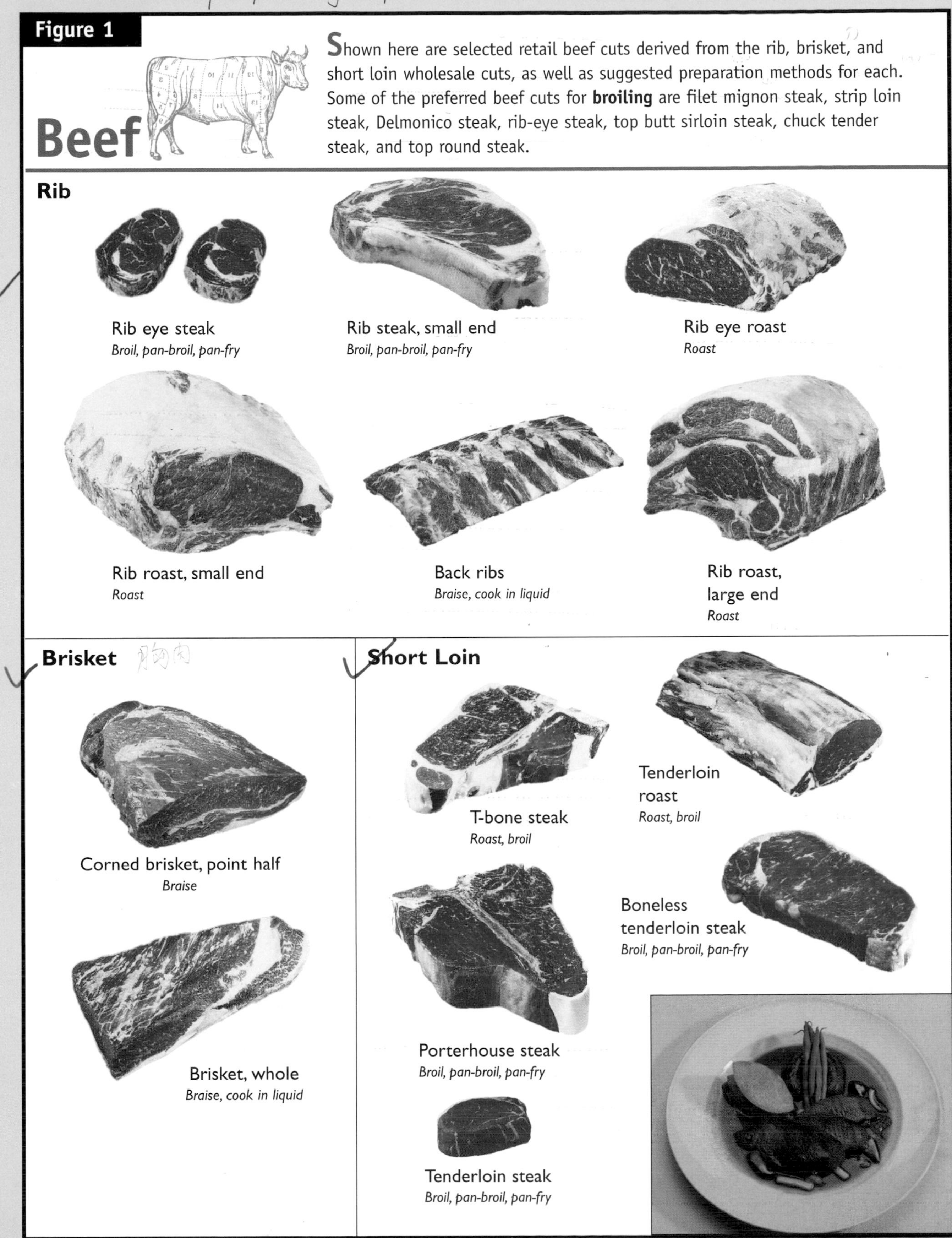

USDA

PhotoDisc

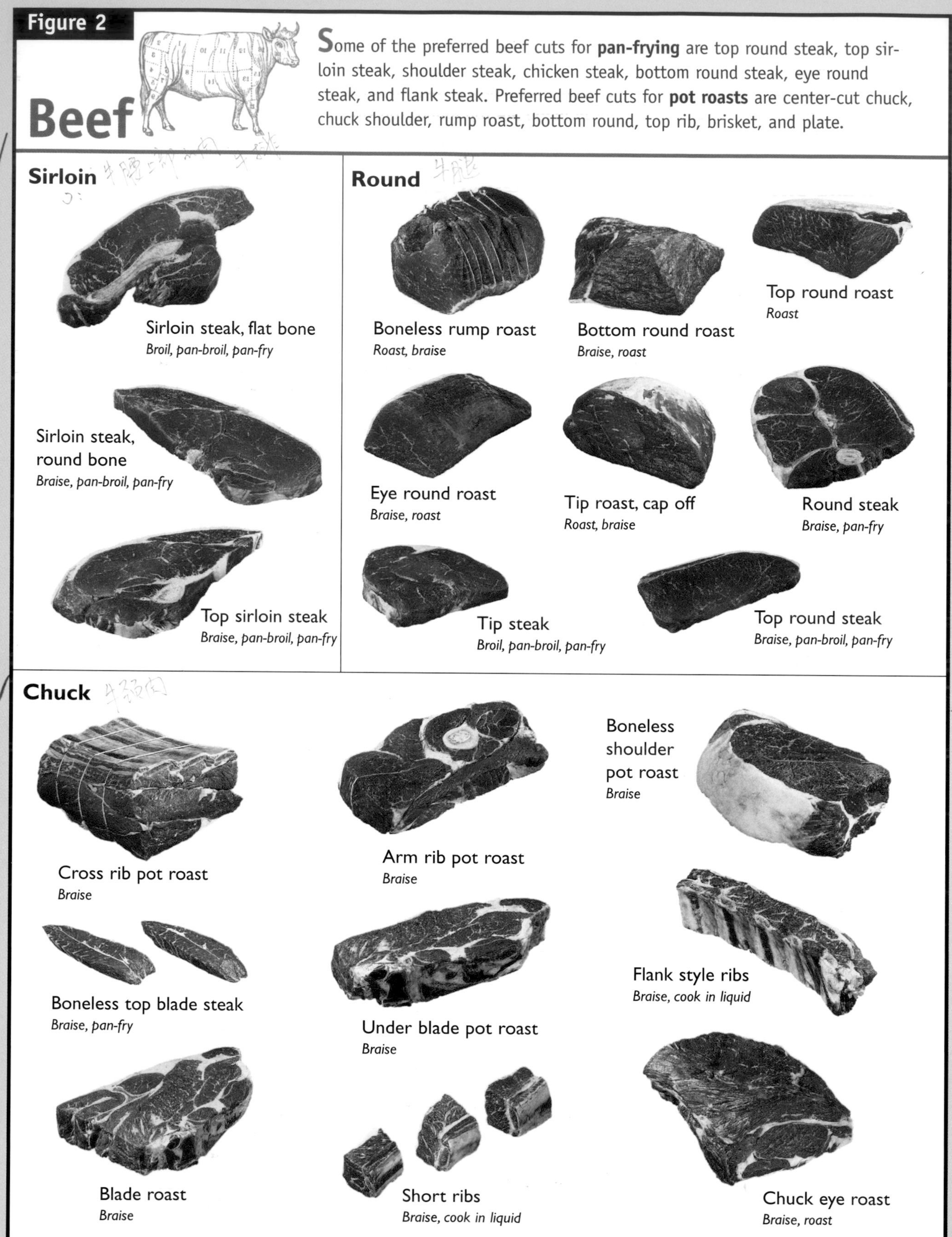
Figure 2
Beef
Some of the preferred beef cuts for **pan-frying** are top round steak, top sirloin steak, shoulder steak, chicken steak, bottom round steak, eye round steak, and flank steak. Preferred beef cuts for **pot roasts** are center-cut chuck, chuck shoulder, rump roast, bottom round, top rib, brisket, and plate.
Sirloin
Sirloin steak, flat bone
Broil, pan-broil, pan-fry
Sirloin steak, round bone
Braise, pan-broil, pan-fry
Top sirloin steak
Braise, pan-broil, pan-fry
Round
Boneless rump roast
Roast, braise
Bottom round roast
Braise, roast
Top round roast
Roast
Eye round roast
Braise, roast
Tip roast, cap off
Roast, braise
Round steak
Braise, pan-fry
Tip steak
Broil, pan-broil, pan-fry
Top round steak
Braise, pan-broil, pan-fry
Chuck
Cross rib pot roast
Braise
Arm rib pot roast
Braise
Boneless shoulder pot roast
Braise
Boneless top blade steak
Braise, pan-fry
Under blade pot roast
Braise
Flank style ribs
Braise, cook in liquid
Blade roast
Braise
Short ribs
Braise, cook in liquid
Chuck eye roast
Braise, roast

USDA

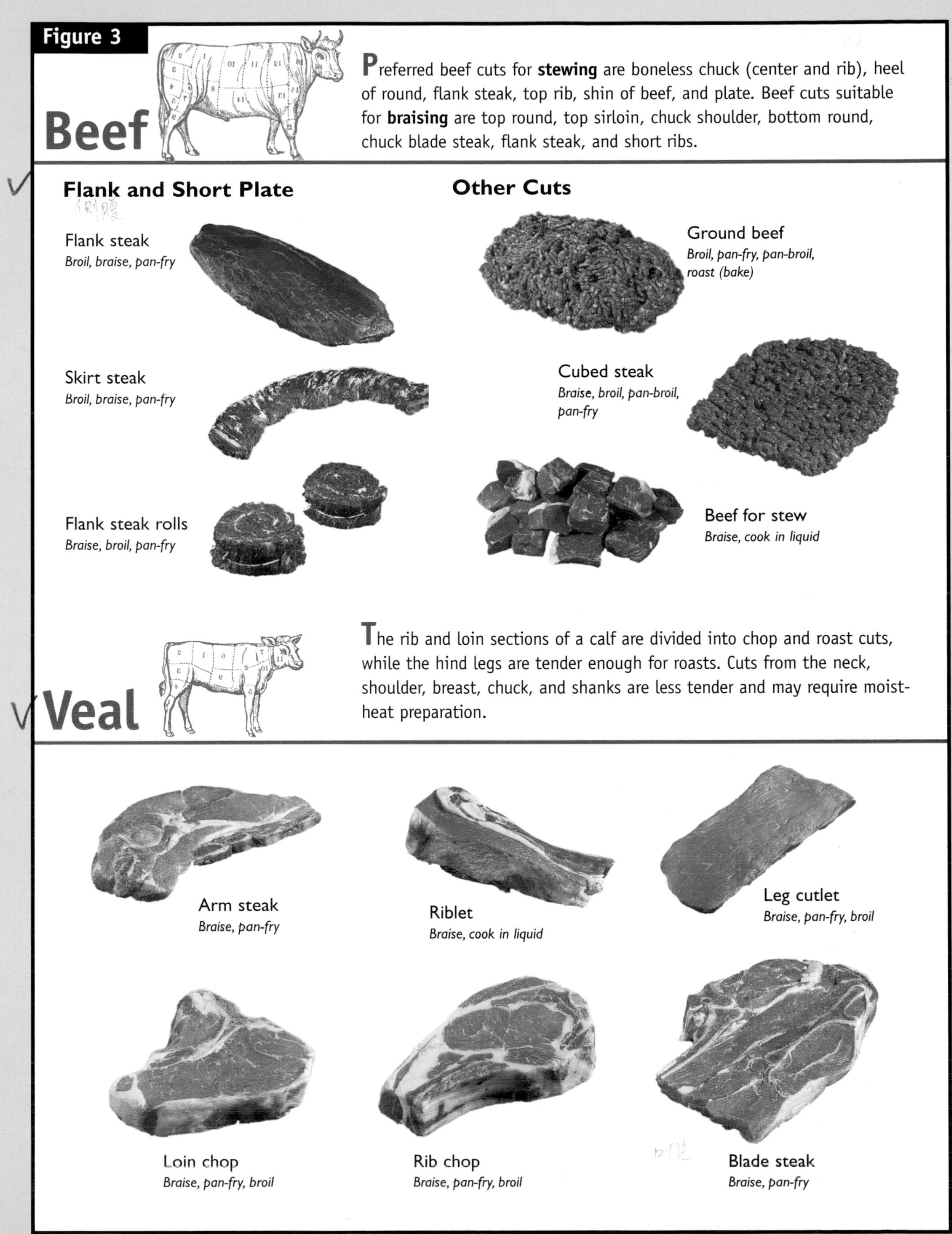
Figure 3
Beef
Preferred beef cuts for **stewing** are boneless chuck (center and rib), heel of round, flank steak, top rib, shin of beef, and plate. Beef cuts suitable for **braising** are top round, top sirloin, chuck shoulder, bottom round, chuck blade steak, flank steak, and short ribs.
Flank and Short Plate
Flank steak
Broil, braise, pan-fry
Skirt steak
Broil, braise, pan-fry
Flank steak rolls
Braise, broil, pan-fry
Other Cuts
Ground beef
Broil, pan-fry, pan-broil, roast (bake)
Cubed steak
Braise, broil, pan-broil, pan-fry
Beef for stew
Braise, cook in liquid
Veal
The rib and loin sections of a calf are divided into chop and roast cuts, while the hind legs are tender enough for roasts. Cuts from the neck, shoulder, breast, chuck, and shanks are less tender and may require moist-heat preparation.
Arm steak
Braise, pan-fry
Riblet
Braise, cook in liquid
Leg cutlet
Braise, pan-fry, broil
Loin chop
Braise, pan-fry, broil
Rib chop
Braise, pan-fry, broil
Blade steak
Braise, pan-fry

USDA

Figure 4

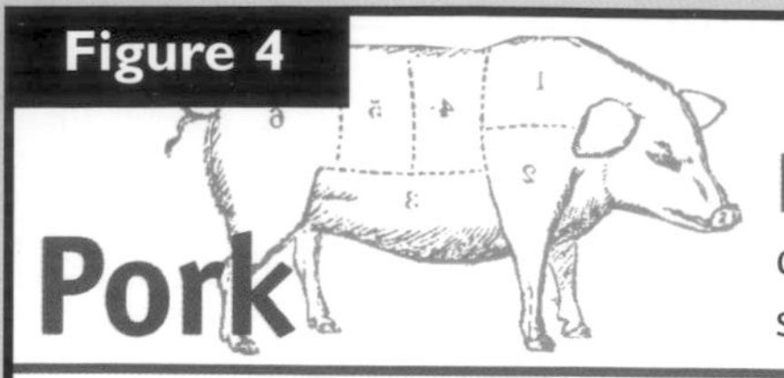

Pork

Preferred pork steaks and chops suitable for **pan-frying** are center-cut loin chop, center-cut rib chop, loin end chop, fresh ham steak, shoulder arm steak, and blade pork steak.

Chops

Sirloin chop

Center-cut loin chop

Blade steak

Chops are one of the most familiar pork cuts. Chops can be prepared by pan-broiling, grilling, broiling, roasting, sautéing, or braising.

Thin chops (3/8-inch) are best quickly sautéed. Thicker chops (3/4-inch to 1 1/2 inches) can be grilled, roasted, braised, or pan-broiled.

Pork rib chop

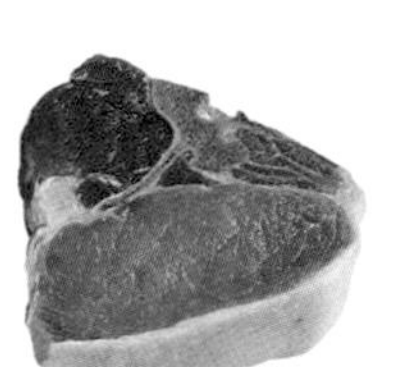

Pork loin chop

Boneless pork sirloin chop

Tenderloin

Pork tenderloin

Pork tenderloins are among the leanest cuts of pork. A pork tenderloin has only 4.1 grams of fat and 141 calories per 3-ounce roasted, trimmed serving.

Ribs

Back ribs

Ribs are commonly used for barbecue meals. Slow-roasting or braising yields tender, flavorful results.

PhotoDisc

Roasts

A roast is a large cut of pork from the loin, leg, shoulder, or tenderloin. It can be roasted in the oven, barbecued over indirect heat, or braised in the oven.

Boneless blade roast

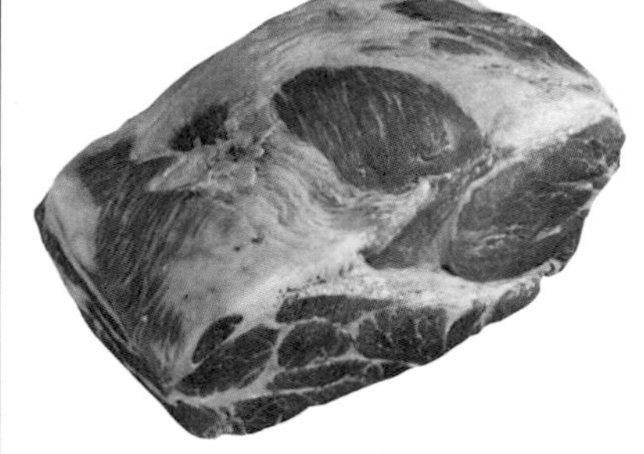

Bone-in blade roast

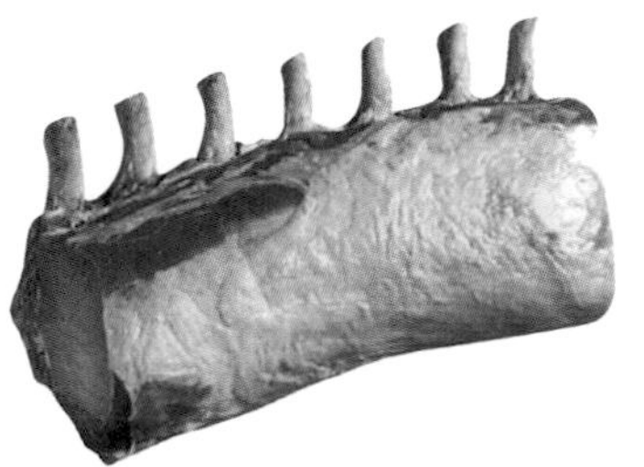

Center rib roast (rack of pork leg)

USDA

Figure 5
Lamb
Lamb is traditional in Middle Eastern and Navajo cuisine. In North America, it is often served with mint sauce or jelly.
METHODS FOR COOKING LAMB
PAN-FRYING (if cut thin enough)
Loin chops
Ground lamb
Rib chop
Sirloin chops
BRAISING, STEWING, AND MOIST COOKING
Shanks
Center leg
Shoulder chops
Half of leg
ROASTING
Boneless lamb shoulder
Boneless loin roast
PhotoDisc
USDA
PhotoDisc

Figure 6

Lower-Fat Meats

More people are achieving the goal of deriving less than 30 percent of their calories from fat. Choosing lean meat cuts and following the tips for reduced fat cooking listed below are some of the steps that can be taken toward achieving a healthy, balanced diet.

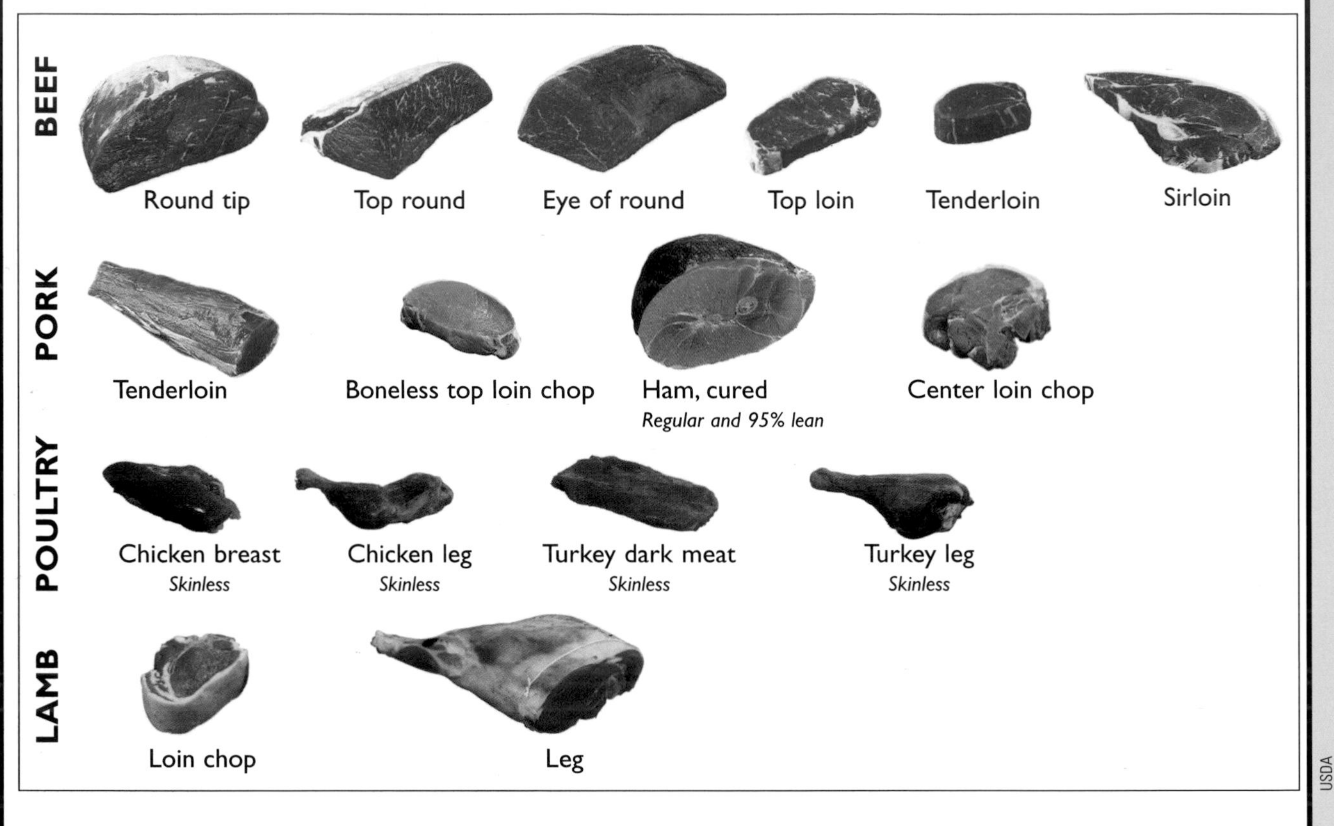

USDA

TIPS FOR REDUCED FAT COOKING

- ✔ Choose 3-ounce servings (for a total of 6 ounces per day). Start with 4 ounces of raw meat to end up with a 3-ounce cooked serving. This will account for cooking losses.
- ✔ Look for beef labeled with the "USDA Select" grade. It's lower in fat and calories than "Choice" or "Prime." Marbling (the flecks of fat in the lean) makes the difference.
- ✔ Use the "loin/round" rule of thumb for beef and "loin/leg" for pork, lamb, and veal. Cuts with these words on the label will be lean choices.
- ✔ Tenderize lean cuts of meat by cooking them slowly in liquid or marinating them before cooking. Pounding, grinding, and slicing across the grain can also help.
- ✔ Keep your meat selections lean. Trim all visible fat and let the remainder drip off during cooking. When you prepare meat, broil, grill, bake, roast on a rack, or microwave. Buy skinless poultry or remove skin before cooking and you will reduce fat content by about half.
- ✔ Remove fat from stews, soups, and casseroles by chilling them and skimming the hardened fat from the top. If you're pressed for time, use a baster to remove it.
- ✔ Don't fry. The batter or breading on fried chicken, for example, acts like a sponge—soaking up fat. And after frying, you're less likely to remove the coating and skin before you eat the meat. Also skip the heavy gravies and rich sauces. Even the butter or margarine you use on broiled food makes the fat add up fast.

Figure 7

Seafood

Consuming fish twice a week, especially those high in omega-3 fatty acids, has been reported to lower the risk for heart disease. Most fish—except mackerel, shark, herring, and eel—also contain fewer than 160 calories (kcal) per three-ounce cooked serving.

Lois Frank

PhotoDisc

Lois Frank

Figure 8

Onions

The type of onion chosen depends on how it will be used in food preparation. Yellow onions are all-purpose, white onions are the most pungent, red onions lend themselves to certain salads, the smaller pearl onions are preferred for soups and casseroles, and vidalias yield a sweeter flavor.

VARIETIES OF ONIONS

WHITE
- **Pungent odor**
- **Sharp flavor**

YELLOW
- **All-purpose**
- **Medium-strong flavor**

RED
- **Best used raw**
- **Tangy flavor**

Lois Frank

VIDALIA
- **Salads**
- **Sweet flavor**

PEARL ONIONS
- **Soups and casseroles**
- **Regular onions whose growth has been stunted**

Vincent Lee
PhotoDisc
PhotoDisc
Vincent Lee

Figure 10

Squash

Summer squashes are harvested in the summer, usually elongated, and can be left unpeeled and cooked whole, sliced, cubed, or grated. Winter squashes, harvested in the fall, usually have hard rinds that are cut in half to remove their fibrous matter and seeds before being baked, broiled, or steamed.

SOME POPULAR VARIETIES OF SQUASH

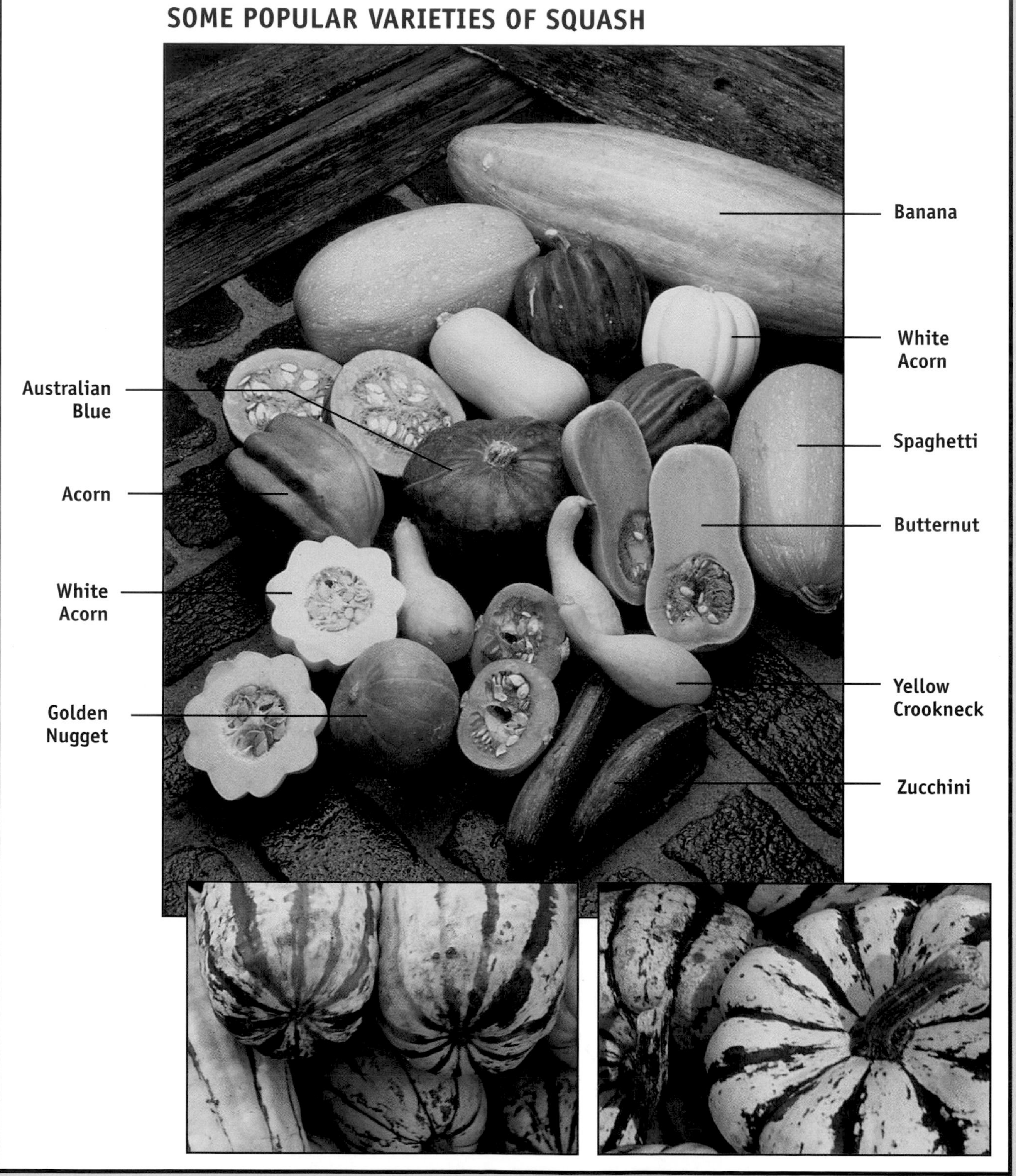

Vincent Lee

Figure 11
Melons
These round to oblong fruits grow on vines. The skin on melons is actually a rind that can be smooth, netted, ridged, wrinkled, or warty. Inside, the edible pulp varies in color and can be white, yellow, pink, green, or red.
SOME VARIETIES OF MELON
Persian
Santa Claus
Casaba
Crenshaw
Juan Canary
Vincent Lee

As the global economy has expanded and transportation methods have improved, the selection of available fruits and vegetables includes produce from around the world.

Exotic Vegetables and Fruits

Figure 12 VEGETABLES

Chinese long beans
- Grow up to 18" long
- Steam or stir-fry

Jerusalem artichoke
- Root of sunflower plant
- Nutty, sweet, mild flavor

Breadfruit
- Not used to make bread
- Used like a potato

Kohlrabi
- Sweeter and crisper than turnips
- Flavor of broccoli stems

Jicama
- Pronounced "hee-ka-ma"
- Sweet, starchy taste

Belgian endive
- Mild, bitter flavor
- Used in salads or soups

Calabaza
- Variety of squash
- Dark orange flesh

Chayote
- Pronounced "chy-o-tay"
- Flavor similar to zucchi-ni/cucumber

Figure 13 FRUITS

Kumquats
- Tart pulp

Cherimoya
- Sweet, custard-like flavor

Plantains
- Served as vegetable
- Starchy, no banana flavor

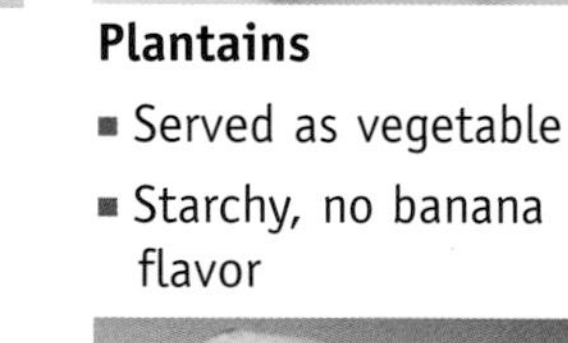

Passion fruit
- Lemony, tart flavor
- Many small black seeds

Lychee
- Grape-like flesh

Red banana
- Maroon when ripe
- Tangy-sweet flavor

Pummelos
- Largest of citrus fruits

Guavas
- High vitamin C content

Figure 14

Apples

Over 7,500 varieties of apples are grown worldwide, but only about 18–25 comprise the majority of the North American commercial crop. While many apples can be used for both eating and cooking, tart varieties with a high acid content and firm texture are best for baking.

Some Apple Favorites

Red Delicious

- Bright to dark red, sometimes striped
- Favorite eating apple
- Mildly sweet, juicy
- Available year-round

Winesap

- Dark red
- Appropriate for cider, snacking, and cooking
- Spicy, slightly tart
- Available October to August

Rome Beauty

- Brilliant red, round
- Great for baked apples; holds shape well when cooked
- Available October to June

Golden Delicious

- Yellow-green
- All-purpose apple for baking, salads, and fresh eating; flesh stays white longer than other apples
- Mellow, sweet
- Available year-round

Criterion

- Sweet, yellow, often with red blush
- Wonderful eaten fresh, in salads, or baked; flesh stays white longer than other apples
- Available October to Spring

Gala

- Yellow to red
- Appropriate for cider, snacking, and cooking
- Spicy, slightly tart
- Available October to August

Granny Smith

- Green
- Excellent for cooking, salads, fresh eating
- Tart, crisp, juicy
- Available year-round

Fuji

- Ranges from yellow-green with red highlights to very red
- Excellent for eating or applesauce
- Sweet, spicy, crisp
- Available year-round

Washington State Apple Commission

Figure 15

Greens

Iceberg, butterhead, romaine, and loose-leaf lettuce are the greens most commonly used in salads, but a variety of other greens are also available.

When selecting greens, look for clean, crisp, tender leaves free of "tipburn"—the ragged brown borders that can appear on a leaf's edge.

Bibb Lettuce (Butterhead)

Boston Lettuce

Belgian Endive

Green Cabbage

Radicchio

Chicory

Savoy

Escarole

Watercress

Red Leaf Lettuce

Swiss Chard

Green Leaf Lettuce

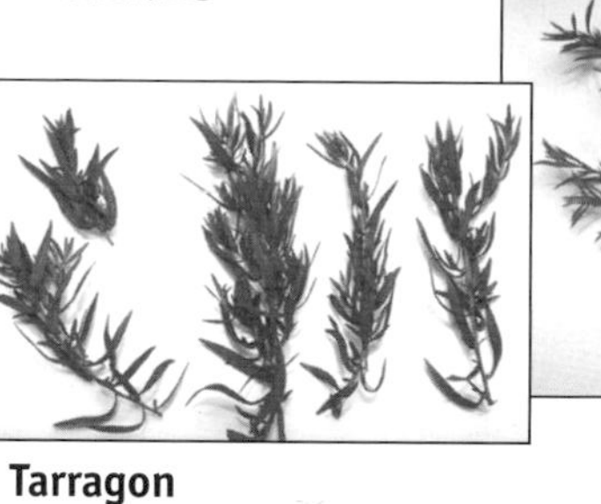
Tarragon

Savory

Cilantro

Flat Parsley

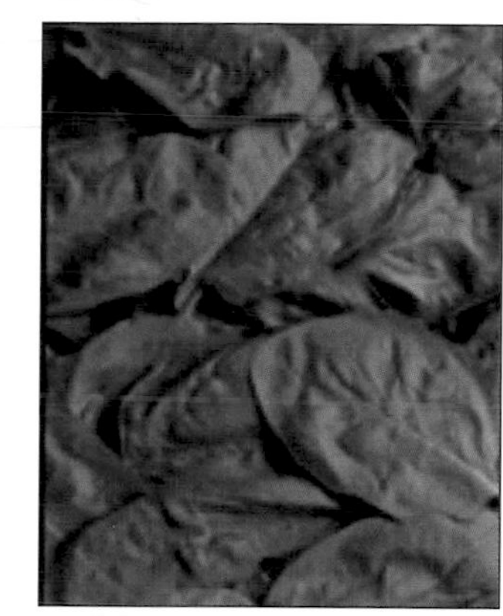
Spinach

Iceberg lettuce

Romaine lettuce

Figure 16 EDIBLE FLOWERS

One of the newest trends in gourmet produce is edible flowers. A colorful, peppery-tasting addition to salads can be made by adding a sprinkle of nasturtium flowers or calendula petals. Daylily, squash, and pumpkin blossoms are delicious dipped in tempura batter and quickly deep-fried. Lavender and many geranium blossoms add a perfumy, herbal scent to beverages and desserts. Viola, pansy, and violet blossoms can be candied and used as edible decorations for cakes and other desserts.

Figure 17

Desserts

Desserts not only satisfy the sweet tooth but are sometimes identified with certain meals or occasions—fortune cookies following a Chinese meal, marshmallow eggs at Easter, birthday cakes, pumpkin pie at Thanksgiving, Christmas cookies and fruitcakes during Christmas, and complimentary mints at some restaurants.

COOKIES

Bar	**Dropped**	**Pressed**	**Rolled**	**Molded**
Brownies	Chocolate Chip	Tea	Sugar	Peanut Butter
	Oatmeal Raisin	Lady Fingers		Shortbread
		Coconut Macaroons		

PIES

Cream
Fruit
Ice Cream
Chiffon
Custard
Meringue

CAKES

Shortened	**Unshortened**	**Chiffon**
White	Angel Food	Lemon Chiffon
Yellow	Sponge	Chocolate Chiffon
Chocolate		
Spice		
Fruit		

CANDY

PASTRIES

Blitz/Puff Pastry	**Phyllo**
Napoleons	Baklava
Tart Shells	
Strudel	**French**
Danish	**Pâte à Choux**
Eclairs	Cream Puffs

FROZEN DESSERTS

Ice Cream
Imitation Ice Cream
Sherbet
Sorbets
Water Ices
Frozen Yogurt
Still-Frozen
—Mousses
—Bombes
—Parfaits

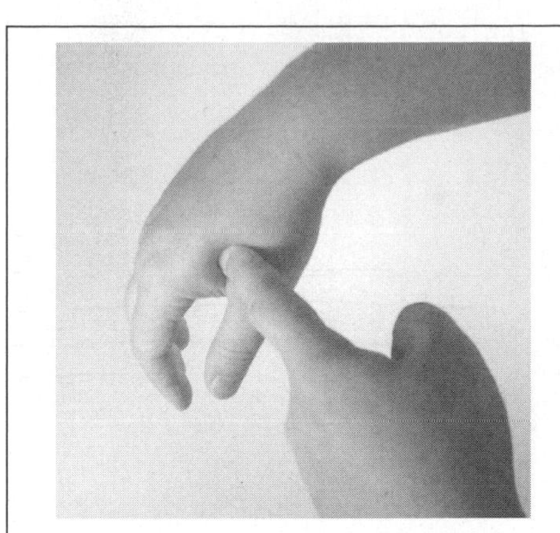

Rare: Shake, dangle, and relax right hand; pressing the area between thumb and index finger feels similar to rare steak—soft and yielding to slight pressure.

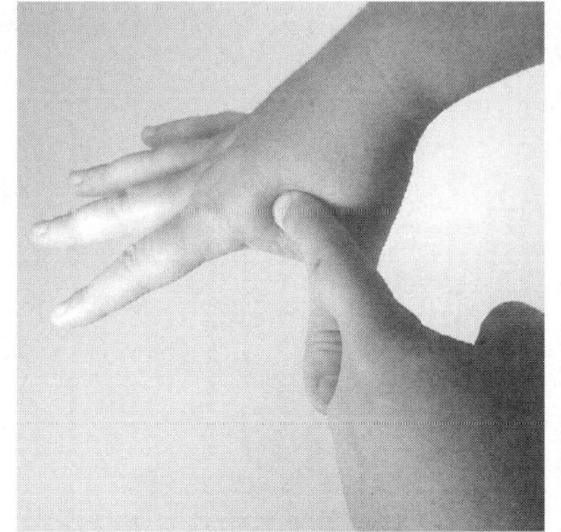

Medium: Stretch out the right hand and tense the fingers; the springy firmness is similar to the resistance felt in medium-cooked meats.

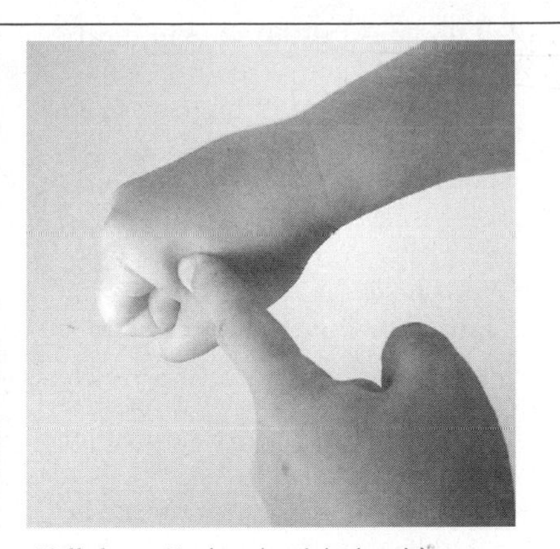

Well done: Harden the right hand into a tight ball; this hard and unyielding feeling with all the springiness gone is how well-done meat feels.

FIGURE 14-18 **Touch as a test for doneness.**

slightly before their final desired temperature is reached and allowed to stand for 15 to 30 minutes in order for carryover cooking to occur. This will also make carving easier and result in a more evenly juicy roast.

Broiling and Grilling. Smaller cuts of tender meat ranging from 1 to 3 inches in thickness can be broiled or grilled. High temperatures and short heating times will keep the meat tender. Broiling and grilling times are based primarily on the meat's thickness and its distance from the heat (Table 14-9). Ovens, whether electric or gas, need at least 15 minutes to reach the desired temperature, while charcoal or wood fires need at least 25 minutes to burn down to the required heat. Beef retail cuts suitable for broiling include the following steaks in descending order of tenderness: filet mignon, strip loin, delmonico, rib eye, top butt sirloin, chuck tender, and top round. A very light layer of oil on the meat will keep it from sticking to the grill, while using a marinade, spice rub, or adding sauces during basting will yield more flavor.

Tips for Broiling/Grilling. The goal in either broiling or grilling is to simultaneously heat the inside of the meat while achieving just the right degree of browning on the exterior. The thickness of the cut and the desired

FIGURE 14-19 **Examples of tender roast cuts.**

TABLE 14-9
Time/Weight Chart for Broiling Sirloin Steak

Beef Cut	*Approximate Thickness (inches)*	*Approximate Weight (pounds)*	*Distance From Heat (inches)*	*Approximate Cooking Time (total minutes)*		
				Rare	*Medium*	*Well*
Sirloin Steak	¾	1¼ to 1¾	2 to 3	10	15	—
	1	1½ to 3	3 to 4	16	21	—
	1½	2¼ to 4	4 to 5	21	25	—

level of doneness dictate the intensity of the heat, which is controlled by altering the distance of the meat from the heat source, from 2 inches for cuts less than 1 inch thick, to up to 5 inches for thicker cuts. When broiling thicker steaks or those to be well-done, the broiler rack in an electric oven should be lowered and the door left open to prevent steam from accumulating, thereby preventing the meat from browning. Gas broiler doors are left closed.

The oven, broiler, or grill should be preheated. Then the meat should be placed under the broiler or over the coals and heated until one side is brown. Tongs should be used to turn the meat, but if a fork is used, it is best inserted into the fat trim to avoid letting the juices escape. The second side is heated to the desired stage of doneness. When heating is complete, remove and serve immediately. One of the benefits of using a grill is that attractive, appetizing grill marks can be made by turning the meat over according to the pattern depicted in Figure 14-20.

FIGURE 14-20 Technique for making grill marks: Rotate clockwise a quarter of a turn.

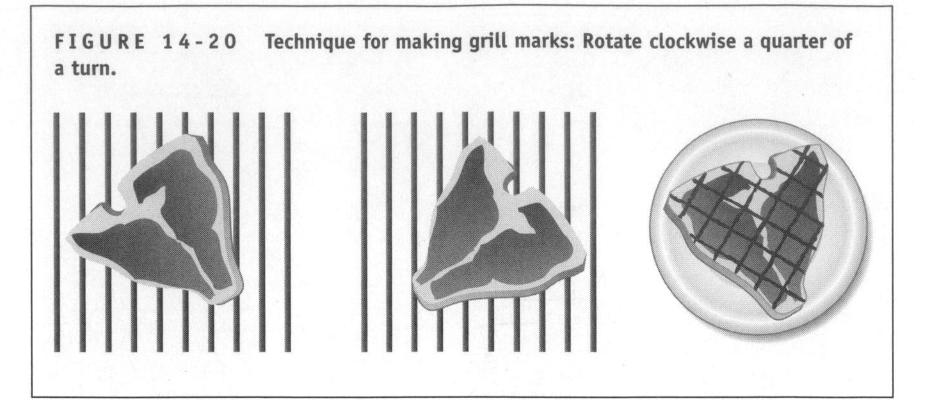

Pan-Broiling. Very thin cuts of meat, less than ½ inch, can be pan-broiled to achieve a tasty outside crust without overcooking the meat. In this method, heat is applied directly through the hot surface of a heavy pan or flat grill (Figure 14-21). Thin, tender cuts of beef steaks, lamb chops, and ground-beef patties are perfect for pan-broiling. Place the meat on the hot surface of the preheated pan with no added fat or oil. Any drippings should be drained during heating to prevent frying. The meat can be seasoned before, during, or after placing it on the pan.

Frying. Sautéing, pan-frying, and deep-frying are suitable for tender, small pieces of meat that are low in fat or that have a breaded coating.

Sautéing. Sautéing is identical to pan-broiling except that a small amount of fat is heated to the sizzling point before the meat is added. Examples of sautéed meat dishes include liver and onions, veal oscar, veal piccata, and veal cordon bleu. Liver should be salted after it is sautéed or else it will toughen and shrivel.

Stir-frying is an adapted version of sautéing that has become increasingly popular. For stir-frying, thin slices of meat are cooked in an oiled wok or other sloping-sided pan. The meat is stirred constantly over high heat for about three minutes to promote even heating. When the meat is done, it is moved to the side, and chopped vegetables are added to the pan. As soon as they are barely tender, they are mixed with the meat and any desired sauces or flavorings.

Pan-Frying. More fat (but no more than up to ½ inch deep) and lower heating temperatures and times are used in pan-frying than in sautéing. Commonly pan-fried meat cuts are larger and include steaks (Figure 14-22), chops, and sliced pieces of liver. Meats are often seasoned and coated with flour or breading before pan-frying. The fat used in sautéing or in pan-frying should be vegetable oil or clarified butter. The low smoking temperatures of whole butter and margarine make them unsuitable for frying. An alternative to frying steaks and chops in oil is to use a teflon pan or to sprinkle the pan with a thin layer of salt. The pan is heated until a drop of water hisses; the meat is then added, fried, and turned when the underside has reached the desired brownness.

Deep-Frying. Meat, with the exception of chicken-fried steak, is seldom deep-fried. When it is, the meat is usually cut into small pieces and dipped in seasoned flour or cornstarch, placed in a wire basket, submerged in oil preheated to 300° to 360°F (149° to 182°C), and heated until golden brown.

1. Place beef in preheated frying pan.
2. Do not add oil or water. Do not cover.
3. Cook slowly (⅝" to 1" cuts), turning occasionally. For cuts thicker than ½" use medium to medium-low heat. For thinner cuts, use medium-high heat.
4. Pour off excess drippings as they accumulate.
5. Season if desired.

FIGURE 14-21 Pan-broiling.

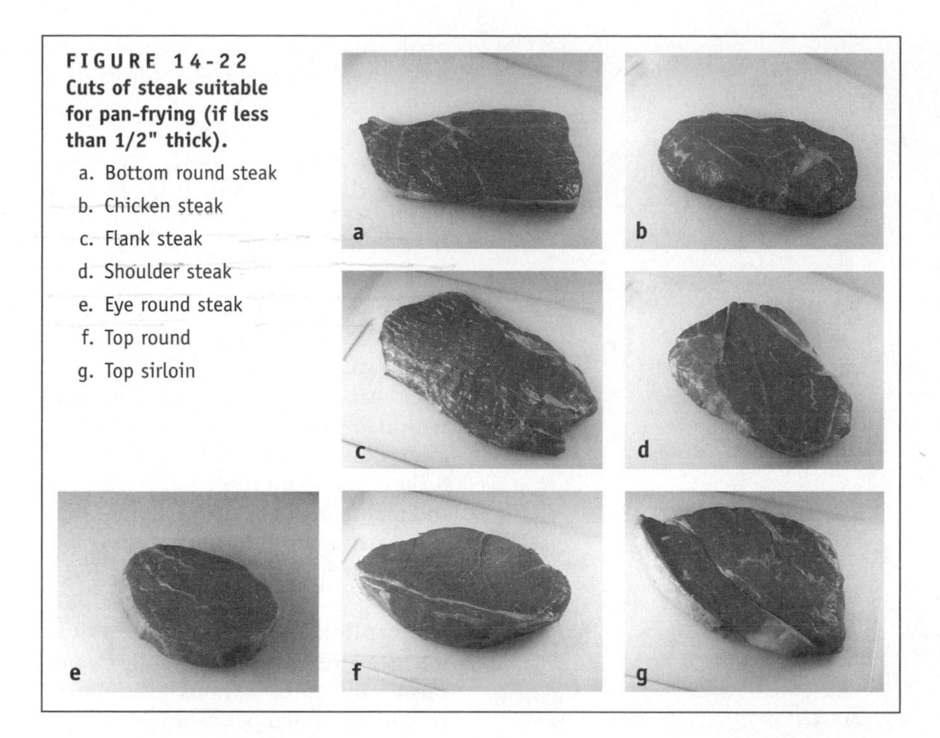

FIGURE 14-22
Cuts of steak suitable for pan-frying (if less than 1/2" thick).
a. Bottom round steak
b. Chicken steak
c. Flank steak
d. Shoulder steak
e. Eye round steak
f. Top round
g. Top sirloin

Moist-Heat Preparation

Less tender cuts of meat, which tend to come from more heavily exercised muscles or older animals, are usually prepared by moist-heat methods such as braising, simmering/stewing, or steaming.

Braising. Braising consists of simmering meat, in a covered pan, in a small amount of water or other liquid. It is ideal for less tender cuts such as beef chuck, round, and flank, because braising breaks down collagen and tenderizes the meat. Some smaller meat cuts such as round steaks, pork and veal chops, and organ meats are also good "braisers." The most common braised meats are pot roasts, which are large cuts of meat cooked whole and served in slices covered with their own cooking liquid. Adding vegetables completes the meal and adds color. Chopped vegetables commonly added to pot roasts include potatoes, carrots, onions, celery, and tomatoes.

Browning Before Braising. Although not necessary, browning the meat prior to adding the liquid improves the final color and flavor. Before browning, the meat should be dried with a paper towel. It is sometimes dredged with seasoned flour before browning. As with any browning, it is essential not to overcrowd the pan and to brown the meat in batches if necessary. After the liquid is added, the pan is covered and the liquid brought to a simmer; boiling must be guarded against because it will toughen the meat. The goal is to simmer the meat until it is tender. Doneness when braising is determined by fork tenderness. The flavor of the braising liquid can be enhanced by the addition of wine, soup stock, marinades, seasonings, or tomato products. Only enough liquid, no more than 1 inch, should be added to produce steam. If too much liquid is used, it can reduce the flavor by sheer dilution.

Simmering or Stewing. Simmered or stewed meat is cooked completely submerged in liquid. The pan is covered, brought to simmering, not boiling, and cooked until the meat is tender. Fricassees are stews in which the meat is first browned in fat. Stews, unlike other simmered meats, are served in their own cooking liquid mixture, thickened or not, as desired, and usually contain vegetables added during the last hour of heating. Cured meats, such as corned beef or tongue and fresh beef brisket cuts, are commonly prepared by stewing. They are not browned first, and the cooking liquid, which has very little flavor, is usually discarded.

Steaming. Steaming exposes food directly to moist heat. Meats can be steamed in a pressure cooker or in a tightly covered pan. They can also be wrapped in aluminum foil or placed in a cooking bag, which is then placed in a heated oven. Since the meat cannot be observed during heating in a pressure cooker, its doneness is determined by timing. Meats also heat very well in a crockery cooker, an electrical appliance that will gently steam meat to extreme tenderness with only a little added liquid. Depending on the size and toughness of the cut, this may take anywhere from six to twelve hours. The long heating time and relatively low temperature may pose food safety concerns, however (see Chapter 3).

Microwaving. Microwave ovens are usually not the best option for cooking meats, except for thawing and reheating leftovers. They decrease juiciness, do not brown, and do not heat sufficiently to kill pathogens such as *Trichinella spiralis.* Microwaved meats do not taste the same as meats cooked by other time-tested methods, primarily because they do not get browned. Brown condiments such as Kitchen Bouquet, Worcestershire sauce, soy sauce, or steak or barbecue sauces can be used to add color to the meat or to cover it up, hiding the fact that the surface appears uncooked. Microwave browning skillets and grills are also available, but the flavor and texture problems remain the same. The power emissions from microwave ovens vary from brand to brand, so the manufacturer's instructions should be followed whenever a microwave is used for preparing meat or meat dishes.

Carving

Meat should not be sliced in just any manner, because the way it is sliced affects its tenderness. The first step in slicing meat is to determine the direction in which the muscle fibers run,

FIGURE 14-23 **Carving across the grain.**

called the grain. This can be seen on the surface of the meat. It may be difficult to find the grain in larger cuts such as roasts, because they consist of parts of several different muscles, each with its own grain. When carving meats, it is important to cut across the grain to increase tenderness (Figure 14-23). Cutting across the grain shortens the muscle fibers into smaller segments, making the meat easier to chew.

Storage of Meats

Meat contains high percentages of water and protein, both ideal for the growth of microorganisms. Consequently, meat should be stored in the refrigerator or freezer.

Refrigerated

Meats are best refrigerated at just above freezing (32°F/0°C), between 32°F and 36°F (0° to 2°C). They do not freeze until the temperature drops to below 28°F (−2°C). The best place to store meats in the refrigerator is in the coldest part. Many refrigerators have such an area or a compartment reserved for meat storage.

Wrapping Meat. Most retail meats are packaged with plastic wrap and can be refrigerated in their original wrap for up to two days. After that time, the store wrapping should be removed and replaced by loosely wrapped plastic wrap, wax paper, or aluminum foil. Leaving the tight store wrapping on meat for more than two days creates moist surfaces, which promote bacterial growth and deterioration of the meat. Exceptions to this general storage guideline are hams and other processed meats that are high in salt. They should not be stored in aluminum foil because the salt's corrosive action on aluminum foil will cause discoloration of the meat. Cured meats are also high in fat, which quickly turns rancid when exposed to oxygen and light. For this reason, ham and other processed meats are best stored in the refrigerator in their original wrappings.

Refrigeration Times. General guidelines suggest that fresh meat should not be stored in the refrigerator longer than three to five days, and that ground meats and variety meats should be cooked within one or two days (see back inside cover). Variety meats are more perishable than regular meat cuts and should be used within a day or two of purchase or frozen immediately. Cooked meat can be kept for about three to four days. If the meat needs to be kept longer than the recommended storage times, it should be frozen.

Controlled-Atmosphere Packaging. One alternative to storing meats for long periods of time at refrigeration temperatures is a new, patented, controlled-atmosphere package (CAP) available only to meat wholesalers. It can extend the shelf life of fresh red meat from the current 2 days to up to 28 days. The process involves using a special package that allows the removal of oxygen and its replacement with a mixture of 70 percent nitrogen and 30 percent carbon dioxide (25).

Frozen

Meats to be frozen should be wrapped tightly in aluminum foil, heavy plastic bags, or freezer paper and stored at or below 0°F (−18°C) (Figure 14-24). It is a good idea to first trim meat of bone and fat and to divide it up into individual servings before wrapping and freezing it. Most beef cuts can be kept frozen for six to twelve months, but ground beef should be frozen for no longer than about three months (see back inside cover). The colder temperatures reached by commercial freezers for at least 20 days at 5°F (−15°C) can kill *T. spiralis*. If not frozen to this degree, pork should always be cooked to the recommended temperature of 160°F (71°C). Wrappers often hide the identity of their contents, so the packages of frozen foods should be labeled and dated. It is better to make more frequent purchases than to freeze meat for extended periods of time, which can reduce its quality.

The texture and flavor of thawed meats will be adversely affected if they are refrozen. Freezer burn, caused by loss of moisture from the frozen food's surface, can result if meat is stored longer than the recommended storage time or when it is wrapped in materials that are not vapor-proof or are punctured. The dehydration of freezer burn causes a discolored surface on the meat that becomes very dry, tough, and somewhat bitter in flavor when cooked.

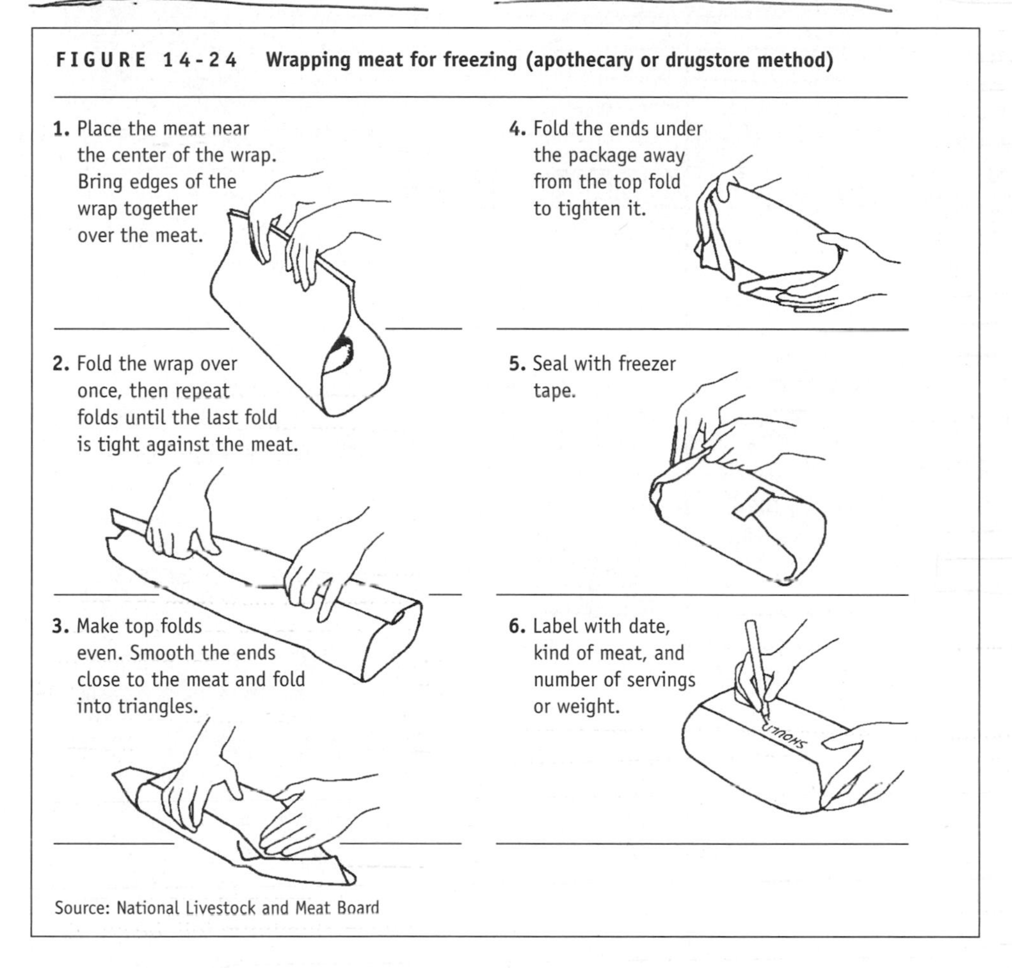

FIGURE 14-24 **Wrapping meat for freezing (apothecary or drugstore method)**

Source: National Livestock and Meat Board

PICTORIAL SUMMARY / 14: Meat

Usually the most expensive item on a menu, meat serves as an important source of complete protein. In North America and Europe, the main sources of meat are herbivores, such as beef cattle, sheep, and swine.

TYPES OF MEATS

Beef. Most beef is supplied by steers, male cattle that are castrated while young so that they will gain weight quickly. Heifers, females that have not borne a calf, are also used for meat.

Veal. Veal comes from male and female calves of beef cattle between the ages of three weeks and three months. These animals are fed a milk-based diet and have their movements restricted for a more flavorful and tender meat.

Lamb. Lamb comes from sheep less than 14 months old; the meat from older animals is sold as mutton.

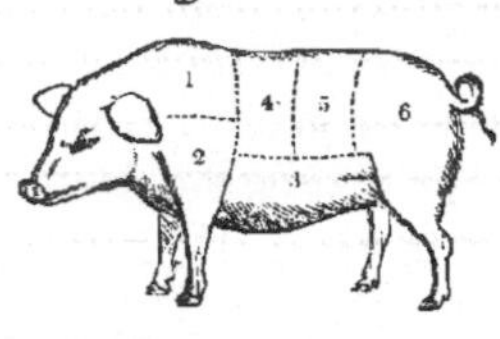

Pork. Most pork comes from young swine of either gender. In the last thirty years, pork has been bred to be leaner and more tender.

PURCHASING MEATS

Meat inspection is mandatory in the United States, but grading is voluntary. There are quality grades for beef, veal, lamb, and mutton. Factors considered in grading are color, grain, surface texture, and fat distribution. Yield grades are ranked from 1 (highest) to 5 (lowest), and indicate the amount of lean meat in proportion to fat, bone, and other inedible parts.

Tenderness in meats is due in part to natural influences such as the cut, marbling, animal age, heredity, diet, and slaughtering conditions. Meats can be artificially treated to make them more tender by aging, adding enzymes, salts, and acids, or subjecting them to mechanical or electrical treatments.

Kosher meats have met standards set by Jewish religious law.

Variety meats include the liver, sweetbreads (thymus), brain, kidney, heart, tongue, tripe (stomach lining), and oxtail of the animal.

Processed meats such as ham and sausage are preserved by curing, smoking, cooking, canning, or drying.

COMPOSITION OF MEATS

Meats consist of muscle, connective tissue, adipose (fatty) tissue, and bone. In meat cuts, the fat deposited in the muscle is visible as white streaks called marbling.

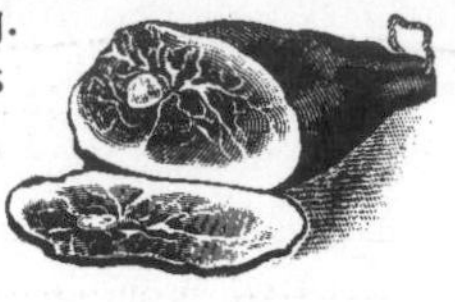

In terms of nutrient composition, meat is primarily water, high-quality protein, fat, some minerals, and B vitamins. Meat is not a good source of carbohydrates, fiber, or vitamin C.

PREPARATION OF MEATS

Meat should be sponged clean of any moisture with paper towels and trimmed of fat before being prepared. Doneness of meats can be determined by a combination of time/weight charts, color changes, internal temperature, and touch. Tender meats are best prepared by dry heat (roasting/baking, broiling, grilling, pan-broiling, and frying), while moist-heat methods (braising, simmering, stewing, and steaming) are best for tougher cuts. Common wholesale and retail cuts of meat are shown below:

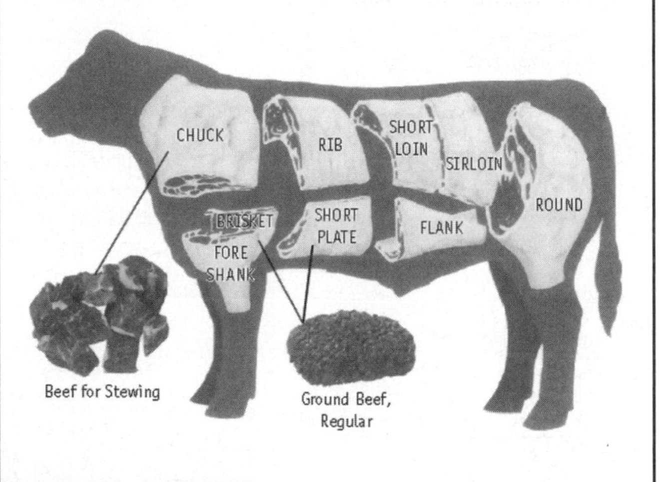

STORAGE OF MEATS

All meats should be refrigerated or frozen according to recommended temperatures. They should be held in the refrigerator no longer than the suggested maximum times, usually three to five days, although ground and variety meats will last only one or two days. Most meats can be kept frozen for six to twelve months if properly wrapped to avoid freezer burn caused by moisture loss.

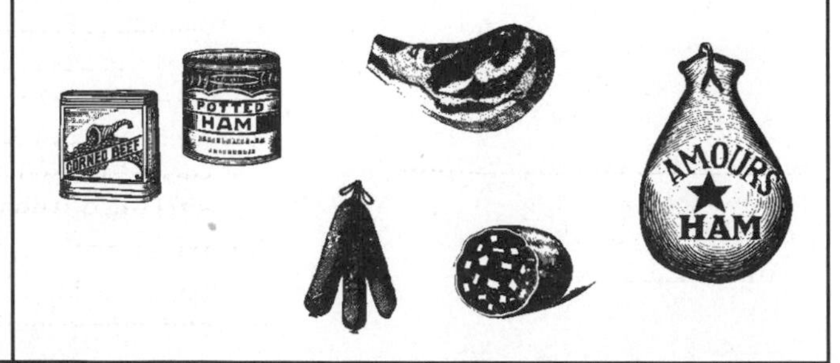

REFERENCES

1. Akinwunmi I, LD Thompson, and CB Ramsey. Marbling, fat trim and doneness effects on sensory attributes, cooking loss and composition of cooked beef steaks. *Journal of Food Science* 58(2):242–244, 1993.
2. Beggs KLH, JA Bowers, and D Brown. Sensory and physical characteristics of reduced-fat turkey frankfurters with modified corn starch and water. *Journal of Food Science* 62(6):1240–1244, 1997.
3. Berry BW. Low fat level effects on sensory, shear, cooking, and chemical properties of ground beef patties. *Journal of Food Science* 57(5):1205–1209, 1992.
4. Berry BW. Fat level, high temperature cooking, and degree of doneness affect sensory, chemical, and physical properties of beef patties. *Journal of Food Science* 59(1):10–14, 1994.
5. Berry BW. Sodium alginate plus modified tapioca starch improves properties of low-fat beef patties. *Journal of Food Science* 62(6):1245–1249, 1997.
6. Best D. Corporations invest in "back to basics" research. *Prepared Foods* 160(5):54–56, 1991.
7. Brewer MS, and FK McKeith. Consumer-rated quality characteristics as related to purchase intent of fresh pork. *Journal of Food Science* 64(1):171–174, 1999.
8. Carter RA, et al. Fabricated beef product: Effect of handling conditions on microbiological, chemical, and sensory properties. *Journal of Food Science* 57(4):841–844, 1992.
9. Cassens RG. Residual nitrite in cured meat. *Food Technology* 51(2):53–55, 1997.
10. Cheah PB, and DA Ledward. Catalytic mechanism of lipid oxidation following high pressure treatment in pork fat and meat. *Journal of Food Science* 62(6):1135–1138, 1997.
11. Correlation of sensory, instrumental and chemical attributes of beef as influenced by meat structure and oxygen exclusion. *Journal of Food Science* 57(1):10–15, 1992.
12. Craig JA, et al. Inhibition of lipid oxidation in meats by inorganic phosphate and ascorbate salts. *Journal of Food Science* 61(5):1062–1067, 1996.
13. DeFreitas Z, et al. Carrageenan effects on salt-soluble meat proteins in model systems. *Journal of Food Science* 62(3):539–543, 1997.
14. Dorsa WJ. New and established carcass decontamination procedures commonly used in beef-processing industry. *Journal of Food Protection* 60(9):1146–1151, 1997.
15. Dowell P, and A Bailey. *Cook's Ingredients.* William Morrow, 1980.
16. Ellert SJ, RW Mandigo, and SS Sumner. Phosphate and modified beef connective tissue effects on reduced-fat, high water-added frankfurters. *Journal of Food Science* 61(5):1106–1011, 1996.
17. Gariepy C, et al. Electrical stimulation and 48 hours aging of bull and steer carcasses. *Journal of Food Science* 57(3):541–544, 1992.
18. Giese J. Developing low-fat meat products. *Food Technology* 46(4):100–108, 1992.
19. Gisslen W. *Professional Cooking.* Wiley, 1998.
20. Haard NF. Foods as cellular systems: Impact on quality and preservation. *Journal of Food Biochemistry* 19:191–238, 1995.
21. Hagen BF, et al. Bacterial proteinase reduces maturation time of dry fermented sausages. *Journal of Food Science* 61(5):1024–1029, 1996.
22. Harris JJ, et al. Evaluation of the tenderness of beef top sirloin steaks. *Journal of Food Science* 57(1):6–9, 1992.
23. Hauge MA, et al. Endpoint temperature, internal cooked color, and expressible juice color relationships in ground beef patties. *Journal of Food Science* 59(3):465–473, 1994.
24. Jiang S, Y Wang, and C Chen. Lysosomal enzyme effects on the postmortem changes in tilapia (*Tilapia nilotica X T. aurea*) muscle myofibrils. *Journal of Food Science* 57(2):277–279, 1992.
25. Katz F. Is it time for changes in meat packaging and handling. *Food Technology* 51(6):99, 1997.
26. Kerler J, and W Grosch. Odorants contributing to warmed-over flavor (WOF) of refrigerated cooked beef. *Journal of Food Science* 61(6):1271–1274, 1996.
27. Kinsman DM, AW Kotula, and BC Breidenstein. *Muscle Foods. Meat, Poultry, and Seafood Technology.* Chapman & Hall, 1994.
28. Kulshrestha SA, and KS Rhee. Precooked reduced-fat beef patties' chemical and sensory quality as affected by sodium ascorbate, lactate and phosphate. *Journal of Food Science* 61(5):1052–1057, 1996.
29. Lonergan SM, et al. Porcine somatotropin (PPST) administration to growing pigs: Effects on adipose tissue composition and processed product characteristics. *Journal of Food Science* 57(2):312–317, 1992.
30. Maruri JL, and DK Larick. Volatile concentration and flavor of beef as influenced by diet. *Journal of Food Science* 57(6):1275–1281, 1992.
31. McWilliams M. *Foods: Experimental Perspectives.* Macmillan, 1997.
32. Miller MF, et al. Consumer acceptability of beef steak tenderness in the home and restaurant. *Journal of Food Science* 60(5):963–965, 1995.
33. Oreskovich DC, et al. Marinade pH affects textural properties of beef. *Journal of Food Science* 57(2):305–311, 1992.

34. Pegg RB, and F Shahidi. Unraveling the chemical identity of meat pigments. *Critical Reviews in Food Science and Nutrition* 37(6):561–589, 1997.
35. Penfield MP, and AM Campbell. *Experimental Food Science.* Academic Press, 1990.
36. Prochaska JF, SC Ricke, and JT Keeton. Meat fermentation: Research opportunities. *Food Technology* 52(9):52–56, 1998.
37. Quinton RD, et al. Acceptability and composition of some acidified meat and vegetable stick products. *Journal of Food Science* 62(6):1250–1254, 1997.
38. Renerre M. Factors involved in the discoloration of beef meat. *International Journal of Food Science and Technology* 25:613–630, 1990.
39. Simmonne AH, NR Green, and DI Bransby. Consumer acceptability and beta-carotene content of beef as related to cattle finishing diets. *Journal of Food Science* 61(6): 1254–1256, 1996.
40. State of the food industry: Meat and poultry. *Food Engineering* 63(6):78, 1991.
41. Watch the water loss. *Food Manufacture* 66(4):39–43, 1991.
42. Yin MC, and WS Cheng. Oxymyoglobin and lipid peroxidation in phophatidycholine liposomes retarded by alpa-tocopherol and beta-carotene. *Journal of Food Science* 62(6):1095–1097, 1997.

WEBSITES

USDA's meat and poultry hot-line about food safety questions:

www.fsis.usda.gov/mph/index.htm

Consumer publications from the North Carolina Cooperative Extension Service on beef, pork, and other meats, additives, freezing, food safety, processing and more:

www.ces.ncsu.edu/depts/foodsci/agentinfo/meat/conspub.html

Meat and poultry labeling terms from the USDA:

www.fsis.usda.gov/oa/pubs/lablterm.htm

15

Poultry

he term "poultry" refers to all domesticated birds raised for their meat. Although chickens are the most popular poultry consumed, other species include turkeys, ducks, geese, guinea fowls, and pigeons (squabs). Game birds such as pheasant, wild duck, and quail are also consumed, but few of them reach the marketplace. Not readily available in all parts of the country yet, but starting to be seen, are emu and ostriches, bred for their lower fat meat.

Despite the variety of poultry, chickens, domesticated by humans for over 4,000 years (10), remain the most common poultry consumed. Chickens are especially useful, since both their meat and their eggs are consumed. The popularity of chicken and turkey continue to increase at the expense of beef (14). In the past 40 years, production of broilers (young chickens) in the United States has increased from about 34 million to over 6 billion (2). Poultry is important to the diet, and the purpose of this chapter is to discuss poultry classification, composition, purchasing, preparation, and storage.

TABLE 15-1
Species and Classes of Poultry*

Species	*Class*	*Sex*	*Age*
Chicken	Cornish game hen	Either	5–6 weeks
	Broiler or fryer	Either	Under 10 weeks
	Roaster	Either	Under 12 weeks
	Capon	Unsexed male	Under 4 months
	Hen, fowl, baking chicken, or stewing chicken	Female	Over 10 months
	Cock or rooster	Male	Over 10 months
Turkey	Fryer-roaster	Either	Under 12 weeks
	Young hen	Female	Under 6 months
	Young tom	Male	Under 6 months
	Yearling hen	Female	Under 15 months
	Yearling tom	Male	Under 15 months
	Mature or old	Either	Over 15 months
Duck	Duckling	Either	Under 8 weeks
	Roaster duckling	Either	Under 16 weeks
	Mature or old	Either	Over 6 months
Goose	Young	Either	
	Mature or old	Either	
Guinea	Young	Either	
	Mature or old	Either	
Pigeon	Squab	Either	
	Pigeon	Either	

*The different species represent "kinds" of poultry, while class is dependent on the bird's sex and age.

Classification of Poultry

Ready-to-eat poultry is classified according to age and gender (Table 15-1). Classifications vary from species to species, with chickens classified as broilers, fryers, etc., and turkeys as toms and hens.

In the past, there was a "stewing hen" classification in the chicken category, but such a designation is now rare. Younger poultry are usually preferred because they are more tender and have less fat than older birds.

Chickens

Chickens sold on the market may be male or female, and differ in the age at which they are slaughtered and their weight. The younger chickens coming to market are classed as broilers/fryers, roasters, capons, and Cornish game hens.

Broilers/Fryers. Broilers and/or fryers are chickens of either sex, slaughtered under ten weeks of age, and weighing 3 to 5 pounds. They can be used not just for broiling and frying, as the names imply, but in any other way desired. At the market, these chickens will have soft skin, tender meat, and a flexible breastbone.

Roasters. Roasters are older and larger than broilers/fryers. These chickens are of either sex, usually processed at nine to eleven weeks of age, and weighing 6 to 8 pounds. The breastbone is less flexible than in broilers, having become calcified with age.

Capons. Capons are neutered male chickens that usually reach the market under four months of age weighing 12 to 14 pounds. The tenderness and juiciness of the meat is comparable to that of broiler/fryers.

Cornish Game Hens. Cornish game hens are bred by crossing a Cornish hen, a breed of chicken, with one of the other common breeds, such as White Plymouth Rock, New Hampshire, or Barred Plymouth Rock. The hens are slaughtered at five to six weeks, at which point they will weigh not more than 2 pounds. The meat is always very tender.

Mature Chickens. Older adult chickens over ten months of age, both female (hens, fowls, baking chickens, or stewing chickens) and male (cocks or roosters), have outlasted their breeding capabilities. Their meat is tougher, the skin coarser, and the breastbone less flexible. They are best used in stews, soups, and other slow-cooking dishes.

Turkeys

The turkeys bred for their meat today look very different from the *Meleagris gallopavo silvestris* depicted in the familiar old paintings of pilgrims and Native Americans at the first Thanksgiving. Turkeys consumed today are actually descended from the *Meleagris gallopavo* domesticated by the Aztecs of Mexico. Presently, seven standard breeds of turkey exist, but only the broad-breasted white is of commercial significance.

Turkeys are classified as fryer-roasters, hens, and toms. Fryer-roasters are very young turkeys, under 12 weeks old, with a ready-to-cook weight of around 7 pounds. They are seldom found in the markets, however; young hens and toms are more often sold. A young hen will weigh less than a young tom of the same age. Young toms are usually processed at about 17½ weeks of age, while the hens are processed earlier, at 14½ weeks, when they weigh 26 and 14 pounds respectively. The ready-to-cook weight varies from 8 to 15 pounds for a young hen and from 25 to 30 pounds for a young tom.

Other Domestic Poultry

The flesh of ducks and geese is not as widely consumed as that of chickens or turkeys, and is considered a luxury food item by many people. Ducks are usually marketed when they are seven to eight weeks old and weigh 3 to 7 pounds in their ready-to-cook state. Geese are marketed at about eleven weeks of age and have a ready-to-cook weight of 6 to 12 pounds. Other birds such as guinea fowl, squab (young pigeon), quail, and pheasant are also sometimes consumed. Occasionally these birds may be served in restaurants as delicacies or special entrées. The immature version of these birds is preferred for consumption. For example, younger guinea fowl weighing 1¾ to 2½ pounds (live weight) are preferred over mature guinea fowl that are normally 1 pound heavier. Squab are processed just before they leave the nest, or at about 30 days of age.

Composition of Poultry

The composition of poultry (muscle tissue, connective tissue, etc.) is similar to meat (see Chapter 14).

Pigments

Turkeys and chickens have both white and dark meat, the lightness or darkness depending on the amount of myoglobin content in the muscle.

HOW & WHY? ?????????

Why is the breast meat in chicken and turkey whiter than the thigh or drumstick? Higher amounts of the red-pigmented myoglobin are found in muscles that are used more frequently, such as those of the thighs and drumsticks (11, 18). Since domesticated chickens and turkeys do almost no flying, their little-used breast meat is white. Wild birds such as ducks have darker breast meat because they actually use the muscles for flying.

Purchasing Poultry

Inspection

In 1968, the Wholesome Poultry Products Act made inspection of poultry shipped across state lines mandatory. It is also required that poultry sold within a state must meet similar regulations, but these vary slightly from state to state. Poultry is inspected for wholesomeness before and after slaughter by a USDA inspector, who also ensures that the poultry is processed under sanitary conditions. Processing plants are encouraged to follow a Hazard Analysis Critical Control Point (HACCP) plan to minimize the risk of foodborne illness among consumers (8). Poultry that passes inspection is stamped with the USDA inspection mark.

Grading

The grading of poultry is voluntary and is paid for by the producer. Three

NUTRIENT CONTENT

The protein, carbohydrate, and vitamin content of poultry is somewhat similar to meats (see Chapter 14) with the exceptions listed below.

Fat and Cholesterol. Contrary to the popular notion that poultry is always lower in fat and cholesterol, Figure 15-1 shows that, with the exception of a few meat cuts, poultry is very similar to other meats in nutritive value.

FIGURE 15-1 Comparing the calories and fat grams in poultry vs meat (3 oz)

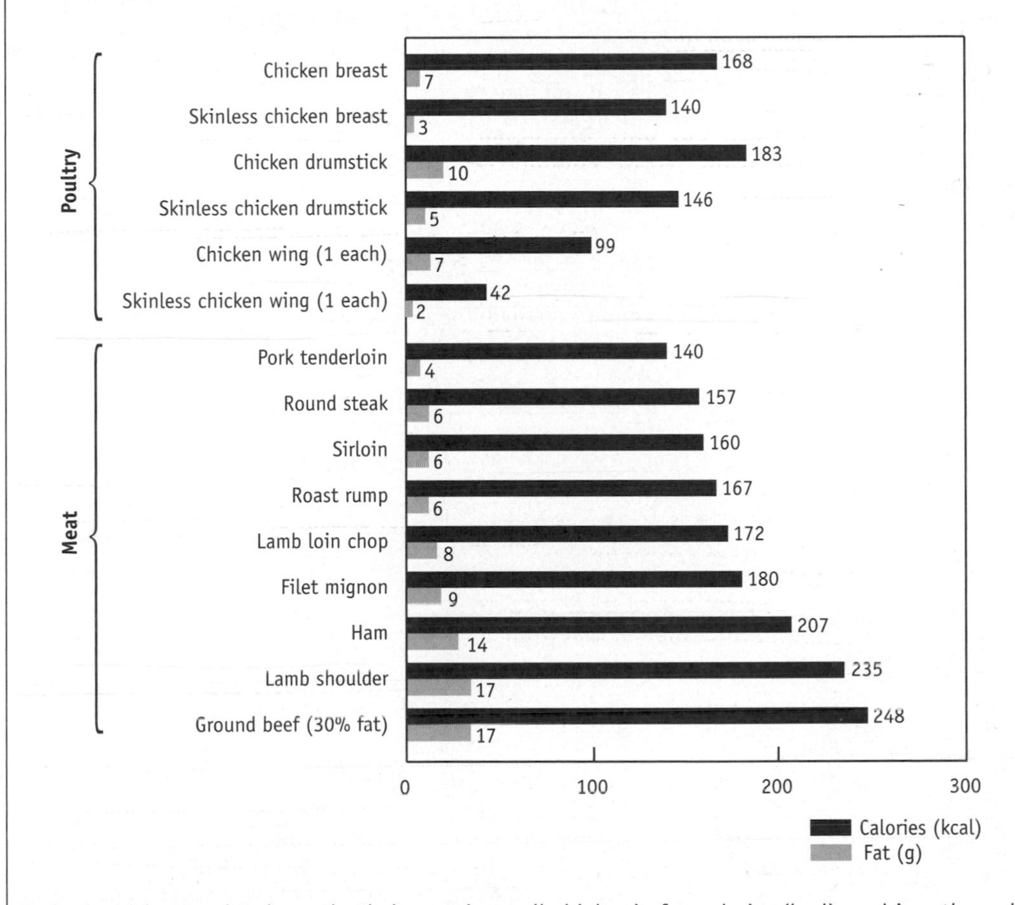

In both chickens and turkeys, the dark meat is usually higher in fat, calories (kcal), and iron than white meat. It is only after removing the skin, about 100 calories (kcal) per ounce, that there is any significant difference in fat content between poultry and lean cuts of meat. Ducks and geese are considerably higher in fat than chickens or turkeys. Emu and ostrich meat are lower in both calories (kcal) and grams of fat, with a 3-ounce serving yielding 93 calories (kcal) per 1.5 grams of fat and 121 calories (kcal) per 2.5 grams of fat respectively.

Minerals. Any processed poultry product (canned, dried, smoked, or self-basting) is higher in sodium than nonprocessed poultry. Processed poultry products are sometimes used as a substitute for the meat in foods like hot dogs, bologna, and hamburgers, and lower-sodium varieties are available for these uses. (9)

grades are used: A, B, and C. Grade A is the best and refers to a chicken that is full-fleshed and meets standards of appearance (Figure 15-2). The criteria used in grading are the conformation (the shape of the carcass), the fleshing (the amount of meat on the bird), the amount and distribution of fat, and freedom from blemishes such as pinfeathers, skin discoloration, broken bones, and skin cuts and tears (5). Poultry parts may also be graded USDA A, B, or C as well. In spite of the claims made by some chicken producers, skin color is not reflective of quality, but rather of the amount of xanthophyll and carotene plant pigments in the bird's diet.

The USDA grade shield shown in Figure 15-2 is used only when the poultry has been USDA graded. Since such grading is not mandatory, some poultry may be marketed under the proprietary "grades" established by

KEY TERMS

Eviscerate To remove the entrails from the body cavity.

individual packing houses, which may or may not match federal standards.

Types and Styles of Poultry

Poultry comes to market in a number of different types and styles. "Type" refers to whether it is fresh, frozen, cooked, sliced, canned, or dehydrated. "Style" describes the degree to which it has been cleaned or processed, i.e., live, dressed, ready-to-cook, or convenience categories. Live birds are rarely bought by the average consumer or restaurant. The other styles are far more prevalent.

- *Dressed.* Dressed birds are those which have had only the blood, feathers, and craw removed. The craw or crop is the pouchlike gullet of a bird where food is stored and softened.
- *Ready-to-cook.* Ready-to-cook poultry is **eviscerated**, free of blood, feathers, head, and feet, and is the style found in the supermarket and in most food service facilities. In ready-to-cook poultry, the internal organs such as the heart, liver, neck, and gizzard (part of the bird's stomach) have been cleaned and had the fat removed, and are frequently put back inside the inner cavity, often in their own giblet bag.
- *Convenience.* For convenience, smaller pieces such as halves, breasts, drumsticks, thighs, and wings of both chicken and turkey are available.

Ground turkey and ground chicken products are also becoming increasingly popular, and are used in a variety of foods ranging from sandwich fillings to frozen entrées. Not all ground poultry products are created equal. Labels should be read carefully, because fat is sometimes added back, which increases the total calorie (kcal) and fat gram count.

Processed Poultry. Convenience is also available to consumers and food manufacturers in the form of processed poultry products. Processed chicken and turkey are commonly used in canned or dried soups, frozen dinners, pot pies, sausages, hot dogs, burgers, and bologna. In addition, larger pieces of processed poultry meat minus the bone are sold as boneless turkey breast, roll, and ham.

U.S. Grade A

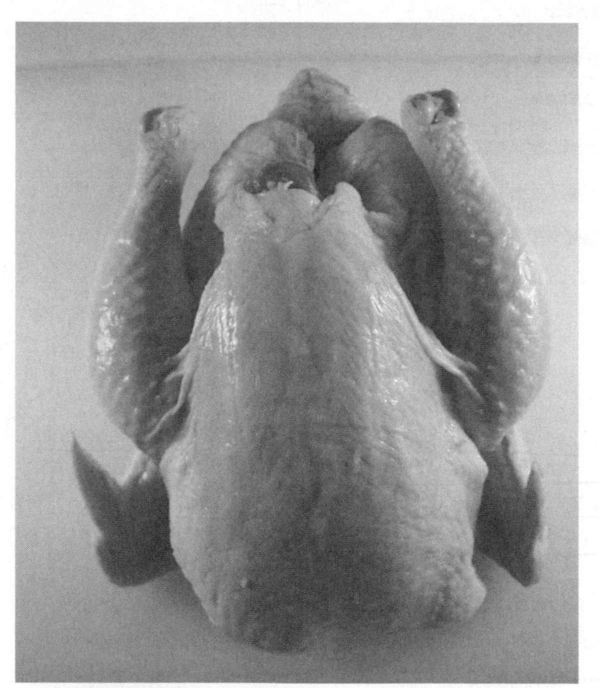

Fully fleshed and meaty; uniform fat covering; well formed; good, clean appearance. This grade is most often seen at retail.

U.S. Grade B

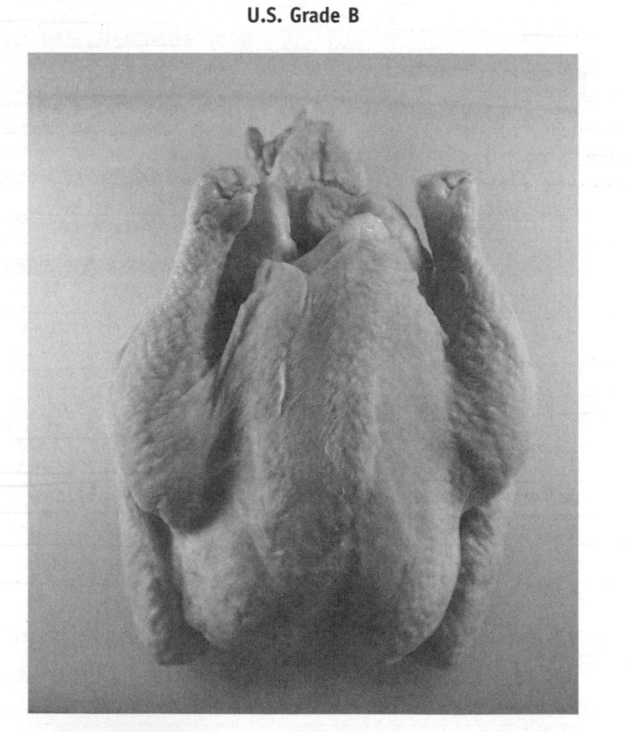

Not quite as meaty as A; may have occasional cut or tear in skin; not as attractive as A.

FIGURE 15-2 USDA grades for poultry.

These meats are made from mechanically deboned poultry in which the bone fragments have been removed. The larger cuts are easy to carve and have a characteristic texture due to binders and other compounds that have been added (Chemist's Corner 15-1).

How Much to Buy

Ready-to-cook poultry contains a good deal of inedible bone and unwanted fat, which must be taken into consideration when deciding how much to buy. A good rule of thumb for most poultry is to buy ½ pound or slightly more per serving. The exception is when purchasing ducks or geese, which have more fat to melt off, resulting in less yield. When purchasing a goose, plan on a bit over ½ pound per serving, and 1 pound for ducks. Turkeys under 16 pounds, which have a higher bone-to-meat ratio, are best purchased at about 1 pound per person.

Common broiler-fryer chickens average 3½ pounds and yield four servings—two breasts, and two leg and thigh pieces. Chickens under 2½ pounds are not economical. Turkeys, especially full-grown toms weighing 18 pounds or more, provide the greatest yield per pound. One of the most economical ways to buy poultry is in its ready-to-cook whole state. Poutry purchased whole can be cut up following the steps illustrated in Figure 15-3.

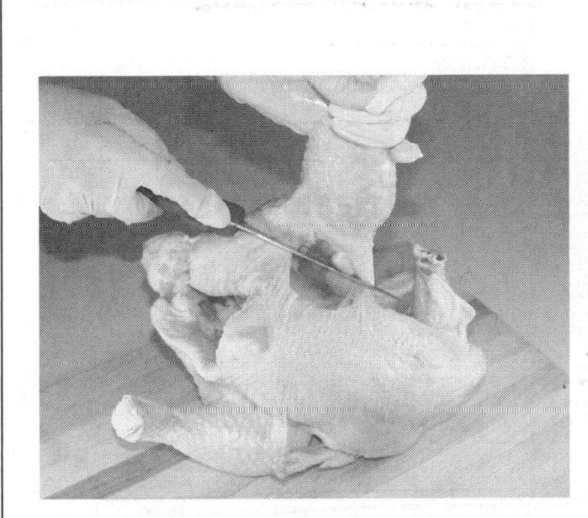

(1) Separate the leg from the breast. Pull the drumstick toward you. Use the tip of a knife to cut through the skin diagonally. Stay close to the thigh, leaving the skin on the breast intact. Often there's a pale strip of fat just under the skin to guide you.

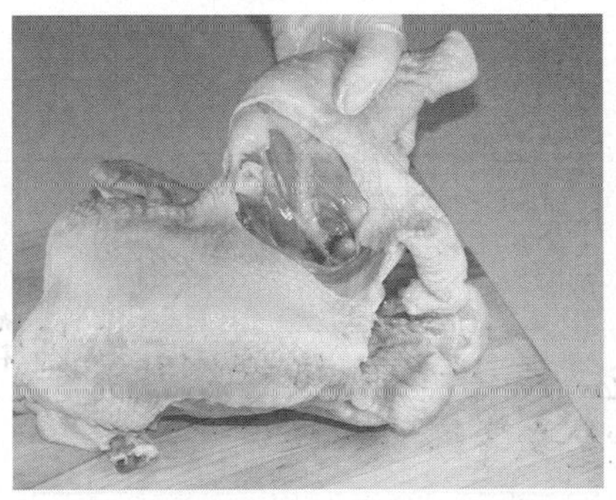

(2) Snap the thigh away from the backbone. Push your thumbs down into the opening between the thigh and the breast and fold the thigh away from the chicken's body until you see the joint snap out of the back. Gently tug the leg away from the carcass.

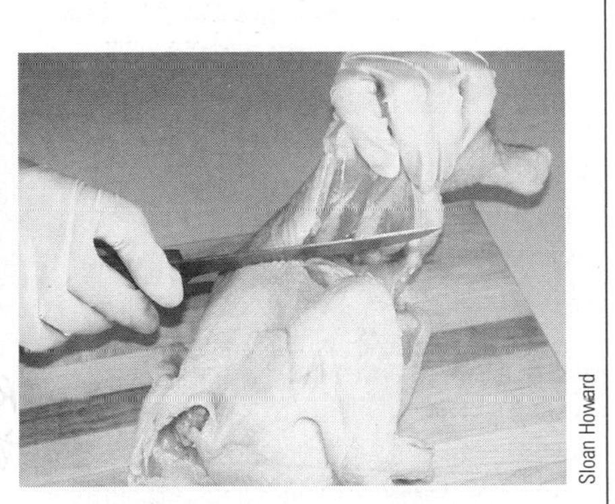

Sloan Howard

(3) Cut the leg away from the chicken. Following the contours of the backbone, trim around the "oyster"—the tender nugget of meat close to the backbone—and leave it attached to the thigh, not the back. Repeat with the other leg.

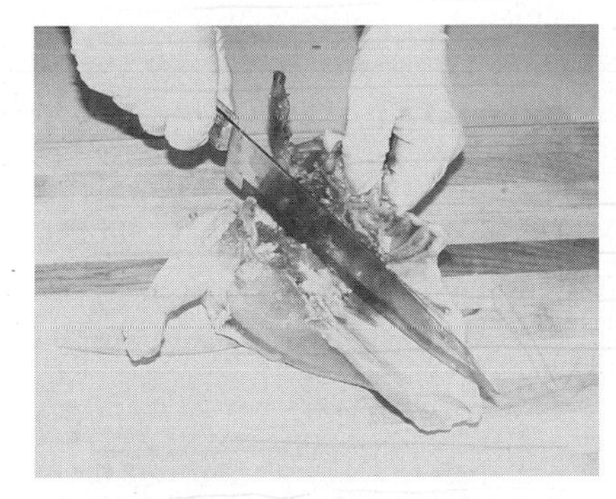

(4) Whack through the ribs with a heavy chef's knife. Holding the chicken with the pointed end of the breast up, use a chopping motion to separate the whole breast from the back.

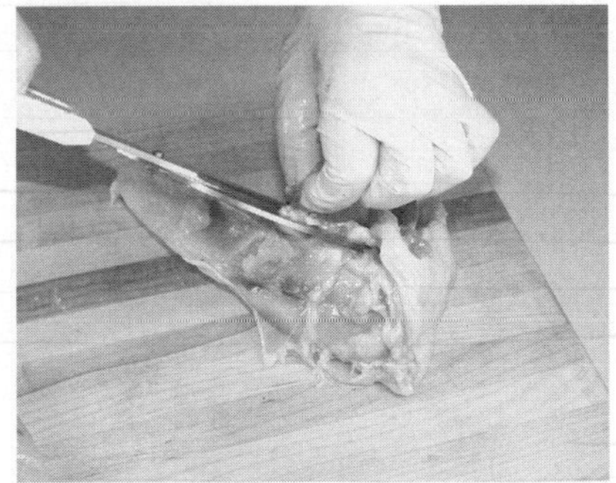

(5) Snap the backbone away from the breast. Hold the breast in one hand and push down on the backbone with the other. With this action, the wishbone is exposed. Cut along the wishbone to fully remove the back. Aim for the point where the wings join the breast, being careful to leave them attached to the breast. Save the back to use for stock.

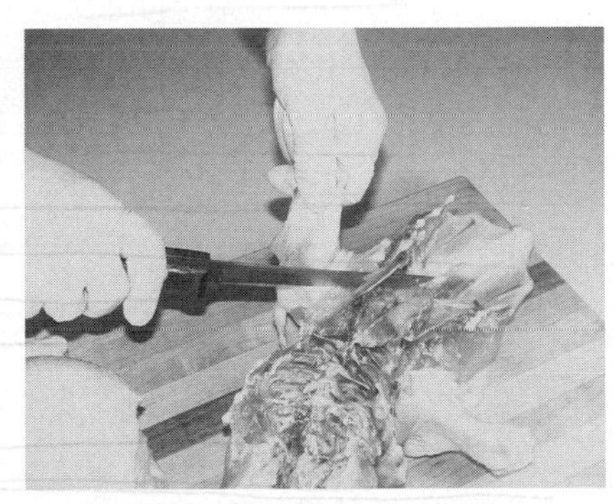

(6) Cut the breast in half. Lay it skin side down and cut through the center of the cartilage. Cut off any pieces of wishbone and rib that remain attached to the breast.

FIGURE 15-3 Cutting up a chicken.

Preparation of Poultry

Throughout the world, chicken is the most widely eaten of all the types of poultry. In Mexico, cooked chicken is shredded to fill tacos, enchiladas, and tamales. The Chinese stir-fry freshly cut-up chicken with vegetables and soy sauce. Chicken Kiev is a Russian specialty consisting of boneless breasts that are stuffed, rolled in a seasoned batter, and deep-fried. Paella, a Cuban favorite, is a combination of chicken with rice, tomatoes, sausage, and shellfish in one dish. In Africa, where peanuts are known as ground nuts, ground nut stew is made by simmering chicken with tomatoes and peanuts. In Japan, chicken may be marinated in a mixture of soy sauce, rice wine, and ginger before being grilled or steamed with cooked rice and egg. The resulting dish, called donburi, is very popular in that country. In India, chicken may be spiced and braised in a curry sauce or marinated in yogurt and spices before being roasted. The French are famous for *coq au vin*, or chicken braised in red wine, and the Italians are known for roasting chicken with rosemary. Some chicken dishes commonly consumed in North America include fried chicken, chicken cordon bleu, chicken and dumplings, chicken à la king, chicken divan, and chicken pies, soups, and salads.

Preparation Safety Tips

As a prelude to preparation, all ready-to-cook poultry should be washed inside and out and then patted dry with paper towels. Dish towels should not be used, because they can become a habitat for microorganisms.

HOW & WHY? ?????????

Why is there so much concern about the safety of poultry in food preparation? About one-fourth of all chickens in the United States carry *Salmonella*, and about half carry *Campylobacter jejuni*. A national survey showed that while only about 4 percent of broilers tested positive for *Salmonella* before processing, the number rose to 36 percent after the carcasses had been subjected to scalding, defeathering, eviscerating, and chilling (6). For this reason, anything that comes in contact with raw poultry (including hands, cutting boards, sinks, utensils, dishes, and counters) should be cleaned and sanitized afterwards.

Thawing Frozen Poultry. Freezing will largely protect against bacterial growth while the poultry is frozen, but precautions should be taken during and after thawing, when any bacteria that are present may begin to grow. The refrigerator is the best place to thaw frozen birds, and its use requires planning ahead. It takes about a day for a 3½ pound chicken and one to five days for a turkey to defrost, depending on its weight (Table 15-2). When the cavity is sufficiently thawed, the package of internal organs should be removed, and the cavity rinsed. Thawing whole poultry at room temperature, in the microwave oven, or under running cold water is not recommended.

Stuffing. After thawing, the bird should be seasoned and/or stuffed and baked immediately. For food safety reasons, the USDA recommends that stuffing be prepared and cooked separately or, if not, at least checked with a meat thermometer to determine if the internal temperature is at least 165°F (74°C). Prestuffed frozen poultry should never be thawed, but should be prepared, according to package directions, directly from the frozen state. The stuffing should be removed from leftover cooked poultry before the bird is refrigerated or frozen.

Changes During Preparation

Properly prepared poultry is tender and juicy, but overcooking causes the flesh to become dry, tough, and stringy. The skin of any poultry, which is primarily fat, can be removed before or after preparation, but if it is left on, it does contribute to flavor and juiciness. Fat that naturally melts off the bird during heating can be used to baste the poultry or to create sauces. Basting adds flavor and helps keep the meat tender and moist. Fat rises to the top of the drippings, so it may be easily removed before the drippings are used for gravy or sauce.

Reheated poultry, especially turkey, has a characteristic "warmed-over" flavor caused by the breakdown of fat (13). Microwave reheating

Chemist's Corner 15-1

Processing Poultry

The texture of processed poultry products is influenced by a variety of factors. First, the physical removal of meat from the bone mechanically causes a redistribution of the collagen fibers and myofibril proteins around the fat globules. This creates a more stable meat emulsion (15). Secondly, a brine mixture containing water, salt, and phosphates is added to improve flavor and cohesiveness. The phosphates in the mixture make the protein more absorbent to water by binding to the calcium and causing the protein fibers to relax. Gums such as carrageenan are then added to absorb water, creating a gel-like texture that prevents water loss during heating and makes slicing easier.

TABLE 15-2

Thawing a Turkey. The rule of thumb is about 24 hours of thawing for every 5 pounds of whole turkey.

Weight	*Thawing Time in Refrigerator (40°F/4°C)*
8–12 lb	1–2 days
12–16 lb	2–3 days
16–20 lb	3–4 days
20–24 lb	4–5 days

results in less of the warmed-over flavor than the conventional methods (4). The other changes that occur during preparation closely parallel those found in meats (see Chapter 14).

Determining Doneness

Poultry should always be heated until well done to enhance flavor and to minimize the risk of foodborne bacterial illnesses. Doneness may be determined by internal temperature, color changes, and/or touch and time/weight tables, each of which is discussed below.

Internal Temperature. The best way to check poultry for doneness is to use a meat thermometer. It should be inserted into the thickest part of the breast, although it can also be inserted in the inner thigh. In either case it should not touch bone or fat. Poultry is sufficiently cooked when the internal temperature reaches 180° to 185°F (82° to 85°C). The pop-up thermometers that some poultry producers place in turkey breasts are not always reliable, so check for other signs of doneness. A thermometer placed in the center of any stuffing must reach a minimum temperature of 165°F (74°C) (Figure 15-4).

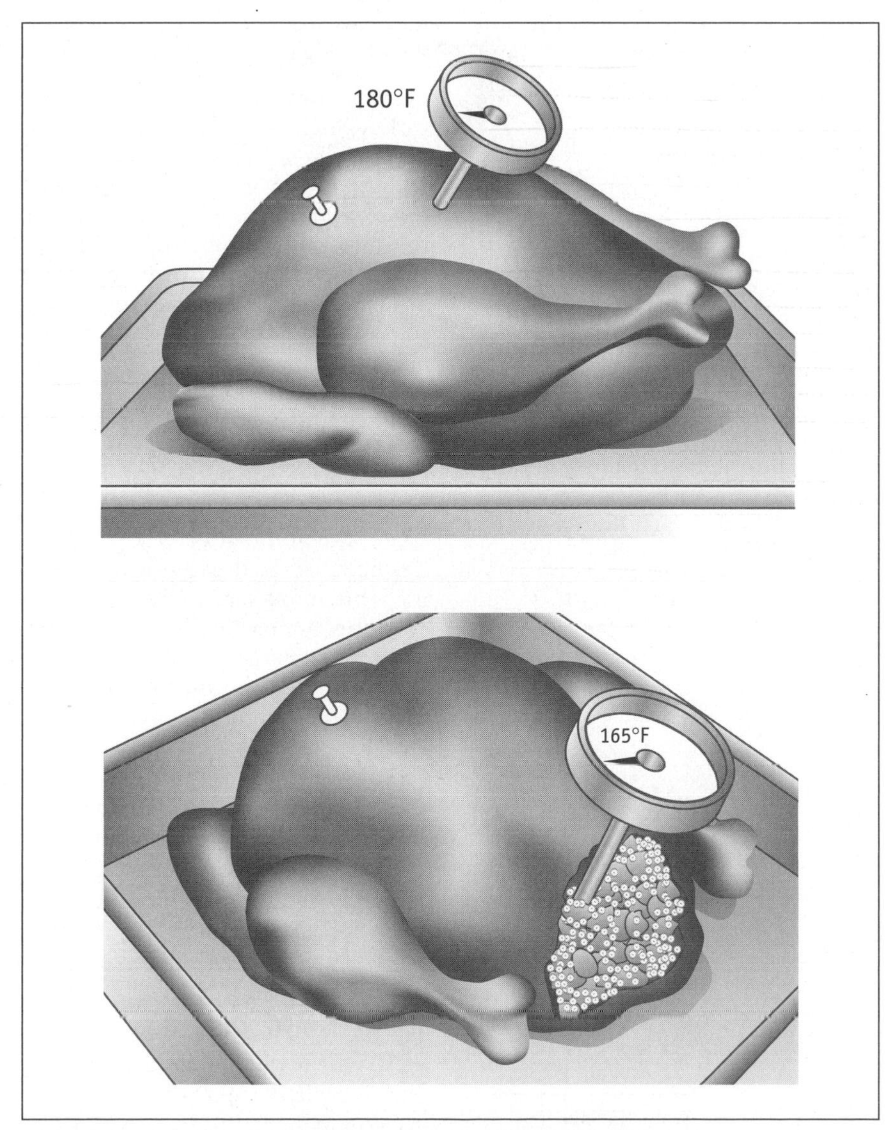

FIGURE 15-4 Internal temperatures for a well-cooked turkey.

Color Change. When oven-roasted chicken or turkey reaches a golden brown color, it is time to test for doneness. The juices coming out of the bird should have turned from pink to clear, and a bit of bone should be showing on the tip of the legs. When a turkey is roasted breast side up, the breast should be covered with metal foil or a bit of cooking oil to keep the breast from over-browning or burning. The foil should be removed 45 minutes to an hour before the end of heating to allow for final browning.

Touch. When pressed firmly with one or two fingers, the well-done bird's flesh will feel firm, not soft. White meat may be firmer than dark meat, in part because certain proteins have a higher gel-forming ability in white muscle than when they are located in the dark muscles (1). Another way to tell whether or not the poultry is done through touch is to wiggle the drumstick—it should move easily in its joint.

Time/Weight Charts. Time/weight charts appear on the packaging of all frozen and many fresh birds. It takes about 1½ hours in a 350°F (177°C) oven to thoroughly cook a 3½-pound chicken. Preparation times for turkeys depend on their weight and are reduced for those roasted in one of the special oven bags (Table 15-3). Although there are time/weight charts for frozen turkeys, it is not recommended that they be cooked from the solidly frozen state, because they may not be heated through enough to destroy microorganisms.

Dry-Heat Preparation

Roasting or Baking. Poultry to be roasted or baked, whether whole or in pieces, should be rinsed and patted dry as described previously. The inside of the cavity of a whole bird is seasoned as desired, and the outside may be coated lightly with vegetable oil to prevent the skin from cracking and allowing moisture loss. Margarine is not recommended for

KEY TERMS

Truss To tie the legs and wings against the body of the bird to prevent them from overcooking before the breast is done.

TABLE 15-3

Time/Weight Chart for Preparing Turkey at 325°F/163°C*

	Cooked in Open Roasting Pan		*Cooked in Oven Cooking Bag*	
Weight (pounds)	*Unstuffed (hours)*	*Stuffed (hours)*	*Unstuffed (hours)*	*Stuffed (hours)*
8–12	2¾–3	3–3½	1¾–2¼	2¼–2¾
12–16	3–4	3½–4¼	2¼–2¾	2¾–3¼
16–20	4–4½	4¼–4¾	2¾–3¼	3¼–3¾
20–24	4½–5	4¾–5¼	3¼–3¾	3¾–4¼

*These times are approximate and should always be used with a properly placed thermometer.

such a coating or for basting because of its low smoking temperature. Seasonings may be added as desired. Although they do not add flavor to the flesh, they make a delightful flavor when the skin has been browned to crispness (Chemist's Corner 15-2). The bird is then placed in an oven set at between 325°F and 350°F (163° to 177°C) and baked for the allotted time:

- 20 to 25 minutes per pound for poultry up to 6 pounds
- 15 to 20 minutes per pound for poultry up to 15 pounds
- 12 to 15 minutes for poultry over 15 pounds

When birds are stuffed, more cooking time must be added, about 5 minutes per pound, to make sure the stuffing is sufficiently heated all the way through to kill microorganisms. A small piece of aluminum foil placed over exposed stuffing in the final stages of baking will prevent it from scorching (Figure 15-5).

Poultry may be **trussed** before roasting. This is usually done with turkeys because of their long preparation time. Figure 15-6 illustrates one method of trussing a bird. Wire clips, which frequently come with a turkey, will hold the legs in place without trussing. The wire clips should be temporarily removed when cleaning the bird prior to preparation, and the wings can be tied up against the breast to prevent their edges from burning.

Chemist's Corner 15-2

Aroma of Roasting Chicken

The classic aroma of roasting chicken comes from volatile compounds such as carbonyls and hydrogen sulfide (12).

Birds to be roasted are placed, usually with the breast up, in a heavy-duty roasting pan on the lowest rack of the oven. The pan should have 2-inch sides; sides higher than 2 inches make basting difficult and prevent the lower portion of the bird from browning. Some cooks claim that the bird is juicier if placed breast down so that it can self-baste. But eventually it must be turned over to brown the breast, and because this task is not easy, most people find that basting a breast-up bird with accumulated pan juices and/or melted butter is quite satisfactory.

Ducks and geese, because of their high fat content, should be placed breast down after having had their skins thoroughly pricked to release excess fat. They are turned breast-up about halfway through heating time. The skin is pricked again at least once during heating to facilitate fat drainage, and pan drippings are removed during roasting. Duck and goose are sometimes preroasted for about 15 minutes and then prepared like chicken. Cornish game hens are roasted the same way as broilers and fryers except that their cooking time is only about a half hour, unless they have been stuffed, which increases their baking time by 15 minutes.

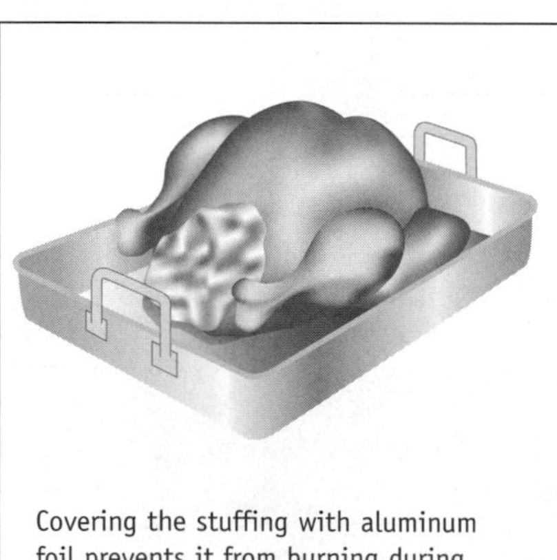

Covering the stuffing with aluminum foil prevents it from burning during the last stage of roasting.

FIGURE 15-5 Protecting stuffing from scorching.

Basting. Basting of chickens and turkeys helps prevent drying of the skin and meat. This involves using a wide spoon or brush or a special tool called a baster to periodically cover the bird with liquid from the drippings, melted butter, or barbecue or other sauce. Any sauce containing sugar (brown or white), such as barbecue sauce, will increase browning, possibly to an undesirable degree, and should be applied toward the end of the cooking process to avoid browning of the sauce. The number and timing of bastings depend on the size of the bird and whether or not it has been covered early in the cooking with an oil-soaked cloth or other covering, but basting once every half hour is usually more than adequate. Basting helps the skin to brown, but to prevent overbrowning, tent the bird with aluminum foil two-thirds through the cooking time.

Duck and goose do not need to be basted; they are so high in fat that they are self-basting. For the same reason, any stuffing for these fowl should be cooked separately, because it would be too soaked with fat if prepared in the cavity of the bird.

Stuffing. Stuffing refers to anything that is placed in the cavity of a bird during cooking. This is usually the familiar breadcrumb or cornbread stuffing; however, other foods such as

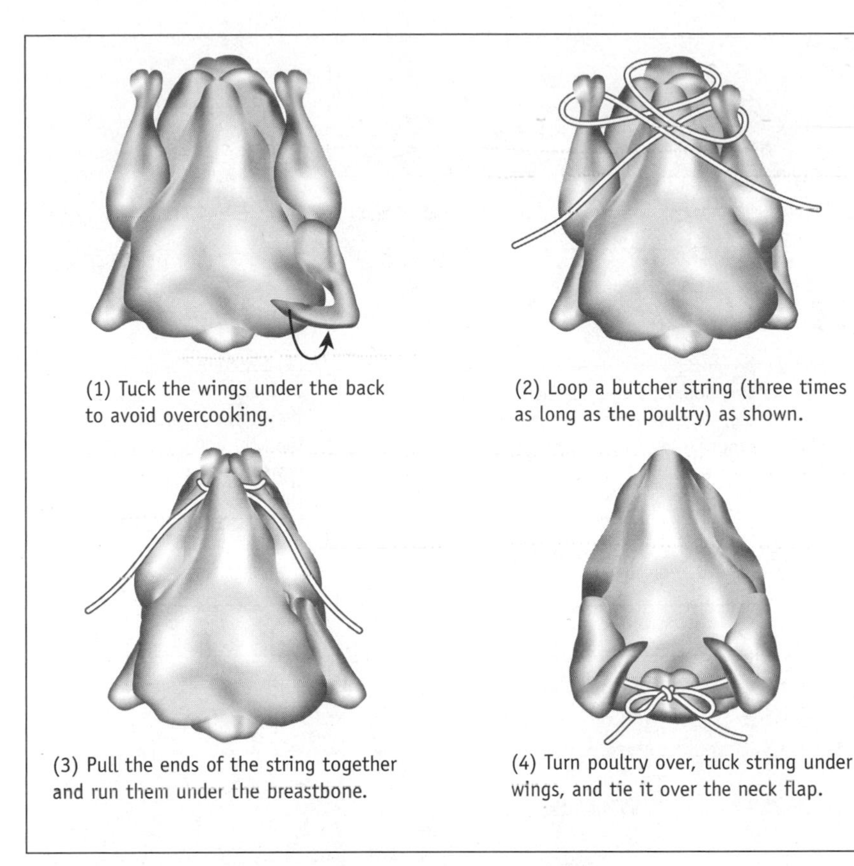

FIGURE 15-6 **Trussing poultry.**

vegetables and meats are sometimes stuffed in the bird's cavity. Dressing is distinguished from stuffing by being heated separately in a casserole or pan and served as a side dish.

The main ingredient of stuffing/dressing is cut cubes of day-old bread, packaged stuffing mixes, or rice. This starch-based foundation absorbs the juices released during cooking, which is why it is important that it be dried or else the dressing will be mushy. Bread cubes (¼ to ½ inch for turkey, smaller cubes for chicken) can be dried by spreading them out on a cookie sheet and baking them on low (275°F/135°C) for 15 minutes or leaving them out overnight. If grains such as rice are to be the main stuffing ingredient, they should be cooked and cooled before being combined with the other ingredients. Added to this bread or grain base is a **mirepoix** (meer-PWAH). Apricot or apple pieces, nuts, mushrooms, oysters, raisins, or other items may also be added to the base, according to personal preference. Liquid such as broth or water is then added to hold the mixture together, and eggs may be included to add cohesiveness. If the stuffing is going into the bird, only enough liquid should be added to make the stuffing barely hold together; if it is too moist, it will not be able to soak up juices.

All the ingredients should be lightly tossed together and then spooned into the poultry cavity. Stuffing should not be packed in, but should fill the cavity only three-quarters full, because it will expand as it cooks. It is important to remember that stuffing a bird increases roasting time, so plan accordingly. The center of the stuffing needs to reach a final temperature of 165°F (74°C) in order to destroy microorganisms.

A stuffed bird should be allowed to stand for only a short time after being removed from the oven and before serving. It should be refrigerated as soon as possible, and all the stuffing should be taken out of the bird's cavity before refrigerating. It cannot be stressed enough that stuffings, particularly those with eggs as an ingredient, are an ideal medium in which microorganisms can grow and flourish.

If stuffing is not used, then an apple, potato, carrot, onion, or celery stalks may be placed in the cavity to absorb off-flavors. The fat and off-flavors absorbed by any such fruits or vegetables during cooking renders them unappetizing to eat, and they are usually discarded.

KEY TERMS

Mirepoix A collection of lightly sautéed, chopped vegetables (a 2:1:1 ratio by weight of onions, celery, and carrots) flavored with spices and herbs (sage, thyme, marjoram, and chopped parsley are the most common).

Carving. Chicken is carved into its breast, leg, thigh, and wing pieces using the technique illustrated in Figure 15-7. Turkey should be allowed to stand for about 20 minutes after it is removed from the oven before carving. This allows the flesh to firm up and makes carving easier. Figure 15-8 demonstrates the carving of a turkey. Carve only what will be used immediately to avoid drying and cooling of the turkey meat pieces.

Broiling or Grilling. Except when cooking a whole bird on a spit over hot coals, only cut-up poultry is used for broiling or grilling. It is frequently marinated or coated with butter and seasonings before being broiled or grilled. In the interest of food safety, marination must take place under refrigeration. A marinade must be fully cooked if it is to be served or used for basting. Failure to heat the marinade to a sufficient temperature to kill the bacteria that remain in it from the raw chicken may cause a foodborne illness. For the same reason, unless it is thoroughly washed in the interim, the plate used to carry the raw poultry to the grill should never be used to carry it back to the table after it is cooked.

Vegetable sprays applied to the pan or grill help to prevent sticking. When an oven broiler is used, the poultry pieces are put skin-side up on a rack in the broiler pan and placed approximately 6 inches below the heat source. The same procedure is used for grilling over coals, except that the skin side goes down. The cooking time varies according to thickness, but in general, chicken takes 20 minutes per side. Turkey pieces are larger and so require longer cooking. Once the skin side is browned, use tongs to turn the poultry pieces over, since the piercing tines of a fork will allow juices to be lost. Sauces are best added during the last 15 minutes of preparation, because

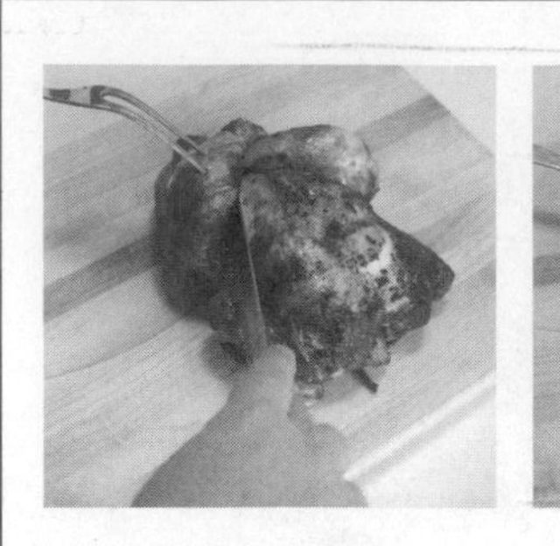

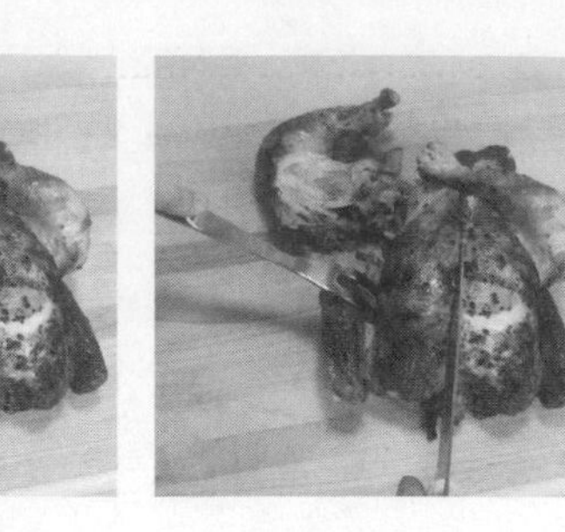

(1) Steady the chicken on a sanitary cutting board. To remove the leg, slice through the skin holding the leg to the breast.

(2) Push the leg down to partially dislodge the joint, cut through the meat between the leg and breast, then cut through the joint.

(3) Separate the breast meat by bracing the chicken with the fork, slicing just inside the keel bone; move the knife downward, pulling/cutting the breast section away from the rib cage.

FIGURE 15-7 Carving roast chicken.

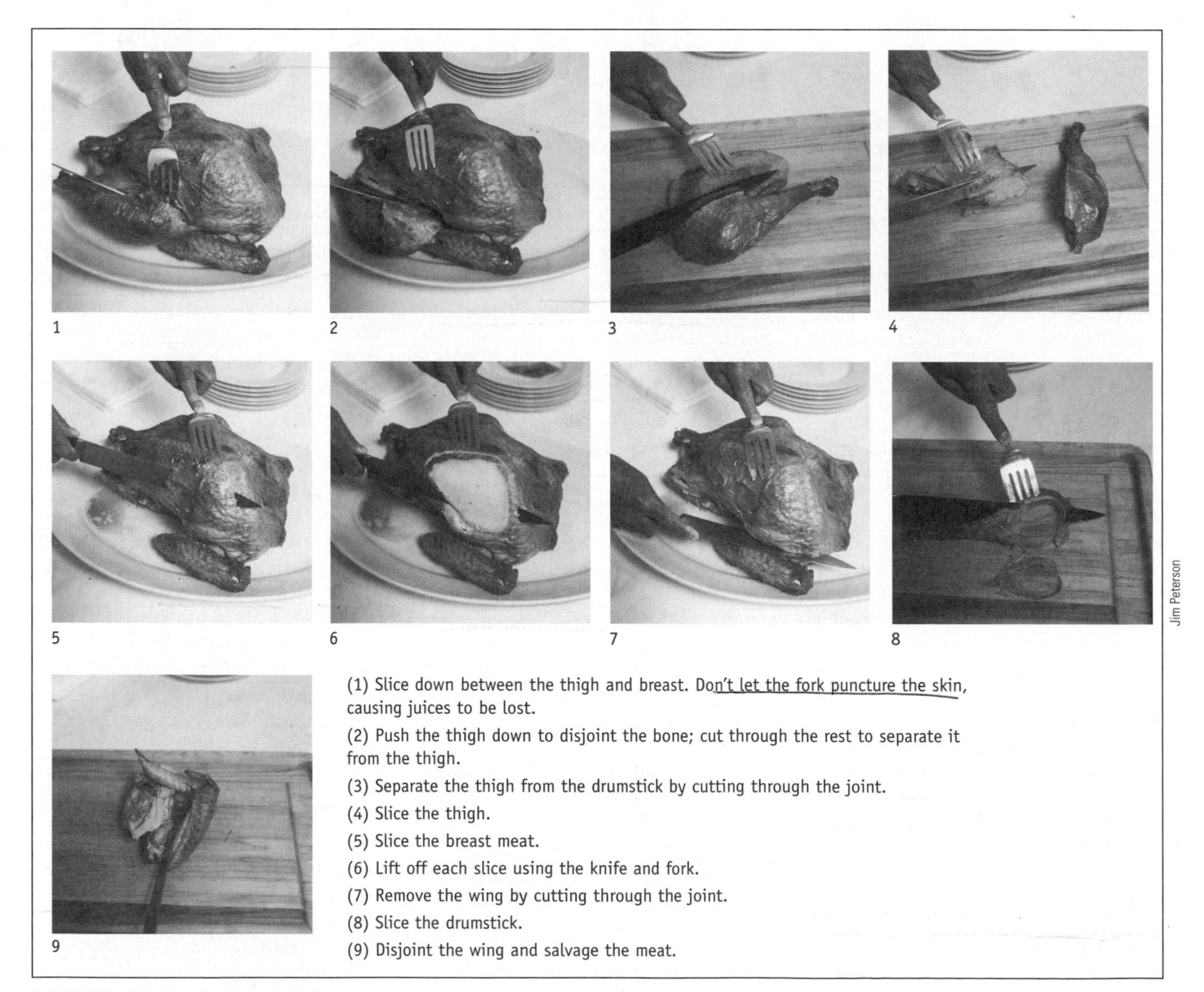

(1) Slice down between the thigh and breast. Don't let the fork puncture the skin, causing juices to be lost.

(2) Push the thigh down to disjoint the bone; cut through the rest to separate it from the thigh.

(3) Separate the thigh from the drumstick by cutting through the joint.

(4) Slice the thigh.

(5) Slice the breast meat.

(6) Lift off each slice using the knife and fork.

(7) Remove the wing by cutting through the joint.

(8) Slice the drumstick.

(9) Disjoint the wing and salvage the meat.

FIGURE 15-8 Carving a turkey.

high heat readily burns sugar, which is the main ingredient of many barbecue sauces.

Frying. Poultry pieces can be sautéed, pan-fried, deep-fried, or stir-fried.

Sautéing. Small poultry pieces are placed in a skillet or pan with a small amount of oil for quick preparation. Pieces must be turned to assure adequate doneness. Sautéing can also be used to brown larger poultry pieces prior to their being baked or braised to completion.

Pan-Frying. Pan-fried chicken pieces are usually breaded or floured before they are fried over high heat in approximately ¼ inch of fat. The breading adds texture and flavor and keeps moisture from being lost from the fried food; it also allows heat to be transmitted to the food without its absorbing as much fat. Fry with the skin side down first; when that side is brown, turn it over with tongs and brown the other side. Lower the heat and turn the pieces occasionally, for 30 to 45 minutes, or until done. If the poultry is placed in the oven following browning, the method of preparation is referred to as "oven fried," even though it is actually baked.

Deep-Frying. Deep-frying poultry pieces (that have been breaded, floured, or battered) involves submerging them completely in oil heated to between 325° and 350°F (160° to 180°C).

Stir-Frying. Stir-frying is lightly frying bite-size pieces of boned chicken while stirring them frequently in a tiny amount of oil. Vegetables, also cut into small pieces, are usually added, along with soy sauce and/or other seasonings.

Moist-Heat Preparation

Braising. Although braising, also called fricasseeing, can be used with any poultry, it is of particular value when it comes to preparing older, tougher birds. The slow, moist heating tenderizes the meat and makes it easier to chew. The chicken or turkey is first cut into pieces and browned in a small amount of oil and/or butter; it may be floured or breaded first. Liquid is added, and the poultry is simmered in a tightly covered pan until tender. The initial browning is important because it helps create a rich flavor and holds in the juices. Desired seasonings are added with the liquid.

Stewing. Any whole or cut-up fresh poultry can be covered in cold salted water and heated to the boiling point, at which point the heat is immediately lowered to simmer. An average 3½ pound chicken usually takes about 2 to 2½ hours. The bones and skin may or may not be removed from the pot, and dumplings, which are made from a dough mixture, can be placed gently on top of the simmering chicken 12 to 15 minutes before the end of preparation time.

Poaching. Chicken pieces can be poached fairly quickly in a small amount of water. The chicken pieces, such as breasts, are placed in a frying pan and covered with 1⅓ cups water. The water is brought to a boil and then reduced to a simmer, and the chicken is cooked about 10 to 15 minutes or until tender.

Microwaving. Microwave ovens do not always heat food deeply or evenly enough, and power levels vary from brand to brand, so it is suggested that stuffed poultry, particularly turkeys, be prepared in the conventional oven. The microwave manufacturers' instructions should be followed for preparing all other poultry. This is equally true when it comes to thawing poultry or any other frozen food. Once thawed in the microwave oven, the poultry should be cooked immediately.

In general, microwave directions call for smaller pieces of poultry rather than whole fowl. If a recipe calls for chicken pieces, a microwave can be handy. The poultry pieces are arranged skin side up, with the thickest portions toward the outside of the dish and any loose flaps of skin tucked under. The dish is covered with wax paper or plastic wrap and cooked on high for about 8 minutes per pound, or according to the manufacturer's directions. Chicken breasts are heated on high for about 10 minutes or until well done. The pieces should be rotated at the 5-minute mark.

Flavor and appearance are enhanced if the pieces are initially covered with browning sauce, barbecue sauce, or some other topping. Cooking is completed when the flesh is firm and fork tender, and the juices run clear instead of pink. However, temperature readings, being more accurate, are recommended. The finished pieces should be left to stand about five minutes before serving. If they are to be used in a salad or other dish, it is best to chill them in the refrigerator for at least two hours. Two boned, skinned, chicken breasts will yield 1 cup of cubed chicken meat.

Storage of Poultry

Precautions should be taken in the handling of poultry, because of the possibility of it being contaminated with bacteria. *Salmonella* is one of the most common causes of foodborne illness.

Some of its major sources include raw poultry, eggs, and stuffing (see Chapter 3 for a full discussion of this bacterium). In 1993, the irradiation of poultry was approved for commercial use in the control of *Salmonella* following several studies that showed it reduces bacterial concentration (3, 16). Irradiated poultry, however, is not sterile and should be handled using the same precautions used for nonirradiated raw poultry. In 1992, the use of trisodium phosphate (TSP), a colorless, odorless, flavorless chemical mixture, also received approval for use by the poultry industry on poultry carcasses to further aid in reducing *Salmonella* contamination.

Refrigerated

Fresh, ready-to-cook poultry can be kept safely in the refrigerator at 40°F (4°C) or below for up to three days (Chemist's Corner 15-3). It should be stored in the vapor-proof wrapping in which it is purchased, since repackaging increases the risk of bacterial contamination. It is best kept in the bottom portion of the refrigerator to prevent its drippings from contaminating other foods. Chickens labeled "fresh" should not have gone below 26°F (−3°C), the temperature at which chickens freeze.

Frozen

Frozen whole poultry can be stored from six to twelve months at 0°F (−18°C), while leftover cooked poultry can be frozen for up to four months. The meat will decline in moistness and eating quality if it is kept frozen beyond the recommended times. Breaded or fried poultry should never be thawed and refrozen.

Thawing. Defrosting is recommended in the refrigerator, where it will take about a day to defrost an average chicken, and one day for each five pounds of turkey. Once defrosted, poultry or any other meat should not be refrozen unless it has been cooked. Stuffing should be refrigerated promptly and consumed within two or three days.

Chemist's Corner 15-3

Oxidation of Cooked Poultry

Dark meat has a higher myoglobin content than white meat. Consequently, it is more easily oxidized because the iron in the myoglobin acts as a metal catalyst to speed up the reaction of the polyunsaturated fatty acids being oxidized. Oxidation of these polyunsaturated fatty acids found naturally in the meat results in disagreeable off-odors (7, 17). As a result, chicken legs with their dark meat cannot be stored as long as chicken breasts.

PICTORIAL SUMMARY / 15: Poultry

Humans have been domesticating chickens for over 4,000 years. These days, the consumption of poultry, especially chicken and turkey, continues to increase in popularity.

CLASSIFICATION OF POULTRY

Poultry, or domesticated birds raised for their meat, includes:

Chicken
- Broilers
- Fryers
- Roasters
- Capons
- Cornish Game Hens

Turkey
- Young Tom
- Young Hen

Duck

Goose

Pigeon

Guinea Fowl

Domesticated birds are classified according to age and weight, and the classifications vary from species to species. Chickens are sold as broilers and/or fryers, roasters, capons, Cornish hens, and stags. The majority of turkeys coming to market are young hens, hens, young toms, and toms.

PURCHASING POULTRY

All poultry scheduled to be transported interstate must have the USDA stamp of approval. For birds sold intrastate, USDA inspection is voluntary. However, strict state inspection guidelines are enforced. These may vary slightly from state to state but are close to federal standards. USDA grade stamps indicate A, B, and C quality, with A being the best. Many processors use their own grading system and stamps.

The least expensive way to buy poultry is to purchase it as a ready-to-cook whole bird. The larger the bird, the more edible meat per pound. For chicken and turkey, approximately ½ pound of whole bird is needed for each serving.

Poultry is available for purchase in the following forms:

- Fresh
- Frozen
- Cooked
- Canned
- Dehydrated
- Live
- Dressed
- Ready-to-cook
- As convenience food

COMPOSITION OF POULTRY

Nutritionally poultry, like meat, is a high-quality protein food. Contrary to the popular notion that poultry is always lower in fat and cholestorol, poultry is very similar to many other meats in nutritive value. Poultry does provide less fat if the skin is removed. The amount of myoglobin determines whether the flesh is white or dark.

PREPARATION OF POULTRY

Poultry can be prepared in any number of ways:

Dry-heat methods
- Roasting
- Baking
- Broiling
- Grilling
- Frying

Moist-heat methods
- Braising
- Stewing
- Poaching

Regardless of the preparation method selected, poultry should always arrive on the table well-done as determined by the combined use of internal temperature, color changes, touch, and time/weight tables. Poultry is sufficently cooked when internal temperature reaches 180° to 185° F (82° to 85°C).

Microwave ovens are not recommended for cooking poultry, except for smaller pieces. Thawing frozen poultry is best done in the refrigerator.

When handling fresh or frozen poultry, cleanliness and personal hygiene are of utmost importance in preventing foodborne illnesses.

Carving a chicken

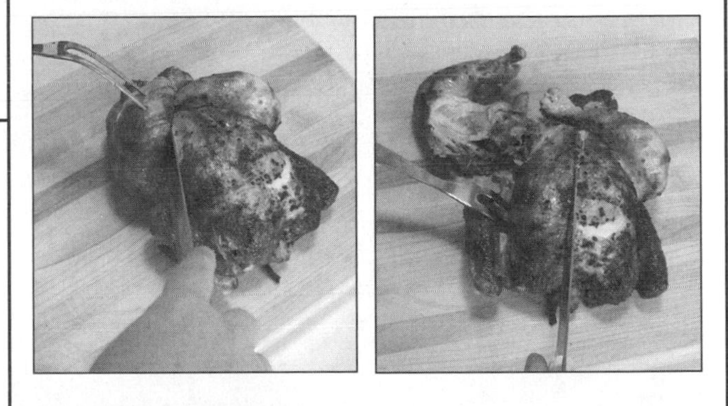

STORAGE OF POULTRY

Fresh poultry will keep in the refrigerator for up to three days, while frozen poultry will keep in the freezer for six to twelve months. All prepared foods should be refrigerated in covered containers and guarded against *Salmonella*.

- Store in the refrigerator a maximum of three days,
- Freeze for a maximum of six to twelve months.

REFERENCES

1. Boyer C, et al. Ionic strength effects on heat-induced gelation of myofibrils and myosin from fast- and slow-twitch rabbit muscles. *Journal of Food Science* 61(6):1143–1148, 1996.
2. Ensminger ME. *Poultry Science.* Interstate Publishers, 1992.
3. Hashim IB, AVA Resurreccion, and KH McWatters. Consumer attitudes toward irradiated poultry. *Food Technology* 50(3):77–80, 1996.
4. Kerler J, and W Grosch. Odorants contributing to warmed-over flavor (WOF) of refrigerated cooked beef. *Journal of Food Science* 61(6):1271–1274, 1996.
5. Kinsman DM, AW Kotula, and BC Breidenstein. *Muscle Foods: Meat, Poultry, and Seafood Technology.* Chapman & Hall, 1994.
6. Li Y, et al. *Salmonella typhimurium* attached to chicken skin reduced using electrical stimulation and inorganic salts. *Journal of Food Science* 59(1):23–24, 1994.
7. Liu G, and YL Xiong. Storage stability of antioxidant-washed myofibrils from chicken white and red muscle. *Journal of Food Science* 61(5):890–894, 1996.
8. McNamara AM. Generic HACCP application in broiler slaughter and processing. *Journal of Food Protection* 60(5):579–604, 1997.
9. Meullenet JF, et al. Textural properties of chicken frankfurters with added collagen fibers. *Journal of Food Science* 59(4):729–733, 1994.
10. Parkhurst CR, and GJ Mountney. *Poultry Meat and Egg Production.* Van Nostrand Reinhold, 1988.
11. Pegg RB, and F Shahidi. Unraveling the chemical identity of meat pigments. *Critical Reviews in Food Science and Nutrition* 37(6):561–589, 1997.
12. Penfield MP, and AM Campbell. *Experimental Food Science.* Academic Press, 1990.
13. Ruenger EL, GA Reineccius, and DR Thompson. Flavor compounds related to the warmed-over flavor of turkey. *Journal of Food Science* 43:1198–1200, 1978.
14. State of the food industry: Meat and poultry. *Food Engineering* 63(6):78, 1991.
15. Tanaka MCY, and M Shimokomaki. Collagen types in mechanically deboned chicken meat. *Journal of Food Biochemistry* 20:215–225, 1996.
16. Thayer DW, et al. Destruction of *Salmonella typhimurium* on chicken wings by gamma radiation. *Journal of Food Science* 57(3):586, 1992.
17. Wettasinghe M, and F Shahidi. Oxidative stability of cooked comminuted lean pork as affected by alkali and alkali-earth halides. *Journal of Food Science* 61(6):1160–1164, 1996.
18. Xiong YL. Myofibrillar protein from different muscle fiber types: Implications of biochemical and functional properties in meat processing. *Critical Reviews in Food Science and Nutrition* 34(3): 293–320, 1994.

WEBSITES

Learn from the USDA about meat from the emu, ostrich, and rhea:

www.fsis.usda.gov/OA/pubs/ratites.htm

The USDA's facts about ground poultry:

www.fsis.usda.gov/oa/pubs/gmdpoul.htm

Consumer publications on poultry from the North Carolina Cooperative Extension Service:

www.ces.ncsu.edu/depts/foodsci/agentinfo/poultry/conspub.html

Fish and Shellfish

Humans were eating fish, shellfish, and sea mammals long before they started cultivating plants or domesticating animals for food. Excavations of Stone Age sites have uncovered fishnets, spears, and fishing hooks made from the upper beaks of birds. Seafood is now the only major food source that is still hunted. Most other food sources are being raised or grown.

At present, there are over 20,000 known species of edible fish, shellfish, and sea mammals. Of these, approximately 250 species are harvested commercially in the United States, with millions of tons annually being served up for the consumption of humans and domesticated animals. This chapter focuses on those species and examines their classification, composition, purchase, preparation, and storage.

Classification of Fish and Shellfish

The staggering variety of creatures harvested from the water makes it difficult to classify them using only one set of criteria. As a result, several categories have arisen in order to distinguish them from each other: vertebrate or invertebrate, salt or fresh water, and lean or fat. Although these classifications are used to separate the identity of different fish, a vertebrate could be in salt or fresh water, and either lean or fat. The Food and Drug Administration has attempted to standardize fish nomenclature by publishing a "Guide to Acceptable Market Names for Food Fish Sold in Interstate Commerce," and requiring that fish be named according to this publication (27). The FDA guide is the recommended way to classify fish and shellfish, but the three common methods mentioned above are now described: (1) vertebrate or invertebrate, (2) salt or fresh water, (3) lean or fat.

Vertebrate or Invertebrate

This classification divides water animals according to the presence or absence of a backbone (Figure 16-1).

Vertebrate. The vertebrate category includes **finfish**, which obtain their oxygen from the water through their gills, and sea mammals, all of which must get their oxygen from above the water's surface.

Finfish. Finfish are found in the fresh water of rivers, lakes, and streams, and the salt water of oceans and seas. The most popular finfish in North America are tuna, cod, Alaska pollack, salmon, catfish, and flounder/sole.

Sea Mammals. Sea mammals include dolphin, whale, and seal, which are consumed in some cultures.

Invertebrate. The invertebrate category includes shellfish, most of which have external skeletons or shells.

Shellfish. Shellfish, which is a commercial rather than a scientific classification, includes the invertebrate **crustaceans** and **mollusks**. Examples

KEY TERMS

Finfish Fish that have fins and internal skeletons.

Crustacean An invertebrate animal with a segmented body covered by an exoskeleton consisting of a hard upper shell and a soft under shell.

Mollusk An invertebrate animal with a soft unsegmented body usually enclosed in a shell.

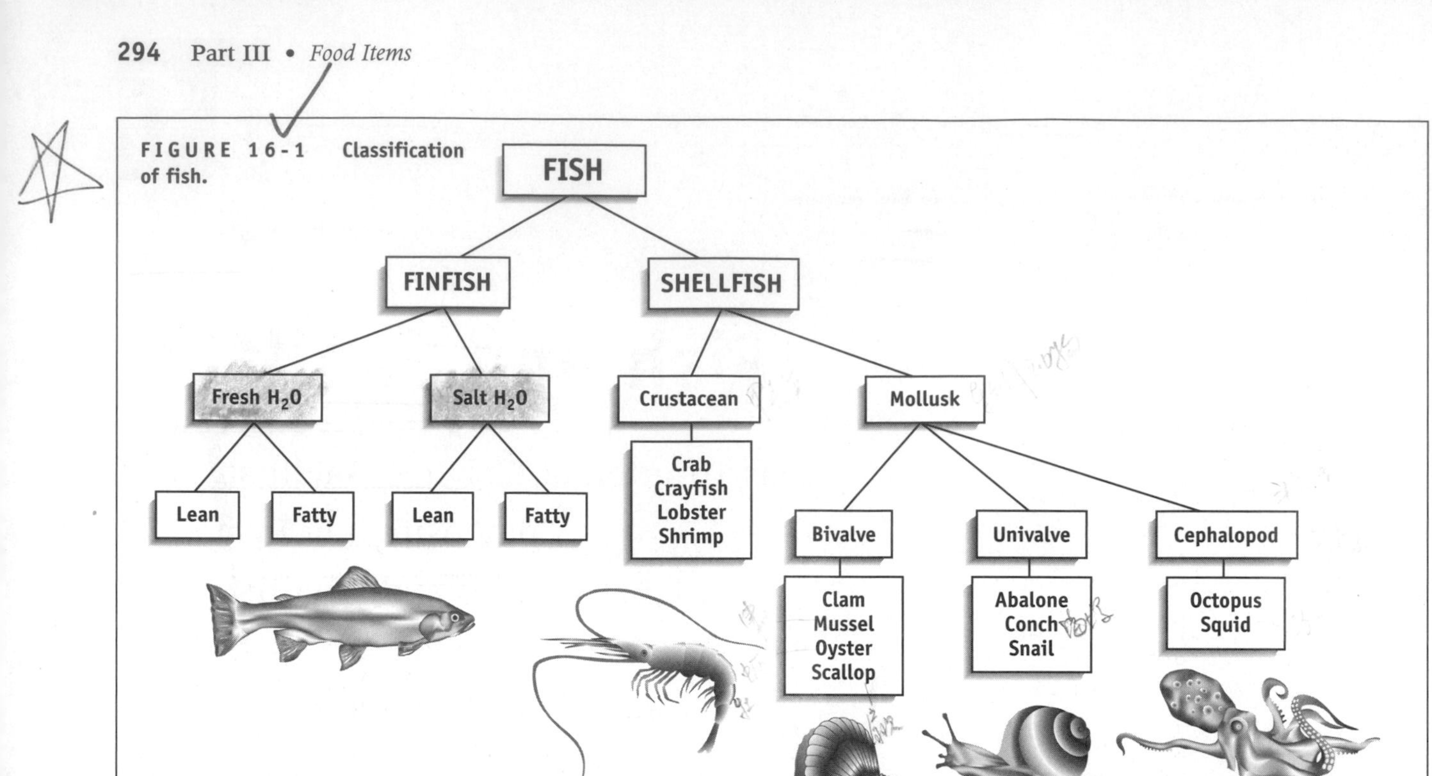

FIGURE 16-1 Classification of fish.

of crustaceans are shrimp, crab, lobster, and crayfish. Mollusks include bivalves, univalves, and cephalopods. Bivalve creatures, including clams, oysters, mussels, and scallops, are contained within two hard shells that are hinged together. The univalves, such as conch and abalone, have only a single hard shell. Cephalopods, which include octopus and squid, have an almost rubbery soft inner shell, which will be familiar to parakeet owners as a cuttlebone.

Salt or Fresh Water

The majority of the fish eaten in the United States are taken from salty waters, but many also come from freshwater lakes, ponds, and streams. Saltwater fish often have a more distinct flavor than freshwater fish. Sole, however, is a very mild-flavored saltwater fish, and is one of several exceptions to the taste generalization (Table 16-1). Some saltwater fish other than sole are halibut, cod, flounder, haddock, mackerel, red snapper, salmon, shark, striped bass, swordfish, and tuna. Catfish, perch, pike, and trout are the most common freshwater varieties.

KEY TERMS

Myotomes Layers of short fibers in fish muscle.

■ ■ ■ ■

Myocommata Large sheets of very thin connective tissue separating the myotomes.

Lean or Fat

Fish are sometimes identified by their fat content, but in this case, "fat" is a relative term. Fish are not very fatty compared to most other meats. Fat content in a 3-ounce cooked portion is less than 2.5 grams in lean fish (less than 5 percent fat) which includes cod, pike, haddock, flounder, sole, whiting, red snapper, halibut, and bass (Table 16-2 on page 296). The same portion of fatty fish (more than 5 percent fat) yields 5 to 10+ grams of fat. Examples include salmon, mackerel, lake trout, tuna, butterfish, whitefish, and herring.

Composition of Fish

Structure of Finfish

Regardless of their classification, fish are usually tender when they come to the table, and three structural factors contribute to this tenderness.

Collagen. When compared with meat or poultry, fish muscle has lower amounts of collagen. The bodies of land animals average 15 percent connective tissue by weight, while fish are only 3 percent collagen (35).

Amino Acid Content. Another reason fish is tender is that there is less of a certain amino acid (hydroxyproline) in the connective tissue, so when fish is cooked, the collagen breaks down more easily at a lower temperature and converts to gelatin.

Muscle Structure. Also unlike mammals and birds, whose muscles are arranged in very long bundles of fibers, the muscles of fish are shorter (less than an inch in length) and are arranged into **myotomes**, which are separated by **myocommata** (Figure

TABLE 16-1
Common Fish and Shellfish Grouped by Flavor and Texture

TEXTURE	*FLAVOR*		
	Mild Flavor	*Moderate Flavor*	*Full Flavor*
Delicate	Cod Crab Flounder Haddock Hake Pollock Scallops Sole	Butterfish Lake perch Whitefish Whiting	Mussels Oysters
Moderate	Crayfish Lobster Pike (walleye) Orange roughy Shrimp Tilapia	Mullet Ocean perch Shad Smelt Surimi products Trout Sea trout (weakfish) Tuna (canned)	Bluefish Mackerel Salmon (canned) Sardines (canned)
Firm	Grouper Halibut Monkfish Sea bass Snapper Squid Tautog (blackfish) Tilefish Wolffish	Catfish Mahi-mahi Octopus Pompano Shark Sturgeon	Clams Marlin Salmon Swordfish Tuna

16-2). This combination of structure and chemistry contributes to the characteristic flaking of prepared fish as the heat softens the collagen in the myocommata.

FIGURE 16-2 **Fish muscle, unlike other meats, is arranged in layers of short fibers (myotomes) separated by very thin sheets (myocommata).**

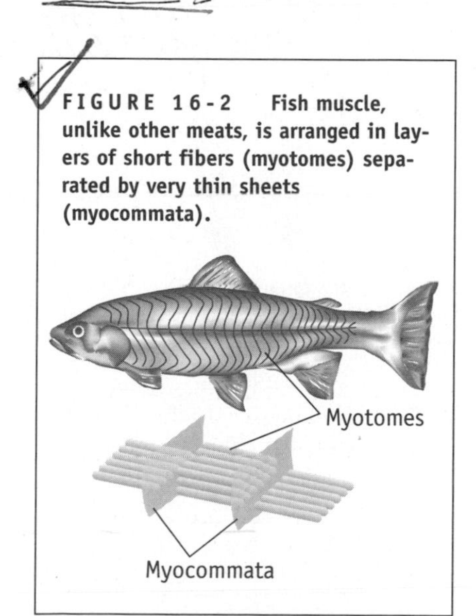

Pigments

When fish flesh is exposed to air during preparation, it will vary in color as a result of the presence of white, pink, or red pigments. The color of a fish's flesh depends on whether that fish relied predominantly on quick or slow movements to stay alive. "Red," or darker meat flesh, such as that seen in salmon, has a higher concentration of the "slow-twitch fibers" needed for long-distance swimming and endurance. "White" meat, like that of the sole, has more "fast-twitch fibers," which are designed for quick bursts of speed of brief duration between long periods on "cruise control." Some fish, such as tuna, are comprised of both fast-twitch and slow-twitch fibers, giving them dark, light, and white meat. A higher fat content will also darken the color of the flesh, as seen in fatty fish like mackerel and tuna.

The concentration of myoglobin contributes to the overall color of fish flesh. The more oxygen required by the muscle, the more myoglobin proteins are necessary, because they carry the oxygen. Unfortunately, a higher myoglobin concentration results in quicker rancidity because the iron in myoglobin accelerates the oxidation of fat found in the muscle (29).

HOW & WHY?

Why do salmon have that characteristic pink/orange hue? Sometimes a specific pigment adds a special hue. For example, a carotenoid pigment, astaxanthin, imparts a characteristic orange-pink color to certain salmon and trout that feed on insects and crustaceans containing this pigment.

Purchasing Fish and Shellfish

Commonly purchased fish and shellfish and their uses are listed in Table 16-4 on page 298. Retailers providing consumers with nutrition information must abide by the nutrition labeling values provided by the FDA for fish and shellfish (24). Fish processors may submit to inspection and grading on a voluntary basis.

Inspection/Grading

Unlike meat and poultry, inspection of finfish is voluntary; inspection, when it occurs, is based on the wholesomeness of the fish and the sanitary conditions of the processing plant. The National Marine Fisheries Service of the U.S. Department of Commerce is responsible for fish inspections, which are paid for by the processor.

Only inspected finfish can be graded. Grading, too, is voluntary and paid for by the processor. Fish products are graded U.S. Grade A, U.S. Grade B, and substandard. Quality grades are based on appearance, texture, uniformity, good flavor, fresh odor, and an absence of defects. Breaded fish products are further evaluated in terms of their breading and bone-to-fish ratio.

TABLE 16-2
Lean vs Fatty Fish: Fat Content of 3-Ounce Cooked Portions of Fish and Shellfish

Lean Fish			
Very low fat—less than 2.5 grams total fat			
Clams	Haddock	Pike (Northern)	Red snapper
Cod	Halibut	Pike (walleye)	Snow crab
Cusk	Northern lobster	Pollock (Atlantic)	Sole
Blue crab	Mahi-mahi	Ocean pout	Squid
Dungeness crab	Monkfish	Orange roughy	Tuna (skipjack)
Flounder	Perch (freshwater)	Scallops	Tuna (yellowfin)
Grouper	Ocean perch	Shrimp	Whiting
Low fat—more than 2.5 grams but less than 5 grams total fat			
Bass (freshwater)	Croaker	Salmon (pink)	Swordfish
Bluefish	Mullet	Shark	Rainbow trout
Blue mussels	Oysters (Eastern)	Smelt	Sea trout
Catfish	Salmon (chum)	Striped bass	Wolffish (ocean catfish)
Fatty Fish			
Moderate fat—more than 5 grams but less than 10 grams total fat			
Butterfish	Salmon (Atlantic)	Lake trout	
Herring	Salmon (coho)	Tuna (bluefin)	
Mackerel (Spanish)	Salmon (sockeye)	Whitefish	
Higher fat—more than 10 grams total fat			
Mackerel (Atlantic)	Salmon (king)		

Labels should be read whenever possible to find out whether or not the fish product has been inspected and graded.

Shellfish Certification. The Department of Commerce also oversees the publication of the Interstate Certified Shellfish Shippers List, which lists department-certified shippers of oysters, clams, mussels, and scallops. Only shellfish from these certified waters, which have been tested and found to be free of excessive levels of various microorganisms, can be sold for consumption. Wholesale containers of shellfish must then be labeled to include the harvester's name, address, and certification number, the date and location of harvest, and the type and quantity of shellfish. Shellfish that have been "shucked," or removed from their shells, must also be tagged with a "sell by date" (for containers under 64 fluid ounces) or "date shucked" (over 64 fluid ounces). These tags are required to be kept by food service operations for at least 90 days upon receipt. If shellfish are not properly tagged or if they are obtained from uncertified waters, the Department of Commerce may report the violation to the Food and Drug Administration, which is the regulatory agency with final jurisdiction over commerce in shellfish.

Selection of Finfish

The criteria for selection of vertebrate and invertebrate seafood are very different and will now be described. Finfish can be purchased fresh or frozen, canned, cured, fabricated, or as fish roe.

Fresh and Frozen Fish. Fish can be purchased fresh or frozen as whole, drawn, dressed, steaks, fillets, and sticks (Figure 16-3).

- *Whole fish.* The body is entirely intact.
- *Drawn fish.* Whole fish that have had their entrails (inner organs) removed.
- *Dressed fish.* The head, tail, fins, and scales have been removed in addition to the entrails.
- *Steaks.* Cut from dressed fish by slicing from the top fin to the bottom fin at a 90 degree angle at varying thicknesses. Steaks contain a portion of the backbone and other bones.
- *Fillets.* Made by slicing the fish lengthwise from front to back to avoid the bones.

FIGURE 16-3 Forms of finfish available for purchase.

Source: Dept. of Fisheries

NUTRIENT CONTENT

Protein. Fish is a high-protein food. In fact, fish is so high in protein, about 18 to 20 percent, that the food industry has devised the means to make protein concentrates by grinding whole fish, including the calcium-rich bones (if consumed), dehydrating it, and removing the fat to take away the fishy flavor. The resulting concentrate of between 70 and 80 percent pure protein is used as an additive in foods such as noodles to increase both their protein quality and their calcium content (28).

Fat and Carbohydrates. As a general rule, finfish are a low-fat food. Most fish are lower in fat than equivalent amounts of beef, pork, lamb, and even poultry. With the exceptions of mackerel, shark, herring, and eel, fish generally contain fewer than 160 calories (kcal) per 3-ounce cooked serving. Most of the calories (kcal) in fish are derived from protein and fat, with few, if any, from carbohydrates. Shellfish contain carbohydrates in the form of glycogen, ranging in concentration from 1 to 3 percent by weight. The fat in fish is generally in low proportions, unless the fish has been fried. It should be noted here, however, that although fish and shellfish are relatively low in fat, squid and some crustaceans such as shrimp contain more than 100 milligrams of cholesterol per 100 grams (23).

Even though fat does contribute to calories (kcal), the fat from fish is a good source of **omega-3 fatty acids** (Table 16-3) (14, 20). The consumption of omega-3 fatty acids has been reported to be related to a decrease in the risk of heart disease (9). It has also been suggested that they play a beneficial role in the alleviation of psoriasis and some inflammatory diseases, such as rheumatoid arthritis and lupus erythematosus (17, 19, 31). The combined benefit of the lower fat content, higher polyunsaturated fat, and omega-3 fatty acids of fish has led to the recommendation that people should eat fish at least twice a week.

Vitamins and Minerals. Fish is also a good source of the B vitamins: thiamin (B_1), riboflavin (B_2), niacin, B_6 (pyridoxine), and B_{12}, although small amounts of these water-soluble vitamins may be lost through decomposition, heating (cooking or canning), and/or extraction in water or salt solutions (15). The higher the fat content, the higher the levels of the fat-soluble vitamins A and D in the fish. Long before vitamin supplements became available, children were given (notoriously awful-tasting) cod liver oil as a dietary supplement of vitamin D to help protect them against rickets. Fish flesh is also a significant source of some minerals. Iodine is found primarily in saltwater fish. Sardines and salmon canned with the bones are good sources of calcium, and fish does contain some iron (12). For those watching their intake of sodium, dried or smoked fish have higher concentrations than the fresh forms.

TABLE 16-3

Fish High in Omega-3 Fatty Acids (3-oz cooked portion)

More than 1.0 gram
Herring
Mackerel (Pacific, jack, Spanish)
Salmon (Atlantic, king, pink)
Tuna (bluefin)
Whitefish
Between .5 and 1.0 gram
Bass (freshwater)
Bluefish
Mackerel (Atlantic)
Salmon (chum, coho, sockeye)
Smelt
Striped bass
Swordfish
Rainbow trout

- *Fish sticks.* Uniform portions cut from fillets or steaks. They can also be made from fish that has been minced, which is then shaped, breaded, and frozen.

Determining Freshness of Fish. Sniffing for aroma may be the safest and easiest method of determining whether or not fish is fresh, but other criteria can be applied in addition to the "sniff test." When selecting whole fish, look for skin that is bright and shiny and eyes that bulge, are jet black, and have translucent corneas (the part surrounding the pupil). The fish should have a "fresh fish" aroma, tight scales, firm flesh, a stiff body, red gills, and a belly free of swelling or gas. The same criteria hold true for drawn fish with the exception of the potential gas-filled belly, which, of course, has been removed.

Rigor Mortis. A stiff body is preferred when selecting a finfish because it is an indication that it is still in rigor mortis, which occurs after slaughter. Flesh that is allowed to go through rigor mortis (stiff to relaxed muscles) has a better texture and flavor. The water-holding capacity of the proteins is increased, which makes the flesh juicier than fish that have not undergone rigor mortis. For these reasons, it is better that handling, packing, processing, and freezing be avoided while fish are in the rigor state (3). It is also recommended that fish, prior to slaughter, not be subjected to excessive stress if possible, because the resulting stronger rigor mortis is detrimental to texture (32).

Rigor mortis in fish can last anywhere from several hours to days, depending on the species, temperature, and condition of the fish when caught. Stiffness is delayed if caught fish are immediately placed on ice and kept chilled. Freshness is extended under these conditions because bacterial spoilage does not occur until after rigor mortis has passed. Freezing fish immediately after capture, rather

KEY TERMS

Omega-3 fatty acids The polyunsaturated fatty acids eicosapentaenoic (EPA) and docosahexaenoic acid (DHA).

than chilling them on ice and allowing rigor mortis to proceed until the muscles relax again, results in a tough-textured flesh. Cooking fish prior to rigor mortis also results in a tough texture.

Phosphate Treatment of Fresh Finfish. The meat of the fish should not be slimy, but this can be tricky to judge, because any slime present may have been produced by the fish having been soaked in a special phosphate-containing solution to prevent moisture loss. This solution increases the pH of the tissue, which denatures the proteins and makes them more capable of binding water. Fishermen frequently treat fish with this solution to cut down on the water loss, which might endanger their weight-based profits. Without this solution, fish that is refrigerated may lose up to 80 percent of its water-binding capacity within five days after harvest. The phosphate-containing solution restores the binding capacity of the muscle proteins and prevents the flesh from becoming dry and stringy. Treatment with phosphates also partially inhibits the oxidation of the natural fats in fish, which can result in "fishy" smells when the phosphates bind with the metal ions that promote oxidation.

Signs of Decay in Fresh Finfish. Other changes that occur in a fish after death is that the eyes flatten and become concave (although this may also be a result of the fish having been picked up by the eye sockets), the pupil turns gray or creamy brown, and the cornea becomes opaque and discolored. In addition, the bright red gills turn a paler brown and as a result are sometimes removed.

When the gills turn brown and the eyes lose their bright look, the fish may be cut up as steaks, fillets, or fish sticks. Steaks and fillets should have a shiny, smooth surface that has no signs of curling at the edges. The pieces should be cut clean with no signs of blood, skin fragments, or loose bone, and they should be firm and free of **gaping** (Figure 16-4). Although gaping is a sign of aging, it may also be a result of rough handling, processing before rigor mortis is complete, the fish having been caught after spawning, or even genetics. Certain fish, like bluefish and hake, are known to gape more easily (10).

How Much Fish to Buy. Part of selecting finfish is knowing how much to buy. A few general guidelines exist. About ⅓ pound of steaks, fillets, or sticks make an appropriate portion per person. Purchases of ½ pound for each person will be required when buying dressed

FIGURE 16-4 **A badly gaping fish fillet.**

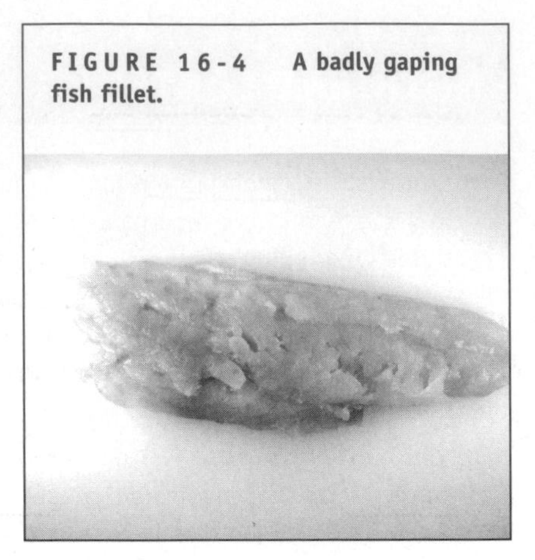

KEY TERMS

Gaping The separation of fish flesh into flakes that occurs as the steak or fillet ages.

TABLE 16-4
Types of Fish and Shellfish and Their Uses

Common Name(s)	*Uses*
American Pollock (Boston Bluefish)	Baked, broiled, pan fried, steamed, or poached.
American Shad (Buck Roe; White Shad)	Baked, broiled, planked, stuffed, or sautéed.
Atlantic Croaker (Croaker; Hardhead)	Baked, broiled, poached, pan fried, or oven fried.
Blue Crab	Steamed (boiled), cakes, patties, deviled, stuffed, casseroles, salads, appetizers.
Carp	Not a favorite of Americans; eaten by Europeans.
Catfish	Baked, broiled, grilled, barbecued, smoked, sautéed or stuffed.
Cod (Codfish; Scrod)	Baked, broiled, poached, fried, or steamed; oven finish or deep fry breaded portions and sticks.
Crayfish (Crawfish)	Like lobster; thick soup, crayfish bisque.
Dungeness Crab	Steamed, baked, broiled, simmered, casseroles, salads, appetizers, cocktails, and sauces.
Geoduck Clam	Steaks, fried, minced as dip or chowder; party snacks.
Flounder (Blackback, Fluke; Summer Flounder, Winter Flounder)	Baked, broiled, poached, fried, steamed; oven finish or deep fry breaded fillets, sticks, or portions.
Haddock (Scrod)	Same as flounder.
Halibut (North Pacific Halibut)	Baked, broiled, poached, fried, or steamed.
Herring (Pacific Sea herring)	Bait, oil, fertilizer; cooked or eaten as kippered herring.
Jonah Crab	Steamed, simmered, or broiled; casseroles, salads, appetizers, cocktails, and sauces.
King Crab	Used interchangeably with other crab meat recipes; casseroles, salads, appetizers, cocktails, and sauces.
Lake Trout	Baked, broiled, poached, fried, steamed, or sautéed.
Lingcod	Broiled, sautéed, baked, poached, or deep fried.
Lobster (Spiny Lobster)	Baked, broiled, or simmered; variety of recipes for use.
Mackerel	Baked, broiled, fried, poached, or steamed.
Menhaden (Pogy; Fatback)	Seldom for human consumption.
Mullet (Black or Striped Mullet)	Deep fried, oven fried, baked, or broiled.
Ocean Perch (Redfish; Rockfish; Rosefish)	Baked, broiled, poached, or steamed fillets; oven fried or deep fried, breaded, raw, or cooked portions.
Ocean Quahog Clam (Mahogany Quahog; Black Quahog)	Deep fried; pan fried patties; deviled clams, Manhattan clam chowder; clam cakes and rolls.
Oysters (Eastern or Atlantic Oyster; Pacific Oyster; Western Oyster)	Steamed, baked, sautéed, or used in variety of dishes.

(Continued)

TABLE 16-4
Types of Fish and Shellfish and Their Uses (Continued)

Common Name(s)	*Uses*
Pacific Cod	Baked, broiled, poached, fried, or steamed; oven finish breaded, cooked portions or sticks, deep fried frozen breaded cuts.
Pompano (Cobblerfish; Butterfish; Palmenta)	Baked, broiled, pan fried, or deep fried.
Porgy (Scup)	Baked, pan fried, or sautéed.
Rainbow Trout	Baked, broiled, pan fried, paoched, or steamed.
Red Crab	Broiled, baked, steamed, sautéed, or served cold; suitable in any recipe for crab.
Red Snapper	Broiled, baked, steamed, or boiled.
Rockfish	Baked, broiled, fried, and in chowders.
Sablefish (Black Cod)	Ready-to-eat; steamed; used in casseroles or salads.
Salmon	Baked, broiled, barbecued, fried, steamed, or poached; variety of recipes and dishes.
Sardines (Atlantic herring)	Ready-to-eat snack; convenience food.
Scallop	Boiled or sautéed, cocktails.
Sea Bass (Striped Bass)	Baked, broiled, pan fried, oven fried, or poached.
Shrimp (Northern Shrimp; North Pacific Shrimp; Southern Shrimp)	Simmered, baked, broiled, fried or oven finish; cocktail; hundreds of uses such as casseroles, salads, and sauces.
Smelt (Whiteball; Eulachon)	Broiled, fried, baked, or prepared in casserole.
Snow Crab (Tanner; Queen)	Interchangeably with other crabmeat recipes.
Sole (Gray Sole; Witch Flounder)	Baked, broiled, fried, steamed, or deep fried.
Spanish Mackerel	Baked, broiled, or smoked.
Squid (Inkfish; Bone Squid; Taw Taw; Calamari; Sea Arrow)	Fried or baked with a stuffing; salads; sauces; combination dishes.
Sturgeon	Specialty item.
Sunray Venus Clam	Chowder, fritters, patties, dips and clam loaf.
Surf Clam	Steamed, fried, or broiled; in chowders, fritters, sauces, dips, or salads.
Swordfish	Baked, broiled, fried, poached, or steamed.
Tuna	As it comes from the can; variety of recipes.
Weakfish (Gray Sea Trout; Squeteagues)	Baked, broiled, sautéed, or pan fried.
Whiting (Frost Fish; Hake; Silver Hake)	Baked, broiled, pan fried, poached, or steamed; portions and sticks oven finished.
Yellow Perch	Baked, broiled, or pan or deep fried.

fish, and ¾ pound per serving for whole or drawn fish.

Caviar. Caviar, which has a mystique surrounding it as a food of the very rich, is really just fish eggs. Its official definition varies according to the country in which it is sold. In the United States and many other countries, caviar is the clean, salted fish eggs of any fish species. The label is required to list the particular type of fish serving as the caviar source. In Europe, caviar is more narrowly defined by law as only the eggs of the Caspian Sea sturgeon.

The most expensive, largest-grained caviar comes from the Beluga sturgeon. These fish can live for over 70 years and may grow to a length of 25 feet (11). Like chicken eggs, roe is very high in cholesterol—about 94 mg per tablespoon. It is also high in salt, but the best caviar is *malassol,* which in Russian means "little salt." To protect the taste of caviar, it is served with a bone or shell spoon, because metal imparts an off-flavor. It is sometimes served on a neutral-tasting bread that has been toasted on one side, with the caviar being gently placed on the untoasted side.

Canned Fish. About half of all fish consumed in the United States is canned. Tuna accounts for 76 percent of canned fish consumption; salmon comes in second at 9 percent, followed by sardines, shrimp, and crab (13). Canning alleviates the problem of the rapid perishability of fish.

Tuna. Six species of tuna are canned and sold in the United States: yellowfish, skipjack, bluefin, Oriental tuna, little tuna, and albacore. "White" canned tuna comes from albacore and is the most expensive. All other tuna is labeled "light meat tuna," although some of it can be quite dark. Canned tuna comes in three different styles: fancy or solid pack (a fillet or whole piece), chunk (large pieces), and flake (fine pieces or grated). Solid pack has the best appearance and is also the most expensive. Tuna may be canned in either water or oil, so buyers should examine labels for nutrition information. The total calories (kcal) can vary drastically, depending on the canning medium. Each tablespoon of vegetable oil added to a can of tuna contains about 100 calories (kcal) and 15 grams of fat.

Salmon. Chinook or King salmon is the most expensive of the canned salmons. In the less expensive ranges are Sockeye (red salmon), Coho (medium red), pink, and chum. Salmon is often packed with the bones, which increases the calcium content, but only if consumed.

Sardines. Sardines are always packed with their bones unless otherwise noted on the label. They come packed in tomato or mustard sauces or in oil. It is even possible presently to find them on the shelves packed in jalapeño sauce or plain water.

Cured Fish. Fish may be cured by drying, salting, or smoking. Curing is one of the oldest ways of preserving fish. Although distinctive tastes and prolonged keeping times are achieved with any of the curing techniques, they can also harden the outer surfaces. Smoked salmon, smoked haddock (finnan haddie), pickled herring, and smoked herring, also known as kippered herring, are some of the more familiar forms of cured fish. Caviar, discussed above, also belongs

KEY TERMS

Surimi Japanese for "minced meat," a fabricated fish product usually made from Alaskan pollack, a deep-sea whitefish, which is skinned, deboned, minced, washed, strained, and shaped into pieces to resemble crab, shrimp, or scallops.

in the cured category, because it is preserved by salting (Table 16-5).

Anchovies. Anchovies are tiny, bony fish that have been cured with salt. They come to the market either salt-packed or oil-packed and in cans as whole fish, fillets, or anchovy paste. Because of their strong flavor, they are usually used as a garnish or in salad dressings and sauces rather than as a food in themselves. The salt-packed anchovies must first be rinsed, but their flavor tends to be superior to the oil-packed variety.

Fabricated Fish. In an attempt to counter the twin problems of the expense and perishability of fish, several fish products, including fish sticks, nuggets, and simulated fillets, have been developed using fabricated fish. Fabricated fish products make use of the less popular species. The fish are mechanically deboned, recovering 60 to 90 percent of the edible meat, and the flesh is then ground, seasoned, shaped, and breaded (16). These products are commonly frozen for sale to the consumer.

The cost of genuine crab meat has led to the introduction in this country of **surimi**, which has been used for centuries in Japan. Over 900 years ago, a Japanese fisherman discovered that fish would last much longer if it were minced, washed, mixed with salt and spices, ground into a paste, and then cooked (35). Today the deboned and minced fish is treated to produce a pure white product with a somewhat elastic chewy texture that "sets" when the mixture forms a translucent, elastic, moist gel (21, 26) (Chemist's Corner 16-1). Part of the treatment consists of adding flavorings such as salt and sugar. Egg whites or starch are included for binding, texture, and flavoring. Red coloring is often added to impart the appearance of cooked crab legs.

Surimi at this stage cannot be consumed until it is cooked, and it is the method of cooking that determines the type of food produced (Figure 16-5). North Americans are most familiar with kamaboko, surimi that has been steamed and shaped into pieces resembling crab, shrimp, or scallops (22, 35). Although the taste may be very similar to crab, the nutritional values are not. The resultant product usually has 75 percent less cholesterol than the original shellfish, but very little, if any, of the omega-3 fatty acids, and it usually has more sodium because of the added salt.

Fish Roe. Fish roe consists of the eggs of vertebrate fish held together by a thin membranous sac. It is available only from female fish during the spawning season and is highly perishable.

Freshwater roe is often breaded and fried, but the surrounding sac must be pierced first or it may explode during frying, causing severe burns. A

TABLE 16-5
Preserved Fish/Roe

Type	*Description*
Anchovy	Tiny, very fatty fish with a powerful flavor, which are cured by having most of their fat content removed by pressure and fermentation.
Arbroath smokies	Haddocks or whiting that are brined.
Bismark herring	German herrings from the Baltic filleted and marinated in vinegar with onion rings for 2–3 days.
Bloaters	First developed at Yarmouth, England, in 1835, bloaters owe their special flavor to the activity of gut enzymes. They are dry-salted and smoked.
Block fillets	Haddock or whiting are brined, dyed to make them a bright yellow, then smoked.
Bombay duck	A well-known Indian condiment made from dried bummaloe. It is used as a condiment with curries.
Glasgow pales	So-called because after brining they are lightly smoked and undyed.
Katsuoboshi	A Japanese fried fish that can also be smoked.
Lutefisk	The reconstituted unsalted cod from Norway known as *stockfish* or *stockfisk*.
Kippers	Good-quality fresh herring, soaked in brine and then smoked.
Matjes herrings	Young Netherlands herring caught in the spring, before they become too fatty.
Migaki-nishin	Japanese dried fish fillets and abalone.
Roes and caviars	Roe is fish eggs. In the United States and many other countries, caviar is defined as the salted roe of any fish species. In Europe, caviar is defined as only the roe from sturgeons (Beluga, Ostra, and Sevruga) originating in the Caspian Sea.
Rollmops	Herring fillets packed in spiced brine.
Sardine	Fish, cooked either by frying in peanut oil or steam cooking.
Smoked eels	Brined, dry-salted, and smoked.
Smoked salmon	Dry-salted with fine salt and smoked. Some curers add brown sugar, saltpeter, and rum.
Smoked sprats	The most famous are the *Kieler Sprotten*—brined sprats from Germany.
Smoked trout	Rainbow and brown trout are brined, speared on rods, smoked, then hot-smoked.

FIGURE 16-5 Surimi products.

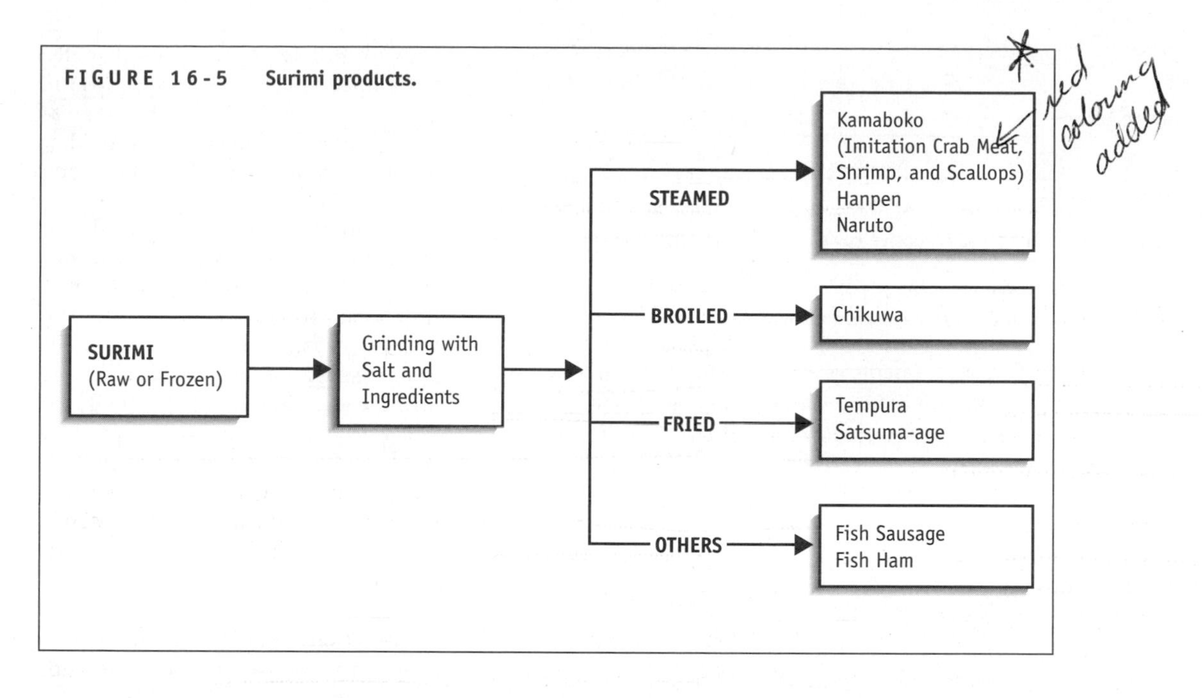

major drawback to fresh fish roe is that it stays fresh for only a day or two at the most; it is usually preserved in a brine solution, which imparts a salty flavor, firms the roe, and extends its usable time. The roe sold in the unrefrigerated section of the supermarket has been pasteurized to extend its shelf life. Fish such as shad and herring from North Atlantic waters are popular roe sources, as are Pacific salmon and whitefish from the Great Lakes. Other roe sources include cod, carp, pike-perch, and gray mullet (Table 16-6) (11).

Selection of Shellfish

The purchaser of shellfish is faced with a number of different forms from which to choose. The first decision is whether to buy them alive or processed.

Purchasing Live Shellfish. Lobsters, crabs, oysters, and clams all may be purchased alive and in their shells. Shellfish are highly perishable, and to maintain their quality, must be kept alive until they are cooked, or in the case of oysters, occasionally consumed raw.

Selecting Live Mollusks. In contrast to crustaceans, which are normally active creatures and easy to tell whether or not they are alive, determining the state of mollusks in their closed shells poses more of a puzzle. Tapping on the shell should cause it to close more tightly; the rule in most cases is that if the shell remains open, the mollusk is dead and should be discarded. The exceptions to this rule are mussels, which ordinarily gape, and longneck (steamer or soft-shell) clams, which normally have a gap in the shells where the "neck" protrudes. Any shells that are broken, have a decaying odor, or float should be discarded.

The "R-Month Rule." An old rule of thumb held that shellfish should be eaten only during the months with the letter "r" in their names, because bacterial illnesses are more common in the warmer months of May through August. This is still a valid guideline, although modern methods of harvesting and storage provide a safer supply of shellfish year round.

Purchasing Processed Shellfish. Shellfish can also be bought cooked in the shell and chilled or frozen. Alternatively, the meat can be removed from the shell and sold fresh, chilled, frozen, canned, salted, smoked, or dried. Shellfish can be sold headless and in their shell, as in the case of shrimp or lobster tail, or they may be shucked, or removed from the shell, with a special knife.

Shucking Shellfish. Shucked shrimp, scallops, oysters, and clams are often breaded and frozen. Shrimp may also be sold with the intestinal tract removed, a form known as "peeled and deveined." Shucking bivalves such

Chemist's Corner 16-1

Surimi

Surimi's "springy" texture is derived in part from gums that are added to help form gels. Another influence is washing, which leaves behind the water-insoluble (myofibrillar) protein that gives surimi its elasticity and gel-forming capacity. Sugars such as sucrose and sorbitol are added as cryoprotectants to protect the myofibrillar proteins from denaturating during freezing (35). Starch granules make the surimi more compact by swelling with water around the protein matrix and filling in the interstitial spaces (37, 38). Improved gel strength is obtained by adding egg whites, which inhibit endogenous protease activity in fish flesh (30). Salt is then added to the surimi to solubulize its protein and produce a firm, elastic gel, and again later during freezing for stabilization (6, 18).

TABLE 16-6
Sources of Caviar and Roe

Sturgeon	
Beluga Sturgeon	The beluga, the largest of the Caspian Sea sturgeon, produces the rarest and most expensive caviar. Beluga eggs are large and gray.
Osetra Sturgeon	Osetra caviar are more available than beluga. The medium-size eggs are gray-brown and have a nutty, meaty taste.
Sevruga Sturgeon	The smallest of the Caspian Sea sturgeon, the sevruga eggs are small and gray, have a stronger, fishier taste than the other Caspian Sea caviars. Sevruga caviar is particularly popular in Russia and Europe.
Other Roe-Producing Fish	
Lumpfish	The lumpfish, caught off Iceland, produces small, colorful (black, red, or yellow) eggs that are popular as a garnish.
Salmon	The large, pinkish eggs of the salmon, from the Pacific Northwest, are also frequently used as a garnish.
Whitefish	The North American Whitefish produces small, golden eggs with a distinct crunch and mild flavor.
American Paddlefish	This denizen of the Mississippi River and its tributaries produces roe that looks like sevruga caviar. The small, gray eggs have a tangy flavor.

as clams and oysters is a somewhat dangerous process (Figure 16-6). The hand holding the bivalve should be protected with a towel or a metal-mesh glove. The hinge is severed as the shells are pried apart, and the empty half of the shell is discarded, while the muscle attachment to the other shell is spliced so the meat can be removed. An average worker can shuck almost 7 pounds in an hour, but automated shucking speeds up the process.

Oysters. Oysters can be bought live in the shell, or shucked and then chilled, frozen, or canned. Live oysters should have tightly closed shells. Any gap between the shells means the oyster is dead and should be discarded. Select shucked oysters that are plump and full-bodied; about one cup is equal to one serving. If the oysters are in their shell, buy half a dozen per person. Three varieties of oyster are commonly available in the United States: eastern oysters from the Atlantic coast, and Olympia (small) and Japanese (large) oysters from the Pacific coast. Oysters in the shell and well refrigerated have a longer shelf life than other mollusks because their shells remain very tightly closed, while other shellfish have a tendency to gape, making them more susceptible to drying out and dying.

Clams. Clams can be bought in the same forms as oysters, and, as with oysters, their shells should be closed tightly and there should be no decaying odor. About six to eight shelled clams are required per serving. Two major east coast types are hard-shell and soft-shell clams, with the meat of hard-shell clams being less tender. Hard-shell clams include cherrystones, which are the most common variety; littlenecks, which are the smallest and most tender; and chowders or quahogs, which are the largest. Soft-shell clams are also known as longnecks or steamers because of the long tube extending from the shell opening. Since soft-shell clams do not completely close, they are very susceptible to drying out

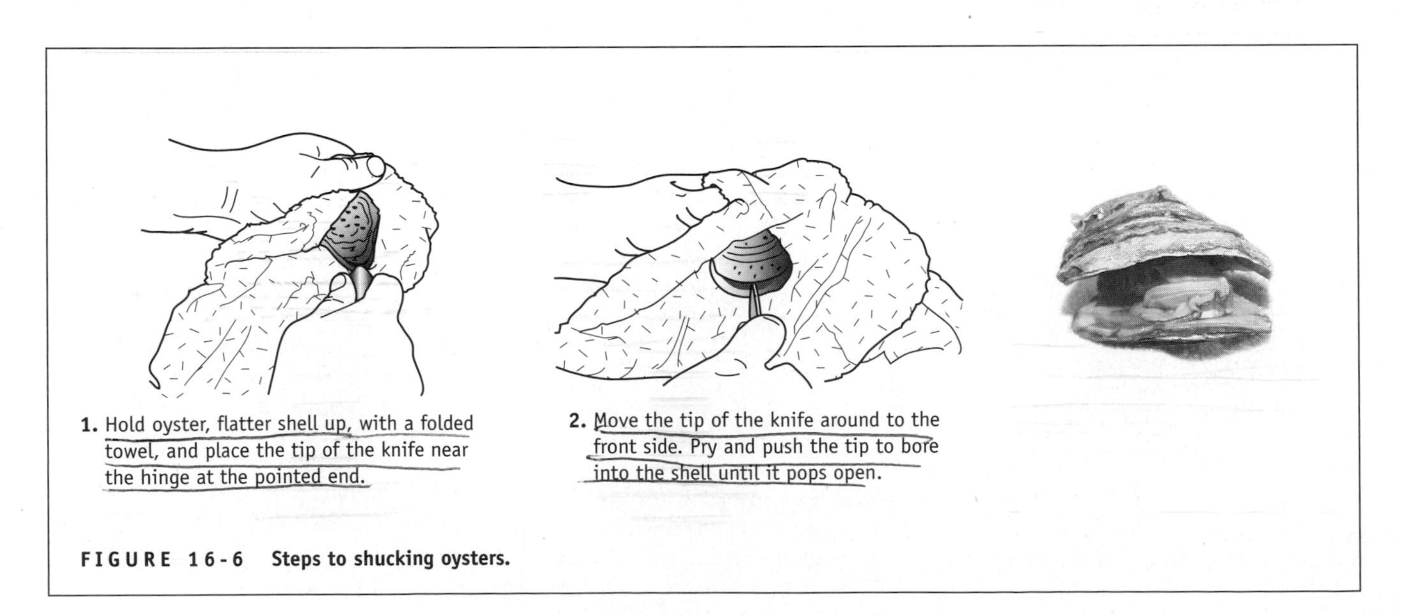

FIGURE 16-6 Steps to shucking oysters.

and dying. A limp neck hanging out of the shell signals that the clam is dead and should be discarded (10). West coast varieties include razor and Pismo clams.

Scallops. In North America, the only part of the scallop that is eaten is the creamy white or tan-colored abductor muscle responsible for opening and closing the shell to move it through the water. Scallops cannot close their shell tightly when taken from the water, so they are usually shucked and then sold fresh, frozen, or canned. This sweet-tasting mollusk varies in diameter from ½ to 2 inches, and about ¼ to ⅓ pound is an adequate portion for one person. The three main varieties of scallops available are bay scallops, which are small, sweet, and delicate; sea scallops, which are larger and not as delicate; and calico scallops, the least expensive and tiniest of all and the blandest in flavor. Calico scallops are often sold cooked or frozen.

Mussels. The black or dark-blue colored shells of common mussels should be scrubbed free of barnacles, but the "beards" or black threads used to attach the shells to solid foundations in the ocean should not be pulled out until the mussels are ready to be cooked, because removing them kills the mussel (10). Mussels are heated in their shells after being purchased live (Figure 16-7), or they are shucked and packed in brine. Extremely hollow or heavy-feeling mussels should be discarded, because they are either dead or filled with sand. Also available are the larger, green-lipped New Zealand mussels whose size makes them ideal for stuffing and baking.

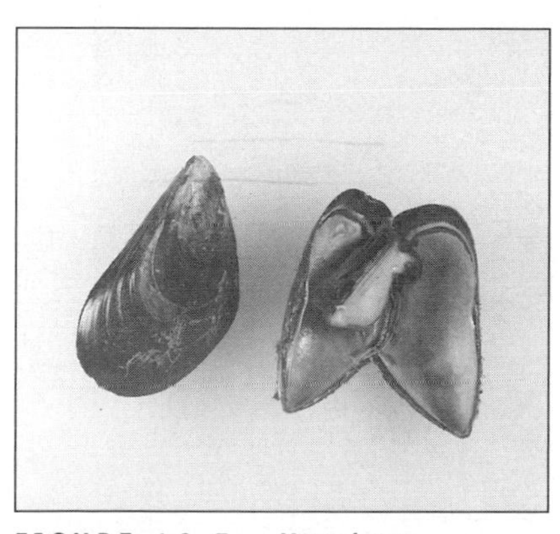

FIGURE 16-7 **Mussels are often steamed open and served in the half-shell.**

Abalone. Abalone is expensive because the supply is limited (10). These large mollusks are found mostly in the waters off California and northern Mexico. Unlike the other mollusks discussed, abalone have only one shell. Most of the animal consists of a massive, muscular foot. Only abalone with meat weighing at least ¼ pound may be legally harvested; some extremely large abalone yield as much as 3 pounds. The strict regulations governing the harvesting of wild abalone have led to farm-raised abalone, which are largely harvested in California and Hawaii (25).

Lobsters. Lobsters are the largest of the crustaceans. They are mainly purchased as Northern (or Maine) lobster, or spiny or rock lobster varieties (Figure 16-8). Gourmet cooks prefer the female lobster for its finer flavor and because it contains "coral," or lobster roe, which is considered a delicacy. When cooked, the roe turns from dark green to red and is often used to color a sauce or served alone as a garnish. Another delicacy, found in both male and female lobsters, is the pale green liver, known as tomalley.

The majority of the meat from a lobster is in the tail, but there is also some in the claws of the Maine lobster. Lobsters are right-handed or left-handed, as indicated by which claw is the larger, and although the larger claw has more meat, that from the smaller claw is sweeter and more tender. Northern lobsters with one or both claws missing are sometimes sold as "culls." They are less expensive and are attractive to the buyer interested only in the tail meat. Unless they are canned or frozen, lobsters must remain alive until cooked, at which point their natural dark blue-gray or greenish color turns deep orange or red.

Shrimp. The tail harbors most of the meat in shrimp. They are sold, headless, in either the raw shell-on (green shrimp), cooked shell-on, or cooked and peeled form. All three forms come both fresh and frozen, but the majority of shrimp are frozen.

Peeling and Cleaning Shrimp. When shrimp are bought in their shells, they must first be peeled. Medium or large shrimp are then deveined, which involves removing the dark-colored "sand vein" that runs along the shrimp's back (Figure 16-9). The "sand vein" is usually left in small shrimp, where it is undetectable.

HOW & WHY? ?????????

Why must the "sand vein" in large shrimp be removed? The "sand vein" is actually the shrimp's intestines. In larger shrimp, it contributes a gritty, muddy taste if it is not removed.

Lois Frank

FIGURE 16-8 **Northern lobster (left), spiny or rock lobster (right).**

KEY TERMS

Prawn A large crustacean that resembles shrimp but is biologically different. Large shrimp are often called by this name.

■ ■ ■ ■

Scampi A crustacean found in Italy and not generally available in North America. The term is often used incorrectly to describe a popular shrimp dish.

After cleaning, the shrimp are dried by pressing them between paper towels to absorb as much moisture as possible. Before they are cooked, shrimp are somewhat grayish green, but they turn dark pink to borderline red when heated.

How Much Shrimp to Buy. Shrimp are available in small, medium, large medium, large, and jumbo sizes (10). They are purchased according to "count per pound," which varies depending on the region, but obviously the smaller the shrimp, the higher the count per pound. Serving size averages ⅓ to ½ pound for headless, unpeeled shrimp or ¼ to ⅓ pound for peeled and deveined shrimp. The largest jumbo-sized shrimp are frequently misnamed **prawns**, but true prawns have lobster-like pincer claws and are otherwise different from shrimp. Another shrimp-related North American misnomer is the use of the word **scampi** for describing large broiled shrimp seasoned with butter and garlic.

Canned Shrimp. Glass-like beads are sometimes found in canned shrimp, but they are completely harmless. They are formed during canning, specifically under the high heats of sterilization. Called struvite crystals, they consist of magnesium-ammonium phosphate compounds that form when the magnesium from sea water combines with the ammonia that is produced during heating of the shellfish's natural protein. Phosphate treatment prevents struvite crystal formation due to the phosphates binding with the magnesium. Struvite crystals can be crushed to a powder by a fingernail or dissolved by boiling for a few minutes in the weak acid of lemon juice or vinegar.

Crab. The majority of meat in a crab is found in its claws and legs. The four top commercially harvested crabs are the blue crab from the Atlantic and Gulf coasts, stone crabs from Florida, Dungeness crabs from the Pacific coast, and, most expensive, king crabs from the northern Pacific waters. Soft-shelled blue crabs are considered a delicacy, particularly on the east coast. These crabs are caught while molting, a process during which they shed their shell and have a soft exterior until the new surface is completely hardened. The process may take several days, during which time the crab is more vulnerable to predators, especially two-legged ones such as birds and humans.

Canned Crab. Canned crab may have a blue tint. This is caused by copper in the crab's blood combining with the ammonia in its flesh. Although the color may appear unappetizing, it is completely harmless.

Crayfish. Referred to as either crayfish, crawdads, or crawfish, these small crustaceans average 4 ounces in weight. Crayfish are similar in appearance to lobsters but smaller, and their first pair of walking legs do not develop into huge, flesh-rich claws. Only their tails serve as a source of meat. They are found mainly as a food source in freshwater streams and ponds of the southeastern United States, especially Louisiana. Crayfish are sold both head-on and tails only, fresh and frozen.

Preparation of Fish and Shellfish

In the preparation of seafood, great care must be taken not to overcook; cooking too long or at too high heat is the most common mistake when preparing fish or shellfish. It results in excessive flaki-

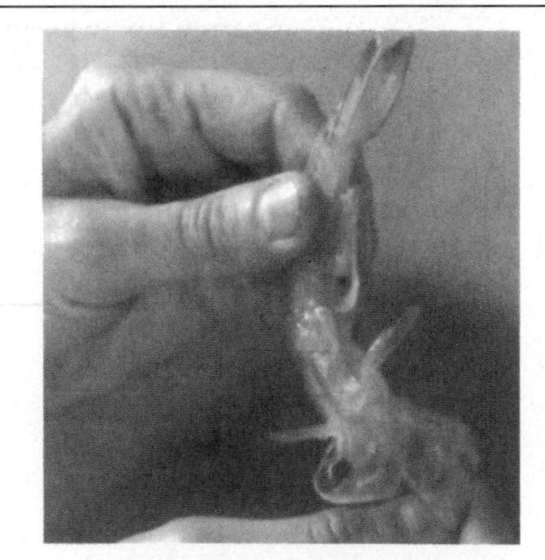

To peel a shrimp, hold it upside down by the tail and remove the legs and shell by clasping all the legs and peeling to the left or right. The last tail segment is usually left on for easy handling and/or appearance.

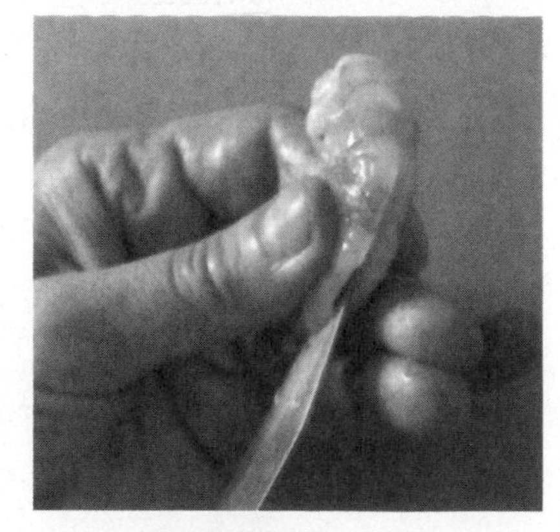

Devein by slitting the shrimp down its back. A shallow cut is sufficient to remove the vein, while a deeper one is used to butterfly the shrimp.

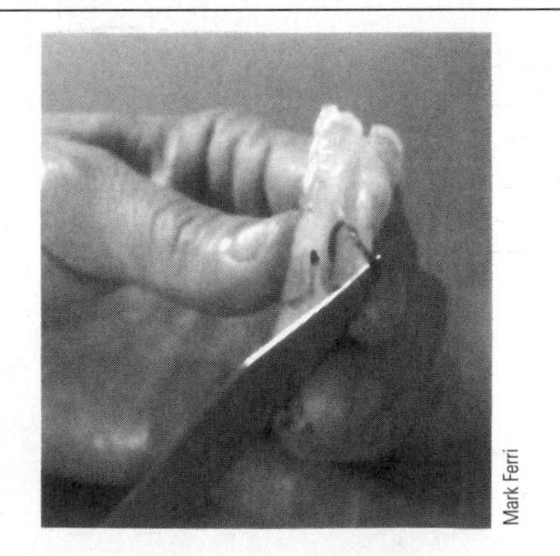

Complete the deveining by lifting out the intestine (black vein) with the tip of a paring knife or by washing the shrimp's back under cold, running water.

Mark Ferri

FIGURE 16-9 Peeling and deveining shrimp.

ness, dryness, and flavor loss in fish and toughness in shellfish. For shellfish, however, the "well done rule" should always be observed, especially when microwaving, because of the increased chance of their carrying foodborne illness. Shellfish are often simmered or steamed; the results with dry heat are harder to guarantee, because the meat may dry out and toughen. Nevertheless, with precautions such as breading taken against drying, shrimp, lobster tails, and half-shelled oysters and clams are fairly commonly baked, broiled, or fried (Color Figure 7). Both temperature and time need to be carefully controlled in either dry- or moist-heat preparation of seafood. The following discussion refers to the preparation of finfish unless shellfish is specifically mentioned.

Dry-Heat Preparation

Baking. Fish to be baked should be rinsed, patted dry with paper towels, and placed in a shallow pan. Season as desired, and place in a moderate oven (350° to 400°F/180° to 200°C). Baking time will vary depending on the shape and thickness of the fish, but a general rule of thumb is to bake up to 10 minutes per inch of thickness measured at the thickest diameter of the fish. Basting with butter or covering the fish with vegetables cuts down on moisture loss. Some prefer to prepare whole or drawn fish with the head and tail left on to help keep juices inside. Additional flavor can be added by filling the cavity with herbs and spices. Leaving the skin on whole or drawn fish seals in the moisture and flavor. Moisture loss may also be prevented by wrapping the fish in foil, parchment paper, grape leaves, or leafy greens. These techniques of enclosing the fish technically results in a moist-heat cooking method because the fish steams in its own juices. The flavor, moisture, color, and texture of baked fish are often enhanced by the addition of sauces (Figure 16-10).

FIGURE 16-10 **Saucing a fish.**

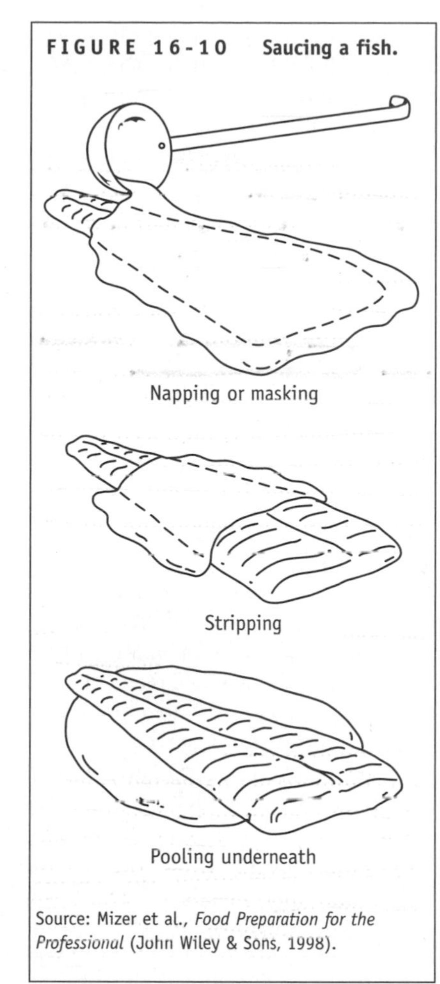

Source: Mizer et al., *Food Preparation for the Professional* (John Wiley & Sons, 1998).

Determining Doneness of Fresh Fish. Fish is done when it flakes easily with the gentle pressure of a fork without falling apart. The opaque look of fish that has been properly prepared is caused by denatured proteins (that unwind and hook together with other proteins) so that water can attach, resulting in a whitish hue. The presence of this "white" mesh results in a moist and tender flesh. Heating much beyond this stage tightens the protein bonds, shrinks the protein mesh, and squeezes out the water, resulting in a tough, dry, unappetizing fish flesh (8). Other signs of doneness include any bone being no longer pink and/or the flesh becoming firm, turning from translucent to opaque, and/or separating from the bone.

Baking Shellfish. Shellfish are often prepared by baking; examples include lobster thermidor, baked soft-shell clams, and oysters Rockefeller. In lobster thermidor, the lobster is split in half and baked. The meat is then extracted and mixed with a seasoned bechamel sauce before being put back in the lobster shell and baked again until golden brown and heated through. If soft-shell clams are to be baked in the oven, they are placed in a pan layered with rock salt and baked at 425°F (218°C) for about 15 minutes or until the shells open. Oysters Rockefeller is made by pouring a spinach mixture over half-shell oysters in a pan layered with rock salt. They are baked at 475°F (246°C) for about 10 minutes and then browned briefly under the broiler.

Broiling. Dressed or filleted finfish or fish steaks are best broiled at 5 inches or less below the heat source. Lean fish should be coated with melted butter, margarine, or oil, but this step can be omitted with most fatty fish. Season the fish as desired, and place it skin-side down on a pan that has been greased to avoid sticking and broil it on one side until tender. Lobsters and large shrimp can also be broiled. Whole lobsters need to be killed and split before broiling, while lobster tails can be broiled whole.

Grilling. Fish can be grilled on an outside grill or in the oven. Grilling is not recommended for delicate fish such as sole, because they may stick to the grill and fall apart easily. Fatty, firm-fleshed fish such as salmon, bluefish, and mackerel that have been drawn or cut into steaks are well suited for grilling. Also, larger shrimp may be put on skewers like kabobs and grilled. A fat coating such as oil or even mayonnaise can be applied to the fish to prevent it from sticking to the grill. The grill itself should be scraped of any residue and lightly oiled to prevent sticking. Steaks are seasoned as desired, and cooked on both sides if thick, but on only one side for thin steaks or fillets. The fish should be about 4 inches from the heat source. When the fish flakes easily, serve it immediately. Drawn fish can be checked for doneness by slipping the tip of a paring knife into the back of the fish and pulling away. It is done if it clings briefly before giving way, but is overdone and dry if cooked to the "flakes easily" stage.

Frying. Lean fish less than ½ inch thick, shrimp, and scallops will sauté nicely in a small amount of butter and/or oil. The fish is seasoned as desired and sautéed over medium heat until it is cooked about three-quarters of the way through, at which time it is turned gently with a spatula and heated until the flesh flakes easily. Shellfish are best sautéed on high heat for a short time. Shrimp and scallops are ideal for this type of

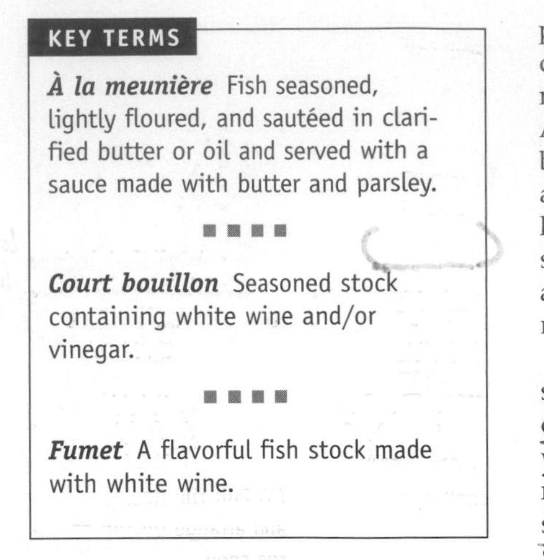
KEY TERMS

À la meunière Fish seasoned, lightly floured, and sautéed in clarified butter or oil and served with a sauce made with butter and parsley.

Court bouillon Seasoned stock containing white wine and/or vinegar.

Fumet A flavorful fish stock made with white wine.

preparation. When done, scallops will be firm and look opaque, and shrimp will be opaque and pink.

Sautéed Fish Variations. Sautéed fish may be prepared **à la meunière** (a-lah-muhn-YAIR). The dish can be served amandine (with almonds), florentine (with spinach), or à la belle (with mushrooms). A variation of this method, but using more fat, is used to prepare trout and other small fish. They are seasoned, breaded or dipped in cornmeal or flour, and panfried until they are golden brown on both sides.

Deep-Fat Fried Fish. Deep-fat frying is a popular method for preparing battered or breaded lean fish and shellfish (shrimp, scallops, clams, and oysters). Whole small fish, shellfish (which must first be shelled), fish fillets, or steaks are dipped in batter or seasoned breading mix before being deep fried in oil until golden brown. The oil is heated to 350°F (180°C) for large fish and around 180°F (82°C) for small seafood such as fish strips, oysters, or clams (25). Fish is always fried alone because it imparts a fishy taste to the oil, which would be picked up by other foods fried in the same oil. Lean fish are preferred because unpleasant oily tastes often occur in fatty fish that are deep fried.

Moist-Heat Preparation

Poaching. Fish is a delicate food suitable for poaching. The lower water temperature of 160° to 180°F (71° to 82°C), which keeps bubbles small and clinging to the sides of the pan, protects the delicate flesh of fish. If a whole, drawn, or dressed fish is being poached, it can be wrapped in cheesecloth to hold it together. The liquid may be a **court bouillon** or a **fumet**. Although fatty fish such as salmon can be poached, best suited for this method are white, lean fish such as cod, pike, haddock, flounder, sole, whiting, red snapper, halibut, and bass. Sole fillets are thin enough to make paupiettes, or rolled fillets (Figure 16-11).

When poaching fish, the water should never be allowed to boil. Boiling causes flavor loss and toughens the fish, while low temperatures retain maximum flavor and moisture. A well-seasoned poaching liquid is also important. Seasonings and/or chopped vegetables such as tomatoes or shallots add flavor, texture, and color. The poaching liquid is often reduced and sometimes thickened for use as a sauce. The fish is placed in the middle of a baking or frying pan and cooking liquid is added until it covers up to an eighth to a quarter of the fish's thickness. Some recipes call for covering the entire fish in liquid, but too much liquid may dilute delicate flavors. On the other hand, too little liquid will evaporate and cause the fish to dry out during cooking.

Fish fillets can be poached in an oven set at 350°F (180°C) or in a pan on top of a range set at poaching temperatures. The pan can be covered to trap more heat and moisture and to prevent volatile flavor compounds from escaping. This technique, when using only a small amount of liquid, is more akin to steaming, another moist-heat method.

FIGURE 16-11 Rolled fish fillet (paupiette): Thin fillets (usually sole) are rolled with the skin side inside so the flesh cooks on the outside.

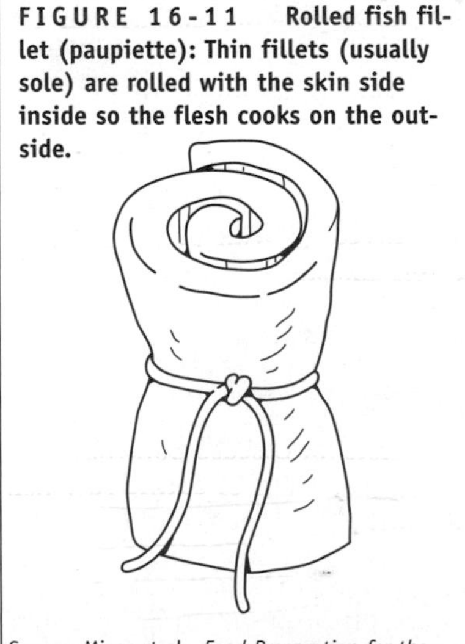

Source: Mizer et al., *Food Preparation for the Professional* (John Wiley & Sons, 1998).

Simmering. Simmering uses slightly higher temperatures than poaching—180°F (82°C) to just under boiling, where gentle bubbles rise but barely break the surface. This method is most often used to cook shrimp, even though the expression "boiled shrimp" is commonly used for the outcome of this process. Shrimp are often simmered and then chilled, shelled, and deveined for shrimp cocktail.

Lobster, crab, and crayfish may also be simmered. The live lobster, crab, or crayfish is killed by inserting it headfirst into boiling water that has been salted with 2 teaspoons per quart. Prior to placing crayfish in the water, the middle tail fin must be grabbed, twisted, and pulled to remove the stomach and intestinal vein. Lobsters will curl their tails when first dropped into the water, which may cause toughening. It is prevented by killing the lobster with the point of a sharp knife inserted directly between the head and the shell. A more expensive technique involves submerging the crustacean in a container of beer or wine, which inebriates it and causes it to relax. Once the shellfish is submerged, the water is brought back to a boil and then immediately reduced to a simmer. Heating time averages 5 minutes per pound for a lobster; a whole crayfish takes less than 7 minutes. When done, the crustacean is immediately removed from the water to prevent further cooking, drained well, and served at once with clarified butter and lemon (Chemist's Corner 16-2). Lobsters are often split in half at restaurants for the diners' convenience.

Steaming. Fish can be steamed in the oven if it is tightly covered in a baking dish, aluminum foil, or parchment paper, or in a pan on top of the range. When fish is wrapped with parchment paper, along with seasonings and aromatic vegetables if desired, and cooked in the oven, this is known as cooking *en papillote*. When the fish is done, the parchment envelope puffs up, turns brown, and provides a dramatic presentation. Each person may then be served a portion still wrapped in its own paper package, making for a novel dining experience. Fish may also be cooked in foil envelopes, although these are generally removed before the fish is served at the table. Regardless of the

Chemist's Corner 16-2

Lemon Juice and Fish Odor

The characteristic smell of fish odor is primarily from trimethylamine, a component of certain phospholipids located in the fat of the fish. Freshly caught fish do not smell until they degenerate. Contributing to decay are bacteria and enzymes that split the trimethylamine from the phospholipid and release it into a form that has a "fishy" odor. Adding acid, such as lemon juice, over cooked fish reduces this odor by converting the unpleasant-smelling liquid trimethylamine into an odorless solid. One way to determine the degree of bacterial deterioration of fish is to measure the amount of trimethylamine (27).

FIGURE 16-12 How to saddleback a lobster tail.

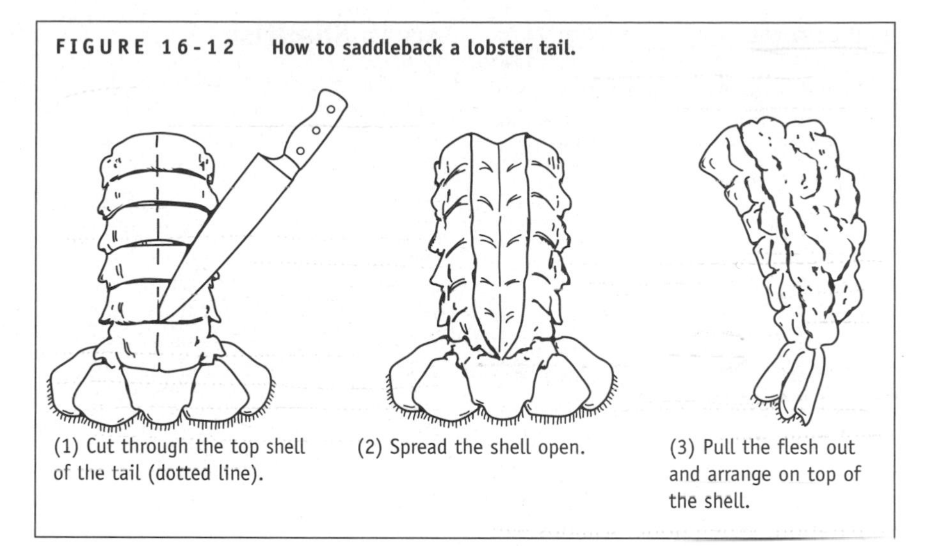

way it is accomplished, steaming heats the fish in its own juices, which locks in the flavor and aroma.

Steaming Shellfish. Steaming can also be used to prepare lobster tails, clams, and mussels. Frozen lobster tails are thawed and "saddlebacked," which involves splitting the tail by cutting through the hard top shell and pulling the meat out so it lies on top (Figure 16-12). The tail is then seasoned and steamed shell-down in a covered pan for a few minutes. Clams and mussels are steamed by placing them in a covered pot with a small amount of liquid on the bottom. Steaming clams or mussels just until the shells open does not kill microorganisms, so it is important to steam them for about 5 minutes or to a temperature of 145°F (63°C). Pressure steaming is not recommended because it tends to toughen both fish and shellfish.

Clambakes Are Underground Steamings. At a clambake, clams are actually steamed rather than baked. A hole a foot deep and 3 feet wide is dug into the sand and lined with smooth, round rocks. This serves as the base of a fire that will be kept going for two or three hours after the rocks and/or embers have been heated hot enough. The embers are raked over the rocks and removed, and soaked seaweed is placed over the rocks to a depth of about 6 inches. Chicken-wire mesh is laid over that to serve as a platform for a layer of hard-shell clams, which are then covered with sweet potatoes, followed by broiler chickens cut into quarters, partially husked corn, and then a layer of soft-shell clams. The whole pile is splashed with a bucket of seawater, covered with a wet tarp, and allowed to "bake," or rather steam, for about an hour. Doneness of the clams is tested by checking to see if their shells have opened. The chickens take longer and thus need to be tested for doneness separately.

Microwaving. Almost any form of fish can be microwaved. If it is commercially frozen, the defrosting instructions on the package should be followed. In general, instructions call for arranging fish fillets or steaks or small fish in a single layer, with thicker portions toward the outside of a microwave-safe dish. Desired seasonings and dots of butter are added before covering with plastic wrap to trap the moisture. Poaching can also be done in the microwave oven.

Raw Fish

The Centers for Disease Control warns about the hazards of eating raw fish or shellfish. This is particularly true for pregnant or nursing women, the very young, the elderly, and anyone with a serious illness or compromised immune system. Not only bacteria and viruses, but parasites as well, may pose a problem. Mollusks are particularly prone to carry contaminants, because they are filter feeders whose usual habitat is in shallow waters, which are more likely to be subject to bacterial, viral, and chemical pollution. Consuming, or even shucking, raw oysters is a potential concern because they may carry *Vibrio vulnificus, V. cholera, V. parahaemolyticus,* Norwalk virus, or hepatitis A (34).

Sashimi and ceviche. Sashimi (raw fish) used in sushi (a rice item), as well as ceviche, should be carefully checked by trained cooks for aniskiasis parasites, which are the width and color of white thread. Ceviche is raw fish that has been prepared by an acid marination, lemon or lime juice based, that denatures the proteins and turns the flesh white. This type of preparation does not involve heating, and thus the fish should still be considered "raw" and treated accordingly. Only heating to 145°F (63°C) for at least one minute or freezing the fish in a commercial freezer to −10°F (14°C) for seven days ensures destruction of aniskiasis parasites.

Storage of Fish and Shellfish

Fish can be purchased fresh, frozen, canned, or cured. Each style has its own storage requirements (back inside cover), but it is important to stress once again that all fresh fish and shellfish are highly perishable and require that precautions be taken to ensure freshness. Although proper preparation helps to destroy microorganisms that occur naturally or are introduced during handling, fish, and especially shellfish, must be stored properly to reduce the risk of foodborne illness.

Fresh Finfish

Fresh fish are best consumed within a day or two of purchase. Fish flesh is much more perishable than animal tissue for several reasons. One of these is that all raw seafood carries some bacteria, which multiply rapidly above 40°F (4°C).

Refrigerated. Fish should be stored in the coldest portion of the refrigerator. It should also be tightly wrapped to prevent odors from coming in contact with other foods (36). Fish bought wrapped in butcher paper should be rewrapped in plastic wrap and then aluminum foil, but prepackaged fish and shellfish can be left in the original package. Any exposure to oxygen increases perishability, because the high levels of polyunsaturated fatty acids in fish can be oxidized into compounds that affect odor and taste (1).

Spoilage Factors. Other factors that can contribute to spoilage are proteolytic enzymes, natural toxins, and contaminants. Proteolytic enzymes break down muscle proteins and provide amino acids for bacterial growth (2). Bacterial enzymes can also break down proteins to amino acids and elevate the levels of histamine, a toxin. Excessive consumption of histamine leads to a foodborne illness known as scombroid poisoning or scombrotoxism (discussed more fully in Chapter 3). Excessive histamine may accumulate in tuna, tuna-like fish, mahimahi, bluefish, and other species that usually have not been chilled immediately after being caught.

Storing Caviar. Caviar is particularly sensitive to oxygen and cannot be left out in the air for more than one hour. Unopened caviar can be stored in the refrigerator for up to three months, but once opened, it should be consumed within three days.

Fresh Shellfish

It is a good practice to eat fresh shellfish the day they are bought. If they must be kept, the storage requirements are varied and depend on the type of shellfish. Most fresh shellfish may be kept alive in cool, salty, wet environments, preferably in the refrigerator. Storing fresh shellfish on ice may kill them if they become submerged in fresh water from the melting ice. Live oysters, clams, and mussels should be well aerated in the refrigerator and not stored in plastic bags or in fresh water, where they will die. Any dead animals, indicated by an open shell or no response when tapped, should be discarded. Crabs, usually sold precooked, should be stored in the coldest part of the refrigerator and used within a day or two. Once cooked, all crustaceans must be refrigerated at temperatures below 40°F (4°C) and consumed within two days.

Frozen

As mentioned in Chapter 7 on Food Preservation, the frozen-foods industry in North America began with fish because of Clarence Birdseye's accidental discovery while ice fishing. Freezing greatly extends the keeping time of fish that, depending on the type, can be stored in the freezer up to nine months. It is absolutely necessary, in order to arrest microbial growth, to freeze fish if they are not cleaned (eviscerated) within 24 hours of being caught (4). Once cleaned, the general rule is that lean fish keep longer than fatty fish. Fish should be stored at 0°F (−18°C) or below and never refrozen once thawed. Prepackaged and frozen fish can stay in the original wrappers but should be kept airtight in order to prevent them from drying out.

Thawing. Fish is best thawed by transferring it from the freezer to the refrigerator one day before preparation; once thawed, it should be cooked immediately. The exceptions are breaded frozen fish, or fish fillets or steaks weighing less than ½ pound; these should not be thawed before cooking because they will become mushy. Frozen, raw shellfish can also be prepared from the frozen state, while frozen precooked shellfish can be used as is after thawing.

Even though it is the most healthful and popular method of preserving fish, freezing tends to cause a reduction in quality, making fish dryer, tougher, less springy, and possibly affecting the flavor (5) (Chemist's Corner 16-3).

Canned and Cured

Canned fish can stay on the shelf for up to twelve months, but any dented, damaged, or bulging cans should be discarded. Unused fish from an opened can should be moved to a covered glass or plastic container and can be stored for up to a week in the refrigerator. Cured fish can be refrigerated, frozen, or canned. Chapter 7 discusses canning and curing in more detail.

Chemist's Corner 16-3

Effect of Freezing on Fish

Freezing fish decreases its quality because the myofibrillar proteins are disrupted (denatured and/or aggregated) (5, 6, 33). The result is a subsequent loss in the muscle proteins' functional properties such as protein solubility, gel-forming ability, and water retention. These properties are important to quality. The tougher texture of frozen fish is thought to be due in part to the enzyme trimethylamine oxide (TMAO) demethylase, which breaks down TMAO to dimethylamine and formaldehyde. Formaldehyde is believed to be a crosslinking agent that is hypothesized to be responsible for a tougher texture (6). Additives can counteract the negative effect of freezing on the quality of fish. Polymerized phosphates improve the texture of frozen fish by increasing water retention, reducing thaw drip, and decreasing cooking losses. Polyphosphates achieve this effect by increasing the binding of phosphates to meat proteins, breaking down actomyosin to actin and myosin, elevating pH, and improving ionic strength (7).

PICTORIAL SUMMARY / 16: Fish and Shellfish

There are over 20,000 known species of edible fish, shellfish, and sea mammals. Of these, approximately 250 species are harvested commercially in the United States, with millions of tons annually being served up for the consumption of humans and domesticated animals.

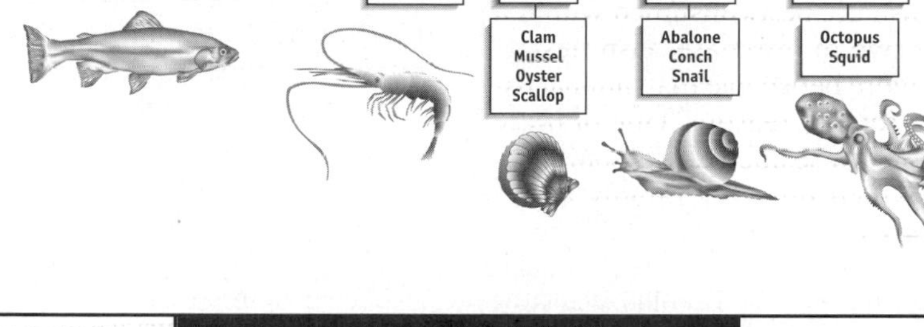

CLASSIFICATION OF FISH AND SHELLFISH

Fish can be classified three different ways:

- Vertebrate or invertebrate: Vertebrate fish or finfish have fins and internal skeletons; invertebrate fish, or shellfish (mollusks and crustaceans), have external skeletons.
- Saltwater or freshwater
- Lean or fatty

COMPOSITION OF FISH

Fish and shellfish are more tender than other flesh foods, and nutritionally, 3 ounces of fish contains fewer calories (kcal) than beef, pork, lamb, or poultry. Fish is high in protein and relatively low in fat. Small amounts of carbohydrate may be present in fish in the form of glycogen. The fat in fish is polyunsaturated, and, depending on the fish, high in omega-3 fatty acids. Fish is also a good source of many B vitamins.

PURCHASING FISH AND SHELLFISH

Inspection of fish is voluntary and is based on the wholesomeness of the fish and the processing plant. Only inspected fish products can be graded U.S. Grade A, U.S. Grade B, and substandard. Grades for shellfish such as shrimp and oysters are based on size. Fish can be purchased fresh or frozen in a variety of market forms, as well as canned, cured, and fabricated (surimi). Fish roe is also sold. Shellfish can be purchased alive, cooked in their shell, or shucked, to be refrigerated, frozen, or canned.

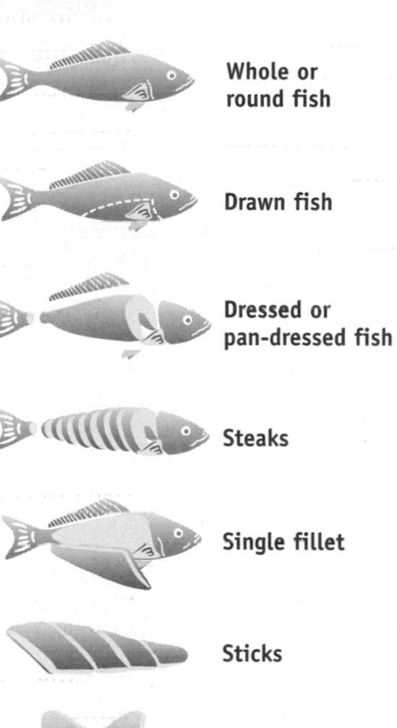

PREPARATION OF FISH AND SHELLFISH

Overcooking is the most common mistake in the preparation of fish, resulting in excessive flakiness, dryness, and flavor loss. Dry heat is the most popular method of preparation, and includes baking, broiling, grilling, and frying. Moist-heat methods include poaching, simmering, steaming and micro-waving. Fish cook quickly and are done when the flesh turns from a translucent to opaque color, is firm to the touch, separates from the bone (if present), the bone is no longer pink, and the flesh is moist and flakes easily at the segments without falling apart.

STORAGE OF FISH AND SHELLFISH

Fresh fish are best consumed within a day or two. If fish is purchased in butcher paper, it should be rewrapped with plastic wrap and aluminum foil. Prepackaged fish and shellfish can stay in the original package. Fish should be frozen at 0°F (-18C) or below and never refrozen once it is thawed. Breaded fish or fish fillets or steaks weighing less than ½ a pound need not be thawed before heating. Most fresh shellfish must be kept alive prior to preparation. Canned fish can stay on the shelf for up to twelve months, but any leftovers should be stored in a glass or plastic container in the refrigerator and used within three days.

REFERENCES

1. Bandarra NM, et al. Seasonal changes in lipid composition of sardine (*Sardina pilchardus*). *Journal of Food Science* 62(1):40–42, 1997.
2. Benjakul S, et al. Physicochemical changes in Pacific whiting muscle proteins during iced storage. *Journal of Food Science* 62(4):729–733, 1997.
3. Berg T, U Erikson, and TS Nordtvedt. Rigor mortis assessment of Atlantic salmon (*Salmo salar*) and effects of stress. *Journal of Food Science* 62(3):439–446, 1997.
4. Bett KL, and CP Dionigi. Detecting seafood off-flavors: Limitations of sensory evaluation. *Food Technology* 51(8):70–79, 1997.
5. Careche M, and ECY Li-Chan. Structural changes in cod myosin after modification with formaldehyde or frozen storage. *Journal of Food Science* 62(4):717–723, 1997.
6. Chang CC, and JM Regenstein. Textural changes and functional properties of cod mince proteins as affected by kidney tissue and cryoprotectants. *Journal of Food Science* 62(2):299–304, 1997.
7. Chang CC, and JM Regenstein. Water uptake, protein solubility, and protcin changes of cod mince stored on ice as affected by polyphosphates. *Journal of Food Science* 62(2):305–309, 1997.
8. Corriher SO. *CookWise*. Morrow, 1997.
9. Daviglus ML, et al. Fish consumption and the 30-year risk of fatal myocardial infarction. *New England Journal of Medicine* 336(336):1046–1053, 1997.
10. Dore I. *Fish and Shellfish Quality Assessment.* Van Nostrand Reinhold, 1991.
11. Dowell P, and A Bailey. *Cook's Ingredients.* Morrow, 1980.
12. Gomez-Baauri JV, and JM Regenstein. Processing and frozen storage effects on the iron content of cod and mackerel. *Journal of Food Science* 57(6):1332–1336, 1992.
13. Grocery edibles. *Progressive Grocer* 71(7):88, 1992.
14. Hepburn FN, J Exler, and JL Wehrauch. Provisional tables on the content of omega-3 fatty acids and other fat components of selected foods. *Journal of the American Dietetic Association* 86:788–793, 1986.
15. Higashi H. Relationship between processing techniques and the amount of vitamins and minerals in processed fish. In: *Fish in Nutrition,* eds. E Heen and R Kreuzer. Fishing News, 1962.
16. Jahncke M, RC Baker, and JM Regenstein. Frozen storage of unwashed cod (*Gadus morhua*) frame mince with and without kidney tissue. *Journal of Food Science* 57(3):575–580, 1992.
17. Kinsella JE. *Seafoods and Fish Oils in Human Health and Disease.* Marcel Dekker, 1987.
18. Konno K, K Yamanodera, and H Kiuchi. Solubilization of fish muscle myosin by sorbitol. *Journal of Food Science* 62(5):980–984, 1997.
19. Kremer JM, et al. Fish-oil fatty acid supplementation in active rheumatoid arthritis: A double-blind, controlled, crossover study. *Annals of Internal Medicine* 106(4):497–503, 1987.
20. Krzynowek J, et al. Factors affecting fat, cholesterol, and omega-3 fatty acids in Maine sardines. *Journal of Food Science* 57(1):63–65, 1992.
21. Lee HG, et al. Transglutaminase effects on low temperature gelation of fish protein sols. *Journal of Food Science* 62(1):20–24, 1997.
22. Ma L, A Grove, and GV Barbosa-Canovas. Viscoelastic characterization of surimi gel: Effects of setting and starch. *Journal of Food Science* 61(6):881–883, 1996.
23. Nettelton JA, and J Exler. Nutrients in wild and farmed fish and shellfish. *Journal of Food Science* 57(2):257–260, 1992.
24. Pennington JAT, and VL Wilkening. Final regulations for the nutrition labeling of raw fruits, vegetables, and fish. *Journal of the American Dietetic Association* 97:1299–1305, 1997.
25. Peterson J. *Fish and Shellfish.* Morrow, 1996.
26. Piyachomkwan K, and MH Penner. Inhibition of Pacific whiting surimi-associated protease by whey protein concentrate. *Journal of Food Biochemistry* 18:341–353, 1995.
27. Potter NN, and JH Hotchkiss. *Food Science.* Chapman & Hall, 1995.
28. Rakosky J. *Protein Additives in Food Service Preparation.* Van Nostrand Reinhold, 1989.
29. Rosell CM, and F Toldra. Effect of myoglobin on the muscle lipase system. *Journal of Food Biochemistry* 20:87–92, 1997.
30. Sareevoravitkul R, BK Simpson, and H Ramaswamy. Effects of crude alpha2 macroglobulin on properties of bluefish (*Pomatomus saltatrix*) gels prepared by high hydrostatic pressure and heat treatment. *Journal of Food Biochemistry* 20:49–63, 1997.
31. Shapiro JA, et al. Diet and rheumatoid arthritis in women: A possible protective effect of fish consumption. *Epidemiology* 7(3):256–263, 1996.
32. Sigholt T, et al. Handling stress and storage temperature affect meat quality of farmed-raised Atlantic salmon (Salmo salar). *Journal of Food Science* 62(4):898–905, 1997.
33. Srinivasan S, et al. Physiochemical changes in prawns (Machrobrachium rosenbergii) subjected to multiple freeze-thaw cycles. *Journal of Food Science* 62(1):123–127, 1997.
34. Tuttle J, S Kellerman, and RV Tauxe. The risks of raw shellfish: What every transplant patient should know. *Journal of Transplant Coordination* 4:60–63, 1994.
35. Vieira ER. *Elementary Food Science.* Chapman & Hall, 1996.

36. Wempe JW, and PM Davidson. Bacteriological profile and shelf life of White Amur (*Ctenopharyngodon idella*). *Journal of Food Science* 57(1):66–68, 1992.
37. Yoon WB, JW Park, and BY Kim. Surimi-starch interactions based on mixture design and regression models. *Journal of Food Science* 62(3):555–560, 1997.
38. Yoon WB, JW Park, and BY Kim. Linear programming in blending various components of surimi seafood. *Journal of Food Science* 62(3):561–564, 1997.

WEBSITES

Consumer publications on seafood from the North Carolina Cooperative Extension Service. Topics include safe handling, shellfish contamination, consuming raw seafood, trace elements in seafood, and proper storage:

www.ces.ncsu.edu/depts/foodsci/agentinfo/sea/conspub.html

The U.S. Food and Drug Administration's Center for Food Safety and Applied Nutrition (CFSAN) website on seafood information and resources:

www.cfsan.fda.gov/seafood1.html

The U.S. Food and Drug Administration's Center for Food Safety and Applied Nutrition (CFSAN) website on foodborne illness and seafood:

Wm.cfsan.fda.gov/~mow.sea-ill.html

Vegetables

Webster's dictionary refers to vegetables as "any plant," but more specifically as those that are edible. The definition of a vegetable has been narrowed in practice to the edible part of a plant in raw or cooked form accompanying the main course of a meal. Vegetables add color, flavor, and texture to meals. The focus of this chapter on vegetables will be on their classification, composition, purchase, preparation, and storage.

Classification of Vegetables

The part of the plant used as a vegetable often serves as a common method of classification. Vegetables may be derived from almost any part of a plant: roots (carrots, beets, turnips, and radishes), bulbs (onions and garlic), stems (celery and asparagus), leaves (spinach and lettuce), seeds (beans, corn, and peas), and even flowers (broccoli and cauliflower) (Figure 17-1). In addition, there are foods that are routinely called vegetables and used as vegetables, but which are actually fruits. Botanically, fruits

FIGURE 17-1 Classification of vegetables.

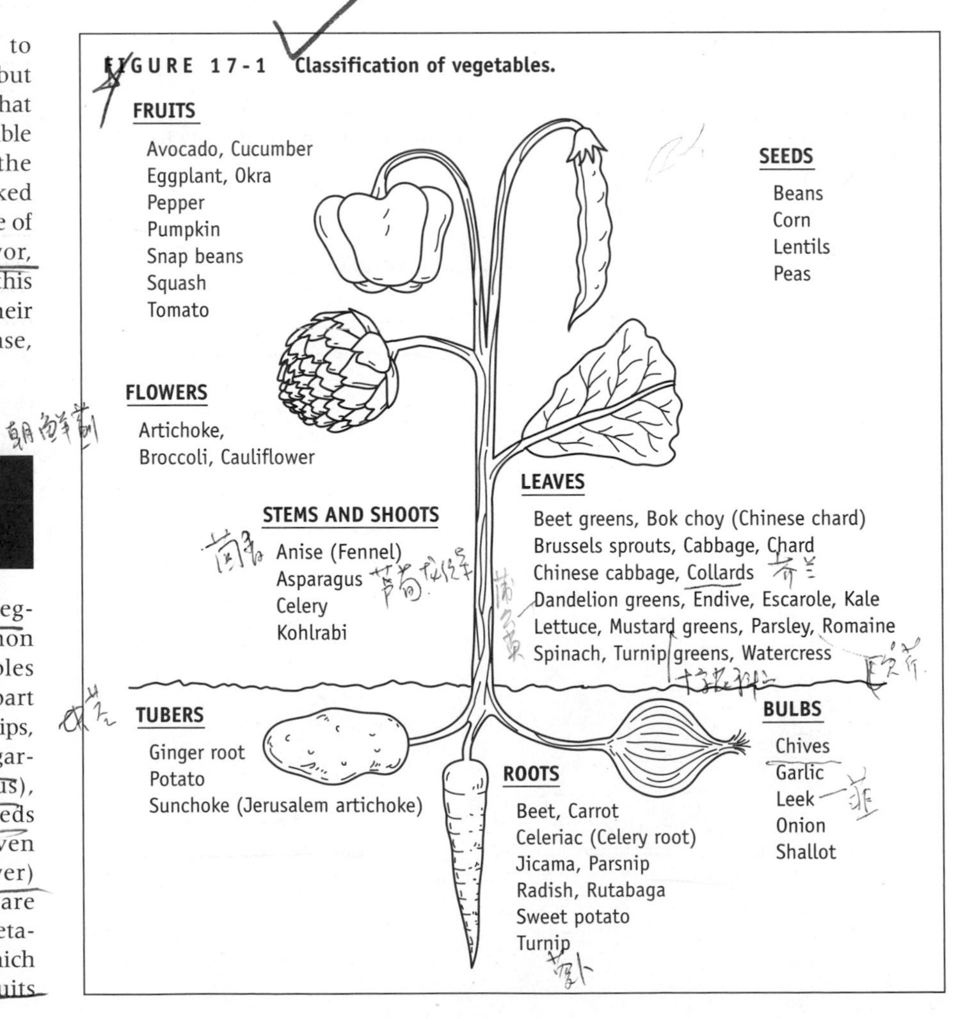

are the part of the plant that contains its seeds, specifically, the mature ovaries of plants. If it comes from a flower, then it is usually a fruit. The fruits most often seen masquerading as vegetables include tomatoes, squash, cucumbers, avocados, okra, eggplant, olives, water chestnuts, and peppers. Although many of the principles discussed in this chapter, such as composition and color, apply to both vegetables and fruits, the main focus is on vegetables.

Composition of Vegetables

Structure of Plant Cells

Cell Wall. Cells are the building blocks of both plant and animal organisms. One of the major differences between the two is that vegetables and other plants lack the skeletal structure that provides support in animal organisms. Instead, each vegetable cell gains its structural support by being surrounded by a sturdy wall (Figure 17-2). Contributing to the strength of these cell walls are several fibrous compounds which are indigestible by humans—cellulose, pectic compounds, hemicellulose, lignin, and gums. Humans are unable to break down fiber because they lack the enzyme necessary to break down cellulose to the glucose molecules that the body can use.

Two such fibers are pectic compounds and hemicellulose, found within and between cell walls where they serve as a type of intra- and intercellular cement, giving firmness and elasticity to the tissues (17, 25). The outer layer of the skin, peel, or rind has a higher proportion of cellulose and hemicellulose than the inner, thinner layers. The surface cells of these protective layers secrete a waxy cutin, a water-impermeable coat that protects the plant.

Lignin is another type of fiber, but it is one of the few non-carbohydrate compounds (polymers of phenolic alcohols). As vegetables mature, their lignin concentrations increase. This is why spinach stems and the inner cores of carrots, asparagus spears, and broccoli become tougher with age and do not soften when heated (43). Other fibrous compounds found in plants include gums, polysaccharides with a unique ability to absorb water and swell to several times their original volume. They are often added to processed foods such as ice cream, candies, and salad dressings to increase their viscosity. Examples of food gums derived from plants include algin, carob bean gum, carrageenan, gum arabic, gum guar, locust bean, gum tragacanth, and gum karaya.

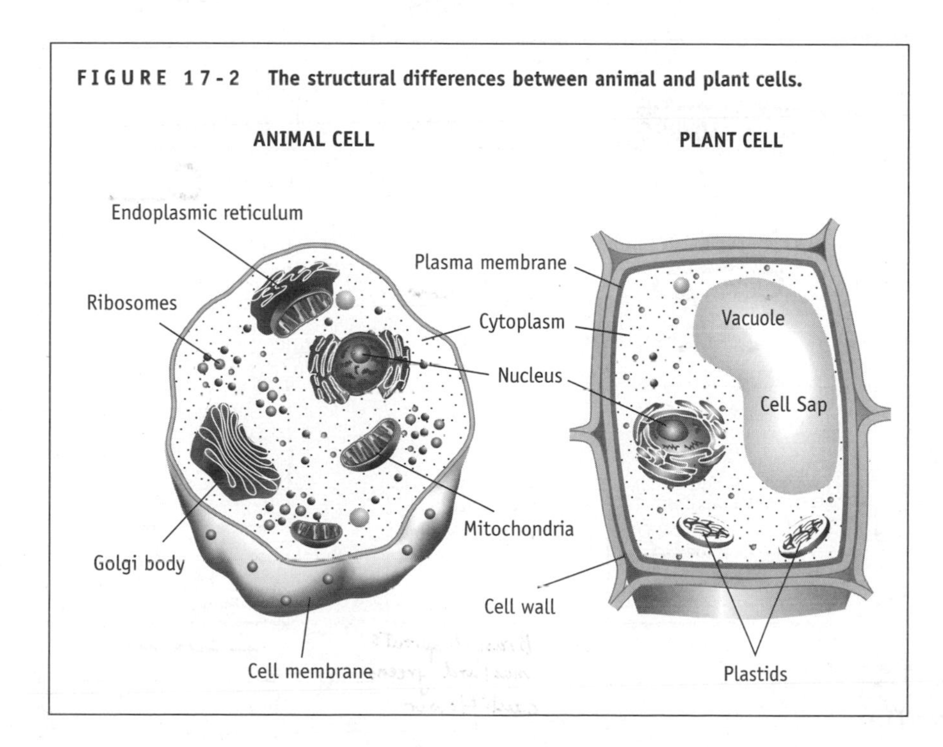

FIGURE 17-2 The structural differences between animal and plant cells.

> **KEY TERMS**
>
> ***Turgor*** The rigid firmness of a plant cell resulting from being filled with water.

Parenchyma Cells. The most common type of cell in vegetables and fruits is the parenchyma cell. Within the jelly-like cytoplasm of these cells are the compounds responsible for the plant's starch content, color, water volume, and flavor. Several of these substances, such as starch and pigments, are stored in organelles called plastids. There are three types of plastids: leucoplasts, chloroplasts, and chromoplasts.

Leucoplasts. Leucoplasts store starch and some water. Starch stored by these parenchyma cells serves as the major digestible portion of the plant. Many vegetables and fruits are consumed for their starch content and resulting source of energy.

Chloroplasts and Chromoplasts. Certain plastids contain pigments that add the color found in many vegetables and fruits. The chloroplast plastids contain the chlorophyll that is essential for carbohydrate synthesis and provides the green color of plants. The orange-yellow colors of certain vegetable and fruits are derived from the chromoplast plastids that contain the carotene or xanthophyll pigments.

Vacuoles. Another function of the parenchyma cells is to store water and other compounds in sacs called vacuoles. The size of the cell is often related to the amount of liquid it holds, which determines the juiciness of the vegetable or fruit. Juiciness ranges from very little in potatoes and bananas to very high in tomatoes and watermelon. The optimum water content provides **turgor** and crunchiness to leafy vegetables like lettuce and spinach. Heat or humidity reduces turgor in plants. Other substances found in the cells' vacuoles besides liquid are the red-blue pigments known as anthocyanins (discussed below) and an array of flavor compounds such as saccharides, salts, and organic acids (17).

Acids. Organic acids found in the cell contribute to its pH and to the food's flavor and acidity. Most vegetables have a pH of about 5.0 to 5.6, with tomatoes slightly lower (pH 4.0 to 4.6) and corn, peas, and potatoes slightly higher (pH 6.1 to 6.3).

Intercellular Air Spaces. Plant cells do not fit tightly next to each other, so intercellular spaces—the spaces *between* cells—fill with air, adding volume and crispness to vegetables and fruits. How close the cells are to each other influences the textural differences between fruits and vegetables in terms of their crispness. For instance, air spaces account for 20 to 25 percent of the volume of an apple, 15 percent of a peach, and only 1 percent of a potato. Without the air, both vegetables and fruits would be soft and flaccid.

tomatoes (35). Light yellow xanthophyll pigments color pineapples.

Heat affects the color of vegetables, most likely because it modifies the pigments' chemical structure. Exposure to oxygen also causes oxidation of pigments and a resulting loss in color. Vegetables containing beta-carotene should not be overheated, because this pigment not only contributes to color but can also be converted to vitamin A; therefore, its destruction would be doubly undesirable (2) (Chemist's Corner 17-1).

Chlorophyll. Chlorophyll, the pigment responsible for the green color of plants (Chemist's Corner 17-2 on page 316), also makes possible the essential process of photosynthesis, in which leaves capture the sun's light energy to convert carbon dioxide and water to carbohydrates. It is not surprising that plants rich in chlorophyll include most of the leafy green vegetables like lettuce, spinach, broccoli, green cabbage, and kale.

In older plants or those picked and exposed to sunlight, chlorophyll is degraded, causing underlying pigments to show. This is why leaves may turn yellow in fresh parsley or broccoli florets left too long on the produce stand (49, 50). The process is similar to what happens in autumn, when the nongreen colors, which have been in the leaves all along but masked by the darker, green chlorophyll, are allowed to show as the chlorophyll diminishes with the changing light and cooler temperatures.

Blanching Enhances Green Color. Chlorophyll may also be hidden by the air between the cells in fresh vegetables. Sometimes the green color can be enhanced by blanching the vegetable, which causes the air to bubble away

> **Chemist's Corner 17-1**
>
> ***Vitamin A Carotenoids***
>
> The carotenoids contributing vitamin A activity include beta-carotene, alpha-carotene, and beta-cryptoxanthin (27). Beta-carotene can be split into two molecules with vitamin A activity (4).

Plant Pigments

Vegetables (and fruits) are clothed in all the colors of the rainbow, and so brighten meals that would otherwise look bland with only meat, dairy products, grains, and bread on the table. The selection of fruits and vegetables is often based upon the way they look, and color is an important attribute of a meal's appearance.

Plant pigments fall into three major groups: carotenoids, chlorophylls, and flavonoids (Figure 17-3). Carotenoids and chlorophylls are found in plastids and are fat soluble. Flavonoid pigments are water soluble, and have a tendency to be lost in cooking water.

Carotenoids. Carotenoids (alpha-, beta-, and gamma-carotenes), along with lycopenes and xanthophylls, account for most of the yellow-orange and some of the red color of fruits and vegetables. Carotenes lend reddish-orange color to carrots and winter squashes. Lycopenes, which are deeper red, provide the bright color of

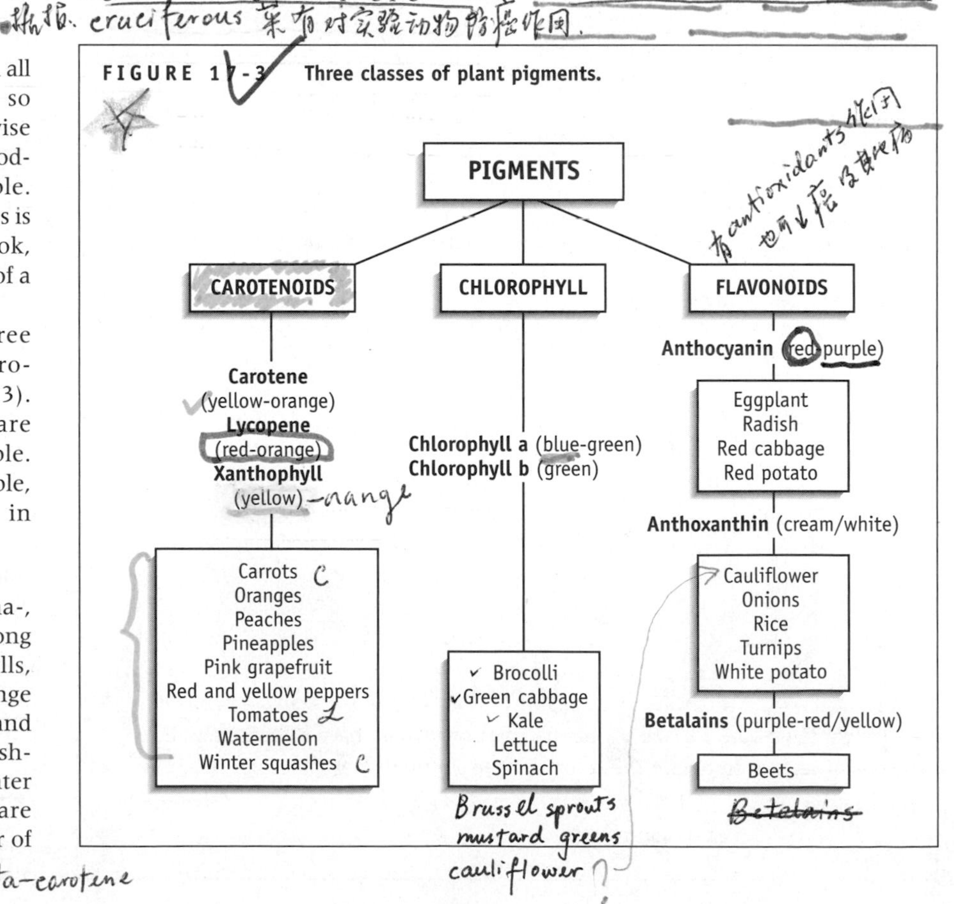

FIGURE 17-3 Three classes of plant pigments.

NUTRIENT CONTENT

With few exceptions, fresh, unprocessed vegetables are naturally low in calories (kcal), cholesterol, sodium, and fat. Most vegetables are good sources of carbohydrate (including fiber), vitamins, and minerals—vitamin C, beta-carotene, certain B vitamins, calcium, and potassium. It has been suggested by researchers that this nutrient combination, in addition to other non-nutritive compounds found in a plant-based diet of vegetables, fruits, whole-grain foods, and **legumes** (especially soy products), contributes to a reduced risk for certain cancers (19, 36).

Calories (kcal). Most vegetables, particularly the leafy, green varieties, are low in calories (kcal). However, the high starch content of vegetables such as beans, beets, peas, corn, and potatoes makes them somewhat higher in calories (kcal).

Cholesterol. Plants contain no cholesterol. For example, peanuts are high in fat but do not contain even a trace of cholesterol. Only animals having a liver are capable of manufacturing cholesterol, so only animal products contain cholesterol. However, cholesterol may be added during processing of a plant-based food item.

Fat. Unprocessed vegetables contain little or no fat. The only vegetable foods high in fat are the processed vegetable oils derived from plant seeds—corn, peanut, rapeseed, cottonseed, safflower, and others. The calorie (kcal) and fat content of selected vegetables are shown in Table 17-1.

TABLE 17-1
Calories (kcal) and Fat Grams in Vegetables

Vegetables	*Serving size**	*Calories (kcal)*	*Fat/grams*
Baked potato	1 medium	57	0.1
Pinto beans		116	0.4
Sweet potato	1 medium	103	0.1
Navy beans		129	0.5
Lima beans		104	0.3
Corn		88	1.1
Peas (green)		67	0.2
Lentils		115	0.4
Most other vegetables		<40	0.4

**Serving size = ½ cup cooked unless otherwise noted.*

Carbohydrates. Some vegetables are so high in complex carbohydrates that they can substitute for grain-based starches. Potatoes, legumes, and corn fall into that category. During the ripening process, vegetable sugars convert to starch for storage. Corn, carrots, and peas taste sweeter when harvested early because their sugars have less time to convert to starches. In fruits, the opposite occurs; starches convert to sugars, which explains why a ripe melon tastes sweeter than an unripe one.

Fiber. Many, but not all, vegetables are rich in dietary fiber. Fiber content varies a good deal among plant foods, as shown in Figure 17-4. For example, iceberg lettuce contains only 1 gram of fiber per cup: one would have to consume 20 to 30 cups of lettuce to obtain the recommended 20 to 30 grams of daily fiber,

(continued)

KEY TERMS

Legumes Members of the plant family *Leguminosae* that are characterized by growing in pods. Vegetable legumes include beans, peas, and lentils.

so that it no longer clouds the colors (6). The brighter green seen in frozen vegetables is caused by blanching them before freezing. Raw vegetable platters prepared by caterers are often more brilliant in color because the vegetables have been blanched and then dipped quickly in cold water to stop the cooking.

Chlorophyll, a fat-soluble pigment, is not as stable as carotenoid pigments. Color changes in green vegetables can be minimized by keeping heating times short, adding a small amount of sugar, and heating the food uncovered for the first few minutes to allow volatile organic acids to escape. Avoid heating green vegetables with the lid on because this causes the natural acids in the vegetables to concentrate and destroy the chlorophyll. A more alkaline cooking medium converts the cholorphyll pigment to a water-soluble form, which is what turns the cooking water green. After blanching, the vegetables are then "shocked" in cold water or ice water to stop carry-over cooking.

HOW & WHY? ?????????

Why do greens become duller in color when cooked? While heating green vegetables causes them at first to turn a sharp, bright green, longer heating periods such as those required for canning cause them to turn a dull olive brown—the color often seen in canned spinach or peas (45). This heat-induced color change occurs when membranes that previously separated acids and pigments are disrupted, allowing acids to come in contact with chloroplasts. Hydrogens from the acids replace the magnesium in the chlorophyll, which results in an olive-brown compound known as pheophytin.

Chemist's Corner 17-2

Types of Chlorophyll

There are two types of chlorophyll, and each produces a specific color in green plants. Chlorophyll a is blue-green, while the more common chlorophyll b is green. These chemical compounds differ in color because of their different structures in that chlorophyll a has a methyl group (-CH_2) attached to one of the carbons on the chlorophyll molecule, while chlorophyll b has an aldehyde group (-CHO).

Flavonoids. Flavonoid pigments include anthocyanins (red-blue), anthoxanthins (creamy to white), and betalains (purplish-red).

Anthocyanin. Most of the red, purple, and blue colors seen in fruits and vegetables derive from anthocyanin. Although numerous fruits contain this pigment, it is found in only a few vegetables—red cabbage, eggplant, radish, and red potato (18). The color of anthocyanins in these foods is affected by several factors, including changes in pH that may occur during simmering. Acidic tap water intensifies the red color of anthocyanins; alkaline water changcs the reddish-blue hue first to an unappetizing blue and then to green. The latter process is sometimes observed in red cabbage and can be prevented by adding acid ingredients such as apple, lemon juice, or vinegar.

HOW & WHY? ?????????

Why does red cabbage sometimes turn purplish-blue when cut with a knife? If a nonstainless metal knife is used to cut red cabbage, the anthocyanin pigment will react with the metal ions in iron, tin, or aluminum and turn to off-colors. Stainless steel and glass are consequently the best choices for the preparation and storage of foods rich in anthocyanins.

NUTRIENT CONTENT *(continued)*

whereas only 1½ to 2 cups of kidney beans would have to be eaten to obtain a similar amount.

Protein. Vegetables lack certain essential amino acids and therefore are not a source of complete protein. The most complete sources of protein in the plant kingdom are the legumes, which nevertheless tend to be low in the essential amino acid methionine (5). Most vegetables average only about 3 percent protein and are far from complete, so various combinations of legumes and cereals serve as a complete protein and act as a dietary staple for much of the world's population (8).

Vitamins and Minerals. Vegetables are usually higher in vitamins and minerals than fruits. Sprouted beans are high in vitamin C and riboflavin (B_2). Dark green, leafy vegetables are good sources of riboflavin (vitamin B_2), carotene, vitamin C, and iron. Though 80 percent of the calcium in an average North American diet is derived from dairy sources, the next most important source is green vegetables such as broccoli. However, compounds called oxalates, found primarily in green vegetables, can bind to the calcium, zinc, or iron in these vegetables and may decrease their absorption (46). Green vegetables are excellent sources of vitamin K and folate, a B vitamin, the need for which almost doubles during pregnancy.

Fruits and vegetables also contain electrolytes such as potassium and sodium, although the amount of sodium in fresh produce is of negligible proportions in the human diet. Canning often increases the sodium content, but lower-sodium versions of many processed foods and beverages are now available on the market.

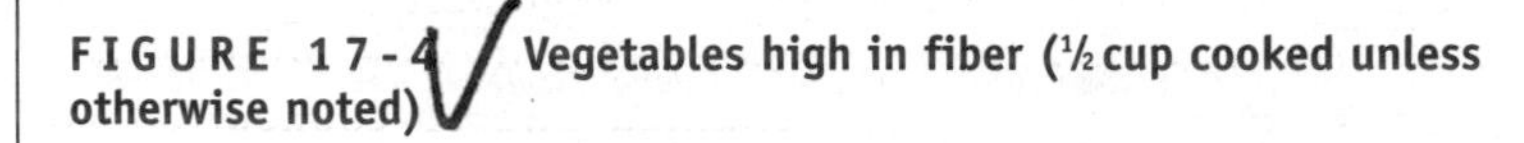

FIGURE 17-4 **Vegetables high in fiber (½ cup cooked unless otherwise noted)**

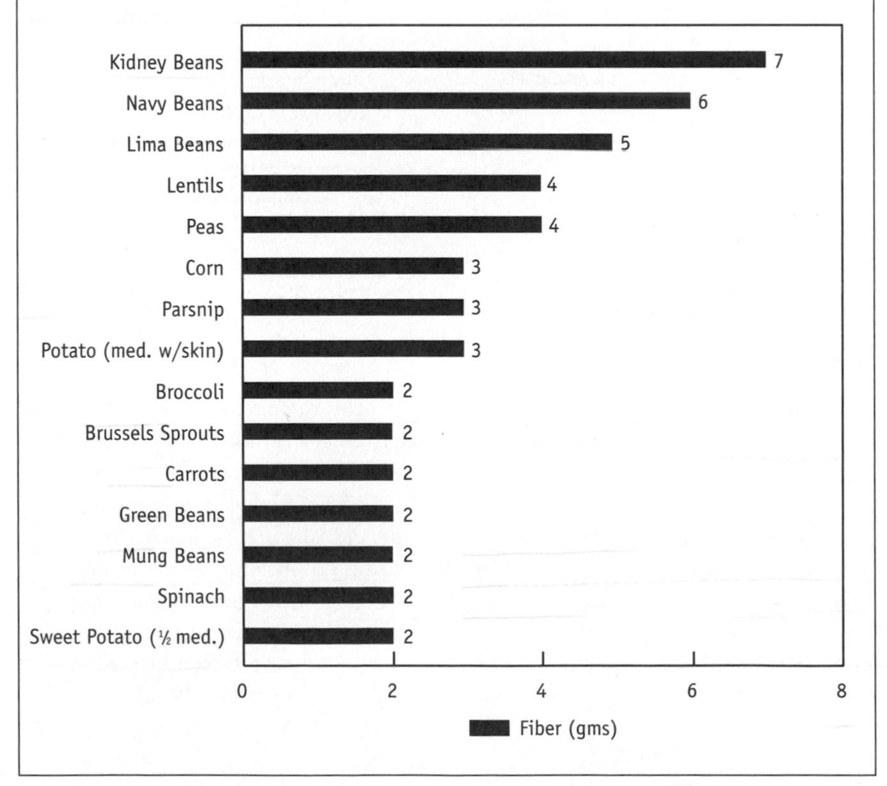

Anthoxanthins. Anthoxanthins are actually a composite of compounds known as flavones, flavonols, and flavonones. They are the reason for the cream or white color of cauliflower, onions, white potatoes, and turnips. Further whitening can be achieved by adding acidic ingredients like cream of tartar or vinegar. Anthoxanthins turn an undesirable yellow color in alkaline water, and can even change to blue-black or red-brown under excessive heating or in the presence of iron or copper.

Betalains. Betalain pigments (red betacyanins and yellow betaxanthins) give beets their deep purplish-red color. It is important to leave beets unpeeled until after they are cooked in order to prevent their rich color from bleeding out into the water. For the same reason, 1 or 2 inches of stem should be left at the top during cooking. Acidic ingredients like vinegar convert the purplish-red hue of beets to a brighter red. In an alkaline medium, the red color shifts to yellow.

Phytochemicals

Research now indicates that fruits and vegetables may carry a vast array of phytochemicals—nonnutritive compounds in plants that possess health-protective benefits (7, 12, 22). Foods that contain phytochemicals, rather than manufactured supplements, are the preferred source. The **cruciferous** vegetables as well as tomatoes, strawberries, pineapples, and green peppers contain phytochemicals that appear to inhibit cancer in laboratory animals. Studies on phytochemicals have prompted the National Cancer Institute to recommend an increased daily intake of fruits and vegetables, particularly cruciferous vegetables (26).

Purchasing Vegetables

Grading Vegetables

Most fresh produce deteriorates too quickly to grade, so most grading is presently voluntary; it is based on ripeness, color, shape, size, uniformity, and freedom from bruises and signs of decay. The only fresh vegetables currently subject to USDA grading are potatoes, carrots, and onions. New labeling laws include USDA grading for the 20 most commonly consumed fresh vegetables (38). USDA grades are shown in Table 17-2.

Selecting Vegetables

Vegetables are from living plants that grow in cycles with the passing seasons. Thus, the season of the year is the most important consideration when selecting vegetables (Appendix H). Color Figure 12 shows some of the exotic vegetables that are available in many stores. Not all vegetables are in season at the same time, and they need to be picked accordingly to maximize quality. Selecting an out-of-season tomato that lacks color and flavor ultimately affects the quality of the resulting meal.

The amount to buy depends on the type of vegetable being purchased. Vegetables, especially leafy ones, tend to lose volume when trimmed and/or cooked, so consumers often purchase slightly more per person to make up for the anticipated waste. Specific tips on selection are discussed for each of the vegetables in alphabetical order (37).

KEY TERMS

Cruciferous A group of indole-containing vegetables named for their cross-shaped blossoms; they are reported to have a protective effect against cancer in laboratory animals. Examples include broccoli, brussels sprouts, cabbage, cauliflower, kale, mustard greens, rutabaga, kohlrabi, and turnips.

Artichoke. An artichoke is the immature flower head of a thistle. The edible "leaves" are actually petals (Figure 17-5). The most common variety is Green Globe. Selection varies according to season, with the best-flavored artichokes available in the winter when the artichokes are heavy, compact, plump, and have bronze-tipped leaves (winter-kissed) from a light frost. Spring artichokes are globe-shaped, green, and have tight leaves. The shape becomes conical in

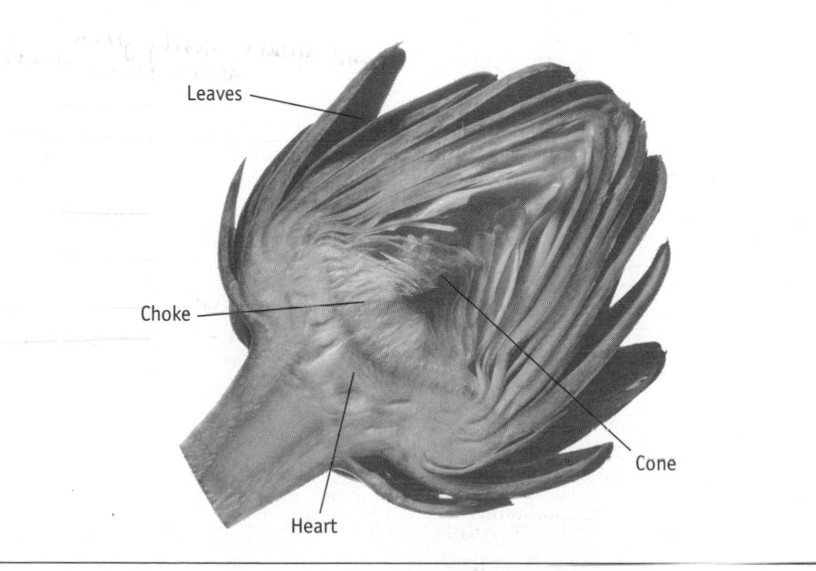

FIGURE 17-5 The major edible portion of an artichoke is the fleshy ends of the "leaves" attached to the plant. The heart is also edible and considered a delicacy. It is reached by lifting out the cone and cutting out the choke (core).

TABLE 17-2
USDA Grades for Vegetables and Fruits—Processed and Fresh

Vegetables and Fruits	*USDA Grade*	*What the Grade Means*
Processed		
Canned and frozen.	U.S. Grade A (or Fancy)	This is the highest grade for canned and frozen vegetables. They are the most tender, succulent, and flavorful and have the best color for type.
	U.S. Grade B (Choice for fruits or Extra Standard for vegetables)	The majority of canned and frozen vegetables. They are not as tender or as well-colored as U.S. Grade A.
	U.S. Grade C (or Standard)	These canned and frozen vegetables are usually more mature, less tender, and not uniform in shape or color. They are an excellent buy when cost, but not appearance, is important. U.S. Grade C vegetables are frequently utilized for soups, stews, and casseroles.
	U.S. Grade D (or Substandard)	Lowest marketable quality.
Dried or dehydrated fruits. Fruit and vegetable juices, (canned and frozen). Also includes preserves, jams, jellies, peanut butter, honey, catsup, and tomato paste.	U.S. Grade A	Very good flavor and color, and few defects.
	U.S. Grade B	Good flavor and color but not as uniform as A.
	U.S. Grade C	Less flavor than B, color not as bright, and more defects.
Fresh		
The grade is more likely to be found without the shield.	U.S. Fancy	Premium quality; only a few fruits and vegetables are packed in this grade, so they are rarely seen by most people. U.S. Fancy is the highest grade used in judging the quality of fresh vegetables and is based on appearance and uniform shape.
	U.S. No. 1	This is the most common grade for fresh vegetables and is given for those that appear fresh and tender, have good color, and are relatively free from bruises and decay.
	U.S. No. 2	Intermediate quality between No. 1 and No. 3.
	U.S. No. 3	Lowest marketable quality.

the summer and fall with the leaves becoming lighter in color and weight. Overmature artichokes are detected by leaves that separate and a woody texture. The edible portion of the artichoke includes the stem, the base (or heart), and the thicker bottom section of the leaves. The furry center, known as the choke, must be removed before consumption. Occasionally baby artichokes are sold; they are tender throughout and may be eaten whole. Mature artichokes are usually trimmed as shown in Figure 17-6 prior to preparation.

Asparagus. Tenderness is the key consideration when selecting asparagus. Tight, compact buds and fresh, firm stalks that break with a crisp snap are desired. Opened buds and flattened or angular stalks are signs of tougher, less sweet asparagus. The size and thickness of the stalk or spear is not related to tenderness. The two main types of asparagus available in North America are the more popular dark green and the less often purchased white or light green.

Beans (Green snap, green, wax, and yellow wax-podded beans). Green (green snap beans) and yellow (wax) are the two most common types of beans. Select sturdy, crisp pods with no sign of wrinkling, pitting, or bulging in the skin due to overmaturity.

Beets. Select beets that are smooth, firm, and free of any sign of dryness. Leaves should be fresh, tender, and clean.

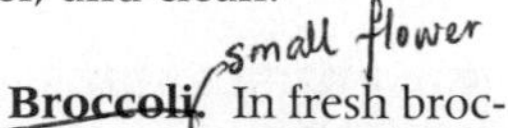

Broccoli. In fresh broccoli, the buds clusters should be firmly closed and the head dark green to purple in color. Stems should be light green and not pale, woody, hard, or dry. The signs of decay in broccoli are subtle, but avoid broccoli with florets that are starting to open or have an uneven, yellow-green color. Calabrese

Trim away both ends of the artichoke.

Scissors are used to cut away the barbs from the ends of leaves.

FIGURE 17-6 Preparing an artichoke.

is the most common variety, but others include Emperor, Premium, Futura, Green Duke, Harvester, Atlantic, Topper 430, Gem, Medium Late 423, and Medium Late 145.

Brussels Sprouts. Brussels sprouts varieties such as Citadel, Lunet, Rampart, and Valiant differ in their bud-bearing characteristics. Select those with compact leaves and clean stalk ends. Brussels sprouts with wilted, yellowish, puffy, or soft leaves are poorer in texture and flavor.

Cabbage. Look for cabbage with leaves firmly attached to the stem. The stem should be solid, hard, and heavy in relation to the head. The cabbage surface is usually smooth (except savoy cabbage), evenly colored, and free from any signs of dehydration. The most common variety is domestic cabbage, but others include Danish, the yellow-colored savoy, and red cabbage. Bok choy is Chinese mustard cabbage; it has large, sturdy, white stems and leaves, and a unique, mild flavor that enhances salads, stir-fry dishes, or soups.

Carrots. The darker the orange color, the more beta-carotene the carrot contains. The surfaces should be smooth, firm, and free from cracks, and the tops should be free from any sprouting green shoots, which would indicate an older carrot. Most carrots are sold by length and range from 2 to 8 inches long. Pre-peeled carrots are also available.

Cauliflower. Cauliflower, like broccoli, is a head of flowers. A fresh cauliflower head will not have started to bloom, spread, or look "ricey." The color ranges from white to creamy white. Lime-green is the color of broccoflower, a hybrid between broccoli and cauliflower. The jacket leaves at the cauliflower's base should be green and crisp. Popular varieties include Early Snowball, Super Snowball, Snowdrift, Pearl, Danish Giant, and Veitch Autumn Giant.

Celery. The first thing to check for in celery is clean, smooth, tender-looking inner stalks. They should snap when broken and have a straight, solid cone in the center. Celery with signs of stringiness or roughness may be tough and should not be purchased. The leaves on top should be fresh, green, and free from signs of dehydration (wilting, color changes) or decay. The green variety is the most popular, although Pascal, a lighter color variety, is growing in popularity because of its special flavor and lack of stringiness and toughness.

Corn. Corn on the cob is relatively easy to select, because the husks, the stem ends, the silk strands on top, and the kernels can all be checked for freshness. Healthy husks and stems are a vibrant green and show no signs of discoloration or drying out. The stem ends are moist and not wilted or decayed. Many stores prefer that shoppers not rip open corn ears, because it causes the kernels to dry out and degrade, but when it is possible, the husk should resist being pulled back and the kernels should be plump, even sized, and ripe-colored. Any indentations in the kernels or large or dark-yellow kernels indicate older, tougher, less flavorful corn. There are more than 200 varieties of corn, but the major type found commercially is Yellow Hybrid Sweet Corn.

Cucumbers. The three types of cucumbers that are available include slicing or table, pickling, and greenhouse. Slicing cucumbers are found in most supermarkets and are best selected based on their firmness and fairly straight-sided oblong shape. The skin should be free from any soft spots or shriveling, and evenly colored, with no yellow evident. Typical supermarket varieties are heavily waxed to protect the fruit. Varieties include Ashley, Cherokee 7, Gemini, Hybrid Ashley, High Mark II, Long Market, Marketer, Palomar, Poinsett, and Straight Eight. Hothouse or greenhouse cucumbers are elongated, seedless cucumbers with a mild flavor and no bitterness. European cucumbers are long and slender and usually come wrapped in plastic. Small- to medium-sized cucumbers are utilized to manufacture pickles (see Chapter 7).

Eggplant. Freshness in eggplants is judged by the condition of the skin. It should be smooth, firm, and glossy, with a deep purple, almost black color. Eggplants are also available in white and striped varieties, and, in these, any sign of yellow indicates aging. Flabbiness, softness, or dark bruise-like discolorations indicate poor quality. Select the heaviest eggplant with at least a 3- to 6-inch diameter. Chinese eggplants (also called Asian or Japanese eggplant) have a thinner skin and are long and thin, averaging 1 or 2 inches in diameter. American varieties include Black Beauty, Florida High Market, Florida Market, Myers Market, New Hampshire, and New York Purple.

Garlic. Roman gladiators ate garlic, believing it would make them stronger and capable of great feats, and even today many people believe it has curative powers against a host of ailments. Garlic is sold in whole heads. Select firm, well-filled cloves with skins that are not overly dry. Avoid those that are sprouting or have dark or rotted areas. Garlic can be peeled and minced by using a garlic press, or the clove can be crushed on a cutting board under the side of a chef's knife or heavy object such as the side of a can. Garlic ranges in color from white to dark maroon, and varieties include Argentina White, California Early, California Late, Mexican Purple, and Mexican White. Elephant garlic, which is much larger, milder, and sweeter than regular garlic, is gaining in popularity, and is capable of reaching 6 inches in diameter and weighing up to 12 ounces.

Ginger Root. This vegetable, often used as a spice in Asian, Indian, Moroccan, and Caribbean dishes, is selected by looking for a smooth, nonwrinkled skin with a fresh sheen and filled-out flesh. Choose ginger root that breaks cleanly with a snap and has the least number of knots.

Greens. Greens are leafy vegetables that include collards, dandelion, kale, mustard greens, beet greens, endive, and Swiss chard. These excellent sources of vitamin A should be fresh, crisp, tender, and free from any drying, insect damage, or decay. Top varieties for each of the greens arc as follows:

- *Collards.* Georgia, Louisiana Sweet, Morris Heading, and Vates
- *Dandelion.* Arlington Thick Leaf, Improved Thick Leaf, and Thick Leaf
- *Kale.* Dwarf Blue Curled Scotch, Dwarf Green Curled Scotch, and Tall Green Curled Scotch
- *Mustard greens.* Florida Broad Leaf, Large Smooth Leaf, Fordhook Fancy, Southern Giant Curled, and Tendergreen
- *Swiss chard.* Burpee's Rhubarb Chard, Dark Green White Ribbed, Fordhook Giant, Giant Lucullus, and Rhubarb Chard

Leeks. This mild-flavored member of the onion family is traditionally used in soups and French dishes. Select leeks with tender leaves that are not too fibrous or coarse. Varieties include American Flag (London Flag), Blue Leaf, Carentan (Winter), and Musselburg.

Legumes. Beans, peas, and lentils serve as excellent sources of fiber, protein, iron, and complex carbohydrates. The single common identifying factor among all legumes is that they grow as seeds within a pod (Figure 17-7). Dried beans have served as a dietary staple since the Bronze Age.

Soybeans are unique in that, compared to other plant sources, they

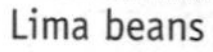

Lima beans

Black beans

Blackeye beans

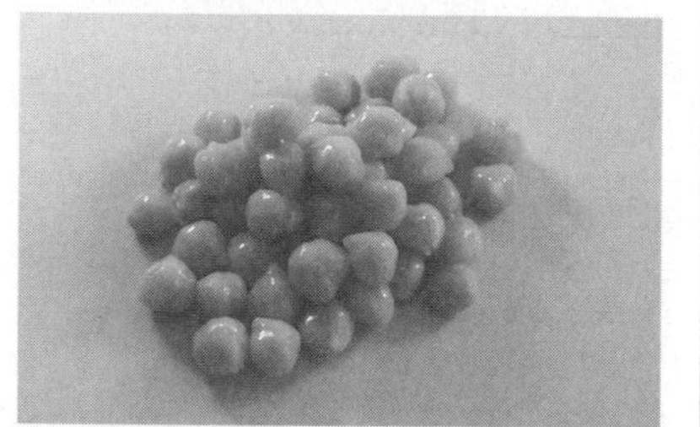

Garbanzo beans

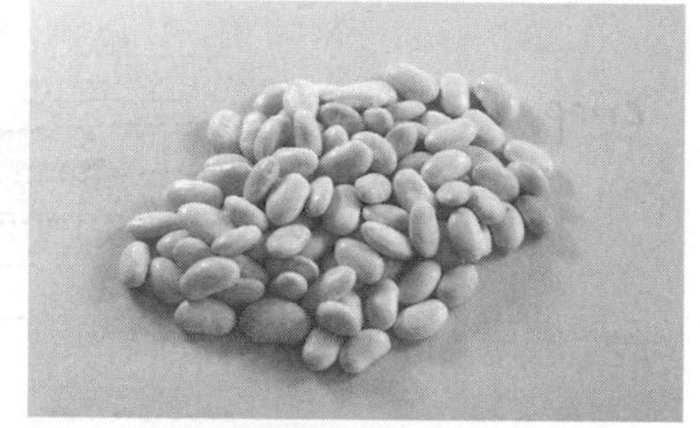

Great Northern beans

Large Lima beans

Navy beans

Pink beans

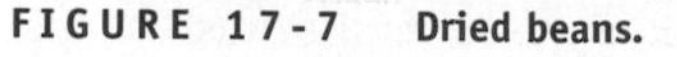

Pinto beans

Red Kidney beans

FIGURE 17-7 Dried beans.

have relatively high protein and fat content, and can be utilized to make textured vegetable protein, soy milk (see Chapter 11), tofu (see Chapter 12), meat analogs, and fermented soybean foods (32).

Textured Vegetable Protein. Textured vegetable protein (TVP) is derived mainly from soybeans, but other plants such as peanuts and cottonseed may be used. In TVP, the plant material has been altered into fibrous, porous granules that rehydrate rapidly. TVP is often used to extend ground meats in order to lower costs and/or reduce fat content (24). The USDA limits the use of TVP in commercially prepared foods to no more than 30 percent of any particular product (41). One cup of TVP is usually rehydrated in ½ cup liquid before being added to hamburgers, hot dogs, chilis, tacos, spaghetti, lasagna, soups, meatloaf, pizza toppings, or other dishes. TVP burns easily, so it should be added only after the meat has been browned. It also retains moisture and contributes to texture by maintaining cohesiveness (24). TVP is available flavored or unflavored, and natural or caramel-colored.

Meat Analogs. Meat analogs are imitations of meat products that are made by blending soy protein with other vegetable proteins, carbohydrates, fats, vitamins, minerals, colorings, and flavors. There are meat analog products that can stand in for breakfast sausages, bacon, ham slices, and chicken or beef chunks. These products contain one-third the fat when compared to the original meats, and no cholesterol; however, depending on how they are made, the protein quality is relatively low, since soy protein lacks sufficient amounts of the essential amino acid methionine. They are also higher in sodium and more costly than the meats they simulate. However, when used with a mixed diet of grains and seeds, these meat analogs provide a balanced selection.

Fermented Soybean Foods. The fermentation of soybeans results in a number of different foods:

- *Miso.* A fermented soybean paste, which is used as a seasoning for soups, stews, dressings, dips, sauces, and gravies (16).
- *Natto.* A condiment made from fermented cooked soybeans held together by a sticky bacterial product.
- *Sufu.* Fermented tofu, also known as bean cake.
- *Tempeh.* A snack or meat alternative entrée made from fermented whole soybeans molded into a cake.
- *Tamari.* A naturally aged and fermented soy sauce.

Lettuce. Lettuces are leafy vegetables that serve as the foundation for salads and as an ingredient in sandwiches or tacos (Table 17-3). Select lettuce with fresh, crispy leaves and no signs of withering or dehydration. Iceberg lettuce can be easily cored by holding the head coredown and rapping it against the counter. This motion forces the core into the head, and it can then be easily twisted and removed. The lettuce head is then rinsed core-up so that the cold running water will spread through the leaves. All moisture should be removed after lettuce is washed to slow decay.

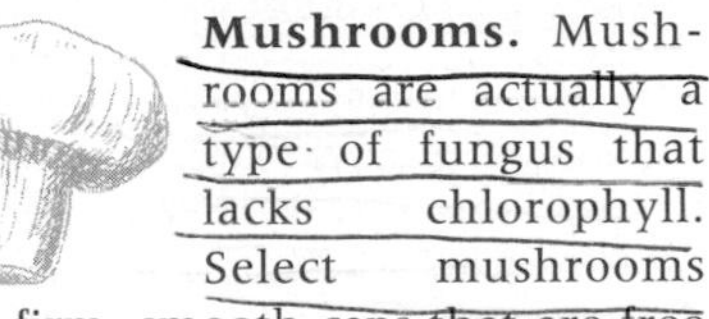

Mushrooms. Mushrooms are actually a type of fungus that lacks chlorophyll. Select mushrooms with firm, smooth caps that are free of slime, mold, wilting, or bruises. The appearance of gills underneath the umbrella can indicate aging, although they are always visible in some varieties. If present, the gills should be clean, not too dark, and unbroken. Once purchased, mushrooms are often not washed, but instead stemmed and cleaned with a damp cloth or paper towel or scraped clean with a paring knife. They are very absorbent, so exposing them to water may ruin their flavor and texture, although many individuals subject them to a quick rinse just prior to preparation. There are quite a few varieties of mushrooms from which to select (Figure 17-8). "Wild" mushrooms should not be picked from fields, forests, or any other location, because even mushroom identification experts have been reported to make mistakes and get extremely ill.

Okra. Okra consists of edible pods filled with seeds. The pods should be young and tender, and range from 1 to 3 inches in length. Fresh pods snap and puncture easily, while older ones are dry, fibrous, have a dull color, and contain hard seeds. Okra pods are usually green, but white and purple varieties are also available. Popular varieties of okra include Clemson Spineless, Dwarf Long Pod, French Market, and Perkins Spineless.

TABLE 17-3
Major Varieties of Lettuce

Variety	Characteristics
Iceberg	Iceberg lettuce accounts for almost 10 percent of the entire produce market. Leaves are crisp, consistently mild in flavor, and have a high moisture content.
Boston or Bibb	Also know as butterhead, this lettuce has a very mild taste and pliable, soft leaves. Varieties include Bibb, Big Boston, May King, and White Boston.
Endive	Curly leaf ends distinguish this lettuce from the rest. The edges tend to be more bitter than the center.
Escarole	These crumpled-looking leaves, relatively fibrous and somewhat bitter tasting, are usually available in the Full Heart Batavian variety and are often served cooked or in soups.
Leaf lettuce	This lettuce has a crunchy texture not unlike iceberg lettuce, but it has more flavor.
Romaine or Cos	The darker green, sometimes reddish-tinged flimsy leaves and stronger flavor differentiate this lettuce from iceberg. The pale, crisp cores or "hearts" of romaine are often sold for Caesar salad.

Shiitakes have a distinct, slightly smoky flavor that adapts well to other strong flavors; they're commonly used in Asian cooking.

Oyster mushrooms have a subtle flavor that's best paired with other simple ingredients.

The Mushroom Council

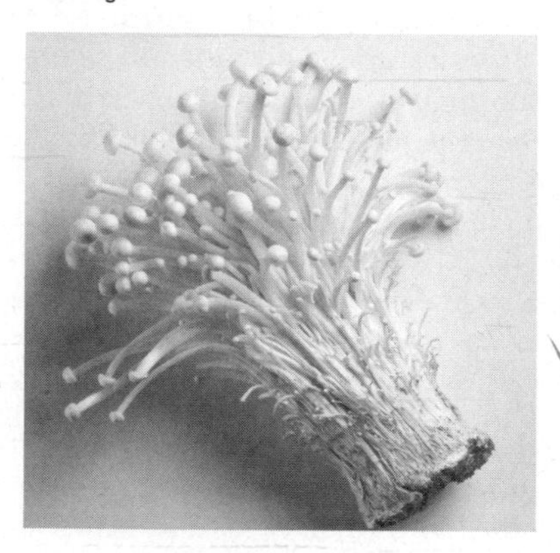

Enoki mushrooms have a mild, pleasant flavor that isn't particular distinctive.

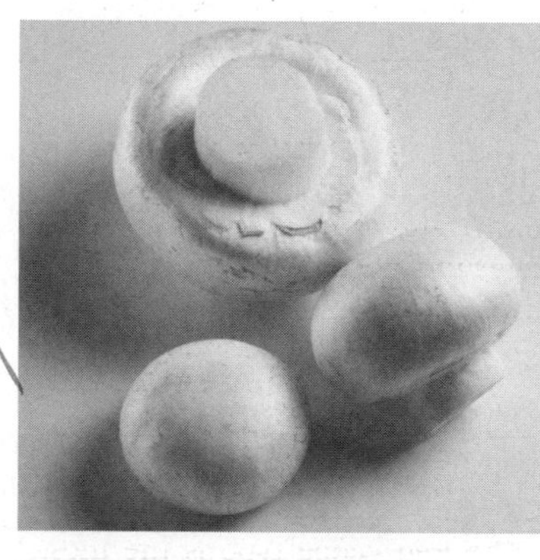

Button mushrooms are the most common variety sold; they are somewhat nutty and creamy when raw and become earthy and rich when cooked.

Portabella mushrooms are simply cremini grown to gargantuan proportions; their flesh is denser and usually more fibrous.

Cremini are a darker variety of the standard button mushroom; they're firmer with a more intense flavor.

FIGURE 17-8 Varieties of cultivated mushrooms.

Onions. The most common onion varieties are yellow, white, red, and green (Color Figure 8). The yellow onion, which is the general all-purpose onion, should feel firm, the skin should be dry and papery, and the stem small. The skin should be free of any green sunburn marks, mold, or blackish decay spots. A typical fresh onion odor emits from well-selected onions. Onion flavor ranges from pungent to sweet, with varieties including Vidalia (small, sweet yellow), Texas 1015 (large yellow), White Boiler (small white), Medium White (large white), and Red. Vidalia onions are known for their distinctive taste, which is sweeter and milder than other onions because of their higher sugar and water content. This type of onion is preferred for onion rings, omelets, pizzas, marmalades, sautéing, or any other dish where a distinctive sweet onion flavor is desired. Only onions from specific counties in the state of Georgia may legally carry the Vidalia name (30). Other types of onions and their relatives are as follows:

- *Red onions.* Commonly used raw in salads.
- *White or "boiling" onions.* Frequently added to stews.
- *Pearl onions.* Tiny onions that have a mild flavor and crisp texture. Pearl onions are often served with peas as a vegetable or as an ingredient in stews.
- *Green onions.* Also known as scallions, green onions are "onions to be" that have been harvested while the bulbs were still small and the leaves still growing. For scallions, tender, succulent, small to medium bulbs with springy leaves that hold their shape are preferred. Green onions are commonly used in salads, in Mexican dishes, for stir-frying, and alone as a garnish.
- *Shallots.* Related to the onion family, although their cloves more closely resemble garlic; they are a common ingredient in sauces.
- *Leeks.* A type of onion.

Peas. The three predominating varieties of peas are green, snow, and sugar snap. Green peas have a bulging pod, snow pea pods are flat, and sugar snap pod diameters range somewhere between green and snow peas. Unlike green peas, snow and sugar snap peas are consumed with their pods. All should be fresh, tender, and sweet. Avoid those that are in an advanced stage of maturity, which is signaled by grayish specks, yellow streaks, a darker color, dryness, and wrinkling.

Peppers, Sweet. Sweet peppers are available in many colors—green, red, yellow, orange, purple, and brown. As peppers ripen, they become sweeter because of their increasing sugar content. Those with a bright, smooth, glossy color and thick, filled-out walls and a good solid weight are recommended, while those that are pitted, dull-looking, or shriveled should be avoided. Bell peppers with a more square or rectangular shape are easier to cut up. California Wonder is the most common variety of pepper, and others include Early Cal-Wonder, Burlington, Chinese Giant, Harris Early Giant, Neapolitan, and Yolo Wonder. Other sweet peppers include the round cherry, pimento, Holland, sweet Hungarian, Cubanelles, Italian, bull horn, and sweet banana wax.

Peppers, Hot (Chili). Chili peppers come in a variety of colors, shapes, and degrees of "heat." Green, red, yellow, and black are the predominant colors. The skins of fresh chilis should always be bright, shiny, firm, and unwrinkled. Several of the more common types of chili peppers (Mexican varieties) include jalapeño, Anaheim, yellow chile, chile de arbol, ancho, pasilla, serrano, and habeñero, which is the hottest of all hot peppers.

Potatoes. Numerous potato varieties are cultivated in South America, but in North America, three basic types are found in the market: Russet/Idaho, white, and red (Color Figure 9). Russets are considered starchy potatoes and are preferred for baking; whites are best suited for roasting with meats or poultry; and reds, known as waxy potatoes, have the least starch and are good for simmering. New potatoes are immature, smaller versions of any mature potato variety. They have a delicate skin and waxy flesh that is ideal for simmering and for salads because they hold their shape. When selecting potatoes, look for those that are firm to the touch, and that have few eyes, good color, unshriveled skin, and a round or oblong shape (irregular shapes result in more waste during peeling). They should all be about the same size to ensure even cooking. Russets should have a net-like texture on their skins with no hint of green coloration, which indicates exposure to sunlight and possible presence of the toxin solanine. The varieties of each of the potatoes are as follows:

- *Russet.* Burbank, Centennial, and Norgold.
- *White.* Available in either long (White Rose) or round (Atlantic, Katahdin, Kennebec, Superior).
- *Red.* Norland, Red LaSoda, and Red Pontiac.
- *Others.* Varieties such as Yukon Gold, a medium-starchy all-purpose potato, and a wide array of small, fingerling potatoes are becoming increasingly available in the market.

Radishes. Radishes are classified by shape: globular, oval, turnip-shaped, oblong, and long. The Red Globe type is the most popular and is selected based on a bright red color, smooth, firm skin, healthy roots, and crisp, white flesh. Avoid those that are cracked open, withered, spongy, and dry or rough looking. The leaves should be fresh and vibrant with no signs of decay. The daikon is a white, sweet, and elongated radish from Asia that is shaped like a carrot. Some markets also sell the large black radish, which is black on the outside and white on the inside.

Rutabagas. Looking much like a turnip, rutabagas are more purple/beige in color, have a yellowish flesh, and have leaves that are thicker and smoother. The three most common varieties are American Purple Top, Laurentian, and the Thomson Strain of Laurentian. Like cucumbers, rutabagas are often sold heavily waxed.

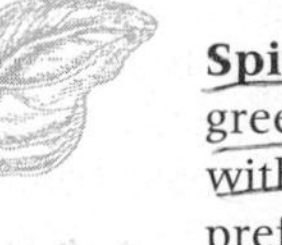

Spinach. Fresh, dark green spinach leaves with good turgor are preferred over those that are wilted, decaying, straggly, slimy, or have large, rough stalks. Varieties include dark green Bloomsdale, Bloomsdale Long Standing, Old Dominion, and Virginia Savoy. Spinach leaves may be wrinkled (savoy), flat, or semi-savoy, which is a combination of wrinkled and flat (sold for fresh and processed purposes). Young, tender leaves are somewhat lighter green and best used for salads, while the darker, more mature leaves are preferred for cooked preparations.

Sprouts. The two dominant types of sprouts on the market are alfalfa and mung bean. The greenish-topped, white-stemmed alfalfa sprouts are selected based on very little breakage of the stems and freedom from wilting or slime. The sprout heads should be full, vibrantly green, and provide an overall nut-like flavor. Bean sprouts grown from mung beans are long, ivory-colored strands of uniform length and with a crisp, snappy texture. Other sprouts, including radish, are also often available.

Squash. There are two categories of squash: Summer or soft-skinned types and winter or hard-shell types (Color Figure 10).

Summer Squash. Summer squash include zucchini, yellow crookneck, yellow straightneck, chayote, cucuzza, scallopini, and sunburst. Look for small squash with a shiny skin and no bruises or scars. The smaller the squash, the more tender its flesh, seeds, and skin. Pits in the skin are caused by chilling, while dull skin or discoloration may indicate old age.

Winter Squash. It is difficult to determine the eating quality of winter squash because of their hard outer rind. Examples of winter squash include acorn, butternut, pumpkin, spaghetti, banana, turban, hubbard, delicata, sweet dumpling, kabocha, and golden nugget. Select those that are dull and heavy for their size; as winter squash age, they become lighter in weight, indicating a dry and stringy texture. When selecting a pumpkin, look for a clean, uncracked surface with a bright orange color. Avoid those that show scarring, excess wrinkling, fading color, or soft spots on the rind.

Sweet Potatoes. There are two types of sweet potato: soft-fleshed, which tend to have a more orange flesh and are soft, moist, and sweet when cooked; and firm-fleshed, which are lighter in color and dry and mealy when cooked. Select sweet potatoes that have large, uniform shapes with no signs of discoloration, wrinkling, or drying. If they are too tapered or pointed at the ends, they may be difficult to peel. Prominent sweet potato varieties include Beauregard, Jewel, Garnet, and Hernandez.

HOW & WHY? ?????????

Why are some sweet potatoes called "yams"? Contrary to popular belief, sweet potatoes and yams are not the same vegetable. True yams are large, yellow to white tubers sold mostly in Latino markets. The "yam" sold in North American markets is actually a sweet potato that was developed in the 1930s by a horticulturist who wanted to distinguish it from other sweet potatoes (37). For many years, the produce and grocery industries have used the term "yam" to describe sweet potatoes with a bright orange color and sweet, moist flesh. The USDA requires that these sweet potatoes described as "yams" must also carry a "sweet potato" label (40).

KEY TERMS

Au gratin Food prepared with a browned or crusted top. A common technique is to cover the food with a bread crumb/sauce mixture and pass it under a broiler.

Scalloped Baked with milk sauce and bread crumbs.

Tomatoes. Tomatoes are often sold unripe because they are easily bruised during transportation (10). The color changes from green to bright red or reddish-orange as tomatoes ripen. Once ripe, they should give to gentle palm pressure; a watery consistency or slippery skin is a sign of being overly ripe. A sweet delicate aroma often accompanies ripeness. Sometimes cutting the tomato open is the only way to determine if it is ripe for consumption.

There are several common types of tomatoes: mature green, which are best for slicing or garnishes; vine-pink/vine-ripe, suited for stuffed tomato entrees or salad wedges; romas, which are elongated in shape, meaty, and flavorful, making them ideal for salads and sauces; and cherry tomatoes, ranging from the size of marbles up to ping-pong-ball size, which are excellent for salads, garnishes, or salad bars. These tomatoes soften at a slower rate than other tomatoes, allowing them to be harvested at a later date after they have had more time to ripen.

Turnips. Soups and stews are the main dishes to which turnips are added. A firm, well-rounded turnip, free from shriveling or rough skin, is preferred. Common varieties include Amber, Golden Ball, Purple Top, Yellow Aberdeen, White Egg, and White Globe.

Preparation of Vegetables

General Guidelines

Vegetables can be prepared by dry-heat methods (baking, roasting, sautéing, or deep-fat frying) or moist-heat methods (simmering, steaming, or microwaving). Serving styles also vary and include plain, buttered, creamed, **au gratin**, glazed, **scalloped**, stuffed, or in soufflés, omelets, and cream soups. Regardless of the cooking method or serving style selected, some general principles governing the handling and preparation of vegetables should be followed:

- *Buying.* Purchase only the freshest possible vegetables in amounts that will be used within a few days.
- *Storage.* Store vegetables immediately at the appropriate temperature and do not leave them out of storage for any length of time unless they are being prepared. Leftovers should be refrigerated immediately and used within three days.
- *Washing.* All vegetables must be thoroughly washed (with a vegetable brush when appropriate) to remove soil, microorganisms, pesticides, and herbicides. Washing should be quick, because most vegetables absorb excess water when soaked. Many root vegetables, except beets and baked potatoes, are peeled of the outer layer that is normally washed.
- *Cooking liquid.* As small an amount of liquid as possible should be used; in many cases, leftover liquids may be saved for stock. Using a microwave minimizes the amount of water used.
- *Timing.* The cooking time should be as short as possible; most vegetables when heated too long will undergo undesirable changes in quality. Vegetables should be served promptly after cooking.

Changes During Heating

When heated, vegetables undergo several changes in texture, flavor, odor, color, and nutrient retention. Understanding these phenomena can help to retain as much of their quality as possible during preparation.

Texture. High temperatures gelatinize starch, decrease bulk by softening cellulose, and cause a reduction in turgor due to water loss. Although this is desirable when baking potatoes or cooking legumes, it is not recommended for most other vegetables; they should be heated until barely cooked. Acids or acidic foods, such as vinegar or tomatoes, should be added toward the end of the cooking time, because they make vegetables more resistant to softening and, by precipitating vegetable pectins, increase required heating time.

To compensate for turgor lost during processing, calcium salts are often added to pickles and canned vegetables to make them firmer (33). The salts combine with pectic substances and become insoluble, firming the food's texture. Other calcium sources include molasses, hard water, and brown sugar. Sometimes the toughness of green beans and other vegetables is actually due to preparing them in hard water (6). Adding alkaline ingredients such as baking soda has the opposite effect of breaking down cellulose and producing a very mushy texture.

Flavor. Vegetables obtain their flavors from an assortment of volatile oils, organic acids, sulfur compounds, mineral salts, carbohydrates, and polyphenolic compounds. In general, to retain these flavor compounds, vegetables should be heated in as little water as possible and for as short a time as will do the job. There are exceptions; long gentle heat yields milder flavors in onions. Garlic flavor also depends on whether it is raw or cooked. Raw garlic is very pungent, yet becomes mellow or sweet when cooked slowly, and even nutty and rich when cooked at length. Only low to medium-low heats should be used for cooking garlic, because it cooks quickly and will become bitter when burned. Most of the substances that cause bitterness in vegetables such as cucumber and eggplant can be eliminated with peeling. Those that lurk under the skin can be drawn out prior to cooking by the technique of **degorging**. Droplets containing the bitter compounds are drawn out by osmotic pressure and can then be soaked up with a paper towel or rinsed off and patted dry.

Odor. Food odors contribute to the perception of flavor, but some odors, such as that from cooked cabbage or onions, may be undesirable. These pungent odors are generated by sulfur compounds present in the *Cruciferae* family (the cruciferous vegetables) and the *Allium* genus (onions, garlic, shallots, leeks, chives) (3).

KEY TERMS

Degorge To peel and slice vegetables, sprinkle them with salt, and allow them to stand at room temperature until droplets containing bitter substances form on the surface; the moisture is then removed.

Garlic, onions, shallots, and leeks are odorless until they are cut or bruised, which allows an enzyme to contact a particular substrate to create a distinctive-smelling sulfur compound (Chemist's Corner 17-3) (31). The strong odors of cooked cabbage and onion can be reduced by shortening the heating time, adding a little vinegar to the cooking water, and/or by removing the lid occasionally during cooking to let volatile organic acids escape.

Color. Both fat- and water-soluble pigments are affected by pH, heating, and the presence of metals

Chemist's Corner 17-3

The Chemistry of Pungent Odors

Pungent odors from vegetables in the *Cruciferae* family and *Allium* genus are released by overheating. The heat triggers enzymes such as myrosinase that release an excess of hydrogen sulfide gas (H_2S) and convert sinigrin, a sulfur compound, to mustard oil (Figure 17-9). Garlic's sulfur compound, alliin, is odorless until the garlic is cut; cutting allows alliinase enzymes to come in contact with the alliin and convert it to thiosulfinates and volatile, odorous disulfides (9). The thiosulfinates actually act as antibiotics, which is probably why warriors rubbed cut onions on their wounds prior to the development of antibiotics. Leeks and scallions contain the same substances, but in lesser concentrations. Pyruvate (pyruvic acid) is also formed and can be measured as a direct equivalent to an onion's pungency (23). The flavor of cabbage is further affected by the sulphur compound (+)-S-methyl-L-cysteine sulfoxide as it converts to dimethyl sulfide. .

Mustard Oil Formation

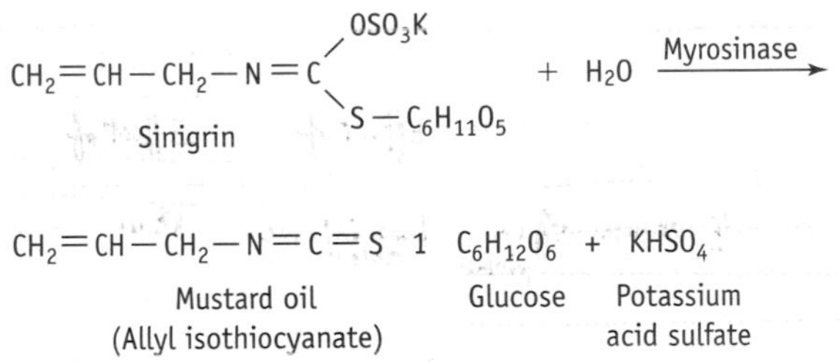

$$CH_2{=}CH{-}CH_2{-}N{=}C(OSO_3K)(S{-}C_6H_{11}O_5) + H_2O \xrightarrow{\text{Myrosinase}}$$

Sinigrin

$$CH_2{=}CH{-}CH_2{-}N{=}C{=}S \; 1 \; C_6H_{12}O_6 + KHSO_4$$

Mustard oil (Allyl isothiocyanate) — Glucose — Potassium acid sulfate

Allicin Formation

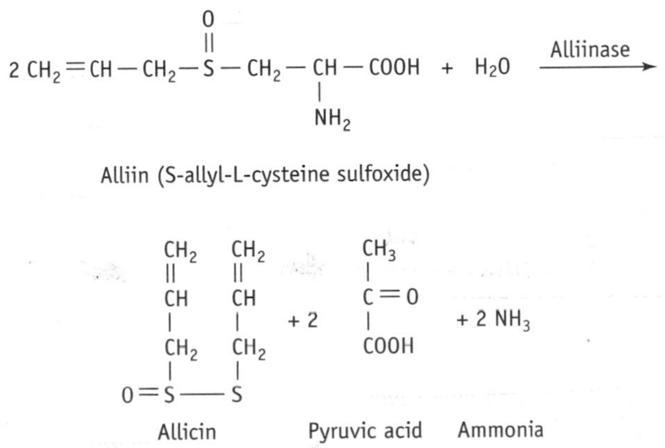

$$2\ CH_2{=}CH{-}CH_2{-}S({=}O){-}CH_2{-}CH(NH_2){-}COOH + H_2O \xrightarrow{\text{Alliinase}}$$

Alliin (S-allyl-L-cysteine sulfoxide)

$$CH_2{=}CH{-}CH_2{-}S({=}O){-}S{-}CH_2{-}CH{=}CH_2 + 2\ CH_3{-}C({=}O){-}COOH + 2\ NH_3$$

Allicin — Pyruvic acid — Ammonia

FIGURE 17-9 The chemistry of pungent odors: Heating vegetables in the *Cruciferae* family (such as cabbage) generates sulfur compounds.

KEY TERMS

Specific gravity The density of a substance compared to another substance (usually water).

(Table 17-4). Undesirable color changes can be prevented in a number of ways (6). Red cabbage, rich in anthocyanins, is prevented from turning blue if cooked with something acidic such as apples or a teaspoon of vinegar or lemon juice. Anthoxanthin-containing foods such as potatoes, rice, cauliflower, and onions are normally cream white in color, but turn yellowish if exposed to hard water. This can be prevented by adding ½ teaspoon of cream of tartar, vinegar, or lemon juice to each gallon of water used (6). Adding baking soda to green vegetables makes them appear greener; however, this is not recommended due to the deleterious effect on B vitamins and texture.

Influences other than pigments on color include the Maillard reaction, the caramelization of sugars, and enzymatic browning (13, 44). Enzymatic browning may be observed in cut-up potatoes that turn pinkish-brown when exposed to oxygen (20). Some potatoes turn dark blue-gray when oxygen reacts with the natural iron content of the potato. This discoloration can be prevented by soaking cut potatoes in water sprinkled with a little lemon or orange juice, or with cream of tartar added (1 teaspoon per quart) to increase acidity.

Nutrient Retention. Careful preparation of vegetables conserves important nutrients. Since leaching is the greatest cause of mineral loss in vegetables, it is important to cook them using as little water as possible. In many instances, it is better to avoid immersing them in water altogether and instead revert to steaming, baking, or microwaving. Other ways to minimize nutrient loss are to leave the skin on whenever possible, to cut vegetables into fewer, larger pieces rather than many smaller pieces, and to cook just to the point of doneness and no further.

Some nutrients may actually increase during food preparation, desirably or not. Frying vegetables increases the fat content of the finished product. Heating also increases the amount of protein available from legumes by destroying the enzymes known as protease inhibitors. Heat softens a food's fiber content and may even increase it as a percentage of weight after some natural water loss due to heating.

Dry-Heat Preparation

Baking. Some vegetables—especially potatoes, winter squash, onions, stuffed green peppers, and tomatoes—can be baked whole at approximately 350°F (177°C). Increasingly popular are roasted vegetables such as peppers, onions, and eggplant, which are generally sprinkled with oil and roasted at 375° to 425°F (191° to 218°C) until tender and well browned.

Potatoes. Potatoes are the most commonly baked vegetable and take approximately 1 hour, depending on their size and the oven temperature. Not all potatoes, however, lend themselves to baking. The starch content, density, and **specific gravity** of a potato determine whether it is a

TABLE 17-4

Pigment Colors Change in the Presence of Acid, Alkali, Heat, and Metals

Pigment	*Original Color*	*Acid*	*Alkali*	*Heat*	*Metal*
Carotenoids					
	Orange-red Yellow-orange	Lighter color	Brownish	Lighter orange	No effect.
Chlorophylls					
Chlorophyll a	Blue-green	Gray-green	Bright green (chlorophyllin)	Initial bright green, then dull olive brown	Copper and zinc help to retain the green color of chlorophyll, but they are not used because of possible toxicity.
Chlorophyll b	Green	Olive brown (pheophytin)	Bright green (chlorophyllin)	Initial bright green, then dull olive brown	See above.
Flavonoids					
Anthocyanins	Red-purple	Red	Blue, purple, green	Dull reddish-brown	Copper, iron, aluminum, and tin change red-purple colors from green to slate blue.
Anthoxanthins	Cream/white	Whiter	Yellowish white	Little change; possible loss of white; yellow hue	Aluminum results in yellowing, while iron darkens the cream/white pigments.
Betalains	Purple-red/yellow	Red	Yellow	Darkens	Iron darkens, aluminum turns betalains a bright yellow.

"boiler" or a "baker." If a potato floats in a saline solution (½ cup salt in 5½ cups water), it is best for simmering. If it sinks to the bottom, it is best for baking. Russet and Idaho varieties, which are "mealy" potatoes, have a high starch content and are good for baking and mashing, because they yield a dry, light, fluffy texture (Chemist's Corner 17-4). When these potatoes are whipped, they readily incorporate air and soak up added milk and butter. Medium-starch potatoes, such as white potatoes, are all-purpose and are suited for both baking and simmering.

Potatoes may be pierced with a fork before baking, or cooked for 20 minutes before piercing them to allow the steam to build up. Once in the oven, it is best to turn them every 20 minutes for even baking. Wrapping them in aluminum foil retains their moisture, resulting in a gummier potato, and produces a softer skin, which may or may not be desirable. Aluminum-foil-wrapped potatoes are steamed, in addition to being baked.

Potatoes can be checked for doneness by squeezing them with an oven glove to see if they give in to pressure and feel soft under their skins. A fluffier potato results if the potato is massaged slightly before being slit open. Once opened, pushing both ends of the potato together plumps up the insides, making them still more fluffy and easier to empty.

Frying. Vegetables can be pan-fried, stir-fried, or deep-fat fried. The potato is the most commonly pan-fried of the vegetables.

Stir-Frying. Stir-frying combines a little oil with the vegetable's natural moisture. For best results, tender, quick-cooking pieces like mushrooms are cut into large, uniform slices, while less tender pieces like carrots and celery are cut to expose the greatest amount of surface area to the heat. Although not necessary, a tight cover over the pan or wok will trap steam and reduce heating time, retaining maximum nutrients, color, and texture, although the vegetables will not brown as readily in the presence of trapped steam.

Deep-Fat Frying. Deep-fat frying continues to be popular despite the associated increase in fat content. This is used especially for French fries, onion rings, and breaded zucchini. Russet and Idaho potatoes are preferred for French-frying and potato chips. Potatoes are selected for producing commercial potato chips based on the amount of sugars (specifically, reducing sugars) they contain: the less sugar, the more desirable the potato. These sugars contribute to excessive Maillard browning during frying and are considered a major quality defect (42). Other vegetables well suited for deep-fat frying are mushrooms, long beans, broccoli florets, okra, and eggplant. An example of the latter is breaded fried eggplant with parmesan.

Moist-Heat Preparation

Simmering. Vegetables should not be boiled, but instead simmered in as little water as possible to avoid nutrient loss and adverse effects on flavor, texture, and color. To simmer, vegetables (with the exception of potatoes) are added to lightly salted boiling water; when the water starts to return to a boil, the heat is immediately lowered to a simmer, and the vegetables are cooked until tender. Mushy textures result from overcooking vegetables or boiling them heavily so they bounce against each other. During simmering, the lid may be left off to allow volatile acids to escape, but leaving it on reduces the heating time. Approximate times for cooking vegetables vary from 5 to 30 minutes, with beets exceeding this general time frame by taking about 30 to 40 minutes for young beets and up to 2 hours for older beets.

Potatoes (peeled or unpeeled) are always started in cold salted water, brought to a boil, and lowered to a simmer. Waxy potatoes such as Round Reds have less starch than others and are more suited for simmering. Their higher moisture content and better ability to hold their shape make them appropriate for such dishes as potato salad, soups, casseroles, and scalloped potatoes.

Simmered vegetables may be served puréed or prepared as baby food. Young, tender vegetables are thoroughly scrubbed, cooked, and puréed. The puréed mixture is then heated to the boiling point, cooled, and served.

Chemist's Corner 17-4

Cream of Tartar and Potatoes

Oxidation of the potato's iron content from ferrous (+2) to ferric (+3) results in a dark pigment that makes the potato appear less white. Acid in the form of cream of tartar keeps mashed potatoes looking whiter by keeping the iron in the reduced ferrous state, which prevents it from being oxidized.

Amount of Water. There are two philosophies on the amount of water to use to cook vegetables properly: one is to use a minimum amount of water to reduce leaching, and the second is to use a large volume of simmering water to assure the vegetables are cooked as fast as possible. Some people feel that the speed of cooking in larger amounts of water far outweighs any nutrients lost in the cooking liquid; however, many professional chefs choose the first option.

Steaming. Steaming sometimes takes a little longer than simmering (about 5 to 10 minutes more), but provides better retention of flavor, texture, and color. For steaming, vegetables are placed in a perforated basket inserted in a pan just above simmering water (Figure 17-10) and cooked, tightly sealed, until just tender. Pressure cookers are not recommended, because their high temperatures make vegetables mushy; legumes such as pinto beans are the exception to this.

FIGURE 17-10 Equipment used to steam food.

Microwaving. One of the best ways to retain a vegetable's texture, color, and nutrient content is to cook it in a microwave oven, since this method requires very little water and is fast enough to minimize loss of quality (Figure 17-11). Manufacturer's instructions should be followed for the specific vegetable being prepared, but general guidelines for microwaving vegetables include adding a minimal amount of water, covering with plastic wrap, and microwaving for about 3 to 10 minutes (rotate or rearrange at half time) until fork tender. A more complete description of microwave preparation is described in Chapter 5.

Frozen vegetables may be cooked in the microwave oven without adding water, because water is already present. Canned vegetables are already cooked and need only to be reheated in the canning liquid and then drained. Home-canned vegetables low in acid concentration (above pH 4.6) such as green beans should not be microwaved, but rather boiled for 10 minutes to prevent possible foodborne illness from botulism.

FIGURE 17-11 Microwaving retains vegetables' texture, color, and nutrient content.

Preparing Legumes

Legumes are best prepared by simmering rather than boiling. One cup (½ pound) of dried beans usually yields 2 to 2½ cups of cooked beans; ½ cup of cooked dried beans or peas may be counted as a 1-ounce serving of meat or as one vegetable serving. Cooked legumes can be eaten plain or mixed and matched with other foods, as in red beans and rice, bean burritos, and tostadas.

There are three methods for preparing dried beans: overnight soak, short soak, and no soak. In the overnight soak method, beans are sorted and thoroughly rinsed and then immersed in water; those that rise to the top or are shriveled are discarded. The remaining beans are soaked in water amounting to three or four times their quantity; i.e., 2 cups of dried beans require 6 to 8 cups of cold water. After soaking for approximately 10 hours, they are either drained and fresh water is added, or they are immediately simmered directly in their soaking liquid until tender. A lid is used but should be shifted slightly to the side to avoid a boil over. During simmering, water is added as needed to compensate for evaporation loss. Soaked beans require a long, slow cooking time to break down their hard-to-digest starches (Table 17-5).

In the short soak method, sorted and rinsed beans are brought to a full boil for 2 minutes, removed from the heat, and allowed to soak in the same *hot* water for one hour. They are then simmered as above in their soaking liquid. Finally, beans can be prepared without soaking, but they take twice the amount of water and double the heating time; they also lose their skins more easily. The no soak method is used for lentils and split peas because their smaller size results in shorter cooking times.

TABLE 17-5
Cooking Time for Beans (after a 10-hour soak)

Beans	*Approx. Cooking Time*
Garbanzo (chick peas)	4 hours
Mung beans	3 hours
Black beans	1½ hours
Soy beans	2 hours
Pinto	2 hours
Soybeans	2 hours
Black-eyed peas	1 hour
Kidney	1½ hours
Lentils, split peas (usually not soaked)	¾ to 1 hour (half the time if soaked)

When cooking legumes, it is important not to use hard water or add salt or acid in the form of tomato products or lemon juice until the beans are well cooked, because these substances inhibit the softening of pectic compounds.

Indigestible Carbohydrates. A pitfall that may be encountered with legumes is that flatulence often occurs. Gases (hydrogen, methane, and carbon dioxide) are produced by intestinal bacteria when they ferment the bean's indigestible carbohydrates, such as raffinose and stachyose. The problem can be minimized by soaking the beans, draining their water, rinsing them well before cooking, and then cooking them properly in fresh water, although this may result in a decrease in nutrients. Enzyme-containing commercial products such as Beano and others are now available on the market and may also be added to the food before consumption to assist in digestion.

Preparing Sprouts

Any whole grain or legume can be sprouted (germinated) into fresh greens. The process of sprouting enhances flavor, nutrient content, palatability, and digestibility. Sprouting releases enzymes such as alpha-amylase that break down the starches into more readily digestible sugars (11).

The sprouting procedure starts with a jar equipped with a mesh top, which can be purchased at a health food store or made by topping a jar with cheesecloth and securing it with a rubber band. The seeds are rinsed, then placed in the jar with three times as much water as seed. An overnight soak in lukewarm water starts the majority of seeds on their way to sprouting, with small seeds needing only four hours and others, particularly seeds with very hard coats, requiring two days. Alfalfa and oat seeds do not need a presoak. Table 17-6 provides a list of seeds and preparation guidelines.

After soaking, the water is drained from the seeds and the jar is placed in a warm, dark place, positioned at an angle to let any water drain, and covered with a towel. Then the seeds must be rinsed at least twice daily, sometimes more often, with lukewarm water. Care must be taken to keep the seeds moist but not wet, because dried-out seeds do not sprout, while too much water causes spoilage. The rinsing process must be carried out gently and evenly so the seeds are not ripped from their tender sprouts.

Once the sprouts reach their full height, they are placed in direct sunlight to develop the chlorophyll, which will turn the leaves green within one day. Most sprouts are ready to eat within three to six days after presoaking. To make sprouts more pleasant to eat, the seed hulls can be removed by filling the sink with cold water, dropping in the sprouts, allowing the hulls to float to the surface, and skimming them off with a strainer.

Sprouts must be stored in the refrigerator and rinsed thoroughly before use. The FDA has warned that sprouts, especially alfalfa sprouts, have been responsible for several foodborne illnesses. As a precaution, some sprout preparers soak seeds in a very weak bleach solution (1 teaspoon per gallon of water) for 30 minutes before sprouting, followed by rinsing with clean water.

Storage of Vegetables

After vegetables are harvested, they are still viable and continue to respire by taking up oxygen and releasing carbon dioxide. This natural respiration contributes to the deterioration of their appearance, texture, flavor, and vitamin content. The faster a vegetable's respiration, the more quickly it deteriorates. Post-harvest respiration rates differ among vegetables, which explains why potatoes (yielding only about 8 ml of carbon dioxide per hour at 59°F/15°C) last longer than cabbage and green beans (yielding 32 and 250 ml of carbon dioxide per hour respectively) (48).

Another factor contributing to produce spoilage is loss of moisture. Vegetables after being picked contain about 70 percent water, but it is then no longer replaced by its root or leaf system. Water loss, resulting in wilting and accelerated decomposition, can be avoided by maintaining a humid atmosphere, spraying with a fine mist, and/or covering the produce with an edible film or coating. Ultimately, the key to properly storing vegetables is to slow their respiration and moisture loss, allowing them to stay fresh longer.

Refrigerated

A cooler temperature is the most important factor in reducing respiration rates, and most fresh vegetables will last at least three days if refrigerated (21). Storage times for various vegetables are ultimately based on their water content. Vegetables with a high percentage of water, such as lettuce, tomatoes, and spinach, have shorter storage times than vegetables with less water content, such as potatoes, carrots, and turnips. Since leaves draw moisture from the rest of the plant, removing the green tops of carrots, radishes, or beets increases their length of storage. Cooked beans will last up to four or five days in the refrigerator, and up to six months in the freezer.

Special Storage Requirements. Some vegetables require special storage treatment. Bean sprouts are best stored in a bowl of cold water in the refrigerator, and the water should be changed frequently. Ginger root should be frozen or stored in an airtight container to trap its moisture. It will keep up to a week at room temperature and for a month in the refrigerator (47). Mushrooms will keep up to five days if refrigerated in a paper bag or basket, but will deteriorate rapidly if exposed to moisture or warm air. Plastic bags are not recommended for mushrooms, which lack the protective skins of vegetables and

TABLE 17-6
Sprout Preparation

Seed	*Quantity*	*Yield*	*Rinses*	*Time*	*Height*
Adzuki	½ C	2 C	4	6 days	1"
Alfalfa (salads)	2 tbs	1 qt	2	4 days	1–2"
Alfalfa (baking)	¼ C	1½ C	2	2 hours	⅛"
Beans (Kidney, Lima, Fava, Green, Pinto)	1 C	4 C	4	6 days	2"
Chia	2 tbs	3 C	1	4 days	1½"
Cress	1 tbs	1½ C	2	4 days	1½"
Fenugreek (salad)	¼ C	4 C	2	5 days	3"
Garbanzo (chick pea)	1 C	3 C	5	4 days	½–1"
Guar beans (soak for 36 hours)	1 C	4 C	4	5 days	2–3"
Lentils	1 C	6 C	2	4 days	1"
Millet	1 C	2 C	3	3 days	¼"
Mung beans	1 C	4 C	4	4 days	2–3"
Peas	1½ C	4 C	2	4 days	½–1"
Radish	1 tbs	2 C	2	5 days	½–1"
Red clover	3 tbs	1 qt	2	5 days	Green
Soybeans	1 C	5 C	8	5 days	½–¾"
Sunflower	1 C	3 C	2	2 days	½"
Wheat	1 C	4 C	3	4 days	½"

fruits, contain 90 percent water, and emit a water vapor that collects inside a plastic bag even if it is vented with small holes. The increased humidity causes brown and black blotches on white mushrooms. Eggplants deteriorate quickly in either warm or cold temperatures, so they should be kept in a cool location (in an aerated plastic bag), where they will normally keep for about a day or two. Asparagus lasts longest if treated like flowers and refrigerated with their stem ends in a jar filled with about 1 inch of water (28).

Maintaining Moisture. An excellent way to retain moisture in vegetables is to store them in the refrigerator's crisper, which is designed for that purpose. If a crisper is not available, plastic bags with tiny holes are a good choice because the holes allow the food to breath. Airtight plastic bags are not suitable for storing vegetables because they cause them to "stew," thus promoting spoilage. Paper towels wrapped around moist vegetables can help to avoid the spoilage caused by dehydration.

Freezing. Fresh vegetables should not be frozen unless they are first blanched; their high water content causes undesirable texture changes when cell membranes burst upon freezing, and certain enzymes cause undesirable browning and deterioration (14, 15).

Dry Storage

Proper storage does not automatically imply refrigeration. Tomatoes (unripe), eggplant, winter squash, tubers (potatoes), dried legumes, and most bulbs (onions) should never be stored in a refrigerator. Tomatoes are picked green while still firm in order to handle the rigors of transportation; however, once at their destination, their ripening can be accelerated by placing them in a paper bag. Keeping them in a bag protects tomatoes from direct sunlight, which softens rather than ripens them. Tomatoes placed in the refrigerator will never reach their optimum flavor and texture. They should be stored in the refrigerator only when fully ripe, to slow spoilage. Dried beans will keep up to a year stored in airtight containers in a dry place, but if kept too long and/or at high humidities (80 percent) and temperatures, dried beans will be hard to cook (6) (Chemist's Corner 17-5).

Storing Potatoes. Potatoes stored in the refrigerator undergo the conversion of their starch to glucose and develop an undesirable waxy consistency when cooked. French fries made from refrigerated potatoes turn an undesirable brown (6). Potatoes should not be exposed to sunlight, which will cause photosynthesis to take place, producing a bitter taste and, eventually, a potentially toxic compound called solanine, which can be seen as a slightly green tint on potato skins and sprouts. Eating large amounts of these green potatoes is not recommended because it may lead to poisoning (39). Potatoes will keep for a couple of weeks at room temperature stored in a basket or bag with holes, or longer in a cool root cellar (45° to 50°F/7° to 10°C). The exception is new potatoes, which are very perishable and should be used within a few days. Onions and potatoes should not be stored next to each other, because they shorten each other's shelf life.

Controlled-Atmosphere Storage. Beyond simple refrigeration is a high-tech method of preservation called controlled-atmosphere storage, available to commercial food companies. This special method slows down the natural respiration of fresh vegetables by reducing the amount of oxygen (below 21 percent) and increasing the amount of carbon dioxide (above 0.03 percent) available to them while in storage (34). Such a contrived environment reduces a vegetable's respiration and/or metabolic rate to such a degree that lettuce can last up to 75 days. Other advanced storage methods include modified-atmosphere packaging (MAP), edible coatings, and plastic shrink- or stretch-wraps (1).

Chemist's Corner 17-5

Legume Storage

It has been known since the third century B.C. that certain changes occur in legume seeds during extended storage, especially at high temperatures and humidity, making them difficult to cook. This hard-to-cook phenomenon is a result of several hypothesized changes at the molecular level. The pectin-cation-phytate theory suggests that during storage, an intracellular enzyme, phytase, hydrolyzes phytin, resulting in the release of divalent cations. Once cooking begins, monovalent cations from the pectin located in the cell wall exchange with these divalent cations to form insoluble pectin. The walls, lined with this insoluble pectin, become extremely strong and make long-stored legumes difficult to cook. Another factor thought to contribute to this phenomenon is the interaction between the protein and starch. The proteins tend to over-coagulate when heated, preventing starch from absorbing enough water necessary for hydration and proper swelling (29).

PICTORIAL SUMMARY / 17: Vegetables

Vegetables in raw or cooked form add color, flavor, and texture to meals as well as enhance a meal's overall nutritional value.

CLASSIFICATION OF VEGETABLES

Vegetables may be derived from almost any part of the plant. Plant parts considered edible include the roots, bulbs, stems, leaves, seeds, and even flowers.

A few vegetables are actually fruits; that is, they are the part of the plant that contain its seeds.

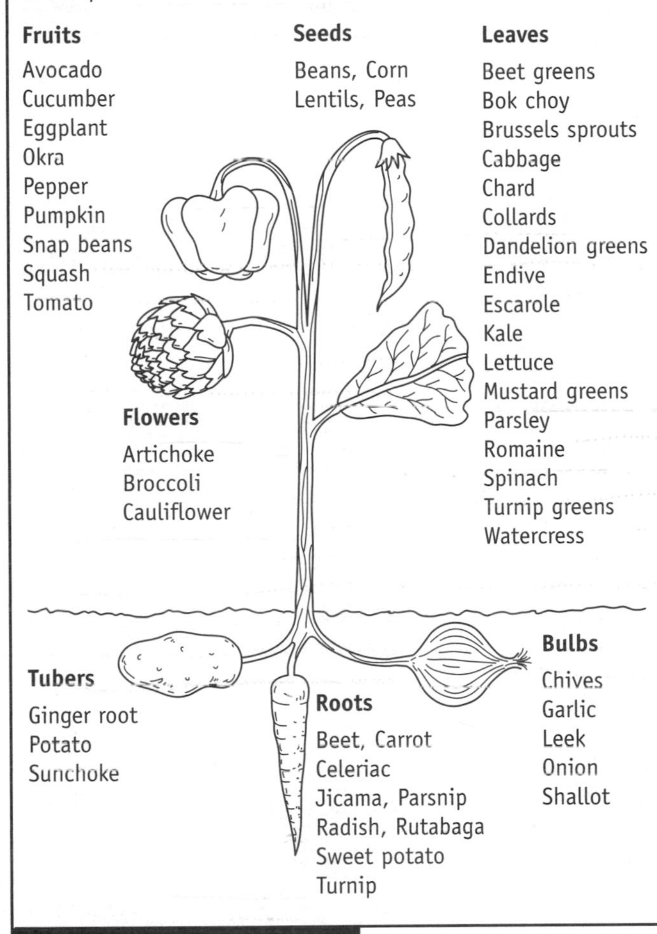

COMPOSITION OF VEGETABLES

Most fresh, unprocessed vegetables are naturally low in calories (kcal), cholesterol, sodium, and fat. Vegetables are usually good sources of carbohydrates (especially fiber), certain vitamins/minerals, and nonnutritive compounds called phytochemicals, which possess health-protective benefits. Vegetables rich in fiber include:

- Kidney beans
- Navy beans
- Lima beans
- Lentils
- Peas
- Corn
- Parsnip
- Potato
- Broccoli
- Brussels sprouts
- Carrots
- Green beans
- Mung beans
- Spinach
- Sweet potato

PURCHASING VEGETABLES

Season, ripeness, freshness, yield, and freedom from bruising or mold are considered when selecting vegetables. Presently, the only USDA graded fresh vegetables are potatoes, carrots, and onions. Grading is voluntary and based on ripeness, color, shape, size, uniformity, and freedom from bruising and decay. Individual selection criteria vary according to the vegetable.

PREPARATION OF VEGETABLES

Because heat affects the vegetable's texture, flavor, color, and nutrient retention, limiting cooking time helps preserve both flavor and appearance. Acidity toughens vegetables, while alkalinity causes excessive softening. Heating vegetables in as little water as possible with the lid on is generally recommended, with a few exceptions, to retain flavor and nutrients. Vegetables may be prepared by either dry-heat (baking, frying) or moist-heat (simmering, steaming, and microwaving) cooking methods. Regardless of the method selected, vegetables should be thoroughly cleaned before preparation to remove soil, bacteria, and pesticide residues.

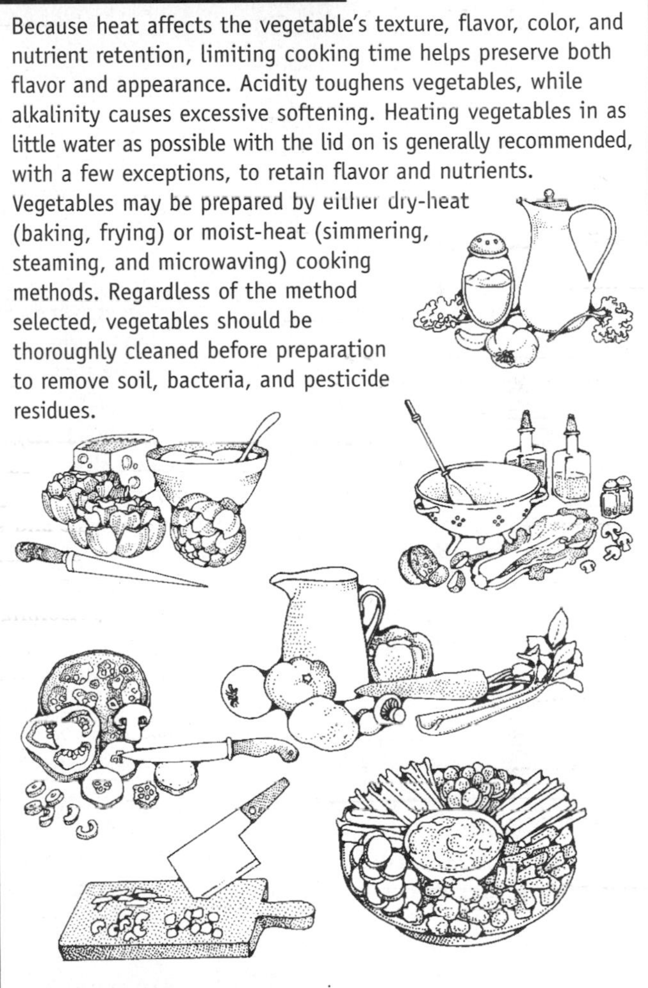

STORAGE OF VEGETABLES

Vegetables continue to respire after harvest, which contributes to the deterioration of their appearance, texture, flavor, and vitamin content. Refrigeration slows this process for most vegetables, except for tubers (potatoes), dried legumes, and most bulbs, which lend themselves to dry storage. Controlled atmosphere storage is a commercial process that extends vegetable shelf-life by reducing oxygen and increasing the carbon dioxide in the surrounding air. Another commercial process that slows down respiration is coating vegetables with a thin, edible coating such as wax.

REFERENCES

1. Avena-Bustillos RD, JM Krochta, and ME Saltveit. Water vapor resistance of red delicious apples and celery sticks coated with edible caseinate-acetylated monoglyceride films. *Journal of Food Science* 62(2):351–354, 1997.
2. Bao B, and KC Chang. Carrot juice color, carotenoids, and nonstarch polysaccharides as affected by processing conditions. *Journal of Food Science* 59(6):1155–1158, 1994.
3. Chin HW, and RC Lindsay. Volatile sulfur compounds formed in disrupted tissues of different cabbage cultivars. *Journal of Food Science* 58(4):835–839, 1993.
4. Clinton SK. Lycopene: Chemistry, biology, and implcations for human health and disease. *Nutrition Reviews* 56(2):35–51, 1998.
5. Coelho RG, and VC Sgarbieri. Nutritional evaluation of bean (*Phaseolus vulgaris*) protein: In vivo versus in vitro procedure. *Journal of Food Biochemistry* 18:297–309, 1995.
6. Corriher SO. *Cookwise: The Hows and Whys of Successful Cooking.* William Morrow, 1997.
7. Craig WJ. Phytochemicals: Guardians of our health. *Journal of the American Dietetic Association* 97(10 Suppl 2):S199–S204, 1997.
8. De Lumen BO, DC Krenz, and MJ Revillesz. Molecular strategies to improve the protein quality of legumes. *Food Technology* 51(5):67–70, 1997.
9. deMan J. *Principles of Food Chemistry.* Van Nostrand Reinhold, 1999.
10. Edan Y. Color and firmness classification of fresh market tomatoes. *Journal of Food Science* 63(4):793–796, 1997.
11. Fahey JW. Underexploited African grain crops: A nutritional resource. *Nutrition Reviews* 56(9):282–285, 1998.
12. Fahey JW, Y Zhang, and P Talalay. Broccoli sprouts: An exceptionally rich source of inducers of enzymes that protect against chemical carcinogens. *Proceedings of the National Academy of Sciences USA* 94(19):10367–10372, 1997.
13. Ferrar PH and JL Walker. Inhibition of diphenol oxidases: A comparative study. *Journal of Food Biochemistry* 20:15–30, 1996.
14. Fuchigami M, N Hyakumoto, and K Miyazaki. Programmed freezing affects texture, pectic composition and electron microscopic structures of carrots. *Journal of Food Science* 60(1):137–141, 1995.
15. Fuchigami M, K Miyazaki, and N Hyakumoto. Frozen carrot texture and pectic components as affected by low-temperature-blanching and quick freezing. *Journal of Food Science* 60(1):132–136, 1995.
16. Funaki J, and M Yano. Purification and characterization of a neutral protease that contributes to the unique flavor and texture of tofu-misozuke. *Journal of Food Biochemistry* 21:191–202, 1997.
17. Garcia-Reverter J, MC Bourne, and A Mulet. Low temperature blanching affects firmness and rehydration of dried cauliflower florets. *Journal of Food Science* 59(6):1181–1183, 1994.
18. Giusti MM, and RE Wrolstad. Characterization of red radish anrhocyanins. *Journal of Food Science* 61(2):322–326, 1996.
19. Goodman MT, et al. Association of soy and fiber consumption with the risk of endometrial cancer. *American Journal of Epidemiology* 146(4):294–306, 1997.
20. Gunes G, and CY Lee. Color of minimally processed potatoes as affected by modified atmosphere packaging and antibrowning agents. *Journal of Food Science* 62(3):572–575, 1997.
21. Haard NE. Foods as cellular systems: Impact on quality and preservation. *Journal of Food Biochemistry* 19:191–238, 1995.
22. Halliwell B. Establishing the significance and optimal intake of dietary antioxidants: The biomarker concept. *Nutrition Reviews* 57(4):104–113, 1999.
23. Hanum T, NK Sinha, and JN Cash. Characertistics of gamma-glutamyl transpeptidase and allinase of onion and their effects on the enhancement of pyruvate formation in onion macerates. *Journal of Food Biochemistry* 19:51–65, 1995.
24. Healthful attributes spur soy consumption. *Prepared Foods* 166(9):107–110, 1997.
25. Kato N, A Teramoto, and M Fuchigami. Pectic substances degradation and texture of carrots as affected by pressurization. *Journal of Food Science* 62(2):359–362, 1997.
26. Kuller LH. Dietary fat and chronic diseases: Epidemiologic overview. *Journal of the American Dietetic Association* 97(Suppl):S9–S15, 1997.
27. Lin SD, and AO Chen. Major carotenoids in juices of ponkan mandarin and liucheng orange. *Journal of Food Biochemistry* 18:273–283, 1995.
28. Lippert S. Asparagus is sweetest in spring. *Fine Cooking* 26:34–39, 1998.
29. Lui K. Storage proteins and hard-to-cook phenomenon in legume seeds. *Food Technology* 51(5):59–61, 1997.
30. McCue N. Vidalia onion: How sweet it is. *Prepared Foods* 165(10):67, 1996.
31. Miller RA, et al. Garlic effects on dough properties. *Journal of Food Science* 62(6):1198–1201, 1997.
32. Murphy PA, et al. Soybean protein composition and tofu quality. *Food Technology* 51(3):86–88, 110, 1997.
33. New process keeps canned veggies crisp. *Food Engineering* 62(6):110, 1990.
34. Ngadi M, et al. Gas concentrations in modified atmosphere bulk vegetable packages as

affected by package orientation and perforation location. *Journal of Food Science* 62(6):1150–1153, 1997.

35. Nguyen ML, and SJ Schwartz. Lycopene: Chemical and biological properties. *Food Technology* 53(2):38–45, 1999.
36. Ohigashi H, A Murakami, and K Koshimizu. An approach to functional food: Cancer preventive potential of vegetables and fruits and their active constituents. *Nutrition Reviews* 51(11):S24–S28, 1996.
37. Packer T. *1997 Produce Availability and Merchandising Guide*. Vance Publishing Group, 1999.
38. Pennington JT, and VL Wilkening. Final regulations for the nutrition labeling of raw fruits, vegetables, and fish. *Journal of the American Dietetic Association* 97:1299–1305, 1997.
39. Phillips BJ, et al. A study of the toxic hazard that might be associated with the consumption of green potato tops. *Food and Chemical Toxicology* 34(5):439–448, 1996.
40. Produce Marketing Association. *Produce Marketing Association Produce Manual*, 1997.
41. Riaz MN. Soybeans as functional foods. *Cereal Foods World* 44(2):88–92, 1999.
42. Rodriguez-Saona LE, RE Wrolstad, and C Pereira. Modeling the contribution of sugars, ascorbic acid, chlorogenic acid and amino acids to non-enzymatic browning of potato chips. *Journal of Food Science* 62(5):1001–1005, 1997.
43. Sanchez-Pineda-Infantas MT, G Cano-Munoz, and JR Hermida-Bun. Blanching, freezing and frozen storage influence texture of white asparagus. *Journal of Food Science* 59(2):821–823, 1994.
44. Sapers GM, et al. Enzymatic browning control in minimally processed mushrooms. *Journal of Food Science* 59(5):1042–1047, 1994.
45. Schwartz SJ, and TV Lorenzo. Chlorophylls in foods. *Critical Reviews in Food Science and Nutrition* 29(1):1–18, 1990.
46. Weaver CM, et al. Calcium bioavailability from high oxalate vegetables: Chinese vegetables, sweet potatoes and rhubarb. *Journal of Food Science* 62(3):524–525, 1997.
47. Wemischner R. Ginger: A fresh flavor that packs some heat. *Fine Cooking* 29:92–93, 1998.
48. Wills R H. Postharvest: *An Introduction to the Physiology and Handling of Fruits and Vegetables*. Van Nostrand Reinhold, 1989.
49. Yamauchi N, and A E Watada. Pigment changes in parsley leaves during storage in controlled or ethylene containing atmosphere. *Journal of Food Science* 58(3):616–618, 1993.
50. Zhuang H, MM Barth, and DF Hildebrand. Packaging influenced total chlorophyll, soluble protein, fatty acid composition and liposygenase activity in broccoli florets. *Journal of Food Science* 59(6):1171–1174, 1994.

WEBSITES

The "why and how" about blanching vegetables from North Dakota State University's Extension Service:
www.ext.nodak.edu/extnews/askext/freezing/4422.htm

Oregon State University's Food Resource website listing numerous foods, including vegetables:
http://food.orst.edu

The Business Newspaper of the Produce Industry, offering produce buyers "The Guide," a publication to assist buyers purchasing produce:
www.thepacker.com

18

Fruits

Imagine bountiful cornucopias overflowing with fruits—nature's desserts available in almost every shape and color of the rainbow. Even in the times of the pharaohs, fruits were transported, sometimes with great difficulty, across continents and seas to eager consumers. Early explorers and migrating peoples carried fruits and their seeds to all parts of the world, so now many are grown in areas far from their original home. For centuries, dates were cultivated in North Africa, but now are also grown in California. Pineapples were once indigenous to South America. Lemons and limes originated in India; oranges, now a daily component in many North American breakfasts, came from southeastern Asia; and the kumquat was brought to North America from China and Japan. The kiwifruit from New Zealand is one of the newest fruits to have spread from its original home to North American markets. The "5 A Day Better Health Program," started in 1991, recommends that people consume at least five servings of vegetables and fruits a day—three vegetables and two fruits (46). This chapter looks at fruits—their classification, composition, purchase, preparation, and storage.

KEY TERMS

Drupes Fruit with seeds encased in a pit. Examples are apricots, cherries, peaches, and plums.

Pomes Fruit with seeds contained in a central core. Examples are apples and pears.

Classification of Fruits

Fruits are the ripened ovaries and adjacent parts of a plant's flowers. They are classified according to the type of flower from which they develop: simple, aggregate, or multiple.

- *Simple fruits*. Develop from one flower and include **drupes, pomes**, and citrus fruits (oranges, grapefruits, lemons, limes, kumquats, and mandarins—tangerines and tangelos).
- *Aggregate fruits*. Develop from several ovaries in one flower. They include blackberries, raspberries, and strawberries.
- *Multiple fruits*. Develop from a cluster of several flowers. Pineapples and figs are two examples.

Classification Exceptions

Sometimes it is difficult to tell the difference between a fruit and a vegetable. For example, is a tomato a fruit or a vegetable? The confusion over the tomato's classification even attracted the attention in 1893 of the U.S. Supreme Court, which ruled it was a vegetable (16). At the time, there was an import tax on vegetables but not fruits, so U.S. tomato growers changed the tomato from a "fruit" to a "vegetable" to protect themselves from foreign tomato growers trying to export to the United States. Despite the Supreme Court ruling, botanists beg to differ; botanically, the tomato grows from a plant's flower, so it is technically a fruit even though it is not sweet. Squash, okra, green beans, and cucumbers, too, are botanically fruits. Nuts are actually fruits, also, but they are seeds instead of fleshy fruits, so they are grouped separately.

Confusion of the opposite sort is stirred up by rhubarb, which is really a vegetable, but is usually treated as a fruit.

Composition of Fruits

The cellular structure and pigments of fruits are similar to those of vegetables, described in Chapter 17. Organic acids, pectic substances, and phenolic compounds are also found in some vegetables, but have more relevance to fruits and are now discussed.

Organic Acids

Natural sugars such as fructose, glucose, and sucrose are the major contributors to the sweetness of fruits, while the tart flavor component is partially due to organic acids located in the cell sap. Acidity varies with the maturity of the plant, usually decreasing as the fruit ripens. These organic acids found in fruits are either volatile or nonvolatile. Volatile acids vaporize during heating, while nonvolatile acids do not, but they can leach out when fruit is cooked in water.

The common organic acids in fruit include citric acid in citrus fruits and tomatoes; malic acid in apples, apricots, cherries, peaches, pears, and strawberries; tartaric acid in grapes; oxalic acid in rhubarb; and benzoic acid in cranberries. Cranberry juice in addition has a unique blend of the organic acids quinic and malic acid, with the ratio of the two so constant that it is used by juice processors to determine cranberry juice authenticity and to calculate the percentage of cranberry juice content in juice drinks (26). The oxalic acid in rhubarb can combine with calcium in the intestine to form calcium oxalate, an insoluble complex that cannot be absorbed. It may also combine with other minerals to form similar compounds.

Acidity of Fruits. Acids cause most fruits to have a pH value below 5.0 (Table 18-1). The tartness of fruits is related in part to their acidic content. For instance, limes, lemons, and cranberries are very tart fruits, with the lowest pH values (around 2.0). The least acidic fruits are more bland and sweet in flavor, and those with a pH above 4.5 most often serve as vegetables.

TABLE 18-1
Average pH Values for Selected Fruits

pH	*Foods*
2.2	Lime juice
2.3	Lemon juice
3.1	Apples, boysenberries, grapefruit, prunes
3.2	Rhubarb
3.3	Apricots, blackberries
3.4	Strawberries
3.5	Orange juice, peaches
3.6	Raspberries, red sour cherries
3.7	Blueberries
3.8	Sweet cherries
3.9	Pears
4.0	Grapes
4.2	Tomatoes
4.6	Bananas, figs
5.1	Cucumbers
5.2	Squash

Pectic Substances

Other compounds frequently found in fruit are pectic substances, of which there are three groups: protopectin, pectin (pectinic acids), and pectic acid. Pectin, a general term describing this group of polysaccharides found in fruits, acts as a cementing substance between cell walls and is partially responsible for the plant's firmness and structure (Chemist's Corner 18-1). It is used commercially to contribute to the gelling of fruit preserves. The string-like pectin molecules bond to each other under the right conditions to

Chemist's Corner 18-1

The Chemical Structure of Pectin

Depending on its source, pectin contains a number of different polysaccharides, but they all consist of methyl pectate, the compound responsible for gelling. Methyl pectate is a molecule that has a high water-holding capacity, with a structure of a long, thread-like chain of repeating galacturonic acid units obtained from the sugar galactose (Figure 18-1).

FIGURE 18-1 **Methyl pectate.**

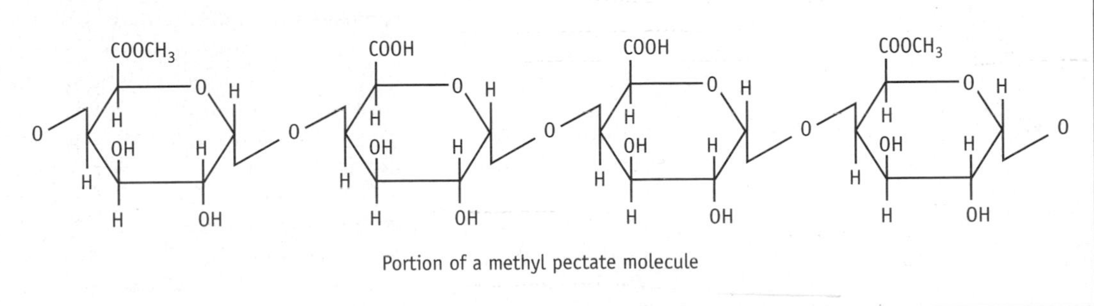

Portion of a methyl pectate molecule

KEY TERMS

Albedo The white, inner rind of citrus fruits, which is rich in pectin and aromatic oils.

FIGURE 18-2 Chemical breakdown of protopectin. Only fruits at the height of ripeness should be used for making fruit spreads without adding pectin.

Protopectin →	Pectin →	Pectic acid
Immature fruit (no gel)	Ripe fruit (gels)	Overripe fruit (no gel)

form a net-like solid structure that is able to trap water and form a gel. Pectin is found between the plant cells and within the cell walls, but not in the juice, so commercial sources of pectin include the pulp (pomace) remaining after apples are pressed for juice, and the spongy **albedo** of citrus fruits.

Use of Pectin by the Food Industry. In addition to its use in jams and jellies, pectin is used in a number of other foods as an emulsifier, stabilizer, thickener, and texturizer. Frozen foods benefit from pectin's ability to improve texture by controlling ice crystal size, preventing loss of syrup during thawing, and improving overall shape. Fruit pieces in yogurt are evenly distributed with the aid of pectin, and diet soft drinks impart more body if pectin is added (48).

Pectin Formation in Ripening Fruit. While fruit is immature, prior to the formation of pectin, its pectic substance is protopectin, a large, insoluble molecule. As the fruit ripens, enzymes convert protopectin to the more water-soluble pectin (6). Enzymes play a key role in the softening of fruits, with the largest influence derived from those enzymes that break down pectin (23). Ripening mechanisms trigger the pectinase enzymes, which break down the pectic substances as the fruit ripens, and the degree of fruit softening is related to how many pectic substances were degraded (29). The stage of ripeness affects pectin concentration, and it is the pectin extracted from ripe fruits that is used to gel jams and jellies.

As fruit continues to ripen and becomes overripe, all the pectin gradually turns to pectic acid (Figure 18-2). Since neither protopectin nor pectic acid can contribute to gelling, only fruit at the height of ripeness should be used for making fruit spreads without added pectin. Heating also converts pectin to pectic acids by hydrolyzing the chemical bonds holding the molecules together, causing the texture of the fruit to become soft and eventually mushy.

Pectic Substances and Juice Cloudiness. When juice is extracted from fruits, pectic substances can sometimes cause it to cloud (28). Although this is desirable in orange juice, where it contributes to body, other juices, such as apple juice, are often more appealing to consumers if they are clear (4, 8). One way to remove the cloudiness is through a clarification process in which enzymes such as pectinases are added to the juice to break down the pectin compounds responsible for juice cloudiness (13, 20). Juice processors can also add enzymes to certain juices to increase juice extraction. Enzymes such as cellulase and hemicellulase break down the cellulose and hemicellulose in cell walls, releasing more juice (7, 11).

Phenolic Compounds

Another group of compounds found in fruits, phenolic compounds, are responsible for the browning and bruising that often occurs in ripening fruit (21, 31). These compounds, also known as tannins, are found predominantly in unripe fruits, giving them a bitter taste and leaving an astringent feeling in the mouth. Fruits containing phenolic compounds include apples, apricots, avocados, bananas, cherries, dates, grapes, nectarines, papayas, peaches, persimmons, pears, and strawberries (44).

Phenolic Compounds and Enzymatic Browning. All of these foods turn brown from enzymatic browning, which, as you learned in Chapter 2, occurs in the presence of three substances: phenolic compounds, found within the cells; polyphenol oxidase enzymes (enzymes that oxidize phenolic compounds), also known as phenolase, catecholase, and tyrosinase; and oxygen, which enters the cells when the food is cut or bruised (Figure 18-3) (54, 56). The polyphen oloxidase enzymes turn the color of the phenolic compounds from clear to brown. These brown compounds, called melanins, though unappetizing, are safe to consume.

Purchasing Fruits

Grading Fruit

Purchasing fruits is based on the individual selection factors for each fruit and whether or not the fruit is graded.

FIGURE 18-3 Enzymatic browning in fruit is catalyzed (caused) by an enzyme (polyphenol oxidase).

Grading fresh produce is difficult, because the quality can change between the time it is graded and the time of purchase. Nevertheless, some fruit producers have fresh fruits graded by the USDA on a voluntary basis. USDA grading is based on size, shape, color, texture, appearance, ripeness, uniformity, and freedom from defects. Nutrient content between the grades does not differ to any great extent. The four grades for fresh fruit are U.S. Fancy, U.S. No. 1, U.S. No. 2, and U.S. No. 3 (49). U.S. Fancy is the best and most expensive. Processed fruits (canned, frozen) are graded differently, but the grading is still voluntary. USDA grades for canned or frozen fruits and vegetables are U.S. Grade A or Fancy, U.S. Grade B or Choice, and U.S. Grade C or Standard.

Selecting Fresh Fruits

Fruit consumption is on the upward trend. Not only does it look and taste good, but people feel that it is good for them (45). According to the FDA, the most frequently consumed fruits, in descending order, are bananas, apples, watermelons, oranges, cantaloupe, and grapes (37). Selection of these and other fruits is based on the grading points mentioned above and on the fruit's peak season (Appendix H). There are many varieties available, and selection tips are different for each fruit (16).

Apples. After bananas, apples are the second most commonly eaten fruit in the United States. There are over 1,000 varieties, but about a dozen constitute 90 percent of commercial apple production in the United States. The most common variety is the Red Delicious, followed by the Golden Delicious, McIntosh, Rome Beauty, Jonathan, York Imperial, Granny Smith, Stayman, Newton Pippin, Winesap, and Cortland. Newer varieties include the Empire, which is a cross between a Red Delicious and McIntosh, the Gala and Braeburn from New Zealand, and the Mutsu, a golden apple, originally from Japan.

Selection is based on whether the apple will be used for cooking or eating fresh (see Color Figure 14). Apples for cooking should be firm, able to hold their shape, and remain flavorful even after heating. The exception is applesauce, for which softer, fleshier apples such as a McIntosh are desired. The Red Delicious and Jonathan varieties are ideal for eating raw, but not for cooking, because they tend to collapse and lose much of their flavor when heated. The Rome Beauty and Newton Pippin are considered best for pies and sauces, but are less acceptable for eating fresh. The remaining varieties can be used for either fresh eating or preparation. When selecting an apple, choose those with skins that are clean and smooth, unbruised, and of good coloration. Avoid apples that lack color; this indicates an immature fruit with less flavor. Conversely, avoid overmature apples, which will be mealy and soft under the skin.

Apricots. Apricots originated in China thousands of years ago, but California is now the main supplier of apricots marketed in the United States. Apricot varieties available include Royal Blenheim, Castlebrite, Improved Flaming Gold, and Katy. The new varieties are easier to ship and maintain their freshness longer than older varieties, which were smaller and softer. For best flavor, select soft, ripe apricots with as much golden orange color as possible. While fresh apricots are only available during the summer, dried and canned apricots are sold year round.

Avocados. This Central American pear-shaped fruit with greenish-purplish skin was once considered exotic, but is now available throughout the year. California varieties such as the Hass, Fuerte, Bacon, and Zutano supply 85 percent of the U.S. market, with Florida providing the remaining 15 percent. The Hass, with its bumpy, dark skin and nutty flavor, is the most commonly sold variety. The others are larger and smoother skinned, but their lower fat content tends to make them less rich tasting.

The selection of avocados is based primarily on touch. If the skin gives slightly to gentle pressure, the fruit is ready for immediate use. Thick-skinned avocados can be tested for ripeness by inserting a toothpick in the stem end; if it moves in and out easily, the avocado is ripe. Most avocados are still hard when they arrive at the produce stand and will continue to ripen. Avoid avocados that are badly bruised or have soft, sunken spots.

Bananas. Bananas are the world's chief tropical fruit. The United States is the world's leading banana importer, and Central America is the major exporter. Bananas ripen best if the bunches are cut while still green. This practice also results in less bruising during handling, shipping, and marketing.

The top two varieties are the bruise-resistant Gros Michel from Martinique, a long banana with a tapered tip, and the more bruise-prone Cavendish, which has a curved shape. Newcomer varieties include the Red Spanish or Red Cuban, Saba, and Manzano. Bananas contain less than 5 percent starch and at least 80 percent sugars (24). When choosing bananas for immediate consumption, pick ones that are firm and greenish yellow to clear yellow in color.

Plantains belong to the banana family, but they are so starchy that they are usually baked, boiled, or fried like potatoes, rather than eaten raw. They average 66 percent starch and only 17 percent sugar when ripe. When plantain skins are black, the fruit is usually ripe and ready to be consumed, but the amount of black on the plantain is a matter of personal preference. Saba and Mazano varieties are ripe when their skins are completely black.

Berries. There are numerous varieties of succulent, delicate berries. They come into season in a succession that starts with the sweetest and ends with the least sweet berries; first come strawberries, then raspberries, blackberries, blueberries, gooseberries, red currants, and finally cranberries.

Strawberries. The short-lived strawberry season starts in April and lasts until sometime in July, but strawberries from

Florida and other parts of the world, such as New Zealand and Mexico, have helped to extend the season. Common varieties include the Chandler, Douglas, Heidi, and Pajaro.

Strawberries do not continue to ripen after they have been picked. For that reason they should ideally come to market when they are bright red and fully ripened. Strawberries are highly perishable because of their high respiration rate and susceptibility to fungal spoilage (43). If they are not to be eaten shortly after purchase, strawberries should be stored, unwashed, in the refrigerator. Rinsing them prior to refrigeration is not recommended because it dilutes the flavor and softens the texture.

Raspberries. Red raspberry varieties include the Willamette, Meeker, Heritage, and Sweetbriar. Munger raspberries are black. Select raspberries that are plump, firm, and free from mold.

Blackberries. Native to North America and Europe, blackberry varieties include Boysen, Cherokee, Logan, Marion, and Olallie. Select blackberries that are plump and have an almost black color that covers the entire berry with no traces of green or red by the stem. Avoid berries with the little caps that indicate immaturity.

Blueberries. Grown in the United States and Canada, blueberries come in Patriot, Bluetta, and Earliblue varieties. Select blueberries that are plump, smooth, and blue-black in color with a light grayish "bloom."

Gooseberries. Most gooseberries, except for the Poorman variety, are so tart that they have to be cooked and sweetened to be eaten; they are primarily used to make pies and preserves. Similar in size and appearance to green seedless grapes, gooseberries are selected based on their smoothness and light green color.

Red Currants. These are also quite tart and generally cooked and sweetened or used primarily as a garnish. Red currants are best selected by picking those firmly attached to the stems.

Cranberries. North America is home to only three native fruits: the Concord grape, the blueberry, and the cranberry. Four varieties—Early Black, Howe, Searles, and McFarlin—supply most of the cranberries consumed. The cranberry season starts in September and ends in November, which may explain why the cranberry is associated with the holidays that occur toward the end of the year. When selecting fresh cranberries, choose those that are firm, plump, and smooth-skinned.

Cherries. Dark red or deep purple cherries are the tastiest. The best choices are Bing, Lambert, Black Tartarian, Chapman, Borlat, Schmidt, and Republican. Tart or sour cherries such as the Montmorency tend to be lighter red in color and are more frequently used for pies. Royal Ann cherries are used to make maraschino and candied cherries. Choose sweet cherries that are plump, firm, and brightly colored with an almost mahogany-looking skin. For tart cherries, expect the skin to be lighter.

NUTRIENT CONTENT

Fruits tend to be low in calories (kcal), fat, and protein. Water content is high, ranging from 70 to 95 percent in fresh fruits.

Fat. The exceptions to the "fruits are low in fat" generalization are coconuts, avocados, and olives, as shown in Table 18-2. The fat in coconut is predominantly saturated, while that in avocado and olives is mostly monounsaturated, but there is no cholesterol in any fruits because they are of plant origin.

TABLE 18-2
Fruits High in Fat

High-Fat Fruit	*Serving Size*	*Calories (kcal)*	*Fat (grams)*
Coconut (shredded)	1 C	466	33
Avocado	1 med.	315	33
Olive	1 C	5	0.6

Carbohydrates. Carbohydrates are the main source of calories (kcal) in fruit. Each fruit serving provides about 60 calories (kcal), derived primarily from 15 grams of carbohydrate. The carbohydrate in fruit starts out in the form of starch, but as ripening takes place, it is converted to sugars such as glucose, fructose, and sucrose. These sugars sweeten the fruit and increase its palatability. If the ripening process is inhibited, which occurs, for example, when bananas are refrigerated, the starch cannot be hydrolyzed to sugar; but when fruits are allowed to ripen normally, their sugar content increases as they mature. Different types of fruit contain different amounts of carbohydrate: dates can run as high as 61 percent carbohydrate, while avocados contain only 1 percent carbohydrate. Canned or frozen fruits with added sugar have higher carbohydrate concentration and thus greater caloric (kcal) content. Food companies are now providing a wide assortment of processed fruit products with lower sugar concentrations.

Vitamins and Fiber. According to the Food Guide Pyramid, it is recommended that people eat at least two to four servings of fruits per day to ensure an adequate intake of vitamin C, beta-carotene, and fiber, all of which have been reported to reduce the risk of cancer (42). Pectin and other soluble fibers may help lower high blood glucose and cholesterol levels in some people (3, 48). Oranges are an excellent source of vitamin C, but fruits containing even more vitamin C than the orange include guava, kiwifruit, papaya, and strawberries as shown in Figure 18-4. The vitamin C content is reduced if the fruit is exposed to air by bruising or cutting, or to alkali or copper, or through cooking or processing. Fruits high in beta-carotene (vitamin A) include mangoes, cantaloupe, apricots, persimmons, and plantain.

(continued)

NUTRIENT CONTENT

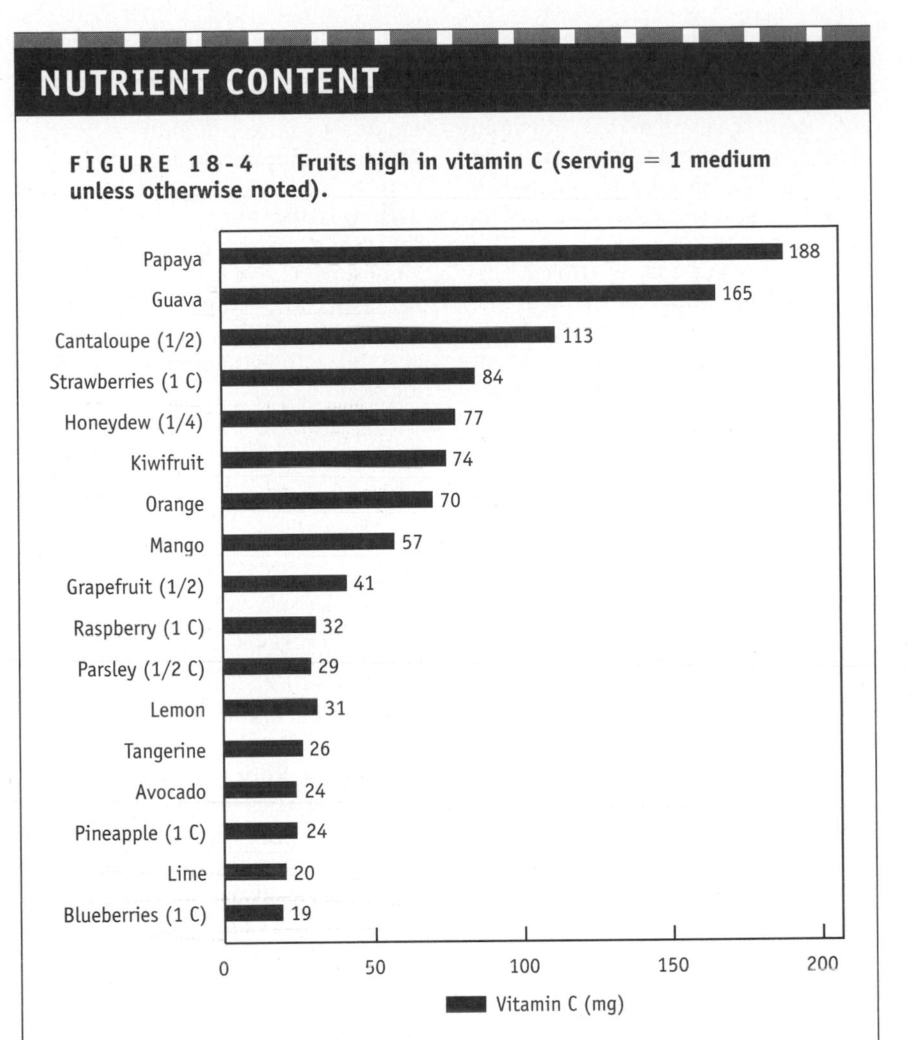

FIGURE 18-4 **Fruits high in vitamin C (serving = 1 medium unless otherwise noted).**

Not all fruits are high in fiber, but some good sources are raisins, pears, apples, raspberries, strawberries, prunes, and oranges. Removing the skin diminishes the number of grams of fiber; paring an apple, for example, removes 0.8 grams of fiber. Fruit juice has less fiber than the whole fruit, with most fruit juices averaging only about half a gram of fiber (27). In addition to being consumed for fiber, prunes and prune juice are often used as a laxative because they contain diphenylisatin, a compound that stimulates intestinal movement.

Minerals. Mineral content is generally low in fruits, but there are some exceptions. Raisins and dried apricots are high in iron, and potassium can be derived from avocados, apricots, bananas, dates, figs, oranges, plums, prunes, melons, and raisins.

Dates. These long, brown, wrinkle-skinned berries from palm trees have been cultivated for centuries in the Middle East and North Africa, where they provided sustenance for many ancient desert nomads (52). Although dates appear to be dried, they are actually fresh. There are two basic classes of dates: soft and semi-dry, although a few are considered dry. Soft dates, such as the Medjool, Khadrawy, and Halawy varieties, contain more moisture, have a very soft, almost caramel-like flesh, and are popular with consumers as a snack. The semi-dry dates are important to food processors. Although there are over 100 varieties, Deglet Noors constitute 95 percent of the market (Figure 18-5). The Zahidi date (semi-dry) has firmer flesh and is smaller than the Deglet Noor. Other varieties of dates include the Barhi, Thoory (the driest date), and Dayri.

FIGURE 18-5 **Deglet Noor dates.**

Most dates at the consumer level are sold packaged, ripe, and ready-to-eat.

Figs. California is the largest producer in the United States of black and green figs. Originating in the Mediterranean basin, they have been cultivated longer than any other fruit in the world (53). The flowers on fig trees are never seen because they develop inside the fruit and produce the crunchy little seeds that give figs their unique texture. These sweet, small, pear-shaped fruits come in a number of varieties, Black Mission being the most common. Others include Brown Turkey, Brunswick, Celeste, and Kadota. Popular dried fig varieties include the Smyrna and Calimyrna. Black figs are selected for eating raw, with green figs primarily for cooking and canning. Dried, pressed figs have for centuries been served as a confection. When buying, avoid dry, hard figs or those with flat sides, splits, or mold.

Grapefruit. Both pink and white/golden grapefruit are available. The pink varieties, which are sweeter, include the popular Ruby Red and Star Ruby, followed by Burgundy Red, Foster Pink, and Marsh Pink. A popular white or golden grapefruit variety is the seedless Marsh. Grapefruit is at its best when the skin is shiny and smooth and both ends are flat. Avoid grapefruit with pointy ends, thick skins, or deep pores.

Grapes. Perhaps no other fruit is used quite as extensively and for as many different purposes as the grape. Grapes are consumed fresh or in the form of wine, or as raisins, juice, jams, or jellies. Most grapes in the United States are grown for the wine industry (41). Grapes are actually berries that grow in clusters, and they come in a wide variety of colors and shapes. Classification is based on color, and the most common commercial table grapes are the red Flame seedless and green Thompson seedless types. Other popular table grapes include Perlette, Lady Finger, Tokay, Cardinal, Catawba, Delaware, and Emperor. The fragile Concord grape is used primarily for making wine, juice, and jelly. Raisins are usually made from Muscat and Thompson seedless grapes. Grapes do not get any riper or sweeter after being picked. Choose grapes that are firm, plump, and well-colored. Avoid those that are soft, wrinkled, moldy, or have bleached areas around the stem. Table grapes should be chilled for the best flavor and texture.

Guavas. There are several varieties of guava, each with its own distinctive flavor, ranging from strawberry-like to pineapple-like to banana-like. The flesh may be white, red, or salmon in color. Seeds may be present or absent, depending on the type of guava. Select guavas whose skins are yellow rather than green, and which yield to pressure and emit a fruity guava aroma. Slightly underripe guavas may be lightly cooked and sweetened.

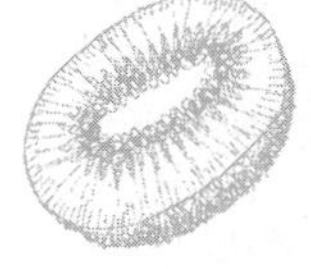

Kiwifruit. New Zealanders changed the name "Chinese Gooseberry" to "kiwifruit" as a salute to their native kiwi bird when they started marketing this fruit to the United States. Grown in both New Zealand and California, kiwifruit are lemon-sized or smaller, oval, with fuzzy, brown exteriors and green interiors sprinkled with a ring of black seeds. They are available year round. When ripe, they yield to gentle pressure. The firmer the fruit, the tarter the flavor. One kiwifruit provides more vitamin C than an orange. Kiwifruit can be peeled, sliced, and eaten, or cut in half and scooped out of its skin with a spoon, but it should not be cooked; heat destroys its color and texture. This fruit is commonly added to fruit salads, but kiwifruit cannot be used in the preparation of gelatin-based foods because an enzyme in kiwifruit, actinidin (a potent proteinase), prevents gelling (19).

Kumquats. These small, orange-type fruits are native to China and are eaten whole without peeling. The Nagami and Meiwa are the two varieties commonly sold in the United States. They are most often found preserved in syrup in the specialty food section of the supermarket.

Lemons. Eureka and Hishon lemons are the two main commercial varieties. Other varieties include Avon, Bearss, Harney, and Villafranco. The Meyer lemon is a cross between a lemon and a mandarin and is prized for its floral, almost sweet flavor. Size is not related to juiciness. Lemons with thin skins and those that are heavy for their size yield more juice than their thicker-skinned, lighter counterparts. Storing lemons in water will produce twice the juice when squeezed.

Limes. Sweet limes are available, but only the acid varieties such as the Tahiti (Persian, Bearss, Idemore, and Pond) and Mexican types are grown in the United States. Key limes differ from the regular limes sold in the supermarket in that they are smaller—closer to cherry tomato size—juicier, have a more potent lime flavor, and are usually more difficult to find and so costlier. When buying, select limes that are heavy for their size and have a thin, glossy skin.

Mandarins. This category of orange-derived citrus fruit includes several familiar fruits, all of which are easily peeled. The most popular are tangerine, tangelos (a tangerine-grapefruit hybrid), and tangors (a tangerine-orange hybrid). Tangerine varieties include the Dancy and Robinson. A common tangelo variety is the Minneola followed by the Nova, Early K, Orlando, and Sampson. The most popular tangor variety is the Temple. Choose mandarins that are firm, full-colored, and covered with a thin, almost oily-feeling skin.

Mangoes. The flavor of the golden-orange flesh of this tropical fruit has been compared to a cross between that of a peach and an apricot. Over 1,000 varieties are available; small varieties include the Haden, Van Dyke, and Atalufo; medium varieties include the Francisque and Tommy Atkins; and the large varieties include the Keitt and Kent. The latter two varieties will stay green even when ripe, but most of the others will pass through a red phase and then turn yellow. Select mangoes that have very little green left and show some red, but are predominantly orange to yellow. The fruit should yield slightly to gentle pressure. Another indication of ripeness is a distinctive mango odor coming from the fruit, especially at the stem end.

Melons. Melons are one of the most difficult fruits to select because of their hard outer rind and subtle color changes. There is a wide assortment of melons and each has its own selection nuances (see Color Figure 11). Melons fall into one of two categories: muskmelons, which include most melons, and gourds, such as the watermelon. Most varieties are sold slightly underripe and will have to ripen a few days at room temperature. When ripe, the stem end becomes slightly indented and yields to pressure. Another indication of ripeness is a distinctive melon odor coming from the fruit, especially at the stem end. Overripe melons will have water-soaked spots, and a rattling sound will occur when the melon is shaken. Selection hints for the different melon varieties follow.

Cantaloupe. Cantaloupe varieties include Topscore, Ambrosia, and Saticoy. Look for a cantaloupe with a rough, rope-like netting over a golden-colored rind. Other signs of ripeness include cracks near the stem end, a cantaloupe aroma, and a springy response to firm pressure.

Honeydew. Ripe honeydews have a waxy-feeling, creamy-white rind and honeydew aroma. For best flavor, choose one weighing at least 5 pounds.

Persian. These melons look like big cantaloupes, but the flesh is firmer, smoother in texture, and a deeper orange. Selection tips are similar to those for a cantaloupe except that more of the rind should show through the webbing.

Watermelon. The condition of the stem and the hollow sound when thumped are the two indicators most often used when selecting a watermelon. Do not try to evaluate a watermelon's ripeness by its green color, but rather look for a creamy or yellowish underside. A dull, smooth rind with a symmetrical shape are signs of a good watermelon. Several varieties found in the market include Calsweet, Triplesweet, and Orchid Sweet (yellow flesh). Figure 18-6 shows how to remove the seeds from a whole watermelon.

FIGURE 18-6 Slicing watermelon.

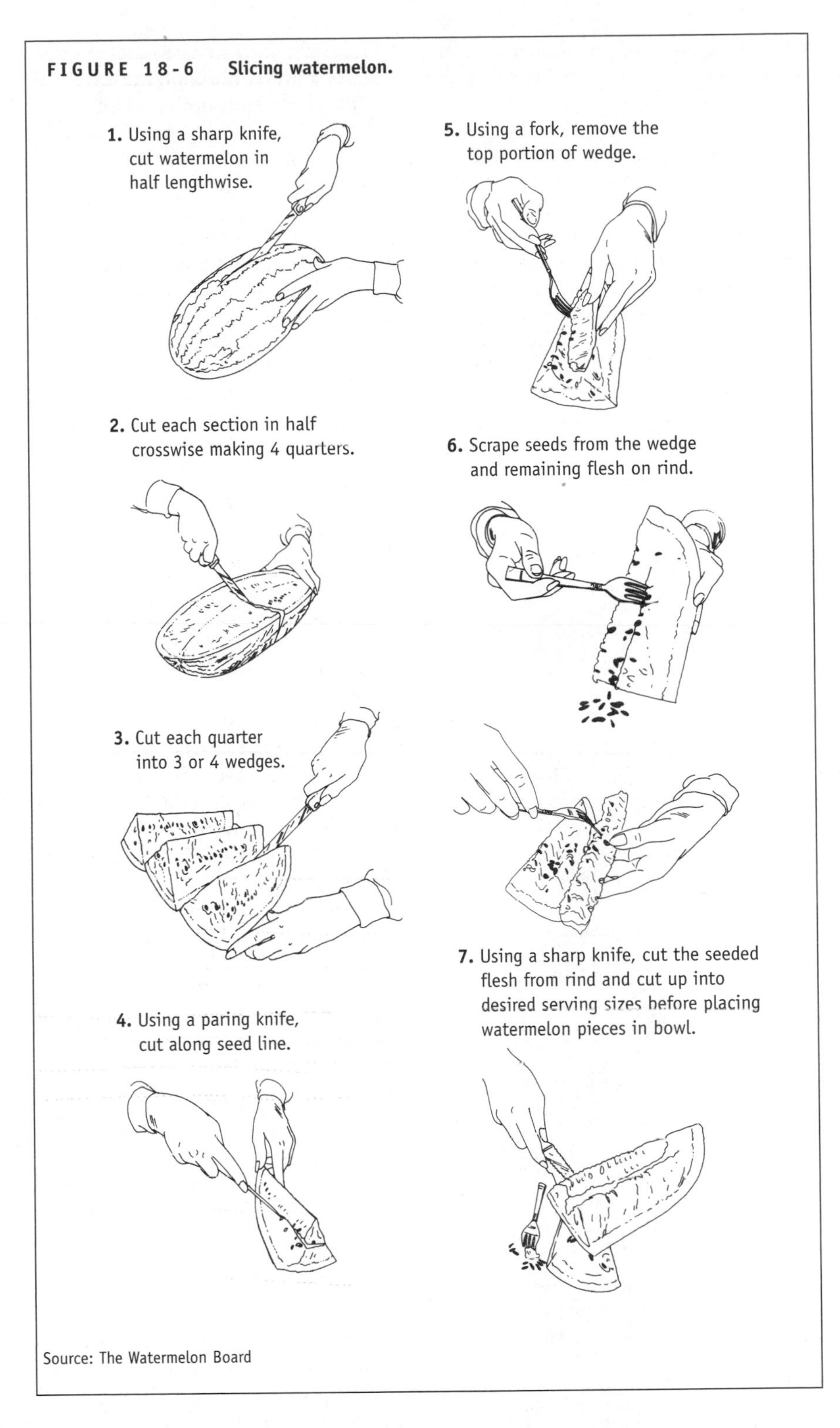

Source: The Watermelon Board

Nectarines. Nectarines have been consumed for over 2,000 years and may soon exceed peaches in popularity. Many children prefer them because they have no fuzz. Almost the entire U.S. supply is grown in California, although some are imported from New Zealand and Chile. The favorite varieties include Firebrite, Red Diamond, Flavortop, Summer Grand, Flamekist, and Red Gold. Select nectarines that yield to soft pressure and that have an orange-yellow background color between the red hues.

Olives. Although olives are botanically considered a fruit, they are often treated as vegetables. The trees on which they grow can live 600 years or more and have one of the longest cultivation histories of any tree in the world (9). Unripe olives are green, and progressively darken the longer they are on the tree. By the time they turn black naturally, the abundance of olive oil they have produced makes them too ripe for consumption. The exception is the dried Greek olive, or salt-cured olive, which is treated with salt to remove the water and looks shriveled and oily. "Black" olives are actually picked green and processed in a curing solution that turns them black, removes their bitterness, and gives them their characteristic olive flavor.

Green olives are fermented rather than subjected to the curing solution used to make black olives.

In North America, the three main types of olives are the California-style black ripe olive, the green, Spanish-style olive, and specialty olives such as Nicoise (neh-swahz), Kalamata (also spelled Calamata), Sicilian, and dried Greek. Most olives are sorted according to size and stored in cans or jars of brine. The variety of olive tree determines if the olive is small, medium, large, extra large, jumbo, colossal, or super colossal (Figure 18-7).

FIGURE 18-7 Olive sizes vary, depending on the variety of olive tree.

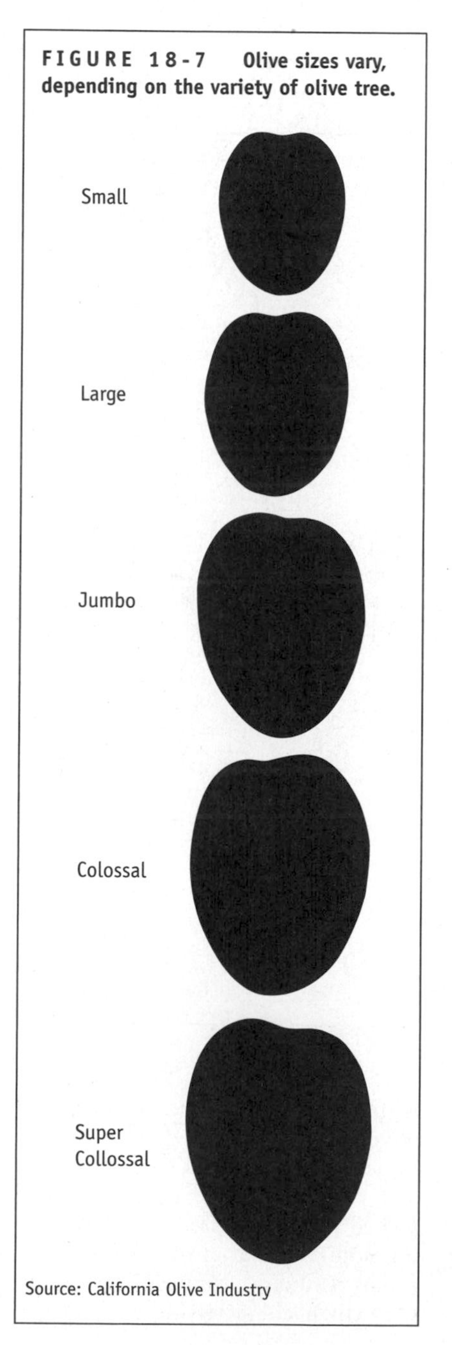

Source: California Olive Industry

Oranges. Oranges, another popular fruit, are grown primarily in Florida and California. Even though the same varieties are grown in each state, the juiciness, color, and texture of their oranges differ because of variations in soil and climate. Thin-skinned, very juicy Florida oranges provide most commercial orange juice. Varieties include the Hamlin, Pineapple, Florida Valencia, and Pope Summer. California oranges are known for their fresh eating qualities and include the Navel and California Valencia varieties. Since California Valencias are thin-skinned, they may also be used for juice. Sour oranges, including Seville and Chinotto, are sometimes sold in specialty markets. Blood oranges with their distinctive deep-red-colored flesh are a wintertime specialty.

The sweetest, juiciest oranges are those picked at the peak of their season, and ideally this is when they come to market. Oranges that feel firm and heavy for their size are the best choice for juiciness. Different varieties are grown throughout the year, and some varieties, inconveniently for the grower, turn orange in the winter before they are ripe. Warmer weather then turns them back to green, so some growers in California expose them to ethylene gas—a gas naturally produced by ripening fruits—to increase the rate of ripening and return their color to orange. In Florida, dye is often used on oranges that do not respond to ethylene.

Papayas. This tropical fruit comes in two major varieties, Hawaiian Kapoho and Sunrise. Ripe papayas are yellow-orange, soft, and prone to damage, so they are shipped to the market in varying degrees of ripeness. The green papayas, which are one-quarter yellow, will ripen in five to seven days; the half-green, half-yellow papayas will be ready in two to four days; and the papayas that are three quarters yellow-orange are at the ideal stage of ripeness. If they are already yellow-orange, they may be refrigerated and kept up to a week. In addition to the color factor, the skin of the papaya should be smooth, unblemished, and free of any dark or mushy spots.

Peaches. Peaches are a very popular fruit. Hundreds of varieties are available, but the newer varieties of yellow freestones tend to dominate the market. They are larger, firmer, and more acidic than the older varieties. Clingstone peaches are still available fresh on the market, but most are used in commercial canning. The flesh of freestone peaches does not stick to the pit, whereas it does in the clingstone variety.

Choose peaches that are already ripe, preferably tree-ripened, and ready to eat. Peaches do not acquire additional sweetness after being picked, although they may become softer and juicier (40). Look for a yellowish or creamy background color. The amount of red is not always a good indication of ripeness, because some of the best peaches are pale-colored. Ripe peaches should yield to gentle pressure. Avoid those that are too firm, have any green background, or are bruised.

Pears. Pears are generally from either the European or Asian type. European pears are juicy and come in four common varieties: Bartletts, the most common, arrive in late summer and are followed by three winter varieties; d'Anjou, Bosc, and Comice. Asian pears differ from European pears in that they vary in color from yellow to yellow-green and brownish-red, and they have a blander flavor and crispier texture that is closer to an apple. There are more than ten varieties of Asian pears, but the top three are Twentieth Century, Shinseiki, and New Kikusui. Skin color is a reliable indicator when selecting Asian pears, which, unlike European pears, are usually sold ripe. European pears can be bought underripe and ripened at room temperatures. Pears bought ripe are best kept refrigerated in plastic bags. Pears are soft-skinned, so occasional surface blemishes should be expected.

Persimmons. Although it is originally from Asia, this bright orange-red fruit is now supplied to the United States primarily from California. Commodore Perry introduced the fruit to North America in 1856 when he sent persimmon seeds from Japan to a friend in the United States (25). Pointy Hachiya persimmons account for 90 percent of the U.S. market. Fuyu is the other main variety. Fully ripe persimmons are slightly transparent, bright orange, and very soft; so soft, in fact, that they are rarely sold ripe on the market, but must be allowed to ripen after purchase.

Pineapple. Rather than a single fruit, a pineapple is really a cluster of fruits of the Ananas tree. The pineapple is native to South America; it was discovered on the island of Guadeloupe by Columbus (32), and got its name from its similarity to a pine cone (18). It is the second most popular tropical fruit after the banana. Unlike bananas, pineapples do not ripen after being picked, and even though the outside color will change from green to yellow and the inside will get juicier, the sweetness does not increase. The best pineapples come fresh off the field after having been allowed to ripen to maturity. Most of the pineapples sold on the market, however, were cut while still green.

Color is not always a reliable indicator of a ripe pineapple. The best way to choose a pineapple from the market is by touch and smell. It should be plump, give slightly to pressure, and have a fresh, sweet, pineapple scent and be topped with fresh green leaves. Avoid fruit with soft spots, brown discoloration, a decaying base, or a fermenting odor. Figure 18-8 shows several techniques for cutting up a pineapple. Like kiwifruit, the enzymes

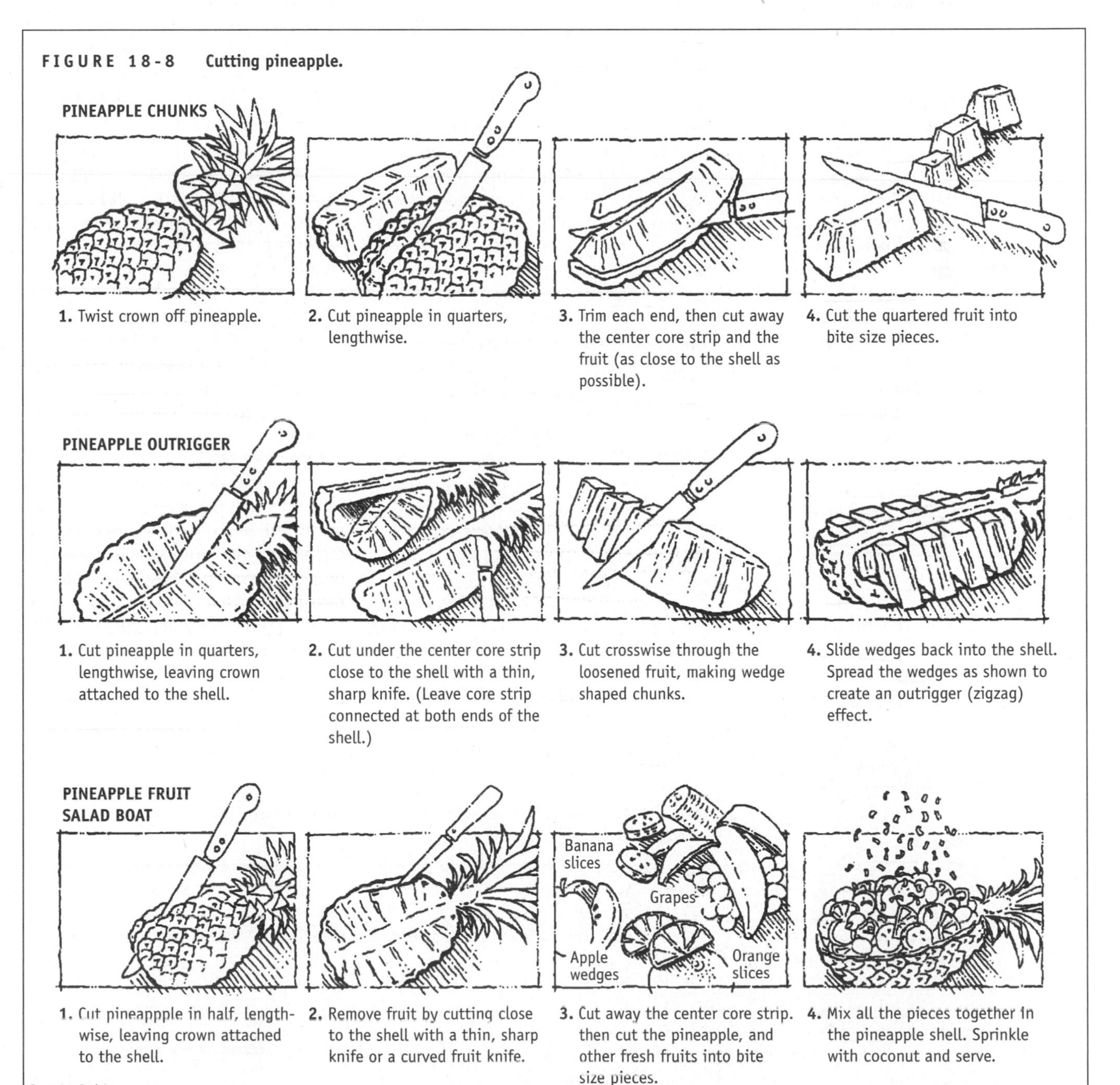

FIGURE 18-8 Cutting pineapple.

Source: Dole

in raw pineapple will prevent gelatin from gelling.

Plums. Plums are available in two types: European plums (blue or purple, yellow, and green) and Japanese plums (yellow or red). Plums, like peaches, come in freestone and clingstone varieties; European plums are usually freestone. Some of the best-tasting varieties of Japanese plums include the Santa Rosa, Laroda, and Elephant Heart. European plums are firmer, milder in taste, and are often dried for prunes. Picked when still immature, plums continue to ripen and should be chosen when they give to gentle pressure and have reached full color.

Pomegranates. The name for these ruby-red, leather-skinned fruits filled with pulp-laden seeds came from the French word for "seeded apple." Choose the largest pomegranates available, since the larger the fruit, the more pulp will be surrounding each seed. Make sure the rind is fresh and not dried out. The bright red seeds may be eaten as is or they may be carefully juiced to extract the tart, bright red juice.

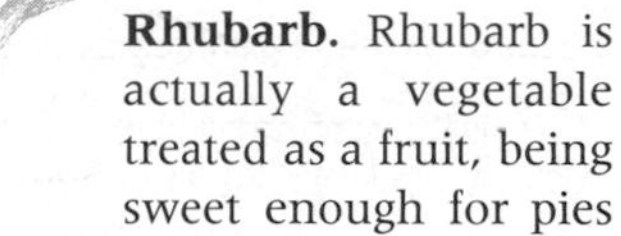

Rhubarb. Rhubarb is actually a vegetable treated as a fruit, being sweet enough for pies once sugar is added. Only the stem of the plant is eaten; the leaves are so high in oxalic acid that they are toxic if eaten in large amounts. Field-grown rhubarb is larger, stringier, and tarter than greenhouse rhubarb. Select rhubarb stalks that are tender, juicy, pink to red in color, and free of fibrous development.

Tangerines—see Mandarins.

Tropical Fruits. As world transportation and marketing outlets have improved, more tropical fruits have found their way into the market (Color Figure 13). Increased consumer awareness of these fruits has led to higher demand for coconut, guava, mango, and papaya. Other tropical fruits gaining in popularity include the atemoya, breadfruit, carambola (star fruit), cherimoya, feijoas, granadilla, jaboticaba, loquat, lychee, mamey, mangosteen, passion fruit, papaw, plumcot, quince, sapodilla, sapote, and tamarind.

Processed Fruits

Not so long ago, fresh fruit was available only in the summer, but refrigerated transportation now makes it possible to import fruits from other countries where growing seasons differ from North America. Canning, freezing, and drying are other ways of making fruit available all year. Home-freezing and home-drying are on the upswing. Conversely, home-canning knowledge is not as widespread as it used to be, but commercial canning remains responsible for providing the largest portion of processed fruits now reaching the market.

Canned. Commercially, more fruits are canned than are frozen or dried. The most commonly canned fruits are applesauce, peaches, pineapple, fruit cocktail, cranberries and cranberry products, and pears (55). The advantages of canned fruits are their convenience, availability, and variety of forms: whole, half, sliced, chunks, crushed, sauce, or juice. Canned fruits have been cooked, which alters their taste and texture and depletes the water-soluble vitamin content if the cooking juices are not used.

Canning Liquids. According to the Code of Federal Regulations, fruits are canned in either their own juice (no sugar added), light syrup (some sugar added, 10 to 14 percent density in fruit cocktail), or heavy syrup (heavier sugar concentration, 18 to 22 percent density in fruit cocktail) (51). The different packing liquids vary in their influence on the flavor and calorie (kcal) content of the fruit (Figure 18-9).

Grading of Canned Fruit. Not all canned fruits are of the same quality. Voluntary USDA grading places the best fruit under U.S. Grade A or Fancy, with less perfect fruit U.S. Grade B or Choice. Choice canned fruit is usually less expensive, and the fruit may be in smaller pieces, which makes it ideal for gelatin salads and fruit mixtures. U.S. Grade C or Standard covers fruits that have uneven pieces and some blemishes, making them best for jams, sauces, and other processed fruit products (49).

Frozen. Some of the most commonly frozen fruits are cherries, strawberries, blueberries, sliced peaches, red and black raspberries, boysenberries, and loganberries. Fruits are simply frozen fresh with or without the addition of sugar or other sweeteners. Freezing a fruit retains its color and taste, but its texture decreases in quality, because the cell membranes have a tendency to rupture as the ice crystals expand during freezing. Thawed fruits are softer and mushier than their fresh counterparts, but still can usually be used the same way as fresh fruit. When used for pies or other desserts, the fruit is usually partially defrosted in the refrigerator or in the microwave and all or some of the juice is drained. The recipe can be altered for sugar content if the fruit has been sweetened.

Signs of Refrozen Fruit. Packages or bags that show signs of juice stains or have a heavy frost clinging to the outside

FIGURE 18-9 Calorie content of canned fruit in water, light syrup, and heavy syrup.

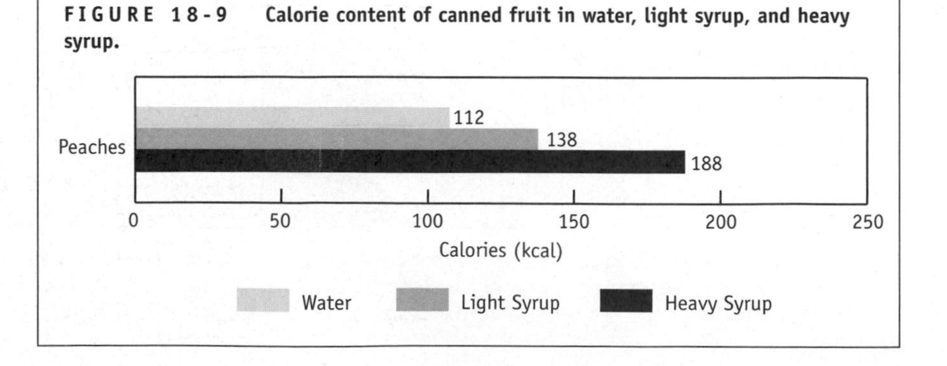

should not be purchased, because these are signs that the fruit has been thawed and refrozen. Refrozen fruit is flaccid and less flavorful. A temperature of 0°F (−18°C) should be maintained at all times during storage and transport.

Dried Fruits

Drying has been used as a method of preserving fruit for thousands of years. Commonly dried fruits, in descending order of popularity, include raisins, processed fruit snacks, prunes, and apricots (14). Dried apple, banana, and pear slices are also popular.

Drying makes these fruits available year round. Dried fruits are often packed in small containers for easy transport in lunch bags or boxes, or by backpackers, campers, and seafarers. These dried fruits contain as much as 70 percent carbohydrates and, with the flavor and nutrients more concentrated, are an alternative to candy or other less nutrient-dense foods. Generally, dried fruit is a good buy if the cost does not exceed five times the price per pound for fresh fruit.

Most fruits are 85 percent water, but drying lowers that amount to below 30 percent. It takes about 5 pounds of grapes to produce 1 pound of raisins. The removal of water during drying increases resistance to microbial spoilage. Drying is also responsible for the loss of volatile substances in fruit, softening of the cellulose, and resulting in a sweeter-tasting fruit.

Rehydrating Dried Fruit. Dried fruits can be rehydrated to some degree by adding ½ cup of water or other liquid, such as juice or liquor as called for in some recipes, for every cup of dried fruit and microwaving on medium high (70 percent) for two minutes.

Fruit Juices

Fruit juices come to market in cans, bottles, and cartons. They arrive fresh, as frozen concentrates, or in powdered forms that often contain added sugar. Concentrated fruit juice has had about three-fourths of its water removed. This may alter flavor somewhat, but the concentration in volume reduces shipping and handling costs.

The juices most commonly sold in the supermarket are apple juice/cider, and orange, grape, cranberry, grapefruit, and prune juices (38). There has been an increase lately in the number of tropical juice mixes on the market. The labels on these mixes can be misleading, because the product can contain anywhere from 10 to 100 percent actual juice. The ratio of fruit juice to water determines the standard for fruit beverages under federal guidelines (Table 18-3). Other ingredients such as vitamins (vitamin C is most prevalent), preservatives, natural or synthetic flavorings, sweeteners, and colorings are frequently added to the juice (see Chapter 29).

Preparation of Fruits

Enzymatic Browning

Certain kinds of fruit, when peeled or sliced, are susceptible to enzymatic browning. Inhibiting enzyme activity can be accomplished by denaturing enzymes, adding acid, lowering the storage temperature, and/or blocking exposure to oxygen through the use of coatings or antioxidants. Each of these methods of inhibiting enzymatic browning is now discussed.

Denaturing the Enzymes. Blanching foods by dipping them briefly in boiling water destroys the enzymes through denaturation. Fruits are not generally blanched, however, because heating already ripe fruits makes them lose texture and flavor.

Acid pH. Polyphenol oxidase enzyme activity is inhibited in the presence of acid. The optimal pH for this enzyme is 7.0, so the high acidic content of fruits such as oranges, lemons, and limes prevents enzymatic browning. Less acidic fruits such as peaches, apples, and bananas can be protected from enzymatic browning by coating them with lemon, lime, or orange juices, or with solutions of cream of tartar or the citric and acetic acids found in vinegar.

TABLE 18-3
Fruit Juice Beverage Names Depend on the Percentage of Actual Juice

Beverage Name	*Fruit Juice (%)*
Juice	Not less than 100
Juice drink	Not less than 50
Nectar	Not less than 30–40
Ade	Not less than 25
Drink	Not less than 10

Cold Temperatures. Cold temperatures reduce the rate of enzyme activity and thereby slow, but never completely inhibit, enzymatic browning (30).

Coating Fruits with Sugar or Water. Cut fruits can be protected from exposure to oxygen by covering them with a light layer of sugar or syrup (33). Submerging fruits in water also blocks oxygen from contacting their surfaces.

Antioxidants. Antioxidants such as ascorbic acid (vitamin C) and sulfur compounds prevent enzymatic browning by using up available oxygen (22). Ascorbic acid and sulfur solutions are available commercially for this purpose.

HOW & WHY? ?????????

Why are some raisins brown and some "golden"? The difference between light- and dark-colored raisins is that the former are often treated with sulfur dioxide, which prevents enzymatic browning (12).

Many dried fruits are treated in a similar manner. However, sulfur dioxide can cause allergic reactions in some people, so its use in salad bars to prevent browning was banned several years ago by the FDA (33). Pineapple juice, which is naturally high in sulfur compounds, can be helpful in inhibiting enzymatic browning when added to fruit salads or compotes.

Changes During Heating

Color. The pigments coloring fruit are the same as those discussed in Chapter 17 on vegetables: carotenoids, chlorophylls, and flavonoids (anthocyanins,

KEY TERMS

Aromatic compound A compound that has a chemical configuration of a hexagon.

■ ■ ■ ■

Essential oil An oily substance that is volatile (easily vaporized), with 100 times the flavoring power of the material from which it originated.

anthoxanthins, and betalains). Fruit becomes more brilliant in color during ripening as chlorophyll breaks down and exposes the underlying pigments. Subsequent color changes in fruits are usually due to a change in pH or a reaction with a metal salt and/or ethylene gas.

pH. Heating fruit may change the pH, which affects the color of some pigments. Sometimes a harmless blue discoloration is seen in baked products surrounding added ingredients such as cherries, cranberries, and walnuts, caused by the interaction of anthocyanins in these foods with the alkaline baking powder or soda in the flour mixture. To prevent this blue color, acid in the form of buttermilk or sour cream can be used to substitute for milk in these baked good products, and the walnuts can be roasted (about 10 minutes at 350°F/177°C), which removes the anthocyanin-containing skins (15).

Metal Salts. Canned fruits or vegetables contain acids, which often react with the tin lining to produce metal salts that change the color of pigments, particularly the red-blue anthocyanins. As a result, canned foods and juices containing anthocyanins are often stored in specially treated cans or glass to prevent food discoloration. Red, purple, or blue juices can also turn bluer when pineapple juice is added if iron has been introduced into the canning process from the pineapple processing equipment.

Ethylene Gas. One of the ways to facilitate the ripening of oranges and tomatoes to their optimum color is by exposing them to ethylene gas. This stimulates respiration, thereby accelerating the ripening process, but it may mean the fruit deteriorates more quickly (35).

Texture. Fruit is often served raw because the texture is usually more desirable than that of cooked fruit. Heat drastically softens fruit, sometimes to the point of mushiness (e.g., strawberries). During heating, several changes occur that contribute to this softeness:

- The conversion of the fruit's protopectin to pectin
- The degradation of cellulose and hemicellulose
- The denaturation of the cell membrane proteins

This last change influences membrane function and the ability of the fruit to maintain its turgor.

Osmosis. During heating, the fruit's osmotic system of selective permeability is replaced with simple diffusion, contributing to the loss of shape in heated fruits. Normally, the cell walls of raw fruits have semipermeable membranes, which allow water, but not solutes such as sugars or minerals, to pass through the membrane. The water moves from a low solute concentration to a higher one, until the solution concentration is equal on both sides of the membrane. This principle is why water is often sprayed on fruits displayed in supermarket stands to keep them crisp and turgid, because the higher solute concentration in the cells draws the water into the fruit, causing cell enlargement and swelling. Fruits continue to look full and fresh when water is sprayed on them in this manner. This is also the reason why lemons stored in water will produce about twice the juice. Sprinkling sugar on fruit produces the opposite effect, causing fruit to become syrupy as water passes out from the cells to the surface where the solute concentration is higher (due to the sugar). This can backfire, however, if sugar is left on fruit too long, producing a soft, mushy texture.

Heating destroys the membrane's ability to prevent the loss or uptake of solutes through the cell wall. Solutes can then freely pass across the membrane until solute concentrations are equal on both sides.

HOW & WHY? ?????????

Why are canned fruits softer and shinier than fresh fruits? Heating fruit in water causes the fruit's natural sugars to move out into the water and water to move into the fruit, resulting in the loss of shape. Fruits become more translucent and shiny when heated as water leaves the cells and the air between cells is expelled.

Sugar may be added to fruit before it is cooked in water to replace the lost sweetness. This also helps the fruit to retain its shape and firmness. For the same reason, fruits are often heated in a syrupy solution of greater than 15 percent sugar. The sugar enters the cells as the water leaves. Sugar further inhibits loss of texture by interfering with the conversion of protopectin to pectin. Calcium salts and acids also help to firm fruit, which is why calcium is added to many canned fruits and vegetables, especially pears and tomatoes (5) (Chemist's Corner 18-2).

Flavor. Sugars and acids contribute to the sweet and sour taste of fruits. The flavors are derived from the chemical configuration of compounds such as sugars, acids, phenolic and **aromatic compounds**, and **essential oils**. The flavor substances and volatile compounds that contribute to aroma can be lost during preparation,

Chemist's Corner 18-2

Calcium in Canned Fruits

Pectin often exists in nature as salts of calcium or magnesium. It is able to do so because of its abundance of carboxyl groups (-COOH). The bridges formed between the free carboxyl groups and the intracellular ions of calcium and magnesium strengthen the tissue of the plant and resist degradation during the heating of canning (36).

which is why fruits are either served raw or heated only for the minimum amount of time.

Zest. The peels of oranges, lemons, and limes serve as reservoirs for potent aromatic oils. The colorful outer layer of these citrus fruits, known as the "zest," often contains more flavor than the fruit's juice (47). The zest can be removed one of three ways: grating with the smallest holes of a grater, a zester tool, or a sharp paring knife (Figure 18-10). Various forms of zest—flat strips, julienned slivers, or finely grated—can be added to a number of dishes. The amount of grated zest and unstrained juice obtained from selected citrus fruits is shown in Table 18-4. Large zest pieces can be strained out of a dish after they have contributed their flavors during cooking, or they can be left in the dish if blanched for two minutes to make them soft and palatable. Finely grated zest is often added to the batter of dessert dishes, where it is small enough to be evenly distributed.

TABLE 18-4

Amount of Zest and Juice Obtained from Citrus

Citrus type	*Grated zest, lightly packed*	*Juice, unstrained*
1 lime	1–1½ tsp	about 3 tbs
1 lemon	½–1 tbs	¼–⅓ C
1 orange	about 1 tbs	about ½ C
1 grapefruit	about 1½ tbs	about 1 C

Dry-Heat Preparation

Fruits add variety to meals, and preparation techniques can make overripe fruits or otherwise unpalatable ones such as rhubarb and green apples appetizing. Before preparation, all fruits must be washed to remove pesticide residues, soil, and microorganisms. Since most fruit spoils faster if it is washed before refrigeration, it is best to rinse it just before use. Cooked fruit, with the exception of applesauce and fruit spreads (jams, jellies, etc.), should be served immediately after it has been prepared.

Baking. Whole apples and some varieties of pears are good for baking. The apples may be left whole, or cored, peeled, and placed in a baking dish. A sauce can be made from brown sugar, water, butter, cinnamon, and nutmeg boiled together and poured over the apples, which are then baked uncovered for approximately 1 hour at 350°F (177°C) with occasional basting. Bananas, plums, peaches, and rhubarb can also be baked successfully. Fruit can be added to many baked goods to enhance their flavor, color, and texture. Cakes, pies, cookies, muffins, scones, and cobblers, as well as pancakes and waffles, all profit from partnerships with fruit.

Broiling. A few fruits, such as grapefruit halves, pineapple halves, and bananas, can be broiled by placing them on a heavy-duty pan, which is put in the oven under the broiler element until the fruit is golden brown. Often a bit of melted butter and sugar is drizzled on the fruit before broiling or browning may not occur.

Frying/Sautéing. A small amount of fat in a frying pan can be used to sauté slices of apple, banana, cherry, and pineapple. Sometimes the fruit is sprinkled with white or brown sugar. Sautéing contributes to a slight increase in the fruit's caloric (kcal) and fat content. Sautéed fruit may be served alongside roasted meats or as a dessert course.

Grater: The smallest holes on a grater can be used to collect zest, which is freed with the use of a pastry brush. Avoid scraping so hard that white pith ends up in the zest.

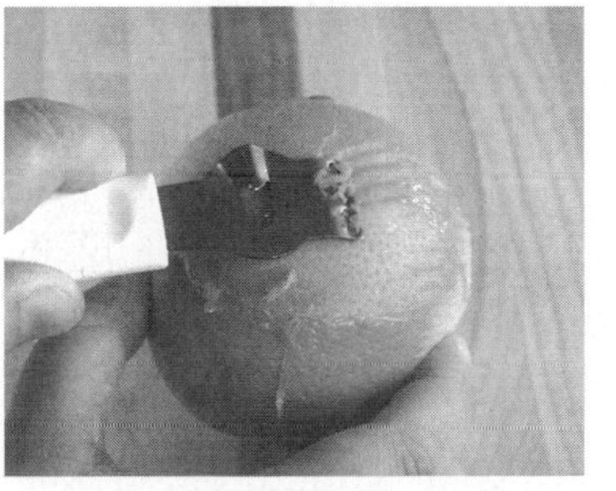

Five-hole zester: The shallow blade of a zester does not go any deeper than the zest, making it easy to collect the flavorful and fragrant peelings.

Source: M. Stevens, *Fine Cooking* 25:76, 1998.

FIGURE 18-10 Scraping the zest off citrus fruits.

Moist-Heat Preparation

Stewing/Poaching. Fresh fruits can be stewed, or better yet, poached. For stewing, fruits are heated either without water (for fruit with a high water content), with very little water or juice, or with added sugar or syrup. Stewing fruits in a sugar syrup helps to maintain their shape.

For poaching, fresh fruit such as pears are gently lowered into boiling water or syrup, the heat is reduced to poaching temperatures, and the fruit is cooked until barely tender. This process does not take long, and the fruit should be removed from the heat as soon as cooking is complete. Very soft fruits are best drained immediately to prevent overcooking, while firmer fruits such as apples and pears are best left to cool in the poaching syrup in order to soak up more flavor and prevent wrinkling.

Applesauce Preparation. Applesauce is relatively simple to prepare from cored and chopped cooking apples. The peels may be removed for a smooth sauce, or left on to add a touch of color and increase the fiber of the applesauce. A pound of apples, about 1 to 1½ cups of water or cider, and a cinnamon stick (optional) is added to a saucepan. The pan is covered, and the mixture is brought to a boil and then reduced to a simmer and cooked for about 10 minutes, until tender. The pan is removed from the heat and the cinnamon stick is discarded. The apples are then pulverized with a potato masher or fork until smooth. Sugar may be added (½ cup per pound of apples). The applesauce should be served immediately or covered and refrigerated. Chunky applesauce is made by adding the sugar before heating and smashing the fruit only slightly.

Preparing Dried Fruit. Dried fruit is usually soaked in water and then simmered in a covered pan, but some dried fruits, such as prunes, have a higher water content and do not need to be soaked before cooking. They may simply be covered with cold water and brought to a simmer. The natural sweetness of dried fruits usually makes it unnecessary to add sugar, but some fluid may be necessary to rehydrate the dried fruits. To add more flavor, some recipes call for juice or liquors instead of water.

Fruit Spreads

Types of Fruit Spreads. The most commonly consumed fruit spreads are fruit preserves, jams, conserves, jellies, marmalades, and butters. Except for bananas and melons, most fruits lend themselves well to these methods of preservation. The type of fruit spread is determined by the amount of added sugar and the form of the fruit-juice, mashed, sieved, whole, halved, or chunks:

- *Preserves.* Made from whole fruit, halves, or chunks.
- *Jams.* Made from ground or mashed whole cooked fruit.
- *Conserves.* Made from a mixture of fruits (usually citrus or at least with some citrus) to which nuts and raisins, but no sugar, are generally added.
- *Jelly.* Made from the juice of cooked fruit, with added sugar and pectin. For a clear jelly, the juice is first strained.
- *Marmalades.* Contain juice with thin slices of fruit and rind, especially citrus fruits.
- *Butters.* These thick and smooth fruit preserves are made from sieved, long-cooked fruit, are usually less sweet than jams and jellies, and do not keep as well as other fruit spread forms.

Ingredients. Fruit spreads gel when pectin molecules bond to each other and form a mesh-like network. Pectin (1 percent), sugar (about 60 percent), and acid (pH 2.8 to 3.4) make the formation of the network possible.

Pectin. Pectin is extracted from fruits to assist in the gelling of fruit spreads (see Pectic Substances under Composition of Fruits). Although pectin is a natural fruit component found in varying degrees in different fruits, additional amounts may need to be added in the manufacture of fruit spreads. At least a 0.5 to 1 percent concentration of pectin needs to be present for fruit to gel, and if the fruit does not contain sufficient pectin, commercial pectin can be added in either powder or liquid form (Chemist's Corner 18-3).

Sugar. Sugar is a natural preservative. Most microorganisms need water to survive and when sugar osmotically pulls the water from their cell(s), they die. In addition to acting as a natural preservative, sugar helps to maintain the firmness of fruits, contributes to their flavor, and makes gelling possible. It is, however, high in calories (kcal). Most jams and jellies average 50 calories (kcal) per tablespoon, but this can be reduced by almost half (30 calories (kcal)/tablespoon) by using alternative sweeteners and a unique type of pectin that relies on calcium ions to create a net-like structure (34). Calcium ions contribute to the creation of crosslinking bridges, which help gel fruit spreads, salad dressings, and diet jellies made without sugar. Also available are calorie-free imitation fruit spreads (1). Another option for lower-sugar spreads is to purchase puréed baby fruits or to blend frozen fruits. Although all fruit spreads should be refrigerated after opening, it is doubly necessary to keep low- or no-sugar spreads refrigerated because there is little or no sugar to prevent microbial growth.

Acid. Acid provides both flavor and the hydrogen ions (H^+) needed to neutralize the negative charges that cause pectin molecules in water to repel each other. The pH range that allows pectin molecules to bind is between 2.8 and 3.4, with an optimum pH of 3.2. All fruit spreads should taste somewhat acidic. In fact, without acid, fruit spreads would be lacking in both flavor and gel strength. Some commercial pectins contain added acid. If the fruit mix is too sweet, 1 tablespoon of lemon juice, 1 tablespoon of vinegar, or ⅛ tablespoon of citric acid can be added for every 8-ounce cup of juice (2).

Fruit. Because of their varying amounts of pectin and acid, some fruits are more suited for making fruit spreads than others. Apples, citrus fruits, cranberries, gooseberries, grapes, loganberries, plums, quinces, and red currants contain sufficient pectin and acid for gelling. Fruits rich in pectin but low in acid are sweet apples (Red and Golden Delicious, Criterion, Gala), unripe figs, and pears. Apricots and strawberries are low in pectin but rich in acid. Peaches

Chemist's Corner 18-3

Pectins and Methoxylation

Commercial pectins are sold according to their degree of methoxylation (DM). High-methoxyl pectins have a DM value of 50 to 80 percent, form gels in the presence of acid and sugar, and are most commonly used for the manufacture of fruit spreads. Low-methoxyl pectins have a 25 to 50 percent DM value and will form gels in the presence of divalent cations such as calcium (39). High-methoxyl pectin makes it possible for fruit spreads to gel through coordinate bonding with Ca^{2+} ions or hydrogen bonding and hydrophobic interactions. In low-methoxyl pectin, gelation occurs due to ionic linkage via calcium bridges between two carboxyl groups belonging to two different chains in close proximity to one another (48).

are low in both pectin and acid. Adjustments for either low pectin or low acid can be made by combining one fruit with another, using commercial pectin, and/or adding lemon juice.

Preparation of Fruit Spreads. To prepare a good fruit spread, the correct balance of pectin, sugar, and acid must be attained. The fruit selected determines the amount of pectin and acid and the overall quality of the spread. Steps for preparing fruit spreads (see Figure 18-11) include:

1. Heat the fruit.
2. Add sugar and possibly pectin and acid.
3. Pour the mixture into sterile containers.
4. Seal the containers properly.

Ripe fruit, whole or cut up, is simmered in a small amount of water to liberate the pectin and juice. Approximately ¼ cup of water is used for every pound of fruit. Too much water will dilute the pectin and prevent gelation. If the fruit is extremely watery, simply crushing it may provide sufficient liquid. Boiling times vary from 10 to 15 minutes for soft fruit to about 20 minutes for hard, sliced fruit. Uncooked or undercooked fruit will not gel, because heat is necessary for the conversion of protopectin to pectin. Sugar and pectin and/or acid is added, after which the mixture is boiled to concentrate the ingredients. Too much sugar creates a gummy product, while too little prevents gelation. Some commercial pectins may be added to the fruit before the sugar, but there will be directions to that effect on the package.

A test to determine if the mixture is done is to dip a cool metal spoon in the boiling fruit, lift it up, and see if it drops from the spoon in flat-sheet fashion. Heating the mixture beyond this stage will cause the pectin molecules to break down and prevent the mixture from gelling; also, the color may get darker because of the caramelization of sugars. When done, the mixture is poured into sterile glass containers to within ¼ inch of the rim and sealed with two-part canning lids according to container directions. Fading of the finished spread's color is prevented by storing the jars in a dark, cool place. Table 18-5 lists possible reasons for gelling problems.

FIGURE 18-11 **Preparing fruit spreads. Heat fruit, sugar, and possibly added pectin and/or acid sufficiently long enough (or it will not gel) before pouring into sterilized jars to set.**

Storage of Fruits

Many types of fruit are picked and shipped to market in an unripe state because the hardship of transportation damages delicate fruits (10). Unripe fruit can be left at room temperature in a paper bag until ripe. Most European pears, for example, are marketed unripe, so it is best to store them outside the refrigerator in a loosely closed paper bag until they give to firm pressure. Commercially, fruit can be stored under controlled-atmosphere storage or with the aid of coatings or films as described in Chapters 7 and 17.

Fruits differ in their respiration rates, and thus in their rates of ripening. A few fruits, known as climacteric, will experience an increased phase of respiratory rate right before becoming fully ripe. These fruits continue to ripen after being harvested. Other fruits and most vegetables are nonclimacteric, and although they continue to respire at the same or

TABLE 18-5
Gelling Problems and Their Causes

Problem	*Possible Causes*
Sugar crystallization (mainly found in grape jelly)	Too much sugar Not enough acid Overcooking Delay in sealing (may be prevented by permitting juice to stand overnight in cold place before making it into jelly)
Weeping	Cause not really known; may occur in jellies made from currants or cranberries high in acid
Cloudiness	Squeezing juice out of bag Starch from apples used to make jelly (avoid pressure on jelly bag when straining)
Failure to gel	Improper balance between pectin, sugar, and acid Lack of pectin or acid in fruit Overcooking
Fermented jelly or mold formation	Jelly glasses not well sterilized Jelly stored in warm, damp place Jelly not completely sealed

even lower rate after being harvested, they are best ripened fully before being harvested (Table 18-6).

Storing Fresh Fruit

Ripe fruit with a high water content is best if consumed within three days of purchase. Grapes spoil quickly, so regardless of whether or not they are ripe, they are stored, unwashed, in plastic bags in the refrigerator and washed just prior to consumption. On the other hand, cooled apples will keep for weeks.

HOW & WHY? ?????????

Why does "one rotten apple spoil the barrel"? An overripe apple stored with good apples will ruin the others by releasing ethylene gas, an agent that speeds up ripening. All bruised, dented, or otherwise damaged apples should therefore be removed before storing the other apples.

Once fruit is ripe, storage time may be increased by placing it in plastic bags punctured with air holes, and then in the refrigerator (back inside cover). An exception is bananas, which are best stored at room temperature, because refrigeration interferes with their ripening process and causes their skin to blacken.

Some storage requirements for other fruits include:

TABLE 18-6
Selected Climacteric and Nonclimacteric Fruits

Climacteric Fruits (continue to ripen after harvest)	*Nonclimacteric Fruits (best ripened before harvest)*
Apple	Blueberry
Apricot	Cherry
Avocado	Citrus fruits (grapefruit, lemon, orange)
Banana	
Breadfruit	
Cantaloupe	Grapes
Guava	Melons
Peach	Olives
Pear	Pineapple
Plum	Strawberry
Tomato	
Tropical fruits (papaya, mango, passion fruit)	

- *Cherries.* Cherries should be arranged in a single layer between paper towels, placed in a plastic bag, and refrigerated.
- *Dates.* Dates will keep for weeks at room temperature, up to a year in the refrigerator, and up to five years in the freezer.
- *Citrus fruit.* Citrus fruit can be stored at a cool room temperature (60° to 70° F/16° to 21° C) or in light refrigeration temperatures (45° to 48°F/7° to 9°C) for two to three weeks.
- *Pineapples.* Uncut ripe pineapples can be kept at room temperature for up three days.
- *Pomegranates.* Ripened pomegranates are kept at room temperature for short periods, but should be refrigerated if storage exceeds several days.
- *Guavas.* Once guavas ripen, they can be stored in the refrigerator for a day or two.

Storing Canned Fruit

Canned fruits keep their quality longer if the cans are stored in a dry place with temperatures under 70°F (21°C). Bulging, dented, leaking, or rusted cans should always be discarded. Bulging may indicate the presence of the anaerobic bacteria that causes deadly botulism (see Chapter 3).

Olives. The storage requirements for olives depends on whether they are in a can or jar. The unused portion from an open can of olives should be stored in the original brine, covered with breathable plastic wrap, and kept in the refrigerator for no more than ten days; possibly harmful toxins develop when such olives are stored in sealed or airtight containers. The brine of olives in a jar has three times the salt concentration and a much lower pH than that found in canned olives, so it does not matter if the storage container is airtight or not.

PICTORIAL SUMMARY / 18: Fruits

Throughout human history, fruits have been prized and their seeds transported wherever man has settled, resulting in the great variety of fruits available today.

CLASSIFICATION OF FRUITS

Fruits are classified according to the type of flowers from which they develop:

Simple fruits develop from one flower. They include:

- Drupes: Fruit with seeds encased in a pit, such as apricots, cherries, peaches, and plums.
- Pomes: Fruit with seeds contained in a central core, such as apples and pears.
- Citrus fruits: Oranges, grapefruits, lemons, limes, kumquats, and mandarins.

Aggregate fruits develop from several ovaries in one flower. They include blackberries, raspberries, and strawberries.

Multiple fruits develop from a cluster of several flowers. They include pineapples and figs.

PURCHASING FRUITS

Fruits are selected based on appearance, size, color, shape, uniformity, and freedom from defects. The four grades for fresh fruit are U.S. Fancy, U.S. No. 1, U.S. No. 2, and U.S. No. 3. Selecting fruits is based on grading factors as well as the variety, whether or not they are in season, and criteria that vary for each individual fruit.

Many fruits are available on the market in summer, but they can be provided throughout the year in various processed forms: canned, frozen, dried, and juice. Refrigerated transport also makes fruits available year-round from countries with different growing seasons. Grades for canned or frozen fruits are U.S. Grade A or Fancy, followed by U.S. Grade B or Choice, and U.S. Grade C or Standard.

COMPOSITION OF FRUITS

The cellular structure and the pigments of fruits are similar to what is found in vegetables. The flavors of fruits are due to combinations of sugars, acids, phenolic and aromatic compounds, and essential oils. The pH in fruits tends to be below pH 5.0. Pectin plays an important role in fruit ripening and the gelling of fruit spreads.

Nutritionally, fruits are low in calories (kcal), fat, and protein. The few fruits that are high in fat are coconut, avocado, and olives. Fruits are high in water and carbohydrates, the latter providing the majority of calories (kcal) found in fruits. Some, but not all, fruits can be excellent sources of vitamin C, beta-carotene, iron, or potassium.

PREPARATION OF FRUITS

Cooking alters the taste, texture, color, and shape of fruit, so fruit is often consumed in its raw state. Cooking fruit adds variety to meals, makes fruit like rhubarb and green apples palatable, and utilizes overripe fruits. Fruits should be prepared using a minimum of water, time, and heat.

Most fruit can be kept beyond the growing season by combining it with sugar in the following forms of fruit spreads:

Preserves: Made from whole fruit, halves or chunks.

Jams: Ground or mashed whole cooked fruit.

Conserves: Made from a mixture of fruits (usually including citrus) to which nuts and raisins, but no sugar, are added.

Jellies: Made from the juice of cooked fruit, with added sugar and pectin.

Marmalades: Juice combined with thin slices of fruit and rind, especially citrus fruits.

Butters: Thick and smooth, made from seived, long-cooked fruit; usually less sweet than jams and jellies, and more perishable than other fruit spreads.

STORAGE OF FRUITS

Most ripe fruit should be stored in the refrigerator in plastic bags punctured with air holes, with the exception of bananas, which will turn brown if refrigerated. Unripe fruit is usually left at room temperature in a paper bag until ripe.

REFERENCES

1. Altschul AM. *Low-Calorie Food Handbook.* Marcel Dekker,1993.
2. American Home Economics Association. *Handbook of Food Preparation.* American Home Economics Association, 1993.
3. Baker RA. Potential dietary benefits of citrus pectin and fiber. *Food Technology* 48(11):133–139, 1994.
4. Baker RA, and RD Cameron. Clouds of citrus juices and juice drinks. *Food Technology* 53(1):64–69, 1999.
5. Baker RA, and RG Hagenmaier. Reduction of fluid loss from grapefruit segments with wax microemulsion coatings. *Journal of Food Science* 62(4):789–792, 1997.
6. Batisse C, B Fils-Lycaon, and M Buret. Pectin changes in ripening cherry fruit. *Journal of Food Science* 59(2):389–393, 1994.
7. Berne S, and CD O'Donnell. Filtration systems and enzymes: A tangled web. *Prepared Foods* 165(9):95–96, 1996.
8. Beveridge T. Electron microscopic characterization of haze in apple juices. *Food Technology* 53(1):44–48, 1999.
9. Boskou D. *History and Characteristics of the Olive Tree.* American Oil Chemists' Society, 1996.
10. Brownleader MD. Molecular aspects of cell wall modification during fruit ripening. *Critical Reviews in Food Science and Nutrition* 39(2):149–164, 1999.
11. Cameron RG, RA Baker, and K Grohmann. Citrus tissue extracts affect juice cloud stability. *Journal of Food Science* 62(2):242–245, 1997.
12. Cannellas J. Storage conditions affect quality of raisins. *Journal of Food Science* 58(4):805–809, 1993.
13. Chen CS, and MC Wu. Kinetic models for thermal inactivation of multiple pectinerases in citrus juices. *Journal of Food Science* 63(5):747–750, 1998.
14. Cohen J. Snack foods. *Progressive Grocer* 70(7):62, 1991.
15. Corriher SO. *Cookwise: The Hows and Whys of Successful Cooking.* Morrow, 1997.
16. Cunningham E. Is a tomato a fruit and a vegetable? *Journal of the American Dietetic Association* 102(6):817, 2002.
17. Elving P, ed. *Fresh Produce.* Lane, 1987.
18. Flath RA. Pineapple. In *Tropical and Subtropical Fruits,* ed. P. Shaw, 157–183. Avi Publishing,1980.
19. Funaki J, and M Yano. Inhibiting the activity of actinidin by oryzacystatin for the application of fresh kiwifruit to gelatin-based foods. *Journal of Food Biochemistry* 19:355–365, 1996.
20. Genovese DB, MP Elustondo, and J Lozano. Color and cloud stabilization in cloudy apple juice by steam heating during crushing. *Journal of Food Science* 62(6):1171–1175, 1997.
21. Groupy P. Enzymatic browning of model solutions and apple phenolic extracts by apple polyphenoloxidase. *Journal of Food Science* 60(3):497–501, 1995.
22. Gunes G, and CY Lee. Color of minimally processed potatoes as affected by modified atmosphere packaging and antibrowning agents. *Journal of Food Science* 62(3):572–575, 1997.
23. Jackman RL, HJ Gibson, and DW Stanley. Tomato polygalacturonase extractability. *Journal of Food Biochemistry* 19:139–152, 1995.
24. Ketiku AO. Chemical composition of unripe (green) and ripe plantain (*Musa paradisiaca*). *Journal of Food Science* 19:139–152, 1973.
25. Knight R. Origin and world importance of tropical and subtropical fruit crops. In *Tropical and Subtropical Fruits,* ed. P. Shaw, 1–120. Avi Publishing, 1980.
26. Kuzminski LN. Cranberry juice and urinary tract infections: Is there a beneficial relationship? *Nutrition Reviews* 54(11):S87–S90, 1996.
27. Lanza E, and R Butrum. A critical review of food fiber analysis and data. *Journal of the American Dietetic Association* 86:732–743, 1986.
28. Lavons JA, RD Bennett, and SH Vannier. Physical/chemical nature of pectin associated with commercial orange juice cloud. *Journal of Food Science* 59(2):399–401, 1994.
29. Lee GH, et al. Ripening characteristics of light-irradiated tomatoes. *Journal of Food Science* 62(1):138–140, 1997.
30. Lozano JE, R Durdis-Biscarri, and A Ibarz-Ribas. Enzymatic browning in apple pulps. *Journal of Food Science* 59(3):564–567, 1994.
31. Lozano-de-Gonzalez PG. Enzymatic browning inhibited in fresh and dried apple rings by pineapple juice. *Journal of Food Science* 58(2):399–404, 1993.
32. Magness JR. How fruit came to America. In *The World in Your Garden.* National Geographic Society, 1959.
33. McEvily AJ, and R Iyengar. Inhibition of enzymatic browning in foods and beverages. *Critical Reviews in Food Science and Nutrition* 32(3):253–273, 1992.
34. McGee H. *On Food and Cooking: The Science and Lore of the Kitchen.* Macmillian,1997.
35. O'Connor-Shaw RE. Shelf life of minimally processed honeydew, kiwifruit, papaya, pineapple and cantaloupe. *Journal of Food Science* 59(6):1202–1206, 1994.
36. Penfield MP, and AM Campbell. *Experimental Food Science.* Academic Press,1990.
37. Pennington JT, and VL Wilkening. Final regulations for the nutrition labeling of raw fruits, vegetables, and fish. *Journal of the American Dietetic Association* 97:1299–1305, 1997.
38. Petreyak RM. Beverages. *Progressive Grocer* 70(7):45–49, 1991.
39. Raphaelides SN, A Ambatzidou, and D Petridis. Sugar composition effects on textural parameters of

peach jam. *Journal of Food Science* 61(5):942–946, 1996.

40. Robertson JA, et al. Relationship of quality characteristics of peaches (cv. Loring) to maturity. *Journal of Food Science* 57(6):1401, 1992.
41. Rommel A, R Wrolstad, and D Heatherbell. Blackberry juice and wine: Processing and storage effects of anthocyanin composition, color and appearance. *Journal of Food Science* 57(2):385–410, 1992.
42. Rouseff RL, and S Nagy. Health and nutritional benefits of citrus fruit components. *Food Technology* 48(11):15–132, 1994.
43. Shamaila M, WD Powrie, and BJ Skura. Sensory evaluation of strawberry fruit stored under modified atmosphere packaging (MAP) by quantitative descriptive analysis. *Journal of Food Science* 57(5):1168–1172, 1992.
44. Siddiq M, et al. Partial purification of polyphenol oxidase from plums (*Prunus domestica* L., cv. Stanley). *Journal of Food Biochemistry* 20:111–123, 1996.
45. Sloan A. Fruit frenzy. *Food Technology* 55(12):14, 2001.
46. Stables G, et al. Changes in vegetable and fruit consumption and awareness among U.S. adults: Results of the 1991 and 1997 5 A Day for Better Health Program surveys. *Journal of the American Dietetic Association* 102(6):809–817, 2002.
47. Stevens M. Pared or grated, citrus zest adds fresh flavor and fragrance. *Fine Cooking* 25:76, 1998.
48. Thakur BR, RK Singh, and AK Handa. Chemistry and uses of pectin. *Critical Reviews in Food Science and Nutrition* 37(1):47–73, 1997.
49. U.S. Department of Agriculture. Agricultural Marketing Service. *How to Buy Canned and Frozen Fruits.* Home and Garden Bull. no. 261. Washington D.C.: GPO, 1994.
50. U.S. Department of Agriculture. Agricultural Marketing Service. *How to Buy Fresh Fruits.* Home and Garden Bull. no. 260. Washington D.C.: GPO, 1994.
51. U.S. Department of Health and Human Services. Food and Drug Administration. *Code of Federal Regulations.* (21 CFR Part 145), 1997.
52. Vandercook CE, S Hasegawa, and VP Maier. Dates. In *Tropical and Subtropical Fruits,* ed. P. Shaw. Avi Publishing, 506–541,1980.
53. Vinson JA. The functional food properties of figs. *Cereal Foods World* 44(2):82–87, 1999.
54. Weemas C, et al. Temperature sensitivity and pressure resistance of mushroom polyphenoloxidase. *Journal of Food Science* 62(2):231–266, 1997.
55. Wold M. Miscellaneous grocery. *Progressive Grocer* 70(7):70, 1991.
56. Zhang X, and WH Flurkey. Phenoloxidases in Portabela mushrooms. *Journal of Food Science* 62(1):97–100, 1997.

WEBSITES

Fruit associations and organizations as compiled by California Rare Fruit Growers, Inc.:

www.crfg.org/related.org.html

Consumer publications by the North Carolina Cooperative Extension Office on safe food handling,drying, jams and jellies, freezing, and more:

www.ces.ncsu.edu/depts/foodsci/agentinfo/fruit/conspub.html

Fruit Facts are a series of publications containing information on individual fruits from avocado to white sapote:

www.crfg.org/bubs/frtfacts.html

19

Soups and Salads

The first impression of a meal is often a soup or salad. The quality of this introductory item often serves as an indication of what is to follow. Starting the meal with an excellent soup or salad is equivalent to putting your best foot forward. It is important to begin a meal with flavor and distinction, rather than place the entire burden of carrying the meal on the main course. The soup or salad serves as the curtain-raiser, and can help introduce the meal in a colorful, entertaining manner.

The versatility of soups and salads allows them to be served not only before a meal, but also during or after, or even as a meal itself. Countless combinations are possible, and this probably led to soups being some of the first meals ever prepared in a pot over an open fire.

KEY TERMS

Stock The foundational thin liquid of many soups, produced when meat, poultry, seafood, and/or their bones, or vegetables are simmered.

White stock The flavored liquid obtained by simmering the bones of beef, veal, chicken, or pork.

Brown stock The stock resulting from browning bones and/or meat prior to simmering them.

Soups

In 16th century France, specialized shops were set up among the ordinary trade shops to serve soup to the workers. A Parisian soup vendor had a Latin inscription above his door that read, "Come to me all of you whose stomachs cry out and I will restore you" (15). The "restorative" services of these soup shops eventually led to them being called "restaurants," and our English word used for the evening meal, "supper," is derived from "soup" (12).

The enormous assortment of soups now available ranges from light soups that serve as appetizers to very heavy soups that can be offered as a main dish. Most soups are served hot, but there are exceptions, the most famous one being vichyssoise (vee-shee-swahz). Many stories exist as to how it originated; one of them was that this soup was invented in 1917 by Louis Diat, chef of the Ritz Carlton in New York, who named it after his native town in France (7). Its legendary status comes from a story in which King Louis XIV of France, suspicious that people were trying to poison him, ordered that all his food be sampled by an official taster. The king's taster felt the need for his own taster, who, in turn, also had a taster, and by the time this "hot" creamed leek and potato soup got to the king, it was cold, and it has been eaten that way ever since (10).

Whatever the temperature, all soups are based on **stock**. To this foundation are added other ingredients, lending each kind of soup its own name and unique characteristics. Stocks are very basic, primarily protein and water with little or no flavor in order to serve as foundations for other added ingredients. The different types of stocks are discussed below, followed by an examination of the three basic categories of soups: (1) clear and thin, (2) thick, and (3) cream.

Stocks

Cracked bones and water are often the main ingredients of meat stock. For this **white stock**, neck and knuckle bones are preferred, because they contain more collagen (which converts to gelatin) and flavorful extracts than any other bone in the animal (10). Although all bones are porous, splitting them open helps to release the gelatin, which imparts a rich thickness and body to the stock. When stock is made from meats, beef and chicken are the two most commonly used. Veal bones contribute the most amount of collagen, making the thickest, most gelatinous stocks, while beef bones contribute the richest, meatiest flavor. Many chefs use both. **Brown stock** has a deeper, caramelized flavor. Browning the bones and meat before adding water has the advantage of discouraging the stock from becoming cloudy. Heating the meat coagulates

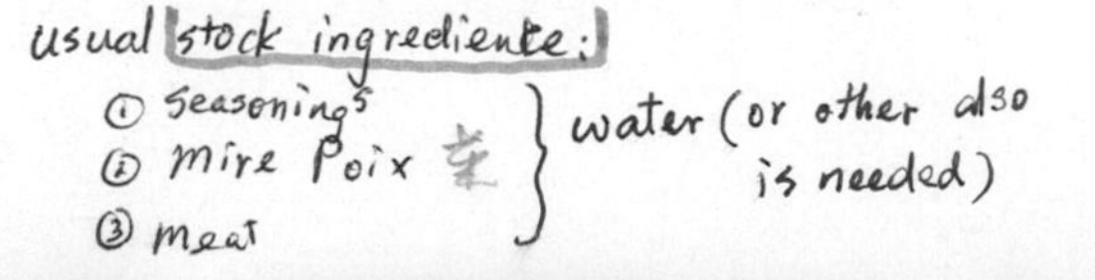

many of the proteins and traps minute particles that could otherwise cloud the stock. Regardless of whether the stock is brown or white, it is important to simmer rather than boil the stock; boiling would cause the particles floating to the top to churn back into the broth, turning it cloudy and less clean tasting (16). Although there is no standard formula for stocks, a general rule is that one pound of bones is required for every pint of water.

Water Is the Main Ingredient. Pure, clean, cold water is the first ingredient of any stock. Spring or distilled water is preferred (although not always used due to cost), because tap water may carry the flavors of chlorine or other substances (11). Flavor is then generated by simmering the water and ingredients for half an hour up to several hours. Boiling should never be allowed, because, in addition to causing clouding, it will toughen the meat and disintegrate any ingredients except bones. If additional water is required during simmering, hot water can be added to avoid cooling the entire stock.

HOW & WHY? ?????????

Why is cold water used to start soups? The water should be cold to start with, because ingredients placed in cold water will transfer their flavor more efficiently to the liquid (9). If hot water is used, the stock will be less flavorful and less clear.

Flavoring Ingredients. A mirepoix (meer-pwah) made up of onions, celery, and carrots is often added to stocks. This standard mirepoix can be modified to produce a white mirepoix, in which leeks replace the carrots in order to create a near-colorless stock (5). Salt or other potent seasonings are not usually added until the last half hour, if at all, in order to prevent their becoming too concentrated as liquids evaporate during cooking. The exception is a **bouquet garni**, which can be added to dishes with a lot of liquid undergoing a long simmer. Standard ingredients of a bouquet garni include parsley, thyme, and a bay leaf, although other aromatics such as cloves, rosemary, sage, and garlic cloves can be added (19). Cracked or ground pepper should never be added at the start of a long heating process because it will turn harsh and acrid in taste (15). Whole peppercorns may be used instead.

Meat Stocks. Meat stock, based on cracked bones and sometimes raw meat, serves as the major ingredient for all meat soups and major meat sauces. Cooked meat will not yield the same flavor, although it is sometimes used. The more mature the meat, the better, because meat becomes more flavorful with age. In general, red meats contain more flavor than white (veal, pork). Some of the cuts most frequently used for meat stock include:

- *Beef.* Oxtail, chuck, shank, bottom round, and short ribs
- *Pork.* Hocks, ham bones, and Boston butt
- *Lamb.* Shank, leg, shoulder

Adding Meat and Bones. The first step in preparing soup stock made with meat and/or bones is to cube or grind the meat, and/or cut the bones into 3-inch sections. This increases the surface area and improves extraction.

To draw out even more flavor, some chefs first soak the meat or bones in a pot filled with cold water, using 2 cups of water to every cup of meat or pound of bones; or add enough water to cover the packed ingredients, plus 1½ inches over. After half an hour to an hour of soaking, heat to the boiling point and then reduce it to a bare simmer for up to 3 to 4 hours. The bones and meat heated in the water will release gelatinous particles as the stock simmers that will cloud and form a scum on top of the liquid. Many chefs skim this scum layer before adding the vegetables.

Adding Vegetables. Vegetables are often not added until the last hour or half hour of preparation, both to prevent their becoming mushy with overcooking and to preserve the desired flavor, which would become too concentrated and bitter with the water evaporation that takes place during longer cooking. Sometimes, however,

KEY TERMS

Bouquet garni A bundle of parsley, thyme, bay leaf, and whole black pepper rolled in a leek and tied together with twine.

vegetables are placed in the stock at the beginning intentionally so they will get mushy and can be puréed to act as a thickener. The size that vegetables are cut also makes a difference—large chunks are best for long-simmering stocks, while small pieces are added near the end of heating time.

Poultry Stocks. The more mature the poultry used to make stock, the more flavorful will be the liquid. Birds that have been free to roam and scavenge, called free range birds, yield the most flavor of all (10). An entire cleaned chicken carcass or an assortment of meaty bones (ideally backs and necks) is placed in enough cold water to cover it by 1½ to 2 inches and brought to a simmer with the lid off. It is cooked at a simmer for about an hour (whole bird) to 4 hours (just bones) until the meat is tender and ready to be cut off in bite-size pieces. During this process, the frothy fat is repeatedly skimmed from the top as necessary. Once the meat is tender, the liquid is strained and returned to the pot before adding the chicken pieces, chopped vegetables, grain (such as rice or barley), and seasoning.

It is then covered and simmered for another hour. The vegetables often include, but are not limited to, onions, celery, parsley, leeks, carrots, parsnips, and turnips. Common seasonings are bay leaf, onion, garlic, pepper, sage, thyme, rosemary, or any other complementary herb. Chicken bouillon is sometimes added if the stock lacks flavor. Another method to add more flavor is to roast the chicken bones and vegetables before adding water.

Fish Stocks. Most fish stocks use the backbones (called frames or racks), heads, and/or tails of lean white fish. Fatty fish are rarely used because of their strong oily flavors. The high gelatin concentration in the heads contributes to flavor and the body of the stock. Fish frames should be thoroughly washed and eviscerated, because remnants and organs, as well as gills and any attached skin, will give the stock an off-flavor (2). The eyes may also be removed. The bones are then combined with all the other ingredients in the liquid. Fish contain less gelatin than meat and poultry, so the plain water usually used to start a soup stock is sometimes replaced by or fortified with chicken or vegetable stock, or bottled, but not canned, clam juice (11). The floating fish frames are occasionally pushed down into the liquid to release the flavorful compounds from the bones. Heating should not exceed about half an hour, because more than that makes the stock too bitter or fishy tasting. For more flavor, chopped fennel leaves and the white portions of leeks are classic additions to fish stock, along with the standard combination of onions, carrots, and celery. The celery may be omitted for a milder flavor, and the carrots may be left out if a golden color is not desired in the stock. Sometimes the fish frames and vegetables are first heated in butter to extract more flavor and create a richer-tasting stock. Any added vegetables must be cut small because of the short cooking time.

Shellfish Stocks. Shellfish stocks are usually made from shrimp, lobster, mussels, or clams, or from the shells of the first two. Leftover shellfish shells can be frozen until the stock is ready to be made.

KEY TERMS

Bouillon A broth made from meat and vegetables and then strained to remove any solid ingredients.

■ ■ ■ ■

Consommé A richly flavored soup stock that has been clarified and made transparent by the use of egg whites.

■ ■ ■ ■

Broth Stock made from meat or meat/bone combinations and some water with little or no flavoring. Since broths are seldom reduced as stocks, they therefore can contain salt.

Vegetable Stocks. Vegetable stocks have advantages over other kinds of stocks in that they are less expensive, less messy, and less time-consuming than their meat, poultry, or fish counterparts. Vegetables that lend themselves well to making stock include carrots, onions, leeks, shallots, garlic, celery, celeriac, parsnips, fennel, and tomatoes. Just as for poultry and beef, the more mature the vegetable, the more flavor in the extract. Vegetable stock takes only about half an hour to cook; any more time will turn the texture of the vegetables into mush and turn the flavor extract bitter. The standard vegetables and seasonings described under poultry stock are used when making vegetable stocks.

Storage of Soup Stocks. Stock should never be put in the refrigerator while it is still hot. The large volume of hot liquid can raise the internal temperature of the refrigerator to the point that the stock will not cool sufficiently within two hours, and may warm everything else in the refrigerator as well. A good way to cool the stock is to place the hot stock pot in a sink full of cold water and ice cubes until it is lukewarm, although the time spent in this procedure should not exceed one hour. After leaving it uncovered for the first half hour and stirring occasionally to cool it, it should then be covered with a lid or an upside down plate to prevent evaporation, which would cause the stock to become too concentrated. Refrigerated stock cools better in shallow pans. If covered, stock lasts up to five days, but is best if used within two days. For future use, stock can be frozen for a couple of months (16). Any animal stock should be considered a potentially hazardous food when it comes to food safety and treated accordingly (see Chapter 3). Simmering stored stock for at least ten minutes before use is a basic safety precaution.

Clear and Thin Soups

Clear soups include bouillons and consommés. Thin soups are somewhat thicker than clear soups.

Boullion. Bouillon is the French word for broth. Bouillons are less gelatinous than stocks, and may be added to poultry and vegetable stocks to add flavor. Traditionally, this type of soup is called bouillon if it is based on beef, and court bouillon or fumet if it is prepared using fish. Court is the French word for "short," and it describes the preparation time of bouillon, which is much shorter than that for stocks.

Consommé. A **consommé** is a perfectly clear beef bouillon. One raw egg white is added for every quart of stock, and the whole mixture is heated to boiling. The egg white will coagulate on the surface of the stock, forming what is referred to as a raft. As the stock simmers, the raft entraps loose particles. The resulting masses of unwanted material can be removed by straining. Since egg whites strip flavor from stock as well as particles, some ground meat and finely chopped vegetables are generally mixed with the egg whites to add flavor back to the stock. Since bouillon is the French word for broth, and because it is so similar to consommé, the three terms are often used interchangeably (15).

Thin Soups. Thin soups have a consistency between a clear liquid soup and a thickened soup. They are **broth** based, but typically contain a garnish of cooked meats, vegetables, and/or starch such as pasta, rice, or barley. Examples include tomato, onion, and noodle soups. Miso soup is an increasingly popular thin soup originating in Japan. There are many variations of this soup, in which miso is used as the primary seasoning agent. Miso itself is made by mixing cooked beans (usually soybeans) with cultured rice or barley (called *koji*), and salt (Figure 19-1). This mixture is then fermented several weeks or as long as three years, depending on the variety of miso. Unpasteurized miso contains potentially beneficial microorganisms.

Thick Soups

Any stock or broth can be thickened with a starch such as cornstarch or with puréed vegetables, or made thick and chunky by adding noodles or grains. Adding starches to soups also reduces the perception of saltiness if

Sarah Chester, South River Miso Co.

FIGURE 19-1 Miso for miso soup is traditionally made by mashing boiled beans (usually soybeans) underfoot like wine grapes, and mixing the mashed beans with cultured grain (called *koji*) and sea salt. This mixture is then fermented in wooden vats for several weeks or as long as three years, depending on the variety of miso. The miso maker in this photo wears special plastic foot coverings.

the components of starch or any added gums bind to sodium (17).

Examples of Thick Soups. There are many varieties of thick soups. Chowders are usually fish-based soups thickened with vegetable pieces, such as potatoes, that have been cooked in milk. Gazpacho is an uncooked, thick, tomato-based soup with added vegetables and Latin American seasonings that is served chilled. Minestrone is an Italian soup that contains macaroni, vegetables, and beans. Egg drop soup is a popular Chinese soup made by dropping a slightly beaten egg into simmering clear stock. In India, mulligatawny soup is popular; it consists of chicken, carrots, green pepper, and apple seasoned with curry, cloves, and mace. Borscht is a Russian/Polish soup that is based on beets and may or may not have vegetables and/or meats added to it (12). Thick soups also include any that are puréed or made from starchy vegetables that have been pulverized and sieved. Potatoes are a common thickener for a range of thickened soups.

Cream Soups

Cream soups and **bisques** are made by adding cream and/or milk to a thickened, flavorful purée made from meats, poultry, fish, and/or vegetables. Common thickeners are white sauce (equal parts of flour + fat combined with liquid), potato starch, or puréed potatoes. Cream is usually the thickening agent in bisques. Cream of chicken is the most popular creamed soup made from an animal source. The most common vegetable creamed soups include asparagus, broccoli, potato, spinach, tomato, celery, cauliflower, and corn.

> **KEY TERMS**
>
> ***Bisque*** Traditionally, a cream soup made from shellfish. Marketers sometimes label creamed vegetable soups as bisques.

Preparation of Cream Soups. In preparing creamed soups, flavorings are added to a thin white sauce, which is then combined with a liquid base that is either milk or white stock. Bechamel is a white sauce often used as a base for traditional cream soups because of its velvety, smooth, creamy consistency. Heavy cream and/or egg yolks may be added after the soup has been heated and seasoned, but these additions increase caloric (kcal) and fat content. If only one vegetable is featured in the cream soup, it helps to add a few aromatic ones such as onion, garlic, leeks, and shallots to enhance the flavor. A garnish of parsley or a sprinkle of chives, paprika, or nutmeg make colorful additions to a bowl of cream soup.

Lower-Fat Cream Soups. Cream soups by definition contain fat, but substitutions can be made to create a lower-fat soup that will still retain most of the characteristics of a cream soup (Figure 19-2). The cream can be replaced by starch in the form of raw rice (½ cup) or potato (thinly sliced) added at the beginning of cooking (16). The starch gelatinizes, creating a thick, smooth consistency. Adding just a touch of cream rounds out the flavor of this starch-based cream soup.

How to Avoid Curdling. One potential problem in making creamed soups is that the acid from ingredients such as tomatoes can cause milk proteins to curdle as the mixture heats. Curdling can be reduced by doing the following:

- Prepare a fat/flour mixture (see Chapter 20) with either the milk or the stock (but cold milk or cream should never be added to hot soup).
- Stir some of the hot soup into the cold dairy product to temper it before adding it to the warm ingredients.

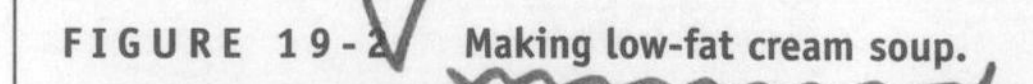

FIGURE 19-2 Making low-fat cream soup.

(1) Sauté aromatic vegetables (onions, garlic, shallots, or leeks) over medium heat. Adding herbs such as thyme, sage, or rosemary improves flavor.

(2) Pour in the liquid (water, broth, milk, or a combination) and heat to a simmer. Use 5 to 6 cups of liquid for every 2 pounds of main vegetable to be added.

(3) Add a thickener in the form of rice (1/2 cup) or raw potato (sliced). Starchy ingredients thicken when they gelatinize, making the soup feel creamy.

(4) The time to add the main vegetable (in even, bite-size pieces) is when the time remaining to cook the thickener is equal to how long it will take to prepare the vegetable.

(5) When the thickener reaches its maximum viscosity, purée the soup in batches using a blender or food processor. Another option is to strain the soup to obtain a finer texture.

(6) Finishing touches include adding a little liquid to thin the purée if it is too thick, and/or adding a small amount of cream to give it a "cream" flavor. Serve hot or cold.

- Do not allow the soup to come to a boil after adding any dairy product, particularly cheese.
- Add acid to milk rather than milk to acid.

Safety Tips for Cream Soups. Because cream soups should not be boiled after the cream or milk has been incorporated, there is more of a chance for bacteria to grow. Therefore, these soups should be treated with the safety steps discussed in Chapter 3 to avoid foodborne illness.

Salads

The Romans are credited with inventing salads, because they brought garden delectables to the table, sprinkled them with salt and seasonings, and added vinegar and oil. The practice continued in the Middle Ages, when salads were a side dish that added greens to the massive amounts of meat served in castles. Salads, which are most often served cold, are now offered as appetizers at the beginning of a meal, as side dishes, as main courses, or even as dessert. Table 19-1 gives a list of the numerous types of salads that are available. Through the years, the concept of salads has evolved and branched out so much that almost any food imaginable can be made into a salad. The most common types of salads are grouped by their main ingredient: leafy greens, vegetables, fruits, proteins, pastas/grains, gelatin, or a combination. Sometimes salads such as tuna, potato, rice, pasta, egg, chicken, and ham are categorized as "cooked," but in this text they are classified by their main ingredient. These will be discussed following the basics of creating a salad. Then salads, dressings, and gelatin foods will be the remaining focus of this chapter.

Salad Ingredients

The salad's base, also called an underliner, is the item that serves as the salad's first layer or foundation. It is usually lettuce or some other green, but the base might also be pasta, rice, cottage cheese, or gelatin. The predominant ingredient or main part of the salad on top of the base is called the body. Topping it all off is the dressing, which adds flavor and moistness. Tart dressings go best with greens or vegetables or protein-based salads, while sweetened dressings are usually reserved for fruit salads.

Garnishes. Sometimes garnishes are added for eye appeal. A few of the more common garnishes are fruit wedges (lemon, orange), hard-cooked eggs (slices, halves, or wedges), tomatoes (cherry, slices, or wedges), mushrooms, pimentos, radishes, olives, pickles, watercress, mint, and parsley.

The flavor, texture, and color of salads can be made more interesting by adding toppings such as croutons, cheese, shredded vegetables (carrots, red cabbage), herbs (Figure 19-3), flowers, seeds, and nuts.

Toasting Nuts. Most nuts can be toasted by spreading them on a cookie sheet, placing them in the oven at 325° to 375°F (162° to 191°C), stirring every few minutes, and removing them after five or ten minutes or when they have turned light brown (20). They should be allowed to cool before being added to the salad.

Principles of Salad Preparation. There are several principles to keep in mind when constructing a salad. The ingredients must be of the finest quality, because fresh dishes cannot hide inferior quality as is sometimes possible in cooked dishes. The more assertively flavored a food, the smaller its role in the salad should be. Color

TABLE 19-1
Types of Salads

Salad Type	*Examples*
Appetizer	Fruit salad Marinated shrimp on lettuce Mandarin orange and avocado slices on watercress Mushrooms, green pepper, and cauliflower Orange and grapefruit sections on butter lettuce
Entrée	Chef's salad Fruit plate with yogurt or cottage cheese Tuna or chicken salad Tomato stuffed with egg salad Spinach salad with chopped egg and bacon
Accompaniment	Mixed vegetable or garden salad Mixed fruit on lettuce Cole slaw Pineapple-carrot salad Fruit in gelatin Three-bean salad
Dessert	Frozen fruit salad Molded fruit salad Mixed fruit salad Melon wedges

Source: Texas Tech University.

FIGURE 19-3 Herbs add interest and flavor to salads.

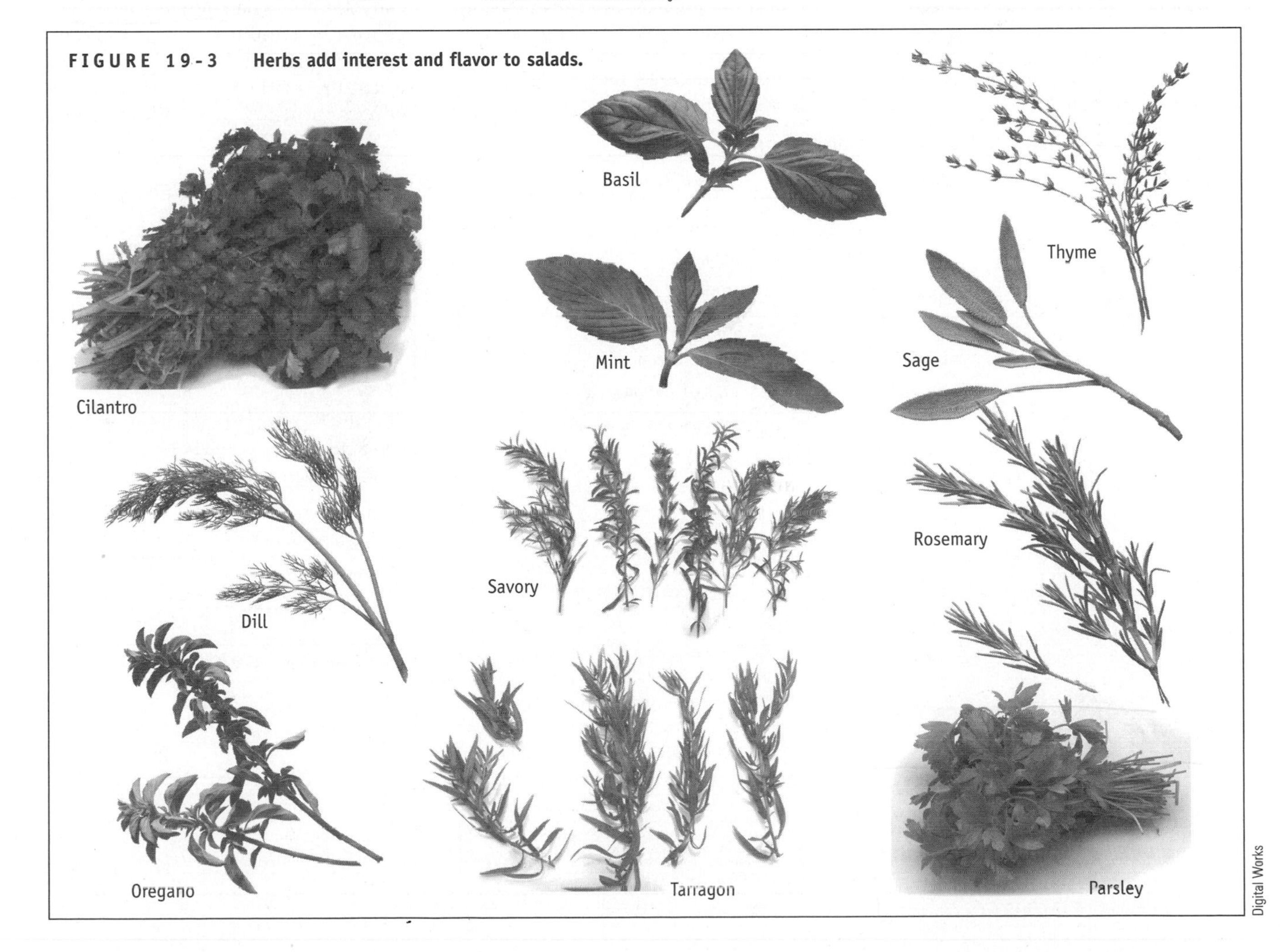

and texture should also be considered when making salads. The ultimate goal is to assemble and arrange ingredients with a good balance of color, texture, and shape within the rim of the chilled salad plate (Table 19-2).

Fat and Fiber in Salads. Another factor to keep in mind when creating a salad is nutrient content. Just because something is called a "salad" does not automatically mean it is low in fat, because many salads contain high-calorie (kcal)/fat ingredients. Some of these include meat, egg yolks, cheese, avocado, coconut, olives, nuts, and oil-based dressings. A chef's salad harbors large amounts of both calories (kcal) and fat even before the dressing is added (Table 19-3). Many dressings contain oil, which averages 120 calories (kcal) and 13 grams of fat per tablespoon. There is also a fair amount of sodium in many of the same ingredients and dressings. Salads are also not necessarily good providers of fiber; iceberg lettuce contains only 1 gram per cup, while the daily recommended amount is 20 to 30 grams. A salad's fiber content may be boosted by topping leafy greens with high-fiber vegetables such as broccoli, cauliflower, carrots, baby corn, beans, and potato pieces with skins.

Leafy Green Salads

In general, when people think of salads, leafy green ones are the most likely to come to mind; examples include garden, tossed spinach, mixed green, and Caesar salads. Greens range in flavor from extremely mild to tangy, bitter, and sharp. Iceberg lettuce is the most mild-flavored and the one most frequently used to fill out the body of salads. Its bland flavor almost necessitates that other, more distinctive-tasting, ingredients be added. Other salad greens include Romaine, Bibb, Boston leaf, and Ruby lettuces, and even spinach (see Color Figure 15). If more assertive flavors are desired, escarole, Chinese (Napa) cabbage, kale, and either green or red cabbage make good choices. Flowers and herbs provide additional flavor and color, but should be added in limited amounts, with the exception of herb salad made only from leafy herbs (Color Figure 16).

Preparation of Green Salads. Constructing a green salad starts with a thorough rinsing of the greens to remove any soil (Figure 19-4). Many greens are grown in the ground, and not adequately removing the soil leaves a gritty feel between bites of leaves. Any water left clinging to greens after washing will dilute the flavor, adversely affect the texture, and accelerate deterioration, so it should be removed either by draining greens in a colander, patting them between paper towels, or using a salad spinner. Next the stems and/or cores must be removed. Greens should be refrigerated for 30 minutes to several hours prior to use in order to promote crispness. Once they are ready for the salad bowl, the greens are torn by hand into bite-size pieces. Hand-tearing the leaves is preferred to cutting with a knife because it reduces bruising and allows more

TABLE 19-2
Adding Color, Flavor, and Texture to Salad Greens

Vegetables	*Fruit*	*Protein Foods*
Artichoke	Apple	Anchovies
Asparagus spears	Avocado	Cheese
Baby corn	Grapefruit	Chicken, turkey
Baby squash	Olives	Egg slices
Broccoli	Orange	Ham, bacon
Carrots	Pimento	Nuts (sunflower, walnut)
Cauliflower	Pineapple	Salmon, sardines, shrimp, tuna
Celeraic (celery root)	Papaya	
Celery	Pears	
Cucumber		
Green pepper		
Heart of palm		
Leek		
Mushroom		
Peas		
Radishes		
Red pepper		
Red onion		
Scallions		
Tomato (cherry, yellow, sun-dried)		

TABLE 19-3
Major Calorie Contributors to Chef's Salad

Salad Ingredients	*Calories*	*Fat (g)*
1 oz beef	75	6
2 oz turkey	80	1
1 oz ham	70	5
1 oz cheese	114	9
1 hard-cooked egg	74	5
3 ripe olives	15	1
Lettuce (2 C)	18	—
3 tomato wedges	19	—
Total	465	27
Salad Dressings (2tbs)	***Calories***	***Fat (g)***
Blue cheese	154	16
French	134	12
Italian	137	16
Russian	151	16
Thousand Island	140	14

Clean greens. Briefly soaking and swirling the greens in a bowl of cool water dislodges grit. Lifting the leaves out with separated fingers is followed by replacing the water in the bowl and repeating the process until no grit remains.

Dry greens with a salad spinner. Water is removed by repeatedly spinning the leaves, draining the bowl, and tossing the leaves between spinnings.

FIGURE 19-4 Cleaning salad greens.

Shredded
Manual–Cut each head in half lengthwise. Then slice crosswise into $^1/_8$-$^1/_2$" strips.
Automatic slicer–Crisp iceberg lettuce is the only lettuce suitable for shredding on slicers or slicing attachments. Use half heads; set the gauge for $^1/_8$-$^1/_2$" strips.

Rafts
Cut each head crosswise into slices about 1" thick.
A head should yield about 3 to 4 rafts.

Chunks
Slice head into rafts. Then cut each raft crosswise and lengthwise into $1^1/_2$" square chunks.

Large Leaves or Cups
Begin with a full head, holding it core side down.
Gently separate leaves and lift them away from the head. For smaller portions, cut heads in half, then lift off leaves.

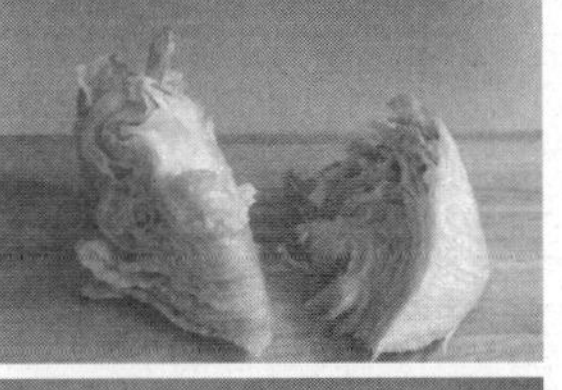

Wedges
Small iceberg lettuce heads cut into quarters yield an average wedge. Wedges should weigh about 4 oz. each. Larger heads should be cut into six or eight wedges to yield similar size portions.

Half Heads
Cut a full head in half lengthwise. Then place cut side down on a board and make two or three parallel cuts partially through the half head. Present over a mound of shredded iceberg lettuce with the cuts spread open and filled.

Full Heads
Select a firm, springy head and core. Remove a slice from the core end, then pull out about a 3" heart from the center of the head and shred. Spread remaining leaves back from the center like a rose. Place shredded lettuce back into the center of the prepared head.

FIGURE 19-5 Methods of preparing iceberg lettuce.

salad dressing to be absorbed into the leaves. Commercial food service units often rely on quicker methods, however, and Figure 19-5 shows several techniques for cutting iceberg lettuce. Cabbage and carrots can also be prepared for salads (Figure 19-6).

Vegetable Salads

Vegetable salads are those whose main ingredients are non-leafy vegetables. These are sometimes composed of a single vegetable, such as a tomato salad, or made up of a combination of vegetables. A few of these are cucumber and tomato, green bean, mushroom, and carrot salads, and coleslaw. Any vegetable is eligible, and vegetables that are raw, cooked, canned, or marinated are all suitable candidates. Vegetables that are commonly used in salads include tomato, cucumber, broccoli, cauliflower, beets, radishes, olives, green bell peppers, carrots, peas, mushrooms, onions, and celery. Tougher vegetables such as carrots and celery will have improved texture when shredded with a knife, grater, or food processor. Potato salad is best prepared with red or white boiling potatoes. They are less starchy and hold their

(a) Removing the cabbage core.

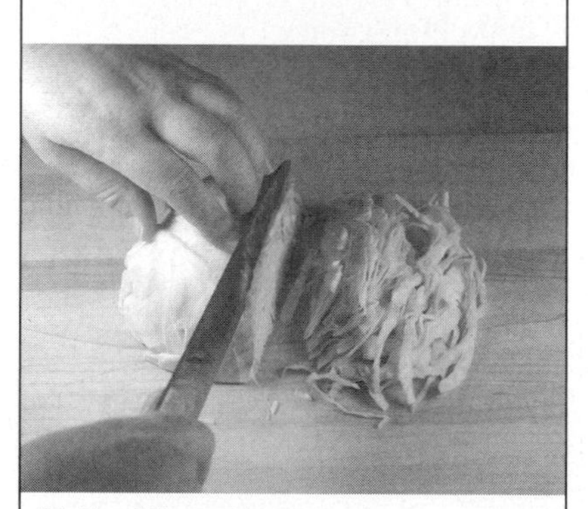
(b) Shredding cabbage.

(c) Grating carrots.

FIGURE 19-6 Preparing various vegetables.

Trimming pineapples.

Cutting out pineapple "eyes."

Coring apples.

Peeling soft-fleshed fruits.

Preparing melon balls.

FIGURE 19-7 Preparing fruit for salad.

shape when boiled better than the Russet or Idaho potatoes usually used for baking.

Fruit Salads

Fruit salads are often served as a dessert or as a luncheon dish, and often combined with a higher-protein food such as cottage cheese or yogurt. Some of the fruits most frequently used for salads are apples, avocados, oranges, bananas, kiwi, pineapple, melons, grapefruit, pears, peaches, cherries, blueberries, strawberries, grapes, and raisins. Acidic fruits should be balanced with mild greens and a light sauce (15). Canned fruit should be thoroughly drained, and all fruit should be cut into pieces no bigger than bite-size (Figure 19-7). Some fruits undergo enzymatic browning when cut, which may be prevented by coating them lightly with citrus or pineapple juice, sugar, or an ascorbic acid (vitamin C) mixture. Fruit salads are often arranged, rather than tossed, in order to prevent their delicate flesh from bruising or damage.

Protein Salads

A salad based on protein foods, such as cold meat, poultry, fish, eggs, or dairy products, helps to make it a main course. All, except the dairy items, must be used in their cooked form. Firm cheeses, such as cheddar or Swiss, are often cubed and added to salads. To quickly prepare chicken breasts specifically for salad, poaching works well (see Chapter 15). Four boned, skinned chicken breasts yield 2 cups of cubed chicken meat, which should be chilled by refrigerating at least two hours prior to placing it in the salad. Some chefs prefer roasted chicken for chicken salads.

Pasta/Grain Salads

A bland background of pasta or grains serving as the body of a salad is a perfect invitation to more colorful and flavorful ingredients, especially if they are acidic or salty. The large number of pasta shapes (elbows, shells, noodles) and varieties of grains (rice, bulgur, wheat, etc.) allow the creation of an almost unlimited array of salads. Chapter 21 discusses the preparation of the grains and pastas themselves.

Salad Dressings

There are as many different kinds of salad dressings as there are kinds of salads. The original dressing was probably just plain salt. The word "salad" is derived from *herba salata,* Latin for salted greens (15). During Roman times, freshly picked greens were salted to enhance their taste, but now there are many different types of dressings, which can be thought of as "sauces for salads" (Figure 19-8).

Oil and Vinegar Dressings. Vinaigrette is the most basic of all dressings. The flavor of the vinaigrette is varied by the type of oil and vinegar used.

FIGURE 19-8
Pouring dressing over salad.

Oils. A vinaigrette may use a flavorful oil such as Extra Virgin olive oil, walnut oil used with a raspberry vinegar, or sesame oil with an oriental rice vinegar. Other salad dressings based on vegetable oil may rely on a neutral-flavored oil such as corn, cottonseed, soybean, or safflower. Winterized oils, often called salad oils, are those that have been treated so they do not cloud when refrigerated. Olive, peanut, walnut, hazelnut, and almond oils are sometimes chosen for their characteristic flavors. A nutty-flavored salad dressing can be made by adding toasted sesame oil to the dressing.

Vinegars. Balancing the oil in a vinaigrette is some form of vinegar, of which the three most common types are wine, cider, and white/distilled (4). Vinegars differ in their base and ingredients, but they are similar in that they are all fermented from alcohol to acetic acid by the bacterium *Acetobacter* (8). The term *vinaigre* is French for sour wine, and the higher the percentage of acetic acid shown on the label in fine print, the more sour and stronger the vinegar. Better-tasting vinegars are slowly fermented naturally, while those fermented with the aid of heat and chemicals often have harsh, metallic tastes (18). Some of the vinegars used in salad dressings include:

- *Wine.* Made from wines such as red, white, cabernet sauvignon, champagne, and sherry. These vary widely in terms of flavor and quality.
- *Cider.* A golden brown vinegar made from fermented apple juice; it has a strong flavor.
- *White/distilled.* A colorless, extremely acidic, almost flavor-free vinegar made from grain alcohol.
- *Herb.* Cider, white, or wine vinegar with herbs added; tarragon is a common herb-flavored vinegar.
- *Fruit.* Cider, white, or wine vinegar with fruits added; raspberry vinegar is increasingly popular.
- *Rice.* Made from rice wine, this vinegar has the lowest acidity, is clear to slightly yellow, and has a mild flavor.
- *Balsamic.* A rare, dark brown vinegar made from a very sweet grape concentrate; it has a slightly sweet taste. It is more costly than other vinegars because it is aged in wooden barrels for over ten years. Commercialized balsamic vinegar available in most grocery stores is actually a red-wine vinegar to which concentrated grape juice and perhaps caramelized sugar have been added.

KEY TERMS

Vinaigrette A salad dressing consisting of only oil, vinegar, and seasoning.

The ratio of oil to vinegar varies among salad dressings. Flavor becomes more bland and less acidic as more oil is added, but shifts to a sharper, acidic taste when vinegar concentration increases. Sometimes lemon juice is added or replaces vinegar when a more lemony flavor is desired.

Emulsified Dressings. Oil and vinegar normally separate if left to stand, and need to be mixed together before being applied to a salad. If the oil and vinegar mixture has been treated to create a permanent emulsion, however, the ingredients will not separate. One such emulsion is mayonnaise, which can be served as is in chicken or potato salad or in coleslaw, or as a base for many other salad dressings, including Chantilly, creamy Roquefort, Green Goddess, Dill, Thousand Island, and Russian.

Other Dressings. There are many other possibilities for salad dressing bases. Little or no oil is found in cooked dressings; these have a viscous consistency because of an added starch thickener. Some dressings are dairy-based, with sour cream or yogurt as the main ingredient. A few dressings are made from fruit juice, and their lower calorie (kcal)/fat content makes them a good alternative for health-conscious consumers.

Adding Dressings to Salads. The best time to add dressing to a salad is just before serving, because the acidity of the dressing will start wilting the leafy green vegetables as soon as the dressing is added. It is best to chill the dressing for at least half an hour to improve its ability to coat the salad

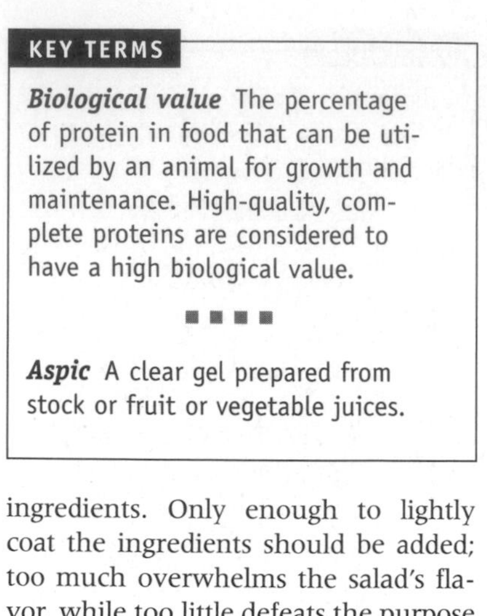
KEY TERMS

Biological value The percentage of protein in food that can be utilized by an animal for growth and maintenance. High-quality, complete proteins are considered to have a high biological value.

Aspic A clear gel prepared from stock or fruit or vegetable juices.

ingredients. Only enough to lightly coat the ingredients should be added; too much overwhelms the salad's flavor, while too little defeats the purpose of adding the dressing, leaving the flavor unenhanced. Salads are best tossed to evenly coat all the ingredients with a taste of dressing. Topping a salad with dressing without mixing is a less desirable way to serve any salad.

Gelatin Salads

Gelatin salads rely on the formation of a gel. Gelatin itself is a mixture of proteins extracted from the collagen found in the bones, hides, and connective tissue of animals. The principle commercial sources of gelatin are pork skins and the bones and hides of cattle (13). These sources of gelatin are sometimes avoided by certain vegetarians or people following kosher guidelines. Gelatin is available in either powder granules or sheets, the latter being used by commercial food manufacturers and institutions (1 teaspoon of gelatin powder is equivalent to 2 sheets). Standard amounts sold to consumers consist of uniform packets containing about 2½ teaspoons (7 grams) of powdered gelatin, which will usually gel about 2 cups of liquid. Gelatin is available in a colorless, almost transparent, odorless, and virtually tasteless state, or flavored, in combination with sugar, coloring, and flavoring agents.

Is Gelatin Nutritious? Even though gelatin is a protein from animal sources, it is low in tryptophan, an essential amino acid. Gelatin protein is therefore considered incomplete and of low **biological value**. A few gelatin manufacturers have added tryptophan to some of their products in an effort to remedy this problem, but gelatin's contribution to dietary protein is nevertheless limited by the small amounts—only 1 tablespoon (7 grams protein, 23 calories/kcal)—required to gel 2 cups of liquid. A serving size is about one-fourth that amount, and so would contribute little in the way of nutrients.

Preparation of a Gel. A gel is prepared by hydrating gelatin granules or sheets, dispersing them by heating and stirring the mixture, and then cooling it. Heating protein solutions decreases the protein's solubility and increases the interaction between dissolved proteins (1). Pieces of foods such as meats, vegetables, or fruits can be added during the cooling period when gelation or thickening takes place. The type of food incorporated into the gel determines whether the gel will be a salad, a dessert, or a main dish. A special type of gel that can be made with gelatin is an **aspic**. Sometimes aspics also have fruit or vegetable juices added, as in a tomato aspic.

Phases of Gel Formation. The three phases of gel formation consist of hydration, dispersion, and gelation.

Phase 1—Hydration. Gelatin can absorb five to ten times its weight when heated in water. Unflavored gelatin will clump unless it is first hydrated in cool or barely warm water before adding it to other ingredients. To separate the granules, the gelatin is sprinkled over the water, in the proportions of 1 tablespoon gelatin to ¼ cup water, and left to stand without stirring for 5 minutes. This initial hydration can be skipped when preparing flavored gelatin because it has been acidulated, or treated. Adding boiling water instantly dissolves this type of gelatin.

Phase 2—Dispersion. After hydration, hot water or a flavorful liquid such as stock or juice is usually added to disperse the protein granules (Figure 19-9). Gelatin disperses at 100°F (38°C), and faster at higher temperatures. When adding hot liquid, it is important to stir, especially along the sides and bottom of the container where the granules may stick, to ensure sufficient dispersion. After the granules have been completely dispersed, cool liquid can be added. Cooling the remaining liquid not only speeds up the gelling process, but prevents the loss of volatile substances that evaporate into the air when hot liquid is used.

Phase 3—Gelation. Refrigeration is crucial to the gelling process. As the liquid containing the gelatin protein cools in the refrigerator, it will convert first to a **sol**, which has a liquid consistency. It is either left in the mixing bowl, or poured into individual serving containers or into a larger mold that may be lightly coated with salad oil for easier unmolding. Molds vary

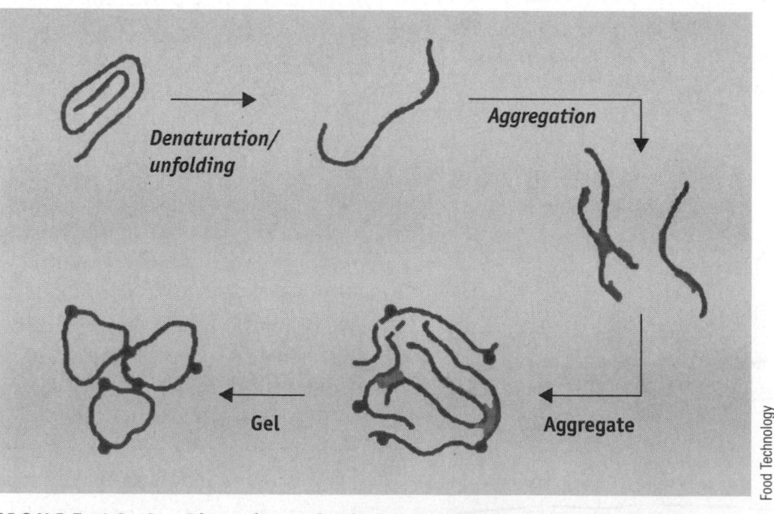

FIGURE 19-9 Dispersion and gelation: Increasing the temperature by adding hot water causes the protein molecules to denature and unfold. Upon cooling, the protein molecules aggregate, forming a gel.

in shape and can include any number of designs—a ring, a fish, a pineapple, or others.

HOW & WHY? ?????????

How does gelatin contribute to gel formation? The long, thin protein strands of the gelatin give it its unique ability to gel. The gelatin protein molecules, each surrounded by water molecules, slowly start to form a network by creating crosslinks (bonds or bridges) among themselves. The water remains trapped within the protein strands, creating an elastic solid (Figure 19-10).

Cooling the sol further (1 to 6 hours) will create a more solid-like gel. The end result is a solid structure that can be used in gelatin-based molded salads (as well as desserts and plain-flavored gelatin). These can be topped with a sauce, a whipped topping, or mayonnaise, depending on the ingredients.

Unmolding a Mold. Once the gel has cooled and set, it is ready to be served in its container(s) or removed from the mold(s) and placed on a serving dish. To unmold a gelatin, first slip the tip of a knife around the top edges of the mold to release some of the vacuum. Dip the mold in lukewarm water up to the upper edge of the container for a few seconds, but not long enough for the gel to start melting. Invert the mold over a plate, or, placing a plate over the mold, invert them together. A few shakes should then release the gel from the mold. The process should be repeated if the gel is still sticking to the container. A few drops of water or oil on the plate will allow the gel to slip to the center. Not all gels are firm after being unmolded; in this case, they should be covered and put back in the refrigerator for further chilling before being served.

FIGURE 19-10 Gelation phase. A gel forms when the protein molecules form a network, trapping water into a more solid mass.

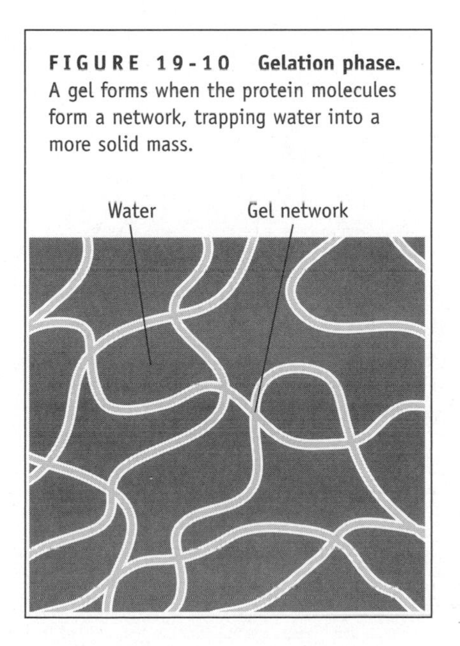

Factors Influencing Gel Formation. Many things can affect gel formation, including the gelatin concentration, temperature, added ingredients, and whether or not the gelatin is whipped.

Gelatin Concentration. Gelatin concentration should be approximately 1 to 2 percent, which is equal to a little under 1 tablespoon of gelatin per 2 cups of liquid (Chemist's Corner 19-1). Adding more gelatin allows the gel to set faster and be firmer, but it can also result in a tough, rubbery texture. Insufficient gelatin will result in a wobbly gel. The amount of gelatin required varies with each dish and its other ingredients. For example, food service organizations use about 2 ounces of unflavored gelatin for every gallon of liquid, but that same amount of water requires 24 ounces of gelatin if it is sweetened (5).

Temperature. The faster the sol cools, the quicker it sets. The cooling process may be accelerated by adding ice cubes or crushed ice to the dispersed gelatin or by placing the container in a larger bowl filled with ice water. The container may also be placed in the freezer for a short time, but should not be allowed to freeze, because this damages the gelatin layers next to the pan. There are some disadvantages to quick cooling: fast-setting sols result in weak gels and are more likely to lose their structures when brought to serving temperature, and there may be more tough lumps formed during gelation.

Sugar. Sugar delays gelling and weakens the gel's structure by competing with water for crosslinking sites. To compensate, more gelatin is added to mixtures containing sugar: approximately 2 tablespoons of gelatin for every half cup of sugar.

Chemist's Corner 19-1

Measuring Gelation Potential

The gelation potential of a particular gel is measured by the "Bloom" test. Bloom is the amount of force required to compress a gel (6.67 percent concentration) a distance of 4 ml (6). The higher the Bloom, the less gelatin is required for the gel to set (13).

Acid. A gel's optimum strength is found between pH 5 and 10. Acids weaken gels by hydrolyzing the gel's network of protein molecules. This is why it is important to avoid adding vinegar or acidic fruits and juices such as lemon or tomato juice to gels. These ingredients can decrease pH below 4, which can prevent gelation or result in a soft gel. This problem is avoided by doubling the concentration of gelatin used in such dishes.

Enzymes. Proteolytic enzymes found in fresh pineapple (bromelin), papaya (papain), kiwi (actinidan), and figs (ficin) prevent and weaken gel formation by hydrolyzing the protein of the gel network. As a result, these foods are often not added to gels. Heat denatures the enzymes in these fruits, so cooked or canned versions pose no threat to the gel's structure.

Salts. Salts strengthen the structure of gels. Natural mineral salts exist in hard water and milk, which is why incorporating these two liquid sources into a gel will cause it to be firmer than usual.

Adding calcium in the form of a salt has long been known to firm gels (14).

KEY TERMS

Sol A colloidal dispersion of a solid dispersed in a liquid (see Chapter 2 for colloidal dispersions).

KEY TERMS

Foam A colloidal dispersion of a gas in a liquid.

Added Solid Ingredients. The addition of solids, such as chopped fruits, vegetables, marshmallows, and nuts, also weakens gel formation. If any of these are acidic or contain proteolytic enzymes, the gel's structure is even further compromised unless it is compensated for by added gelatin. In order to avoid lowering the gelatin concentration, any added solids should be completely free of liquid. If they are not, any liquid present should be compensated for by reducing the amount of other liquids added. It is best to use only a small amount of solid ingredients and to chop them up as finely as possible in order to decrease interference with the protein crosslinks being formed. Solid ingredients should be added just before the sol has the consistency of beaten egg whites and is about to turn into a gel or else the solids will float to the top.

Whipping. A whip or sponge is formed by combining a gel with beaten egg whites or whipped cream. Although not a familiar sight, the consistency should look like raw egg whites in syrup. Gels increase to two to three times their volume when whipped into a **foam** just before they are set. If this is attempted prematurely, it will prevent the structure from forming correctly and the foam will float to the top as the rigid gel sinks to the bottom. Overgelling results in broken-up chunks of gel. Properly formed gels and foams can be used together to create a two-tone effect seen in parfaits.

Storage of Gelatin. Dried gelatin can be stored in a cool, dry place for many months. When stored too long, however, the gelatin becomes less soluble and loses its ability to hydrate; and without hydration, there is no gelation. Once prepared, gels are best served within 24 hours. Beyond this point, the gel continues to form bonds between the gelatin proteins, resulting in an increasingly firm, more compact gel as the spaces between the lattices in the gel network become smaller and smaller. Eventually some of the water will be squeezed out in a process known as syneresis. Most gels will lose their shapes when exposed to temperatures above 80°F (26.5°C) (3).

PICTORIAL SUMMARY / 19: Soups and Salads

As the first act, a soup or a salad often sets the tone for the meal to come. But their versatility also allows them to take center stage as a complete meal in themselves.

SOUPS

Soups are based on a stock, which is the flavorful liquid remaining from cooking bones, meat, and/or vegetables. Stock should be simmered, never boiled, and cooking time varies from half an hour to 12 hours. Hot stock should never be placed in the refrigerator without proper precooling.

The main types of stock are meat, poultry, fish, and vegetable.

Meat/poultry stocks. Simple stocks made from water and bones, cracked to release flavor, are called *white stocks*. If the bones are browned prior to simmering, it is a *brown stock*. If meat is added, it is usually raw. A bouquet garni or mirepoix of onions, celery, and carrots is often added to enhance flavor.

Fish stocks. Fatty fish are rarely used for stock because of their strong and oily flavor. Fish frames must be thoroughly cleaned and heating should not exceed about half an hour to preserve a delicate flavor. Another stock (chicken or vegetable) or bottled clam juice may be added to "fortify" the base. Fennel and leek are classic vegetable additions, along with the standard onions, carrots, and celery.

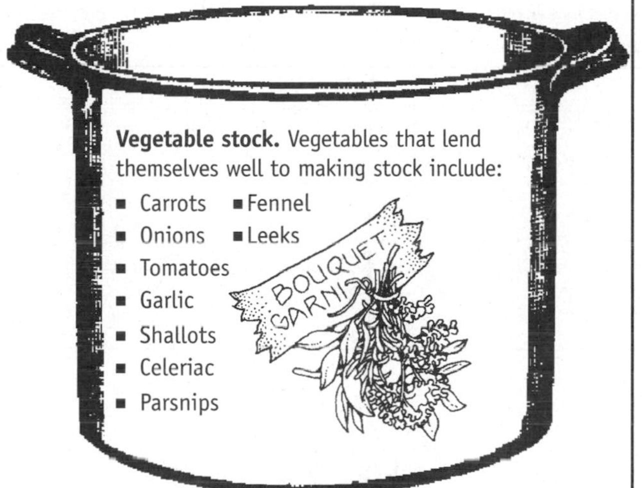

Vegetable stock. Vegetables that lend themselves well to making stock include:

- Carrots
- Fennel
- Onions
- Leeks
- Tomatoes
- Garlic
- Shallots
- Celeriac
- Parsnips

There are three different types of soups:

- **Clear and thin:** Removing the particles found in a stock results in a clear soup such as a bouillon or consommé. Thin soups have a consistency between a clear liquid soup and a thickened soup.
- **Thick:** Any stock can be thickened with starch or puréed vegetables to yield thick soups such as chowders, gazpacho, minestrone, egg drop, mulligatawny, and borscht.
- **Cream:** Adding cream and/or milk plus a thickening ingredient to a purée made from meats, poultry, fish, or vegetables results in a creamed soup, or bisque.

SALADS

A great salad combines the freshest possible ingredients, with a good balance of flavor, color, texture, and shape.

Types of Salad. The most common types of salad are grouped by their main ingredient: leafy green, vegetable, fruit, protein, pasta/grain, gelatin, or a combination. Leafy green salads are the most common, and examples include garden, spinach, mixed green, and Caesar salads.

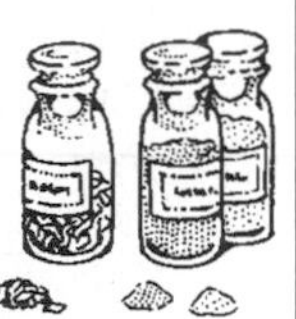

Salad dressings. There are many types of salad dressings, but the most basic are dressings based on oil and vinegar. Emulsified dressings such as mayonnaise do not separate as regular oil and vinegar dressings do. Dressings may also be made with little or no oil, with dairy products such as sour cream or yogurt, or from fruit juice.

Gelatin salads are made from proteins extracted from the collagen found in animal skins, bones, and hides. Gelatin is prepared by hydrating the granules, dispersing the protein by heating and stirring the mixture, and cooling it until it sets. Factors influencing these three stages of gel formation include gelatin concentration, temperature, acid, salt, sugar, enzymes, and added solid ingredients. Dried gelatin can be stored in a cool, dry area for months, but prepared gelatin salads should be served within 24 hours.

Common garnishes:

- Fruit wedges
- Hard-cooked eggs
- Tomatoes
- Mint
- Mushrooms
- Pimentos
- Radishes
- Parsley
- Olives
- Pickles
- Watercress

Salad Toppings

- Croutons
- Herbs
- Cheese
- Seeds, nuts
- Shredded vegetables
- Edible flowers

REFERENCES

1. Aguilera JM. Gelation of whey proteins. *Food Technology* 49(10):83–89, 1995.
2. Alford K. One stock makes three flavorful fish stews. *Fine Cooking* 18:57–60, 1997.
3. American Home Economics Association. *Handbook of Food Preparation*. American Home Economics Association, 1993.
4. Better Homes and Gardens. *Salads*. Better Homes and Gardens Books, 1992.
5. Bocuse P, and F Metz. *The New Professional Chef. The Culinary Institute of America*. Van Nostrand Reinhold, 1996.
6. Carr JM, K Sufferling, and J Poppe. Hydrocolloids and their use in the confectionary industry. *Food Technology* 49(7):41–44, 1995.
7. *The Chef's Companion: A Concise Dictionary of Culinary Terms*. Van Nostrand Reinhold, 1996.
8. Chukwu U, and M Cheryan. Concentration of vinegar by electrodialysis. *Journal of Food Science* 61(6):1223–1226, 1996.
9. Corriher SO. Make grease-free soups full of satisfying flavor. *Fine Cooking* 26:78–79, 1998.
10. DeGouy L P. *The Soup Book*. Dover Publications, 1974.
11. Dragonwagon C. *Soup and Bread Book*. Workman, 1992.
12. Ensminger AH, et al. *Foods and Nutrition Encyclopedia*. CRC Press, 1994.
13. Gelatin Manufactures Institute of America. *Gelatin*. Gelatin Manufactures Institute of America, 1993.
14. Hongsprabhas P, and S Barbut. Protein and salt effects on Ca^{2+}-induced cold gelatin of whey protein isolate. *Journal of Food Science* 62(2):382–385, 1997.
15. Kolpas N. *Soups*. Time-Life Books, 1993.
16. Peterson J. Simple steps to making versatile chicken stock. *Fine Cooking* 19:18–19, 1997.
17. Rosett TR, et al. Thickening agents effects on sodium binding and other taste qualitites of soup systems. *Journal of Food Science* 61(5):1099–1104, 1996.
18. Stevens M. A cook's guide to vinegar. *Fine Cooking* 17:74–76, 1996.
19. Stevens M. Add flavor to soups and stews with a bouquet garni. *Fine Cooking* 17:74–76, 1998.
20. Stevens M. Toast nuts to give them more flavor and crunch. *Fine Cooking* 30:74–75, 1998.

WEBSITES

Steps to preparing soups, soups, and more soups:

http://fp.enter.net/~rburk/soups/soups.htm

Learn the basic ingredients of the major stocks and soups:

http://www.chefdepot.net/sauces.htm

Culinary Connection's website on stocks:

http://terkelowna.tripod.com/whatscooking/id113.html

Starches and Sauces

Starchy foods are the mainstays of diets throughout the world (11). In Ireland in the mid-1800s, the potato was so important to the diet that a widespread potato blight resulted in famine and starvation for many thousands of Irish people. In China, Japan, India, and other Asian countries, rice is the most important staple food, followed by a wide variety of noodles. Certain African countries rely on roots, tubers, and sorghum (a grain) as the main starch.

As a complex carbohydrate, starch provides energy, and a well-balanced diet derives at least 55 to 65 percent of its calories (kcal) from carbohydrates. Starches are made up of glucose molecules synthesized by plants through the process of photosynthesis, so they can be obtained only by consuming foods of plant origin. The glucose is converted to starch by the plant, and then utilized for energy or stored in the seeds, roots, stems, or tubers (enlarged underground stems). When foods from these sources are ingested by mammals, including humans, the digestion process converts the starch back to glucose. Such starches provide 4 calories (kcal) of energy per gram.

The food industry makes widespread use of the thickening capability of starches as well as of their abilities to act as stabilizers, texturizers, water or fat binders, fat substitutes, and emulsification aids (21). Starches contribute to the texture, taste, and appearance of foods such as sauces, gravies, cream soups, Chinese dishes, salad dressings, and desserts, including cream pies, fruit pies, puddings, and tapioca. The preparation of cream soups and pie fillings is covered in Chapters 19 and 26 respectively, while this chapter covers the preparation of sauces with and without starch. First, a look at starch sources, their structure, and characteristics.

Starches as Thickeners

Sources of Starch

The word "starch" is derived from the Germanic root word meaning "stiff," and commercial starch lives up to the original meaning by acting as a thickening or gelling agent in food preparation. Plants serve as the source of starch granules, which are the plant cell's unit for storing starch. Starch granules differ in size and shape, depending on their botanical origin. Cereals such as wheat and rice are common sources of starch. Root starches include potatoes, arrowroot, and cassava (tapioca). Other sources of complex carbohydrates are dried beans, peas, and the sago palm. Potato starch granules are the largest, while corn, tapioca, rice, and taro root starch granules are progressively smaller (Figure 20-1). Some starches derived from plants can be considered food additives and are used in a wide variety of ways.

Cornstarch. The wet milling process is used to derive starch from corn, which is the major source (95 percent) of starch in the United States (27). The dried kernels are softened by soaking them in warm water containing sulfur dioxide. Once softened, the kernels are cracked, any extraneous material is removed, and the cracked kernels are ground and screened, or sifted down to yield starch and protein. The protein is removed and the starch is filtered, washed, dried, and packaged as cornstarch.

Starch in Food Products

Starch serves several purposes in the food industry, including its use as a thickening agent, edible film, and sweetener source (dextrose and syrup) (Figure 20-2).

Thickening Agent. Starch's main use in processed foods is as a thickening and/or gelling agent. Foods that are frequently thickened with starch include soups, sauces, pie fillings, gravies, chili,

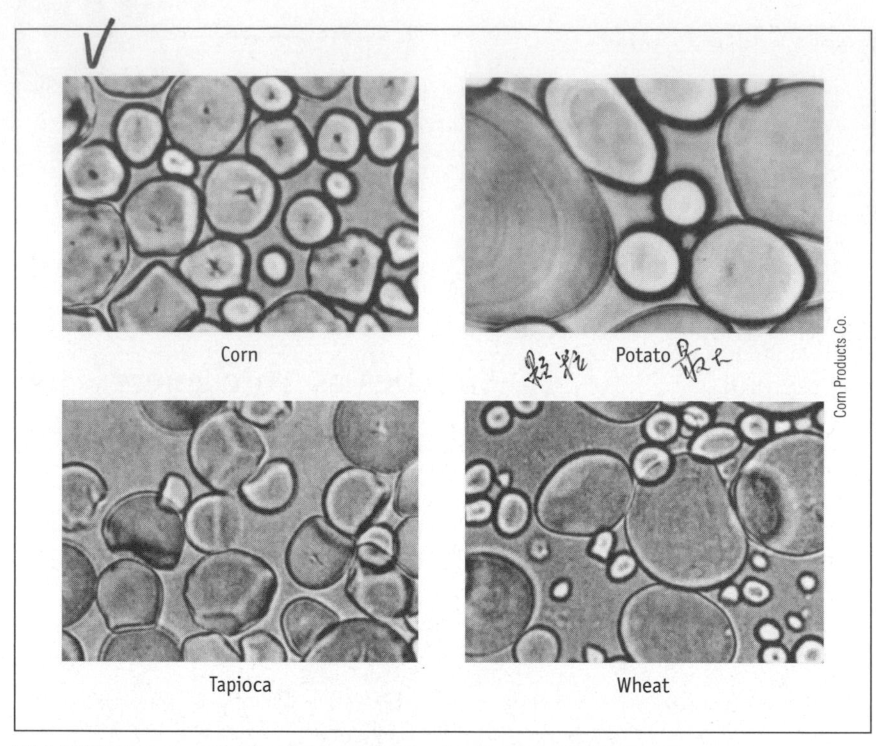

FIGURE 20-1 Starch granules differ in size based on their source.

stews, cream-style corn, cream fillings, custards, fruit pie fillings, whipped toppings, and icings. Certain puddings, candies, gums, and salad dressings are also thickened with starch.

Edible Films. Starch films are used as a protective coating for chewing gums, to bind foods such as meat products and pet foods, and to act as a base on the food for holding substances (such as flavor oils) in chocolates.

FIGURE 20-2 Examples of foods containing starches.

Bakery products
- Cake mixes
- Cream pie fillings
- Cupcakes
- Frostings
- Glazes
- Fruit pie fillings
- Refrigerated cookie doughs

Breadings
- Breaded mushrooms
- Breaded zucchini
- Fried chicken
- Fried fish
- Onion rings

Beverages
- Carbonated sodas
- Fruit juices
- Fruit drink mixes

Canned foods
- Chili
- Chow mein
- Canned pastas
- Cream-style corn

Condiments

Confections
- Candies
- Fillings for chocolates
- Gum

Dairy products
- Cheese powders
- Cheese spreads
- Dips
- Sour creams
- Yogurts

Dressings
- Salad dressings
- Sandwich spreads

Frozen foods
- Ice cream
- Pasta
- Pot pies
- TV dinners

Gravy and sauces

Meat products
- Luncheon meats
- Sausage
- Turkey loafs

Preserves
- Jam
- Jelly

Soups, stews
- Chowders
- Dry soup mixes
- Soups
- Stews

Toppings
- Butterscotch
- Marshmallow

Source: Whistler and BeMiller, *Carbohydrate Chemistry for Food Scientists* (Eagen Press, 1997).

Dextrose. Starch consists of repeating units of glucose, which can be broken down into their individual units for use as a sweetener in the production of confections, wine, and some canned goods. The food industry refers to this glucose derived from starch as dextrose, and measures the degree of conversion from starch to glucose in Dextrose Equivalents (DE) (26) (see Chapter 9).

A hydrolysis product of starch is considered a starch hydrolysate (6). These starch hydrolysates have been used as sweeteners for over 3,000 years, starting most likely with candy made in China with the aid of maltose-bearing syrups derived from rice starch. The most common solid starch hydrolysates used now are dry maltodextrins ($DE < 20$) and corn syrup solids ($DE = 24$ to 48) (25).

Starch Syrups. Over half of all the starch produced in the United States is converted eventually to syrup (11). Corn syrup made from corn starch is added to a large assortment of foods, including soft drinks, canned fruits, jams, jellies, preserves, frozen desserts, confections, frozen fruits, fountain syrups, and many others (8).

Starch Structure

As discussed in Chapter 2, starch is a polysaccharide consisting of long chains of repeating units of glucose molecules linked together either in the form of amylose, which is made up of largely linear molecules, or amylopectin, whose molecules are highly branched (30) (Chemist's Corner 20-1). The amount and proportions of amylose and amylopectin found in starches vary according to its plant source (Table 20-1), but most starch contains about 75 percent amylopectin and 25 percent amylose (14). All starches contain some amylopectin, but a few consist entirely of amylopectin, and these are known as "waxy" starches (5, 31). It is the varying amylose content that causes texture differences in starch-containing foods (24).

TABLE 20-1
Proportion of Amylose and Amylopectin in Various Starch Sources

Starch	*Amylose*	*Amylopectin*
Potato	21	79
Tapioca	17	83
Corn	28	72
Waxy maize	0	100
Wheat	28	72

Source: Cereal Foods World

Starch Characteristics

Starches have the capacity to go through the processes of gelatinization, gel formation, retrogradation, and dextrinization. Though some of these are more useful than others, it is this capacity that makes them so valuable in food preparation. The concentration of amylopectin and amylose in starch determines the degree to which these processes take place (Chemist's Corner 20-2). Starches can also be modified chemically or physically to better serve specific purposes.

Gelatinization

Gelatinization occurs when starch granules are heated in a liquid (1). When the liquid is heated, the hydrogen bonds holding the starch together weaken, allowing water to penetrate the starch molecules, causing them to swell until their peak thickness is reached (13). The swelling of the starch granules increases their size many times over. The increased volume and gumminess associated with gelatinization radically changes the texture of many foods. Pasta, rice, oats, scalloped potatoes, and most sauces, soups, and puddings are very different in consistency before and after cooking, and this is because of gelatinization (10).

KEY TERMS

Gelatinization The increase in volume, viscosity, and translucency of starch granules when they are heated in a liquid.

Factors Influencing Gelatinization. Gelatinization is dependent on a number of factors, including the amount of water, the temperature, timing, stirring, and the presence of acid, sugar, fat, and protein. These factors need to

Chemist's Corner 20-1

Amylose and Amylopectin

The varying ratios of amylose to amylopectin cause the texture differences seen in different starches. Amylose consists of a linear molecule containing 500 to 2,000 glucose units linked through alpha-1,4 linkages. Amylopectin contains from 10^4 to 10^5 glucose units; they are highly branched and are bound with both alpha-1,4 and alpha-D-1,6 linkages (Figure 20-3) (20).

FIGURE 20-3 Two types of starch.

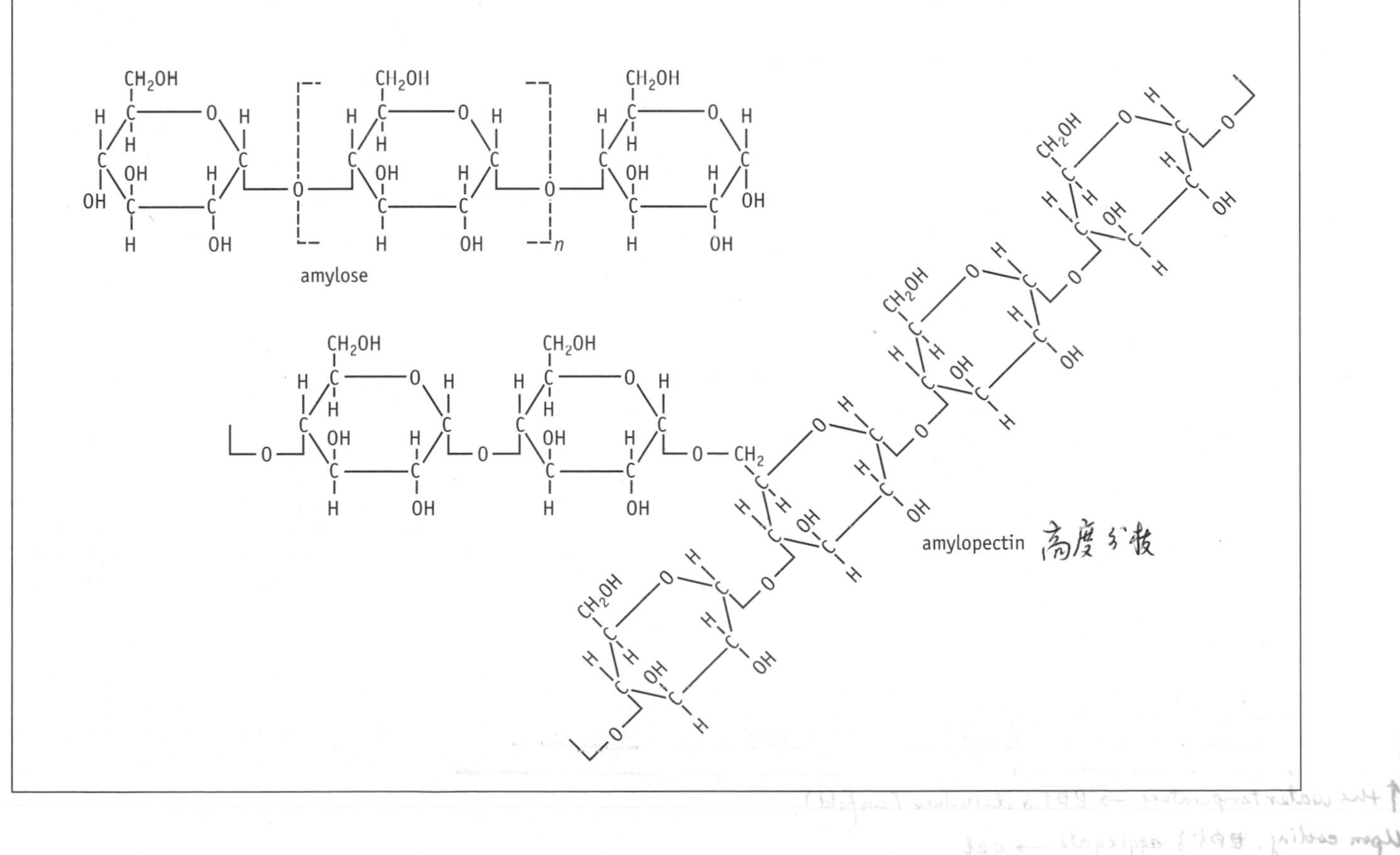

Chemist's Corner 20-2

Testing for Starch

Food scientists can determine if starch granules are either amylose or amylopectin by staining them under a microscope. If they turn blue, they contain more amylose. An amber-violet color indicates more amylopectin (18).

TABLE 20-2

Critical Temperatures of Gelatinization

Starch Source	*Critical Temperature °F (°C)*	*Characteristics of Cooked Starch*
Roots and tubers (potato and tapioca)	133–158 (56–70)	Form viscous, long-bodied, relatively clear pastes; weak gel upon cooling.
Cereal grains (corn; sorghum, rice and wheat)	144–167 (62–75)	Form viscous, short-bodied pastes; set to opaque gel upon cooling.
Waxy hybrids (corn and sorghum)	145–165 (63–74)	Form heavy-bodied, stringy, clear pastes; resistant to gelling upon cooling.
High amylose hybrids (corn)	212–320 (100–160)	Form short-bodied pastes; set to very rigid opaque gel upon cooling.

work in synchrony in order for maximum gelatinization to occur.

Water. Sufficient water must be available for absorption by the starch. The amount of liquid needed for absorption depends on the concentration of amylose and amylopectin in the starch (17). When preparing starchy foods such as grain or pasta, sufficient water is added to cover, and then more to allow for evaporation and a two to three times expansion in volume.

Temperature. Table 20-2 shows that the temperature range within which gelatinization can occur varies according to the type of starch. Most starches gelatinize when heated above 133° to 167°F (56° to 75°C) (8). The temperature at which a particular starch gelatinizes falls generally within a narrow range, often referred to as its transition temperature (28). Larger starch granules, such as those of the potato, gelatinize at lower temperatures, while smaller granules, such as wheat, gelatinize at higher temperatures.

Timing. Heating beyond the gelatinization temperature decreases viscosity (Figure 20-4). Starch granules break apart when continued heating stresses the bonds holding them together.

Stirring. Stirring during the early formation of the starch paste, or the gelatinizing starch mixture, is required in order to assure uniform consistency and to prevent lumps from forming. Continued or too vigorous stirring, however, causes the starch granules to rupture prematurely, resulting in a slippery starch paste with less viscosity.

Acid. A pH below 4.0 decreases the viscosity of a starch gel. Any acidic fruit juices should be added to pie fillings or salad dressings after gelatinization has occurred. In order to maintain viscosity in canning high-acid foods, commercial food processors use chemically altered starches that are resistant to acid breakdown (see "Modified Starches" below).

Sugar. Sugar competes with starch for available water, delays the onset of gelatinization, and increases the required temperature (12). Adding too much sugar inhibits complete gelatinization and results in a thick, runny paste. Sugars differ in their ability to delay gelatinization, with the following sugars having the greatest impact (in order from least to greatest): fructose, glucose, lactose, and sucrose. Other factors contributing to the delayed gelatinization caused by sugars include reduced granular swelling and starch-sugar and starch-water interactions.

Fat/Protein. Fat or protein delays gelatinization by coating the starch and preventing it from absorbing water.

Gel Formation

Gelatinization must occur before the next step, gel formation, which is the same as gelation (see Chapter 19) (Chemist's Corner 20-3). A fluid starch paste is a sol, while a semi-solid one is known as a gel. Not all starches will gel, but among those that do, the gel forms after the gelatinized sol has been cooled, usually to below 100°F (38°C) (1). Gel formation is dependent on the presence of a sufficient level of amylose molecules, because amylose will gel and amylopectin will not. The linear amylose molecules form strong bonds, while the highly branched amylopectin molecules form bonds that are too weak to contribute to rigidity (34). The bonds that form between the amylose molecules create a three-dimensional network that traps water and increases the rigidity of the starch mass (2).

Regular cornstarch contains large amounts of amylose, which makes it a good gelling agent. High-amylose starches form more opaque gels than low-amylose starches such as potato and tapioca, which form clearer gels. Potato starches are often used to thicken foods such as soups and sauces in which gel formation during cooling would be undesirable (32). Waxy hybrids, a cross between corn and sorghum, contain very little amylose and do not gel (19). Commercially, regular cornstarch is most often used to thicken puddings and fruit juices, while waxy cornstarch is preferred for the manufacture of canned and frozen food products. Waxy maize starches are used in many frozen foods because other starches lose some of their thickening power when frozen.

Retrogradation

As the gel cools, bonds continue to form between amylose molecules, and **retrogradation** occurs (9, 16). This retrogradation is accelerated by freezing, so the starches used in frozen food products usually come from low-amy-

FIGURE 20-4 Gelatinization: Starch granules heated in liquid initially swell, increasing viscosity. Heating beyond the maximum obtainable volume ruptures the starch granules, resulting in less viscosity.

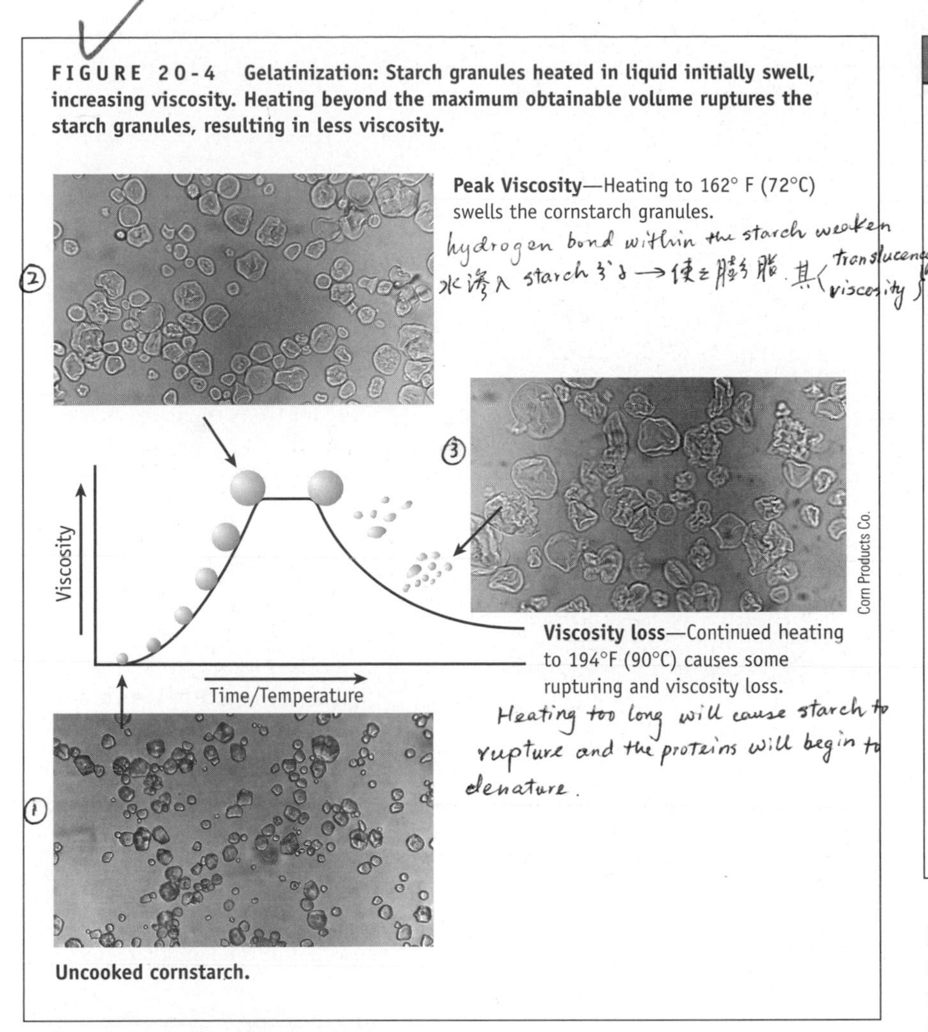

Uncooked cornstarch.

Chemist's Corner 20-3

Gelatinization Under a Microscope

The process of gelatinization can also be detected by looking at the starch granules under a microscope. Starch molecules are stored within the cell's starch granules, and the high degree of order within these granules is depicted by their birefringent nature, which enables them to refract light in two directions. Figure 20-5 shows this birefringence depicted by "dark crosses" under a microscope providing polarized light. Once gelatinization occurs, this birefringence disappears, indicating a loss of crystallinity (22).

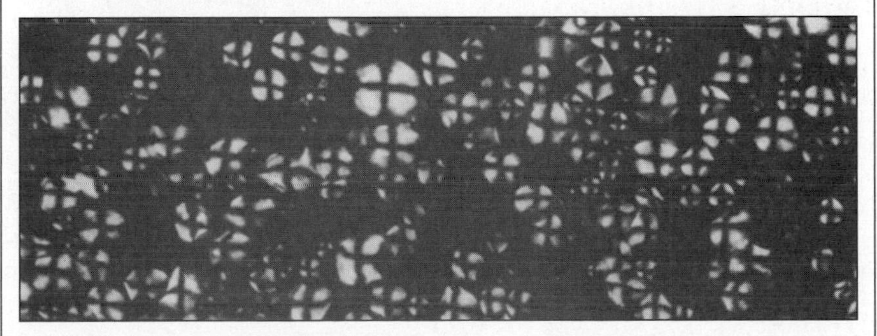

FIGURE 20-5 Cornstarch molecules.

Chemist's Corner 20-4

Resistant Starch

A small amount of starch, called resistant starch, is not digested in the small intestine and therefore does not contribute calories (kcals) (35). The three types of resistant starch include physically inaccessible starch, resistant starch granules, and retrograded starch (4). Physically inaccessible starch includes those starch granules trapped in the food that are prevented from gelatinizing. Resistant starch granules are those that are indigestible because of their chemical configuration. Retrograded starch occurs during processing, in which the heating and subsequent cooling of a starch renders the molecules of amylose and amylopectin inaccessible to enzymatic hydrolysis (3).

lose sources such as waxy corn or sorghum. The best way to prevent retrogradation is to use the gelled food as soon as possible (Chemist's Corner 20-4).

Dextrinization

Another process characteristic of starches is **dextrinization**. The result is an increase in sweetness. A side effect is that dextrinized starches lose much of their thickening power because they

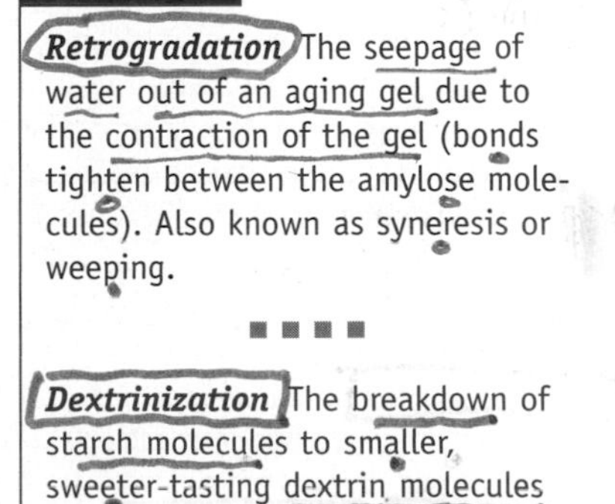
KEY TERMS

Retrogradation The seepage of water out of an aging gel due to the contraction of the gel (bonds tighten between the amylose molecules). Also known as syneresis or weeping.

Dextrinization The breakdown of starch molecules to smaller, sweeter-tasting dextrin molecules in the presence of dry heat.

have been broken down into smaller units, so more flour is required to thicken gravy if the flour has been browned in the gravy-making process. The traditional darkened flour used in Louisiana gumbo has almost no thickening power left, although it adds tremendous flavor to the stew.

FIGURE 20-6 Dextrinization: The formation of dextrins from amylopectin. Each "G" represents a glucose molecule.

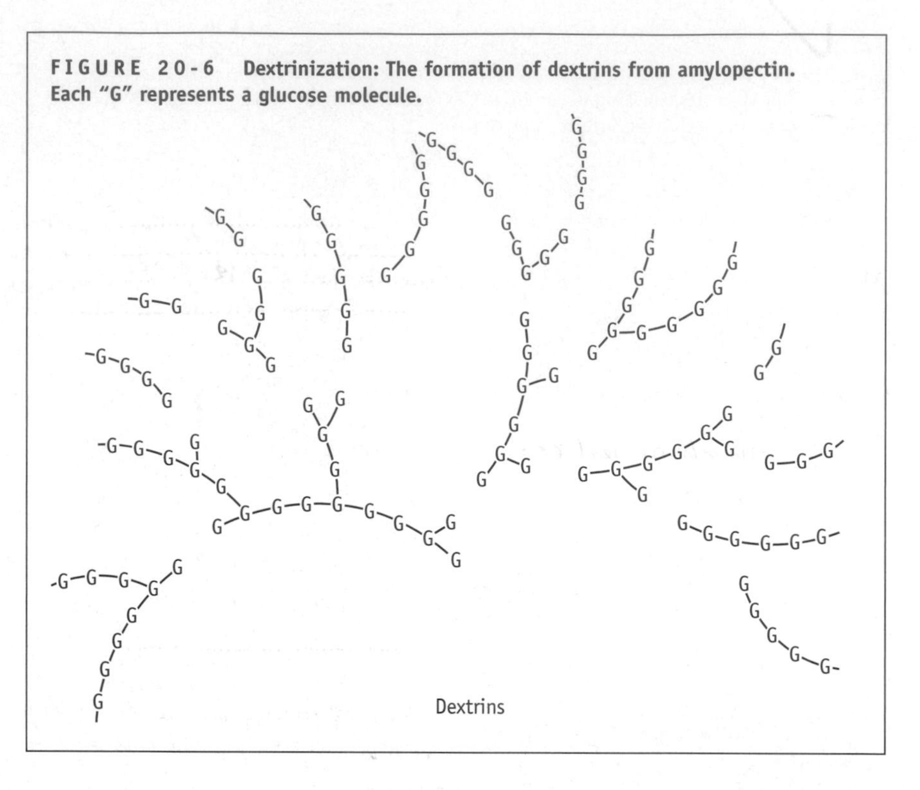

HOW & WHY? ?????????

Why does toasted bread taste sweeter than untoasted bread? Toasting or browning breaks down amylose and amylopectin (Figure 20-6), and the resulting dextrins cause toast to taste noticeably sweeter than the original bread. Gravies and cooked commercial breakfast cereals also taste sweeter than their unprocessed ingredients because of this process.

Modified Starches

Some starches are sold only to food service operations and food companies and are not encountered at the retail level. These starches have been altered to yield a wide variety of **modified starches**, extending their usefulness in food processing (7, 33). The modifications may affect the starch's gelatinization, heating times, freezing stability, cold water solubility, or viscosity. Three types of modified starches include crosslinked starch, oxidized starch, and instant or pregelatinized starch.

Crosslinked Starch. One of the more commonly used modified starches, crosslinked starch has been treated chemically to link the starch molecules together with crossbridges (Figure 20-7). Crosslinking makes a starch more heat resistant. This type of starch is less likely to lose viscosity when exposed to heat (15, 29), making it ideal for use in cooked or canned foods such as pizza, spaghetti, cheese, barbecue sauces, pie fillings, bakery glazes, and puddings (18). Incorporating crosslinked waxy cornstarch into a pie filling significantly improves its appearance and texture when compared to one that utilizes unmodified cornstarch (Figure 20-8).

KEY TERMS

Modified starch A starch that has been chemically or physically modified to create unique functional characteristics and increase the starch's usefulness.

Oxidized Starch. Starches may also be modified by exposing them to chemical oxidizers. Oxidized starches become less viscous than crosslinked starches, but they are clearer and more appropriate for use as emulsion stabilizers and thickeners. The more powder-like consistency of oxidized starches also makes them ideal for dusting foods such as marshmallows and chewing gum (18).

FIGURE 20-7 Crosslinking within the starch granule of a modified starch.

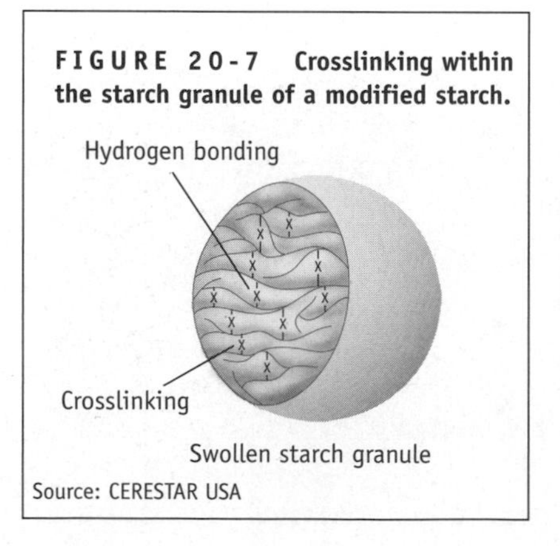

Source: CERESTAR USA

Instant or Pregelatinized Starches. These starches do not have to be heated in water to expand and gel (23). They have already been cooked and dried, so when cold water is added they absorb it immediately and expand

Pie made with unmodified cornstarch exhibits syneresis.

Food Technology

Pie made with modified starch (crosslinked waxy cornstarch) exhibits good appearance.

FIGURE 20-8 Pies made with and without modified cornstarch.

on the spot. Instant starches make it possible for the food industry to sell instant dry-mix puddings, gravies, and sauces. The dry-mix puddings made from instant pregelatinized starch, listed as "modified cornstarch" on the ingredient label, are easily prepared by adding cold milk and mixing briefly. Mixes that require heat for preparation contain plain "cornstarch."

Sauces

Since so many sauces contain starch, it is appropriate to discuss sauces in conjunction with starches. The purpose of a sauce is to enhance a food's flavor, texture, moisture, and appearance. The sauce itself should have a good consistency, a distinctive flavor, and a shine characteristic of that particular sauce. The number of different sauces, with wide variations in consistency, temperature, seasoning, and richness, makes categorizing them difficult. For purposes of our discussion, sauces will be distinguished by whether they are thickened or unthickened, and by their major unique ingredients. The purpose of this section on sauces is to provide an overview of sauces along with their functions, discuss their basic ingredients and preparation methods, and finish with safe storage since they often contain animal products (meat stock, milk/cream, and eggs).

Functions of Sauces in Foods

Sauces perform a myriad of functions in foods, not limited to adding moistness, flavor, texture, body (especially to soups), and appearance through their rich color and shine. Fish, meats, and vegetables are much more appetizing and appealing when accompanied by sauces. Despite their positive benefits to the palate, sauces are often not prepared because of their delicate nature, apparent mysterious preparation techniques, and extra required time and effort. However, a basic understanding of sauces will eliminate the mystery and open the opportunities of them being used more in foods.

Types of Sauces. The major sauces used in food preparation are thickened sauces, including cheese sauce, white sauce, and some gravies; and unthickened sauces, including other gravies, hollandaise, butter, fruit, barbecue, tartar, and tomato sauces (Table 20-3). More frequently, condiments such as mayonnaise, relish, chutney, salsa, ketchup, and mustard serve the same purpose, although they are not generally referred to as sauces.

Whether thickened or unthickened, sauces can be grouped into mother sauces and small sauces.

Mother Sauces. There are five groups of **mother sauces**, also known as grand, leading, or major sauces:

- Béchamel, or white, sauce
- Espagnole, or brown, sauce
- Hollandaise sauce
- Tomato sauce
- Velouté sauce

Small Sauces. With the exception of the tomato and hollandaise sauces, the mother sauces are not usually used as sauces by themselves, but rather serve as the base for many other secondary or **small sauces**. Examples of small sauces include cheese, cream, curry, mushroom, and shrimp sauces (Table 20-4 on page 377).

Thickened Sauces

Ingredients of Thickened Sauces. Thickened sauces rely predominantly on the gelatinization of starches for their smooth texture. The three ingredients that serve as the foundation of a thickened sauce are a liquid, a thickening agent, and seasonings and/or flavorings.

Liquid. Any liquid can serve as the body of the sauce, but the most common are:

- White stock from chicken, veal, or fish (velouté)
- Brown stock from beef or veal (espagnole)
- Milk (béchamel)
- Clarified butter (hollandaise)
- Tomato juice or purée (tomato sauce)

KEY TERMS

Mother sauce A sauce that serves as the springboard from which other sauces are prepared.

Small sauce A secondary sauce created when a flavor is added to a mother sauce.

Glaze A flavoring obtained from soup stock that has been concentrated by evaporation until it attains a syrupy consistency with a highly concentrated flavor.

Thickening Agent. A thickening agent, usually starch, may be added to the liquid to make it more viscous. Like gravies, sauces can be made with or without a thickening agent. Wheat flour is most frequently used as the thickening agent in North America and Europe, while cornstarch is preferred for Chinese food. The 10 percent by weight protein content of wheat starch creates a cloudier appearance and leaves a more floury taste than the purer starch content of cornstarch, which yields a glossy, translucent sheen. Arrowroot gives a result that is even clearer than cornstarch, but its cost limits its use. Pregelatinized or instant starches speed up the preparation process because they thicken immediately in cold water and do not have to be heated.

Seasonings/Flavorings. The most common seasonings and flavorings for starch-based sauces are salt, black pepper, white pepper, lemon juice, cayenne, herbs, and wine. Any acids in the form of vinegar, wine, tomato products, or lemon are usually added after gelatinization, because acid can break down the starch.

Many restaurants use **glazes** as flavorings for sauces. The word "glaze" is from the French *glace* for glossy. This highly flavored concentrate, which congeals when refrigerated into a shiny, rubbery mass, may be obtained from either meat (3), chicken, or fish stock. Since glazes take up much less shelf space than the stocks from which

TABLE 20-3
Thickened and Unthickened Sauces

THICKENED SAUCES	*Basic Ingredients*	*Uses*
Cheese sauce	1. White sauce with cheese added. 2. Melted cheese with cream or milk added.	Used for egg dishes, fondues, pasta dishes, and vegetables.
Custard sauce	Eggs, milk, flavoring.	Used in a variety of desserts.
Gravy	Meat stock or juices (extracted from meat during the cooking process) added to flour and possibly liquid.	Pan gravy is served with the meat from which it was derived or with mashed potatoes, vegetables, or breads.
White sauce	Butter, flour, milk, salt, and pepper.	Basic sauce that can be as thin or as thick as desired. Can have added ingredients. It can be used for fish, poultry, and vegetables.
Variations of white sauce	Variations of white sauce are used for caramel, butterscotch, lemon, honey, and other sauces.	These sauces are used for ice creams, steamed puddings, cottage pudding, soufflés, and bread puddings.
UNTHICKENED SAUCES	***Basic Ingredients***	***Uses***
Barbecue sauce	Ketchup, onion, vinegar, Worcestershire sauce, and other seasonings. Green pepper optional.	For steaks, burgers, ribs, barbecued ribs, etc.
Butter sauce	Butter, lemon juice, chopped parsley.	A sauce for fish dishes.
Chocolate sauce	Chocolate, water, and/or milk.	Used for cakes, ice cream, and parfaits.
Fruit sauces	Mostly mashed berries and sugar. Optional: egg white, butter. Added ingredients: sugar, water, cornstarch, etc. (Adding cornstarch makes these thickened sauces)	Fruit sauces are usually not cooked. They are served over ice cream, cottage cake, cheesecake, or in parfaits, etc.
	Apple	For game, goose, pork
	Cranberry	For turkey
	Currant jelly	For game
	Orange with or without pineapple	For ham or tongue
	Orange-currant	For chicken, duck, ham, and lamb
	Raisin	For ham or tongue
Gravy	"Just the juices" extracted from meat during cooking (*au jus*).	Over meats, especially roast beef.
Hard sauce	Butter, powdered sugar, vanilla.	Creamed butter, sugar, and vanilla or other flavoring. This sauce can be used with any hot pudding, especially steamed plum pudding at Christmastime.
Hollandaise sauce	Butter, egg yolks, boiling water, seasoning, lemon juice.	An emulsion similar to mayonnaise, which is beaten vigorously until smooth and creamy. This aristocrat of sauces is used mostly for fish, asparagus, broccoli, and cauliflower.
Tartar sauce	Mayonnaise, capers, olives, parsley, chopped pickles.	A favorite sauce for fish.
Tomato sauce and ketchup	Tomatoes, onion, cloves, flour, fat, salt, and pepper.	Tomato sauces have a wide variety of uses with pasta dishes and meats—especially hamburgers, hot dogs, and Mexican dishes.

they originate and last for months when refrigerated, they are highly convenient to use.

KEY TERMS

Roux A thickener made by cooking equal parts of flour and fat.

Thickeners. One of the first steps in preparing a sauce is to add a starch thickener in the form of either a roux (roo), beurre manié (burr-mahn-*yay*), or slurry.

Roux. Wheat flour is usually used in preparing a **roux**, rather than cornstarch, potato starch, or arrowroot. Hot liquid is gradually added to the cooked flour and butter, and this combination is cooked until it reaches the desired consistency, depending on what kind of sauce is being prepared.

There are three types of roux that serve as the foundation in making thickened sauces: white, blond, and brown. Variations in the heating times of the fat-flour combination cause the differ-

TABLE 20-4
Selected Small Sauces Derived from Mother Sauces

Mother Sauces	*Base Ingredient*	*Small Sauces*
Béchamel (white) sauce	Milk	Cheddar cheese Cream Mornay Mustard
Espagnole (brown) sauce	Brown stock	Mushroom
Hollandaise sauce	Butter	Maltaise Mousseline
Tomato sauce	Tomato	Creole Portuguese Spanish
Velouté sauce	White stock Chicken stock Fish stock	Curry Mushroom Herb

KEY TERMS

Beurre manié (pronounced burr mahn-YAY) A thickener that is a soft paste made from equal parts of soft butter and flour blended together.

Slurry A thickener made by combining starch and a cool liquid.

ences in the colors and flavors, with the most cooking producing the brown roux. As the roux cooks, it becomes darker and its starchy taste lessens, but its ability to thicken is also reduced as the starch molecules are broken down by heat (27). Thus, the darker the roux, the more of it will be needed to add to the liquid for thickening purposes.

Beurre manié. In a **beurre manié**, the butter and flour are not cooked (Figure 20-9). Also unlike a roux, which is added to a sauce at the beginning, a beurre manié is whisked in, bit by bit, to a simmering sauce until it reaches its desired thickness just before serving. Since a beurre manié is not cooked, it should only be used in small amounts to prevent the taste of the sauce from becoming starchy and unpleasant. A pot of stew (about 3 quarts) generally takes about 2 to 3 tablespoons of butter in an equal amount of flour. Once cooked, the stew needs to be set aside off the heat because extended simmering can also bring out the floury taste.

Slurry. The third type of thickener, **slurry**, is made by gradually mixing cold water, which will not cause the starch granules to expand, into either cornstarch or flour to make a fairly thin liquid. This slurry may then be mixed gradually into a simmering liquid sauce base. Under the simmering heat, the starch granules then expand and the sauce thickens. Typically, pan gravies are made with a slurry of flour and water. Slurry-thickened sauces are inferior to roux-thickened sauces, because a slurry can leave behind a starchy taste and be less stable than other sauces.

Preparing a Sauce from a Roux. The most common sauces that depend on the formation of a roux are white sauces and thickened gravies. A basic white sauce is a combination of butter, flour, milk, and seasonings. Whether it is a thin, medium, thick, or very thick white sauce depends on the amount of fat and flour added (Table 20-5). Veal, chicken, or fish stock can be substituted for the milk, and each will add its own distinctive flavor. Thickened gravy is easily created by adding a roux to the pan drippings of a roast.

Combining the Liquid and Roux. Once the roux has been prepared by melting the fat, mixing in the flour, and gently heating the mixture until smooth and cooked to the desired

FIGURE 20-9 Creating a beurre manié.
1. Equal parts butter and flour are used to make a beurre manié.
2. The blended beurre manié is whisked into the dish to be thickened.

Scott Phillips, *Fine Cooking*

TABLE 20-5
White Sauce Ingredient Proportions

Sauce	*Fat*	*Flour*	*Liquid*	*Salt*	*Pepper*
Thin	1 tbs	1 tbs	1 C	¼ tsp	dash
Medium	2 tbs	2 tbs	"	"	"
Thick	3 tbs	3 tbs	"	"	"
Very thick	4 tbs	4 tbs	"	"	"

KEY TERMS

Au jus Served with its own natural juices; a term usually used in reference to roasts.

degree of doneness (white, blond, brown), the next step in sauce-making is to combine the liquid with the roux. The amount of liquid added to the roux depends on the desired thickness of the sauce. The temperature of the liquid also makes a difference. It is preferable to add room-temperature or warm, not hot, liquid into a moderately hot or warm roux. Neither the roux nor the liquid should be excessively hot. If the roux is cold, the liquid may be fairly hot, but cold liquid should never be added to a cold roux or the fat in the roux will solidify. Whatever the temperatures of the various elements, the key is to add the liquid gradually and to use a whisk to blend it into the roux.

Heating the Sauce. The next step in sauce preparation is to heat the sauce just to the boiling point and then reduce it to a simmer until its maximum thickness is reached. The brief, initial boiling for one or two minutes removes the starch flavor. If the temperature is not reduced to a medium or low heat soon enough, the result will be a burned roux that is gritty and has an off-taste. Simmering the mixture for up to half an hour creates a velvety, smooth texture and further reduces the starchy taste of the flour.

Adding Seasonings/Flavorings. The finishing step in thickened sauce preparation is the addition of seasonings and other flavorings by incorporating other ingredients, such as cream, egg yolks, and additional butter.

Cheese Sauce. Adding cheese to a basic white sauce creates a cheese sauce. Several problems may occur in commercial cheese sauce production, including but not limited to poor texture, oiling off, syneresis, and oil streaking. Table 20-6 suggests a number of modifications to remedy these problems.

Preventing Lumps. The achievement of a smooth, lump-free sauce depends on the principle of separating the starch granules from each other to prevent them from being trapped in a ball surrounded by a film of gelatinized starch. Some ways to avoid lumps include:

- The fat and flour in a roux should be blended until smooth before adding the liquid. If the flour is coated with fat in this way, it will not form lumps when it contacts the liquid.
- A small amount of sugar may be added to separate the granules, but care must be taken: too much sugar will cause the sauce to be irreversibly runny.
- A small amount of the starch (2 tablespoons) may be vigorously mixed with cold water in an enclosed jar before incorporating it into the rest of the liquid to be added to the roux.

Unthickened Sauces

Sauces prepared without a starch or any other thickening agent are considered to be unthickened. Briefly discussed below are examples of unthickened sauces such as some gravies, and butter, fruit, tartar, barbecue, and tomato sauces. Some salad dressings also fall into this category of unthickened sauces; they were discussed in Chapter 19.

Gravy. Gravy is made from the juices or drippings remaining in the pan after meat or poultry is cooked. The drippings can be served thickened and with added seasonings as described previously, or unthickened, which is commonly referred to as **au jus** (oh zhue), meaning "in its own juice." Roast beef is frequently served au jus. The basic method for making gravy from the pan drippings of roasted meat or poultry involves the five steps of degreasing, deglazing, reduction, straining, and seasoning (Figure 20-10).

Step 1—Degreasing. Degreasing consists of separating the liquid from the fat. After cooking, the meat or poultry is removed from the pan so the liquids and residues on the bottom of the pan can be separated from the fat. There are several possible methods of degreasing the sauce:

- Refrigerating the mixture so the fat can rise to the surface, harden, and be removed.
- Tipping the pan so the fat can be skimmed or spooned off from the pan juices.

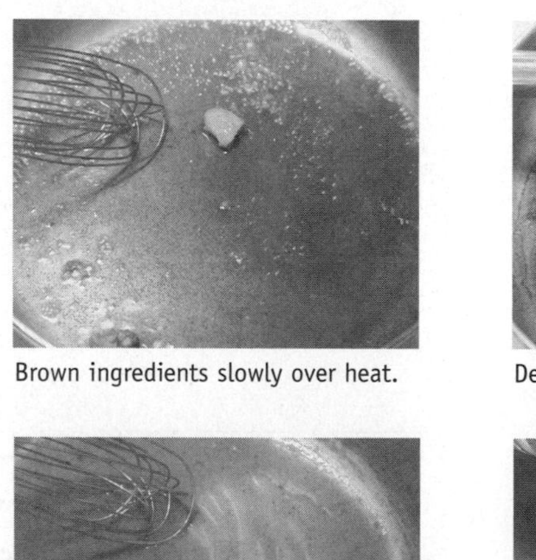

Brown ingredients slowly over heat.

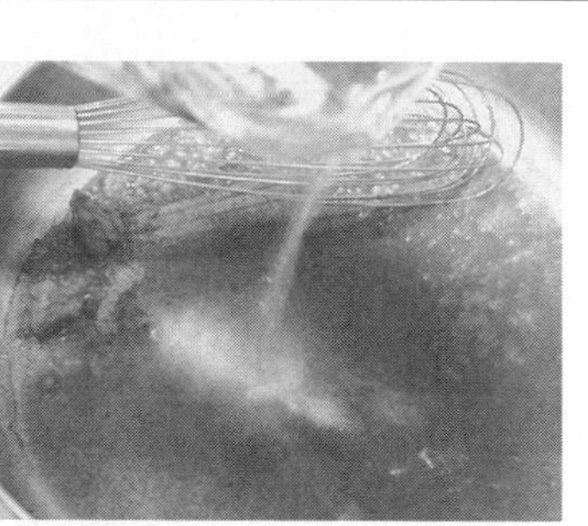

Deglaze with water one cup at a time.

Further reduce to concentrate the sauce.

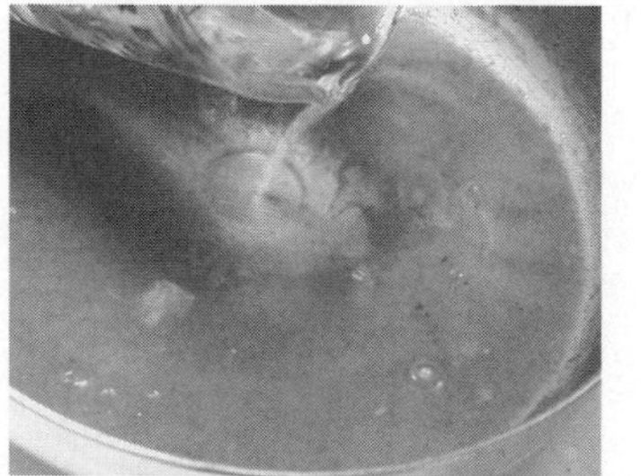

Add more water, deglaze, and reduce again.

FIGURE 20-10 Preparing gravy through deglazing and reducing.

- Using special utensils that permit the fat and liquid to separate so that only the liquid is poured off (Figure 20-11).

Step 2—Deglazing. The next step in making gravy is to **deglaze** the pan. The liquid used may be water, but stock, wine, beer, milk, cream, tomato juice, or vegetable juice may also be suitable, depending on the kind of taste result that is preferred.

FIGURE 20-11 Fat separator allows fat to rise to the top so remaining liquid can be poured out separately.

KEY TERMS

Deglaze Adding liquid to pan drippings and simmering/stirring to dissolve and loosen cooked-on particles sticking to the bottom of the pan.

TABLE 20-6
Troubleshooting Problems in Commercial Cheese Sauce Production

Symptom	*Causes*	*Changes to Make*
Poor texture	Viscosity too thick	Decrease level of thickening agents and check thickening agent selections. Reduce heat exposure during processing. Minimize evaporation losses during processing.
	Viscosity too thin	Increase level of thickening agent and check thickening agent selections. Increase heat exposure during processing.
	Lumpiness or gumminess	Decrease level of thickening agents or whey protein concentrate. Increase blending time for uniform dispersion and hydration of thickening agents.
	Graininess	Ensure uniform blending. Raise pH to avoid protein precipitation.
	Grittiness or sandiness	Reduce level of salt in formulation. Reduce cooking to decrease extent of protein denaturation. Decrease lactose content to control crystallization.
Oiling off	Emulsion breakdown	Increase levels of emulsifiers or emulsifying salts. Use less heat. Increase blending time to disperse ingredients more uniformly.
Syneresis, weeping	Colloidal system breakdown	Increase stabilizer level and check stabilizer selection. Use more casein or cheese or less whey to bind water. Use less heat to decrease extent of protein denaturation. Ensure proper blending to hydrate stabilizer. Check mineral balance.
Oil streaking	Emulsion breakdown	Increase levels of emulsifiers or emulsifying salts, use less heat, or increase blending action to disperse ingredients more uniformly.
Color streaking	Uneven color distribution	Increase blending to disperse ingredients more uniformly.
	Poor color solubility	Select color with solubility properties to match the system.
Skin formation	Air exposure and evaporation	Stir and cover during usage and cover during storage.
Rancid flavor	Lipase activity	Check incoming ingredients for off-flavors. Avoid mixing raw milk with pasteurized milk product.
Oxidized flavor	Oxidation	Minimize exposure to light and maintain cool temperatures during product storage. Check incoming ingredients for off-flavors.
Bitter flavor	Protein breakdown to bitter peptides	Use milder varieties or less aged cheeses. Check incoming ingredients for bitter notes. Check age of cheese sauce.
Improper flavor intensity	Flavor too strong	Use mild-flavored or less aged cheeses. Decrease level of flavoring agents. Decrease amount of enzyzme-modified cheeses in blend.
	Too little flavor	Use strong-flavored or more aged cheese. Add more flavoring ingredients. Use enzyme-modified cheese in blend.

KEY TERMS

Reduction To simmer or boil a liquid until the volume is reduced through evaporation, leaving a thicker, more concentrated, flavorful mass.

Step 3—Reduction. After deglazing comes **reduction**, which concentrates the volume and flavor of the contents in the pan.

Step 4—Straining. At this point, unthickened gravies are often strained through a cheesecloth, strainer, sieve, or china cap to remove large particles.

Step 5—Seasoning. Seasoning is the last step, because if it is done before reduction, the result might be an overly seasoned or salty sauce.

Hollandaise Sauce. Egg yolks added to a base of butter are the main constituents of a hollandaise sauce. A small amount of water, salt, vinegar (or lemon juice), and cayenne are also incorporated. The key to this sauce is to heat the ingredients without curdling the egg yolks, and this is most easily accomplished by using indirect heat, usually in a double boiler. Hollandaise sauce is particularly susceptible to contamination because after the egg is added, it is not cooked to a temperature high enough (140°F/60°C) to kill bacteria.

Barbecue Sauce. Barbecue sauces consist of tomato products and water, juice, or some other liquid as a base. Sugar, sautéed onions, vinegar, and seasonings or flavorings such as chili powder, pepper, salt, and dry mustard distinguish one barbecue sauce from another. All the ingredients are combined in a saucepan and simmered for at least 20 minutes.

Butter Sauce. The simplest butter sauce is plain melted butter, but many variations are possible. These variations, such as clarified, brown, and black are described more fully in Chapter 10.

Fruit Sauce. Sauces based on sugar and fruit are relatively easy to prepare. Some fruit sauces are thickened by adding cornstarch or flour and briefly heating the concoction until the starch gelatinizes.

Tartar Sauce. This sauce is prepared by adding chopped pickles, onions, capers, parsley, mustard, or shallots, or some combination of these, to a mayonnaise base. No cooking is required.

Tomato Sauce. As one of the mother sauces, tomato sauce is a popular base for many foods. Spaghetti sauce is one of the most common and familiar tomato sauces. Beef or chicken stock serves as the base and any of a number of ingredients are added—chopped onion, tomato purée, tomato paste, canned or fresh tomatoes, cooked ground beef or poultry, and Italian herbs such as basil, oregano, sage, and marjoram.

Storage of Starches and Sauces

The quality of dry starches deteriorates with improper storage. Like any other grain product, they should be kept in airtight containers and stored in a cool, dry place away from moisture, oxygen, light, and pests. Many foods made with starches contain eggs, milk, cream, or other dairy products, all of which make them prone to bacterial contamination and thus to foodborne illness. Sauces made with these ingredients should be kept out of the temperature danger zone. Thickened sauces should also be prepared, served, and stored with caution. These products should be stored in the refrigerator and never left to sit for long at room temperature.

PICTORIAL SUMMARY / 20: Starches and Sauces

As a complex carbohydrate, starch, in the form of potatoes, rice, noodles, and sorghum is a mainstay of diets throughout the world. In food preparation, starch contributes to the texture, taste, and appearance of many foods such as sauces, gravies, cream soups, salad dressings, and desserts.

STARCH CHARACTERISTICS

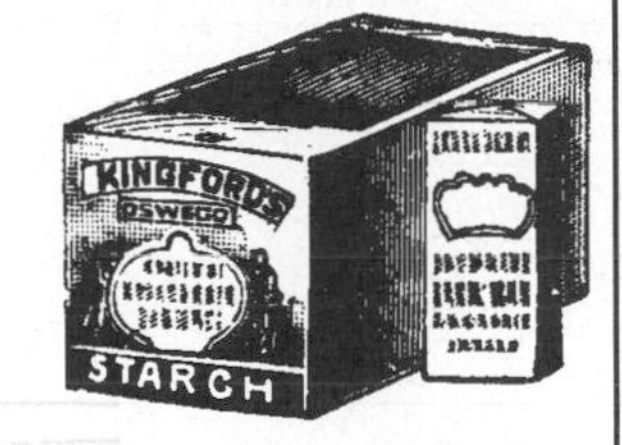

Starches are derived from the seeds and roots of various plants. The size and shape of starch granules differ according to their source. Starches provide 4 calories (kcal) per gram and are a good source of energy. Starches are used in many food products as thickening agents, edible films, and sweeteners (dextrose and syrups).

STARCHES AS THICKENERS

When starch granules are heated in a liquid they absorb the liquid, swell, and increase in viscosity and translucency —a process called gelatinization.

Factors affecting gelatinization include:

- **Water:** Sufficient for absorption by the starch.
- **Temperature:** Transition temperatures for various starches differ, but most are 133° to 167°F (56° to 75°C).
- **Timing:** Heating beyond the gelatinization temperature decreases viscosity.
- **Acidity:** A ph below 4.0 decreases the viscosity of a starch gel. Acidic fruit juices should be added after gelatinization.
- **Fat/Protein:** These delay gelatinization by coating the starch and preventing water absorption.

SAUCES

Starches are used to thicken sauces and gravies. The basic ingredients of most sauces are liquid, optional thickening agent, and seasonings and/or flavorings.

The wide array of sauces makes it difficult to classify them, especially as they differ so much in consistency, temperature, seasoning, and richness.

Sauces used in food preparation:

Thickened	Unthickened	
■ Cheese	■ Hollandaise	■ Barbeque
■ White sauce	■ Butter	■ Tomato
■ Some gravies	■ Fruit	■ Some gravies
	■ Tartar	

The five groups of mother sauces:

- Velouté
- Espagnole/brown
- Béchamel/white

These three serve as a base for secondary sauces

- Hollandaise
- Tomato

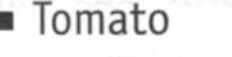

STORAGE OF STARCHES AND SAUCES

Dry starches are stored like any other grain product—in an airtight container in a cool, dry place away from moisture, oxygen, light, and pests. Foods and sauces made with starches are particularly prone to microbial contamination when they contain eggs, milk, cream, or any other dairy product, and should be stored in the refrigerator.

REFERENCES

1. Aguilera JM, and E Rojas. Rheological, thermal and microstructural properties in whey protein cassava starch gels. *Journal of Food Science* 61(5):963–966, 1996.
2. Bora PS, CJ Brekke, and JR Powers. Heat induced gelation of pea (*Pisum sativum*) mixed globulins, vicilin and legumin. *Journal of Food Science* 59(3):594–596, 1994.
3. Brown I. Complex carbohydrates and resistant starch. *Nutrition Reviews* 51(11):S115–S119, 1996.
4. Cairns P, et al. Physiochemical studies on resistant starch in vitro and in vivo. *Journal of Cereal Science* 23:265–275, 1996.
5. Carriere CJ. Evaluation of the entanglement molecular weights of maize starches from solution rheological measurements. *Cereal Chemistry* 75(3):360–364, 1998.
6. Chronakis IS. On the molecular characteristics, compositional properties, and structural-functional mechanisms of maltodextrins: A review. Critical Reviews in *Food Science and Nutrition* 38(7):599–637, 1998.
7. Committee on Codex Specifications. *Food Chemicals Codex.* National Academy Press, 1993.
8. Ensminger AH, et al. *Foods and Nutrition Encyclopedia.* CRC Press, 1994.
9. Fan J, and BP Marks. Retrogradation kinetics of rice flours as influenced by cultivar. *Cereal Chemistry* 75(1):153–155, 1998.
10. Galvez FCF, AVA Resurreccion, and GO Ware. Process variables, gelatinized starch and moisture effects on physical properties of mungbean noodles. *Journal of Food Science* 59(2):378–381, 1994.
11. Guzman-Maldonado H, and O Paredes-Lopez. Amyloytic enzymes and products derived from starch: A review. *Critical Reviews in Food Science and Nutrition* 35(5):373–403, 1995.
12. Hoover R, and N Senanayake. Effect of sugars on the thermal and retrogradation properties of oat starches. *Journal of Food Biochemistry* 20:65–83, 1996.
13. Hoover R, and T Vasanthan. Effect of heat-moisture treatment on the structure and physiochemical properties of cereal, legume and tuber starches. *Carbohydrate Research* 252:133–153, 1994.
14. James JA, and BH Lee. Glucoamylases: Microbial sources, industrial applications and molecular biology: A review. *Journal of Food Biochemistry* 21:1–52, 1997.
15. Kiribuchi-Otobe C, et al. Wheat mutant with waxy starch showing stable hot paste viscosity. *Cereal Chemistry* 75(5):671–672, 1998.
16. Klucinec JD, and DB Thompson. Amylose and amylopectin interact in retrogradation of dispersed high-amylose starches. *Cereal Chemistry* 76(2):282–291, 1999.
17. Leslie RB, et al. Water diffusivity in starch-based systems. *Advances in Experimental Medicine and Biology* 302:365–390, 1991.
18. Luallen TE. Starch as a functional ingredient. *Food Technology* 39(1):59–63, 1985.
19. Marrs WM. Blending starches, gums provides texture and processing benefits. *Prepared Foods* 166(10):63–66, 1997.
20. Mauro DJ. An update on starch. *Cereal Foods World* 41(10):776–780, 1996.
21. McCue N. Starches build flavor and functionality. *Prepared Foods* 166(2):73, 1997.
22. Penfield MP, and AM Campbell. *Experimental Food Science.* Academic Press, 1990.
23. Radosavljevic M, J Jane, and LA Johnson. Isolation of amaranth starch by diluted alkaline-protease treatment. *Cereal Chemistry* 75(2):212–216, 1998.
24. Raeker MO, et al. Granule size distribution and chemical composition of starches from 12 soft wheat cultivars. *Cereal Chemistry* 75(5):721–728, 1998.
25. Schenck FW. Solid starch hydrolysates. *Cereal World Foods* 41(5):388–390, 1996.
26. Singh N, and M Cheryan. Microfiltration for clarification of cornstarch hydrolysates. *Cereal Foods World* 42(1):21–24, 1997.
27. Stevens M. Making and using a roux as a classic thickener. *Fine Cooking* 19:72, 1997.
28. Vodovotz Y, and P Chinachoti. Thermal transitions in gelatinized wheat starch at different moisture contents by dynamic mechanical analysis. *Journal of Food Science* 61(5):932–937, 1996.
29. Wang LZ. Starches and starch derivative in expanded snacks. *Cereal Foods World* 42(9):743–745, 1997.
30. Wang LZ, and PJ White. Structure and properties of amylose, amylopectin, and intermediate materials of oat starches. *Cereal Chemistry* 71(3):263–268, 1994.
31. Whistler RL, and JN BeMiller. *Carbohydrate Chemistry for Food Scientists.* Eagen Press, 1997.
32. Wiesenborn DP, et al. Potato starch paste behavior as related to some physical/chemical properties. *Journal of Food Science* 59(3):644–648, 1994.
33. Yamin FF, et al. Thermal properties of starch in corn variants isolated after chemical mutagenesis of inbred line B731. *Cereal Chemistry* 76(2):175–181, 1999.
34. Yook C, UH Pek, and KH Park. Gelatinization and retrogradation characteristics of hydroxypropylated and cross-linked rices. *Journal of Food Science* 58(2):405–407, 1993.
35. Yve P, and S Waring. Resistant starch in food applications. *Cereal Foods World* 43(9):690–695, 1998.

WEBSITES

The National Starch and Chemical Company provides an excellent dictionary on food starch terms:

www.foodstarch.com

Lots of technical articles on food starch:

www.foodstarch.com/products_services/pns_techart.asp

See what starch looks like under the microscope:

http://distans.livsteck.lth.se:2800/microscopy/f-starch.htm

Cereal Grains and Pastas

Cereal grains are seeds from the grass family *Gramineae*. These seeds and their products may indeed be regarded now, as in ancient times, as "the staff of life." According to archaeological evidence, wheat and barley were being cultivated in the Fertile Crescent around 8000 B.C., and rice was cultivated in Thailand as early as 4500 B.C. (14). The importance of grains to the well-being of humans is so great that breads are religious symbols throughout the world. In Christianity, many denominations symbolize the body of Christ with a form of bread, and during Judaism's Passover celebration, unleavened bread is served to symbolize the Jews' hasty exodus from Egypt, when there was no time to wait for bread to rise.

Grains are the world's major food crops, and there are numerous varieties (Figure 21-1). The most common grains are wheat, rice, corn, and barley, which together account for 85 percent of the world's production of grains. The remaining 15 percent consists of millet, sorghum, oats, and rye (Figure 21-2).

The term "cereal" is derived from Ceres, the Roman goddess of agriculture, and refers to grains in general, not just what may be poured out of a box at breakfast. Although breakfast cereals are a common manner in which grains are consumed, they also find their way to the table as flour in baked goods and pastas, alone as vegetables (corn, hominy), as alcohol, and indirectly through the meat of animals consuming grains.

Composition of Cereal Grains

Structure

All grasses have individual kernels or grains, called caryopses, which are similar in structure. Each caryopsis has a protective outer **husk**, a **bran** covering, a starchy **endosperm**, and a **germ** (Figure 21-3 on page 385).

Husk. The husk, also called the chaff, protects the grain from frost, wind, rain, extreme temperatures, insects,

> **KEY TERMS**
>
> ***Husk*** The rough outer covering protecting the grain.
>
> ***Bran*** The hard outer covering just under the husk that protects the grain's soft endosperm.
>
> ***Endosperm*** The largest portion of the grain, containing all of the grain's starch.
>
> ***Germ*** The smallest portion of the grain, and the embryo for a future plant.

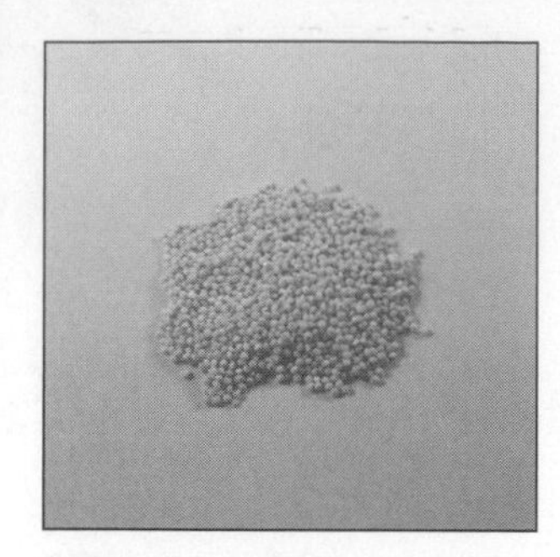

Millet. The seed of an annual, gluten-free grass which is widely eaten as a cereal in Africa and Asia. It is also used as a source of starch.

Corn. Indigenous to Mexico, corn is one of the most important cereals in the form of grain, meal, and flour. It is used to make corn bread and hominy, and is also an important source of starch and of cooking oil.

Wheat. Thought to have first been cultivated in the Nile region, it is the source of the highest-quality bread and baking flours. There are many different varieties: the durum wheat type is best known for making pasta.

Oats. Native to Central Europe, oats are used to make oatmeal and flour, and are often added to cakes and cookies.

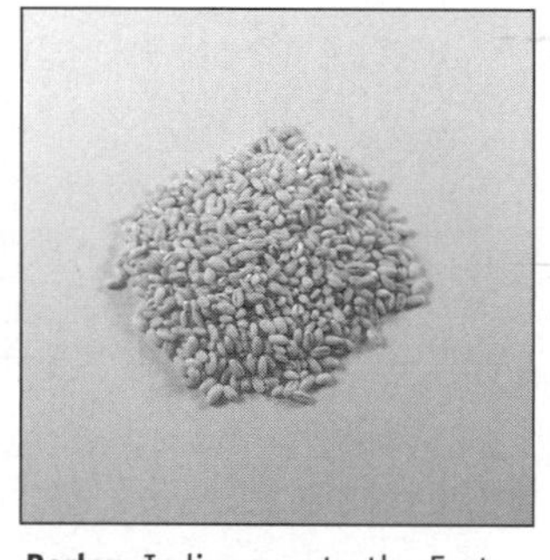

Barley. Indigenous to the East, barley is used for making malt liquor, as a side dish similar to rice, and also in soups.

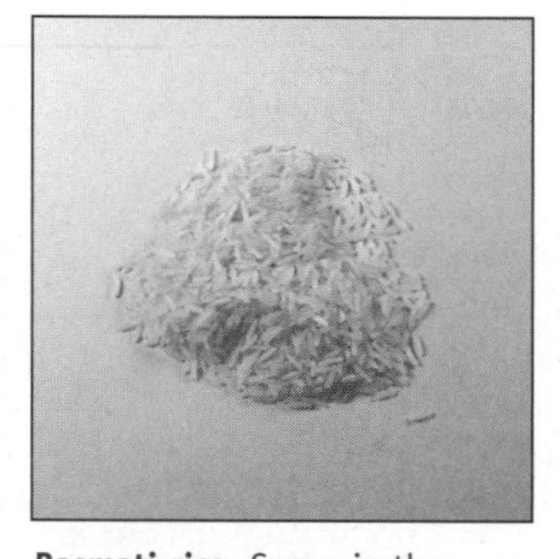

Basmati rice. Grown in the foothills of the Himalayas, this narrow long-grain rice is one of the finest. It should be soaked before cooking, and is the best rice to eat with Indian food.

FIGURE 21-1 Several of the many varieties of grain.

and other potentially damaging environmental factors. Husks are not usually consumed but are sometimes processed into fiber supplements.

Bran. Bran is about 14.5 percent of the grain by weight and is an excellent source of fiber and minerals. Just beneath the bran, there is a less fibrous coating called the aleurone layer, which contains protein, phosphorus, thiamin and other B vitamins, and some fat. Both the bran and the aleurone layer are removed from grains as they are processed into white flour.

Endosperm. The endosperm comprises about 83 percent of the grain. The starch in the endosperm makes grains, especially whole grains with their added fiber from the bran, excellent sources of complex carbohydrates. The endosperm serves as the basis for all flours, which are made by separating the endosperm from the husk, bran, and germ, and then milling it into fine powder. In whole-grain flours, the bran and germ are also milled into the flour, thus the name "whole grain."

Germ. The germ (or embryo) is found at the base of the kernel and accounts for only 2.5 percent of the grain's weight. Rich in fat, and with some incomplete protein, vitamins, and minerals, the germs are collected separately and sold as wheat germ, an excellent source of B vitamins and vitamin E. The fat content found in the germ makes it susceptible to spoilage, which is why wheat germ should be refrigerated. Also, because whole-grain flours contain the germ, they have a much shorter shelf life than pure white flours.

Uses of Cereal Grains

Most of the cereal grains produced are used for flour, pasta, and breakfast cereals. They are also used in the production of alcoholic beverages and animal feeds.

Flour

Flour is the fine powder obtained from crushing the endosperm of the grain. In the case of whole-grain flours, the bran and germ are also milled into the flour. Although any grain can be used to make flour, wheat flour is the predominant choice because it provides the protein structure that facilitates the rising of baked goods (see Chapter 22). Flour is used to make breads and an assortment of other baked products, such as biscuits, rolls, crackers, pretzels, chips, cookies, cakes, and pastries. It also plays an important role as a thickener (see Chapter 20).

Pasta

Pasta is thought to have originated in China. It comes in a variety of shapes, and is sold in both dried and fresh forms.

Breakfast Cereal

During the mid- to late 1800s, a vegetarian craze hit the United States. Among its advocates were the Seventh-Day Adventists, who hired Dr. John Harvey Kellogg to be the director of a health sanitarium. Dr. Kellogg instituted many health-promoting innovations, among them the use of a vegetarian substitute, in the form of dry cereal, for the traditional breakfast of ham or sausage and eggs. Charles William Post, another Seventh-Day Adventist, visited the sanitarium, returned home with the dry breakfast cereal concept, and started a cereal company. At that point, Dr. Kellogg's brother, William Keith Kellogg, seeing the financial success experienced by Post, introduced Kellogg's Corn Flakes to the general public, and the dry cereal industry boom began.

Now, over 75 percent of breakfast cereals are the ready-to-eat type made from wheat, corn, or oats, and it is probable that neither of their originators could have foreseen the emergence of cereals with bits of chocolate or tiny colored marshmallows in the box. New cereal products are constantly being introduced into the market by food companies. Even if they capture only 1 percent of the market, that translates into $90 million (18). The varying shapes of ready-to-eat cereal available to consumers depend on whether they have been extruded (pressed through a die), flaked, granulated, puffed, rolled, or shredded (Figure 21-4). Extrusion cooking is popular for ready-to-eat cereals, after which they are shaped, toasted, flavored, colored, and sometimes enriched or fortified before being packaged and shipped to the consumer (6, 41). Fortified cereals often provide a substantial percentage of a person's RDA for vitamins and minerals.

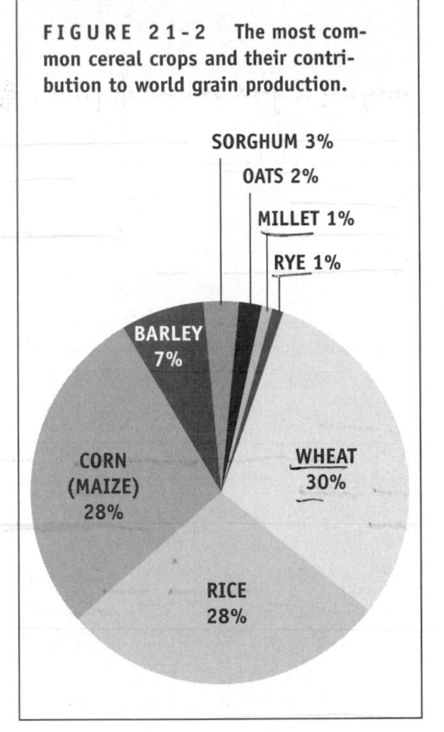

FIGURE 21-2 The most common cereal crops and their contribution to world grain production.

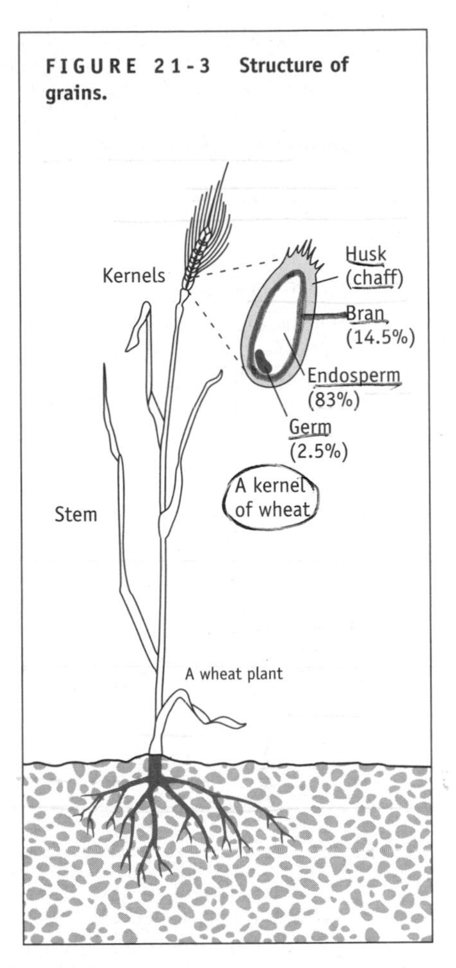

FIGURE 21-3 Structure of grains.

Source: Sizer and Whitney, *Nutrition Concepts and Controversies* (Wadsworth, 1997).

FIGURE 21-4 Processing grains into cereal.

Alcoholic Beverages

Various grains are used to make alcoholic beverages. For example, rice is used to make sake, a Japanese rice wine; and barley, rye, or corn may be used in brewing beer or distilling liquor.

Animal Feeds

Grains are important in the manufacture of livestock and pet feed. Many dried dog foods list "ground yellow corn" as the first ingredient on their labels.

Types of Cereal Grains

The three dominant grains grown within the last 5,000 to 8,000 years of human history appear to have been wheat, barley, and rice. The grains of primary importance in the world now in descending order are wheat, rice, corn, barley, millet, sorghum, oats, and rye.

Wheat

More wheat is now produced throughout the world than any other grain. The country producing the most wheat is Russia, followed by the United States. About 75 percent of all of the harvested wheat is made into flour, while the remaining 25 percent is used for cereals, pasta products, animal feed, wheat germ, and wheat germ oil.

Classification of Wheat. Wheats can be classified in a number of different ways, according to their species, growing season, texture, or color. There are fourteen different species of wheat, and each of them has a number of different varieties. As a result, there are over 30,000 varieties throughout the world, but only three species—common, club, and durum—account for almost 90 percent of all the wheat grown in the world (Figure 21-5). Among the lesser known wheats are spelt, emmer, and einkorn, referred to as "ancient" wheats because they were some of the earliest cultivated wheats (1).

The two major types of wheat defined by their growing season are winter wheat (hard) and spring wheat (soft). Hard and soft wheats also differ in their protein content, which makes wheat flour more suited than any other grain for a variety of different baking purposes. Hard wheats, of which durum is the hardest and highest in protein, are more suitable for bread and pasta production (16). The majority of wheat grown in the United States is soft, or common, wheat. Soft

NUTRIENT CONTENT

A cup of cooked cereal, grain, or pasta contains about 160 calories (kcal), 30 grams of carbohydrate, 6 grams of protein, some vitamins and minerals, and a trace of fat. Whole-grain products provide additional fiber.

Carbohydrate. Grains are an excellent source of complex carbohydrates. At least six to eleven servings a day of grains, in the form of breads, cereals, rice, or pasta, fulfill the Food Guide Pyramid's dietary recommendation that a large portion of calories (kcal) be derived from carbohydrates. Consuming grains at this level may also help meet the American Dietetic Association's recommendation that at least 20 to 35 grams of fiber be consumed each day.

Protein. Although cereals supply almost half of the dietary protein worldwide (15), their protein is incomplete because grains are low in the essential amino acid lysine. They do, however, have adequate amounts of the amino acid methionine. For this reason they are often paired with legumes, which lack methionine, to achieve protein complementation.

Fat. Cereals are very low in fat and contain no cholesterol. A slice of bread averages only 100 calories (kcal), but spreading 1 tablespoon of butter, margarine, mayonnaise, or peanut butter on the bread adds another 100 calories (kcal) and about 10 grams of fat.

Vitamins and Minerals. Unfortunately, refined grains are also low in many vitamins and minerals, because the milling process that removes the husk, bran, and germ also removes nutrients. Certain vitamins are lost during milling: 77 percent of the thiamin (B_1), 80 percent of the riboflavin (B_2), 81 percent of the niacin, 72 percent of vitamin B_6, 50 percent of pantothenic acid, and 67 percent of folate. About 86 percent of vitamin E is also lost, but is not replaced by enrichment (33). Grains that are enriched have had certain nutrients added back to levels established by federal standards. The term "enriched baked products" refers to those that have had specific nutrients replaced: thiamin (B_1), riboflavin (B_2), niacin, folate, and iron.

A further difficulty in obtaining minerals from grains is that some minerals may not be absorbed as efficiently in the intestinal tract because they bind to a compound in grains called phytate (phytic acid). A few researchers, however, question phytate's overall effect on mineral metabolism (30, 42). Yeast, commonly used to leaven bread during bread-making, breaks down the phytate and largely prevents it from interfering with mineral absorption (36).

Fiber. Whole-grain products are a good source of soluble fiber, which has been shown to reduce high blood cholesterol and help stabilize high blood glucose; and of insoluble fiber, which may help to reduce the risk of colon cancer (4, 23). Whole grains also contain phytochemicals, lignans, and phytoestrogens, compounds that are structurally similar to estrogenic substances and may have a protective effect against certain types of cancer (12). Following the wave of consumer interest in fiber, food companies have introduced high-fiber cereals that incorporate psyllium, a fiber that has been reported to help lower high blood cholesterol (29). Other fibers that have the same effect are oat and rice bran (7, 21). Other fibers that are natural gums have been added to foods such as breads, cookies, and crackers for their functional properties (40). They make such products feel smoother in the mouth and cause them to hold their shape better in the package.

FIGURE 21-5 The three most common types of wheat.

wheat produces a flour that is lower in protein and is ideal for cakes, cookies, crackers, and pastries.

Wheat may also be distinguished as being red or white, with various shades of yellows and ambers in between. Once it is milled, however, all flours appear white and lose many of their distinguishing color characteristics.

Forms of Wheat. Flour is just one of the many possible forms of wheat. Other forms on the market include wheat berries, cracked wheat, rolled wheat, bulgur, farina, wheat germ, and wheat bran.

Wheat Berries. The simplest form of wheat is wheat berries, or groats, which are whole wheat kernels that have not been processed or milled. These take the longest time to cook compared to other wheat forms.

Cracked Wheat. This consists of wheat berries that are ground until they crack. It is available in coarse, medium, or fine grinds. Cracking wheat reduces its cooking time from 1 hour to about 15 minutes.

Rolled Wheat. Flattening wheat berries between rollers produces rolled wheat, a product similar to rolled oats, but different from the extruded wheat flakes used in breakfast cereals.

Bulgur. Bulgur is a type of wheat that has been partially steamed, dried, and then cracked to produce a more pronounced flavor. Bulgur is a common ingredient in tabbouleh, a dish originally from Lebanon that is a salad of bulgur, vegetables, and herbs.

Farina. Farina (which may be familiar as the product known as Cream of Wheat) is made by granulating the endosperm of the wheat into a fine consistency.

Wheat Germ. The germ of wheat is a good source of vitamin E, unless it has been defatted, and a good source of some B vitamins and fiber. Wheat germ contains polyunsaturated fat and, unless it has been defatted, it will become rancid if not refrigerated.

Wheat Bran. Wheat bran is a source of insoluble fiber, specifically indigestible cellulose. The various fibers, along with their chemical compositions, are described in Chapter 2.

Rice

Over half the world's population relies on rice as a staple food (5). In Asia, where 94 percent of the world's rice is produced, rice is so important that it is a symbol of life and fertility. This is why rice is sometimes thrown at the bride and groom at a wedding. Per capita yearly consumption in the Orient, where people may eat rice three times a day, averages 200 to 400 pounds, compared to only about 10 pounds in the United States (14). China and India lead the world in rice production.

There is some historical evidence that Alexander the Great may have brought rice to Greece after discovering it while invading China. The Moors carried rice to Spain, whence it was transported to the West Indies and South America by Spanish ships in the early 1600s. It is believed that rice was introduced into the southeastern United States during the mid-1600s when a ship from Madagascar was damaged by storms and its captain and crew took refuge in South Carolina (14).

Classification of Rice. Rice is classified according to its mode of cultivation, grain length, and texture. Over 90 percent of all rice is grown with its roots submerged in water, and is known as lowland, wet, or irrigated rice. It is not necessary for water to cover the base of the rice plant, but this method protects it from insects and weeds, which results in a better yield. Highland, hill, or dry rice is grown in areas of plentiful rain, where the hilly terrain prevents flooding. The length in relationship to the width of the grain determines if rice is considered to be long-grain, medium-grain, or short-grain (Figure 21-6). Most long-grain rices, which are

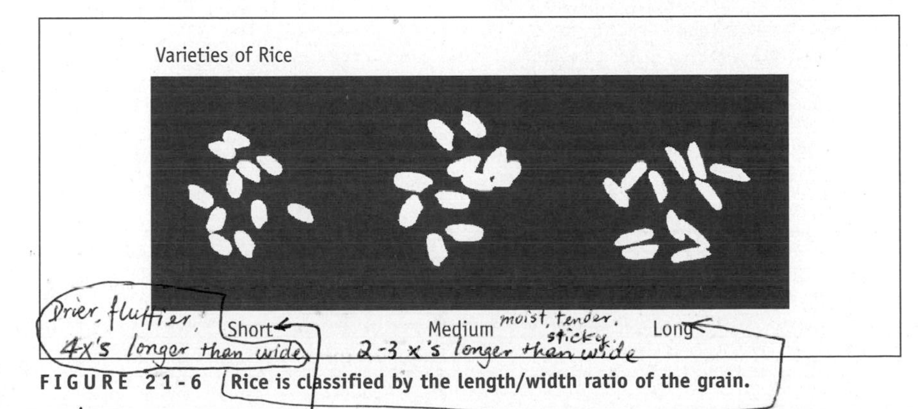

FIGURE 21-6 Rice is classified by the length/width ratio of the grain.

about four times longer than they are wide, cook to a drier, fluffier consistency, which allows the grains to separate. Medium- and short-grain rices have less amylose, which makes them stickier when cooked. The very starchy short-grain rice known as Arborio gives Italian risotto its characteristically creamy texture. It is ideal for rice pudding (13). The sticky short-grain rice with an oval shape is also preferred by the Chinese and Japanese because of its easier handling with chopsticks and preparation of sushi (Figure 21-7). Dry, long-grain rice is very difficult to eat with chopsticks. Glutinous, or waxy, rice is especially sticky when cooked. The rice varieties consumed in North America are typically nonglutinous and include long-, medium-, and short-grain rice. The differences in stickiness among the rices is due in part to the fact that long-grain rices contain more amylose, while short-grain rices are higher in amylopectin.

Forms of Rice. Rice comes in various forms, depending on its type and the way it is processed. These are white, converted, instant, brown, wild, glutinous, specialty rice bran, and wild rice. Although there are more than 40,000 different varieties of rice worldwide, white rice is the most common (31).

White Rice. White rice has been milled and polished to remove the husk, bran, and germ. By removing the bran layer and germ, all of the fiber and most of the B vitamins and iron are eliminated, although some of these are replaced in enriched grains.

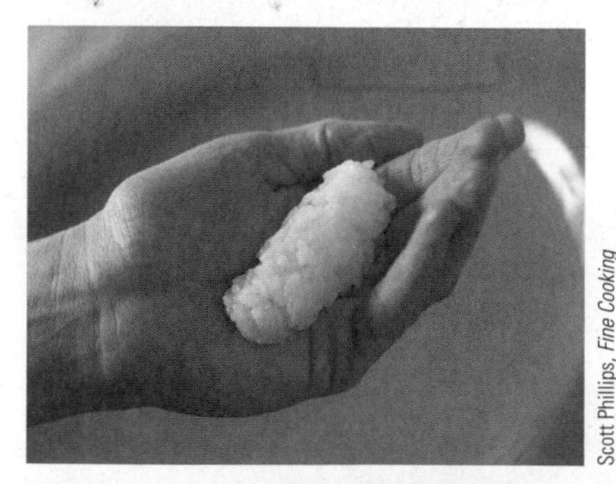

Scott Phillips, Fine Cooking

FIGURE 21-7 The sticky starch of short-grain rice makes it better suited than other rices for sushi and being picked up by chopsticks.

The exception is riboflavin (B_2), a B vitamin that is not normally added to rice because it turns the grains slightly yellow. The surface of white rice becomes smooth when it is placed in machines equipped with revolving bands of felt or leather, which rub off the grain's bran and some of the endosperm.

Converted Rice. Converted or parboiled rice is long-grain rice that has been soaked, steamed under pressure, and dried before milling. It accounts for about 20 percent of all the rice sold in North America. Converted rice is commonly used in food service establishments because the grains stay firm and separate.

Instant Rice. Instant rice is rice that has been cooked and then dehydrated, so it takes only a few minutes to prepare. The texture is inferior, however, and the grains have a tendency to split during cooking and become dry.

Brown Rice. Brown rice has had only the hull removed, leaving the bran and germ intact. The presence of the outer covering of bran results in more fiber, but longer cooking times are required and the result is a tougher texture compared to white rice.

Glutinous Rice. Also known as sweet rice, glutinous rice is slightly sweeter, stickier, and more translucent than regular white rice when heated. This rice is easily shaped and molded and is thus preferred for preparing rice dumplings, rice cakes, and sushi.

Specialty Rice. A number of long-grain rice varieties known as specialty rices have nuttier tastes, separate easily, and are more expensive. These specialty rices are not enriched. Examples include Basmati, Jasmine, Texmati, Wehani, and Wild Pecan/popcorn rice.

Rice Bran. Rice bran is just as effective in lowering high blood cholesterol as oat bran, but the bran from rice has a slightly different taste and a shorter shelf life than that from oats.

Wild Rice. Wild rice is not really rice, or even a grain, but rather a reed-like water plant (*Zizania aquatica*). It is harvested from areas in the Great Lakes region and in Canada where it grows wild, although increasingly it is deliberately cultivated. This food, historically consumed by Native Americans, has a nutty flavor and is often used to stuff game and poultry (28). It may be mixed with brown or white rice for commercial purposes, and it contains twice the amount of protein and more B vitamins than white rice.

Corn

Corn is native to the Americas, where fossilized corn pollen grains found near Mexico City have been estimated to be over 80,000 years old. Corn plays a very important role in the religious life of many Native American peoples, especially the Hopi in Arizona, among whom life from birth to death revolves around corn. Over 50 percent of the world's corn is grown in the United States. Outside the United States, corn is referred to as "maize" worldwide.

Classification of Corn. Corn is classified according to its kernel type—dent, sweet, flint, popcorn, flour, and pod (Figure 21-8)—and by its color. Yellow and white predominate, but there is also red, pink, blue, and black corn as well as corn with bands or stripes.

Dent Corn. Dent corn, so named because each corn kernel has a small dent, accounts for over 95 percent of all corn grown in the United States. Although almost half is sold as livestock feed, the rest is stored as a buffer against the next year's crop, exported, or used by food manufacturers in the production of corn syrup, alcohol, starch, and canned or other processed corn.

Sweet Corn. Most canned corn is derived from sweet corn, which can be either yellow or white. Sweet corn tastes best before the milky fluid in fresh corn kernels has had a chance to harden.

Flint Corn. Colonial settlers were taught to grow flint corn by the Native Americans. Grinding the extremely hard corn kernels makes a good quality cornmeal.

Popcorn. A special variety of corn with thick-walled kernels is used to make popcorn. During popping, the mois-

FIGURE 21-8 Corn is classified by its kernel type.

ture inside the kernels heats into steam that builds up to 135 pounds of pressure per square inch, finally bursting the kernel open and exposing the puffy starch.

HOW & WHY? ?????????

Why don't some popcorn kernels pop? Kernels that do not pop are believed to contain leaks that allow the steam to escape. Making sure the popcorn is stored in an airtight container, preferably refrigerated, reduces the number of UPKs—unpopped popcorn kernels.

Flour Corn. This white or blue corn is common among Native Americans of the southwest who have been using its flour as a staple for over 1,000 years. Blue corn chips are sold commercially on the same shelf as the standard potato, corn, and tortilla chips.

Pod Corn. Pod corn is a noncommercial corn in which each kernel is encapsulated in a separate husk, and the entire ear is wrapped in a large husk.

Forms of Corn. The bulk of corn grown in the United States is used for livestock feed. The remainder is used for human food, to make alcohol, and for its seed (32). Corn is consumed in its natural state on the cob, as kernels, or as processed foods, including hominy, hominy grits, cornmeal, cornstarch, corn syrup, and corn oil.

Corn on the Cob and *Kernel Corn.* Corn on the cob is eaten directly off the cob. Corn kernels removed from the cob can be used as a vegetable in a wide variety of dishes. Fresh corn varieties include yellow and white corn, and a hybrid corn that contains both yellow and white kernels.

Hominy. The endosperm of this white corn is enlarged because it has been soaked in lye and dried (Figure 21-9). Hominy is enlarged kernels of hulled corn with the bran and germ removed. It is available in canned or frozen form and is particularly popular in the southeastern region of the United States.

FIGURE 21-9 Hominy is processed from the endosperm of the corn kernel.

Hominy Grits. Grits are coarsely ground grain. Hominy grits are obtained by grinding dried hominy into small uniform particles that are boiled as a breakfast dish in the southern United States.

Cornmeal. Meal is the coarsely ground version of any grain, so cornmeal is coarsely ground corn. The original corn can be whole or degerminated and either yellow or blue-white in color.

Bread made from cornmeal has a shorter shelf life than wheat bread because cornmeal has a higher fat content than wheat flour.

Cornstarch. Cornstarch is finely ground corn endosperm. It is used to thicken gravies, puddings, and sauces because of its high starch content (see Chapter 20).

Corn Syrup. Corn syrup is a viscous liquid consisting of fructose, glucose, and other sugars (maltose, dextrins), obtained by treating cornstarch with selected enzymes. Corn syrup comes in light and dark, with the dark being the stronger flavored of the two.

Corn Oil. Corn oil is extracted from the germ of the corn kernel.

Barley

Barley was one of the first grains cultivated by humans (20). It is a hardy plant that grows in a wide range of climates. The heavy "beards" growing on the grain made it a symbol of male potency for the ancient Chinese. Early Greek athletes believed that their sports ability was enhanced when they ate barley mush during training. Roman gladiators were called *hordearli*, meaning "eaters of barley." Spaniards introduced barley to South America in the mid-1500s, and from there it spread northward. Barley is now used primarily as a malt (see below), in cereals and soups, for livestock feed, in the manufacture of beers and whiskey, and in soups, salads, and stews (19).

Forms of Barley. Used for thousands of years in the production of certain foods and beverages, barley is available in various forms: hulled, pot, pearled, flaked, barley grits, and malt.

Hulled Barley. Barley is enclosed by a tough hull that can be removed.

Pot Barley. Pot barley has had the hull and some of the bran removed from the grain in the process called "pearling" (27).

Pearled Barley. In pearled barley, additional pearlings remove more of the bran, germ, and part of the endosperm, resulting in tiny grains with a "polished" pearl-like color. It is used to make barley flour, grits, and flakes, which can then be incorporated into breakfast cereals, breads, soups, cookies, and crackers. It is also cooked as is and added to soups, salads, and stews.

Flaked Barley. Flaked barley is rolled/pressed and used primarily as a hot cereal.

Barley Grits. Barley grits are barley grains that have been toasted and cracked into particles.

Malt. Malt is germinated (sprouted) barley that has been gently dried to stop the growth of germinating roots while leaving intact the enyzmes that contribute to flavor and color. The enzymes are important for converting the starches to sugars such as maltose (Chemist's Corner 21-1). The barley grain is well suited for the production of the malt required in some alcoholic beverages, because the aleurone layer contains the enzymes needed to convert some of the endosperm starch to maltose. Also, the aleurone layer of barley is three to four cells thick, as compared to the single-cell layer in other cereal grains. About 4 percent of all barley is utilized in the production of malt, which can be purchased in dry or liquid form for use as food flavorings, as color additives in caramel, and as optional ingredients in baked products. Common foods to which malt is added include milk shakes, cereals, waffles, pancakes, ice cream, baked foods (cookies, crackers, pretzels, pizza, bagels), imitation coffees, and confections (19).

Millet

One of the smallest among the cereal grains, millet contains insufficient amounts of gluten, the protein required for bread to rise properly (see Chapter 22), but it is used in many countries throughout the world to make unleavened bread as well as beer. Millet is less common in the United States, where it is used primarily for bird seed, but it is grown widely in Russia, China, Japan, India, Africa, and southern Europe. The varieties of millet include common (also known as Indian, Proso, Hog, and Broomcorn), finger, foxtail, Japanese, little, and pearl or bulrush millet.

Sorghum

Sorghum is a cereal grain of major importance in Africa and certain parts of Asia (38), Central America, Pakistan, and India. It is consumed in the form of food (usually porridge), alcoholic beverages, or livestock feed. Cushiony puffed sorghum grains can also be used as a biodegradable packing material.

Oats

Oats are usually eaten as a hot or cold cereal or in breads, muffins, and cookies. A low protein content limits the use of oats for bread-making. Approximately 85 percent of the oats grown in the United States is used for livestock feed.

Forms of Oats. Oats are available in a variety of forms: groats, steel-cut, rolled, and bran (Figure 21-10).

Oat Groats. Oat groats are whole oats without the husks. They can be prepared like rice or used as an ingredient in other foods.

Steel-Cut Oats. Steel-cut oats are groats that have been cut lengthwise and then placed in cans or packages. They have a chewy texture, and are often imported from Scotland and Ireland, where they are more frequently consumed.

Rolled Oats. Rolled oats have been heated and pressed flat with steel rollers. They come to market as regular old-fashioned, quick-cooking, or instant oatmeal. Instant oats have been steamed to pregelatinize the starch, allowing immediate hydration and making them, like all other pregelatinized cereals, the quickest to prepare. Rolled oats are often incorporated into granola, which is a mixture of toasted oats, nuts, dried fruits, and sweeteners. Muesli, the European version of granola, differs in that the oats have not been toasted and it usually contains less fat.

Oat Bran. This is the bran isolated from oats; it has been reported to have a lowering effect on high blood cholesterol (21).

Rye

Rye is second only to wheat as a grain used in bread-making, but in the United States, rye is used primarily to produce rye crackers and whiskey. Compared to wheat, rye usually contains less protein (but still contains gluten) and starch, but more free sugars and dietary fibers (3). As a result, the loaf volumes of rye bread are half of that obtained for wheat products; however, rye breads are richer in flavor and aroma, and have a longer shelf life (Chemist's Corner 21-2) (39). Rye thrives in cold, wet weather and is widely grown and consumed in Scandinavia, eastern Europe, and Russia. During the Middle Ages in parts of Europe, only the nobility were allowed to consume wheat bread, leaving the

Chemist's Corner 21-1

Malt Production

Malt is derived from maltose, the monosaccharide naturally found in barley grain. Maltose is a part of the grain's starch, which first must be broken down to the smaller molecules of dextrins and maltose. Enzymes (alpha- and beta-amylase) found in the barley grain convert this starch to dextrin and maltose, resulting in a liquefied mass known as wort. The wort is evaporated under a vacuum to convert it into a syrup (80 percent solids), and the temperatures at which this occurs determine what type of malt extract is produced. These malt syrups can then be dried into "malts" that have the same color, flavor, and sweetness of their liquid counterparts without the enzyme activity (19).

FIGURE 21-10 Varieties of oats.

peasants to subsist on a type of rye bread. Some historians believe that ergot poisoning (Holy Fire or Saint Anthony's Fire), a foodborne illness caused by a fungus that grows on grains, especially rye, was a contributing factor to the French Revolution: the peasants were consuming ergot-prone rye, while the ruling classes were eating wheat bread free of the taint of ergot poisoning. Ergot poisoning can cause delusions, hallucinations, and erratic behavior, and it has been suggested that it may have been a factor in the Salem witch trials as well (14).

Other Grains

Triticale. Triticale is a relatively new hybrid grain that was developed in 1875 by a Scottish botanist who crossed wheat (*Triticum*) with rye (*Secale*) to create a grain with the bread-making quality of wheat and the climate hardiness of rye. Triticale, one of the few fertile hybrids, has a higher protein content and more of the essential amino acid lysine than wheat. Lysine is the most common limiting amino acid in grains. The

Chemist's Corner 21-2

Rye Proteins and Starches

The lower volume and sweeter taste of rye breads compared to wheat breads are due to the unique characteristics of rye's protein and carbohydrate content. Rye proteins are more water soluble than wheat proteins and exhibit both the properties of a viscous liquid and an elastic solid (37). Starches also play a key role in baked crumb structure, firmness during slicing, and chewiness. This contributes to the compact and moist nature of rye bread. The starch granules found in rye bread more easily disrupt and gelatinize at lower temperatures than wheat starch, which allows them to be degraded by alpha-amylase enzymes for a longer period of time during baking. Also contributing to the stickiness, moistness, and sweeter taste of rye bread is the ability of the pentosans (water-soluble, five-carbon carbohydrates such as D-xylose and L-arabinose) in rye to retain water (39).

flour from triticale produces excellent loaf volume, but "sticky dough" problems prevent it from being widely used (27).

Buckwheat. Buckwheat is not related to wheat and is not, strictly speaking, a true grain. It is, in fact, the fruit of a leafy plant that is related to rhubarb. It is, however, generally categorized as a grain because of its use as a flour and cereal. The nut-like flavor of buckwheat flour makes it a popular addition to pancake flour and some breakfast cereals and crepes. The roasted, cracked, and granulated buckwheat is known as kasha, a dish made of buckwheat groats, popular in Eastern European countries.

New Waves of Grain. Some new grains have entered the market, including amaranth, kamut, and quinoa.

Amaranth. Amaranth is a high-protein grain that is becoming more available in the market. It yields a flour that imparts moistness and a nutty flavor to baked products. The grain is so small that an expression in Africa states, "Some things are not worth an amaranth (seed)" (22).

Kamut. Kamut is a grain nutritionally superior to wheat that is lower in protein and has a buttery flavor. However, it does contain gluten and is avoided in a gluten-free diet.

Quinoa. A staple food of the Andean culture since 3000 B.C. and sometimes referred to as "Inca rice," quinoa is a very good source of plant protein (13 percent) and is popular among vegetarians (10). The tiny, seed-like kernels must be rinsed before cooking to remove the naturally bitter coating, leaving a rather mild, pleasant-flavored grain that is very quick-cooking.

Preparation of Cereal Grains

Cereals in their natural form are nearly indigestible. The hard outer covering of the seeds of these grasses prevents their immediate consumption and can even break a tooth. This barrier is overcome by heating the grain in water, which softens the outer covering and makes the starchy endosperm digestible. Cooking also gelatinizes the starch, which improves the flavor and texture of cereals as the grains soften and expand. Gelatinization occurs when heated starch molecules absorb water and expand—a phenomenom seen when grains expand to two or three times their volume when cooked. The degree of expansion depends on the type of grain, and is partially caused by the escape of amylose and amylopectin from the starch granule. The desired results in prepared grains are most commonly achieved by moist-heat methods: boiling, simmering, microwaving, and baking in the presence of liquid.

Moist-Heat Preparation

Boiling/Simmering. The type of grain dictates the amount of water to be added and the intensity or duration of heating. Table 21-1 lists recommended ratios of water to grain, cooking time, and yield for each of the different grain types. The two most important factors in grain preparation are the amount of water used and the exposure to heat.

Cooking the Grain. The most common preparation method for grains such as rice is to place water in an uncovered saucepan, lightly salt it, and heat to boiling. The salt, added in the ratio of ¼ teaspoon per cup of uncooked grain, provides flavor. The grain is then added to the boiling water. Adding the grain to hot water results in a fluffier product; adding it to cold water yields a stickier grain. The grain is stirred only as much as it takes to disperse it and the salt evenly in the water. To avoid stirring altogether, the grain can be poured in a zig-zag fashion over the entire surface of the boiling water for a more even distribution. The pan is then covered, and the water is brought back to a boil. The heat is then immediately reduced, and the contents are allowed to simmer (covered) for the remainder of the preparation time. Most grains should be stirred as little as possible while cooking. Stirring can cause a gummier texture because it makes the starch granules rupture prematurely. The exception to this rule is risotto, an Italian rice dish prepared by constant stirring.

Determining Doneness. After the minimum amount of covered cooking time has passed, the grains are tested for doneness by tasting. The grain should be tender but have a slightly resistant core. Undercooked grains are difficult to chew and have a starchy, raw flavor. Overcooked grains may form a mushy, formless mash. Too much water contributes to stickiness, sogginess, and loss of nutrients, but insufficient water causes dry, toughened textures, and may even allow the grain to burn.

Standing Time. Once cooked, it helps to let the grain stand for 10 to 15 minutes. This allows steam to further separate the granules, creating a light, airy texture. To further this goal, after removing the saucepan from the heat, a fork can be used to fluff the grain by gently and quickly forming a pyramid with the grain in the pan. The fork handle is inserted into the pile in four places, moving it back and forth each time to create a ¼-inch tunnel for steam to escape from the pyramid. Placing a paper towel flat over the pan and then covering the pan with its lid allows the paper towel to absorb the rising steam. A modified version of this method is often used in the Middle East to ensure a light and fluffy grain.

Sauteeing and Baking. Although they are not as frequently used as the boiling method, there are two other procedures for preparing grains, both intended to add flavor to the finished dish. These are sautéing, referred to as the pilaf method, and baking. Grains such as rice and bulgur can be first sautéed in fat (1 tablespoon per cup of uncooked grain), after which boiling chicken broth or other stock, instead of water, is poured over the grain. It is then covered and simmered until done. Cereal grains can also be prepared in a casserole dish and baked in the oven if sufficient liquid is provided. The grain, usually rice, may already be cooked prior to being included with the rest of the ingredients that will be baked, but if not, boiling water is added to the dish. The casserole is usually covered to take advantage of the steam that further aids in cooking its contents. Baking times vary according to the grain used, but average 20 to 30 minutes.

Adding Seasonings. Regardless of the preparation method, grains are usually bland unless seasoning ingredients are added at the beginning of the cooking process. Chicken- or beef-flavored soup stock or bouillon can be used instead of water, giving the grain a slight chicken or beef flavor.

Factors Influencing Grain Cooking. The form of grain, the presence of the bran or hull, the pH of the water, and the desired tenderness are factors that influence the amount of water to be used, the heat intensity, and the cooking time. Any reduction in particle size through cracking, rolling, cutting, or flaking decreases the heating time. For example, cracked wheat cooks in 15 minutes, while whole wheat berries may take over an hour. Removing the bran or hull also drastically cuts heating time. Brown rice takes about twice as long to cook as white rice, while barley with its hull intact takes about 35 minutes longer than pearl barley. A more alkaline environment, such as the one created by using hard tap water, causes grains to cook at a faster rate and possibly to overcook because of increased breakdown of the cellulose.

Hot Breakfast Cereals. Hot breakfast cereals account for 7 percent of all breakfast cereals. The most commonly consumed hot breakfast cereals are oatmeal, farina, hominy or corn grits, cream of rice, and bulgur. They are available in three forms: regular, quick-cooking, and instant. Most people like cereals prepared to a thick yet moist consistency, which can be varied by using more or less water. To prevent clumping, particularly when preparing the more finely granulated cereals such as farina and grits, it is best to sprinkle the cereal slowly over the boiling water while stirring so that each grain has a chance to be surrounded with hot liquid. Another way to prevent clumping is by mixing the cereal with cold water to make a

TABLE 21-1
Grain Heating Times

Per ½ C Uncooked	*Liquid (Cups)*	*Cooking Times*	*Yield (Cups)*
WHEAT			
Whole berries	1½	1 hour, 10 minutes	1¾
Cracked wheat	1	15 minutes	1
Bulgur	1	15 minutes	1½
Wheat flakes	2	53 minutes	1
RICE			
Brown, long-grain	1	25–30 minutes	1½
Brown, short-grain	1	40 minutes	1½
Brown, instant	½	5 minutes	1¾
Brown, quick	¾	10 minutes	1
White, long-grain	1	20 minutes	1¾
White, instant	½	5 minutes (let stand 5 minutes)	1
Converted	1½	31 minutes	2
Arborio	¾	15 minutes	1½
Basmati, white	½	15 minutes	2
Jasmine	1	15–20 minutes	1¾
Texmati long-grain, brown	1 (plus 2 tbs)	40 minutes	2
Texmati long-grain, white	¾ (plus 2 tbs)	15 minutes	1¾
Wehani	1	40 minutes	1¼
Wild	2	50 minutes	2
Wild pecan	1	20 minutes	1½
BARLEY			
Hulled	2	1 hour, 40 minutes	1¾
Pearl	1½	55 minutes	2
Quick	1	10–12 minutes (let stand 5 minutes)	1½
Grits	⅓	Let stand 2–3 minutes	⅓
Flakes	1½	30 minutes	1
OATS			
Groats	1	6 minutes (let stand 45 minutes)	1¼
Steel-cut	2	20 minutes	1
Old-fashioned rolled	1	5 minutes	1
Rolled, quick	1	1 minute (let stand 3–5 minutes)	1
Oat bran	1	6 minutes	1
RYE			
Rye berries	1½	1 hour, 55 minutes	1½
Rye flakes	1½	1 hour, 5 minutes	1¼
BUCKWHEAT			
Groats, unroasted, whole	1	15 minutes	1¾
Groats, roasted, whole	1	13 minutes	1½
Kasha	2½	12 minutes	2

slurry before adding it to the hot water. Once all the cereal is added, maintain a slow boil and stir occasionally until the grains are translucent. There are virtually no clumping problems with instant breakfast cereals, which simply require the addition of boiling water and a bit of stirring. One cup of uncooked cereal yields about four to six servings when cooked.

Microwaving. Heating times are not significantly reduced with a microwave oven. Follow the manufacturer's directions for the amounts of required water, grain, and heating times. Add the grain to the boiling water, cover with a lid, heat according to the recommended time, stir at the halfway point, and let stand for five minutes before serving. Microwave preparation reduces cleanup because the same dish can be used for cooking and serving. The microwave oven is ideal for reheating because it yields a near "fresh cooked" flavor and texture, while reducing the tendency of grains to dry or overcook. It is also ideal for preparing instant hot cereals because single servings can be prepared directly in a microwave-safe serving bowl in a one-step process.

Storage of Cereal Grains

Dry

Dry grains, freed of their bran and germ, are best kept in airtight wrappings or containers in a cool, dry area free of rodents, insects, and other pests. Moisture is the biggest contributor to the deterioration of grains. The relative humidity in the environment determines the grain's moisture content, as grains take up moisture until an equilibrium is reached with the atmosphere's water vapor. In practice, this means that a relative humidity of 70 percent or less is considered safe; microbial growth will occur above 75 percent relative humidity, resulting in extensive grain deterioration (15). Thus, once opened, grain packages should be tightly resealed or the grain placed in another airtight container that will protect it from air or animal invasion. Most grains, when stored properly, will keep for six to twelve months.

Refrigerated

Whole grains should be refrigerated in airtight containers to retard rancidity and prevent mold growth, which can be caused by moisture. Usually, only whole or cooked grains are refrigerated.

Cooked grains will keep up to a week if they are tightly covered. The best way to reheat grains is in a microwave oven or in a covered saucepan on top of the range with about 2 tablespoons of water added for each cup of grain.

Frozen

Cooked whole grains can be frozen for future use if they are tightly wrapped or placed in airtight containers. Uncooked grains should not be frozen, because freezing alters the protein structure in such a way that any baked products made from such grain will not rise as high.

Pastas

The Chinese developed noodles as early as 5000 B.C., and Marco Polo, one of the first Europeans to reach the Orient, is credited with bringing them back from his travels and introducing pasta to Italy and the rest of Europe in the late 1200s. There is some evidence, however, of pasta having been made in Italy as early as 400 B.C. (14). Much later, pasta was introduced to North America in the late 1700s by Thomas Jefferson after he visited Naples while serving as the American ambassador to France. Pasta now serves as a popular source of complex carbohydrates in many parts of the world.

Pasta, meaning "paste" in Italian, is made predominantly from flour starch and water. It is usually made from semolina, a flour derived from durum wheat, although other flours are sometimes used; flavorings and colorings can also be added (9). The highest-quality pastas are made from the higher-protein wheats. Durum flour's higher protein content makes it best suited to withstand the pressures of mechanical kneading and manipulation during commercial pasta production, as well as the heat during preparation (26). It is the protein in durum wheat flour that gives pasta its elasticity and helps it maintain its shape during cooking. Durum wheat is also higher in carotenoid pigments, which contribute to pasta's rich, golden color (11). Also, many pastas contain egg yolks, which further enhances their yellow color. Most pasta manufactured in North America is enriched with several B vitamins and iron.

Types of Pasta

The shapes by which pasta is identified are formed by placing the freshly made pasta dough in a cylinder and forcing it through holes in small discs (dies). Pasta dough is best extruded at 115°F (46°C), because temperatures higher than 140°F (60°C) will denature the protein and reduce the pasta's quality. Once pressed through the appropriate die, pasta is cut and then dried until the moisture level drops from 31 percent to 10 or 12 percent (34).

The type of disc used determines what kind of pasta is produced. Depending on the selected disc, pasta can be called spaghetti ("little strings"), linguine ("little tongues"), vermicelli ("little worms"), rigatoni ("grooved"), or fettuccine, capellini, cannelloni, tortellini, lasagna, ravioli, macaroni, noodles, won-ton wrappers, and others (Figure 21-11). There are over 600 pasta shapes now in existence, but only about 150 are available in North America (25). "Long goods" such as spaghetti and linguini are most commonly consumed (41 percent), followed by "short goods" such as macaroni (31 percent) and noodles (15 percent), and specialty items such as lasagna and manicotti (13 percent) (17).

Pasta Nomenclature

Pasta, or "alimentary (nourishing) paste," is made by combining water with semolina flour and/or farina. Macaroni is a generic term for all types of dried pasta. Prior to being dried in various shapes, optional ingredients may be added, such as vegetable purées, as in spinach pasta, and a variety of seasonings. Different types of pasta vary not only in shape, but also in their ingredients, which influences the following pasta nomenclature.

Noodles. If eggs are added (at least 5.5 percent egg by weight), the pasta product is referred to as "noodles," although

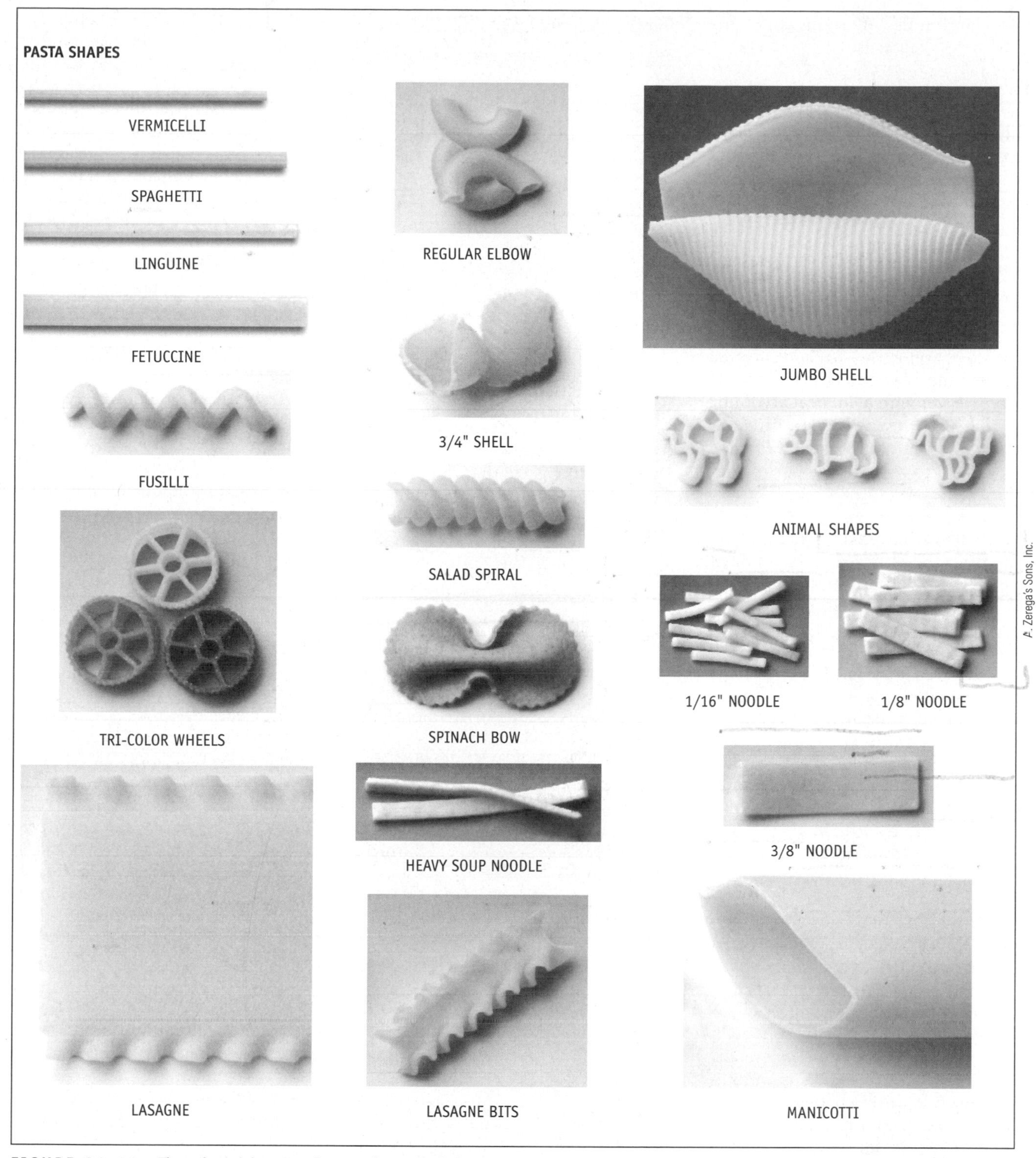

FIGURE 21-11 There is a rich array of pasta shapes and sizes.

eggless noodles are available on the market.

Asian Noodles. Asian noodles are often made from flours other than the standard semolina flour or farina. This and the fact that they rarely contain eggs is why they are sometimes called "imitation noodles" in the West. Asian noodles may be made from rice, mung bean, taro, yam, corn, buckwheat, or potato flours. Examples of Asian noodles are rice, ramen, soba, and bean thread noodles (25). Ramen are instant Japanese noodles that have been previously dehydrated by frying, which makes them extremely porous and much more likely to absorb water than regular noodles. Adding water to ramen instantly rehydrates them, making them popular for use in soups and luncheon noodle meals.

Whole Wheat. Pasta made from whole-wheat flour is slightly higher in nutrients and fiber than standard pasta, but it has a tougher texture, a stronger taste, and tends to disintegrate if cooked too long.

Flavored. Vegetable purées made from spinach, tomatoes, or beets can be added to pasta to alter its color and flavor. A newer version of flavored pasta incorporates herbs and spices such as basil, garlic, parsley, and red pepper into the pasta dough.

Fanciful. Unusual shapes such as dinosaurs and turtles have been developed to appeal to youthful consumers (17).

High-Protein Pasta. Adding soy flour, wheat germ, or dairy products yields high-protein pasta products that contain 20 to 100 percent more protein than standard pasta (25).

Fresh. Fresh pasta, found in the refrigerated section of the supermarket, has a higher moisture content, which gives it a softer consistency and shorter cooking time.

Couscous. Couscous looks like a grain, but it is actually "Moroccan pasta" made from semolina that has been cooked, dried, and pulverized into small, rough particles the size of rice grains. Traditionally, lamb stew and many Middle Eastern dishes were served over a bed of couscous. Middle Eastern or Israeli couscous is larger than regular couscous and resembles small balls of pasta. It is usually toasted before preparation and retains a firm bite after cooking. In another part of the world, people in many African countries consume their couscous with milk for breakfast or as a starch with lunch or dinner (2).

KEY TERMS

Al dente Meaning "to the tooth" in Italian, it refers to pasta that is tender, yet firm enough to offer some resistance to the teeth.

Preparation of Pasta

Moist-Heat Preparation

Pasta is easy to prepare by boiling or simmering (Figure 21-12). It is dropped into lightly salted boiling water, stirred to keep the strands or pieces from sticking together, and heated until it reaches the **al dente** (ahl-den-tay) stage. Properly prepared pasta is not excessively sticky, and once it is done, it is drained and ready to serve.

During heating, the majority of pastas expand to two or three times their original size. Flat noodles are an exception in that they increase in size only by half. The pasta-to-cooking-water ratio should ensure that the pasta is heated in plenty of water: 4 quarts of water per pound of pasta is recommended. The water should remain at boiling through the entire cooking period, and both undercooking and overcooking should be avoided. Undercooked pasta is identified by a white core of stiff, ungelatinized starch in the center, while overcooked pasta is mushy and limp in consistency and very bland.

Water with an alkaline pH tends to cause stickiness (24). Oil may be added to the water to prevent the pasta pieces from sticking together, both during boiling and when the product is allowed to cool. Pasta may be rinsed with water after cooking to prevent sticking, but this may remove any B vitamins added during enrichment that have not already been lost in the cooking water.

Stickiness is influenced by the amount of water on the pasta's surface, the degree of force used to press the pasta through the disc, and most importantly, by the age of the pasta. Older pasta is not recommended because water is lost from the surface

1. In a large, heavy pot, bring 4 quarts of water for each pound of pasta to a full rolling boil. Add salt to taste.

2. For long pasta, add to boiling water in batches, pushing batch down as pasta softens; this avoids breaking. Short pasta can be added all at once.

3. Stir, carefully separating pasta pieces. Cover, return to boil, then immediately remove lid. Stir gently while cooking to prevent pieces from sticking together.

4. Cook pasta until tender but slightly firm (al dente). Taste to test doneness. Drain pasta, but do not rinse because starchy surface holds sauce better.

FIGURE 21-12 Preparing pasta.

NUTRIENT CONTENT

Pasta is high in complex carbohydrates and naturally low in protein, fat, and cholesterol unless it is processed with eggs or fat. Naturally low-fat pasta may be beneficial to health, particularly of diabetics as well as people interested in lowering their dietary fat and blood cholesterol levels (8, 35). Pasta, which used to have the reputation of being "fattening," only recently became popular when it was recognized as an excellent, low-fat, low-calorie (kcal) source of complex carbohydrates. Nevertheless, noodles, which are pastas made with eggs, have more fat and cholesterol than non-egg-containing pastas, and sauces made from butter, oil, cream, or meat will also add calories (kcal) and fat. When pastas are stuffed with cheese or meat, as in tortellini and ravioli, the fat and calorie (kcal) content soars even higher.

Traditional ramen, the Japanese noodle sold as an instant noodle or soup dish, is made from wheat noodles that have been deep-fried in highly saturated lard or palm oil and then dried, but lower-fat varieties are also available. Canned chow mein noodles have been fried in oil or fat and are high in calories (kcal)—each cup contains about 240 calories (kcal) and 14 grams of fat (25).

of pasta as it ages, causing a simultaneous loss in lubrication and increased stickiness (34).

For most regular pasta, heating time is approximately 10 minutes, but fresh pasta and Asian noodles can take as little as 3 minutes. If cooked pasta needs to be held before serving, it is placed in a colander over a pot of hot, steaming water, or it can be cooled and later briefly dropped in simmering water to be reheated, then drained and served. Pasta should be slightly undercooked if it is to be stir-fried, baked, or added to soup, stews, or casseroles. Figure 21-13 lists the amount of pasta to use for a 1-cup serving.

Microwaving. Like grains, pasta can be prepared in a microwave oven, but again, there is no significant savings in time compared to the conventional method. Follow the manufacturer's instructions for the recommended amounts of water and pasta and the required heating times. The pasta is stirred into very hot water, covered with a casserole lid or plastic wrap, and heated for the recommended amount of time. It should be stirred halfway through the cooking period, drained immediately when done, and served. Many new microwavable single-serve pasta entrees are being marketed.

Long pasta (spaghetti, linguine, fettuccine): Use 2 oz. uncooked, or a bunch 1/2-inch in diameter.

Short pasta (penne, shells, rigatoni): Use 2 oz. uncooked, or just over 1/2 cup.

FIGURE 21-13 How much pasta to cook for a 1-cup serving.

Storage of Pasta

The storage of pasta depends on whether it is dried, fresh, or cooked. Dried pasta should be tightly wrapped and stored in a cool, dry place. Fresh pasta should be kept in the refrigerator until the "use by" date. It will be at its best for about a week and will keep in the freezer for about a month. Fresh pasta stored in modified-atmosphere packages may last up to 120 days, but there is an increased risk of microbial contamination because of the long storage time (16). The additional ingredients often found in fresh Asian noodles reduce their keeping time in the refrigerator to two days. Cooked pasta will keep for two to three days in the refrigerator and is easily reheated in the microwave oven under vented plastic wrap, or by placing it in a pot of boiling water for half a minute.

PICTORIAL SUMMARY / 21: Cereal Grains and Pastas

Cereal grains are seeds from the grass family. Serving as the world's major food crop, they have long been regarded as "the staff of life."

COMPOSITION OF CEREAL GRAINS

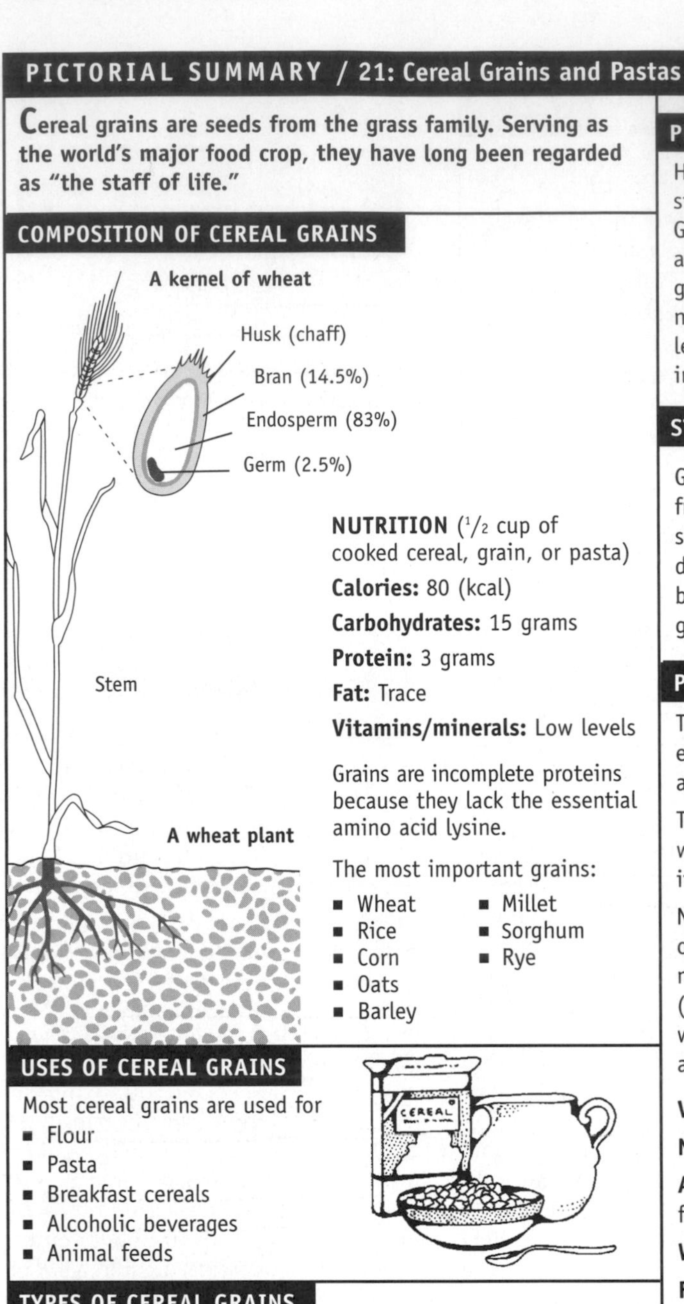

NUTRITION (1/2 cup of cooked cereal, grain, or pasta)

Calories: 80 (kcal)

Carbohydrates: 15 grams

Protein: 3 grams

Fat: Trace

Vitamins/minerals: Low levels

Grains are incomplete proteins because they lack the essential amino acid lysine.

The most important grains:

- Wheat
- Rice
- Corn
- Oats
- Barley
- Millet
- Sorghum
- Rye

USES OF CEREAL GRAINS

Most cereal grains are used for

- Flour
- Pasta
- Breakfast cereals
- Alcoholic beverages
- Animal feeds

TYPES OF CEREAL GRAINS

Wheat: Available as wheat berries, cracked and rolled wheat, bulgur, farina, wheat germ, and wheat bran.

Rice: Available as white, converted, instant, brown, wild, glutinous, specialty rices, or rice bran.

Corn: Can be classified according to its kernal type: dent, sweet, flint, popcorn, flour, and pod. Can be consumed as corn on the cob or kernal corn, hominy, corn or hominy grits, cornmeal, cornstarch, corn syrup, and corn oil.

Barley: Primarily used for the manufacture of beer and whiskey, and for livestock feed.

Sorghum: Major cereal grain in Africa and parts of Asia.

Oats: Available as groats, steel-cut, rolled, and bran.

Rye: Second only to wheat for bread-making.

PREPARATION OF CEREAL GRAINS

Heating dried cereals in water softens and gelatinizes their starch, creates an edible texture, and improves flavor. Gelatinization occurs when the starch molecules absorb water and expand. Although rice is the most commonly prepared grain in the world, all grains can be cooked in a similar manner, with just slight variations in the amount of water and length of heating. Cooking time decreases with any reduction in particle size due to cracking, rolling, cutting, or flaking.

STORAGE OF CEREAL GRAINS

Grains are best kept in airtight containers in a cool, dry area free from rodents, insects, and other pests. Whole grains should be refrigerated to retard the rancidity that can occur due to their fat content. It is important to keep grains from becoming damp and subject to mold growth. Cooked whole grains can be refrigerated or frozen for future use.

PASTAS

There are over 600 pasta shapes now in existence, although only about 150 are available in North America.

The higher protein content of durum wheat flour, also called semolina, makes it best suited for pasta production.

Nutritionally, pasta is an excellent source of complex carbohydrates. Although it is naturally low in fat, the fat and calorie (kcal) count goes up when it is served with sauces made from butter, oil, cream, and/or meat.

Various pastas available:

Noodles: Eggs are added (at least 5.5 percent egg by weight).

Asian noodles: These rarely contain egg and are often made from flours other than semolina.

Whole wheat: Made from whole-wheat flour.

Flavored: Vegetable purées are added for flavor and color.

High-protein: Added soy flour, wheat germ, or dairy products contain 20 to 100 percent more protein.

Fresh: Has a higher moisture content and faster cooking time.

Couscous: Made from semolina that has been cooked, dried, and pulverized. Popular in Africa and the Middle East.

PREPARATION OF PASTA

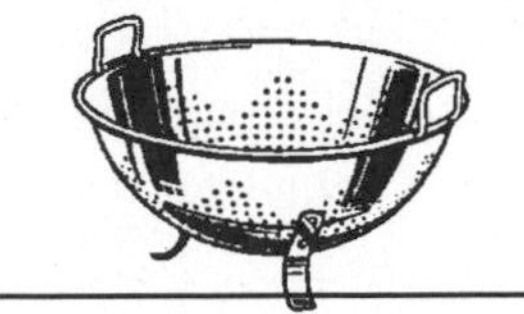

Pasta is best prepared by boiling it in plenty of water and avoiding both under- and over-cooking.

STORAGE OF PASTA

Dried pasta should be stored in an airtight container and placed in a cool dry place.

REFERENCES

1. Abdel-aal ESM, FW Sosulski, and P Huel. Origins, characteristics, and potentials of ancient wheats. *Cereal Foods World* 43(9):708–715, 1998.
2. Aboubacar A, and BR Hamaker. Physiochemical properties of flours that relate to sorghum couscous quality. *Cereal Chemistry* 76(2):308–313, 1999.
3. Aman P, M Nilsson, and R Andersson. Positive health effects of rye. *Cereal Foods World* 42(8):684–688, 1997.
4. Andlauer W, and P Furst. Does cereal reduce the risk of colon cancer? *Cereal Chemistry* 44(2):76–77, 1999.
5. Barton FE, et al. Optimal geometries for the development of rice quality spectroscopic chemometric models. *Cereal Chemistry* 75(3):315–319, 1998.
6. Batterman-Azcona SJ, JW Lawton, and BB Hamaker. Effect of specific mechanical energy on protein bodies and alpha-zeins in cornflour extrudates. *Cereal Chemistry* 76(2):316–320, 1999.
7. Bell S, et al. Effect of beta-glucan from oats and yeast on serum lipids. *Critical Reviews in Food Science and Nutrition* 39(2):189–202, 1999.
8. Best D. Technology fights the fat factor. *Prepared Foods* 160(2):48–49, 1991.
9. Boyacioglu MH and BL D'Appolonia. Durum wheat and bread products. *Cereal Foods World* 39(3):168–174, 1994.
10. Brinegar C. The seed storage proteins of quinoa. In *Food Proteins and Lipids*, ed. S Damodaram, 109–115. Plenum, 1997.
11. Cole ME, DE Johnson, and MB Stone. Color of pregelatinized pasta as influenced by wheat type and selected additives. *Journal of Food Science* 56(2):488–493, 1991.
12. Craig WJ. Guardians of our health. *Journal of the American Dietetic Association* 97(10 Suppl 2):S199–S204, 1997.
13. Dodge A. Making the Creamiest Rice Pudding. *Fine Cooking* 43:73, 2001.
14. Ensminger AH, et al. *Foods and Nutrition Encyclopedia*. CRC Press,1994.
15. Eskin NAM. *Biochemistry of Foods*. Academic Press,1990.
16. Feillet P, et al. Past and future trends of academic research on pasta and durum wheat. *Cereal Foods World* 4(4):205–212, 1996.
17. Giese J. Pasta: New twists on an old product. *Food Technology* 46(2):118–126, 1992.
18. Grieder J. State of the industry report: Breakfast cereals in the U.S. *Cereal Foods World* 41(6):484–487, 1996.
19. Hickenbottom JW. Processing types, and uses of barley malt extracts and syrups. *Cereal Foods World* 41(10):788–790, 1996.
20. Jadhav SJ, et al. Barley: Chemistry and value-added processing. *Critical Reviews in Food Science and Nutrition* 38(2):123–171, 1998.
21. Kahlon TS, RH Edwards, and FI Chow. Effect of extrusion on hypocholesterolemic properties of rice, oat, corn, and wheat bran diets in hamsters. *Cereal Chemistry* 75(6):897–903, 1998.
22. Lehman JW. Case history of grain amaranth as an alternative crop. *Cereal Foods World* 41(5):399–411, 1996.
23. Le Marchand L, et al. Dietary fiber and colorectal cancer risk. *Epidemiology* 8(6):658–665, 1997.
24. Malcolmson LJ and RR Matsuo. Effects of cooking water composition on stickiness and cooking loss of spaghetti. *Cereal Chemistry* 70(3):272–275, 1993.
25. Margen S, et al., eds. *The Wellness Encyclopedia of Food and Nutrition*. Rebus,1999.
26. Mariani BM, MG D'egidio, and P Novaro. Durum wheat quality evaluation: Influence of genotype and environment. *Cereal Chemistry* 72(2):194–197, 1995.
27. Nelson KA and SJ Eilertson. Cereal grain speciality products: Brief overview. *Cereal Foods World* 41(5):383–385, 1996.
28. Oelke EA, Porter RA, Grombacher AW, and Addis PB. Wild rice: New interest in an old crop. *Cereal Foods World* 42(4):235–247, 1997.
29. Olson BH, et al. Psyllium-enriched cereals lower blood total cholesterol and LDL cholesterol, but not HDL cholesterol, in hypercholesterolemic adults: Results of meta-analysis. *Journal of Nutrition* 127(10):1973–1980, 1997.
30. Prasad AS, et al. Biochemical studies on dwarfism, hypogonadims and anemia. *Archives of Internal Medicine* 111:407–428, 1963.
31. Pszczola DE. Rice: Not just for throwing. *Food Technology* 55(2):53–57, 2001.
32. Rhoades RE, Corn, the golden grain. *National Geographic*, 138(6):92, 1993.
33. Schroeder HA. Losses of vitamins and trace minerals resulting from processing and preserving foods. *American Journal of Clinical Nutrition* 24:562–573, 1971.
34. Smewing J. Analyzing the texture of pasta for quality control. *Cereal Foods World* 42(1):8–12, 1997.
35. Temelli F. Extraction and functional propeties of barley beta-glucan as affected by temperature and pH. *Journal of Cereal Science* 62(6):1194–1197, 1997.
36. Turk M, NG Carlsson, and AS Sandberg. Reduction in the levels of phytate during wholemeal bread making: Effect of yeast and wheat phytases. *Journal of Cereal Science* 23:257–264, 1996.
37. Walker CE and JL Hazelton. Dough rheological tests. *Cereal Foods World* 41(1):23–28, 1996.
38. Weaver CA, BR Hamaker, and JD Axtell. Discovery of grain sorghum germ plasm with high uncooked and cooked in vitro protein digestibilities. *Cereal Chemistry* 75(5):665–670, 1998.

39. Weipert D. Processing performance of rye as compared to wheat. *Cereal Foods World* 42(8):706–712, 1997.
40. Whistler RL, and JN BeMiller. *Carbohydrate Chemistry for Food Scientists.* Eagen Press, 1997.
41. Yeh A, and YM Jaw. Effects of feed rate and screw speed on operating characteristics and extrudate properties during single-screw extrusion cooking of rice flour. *Cereal Chemistry* 76(2):236–242, 1999.
42. Zhou JR, and JW Erdman. Phytic acid in health and disease. *Critical Reviews in Food Science and Nutrition* 35(6):495–508, 1995.

WEBSITES

A website encyclopedia listing many foods, including grains:

www.foodsubs.com/FGGrains.html

The National Institute of Diabetes and Digestive and Kidney Diseases offers a four-star site on numerous digestive conditions. Click on "Celiac Disease," for more information on the condition where people are allergic to the gluten in wheat, oat, rye, and barley:

www.niddk.nih.gov/health/digest/digest.htm

The website for the American Association of Cereal Chemists (AACC), a nonprofit international organization specializing in the use of cereal grains in foods. They publish the scientific journals entitled, *Cereal Chemistry* and *Cereal Foods World*:

www.aaccnet.org

Flours and Flour Mixtures

For over 99 percent of the 2 million years humans have been on earth, they "made their living" as hunter-gatherers. It has only been within the last 1 percent of that time that the shift to domesticating plants and animals has occurred (54). About 10,000 years ago, humans are believed to have started eating a crude form of bread—a baked mixture of flour and water. In the time since humans first discovered that grains could be crushed for their starchy insides and mixed with other ingredients to provide more palatable nourishment, there has been a huge increase in the number and variety of baked goods. These range from basic staples, such as **yeast breads** and **quick breads**, to the crusty breads enjoyed by the French and the chewy bagels that have become so widely popular in recent years, to desserts such as cakes, cookies, and pastries. No matter what their outward appearance and taste, the foundation of all these baked products consists of a flour mixture.

The simplest flour mixture is one made from flour and water. Other ingredients that may be added include milk, fat, eggs, sugar, salt, flavoring, and leavening agents. Commercial manufacturers of baked products may also add certain additives. The ingredients of a flour mixture may be divided into categories as dry—flour, leavening agents, sugar, and salt/flavoring; or liquid—water, milk, fat, and eggs. The types and proportions of these ingredients determine the structure, volume, taste, texture, appearance, and nutrient value of the finished baked product.

The principles of measuring the ingredients of a flour mixture and the descriptions of various mixing methods were discussed in Chapter 5. The purpose of this chapter is to review each of the basic ingredients in a flour mixture and its specific functions, along with preparation and storage guidelines for flour mixtures.

KEY TERMS

Yeast bread Bread made with yeast, which produces carbon dioxide gas through the process of fermentation, causing the bread to rise.

Quick bread Bread leavened with air, steam, and/or carbon dioxide from baking soda or baking powder.

Crumb The texture of a baked product's interior.

Flours

Flours provide structure, texture, and flavor to baked products (Figure 22-1). Starch is one of the compounds in flour that strengthens the baked item through gelatinization, and is one of the factors that contributes to **crumb**. Crumb is partially created during baking by the number and size of air cells produced, the degree of starch gelatinization, and the amount of protein coagulation (27, 47). A fine crumb is delicate with small, densely packed air bubbles, while a coarse crumb has large, often irregular air holes. A secondary

Baker Technology

FIGURE 22-1 Flour from grains is the starting point of all baked products.

KEY TERMS

Gluten The protein portion of wheat flour with the elastic characteristics necessary for the structure of most baked products.

Knead To work the dough into an elastic mass by pushing, stretching, and folding it.

function of the starch in flour is that it can be partially broken down by enzymes (amylases) into dextrin, malt, and glucose. These compounds add a slight sensation of sweetness, darken the crust color, and improve fermentation, making the mixture lighter in texture. The heat of baking, in the presence of moisture from vaporization, causes the dextrins to coat the crust with a shiny layer (47).

Gluten

Other components in flour that play an important role in the structure of bakery items are the proteins that form **gluten.** These proteins, as well as those from eggs, contribute to the firming of the flour mixture, while sugar and fat act as tenderizing agents. The ability of a baked product to rise is directly related to its protein content (3). Because wheat flour has the highest concentration of the proteins that form gluten, it yields baked products with light, airy textures and is, therefore, most often preferred over other flours for baking.

The Purpose of Gluten Formation. When flour is mixed with water, an elastic network forms when two types of proteins in flour, gliadin and glutenin, combine to yield the protein complex gluten (Figure 22-2). Gluten is both elastic and plastic. Its ability to expand with the inner pressure of gases such as air, steam, or carbon dioxide results from the combination of glutenin's elasticity and gliadin's fluidity and stickiness (50) (Figure 22-3). Bread dough rises as the gas resulting from the yeast or other leaveners, as well as the air bubbles entrapped in small pockets by **kneading**, expands and stretches the gluten

FIGURE 22-2 Gluten is formed from the combination of two wheat flour proteins: gliadin and glutenin.

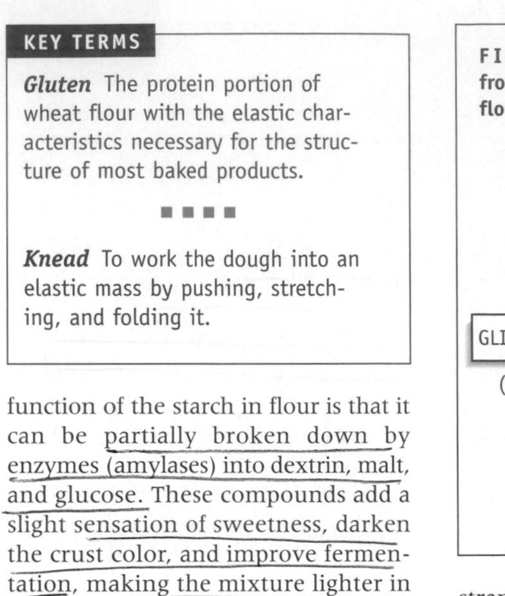

strands upward and outward. Then, during the temperature rises of baking, steam along with the expanding gases of carbon dioxide and ethanol cause the gluten to expand further. The baked product's structure sets when the heat from baking coagulates the proteins and gelatinizes the starch. If the oven door is opened frequently or if baking is stopped prematurely before these processes can occur and set the structure, the steam may be released from the gluten complex, causing the baked product to partially collapse.

Separating Gluten from Flour. Gluten forms such a solid structure that it can be physically separated from the flour mixture by kneading a handful of dough under cold running water (59). This washes away the water-soluble proteins and frees the starch, which is lost down the drain, leaving a rubbery gluten mass. The gluten balls derived from cake, all-purpose, and bread flours are shown in Figure 22-5, and the larger, darker forms behind the gluten balls are the same items baked.

Steps to Gluten Formation. The main purpose of combining the ingredients of a flour mixture prior to baking is to encourage the development of the gluten, which will contribute to the baked product's structural strength. The two major steps of gluten formation involve hydrating the flour mixture and kneading the dough.

Hydration. Hydration of the flour proteins is the first step in gluten formation. Gliadin and glutenin form gluten, while the remainder of the flour proteins consisting largely of albumin and globulin become part of the dough or batter (34) (Chemist's Corner 22-1). The greater the protein content in the flour, the more water will be absorbed, in part because gliadin and glutenin absorb about twice their weight in water. The water helps to draw out the gluten-forming proteins from the crushed endosperm cells, and most doughs are 40 percent water by weight. Once hydrated, the

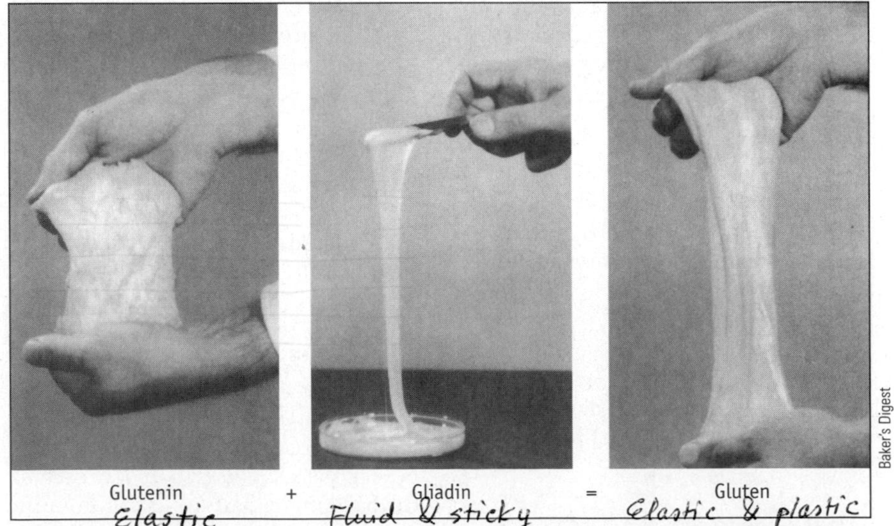

FIGURE 22-3 Glutenin's elasticity (left), plus gliadin's fluidity and stickiness (center) combine to form gluten's consistency (right), which is ideal for preparing baked products.

Chemist's Corner 22-1

Classifying Plant Proteins

All plant proteins are classified according to the Osborne system, which identifies proteins based on their solubility (ability to dissolve) in different solvents (Figure 22-4). The four major classification groups are albumin, globulins, prolamines, and glutelins. The albumin fraction generally consists of enzymes. Common enyzmes in flour include proteolytic enzymes; oxidative enzymes, which help bleach the flour; and phytase, which breaks down phytic acid (39). Prolamines and glutelins are the major storage proteins in most cereal grains (41). Prolamines have a high concentration of proline and lack the essential amino acid lysine (55). The main prolamine fractions for wheat and corn are gliadin and zein respectively, while glutenin is from the glutelin group (17).

FIGURE 22-4 The Osborne classification of plant proteins is based on the solubility of proteins in different solvents.

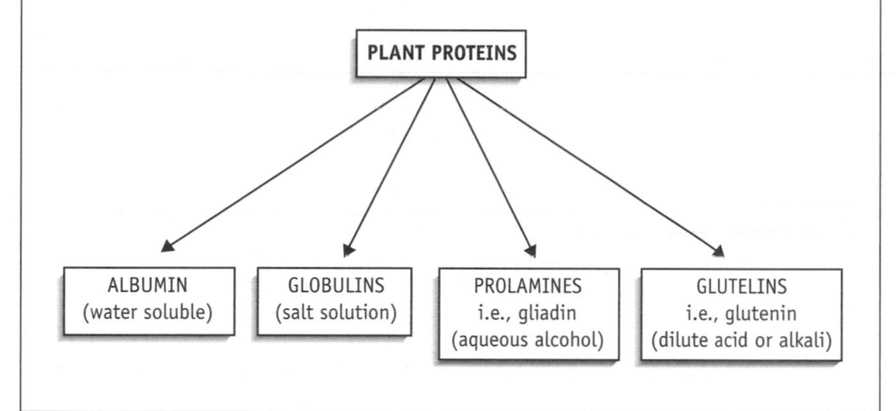

two proteins start to form gluten's complex, intertwined network, which is filled with water in its inner spaces.

Kneading. Kneading is used extensively in bread-making and briefly for biscuits and pastries. It alternately compresses and stretches the dough to increase gluten strength. Kneading also evenly distributes the yeasts to all their sources of food throughout the dough mass, redistributes the air bubbles, and warms the dough, increasing fermentation and carbon dioxide gas production. During kneading, the dough changes from a sticky mass to a smooth, stretchable consistency that is easily molded, yet springs back to light pressure (Figure 22-6). The "net" formed by the gluten stretches as the

FIGURE 22-5 The proportion of gluten derived from cake, all-purpose, and bread doughs. The ball in front is the gluten carefully extracted by running the dough through cold water to remove the starch. It was then baked to reveal the ball in back.

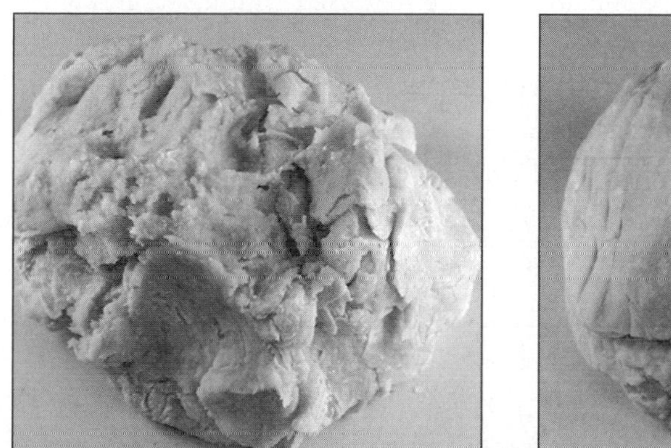

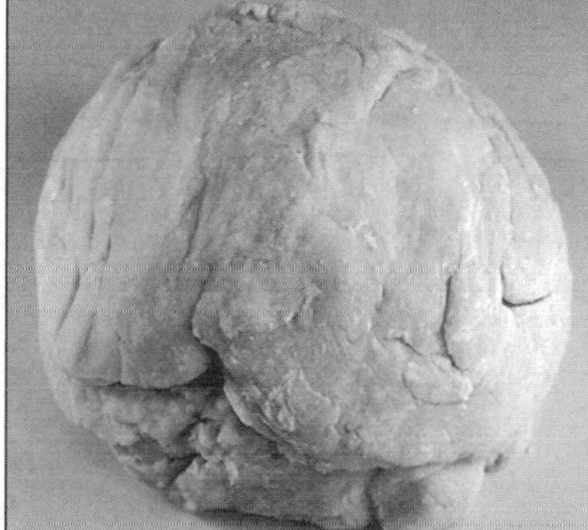

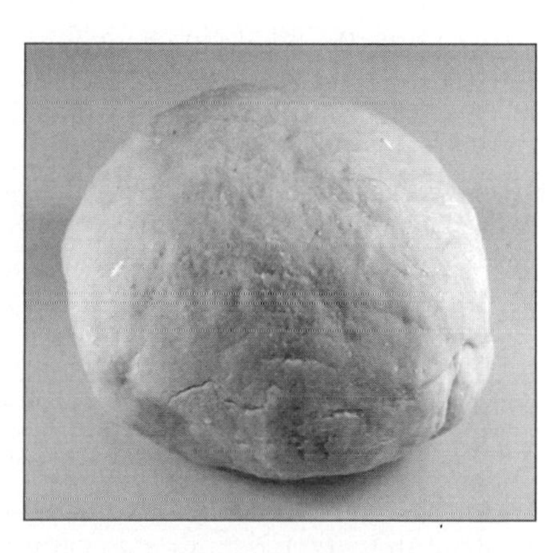

FIGURE 22-6 Dough is developed by kneading.

gases, air, steam, and carbon dioxide rise, but its elastic properties hold the general shape of the baked product.

HOW & WHY? ?????????

How does kneading change the dough? At the molecular level, kneading realigns protein molecules so they run roughly in the same direction and are more likely to form the crosslinks that make the dough stiffer (Figure 22-7). Sulfur-to-sulfur, or disulfide, bonds also help link the gluten proteins together. Kneading physically breaks the bonds between sulfur atoms, creates new ones, and allows the gluten molecule to stretch out in length (39).

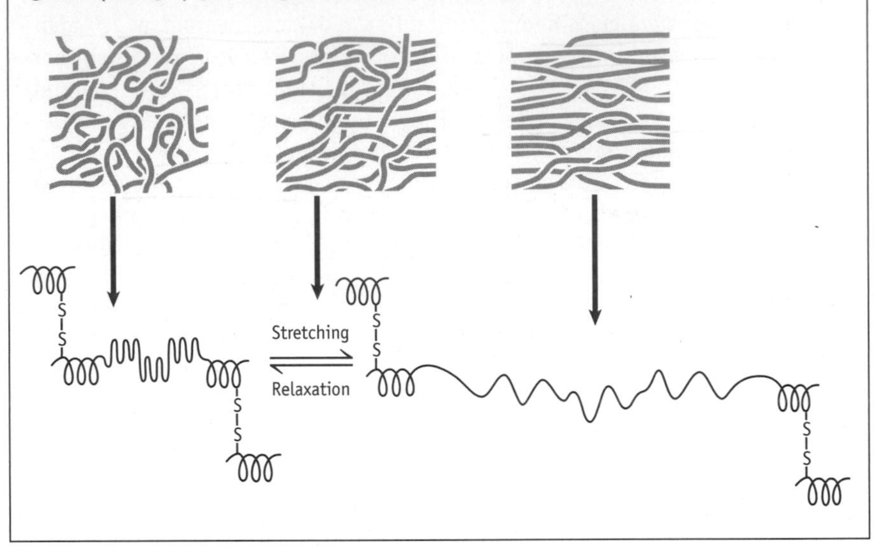

FIGURE 22-7 Kneading realigns protein molecules. The physical pressure of kneading stretches out the gluten molecules crosslinked by sulfide bonds. Sheets of gluten (far right) give dough a smooth, fine texture.

The lipids in flour also play a role in kneading, although their exact role is unknown. Lipids make up only about 1 to 1.5 percent of flour by weight, but without them, the dough will not rise. It has been suggested that lipids serve as an inner core surrounded by gluten and water (Figure 22-8). It has also been postulated that certain lipids link with flour proteins and gelatinized starch (42).

When kneading dough, it is important to avoid excess kneading. Too much kneading will break the gluten strands—the disulfide bonds between the gluten strands pull apart—resulting in a sticky, lumpy dough with little elasticity (9). Although it is difficult to over-manipulate a dough manually, it is fairly easy to do so when using a food processor, standing mixer, or food service dough developer.

Dried Gluten. Dried gluten, sold as vital wheat gluten, is often used to increase the protein content of flours, cereals, meat analogs, pasta, and other bread products. It can also be used to manufacture simulated meat products for vegetarians. Adding vital wheat gluten to regular meat enhances its ability to be sliced and reduces cooking losses during processing. Increasing the gluten content of a flour will also contribute to improved texture and higher volumes during baking (11). Vital wheat gluten is added to some hamburger and hot dog buns to strengthen their structure so that they may hold heavy condiments and prevent the hot dog buns from breaking at their hinges (32).

Celiac Disease. Individuals with celiac disease need to avoid gluten because their immune systems react negatively to this protein. If they consume gluten or any foods that contain gluten, such as those containing wheat, oat, rye, or barley, their small intestines become damaged and they suffer from abdominal bloating and pain, chronic diarrhea, fatigue, weight loss, and failure-to-thrive in infants. The malabsorption then results in malnutrition. In the United States, the time between the symptoms' first appearance and diagnosis averages 10 years because celiac disease mimics many other diseases; but in Italy, where celiac disease is more common, the diagnosis takes less than a month. Celiac disease is genetically transmitted and tends to run in families of European descent, but it is also known to be triggered by surgery, pregnancy, viral infection, or severe stress.

Milling

Milling, or grinding, is the process in which the grain kernel is first freed of its bran and germ and then has its endosperm ground into a fine powder known as flour. Any grain can be milled, but wheat flour dominates the market. Other flours include rye, corn, buckwheat, and triticale flours, and flours made from legumes such as soybeans and peanuts.

Originally, a mortar and pestle were used to grind grain into whole-grain flour. In Mesopotamia sometime around 800 B.C., equipment was developed that used grinding stones moving in circles (Figure 22-9). Eventually

FIGURE 22-8 The role of lipids in gluten development. One theory is that they slide more easily past each other. Ultimately, this contributes to proper dough expansion.

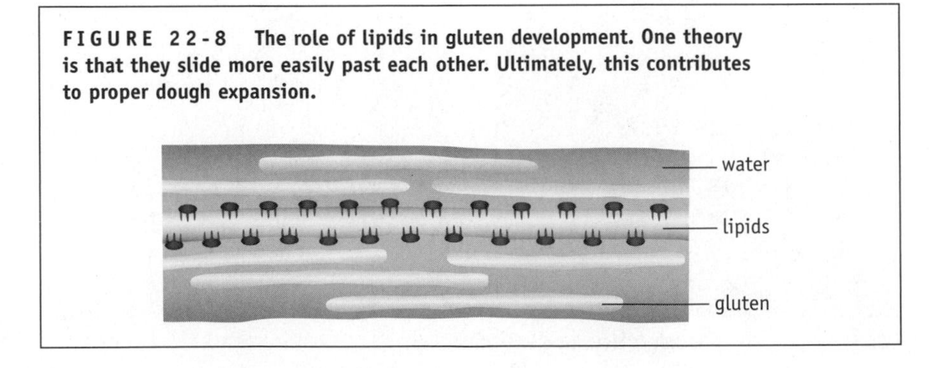

FIGURE 22-9 Different types of stone mills used for grinding grain.

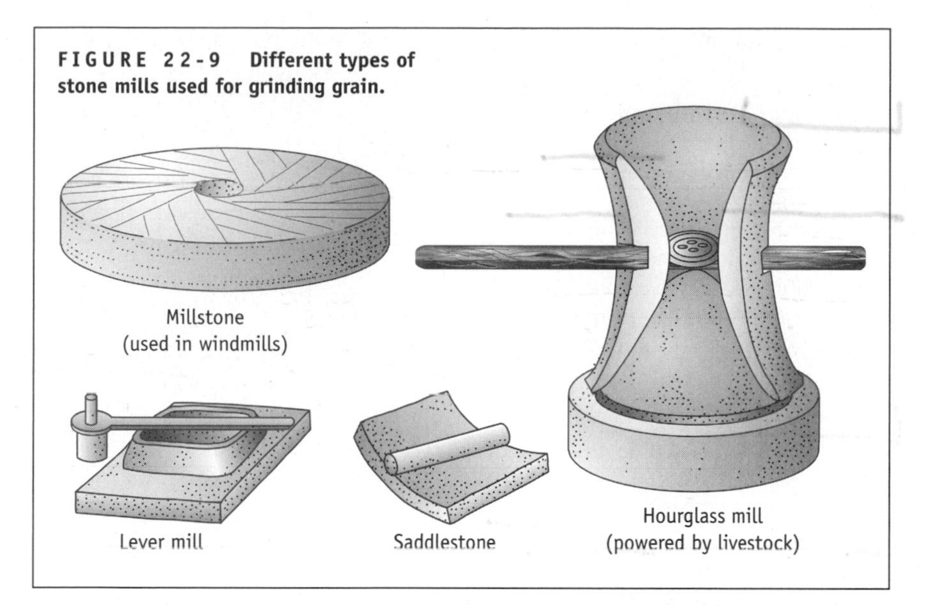

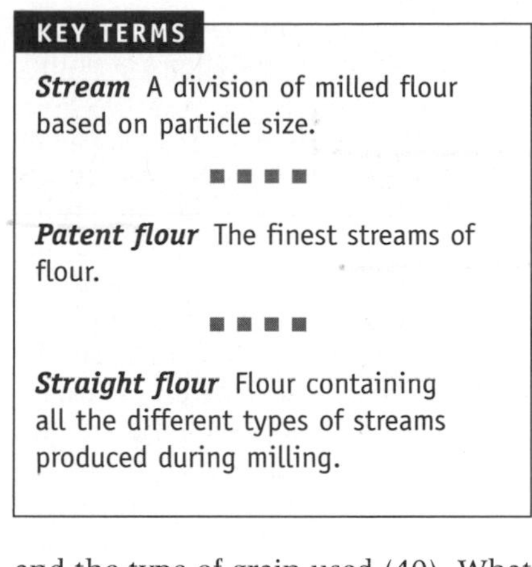
KEY TERMS

Stream A division of milled flour based on particle size.

Patent flour The finest streams of flour.

Straight flour Flour containing all the different types of streams produced during milling.

mills, using flowing water for power, became capable of moving one circular stone over another stone that was held stationary.

Five Steps of Milling. Milling today consists of five basic steps: breaking, purifying, reducing, sifting, and classifying.

Step 1—Breaking. In breaking, special machinery equipped with break rollers removes the bran and germ layers from the grain's endosperm (Figure 22-10). Each succeeding pair of steel break rollers is set closer together to remove more of the bran and germ. The result is called break flour. Break flour still has some bran attached to the endosperm layer.

Step 2—Purifying. To further remove this, the flour is moved through containers where blowing air currents remove any remaining bran. The endosperms, thus freed, or purified, from the whole grain and now free of bran, are known as middlings.

FIGURE 22-10 Machine milling: Break rollers are used to separate the grain's endosperm from the bran and germ.

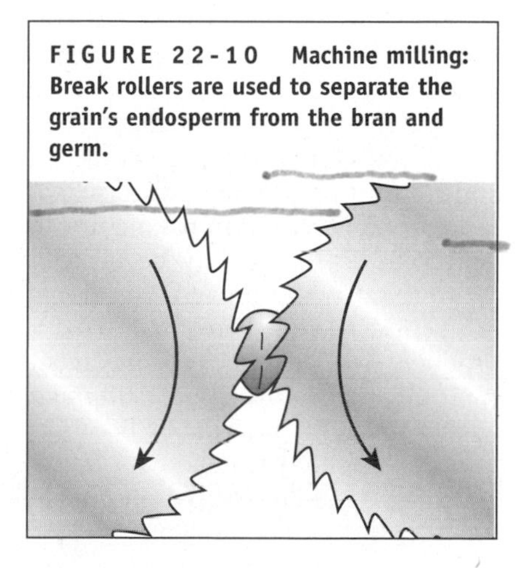

Step 3—Reducing. At the reducing stage, ten to fifteen smooth-surfaced reduction rollers grind the middlings into flour. To prevent the starch granules from being damaged, the spaces between the rollers are progressively reduced.

Step 4—Sifting. The flour is then sifted in **streams**, and this determines how the flour is classified.

Step 5—Classifying. Flour streams range from fine, or first break, to coarse, or clear, and the types of flour obtained from them range from **patent flour** to **straight flour.** Patent flour is divided into long, medium, or short patents depending on how much of the total endosperm was milled. Short patents come from the center of the endosperm, are high in starch, and are best for making pastry flour. Long patents contain more protein from the outer areas of the endosperm and are preferred for the production of bread flours. Clear flour, which is left over from patent flour, is further subdivided into fancy clear, first clear, and second clear. The coarser clear flours are primarily used for all-purpose flour. The quality and grade of the final flour is determined by which streams were combined in its production. The protein content varies according to the origin of the stream, and the type of grain used (40). What follows is a discussion of flours from various wheat and nonwheat grains.

Types of Wheat Flour

The gluten-forming properties of wheat make it the most commonly milled grain and the major source of flour for bread-making (30). The only other flour that comes close to wheat for producing baked goods is rye, but even then it has to be mixed with about one-fourth wheat flour. The different types of wheat differ in their protein content. Soft wheats have the least protein and highest starch content, making them ideal for the tender, fine crumb of cakes and pastries, while hard wheats, with their higher protein content, are preferred for making yeast bread (Table 22-1) (25). Durum wheat, with the most protein, is milled into semolina flour, which is used primarily for pasta production. Flours are varied by using specific types of wheats for their particular qualities, by blending them in different combinations, and by using

TABLE 22-1
Percent Protein of Various Wheat Flours*

Flour	*Protein %*
Gluten	41
Whole wheat (hard)	14
Durum wheat (hard)	13
Bread (hard)	11
All-purpose	10
Pastry (soft)	9
Cake (soft)	8

*As protein increases, so does the "hardness" of the flour.

selected streams. The result can be any of a number of different types of flours.

Whole-Wheat Flour. Whole-wheat flour, also known as graham or entire-wheat flour, is made from the entire wheat kernel, including the bran, germ, and endosperm. One drawback to using whole wheat-flour alone in bread-making is that the bran's coarse granules cut the gluten strands, decreasing the final volume of the bread. The flour is finely ground to reduce this effect and is often combined, usually half and half, with white flour for the same reason. A whole-wheat baked item with no added white flour will be dense and heavy. Whole-wheat flour contains fat from the germ so it requires refrigeration to prevent rancidity.

White Flour. White flour is made using only the endosperm of the wheat grain.

Bread Flour. Most bread flour is a long patent white flour made primarily from hard, winter wheat. The higher gluten content of this flour makes it ideal for making yeast breads and hard rolls that require elastic gluten for multiple rising periods.

Durum Flour (Semolina). Durum flour is made from hard winter durum wheat. Its high protein content makes it best suited for the manufacture of the semolina flour used for pasta products.

All-Purpose Flour. All-purpose flour, also known as family-type flour, contains less protein than bread flour. The protein content of all-purpose flour averages about 11 percent, while bread flours that need strong gluten development have between 12 to 14 percent protein. Blending hard and soft wheat flours yields a flour that can be used for all purposes—breads, cakes, or pastry. Despite what appears as a uniform product, the actual protein content of all-purpose flour may differ from region to region and by brand. Regional preferences have influenced manufacturers to place more protein in flour sold in the northern United States, where yeast breads are more popular, compared to the South, where biscuits requiring a softer flour are more routinely consumed.

Pastry Flour. Cream-colored pastry flour is derived from soft wheat with short to medium patents. Its lower protein content of about 9 percent is preferred by commercial and professional bakers for preparing pastries, some cookies, sweet yeast doughs, biscuits, and muffins. Some people substitute a combination of all-purpose and cake flours for pastry flour.

Cake Flour. Soft, extra-short patent wheat flours are used to make cake flour. It is pure white and has a very fine, silky, soft texture. Its lower protein content of only 8 percent and small particle size compared to all-purpose flour result in less gluten being formed, which gives cakes a fine grain, a delicate structure, and a velvety texture. Cake flour is treated with chlorine gas, which lowers the pH from about 6.0 to 5.0, bleaches the plant pigments, and improves baking quality (20). All-purpose flour can be used to make cakes, but they may have a slightly coarser texture and lower volume (60). Cake flour has the same amount of protein as pastry flour, but more starch.

Gluten Flour. Gluten flour is made from wheat flour that has been milled in such a way to retain the gluten. A small amount of this flour is used in combination with other flours (1 tablespoon for every cup of other flour) to help heavy breads rise more readily. The extra protein binds with water and contributes to a moister bread, making it appear more "fresh" to the consumer. Vegetarians often use gluten flour to make seitan, a protein food with a springy texture and able to sponge up flavors from gravies or broths. Two factors to consider when using gluten are (1) all gluten products need to be refrigerated or frozen because they contain natural fats that can become rancid, and (2) gluten causes digestive problems (abdominal bloating, pain, diarrhea) for people with celiac disease.

Types of Non-Wheat Flour

Although most baked products are made primarily of flour from wheat, there are other, non-wheat flour sources. These include rice, rye, cornmeal, buckwheat, triticale, soybeans, potatoes, and even peanuts.

Rice Flour. Naturally gluten-free rice flour is popular in Asian cultures where it is used to make a variety of food products.

Rye Flour. The lower gluten potential of rye flour contributes to a very compact bread such as pumpernickel. In addition, rye flour contains a high concentration of water-soluble carbohydrates called pentosans, which gives the flour a high water-binding capacity (56). As a result, the gases in rye bread do not expand very well (39). Wheat flours are often added to rye flour in a 1:4 ratio to create a more porous, lighter bread. The pentosans also are responsible for the characteristic stickiness of rye bread.

Cornmeal Flour. Corn's chief protein is zein, which is incapable of mimicking the elastic or plastic properties of flours with higher gluten-forming potential (13). Cornbread and corn muffins do not rise to any great extent and tend to have a crumbly texture unless the cornmeal is combined with all-purpose flour. Cornmeal flour is available from yellow, white, and even blue varieties of corn. Masa farina is a finely ground cornmeal flour made from corn that was presoaked in lime or lye. It is primarily used to make corn tortillas and tamales.

Soy Flour. Soy flour is higher in protein than other flours because its source, the soybean, is actually a legume. It has more of the amino acid lysine than wheat, and is sometimes added to wheat flour to improve its protein profile (18, 37). Soybean flour has a low gluten capacity, however, and so must be combined with wheat flour when it is used in baking. Up to 3 percent soy flour may be used in commercial white breads. Soy flours are generally added to many bakery products at levels of 3 to 6 percent (percent total formulation), although adding about 20 percent soy flour has been reported to result in a texture and color similar to whole-wheat bread (32).

Buckwheat Flour. Buckwheat flour contains more starch and less protein, the flour. than wheat flour and is used primarily in pancake and waffle mixes as well as in crêpes and blinis.

Triticale Flour. This is a flour obtained from triticale, a hybrid of wheat and rye grains. Triticale flour is best used in a 1:3 ratio with white flour in the production of breads.

Potato Flour. Cooked potatoes that have been dried and ground yield potato flour. The starch in potato flour increases loaf volume, which is why the liquid left over from cooked potatoes is sometimes used in homemade breads.

Treated Flours

Wheat flours can be treated to improve their functional properties, resulting in aged, bleached, phosphated, self-rising, instant, or enriched flours. Many of these are used in the production of baked products.

Aged Flour. Freshly milled flour is not white, and the resulting baked products are of poor quality. Historically, this problem was corrected by storing the flour for several months so it could "age" and become naturally bleached by the oxygen in the air. Natural aging is expensive, however, because of the required storage space, increased labor costs, and the risk of pest infestation.

Bleached Flour. To bypass the costs involved in aging flour, all-purpose flour can be bleached by exposing it to chlorine gas or benzoyl peroxide. Artificial bleaching, used by the food industry since the 1920s, oxidizes the flour's carotenoid pigments and improves the condition of the gluten. The oxidation occurring during bleaching not only lightens the flour's color, but also increases the number of disulfide bonds between the protein chains, improving the strength and elasticity of the gluten (22). Oxidizing agents improve the volume, texture, and crumb structure of baked products (43). Flours whitened by this method are labeled "bleached." The bleaching agents evaporate and do not leave residues or alter the nutrient value of the flour. Cake flour is always bleached with chlorine gas, which not only whitens the color, but creates a very tender, fluffy baked product. All-purpose flours may or may not be bleached. Semolina flour is never bleached because its yellow hue contributes to the color of pasta (45).

Phosphated Flour. Flour may be leavened by baking soda instead of baking powder if an acid, specifically monocalcium phosphate (no more than 0.75 percent), has been added. Flour thus phosphated also has the advantage of increased calcium content, with 68 to 165 mg of calcium versus 18 mg per sifted cup of unphosphated flour (16).

Self-Rising Flour. Self-rising flour is all-purpose flour with the leavening agent and salt already added. The leavening agent in self-rising flour is a baking powder (baking soda combined with a salt, monocalcium phosphate), which also contributes a significant amount of dietary calcium. One cup of self-rising flour contains about 1½ teaspoons baking powder and ½ teaspoon salt. These amounts should be added to all-purpose flour when the recipe calls for self-rising flour.

Instant or Agglomerated Flour. Instant flour mixes easily with water, readily gelatinizing without lumps, which makes it ideal for powdered soups, sauces, and gravies. This kind of flour may create a coarse texture in baked products, so it is not recommended for that use. Instant flour is created by passing flour through jets of steam, which wet it and allow it to stick together, or agglomerate, in very small particles. Heated chambers dry the uniformly sized particles so they flow freely and do not need to be sifted. When using it in recipes, 1 cup of all-purpose flour equals 1 cup minus 2 tablespoons of instant flour.

Enriched Flour. Enriched flour is white flour to which the B vitamins thiamin (B_1), riboflavin (B_2), niacin, and folate as well as the mineral iron have been added in order to reach levels established by federal standards. Calcium is an optional enrichment nutrient. Fiber and other nutrients, such as vitamin E, that are lost with the removal of the bran and germ are not replaced (58).

Flour Mixture Ingredients

Flour mixture ingredients may include leavening agents, sugar, salt, liquid, fat, eggs, and in some cases, commercial additives.

Leavening Agents

The presence of a leavener causes the flour mixture to rise. Leaveners may be physical, biological, or chemical:

- *Physical leaveners:* air and steam
- *Biological leaveners:* yeast and bacteria
- *Chemical leaveners:* baking powder and baking soda

Although flour mixtures can rise with the physical help of air and steam, most of the leavening is accomplished by carbon dioxide gas produced from either biological or chemical sources (23). The major biological sources of carbon dioxide are yeasts. Chemical leaveners, such as **baking soda** and **baking powder**, yield carbon dioxide when an alkali reacts with an acid in the presence of a liquid. In order to do its job, baking soda must have an acid ingredient added to the flour mixture, while baking powder has already had the acid incorporated. Acid ingredients are often added in the form of buttermilk (lactic acid), sour cream, choco-

KEY TERMS

Baking soda A white chemical leavening powder consisting of sodium bicarbonate.

Baking powder A chemical leavener consisting of a mixture of baking soda, acid(s), and an inert filler such as cornstarch.

late, brown sugar (contains molasses), or molasses (aconitic acid).

The type of food is the primary determinant of what type of leavening agent will be used. Yeast breads are usually leavened with biological agents such as yeast. Quick breads, as well as cakes, cookies, and pastries, are leavened physically with air, steam, or carbon dioxide generated by baking soda, and/or baking powder. Regardless of the source, leavening changes the baked product's volume, crumb, texture, and ultimately, its flavor (Figure 22-11).

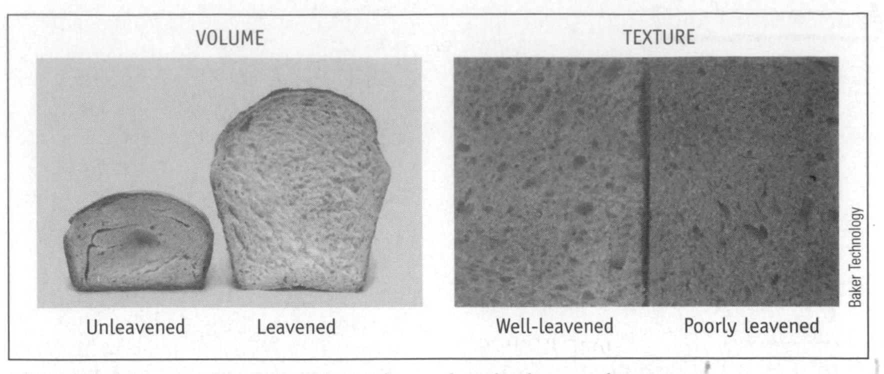

FIGURE 22-11 **The role of leavening on bread volume and texture.**

Air and Steam. Air and steam are physical agents that help dough to rise.

Air. Air is incorporated into almost all flour mixtures during mixing, during the creaming of fat and sugar, by sifting dry ingredients, or by using whipped egg whites (Figure 22-12).

Steam. Water incorporated into flour mixtures produces steam when heated, expanding to 1,600 times its original volume. Steam, either from liquid or from other ingredients such as egg whites, is the primary leavening agent for pie crusts, pastry, cream puffs, and popovers.

Yeast. The ability of yeasts (*Saccharomyces cerevisiae*), which are naturally found in air, water, and living organisms, to produce carbon dioxide through fermentation was probably discovered by accident. An Egyptian baker is reputed to have set a flour-and-water dough aside in a warm place, where it became contaminated with yeasts from the air (16). The yeasts started to feed off the available sugar in the bread, producing carbon dioxide and water through the process of fermentation. This same scenario in the absence of oxygen results in the yeast producing ethyl alcohol, or ethanol (Figure 22-13).

HOW & WHY? ?????????

If yeasts need sugar to ferment, how do they ferment and multiply in a flour mixture? The yeasts feed off saccharides derived from either the flour or sugar added to the dough. Flour contains about 1.5 percent sugar (dry weight basis), of which the glucose is quickly fermented. In addition, an enzyme in yeast (yeast invertase) hydrolyzes starch to glucose and fructose, while flour amylases break down some starch to maltose, which is then transported into the yeast cell where it is hydrolyzed to glucose. Sugar (sucrose) added to the flour mixture is also broken down by yeast invertase to glucose and fructose.

Breadmakers now add yeast directly to flour mixtures and control their growth through hydration, temperature, and salt concentration. During bread-making, yeasts that have been alive but dormant become activated when they are hydrated in water at an optimum temperature, which varies according to yeast type. Once activated by the warm water, they are added to the flour mixture. After they are mixed in, the dough is kneaded and allowed to sit in a warm place as carbon dioxide

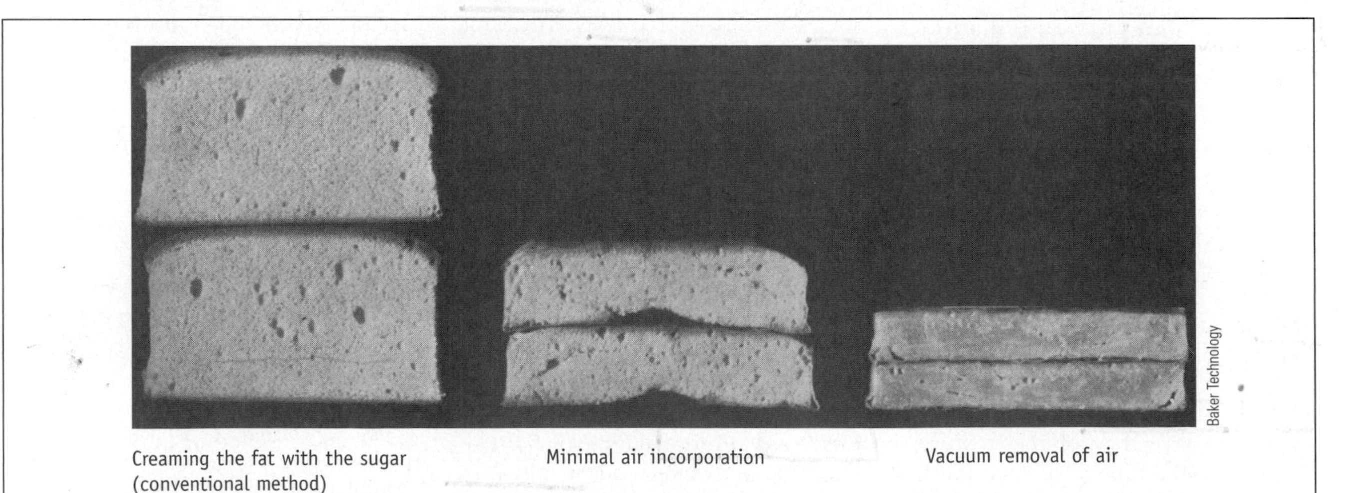

FIGURE 22-12 **The role of air as a leavening agent: three pound cakes made without baking powder and subjected to three different degrees of air incorporation.**

FIGURE 22-13 Fermentation process.

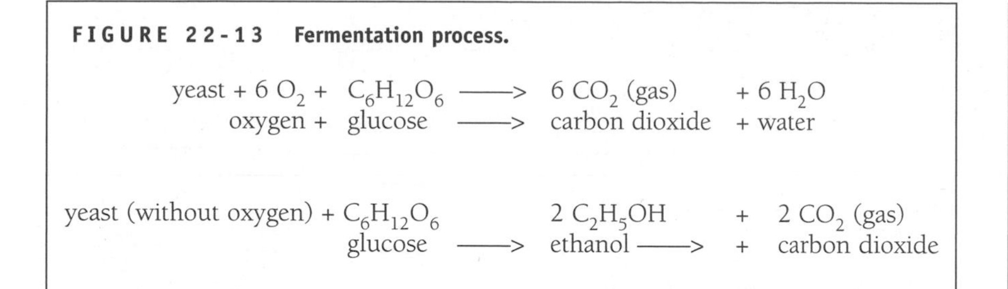

generated by the growing yeast makes the dough rise. Yeasts multiply best at temperatures of 68° to 81°F (20° to 27°C), while fermentation is optimum at 81° to 100°F (27° to 38°C), specifically 95°F (35°C) (Figure 22-14). As the yeasts multiply and ferment in this warm environment, a small amount of sugar is sometimes added to serve as a food source for them. Adding too much sugar or salt, however, pulls water osmotically from the yeast cells and can literally dry them to death.

Yeasts are available in several forms, and are classified on the basis of their activity. Active yeasts include baker's yeast, brewer's yeast, and yeasts for alcoholic beverages. Inactive yeasts, such as dried brewer's yeast and primary-grown yeasts, are used primarily for their nutritional value and contribution to flavor (29).

HOW & WHY? ?????????

How do brewer's, baker's, and nutritional yeast differ? Brewer's yeast is a brewery by-product that is used as a nutrient supplement. It is a source of B-complex vitamins, amino acids, chromium, calcium, phosphorus, potassium, and other nutrients. Brewer's yeast has a bitter taste and is made from the pulverized cells of *Saccharomyces cerevisiae* dried at high temperatures, so the yeast are killed and do not contribute to leavening.

Baker's yeast is used for leavening breads and other baked goods. It is dried at low temperatures so the yeast are not killed. It can be the same species as brewer's yeast (*S. cerevisiae*), or *Candida milleri* and *Lactobacillus sanfrancisco* for some sourdough breads.

Nutritional yeast is grown specifically for a supplement. Available in powder or flake form, nutritional yeast is grown on a molasses solution so it lacks the bitter taste of other yeasts—it actually has sweet, nutty, cheesy flavor. It is the same species as brewer's yeast, dried at high temperatures so it is no longer alive. Nutritional yeast has a similar nutrient composition to brewer's yeast, but more than 3 tablespoons per day may raise uric acid levels.

In this chapter, the focus is primarily on baker's yeasts, which are sold as dry, fresh (compressed), or instant yeast.

Dry (Active) Yeast. This is the most widely available type of yeast sold in small packets. This porous and free-flowing type of yeast can be stored at room temperature, but keeps longer when refrigerated or frozen. It becomes inactive if not used before the expiration date. Water warmed to 115°F (46°C) is ideal for rehydrating and activating the dry yeast. Once hydrated, it should not be exposed to temperatures below 100°F (38°C), which lowers its activity and results in a sticky dough (Chemist's Corner 22-2), or above 140°F (60°C), which will kill it. In general, one package (¼ ounce or 2¼ teaspoons) of active dry yeast is enough to leaven 4 to 6 cups of flour. It contains less moisture (8 percent) than compressed yeast and is therefore less susceptible to deterioration.

Fresh Yeast. Also called compressed or cake yeast, this type of yeast is sold as a semi-solid cake with about 70 percent moisture. It therefore has a short shelf life and develops mold easily, so refrigeration is required. Compressed yeast is not, consequently, a favorite of consumers, but it is preferred by professionals because it dissolves easily without excessively warm temperatures and produces more consistent results. Packs that are brownish (instead of cream-white), dried out, or have a bad odor should be discarded (Figure 22-15). Compressed yeast is reactivated by the addition of warm water (85°F/29°C).

FIGURE 22-14 Yeast activity related to temperature.

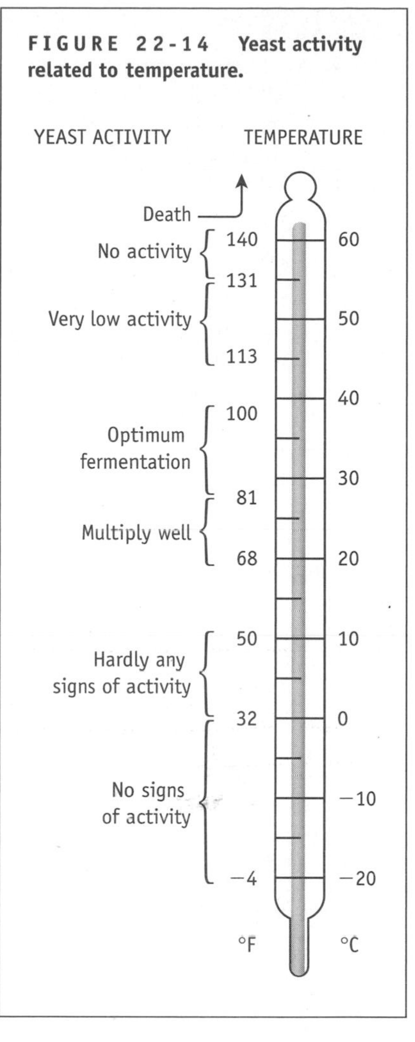

Instant, Quick-Rising or Fast-Acting Yeast. Primarily for commercial bakers, this yeast strain reproduces more quickly

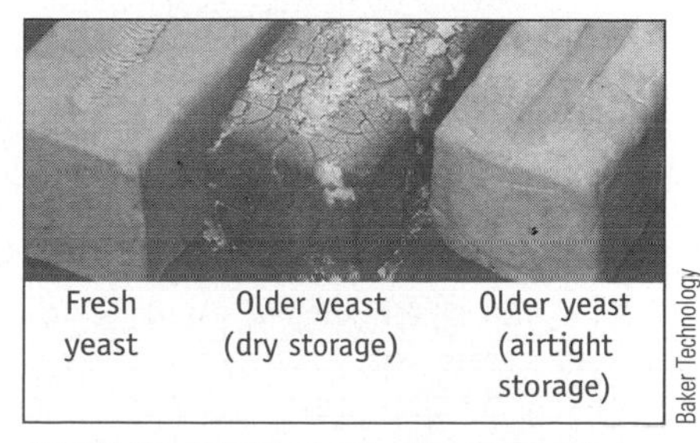

FIGURE 22-15 The freshness of fresh yeast is tested with a fingernail scratch on the surface.

KEY TERMS

Starter A culture of microorganisms, usually bacteria and/or yeasts, used in the production of certain foods such as sourdough bread, cheese, and alcoholic beverages.

than the traditional yeasts. Breads rise twice as fast with this yeast, eliminating the need for a second rising. However, there is less time for flavor to develop. Another drawback is that this type of yeast is extremely sensitive to temperature and moisture, and requires an activation temperature between 125° and 130°F (52° to 54°C).

Bacteria. Harmless bacteria that generate carbon dioxide are used as leavening agents in sourdough and salt-rising breads. These baked products depend on a **starter**. Two types of fermentation—one from bacteria and one from yeasts—contribute to the production of carbon dioxide. The bacteria also contribute a desirable, slightly sour flavor to certain baked products, which has become more and more popular in making the European-style rustic hearth breads. A sourdough starter must be kept alive and fed additional flour in order to keep on contributing to the leavening of breads.

Baking Soda. Baking soda chemically yields carbon dioxide in the presence of moisture and an acid. Up to ¼ teaspoon of baking soda is required for each cup of flour to be leavened. Baking soda is not typically used by itself as a leavening agent, because so much would be required that the flavor and appearance of the product would be adversely affected. Baking soda is only used when the flour mixture includes acid ingredients—lemon, vinegar, buttermilk, yogurt, molasses, brown sugar, cocoa, chocolate, citrus fruits, cream of tartar (¾ teaspoon per ¼ teaspoon baking soda), or sour milk (made by combining 1 cup milk with 1 tablespoon lemon juice or vinegar, or 1¾ teaspoon cream of tartar) (Chemist's Corner 22-3). The acid reacts with the baking soda and creates an intermediate compound, carbonic acid, that reacts to give off carbon dioxide and water (Figure 22-17). The immediate leavening effect caused by this reaction necessitates that any baked product prepared with baking soda be placed in the oven as soon as possible after it is mixed.

Chemist's Corner 22-2

Cold Water and Yeast

Hydrating yeast with water that is too cool results in a sticky dough. Cold water slows cell membrane recovery in yeast and allows cell constituents to leach out. Sticky yeast dough can be caused by glutathione, which is naturally released from the yeast through their cell walls and into the dough when they are hydrated below 100°F/38°C. Glutathione is a natural reducing agent that disrupts the disulfide bonds between and within the protein molecules. The proteins quickly unfold, resulting in a softer, stickier dough. This problem does not happen with fresh yeast, which disperses in cold water easily. Rather it tends to happen with dry yeast, where optimum water temperature for cell restoration is 104°F/40°C.

FIGURE 22-16 Glutathione released from yeast cells results in sticky dough.

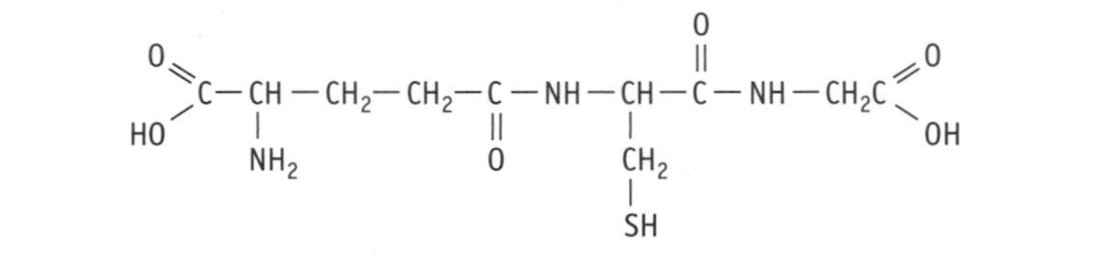

Baking Powder. When baking powder is used, it is not necessary to add an acidic ingredient to the flour mixture in order to produce carbon dioxide because the acid has already been added. The inert filler in baking powder absorbs any excess moisture in the air, which would otherwise cake the powder and/or reduce its potency.

Acid in Baking Powder. Baking powder is a kind of enhanced baking soda that can be made by combining ¼ teaspoon of baking soda with ½ teaspoon of cream of tartar. In the past, cream of tartar was derived from the sediment that collected on the sides of wine casks, and was the most widely used acid in the manufacture of commercial baking powders (31). When a liquid is added, the acid reacts with the alkaline baking soda to release carbon dioxide gas. The type of acid determines the speed with which carbon dioxide is produced. Federal standards require that all baking powders must be formulated to be able to produce at least 12 percent carbon dioxide when water and heat are applied, yet many yield about 14 percent to allow for losses during storage (39). A non-laboratory method to test the potency of baking powder can be conducted by mixing 1 teaspoon of baking powder with ⅓ cup of hot water to see if it causes bubbling.

Types of Baking Powder. The two main types of baking powder are fast, or single-acting, powder and slow, or double-acting, powder. Fast/single-acting powder is available only to commercial bakers. A flour mixture made with fast/single-acting baking powder should be handled quickly and efficiently and placed in the oven as soon as possible, because it starts to produce carbon dioxide as soon as water is added. Any delay allows car-

Chemist's Corner 22-3

Acid and Baking Soda

Organic acids give up protons to react with the sodium bicarbonate, eventually yielding carbon dioxide. Acids contribute to the crumb structure formation by providing small gas cell nuclei during mixing that serve as starting points for the carbon dioxide that will be produced (23).

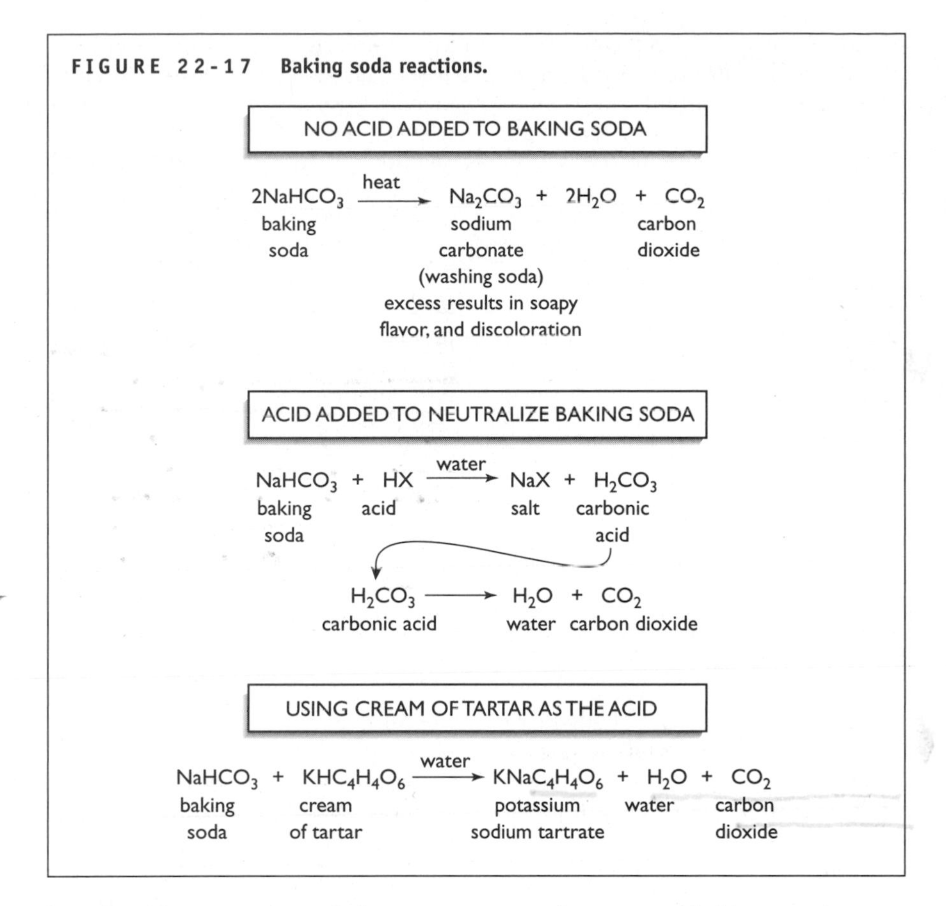

FIGURE 22-17 Baking soda reactions.

NO ACID ADDED TO BAKING SODA

$$2NaHCO_3 \xrightarrow{\text{heat}} Na_2CO_3 + 2H_2O + CO_2$$

2NaHCO₃ = baking soda; Na₂CO₃ = sodium carbonate (washing soda) excess results in soapy flavor, and discoloration; CO₂ = carbon dioxide

ACID ADDED TO NEUTRALIZE BAKING SODA

$$NaHCO_3 + HX \xrightarrow{\text{water}} NaX + H_2CO_3$$

baking soda + acid → salt + carbonic acid

$$H_2CO_3 \longrightarrow H_2O + CO_2$$

carbonic acid → water + carbon dioxide

USING CREAM OF TARTAR AS THE ACID

$$NaHCO_3 + KHC_4H_4O_6 \xrightarrow{\text{water}} KNaC_4H_4O_6 + H_2O + CO_2$$

baking soda + cream of tartar → potassium sodium tartrate + water + carbon dioxide

bon dioxide to escape and decreases the ability of the mixture to rise. Approximately 1½ to 2 teaspoons of single-acting powder are required for every cup of flour.

Many commercial bakers use double-acting baking powder (sodium aluminum sulfate (SAS)-phosphate powder), which reacts twice: once when it is moistened, and again during heating (Figure 22-18). Approximately 1 to 1½ teaspoons of double-acting baking powder are required for every cup of flour.

Other types of baking powder may be used. Potassium bicarbonate is available for people on low-sodium diets. Commercial bakers use baking powders that are specifically designed for certain baked products. Ammonium bicarbonate, for example, is used for cookies that require very little water and have a high surface area; this allows the ammonia to completely evaporate, preventing a bitter flavor (23).

Too Much/Too Little Leavening. Excess leavening results in a baked product that falls, and has a low volume and/or a coarse texture. Adding more than the required amount of SAS-phosphate powder can result in a bitter taste. Too much baking soda, or pockets of baking soda created by inadequate mixing with the dry ingredients, will cause the production of residues of sodium carbonate, resulting in a soapy flavor and discolored brown or yellow spots. This excess alkalinity can also affect the flavor of chocolate and turn its color slightly reddish, although the color change is desirable when preparing devil's food cake (Chemist's Corner 22-4).

Too little leavening results in a compact, heavy baked product. Unfortunately, this can also occur when the baking soda or powder has lost its potency through being exposed to moisture or stored for over six months (31).

Too Much/Too Little Flour. Too much flour results in a lower volume, an increased number of "tunnels," and a drier, tougher crumb. A baked product made with insufficient flour often has a course texture and a weak, possibly collapsible, structure.

Sugar

Aside from contributing sweetness, sugar also influences the volume, moistness, tenderness, color, appearance, and caloric (kcal) content of baked products.

Functions of Sugar. The obvious sweetening role of sugar is apparent in baked products such as cakes, cookies, sweet breads, doughnuts,

FIGURE 22-18 Baking powder reactions, specifically, double-acting SAS-phosphate.

Step 1.
$$2Na_2SO_4Al_2(SO_4)_3 + 6H_2O \xrightarrow{\text{heat}} Na_2SO_4 + 2Al(OH)_3 + H_2SO_4$$

SAS (sodium aluminum sulfate) + water → sodium sulfate + aluminum hydroxide + sulfuric acid

Step 2.
$$3H_2SO_4 + 6NaHCO_3 \longrightarrow 6H_2O + 3Na_2SO_4 + 6CO_2$$

sulfuric acid + baking soda (sodium bicarbonate) → water + sodium sulfate + carbon dioxide

Chemist's Corner 22-4

The Red in Chocolate Cake

The reddish hue of chocolate cake resulting from excess sodium bicarbonate is due to the oxidation of a polyphenol in cocoa to phlobaphene. The higher the pH in chocolate cakes, the redder the color: brown (pH = 6.0 to 7.0), mahogany (pH = 7.0 to 7.5), and reddish (pH = 7.5+) (35).

coffee cake, and sweet rolls. Many breakfast cereals contain sugar for sweetening power, but also to act as a protective coating (2). Other functions of sugar in flour mixtures include the following:

- Increases the volume of cakes and cookies by the incorporation of air into the fat during creaming (especially true for granulated sugars).
- Contributes to volume by providing food for the yeast—although too much sugar (over 12 percent) results in a proportional decrease in volume (Figure 22-19).
- Raises the temperature at which gelatinization and coagulation occur, which gives the gluten more time to stretch, thereby further increasing the volume of the baked product and contributing to a finer, more even texture (4).
- The hygroscopic, or water-retaining, nature of sugar increases moistness and tenderness and also helps delay staling, thus improving the shelf life of baked products.
- Contributes to tenderness by competing with starch for the available water necessary for the hydration of flour proteins and eventual gluten development. The crust is initially very crisp, but becomes softer as sugar attracts moisture from the air or crumb (47).
- Helps to brown the outer crust of baked products through caramelization and the Maillard reaction (7, 46).

Types of Sugar. Different types of sugars have different weights per cup, which should be taken into consideration when making substitutions. As discussed in Chapter 9, alternatives to sugar are available and can sometimes serve as substitutes for sugar in baked products. Although no code of federal regulation exists specifying the amounts of honey to be used in baked products, the FDA states that honey breads and rolls should contain at least 8 percent honey (based on flour weight) and impart a noticeable flavor (1).

Too Much/Too Little Sugar. Baked products made with too much sugar may fall, have a lower volume, a coarse grain, a gummy texture, and an excessively browned crust (Figure 22-20). Dough is considered sweet if it contains more than ½ cup of sugar per 3½ cups of flour (1¼ ounces sugar per cup of flour). Too little sugar results in dryness, reduced browning, lower volume, and less tenderness.

Salt/Flavoring

Small amounts of salt are added to flour mixtures for flavoring, for producing a firmer dough, for improving the volume, texture, and evenness of cell structure, and to prolong shelf-life (47). Baked products made without salt tend to be bland, so flavor is one of the most important reasons for adding salt. It also plays a large role in firming the dough by adjusting the solubility and swelling capacity of the gluten (Figure 22-21) (19). Bakers often prefer flaky sea salt over granular table salt because the greater surface area of the more fragile flakes allows for greater distribution. Some people like the ease of adding the salt to the flour, while others dissolve it in the liquid for greater dispersion; however, the latter increases the risk of greater gluten formation and resulting toughness. Slightly less gluten is formed when salt is added to the flour mixture because fat may cover up some of the granules (10).

Salt Controls Yeast Growth. In the production of yeast breads, salt helps control yeast growth. Without salt, fermentation would be too rapid and result in a sticky, difficult-to-handle dough (38). Too much salt, however, inhibits yeast activity, reducing the amount of carbon dioxide gas produced and decreasing the volume of the loaf (Figure 22-22) (15). There are now available special formulations for salt-free breads for salt-sensitive consumers.

Flavorings. Because the basic ingredients of all baked products are the same, all would taste very similar without variations in added flavorings. Flavor extracts, cocoa, melted baking chocolate, fruits, spices, nuts, and other flavorings, seemingly limited only by the baker's imagination, may be added to vary the taste experience.

Too Much/Too Little Salt. Excess salt produces a firm dough with a low volume (because of partially inhibited fermentation), dense cells, and a too salty taste (47). Too little salt produces a flowing, sticky dough with a low volume, uneven cell structure, lack of color, and a bland taste (35).

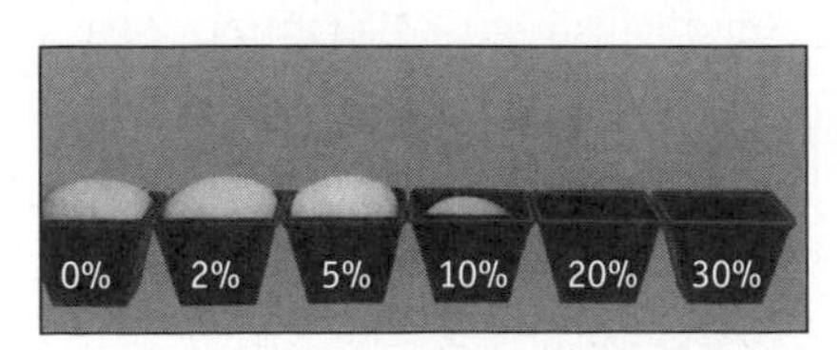

FIGURE 22-19 Sugar's influence on the volume of loaf bread. Source: Schunemann and Treu, *Baking: The Art and Science*. (Baker Technology, 1988).

FIGURE 22-20 Excess sugar decreases cake volume and results in a gummy texture.

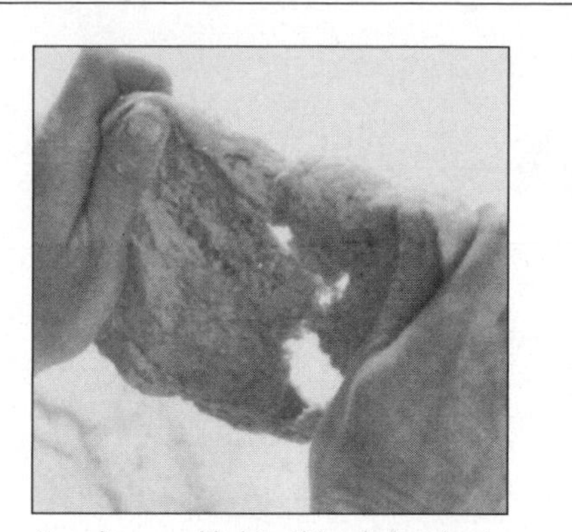
No salt was added to this whole-wheat dough kneaded for five minutes.

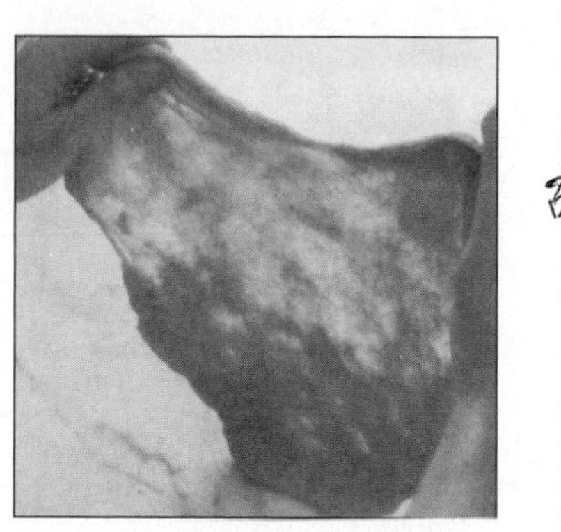
The same dough made with salt is much more pliable, which aids in gluten formation.

Scott Phillips/Fine Cooking Magazine

FIGURE 22-21 Salt makes dough more elastic.

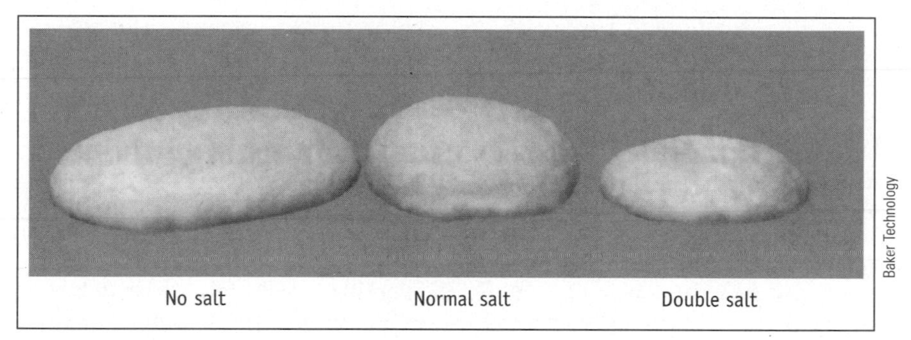

Baker Technology

FIGURE 22-22 Salt influences dough firmness.

Liquid

Liquid in some form is required in flour mixtures to hydrate the flour and to gelatinize the starch. The water in the liquid also allows gluten to be formed (44), acts as a solvent for the dry ingredients, activates the yeast, provides steam for leavening, and allows baking powder or soda to react and produce carbon dioxide gas.

Milk. It is not necessary to include milk in a flour mixture, but it is usually recommended over water, because it improves the overall quality of the baked product. In addition to contributing water, milk adds flavor and nutrients (complete protein, B vitamins, and calcium), and contains certain compounds that help produce a velvety texture, a creamy white crumb, and a browner crust. Doughs made with milk are easier to shape, less sticky and heavy, and retain their shape better. They also tend to expand during fermentation without over- or under-development, and retain gas better, which results in a higher volume. These positive attributes are the result of the presence of milk fat and a natural emulsifier, lecithin.

The lactose in milk participates in the Maillard reaction, resulting in a browner crust. Fresh fluid milk, buttermilk, nonfat dried milk, canned milks, yogurt, and sour cream can all be used at various times in baked goods, depending on the desired end product.

Too Much/Too Little Liquid. Excess liquid may result in a very moist baked item with low volume. Too little liquid may produce a dry baked product that is low in volume and stales quickly.

Fat

Fat incorporated into the flour mixture physically interferes with the development of gluten, creating a more tender crumb (51). In fact, the higher the fat content, the shorter the gluten strands, and the softer, more easily handled, and more pliable the dough. It is for this reason that fat is sometimes referred to as "shortening" (52). To maintain the correct consistency when fat is added to a flour mixture, the liquid is reduced by an amount equal to one-fifth the amount of added fat. Although fat makes the dough softer and able to rise higher, adding too much of it (over 20 percent of the flour's weight) makes the baked product too "short" and results in a lower volume because of an insufficient gluten structure and tearing of the crumb (47).

Functions of Fat. Fat performs many functions in baked goods. It acts as a tenderizer and adds volume, structure, flakiness, flavor, color, and a resistance to staling; it also plays a role in heat transfer (Figure 22-23).

Fat Improves Volume. Fat increases the volume of the baked product as fat particles (crystals) melt during baking, making the batter more fluid and prone to expansion (6, 8). The role of fat in improving loaf volume is also derived from its ability to stabilize large numbers of small air bubbles by adhering to their surfaces. This allows the bubbles to expand during baking without rupturing. Fat also contributes to volume because the creaming together of fats and sugars traps some air, which acts as a leavening agent during heating. The volume of

Baker Technology

FIGURE 22-23 Fat influences the volume and texture of loaf bread.

baked breads increases by about 15 to 25 percent using 5 percent shortening (3 to 5 percent is generally used) (52).

Fat Improves Strength, Crumb, and Flakiness. In the process of adding volume, fats provide strength to the baked product's structure. As a result, it is more resistant to shocks that otherwise might cause it to collapse during handling. The appearance of the crumb is also affected by fat concentration. Lower-fat baked products have a fine, velvety crumb, while those higher in fat lose this characteristic. The flakiness in pie crusts and pastries is entirely dependent on fat being incorporated into the flour (see Chapter 26).

Fat Improves Flavor and Color. Fat is an important contributor to flavor. The moister crumb of many baked products and the smooth mouthfeel of many fillings used in cookies or pastries is heavily dependent on the presence of fats. The color is also influenced by the fat content and the type of fat used. Croissants owe much of their flavor and color to their large butter component.

Fat Delays Staling. Another benefit of the inclusion of fats in flour mixtures is that they delay staling in the final baked product (52). Emulsifiers play a role here, also. The dry, hard, resistant, breakable crumb and crust of staled bakery items is not as evident in baked products made with fat and/or emulsifiers, because these ingredients interfere with the main mechanism of staling—recrystallization of starch molecules, specifically amylopectin. Starch gelatinizes in freshly baked products, but during cooling the randomly distributed amylose and amylopectin recrystallize. The amylose recrystallizes quickly, enabling the bread or other baked product to be sliced, while amylopectin takes several days and firms the crumb in the process (Figure 22-24).

FIGURE 22-24 How bread stales.

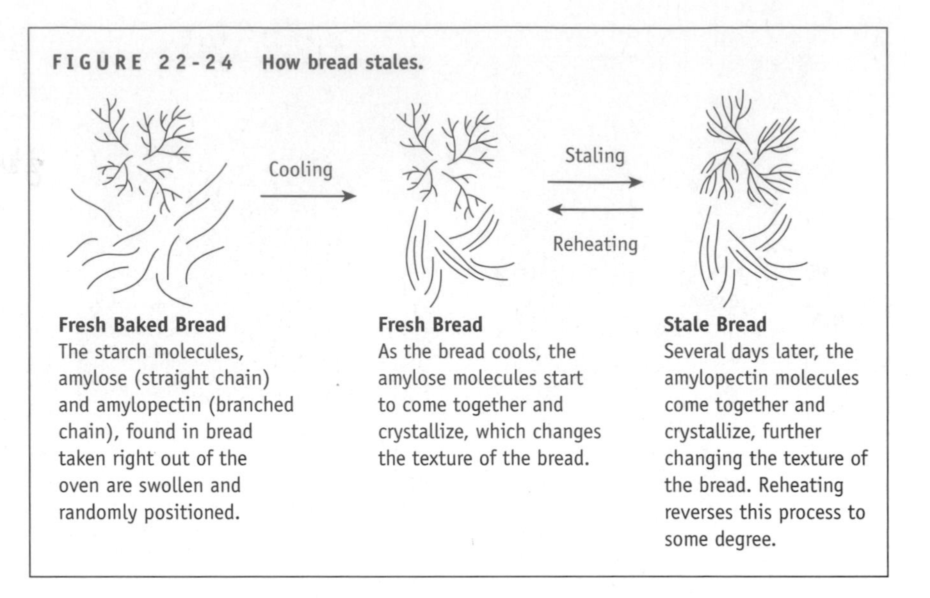

Types of Fat Used in Baked Goods. The fats most commonly used in baking are shortenings, unsalted butter, and margarine. Oil and lard are sometimes used. Shortenings have a much wider useful temperature range than either butter or margarine and have the added benefit of containing their own emulsifiers (about 3 percent) and 10 to 12 percent gas by volume (53). They are also less expensive. The emulsifiers in the shortening distribute the fat more finely throughout the batter, while the gas contributes to volume (Figure 22-25). Cakes made with hydrogenated vegetable shortenings rather than other fats have a higher volume and a finer, more even grain, but lack the flavor that butter provides. Emulsifiers by themselves increase volume, produce a more even, finer pore, and improve the shelf life of baked goods (36). Oil coats flour too thoroughly and prevents adequate gluten development and is, consequently, not often used in baked products, although in some cases oil is added to cake and quick bread recipes specifically for its tenderizing properties.

Temperature of Fat. Except when it is to be used in making pie crusts and pastries, fat should be at room temperature for baking. A completely melted fat does not incorporate air well, and a cold fat does not disperse evenly with the other ingredients. It is important to distribute the fat throughout the batter so it can make its contribution to tenderness and volume.

Lower-Fat Alternatives. The functional roles of fats in flour mixtures cannot be denied, but its elevation of the total fat gram count and calories

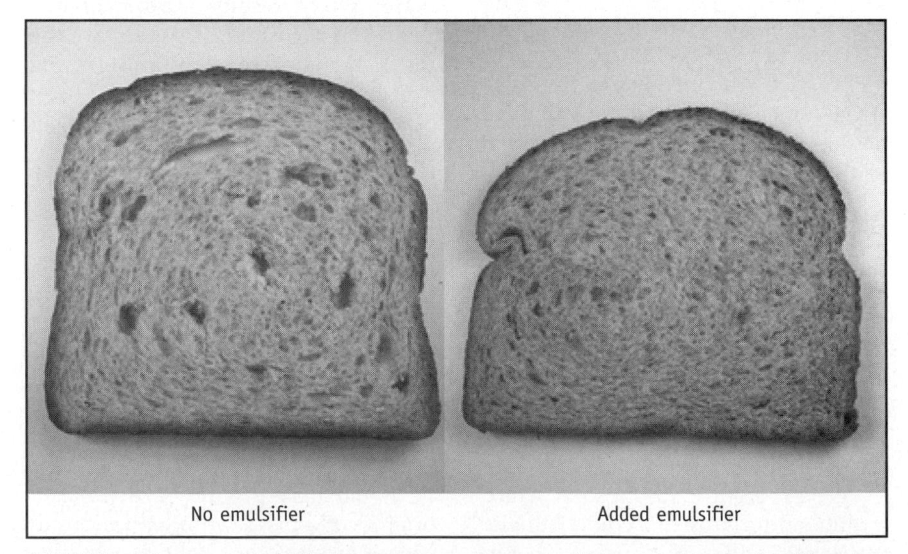

FIGURE 22-25 Emulsifier's influence on the pore pattern and volume of loaf bread.

(kcal) has led to the development and use of several commercial flour blends that incorporate a fat substitute (5). Ingredients recommended for yielding a fat-like effect are various starches, gums, maltodextrins, beta-glucan, and both soluble and insoluble fibers (48). Although a number of baked goods, such as cakes, cookies, and muffins, are available in lower-fat or fat-free varieties, these do not necessarily have fewer calories (kcal). They may even have more calories (kcal) because the fat has been replaced by a large quantity of sugar (24). Chapter 10 discusses fat alternatives in more detail, but the bottom line is that fat-free doughs perform poorly, and the fat-reduction technology to achieve such a dough has not yet evolved to replace the functional properties of fat. Even if achieved, transferring the fat-free technology from one product to another is not automatic, because what works for cakes may not necessarily be correct for tortillas.

Too Much/Too Little Fat. Excess fat makes a batter too fluid, weakens its structure, and decreases the volume of the finished product. Too little fat makes a batter resistant to expansion during leavening and results in a tougher crumb.

Eggs

Eggs are added to some flour mixtures to enhance their structural integrity, or for their contributions to leavening, color, flavor, and/or nutrient content (Figure 22-26). The delicate structure and fine crumb of many baked goods are fortified by the coagulation of egg proteins during baking. Air may be incorporated by adding beaten whole eggs. This reaches a different level of sophistication when egg whites are whipped into a foam and incorporated into cakes such as angel food cake, meringues, cookies, or other baked products. Eggs also contribute to leavening as their liquid turns to steam when heated. Egg yolks act as emulsifiers and add flavor, nutrients, and color in the form of a yellower crumb and a browner crust. The emulsifiers, along with the fat content in an egg, delay staling and improve the shelf life of baked products. Adding eggs to a cake batter increases its nutritional content by supplying protein, the fat-soluble vitamins (A, D, E, and K), B vitamins, cholesterol, and fat. The appearance of many baked products may be enhanced with a shiny glaze made of egg white and sugar.

FIGURE 22-26 The influence of eggs on the volume and texture of loaf bread.

Too Much/Too Little Egg. Excess egg causes a tough, rubbery texture in the baked product. Too little egg causes insufficient volume, and inferior structural strength, color, flavor, and nutrient content.

Commercial Additives

The baking and milling industries often add commercial additives to the flour mixture to improve commercial production and the quality of the final baked product (49). For example, malt, which is derived from germinated barley grains, is one of the oldest additives used by the baking industry. Other additives include antioxidants, mold inhibitors, and dough conditioners. The fat ingredients are often protected from rancidity by added antioxidants such as BHA (butylated hydroxyanisole), BHT (butylated hydroxytoluene), vitamin E, and ascorbic acid (vitamin C). In addition to rancidity, breads and other baked products are susceptible to molds. Adding mold inhibitors such as calcium or sodium propionate extends the shelf life of breads.

Dough Conditioners. Bakers noticed that bread products did not rise as well if freshly milled flour was used. Better baked products were obtained by using flour that had aged or yellowed for several weeks.

HOW & WHY? ?????????

Why does flour have to be aged? Aging increases the exposure to air (oxygen), which improves the strength of the resulting dough and results in faster mixing and reduced staling of the finished baked product.

To speed the aging process, chemical maturing agents were added to the flour. In addition, dough conditioners were developed to improve the effectiveness of the flour mixture. These include reducing agents, oxidizing agents, emulsifiers, and enzymes.

Maturing Agents. Flour can be chemically aged by the addition of the powdered maturing agents potassium bromate, ascorbic acid, and azodicarbonamide (47). For example, potassium bromate is known for providing improved gas retention, increased loaf volume, better dough-handling properties, and an improved crumb and texture (14).

Reducing Agents. Reducing agents encourage gluten development and thereby shorten mixing time. They act opposite to oxidizing agents and actually disrupt the disulfide bonds, resulting in "disconnected" proteins that need less mixing.

Oxidizing Agents. Benzoyl peroxide, an oxidizing agent already mentioned in the discussion of bleached flour above, acts to whiten the flour and improve the functionality of the resulting dough. Oxidizers work by making more sulfur available for the disulfide bonds between proteins in gluten, which strengthens the dough. Oxidizers rebuild the bonds, preventing the dough from becoming too soft and sticky.

Emulsifiers. Common emulsifiers added include lecithin, monoglyceride, and diglyceride, which function to disperse fat ingredients more evenly through the flour mixture. Gas generated by yeast is more easily trapped, reducing

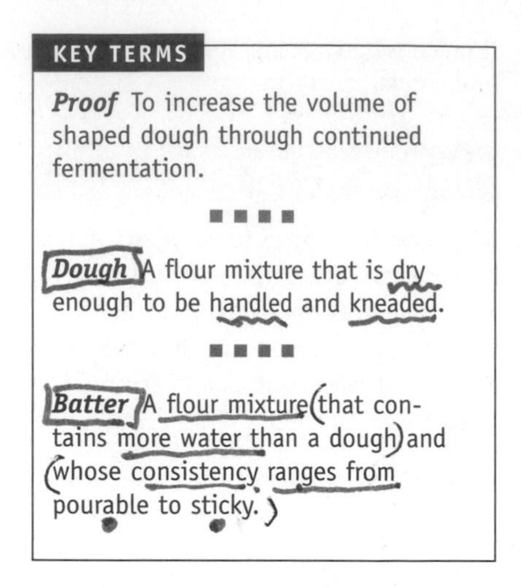
KEY TERMS

Proof To increase the volume of shaped dough through continued fermentation.

Dough A flour mixture that is dry enough to be handled and kneaded.

Batter A flour mixture that contains more water than a dough and whose consistency ranges from pourable to sticky.

proofing time and creating a more evenly textured baked product.

Enzymes. Enzyme supplementation is available in three types to assist baked products: amylase, which converts starch to sugar; protease, which breaks down protein; and lipooxygenase, which bleaches the flour and strengthens the dough. Lack of natural amylase enzymes in the flour will result in baked products with less than desirable volume, texture, and color (Chemist's Corner 22-5) (17). Formerly, amylase was obtained from malted wheat or barley flour, but now is produced by fungal or bacterial fermentation. Loaf volume increases in breads made from doughs where amylase has been added. This occurs because amylase converts some of the flour starch to sugars for the yeast, which increases gas production, and it also delays starch gelatinization during baking, resulting in more oven spring. Proteases have to be added carefully because too much will break down the gluten strands, but this helps soften strong flours with higher protein contents. Lipoxygenase enzyme increases gluten strength and makes a whiter loaf by releasing oxidizers that bleach natural flour pigments.

Preparation of Baked Goods

Doughs and Batters

Flour mixtures are either **doughs** or **batters**, depending on their flour-to-liquid ratio (Figure 22-27).

Doughs. Doughs are divided by their moisture content into stiff or soft doughs.

Stiff/Firm and Soft Doughs. A stiff dough is created by adding only ⅛ cup of liquid per cup of flour, while a soft dough is made more pliable by adding ⅓ cup of liquid per cup of flour. The handling styles for these doughs also vary; some soft doughs, such as those prepared for yeast breads, rolls, and biscuits, are kneaded, while stiff doughs are used to make pasta, pie dough, and pastry. Cookie dough, generally a soft or stiff dough, is either shaped, rolled, or dropped, while pasta dough is rolled or otherwise shaped.

Refrigerated Doughs. In addition to their ability to be handled, Doughs also differ from batters in that they can be refrigerated with little effect on their overall quality. This characteristic has been profitably exploited by food companies that sell convenient pop-open packages of dough for baked products such as rolls, cookies, croissants, and bread sticks.

Batters. Batters, like doughs, are also classified according to their moisture content, and may be either pour or drop batters (see Figure 22-27).

Pour Batters. Pour batters, which average from ⅔ to 1 cup of liquid per cup of flour, are liquid enough to be

Chemist's Corner 22-5

Commercial Enzyme Additives

Enzyme supplementation lengthens shelf life and improves bread quality. Adding alpha-amylase to flours enhances to the conversion of starch to maltose, which aids yeast fermentation and retards staling (12, 33). The retrogradation thought responsible for staling is partially inhibited when alpha-amylase breaks down the size of the amylopectin molecule. Adding too much alpha-amylase, however, creates too many smaller molecules (dextrins) and results in a sticky dough. Proteases, which act on proteins, when added to bread formulations improve loaf uniformity and grain texture, and yield a softer crumb. The breakdown of proteins catalyzed by the proteases makes the dough more elastic and easier to handle. Proteases are also thought to improve crust color and flavor of baked products by hydrolyzing gluten, which releases more amino acids to participate in the Maillard reaction (28).

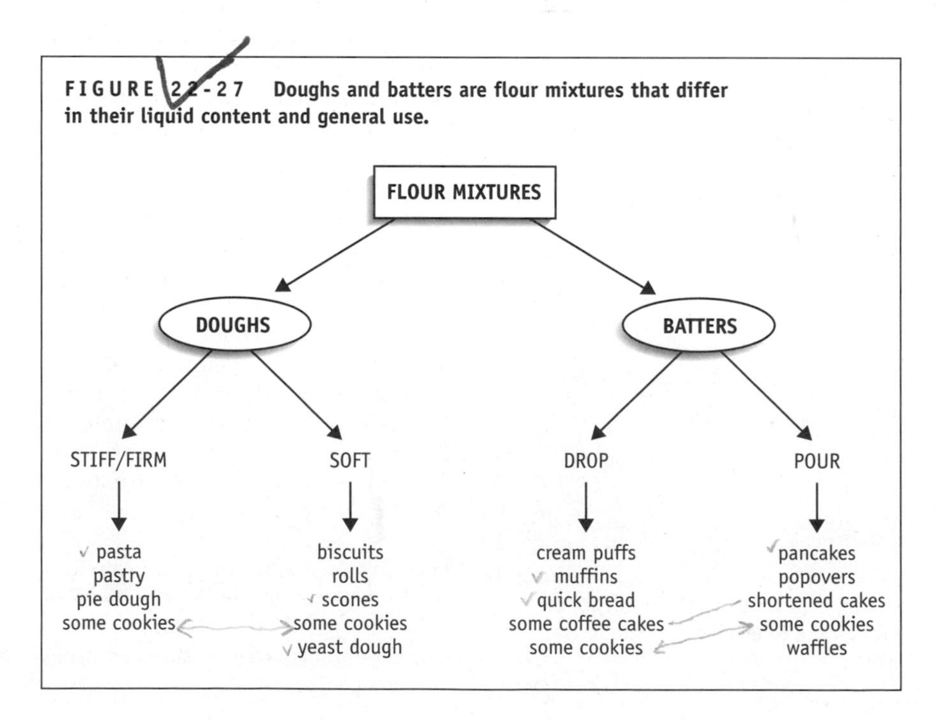

FIGURE 22-27 Doughs and batters are flour mixtures that differ in their liquid content and general use.

poured, piped (emptied from a piping bag), or spread (Figure 22-28). Common foods prepared from pour batters include pancakes, waffles, popovers, cookies, and cakes.

Drop Batters. Drop batters are too thick to be poured, so they are "dropped." Less liquid, ½ to ¾ cup per cup of flour, forms drop batters that are thick enough to be dropped or pushed onto a baking sheet. Examples of foods made from drop batters include muffins, quick tea breads, cream puffs, and some coffee cakes. In general, batters are best used immediately, although some "refrigerator" batters may be kept for a short period in the refrigerator.

Changes During Heating

After the ingredients have been mixed and, if necessary, kneaded to develop gluten, the flour mixture is ready for baking, during which several changes take place. When heat is applied, the gases, such as steam, carbon dioxide, and air, almost immediately expand, creating a pressure that stretches the intricate, elastic network of gluten. The baked product rises in this fashion until the heat melts the fat, gelatinizes the starches, coagulates the flour, eggs, and/or milk proteins, and browns the outer surfaces by the caramelization of sugars, the dextrinization of starch, and/or the Maillard reaction. Once all this has been accomplished, the structure of the baked product is set and cooking is completed.

High-Altitude Adjustments. When preparing baked products at high altitudes, several adjustments are necessary. The lower atmospheric pressure lowers boiling temperatures and raises the amount of water lost through evaporation. As a result, the leavening gases in baked products meet less resistance and tend to over-expand, and thereby collapse more easily. Recipes can be modified by using less leavening, fat, and sugar, while adding a bit more flour and liquid, and baking at higher temperatures in order to set the batter before over-expanding can cause the cells to collapse.

Storage of Flour and Flour Mixtures

Flour should be stored in pest-proof containers and kept in a cool, dry place. White flour will keep in such conditions for about a year, while whole-grain flours, which still contain the fat-rich germ and can therefore turn rancid, should be refrigerated and can be held for only about three months (47).

Dry Storage

Keeping the flour dry is important because moisture attracts insects. Even if it is kept completely dry, flour is still an attractive food source, and insects, beetles, and other pests can bore through paper, plastic bags, and even cardboard to get to it. Only metal, glass, or hard plastic, airtight containers keep pests out. In addition, some flours contain a chemical leavening agent that is triggered by moisture to release carbon dioxide. Figure 22-29 shows how the compression test can be used to determine if a flour is adequately dry.

Cool Storage Temperatures Required. Flour should be kept cool to prevent the activation of its natural enzymes, which can cause it to deteriorate if it is stored too long. At first it is advantageous for flours to age, because this allows the amylase to break down starch into the smaller sugars, dextrin, malt, and glucose that benefit the baked product (malt flour is often added to bypass this "maturing" stage). Storing the flour beyond its recommended storage time, however, results in a higher sugar content, which excessively browns the crust of white bread. Some flour also contains proteases, which break down its protein to the point where the gluten-forming ability is compromised, but this is a problem only in low-gluten flours. White flour is not affected by its own proteases because it has enough initial protein.

FIGURE 22-28 Batters can be spread, poured, or piped.

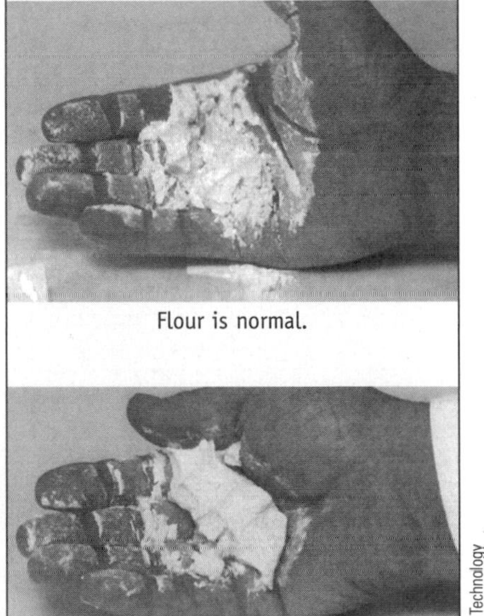

FIGURE 22-29 Flour compression test.

Another enzyme, lipase, breaks down any small fat component that may be present, resulting in rancid off-odors and flavors. In addition to rancidity, another drawback of fat deterioration is that the resulting baked product is usually lower in volume and exhibits large pores (47).

Frozen

Kneaded flour mixtures can be frozen; after defrosting, they are ready to be shaped and baked. The dough may require being kneaded about ten times after defrosting to improve its quality, because frozen dough loses some of its originally retained gases while in the freezer (21). Extended frozen storage can lead to a gradual loss of dough strength, which is why frozen doughs have a relatively short shelf life (26, 57).

The storage guidelines for breads and other baked products made from flour mixtures, such as cakes, cookies, pies, and pastries, are discussed in their respective chapters.

PICTORIAL SUMMARY / 22: Flours and Flour Mixtures

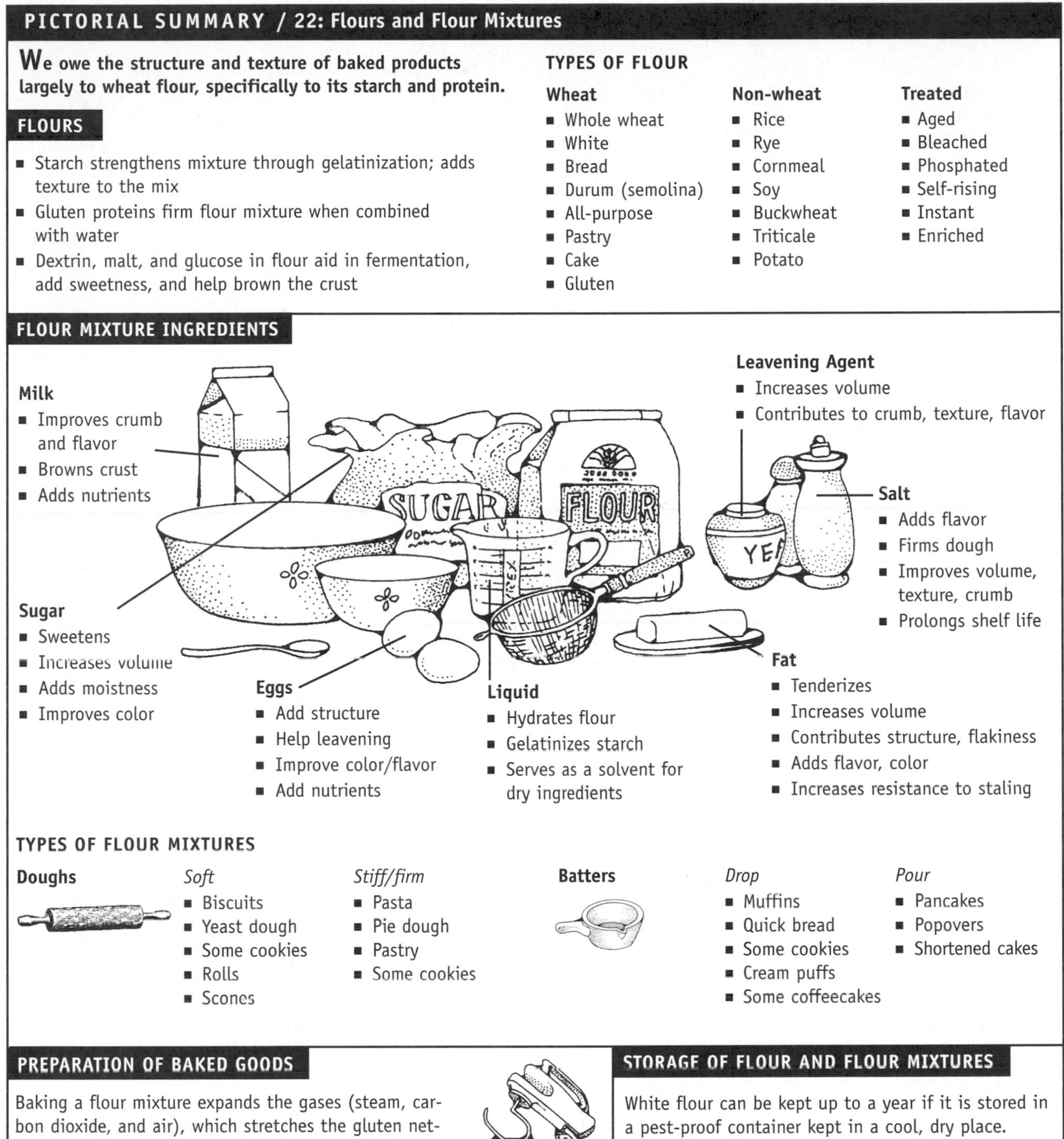

We owe the structure and texture of baked products largely to wheat flour, specifically to its starch and protein.

FLOURS

- Starch strengthens mixture through gelatinization; adds texture to the mix
- Gluten proteins firm flour mixture when combined with water
- Dextrin, malt, and glucose in flour aid in fermentation, add sweetness, and help brown the crust

TYPES OF FLOUR

Wheat
- Whole wheat
- White
- Bread
- Durum (semolina)
- All-purpose
- Pastry
- Cake
- Gluten

Non-wheat
- Rice
- Rye
- Cornmeal
- Soy
- Buckwheat
- Triticale
- Potato

Treated
- Aged
- Bleached
- Phosphated
- Self-rising
- Instant
- Enriched

FLOUR MIXTURE INGREDIENTS

Milk
- Improves crumb and flavor
- Browns crust
- Adds nutrients

Sugar
- Sweetens
- Increases volume
- Adds moistness
- Improves color

Eggs
- Add structure
- Help leavening
- Improve color/flavor
- Add nutrients

Liquid
- Hydrates flour
- Gelatinizes starch
- Serves as a solvent for dry ingredients

Leavening Agent
- Increases volume
- Contributes to crumb, texture, flavor

Salt
- Adds flavor
- Firms dough
- Improves volume, texture, crumb
- Prolongs shelf life

Fat
- Tenderizes
- Increases volume
- Contributes structure, flakiness
- Adds flavor, color
- Increases resistance to staling

TYPES OF FLOUR MIXTURES

Doughs

Soft
- Biscuits
- Yeast dough
- Some cookies
- Rolls
- Scones

Stiff/firm
- Pasta
- Pie dough
- Pastry
- Some cookies

Batters

Drop
- Muffins
- Quick bread
- Some cookies
- Cream puffs
- Some coffeecakes

Pour
- Pancakes
- Popovers
- Shortened cakes

PREPARATION OF BAKED GOODS

Baking a flour mixture expands the gases (steam, carbon dioxide, and air), which stretches the gluten network and causes the baked product to rise. During this time the fat melts, starches gelatinize, proteins coagulate (from the flour, eggs, and/or milk proteins), and outer surfaces brown by the caramelization of sugars, the dextrinization of starch, and/or the Maillard reaction. Heat ultimately sets the structure of the baked product.

STORAGE OF FLOUR AND FLOUR MIXTURES

White flour can be kept up to a year if it is stored in a pest-proof container kept in a cool, dry place. Whole-grain flours, which contain the germ, need to be refrigerated and will keep up to three months. Any flour or dry flour mixture containing chemical leavening agents should be protected from moisture.

REFERENCES

1. Addo K. Effects of honey type and level on the baking properties of frozen wheat flour doughs. *Cereal Foods World* 42(1):36–40, 1997.
2. Alexander RJ. Sweeteners used in cereal products. *Cereal Foods World* 42(10):835–836, 1997.
3. Autio K, T Parkkonen, and M Fabritius. Observing structural differences in wheat and rye breads. *Cereal Foods World* 42(8):702–705, 1997.
4. Bean MM, and WT Yamazaki. Wheat starch gelatinization. 1. Sucrose: Microscopy and viscosity effects. *Cereal Chemistry* 55:936–944, 1978.
5. Blends reduce fat in bakery products. *Food Technology* 48(6): 168–170, 1994.
6. Brooker BE. The role of fat in the stabilization of gas cells in bread dough. *Journal of Cereal Science* 24:187–198, 1996.
7. Camire ME, A Camire, and K Krumhar. Chemical and nutritional changes in foods during extrusion. *Critical Reviews in Food Science and Nutrition* 29(1):35–57, 1990.
8. Castello P, et al. Effect of exogenous lipase on dough lipids during mixing of wheat flours. *Cereal Chemistry* 75(5):595–601, 1998.
9. Chen WZ, and RC Hoseney. Wheat flour compound that produces sticky dough: Isolation and identification. *Journal of Food Science* 60(3):434–437, 1995.
10. Corriher SO. Salt makes bread doughs strong, but it can make pastries tough. *Fine Cooking* 25:78–79, 1998.
11. Czuchajowska Z, and S Smolinski. Instrumental measurements of raw and cooked gluten texture. *Cereal Foods World* 42(7):526–532, 1997.
12. Delcour JA, et al. Partial purification of a water-extractable rye (*Secale cereale*) protein capable of improving the quality of wheat bread. *Cereal Chemistry* 75(4):403–407, 1998.
13. Dombrink-Kurtzman MA, and JA Bietz. Zein composition in hard and soft endosperm of maize. *Cereal Chemistry* 70(1):105–108, 1993.
14. Dupuis B. The chemistry and toxicology of potassium bromate. *Cereal Foods World* 42(3):171–183, 1997.
15. Eliasson AC, and K Larsson. *Cereals in Breadmaking*. Marcel Dekker, 1993.
16. Ensminger AH, et al. *Foods and Nutrition Encyclopedia*. CRC Press, 1994.
17. Eskin NAM. *Biochemistry of Foods*. Academic Press, 1990.
18. Faller JY, BP Klein, and FJ Faller. Characterization of corn-soy breakfast cereals by generalized procrustes analyses. *Cereal Chemistry* 75(6):904–908, 1998.
19. Fu BX, HD Sapirstein, and W Bushuk. Salt-induced disaggregation/solubilization of gliadin and glutenin proteins in water. *Journal of Cereal Science* 24:241–246, 1996.
20. Fustier P, and P Gelinas. Combining flour heating and chlorination to improve cake texture. *Cereal Chemistry* 75(4): 568–570, 1998.
21. Gelinas P, I Deaudelin, and M Grenier. Frozen dough: Effects of dough shape, water content, and sheeting-molding conditions. *Cereal Foods World* 40(3):124–126, 1995.
22. Gerrard JA, et al. Dough properties and crumb strength of white bread as affected by microbial transglutaminase. *Journal of Food Science* 63(3):472–475, 1998.
23. Heidolph BB. Designing chemical leavening systems. *Cereal Foods World* 41(3):118–126, 1996.
24. Hippleheuser AL, LA Landberg, and FL Turnak. A system approach to formulating a low-fat muffin. *Food Technology* 49(3):92–96, 1995.
25. Hoseney RC, and DE Rogers. The formation and properties of wheat flour doughs. *Critical Reviews in Food Science and Nutrition* 29(2):73–93, 1990.
26. Inoue Y, et al. Studies on frozen doughs. III. Some factors involved in dough weakening during frozen storage and thaw-freeze cycles. *Cereal Chemistry* 7(2):118–121, 1994.
27. Jacobson KA. Whey protein concentrates as functional ingredients in baked goods. *Cereal Foods World* 42(3):138–141, 1997.
28. James J, and BK Simpson. Application of enzymes in food processing. *Critical Reviews in Food Science and Nutrition* 36(5): 437–463, 1996.
29. Labuda I, C Stegmann, and R Huang. Yeasts and their role in flavor formation. *Cereal Foods World* 42(10):797–799, 1997.
30. Leonard AL, F Cisneros, and JL Kokini. Use of the rubber elasticity theory to characterize the viscoelastic properties of wheat flour doughs. *Cereal Chemistry* 76(2):243–248, 1999.
31. Malgieris N. *Nick Malgieri's Perfect Pastry*. Macmillan, 1998.
32. Mannie E. Issues and answers in formulating with proteins. *Prepared Foods* 166(6):93–96, 1997.
33. Mannie E. Tapping the power of enzymes. *Prepared Foods* 167(5): 123–124, 1998.
34. Marcone MF, and RY Yada. A proposed mechanism for the cryoaggregation of the seed storage globulin and its polymerized form from *Triticum aestivum*. *Journal of Food Biochemistry* 18:147–163, 1995.
35. McWilliams M. *Foods: Experimental Perspectives*. Macmillan, 1997.
36. Mettler E, and W Seibel. Effects of emulsifiers and hydrocolloids on whole wheat bread quality: A response surface methodology study. *Cereal Chemistry* 70(4): 373–377, 1993.
37. Myers DJ. Industrial applications for soy protein and potential for increased utilization. *Cereal Foods World* 38(5):355–359, 1993.

38. Niman S. Using one of the oldest food ingredients—salt. *Cereal Foods World* 41(9):728–731, 1996.
39. Penfield MP, and AD Campbell. *Experimental Food Science*. Academic Press, 1990.
40. Petrofsky KE, and RC Hoseney. Rheological properties of dough made with starch and gluten from several cereal sources. *Cereal Chemistry* 72(1):53–58, 1995.
41. Phillips RD. Nutritional quality of cereal and legume storage proteins. *Food Technology* 51(5): 62–66, 1997.
42. Pomeranz Y. Molecular approach to breadmaking: An update and new perspectives. *Bakers Digest* 57(4):72, 1983.
43. Poulson C, and PB Hostrup. Purification and characterization of a hexose oxidase with excellent strengthening effects in bread. *Cereal Chemistry* 75(1): 51–57, 1998.
44. Ruan RR, et al. Study of water in dough using nuclear magnetic resonance. *Cereal Chemistry* 76(2):231–235, 1999.
45. Samson MF, and MH Morel. Heat denaturation of durum wheat semolina beta-amylase effects of chemical factors and pasta processing conditions. *Journal of Food Science* 60(6):1313–1319, 1995.
46. Sapers GM. Browning of foods: Control by sulfites, antioxidants, and other means. *Food Technology* 47(10):75–83, 1993.
47. Schunemann C, and G Treu. *Baking: The Art and Science. A Practical Handbook for the Baking Industry*. Baker Tech, 1988.
48. Shukla TP. Baking fat-free tortillas. *Cereal Foods World* 42(3): 142–143, 1997.
49. Si JQ. Synergistic effect of enzymes for breadmaking. *Cereal Foods World* 42(10):802–807, 1997.
50. Sontag-Strohm T, PI Payne, and H Salovaara. Effect of alleic variation of glutenin subunits and gliadins on baking quality in the progeny of two biotypes of bread wheat cv. ulla. *Cereal Foods World* 21:115–124, 1996.
51. Stauffer CE. *Fats and Oils: Practical Guides for the Food Industry*. Eagen Press, 1996.
52. Stauffer CE. Fats and oils in bakery products. *Cereal Foods World* 43(3):123–126, 1998.
53. Thompson SW, and JE Gannon. Observations on the influence of texturation, occluded gas content, and emulsifier content on shortening performance in cake making. *Cereal Chemistry* 33:181–189, 1956.
54. Varughese G, WH Pfeiffer, and RJ Pena. Triticale: A successful alternative crop (part 1). *Cereal Foods World* 41(6):474–482, 1996.
55. Vasco-Mendez NL, and O Paredes-Lopez. Antigenic homology between amaranth glutelins and other storage proteins. *Journal of Food Biochemistry* 18:227–238, 1995.
56. Wang L, RA Miller, and RC Hoseney. Effects of (1->3)(1->4)-beta-D-glucans of wheat flour on breadmaking. *Cereal Chemistry* 75(1):629–633, 1998.
57. Wang Z, and J Ponte. Improving frozen dough qualitites with the addition of vital wheat gluten. *Cereal Foods World* 39(7):500–503, 1994.
58. Wennermark B, and M Jagerstad. Breadmaking and storage of various wheat fractions affect vitamin E. *Journal of Food Science* 57(5): 1205, 1992.
59. Wrigley CW. Wheat proteins. *Cereal Foods World* 39(2):109–110, 1994.
60. Yamazaki WT, and DH Donelson. The relationship between flour particle size and cake volume potential among Eastern soft wheats. *Cereal Chemistry* 49:649–653, 1972.

WEBSITES

This website contains information on types of flour (for example, "What is organic flour?"), history of flour milling, and a flour milling chart:

www.flour.com

Want to know what flours to substitute for wheat flour?

www.ag.uiuc.edu/~robsond/solutions/nutrition/docs/janan111.html

The Wheat Foods Council answers questions about wheat flours:

www.wheatfoods.org/grain_info/flour.html

23

Quick Breads

Archaeological evidence suggests that the first bread eaten by human ancestors was a quick bread, made by mixing flour with water and baking the dough on hot stones. Quick breads are called "quick" because they are baked immediately after the ingredients have been mixed. There is no waiting, as in yeast breads, for leavening to take place through the slow fermentation of yeast. These breads are leavened during baking instead, with air, steam, and/or carbon dioxide produced through the action of baking soda or baking powder.

As seen in the previous chapter on flour mixtures, the basic bread ingredients are flour, liquid, salt, and leavening agent. All-purpose flour is most commonly utilized in quick breads, but various other grain flours can be added that contribute to flavor, color, and texture. Occasionally, cake flour is used in the sweeter quick breads such as coffee cakes. Milk is the most frequently added liquid. Quick breads may also contain added fat, eggs, and sugar, making them sweeter than breads leavened with yeast. If fat is added, it is usually in the form of butter or margarine. The proportions of these various ingredients differ depending on the specific quick bread, but Table 23-1 lists some general guidelines.

The focus of this chapter is on the description and/or preparation of various quick breads including pancakes, crepes, waffles, popovers, biscuits, dumplings, muffins, scones, tea breads, coffee cakes, dumplings, and "unleavened," actually steam-leavened, breads such as tortillas, and chapatis, crisp flat breads, and matzo.

Preparation of Quick Breads

The two most important considerations when preparing quick breads are the consistency of the batter and the cooking temperature. In order to avoid undesirable gluten development, batters are mixed only to the point of moistening the dry ingredients. Thinner pour batters are used to make pancakes, crepes, waffles, and popovers. Thicker drop batters are used to produce quick breads such as muffins, Boston brown bread, corn bread, hushpuppies, quick tea bread, some coffee cakes, and dumplings. (Even though quick tea bread and coffee cake batters are poured from the mixing bowl into baking pans, they are still called drop batters.) Doughs contain more flour than drop batters and are kneaded briefly. They serve as the basis for unleavened breads, biscuits, and scones.

The Muffin Method

The muffin method is the basic method of preparing many quick breads. It consists of three steps:

1. Sift the dry ingredients together.
2. In a separate bowl, combine the moist ingredients.
3. Stir the dry and moist ingredients together with only a few strokes, until the dry ingredients are just moistened but still lumpy.

TABLE 23-1
Quick Bread Ingredient Proportions

Quick Bread	Flour	Liquid*	Fat	Sugar	Eggs	Baking Powder	Salt
Biscuits	1 C	⅓ C	2½ tbs	0	0	2 tsp	½ tsp
Cream Puffs	1 C	1 C	½ C	0	4	0	¼ tsp
Muffins	1 C	½ C	1–2 tbs	1–2 tbs	½	2 tsp	½ tsp
Pancakes	1 C	1 C	1 tbs	1 tsp	1	1–2 tsp	¼–½ tsp
Popovers	1 C	1 C	0–1 tbs	0	2	0	½ tsp
Waffles	1 C	⅝ C	¼ C	1 tsp	1	1–2 tsp	¼ tsp

*Liquid is milk, except for boiling water in cream puffs.

When kneading is called for, it is very brief, approximately ten strokes. Overkneading creates too much gluten, which causes the finished bread to be dense and heavy. Unlike yeast breads, quick breads are baked immediately after the batter is mixed. For most quick breads the pans are greased, filled two-thirds full, and baked at between 350° and 450°F (177° to 232°C), depending on the type of bread. The bread is done when it is brown and passes the toothpick test, which consists of inserting a wooden toothpick, straight up and down, into the center, immediately withdrawing it, and checking to see if it is completely clean of batter. If there is no batter clinging to the toothpick, the bread is done.

Varieties of Quick Breads

Quick breads can be categorized according to the flour mixture from which they originate: batter (pour or drop) or dough.

Pour Batters

Pancakes. Pancakes are made from a pour batter, the consistency of which is dependent upon the proportions of ingredients. The mixing technique and the griddle temperature are the key factors affecting pancake quality. The muffin method is used for mixing the liquid and dry ingredients. Too much stirring will result in dense, heavy pancakes, because it develops gluten and causes the carbon dioxide gas in the batter to escape.

To cook pancakes, a griddle or frying pan is heated and then tested by flicking a few drops of cold water onto the surface. If the water pops up and "dances" across the surface of the griddle, the temperature is just right; it is too cold if the water droplets stay on the surface and boil and too hot if the water drops vanish instantly in a whisk of evaporation. The griddle should be lightly greased, although some griddles are specially coated so that adding a film of fat is unnecessary. For each standard sized pancake, ¼ cup of batter is gently poured onto the griddle. When bubbles start to appear over most of the pancake's surface, the underside should be a delicate brown, and it is ready to turn over. Any additions, such as fruit (e.g., blueberries) or nuts, are added before turning. The second side will usually not brown as evenly as the first, which always serves as the presentation side. For best results, turn the pancakes only once and do not press down on them with the spatula. Such pressure will result in a too-flat, heavy product.

Crepes. Crêpe is the French word for a thin pancake used to wrap other ingredients. The fillings in crepes may be quite sweet—syrups, creams, or fruit—in which case the crepes are classified as a dessert (crepes Suzette, crepes Jacques, crepes Empire, and crepes Soufflé) or as a sweet-roll type breakfast item (5). Blintzes are sweet crepes that have been filled and then sautéed. Crepes can also be filled with nonsweet preparations such as chicken, meat, seafood, and vegetable combinations.

Lacking the baking soda and/or powder used in pancakes, crepes are much thinner. Crepe batters are best made in advance and allowed to sit in the refrigerator for several hours or overnight to allow the flour to absorb all the liquid. The batter is then thinned with milk. A special crepe pan, essentially a smaller diameter (6 inches) frying pan, is often used, although it is not required. The pan is heated until hot, brushed lightly with melted butter, and filled with a thin layer of batter that is then spread quickly by tilting and rotating the pan. Care should be taken not to let batter slip up the sides. If the batter is too thick to pour, slightly more milk should be added; conversely, if large bubbles form as the crepe cooks, the batter is too thin and needs more flour. The crepe is turned with a flat spatula when it is golden-brown on the bottom, after which it is heated for another minute, and then gently slid from pan to plate. Crepes are usually stacked on top of each other to prevent them from drying out while any further ones are made.

Waffles. Waffles are made from a pour batter that contains more fat than a pancake batter. Folding beaten egg whites into waffle batter adds extra crispness and lightness. A waffle iron should be greased lightly with a vegetable spray. Most have an indicator that tells when the waffle iron has reached the proper temperature. Once that temperature is reached, the batter is poured into the middle of the waffle iron for even distribution. Pouring the batter from a pitcher, ladle, or measuring cup allows proper monitoring of the batter flow. When the waffle iron is two-thirds full, the cover is closed and the waffle is cooked for approximately 5 minutes, or until steam has stopped escaping from the waffle iron. If the lid offers any resistance beyond a momentary balkiness to opening, the waffle is not quite ready. Waffles are done when they are golden brown, crisp, and tender. Waffle irons should never be submerged in water for cleaning. They are brushed free of crust and crumbs, and the outside is wiped clean with a damp cloth.

Popovers. A popover is a puffy bread product that looks like an oversized, tall muffin, but with a consistency more like a warm roll. A popover has very thin, moist sides and a hollow center. The name of this quick bread comes from the fact that during baking the batter expands to such a degree that it "pops over" the sides of the container. Popovers can be plain or flavored with herbs, spices, or cheese. They are made from one of the thinnest of all quick bread batters: their ratio of liquid to flour is 1:1.

HOW & WHY? ?????????

Why do popover batters contain so much liquid? The high liquid concentration of this pour batter allows the popovers to be leavened with steam, creating the characteristic large cavity in the center. The large amount of liquid in popover flour mixtures also allows gelatinization of the starch during baking, which is indicated in part by the soft, gel-like texture in the inside walls of a baked popover.

Gluten does not readily develop when there is so much liquid, so popovers rely on protein coagulation

NUTRIENT CONTENT

Quick breads consist primarily of carbohydrate, with varying amounts of fat depending on the type of quick bread, as shown in Table 23-2. The fat content of quick breads influences their calorie (kcal) count, with croissants having the highest fat and calorie count (12 grams/223 kcal). Both fat and sugar contents are somewhat higher in quick breads than in yeast breads, although modifying the ingredients can produce quick breads with lower fat and calorie (kcal) contents. The nutritional profile of certain quick breads can be altered to suit special needs by (6, 7):

FIGURE 23-1 Low-fat tortillas.

- Substituting whole-wheat flour, bran, oatmeal, or wheat germ for about one-fourth of the white flour to increase fiber content
- Replacing hydrogenated vegetable shortening with non-hydrogenated margarine or vegetable oils to decrease the saturated fat content
- In a few commercial cases, adding starches to reduce fat content
- Reducing sugar by 25 percent by using dried fruit purées
- Substituting two egg whites for each egg
- For commercial bakers, using flour blends made with Simplesse fat substitute (3)

Removing the fat from quick breads can cause several problems in the final product. For instance, low-fat tortillas may have poor dough-mixing properties due to a tough dough, chewy texture, cracking during storage, and decreased shelf life (11). Adding a carbohydrate-based fat replacer (hydrolyzed rice flour) that is hydrated with water mimics the texture and mouthfeel of fat while improving the dough's pliability and rollability and the freshness of the final tortilla (Figure 23-1).

TABLE 23-2

Calorie (kcal), Fat, and Fiber Content of Various Quick Breads Compared to Whole Wheat Bread

Bread (55 gm)	*Calories (kcal)*	*Fat (gm)*	*Fiber (gm)*
Whole-wheat bread	135	2	3
Banana bread	186	6	1
Biscuit	200	9	1
Corn bread	149	4	2
Croissant	223	12	1
Crepe	128	7	—
Pancake	150	8	1
Waffle	159	8	1

and starch gelatinization for structure. It is for this reason that at least two large eggs per cup of flour are added to a popover flour mixture (9). Lack of sufficient eggs in the mixture will result in a soggy, compact popover that will fail to expand properly. Egg proteins and starch both provide structural strength, while any added fat, either provided directly or obtained through the egg yolks, helps to make the popovers tender.

It is important not to add too much fat to the flour mixture, because it will weaken the popover structure, allowing steam to escape and hindering the formation of the cavity. Too much water will have the same effect. Another crucial factor in the formation of the cavity is the depth of the pans, which should be twice as deep as those used to make muffins. Although muffin pans or custard cups can be used, special popover pans are best. These pans enable the steam to collect in such a way that if the ingredients are prepared and baked properly, they literally pop up from the baking cup. The individual cups should be heavily greased and the oven preheated while the batter is being made.

The muffin method is used to combine the ingredients; however, the batter is beaten until it is smooth and free of lumps. In the case of popovers, this additional beating is desirable because it gives the finished product its characteristic chewy texture. Overmixing is not a problem because of the large amount of water, which prevents excess gluten formation. When the batter is ready, the preheated cups are filled three-quarters full of batter and immediately placed in a hot (450°F/232°C) oven for about 15 minutes. The temperature is then lowered to 350°F (177°C) and the popovers are baked an additional 20 minutes, or until the proteins coagulate and the structure is properly set. The initial high temperature setting and the preheated muffin tins creates leavening by converting the water into steam. Opening the oven or removing the popovers too soon can cause them to collapse. Once baked and out of the oven, soggy popovers are prevented by making a small slit in the top to allow this steam to escape.

Drop Batters

Muffins. Muffins are one of the easiest quick breads to prepare. They usually contain flour, liquid, fat, egg, sugar, salt, a leavening agent, and a flavoring ingredient that may also add texture. The general ratio of flour to liquid is usually 2:1. Muffins can be quite high in fat and sugar, with honey, molasses, brown sugar, or syrup sometimes taking the place of sugar. White flour can be partially replaced by other cereal products, such as cornmeal, oatmeal, or whole-wheat flour.

Avoid Overmixing. One of the keys to making a good muffin is to utilize the muffin method in mixing the ingredients, just barely moistening the dry ingredients. Some small lumps in the batter are desirable; a smooth batter means that overmixing has occurred, which develops too much gluten. Overmixing creates a muffin with a smooth, peaked top and an interior that is tough and riddled with tunnels (Figure 23-2). The formation of tunnels occurs as the protein in the exterior portion of the muffin coagulates with the increased temperatures of baking, while the interior is still expanding. If there is too much gluten, it continues to stretch with the pressure of the expanding gases. The tunnels form when they are finally coagulated with the heat moving toward the center of the muffin. The likelihood of tunnels is reduced by utilizing whole-grain flours or flours other than all-purpose. The bran from whole-grain flours cuts gluten and interferes with its development, while the lesser amount of protein in most other flours decreases their potential for gluten development (9).

Avoid Undermixing. While both avoiding overmixing and utilizing the appropriate flour help to avoid tunnels, undermixing leaves lumps that are too large, indicating insufficient gluten development and resulting in a crumbly muffin that falls apart. Insufficient mixing also leaves the baking powder incompletely moistened, which results in a low-volume muffin. It is also important to beat the eggs separately before adding them to the liquid ingredients; insufficiently beaten eggs can concentrate on the outside of the muffin, where their coagulation results in an undesirable shiny, waxy appearance (9). A summary of these and other potential problems with muffins and their causes are shown in Table 23-3.

Added Ingredients. Fruits or nuts may be added to the batter for flavor enhancement. To prevent their sinking to the bottom of the muffins, gently roll the fruit and nut pieces between lightly floured hands or thoroughly toss them in a bag containing some flour.

Baking. Once mixed, the batter and all its ingredients are ready for the muffin pan. Only the bottoms of the individual cups should be greased, because ungreased sides give the dough traction as it rises and allow it to rise higher. The muffin cups are filled two-thirds full, and the batter is baked at about 400°F (204°C) for about 20 to 25 minutes. The muffins will slide out of their individual cups more easily if the hot pan is first placed on a wet towel.

Muffin Breads. Changes in the basic ingredients of a muffin recipe result in a variety of other quick breads, such as Boston brown bread, corn bread, hushpuppies, and a variety of tea breads.

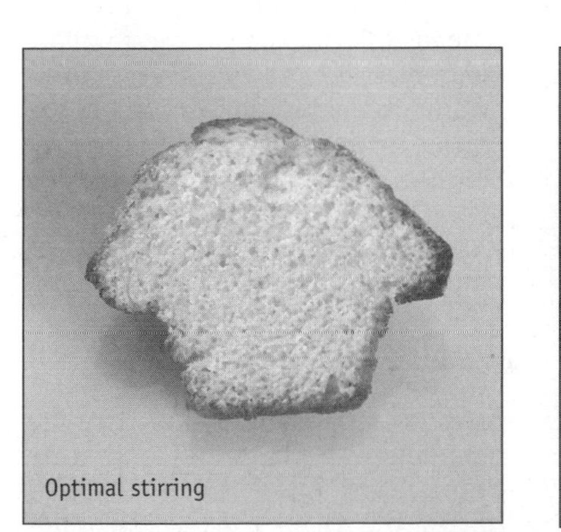

FIGURE 23-2 **Overstirring muffin batter results in tunnels.**

TABLE 23-3
Quick Bread Problems and Their Causes

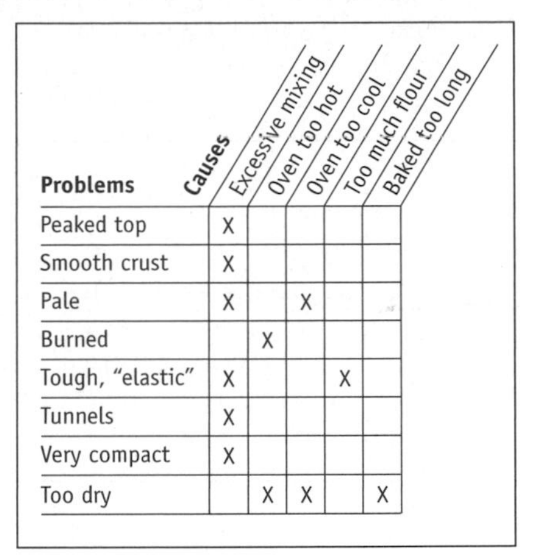

Problems	Excessive mixing	Oven too hot	Oven too cool	Too much flour	Baked too long
Peaked top	X				
Smooth crust	X				
Pale	X		X		
Burned		X			
Tough, "elastic"	X			X	
Tunnels	X				
Very compact	X				
Too dry		X	X		X

Boston Brown Bread. Unlike most other breads, Boston brown bread is made with rye and graham flours, and is steamed instead of baked. The mixture is placed in containers (coffee cans or round cylindrical baking pans are an option), covered, and placed in a pan of simmering water. The containers are then covered and allowed to steam for 2 to 3 hours, depending on the size of the container. Following steaming, the covering is removed and the bread is placed in a moderate (350°F/177°C) oven for about 15 minutes to allow the top to dry out.

Corn Bread. Corn bread flour mixtures are similar to a muffin batter, but include a combination of cornmeal (1¼ cup) and all-purpose flour (¾ cup). Sometimes small chunks of cheddar cheese, onions, chili peppers, corn kernels, and/or cooked meats such as ham or bacon are added. The muffin method is used to combine the ingredients, which can then be used to prepare corn bread, muffins, or sticks. To obtain the characteristic brown, crunchy crust and slightly gritty textured crumb, the pan should first be greased and heated to 425°F (218°C) before filling it with the batter. Baking times average 15 minutes for sticks and about 20 minutes for corn bread or muffins.

Hushpuppies. Hushpuppies are a variation of corn bread and are made from a mixed, stone-ground cornmeal drop batter. They are shaped into small round or oblong balls and deep-fried in a skillet. This bread is usually made to accompany pan-fried fish, especially trout or catfish (4). The name is thought to have come from fishermen who threw them to their hungry, whining dogs to "hush" them up. Another possibility is that hushpuppies got their name from cowboys using the bits of bread to quiet their dogs out on the range.

Tea Breads. Tea breads are similar to muffins, but they are baked in a loaf pan. These are the sweetest of all quick breads and are frequently sliced and served, as the name implies, at teatime or as dessert. Those most commonly encountered are banana, zucchini, nut, carrot, cranberry, and blueberry breads. Normally, these breads are made from a drop batter that is placed in loaf pans and baked at 350°F (176°C) for about 1 hour, but smaller containers can be used and the baking time reduced accordingly. Most tea breads keep well if they are wrapped tightly and stored in the refrigerator.

Coffee Cakes. Coffee cakes usually contain nuts and/or raisins and are often topped with a brown sugar and butter or other sweet topping. They may be mixed by the muffin method or by following the individual recipe instructions. Unlike tea breads, they are best served immediately after baking. Coffee cakes may also be made from yeast doughs; these are not considered to be quick breads.

Dumplings. A dumpling is a small ball of flour (about an inch in diameter) combined with a few other ingredients. Dumplings are simmered briefly (5 to 20 minutes) in water, stock, or gravy and are commonly used as an ingredient in soups or stews. Plenty of liquid should be used to prevent them from becoming overcrowded in the pan or they will stick together and cook unevenly. Overcooking should also be avoided, because dumplings are usually bound together by eggs, the protein of which will toughen when it is exposed too long to heat.

Doughs

Some quick breads are made from doughs that are briefly kneaded. The dough may be unleavened, producing a "flat" bread, or include leavening and used to make biscuits or scones.

Unleavened Breads. Unleavened breads are the world's oldest breads and the easiest to prepare (1). Sometimes they are called "flat" breads, but that is an ambiguous term, because some breads leavened with yeast, such as pizza, pita, and English muffins, also appear flat (12). True "unleavened" breads are leavened by steam rather than some other agent. Flat breads can be steamed, oven- or skillet-baked, fried, grilled, or baked underneath the hot desert sand. Throughout the world, unleavened breads appear in different forms: tortillas in Mexico; chapatis in India, Pakistan, and Iran; crisp flat breads in Sweden and other Scandinavian countries; and matzo in Israel. Flat breads are actually a major form of wheat consumption in many Middle Eastern and North African countries (10).

Tortillas. Mexico is the home of the tortilla, an unleavened, circular bread ranging in diameter from 4 to 14 inches. Tortillas may be prepared using either wheat flour or cornmeal.

- *Flour Tortillas.* In contrast to corn tortillas, the more pliable, softer flour tortillas are made from white flour, vegetable shortening or lard, water, and salt. After mixing, the dough is shaped into a circle, rolled or pressed into a flat, thin tortilla, and heated 30 to 60 seconds on each side on a hot griddle at 356° to 410°F (180° to 210°C) (8).
- *Corn Tortillas.* Corn tortillas have been a staple food in Mexico and Central America for many centuries, and sales of tortilla and tortilla chips exceed $5 billion (2). The tougher-textured corn tortillas are prepared from masa, a cornmeal made from corn treated with lime, an alkaline solution.

HOW & WHY? ?????????

Why is lime added to corn when making cornmeal? Whole corn has a hard outer husk, which is softened by boiling it in lime (calcium hydroxide) water, holding it for about 45 minutes, and steeping it at room temperature for an additional 12 to 18 hours (8). This cooked maize (nixtamal) is then cooled, washed, and ground into cornmeal. Water and some salt are then added to yield the final product called masa. The alkaline solution also increases the nutrient content of tortillas made from masa, adding calcium and simultaneously releasing niacin, a B vitamin, from the corn (13).

Commercially, tortillas are used to make nachos (deep-fried wedges of corn tortillas), tacos, enchiladas, burritos, and tostadas. Strips of tortillas can also be used to thicken traditional soups and stews.

Chapatis. Common in India, Pakistan, and Iran, chapatis are one of the bread staples in these countries. They are prepared by mixing whole-wheat flour, water, clarified butter (known as *ghee* in Indian cuisine), and salt. This mixture is kneaded and rolled into very thin 6-inch circles before being heated on a hot griddle until the crust browns and starts to blister.

Crisp Flat Breads. Scandanavian countries are famous for their crisp flat breads made without any added leavening. The degree of "flatness" varies from paper thin for Norwegian *lefse* to the thicker crisp breads of Norway, Sweden, and Finland. Rye and wheat are the two most common flours used in their preparations, and the color varies from very light to brown.

Matzo. Because of their rapid exodus from Egypt in Biblical times, the Israelites did not have time to allow their bread to rise, so Passover is commemorated in part by consuming the unleavened bread known as matzo. Matzo meal is made from the crumbs of matzo and is used for breading and stuffing, for dumplings in Passover soup, in cakes, and as a binding agent.

Biscuits. Biscuits, which are relatively quick to prepare, rely on fat for shortening power, and on just the right amount of kneading to increase gluten formation. Formerly, lard was usually the fat of choice, but butter may also be used, and margarine or hydrogenated vegetable shortening are common choices. Ingredients for biscuits include flour, fat, milk, baking powder, and salt. The dough is much less sticky than muffin batter because the ratio of flour to milk is 3:1 (9).

The dry ingredients are first mixed together, and then the fat is cut in with a pastry blender or two knives until particles the size of rice grains are formed, and then milk or another liquid is added to an indentation, or well, made in the center of the flour mixture. Once the liquid is mixed in, the dough is then formed into a ball and placed on a lightly floured bread board where it is kneaded briefly, for about half a minute, until the stickiness disappears. The lack of liquid makes kneading necessary for gluten development. Overkneading causes toughening because of excess gluten formation, and allows carbon dioxide gas to escape, resulting in a compact, less tender biscuit.

Once kneaded, the dough is rolled out on a floured surface to the desired thickness: ¼ inch for plain biscuits, ½ inch or less for tea biscuits, and 1 inch or more for shortcake. It is then cut into round shapes using a sharp-edged cutter that has been very lightly dipped in flour. The dough rounds are placed on an ungreased cookie sheet 1½ inches apart for crisp sides. Inadequate spacing between the dough pieces will result in uneven or caved-in sides (Figure 23-3). Lightly brushing the tops with a thin coat of butter or margarine improves browning. Biscuits are baked at 425°F (218°C) for about 10 to 20 minutes or until the surface turns a golden brown color.

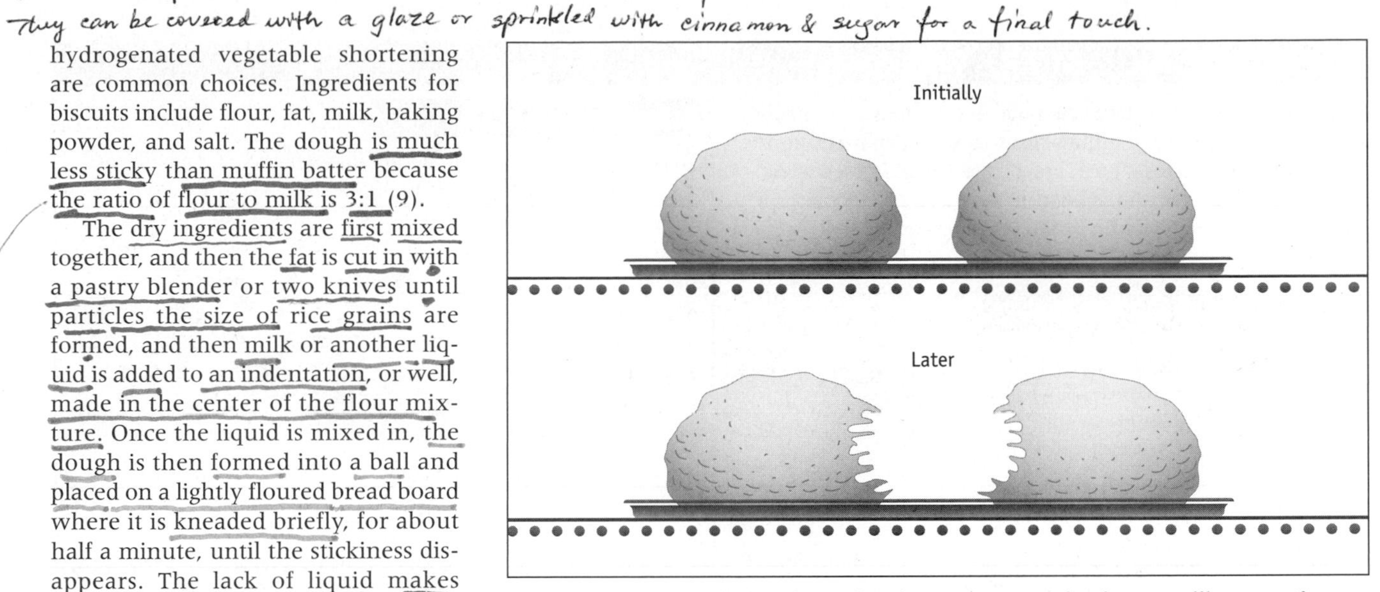

FIGURE 23-3 Baked products placed next to one another in oven will tear on the sides that are touching.

Buttermilk, whole-wheat flour, cheese, herbs, vegetables, and other ingredients can be added or substituted to produce variations of the basic biscuit. Buttermilk or any other acidic ingredient promotes the formation of biscuits with a white interior, while more alkaline ingredients, such as high levels of baking powder, contribute to a more creamy color.

Scones. Scones contain eggs and milk or cream, which makes them richer than ordinary biscuits. Their flavor also differs from biscuits because they often contain pieces of dried fruits such as raisins, cranberries, blueberries, apples, apricots, currants, or cherries. They are baked at about the same temperature and for the same length of time as biscuits.

The eggs are beaten separately and combined with the cream. Fat is cut into the dry ingredients (flour, baking powder, sugar, and salt), a well is made, and the remaining liquid ingredients are poured into it. The dry and liquid ingredients are combined with a minimum number of strokes, the dough is placed on a floured board, patted down to ¾-inch thickness, and cut into diamonds or other shapes. To produce a sheen, the tops are brushed with 2 tablespoons of the egg-cream mixture that has been set aside, and, if desired, the scones are then sprinkled lightly with sugar.

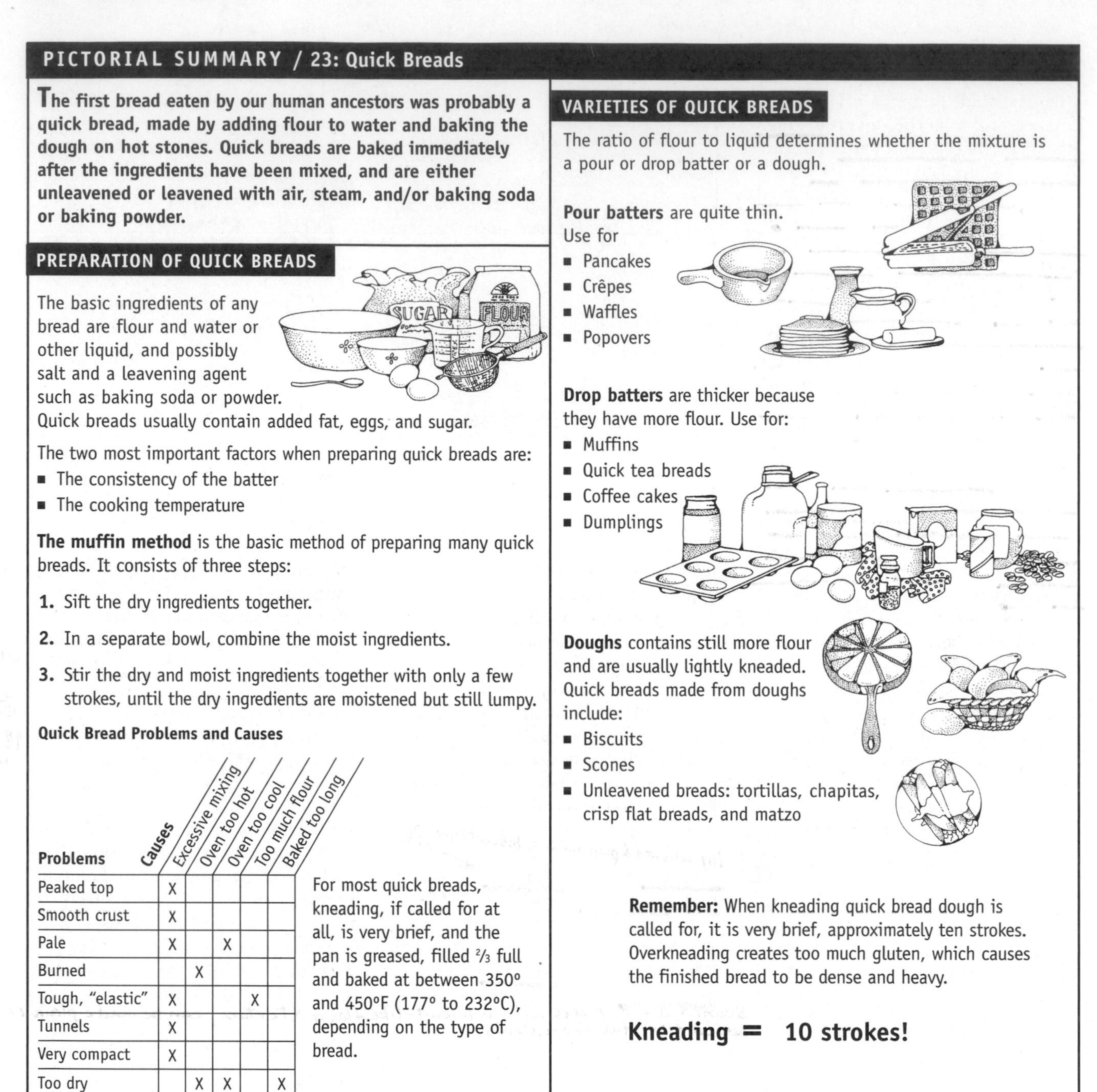

PICTORIAL SUMMARY / 23: Quick Breads

The first bread eaten by our human ancestors was probably a quick bread, made by adding flour to water and baking the dough on hot stones. Quick breads are baked immediately after the ingredients have been mixed, and are either unleavened or leavened with air, steam, and/or baking soda or baking powder.

PREPARATION OF QUICK BREADS

The basic ingredients of any bread are flour and water or other liquid, and possibly salt and a leavening agent such as baking soda or powder. Quick breads usually contain added fat, eggs, and sugar.

The two most important factors when preparing quick breads are:

- The consistency of the batter
- The cooking temperature

The muffin method is the basic method of preparing many quick breads. It consists of three steps:

1. Sift the dry ingredients together.
2. In a separate bowl, combine the moist ingredients.
3. Stir the dry and moist ingredients together with only a few strokes, until the dry ingredients are moistened but still lumpy.

Quick Bread Problems and Causes

Problems / Causes	Excessive mixing	Oven too hot	Oven too cool	Too much flour	Baked too long
Peaked top	X				
Smooth crust	X				
Pale	X		X		
Burned		X			
Tough, "elastic"	X			X	
Tunnels	X				
Very compact	X				
Too dry		X	X		X

For most quick breads, kneading, if called for at all, is very brief, and the pan is greased, filled ⅔ full and baked at between 350° and 450°F (177° to 232°C), depending on the type of bread.

VARIETIES OF QUICK BREADS

The ratio of flour to liquid determines whether the mixture is a pour or drop batter or a dough.

Pour batters are quite thin. Use for

- Pancakes
- Crêpes
- Waffles
- Popovers

Drop batters are thicker because they have more flour. Use for:

- Muffins
- Quick tea breads
- Coffee cakes
- Dumplings

Doughs contains still more flour and are usually lightly kneaded. Quick breads made from doughs include:

- Biscuits
- Scones
- Unleavened breads: tortillas, chapitas, crisp flat breads, and matzo

Remember: When kneading quick bread dough is called for, it is very brief, approximately ten strokes. Overkneading creates too much gluten, which causes the finished bread to be dense and heavy.

Kneading = 10 strokes!

REFERENCES

1. Alford J, and N Duguid. *Flatbreads and Flavors.* William Morrow, 1995.
2. Almcida-Dominguez IID, GG Ordonez-Duran, and NG Almedia. Influence of kernel damage on corn nutrient composition, dry matter losses, and processibility during alkaline cooking. *Cereal Chemistry* 75(1):124–128, 1998.
3. Blends reducc fat in bakcry products. *Food Technology* 48(6): 168–170, 1994.
4. Bocuse P, and F Metz. *The New Professional Chef. The Culinary Institute of America.* Van Nostrand Reinhold, 1996.
5. Friberg B. *The Professional Pastry Chef.* Van Nostrand Reinhold, 1996.
6. Hippleheuser AL, LA Landberg, and FL Turnak. A system approach to formulating a low-fat muffin. *Food Technology* 49(3):92–96, 1995.
7. Hogbin M, and L Fulton. Eating quality of biscuits and pastry prepared at reduced fat levels. *Journal of the American Dietetic Association* 92(8):993–995, 1992.
8. Martinez-Flores HE, et al. Tortillias from extruded masa as related to corn genotype and milling process. *Journal of Food Science* 63(8):130–133, 1998.
9. McWilliams M. *Foods: Experimental Perspectives.* Macmillan, 1997.
10. Paulley FG, PC Williams, and KR Preston. Effects of ingredients and processing conditions on bread quality when baking Syrian two-layered flat bread from Canadian wheat in a traveling oven. *Cereal Foods World* 43(2): 91–95, 1998.
11. Pszczola DE. Lookin' good: Improving the appearance of food products. *Food Technology* 51(11): 39–44, 1997.
12. Qarooni J. Wheat characteristics for flat breads: Hard or soft, white or red? *Cereal Foods World* 41(5): 391–395, 1996.
13. Serna S, M Gomez, and L Rooney. Technology, chemistry, and nutritional value of alkaline-cooked corn products. In *Advances in Cereal Chemistry,* ed. Y Pomeranz. American Association of Cereal Chemists, 1990.

WEBSITES

View various interesting flat breads at this website:

www.foodsubs.com/flatbread.html

Want low-fat versions of flatbreads? Here they are at:

www.fatfree.com/recipes/breads-quick/

Step-by-step instructions on how to prepare quick breads by Quaker Oats:

www.quakeroatmeal.com/kitchen/classes/class.cfm?CategoryID=4&SectionID=1

Yeast Breads

When breads are leavened with carbon dioxide produced by baker's yeast, they are known as yeast breads. Baker's yeast, or *Saccharomyces cerevisiae,* is a one-celled fungus that multiplies rapidly at the right temperature and in the presence of a small amount of sugar and moisture. Yeast breads have been around since long before the birth of Christ, and no one knows exactly how it happened that people came to realize that naturally-occurring, airborne yeast could be used as a leavening agent. The ancient Egyptians knew the technique and preferred the light airy texture of yeast-leavened bread, and the Romans adopted this procedure, after which it became known throughout the Roman Empire. The basic principles for preparing yeast breads have not changed in the many intervening centuries, and they are the subject of this chapter.

Preparation of Yeast Breads

Yeast bread is prepared by mixing the ingredients into a dense, pliable dough that is kneaded, allowed to rise by fermentation, and then cooked, typically by baking, but sometimes by steaming or frying. Normally, the preparation of yeast bread is at least a two-and-a-half to three-hour operation, which is one reason why many people buy their own bread. Sourdough breads take even longer, requiring about eight to ten hours to make and fully develop their flavor. Regardless of the time involved, most yeast breads are prepared with the same basic ingredients.

Ingredients

The fundamental ingredients of any yeast bread are flour, liquid, sugar, salt, and yeast; fat and/or eggs are optional. Wheat is the most commonly used flour for making breads; rye is the next most common, followed by flours such as oat, barley, cornmeal, rice, and others. Because flours other than wheat have less gluten-forming ability, a certain amount of wheat flour must usually be added to these flours in order to improve their baking quality (5, 17).

Mixing Methods

The four best-known methods for mixing yeast breads are the straight dough, sponge, batter, and rapid mix methods. Regardless of the mixing method used, all ingredients should be brought to room temperature prior to mixing in order to obtain the desired dough consistency. It is the ingredients, their amounts and types, along with how much the dough is mixed, that determine whether it is too soft, too firm, or just right. It is important for the dough to reach the desired degree of cohesion, because this influences the dough's handling characteristics and the final quality of the baked item (Table 24-1).

TABLE 24-1
Excessively Firm or Soft Dough Problems

Excessively Firm	*Excessively Soft*
Difficult to mix	Tend to "flow"
Difficult to weigh	Sticky to work with
Slower fermentation	Dough pieces easily lose their shape
Difficult to shape	Finished product remains fairly flat
Poor symmetry in baked goods	Decreased volume
Pale crust	Crumb shows uneven, large cells
Crumb shows dense cells and often has a streaky texture	Deep brown crust

Straight Dough Method. The straight dough method consists of placing all the ingredients at once into a bowl, where they are mixed. For automatic mixing, various dough attachments are available (Figure 24-1). Whether by hand or by machine, the dough is kneaded to develop the gluten and then allowed to rise once or twice before being shaped into a loaf or other form.

Sponge Method. Combining the yeast with water and slightly over one-third of the flour creates a foamy, bubbly mixture that looks like a sponge. This is allowed to ferment in a warm place for a half hour to an hour in order to become foamy and spongy, after which all the remaining ingredients except salt are added to the mixture. Salt inhibits the yeast, so it is added last, after yeast activity is well under way (3).

Batter Method. The batter method is the simplest of all the mixing techniques and requires no kneading after the ingredients have been mixed. It is a good method for the rapid production of the kinds of bread products required in food service operations. Once the ingredients are combined they are beaten by hand, or by electric mixer or dough hook, to develop the gluten. The batter is ready when it no longer sticks to the sides of the bowl, but is still sticky itself. The batter method saves time and is often used for preparing rolls and hot dog and hamburger buns, though it may result in a bread with a more coarse and porous texture.

Rapid Mix. The rapid mix method differs from the others in that it is used primarily with bread-making machines. Millions of North Americans own machines that make bread-making easier. Inside the bread machine is a nonstick pan with a kneading paddle, usually located on the bottom. The machine kneads the bread, allows it to rise, then bakes it in the same pan. For those who prefer more "hands-on" bread-making, the dough mixture can be beaten by hand; alternatively, it can be mixed in the bread machine, then shaped by hand. Ready-to-use mixtures often come with these machines, but if using a recipe, the simplified process consists of placing warmed water, bread flour, yeast, and salt into the container, closing the lid, and pressing a button. Fresh bread can be completed within two to four hours.

Kneading

Kneading develops the dough's gluten to its maximum potential (see Chapter 22). This step of bread-making involves physically handling the dough until it achieves a smooth, soft, nonsticky surface and springs back when pressed gently. However, some traditional bread doughs do remain somewhat sticky after kneading. Kneading is not required if the batter method is used.

A kneading surface should be covered with a fine layer of flour, both to prevent sticking and to allow some flour to be kneaded into the dough. Kneading in too much flour should be avoided, however, because it will slow fermentation time, leaving the final product dry and streaked or heavy. Hands should be lightly floured prior to lifting or tipping the dough gently out of the bowl. The ball of dough is placed on the floured surface and the farthest edge of it is lifted up and folded toward the nearest edge, as shown in Figure 24-2. The fold is then pressed with the heels of both hands in a single rhythmic, rocking motion that pushes the dough fold, first down, and then away. Short presses that are neither too heavy nor too light are best. The mass is turned a quarter of a turn and the process is repeated. More flour may be necessary as kneading continues. To determine if more flour is required, hit the dough ball with an open hand, count to ten, and lift the hand off. If the hand sticks to the dough, it needs more flour (5). If the dough is too firm and inelastic, additional water may be required.

FIGURE 24-2 Kneading dough.

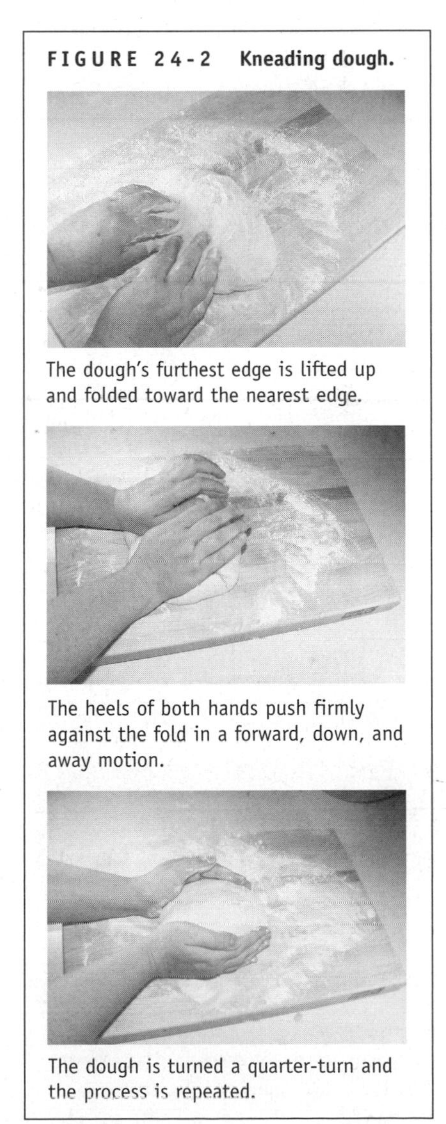

The dough's furthest edge is lifted up and folded toward the nearest edge.

The heels of both hands push firmly against the fold in a forward, down, and away motion.

The dough is turned a quarter-turn and the process is repeated.

FIGURE 24-1 Types of dough mixers.

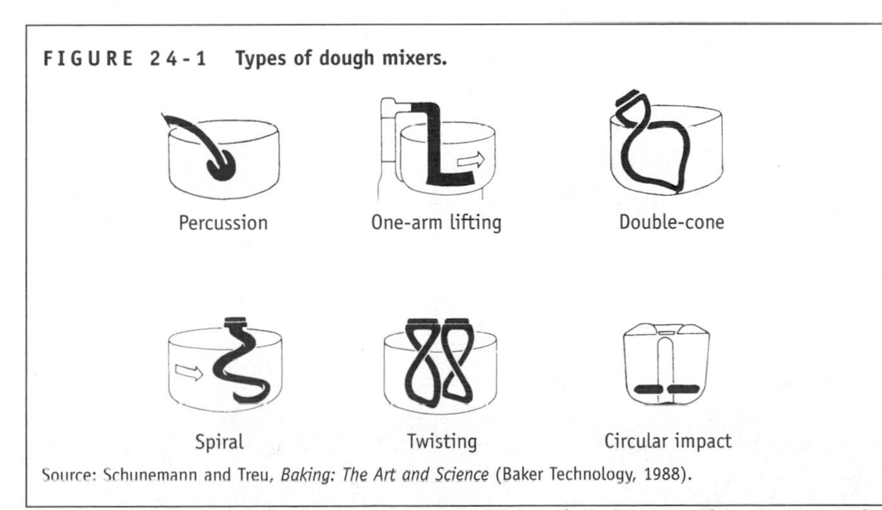

Source: Schunemann and Treu, *Baking: The Art and Science* (Baker Technology, 1988).

Observing an experienced bread baker helps in learning to determine the right amount of pressure and stretching and the length of kneading required to yield a dough ready for rising. In about 10 minutes, the dough usually signals it is completely kneaded by the development of a shinier surface that tends to "push back" when two fingers are gently pressed against the dough to test its gluten strength. Another way to determine if kneading is complete is to stretch some of the dough into a "gluten window" (Figure 24-3). Avoid excessively kneading the dough, because it results in a baked product with a coarse texture.

The effort and time involved in kneading can be eliminated by using a food processor or an electric mixer with a dough hook attachment. It is important, however, to first develop a skill for recognizing the correct amount of kneading, because it is very easy to overknead a dough using a food processor or an electric mixer. On the other hand, it is very difficult to overwork a dough by hand.

NUTRIENT CONTENT

Breads are an excellent source of complex carbohydrates. They may also, depending on the type of bread, contribute fiber to the diet. A slice of whole-grain bread is similar in nutrition to enriched white bread; each contributes calories (kcal) (80), protein (3 grams), carbohydrate (12 grams), fat (1 gram), and a few vitamins and minerals. The major difference between the two is that as much as 3 grams of fiber are found in a slice of whole-grain bread, while there is only 1 gram of fiber in a slice of white bread. At least 55 to 60 percent of daily calories (kcal) should come from carbohydrates, which ideally should contribute a minimum of 20 to 30 grams of fiber.

Although many yeast breads contain very little if any fat (about 1 gram per slice), a few exceptions are croissants (12 grams of fat per 55-gram croissant), submarine rolls (4 grams of fat per 135-gram slice), and hot dog/hamburger rolls (2 grams of fat per 40-gram roll). Several baked products use fat substitutes to lower their fat content (22, 24). Another change being explored is the addition of beta-carotene, a form of vitamin A from plant sources, to certain baked items (25). The retail market now offers a wide variety of specialty breads, including those that are low in calories (kcal), low in fat, high in calcium, high in fiber, or a combination of these.

In the case of pizza, a popular bread-based food, the size and thickness of the slice and its crust determine its basic caloric (kcal) content, but the other ingredients, especially cheese and meats such as pepperoni and sausage, increase the fat content. Pizza slices vary considerably among restaurants, but generally, a large, thick slice of pizza provides about 400 calories (kcal) and 11 grams of fat, compared to a small, thin slice of pizza with only 130 calories (kcal) and 4 grams of fat. Serving size has sometimes been overlooked in perceptions of the nutrient content of pizza. Large servings have led many people to believe that pizza is always high in fat and calories (kcal), but this is not necessarily the case. Three small, thin pieces or one large, thick slice of pizza fall under 400 calories (kcal) and 15 grams of fat; but three large, thick slices will add up to 1,221 calories (kcal) and 33 grams of fat.

Fermentation—First Rising

After kneading, the dough's surface is greased by rolling its sides gently in a lightly greased bowl; this prevents it from drying out as it rises. It is then placed in the bowl, covered with a clean, moist dish or paper towel or plastic wrap, and allowed to rise in an undisturbed, warm location. The covering over the bowl also helps to maintain humidity and prevent drying. For approximately the next hour, the bowl must be left in a preferably humid place that is slightly warmer than room temperature; approximately 85°F (30°C) is optimum. In these conditions, the dough can rise properly. Some traditional breads are left to rise more slowly in cooler (68°F/20°C) environments.

Changes During Fermentation. As the yeast ferments, the dough will double in size as carbon dioxide is produced by the yeast and as enzyme and pH changes take effect (Figure 24-4) (Chemist's Corner 24-1).

Scott Phillips/Fine Cooking

FIGURE 24-3 Testing yeast dough for doneness. When stretched, a correctly kneaded dough shows a thin, translucent window.

FIGURE 24-4 First fermentation rise for dough. Shaped dough placed in a humid, warm, undisturbed place for rising (left); risen dough ready for punching (right).

Chemist's Corner 24-1

Enzymes in Yeast Dough

During fermentation, several enzymes naturally present in the flour eventually influence the flavor, color, and texture of the bread. One of these enzymes, alpha-amylase, starts to degrade damaged starch granules found in flour to sugars, specifically dextrins, which are then converted by beta-amylase to maltose (23). The most common commercial sources for alpha-amylase are fungal (*Aspergillus oryzae*), bacterial (*Bacillus subtilis*), and cereal (malted barley and wheat) (8). Proteases break down the protein into peptides and amino acids. Professional bakers sometimes add enzymes. Caution must be used here, however, because an excess of protease enzymes will cause the bread to have decreased volume and an interior texture. Another enzyme now being used in the baking industry is maltogenic alpha-amylase, which has an anti-staling effect in baked products (27)

Dough becomes more acidic (pH of 6.0 to 5.5–5.0) during fermentation because of the formation of carbonic acid, which is created when carbon dioxide combines with water; and because of the lactic and acetic acids produced by the yeast. The increased acidity may improve gluten's ability to hydrate with water. Acids also improve flavor, extend shelf life by inhibiting staling and mold growth, and reduce the stickiness of the dough. Alcohol is a by-product produced during fermentation, but it evaporates during baking. The amount of time it takes for the dough to rise will depend upon the type and concentration of yeast, the amount of available sugars, the temperature of its environment, salt concentration, and the mixing method: anywhere from three-quarters of an hour to two hours, or even overnight in the case of some European-style breads.

Optimal Fermentation Temperatures. Yeasts are very sensitive to temperature extremes: becoming activated at 68 to 100°F (20 to 38°C), slowing down below 50°F (10°C), and dying if exposed to temperatures at or above 140°F (60°C). Professional bakers use **proof boxes** to optimize fermentation conditions. At home, the dough can be placed in a warm corner free of drafts, in a closed oven with a bowl of hot water on the shelf below, or in the oven after it has been heated for a minute or so, until the warmth is just beginning to be felt, and then turned off.

Avoid Overfermentation. The first rise is completed when the dough has approximately doubled in size and two fingers pushed into the dough near the edge leave an indentation. As the dough rises during fermentation, the gluten stretches and becomes weaker; thus, rising should not be allowed to continue too long or the expanding dough will collapse. Allowing the dough to rise too high can also cause a coarse grain and a sour odor from the excess acid production. Overfermentation can also affect color; most of the sugar is then used by the yeast and is unavailable to interact with flour proteins to create the desired browning of the crust through the Maillard reaction. Similarly, since the Maillard reaction also contributes to the sweet, aromatic, and roasted flavors in baked products, over-fermentation affects flavor (21, 29).

Punching Down—Second Rising

For most homemade and finer-textured breads, once the dough has risen to double its size, it is punched down and left to rise a second time (Figure 24-5).

HOW & WHY?

Why does dough have to be punched down after the first rising? Punching down evens out temperatures, redistributes sugars, yeast, and gluten, breaks large air bubbles into smaller cells, and lets excess carbon dioxide gas escape. The bread would have large holes without the release of excess carbon dioxide.

KEY TERMS

Proof box A large, specially designed container that maintains optimal temperatures and humidity for the fermentation and rising of dough.

The dough can either be punched down directly in the bowl, or placed once again on a lightly floured surface and gently pushed down in the center with a clenched fist, followed by about four kneading motions.

After punching, the dough is sometimes allowed to rise again to double its size, but no more, or the gluten will overstretch and cause the bread to fall. Some baked products are allowed to rise only once before shaping. When there is a second rising, it takes about half the time of the first. The completion of the second rising is signaled strictly by whether or not the dough has doubled its size and not by the finger indentation test.

Shaping

After the bread has risen, it is ready for shaping. Breads can take on a wide variety of shapes. Figure 24-6 provides instructions for shaping rolls, braids, and other, more unusual bread shapes. The basic loaf of bread is shaped by first dividing the dough into the desired number of portions so the pan(s) will be at least half, but no more than two-thirds, full of dough. The dough is shaped into an oblong, roughly rectangular, mound as long and as wide as the loaf pan. One third of each end is

FIGURE 24-5 Punching down the dough allows excess gas to escape and redistributes the ingredients.

KEY TERMS

Proof To increase the volume of shaped dough through continued fermentation.

then folded under the mound. All the edges are pinched together to seal the seams, and the dough is then placed in the pan so the sealed edges are on the bottom and the dough is touching all four sides of the pan. The bottom and sides of the pan are sometimes greased so the loaf can be easily removed, but some bakers leave the sides ungreased so the bread will have more traction during rising.

Proofing

Proofing is the final rising that occurs in the pan or on a baking sheet, and it has an important effect on the quality of the finished bread. It is facilitated by placing the shaped dough in a warm, humid, and undisturbed environment. The pan is covered with a cloth, and the dough is usually allowed to double in size and take on the shape of the bread pan. The amount of rising allowed during proofing varies with the bread type, however. One of the purposes of proofing is to create a dough that is adequately aerated. If the dough expands beyond what is recommended, it leads to overextension of the gluten, which causes the cell walls to break and collapse, the fermentation gas to escape, and, ultimately, a low volume in the finished product (Figure 24-7). Temperature is as important as timing; doughs that are too cool ferment too slowly, while those that are too hot produce breads that have a small volume, large cells, a pale crust, and a reduced shelf life. Humidity also plays an important role, as shown by the bread in Figure 24-8 that was proofed under conditions that were too dry. While proofing is taking place, the

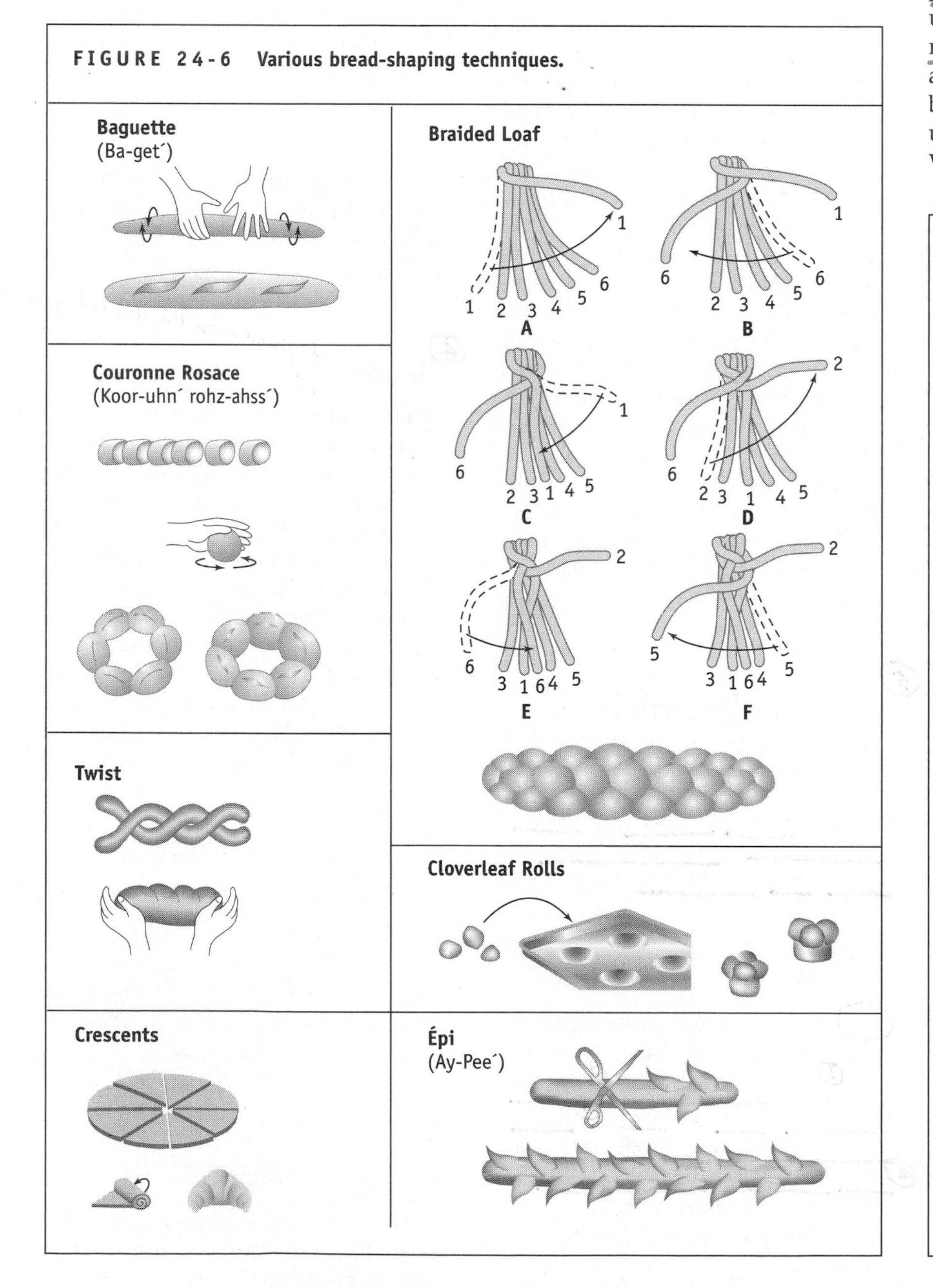

FIGURE 24-6 Various bread-shaping techniques.

FIGURE 24-7 Overproofing produces a low-volume bread due to collapsing cells and the escape of fermentation gases.

Optimum proofing

Overproofing

Crust faults due to drafty, final proofing

Uneven, rough crust

Baker Technology

FIGURE 24-8 Bread proofed in conditions that were too dry; lack of humidity results in crust problems.

oven may be preheated in anticipation of the next step—decorating and baking the bread.

Decorating

In preparation for baking, the bread dough may be decorated with sesame, caraway, or poppy seeds by brushing the top of the loaf with a thin layer of egg white and sprinkling the seeds on top. Many types of bread are **scored** just before baking to allow them to rise evenly without tearing the crust. Another decorative touch, which also adds flavor, is to score the top of the bread with a sharp knife and pour a bit of melted butter into the slashes. Although scoring serves the dual purpose of decoration and allowing excess steam to escape, neither is strictly necessary. Milk brushed on the surface of the loaf before or during baking will give crusts a golden brown color due to the caramelization of the milk. The technique of brushing loaves with water or introducing steam into the oven will give loaves a crispier crust.

Baking

A standard loaf of bread will bake in about 45 minutes. It is usually heated at 400°F (205°C) for the first 10 to 15 minutes, and then at 350°F (177°C) for the remaining 30 minutes. Temperatures and time will vary for different types of bread. The initial hot temperature contributes to **oven spring**. The initial increase in fermentation, enzyme activity, and softening of the ingredients also contribute to oven spring. If the initial temperature is not hot enough, the protein will not congeal properly and the yeast will continue to ferment, causing the dough to spill over the sides. This can also occur when the dough has been overfermented prior to baking. Underfermented dough, however, may not have sufficient oven spring.

Changes During Baking. Baking changes the appearance, texture, flavor, and aroma of the dough as well as its structure. Once the dough is placed in the oven, the hot temperature kills the yeast, thereby stopping fermentation. It also inactivates enzymes, vaporizes the alcohol, and coagulates the protein that firms the dough.

Protein coagulates and starch granules start to swell and gelatinize at about 140°F (60°C), but because of the limited water content, gelatinization does not proceed as far as it might have. If the oven temperature is too high, the proteins will coagulate too soon, resulting in low volume; and conversely, low oven temperatures will cause the structure to collapse.

If baked correctly, bread right out of the oven has an aroma all its own, but the wonderful smell is rapidly lost; this is believed to be the result of the evaporation into the air of volatile compounds (Chemist's Corner 24-2). Other components of flavor are developed in part by the breakdown of starch to dextrins, and by some caramelization, but primarily from the browning of the crust in the Maillard reaction. Substrates for the Maillard reaction in breads can be derived from any added milk, sugar, and/or egg.

> **KEY TERMS**
>
> ***Score*** The technique of taking a sharp knife or a special blade called a lame and creating 1/4- to 1/2-inch deep slashes on the risen dough's top surface just prior to baking.
>
> ***Oven spring*** The quick expansion of dough during the first ten minutes of baking, caused by expanding gases.
>
> ***Crumb*** The cell structure appearing when a baked product is sliced. Evaluation is based on cell size (called "open" if medium to large, or "closed" if small), cell shape, and cell thickness (thin walls occur in fine crumb, while thick walls predominate in a coarse crumb).

The changes associated with baking require even heat exposure, so a minimum distance between baked items is necessary: at least 1 inch between rolls and about 3 inches between loaves (26).

Crumb Development. During baking, the dough converts from an elastic, undefined mass into a set structure with a defined bread **crumb** (11). The formation of a desirable crumb is dependent on gases produced during fermentation and proofing and air introduced during mixing and kneading.

> **HOW & WHY? ?????????**
>
> **How does crumb form in a loaf of bread?** The carbon dioxide produced during fermentation, proofing, and oven spring becomes many small bubbles entrapped in the gluten network (Figure 24-9). When the bread sets and the gas escapes, all that remains are the pores, or cells, forming the crumb of the baked product.

Numerous small or medium cells result in a baked product with a fine and

FIGURE 24-9 Fermentation gases produce the pores of the baked bread.

tender crumb, a large volume, and a longer shelf life. These cells multiply and enlarge during fermentation and proofing; during baking they will enlarge further, but will not multiply. Yeasts are not entirely responsible for cell number and size. Air incorporated into the flour mixture during mixing, punching, and kneading also contributes to the number of cells. This is why using sifted flour in baked products results in larger volumes than unsifted flour (26).

Problems with Texture. An overfermented dough, with its large cells, gives the baked item a moth-eaten appearance and a coarse texture. Underfermented dough, in which the carbon dioxide was not properly distributed, results in a very dense loaf with thick cells, low volume, and a tough crust (Figure 24-10). Distribution of the carbon dioxide throughout the dough to create fine cells is accomplished through mixing, punching, kneading, and shaping. Bakery products in which the cells are either abnormally small or large have shorter shelf lives (26).

Testing for Doneness. Bread can be tested for doneness by inverting the pan with one gloved hand, allowing the bread to drop into the other gloved hand so the bottom of the loaf is facing up, and tapping the bottom. If it rings hollow, it is done; if it does not, it is placed back in the pan and returned to the oven for an additional 5 to 10 minutes. Another method consists of combining the hollow sound technique with an instant-read thermometer, which should read above 195°F (91°C) when inserted into the bread's interior. If the bread is not done but the top is already golden-brown, it is covered with aluminum foil or otherwise protected to prevent further browning. Once it is done, the baked bread is set out on wire racks for cooling.

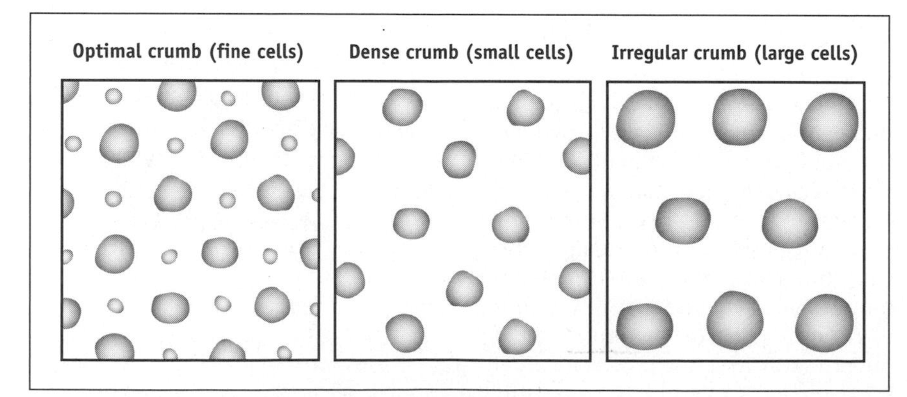

FIGURE 24-10 Crumb quality depends on cell size.

Chemist's Corner 24-2

Bread Flavors and Aromas

The contribution to flavor by fermenting yeasts is due to their formation of volatile and nonvolatile products. Nonvolatile compounds such as amino acids and saccharides are important for generation of flavors via the Maillard reaction. Volatile substances contributing to the odor of fermenting dough include isobutanol, amylalcohol, isoamyl alcohol, phenylethylalcohol, esters, lactones, and organic acids (14). Many of the alcohols formed during fermentation, such as ethanol, n-propanol, isobutyl alcohol, and others, are lost into the air during baking (6). The odor of baked crust is obtained in part from the formation of 2,5-dimethyl-4- hydroxy-3(2H)-furanone, which results from the thermal reaction between certain amino acids and sugars such as glucose, fructose, and rhamnose. Organic acids and carbonyl compounds also contribute to crust flavor. The unique flavor of French bread is thought to be due to its high level of acetic acid.

A well-prepared bread product has certain characteristics: optimum volume, color, and flavor; a symmetrical shape with closed seams; a porous, pliable, firm, and even crumb; and a golden-brown, crispy crust. Major problems that may interfere with the attainment of these attributes, and their causes, are summarized in Table 24-2.

Microwave Preparation. The use of a microwave for baking bread is not highly recommended; because no dry heat is available for dextrinization, the Maillard reaction, or caramelization, the crust will not brown. If a microwave oven is used, breads should be placed in a conventional oven during the last few minutes of baking to yield at least a partial crust. Individual microwave manufacturers' instructions should be followed.

TABLE 24-2
Problems with Yeast Breads and Their Causes

Problems / *Causes*	Improper mixing	Insufficient salt	Too much salt	Dough weight too much for pan	Dough weight too light for pan	Insufficient yeast	Dough proofed too much	Dough underproofed	Dough temperature too high	Dough temperature too low	Dough too stiff	Oven temperature too high	Proof box too hot	Green flour	Dough chilled	Too much sugar	Insufficient sugar	Dough too young	Dough too old	Improper molding	Insufficient shortening	Oven temperature too low	Over-baked	Dough too soft	Under-baked
Volume too low	X		X		X	X		X			X	X			X			X	X						
Volume too high		X		X			X												X			X			
Crust too thick		X	X				X																		
Dry crumb	X										X											X			
Moist crumb																								X	X
Streaky crumb																				X					
Gray crumb								X	X				X												
Lack of shred							X											X	X						
Coarse crumb	X				X		X			X								X	X	X					
Poor texture, crumbly							X						X					X	X			X			
Crust color too pale										X			X				X		X						
Crust color too dark										X		X				X		X							
Crust blisters														X				X	X	X					
Shelling of top crust								X			X			X			X								
Air pockets								X	X			X											X	X	
Poor taste and flavor		X							X										X						
Poor keeping qualities		X					X		X		X						X		X		X	X			

High-Altitude Adjustments. Altitudes above 3,000 feet require changes in recipe ingredient measurements, preparation time, and oven temperatures. Yeast breads rise faster at higher altitudes because there is less atmospheric pressure, so less leavening agent will be needed. Liquid also evaporates more quickly at higher altitudes, so more may be needed. Yeast bread should not be allowed to more than double in volume. Baking temperatures are increased slightly, about 10° to 15°F (6° to 8°C), to help coagulate the proteins and prevent the gases from overexpanding the structure.

can be made from modifying the standard yeast bread recipe

Varieties of Yeast Breads

The simplest yeast bread is made from flour, water, and yeast, but this basic formula has evolved into more complicated varieties that include loaf breads (white, whole-wheat, sourdough, and malt bread), rolls, pita bread, bagels, English muffins, pizza crust, raised doughnuts, and specialty breads.

Loaf Breads

Wheat (White) Bread. Wheat, or white, bread is made with all-purpose flour, milk, water, and small amounts of sugar, salt, and yeast. Fat is an optional ingredient. Almost all of the white bread in the United States has been enriched.

Whole-Wheat Bread. The bran in whole-wheat bread increases the fiber content from 1 to 3 grams per slice, but the sharp edges of the bran cut the strands of gluten in the dough, resulting in a shortened kneading time and a lower-volume loaf. If the bread is designated as "wheat bread," it has been made from a combination of whole wheat and white flours, which produces a loaf with more volume and a lighter texture.

Sourdough Bread. Sourdough bread is made with a starter, which consists of both yeast and lactic acid bacteria, *Lactobacillus plantarum* (10). The lower pH of sourdough bread (4.0 to 4.8 compared to 5.1 to 5.4 pH of regular breads) provides its characteristic texture and taste. Two cups of sourdough starter are equivalent to one small package of yeast.

HOW & WHY? ?????????

Why is sourdough bread "sour"? The *Lactobacillus plantarum* strain of bacteria that is widely used as a starter produces a lactic acid that results in a pleasantly sour taste (9, 28). The lower pH also makes sourdough bread less sweet by inhibiting the amylase enzymes in the flour from breaking down starch into the sweeter-tasting maltose.

Malt Breads. The malt in malt breads makes them sweeter, stickier, and heavier than other breads. The added malt also contains enzymes that convert starch to sugars and digest some of the gluten. A weakened gluten structure can be avoided by using an inactive malt derivative and molasses, but the resultant bread product is then considered to be an imitation malt bread (10).

Rolls

Rolls, including Kaiser, submarine, sandwich, and others, contain the same ingredients used to make loaf bread. Some may contain additional fat, sugar, and/or eggs. The many available types of rolls vary not only in their ingredients, but in their shapes as well (Figure 24-11). Some professional bakers weigh the individual dough allotments for each kind of roll to ensure uniform proportions and appearance.

Pita Bread

This circular Middle Eastern bread with a large hollow center is also known as "pocket bread." It is prepared by flattening dough into thin circles about 9 inches in diameter and

FIGURE 24-11 There are many varieties of both soft and hard rolls.

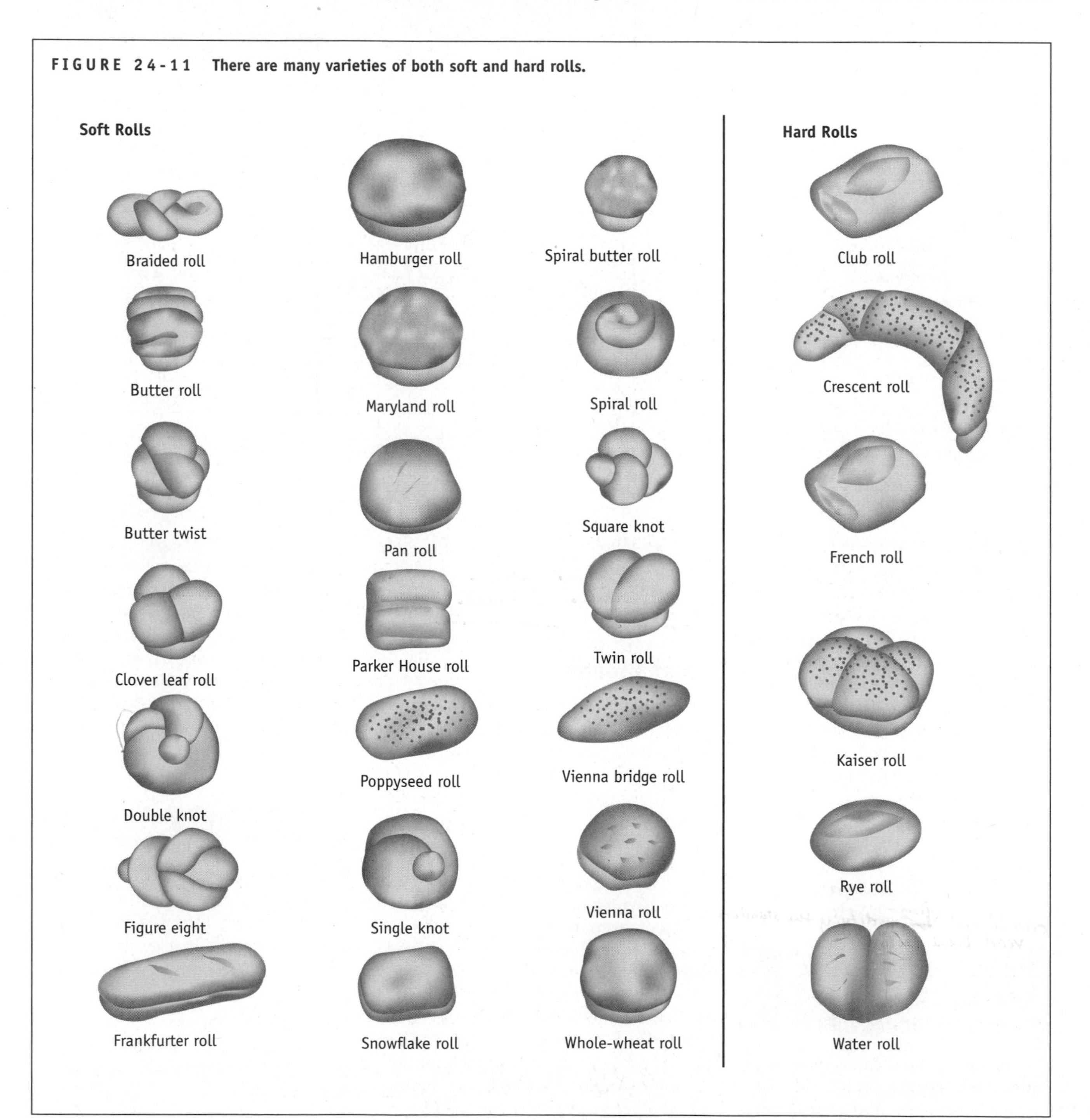

King Arthur Flour

FIGURE 24-12 **A bagel cutter.**

less than ¼ inch thick, and then baking them briefly in a very hot oven (500°F/260°C) until they form the characteristic two symmetrical halves (1). The moisture in the dough heats up so quickly under these high heats that it does not have time to escape through the pores. Instead the dough balloons, and the bread bakes in that position, collapsing again upon removal from the oven. The entire process takes less than a minute. Another method is to cook the dough on a lightly greased griddle for 15 to 20 seconds, gently turn it over, and continue heating for about 1 minute until bubbles appear, and then to turn it back over until it balloons. Pita bread is handy in making sandwiches because it can be filled with a variety of foods, ranging from standard sandwich ingredients to vegetables and cooked meats. Almost any pita sandwich may be topped with lettuce, tomato, and onions.

Bagels

Bagels are made by adding egg whites to the standard white bread formula. The dough is cut into 3-inch rounds with a ¾-inch center hole using a special bagel cutter (Figure 24-12). After the bagels rise, they are boiled in water, cooled, possibly brushed with an egg yolk and water mix, and then baked.

HOW & WHY? ?????????

Why are bagels boiled before baking?

Boiling bagels creates a moist surface that does not brown as readily, so the crust remains lighter and less crunchy. The moisture also causes the starch to gelatinize into a thin, transparent coating that gives bagels their unique, soft, smooth crust texture.

English Muffins

English muffins, unlike other breads, are baked on a greased griddle or pan. These yeast breads, usually about 4 inches in diameter when baked, are made by combining water or milk, sugar, salt, flour, yeast, and a little butter. After the first rising, they are shaped, with or without the use of muffin rings. If muffin rings are used, they are greased, dusted with cornmeal, and filled up to ½ inch with the dough. If rings are not available, the dough is placed on a board that has been lightly sprinkled with cornmeal, where it is pressed down to a thickness of ½ inch, cut into circles about 3 inches in diameter, and moved to a greased cookie sheet. English muffins undergo their second rising before being transferred by spatula to the greased griddle or to a pan where they will be heated until lightly browned.

Pizza Crust

Pizza crust is made with yeast and hard wheat flour. It is allowed to rise only once, for two hours, and then baked in a very hot oven at 400°F (205°C) for about 25 minutes.

Raised Doughnuts

Raised doughnuts are made with bread flour and leavened with yeast. They are typically deep-fried in fat or baked in an oven. Cake doughnuts, on the other hand, are made with cake flour and leavened with baking powder.

Specialty Breads

Specialty breads differ from regular varieties either in their method of preparation or in their ingredients. Some breads that are made differently include:

- Wood-burning oven breads that are often referred to as country loaves or hearth breads
- Steamed bread that is processed with 15 to 20 percent steamed rye flour
- Stonemason bread containing grain that has been treated with moisture and then crushed or cut

Pumpernickel and rye-meal breads lack a crust because they are cooked under continuous steam, which breaks down the starch to sugar and imparts a sweet flavor and dark-colored crumb. Swedish rye crackers are baked in flat sheets. The unique flavor and greater mold resistance of rye bread is due to lactic acid fermentation in which the lactic acid bacteria acidify the dough (2). Rye breads also have a much harder crumb than wheat breads due to the smaller number of pores and greater concentration of large particles.

The use of unique ingredients can also result in a specialty bread. Examples include breads made from flours other than wheat or rye, such as oat, barley, corn, rice, and others, to make three-, seven-, or nine-grain breads; breads prepared with extra dairy items such as butter, milk, buttermilk, whey, and yogurt; and those made with vegetable ingredients such as wheat germ, soy, bran, sesame, linseed, or spices (pepper or paprika). One item to avoid adding to bread dough formulations is garlic, because it causes low volume and decreases the dough strength to the point that it often breaks down during mixing. One theory behind this phenomenon is that allicin, a unique flavor compound in garlic, may interfere with the dough's disulfide bonds (18).

Specialty breads also include those that have had their nutrient value altered for people with health conditions that require dietary modifications (7, 20). Graham bread is made without salt or yeast; other breads reveal their differences in their names—low fat, low calorie, low sodium, and high fiber. Gluten-free bread is produced for people with celiac disease, or gluten-sensitive enteropathy. "Gluten-free" means without gliadin protein. A bread that is gluten-free may still contain other proteins from flour and any dairy, eggs, or other protein-containing ingredient added to the flour mixture.

Storage of Yeast Breads

Fresh

Freshly baked wheat bread is best when consumed within one or two days, while some rye breads can last an average of seven days. When just out of the oven, bread can be kept warm by placing aluminum foil under the cloth in the bread basket. If it is not going to be consumed right away, it should be completely cooled before being wrapped and stored in a dry, cool place at room temperature. Unfortunately, staling starts as soon as the bread leaves the oven (Chemist's Corner 24-3). This unavoidable part of bread preparation is thought to be responsible for an estimated loss of 3 to 5 percent of all baked breads sold in the United States (27). Staling is defined as the deleterious changes in the crust and crumb during storage that result in firmness, crumbliness, and decreased bread quality (13, 15). These undesirable changes are thought to be due primarily to retrogradation (crystallization) of the starch molecules released during gelatinization (4). Moisture levels also play an important role in the staling of bread as water moves from the center of the loaf toward the crust.

> **Chemist's Corner 24-3**
>
> ***The Chemistry of Staling***
>
> Some of the theories as to why staling occurs include possible aggregating amylose molecules (12), recrystallization of the amylopectin (13), and/or a transfer of moisture from the gluten to the starch.

Preventing Staling. Staling is best prevented by keeping the bread away from air. Several techniques include wrapping breads in plastic or paper bags, adding moisture retainers such as fat or sugar (nonfat French bread stales quickly), and/or freezing. The staling caused by retrogradation is reversible when the bread is warmed but returns upon cooling. Reheating the bread in an oven at 125° to 145°F (52° to 63°C) for a few minutes recreates many of the characteristics of fresh bread, especially if a damp cloth or paper towel is placed over the bread during reheating. This method is not recommended for microwave ovens, which cause the bread to become tougher, rubbery, and more difficult to chew (19).

Anti-Staling Additives. Commercial bakers add mono- and diglycerides or fat to bread doughs to help prevent staling (16), and sodium or calcium propionate to retard mold and the bacteria that cause **rope**. Breads affected with rope smell like overripe melons. These bacteria, which live in the soil, can contaminate grains and the flour made from them. The spores survive baking, germinate, and feed off the carbohydrates they obtain by decomposing the bread's starch. The end result is a bread that appears fine on the outside, but is internally mushy or stringy to the point that it can be pulled into "ropes." This condition may mistakenly be perceived as a failure to make sure the bread is thoroughly baked, but consuming this type of contaminated bread can cause vomiting and diarrhea. Rope contamination is more apt to occur in the summer months (26).

> **KEY TERMS**
>
> ***Rope*** The sticky, moist texture of breads resulting from contamination by *Bacillus mesentericus* bacteria.

Refrigerated

Bread should be refrigerated immediately in the warm temperatures and moist humidities of tropical regions. In less humid areas, refrigerating bread is not recommended because it speeds staling.

Frozen

Freezing is one of the best ways to maintain some of the texture and flavor of freshly baked bread. Most breads can be frozen for two or three months. The bread should be wrapped in heavy-duty aluminum foil and dated. Frozen bread should be removed from the freezer and thawed at room temperature in the wrapper. Thawing in a hot oven results in a soggy, flavorless loaf. Home-baked bread can get back its freshly baked flavor if the top portion of the foil covering the thawed bread is opened and it is placed in a preheated 250° to 300°F (121° to 149°C) oven for about 10 minutes. The foil will keep the loaf warm during slicing and serving.

Unbaked bread dough can be frozen for up to two weeks by first shaping the dough and then wrapping it in freezer paper or foil. It should thaw and rise to double its height before being baked. It can also be placed overnight in the refrigerator and then allowed to rise for two hours.

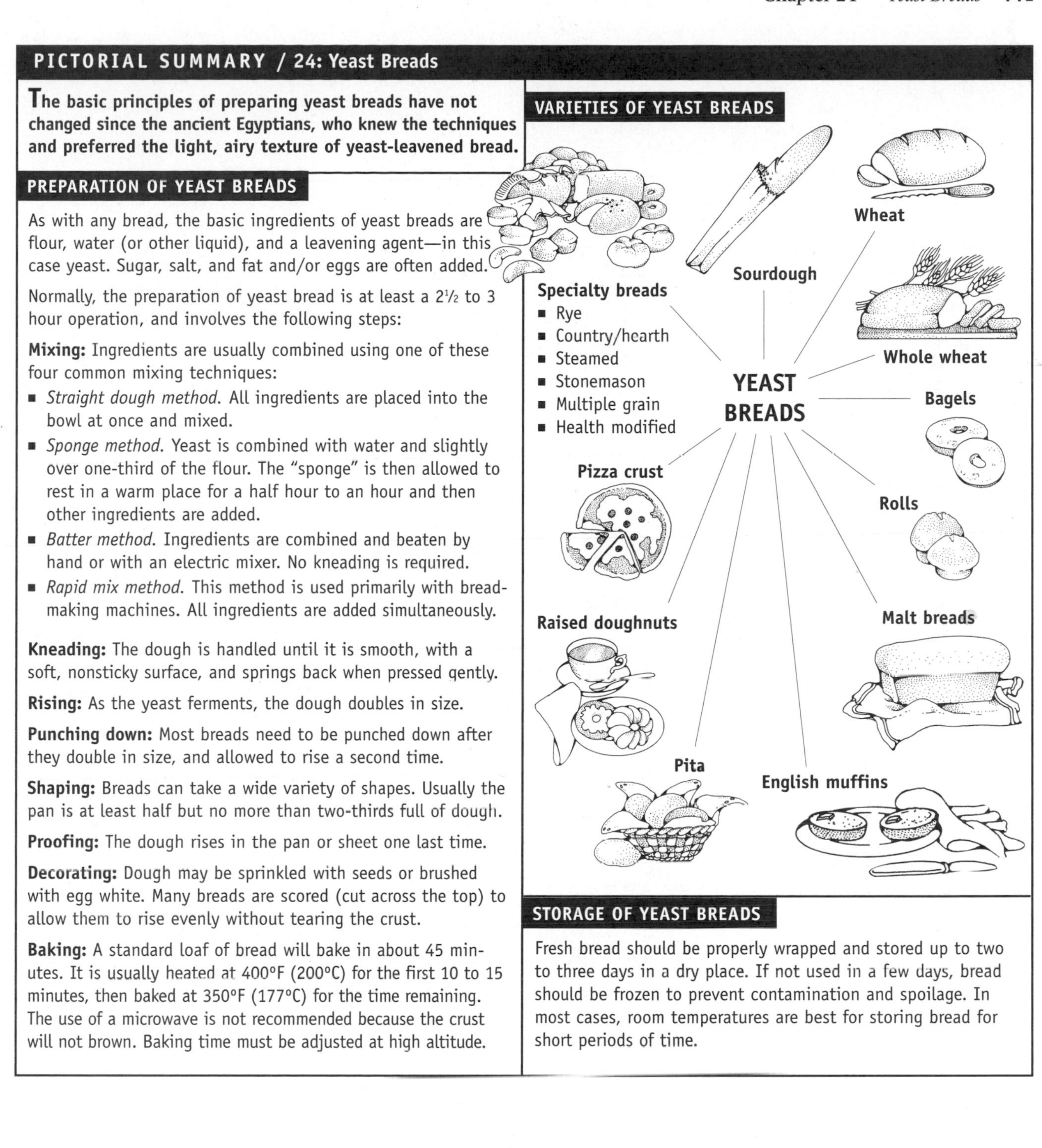

PICTORIAL SUMMARY / 24: Yeast Breads

The basic principles of preparing yeast breads have not changed since the ancient Egyptians, who knew the techniques and preferred the light, airy texture of yeast-leavened bread.

PREPARATION OF YEAST BREADS

As with any bread, the basic ingredients of yeast breads are flour, water (or other liquid), and a leavening agent—in this case yeast. Sugar, salt, and fat and/or eggs are often added.

Normally, the preparation of yeast bread is at least a 2½ to 3 hour operation, and involves the following steps:

Mixing: Ingredients are usually combined using one of these four common mixing techniques:

- *Straight dough method.* All ingredients are placed into the bowl at once and mixed.
- *Sponge method.* Yeast is combined with water and slightly over one-third of the flour. The "sponge" is then allowed to rest in a warm place for a half hour to an hour and then other ingredients are added.
- *Batter method.* Ingredients are combined and beaten by hand or with an electric mixer. No kneading is required.
- *Rapid mix method.* This method is used primarily with bread-making machines. All ingredients are added simultaneously.

Kneading: The dough is handled until it is smooth, with a soft, nonsticky surface, and springs back when pressed gently.

Rising: As the yeast ferments, the dough doubles in size.

Punching down: Most breads need to be punched down after they double in size, and allowed to rise a second time.

Shaping: Breads can take a wide variety of shapes. Usually the pan is at least half but no more than two-thirds full of dough.

Proofing: The dough rises in the pan or sheet one last time.

Decorating: Dough may be sprinkled with seeds or brushed with egg white. Many breads are scored (cut across the top) to allow them to rise evenly without tearing the crust.

Baking: A standard loaf of bread will bake in about 45 minutes. It is usually heated at 400°F (200°C) for the first 10 to 15 minutes, then baked at 350°F (177°C) for the time remaining. The use of a microwave is not recommended because the crust will not brown. Baking time must be adjusted at high altitude.

VARIETIES OF YEAST BREADS

STORAGE OF YEAST BREADS

Fresh bread should be properly wrapped and stored up to two to three days in a dry place. If not used in a few days, bread should be frozen to prevent contamination and spoilage. In most cases, room temperatures are best for storing bread for short periods of time.

REFERENCES

1. Alford J, and N. Duguid. *Flatbreads and Flavors.* William Morrow, 1995.
2. Autio K, T Parkkonen, and M Fabritius. Observing structural differences in wheat and rye breads. *Cereal Foods World* 42(8):702–705, 1997.
3. Bocuse P, and F Metz. *The New Professional Chef. The Culinary Institute of America.* Van Nostrand Reinhold, 1996.
4. Cauvain S. Stalling staling of bakery products. *Prepared Foods* 168(3):69–75, 1999.
5. Clayton B. *Bernard Clayton's New Complete Book of Breads.* Simon and Schuster, 1987.
6. deMan JM. *Principles of Food Chemistry.* Van Nostrand Reinhold, 1999.
7. Ensminger AH, et al. *Foods and Nutrition Encyclopedia.* CRC Press, 1994.
8. Eskin NAM. *Biochemistry of Foods.* CRC Press, 1990.
9. Esteve CC, C Benedito de Barber, and MA Martinez-Anaya. Microbial sour doughs influence acidification properties and breadmaking potential of wheat dough. *Journal of Food Science* 59(3):629–633, 1994.
10. Gobbetti M, et al. Free D- and L-amino acid evolution during sourdough fermentation and baking. *Journal of Food Science* 59(4):881–884, 1994.
11. Hayman D, RC Hoseney, and JM Faubion. Bread crumb grain development during baking. *Cereal Chemistry* 75(4):577–580, 1998.
12. Huang JJ, and PJ White. Monoglyceride interaction in a model system. *Cereal Chemistry* 70(1):42–47, 1993.
13. Keetals CJAM, et al. Structure and mechanics of starch bread. *Journal of Cereal Science* 24: 15–26, 1996.
14. Labuda I, C Stegmann, and R Huang. Yeasts and their role in flavor formation. *Cereal Foods World* 42(10):797–799, 1997.
15. Lahtinen S, et al. Factors affecting cake firmness and cake moisture content as evaluated by response surface methodology. *Cereal Chemistry* 75(4):547–550, 1998.
16. Mettler E, and W Seibel. Effects of emulsifiers and hydrocolloids on whole wheat bread quality: A response surface methodology study. *Cereal Chemistry* 70(4): 373–377, 1993.
17. Meuser F, JM Brummer, and W Seibel. Bread varieties in Central Europe. *Cereal Foods World* 39(4):222–230, 1994.
18. Miller RA, et al. Garlic effects on dough properties. *Journal of Food Science* 62(6):1198–1201, 1997.
19. Miller RA, and RC Hoseney. Method to measure microwave-induced toughness of bread. *Journal of Food Science* 62(6): 1202–1204, 1997.
20. Nordlee JJA, and SL Taylor. Immunological analysis of food allergens and other food proteins. *Food Technology* 49(2):129–132, 1995.
21. Partnership yields new Maillard flavor systems for microwave foods. *Food Engineering* 65(6): 36–37, 1993.
22. Potato maltodextrin gels low-calorie baking opportunities. *Prepared Foods* 159(11):88, 1990.
23. Potus J, A Piffait, and R Drapron. Influence of dough-making conditions on the concentration of individual sugars and their utilization during fermentation. *Cereal Chemistry* 71(5):505–508, 1995.
24. Reduced-fat bakery foods: Meeting the taste challenge. *Prepared Foods* 162(8):79–80, 1993.
25. Rogers DE, et al. Stability and nutrient contribution of betacarotene added to selected bakery products. *Cereal Chemistry* 70(5):558–561, 1993.
26. Schunemann C, and G Treu. *Baking: The Art and Science. A Practical Handbook for the Baking Industry.* Baker Tech, 1988.
27. Si JQ. Synergistic effect of enzymes for breadmaking. *Cereal Foods World* 42(10):802–807, 1997.
28. Wehrle K, and EK Arendt. Rheological changes in wheat sourdough during controlled and spontaneous fermentation. *Cereal Chemistry* 75(6):882–886, 1998.
29. Whitefield FB. Volatiles from interactions of Maillard reactions and lipids. *Critical Reviews in Food Science and Nutrition* 31(1/2):1–58, 1992.

WEBSITES

Descriptions of various doughs and breads:

www.foodsubs.com/Dough.html
and
www.foodsubs.com/bread.html

Generous information on yeast for the nonspecialist to the scientist:

http://genome-www.stanford.edu/Saccharomyces/VL-yeast.html

Step-by-step instructions on how to prepare yeast breads by the Quaker Oats Company:

www.quakeroatmeal.com/kitchen/classes/class.cfm?CategoryID=6&SectionID=15

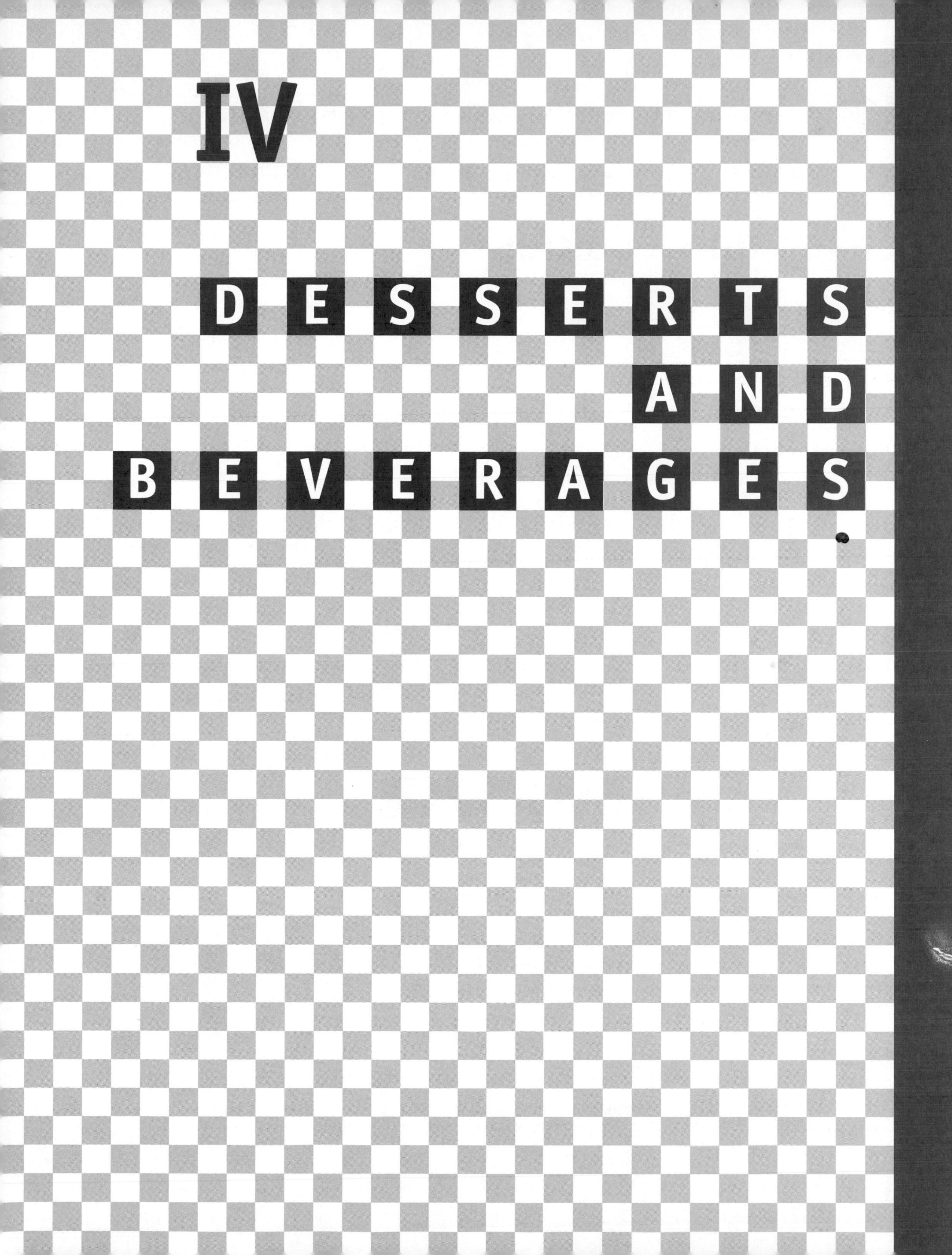

IV

DESSERTS AND BEVERAGES

Cakes and Cookies

A basic flour mixture serves as the foundation for quick and yeast breads, but it is also the basis for their sweeter cousins, cakes and cookies. Cakes are essentially sweetened breads, while cookies can be considered to be "little cakes." Ingredients can be combined in a number of different ways and styles, creating confections ranging from simple sugar cookies to elegantly decorated, many tiered wedding cakes. In the not-too distant past, such items were put together "from scratch," but now the vast majority of cakes and cookies are made from packaged mixes that come ready to be combined with liquid ingredients. To make the process even easier, there are cookies in the supermarket's refrigerator section that are sold ready to bake. This chapter discusses the different types, the nutrient content, and the preparation and storage of cakes and cookies, whose ingredients, along with their specific functions, were discussed in Chapter 22 (Flours and Flour Mixtures).

KEY TERMS

Shortened cake A cake made with fat.

Types of Cakes

Cakes are classified according to whether or not they contain fat. The majority of cakes are either shortened or unshortened; a third category is chiffon cakes.

Shortened Cakes

Shortened cakes, also called butter or conventional cakes, are usually leavened with baking powder or baking soda, although steam generated from the liquid ingredients and air incorporated during the mixing process also contribute to leavening. Examples of shortened cakes include the standard plain white, yellow, chocolate, spice, and fruit cakes. Pound cake is a compact, shortened cake leavened only by air and steam (Figure 25-1). In contrast to other shortened cakes, pound cake contains equal amounts of fat and sugar in addition to cake flour (or, less commonly, all-purpose flour), large amounts of egg, and flavoring. Pound cake derives its name from the original recipe, which called for 1 pound each of butter, sugar, flour, and eggs. Grinding the sugar with a food processor makes the pound cake even lighter and smoother.

Baker Technology

FIGURE 25-1 Pound cakes are leavened only by steam and air.

Unshortened Cakes

Unshortened cakes are also known as sponge or foam cakes. The term "sponge" in food preparation is frequently used to denote foods made with beaten egg whites. The light, delicate structures of angel food and sponge cakes rely on steam and air from foamed, or beaten, eggs as the major leavening agent. Angel food cakes are made with beaten egg whites, while sponge cakes are made with whole eggs, which contribute to the latter's rich, yellow color.

Chiffon Cakes

Chiffon cakes are a hybrid of shortened and unshortened cakes. Fat, usually from vegetable oil and egg yolks, is combined with foamed egg whites, cake flour, and leavening agents. Common examples include lemon or chocolate chiffon cakes.

Preparation of Cakes

As discussed earlier, the ingredients used to make shortened and unshortened cakes differ, as does the mixing of the batters. Different mixing methods also result in different cakes. Some of the most common ways of combining the ingredients of a shortened cake are the conventional, conventional sponge, pastry-blend, single-stage (quick-mix), and muffin mix methods. These mixing methods apply to other foods besides cakes, and are discussed in Chapter 6.

Overall, the flour mixtures that produce cakes and cookies are very similar to those used to make breads. The significant differences are that they are sweeter and often have added flavorings not usually used in breads.

Ingredients

Cakes have a higher proportion of sugar, milk, and fat to flour than do breads, and the flour used is usually cake flour. Both flour and eggs contain the proteins that contribute strength and structure to cakes (5). Fat and sugar have the opposite effect, softening the cake's structure by providing moisture and tenderness. Too much flour and eggs may make the cake tough and/or dry, while too much fat and sugar may weaken the cake to the point where it does not set. Ultimately, the goal is to create cakes that have the strength to hold together, but are still tender and moist.

Flour. The chlorination of cake flour has been reported to break bonds (hydrogen and peptide) within and between flour proteins, and this has been theorized to result in improved dispersion of ingredients, increased swelling of the starch granules, and improved baking quality (10). Cake flour provides structure to cakes when its starch gelatinizes and its proteins form gluten, but the structural strengthening effect of cake flour and egg is balanced by the tenderizing effect of the sugar and fat ingredients.

Sugar. Sugar's tenderizing effect is thought to be due to its dilution of the flour proteins and delaying of starch gelatinization. For many years, the weight of the sugar in cake mixtures could not exceed that of flour because higher proportions of sugar would interfere with the gelatinization of starch and the hydration of proteins. Now, high-sugar (high-ratio) cake mixes with a sugar-to-flour ratio ranging from 1.25:1 to 1.40:1 are common due to improvements in cake flour and shortenings. The extra sugar results in cakes with greater moisture content, which improves their shelf lives. Sugar also functions to sweeten, to increase volume, and to brown the crust.

Three Formulas for "High-Ratio" Cakes. There are three basic formulas for preparing the sweeter "high-ratio" cakes that contain more sugar than flour (5). Following these ingredient proportions will ensure a high-ratio cake that is not too dry or moist.

1. The sugar should weigh the same as or slightly more than the flour. It is the *weight* and not the volume that counts. Remember, 1 cup of sugar (7 ounces by weight) weighs more than 1 cup of flour (4½ ounces by weight).
2. Eggs should weigh almost as much as or slightly more than the fat. Since one large egg weighs about 1¾ ounces, a recipe with 4 ounces of butter would call for two large eggs (3½ ounces).
3. The liquid ingredients (including eggs) should weigh the same as or more than the sugar.

KEY TERMS

Unshortened cake A cake made without added fat.

■ ■ ■ ■

Chiffon cake A cake made by combining the characteristics found in both shortened and unshortened cakes.

Fats. Fats such as butter and shortening also contribute to tenderness, volume, moistness, and flavor. These attributes are best achieved by fats other than vegetable oil, which does not entrap air during creaming. Vegetable oils are generally not used (except for tea breads like carrot cake and commercial cake mixes) because of the resulting decreased volume and harsh crumb (9). However, some people add olive oil, which contains natural emulsifiers, to make cakes more tender and moist (11). Oil coats the flour proteins, preventing them from adhering to water; this reduces gluten formation and leaves more moisture in the batter. The key is not too add too much oil or the cake becomes too heavy and compact. Air bubbles are not as easily incorporated into oil as they are butter, so the only leavening agent that remains is from the chemical leaveners like baking soda, or physical leavening from whipping air into the batter, especially the egg whites.

Eggs. Eggs are added to help strengthen the structure, as well as to increase leavening, to act as emulsifiers, and to add color and flavor.

Milk. Milk is usually used as the main liquid in cake preparation. It hydrates the dry ingredients, dissolves the sugar and salt (Chemist's Corner 25-1), provides steam for leavening,

Chemist's Corner 25-1

The Two Phases of Cake Batter

A cake batter consists of two phases: a continuous, aqueous phase, holding the dissolved solutes of sugar, salt, and leavening salts; and the dispersed particles too large to go into solution—the colloidal proteins and the suspended starch granules, fat globules, and gas cells (10).

and allows baking soda or powder to react and produce carbon dioxide gas.

Leavening Agent. Both cakes and cookies are leavened with gas produced by either baking soda, baking powder, air, and/or steam. The amount of chemical leavening agent used is dependent on how much flour is used. For every cup of flour, high-ratio cakes use 1 teaspoon of baking powder or ¼ teaspoon of baking soda (5).

Additional Ingredients and Other Factors. Salt is important because it is a flavor enhancer. Also, flavors such as vanilla, chocolate, spices, fruits, and nuts are commonly incorporated into the basic flour mixture. Surfactants are often added to commercial cake mixes (Chemist's Corner 25-2).

Chemist's Corner 25-2

Surfactants in Cake Batters

Commercial cake mixes often contain surfactants in their shortenings. These compounds improve texture and flavor, and aid in the emulsification of ingredients and the incorporation of air into the batter, which improves volume. Examples of surfactants include monoglycerides, diglycerides, polysorbate 60, sorbitol—fatty acid esters, glycerol-lactic acid esters, and propylene glycol—fatty acid esters. Batter viscosity and stability can be improved by adding hydrophilic colloids such as gums and carboxymethyl cellulose (10).

NUTRIENT CONTENT

Cakes. Cakes, which consist of flour mixtures with extra sugar and sometimes fat, can be high in carbohydrates and fat, contributing to a general figure of about 200 calories (kcal) and 8 grams of fat per slice. Cakes slices, however, will vary depending on the type, with angel food cake being the lowest in fat and calories, and carrot cake being the highest, as shown in Table 25-1. Protein is present in the flour, but not in large quantities. The total protein is more dependent on other ingredients such as milk, eggs, and occasionally, nuts.

Any nutrient modification of cake mixtures usually focuses on the fat content. Fat is not always easy to replace, but fine tuning the ingredients can reduce fat content somewhat (2). Fat's function as a moistener can be partially fulfilled by substituting yogurt, nonfat sour cream, applesauce, etc. The flavor lost by the removal of the fat can be replaced, in part, by adding more vanilla or another flavor extract. When baking from scratch, the fat in chocolate items can be reduced by using unsweetened cocoa powder, which contains only 6 grams of fat per ounce compared to the 15 grams found in the same amount of unsweetened chocolate. Food companies have more flexibility than the consumer in reducing the fat content of cakes, because the thicker, richer consistency often provided by fat can be partially replaced by using the commercial fat substitutes available to the food industry, but not, as yet, to the consumer (14). There are now several commercial cake mixes on the market that are 94 percent by weight fat-free.

TABLE 25-1

Fat and Calorie (kcal) Content of Cakes

Cake	*Size (round pan)*	*Fat (g)*	*Calorie (kcal)*
Angel food	1/12 of 10″	0.1	129
Carrot	⅛ of 9″	11	239
Chocolate with icing	⅛ of 9″	9	253
Coffee	⅛ of 9″	5	155
Fruit	¼ of 8″	5	155
Sponge	½ of 9″	2	188
Pound	1/10 of loaf	6	113
White	⅛ of 9″	10	266
Yellow	⅛ of 9″	7	221
Cupcake with icing	1 standard	7	154

In addition to ingredients and mixing methods, four other factors to consider when baking cakes are:

- The type of pans to use and their treatment
- Timing
- Temperature
- Testing for doneness

These factors vary depending on whether the cake is shortened or unshortened.

Preparing Shortened Cakes

Shortened cakes are the most commonly prepared cakes, especially for birthday and wedding celebrations. They can be made from scratch or purchased as a boxed mix in the supermarket.

Conventional round cake pans are typically used, but large rectangular pans create large surfaces that can be covered with a limitless number of icing decorations or messages.

Type and Treatment of the Pans. Pan characteristics affect cake quality, so it is important to select the best pan for the job and to prepare it properly. Dull, rough-surfaced pans are best for baking cakes because they absorb heat more readily, resulting in the cake baking more quickly and having a larger volume, a finer grain, and a more velvety texture. Crumb formation is partially dependent on the degree of rising that occurs when the cake batter is first placed in the oven, and rapid heat absorption plays a role in this. On the other hand, shiny surfaces reflect heat, which causes the cake to take longer to bake, resulting in a coarser grain and lower volume. The weight, or gauge, of the pan's metal also affects quality: the heavier the pan, the better. When using glass pans for baking cakes, baking temperatures should be lowered by 25°F (14°C), because glass pans lead to shrunken corners from overcooking and cause the exterior crust to brown readily—a condition desirable for breads, but less so for cakes.

Pan Preparation. The pans are prepared prior to mixing the batter. The bottom is greased, but the sides generally are not; the ungreased sides provide traction, allowing the rising mixture to reach its full volume. Dusting the pan's greased surface with cake flour, or cocoa if it is a chocolate cake, makes later removal of the cake from the pan easier. After dusting, if the pan is turned upside down and tapped to remove excess cake flour, it will prevent the cake from having a mottled bottom. Waxed or parchment paper may also be placed in the bottom before greasing to allow for easier cake removal.

Temperature/Timing. The timing of pouring the cake batter and getting it into a properly heated oven is another important factor in cake quality. If a batter is allowed to stand too long after pouring, carbon dioxide and air will escape. This reduces the volume and increases the coarseness of the cake's cells. Immediately after mixing, the pan is filled with cake batter between half and two-thirds full. The cake should be placed immediately in a preheated oven to ensure proper leavening. In order to avoid uneven baking and burning from hot spots, pans should not be allowed to touch each other or the sides of the oven, nor should they be placed directly above or below each other (Figure 25-2).

Most shortened cakes are baked at 325° to 350°F (163° to 177°C). In general, layered cakes (8-, 9-, or 10-inch round pans) bake in about half an hour, while the thicker loaf cakes take three-quarters of an hour to an hour to bake. Cupcakes normally take only about 20 minutes because of the small quantity of batter in each cup.

Changes During Baking. Heat plays several roles during cake baking. It increases volume by expanding air, steam, and carbon dioxide. It sets the structure by coagulating protein and gelatinizing starch. The heat flows from the edges toward the center of the pan, so cakes become rounded on the top as their interiors continue to rise after the outside portions of the cake have started to set. In addition, heat browns the crust via the Maillard reaction and the caramelization of sugars. If the oven temperature is too low, leavening gas is lost from the batter before there is a chance for coagulation of the proteins and gelatinization of the starch. As a result, a cake is produced with low volume, thickened cells, and possibly an indentation in the center. Excessively high temperatures create a crust before the cake can rise, resulting in a hump formed as the interior of the cake continues to rise (10).

Testing for Doneness. When cakes are nearing doneness, they start to "wrinkle" at the pan edges. They should be removed from the oven before a gap forms between the cake and the pan. To test for doneness in shortened cakes, one method is to insert a cake tester or toothpick in the center of the cake. If it comes out clean, with no batter clinging to it, the cake is done. If a moister cake is desired, it is best to remove it from the oven while a few crumbs are still sticking to the toothpick. Another way to test for doneness is to touch the top of the cake lightly with a finger. If it springs back, the cake is done (Figure 25-3). Testing for doneness should be reserved until close to the end of heating time, and done as infrequently as possible to avoid drafts from the open oven door and to keep from making unnecessary holes with the tester. Every time the oven door is opened, the oven temperature drops, drafts occur, and baking time is prolonged. The drafts may even cause the cake to fall.

Cooling. Once the cake is done, it should be removed gently from the oven and allowed to cool on a rack for 5 or 10 minutes. The warmer interior of the cake needs a chance to become firm, so any drastic, sudden movement or lack of adequate cooling jeopardizes its structure. The rack allows even air circulation under the cake; this prevents condensation and sogginess (Figure 25-4). Shortened cakes

FIGURE 25-2 Pan placement: Adequate air circulation is achieved by placing the pans on the middle rack and at least 1 inch from each other and the oven sides.

FIGURE 25-3 Testing a cake for doneness: The cake is done baking if a light touch to the center leaves no imprint or if a wooden toothpick inserted in the center comes out clean.

FIGURE 25-4 Cakes should be cooled on a rack. This allows them to set and prevents sogginess.

are usually removed from the pan before they are completely cooled, usually about 10 to 15 minutes after being taken out of the oven. Once the cake has cooled, a spatula or knife can be inserted and moved around the edges of the pan before inverting the cake onto a cake plate. Cakes should have a fine crumb, a tender texture, optimum volume, and a lightly browned, delicate crust. Common cake faults and their contributing causes are listed in Table 25-2.

High-Altitude Adjustments. As in bread preparation, cake ingredients must be modified at altitudes higher than 3,000 feet. The lower atmospheric pressure at higher elevations reduces the need for baking powder or soda. Also at higher altitudes, water evaporates more quickly and the concentration of sugar increases. Structural strength can be improved by adding 1 to 2 tablespoons of cake flour, increasing the amount of water, and reducing baking powder and sugar quantities (Table 25-3). Increasing the baking temperature 10° to 15°F (6° to 8°C) increases the rate at which the cake sets by speeding the coagulation of the protein and the gelatinization of the starch. Instructions on cake mix boxes provide altitude adjustment instructions.

Microwave Preparation. It is possible to prepare cakes in the microwave oven, but the process has not yet been perfected. All cakes are better prepared in a conventional oven. Still, microwavable cake mixes are available and some of these contain special ingredients, such as xanthan gum, to increase moisture retention (15). Microwaved cakes are usually cooked in about 10 minutes and rise higher than conventional cakes. They lack the characteristic browning and crust formation expected from conventional cakes. The lack of crust is often masked with frosting; however, it is more difficult to hide the soft, uneven tops that may occur without crust formation.

Packaged microwave cake mixes usually include a microwave-safe baking pan. If not, round pans are preferred over square or rectangular ones because the corners of cakes baked in the latter are prone to burning. Two or more pans filled with cake mix can be cooked at the same time as long as the air can circulate; otherwise they are prepared one at a time or baked entirely in a 16-cup fluted pan. Some cake mixes on the market can be baked in either conventional or microwave ovens. As always, follow the manufacturer's instructions.

Preparing Unshortened Cakes

Unshortened cakes rely on foam formation for their structure, making their preparation slightly more involved than that for conventional, shortened cakes. The preparation of angel food, sponge, and chiffon cakes (hybrids of

TABLE 25-2
Cake Problems and Their Causes

Problems / *Causes*	*Over mixing*	*Under mixing*	*Batter too firm*	*Too much flour*	*Too much air*	*Too much sugar*	*Too little sugar*	*Too much fat*	*Too little fat*	*Too much egg*	*Too much liquid*	*Too little liquid*	*Too little salt*	*Over baking*	*Under baking*	*Too high temperature*	*Too low temperature*	*Too little or old leavening*	*Too much leavening*	*Uneven oven heat*	*Too much flour*	*Oven door opened too early*
Decreased volume	X	X						X			X						X	X				
Increased volume					X																	
Too brown						X								X		X						
Too pale	X						X								X		X					X
Falls in center		X				X		X			X				X		X	X	X			X
Uneven shape																				X		
Peaked	X			X					X							X		X			X	
Cracks on top	X		X													X			X			X
Tunnels	X		X					X		X						X		X	X			
Tough	X			X			X		X					X				X			X	
Dry/crumbly		X		X			X		X			X			X	X			X		X	
Soggy		X				X		X			X				X				X			
Heavy	X					X		X			X				X			X				
Shrinks	X															X		X				
Flat taste													X									

TABLE 25-3
Altitude Adjustments for Shortened Cakes

Ingredient Adjusted	*3,000 Feet*	*5,000 Feet*	*7,000 Feet*
Baking powder			
For each teaspoon, decrease	⅛ tsp	⅛–¼ tsp	¼ tsp
Sugar			
For each cup, decrease	0–1 tbs	0–2 tbs	1–3 tbs
Liquid			
For each cup, add	1–2 tbs	2–4 tbs	3–4 tbs

shortened and unshortened cakes) are discussed below.

Angel Food Cake. The three basic ingredients of angel food cake are egg whites, sugar, and cake flour. Small amounts of cream of tartar, salt, and flavoring are also added. The tender texture of an angel food cake is partially due to the use of cake flour; the majority of leavening is due to steam produced by the evaporation of liquid from egg whites (3). All angel food cake ingredients should be at room temperature; however, for food safety reasons, if eggs are too cold, they may be dipped briefly in warm water and used immediately. A stable egg-white foam, for which the eggs must be at room temperature, is necessary to form the cake's basic structure. Sugar contributes further to the stability of the foam by the incorporation of air into the mixture as it is added, allowing the formation of small air bubbles. It has a tenderizing effect on the cake by interfering with gluten development, and it raises the coagulation temperature of the egg proteins and the gelatinization temperature of starch. Too much sugar, however, will cause the cake to collapse.

HOW & WHY? ?????????

Why is cream of tartar added to angel food cakes? Cream of tartar is an acid that strengthens the egg-white foam by denaturing the protein, thus stabilizing the air cell structure. The acidic environment created by the cream of tartar helps increase tenderness and contributes to the white color of angel food cake by whitening the flour's natural yellowish, anthoxanthin pigments.

Mixing Technique. Proper mixing technique must be followed carefully when making angel food cake. The sugar is added after the egg whites have began to foam. It should be added very gradually or it will pull water from the egg whites, creating a syrupy foam and producing a low-volume cake. Salt and flavoring are also added at this time. The cake flour is sifted gradually over the egg-white foam to prevent its weight from collapsing the air cells. The ingredients must be thoroughly blended, while avoiding overmanipulation, which would reduce tenderness and volume (Figure 25-5).

Temperature/Timing. The prepared batter is poured quickly into an ungreased tube pan, a spatula is run through the batter, sealing it to the sides of the pan, and then the pan is placed in the lower third of a preheated moderate (350°F/177°C) oven. Too cool an oven will cause a low-volume cake because the sugar will absorb liquid from the egg whites, turn syrupy, weep out of the batter, and disrupt the air cells. An oven that is too hot will set the cake's exterior before the cake has been fully expanded and baked through, resulting in a low-volume, dense cake (3). During baking, the light batter rises and relies on the additional support of the central tube (Figure 25-6). As the cake bakes, the proteins coagulate, stabilizing the air cells; water evaporates from the fluid mixture to create a more rigid structure; starch gelatinizes, further contributing to structure; and browning on the surface occurs due to the Maillard reaction (1). Baking time is approximately 45 minutes.

When the cake is done, it is inverted in its pan and allowed to stand for about 1½ hours in this position to stretch and strengthen its structure. For serving, it is best to cut or separate angel food cake into pieces by using the special divider made for that purpose (it looks like a large comb), or by using a sawing motion with a serrated knife

Under-manipulation: uneven grain, course texture, low volume

Optimal folding of ingredients: fine grain, light spongy texture, high volume

Over-manipulation: compact grain, low volume

FIGURE 25-5 Under- and overmixing angel food cake.

FIGURE 25-6 The pan for angel food cake.

FIGURE 25-7 Cutting angel food cake. Unlike other cakes, the delicate structure of angel food cakes requires that they be cut with a "comb" or serrated knife.

(Figure 25-7). Conventional knives will crush the delicate cake.

The few ingredients and minimal number of steps needed for the preparation of angel food cake may make it seem simple, but the delicacy of the egg-white foam can make it a tricky procedure. Chapter 13 on Eggs describes how to prepare an egg-white foam and explains the many factors that influence its formation.

Sponge Cake. Sponge cake is similar to angel food cake, except that there are two foams—an egg-white foam and an egg-yolk foam—and lemon juice often replaces the cream of tartar as the acid ingredient. There are three possible methods for preparing sponge cake.

Method 1. In the first method, the egg whites and yolks are first separated and each is beaten separately. Sugar, and possibly vanilla extract, are beaten into the whipped yolks. The cake flour is then folded in, followed by the gentle folding in of the egg-white foam.

Method 2. Another method, called the syrup or meringue method, creates a finer-textured end product. A syrup is made by cooking two parts sugar with one part water and boiling the mixture until it has reached the soft-ball stage (238° to 240°F/109° to 116°C) (see Chapter 27). Egg whites are beaten with cream of tartar until they are stiff but have not yet formed dry peaks. Then, as the egg-white foam is beaten constantly, the hot syrup is poured into it in a fine stream. When this meringue is completed, beaten egg yolks are combined with lemon juice and folded into the mixture, and, finally, cake flour is sifted over and lightly folded into the other ingredients.

Method 3. In the third method, whole eggs are beaten until foamy and pale yellow in color, then a small amount of cream of tartar or lemon juice is added. The mixture is beaten until stiff before the sugar is added in 2-tablespoon increments. A presifted flour-and-salt combination is then sifted into the egg mixture and folded into the batter.

Chiffon Cake. Chiffon cakes are more tender than either angel food or sponge cakes because of added vegetable oil. A chiffon cake is prepared by folding whipped egg whites into a mixture of cake flour, sugar, beaten egg yolks, and oil. The batter is poured into an ungreased tube pan and baked in a preheated oven of 325°F (163°C). When baking is completed, the pan is inverted over a cooling rack and allowed to cool for at least 20 minutes before being turned out onto a cake plate.

Type and Treatment of the Pans. When preparing unshortened cakes, pans are left ungreased in order to provide traction so the batter can achieve optimum volume; this also prevents angel food cake from falling out of its pan while it is cooling upside down. Tube pans provide structure, help set the delicate cake by allowing heat to reach a greater surface area, and allow for easier cake removal.

Temperature/Timing. Unshortened cakes are baked in a moderate 350° to 375°F (177° to 191°C) oven for approximately three-quarters of an hour to an hour. Excessively high temperatures toughen the cakes' delicate structure, coagulate the top before the air and steam have had sufficient time to accomplish leavening, and may burn the crust.

Testing for Doneness. The unshortened cake is cooked when the surface is lightly brown and springs back when touched. Using a toothpick to test for doneness does not work for unshortened cakes.

Storage of Cakes

Cakes stale fairly quickly. Staling can be prevented to some degree by keeping them covered. Placing half an apple in the cake box also seems to extend the shelf life of a cake. Frosting the cake as soon as it cools is another method to slow down moisture loss. The amount and type of sweetener used in the preparation of a cake affects its ability to be stored. As mentioned before, high-sugar cake mixes have longer shelf lives than lower-sugar cakes. Substituting honey for part of the sugar in cake contributes to even more moisture retention; it contains fructose, which is extremely hygroscopic, or water-loving. However, the fructose and glucose in honey may result in excess browning due to the Maillard reaction.

Freezing is another method to deter staling. Cakes that have been wrapped airtight can be frozen unfrosted, or with one of the types of frostings that freezes well. Fruit cakes should not be frozen, because defrosting will cause the fillings to run, resulting in soggy cakes. Frozen cakes keep up to three months if frosted, and up to six months without frosting.

Types of Cookies

Cookies contain many of the same ingredients as cakes except that the proportion of water is low, while sugar and fat are high.

HOW & WHY? ?????????

Why are cookies crispier than cakes?

Sugar, starch, and flour proteins all compete for the small amount of available water. Cookies are crisp and tender because very little to no starch gelatinization or gluten formation occurs. Gluten development is also hindered by the high fat concentrations found in cookies.

If the recipe does call for a higher amount of water than usual, a more cake-like cookie is produced. However, many commercial cookies are baked with very little water (3 to 4 percent) to create a crisp cookie and one in which the sucrose will not dissolve but rather contribute to the cookie's shiny, glassy appearance (10).

There are hundreds of different cookie recipes, and this enormous variety is possible because a wide range of flavoring agents may be added (Figure 25-8). Chocolate chips, nuts, coconut, fruit, marshmallows, peanut butter, and many other ingredients find their way into cookies. The seemingly endless assortment of cookies makes it difficult to categorize them, and not all fit neatly into one classification. In general, however, the fluidity of the batter or dough determines which of the following six categories—bar, dropped, pressed, molded, rolled, or icebox/refrigerator—cookies fall into.

Bar Cookies. The most fluid of cookie batters are used to make bar cookies such as brownies. The batter is baked in a pan instead of on a baking sheet and cut into individual pieces, or bars, with a knife or spatula.

Dropped Cookies. Dropped cookie batter is literally dropped from a spoon or portion control scoop onto the baking sheet. The batter contains just enough flour so the cookie will not spread out like a pancake when it is dropped on the baking sheet. Figure 25-9 shows the desired spread of cookies during baking. Both spread and surface cracking are used to determine the baking quality of cookies (10). Chocolate chip, oatmeal raisin, and meringue cookies are examples of dropped cookies.

Pressed Cookies. The flour mixture for pressed cookies is viscous enough to be stuffed into a pastry bag or cookie press and forced out through cookie dies. Examples of pressed cookies include tea cookies, lady fingers, and coconut macaroons.

Molded Cookies. This dough is heavy enough to be formed or molded into balls, bars, or other shapes before being placed on the baking sheet. Molded and other cookies sometimes have powdered sugar sprinkled on top, which can be applied evenly by using a mesh tea strainer. Peanut butter cookies are an example of molded cookies.

Craig Aurness/CORBIS

FIGURE 25-8 Cookie varieties are limited only by the imagination.

Rolled Cookies. These come from a slightly heavier dough than molded cookies. The dough is rolled out on a lightly floured board and cut into the desired shape. Using too much flour on the cutting board leads to hard-textured cookies. Powdered sugar may be used in place of flour on the cutting board, but obviously it will make the cookies slightly sweeter. Any leftover dough from the initial cuttings of cookies can be salvaged to make more cookies by rerolling it into a ball, spreading it out, and cutting it again without concern about additional gluten development. The most common type of rolled cookies are sugar cookies. Shortbread cookies are also rolled.

Icebox/Refrigerator Cookies. The same kind of dough used for rolled cookies can be formed into a cylinder,

FIGURE 25-9 Desired spread of a dropped cookie.

NUTRIENT CONTENT

Cookies. The high sugar, fat, egg, and flavoring concentration contributes to the higher calorie (kcal) and fat intake of cookies. Standard cream cookie filling may contain up to 40 percent fat, but glycerin-based fat-free fillings are available (7). Cookies average approximately 50 calories (kcal) each even though the food industry has marketed many new products that are lower in fat (13) or even fat-free.

Sugar content is of concern to people with diabetes or hypoglycemia, but certain cookies are now available for such people (12). Research continues toward the development of a cookie dough containing no added simple sugars (4). One problem of reducing the amount of sucrose in cookies is that it diminishes the bulk properties contributed by sugar. Commercial cookie manufacturers get around this by using bulking agents such as maltodextrins, sugar alcohols, polydextrose, cellulose, and insoluble fiber compounds (4).

Chemist's Corner 25-3

Cake Flour Use in Cookies

Although cake flour is lower in protein than hard wheat flour, it is chlorinated, which reduces cookie spread. Soft wheat flour is preferred for puffier, softer cookies because it blends more easily with whipped egg-whites (if included). In addition, it takes up less water than a higher-protein, hard wheat flour, leaving more water to dissolve the sugar. The sugar syrup that forms with this water makes the dough more relaxed and spreadable (10).

wrapped, and placed in the refrigerator to harden. The chilled dough is then sliced into thin cookies for baking. The commercially prepared cookie doughs sold in the refrigerator section of the supermarket are of the icebox/refrigerator type.

Preparation of Cookies

Mixing Methods

The type of cookie to be prepared determines the mixing method, but for most types the conventional cake method is used. The degree of gluten development is not as important for cookies as it is for cakes, which need a soft, delicate crumb, so all-purpose flour, rather than cake flour, is usually used for cookies. Cake flour is used, however, if a puffy, soft cookie is desired (Table 25-4)(Chemist's Corner 25-3). Lower-protein flours such as cake flour also do not absorb as much water as those that are higher in protein, so more water is available for steam generation, resulting in a more puffed cookie. Also, the chlorination of cake flour tends to inhibit the spread of cookie dough, so flat cookies are less likely to occur.

Once the ingredients are chosen based on whether a flat or puffy cookie is desired, they are usually just barely mixed together until moistened, because development of gluten is not necessary, and the cookies are ready for baking unless chilling is required. Overmixing will cause the cookies to be hard and tough due to the addition of too much air, which facilitates the formation of a protein foam.

Baking Cookies

Type and Treatment of the Pan. Cookie baking sheets are preferred for all except bar cookies. Their low or nonexistent sides allow hot air to circulate and bake the cookies evenly. A shiny top surface and a dull bottom allow even browning. Pans are usually greased for dropped, bar, or rolled cookies, but not for pressed, molded, or icebox/refrigerator cookies. Cookies should be placed far enough apart on the baking sheet so they will not touch or flow together during baking. To prevent spreading, cookie dough should be placed on cool cooking sheets or on a sheet of aluminum foil or parchment. Cookies bake better if the pan is placed in the middle or top rack in the oven with at least 2 inches between the pan and the oven wall. Better air circulation can be achieved with unrimmed cookie sheets that have one or two raised edges. Burned cookie bottoms may be prevented by inserting one pan into another, leaving a pocket of air between them (8).

TABLE 25-4
Change the Ingredients, Change Cookie Character

Ingredient	*Result*
For puffy, soft, pale cookies:	
Cake flour (low protein, acid)	More steam and puff; less browning
Shortening (high melting point)	Less spread
All brown sugar (hygroscopic, acid)	Soft and moist; less spread when used with egg
Egg	Moisture for puff; less spread with acidic ingredients
For thin, crisp cookies:	
All-purpose flour (high protein)	Browning
Butter (protein)	More spread; browning
Baking soda (alkali)	Browning
Corn syrup (glucose)	Browning; crisp
White sugar (sucrose)	Crisp
No egg	No puff; more spread

Source: *Fine Cooking*, 1998

Temperature/Timing. Hotter temperatures of up to 375°F (191°C) are used in the baking of cookies. Exceptions include meringue or sponge cookies, which bake at about 225°F/107°C. Higher temperatures help prevent the dough from spreading and facilitate browning, but if the temperature is too high, excessive drying and browning will result. Baking times average between 10 and 30 minutes, depending on the type of cookie.

Testing for Doneness. Cookies are done when the browning is complete and the centers are cooked. The easiest and most tasty way to determine doneness is to split a sample cookie open and do a taste test. Once done, cookies should be removed immediately from the pan and placed on a cooling rack. They should have a crisp or chewy texture, a uniform shape, even browning, and good flavor. The flexibility of a cookie often observed after being taken out of an oven soon disappears, in part due to the sucrose crystallizing out of solution (10). Cookies burned on the bottom can be salvaged by rubbing them on the coarse side of a grater. Further excess browning can be prevented by double-panning, or using two cookies sheets of the same size, one placed on top of the other. Poor characteristics in prepared cookies and their causes are explained in Table 25-5.

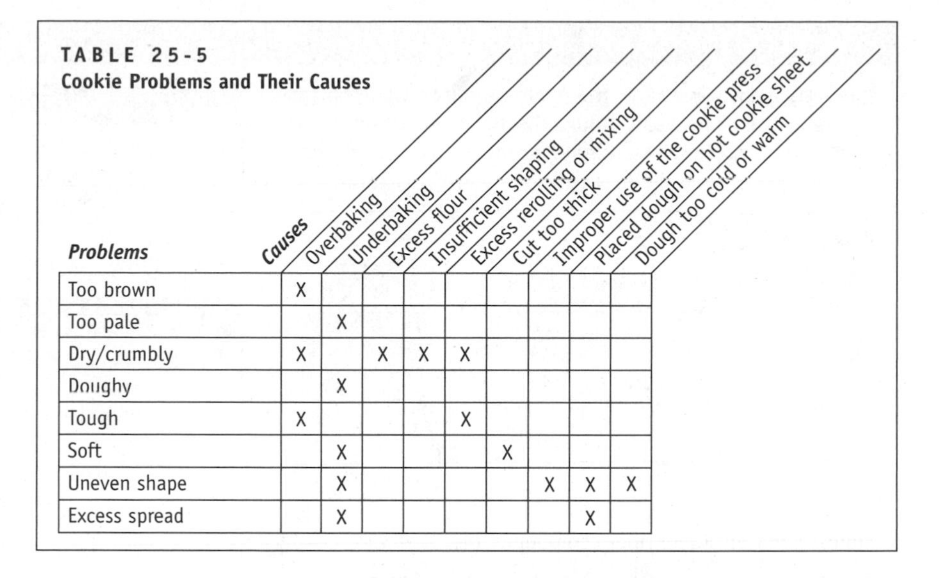

TABLE 25-5
Cookie Problems and Their Causes

Problems / Causes	Overbaking	Underbaking	Excess flour	Insufficient shaping	Excess rerolling or mixing	Cut too thick	Improper use of the cookie press	Placed dough on hot cookie sheet	Dough too cold or warm
Too brown	X								
Too pale		X							
Dry/crumbly	X		X	X	X				
Doughy		X							
Tough	X				X				
Soft		X				X			
Uneven shape		X					X	X	X
Excess spread		X						X	

High-Altitude Adjustments. When baking at higher altitudes, a slightly higher temperature may be needed. A decrease in the quantity of baking powder and sugar or more flour may also be necessary.

Microwave Preparation. Heating time for cookies is so short that using a microwave oven is often impractical and may be detrimental to cookie quality. There are many prepared cookie mixes that come with microwave instructions, however.

Storage of Cookies

Airtight containers are best for maintaining cookie freshness. As soon as they are cool, cookies are transferred to a flat dish or plate and covered with plastic wrap or metal foil. They may also be arranged in layers in a covered cookie jar or plastic zipper bag. Some commercially packaged cookies can keep for months. Most cookies (except those with fresh fruit fillings) are ideal for freezing because their relatively low moisture content means fewer ice crystals will form (6).

PICTORIAL SUMMARY / 25: Cakes and Cookies

The simplest cookies and the most elaborate wedding cakes are really the same basic flour mixture as quick and yeast breads—but made with a higher proportion of sugar, fluid, and fat. And what a difference those additions can make!

TYPES OF CAKES

Shortened cakes: Also called butter or conventional cakes, these are made with fat. Examples are white, yellow, chocolate, and spice cakes.

Unshortened cakes: Also called sponge or foam cakes, these are made without fat, and rely on whipped eggwhite foam for structure. Angel food and sponge cake are two examples.

Chiffon cakes: These combine the characteristics of both shortened and unshortened cakes. Examples: lemon and chocolate chiffon cakes.

PREPARATION OF CAKES

Compared to breads, cakes have a higher proportion of sugar, milk, and fat to flour. Structural strength is provided by flour and eggs, with the latter also contributing to color, flavor, and leavening, and acting as emulsifiers. Fats and sugar add tenderness and volume, while sugar in addition contributes to sweetening and browning of the crust. Cakes are leavened by baking soda or powder, air and/or steam, but a true pound cake is leavened only with air and steam. Flavorings provide variety in cakes.

Nutritionally, an average cake serving yields about 8 grams of fat and 200 calories (kcal) per serving. Per slice, angel food cake is the lowest in fat and calories (0.1 grams/143 calories (kcal)), while carrot cake is the highest (12 grams/283 calories (kcal)).

Four considerations in baking cakes are the type and treatment of the pans, timing, temperature, and testing for doneness. Commonly used mixing methods include conventional, conventional sponge, pastry-blend, quick-mix, and muffin mix. Regardless of the type of mixing method used, too much or little stirring can cause problems, and special measures are taken at altitudes higher than 3,000 feet.

TYPES OF COOKIES

In general, the fluidity of the batter or dough determines the type of cookie: bar, dropped, pressed, molded, rolled, and icebox/refrigerator cookies. All contain flour, sugar, salt, fat, and liquid. Cookies are usually higher in sugar and fat than cakes, although low-fat and nonfat cookies are available.

PREPARATION OF COOKIES

The type of cookie determines the mixing method. All-purpose flour instead of cake flour is usually used because the degree of gluten development does not dramatically affect cookies. With the exception of bar cookies, which are baked in a pan, baking sheets are the preferred pan for baking cookies.

STORAGE OF CAKES

Cakes do not retain their freshness for very long and should be protected from any exposure to air. Iced cakes can be frozen and will keep their quality for up to three months, while unfrosted cakes can be kept frozen for up to six months.

STORAGE OF COOKIES

Like cakes, cookies need to have their exposure to air minimized. Airtight containers or wrappings are crucial to maintaining the freshness of home-baked cookies. Most cookies freeze well due to their low moisture content.

REFERENCES

1. Arunepanlop B, et al. Partial replacement of egg white proteins with whey proteins in angel food cakes. *Journal of Food Science* 61(5):1085–1093, 1996.
2. Baggett N. Chocolate quartet: Classic cakes for an enlightened repertoire. *Eating Well* 4(1):67–73, 1993.
3. Braker F. When it comes to angel food cakes. *Fine Cooking* 22:43–45, 1997.
4. Bullock LM, et al. Replacement of simple sugars in cookie dough. *Food Technology* 46(1):82–86, 1992.
5. Corriher S. For great cakes, get the ratios right. *Fine Cooking* 42:78, 2001.
6. Klivans E. Freezing cookies. *Fine Cooking* 32:13, 1999.
7. LaBell F. A less-filling filling. *Prepared Foods* 162(12):67, 1993.
8. Middleton S. Pros pick the best baking sheets. *Fine Cooking* 26:55–57, 1998.
9. Murano PS, and JM Johnson. Volume and sensory properties of yellow cakes as affected by high fructose corn syrup and corn oil. *Journal of Food Science* 63:1088–1092, 1998.
10. Penfield MP, and AM Campbell. *Experimental Food Science.* Academic Press, 1990.
11. Revsin L. Old-Fashioned cakes with a subtle twist. *Fine Cooking* 43:59–63, 2001.
12. Sangronis E, and M Sancio. Development and characterization of rice bran cookies. *Acta Cientifica Venezolana* 41(3):199–202, 1990.
13. Sigman-Grant M. Can you have your low-fat cake and eat it too? The role of fat-modified products. *Journal of the American Dietetic Association* 97(7):S76–S81, 1997.
14. Smith P. "Lite" cakes: A matter of formula manipulation. *Bakers Digest* March(13):31, 1984.
15. Xanthan gum eliminates problems in microwave cakes. *Food Engineering* 61(9):54, 1989.

WEBSITES

View various cakes cookies at this website:

www.foodsubs.com/cakes.html and **www. foodsubs.com/ cookies.html**

An encyclopedia of baking with lots of troubleshooting tables:

www.bakingbusiness.com/ refbook.asp

Helpful baking tips and information:

www.joyofbaking.com/cakes .html

Pies and Pastries

Pastry is essentially a variety of bread that, in its many forms, is characteristically flaky, tender, crisp, and lightly browned. This delicate combination is not always easy to achieve. Unlike many other food preparations, pastries are made with precisely measured ingredients, in a time/temperature-sensitive manner, and with an artistic touch. These labor-intensive pie shells and desserts are the true test of a food preparer's skill.

Types of Pastry

All pastries fall into one of two basic types: plain pastry and puff pastry. **Plain pastry** is also known as pie pastry, and as the name implies, it is used for pie crusts. **Puff pastry** is primarily found in desserts such as Napoleons, turnovers, patty shells, tarts, and cream horns (Figure 26-1). Puff pastry also differs from the pie variety in that it can increase up to eight times its original size. This is accomplished by a series of folding, rolling, and turning manipulations that can create up to as many as 1,000 alternating layers of fat and dough (1). As might be imagined, the necessity of folding many layers together, even when there are considerably fewer than 1,000, makes puff pastry very time-consuming to produce.

KEY TERMS

Plain pastry Pastry made for producing pie crusts, quiches, and main-dish pies.

Puff pastry A delicate pastry that puffs up in size during baking due to numerous alternating layers of fat and flour.

FIGURE 26-1 An assortment of puff pastries.

Types of Puff Pastry

The vast number of different folding techniques, which may be used in preparing many different recipes, results in a wide assortment of puff pastry variations, including:

- Blitz or quick puff pastry
- Strudel
- French pastries
- Phyllo (fee-low)
- Danish pastries
- Pâte à choux (pot-a-shoe)

Blitz/Strudel/French Pastries. Although quicker and easier to prepare than regular puff pastry, blitz pastry does not rise as high. This pastry combines the mixing technique of plain pastry with the rolling and folding technique of puff pastry. Blitz pastry is usually used to prepare cream-filled pastries like Napoleons and tart shells. Cream-filled pastries are often referred to as French pastries, while the term "strudel" describes the Hungarian version of puff pastry.

Phyllo Pastry. Phyllo, which can also be spelled filo or fillo, is the Greek or Near Eastern version of puff pastry made of very thin sheets of dough. The term "phyllo" means "leaf" in Greek, which describes the characteristic paper-thin sheets of dough that can be rolled or folded into a variety of shapes. Baklava is a very sweet, heavy Greek dessert shaped into triangles consisting of multiple layers of phyllo dough that have been soaked in honey and topped with nuts.

Danish Pastry. Danish pastries are sweet rolls made with yeast—a very crisp pastry separated into fine layers. Danish pastries often have sweet fillings or toppings.

Pâte à Choux Pastry. Cream puffs and éclairs are the most common pastries made from pâte à choux dough, or rather choux paste as it is sometimes called, because it is more like a thick paste. Pâte à choux is French for "cabbage paste," referring to the resemblance of a cream puff to a cabbage head (4).

Preparation of Pastry

Pastries are the most delicate of all baked products, and their success depends on the right proportions of ingredients and the correct preparation technique. It takes skill to correctly distribute the fat and develop the gluten to the point of creating a crust that is flaky, tender, and crisp. Each type of pastry, especially the various types of puff pastries, has its own ingredient requirements and unique instructions for mixing, rolling, filling, and baking, but the general guidelines are discussed below.

Ingredients of Pastry

Most pastry flour mixtures usually contain at least four ingredients: flour, fat, liquid, and salt. Eggs and sugar, the latter sometimes added for its flavor and browning properties, are optional. Except for croissant, Danish, and brioche doughs, which make use of yeast, the leavening agents for most pastry doughs are steam and air. The type and quantity of each of these ingredients determine the final quality of the pastry product. Table 26-1 compares the ingredients of these pastry doughs against each other and against bread dough. Pie and pastry doughs contain more fat than other baked products. In fact, they have 50 percent as much fat as they do flour, by weight, and may have as much as an equal amount of fat and flour. Compare this to bread doughs, which usually contain only about 12 percent fat. Although the basic ingredients of pastries do not vary much, there is a wide assortment of pastries, partially because the flour mixtures can be handled so many different ways. Among all the doughs listed in Table 26-1, short dough is most commonly used in the preparation of pastries (4).

Flour. Pastry depends heavily on the type of flour used, its amount, and how it is handled. Pastry flour and the unbleached all-purpose variety are popular. There is less protein in pastry flour, and it is preferred by professional bakers, but the easy availability of all-purpose flour makes it more popular with the general public. Cake flour, which is even lower in protein than pastry flour, is sometimes used. If all-purpose flour is used, however, the gluten formation increases, which may result in a tougher pastry, so additional fat is sometimes necessary to increase tenderness.

Too much of any type of flour will toughen pastry. The flakiness of pastry is attributed in part to its limited gluten formation achieved by using a very small amount of chilled water, coating the flour with fat, and chilling the dough briefly before handling. Other techniques that can contribute to tenderness include adding acid in the form of unflavored vinegar, lemon juice, yogurt, or sour cream to break apart the long gluten strands; handling the dough minimally (the warmth of the hands can melt the fat); and using as little water as possible because it facilitates gluten formation (2, 9).

Fat. The proportion of fat is probably the most important determinant of quality in pastry, especially in creating flakiness. Large amounts of fat are required to produce a flaky crust (Figure 26-2), with the proportion varying depending on the type of pastry. The specific factors influencing flakiness and tenderness are discussed below.

Flakiness. Fat—the size of its particles, its firmness, and how evenly it is spread—is the major contributor to flakiness. When it is cut in small, cold

FIGURE 26-2 The role of fat in producing flaky pastry.

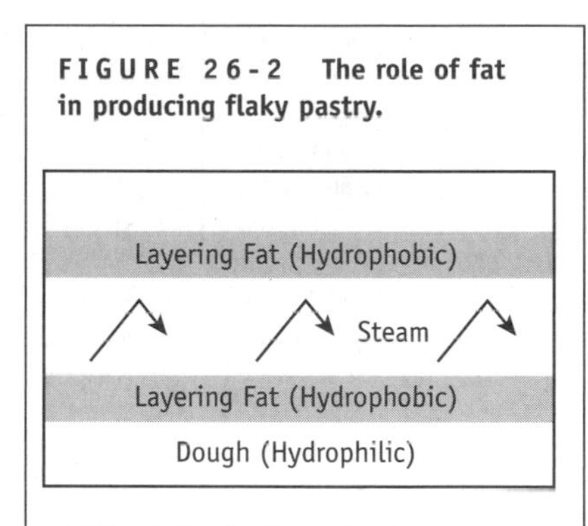

- Water in the dough turns to steam upon baking
- The layering fat creates an impervious layer
- The steam stays inside each dough layer, forcing it to expand because of the pressure it develops underneath each impervious fat layer
- Fat melts and is absorbed by the dough, improving the eating properties of the pastry

Source: Cereal Foods World

TABLE 26-1
Pastry Dough Formulas Compared to Bread Dough (Based on 1 Pound of Flour)

	Basic Bread Dough	*Brioche Dough*	*Croissant Dough*	*Danish Dough*	*Pâte à Choux*	*Pie Dough*	*Puff Pastry*	*Quick Puff Pastry*	*Short Dough*
Flour	1 lb (455 g)	1 lb (455 g)	1 lb (455 g)	1 lb (455 g)	1 lb (455 g)	1 lb (455 g)	1 lb (455 g)	1 lb (455 g)	1 lb (455 g)
Butter (Margarine in Danish)	2 oz (55 g)	5 oz (140 g)	11½ oz (325 g)	17 oz (485 g)	6½ oz (185 g)	10 oz (285 g)	1 lb (455 g)	1 lb (455 g)	14 oz (400 g)
Yeast	1 oz (30 g)	1 oz (30 g)	1¼ oz (37 g)	1½ oz (40 g)	0	0	0	0	0
Sugar	1 oz (30 g)	2½ oz (70 g)	7 tsp (35 g)	1½ oz (40 g)	0	0	0	0	6 oz (170 g)
Eggs	0	4	0	2	14	0	0	0	1
Water	0	0	0	0	3⅓ C (800 ml)	3 oz (90 ml)	6½ oz	3 oz (90 ml)	0
Milk	8 oz (240 ml)	1½ oz (45 ml)	9 oz (270 ml)	6½ oz (195 ml)	0	0	0	0	0
Salt	2¼ tsp (11.25 g)	2 tsp (10 g)	2½ tsp (12.5 g)	⅔ tsp (3 g)	1¼ tsp (6.25 g)	2½ tsp (12.5 g)	2½ tsp (12.5 g)	2½ tsp (12.5 g)	0
Other	0	0	few drops lemon juice	⅔ tsp cardamon	0	3 oz (85 g) lard	1½ tsp (7.5 ml) lemon juice	0	1 tsp (5 ml) vanilla extract
Mixing/ Production Method	Dissolve the yeast in warm milk. Form the dough by combining the flour, sugar, and salt. Mix in the butter. The dough is kneaded, proofed, formed, and possibly proofed again.	Dissolve the yeast in warm milk. Add the sugar, eggs, and salt. Add the flour. Add the butter. The dough is kneaded before being proofed in the refrigerator, shaped, and proofed again.	Form butter block by mixing together butter, some flour, and lemon juice. Form the dough by dissolving the yeast in milk and kneading in the remaining ingredients with the dough hook.	Form margarine block. Form the dough by dissolving the yeast in the cold milk and eggs and adding the remaining ingredients.	Boil a mixture of the water, butter, and salt. Stir in the flour and cook the roux for a few minutes. Cool slightly before adding the eggs.	Combine flour and salt. Add butter and lard into flour mixture by cutting them into marble-size pieces. Add water and mix just until the dough holds together.	Form butter block by mixing together most of the butter, 1/3 of the flour, and lemon juice into a block. Form dough by mixing the remaining ingredients.	Combine flour and salt. Cut butter into large pieces and add to flour mixture without thorough mixing. Add water just until dough holds together.	Mix for a few minutes at low speed: butter (soft), sugar, egg, and vanilla. Mix in flour, but until just incorporated.

particles and incorporated into the dough without being creamed or absorbed by the flour, fat melts during baking, leaving empty spaces where steam may collect to leaven and lift the layers of dough. These pockets vacated by fat create the characteristic flakiness of pastry (Figure 26-3). The more numerous the layers, the more flaky the pastry.

To maximize flakiness, cold fat is cut into chilled flour to form a fat/flour mixture. Keeping the fat cold increases the flakiness of pastry in two ways: (1) less fat is absorbed by the flour, and (2) more pea-size balls of fat are dispersed and surrounded with flour to become pockets of air during baking (Figure 26-4).

The type of fat used in making the pastry also affects flakiness. Firmer, plastic, 100 percent fats such as hydrogenated shortenings and lard produce the flakiest pastries, although lard is being used less often because of its tendency to go rancid and its high saturated fat content. Shortening is softer and more pliable than cold butter, so it coats the flour more easily and can be rolled out even when refrigerated. The fat is added to the flour before any liquid to protect the flour and its proteins from water, which would increase the formation of gluten (2). Butter and margarine may be used in pastry-making, but their water content causes increased gluten forma-

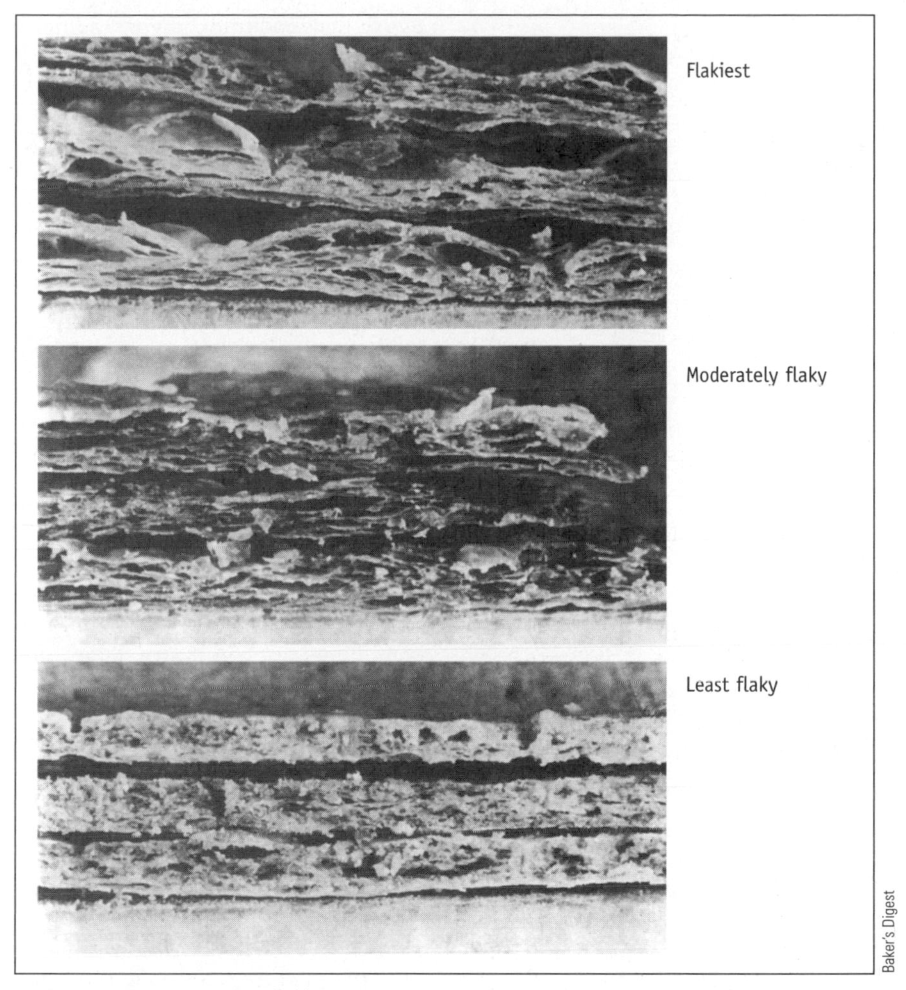

FIGURE 26-3 **Pastry flakiness.**

Tender pastry occurs when fat melts more into flour.

Flaky pastry is achieved with cold fat in pea-size balls.

FIGURE 26-4 **How fat is added influences pastry flakiness.**

tion. Butters vary in water content, with the best high-quality butters having the lowest water content, making them the number-one choice for the finest pastries. Because of butter's desirable flavor, it is often selected to prepare pastries; unlike dough containing margarine, however, when a butter-containing dough is refrigerated it tends to harden and requires more effort to roll out (4). The food industry has combined the positive aspects of both butter and hydrogenated shortenings by producing a butter-flavored hydrogenated shortening. Oil is the least desirable fat for making flaky pastry because it coats each flour particle, resulting in an extremely tender but mealy (grainy) texture (5).

Tenderness. Tenderness differs from flakiness, and in fact, factors that contribute to one may detract from the other (Table 26-2). Tenderness is described as the ease with which pastry gives way to the tooth. The major influence on tenderness is the concentration and distribution of gluten. As gluten development is inhibited by the fat coating the flour during baking, increased tenderness will occur. To achieve maximum tenderness in pastry, the fat must be thoroughly combined with the flour. For most pastries, including pie crusts, the ideal is a combination of tenderness and flakiness achieved by having some fat absorbed by the flour and leaving some fat in pea-sized pieces to melt and let off steam during baking. Tough, dry, or flat-flavored pastry may result from excess gluten formation caused by not cutting enough fat into the flour; by using more flour than necessary, especially during rolling; by adding too much water; or by excessive manipulation of the dough. It is also important to work quickly: the greater the amount of time allowed between adding water to the dough and baking the pastry, the less tender the pastry will be, because the gluten will have more opportunity to be hydrated by the water (7).

On the other hand, too little gluten development results in a pastry that is too tender and crumbly. This occurs when fat is cut into too small pieces; when oil, rather than shortening, is used as the fat; when too little water is added; or when the dough is undermanipulated. It may also happen when conditions are too warm during handling, causing fat to melt and coat the flour, inhibiting gluten development. When using butter, the goal is to work quickly and to put the bowl back in the freezer for about 5 minutes whenever the butter starts to soften (2).

Fat Increases Calories (kcals). With the delicious tenderness and flakiness provided by fat comes a high fat-gram and calorie (kcal) count. Pies, for example, average about 14 grams of fat and 300 calories (kcal) per serving (⅛ of a 9-inch pie). Among pies, some of the lowest and highest in terms of fat and calories (kcal) are strawberry pie (9 fat grams/228 calories (kcal)), and pecan pie (27 fat grams/503 calories (kcal)).

TABLE 26-2
Making a Tender, Flaky Pastry

To Make Tender Pastry	Why
Blend soft fat into the flour before adding any liquid.	Fat coats the proteins and prevents them from forming gluten.
Instead of water, use an ingredient that is part fat, like sour cream, cream, or egg yolks.	Gluten can't form without water, and the additional fat contributes to tenderness.
Add acid to the dough in the form of lemon juice, vinegar, or sour cream.	Acid breaks long gluten strands.
To Make Flaky Pastry	**Why**
Keep the fat cold and in large pieces (pea-sized).	Large, cold pieces will remain firm in the oven long enough to create flakes.
Flatten large pieces of cold fat.	Chunky pieces will make holes in the crust rather than act as spacers.

Source: Corriher, *Fine Cooking Magazine*

Liquid. The liquid component of pastry dough is important for leavening, hydration, and the crispiness of the crust. Pastry is leavened by steam, so liquid is necessary for that purpose. Liquid is also needed to hydrate the proteins so gluten can develop, and to dissolve the salt. The pastry structure is set primarily by the coagulation of flour proteins during baking. As little liquid as possible should be added when making pastry, because too much water will cause shrinkage and a tougher crust from excess gluten development (7). Too little water results in a crumbly crust.

More liquid may be needed at higher altitudes where water evaporates more quickly. Vinegar or lemon juice may be added to the cold liquid, on the principle that acid helps to inhibit gluten formation (3).

HOW & WHY?

Why is pastry so crisp? A crisp crust is created, in part, by water evaporation. In addition, the very little water used in pastry preparation inhibits starch gelatinization, which would make the pastry less crisp (8). Other factors influencing crispness are baking time and temperature; thickness of the rolled dough; whether the rolled dough is baked immediately or refrigerated or frozen first; and the moisture content of pie fillings.

Eggs. Eggs add color, flavor, and richness to pastry dough. Sometimes the egg yolk alone is used, because the water and protein content of the white contributes to toughness. Egg yolk can also be mixed with a minute amount of water and then brushed over the pastry to produce a golden crust.

Salt. In pastry, the only function of salt is to add flavor. It can be omitted, but the crust will lack flavor. Sometimes sugar is added for the same reason, and it has the additional benefit of browning the crust.

Mixing

Mixing Plain (Pie) Pastry. The classic pastry method is used to mix the ingredients of plain pastry. Flour and salt are first sifted together and then chilled. Cold fat is cut into the chilled flour and salt mixture using a pastry blender or forks, or by criss-crossing two knives together until the particles of the mixture are reduced to the size of peas. This can also be achieved in a food processor. Next, cold or ice water is sprinkled evenly, one tablespoon at a time, over the flour. After each addition, the mixture is tossed lightly with a fork or pastry blender until the flour is just moistened (Figure 26-5). Only enough water is used to moisten all the flour and make the particles hold together. Overmixing or adding water beyond the "just moistened" stage results in a tough pastry. When the

1. The flour and salt are mixed together with a pastry blender and the fat is cut into the size of tiny peas.

2. Water is sprinkled into the flour, a tablespoon at a time. A fork is used to mix it with the flour until all the flour is moistened.

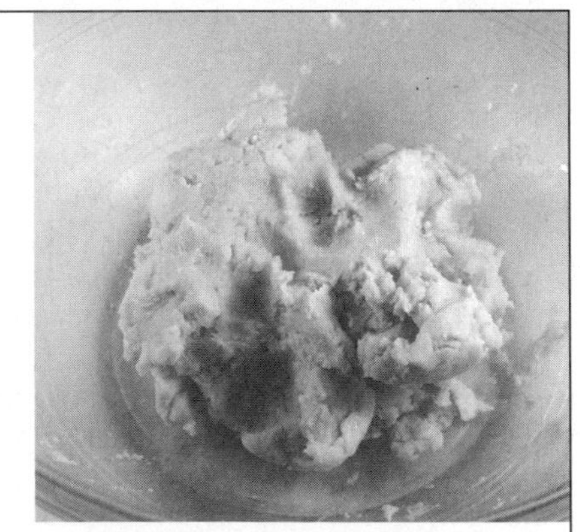

3. The dough is mixed thoroughly until the sides of the bowl are clean, indicating a correct amount of gluten development.

FIGURE 26-5 Mixing pastry ingredients.

dough no longer clings to the side of the bowl, it is pressed lightly into a flat disc, wrapped in waxed paper or plastic wrap, and refrigerated for about 15 minutes to chill the fat. It is then ready for rolling.

Mixing Puff Pastry. There are two separate mixings in the preparation of puff pastry. The first mixing is for the fat component, or butter block, consisting of fat, flour, salt, and perhaps an acid. These ingredients are mixed together by hand and shaped into a ¼-inch thick, 12-inch-square block. The second mixing is for the dough, made from flour, salt, water, and often a little fat. Puff pastry usually contains both cake and all-purpose flours because its dough is manipulated more than that of plain pastry, and the inclusion of cake flour lessens the likelihood that excessive gluten will form. The flours are sifted together and chunks of butter, if used, are cut into the flour with the fingertips until the mixture forms particles the size of coarse crumbs. An electric mixer might be used as an alternative to the hands. The mixture is formed into a mound with a well in the center into which the water and salt are poured. The flour or flour-fat combination is integrated into the water with both hands, and more water is added, if needed, to make a dough that is sticky but manageable. Both the butter block and the dough are refrigerated for 30 minutes. Refrigeration keeps the fat firm and prevents it from being absorbed by the flour. After refrigeration, the dough and the butter block are ready to be folded and rolled together.

Cream Puffs and Eclairs. A variant of this method is used for pâte à choux pastry dough. This flour mixture, used to prepare cream puffs and éclairs, may be referred to as a paste rather than a dough. It has the highest proportion of liquid of any of the pastry doughs. This contributes to the very light weight of these pastries. The large cell formation that occurs, leaving a hollow middle, is produced by steam leavening during baking.

The liquid, usually water, is combined with the fat and salt in a saucepan and heated to boiling (Figure 26-6). As the mixture boils, the flour is added (often all at once) and constantly stirred to keep lumps from forming. The paste is then cooked and stirred for about 3 minutes until it is sufficiently dried out to pull away from the sides of the pan. The paste is then transferred to a bowl, where it must be allowed to cool only slightly before eggs are added, one or two at a time. When the dough is glossy and soft enough, it is put in a pastry bag, which is used to extrude dough onto the baking sheet into small rounds for cream puffs or long, narrow shapes for éclairs. During baking, pâte à choux expands in such a way as to leave a hollow center. This will be filled with whipped cream for cream puffs or cream filling for éclairs. The top is covered with a chocolate syrup. Specific problems and their causes in preparing pâte à choux pastry are listed in Table 26-3.

1. Pâte à choux in progress

2. Ready to come off the heat

3. Add eggs one at a time

4. Mixing the dough

5. Pipe dough onto sheet pan

FIGURE 26-6 Preparing a pâte à choux.

Rolling

Both plain and puff pastry dough must be rolled, with minimum hand contact, in order to spread the fat and gluten in fine sheets layered on top of each other in the process called **lamination**. Over-manipulating the dough by rolling it too much, too hard, or too often will decrease the flakiness, tenderness, and crispness of the pastry.

KEY TERMS

Lamination The arrangement of alternating layers of fat and flour in rolled pastry dough. During baking, the fat melts and leaves empty spaces for steam to lift the layers of flour, resulting in a flaky pastry.

TABLE 26-3
Choux Pastry Problems and Their Causes

Problem	*Cause*
Volume of the pastries too small.	Insufficient roux formation; starch insufficiently gelatanized.
Volume of the pastries too small.	Batter too firm, not enough eggs added.
Contours of the pastries are diluted, they expand too much laterally.	Batter too soft, eggs added too quickly or too many eggs added.
Small volume, thick crust, dense cells.	Lack of steam during baking.

Chilling the Dough. The first step in rolling pastry dough is to let the dough chill in the refrigerator for a set amount of time (minutes, hours, or days) to make it easier to handle and to keep the fat from melting into the flour. Properly wrapped dough can be refrigerated up to four days or be frozen for up to six months (9). Chilling the dough in the refrigerator allows the flour more time to rehydrate and gives the gluten strands an opportunity to relax so that during baking they can expand at the same rate as the gases. Once it is taken out of the refrigerator, the cooled dough is allowed to sit at room temperature until it is malleable before it is rolled.

Rolling Surface. Only the amount of dough needed for one crust is rolled. A cold surface is best for rolling out the dough, which is why marble rolling boards are often used. Any other surface can be cooled by placing an ice water-filled roasting pan on it for a few minutes and then drying it well (9). The rolling surface is prepared by sprinkling it lightly with flour. Another option is to rub a minimal amount of flour into a pastry cloth covering a cutting board. The rolling pin can also be covered in cloth and/or lightly floured to prevent sticking, but the amount of added flour should be kept to a minimum.

KEY TERMS

Streusel topping A crunchy, flavorful topping that can be strewn over the top of pies; it is made by combining flour, butter or margarine, brown sugar, and possibly spices (cinnamon) and chopped nuts (pecans, walnuts, or almonds).

General Guidelines. Next, the dough is placed on the rolling surface and flattened slightly on top with a hand or the rolling pin to provide a starting point. Short strokes are made with the rolling pin from the center outward in a circle until the dough is flattened to ⅛ inch thick. Less pressure is used at the end of each roll to avoid excessive thinning of the edges. Lifting the edge of the dough intermittently and dusting the surface with flour as needed will help prevent sticking. It is best to roll as lightly and as little as possible and to avoid rolling repeatedly in one area to obtain the desired thickness. This will toughen and shrink the pastry by creating too much gluten.

Rolling the pastry too thin will make it too weak to hold fillings and may cause it to become too brown or to burn during baking. Pastry dough rolled too thick and used as a top crust may form a raised dome when baked.

Rolling Plain Pastry. For plain pastry, the dough is rolled out in a circle 1 to 2 inches larger than the bottom of the pan. The wider diameter allows sufficient dough to cover the sides of the pie pan. Once the dough is rolled, the simplest way to transfer it to the pan is to fold it in half and then in quarters to form a wedge shape (Figure 26-7). Rolling the dough on wax paper is an option that makes it easier to pick up, since only the edges of the wax paper need be lifted to accomplish the folding. The wedge is placed in the pie pan with the point in the center and the outside edge on the pan rim and then unfolded so it covers the bottom and sides of the pan. It should be large enough to cover the pan without stretching. Stretching dough to fill a pie pan may cause it to shrink back during baking. It is then pressed gently to eliminate any air bubbles, and the dough is squeezed together to patch up any tears or holes. Bottom crusts to be baked empty and filled later must be pricked with a fork so air can escape. Bottom crusts are not pricked if the filling is to be cooked in them, because pricking allows the filling to leak out.

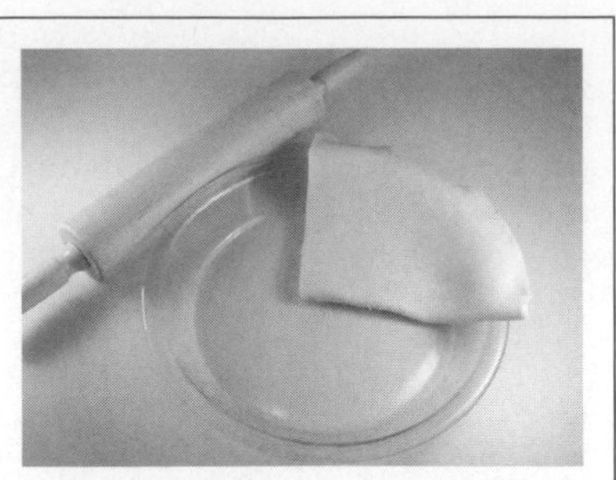

Method 1: Fold the rolled-out circular pie dough in half and in half again for easy transfer; unfold in pie plate.

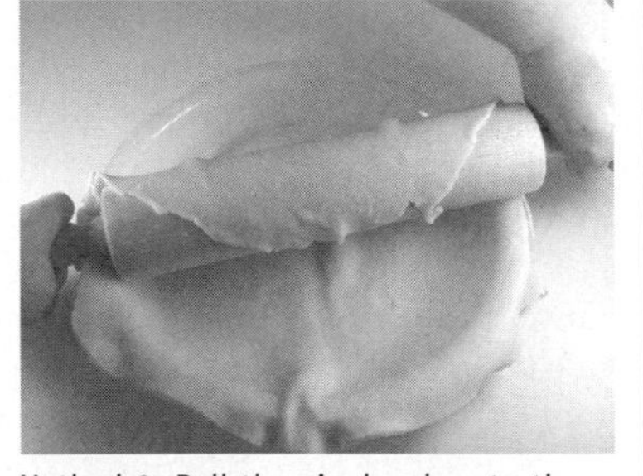

Method 2: Roll the pie dough onto the rolling pin, place on edge of pie plate, and gently unroll.

FIGURE 26-7 Transferring a rolled pie crust to a pie pan.

The top crust is placed over the filling using the wedge procedure, or by lifting the wax paper and dough together, gently turning it over onto the filling, and slowly peeling off the wax paper. Top crusts are usually pricked with a fork, slashed with a thin knife, or opened with picturesque designs to create vents that allow the steam to escape during baking (Figure 26-8). This prevents the contents from boiling over. A decorative open latticework crust can also be made by cutting strips of rolled dough and arranging them on top of the filling. Another option, especially with one-crust apple pies, is to sprinkle the filling with a **streusel topping**. Adding

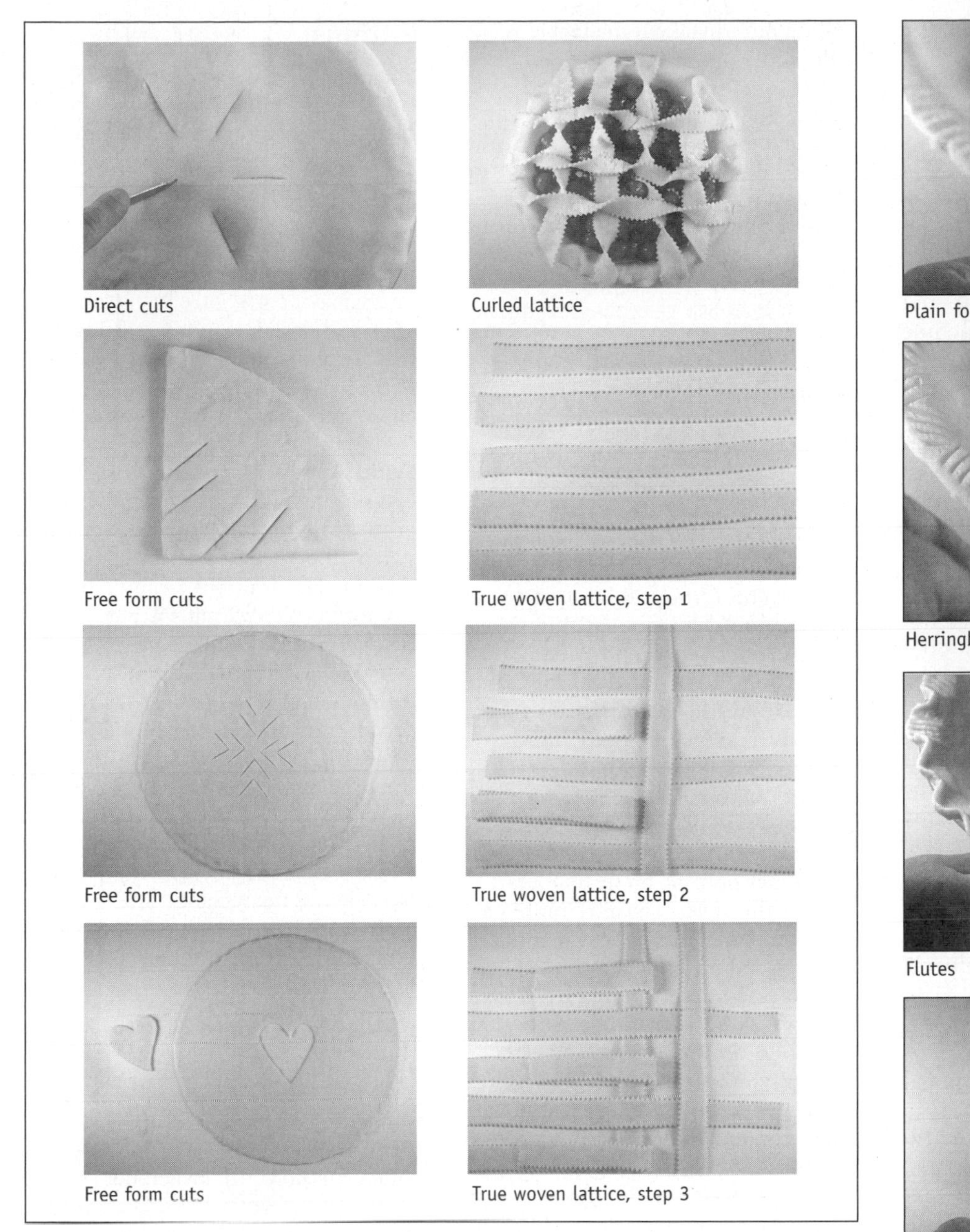

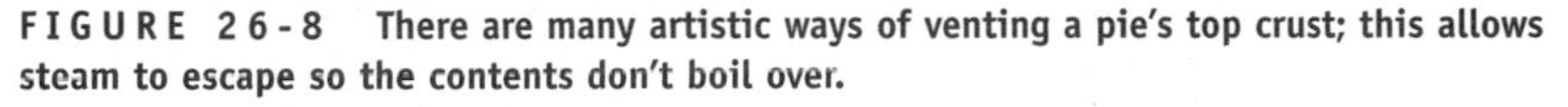
FIGURE 26-8 There are many artistic ways of venting a pie's top crust; this allows steam to escape so the contents don't boil over.

a crimped edge gives pies a finished look and helps to seal top and bottom crusts together in a two-crust pie. The crust edges can be pinched together with fingers, a utensil handle, or the prongs of a fork to make various designs (Figure 26-9).

HOW & WHY? ?????????

How do you avoid a soggy bottom crust? To avoid a soggy bottom crust, it is important to fully thicken the filling before adding it into the pie shell. Another method is to prebake the bottom crust on the lower shelf of an oven set at 450°F (232°C) for at least 3 minutes, and then move it to the center shelf where it is baked at 350°F (175°C) for the remaining three-quarters of an hour or until the filling is ready. The baked pastry is then chilled before lightly coating the bottom crust with cold fat and adding the filling. Another method to prevent a soggy crust is to prepare a

Plain fork

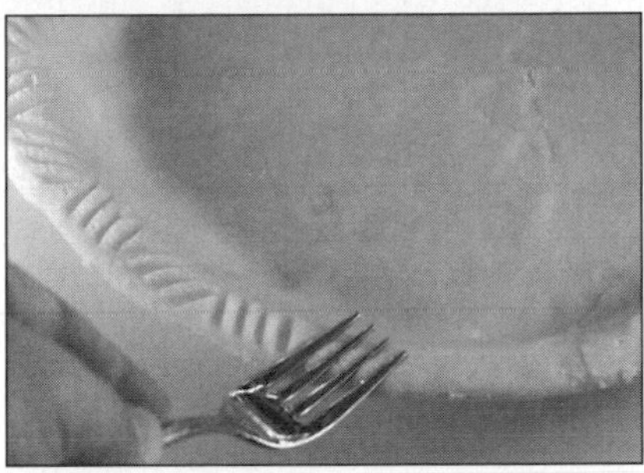
Herringbone

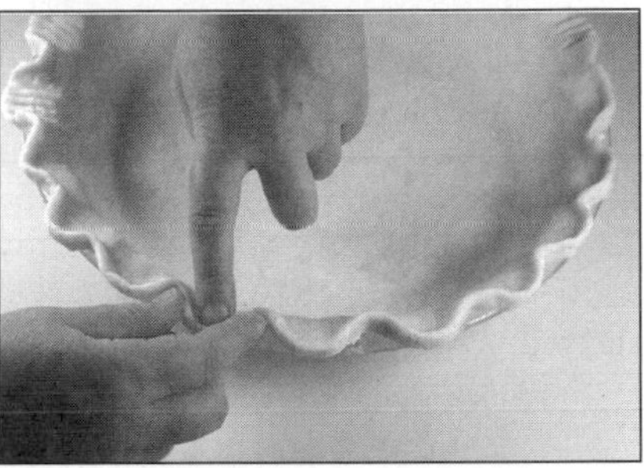
Flutes

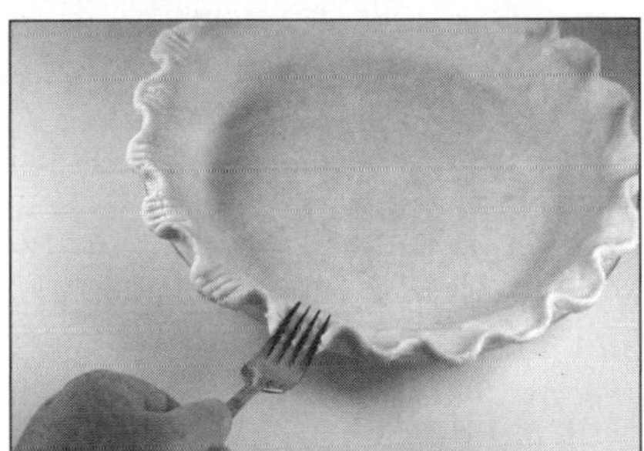
Scallops

Rope

FIGURE 26-9 Decorative pie crust edges.

mealy crust, which is more resistant to fluid absorption than a flaky crust. Using vegetable oil, completely melting the fat, or cutting up the fat to resemble cornmeal results in a mealy pastry.

Alternative Pie Crusts. Flour is not the only ingredient that can be used to create pie crusts. Other options, not strictly considered pastry, include crushed graham crackers, cereal flakes, granola, and cookies. All can be made into a cohesive mass by mixing them with a little fat and sugar. This type of crumb crust is then shaped by pressing it against the sides of the pie pan. It is then baked or left unbaked, depending on what the individual recipe indicates.

Rolling Puff Pastry. Puff pastry, which is higher in volume and fat than other pastries, relies on first folding the fat block and dough together before rolling. Repeated folding creates numerous layers of alternating fat and dough. If the fat does not stay as a separate layer, the pastry will have fewer layers and a gummy texture (4). These numerous folds contribute to the puffing up that occurs when the steam generated during baking forces the layers apart.

Laminating Puff Pastry Dough. To begin, the chilled dough is rolled into a ¼-inch-thick rectangle. The chilled fat, which is in a block two-thirds the size of the dough rectangle, is placed on top of the dough. Then the dough and the butter are folded together (Figure 26-10). The package of dough is then turned 90 degrees, so the fold is running vertically, and rolled out to its original size. The fat must be at the right temperature; if it is too warm it will flow into the dough, and if it is too cold it will damage the dough (10). If butter is used, it can be partially softened by pounding it with a rolling pin to make it malleable while still cool (6). The folded, layered dough is then chilled, folded, and rolled again. The process is repeated two or more times before it is ready to be cut. Figure 26-11 shows the results of insufficient or excessive folding of puff pastry dough. Cutting must be done with a sharp knife or the layers will press together, drastically decreasing volume during baking.

The several extra layers of fat in puff pastry dough make its resulting baked products extremely tender, flaky, delicious, and, of course, rich and high in calories (kcal).

Frozen Rolled Puff Pastry. Puff pastry that has already been rolled is available in the frozen-food section at supermarkets. It is sold as thin sheets of phyllo dough, approximately 12 inches wide and 16 to 20 inches long, that are rolled and packaged into long, thin cardboard boxes. Defrosting the phyllo dough slowly results in the best pastry texture. The dough should be allowed to defrost in the refrigerator for about 12 hours and for an additional hour at room temperature so that the sheets separate more easily. The phyllo dough, which contains no eggs and is extremely thin, should be protected from drying out by covering it with a layer of plastic wrap and then a clean, moist towel (11). The phyllo dough can be brushed with melted butter to keep it supple, then cut and rolled or shaped into the desired pastry (Figure 26-12). Brushing each phyllo dough sheet with a thin layer of fat keeps it supple; too much, however, will weaken the dough, and too little will result in a thick, heavy pastry (11). Tearing of the delicate sheets is not uncommon during handling; this can be remedied by covering the rip with another sheet of phyllo dough. Phyllo dough that is ripping excessively may have already been accidentally defrosted at the supermarket, or have dried out with age. Neither problem can be deciphered by looking at the frozen phyllo dough, so it is best to purchase such dough at a store with a high turnover rate, such as a market specializing in Middle Eastern foods.

KEY TERMS

Mealy A pastry with a grainy or less flaky texture, created by coating all of the flour with fat.

Fillings

The list of possible pie and pastry fillings runs the gamut from fruits to nuts. Main-dish pie options include quiches with egg, vegetable, and meat fillings. Then there are desserts such as fruit pies, cream pies, ice cream pies, chiffon pies, custard pies, and meringue pies. There are one-crust versions of pies known as tarts (Figure 26-13 on page 466), and smaller versions, called tartlets, that are produced as individual servings. It is obvious that there is a pie to fit almost any taste at almost any meal.

Fruit Fillings. Successful fruit pie fillings depend on the proper combination of fruit, fruit juice, sweetener, and starch thickener. Gelatin is sometimes used to glaze fruit pies or tarts that are not to be baked. Fruit for pies can be fresh, frozen, cooked, canned, or even dehydrated if it is soaked and simmered before being used. Sugar is usually added, but too much will draw water out of the fruit, causing it to become shriveled and tough. The usual thickening agent is cornstarch or tapioca, but modified starches such as waxy maize create a clear gel and are best for fruit pies. Flour is not recommended because it has a tendency to cloud the filling.

There are three methods of preparing fruit pies:

- The old-fashioned method
- The cooked-juice method
- The cooked-fruit method

Old-Fashioned Method. In the old-fashioned method, the sweetener, spices, and starch thickener are mixed together and then combined with the fruit. This filling is then placed in the unbaked pie shell. Thin slices of butter are sometimes strewn across the top of the filling before it is covered with a top pie crust and baked.

Cooked-Juice Method. The cooked-juice method allows the uncooked fruit to retain more of its shape, flavor, and texture. Most frozen or canned fruit pies and berry pies can be prepared by this method. The juice is drained from the fruit and water is added if needed. The liquid is brought to a boil, and a mixture of starch dissolved in cold water is stirred into the boiling liquid, followed

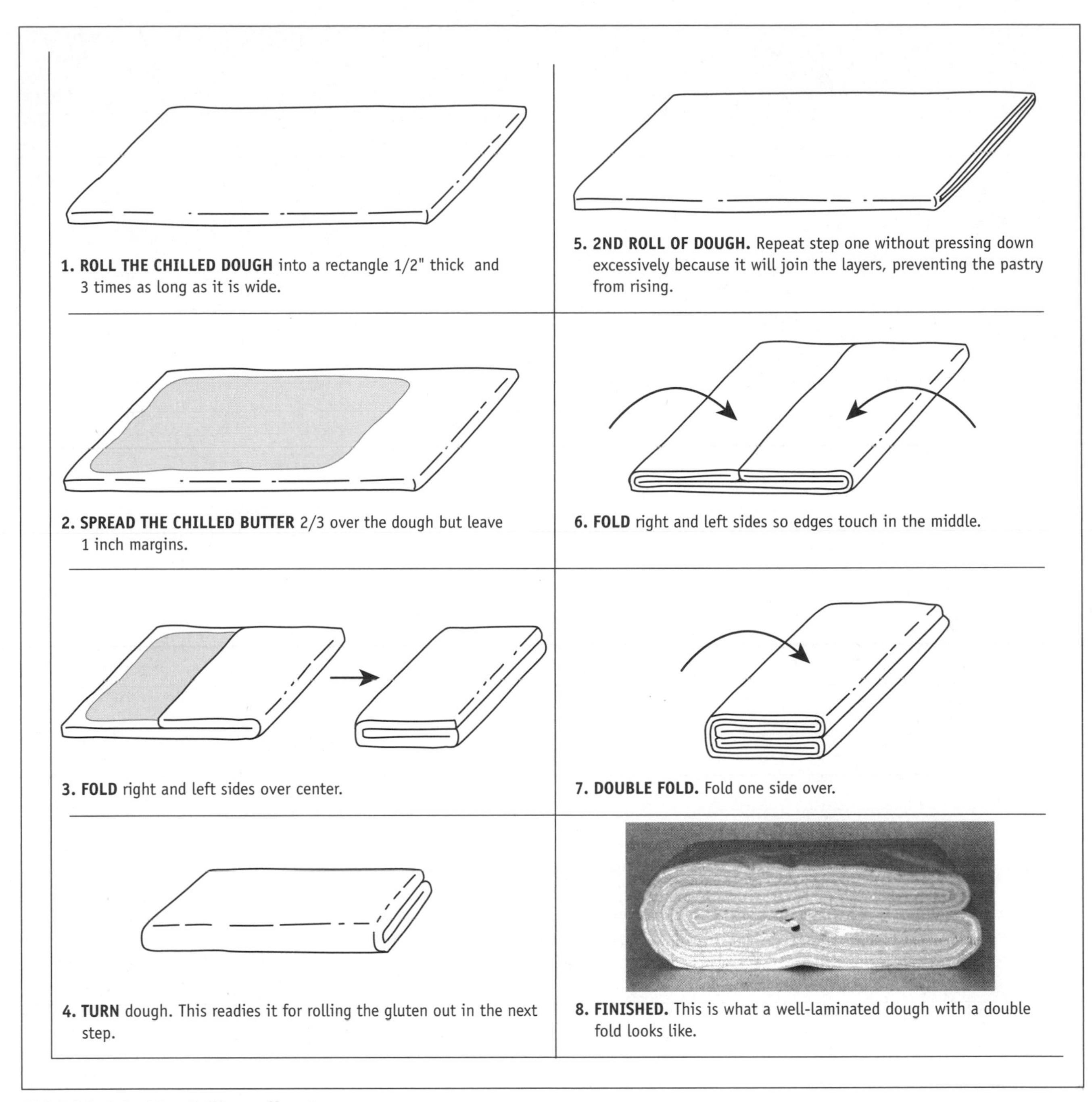

FIGURE 26-10 **Rolling puff pastry.**

by the sweetener and spices. When the mixture is thick and clear, it is cooled slightly and poured over fruit that has been previously arranged in a prebaked crust. The pie is then refrigerated.

Cooked-Fruit Method. If not enough juice is available from the selected fruit, the other option is the cooked-fruit method. This method is the same as the cooked-juice method except

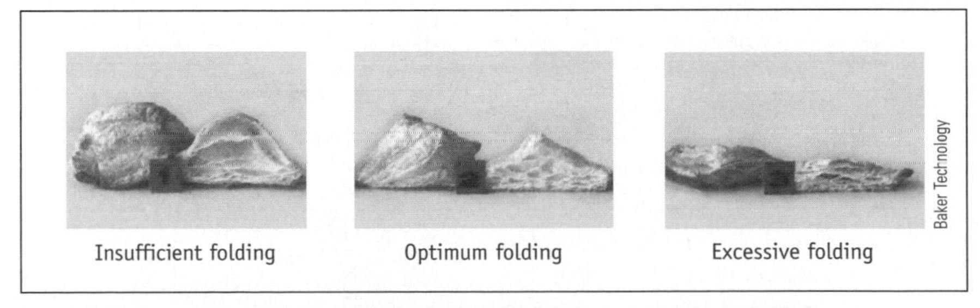

FIGURE 26-11 **Puff pastry that was insufficiently or excessively folded.**

1. Cover phyllo dough so it won't dry out. Work with one sheet at a time; cover the rest with plastic and a moistened kitchen towel.

2. To keep the dough supple, brush the dough with melted butter. Begin by buttering the edges, which dry out faster, and then work toward the center.

3. Use a sharp knife or scissors to cut the dough to size. Buttering and layering the sheets before cutting is faster and keeps the phyllo more pliable.

4. Repair tears by pasting on a piece of phyllo from an extra sheet with melted butter.

5. Versatile phyllo dough can be made into many shapes and will hold sweet or savory fillings. Shown here: folded triangle and a molded cup.

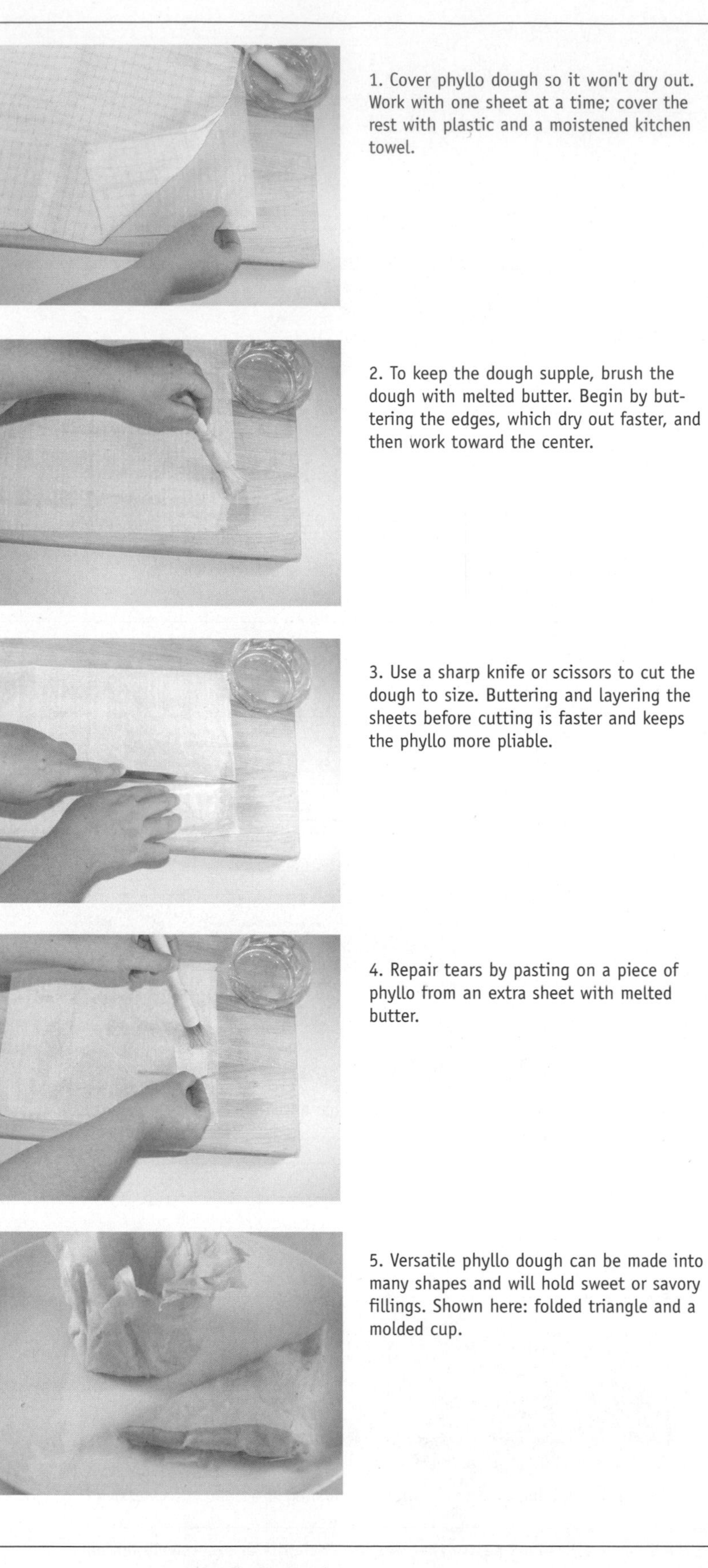

FIGURE 26-12 Handling phyllo dough.

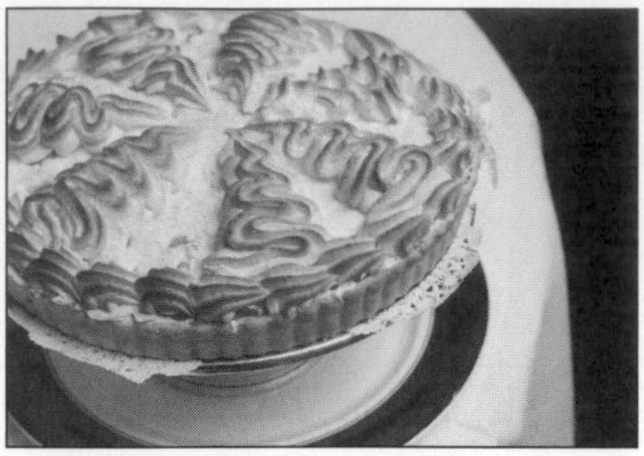

PhotoDisc

FIGURE 26-13 Tarts are one-crust versions of pies.

the fruit is added to the thickened juice and the mixture brought back to a boil. The results are poured directly into a prebaked pie shell. No further baking is required unless it is a two-crust pie.

Cream Fillings. Cream pie fillings are made by heating a mixture of milk, sugar, flavoring, cornstarch, and often egg until the starch gelatinizes, then pouring this mixture into a one-crust prebaked pie shell. (The principles of starch gelatinization are covered in Chapter 20.) Whole, low-fat, or nonfat milk, light or heavy cream, or evaporated milk may be used. Cream pies such as chocolate, banana, or coconut are often topped with a whipped topping or meringue.

Custard Fillings. Custard pies are one-crust pies filled with a milk and egg filling. Pumpkin and pecan pies are examples of custard pies. Cheesecake, in spite of its name, is really a custard pie, because it relies on the coagulation of eggs for structure. The mixture of custard pies is thickened by the eggs; an average custard pie contains two eggs per cup of milk. During baking, the egg proteins coagulate and set the filling. Since the egg-based custard filling coagulates much more quickly than the crust can bake, the two are often baked separately, with the cooled filling being placed in the cooled shell right before serving. Another method is to start the custard pie on the bottom rack of a hot 425° to 450°F (220° to 30°C) oven for the first 10 minutes. Then, after the crust sets, to ensure that it will be crisp, the oven temperature is lowered to 325° to 350°F (165° to 175°C) where it remains for the rest of the time, usually about 25 to 40 minutes.

Chiffon Pies. The use of gelatin in their preparation distinguishes chiffon pies from other types. The foam structure of chiffon pies imparts a light, fluffy texture. Egg whites or whipped cream are often an ingredient in these delicate pies. Only the shell is baked, so nondairy whipped toppings or rehydrated dried egg-white powder whipped into a foam can serve as substitutes for raw whipped egg whites to eliminate the risk of *Salmonella* contamination sometimes posed by raw eggs.

Meringue Pies. Meringue pies are simply cream or chiffon pies covered with meringue (see Chapter 13). The meringue should be spread to cover all parts of the crust edge so it will not shrink back during baking. The meringue is swirled on top and quickly browned in the oven. For serving, an easy way to cut through the meringue topping is to butter or oil the knife blade.

Pastry Fillings. There are a number of possible different fillings for pastries, including custards, creams, Bavarians, mousses, jams, fresh or cooked fruits, chocolate, and even cheese, cooked poultry, fish, or ham. The preparations of these fillings are discussed in the respective chapters devoted to their food type. Cones and pastry bags are frequently used to fill pastries; they are usually made from canvas, paper, or some other material (Figure 26-14).

Baking

Whether using a conventional oven, convection oven, or microwave oven, the type of pan, the temperature, the time, and the method of testing for doneness are important to the finished pastry product.

Pans. Either Pyrex glass pans or pans with dull finishes are best for pies because they help absorb the heat. Shiny metal pans, which deflect heat, and thick metal pans, which take too long to heat, are not recommended (9). Pie crusts are usually baked in 8- or 9-inch pans with sloping sides and an edge designed to catch escaping juices. The size of the pan determines the number of pie slices.

8-inch pan = 4 slices of pie
9-inch pan = 6 slices of pie
10-inch pan = 8 slices of pie

Puff pastries are baked on cookie sheets or specially prepared pans. Pies and puff pastries should be baked in the center of the oven, making sure the pan or cookie sheet does not touch the sides. Custard-based pies, however, should be placed on the bottom rack so the crust will bake quickly and not have time to get soggy. An extra crispy, flaky bottom crust can be obtained by placing a cookie sheet on the rack prior to heating the oven. The direct contact of the bottom of the pie pan and the cookie sheet delivers more direct heat.

Temperature/Timing. A preheated hot oven at approximately 425° to 450°F (218° to 232°C) is best for baking pies and puff pastries. Slower baking would contribute to shrinking. The high heat helps to generate steam, melt the fat, and set the gluten. Adjustments

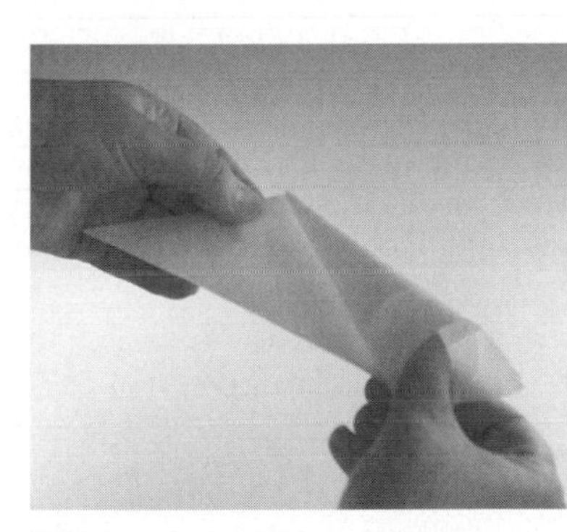

Making a pivot point

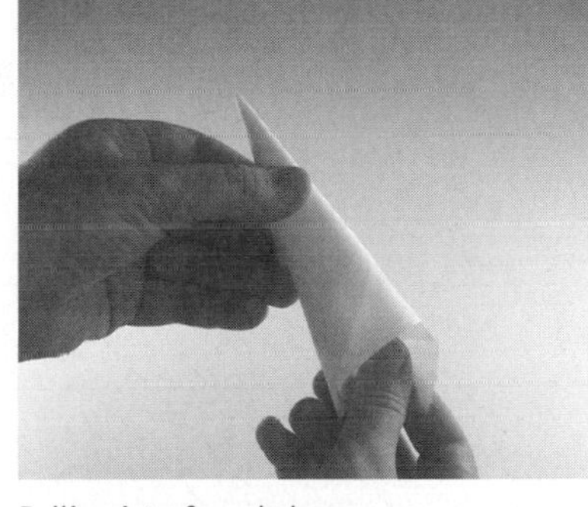

Rolling into funnel shape

Sealing the filled cone

Creating small opening

Filling pastry bag

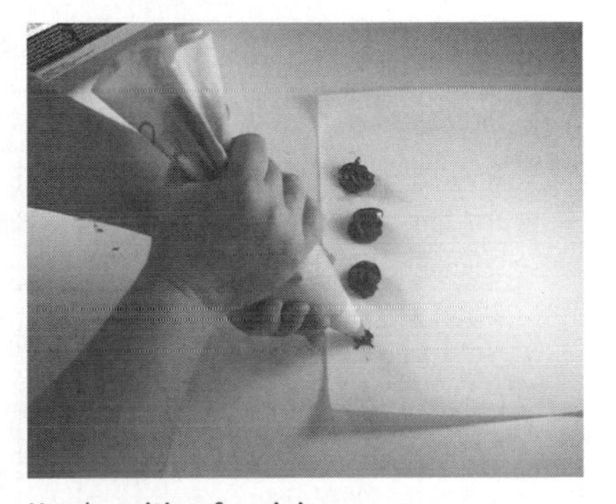

Hand position for piping

FIGURE 26-14 **Parchment cones and pastry bags.**

KEY TERMS

Blind bake To bake an unfilled pie crust.

for temperature are sometimes made, depending on the type of filling.

Pie shells **blind baked** tend to bubble, lose their shape, and brown unevenly. These faults can be prevented by pricking the pie dough with a fork and/or weighting it down with beans, which are removed a few minutes before the baking is completed. Aluminum foil placed around the edge of the pie crust can prevent it from overbrowning or burning (Figure 26-15).

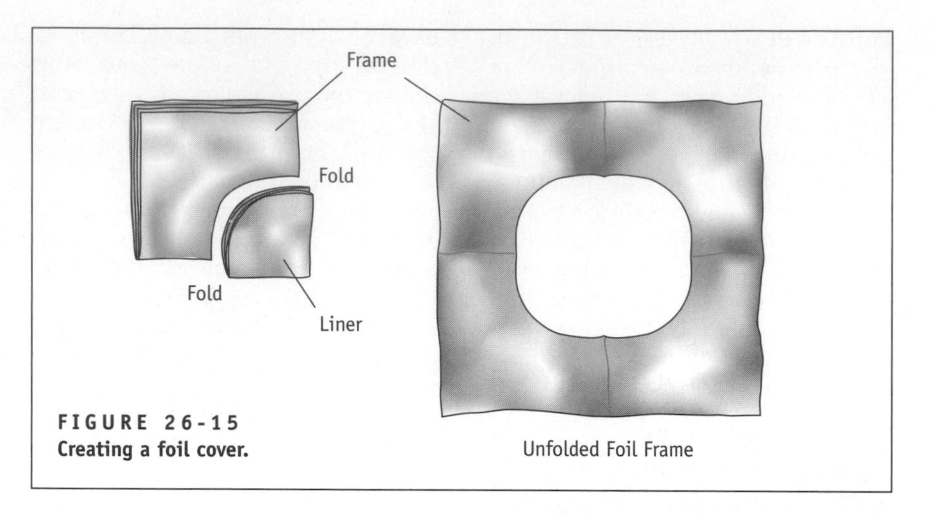

FIGURE 26-15
Creating a foil cover.

Fillings should not be added until immediately before baking. It is important not to prick the bottom crust of pies to be baked already filled, because the juices will run through. The bottom crust can be moisture-proofed by several methods: brushing beaten egg glaze or fruit preserves on the pie dough and then placing it in the oven for a short prebaking; placing a very thin layer of melted butter or flour on the bottom crust; or making sure the filling is very hot as it is poured into the crust.

Several decorative touches can be added just before baking. One of these is accomplished by brushing the top of a two-crust pie lightly with water and sprinkling it lightly but evenly with granulated sugar or coarse decorating sugar. For a shiny or glazed look, the top can be brushed with milk and beaten egg yolk combined with an equal amount of water. Be aware, however, that although glazes make pastry crusts appealing, they also make them tough.

Testing for Doneness. A lightly, delicately browned crust signals that baking is complete. Adding a thin layer of butter on the crust helps it to brown. Once done, all pies should be cooled in the pan on a wire rack to prevent moisture condensation, which results in a soggy lower crust. Table 26-4 lists several reasons why the baked pastry's appearance, tenderness, texture, or flavor may not be up to par.

Storage of Pastry

Pastries are best consumed while fresh, but most keep longer when refrigerated, and some can even be frozen. Pastry doughs freeze up to six months. Unbaked pies will last about four months in the freezer, while baked berry pies can be frozen for six to eight months.

Not all pies can be frozen, especially if they contain egg and milk products, which may separate out. Custard and cream pies are frozen commercially, but freezing is not recommended for home recipes. As with all foods containing milk and/or eggs, custard, cream, and meringue pies should be kept refrigerated to avoid the growth of bacteria and resulting foodborne illness.

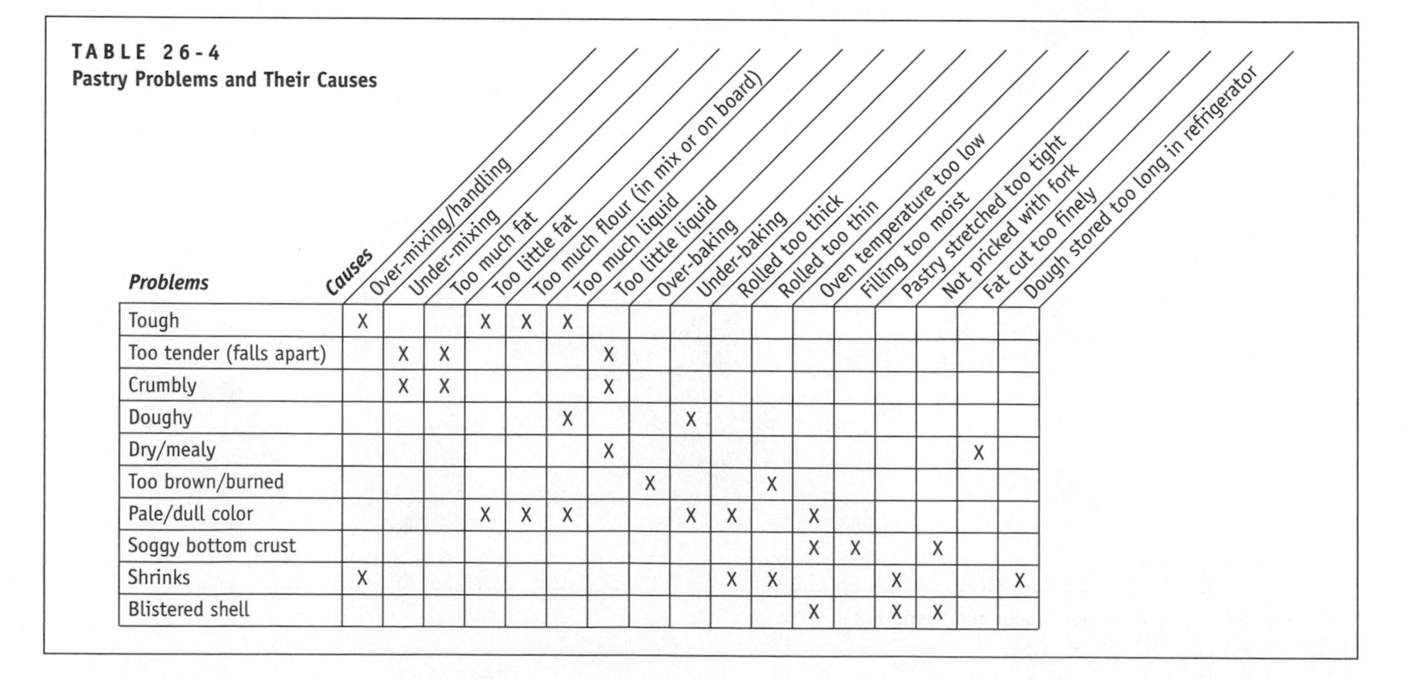

TABLE 26-4
Pastry Problems and Their Causes

Problems / Causes	Over-mixing/handling	Under-mixing	Too much fat	Too little fat	Too much flour (in mix or on board)	Too much liquid	Too little liquid	Over-baking	Under-baking	Rolled too thick	Rolled too thin	Oven temperature too low	Filling too moist	Pastry stretched too tight	Not pricked with fork	Fat cut too finely	Dough stored too long in refrigerator
Tough	X			X	X	X											
Too tender (falls apart)		X	X				X										
Crumbly		X	X				X										
Doughy						X			X								
Dry/mealy							X									X	
Too brown/burned								X			X						
Pale/dull color				X	X	X			X	X		X					
Soggy bottom crust												X	X		X		
Shrinks	X									X	X			X			X
Blistered shell												X		X	X		

PICTORIAL SUMMARY / 26: Pies and Pastries

Pastry is essentially a bread that is flaky, tender, crisp, and lightly browned. But this delicate combination is not always easy to achieve, and pie crusts and pastries are the true test of a food preparer's skill.

TYPES OF PASTRIES

The two basic types of pastry are plain and puff.

Plain (or pie) pastry

- Pie crusts
- Tarts
- Quiches
- Main-dish pies

Puff pastry

- Blitz or quick puff pastry
- Phyllo
- Strudel
- French (cream-filled)
- Danish (sweet rolls)
- Pâte à choux (cream puffs, eclairs)

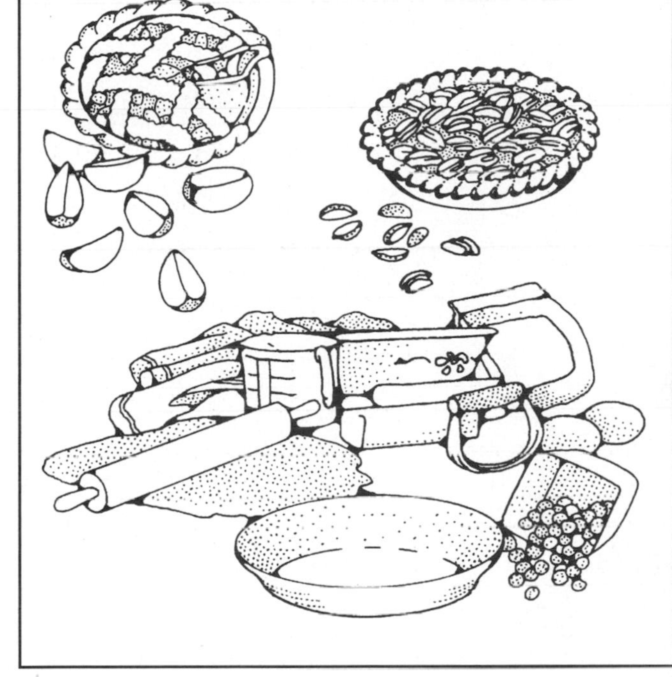

PREPARATION OF PASTRY

The main ingredients in pastry are flour, fat, liquid, and salt (egg is optional). Either pastry or all-purpose flour can be used. Fat is probably the most important ingredient, because it contributes the most to a pastry's flakiness and tenderness as well as its taste. It also, however, contributes to making most pastries high in fat and calories (kcal). Liquid is needed to develop gluten and leaven the pastry with steam, and salt is needed for flavor.

Plain and puff pastries are the most delicate of all baked products, and their making requires skill to correctly distribute the fat and develop gluten to the point of creating a crust that is flaky, tender, and crisp. Mixing, rolling, type of filling (fruit, cream, custard, chiffon, or quiche ingredients), decorations (sugary, shiny, glazed, or frosted surface), and baking all contribute to the success of the finished product.

STORAGE OF PASTRY

When held for later consumption, pastry and pastry products will keep longer when refrigerated. Pastries vary in how successfully they may be frozen. Custard, cream, and meringue-topped pies should not be frozen, but need to be refrigerated to avoid bacterial growth and foodborne illness. Frozen pastry crusts should go as quickly as possible from the freezer into a preheated oven.

REFERENCES

1. Amendola J. *The Baker's Manual.* Van Nostrand Reinhold, 1997.
2. Corriher S. The secrets of tender, flaky pie crust. *Fine Cooking* 17:78–79, 1996.
3. Dodge AJ. Baking classic American pies. *Fine Cooking* 29:73–77, 1998.
4. Friberg B. *The Professional Pastry Chef.* Van Nostrand Reinhold, 1996.
5. Kazier H, and B Dyer. Reduced-fat pastry margarine for laminated dough in puff, Danish, and croissant applications. *Cereal Foods World* 40(5):363–365, 1995.
6. Malgieris N. *Nick Malgieri's Perfect Pastry.* Macmillan, 1989.
7. McWilliams M. *Foods: Experimental Perspectives.* Macmillan, 1997.
8. Penfield MP, and AM Campbell. *Experimental Food Science.* Academic Press, 1990.
9. Purdy SG. *As Easy As Pie.* Macmillan, 1990.
10. Schunemann C, and G Treu. *Baking: The Art and Science. A Practical Handbook for the Baking Industry.* Baker Tech, 1988.
11. Stevens M. Baking flaky pastries with phyllo dough. *Fine Cooking* 20:18–19, 1997.

WEBSITES

Lots of tips and information from *Fine Cooking* magazine:

http://www.taunton.com/finecooking/pages/fc_feat_desspastry.asp

A resource website for information on baking pies:

http://www.bettycrocker.com/allabout/baking/aab_pie.asp

Information on baking tarts:

http://www.joyofbaking.com/tarts.html

27

Candy

In many minds, love and sweets have a close association that has woven itself into everyday language through such terms of endearment as "sweetheart," "honey," and "sugar." There are few who disapprove of love, or dislike sweets such as candy. This is nowhere more true than in the United States, which produces more candy than any other country in the world. Despite the high production of confections in the United States, people from Switzerland and Britain actually consume the most candy, with an average intake of almost 30 pounds of confectioneries per person each year (5). Americans are not far behind, however, with a new confectionery-eating high of 24 pounds per person (26). A little over half of this is in the form of chocolate, with the rest coming from all other types of sugar confectioneries (9).

In pursuit of the gratification of the sweet tooth, candy-making has developed into a many-faceted art. No one really knows how candies first got their start, but evidence points to the ancient Arabs, who used refined sugar to make medicinal lozenges. One of the factors supporting the role of Arabs as the first makers of confections is that the term "candy" is derived from *qand*, the Arabic word for sugar (2). Sugar (specifically sucrose), along with its close relative, corn syrup, are the foundational ingredients of almost all candies and essential to the process of confectionery production. Sugar is mixed with liquid and other ingredients into a solution that is then heated to concentrate its sweet taste. The type and concentration of ingredients, the temperature to which the mixture is heated, the degree to which it is then cooled, and any stirring or manipulation that follows determine the type of candy produced. This chapter discusses the classification of candies and their basic preparation methods.

Classification of Candies

There are thousands of different candies, but they can be classified according to their ingredients and/or preparation method (Table 27-1). One way of categorizing candies according to ingredients is to divide them into candies termed syrup-phase(or sugar-phase) and fat-phase. Most candies are syrup-phase, meaning they are made from a **simple syrup** mixture. Examples of some syrup-phase candies are hard candy, fondants, marshmallows, nougats, jelly beans, gums, caramels, and fudges. These candies are basically sugar with added flavorings. When chocolate or nut pastes such as peanut butter are used, the candy is considered fat-phase. A combination of both fat- and syrup-phases is found in candies such as chocolate-covered candy bars.

Another way to classify confectioneries is based on the method of preparation, which determines whether the candy will be crystalline or noncrystalline in nature. **Crystalline** candies, which are soft, smooth, and creamy, include fudge, fondant, and divinity. **Noncrystalline**, or amorphous (without form), candies include caramel, toffee, taffy, hard candy, and gummy candies like jelly beans, gummy bears, fruit slices, and gum drops. The difference in texture between the two types depends on how the candy's ingredients are combined and/or manipulated.

KEY TERMS

Simple syrup A basic mixture of boiled sugar and water.

Crystalline candy Candies formed from sugar solutions yielding many fine, small crystals.

Noncrystalline candy Candies formed from sugar solutions that did not crystallize.

TABLE 27-1
Candy Classifications

Class	*Comments*
Candied or crystallized fruits	Certain fruits (cherries, citrus peels, plums, etc.) are prepared and dropped into a syrup at 234°F (112°C). They are simmered until clear, spread on a screen, and dried until no longer sticky.
Caramels	Caramels contain sugar, corn syrup, milk, cream, and butter; and are cooked to stiff-ball stage of 246°F (119°C). After pouring into a buttered pan and cooling, they are turned out and cut into squares.
Chewing gums	Although chewing gum is not eaten, it is chewed; and, because of the sugar content in it, can be classed as a candy.
Chocolates	Chocolate liquor is the fundamental ingredient of most chocolate products. Other ingredients can include cocoa butter, sugar, milk products, and flavor.
Fondants	For fondants, the syrup is cooked to the soft-ball stage (238°F, or 114°C), poured into large platters, cooled until lukewarm, then stirred and kneaded until smooth.
Fudges	Fudge is made by gently boiling sugar and corn syrup with milk to 238°F (114°C), adding butter, cooling, then beating until it holds its shape. Spread in buttered pan and, when hard, cut into squares.
Glacé fruits and nuts	Similar to candied fruits, but the fruit is not cooked. The fruit or nut is dipped into the syrup, which has been cooked to 300°F (149°C) and dried on a rack.
Hard candies	Hard candies are the simplest form of candy—they are made mainly from sugar and syrup and are usually boiled to 300°F (149°C). They come in all shapes, sizes, colors, and flavors.
Jellies	The main ingredients for jellies (jelly beans, gum drops, etc.) are sugar, corn syrup, and a jellying agent such as gelatin, natural gums, pectin, or starch.
Licorice	Licorice sticks are made with flour, molasses, sugar, and corn syrup, and flavored with licorice extract.
Marshmallows	Marshmallows are made by whipping a combination of sugar, corn syrup, gelatin, and/or egg whites. This makes a light fluffy-textured candy that can be served plain or as a filling for Easter eggs.
Marzipan	Egg whites are beaten and mixed with almond paste plus sugar. It is not cooked. After standing 24 hours, it can be shaped, stuffed, or dipped.
Nougats	The syrup is cooked to 246–250°F (119–121°C) and then poured over the well-beaten egg whites, or gelatin, or both. Fat is added and it is beaten until cool.
Peanut brittle	The syrup is cooked to 300°F (149°C). It is then poured over the nuts spread out on a pan, allowed to cool, and then broken into pieces.
Popcorn balls	The sugar, corn syrup, and water is cooked to the medium-crack stage (280°F, or 138°C), flavoring added, then poured over the popcorn, and stirred. Then, with buttered hands, the mixture is shaped into balls.
Spun sugar	The syrup is boiled to 310°F (154°C). Then the pan is put into cold water to stop cooking, and back into a warm pan before it crystallizes. Using a wooden spoon, the threads of syrup are wrapped across the greased handles of two wooden spoons that have been anchored into a drawer.
Taffy	Similar to a caramel mixture except more concentrated and pulled to incorporate air.
Toffee	Toffee is hard caramel.

Preparation of Candy

When it comes to preparation, candies are "temperamental." Confectionery production is highly sensitive to timing, temperature, and the skill of the preparer, making it an art successfully executed only with practice and patience. The obscure aura surrounding candy-making is further blurred by the fact that there are even more recipes than the number of candies available. For example, there are over 1,000 different formulas for making marshmallows alone (2). Despite the nebulous state of producing confectioneries, this chapter focuses on their preparation based on whether they are crystalline, noncrystalline, or chocolate. Frostings, although not specifically called candies, are also covered because their main ingredient is sugar and they are prepared from a syrup-like mixture. A brief summary of confectionery preparation follows in order to put the various types of candies in perspective.

Steps to Confectionary Preparation

The preparation of many, but not all, confections can be generally summarized by four basic steps (18):

Step 1: Creating a syrup solution

Step 2: Concentrating the contents of this mixture via heating and evaporation

Step 3: Cooling

Step 4: Beating

These steps and how they differ in the preparation of crystalline and noncrystalline candies are now discussed.

The preparation of many crystalline and noncrystalline candies starts out the same—a syrup or sugar solution is heated to melt the sugar, evaporate the liquid, and concentrate the sugar. The formation of sugar crystals from this syrup solution is the basis of crystalline candies, while the goal in preparing noncrystalline candies is to inhibit their formation. This is achieved in part by how their sugar solutions are cooled and beaten. Rapid cooling and beating results in crystalline candies, while slow cooling without agitation forms noncrystalline candies (1).

Crystalline Candies

The goal in preparing crystalline candies is to develop numerous, very fine nuclei in the syrup solution, which will serve as the basis of the sugar crystals. Confectioners generate small nuclei by (1) controlling the form and content of sugar, (2) controlling the temperature, and (3) stirring correctly. As the solution cools, the sugar hardens out, or crystallizes, creating a candy. Crystals are a compilation of loosely packed sugar molecules organized around **nuclei**. The size of the sugar crystals is determined by the rate or speed of nuclei formation. If the nuclei appear slowly in the syrup solution, there is more time for the sugar molecules to aggregate around the nuclei and become large.

Candies Start with a Syrup Solution. The sugars added to the sugar solution vary, but often glucose, invert sugar (a mixture of glucose and fructose), or corn syrup is added to the sucrose to make it more soluble and less likely to form large crystals. This same functional characteristic can be a problem, however, because too many monosaccharides may make the syrup so runny that it never crystallizes. To derive the benefits of invert sugar, it is usually purchased commercially or made by adding an acid such as cream of tartar to sucrose (see Chapter 9). When added to chewy candies, invert sugar's hygroscopic nature prevents them from drying out. The benefit of adding corn syrup is that, like invert sugar, it contributes to chewiness but also adds viscosity, slows the dissolving rate of candies in the mouth, and strengthens the structure of sugar crystals so they are less likely to be affected by temperature or mechanical shock (19).

In the confectionery business, glucose is referred to as dextrose, and fructose as levulose (19). Both of these sugars crystallize more slowly than sucrose and, when combined, are sweeter than sucrose. Other foods, such as chocolate, fat, milk, cream, and eggs, are often added and help to interfere with large crystal formation.

Heating the Syrup. Temperature is of major importance in confectionery production, influencing crystallization at all stages of heating and cooling. The syrup mixture is heated to supersaturation to increase the amount of sugar that can be added to the solution. (For more on saturated and supersaturated solutions, see "Crystallization" in Chapter 9.) Moisture escapes from the candy mixture through evaporation, leaving the solids, primarily sugar, behind to form into a hard mass. As the concentration increases, so does the boiling point, until eventually it surpasses the normal boiling point of pure water. The temperature of a syrup solution reflects its concentration, so reaching a particular candy's final temperature is crucial (Table 27-2). For example, a syrup heated to a high temperature will result in a harder candy than one heated to a lower temperature. Lower temperatures evaporate less water, and the more water a candy mixture retains, the softer its consistency. Even the weather is a factor when making the syrup solution.

> **KEY TERMS**
>
> ***Nuclei*** Small aggregates of molecules serving as the starting point of crystal formation.

> **HOW & WHY? ?????????**
>
> **Why is the weather considered by confectionary makers when preparing candy?** Simple syrups can absorb moisture from the air during a humid day, causing a dilution of the syrup mixture and making the candy softer than desired. As a result, cool, clear days when there is little humidity in the air are preferred for candy-making (1).

Over- and Underheating. One problem sometimes encountered in confectionery cooking is overheating the solution, resulting in a too-hard, sometimes excessively brittle candy. Overheating can also cause color and flavor changes. On the other hand, too cold a temperature results in a too-soft, runny consistency.

There are two ways to determine the correct final temperature during candy making: (1) using a candy thermometer, and (2) the cold-water test.

Thermometer Test. Candy thermometers have been specifically designed for the high heats that syrup mixtures reach, and have a clip on the side to hold them inside the pan. Home-use candy thermometers have a range of about 100° to 320°F (38° to 160°C). Commercial models are longer to fit larger pans, and have a broader temperature range of 60° to 360°F (16° to 182°C) (20). When using a candy thermometer, the bulb should be completely immersed in the mixture without touching the bottom of the pan. The top of the mercury line should be read by bringing the eyes level with the thermometer, not the thermometer to eye level.

Before each use, and especially at high altitudes, it is important to check the candy thermometer for accuracy.

TABLE 27-2

Candy Syrup Temperatures and Doneness Tests

Candy	*Doneness Test*	*Final Temperature of Syrup Begins At** *F°*	*C°*	*Description of Test*†
Jelly		220	105	Syrup runs off spoon in drops that merge to form a sheet.
Syrup	Thread	230	110	Syrup spirals a 2-inch thread when dropped from spoon.
Fondant Fudge Panocha	Soft ball	234	112	Syrup forms a soft ball that flattens out between fingers.
Caramels Nougat	Firm ball	244	118	Syrup forms a firm ball that retains its shape when held between fingers.
Divinity Marshmallows Popcorn balls	Hard ball	250	121	Syrup forms a ball that is hard enough to hold its shape, yet plastic enough to roll out.
Butterscotch Hard candies Toffees	Soft crack	270	132	Syrup separates into threads that are hard and brittle under water, but become soft and sticky when removed from the water.
Brittle Glacé Some hard candies	Hard crack	300	149	Syrup separates into threads that are hard and brittle, but do not stick to fingers.
Caramel	Light brown liquid	338	170	No cold water test. The sugar liquifies and becomes light brown.

*For each increase of 500 feet in elevation, cook the syrup to a temperature 1°F lower than temperature called for at sea level. If readings are taken in Celcius (Centigrade), for each 900 feet of elevation, cook the syrup to a temperature 1°C lower than called for at sea level.

†Remove pot from heat while testing to avoid overcooking. Drop an eyedrop amount of syrup into fresh cold water, which is replaced with each test.

This is accomplished by determining the boiling point of water on the thermometer, which should be 212°F (100°C) at sea level, and 1°F lower for each 500-foot increase in elevation above sea level (1°C lower for each 900 feet in elevation). If the boiling point of the thermometer is below the standard 212°F (100°C), then the number of degrees below this point is subtracted from the recipe's cooking temperature. Once the thermometer has been used to make candy, it helps to immerse it immediately in very hot water to make it easier to clean and to prevent breakage (20).

Cold Water Test. The older and still viable method of testing the temperature of candy mixtures is the cold-water test, which measures the syrup's consistency. This test is carried out by placing a very small amount of the candy mixture, about an eye-drop or ¼ teaspoon, into a cup of very cold (not ice) water, after which the drop is observed for softness or firmness and possibly double-checked by pinching it between two fingers. As the mixture cooks, drops when cooled at successive stages will form first a soft, then a firm, and finally a hard ball. Since the mixture can move from one stage to the next fairly rapidly, the saucepan should be removed from the heat source before doing the test. Experience is the best teacher in learning the precise look and feel of the candy mixture.

NUTRIENT CONTENT

Sugar, at about 45 calories (kcal) per tablespoon, is the main ingredient of candies. Add a little chocolate at 146 calories (kcal) and 9 grams of fat per ounce, and a chocolate-covered candy bar easily contributes between 200 to 300 calories (kcal) and 10 grams of fat to a person's diet. Hard candies, jelly beans, gum drops, and taffy have no fat, but still contain about 100 calories (kcal) per ounce.

The Food and Drug Administration does not allow the use of alternative sweeteners in confectioneries, so reducing calories (kcal) in candies is not easy, even though consumer demand for such products continues to increase (3, 8). Since the 1960s, however, certain gums and breath mints have been made with alternative sweeteners (25), and the FDA policy may soon be amended or revoked in response to numerous requests (16).

Despite this hurdle, one option for reducing calories (kcal) in confectioneries is to use polydextrose, a low-calorie (kcal) bulking agent that is allowed in hard and soft candies and in frostings (22). Another option is the sugar alcohols sorbitol and xylitol, sometimes found in mints and chewing gums (14). These sugar alcohols used by the confectionery industry are not carcinogenic, unlike some other alternative sweeteners (22).

Avoid Vigorous Stirring. Stirring, or agitation, is another factor governing crystal formation. It is important to avoid vigorous boiling or stirring when heating the solution to its final temperature. Syrup splashing onto the sides of the pan can prematurely **seed** the solution and initiate crystal formation. One sugar crystal falling into the mixture from the side can start a chain reaction, resulting in a large sugar mass. This can be prevented by covering the boiling syrup solution with a lid, creating steam that returns the syrup on the sides of the pan back into the solution before it can form crystals. The pan must then be uncovered later in cooking to allow the water to evaporate and the temperature to rise. To prevent crystals from forming during this final cooking stage, the sides of the pan may be wiped often with a wet pastry brush or a damp paper towel or cheesecloth. If this is not done, large crystals, instead of the desired small ones, will likely form.

Cooling and Beating. Although agitating the mixture can result in some premature crystal formation, the crystallization desired for candy-making begins during the cooling of a supersaturated solution. After the correct temperature is reached, the solution is cooled immediately, without any additional movement, not even to stir in flavoring or move the thermometer or spoon in the mixture. As the mixture cools, it becomes supersaturated, which allows for the formation of nuclei. Immediate cooling also prevents further evaporation of the water.

Cooling Is Crucial. Cooling at the correct rate is crucial for candy production. Sugar molecules move rapidly in a hot solution, but drastically slow down as the temperature drops. Small crystals are less likely to aggregate in a hot solution where molecules are rapidly moving. At the same time, the small crystals that do occur grow large because the greater frequency of contact among sugar molecules encourages their growth. The syrup mixtures for crystalline candies such as fudge and fondant are quickly cooled to slow molecular movement, and stirred to form small crystals by fostering aggregation of the molecules. Figure 27-1 shows that starting off with a greater number of small nuclei results in smaller crystals.

Stirring During Cooling. Stirring the mixture after it cools promotes the formation of numerous small crystals that contribute to a smoother consistency in the candy, but this is done only after the desired cooler temperature has been reached. Then the crystallization is initiated by beating the mixture rapidly until its shiny, glossy appearance turns dull. If it is cooled too long before beating, the development of the finer crystals necessary to produce a smooth, crystalline candy will be inhibited.

Types of Crystalline Candies. In the preparation of crystalline candies, small crystals form when many nuclei quickly appear, leaving less time for the sugar molecules to collect around the nuclei, along with less sugar per nuclei because of the higher dilution. The smooth, creamy texture of fondant, fudge, and divinity depends on the formation of numerous, small sugar crystals (Figure 27-2). A discussion of these crystalline candies and their preparation follows.

> **KEY TERMS**
>
> ***Seed*** To create nuclei or starting points from which additional crystals can form.

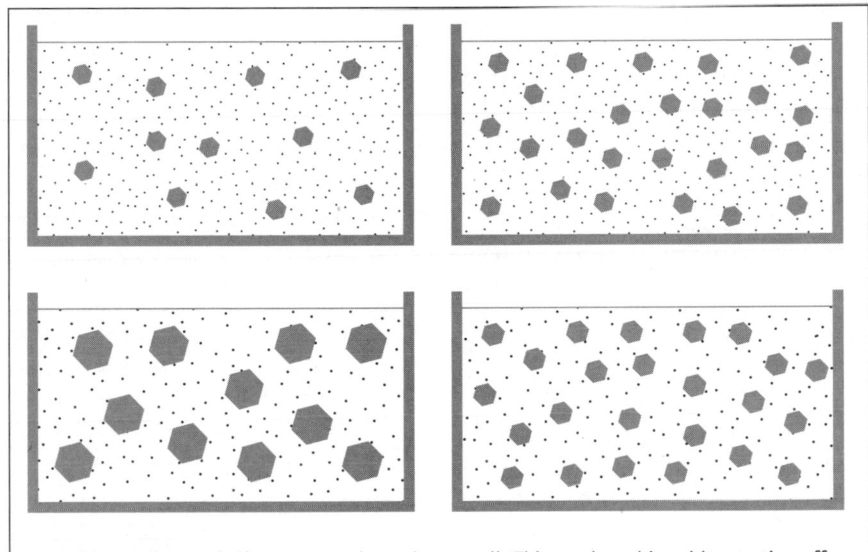

Smoother candies result if sugar crystals are kept small. This can be achieved by starting off with more (top right) rather than less (top left) seed crystals so there are more sites for the remaining sugar to crystallize on. A greater number of starting points results in smaller crystals (bottom right).

FIGURE 27-1 Correct cooling is critical to the formation of many small crystals.

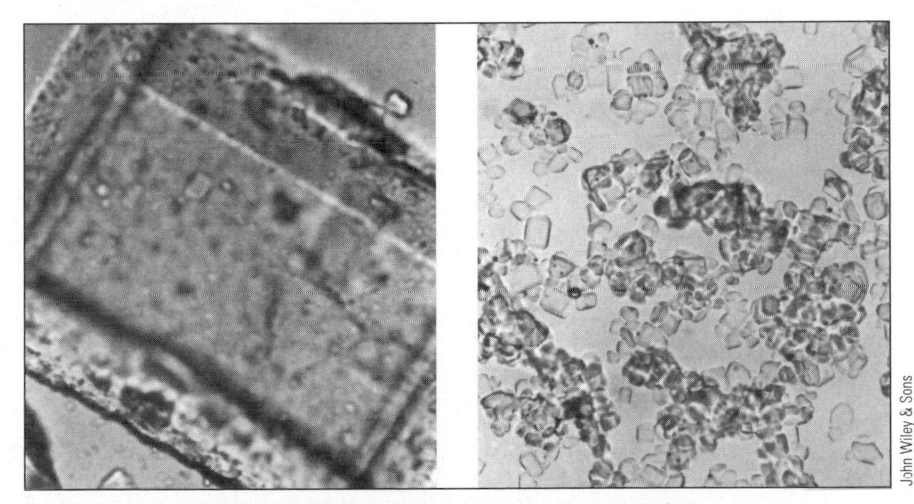

John Wiley & Sons

FIGURE 27-2 Sugar crystal size changes during candy preparation.

Fondant. Fondant is a crystalline candy made from a sugar syrup that has been crystallized into a creamy, white paste. It serves as the soft, rich filling of many chocolates, mints, and candy bars (Chemist's Corner 27-1). Fondant is also used as an icing to glaze and decorate pastries, sealing baked products from air and the loss of moisture (6). Professional bakers buy it ready-made or in powder form, but large amounts can also be prepared by hand.

Fondant preparation starts with sugar, preferably glucose, which is combined with half as much water and a little cream of tartar or corn syrup. The mixture is heated in a covered saucepan to a rapid boil. The solution is then cooked to the soft-ball stage, or 234° to 240°F (112° to 115°C). After reaching the final temperature, the mixture is poured onto a flat marble slab and cooled to about 120°F (49°C). The fondant should be poured, with no scraping or shaking, over a wide surface for even cooling. When the proper temperature is reached, a spatula or wooden spoon is used to mix the candy until it becomes white and creamy. The color and consistency come from the formation of fine, uniform crystals (13). At this point, ½ teaspoon of vanilla or other flavor extracts or colorings may be added. The mixture is then kneaded until smooth. The end product should be stored overnight or even several days in a tightly covered jar before being soft enough to use (6). It can keep up to several weeks on the shelf if kept in an airtight container in a cool, dry location away from high humidity, and up to several months once it is surrounded by chocolate.

Chemist's Corner 27-1

Liquid Fondant

The fondant of cream-filled centers is often prepared with an invertase enzyme that slowly hydrolyzes the sucrose to invert sugar. The higher solubility of this invert sugar compared to sucrose makes it melt under the chocolate layer, turning the center into a creamy liquid (19).

Fudge. Fudge is the most popular of all crystalline candies. It is made by adding chocolate, cream, milk, and butter to a simple syrup. Table 27-3 lists the functions of dairy ingredients in the manufacture of fudge and other candies. After the ingredients have been heated to the soft-ball stage (234° to 240°F/112° to 115°C), the mixture must be cooled to about 120°F (49°C). The fudge is then beaten until the gloss dulls and the texture turns creamy smooth. Beating fudge too long hardens it, but this can be corrected with recooking (Figure 27-3). After cooking, the fudge should be poured at once into a greased pan. Covering it will prevent its drying, as will waiting to cut it into squares until serving time. Panocha or penuche is similar to fudge except it is made with brown, instead of white, sugar and contains no chocolate or cocoa.

Divinity. Divinity is prepared like fondant and fudge up to the point of beating. When the mixture has reached 250° to 266°F (121° to 130°C) (the hard-ball stage), it is poured slowly over stiffly beaten egg whites, with continuous beating. The beating continues until the candy holds its shape when dropped from a spoon. At this point, it can be poured into a buttered pan or dropped by spoonfuls onto a buttered surface.

Noncrystalline Candies

The goal in preparing noncrystalline candies is to make sure the sugar does not crystallize. Two major methods are

TABLE 27-3
Functions of Dairy Ingredients in Confections

Dairy Ingredient	*Function in Confections*
Milk fat	Gives rich flavor, texture, and mouthfeel.
	Provides lecithin and monoglycerides, which aid in emulsification.
	Precursors of key flavor compounds formed by cooking.
	Acts as a flavor carrier.
	Provides a moisture barrier.
	Inhibits bloom defect.
	Oxidation and/or hydrolysis produces distinctive desirable flavor compounds (however, under certain conditions of processing and storage, off-flavors [rancid, fishy, sunlight, stale] may develop).
Milk proteins	Aid in emulsification and miscibility of various ingredients.
	Assist in whipping and foam formation.
	Absorb moisture due to water-binding capacity.
	With heat, can yield protein fiber network that gives rigidity, shape, texture, and chewability. Furthermore, impede movement of water molecules, leading to retardation of sugar crystallization.
	Contribute to flavor and color development.
	Improve nutritional profile.
Lactose	Imparts chewiness and graininess.
	Causes sandiness defects under certain conditions, e.g. at high lactose concentrations.
	Acts as a flavor carrier.
	Polymerizes to form a flexible and extensible matrix (textural effect).
	Caramelizes to generate color and flavor compounds in caramels and toffees.
	Provides reducing sugar moiety that can participate in Maillard browning, which affects color and flavor.
	Acts as a precursor of pyrazines (nutty flavor), maltol and isomaltol (caramel flavor), furfurol and furfuraldehyde (cooked cereal).
	Contributes to dark color and bitter taste defects under certain conditions of overcooking.

Adequately beaten fudge is smooth and the gloss has just dulled.

Overbeaten fudge becomes hard and loses its gloss.

FIGURE 27-3 Beating fudge.

used to inhibit crystallization: creating very concentrated sugar solutions, and/or adding large amounts of **interfering agents** to block the sugar molecules from clustering together. The high viscosity inhibits crystallization by impeding molecular movement.

Concentrating the Sugar Solution. Using high temperatures to evaporate much of the water results in a very concentrated syrup, and the degree of evaporation determines the percentage of moisture in noncrystalline candies. Sugar-syrup-based candies containing 2 percent moisture or less are hard candies, sour balls, butterscotch, and nut brittles. A higher moisture content of 8 to 15 percent is found in caramel and taffy, and an even higher level of 15 to 22 percent moisture is contained in marshmallows, gum drops, and jelly beans (19).

Interfering Agents. Interfering agents may inhibit crystallization in several ways:

- By inhibiting nuclei formation
- By physically coating the crystals, which prevents their growth
- By decreasing water activity, which leaves less water available in which the sugar can dissolve

The two main interfering agents used in confectionery production are corn syrup and cream of tartar, but certain other ingredients added in large amounts can also contribute to preventing crystallization. Corn syrup inhibits crystallization because its glucose chains form a complex that impedes the movement of the sugar and water molecules. The inhibiting action of cream of tartar works by forming invert sugar from sucrose, which then acts similarly to corn syrup by getting in the way of the sucrose molecules. Both interfering compounds provide more time for cooling and/or beating, which promote small crystal formation (1).

Types of Noncrystalline Candies. Noncrystalline candies vary in their degree of sugar solution concentration and the types of interfering substances added (Chemist's Corner 27-2). The preparation of a few of these noncrystalline candies are now discussed in more detail.

KEY TERMS

Interfering agent A substance added to the sugar syrup to prevent the formation of large crystals, resulting in a candy with a waxy, chewy texture.

Hard and Brittle Candy. All but 1 percent of the moisture from simple syrup is removed when making hard candies (4). Flavorings and colorings are then added, with centers or fillings being optional. Brittle candies such as peanut brittle and toffee have baking soda added. The syrup to make brittle is heated to the high temperatures of caramelization and then spread out on a hard surface to cool (Figure 27-4). The bubbles in these candies come from the carbon dioxide gas produced by the baking soda.

Caramels. To inhibit the formation of crystals when making caramel, large amounts of interfering substances such as fat, cocoa butter, concentrated milk products, and/or corn syrup are added to the sugar syrup. The result is a candy with a waxy, chewy texture. The dairy products contribute protein, calcium, and flavor. The amount of fat added to caramels is high—about

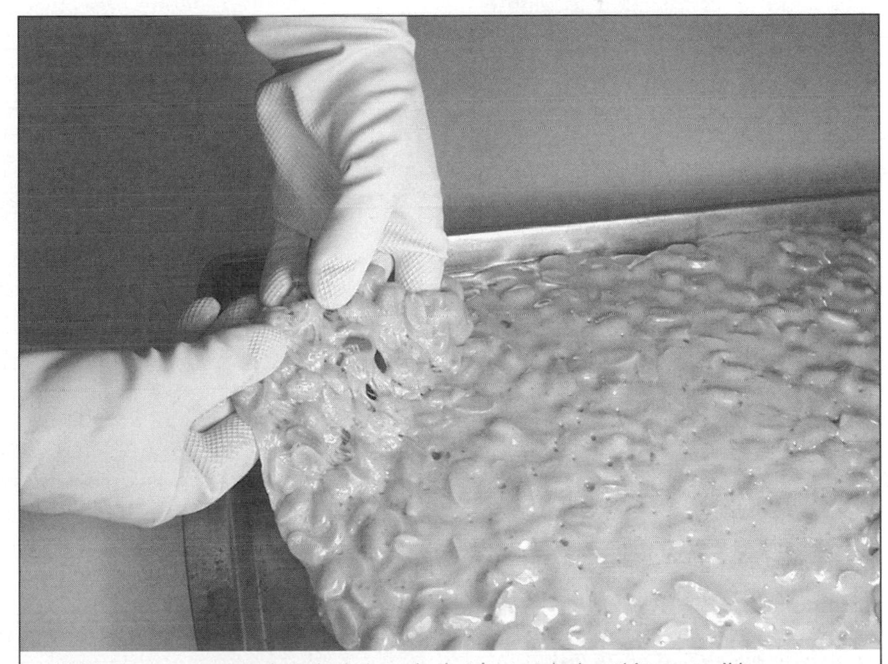

Thick gloves protect against the hot candy that is stretched as thin as possible.

FIGURE 27-4 Spreading peanut brittle to cool.

Chemist's Corner 27-2

Citric Acid in Candies

Citric acid is added to some candies for flavor. The only problem is that the acid, being very hygroscopic (draws moisture from the air), turns the candy somewhat liquid so that it sticks to the wrapper. The confectionery industry has bypassed this side effect of citric acid by encapsulating it in a film or coating (17).

¼ cup of fat for every cup of sugar. The characteristic color of caramel candies, ironically, is not from caramelization, but from the Maillard reaction, which occurs as sugars interact with the milk proteins (9). The color starts to develop when the temperature reaches about 325°F (163°C); however, if it is allowed to become too dark, a bitter taste results (13).

Taffy. Taffy is very similar to caramel except that it is made from a more concentrated solution, which makes it firmer. The other difference is that, once solidified, the syrup mass is pulled to aerate the mixture by incorporating air bubbles. This pulling transforms the mass into a candy that is lighter, chewier, and paler colored than caramel. Different flavorings can be added to create varied types of taffy.

Aerated Candies. Corn syrup serves as the foundation for aerated candies such as marshmallows, jelly beans, and gum drops. Candy can be aerated physically (pulling), chemically (adding sodium bicarbonate), or by the addition of foams for structure; marshmallows, for example, incorporate an egg-white foam (1). Gummy textures are formed by adding gelling agents such as starch, gelatin, pectin, and gums to the sugar. Gelatin can be used in confectionery production to stabilize marshmallows and other aerated products, to add elasticity to gums and jelly beans, and to soften and bind water in caramels and other chewable sweets (21).

KEY TERMS

Tempering To heat and cool chocolate to specific temperatures, making it more resistant to melting and resulting in a smooth, glossy, hard finish.

Chocolate

Chocolate, derived from the tropical cocoa or cacao tree (Figure 27-5), is the chief ingredient for many different types of candies. Warm, moist climates near the equator provide the best environment for the growth of cacao trees. Botanists believe the tree originated in the Amazon-Orinoco river basin in South America; the crop is now cultivated primarily in West Africa and Brazil.

The tree's cacao beans played an important role in the traditions, religion, and legends of the Aztecs. Historians suggest that it was the Aztecs who first came up with chocolate beverages, and who believed that the cocoa tree had a divine origin, leading to its botanical name, *Theobroma*, meaning "food of the gods." The Aztec's term for their drink, "bitter water" or "cocoa water," was *xocolatl*, which the Spanish later converted to "chocolate." Hernando Cortez brought the cocoa beans from Central America back to Spain in 1758, and from there they quickly spread to other parts of Europe. The popularity of chocolate has led to a hypothesis that for some people its addictive-like quality and resulting "chocolate binges," especially as they relate to depression, may be parallel with other addictions in general (24).

Chocolate Manufacturer's Assn.

FIGURE 27-5 Cacao tree with pods (fruit).

Obtaining Chocolate Liquor from Cocoa Beans. The first step in chocolate production is the manufacture of chocolate liquor, the basic ingredient for all chocolate products. After being picked, cocoa beans are scooped out of the pods in which they are packed and heaped into large piles, covered with banana leaves, and allowed to ferment. The fermentation alters the seed coat, making it easy to remove, and modifies the beans' flavor and color (10). The beans are then dried to 7 percent moisture to give them good keeping quality before being shipped to a chocolate manufacturer.

Upon arriving at the manufacturing plant, the beans are blended into various combinations to obtain specific flavors and colors, which are further developed by roasting. The hull and the germ of the cocoa beans are then removed, and what remains is called the nibs. These nibs, containing 54 percent cocoa butter, are ground very fine under heat to yield chocolate liquor (Figure 27-6). The cocoa butter in this chocolate liquor has a melting point just below body temperature, and is responsible for chocolate's melt-in-the-mouth appeal and brittle snap at room temperature (6, 19).

Conching. The next step in manufacturing good-quality chocolate is conching, during which the chocolate's characteristic flavor and consistency are developed. Warmed chocolate (usually between 70° to 160°F/21° to 71°C) is kneaded ("conched") and aerated by machines to increase its smoothness, viscosity, and flavor (19). Several ingredients may be added during this time. These include flavorings, additional cocoa butter, and lecithin, which increases viscosity and causes the chocolate to develop a velvety smooth coating. Conching complete, the chocolate may be poured into blocks and cooled before being packaged, where it can be stored without any major problems for up to one year (5).

Tempering. Chocolate may also undergo **tempering** before it is formed and packaged. The three basic

FIGURE 27-6 The process used to manufacture chocolate.

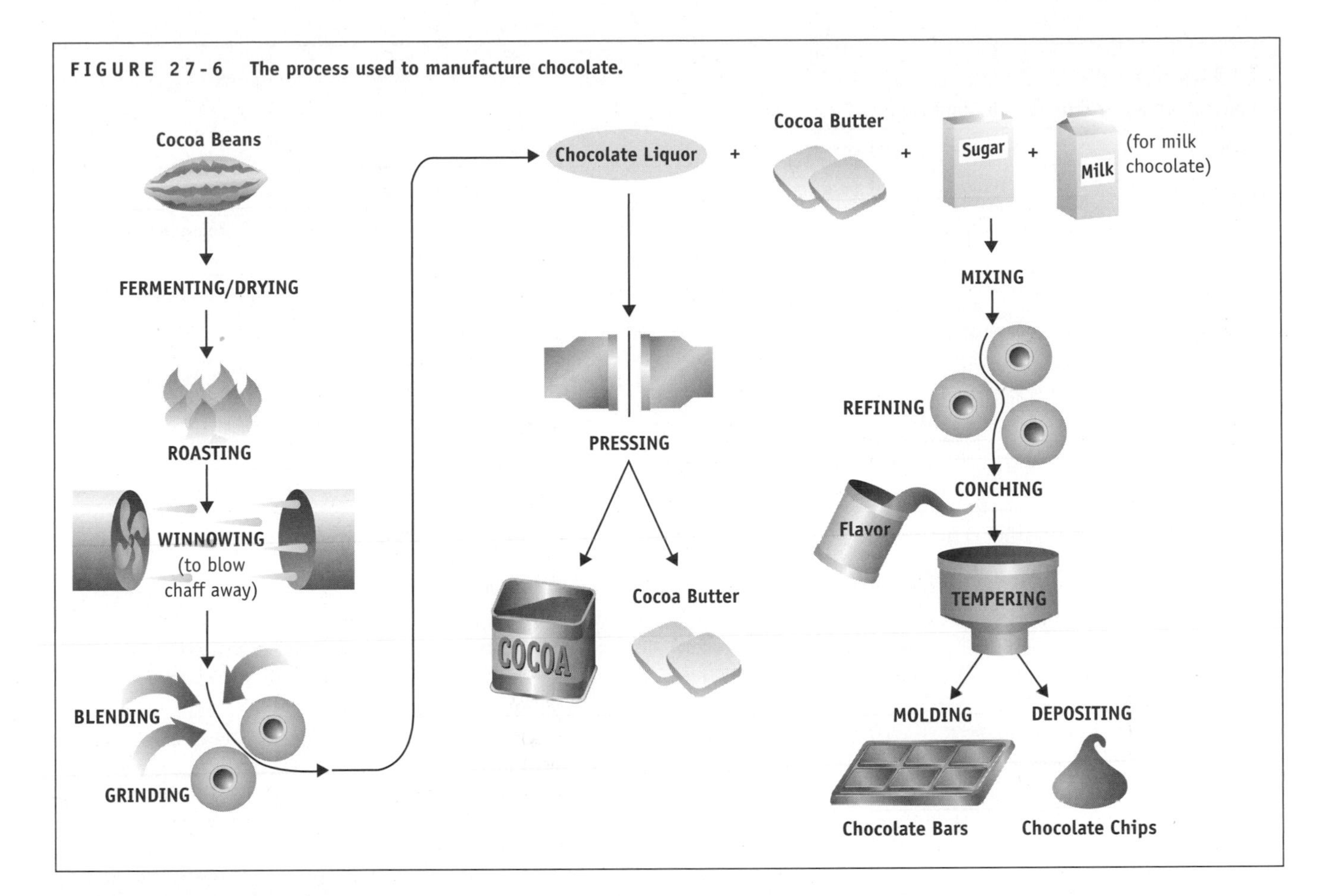

Chemist's Corner 27-3

Tempering and Chocolate Melting Point

Tempering helps the fatty acids in cocoa butter change to the stable beta form. Table 27-4 shows that fatty acids in the beta form have the highest melting points. One of the reasons to avoid rapid cooling is that fatty acids convert to the unstable alpha phase. However, if this is followed by slow heating, the crystals convert from alpha to beta prime, and finally to the more stable beta form (23).

TABLE 27-4

Polymorphism and Melting Points of Cocoa Butter

Fatty Acid Crystal Form Phase	*Melting Point °F*	*°C*
α	73°	23.3°
β′	82°	27.5°
β	93°	33.8°

steps of tempering a chocolate are melting, cooling, and rewarming (12). Proper tempering is what gives high-quality chocolate that "snap" when bitten into or broken in half (Chemist's Corner 27-3). Tempered chocolate is often used to **enrobe** candies and other foods such as ice cream bars, fruits like raisins and strawberries, cookies, nuts, and doughnuts (23). Untempered or improperly tempered chocolates may exhibit gray surfaces, or bloom, on the surface of hardened chocolate.

HOW & WHY? ?????????

Why does a grayish film sometimes form on chocolate? The grayish bloom, which usually occurs after the chocolate has been stored or allowed to get warm, is caused by one of two things: fat crystallizing onto the surface, or moisture in the air reacting with the sugar on the surface (11, 19). Fat bloom may be caused by improper crystallization, incorrect temperatures in heating or cooling, insufficient stirring, or inadequate conditions where the chocolate is prepared, such as less than optimal room temperatures, humidity, or drafts. It also may occur when the soft fat centers or added fats are incompatible with cocoa. Sugar bloom occurs when chocolate is stored under damp conditions (sugar in the chocolate absorbs water) and then exposed to a drier environment (moisture on the surface evaporates, causing sugar to recrystallize on the surface) (1).

Fat bloom can be distinguished from sugar bloom by its greasy feeling. Bloom and other problems of chocolate production along with several solutions are listed in Table 27-5.

KEY TERMS

Enrobe To coat food with melted chocolate that hardens to form a solid casing.

TABLE 27-5
Confectionary Problems and Their Causes

Problem	*Causes*	*Solution*
Soapy, rancid flavor	Lipase activity	Check incoming ingredients for off-flavors.
Burned, medicinal, fishy, or chemical taste	Oxidation of oils or fats during product storage.	Minimize exposure to light and maintain cool temperatures. Check incoming ingredients for off-flavors.
Scorched or burned taste	Overheated sugar and milk solids	Control temperatures and times during heating and cooling.
Lack of brown color and flavor notes	Insufficient Maillard browning	Increase amount of milk products.
		Use a slower cooking process.
	Too little caramelization	Increase sugar content.
		Cook at higher temperatures and/or longer cooking times.
Poor caramel texture	Too thick or poor elasticity	Decrease milk protein level.
	Too thin	Increase milk protein level to increase viscosity.
Gritty or grainy texture	Change in sugar crystal type	Adjust balance or levels of sugar types for proper crystallization.
	Large sugar crystals formed	Reduce lactose content.
		Check seeding and agitation during heating and cooling.
	Sugar crystals formed on product surface	Optimize storage conditions (humidity and surface temperature) to prevent migration of fat and water to surface.
Bloom (white, dusty appearance)	Fats not in proper phase	Check melting temperatures and cooling temperatures and times.
		Check amount and type of seed materials to ensure proper tempering.
		Increase tempering time.
	Incompatible fats	Use fats with melting points that are appropriate for the processing conditions.
		Avoid lauric fats in coatings.
		Include milkfat in formula.
	Improper storage temperatures	Store at recommended temperatures (80–86°F, 27–30°C).

Source: Chandon, *Dairy-Based Ingredients* (Eagan Press, 1997).

Factors Affecting Tempering. The correct chocolate, the temperature, and the timing are crucial factors in tempering. A certain amount of fat is needed to produce a smooth, glossy coating, so using the right chocolate is imperative. Baking chocolate and cocoa, for example, are too low in fat for dipping candies and will not respond to tempering. The weather also affects chocolate consistency; too much humidity can interfere with the solidification of the chocolate. Clear, cool days not exceeding 70°F (21°C) are preferred (20). A minimum of 1½ pounds of chocolate is required for tempering, any lesser amount being too unpredictable with the rapid temperature changes (13). If coating candies, 1½ pounds of chocolate will cover about 90 ¾-inch centers (20). Prior to heating, the chocolate should be chopped or grated to allow for faster, more uniform melting. Chopping is easier if the chocolate is at room temperature and a clean, dry cutting board is used.

The three most common tempering methods are tabliering, seeding, and the cold-water method.

Tabliering Method. Melting the chocolate pieces is the first step in the tabliering technique. It is never done over direct heat; a double-boiler is used in order to keep temperatures from escalating too high. A bowl placed over a saucepan of simmering water might also be used. The bowl should not touch the water and should be large enough to prevent steam escaping from the saucepan from condensing and coming in contact with the chocolate. In either case, the chocolate is stirred constantly to promote melting and prevent burning. The spoon or spatula should be wooden or plastic because these materials are poor conductors of heat. A metal utensil will conduct heat away from surrounding areas of the chocolate, cooling them and making them more saturated.

For the same reason, as well as to prevent lumping, cold liquids in the form of milk, cream, or liqueur should never be directly added. Melting the chocolate at too high a temperature or too quickly can scorch it or cause the cocoa butter to separate out, forming undesirable lumps.

When the temperature reaches about 115° to 120°F (46° to 49°C), it is ideally kept there for about 30 minutes with continued stirring. The temperature should not be allowed to exceed 125°F (52°C). The chocolate is then cooled by removing it from the heat and placing one-third of it on a marble slab (Figure 27-7). It is scraped back and forth until it cools to just under 80°F (26°C), at which point it will crystallize and thicken, signaling that it is to be added back to the other two-thirds. Then it is again warmed slowly to 85° to 90°F (29° to 32°C), where it reaches a consistency that can be worked. The entire process should be repeated if the temperature is allowed to rise above 91°F (33°C) (13).

Seeding Method. Slight variations of the tabliering method have resulted in two other tempering techniques, the seeding and cold-water methods. In seeding, already tempered chocolate is melted in the double boiler, but then removed from the heat source, at which time more unmelted chocolate pieces are added. The chocolate is cooled to 80°F (26°C) and stirred for 2 minutes at this temperature before being warmed again to 85° to 90°F (29° to 32°C) (6).

Cold Water Method. The cold water method consists of cooling the heated chocolate by placing the bowl of chocolate in a larger bowl of ice water instead of pouring it onto a slab. The chocolate must be stirred as the bowl is lowered into the ice water, because it starts to set immediately. As in the other methods, the chocolate is then rewarmed to 85° to 90°F (29° to 32°C) so that it can later be cooled and shaped.

Tempering for Chocolate Dipping. A slight variation is made in the tempering process if the chocolate is to be used for dipping fruits, nuts, or other items (Figure 27-8). After it reaches its maximum temperature, it is cooled to about 83° to 85°F (28° to 29°C), the ideal temperature for dipping. Items must be dipped quickly, because the temperature range in which the chocolate stays at the right consistency is extremely narrow. It is kept from getting too cool by leaving it in the double boiler on reduced heat. After the items have been dipped, they are held up for a moment in the air, allowing the excess to drip off; then they are placed on wax paper in such a way that the hardening chocolate does not form an unsightly line or an undesirably large base. A broad base is caused by using too much chocolate or having the temperature too high.

FIGURE 27-7 Further cooling and manipulation of chocolate.

Working chocolate on marble helps to keep it cool.

Shaping chocolate in a plastic-lined frame.

FIGURE 27-8 Tempering chocolate for dipping.

Chocolate chopped for uniform melting.

Checking chocolate temperature.

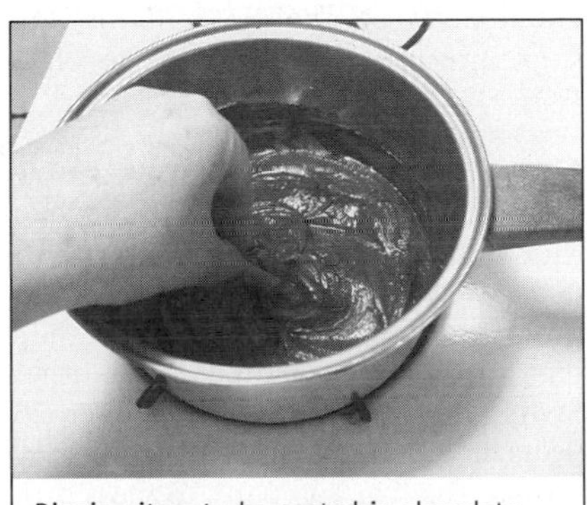

Dipping item to be coated in chocolate.

KEY TERMS

Nontempered coating A coating resembling chocolate that is not subject to bloom because it is made with fats other than cocoa butter.

Coatings. The inconvenience of tempering and the risk of bloom can be entirely avoided by using **nontempered coatings**. Known as "coating chocolate," "chocolate icing," "confectionery coating," or "non-temp chocolate," these products are not true chocolates. The cocoa butter in non-temp chocolates has been replaced with other fats, such as hydrogenated palm kernel, lecithin, soy, or cottonseed oils. Nontempered coatings are available in white, milk, and dark chocolate and are easier to use than tempered chocolate, but they fall short of real chocolate in flavor and texture. They are accordingly reserved for use as glazes or decorations rather than as fillings or major ingredients (6). Their major benefit is that items can be dipped in them without the preparer having to follow the detailed, time- and temperature-dependent steps of tempering.

Types of Chocolate Products. The chocolate products obtained from the chocolate manufacturing process differ based on how much cocoa butter and other ingredients are added to the chocolate liquor.

Baking Chocolate. Baking chocolate is chocolate liquor that has cooled and solidified into cakes. It is also sold as bitter, unsweetened baking chocolate, but in any form it must have a chocolate liquor content of at least 35 percent.

Other possible ingredients include sugar, cocoa butter, lecithin, and flavoring (13). Lecithin is often used in confections because it can replace some of the cocoa butter in chocolate bars (21).

Cocoa. Chocolate liquor with most of the cocoa butter removed results in cocoa. This reduces the fat content in cocoa to 10 to 24 percent, compared to 38 percent in eating chocolate and 50 percent in unsweetened baking chocolate (15). The lower fat proportion gives cocoa a very concentrated chocolate flavor. Cocoa is made by pressing much of the cocoa butter out of the heated liquor as it hardens into cakes. The cakes are then ground into cocoa powder. Two types of cocoa are available: natural cocoa, which is slightly acidic, and the less bitter cocoa, which may be called alkalized, Dutch-processed, or Dutch cocoa. A Dutchman started one of the first hard chocolate businesses by inventing the process of "Dutching," which consists of treating crushed cocoa beans or chocolate liquor with an alkali (usually potassium carbonate or sodium carbonate) to raise its pH. The degree of alkalization in the cocoa determines its color, which ranges from deep reddish brown to charcoal black. As the color darkens, the chocolate's flavor becomes more bitter and astringent. Europeans prefer the darker, reddish-colored Dutch cocoa for their recipes, while North Americans predominantly use natural cocoa (15).

Semi-Sweet or Sweet Chocolate. Granulated sugar and extra cocoa butter are added to the chocolate liquor to produce the sweeter, smoother taste of semi-sweet or sweet chocolate. Semi-sweet chocolate must contain at least 15 percent chocolate liquor, while equal parts of chocolate liquor and sugar are present in sweet chocolate (13).

Milk Chocolate. Milk chocolate candy bars were first produced in 1875 when a Swiss manufacturer added condensed milk to chocolate liquor and other ingredients. Condensed milk had just been invented by another Swiss, Henri Nestlé. A little over 25 years later, in 1903, Milton Hershey formed the first company to mass produce milk chocolate in the United States, which later became the largest chocolate manufacturing plant in the world. The name of the Pennsylvania town where the factory was built was eventually changed to Hershey (6).

Milk chocolate has a lighter color and sweeter, milder flavor than other chocolates. It is made like semi-sweet and sweet chocolate except that it contains less chocolate liquor (10 percent minimum), more granulated sugar, cocoa butter, and dried whole-milk solids (12 percent minimum) (9). The milk is added in powder form to prevent the chocolate's consistency from becoming too liquid. At the molecular level, milk chocolate is a suspension in which the continuous phase is the cocoa butter and milk fat, while the dispersed phase consists of the cocoa, sugar, and nonfat milk solids (7).

Imitation Chocolate. Some or all of the cocoa fat is replaced with vegetable fat in imitation chocolates. These less costly versions of chocolate, which are also less likely to melt during the warmer summer months, are sometimes used to coat candies, ice cream bars, and crackers.

White Chocolate. Technically, white or ivory chocolate is not chocolate, nor is it recognized as such by the FDA, because it contains no chocolate liquor or cocoa. Its basic ingredients are sugar, cocoa butter, milk, natural flavor, lecithin, and vanillin. If the cocoa butter is removed from white chocolate, the product is referred to as "white coating" or "confectionery coating."

Frostings

Frostings or icings, either uncooked or cooked, are spread on baked products such as cakes, cupcakes, and pastries. In addition to improving the appearance, flavor, and texture of baked dessert goods, frostings retain moisture and increase shelf life. Some of the more popular types of frostings are flat (also known as simple), decorating, buttercream, and fondant (previously discussed above).

Flat Frostings. Uncooked flat or ornamental frostings are the simplest to prepare. Flat frostings are made by beating powdered sugar (10X) with water, milk, or cream, and a flavoring. Powdered sugar is used rather than granulated because it dissolves instantly in unheated liquids. Unfortunately, the presence of cornstarch (about 3 percent) in powdered sugar gives uncooked frostings a slight raw taste. Also, unless a fat is added, they have a tendency to become dry and crack. Flat frostings are most frequently used on coffee cakes and sweet rolls.

Decorating Frosting. Decorating frosting, also called royal or ornamental frosting, is used for decorating cakes. It is similar to flat frosting, but beaten egg whites are added to impart a firmer structure to the icing when it dries.

Cooked Frosting. Cooked frostings are often made with granulated sugar that is dissolved in heated water; additional ingredients may include egg whites, cream of tartar, or corn syrup. If confectioners' sugar is used, heating allows the formation of a sugar syrup that will serve as the frosting's foundation. A common cooked frosting, made with either granulated or confectioners' sugar, is buttercream frosting. This frosting is used primarily as a filling and icing for pastries and cakes, and frequently to create rose or leaf decorations.

As the name implies, buttercream frostings have butter (preferably sweet, unsalted) added to make them creamier, easier to spread, more flavorful, and less likely to crack. They are made by combining butter and/or shortening with a boiled sugar syrup. Modifications of buttercream frostings include those that are meringue-based, which are made by beating soft butter into whipped egg whites sweetened with a sugar syrup. A very rich, delicate cream called French or Italian buttercream is prepared slightly differently by beating soft butter into whipped egg yolks (or whole eggs) mixed with the hot sugar syrup.

Storage of Candy

Storage requirements vary depending on the candy. Those lowest in water content, such as hard candies and brittle, keep indefinitely if properly wrapped, because they do not support the growth of microorganisms. Extended storage and subsequent exposure to moisture in the air, however, can cause their surfaces to become gummy. Candies such as fudge and fondant, which have a higher moisture content, get softer and smoother in texture if left in an airtight container, but after about a day, a graininess develops as the sugar crystals become larger. Ingredients other than sugar, such as fat or milk products, are subject to rancidity, which results in off-flavors and odors; however, this degradation can be delayed by refrigeration or freezing. Ingredients that improve shelf life include humectants, which act to hold moisture. Common humectants incorporated into candies include glycerin (glycerol), sorbitol, pectins, and gums (19).

Shelf Life of Chocolate

Chocolates can stay on the shelf for over a year as long as they are not subjected to wide fluctuations in temperature and humidity. The exceptions are milk and white chocolates, which have a slightly shorter shelf life of about eight months to a year due to their milk content.

In order to reach these long storage life spans, chocolate should be properly wrapped and stored in a cool, dark place, the ideal temperature being 65°F (18°C) and the humidity 50 percent (13). The slightest melting may result in fat bloom as the cocoa butter crystals form a grayish or whitish film on the chocolate's surface. Sugar bloom, which is rougher in appearance and texture, can result under conditions of high humidity.

PICTORIAL SUMMARY / 27: Candy

The United States produces more candy than any other country in the world. Americans eat an astonishing 24 pounds of confectioneries per person per year, with a little over half of this in the form of chocolate.

CLASSIFICATION OF CANDIES

There are thousands of different candies, but they can be classifed according to their ingredients and/or preparation method.

Classification by ingredients:

Syrup-phase (or sugar-phase): Most candies fall into this category; they are made from a simple syrup, which is a mixture of boiled sugar and water. Composed of sugar and flavorings, syrup-phase candies include:

- Hard candy
- Nougats
- Caramels
- Marshmallows
- Jelly beans
- Gums
- Fudges
- Fondants

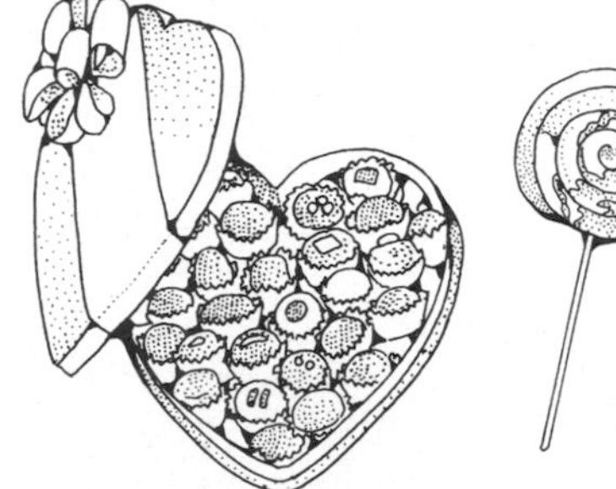

Fat-phase: When chocolate and/or peanut butter are added, the candy is considered fat-phase.

Combination of both fat- and syrup-phases can be found, for example, in a chocolate-covered candy bar.

Classification by method of preparation:

Crystalline candies, which are soft, smooth, and creamy, are formed from sugar solutions yielding many fine, small crystals. They include fudge, fondant, and divinity.

Noncrystalline candies are formed from sugar solutions that did not crystallize, and are amorphous, or without form. They include caramel, toffee, hard candy, and gummy candies.

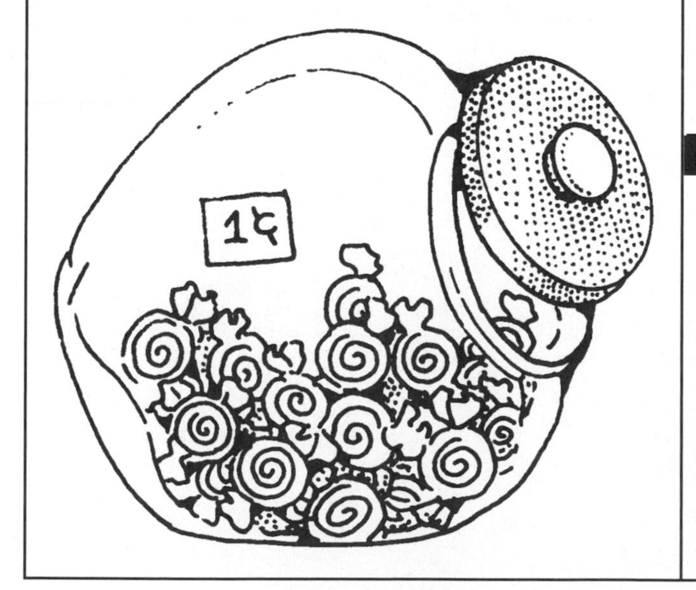

PREPARATION OF CANDY

The art of confectionary production, highly sensitive to timing, temperature, and skill, requires patience and practice.

The preparation of many, but not all, confections can be generally summarized by four basic steps:

1. Creating a syrup solution.
2. Heating this mixture to concentrate the contents through evaporation.

—These steps are basically the same for crystalline and noncrystalline candies.

3. Cooling
4. Beating

— *Crystalline candies:* Goal is to form sugar crystals, so rapid cooling and beating are used.

Noncrystalline candies: Goal is to avoid the formation of sugar crystals, so cool slowly, without agitation.

Chocolate from the tropical cacao tree is the main ingredient of many different types of candies. Cocoa beans are ground very fine and heated to yield chocolate liquor, the basic ingredient for all chocolate products—cocoa, baking chocolate, milk chocolate, semi-sweet chocolate, and sweet chocolate. Other ingredients can include cocoa butter, sugar, milk products, and flavorings.

Frostings, either cooked or uncooked, are spread on baked dessert products to improve their appearance, flavor, texture, and shelf life. The most common types of prepared frostings are:

- Flat (simple)
- Decorating
  - –Royal
  - –Ornamental
- Buttercream
  - –Meringue
  - –French
  - –Italian
- Fondant

There are over 1,000 different formulas for making marshmallows.

STORAGE OF CANDY

Candies lowest in water content (hard candies/brittle) keep the longest if properly wrapped; however, extended storage times or exposure to moisture can cause their surfaces to become gummy. Fat or milk added to candies makes them prone to rancidity, but refrigeration or freezing delays this process. Humectant added to candies improves their shelf life.

Chocolates can stay on the shelf for over a year, except for milk and white chocolates, which last about eight months to a year. All chocolate should be properly wrapped and stored in a cool (65°F/18°C, humidity 50 percent) dark place. The unsightly grayish or whitish film on a chocolate's surface is bloom resulting from either fat or sugar crystallizing on the surface.

REFERENCES

1. Alexander RJ. *Sweetners: Nutritive.* Eagen Press, 1998.
2. Alikonis JJ. *Candy Technology.* Avi Publishing, 1979.
3. Altschul AM. *Low-Calorie Food Handbook.* Marcel Dekker, 1993.
4. Chin M, and D Frick. Formulating a color delivery system for hard candy. *Food Technology* 49(7):56–61, 1995.
5. Ensminger AH, et al. *Foods and Nutrition Encyclopedia.* CRC Press, 1994.
6. Friberg B. *The Professional Pastry Chef.* Van Nostrand Reinhold, 1996.
7. Full NA, et al. Physical and sensory properties of milk chocolate formulated with anhydrous milk fat fractions. *Journal of Food Science* 61(5):1068–1072, 1996.
8. Izzo M, C Stahl, and M Tuazon. Using celluose gel and carrageenan to lower fat and calories in confections. *Food Technology* 49(7):45–50, 1995.
9. Jeffrey MS. Key functional properties of sucrose in chocolate and sugar confectionery. *Food Technology* 47(1):141–144, 1993.
10. LaBell F. Cocoa: A bean of many flavors. *Prepared Foods* 166(8):81, 1997.
11. Loisel C, et al. Fat bloom and chocolate structure studied by mercury porosimetry. *Journal of Food Science* 62(4):781–788, 1997.
12. Loisel C, et al. Tempering of chocolate in a scraped surface heat. *Journal of Food Science* 62(4):773–779, 1997.
13. Malgieris N. *Nick Malgieri's Perfect Pastry.* Macmillan, 1989.
14. Masalin K. Caries-risk-reducing effects of xylitol-containing chewing gum and tablets in confectionery workers in Finland. *Community Dental Health* 9(1): 3–10, 1992.
15. Medrich A. Rediscovering cocoa. *Fine Cooking* 17:68–73, 1996.
16. Nabors LO. Intense sweeteners: Acesulfame K, alitame, aspartame, saccharin, sucralose. *The Manufacturing Confectioner* 70(11):65, 1990.
17. Ohr LM. A means to various ends. *Prepared Foods* 166(11):67, 1997.
18. Penfield MP, and AM Campbell. *Experimental Food Science.* Academic Press, 1990.
19. Potter NN, and JH Hotchkiss. *Food Science.* Chapman & Hall, 1995.
20. Pritchard A. *Anita Pritchard's Complete Candy Book.* Harmony Books, 1978.
21. Pszczola DE. Ingredient developments for confections. *Food Technology* 51(9):70, 1997.
22. Shinsato E. Confectionery ingredient update. *Cereal Foods World* 41(5):372–375, 1996.
23. Stauffer CE. *Fats and Oils: Practical Guides for the Food Industry.* Eagen Press, 1996.
24. Tuomisto T, et al. Psychological and physiological characteristics of sweet food "addiction." *International Journal of Eating Disorders* 25(2):169–175, 1999.
25. Vink W. Applications in confectionery: Tableted confections. *The Manufacturing Confectioner* 70(11):77, 1990.
26. U.S. confectionary consumption. *The Manufacturing Confectioner* 74(9):73–78, 1993.

WEBSITES

A website on everything "candy":
www.candyusa.org

Tour the world's largest chocolate factory via the web:
www.hersheys.com/tour/pic_text/intro.htm

The official site for the United States Chocolate Industry. It includes links to the Chocolate Manufacturer's Association, the American Chocolate Institute, and the World Cocoa Foundation:
www.chocolateandcocoa.org

28

Frozen Desserts

There is no denying that frozen desserts have maintained their popularity throughout the ages. In 62 A.D., the Roman Emperor Nero sent slaves to the mountains to retrieve ice, which was then flavored with nectar, fruit pulp, and honey to be enjoyed by him and his court. Marco Polo brought the formula for water ices to Europe from Asia where they had been enjoyed for at least a thousand years. George Washington and Thomas Jefferson were both reputed to have made ice cream in their homes, and Dolly Madison is known to have served ice cream at the White house in 1812 (17). By 1840, ice cream had moved from palaces and mansions to the streets of America's largest cities. The ice cream scoop was invented 66 years later, and by 1921 ice cream was served to immigrants arriving on Ellis Island as part of their first American meal. Now American consumption of ice cream and other frozen desserts is higher than ever.

KEY TERMS

Stabilizer A compound such as vegetable gum that attracts water and interferes with frozen crystal formation, resulting in a smoother consistency in frozen desserts.

Types of Frozen Desserts

Ice cream and other commercially frozen desserts, from simple water ices to elaborate ice cream cakes, are probably the most commonly consumed desserts in North America. Newer frozen desserts or modifications of old ones appear on the market regularly in the form of pies, mousses, cakes, parfaits, pudding sticks, frozen yogurt, popsicles, and new flavors of ice cream.

What makes one frozen dessert different from another? It is the ingredients, especially the type and proportion of fat (milk fat) and milk solids-not-fat (MSNF), and the way in which these and other ingredients are combined. Common ingredients other than milk fat and MSNF include sugar, **stabilizers** (gums), emulsifiers, water, air, and flavorings. Ice cream, imitation ice cream, sherbet, sorbets, water ices, frozen yogurt, and still-frozen desserts, their ingredients and preparation, are the subjects of this chapter.

Ice Cream

Americans consume more ice cream than any other nation in the world, undeterred by the fact that it is the frozen dessert with the highest fat content (Figure 28-1). The many variations of ice cream are listed in Table 28-1. Ice cream is a food prepared by simultaneously stirring and freezing a pasteurized mix of dairy (milk, cream, butterfat, etc.) and nondairy ingredients (sweeteners, stabilizers, emulsifiers, possibly eggs, colors, and flavors) (Chemist's Corner 28-1).

Contents of Ice Cream. By law, ice cream must contain at least 10 percent milk fat and 20 percent milk solids-not-fat. (15). Only 8 percent MSNF is required for bulky flavors like chocolate or fruit ice cream (11). Milk solids-not-fat influence ice cream's flavor, **body**, texture, dis-

PhotoDisc

FIGURE 28-1 Ice cream is the most popular of the frozen desserts—and also the highest in fat.

TABLE 28-1
Frozen Desserts

Dessert	*Description*
Bombe	A decorative ice cream, in two or more different flavors, layered in a mold.
Frozen custard	Ice cream that contains 14% egg yolk solids. Also known as French ice cream.
Frozen yogurt	Frozen dessert made from a cultured diary product with added sweeteners and flavors.
Gelato	Frozen dessert with intense flavor and colors. It contains sugar, milk, cream, egg yolks, and flavoring.
Ice cream	Frozen mixture of dairy ingredients containing at least 10% milk fat and weighing at least 4.5 pounds to the gallon. Marketing terms include regular ice cream, economy ice cream (lower price), premium ice cream (higher fat content), and super premium (high fat content—at least 12% milk fat—and the best ingredients).
Ice cream cone	Born at the 1904 St. Louis World's Fair when a vendor selling ice cream ran out of dishes and created a "cup" from waffles being sold in a nearby stand.
Light ice cream	Contains 50% less total fat or 33% fewer calories than regular ice cream.
Low-fat ice cream	Contains no more than 3 grams of total fat per serving.
Neapolitan	Alternating lengthwise layers of two to four ice cream flavors.
Nonfat ice cream	Contains no more than 0.5 grams of total fat per serving.
Mellorine	A product made without milk fat. Other fats from either animal or vegetable sources can be used as long as fat content is not less than 6%.
Mousse	A French term meaning "foam" or "froth" and used to describe an airy, rich dish that can be hot, cold, or frozen. The dish can be a savory main meal (meat, fish, shellfish, cheese, or vegetables) or a dessert that is cold or frozen. The fluffy texture is from either whipped cream or egg whites and its structure is often strengthened by adding gelatin.
Parfait	Alternating layers of ice cream and fruit or syrup in a tall, slender glass.
Reduced-fat ice cream	Contains at least 25% less total fat than regular ice cream.
Scoop ice cream	The two main types of scoop ice cream are uncooked and cooked. Both contain cream, sugar, and flavoring, but cooked ice cream adds at least 1.4% egg yolk solids. Thickening agents are sometimes substituted for eggs.
Sherbet	Frozen dessert containing less than 2% milk fat and often more sugar than ice cream.
Sorbet	Similar to sherbet, but it contains no dairy, fat, egg, or gelatin ingredients.
Still-frozen desserts	Frozen desserts not stirred during freezing—mousses, bombes, and parfaits.
Sundae	Ice cream topped with syrup, nuts, whipped cream, and a cherry.
Water ice (glacé)	Frozen dessert made from sweetened water and fruit juice.

pensing qualities, shelf life, and nutritive value, specifically in terms of protein, B vitamins, vitamin A, and calcium (1). The milk fat in frozen desserts comes primarily from milk products such as cream, butter (unsalted), sweetened condensed milk, or whole milk. The egg yolk that is used in some ice cream may be an additional source of fat and also provides cholesterol, as do cream and butterfat. Expensive ice cream brands labeled "premium" or "super premium" are denser, smoother, richer, and higher in calories (kcal) because they use heavier cream than that used in standard ice cream. Federal and state regulations determine the minimum levels of fat and MSNF for many frozen desserts, and at least 14 percent fat must be present in premium ice creams (9, 16). Since milk fat is the most expensive ingredient, the more that is used, the higher the frozen dessert's cost. Other ingredients usually found in ice cream include eggs, sugar, stabilizers (gums, gelatin), flavoring, and coloring. When real cocoa or other forms of chocolate are added, the fat content rises even further.

Chemist's Corner 28-1

The Chemistry of Ice Cream

Ice cream is a colloid food foam consisting of frozen ice crystals, air bubbles surrounded with fat globules and coated with an emulsified protein layer, and an unfrozen liquid phase containing sugars and salts in solution (8).

KEY TERMS

Body The consistency of frozen desserts as measured by their firmness, richness, viscosity, and resistance to melting.

Low-Fat Ice Cream. Low-fat ice cream, previously called ice milk, usually contains more sugar than milk (by weight) and less than 7 percent fat. Low-fat ice cream was first introduced during the Depression as a lower-cost (and not very popular, according to some reports) alternative to ice cream (11). Until recently, the milk fat content of any frozen dairy product sold as ice cream had to be by law at least 10 percent of its weight. At present, about one-third of the nonfat milk sold commercially in the United States is used in the manufacture of ice cream and related products. Aiming to encourage lower-fat eating patterns, the FDA modified its restrictions and introduced the following definitions for commercial low-fat ice cream based on a ½-cup serving (3):

- Reduced-fat ice cream: less than 7 grams of milk fat
- Light or low-fat ice cream: less than 3 grams of milk fat
- Nonfat ice cream: less than 0.5 gram of milk fat

Imitation Ice Cream. Replacing the milk fat and milk solids-not-fat in ice cream with other ingredients results in imitation ice cream. Mellorine is a frozen dessert similar to ice cream except that the milk fat has been replaced with vegetable fat. Another type of imitation ice cream contains soybean products such as tofu, soy protein, or soy isolate, with the milk fat usually replaced by corn oil. Parevine is a frozen dessert free of any milk fat or MSNF, which meets the kosher requirements set by Jewish food laws (14). An example of such desserts would be those made with tofu. Frozen desserts made for diabetics usually replace the sucrose with sorbitol.

Frozen Yogurt

Frozen yogurt sales crept along for two decades after the product was introduced, but now frozen yogurt consumption is at an all-time high (15). The increasing demand for frozen yogurt is owed in part to its taste, texture, lower fat content, "good-for-you" image, and effective marketing strategies (11, 15). Perhaps unknown to consumers is that full-fat frozen yogurt has almost as many calories (kcal) as regular ice cream—both contain about 280 calories (kcal) per 8-ounce cup. Nonfat yogurt, on the other hand, averages 180 calories (kcal) per 8-ounce cup and zero grams of fat. Decreasing the fat can compromise the taste and texture of the finished product, but improved food technology is overcoming this problem. Most frozen yogurt contains no live yogurt cultures, although some manufacturers ensure that live and active cultures are present in their frozen yogurt.

Sherbet

Sherbets are even lower in fat content than lower-fat ice cream. By law, they must contain less than 2 percent milk fat. Sherbets are often made with egg whites and/or gelatin to give them a creamy consistency. To compensate for the body lost because there is less fat, more sugar is added to the basic milk and fruit juice mixture. This makes sherbet's caloric (kcal) content similar to, and sometimes higher than, ice cream (Figure 28-2) even though sugar provides 4 calories (kcal)/gram, while fat contributes 9 calories (kcals)/gram.

Sorbet

Sorbets are made without fat, eggs, gelatin, or dairy products (milk solids-not-fat) and as such have a harder consistency than either ice cream or sherbet. They consist of puréed fruit or fruit flavoring and a sugar syrup made of equal amounts of sugar and water simmered together. Countless flavors are possible and include any of the fruits or combination thereof.

Water Ices

Water ices (glacés), made from a base of sweetened water and fruit juice, lack both fat and milk solids-not-fat. They may contain gelatin, vegetable gums, egg whites, flavorings, and/or

FIGURE 28-2 Calories and fat in frozen desserts (per cup).

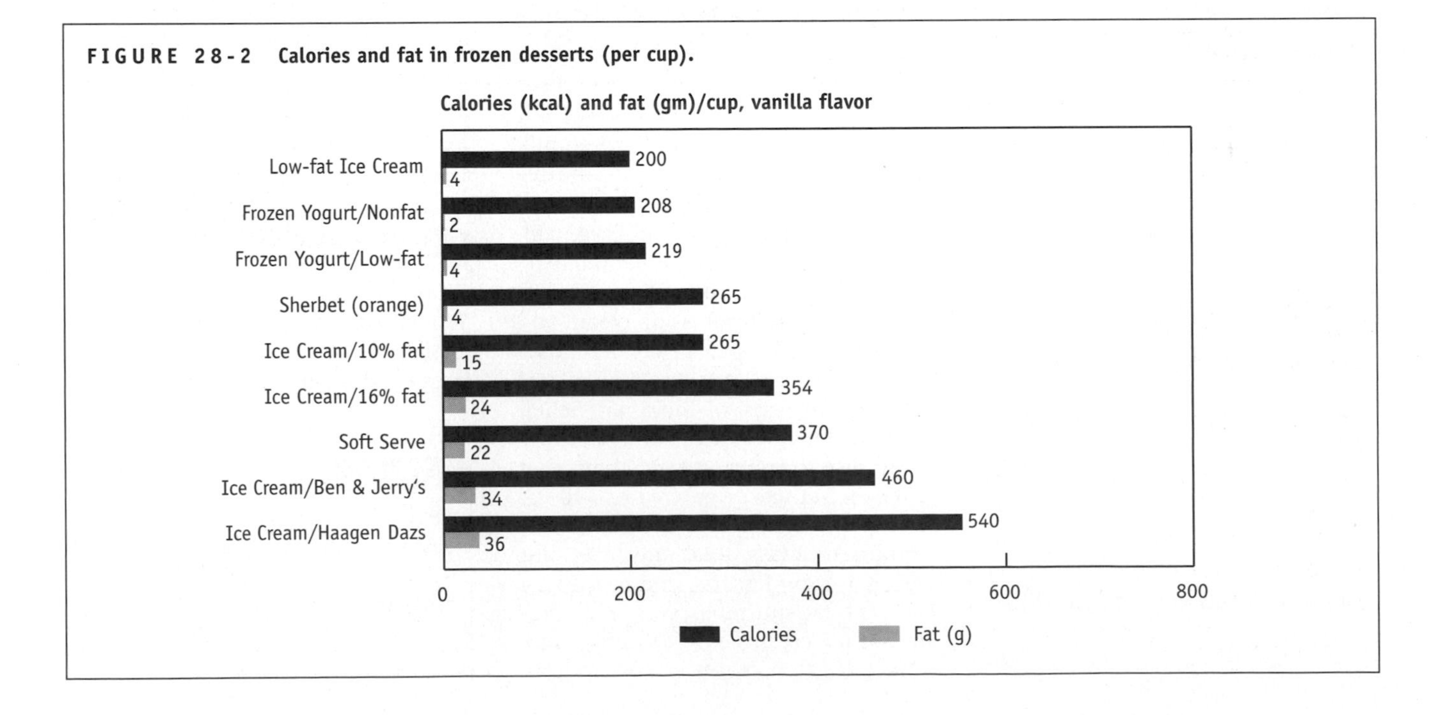

colorings. Popsicles are water ices that evolved from frozen lemonade on a stick (13). Granites from Italy are flavored ices that contain fruit and are quickly frozen to promote the formation of the large ice crystals that give them a rough texture. They are generally stirred a few times during freezing to further promote large crystalline formation. The term "granites" is often used interchangeably with "granitas," which are technically frozen fruit slushes or beverages and not frozen desserts. Though made from a similar mixture, they are different in that they are frozen solid. Another type of ice dessert is frappé, which is crushed and served as a slush after it is frozen.

Still-Frozen Desserts

A still-frozen dessert is one that is not stirred during freezing. Examples include mousses, bombes, and parfaits. The light, airy, smooth, velvety texture of these desserts comes from the incorporation of whipped egg whites or whipped cream into the mixture. The foam structure acts as a physical barrier preventing large ice crystals from forming.

Preparation of Frozen Desserts

The preparation of frozen desserts can be very time consuming. As a result, the convenience and wide availability of commercial frozen desserts results in food service establishments and individuals not routinely preparing ice cream or other frozen desserts. Some frozen desserts, however, do lend themselves to preparation on certain occasions, and this section discusses that option.

Factors Affecting Quality

The structure of frozen desserts depends on the crystallization of water from a sugar mixture (16). These crystals in many frozen desserts are made by either churning a mixture while in the process of freezing, or placing it in a mold where it is allowed to freeze (2). In preparing frozen desserts, the three general factors crucial to their quality are flavor, texture, and body.

Flavor. It is important to remember when making frozen desserts that cold temperatures mute flavors, so the mixture must be boldly flavored before freezing. The number of possible flavors that may be added to the foundation mixture of frozen desserts is almost limitless. When it comes to ice cream, however, vanilla is the most popular, and chocolate ranks second (18). Other flavor choices include, but are certainly not limited to, butter pecan, strawberry, Neopolitan, chocolate chip, cookies and cream, cherry, coffee, peach, pineapple, orange, raspberry, coffee, and cappuccino. Flavor may be further enhanced by the addition of nuts, candies, cookies, and other ingredients.

Texture. A smooth texture is preferred in most frozen desserts, with the exception of frozen ices and granites. All are crystalline products whose textures are dependent on the formation of ice crystals. The smaller and more evenly distributed these crystals, the smoother the texture, so larger crystals, with their coarser texture, are to be avoided. Ice crystal size is directly related to the number of nuclei. The greater the number of nuclei, the smaller the crystal size and the smoother the consistency (4). Other factors determining texture are the frozen dessert's content of milk solids-not-fat, fat, sugar, and other substances. About 50 percent of the total solids in ice cream and many other frozen desserts consists of dairy products (3).

HOW & WHY? ?????????

Why are MSNF sometimes added to frozen desserts? One reason MSNF or fat is added to frozen desserts is to help produce a smooth texture (9). The addition of MSNF results in smaller ice crystals, tinier air cells, and thinner air cell walls (1). Fat assists in separating water molecules and coating the frozen crystals to create the sensation of smoothness (17). Use of excess MSNF, however, produces a sandy texture from the precipitation of lactose, the least soluble of sugars (15).

Sugar lowers the freezing temperature of water, so lower-fat ice cream, sherbet, and water ices, which all contain more sugar and less fat, feel colder on the tongue than ice cream.

Monosaccharides lower the freezing point even more than sucrose; consequently, corn syrup and dextrose (glucose) are commonly added to commercial frozen desserts (16). Although the freezing point is lowered with added sugar, the sugar and other ingredients prevent the frozen dessert from freezing into a solid block of ice.

Other factors contributing to texture are air cells, emulsifiers, and the type of treatment. Ice cream is a foam of air cells, each surrounded by a layer of fat coated with emulsified protein films, and a network of ice crystals (Figure 28-3). The air cells, of which there are millions in every bite of ice cream, make the frozen mixture light and airy.

NUTRIENT CONTENT

Fat content is often the factor that differentiates frozen desserts from one another. In order to meet the demand for lower-fat products, food companies have provided an army of reduced-fat frozen dessert products that have challenged ice cream's historical share of the market. The new products use skim milk, fat substitutes, emulsifiers, and stabilizers. Overall, consumers taking a product acceptance survey reported a higher preference for lower-fat ice cream and frozen yogurt than for other kinds of frozen low-fat dairy desserts, which tend to have bland or off-flavors (11). Fat and sugar are difficult to replace in frozen desserts, which rely on these ingredients to provide smooth mouthfeel and body. The functions and replacement of sugar and fat are discussed in Chapters 9 and 10 respectively.

FIGURE 28-3 The structure of ice cream: Ice cream is a foam of air bubbles trapped in frozen liquid. The liquid contains the dissolved sugar and milk solids.

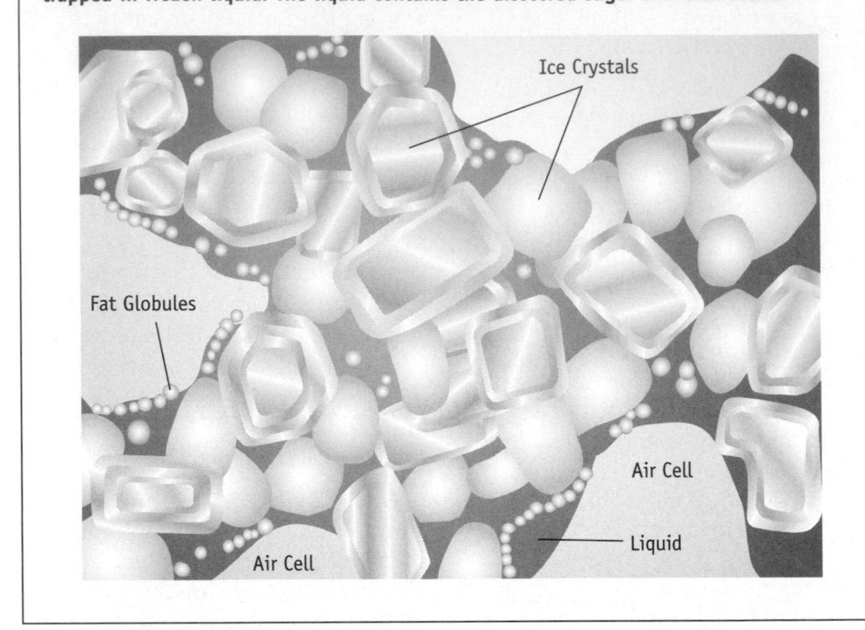

Emulsifiers such as mono- and diglycerides, egg yolk, and/or lecithin are often added to create a smoother texture by stabilizing the dispersion of air in the foam. Rapid freezing and correct storage temperatures also contribute to the smooth texture of frozen desserts. Commercial ice creams are aged, which consists of holding them at refrigerator temperatures for a given period of time. Glycerol monostearate (GMS) may be added to destabilize the emulsion by displacing protein from the fat globules, allowing them to partially aggregate. This results in a smooth texture and more resistance to melting (6).

Body. Commercial ice cream has more body than homemade ice cream because of the added stabilizers (no more than 0.5 percent). These include vegetable gums such as sodium carboxymethylcellulose (CMC), carrageenan, guar gum, agar, acacia, alginate, furcelleran, karaya, locust bean, tragacanth, and gelatin (22). These food additives (see Chapter 8) are primarily of natural origin, although a few are chemically modified. The food industry uses gums because they contribute to body, resistance to melting, viscosity, and reduced ice crystal formation during storage (12). Stabilizers are common ingredients in frozen desserts because they attach to any water that is freed from the melting ice crystals during an increase in temperature such as the opening of a freezer door. The water bound to gums cannot attach to existing ice crystals and make them larger, which is detrimental to quality (21).

Ice Cream Shrinks as It Ages. The volume of ice cream shrinks as it ages due to the collapsing films around the air cells (Figure 28-4). Another cause of shrinkage is mechanical compaction caused by a dipper being pressed against ice cream during scooping (17).

Overrun. Body is influenced to a large degree by **overrun**. The numerous small air cells created by overrun prevent the ice cream from being too hard, dense, or cold. Commercial ice cream usually contains an overrun of 70 to 100 percent (17). A 100 percent overrun means that the volume of air is equal to the volume of the mix before it was churned. Too much overrun damages body by creating a foamy texture, but too little results in heavy, compact ice cream with a coarse texture. Excessive overruns are prevented by federal standards that protect customers by setting the minimum weight for ice cream at 41.2 pounds per gallon (17). Ice cream with lower overrun and higher fat content is classified as premium ice cream, while very low overruns and high fat content—at least 12 percent milk fat—and the best ingredients constitute super premium ice cream. Overruns greater than 150 percent are possible if more stabilizers and emulsifiers are added, but the product is classified as a "frozen dairy dessert" instead of ice cream.

Mixing and Freezing

Ice Cream. Commercial mixes are used by the food industry to produce many different kinds of ice cream (Figure 28-5). Cream is the main

KEY TERMS

Overrun The volume over and above the volume of the original frozen dessert mix, caused by the incorporation of air during freezing.

FIGURE 28-4 Why ice cream shrinks as it ages.

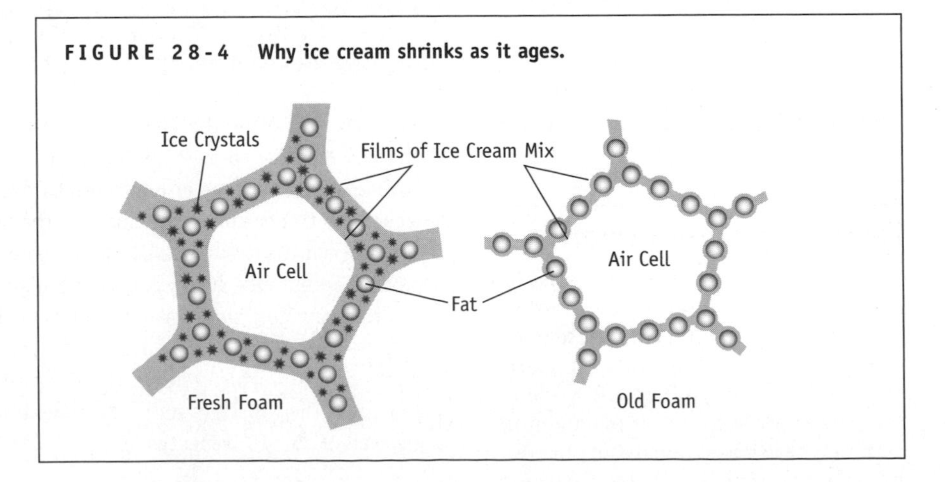

A complete mix comes all ready to just add water and blend. It can be combined with hundreds of flavors to produce an endless variety of ice cream desserts.

Reconstituted mix is transferred to the freezer equipment.

Ready to serve.

FIGURE 28-5 Making a frozen dessert from a commercial mix.

ingredient; this provides a smooth mouthfeel. In addition, sugar and flavorings such as vanilla bean, chocolate, strawberry, mint, pistachio, candy, cookie pieces, brownie chunks, and caramel or chocolate sauce are added.

Cooked vs Uncooked Ice Cream. Eggs are added to cooked ice cream. The cream is heated to scalding, and the eggs and sugar are beaten into a smooth paste and then added to the cream. The mixture is then heated until it thickens while being stirred constantly with a whisk or wooden spoon. The mixture should be heated to at least 180°F (82°C) to protect against the growth of any *Salmonella* that may be present. The mixture is removed from the heat before the flavoring agents are added. The addition of flavorings may be delayed until after the ice cream has been churned (see below) and is still soft, to prevent the breakage of ingredients such as nuts and chocolate chunks.

Whether cooked or uncooked, if fruit is used for flavoring, it should first be combined with a sugar syrup or liqueur to eliminate the unpleasant effect of solidly frozen fruit chunks on the texture. The acidity of fruit flavors is balanced by sugar; thus, double the normal amount of sugar is usually added to frozen desserts such as fruit sherbets (19). The ice cream mixture is cooled for several hours or overnight in the refrigerator before it is processed into a frozen product. Uncooked ice cream can bypass the refrigeration stage. Care should be taken to ensure that only pasteurized products are used in either cooked or uncooked ice cream to reduce the risk of foodborne illness.

The consistency of uncooked ice cream is closer to that of sherbet, which has a grainier consistency. This can be compensated for to some degree by using thick fruit pulp (7). Food safety guidelines dictate that uncooked ice cream, commercial or homemade, should never be prepared with raw eggs because of the risk of *Salmonella* foodborne illness.

Heating and Aging. Commercial preparation of ice cream consists of combining the liquid ingredients in a mixing vat where they are heated to 104°F (43°C), warm enough to dissolve the added sugar and other dry ingredients. The ice cream mix is pasteurized and homogenized; the latter improves the overall texture and body of ice cream. Commercial mixes are then aged from 3 to 24 hours at 40°F (4.4°C) in vats. During this stage, several beneficial changes take place: the fat solidifies, and there is a swelling of milk proteins, gelatin, and other stabilizers, which increases the viscosity of the mix. Smoother texture, improved body, and resistance to melting also result from the aging process (17).

Churning and Freezing. After being aged, ice cream mixes are ready to be frozen. During freezing, the ingredients are churned to promote the formation of numerous small nuclei necessary for a smooth, velvety texture, and the incorporation of air to increase volume. The two major noncommercial methods for mixing ice cream ingredients are the electric mixer method (ice cream machine) and the old-fashioned hand-cranking method. In both methods, the ice cream freezer canister is filled only two-thirds full to allow for the natural expansion that occurs with freezing (water expands when frozen) and the incorporation of air during churning. This container, equipped with a dasher for rotating the mixture, is surrounded with a combination of rock salt and crushed ice (1:4–6 ratio, with the ratio depending on the dessert) to promote freezing to temperatures below 32°F (0°C) during churning (Figure 28-6) (Chemist's Corner 28-2). The dasher should be turning before the mixture is added.

HOW & WHY? ?????????

Why is rock salt added to ice cream machines? Salting the ice surrounding a churning ice cream canister speeds the melting of the ice, which absorbs energy (heat) away from the contents of the canister. It keeps the canister and its contents cooler. The goal in adding an ice and salt mixture is to remove sufficient heat from the ingredients in the canister so that they start to freeze. The melting ice signals this absorption of heat.

It is best to fill the space around the canister half full with crushed ice and then add the salt to prevent it from

FIGURE 28-6 Ice cream freezer.

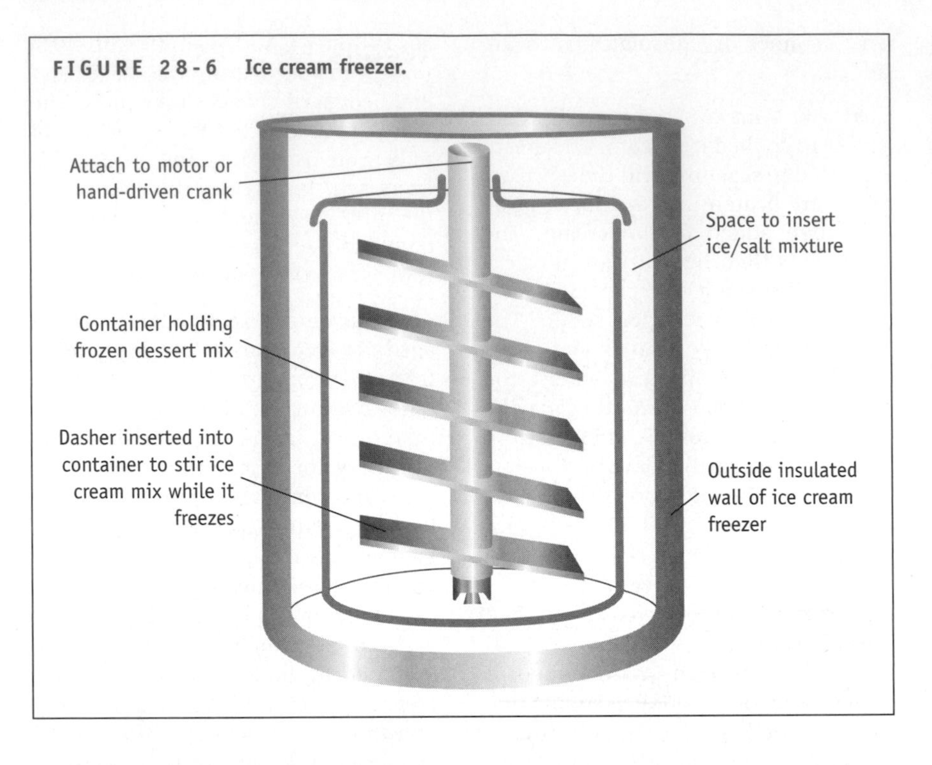

aggregating at the bottom. The remaining salt and ice can be mixed together and then added, or layered alternately on top of the ice around the canister. Crushed rather than cubed ice is used because of its greater surface area, and rock salt is preferred over table salt, which tends to cake, is more expensive, and dissolves too quickly. To maintain the cooler temperatures, it is important not to discard the liquid that results from the salt melting the ice. The cooler temperature of the outside container necessitates the constant churning of the ice cream mixture in order to prevent large ice crystals from forming on the perimeter. The sharp edge of the revolving dasher moves ice crystals forming on the container's wall toward the center to promote uniform crystal size formation.

Chemist's Corner 28-2

Salt's Effect on Freezing

The freezing point of water decreases 6.7°F (3.72°C) for each gram molecular weight (58 grams) of sodium chloride (NaCl) added to 1 liter of water. Every gram of melting ice requires 80 calories in energy, which is absorbed by the ice-salt mixture around the ice cream canister (14).

Speed of Churning/Freezing. The number of revolutions per minute of the mixer is low at first, but is increased near the end to facilitate the formation of small crystals. This is based on the principle that slow freezing results in large crystals, while quick freezing creates tiny ice crystals (4). The speed of the dasher also determines how much air is incorporated into the mixture. A faster dasher introduces more air, creating a lighter ice cream. Slowing down the dasher traps less air, yielding a denser ice cream (10). The ice cream is finished churning in about 20 minutes, when the dasher becomes difficult to rotate.

Storing and Hardening. The dasher is then carefully removed and the ice cream is packed down and put in the household or institution freezer for about four to six hours before serving. Ice cream may be served immediately after preparation, but its texture will be somewhat soft. If the ice cream is too grainy, a common fault, it may be because of one of several preparation errors: filling the container more than two-thirds full; using too much salt, which freezes the ingredients too rapidly and does not allow enough time for the formation of small crystals; and/or churning too rapidly. Freezing will also have an impact on the ice cream's texture. The initial churning freezes about 33 to 67 percent of the water, while the second freezing during storage, called the hardening phase, freezes an additional 23 to 57 percent of the water (16).

Frozen Yogurt. An ice cream freezer is used to prepare frozen yogurt. The process is similar to that for making ice cream, except that the ingredients are yogurt based. Emulsifiers may also be added.

Sherbet. The procedure for preparing sherbet is again similar to preparing ice cream, but the ingredients are syrup based rather than milk based. No egg yolks are used. The sugar concentration is key to determining the final flavor of the sherbet. This can be fine-tuned either by tasting or by using an instrument that measures sugar concentration (Chemist's Corner 28-3).

Sorbet. Sorbets are relatively easy to prepare. A purée of either fresh or frozen fruits is combined with a sugar syrup and then frozen in an ice cream maker. Fruits that are beyond ripe, but not rotten, are best because they yield the most intense flavors. A sweet fruit should be the primary choice and may be balanced by one that is more tart, such as strawberries or plums (5). Herbs such as rosemary, thyme, or basil may also be added as a secondary flavor (7). If the fruit is less than ripe, it can be poached in the

Chemist's Corner 28-3

Measuring Sugar Concentration

A saccharometer, also known as a syrup-density meter, hydrometer, or Baumé hydrometer, measures the sugar concentration of a liquid. The calibration of the hollow glass tube ranges from 0° to 50° and is read as "degrees of Baumé." The recommended range for sherbets, sorbets, and ices is about 16° to 20° Baumé (7).

sugar syrup to soften the flesh and make it easier to purée. Poaching is always recommended for pineapple and kiwis, no matter how ripe, in order to destroy certain enzymes that will interfere with freezing.

Sorbet Syrup Preparation. The syrup of equal parts sugar and water is simmered prior to adding the fruit purée. It is best if the syrup is still warm when the purée is added in order to bring out the most flavors from the fruit. Chilling the mixture before churning speeds up the freezing process, but the mixture should first be tasted for any necessary flavor adjustments, because once it is frozen the flavors will be less intense. In its liquid state, the sorbet mixture should have the desired flavor but be too potent to drink. Tart citrus flavors such as lemon, grapefruit, or lime can be added if the mixture is too sweet.

Churning Sorbets. The final stage of sorbet preparation consists of placing the mixture in an ice cream freezer that incorporates air into the fruit syrup, creating a light, airy texture. After it thickens, the sorbet should be transferred to an airtight container and frozen for at least a couple of hours before consumption, although it may be consumed immediately. Sorbet made this way will keep for up to two weeks (5). The use of stabilizers such as gelatin, pectin, agar-agar, or gum tragacanth improves the shelf life of sorbets by reducing the tendency of their frozen liquids to separate from the ice crystals (7).

Water Ices. Water ices are made by combining one part sugar with four parts liquid and flavoring. This flavored syrup is then placed in a mold, covered with aluminum foil, and frozen. Some frozen ices, such as granites, are stirred periodically during freezing to promote large ice crystal formation. Too much sugar will prevent the mixture from freezing, while too little results in an unappetizing hard-frozen mass.

Still-Frozen Desserts. Still-frozen desserts are frozen without stirring, usually in molds. The mixtures for these vary tremendously in their ingredients and preparation. Their light, airy texture and high volume are derived from whipped cream or egg-white foams that are folded into the other ingredients. The smoothness of still-frozen desserts is often the result of emulsifying agents having been added to prevent the formation of large crystals during freezing. Examples of such emulsifiers include eggs, gelatin, corn syrup, and/or cornstarch. It is best to remove still-frozen desserts from the freezer a half hour prior to serving, but care should be taken with regard to food safety because many contain eggs or cream.

Storage of Frozen Desserts

Ice cream is best stored at temperatures of 0°F (−18°C) or below for one to two months. A thin, plastic film is sometimes used inside the carton to cover commercial ice cream. If this is lacking, a sheet of wax paper can be pressed against the ice cream before resealing the carton or the entire carton can be placed in an airtight plastic bag. This prevents absorption of other food odors in the freezer and exposure to moisture buildup, which can promote the formation of large crystals. Ice crystals do not remain the same size as when they formed, but slowly enlarge, so the consistency of frozen desserts rarely stays the same throughout their storage life.

Texture Changes

The gradual change in texture often observed in ice cream, lower-fat ice cream, and sherbet results from transportation from the store, repeated removal of the frozen dessert from the freezer, and/or freezer temperature fluctuations. During any one of these states, some of the ice crystals melt into water, which attaches to neighboring crystals. Consequently, the crystals get larger and larger, creating a coarse texture (20). Over time, the number of ice crystals decreases while the average ice crystal size increases. Thawing frozen desserts, even partially, and then refreezing them has a negative effect on their quality (Figure 28-7). It is best, therefore, to keep them at freezing temperatures and protect them as much as possible from changes in temperature.

Scooping Frozen Desserts

In spite of the hazard to its quality, slightly warmer ice cream is preferred during serving because it is easier to scoop and more flavorful. A half gallon can be easily sliced or scooped when left in the refrigerator for about 15 minutes. Scooping downward into the container to gather a compact ball of ice cream or sherbet results in fewer servings per tub than skimming the dipper over the surface. Dipping the dipper in cool water before scooping also helps to keep the ice cream from sticking to the dipper.

FIGURE 28-7 **Ice cream quality is best when first purchased because numerous small crystals are present (left). Any increase in temperature causes the smallest crystals to melt (center). The extra water is taken up by the remaining crystals, making them larger and the ice cream more grainy (right).**

PICTORIAL SUMMARY / 28: Frozen Desserts

From 62 A.D., when Roman Emperor Nero sent runners to the mountains to fetch ice to be flavored with nectar, fruit pulp, and honey for him and his court, frozen desserts have maintained their popularity. Today, commercially frozen treats are probably the most commonly consumed desserts in North America.

TYPES OF FROZEN DESSERTS

Federal and state regulations determine the minimum levels of fat and milk solids-not fat (MSNF) for many frozen desserts.

What makes one frozen dessert different from another? The ingredients, proportions, and the way they are prepared:

Ice cream: Ice cream is prepared by simultaneously stirring and freezing a pasteurized mix of dairy (milk, cream, butterfat, etc.) and non-dairy (sweeteners, stabilizers, emulsifiers, possibly egg, colors, and flavors). By law ice cream must contain at least 10 percent milk fat and 20 percent milk solids-not-fat, with premium ice creams containing at least 14 percent.

Low-fat ice cream (ice milk): Lower-fat ice cream usually contains more sugar than milk and less than 7 percent fat.

- *Reduced-fat ice cream:* Less than seven grams of milk fat.
- *Light or low-fat ice cream:* Less than 3 grams of milk fat.
- *Nonfat ice cream:* Less than 0.5 grams of milk fat.

Imitation ice cream: Milk fat is replaced with other ingredients such as vegetable fat or corn oil. Other imitation ice cream products contain soy products.

Frozen yogurt: Full-fat frozen yogurt has almost as many calories (kcal) as regular ice cream (about 280 kcal per 8 oz.) but nonfat yogurt averages only 180 kcal per 8 oz., plus zero grams of fat.

Sherbet: Sherbets must contain less than 2 percent milk fat, and their creamy consistency often comes from egg whites and/or gelatin. More sugar is added so their caloric value is actually very similar to regular ice cream.

Sorbet: Sorbets consist of puréed fruit or fruit flavorings and a sugar syrup. Sorbets have a harder consistency than ice cream or sherbet.

Water ices (glacés): Water ices, made from a base of sweetened water and fruit juice, lack both milk fat and milk solids-not-fat. Examples: popsicles, granites, and frappés.

Still-frozen desserts: These desserts are not stirred while freezing. Examples include mousses, bombes, and parfaits. The use of whipped egg whites or whipped cream results in an airy, velvety texture, without large ice crystals.

Ice cream and other frozen desserts continue to be popular.

PREPARATION OF FROZEN DESSERTS

The first step in preparing a frozen dessert is to combine the ingredients, which may or may not be cooked. Frozen dessert mixtures are then either churned using an ice cream freezer (electric or hand-cranked) until they are frozen (ice cream, sherbet, sorbet, frozen yogurt), or placed in a mold (sorbet, water ices, still-frozen desserts) where they are allowed to freeze.

Flavor, texture, and body all contribute to the quality of frozen desserts. The texture of frozen desserts is based on their crystalline nature—the smaller and more evenly distributed these crystals are, the smoother the texture. Larger crystals yield a coarser texture. Other factors contributing to texture are fat, milk solids-not-fat, air cells, emulsifiers, and processing. Body is defined as the product's combined firmness, richness, and viscosity.

Among the different flavors of ice cream . . .

. . . **vanilla** is the most popular, followed by **chocolate**.

STORAGE OF FROZEN DESSERTS

Ice cream is best stored at temperatures of 0°F (-18°C) or below for one to two months, and is best if protected as much as possible from fluctuations in temperature.

REFERENCES

1. American Dairy Institute. *Nonfat Dry Milk in Frozen Dairy Desserts.* American Dairy Institute, 1988.
2. Bocuse P, and F Metz. *The New Professional Chef. The Culinary Institute of America.* Van Nostrand Reinhold, 1996.
3. Chandan R. *Dairy-Based Ingredients.* Eagen Press, 1997.
4. deMan JM. *Principles of Food Chemistry.* Van Nostrand Reinhold, 1990.
5. Deville D. Cool sorbets, intensely flavored. *Fine Cooking* 16:67–71, 1996.
6. Euston SE, et al. Oil-in-water emulsions stabilized by sodium caseinate or whey isolate as influenced by glycerol monostearate. *Journal of Food Science* 61(5):916–920, 1996.
7. Friberg B. *The Professional Pastry Chef.* Van Nostrand Reinhold, 1996.
8. Goff HD. Partial coalescence and structure formation in dairy emulsions. In *Food Proteins and Lipids,* ed. S Damodaram. Plenum, 1997.
9. Guinard JY, et al. Sugar and effects on sensory properties of ice cream. *Journal of Food Science* 62(5): 1087–1094, 1997.
10. Jay S. What to look for in an ice cream machine. *Journal of Food Science* 31:50–53, 1999.
11. Low-fat frozen desserts: Better for you than ice cream? *Consumer Reports* 57(8):483, 1992.
12. Martin DR, et al. Diffusion of aqueous sugar solutions as affected by locust bean gum studied by NMR. *Journal of Food Science* 64(1):46–69, 1999.
13. McGee H. *On Food and Cooking: The Science and Lore of the Kitchen.* Macmillan, 1997.
14. McWilliams M. *Foods: Experimental Perspectives.* Macmillan, 1997.
15. O'Donnell CD. Dairy derivatives that deliver. *Prepared Foods* 166(6):129, 1997.
16. Penfield MP, and AM Campbell. *Experimental Food Science.* Academic Press, 1990.
17. Potter NN, and JH Hotchkiss. *Food Science.* Chapman & Hall, 1995.
18. Pszczola DE. 31 ingredient developments for frozen desserts. *Food Technology* 56(10):46–65, 2002.
19. Smith D. Sugar in dairy products. In *Sugar: A User's Guide to Sucrose,* eds. N Pennington and C Baker. Van Nostrand Reinhold, 1990.
20. Sutton RL, D Cooke, and A Russell. Recrystallization sugar/stabilizer solutions as affected by molecular structure. *Journal of Food Science* 62(6):1145–1149, 1997.
21. Sutton RL, and J Wilcox. Recrystallization in model ice cream solutions as affected by stabilizer concentration. *Journal of Food Science* 63(1):9–11, 1998.
22. Whistler RL, and JN BeMiller. *Carbohydrate Chemistry for Food Scientists.* Eagen Press, 1997.

WEBSITES

Instructions and tips on preparing frozen desserts:

http://www.essetti.com/applncs/frznrecipes.htm

Prepare various frozen desserts using this website:

http://www.bartleby.com/87/0026.html

Nutrition information from Nutrition Action on various frozen desserts:

http://www.cspinet.org/nah/6_98des.htm

29

Beverages

All living organisms require water for survival. This essential nutrient can be absorbed by some through cell membranes or walls, through skin by others, and in higher organisms consumed in the form of foods and/or directly by drinking. Human bodies are approximately 65 to 70 percent water, so fluids must be consumed daily to maintain that supply. Helping to meet this need are foods such as fruits, vegetables, milk, meat, and eggs, which all contain approximately 70 percent water. Beverages, however, contribute a greater share to the human requirement for water, and they do this in sometimes mundane, sometimes delicious ways. Most beverages used for obtaining fluids fall into one of the following seven categories: water, carbonated beverages, New Age beverages, fruit/vegetable beverages, aromatic beverages (coffee and tea), dairy beverages (see Chapter 11), and alcoholic beverages (21). This chapter discusses these beverages, specifically their composition, processing, preparation, and storage.

KEY TERMS

Mineral water Water from natural springs having a strong taste or odor due to small amounts of salts of calcium, magnesium, and sodium (sodium bicarbonate, sodium carbonate, sodium chloride), and sometimes iron or hydrogen sulfide.

Water

Water is so vital to human life that its source determined where early humans lived—near streams, rivers, lakes, and springs. A person can survive up to 40 days without food, but only about 7 days without fresh water or water-based liquids. Water contains no calories (kcal) or vitamins. Its predominant mineral content, usually either calcium or sodium, varies according to region and determines whether the water is hard or soft (see Chapter 2). In some locations, the water is so high in calcium that it serves as a significant source of daily dietary calcium (19). The wide number of plain water variations available on the market are now discussed in more detail.

Types of Water

There are now so many different types of water available that bottled water is not just a trend, it is a gusher (30). So much so that both Coca-Cola and PepsiCo have entered the bottled water market. Bottled water can be made from various types of water—mineral, deionized, distilled and sparkling.

Mineral Water. Unless it has been distilled or deionized, all water is **mineral water** because it naturally contains dissolved mineral salts. Natural springs tend to carry higher concentrations of minerals, such as sodium chloride, sodium bicarbonate, sodium carbonate, salts of calcium and magnesium, and sometimes iron or hydrogen sulfide, and these give mineral water its characteristic taste. Bottled water labeled "natural" has an unaltered mineral content. Examples of bottled mineral water from springs, some of which are flat and some "sparkling" (i.e., carbonated), include Perrier (Figure 29-1), Evian, Vichy, Contrexeville, and Apollinaris. Bottled water may be labeled **spring water**, **well water**, or **artesian water**.

Deionized Water. Deionized water is **purified water** that has had all of its mineral content removed. The deionization process removes all the mineral salts (via ion exchange), leaving only pure water. Use of deionized water can make beverages taste less flavorful.

Distilled Water. Distilled water is also purified water, but it is formed by converting water into steam and then condensing it in another, cooler container. The purpose of distillation is to separate a liquid from other particles or liquids. The steam leaves behind the dissolved particles and/or other liquids with higher boiling temperatures. The steam funnels upward into another chamber, where it is collected as it condenses back into a liquid. This process not only removes dissolved minerals, but destroys pathogens. Sometimes normal concentrations of minerals are added back into the water (21).

Distilled water is preferred for making ice cubes because they will be clear rather than cloudy, as they are when made from tap water. The cloudiness is caused by dissolved air bubbles, which are removed by the process of distillation. The same result can be obtained by boiling tap water for several minutes. Whether the water is distilled or boiled, the resultant ice cubes last longer because they are denser. Ice cubes can also pick up off-flavors from being held in the freezer too long with other foods. Rinsing the ice cubes before putting them in beverages promotes the elimination of any possible freezer odors (26).

Sparkling Water. Carbon dioxide gas contributes the fizz and bubbles to sparkling water, which is available naturally from underground sources (14). Tap water that has been commercially filtered and carbonated is called seltzer. Club soda (soda water or plain soda) is somewhat different from carbonated water in that, in addition to having been filtered and carbonated, it has had mineral salts such as bicarbonates, citrates, and phosphates of sodium added. Sometimes club sodas, flavored club sodas, and seltzers serve as a source of sodium (10).

KEY TERMS

Spring water Water that, according to the FDA requirements, flows from its source without being pumped and contains at least 250 parts per million of dissolved solids.

Well water Water pumped from an aquifer, an underground source of water.

Artesian water Water that has surfaced on its own from an aquifer, rather than being pumped.

Purified water Water that has undergone deionization, distillation, reverse osmosis, or any other method that removes minerals, chemicals, and flavor.

Distilled water Water that has been purified through distillation to remove minerals, pathogens, and other substances.

FIGURE 29-1
Carbonated mineral spring water.

Carbonated Beverages

Carbonated soft drinks, sodas, or "pops" are the most widely consumed beverage in North America, even more so than tap water (34). Cola drinks are the most commonly purchased soft drinks, followed by lemon-lime, orange, ginger ale, root beer, and grape drinks (57). What are people actually buying when they purchase soft drinks? About 90 to 98 percent is sparkling water, 10 percent is sugar, and the remaining small percentage consists of additional flavors, colors, acids, and preservatives (Chemist's Corner 29-1).

Chemist's Corner 29-1

The Water in Soft Drinks

The water used in soft drinks is relatively chemically pure compared to that obtained from municipal drinking water. Tap water cannot be used in soda production because the alkalinity is too high, causing it to neutralize the acids; minerals such as iron and manganese interfere with coloring and flavoring agents; and residual chlorine is detrimental to flavor. Bottling plants treat water to be used for sodas by deionization, chemically precipitating the minerals, using activated charcoal to removed undesirable flavors, odors, and chlorine, deaeration to remove oxygen, and a final filtration to remove any remaining compounds passing the carbon filter (43).

Early Soft Drinks

The major ingredient in soft drinks, sparkling water, was created in 1772 by the chemist Joseph Priestley, who added carbon dioxide, a colorless, nontoxic gas, to water. One of the early methods of creating carbon dioxide consisted of acidifying baking soda (sodium bicarbonate), which is where the term "soda" originates. The word "soft" was originally used to distinguish these beverages from so-called "hard" alcoholic drinks (21). In addition to providing sparkling bubbles, carbon dioxide produces a tingling mouthfeel, an effervescent appearance, and a preservative effect to the beverage.

In the 1800s, another chemist, Benjamin Silliman, began selling carbonated soda water to the public (21). Flavored soda water became popular after 1830, and in the late 19th century a druggist invented a flavored soda by adding an extract from the African kola nut. Coca-Cola was later invented in 1886 by a pharmacist, John Pemberton, as a hangover remedy (38). The list of Coca-Cola ingredients has been guarded for over 100 years, but one researcher claims the original Coca-Cola recipe included citrate, caffeine, extract of vanilla, seven flavoring oils, fluid extract of coca (cocaine), citric acid, lime juice, sugar, water, caramel, and alcohol (52). The use of cocaine became illegal in the 1930s. Another pharmacist creating a soft drink was Charles Hires, who added a unique extract to sparkling water and introduced root beer.

Soft Drink Processing

Soft drinks today are carbonated by placing a syrup/water mixture in a pressurized carbon dioxide container known as a carbo-cooler. After interacting with the carbon dioxide and cooling off, the newly carbonated syrup water is pumped to a filler, which then allocates a set amount of the soft drink into a sterile container (23). Added sweeteners, usually in the form of corn syrup, sucrose, or monosaccharides, contribute flavor and increase density, viscosity, mouthfeel, and calorie (kcal) content. Sucrose was originally used to sweeten soft drinks, but was gradually replaced by the less expensive corn syrups. The fructose in corn syrup is sweeter than sucrose, so less is required.

Various acids also contribute to the flavor of soft drinks: citric acid from citrus fruits contributes tartness, usually to fruit drinks; phosphoric acid, usually used in cola drinks and root beers, is flat and sour; and malic acid from apples is used for its long-lasting flavor (43). Preservatives added to soft drinks usually consist of benzoic and sorbic acids and their calcium, potassium, and sodium salts (23).

Diet Soft Drinks. Diet soft drinks utilize alternative sweeteners such as aspartame, saccharin, acesulfame-K, or sucralose. The result is an increase in water content from 90 to 98 percent, and a decrease in sugar to zero, which has a detrimental effect on mouthfeel (23, 24). Bulking agents such as carboxymethyl cellulose or pectin are often added to counter this side effect (43). Diet soft drinks were introduced in the 1950s and have steadily grown in popularity, with diet colas now accounting for at least one-third of the soft drink market (11, 41).

Fruit and Vegetable Beverages

Most fruits and vegetables contain very high proportions of water, which makes them easy to squeeze or pulverize into a juice. Two exceptions to this rule are bananas and avocados, which hold very little water. Orange juice sales lead the fruit juice market, followed by apple and grape juices (7). Although not a major threat to the market share of these traditional juices, tropical juices such as guava, mango, and kiwi are growing in popularity (37). Depending on the fruit or vegetable from which it is made, the juice may provide some vitamin A and/or C. The amount of these nutrients from juices compared to soft drinks is shown in Table 29-1.

Fruit/Vegetable Juice Processing

Extraction of juices is the first step in manufacturing fruit and vegetable juices. Other steps of juice production include clarification, deaeration (removing the air), pasteurization, determining concentration/additions, and packaging (bottle, can, or freezing) (43).

Juice Extraction. Prior to juicing fruits or vegetables, most of them must be thoroughly washed, freed of any bruises, seeds, and mold, and often peeled and pitted. Citrus fruits are generally cut in half and squeezed either mechanically or by hand. Although the pithy white parts of oranges and grapefruits containing bioflavonoids and vitamin C may remain, it is important to remove the skins in order to remove bitter-tasting and possibly toxic compounds. The skins of tropical fruits are also usually peeled, because they may have been grown in a country allowing questionable pesticides. The skins of waxed produce and the leaves of carrots and rhubarb also must be removed. Cutting the fruits and vegetables into slices or chunks allows them to be placed in a juicer, where they can be processed into a beverage.

TABLE 29-1

Nutrient Comparison Between Various Fruits and Vegetable Beverages vs Soda

Beverage (8 oz/240 ml)	*Calories*	*Vitamin A (RE)**	*Vitamin C (mg)*
Soda			
Soda	150	0	0
Juice			
Apple, canned/bottled	117	–	2
Apricot nectar, canned	141	332	2
Carrot, canned	95	6327	21
Cranberry cocktail, bottled	144	1	90
Grape, canned/bottled	155	2	–
Grapefruit, fresh	96	3	94
Lemon, fresh	60	5	112
Lime, fresh	66	3	72
Orange, fresh	111	50	124
Papaya nectar, canned	142	28	8
Passion fruit (yellow)	149	595	45
Peach nectar, canned	134	64	13
Pineapple canned	157	–	26
Prune, canned	182	1	10
Raspberry, bottled	98	24	36
Tangerine, canned	124	104	55
Tomato, canned	42	136	45
Vegetable (V8), canned	46	283	67
Reference Daily Intake (RDI)		5000	60

*RE = Retinol equivalents—a measure of vitamin A activity based on an older estimation of how much vitamin A the body derives from carotenoids.

Different types of juice extractors result in varying amounts of pulp; some strain out most of the pulp, while others produce a thick, pulp-rich beverage. The most traditional commercial method of extracting juices is through the use of presses (7). Fresh fruit and vegetable juices taste best and are most nutritious when they are consumed immediately after preparation rather than stored for future use (12).

Clarification. The natural pressing of plant material leaves a semifluid mass of cell wall material (cellulose and pectin). The small particles of suspended pulp remaining in pressed juice are called haze or cloud (6). Many consumers prefer crystal-clear fruit juice, so these particles are removed.

HOW & WHY? ?????????

How is the pulp removed from juice?

Fine filters or high-speed centrifuges are used to remove the haze. The latter separate contents based on differences in density. Any minute pulp and colloidal particles still remaining can be removed by adding commercial enzymes (Chemist's Corner 29-2), which break down the pectic substances that settle to the bottom of the juice, followed by further filtering or centrifugation (2). These enzymes, often called "liquification" enzymes, are often added in the clarification of apple, cranberry, and grape juices (5, 7).

In tomato juice, the suspended pulp, which contributes to viscosity, is desired. If fresh tomato juice is left to stand, however, a natural enzyme called pectin methyl esterase breaks down the pectin, resulting in a progressively thinner tomato juice over time. Tomato juice manufacturers can use a process called the hot-break process to inactivate the enzymes by quickly heating the product to 180°F (82°C). Allowing the enzyme activity to proceed and produce a tomato juice with a thinner consistency is called the cold-break process (43).

Deaeration. Entrapped air in juices is removed by deaeration, a process that reduces undesirable changes due to oxygen. Juices are deaerated by spraying them into a vacuum deaerator. Deaeration improves shelf life, maintains flavor, and reduces the breakdown (oxidation) of vitamin C (43).

Pasteurization. The high heats of pasteurization inactivate enzymes and destroy microorganisms that can cause foodborne illness. Some juices, especially fresh juices, are not pasteurized, which raises concerns when *Escherichia coli* outbreaks from nonpasteurized juices are reported (see Chapter 3). Pasteurizing juices diminishes their fresh-squeezed flavor, which can be partially replaced by adding back essence oils and folded peels from the original juice (32). Alternatives to pasteurizing juices currently being studied include irradiation, hydrostatic pressure, ultrasound, high-intensity pulsed electrical fields, and oscillating magnetic fields (36) (see Chapter 7).

Concentration/Additions. Blends of fruit flavors and juices are common and range from pure fruit juices to highly diluted, artificially flavored and colored "drinks," which may contain little of the actual juice or its nutrients. The latter types of drinks may also be carbonated, and although they may be less thick and sweet than regular fruit juice, some are higher in sugar. Some juices are fortified with calcium to appeal to consumers interested in lowering their risk of osteoporosis (50).

The leading beverages to which fruit juice has been added include sparkling water, noncarbonated drinks, teas, and

Chemist's Corner 29-2

Enzymes Reduce Haze

Cellulase and pectinases are added to juices to hydrolyze the cellulose and pectin respectively (5). Amylase enzymes are added to eliminate starch haze caused by the presence of glucose polymers (6).

wine coolers (44). The sport or isotonic beverages (discussed below) designed to maintain electrolyte balance are noncarbonated drinks that sometimes contain fruit juice. There are also "texturally modified" beverages that use real fruit pulp or pieces and other components to enhance mouthfeel (45).

New Age Beverages

New Age beverages are the latest breakthrough in the manufacture and sale of beverages. The term was coined in the mid-1990s to represent all beverages that did not fit into the traditional categorization of drinks, and includes, but is not limited to, beverages identified as sport or isotonic, herbal, neutraceutical, energy, smart, and fun (28).

Sport or Isotonic Beverages. Physical performance is often enhanced by sport or isotonic beverages that prevent dehydration, replace important electrolytes, and provide extra energy in the form of carbohydrates. Carbohydrates are available in many forms, but usually include sucrose, glucose, and maltodextrin. Adequate hydration, according to the American College of Sports Medicine, is achieved by most people if they drink about 17 ounces (500 ml) of fluid approximately two hours before exercise and at regular intervals during physical exertion (18). However, physical performance lasting under one hour does not benefit from the carbohydrates in isotonic drinks. Another benefit of sport or isotonic drinks is their electrolyte content of potassium and sodium in the form of monopotassium phosphate, potassium chloride, sodium chloride, and sodium citrate.

Herbal Beverages. Adding an herb to a drink results in an herbal beverage. In the United States, beverages are now sold containing popular herbal supplements (9). Examples include ginseng ("boosts energy"), ginko ("sharpens the mind"), echinacea ("boosts immune system"), kava ("alleviates stress"), and St. John's wort ("herbal antidepressant"). Additional supplements may include vitamins, minerals, bee pollen, chromium, carnitine, picolinate, and selenium, to name a few (3, 16).

Nutraceutical Beverages. Drinks delivering health benefits are described as nutraceutical or functional beverages. Japan was selling functional beverages 20 years ago, elixers aimed at countering stress, fatigue, and alcohol (27). Ingredients include green tea, antioxidants, soy, fiber, probiotics, phytochemicals, vitamins, minerals, and even oxygen (48, 55). Despite the concern with overblown claims, these nutraceutical beverages continue to appear on the market. One of the oldest nutraceutical beverages is cranberry juice, recommended to reduce or treat urinary tract infections (60). Normally, the bladder is sterile, but bacteria like *Escherichia coli* can cause an infection that can enter by the urinary tract, pass the bladder, and end up in the kidneys, where it can cause serious damage if not treated. The unique blend of organic acids in cranberry juice appears to have an inhibitory effect on potentially harmful bacteria. Another plant compound serving as the foundation of a nutraceutical drink is inulin. This carbohydrate-based fiber promotes the growth of beneficial bacteria in the large intestine, which are thought to inhibit the risk of colon cancer (49).

Energy Beverages. Energy drinks are either high in carbohydrates or added stimulants such as caffeine. Combining both compounds is an orange juice containing high levels of caffeine.

Smart Beverages. Certain compounds claimed to boost brain powers are added to beverages sold as "smart" drinks. Although there is very little to no research supporting such claims, some of these substances said to stimulate mental activities that are added to smart drinks include amino acids such as choline, L-cysteine, taurine, and phenylalanine (28).

Fun Beverages. Eye-catching, "fun" beverages are some of the latest New Age beverages to appear on the market. One example suspends minicapsules within liquid to create a visually distinctive beverage that looks like the inside of a lava lamp (46).

Coffee

Coffee is one of the two most common aromatic beverages, the other being tea. An aromatic beverage is defined as one generating a pleasant odor from plants or spices. Coffee consists primarily of water with some additional compounds extracted from coffee beans during brewing. Scandinavians consume more coffee overall than any other nation in the world, with Americans second in consumption (51).

Coffee was probably discovered in Ethiopia around the 3rd century A.D., when an Arabian goat herder named Kaldi noticed that his goats became particularly frolicsome after eating certain berries. The berries, of course, turned out to be from the coffee plant. Although many varieties of this plant are grown, only two, *Coffea arabica* and *Coffea robusta*, constitute 99 percent of the commercial crop. A third variety is *Coffea liberica* (21). Brazil and Columbia are the world's top coffee-producing countries, but other countries with tropical rain forest climates conducive to coffee production include Indonesia, Mexico, Central America, and several African countries. Knowing where a coffee is grown is important to coffee producers, because the flavor and quality of coffee vary according to the region where the particular coffee plant is grown.

Coffee Processing

Coffee is made from beans that are picked, partially dried, processed to remove the hull, and either roasted and ground immediately, or exported to a processing plant for roasting and grinding. A small percentage arrive on the market as whole beans. Most imported beans are shipped green, because they can be stored in this state with little loss of quality. Coffee beans can also be processed to remove the caffeine, and liquid coffee extract can be used to produce instant coffee.

Removing the Hull. Coffee plants, which are evergreen, generate numerous white flowers that yield a cherry-like fruit containing two seeds, the coffee "beans," enclosed in a tough skin. Coffee beans are usually hand-picked and then subjected to one of two treatments to remove their hulls and free their greenish-blue berries. Countries with a limited water supply, such as those in the Middle East, historically have used the dry method where they sun-dry the beans. In the wet method, the berries are washed and then hulled by machine and partially fermented. Coffee processed by the wet method is more expensive, but experts agree that it has a better flavor than coffee processed by the dry method. Whichever method is used, the beans are then graded for size, type, and quality. It takes about 2,000 hulled coffee beans to produce 1 pound of coffee.

Roasting. Roasting changes the beans' physical and chemical makeup and develops their flavor, aroma, and appearance. Physically, the beans dehydrate, becoming lighter and more porous as their surface area is increased.

HOW & WHY? ?????????

Why do roasted coffee beans have an oily coating? The oily surface of roasted coffee beans is due to the breakdown (hydrolysis) of fats (fatty acids released from triglycerides) that occurs during the heat of roasting.

The beans also darken from green to brown during roasting. Unroasted green coffee beans would have little, if any, flavor and aroma.

Roasting time determines whether a coffee is classified as a light, medium, or dark roast. The darkest roasts, in ascending order, are French (New Orleans), Spanish (Cuban), and Italian (Espresso) roasts. Europeans prefer darker roasts, while Americans gravitate toward medium-roast coffees.

Grinding. Once roasted, the beans are ground to create more surface area from which the hot water can extract the compounds that contribute to flavor, aroma, and appearance. The type of grind—fine, drip, or regular—depends on the equipment that will be used to brew the coffee. Coffee beans ground too fine for their equipment deliver a bitter cup of coffee, while coarse grinds produce a weak, flavorless coffee. Generally, fine grind is used for vacuum and pressure pots, drip grind for drip pots, and regular for percolators. Particle size does not differ very much among these three main grinds, although fine has a greater number of smaller particles, while regular has the largest.

Decaffeination. Currently, over 20 percent of the coffee consumed is decaffeinated. Caffeine is usually removed by soaking green coffee beans in steam or water and extracting the caffeine from the water with a solvent such as methylene chloride or ethyl acetate. When used in large amounts, methylene chloride has been shown to cause cancer in laboratory animals, so the FDA limits its concentration to 10 parts per million. Ethyl acetate occurs naturally in fruits and vegetables, but some manufacturers choose nonchemical decaffeination methods, such as filtering the soaking water with activated charcoal. Such coffee is labeled as "water-processed" decaffeinated coffee. Regardless of which method is used, not all the caffeine is removed. Most decaffeinated coffees are about 97 percent caffeine free. Scientists are working to produce caffeine-free coffee beans through genetic engineering (22).

Instant Coffee. There are two major methods for manufacturing instant coffee, spray-drying and freeze-drying. In the first method, a strong extract of coffee is sprayed through a jet of hot air, evaporating the water and leaving dried coffee particles. Freeze-drying is accomplished by freezing coffee concentrate into a solid mass and breaking the mass into small particles, which are then heat-dried in a vacuum.

KEY TERMS

Polyphenol An organic compound with two or more phenols—carbon atoms structured into an aromatic ring with one or more hydroxyl (-OH) groups.

Composition of Coffee

Coffee's aroma, taste, and stimulating qualities are derived from substances extracted from the beans when hot water is poured over them. These substances are almost completely depleted from the ground coffee in this way, so the grounds should never be reused. There are more than eighteen classes of flavor compounds in ground coffee, but to simplify the discussion, only the volatile compounds, bitter substances, and methylxanthines will be reviewed.

Volatile Compounds. The roasting of green coffee beans creates many volatile substances, which give coffee its characteristic fragrance (53). All of the compounds responsible for the aroma of coffee have not been deciphered yet, but over 600 volatile compounds have been identified so far (33). Heat vaporizes these substances into the air, where they signal to the human nose that the coffee is ready.

HOW & WHY? ?????????

Why does coffee lose its flavor as it cools or stands? The instability of some of the volatile compounds causes the flavor of roasted coffee to deteriorate quickly as it cools. In addition, when coffee is kept hot too long, evaporation causes the loss of many of these compounds and their accompanying flavors.

Bitter Substances. The bitter, slightly sour taste of coffee is partially due to its organic acids. Chlorogenic acid, which constitutes about 4 percent of roasted coffee beans, contributes to over half of the acid found in a cup of coffee. Two other components contributing to the bitter taste of coffee are the **polyphenol** compounds and caffeine. Coffee should never be

KEY TERMS

Methylxanthine A compound that stimulates the central nervous system.

boiled, because temperatures at or above boiling increase the solubility of these compounds, resulting in coffee that has moved from pleasantly bitter to unpalatable.

Methylxanthines. Coffee contains two substances, caffeine and theobromine, which belong to a group of compounds called **methylxanthines** (Figure 29-2). Methylxanthines can have either positive or negative effects, depending upon the individual. Possible side effects of caffeine, other than alertness and increased exercise performance, include temporarily increased heartbeat, metabolism, and stomach acid; sleep disturbance; and dilation and/or constriction of certain blood vessels. It also has diuretic effects, causing increased urination and increased loss of calcium through the urine. In addition, withdrawal symptoms, including headache, fatigue, moodiness, depression, and anxiety, often appear after abrupt cessation of habitual caffeine use (59). Caffeine is also present in over 60 other plants, including tea leaves, cocoa beans, and the kola nut, but coffee and tea contain the highest concentration of caffeine. Theobromine is slightly less stimulating than caffeine; it is also found in cocoa and chocolates.

The amount of caffeine in a cup of coffee depends on which brewing method is used. Brewed drip coffees have the most caffeine, about 132 to 180 mg per cup, while decaffeinated coffees have the least, about 3 mg per cup. In comparison, tea contains about 40 mg of caffeine per cup. New caffeinated beverages on the horizon include a combination of a cola drink and coffee, a citrus-flavored drink with caffeine, and orange juice mixed with caffeine in a bid to replace the traditional morning coffee. Caffeine sells soft drinks, as evidenced by many of the top-selling soft drinks containing caffeine. Countering the caffeine-loaded beverages are caffeine-free sodas, decaffeinated coffee and tea, herbal tea, and caffeine-free chocolate and cocoa substitutes.

FIGURE 29-2 Methylxanthines: Compounds found in certain foods and beverages that stimulate the central nervous system.

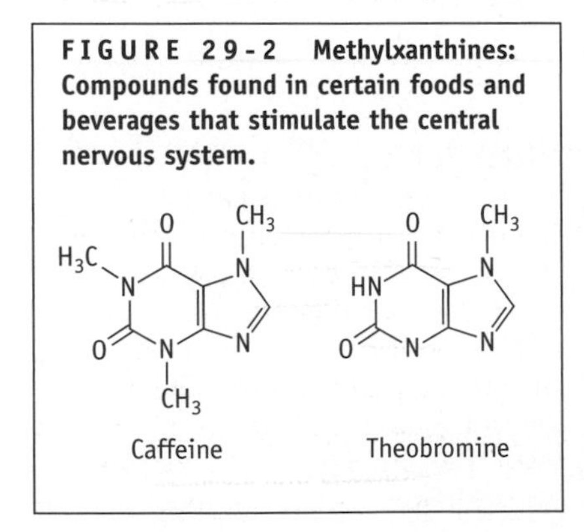

Preparation of Coffee

Many coffee-drinking consumers judge an establishment by its coffee. The key to making a good cup is to extract the compounds that contribute to good flavor and aroma, while simultaneously limiting the extraction of bitter substances. The elements that go into an enjoyable cup of coffee are quality, freshness, water-to-coffee ratio, water, temperature, brewing time, brewing equipment, and holding time.

Coffee Quality. *Coffea arabica* produces the finest coffee, while *Coffea robusta* yields a strong but inferior quality coffee. However, almost all coffee sold in the United States is blended and known by a brand name rather than its variety.

Blended Coffees. Blending different varieties of beans allows the best characteristics of each to be fully expressed. Coffee beans found in nature rarely possess the taste, aroma, body, and consistency of blended varieties, so the beans chosen for blends are those that will contribute the best qualities derived from various beans (21). Sometimes a vegetable called chicory is added to blends to reduce their cost. This root is dried, roasted, and ground to look like coffee, but its use produces a heavier, more bitter, and darker cup of coffee.

Some desired qualities and the coffees that contribute these qualities to a blend include:

- Richness/body (Sumatran, Java)
- Sweetness (Venezuelan, Haitian)
- Extra sweetness (Mature Java, Mysore)
- Flavor/aroma (Sumatran, Colombian, Celebes)
- Brightness/acidity (Costa Rican, Central American coffees)

Unblended Coffees. Lately the demand for unblended coffees such as Mocha, Java, and Kona has increased. Many specialty coffee roasters are also offering popular "estate" coffees, or coffees grown in specific locations, resulting in certain characteristic flavors.

Factors Influencing Coffee Quality. The quality of coffee is also influenced by whether the coffee is ground, instant, decaffeinated, or espresso. Instant coffee is convenient, less expensive, and constitutes about one-fifth of all coffee sold. Its quality, unless it is decaffeinated, is inferior to freshly brewed coffee, which limits its use in restaurants. Quality of decaffeinated coffees varies, ranging from detectable in difference from regular coffee to no discernable difference. Espresso coffee is made from beans that are roasted until they are almost black, and twice the normal amount of ground coffee is used. Espresso is Italian for "pressed out," describing how high-pressure steam is used to force water through the ground coffee, yielding a very dark, very strong coffee, usually served in small cups. Espresso can be used as the basis for a number of drinks. Caffé mocha is espresso and hot chocolate mixed together in a tall glass, possibly topped with whipped cream and sprinkled with cocoa powder. Caffé latte is primarily steamed milk with espresso poured into it, while cappuccino consists of espresso with foamed milk on top.

Coffee Variations. Coffee can be combined with other ingredients to yield many different and flavorful products:

- *Specialty coffees.* Blends of coffee flavored with chocolate, sugar, almonds, and even liquor. They are higher in price but preferred by certain coffee connoisseurs.
- *Imitation coffee.* A cheaper version of coffee made from roasted and ground grains. Molasses is often added to cover the grainy taste and aroma.

- *Café au lait.* A double-strength coffee mixed with equal parts of hot milk.
- *Flavored coffees.* Coffees containing flavored syrups made from sugar, water, natural flavor, and a preservative.
- *Iced coffee.* A popular beverage in Japan and increasingly more common in North America.

Coffee Freshness. Coffee flavor and aroma begin to deteriorate as soon as the coffee beans are roasted. Deterioration and staling become even more rapid once the coffee is ground, so it is usually packaged in vacuum-packed cans. The fresher the bean, the better the coffee, and for that reason, many people buy whole beans and grind them on a daily basis for ultimate freshness. Coffee, whether ground or in bean form, is often stored in the freezer to maintain its freshness.

Water-to-Coffee Ratio. The normal ratio of coffee to water ranges from 1 to 3 tablespoons of coffee (1 tbs = weak, 2 tbs = medium, 3 tbs = strong) for each 6-ounce cup, or about 1 pound of coffee for every 1¾ to 2 gallons of water. Measurements for coffee pots are based on an average 6-ounce coffee cup. Personal preferences and geographical locations dictate the strength of the dilution.

Water Type. Fresh, cold, tap water is usually best for making coffee. Hot water has a tendency to produce a flat, stale cup of coffee. Deionized or distilled water tends to produce a sour cup of coffee. Water that is naturally soft is best, although chemically softened water is not recommended, because it filters more slowly and the sodium ions may combine with the fatty acids of the beans to form soaps (58).

Water Temperature. The water to be used for coffee should not be boiled because the dissolved oxygen will escape, resulting in flat-tasting coffee. Boiling also causes a loss of volatile compounds responsible for flavor; thus the aroma of boiling coffee smells good, but the coffee itself lacks flavor. It is best to warm the water to just below boiling temperatures (190° to 200°F/88° to 93°C) in order to extract just the right amounts of substances from the coffee grind. At these temperatures, about 20 percent of the weight of the grounds is extracted, which makes the best cup of coffee. Lower temperatures do not extract sufficient flavor, while higher temperatures extract too many bitter compounds.

Brewing Time. Brewing time usually averages about 5 minutes, depending on the type of grind. A fine grind requires less time for extraction than a coarse grind. The fineness or coarseness of grind selected is often determined by the brewing equipment used (Table 29-2).

Brewing Equipment. Different types of coffee-making equipment are shown in Figure 29-3. Materials used to make coffee equipment should be stainless steel, glass, earthenware, or enamelware; uncoated metals leave an undesirable aftertaste. After use, the equipment must be washed to remove the natural oils that accumulate and may become rancid. Thorough rinsing removes any soap residue. The basic styles of brewing equipment rely on drip, vacuum, percolator, filter, and steeping mechanisms.

Drip Coffee Makers. Automatic or nonautomatic drip coffee makers are probably the most popular devices for making coffee. They consist of a brewing basket with a perforated bottom, a filter that goes in the basket, and a coffee pot to catch the dripping coffee that results from the heated water passing through the coffee grounds.

Vacuum Coffee Maker. The vacuum coffee maker looks something like an hourglass. The cold water goes in the lower portion, and fine coffee grounds go in the top. The seal between the two is tight, creating a closed system that allows steam to build up in the bottom. The pressure from the steam forces the water through the funnel to the top, where it comes in contact with the coffee grounds. The heat is then reduced to allow the water to stay on top for 1 to 4 minutes. Removing the pot from the heat source cools the water and creates a vacuum that pulls the coffee down. Two potential problems with vacuum coffee makers are that the coffee may boil, and it may not be hot enough when served.

Percolator. In a percolator, a brewing basket is placed on the top of a tube whose base rests on the bottom of the pot. A tight lid with a glass dome is placed on the pot. Heating forces the water up the tube into the dome and down over the coffee. Electric percolators are considered best, because both the heat and timing are automatically regulated. Stove-top percolators must be timed carefully, and coffee quality is harder to control. Over-percolating by either method results in flavor loss. When the coffee is ready to serve, the grounds should be removed to prevent them from absorbing the aroma of the coffee.

Filter. The filter method is relatively easy, consisting only of two parts. A coffee filter is placed in a filter holder, which is then set on top of an empty coffee pot or coffee cup. Coffee grounds are placed in the filter, and enough hot water is poured through to match the number of cups desired.

TABLE 29-2
Brewing Time Increases with Larger Grind Sizes

Grind	Brewing Time (Minutes)	Usual Equipment
Very fine	1–2	Expresso, Filter
Pulverized	2–3	Turkish coffeepot
Fine	2–4	Vacuum
Drip	4–6	Drip pot
Regular	6–8	Percolator/steeped
Coarse	6–8	Percolators with aluminum filters

FIGURE 29-3 Coffee brewing equipment.

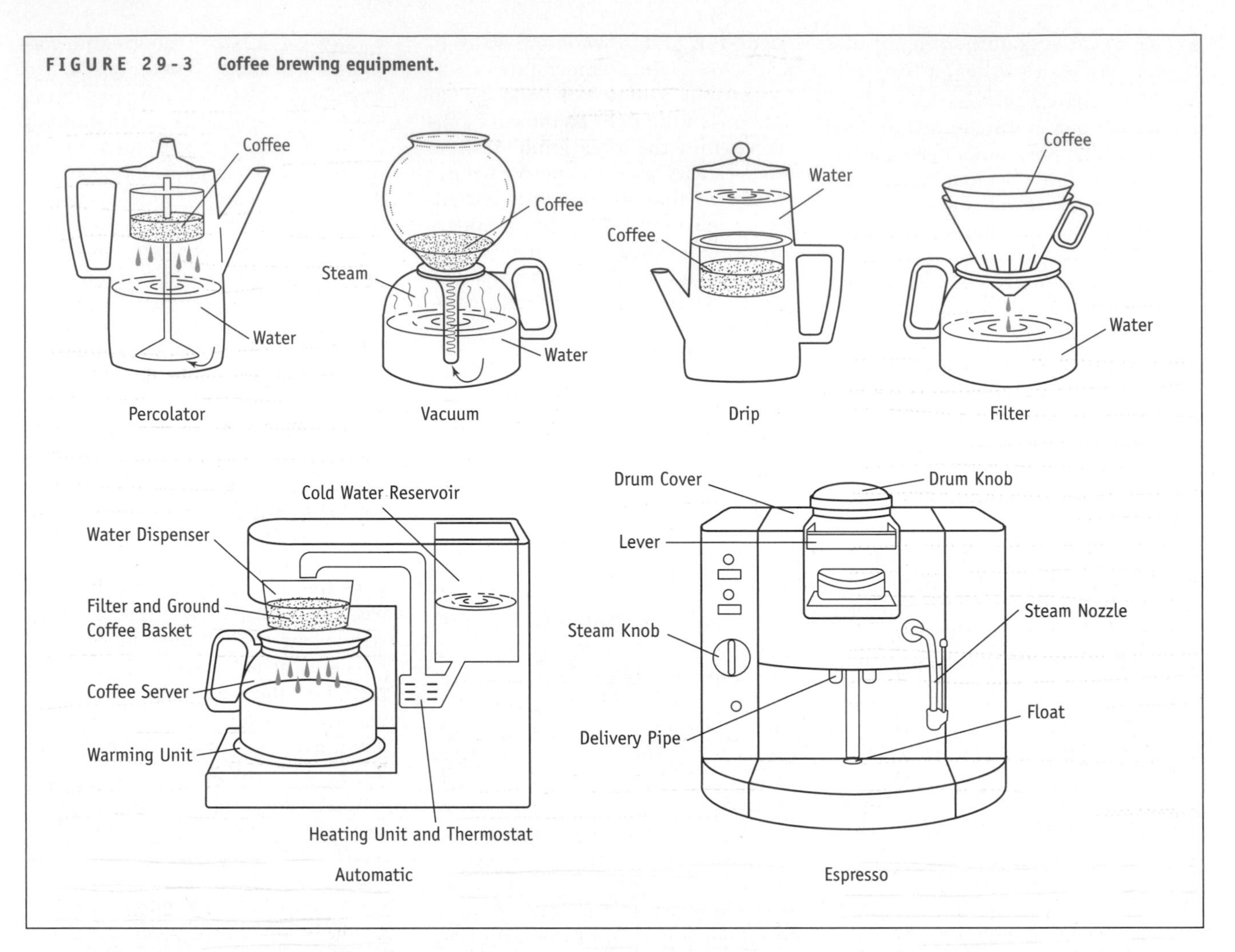

Steeping Method. The steeping method requires only a pot and a lid. It is a straightforward method ideal for use on a picnic or camping trip. Regular ground coffee is heated in water to just below the boiling point and held for several minutes. The mixture is then filtered to remove the larger grounds, while the remaining fine particles sink to the bottom. Large quantities, known as "camp coffee," can be steeped by wrapping and tying up the grounds in cheesecloth.

Holding Time. For the best flavor, piping hot coffee should sit for 3 to 5 minutes before being sipped in order to allow the flavor to mellow. Coffee can be held at 185° to 195°F (85° to 88°C), but the longer it is held, the more the flavor deteriorates. Coffee should never be held for over an hour or the loss of flavor will be substantial. Higher temperatures will also negatively affect the quality of the coffee. Reheating is not recommended because it causes coffee to lose much of its flavor and aroma due to volatile compounds being lost, destroyed, or altered.

Storage of Coffee

Whole, roasted coffee beans should be transferred from their paper bag to a tightly sealed glass or metal container and stored in the refrigerator, freezer, or other cool, dry place. Properly stored, they will retain their freshness for up to three weeks. Whole coffee beans can be frozen for several months, and there is no need to thaw the beans before grinding. For the freshest taste and aroma, they should be ground just before brewing. Coffee beans contain sufficient fat to make them good candidates for rancidity with improper or over-long storage; this will also cause them to lose volatile essential oils. Most ground coffee on the market is vacuum-packaged, but once opened, it starts to lose its freshness. For maximum keeping time, ground coffee should be protected against any exposure to air or moisture by being stored in an airtight container in the refrigerator or freezer.

Tea

Despite the popularity of coffee, its consumption has been steadily decreasing over the past several years as consumption of various kinds of teas and other alternative beverages has increased (15). Tea remains one of the most popular beverages in the world with about one-third of Americans drinking tea every day (29).

Tea's origins can be traced to ancient China. Legend has it that around 2737

B.C., Chinese emperor Shen Nung was boiling water when some wild tea leaves fell into the pot. The aroma was so pleasant that he tasted the brew and declared it good (13). The plant from which the leaves fell was an evergreen shrub, the *Camellia sinensis.* It is believed to have been first grown in India. In the 16th century, traders from Europe carried tea leaves and the knowledge of how to make the brew home from the Far East, and by the 18th century it had become the national drink of England. A tea tax increasing the cost of tea imported by American colonists is said to have triggered the Boston Tea Party of 1773 and the start of the American Revolutionary War (21).

Tea comes from many regions throughout the world. The United States imports most of its tea from Sri Lanka (Ceylon) and India, but other main tea-producing countries include China, Indonesia, and Japan. As with coffee, the country of origin and its climate influence the flavor characteristics of each tea variety, but the method of processing is even more important.

Tea Processing

Only the smallest and most tender leaves are picked (Figure 29-4). About 1,000 leaves are needed to make 1 pound of manufactured tea. Processing consists of withering, rolling, oxidizing, and firing. Withering tea leaves is accomplished by spreading them out in thin layers to expose them to warm air for 6 to 18 hours to reduce their moisture content to 55 to 65 percent. Rolling disrupts cell structure within the tea leaves, allowing oxidation to occur. Oxidizing, also called "fermenting," is the process whereby the natural enzymes in the leaves, released by rolling, cause the leaves to become darker. Firing dries the leaves by passing them through a hot dryer for 20 minutes, inactivating the enzymes and decreasing their leaf moisture (43).

Types of Tea. Three categories of tea that result from different processing techniques are black, green, and oolong (Figure 29-5). The majority, about 77 percent, of commercially produced tea is black tea, the predominant choice in North America, Great Britain, Europe, and some Asian countries. People in China, Japan, Korea, and India prefer green tea, which accounts for 21 percent of the produced tea. Oolong tea, which is consumed primarily in southeastern China, accounts for the remaining 2 percent of tea production (1). These teas may be further distinguished by added flavorings. In addition, various herbal teas are available on the market.

Black Tea. The leaves are allowed to wither and dry under hot blown air (80° to 200°F/27° to 93°C) before being rolled to break the membranes between the cells. Enzymes, naturally occurring in the leaves, then oxidize the polyphenolic compounds in the cells and create changes in color, taste, and aroma.

Green Tea. For green tea, the leaves are heated and steamed before they are rolled in order to inactivate the enzymes, thus preventing oxidation and preserving the lighter color.

Oolong Tea. Oolong tea is processed to be somewhere between black and green tea. It is only partially oxidized

Tea Board of India

FIGURE 29-4 **Harvesting tea leaves.**

FIGURE 29-5 Three major types of tea based on processing technique.

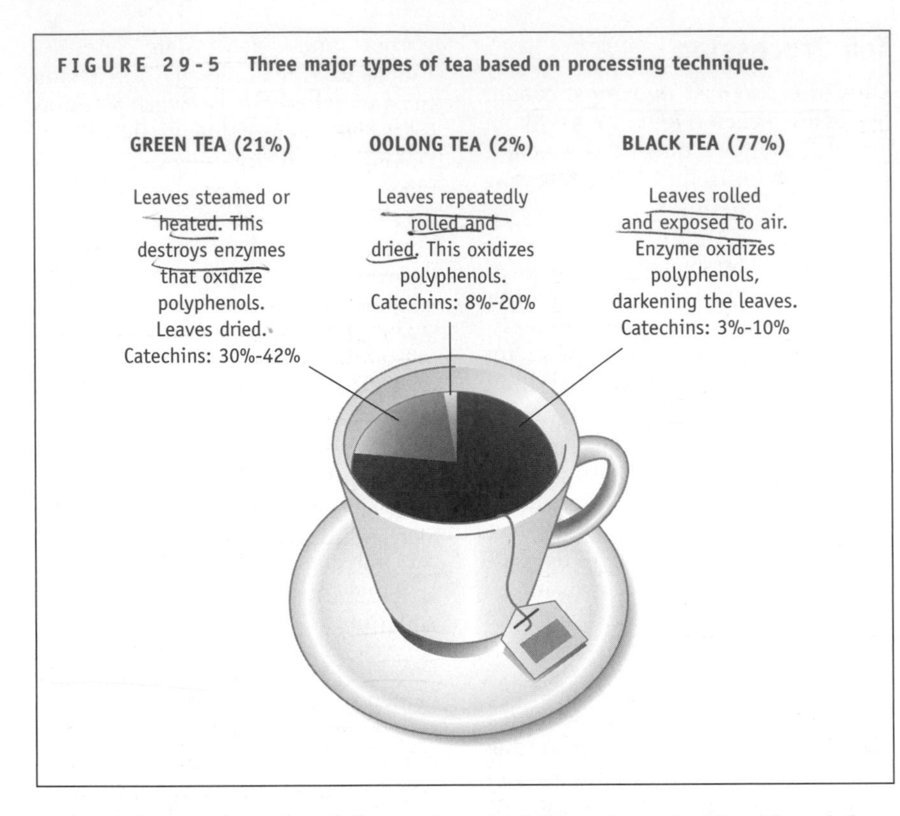

and is often perfumed and flavored with jasmine flowers.

Flavored Teas. These are teas flavored with natural oils, spices, and dried flowers or fruits. Almond, orange peel, cinnamon, and lemon peel are a few such teas.

Herbal Teas. A wide variety of leaves, flowers, barks, and/or fruits of plants other than the tea plant are used to make herbal teas, which are sometimes called "infusion." Technically these are not teas, because they do not contain tea leaves. They take their name from the fact that they are brewed in a similar fashion to tea. Examples of herbal teas include chamomile, ginseng, spearmint, rose hip, and raspberry. These and other herbal teas have had health and therapeutic benefits claimed for them, but research is still inconclusive. On the other hand, it has been reported that sassafras tea contains the carcinogen safrole; that comfrey in large amounts causes liver damage; that lobelia can cause vomiting; and that woodruff, tonka beans, and melilot are anticoagulants that can result in bleeding (52, 56).

KEY TERMS

Catechins Flavonoid pigments that are a subgroup of the flavonol pigments.

■ ■ ■ ■

Tannins Polymers of various flavonoid compounds, of which some of the larger ones yield reddish and brown pigments.

Grades of Tea

The grade of a tea is decided primarily on leaf size. The leaves are separated by screens with different sizes of holes. Quality decreases as the size of the leaf increases. The largest leaves are generally packaged as loose tea and are named orange pekoe (peck-oh), pekoe, and pekoe souchong. Smaller or broken leaves are usually used in tea bags and called broken orange pekoe, broken orange pekoe fannings, and fannings (21). Tea bags became popular when an American merchant sent tea leaf samples wrapped in silk bags to customers, who took advantage of this to brew individual cups of tea. The leaves in tea bags are usually blended to yield a characteristic flavor associated with a given brand.

Composition of Tea

Over 300 compounds have been found in black tea, and about 30 in green tea. Some of the major constituents, other than water, in a cup of tea are polyphenolic compounds (such as **catechins** and flavonols), methylxanthines, and volatile compounds. Catechins tend to be found in higher concentrations in young tea leaves than in older leaves (20). The concentration of polyphenols also varies depending on whether the tea is black, green, or oolong. Polyphenolic compounds and/or their oxidative products are largely responsible for the flavor and strong astringency of tea. Green tea has the greatest amount, which gives this tea its metallic taste. Phenolic compounds are believed to have possible antioxidant and anticarcinogenic properties (17, 40). On the other hand, polyphenols bind iron and prevent some of it from being absorbed by the intestine if tea is consumed with a meal containing iron (8). Methylxanthines, such as caffeine and theobromine, are found in tea, but in smaller quantities than are found in a cup of coffee. Green tea is an excellent source of fluoride, and a moderate source of folate, a B vitamin. Other compounds in tea, the aflavins, help decrease the bitterness of the caffeine, while **tannins** contribute to its astringency and characteristic reddish-brown color. Very few nutrients, other than water, fluoride, and folate (B vitamin), are found in tea.

Preparation of Tea

The goal in making tea, as in coffee-making, is to extract the compounds responsible for good flavor and aroma, while limiting the extraction or development of bitter substances.

Fresh tea may be consumed hot or iced, but the brewing process is the same for both.

Brewing Tea. Tea is easy to prepare. Water is heated to just below the boiling point (212°F/100°C) and poured over a tea bag or 1 teaspoon of loose black tea per 6-ounce cup and allowed to steep for 3 to 5 minutes. Then the tea or bag is removed, and the beverage is served hot or with ice (23). In any case, it is important to avoid boiling the water or steeping too long, both of which will extract more of the bitter compounds, increase the degree of astringency, and elevate caffeine concentration. The caffeine content of tea can be reduced by 50 to 75 percent if the tea bags are soaked in cold water overnight. If the opposite is preferred, stronger teas can be prepared by increasing the amount of tea leaves rather than the brewing time, because the longer the tea steeps, the more bitter it will become. The addition of milk decreases the bitterness of tea because milk proteins bind the astringent tannic acids (26). Using glass, pottery, china, or stainless steel teapots will keep metallic tastes to a minimum. The teapot should always be covered to prevent the escape of volatile compounds.

Iced Tea. The story goes that a vendor selling hot tea on a blistering, humid day at the 1904 World's Fair in St. Louis was not having much luck until he dropped a couple of ice cubes into the brew. Iced tea now accounts for more than half the tea consumed in the United States. It starts out hot, using double the amount of tea, and it can be poured over ice cubes immediately or allowed to cool before the ice is added. Sun tea is a popular means of creating iced tea. This is made by placing eight to ten regular-sized tea bags in a 1-gallon jar of cold water and setting it out in the sun for two to four hours. The bags are then removed and the brew refrigerated. Overnight tea can be made by placing six regular-sized tea bags in 1 quart of cold water, covering, and placing it in the refrigerator for at least eight hours. Fruit-flavored and herbal teas also make good iced teas. Flavored ice cubes made of citrus, cranberry, or other fruit juice add a special taste to iced teas. Sugar and lemon are often added to iced tea; mint less often. Lemon combats the astringency and lightens the tea by bleaching the tannins. It also prevents the tea from clouding by creating an acid environment, which inhibits the precipitation of compounds formed between caffeine and the aflavins. "Ready-to-drink" iced teas were introduced in 1991 and are available unsweetened or sweetened.

Instant Tea. Instant iced tea can be made by mixing soluble tea powders with water and ice. Instant tea powder is manufactured by dehydrating a strong concentration of tea and sometimes mixing it with sugar and other flavorings.

Microwaving. A single serving of tea can be made in the microwave oven by placing one tea bag in a 6-ounce microwave safe cup filled with cold water, microwaving on high for 30 seconds without boiling, and letting it stand for 30 seconds before removing the tea bag. Iced tea can be made the same way using a 12-ounce glass and ½ cup of cold water. It is allowed to cool a few minutes before adding ice.

Storage of Tea

Proper storage is important in maintaining the quality of tea. It keeps longer than coffee, but when it is stored improperly, it is susceptible to oxidation as well as to the loss of volatile compounds. Tea should be stored in airtight containers at temperatures below 85°F (30°C).

Dairy Beverages

Milk-based beverages contain more nutrients, in the form of protein, calcium, and B vitamins, than most other beverages. The choices are many, beginning with whole milk and moving to reduced fat (2 percent), low fat (1 percent), fat-free (nonfat), and reconstituted nonfat dried milk (NFDM). Any of these can be used to make hot chocolate, milk shakes, and chocolate or other flavored milk drinks. Some drinks, known as "nutritional beverages," formulated to enhance dietary intake, may be made with a milk base (45). Fermented or cultured dairy beverages such as kefir, acidophilus, and buttermilk are also available. These and a variety of other dairy products are explained more fully in Chapters 11 and 28. Smoothies are a unique way of blending dairy and fruit flavors, are often perceived as "healthy," and are becoming increasingly available (47).

Alcoholic Beverages

Alcoholic beverages fall into three major categories: beer, wine, and spirits. Figure 29-6 shows that, although each of these drinks differs in its volume of serving size, each contains the same amount of alcohol—1 ounce. The differences among alcoholic beverages stem from how they are manufactured. Beers are fermented by the action of yeast on grain, primarily barley, while wines are derived from yeast fermenting the sugars derived from fruit, primarily grapes. Spirits, or hard alcohols, contain more alcohol and are so called because they contain (via distillation) the "spirit or soul" of fermented grain or fruit mixtures. Over 50 years ago, apple cider was

FIGURE 29-6 All these drinks contain the same amount of alcohol—1 ounce.

KEY TERMS

Hops The dried fruit of the *Humulus lupulus* plant, which grows in the Pacific Northwest of the United States.

considered an alcoholic beverage, but the term, for which there is no legal definition, is now used to describe fresh-pressed apple juice with a darker color and less clarity than regular apple juice because of its suspended solids (54). However, "hard cider" is often used to describe an alcoholic beverage.

Calorie (Kcal) Content

Regardless of their source, alcoholic beverages have little to offer in the way of nutrients. Alcohol does provide calories (9 calories/kcal per gram), however, and lower-calorie (kcal) alcoholic drinks are appearing on the market as weight-conscious consumers become a growing segment of the population. Mild vodkas and whiskeys, with half the alcohol, are available in Europe (42). In North America, it is more common to find "lite" beers and wine coolers. The term "lite" refers to lower-calorie (kcal) beer. The word "light" is used to refer to beer that is lighter in color than the "dark" beers, so some confusion is possible between the two terms. Normally, beer averages about 150 calories (kcal), while "lite" beers contain about 100 calories (kcal) per 12 ounces. A martini made with vodka or gin will average 150 calories (kcal), while a gimlet can run as high as 350. A number of nonalcoholic beverages are available that have fewer calories than their original counterparts (4).

Beer

Baere is the German word for barley and the root of the word "beer." Barley is the principle grain used in the production of beer, although other grains can serve as a source of carbohydrates, which are eventually fermented by yeast to alcohol. Different types of beer result from brewers using different grains and processing techniques.

Malt Production. Regardless of the grain, the yeast cannot utilize the starch unless it is first converted to simple sugars or malt. Malt is an extract of the grain's sugars (maltose, maltotriose, and glucose) and the compound giving beer its characteristic color and flavor. The darker the malt, the darker the beer.

The natural process of germination activates starch-breaking enzymes that supply sugars, such as malt, to the growing seedlings. In order to produce malt, beer manufacturers must first allow the barley to germinate by soaking the grain in water, a production step known as steeping (31). Germination is then halted by drying the grain, a process called kilning, which results in malt. The barley malt is made further available to the yeasts by the next step, called mashing, in which malted barley is mixed with water and heated to gelatinize its starches. Mashing converts the nonfermentable starches into simpler sugars. As a result, the liquid fraction resulting from mashing, called wort, is very high in sugars capable of being fermented by yeast (43).

Brewing Beer. The next stage in beer production, brewing, is such a critical step that the entire process of manufacturing beer is often referred to as "brewing." Extracting plant materials by pouring hot water over them describes the brewing of coffee, tea, and beer. Beer brewing further consists of adding **hops** to the wort and boiling the mixture.

HOW & WHY? ?????????

Why are hops important to beer brewing? Hops contribute a number of substances to beer flavor: essential oils (responsible for aroma and flavor), bitter resins, and the tannins that contribute to color and astringency. The bitterness of hops is lost during storage when the resins are broken down (oxidized and polymerized) (31).

Brewing also concentrates and sanitizes the wort, inactivates enzymes, precipitates proteins contributing to haze, and caramelizes the sugars.

Adding Yeast. After cooling, this mixture is ready to be innoculated with *Saccharomyces carlsbergensis* yeast. This ferments for approximately nine days to produce a beverage containing about 4.6 percent alcohol by volume (31), though the alcoholic content of beer can range from 2 to 8 percent (21). Yeast is also important in determining the dryness or sweetness of beers. The degree to which sugar is converted to alcohol is defined as attenuation, and high attenuation yields dry (nonsweet) beers, while low attenuation produces sweet beers. To stop the production of alcohol from yeast, the temperature is quickly lowered to 32°F (0°C), and the beer is filtered to remove the yeast and other suspended particles.

Lagering. The next step, called lagering, involves storing the beer in tanks for several weeks to months, allowing the development of flavor, body, and the settling of particles. The particles, traces of degraded proteins and tannins, will cause an unsightly haze in chilled beer, so they are filtered out to give the beer a clear appearance (39). Enzymes may be added to further degrade the proteins, or earths or clays may be utilized to adsorb these materials (43).

Draft Beer and Pasteurized Beer. The yeast in a bottle of beer brought back to room temperature will continue to ferment up to a certain concentration of alcohol, creating undesirable pressure within the sealed bottle. To avoid this, most beer is either cold filtered (draft) or pasteurized to completely halt fermentation. Filtering removes the yeast, while pasteurization kills it. Some of the off-flavors in canned and bottled beer result from the heat of pasteurization. Draft beer is filtered, packaged in kegs, and refrigerated, resulting in a better flavor than pasteurized beer. Beer that has been filtered and then refrigerated is identified as having undergone "cold pasteurization."

Home-Brewing. In addition to differing in whether they have been cold filtered (draft) or pasteurized, beers differ in where they have been produced—at home, or by a small or large commercial brewer. Home-brewing has become popular, and brew pubs and small brewing companies are enjoying widespread patronage (15). These small brewers produce much of the regional and specialty beer: ales, pale and bitter, light and dark, porter, bock, stout, and flavored beers.

Wine

Wines are made from the fermented juice of fruits, usually grape (Vitus vinifera). The sugar in grapes contributes to yeast fermentation, while their acidic concentration (pH=3) discourages the growth of most other microorganisms. The wines produced from year to year start with the growth of grapes, their harvesting, the extraction of their juices, and finally fermentation (Figure 29-7). Yeasts produce many of the flavors found in wine, so selecting a particular wine yeast strain is very important. Yeasts produce certain compounds (higher alcohols, acids, and esters) that all contribute to flavor quality (25).

HOW & WHY? ?????????

Why are wines described as "sweet" or "dry"? The degree to which sugars are fermented to alcohol determines whether the wine is sweet or dry. Sweet wines contain higher concentrations of unfermented sugars than dry wines; almost all of the carbohydrate in dry wines has been converted to alcohol.

The sweetness and/or dryness of a wine does not always correspond to alcohol content. A dry wine containing 14 percent alcohol can be made sweeter by adding juice or sugar. Sweet wines are not always lower in alcohol because distilled spirits may have been added. In general, wines average 11 percent alcohol.

Racking. After fermentation, the wines undergo "racking" in which they are allowed to stand to settle out the yeast cells and finely suspended material. The wine is then drawn out, leaving behind the sediment or "lees." Racking was classically done in large 50- to 1,000-gallon barrels, but is now often achieved with large tanks.

Aging. Aging occurs next, in which the wine is stored in casks (often made of oak) or tanks for several months or years to allow the remaining traces of sugar to ferment and the further development of flavor (Figure 29-8). Most of the wines with a final alcohol concentration under 17 percent are filtered (cold pasteurized), or heat pasteurized just before bottling (Chemist's Corner 29-3). The exception is sparkling wine, in which the carbon dioxide content acts as a partial preservative. Sulfur dioxide (SO_2) is sometimes added to wines to inhibit microorganisms and the enzymatic browning that occurs when the phenolic compounds found in grapes are oxidized by phenolase enzymes (43).

Chemist's Corner 29-3

Filtering Wines

Tartrates naturally found in grape juice tend to crystallize as salts of tartaric acid. If these salts are not removed from wine by ion exchange treatments or a combination of chilling and filtering, the salts will slowly appear as glasslike crystals in the bottom of stored wine bottles (43)

The Wine Institute

FIGURE 29-7 **Grapes arrive at the winery in gondolas from the vineyard, ready to be crushed into juice for fermentation. Gondolas are usually pulled by tractor through the vineyard, where grapes are picked either by hand or by mechanical harvester.**

The Wine Institute

FIGURE 29-8 **Fifty-gallon oak barrels age wine and impart flavor and complexity. Wine may be aged in oak barrels on an average of one to three years. Many fine wines will also benefit from further aging once in the bottle.**

KEY TERMS

Congener Alcohol by-product such as methanol or wood alcohol.

Vintage The year in which a wine was bottled; especially, an exceptionally fine wine from a year with a good crop.

Evaluating Wines. Appearance or clarity, mouthfeel, taste, and aroma are characteristics used to judge various wines. In terms of appearance, wine may be clouded by bacteria or trace metals. The wine's astringency and body (viscosity) contribute to the mouthfeel. Body is the weight of the wine in the mouth—light, medium, or heavy bodied. The more alcohol, the fuller the body (35). Another way to gauge alcohol content is to swirl the wine in a glass and watch how the liquid streams back down the sides. Thicker rivulets, also called legs or tears, often indicate a higher alcohol level (35). The major taste difference in wines is their balance between sourness and sweetness. The longer the taste remains in the mouth after being swallowed, the longer the "finish" and the better the wine. Aroma, also known as bouquet or "nose," comes from over 200 volatile compounds, among them **congeners**, which contribute to aroma as well as to hangovers (38). Aromatic wines such as vermouth are flavored with the flowers, leaves, roots, or bark of plants. The aromas and tastes of wine are more easily detected if the wine is aerated with oxygen by swirling the glass. A dry, crumbly cork means that the wine was improperly stored (35).

Selecting a Wine. Choosing a wine can be confusing, because over 10,000 different strains of Vitis vinifera exist, and the names of vineyards or regions on the labels of wines are unfamiliar to most consumers.

Vintage. Even if a good wine is selected, the quality will vary depending on its **vintage**. If a particular growing year results in an especially good wine, it is known as a "vintage year."

Types of Wines. Table 29-3 lists some of the common types of wines. Some of these wines are named after their grape of origin—such as Cabernet

TABLE 29-3
Wines

Product	*Description*	*Uses*
Apple wine (hard cider)	Apple cider or juice that has been allowed to ferment	Served cold or hot as a beverage. Distilled to make apple brandies such as applejack (an American product) or Calvados (from Normandy, France)
Aromatic wine	A fortified wine flavored with one or more aromatic plant parts such as bark, flowers, leaves, roots, etc. Vermouth is one of the best-known types of aromatic wines.	An aperitif (drink served before a meal to stimulate the appetite) that is best when poured over ice. Mixer for cocktails and similar drinks.
Bordeaux	A wine produced in the Bordeaux region of France.	During meals or with the dessert. Good when chilled slightly and served in elongated Bordeaux glasses.
Brandy	Distillate from a wine (hence, the characteristics of each product stem from those of the original wine, the type of distillation and the aging process).	After-dinner drink. In desserts and other dishes.
Burgundy	A wine produced in the Burgundy region of France. May be red, white, or sparkling.	During meals or with the dessert. Good when chilled slightly.
Cabernet	Wine made from the Cabernet Sauvignon grape, which was brought from Bordeaux, France to California.	After-dinner drink.
Chablis	An excellent dry white wine (with a green-gold tint) from the French town of Chablis. However, the name is sometimes applied to similar dry, white wines made elsewhere.	Also good with fish, hors d'oeuvre, seafood, and shellfish. The best wine for serving with oysters.
Champagne	A sparkling wine that is made by allowing wine from Pinot grapes to undergo a second fermentation after a small amount of sugar has been added to the bottle. In France, the name Champagne is limited to the sparkling wines produced in the province of Champagne.	An aperitif that is served chilled. However, it may also be served at any time during any meal. A tulip-shaped glass helps to retain the bubbles.
Chianti	Red wine from the Tuscany region of Italy that is often sold in a round-bottom flask placed in a straw basket. However, the best wine comes in tall bottles that can be binned for aging.	With meals, particularly when Italian meat or pasta dishes are served.
Claret	A dry, red Bordeaux wine made from Cabernet Sauvignon grapes.	With or after a meal.
Cognac	Brandy that is double-distilled from wine made in the Charente district of France.	After-dinner drink. In desserts and other dishes.

TABLE 29-3 (continued)
Wines

Product	*Description*	*Uses*
Dry wines	A wine that is not sweet or sweetened. (In other words, all or most of the natural sugar content has been converted to alcohol.)	With or after a meal.
Fortified wines	Wines that have had their natural alcohol content increased by the addition of a brandy.	With dessert or after dinner. Should be served in small, narrow glasses.
Honey wine (mead)	An ancient type of wine that was made from fermented honey flavored with herbs.	With meals.
Light wines	A wine that has a low alcoholic content.	With meals.
Madeira	One of the wines made on the island of Madeira. Madeira wines are the longest-lived (they keep for many years without deterioration) of any of the wines.	Depending upon the type of wine, it may be served at various parts of the meal.
May wine	A light, white Rhine wine that is flavored with the herb woodruff.	Served chilled in a punch bowl with pieces of fresh fruit floating on top.
Moselle wines	Light wines (the alcohol content is usually about 10% or less) made in the valley of the Moselle River in Germany, which lies to the west of the Rhine.	With lunch or dinner.
Mulled wines	Heated, sweetened, spiced wine served in a cup.	Served during the winter holidays.
Muscatel	A sweet, fortified wine made from Muscat grapes.	Served with dessert.
Perry (pear wine)	Light wine made from pear juice.	With meals.
Pinot	Wine made from Pinot grapes.	Starting material for making champagne. Served with meals.
Port	The type of fortified wine that originated in the town of Oporto in Portugal.	With dessert or after dinner.
Pulque	Fermented juice of the agave plant that grows in Mexico and in southwestern U.S. A common drink in Mexico.	Used to make a distilled liquor, or used shortly after its preparation because it does not keep well.
Red wines	Wines produced from dark-colored grapes that are fermented together with their skins (which contain most of the color pigments).	Served at meals featuring beef or lamb dishes, other than stews flavored with wine. (In the latter cases, wine is served after the meal.)
Resinated (Greek wines)	Greek wines that contain a resin which imparts a pinelike flavor.	Best served with mild-flavored main dishes made from fish, pork, or poultry.
Rice wine (sake)	A Japanese wine made from fermented white rice. Although made from grain and sometimes referred to as beer, sake's alcohol content is similar to wine.	With meals at Japanese restaurants. May be served hot.
Riesling	White wine made from the Riesling grape, which is considered to be the finest wine grape grown in Germany.	With meals.
Rhine wines	Wines vary from grapes grown in the Rhine River Valley of Germany. (The wines range from dry and light to rich and sweet.)	Depends upon the characteristics of the wine.
Rose wines	Rose-colored wines produced by fermenting dark-colored grapes. The best rose wines are made from grenache grapes.	With cold foods and light meals, or when either a red or white wine may be used.
Sauternes	Wines made in the Sauterne district of Bordeaux, France from grapes withered somewhat by a *Botrytis* mold that is also called "noble rot."	Should be served cold at the end of a meal; preferably, at which no other wine has been served. Serve in small, narrow glasses.
Sherry	A fortified wine made by a process similar to the one developed in Jerez de la Frontera, Spain. (Sherries range from pale-colored dry wines, to rich, sweet ones.)	Depends upon the characteristics of the particular wine.
Sparkling wines	Wines that are bubbly with carbon dioxide gas by virtue of having undergone a second fermentation initiated by the addition of a small amount of sugar.	Accompaniments to any part of a meal.

Continued

TABLE 29-3 (continued)
Wines

Product	Description	Uses
Sweet wines	Fortified wines that contain considerable amounts of unfermented sugars. (The addition of extra alcohol prevents the fermentation of the sugars which are present.)	Served as dessert. Should be served in a small, narrow glass.
Table wines	Unfortified wines of low to moderate alcoholic content. (They usually contain 14% or less alcohol.)	Served with meals.
Tokay	A rich white dessert wine made in Hungary that comes in dry and sweet varieties.	At meals or with desserts, depending upon whether the dry or sweet variety is served.
Vermouth	A fortified wine that is flavored with a variety of aromatic herbs and comes in dry and sweet varieties.	Preparation of Martinis or other cocktails. Sweet Italian vermouth is often served on ice as an aperitif.
White wines	Made by fermenting grapes separated from their skins in order to keep the content of colored pigments low.	Served at meals featuring fish, pork, poultry, seafood, shellfish, or other bland-flavored items.
Zinfandel	A red wine made from Zinfandel grapes grown in California.	At meals featuring beef or lamb dishes other than stews that contain wines.

Sauvignon, Concord, Muscatel, Pinot, Riesling, Zinfandel. Others are named for their region of origin (more common with French wines)—Bordeaux, Burgundy, Chablis, Madeira, Moselle, Port, Rhine, Sauterne. Wines may also be classified based on certain characteristics such as sweetness or dryness, carbonation, or color. As a result, wines are sometimes categorized as appetizer wines, sweet dessert wines, sparkling wines, and red or white table wines. Champagne, named after a region in France, is a wine that has been carbonated—either naturally in the bottle, a method known as the "champagne method," or mechanically by adding carbon dioxide to the wine.

The Colors of Wine. Wines may be white, red, or pink. White wines can be made from lightly colored (white or green) grapes; or they can be made from red grapes by one of two methods (Chemist's Corner 29-4). The first method is to remove the skins, pulp, and seeds from the red grapes prior to pressing, and the second is to gently press the red grapes and collect the juice early enough before the pigments from the skin have time to release. Red wines are produced by leaving the skins, pulp, and seeds in the juice. Pink wines are made by adding a small amount of red wine to white wine (43). The traditional guideline has been that red wine goes with red meat, and white wine with white meat, but in reality, the decision should be made entirely on the basis of which wine one prefers.

KEY TERMS

Proof Alcoholic strength indicated by a number that is twice the percent by volume of alcohol present.

Spirits

Distilled beverages are often called spirits, because they embody the "spirit" of the fermented mixture. They are also often referred to as "hard" because they contain more alcohol than beer or wine. The boiling point of alcohol (173°F/ 78°C) is almost 40° lower than that of water, so alcohol can be easily vaporized, cooled, condensed, and collected. Without this distillation, the alcoholic concentration of most alcoholic drinks would not exceed 15 percent, because that is the highest alcohol concentration that yeast can tolerate.

Proof. The amount of alcohol in spirits is called its **proof**. An 80-proof liquor contains 40 percent alcohol by volume.

Common Spirits. Table 29-4 lists some of the common distilled spirits. Gin and most whiskeys are derived from fermented grains. Vodka is made from either grains or potatoes. Liqueurs, which are sweet, syrupy, flavored liquors, are produced by steeping herbs or fruits in strong spirits before distillation. Rum is made from products of sugar cane. Tequila, a less common liquor, comes from the fermented sap of the Mexican-grown mescal plant. Spirits can also be produced by distilling the alcohol in wines. This process results in such beverages as brandy, which is distilled wine; cognac, which is double distilled brandy; and fortified wines, to which brandy is added to double the alcohol content.

Chemist's Corner 29-4

White Wine from Red Grapes

If white wines are made with red grapes, any remaining pigments can be removed with ion exchange, anthocyanase enzymes, or activated charcoal treatments (43).

TABLE 29-4
Spirits

Type of Liquor	*Production*
Brandy	Brandies are distilled from either a wine or a fermented mash of fruit. They are often aged for 2 or more years to mellow the harshly flavored constituents common to distilled liquors.
Gin	A fermented grain mixture is distilled to yield a strongly alcoholic mixture, which is then either redistilled or mixed with flavoring derived from juniper berries or other botanical substances.
Liqueur or cordial	A distilled liquor is either mixed or redistilled with one or more flavoring materials such as fruits, fruit peels, herbs, spices, flowers, cocoa, coffee, or roots. Liqueurs may contain from 2½% (minimum) to 35% added sugar.
Rum	Rum is distilled from a fermented sugarcane product. The stronger-flavored rums are usually aged for 3 or more years. Sometimes the distillation residue is added to subsequent rums to produce a strongly flavored product.
Vodka	Produced from starchy materials such as potatoes and/or grains (the starch is converted to sugar by enzymes). The distillation process used for vodka yields a product high in alcohol and low in congeners. The distillate is usually run through charcoal to remove any unwanted components.
Whiskey:	
All types	Produced from malted and unmalted grains. The enzymes in the former convert the starch in the latter to sugar. Distillation of the fermented grain mash is conducted so as to yield a product rich in congeners. The newly distilled liquor is usually aged in wooden barrels. Various whiskeys are often blended.
Bourbon	Made from both malted or unmalted grain, mainly corn. The distillation mixture is a sour mash containing about 1/4 old mash (previously fermented) and 3/4 new mash. Aging is done in charred barrels.
Canadian	Produced in Canada from malted barley and unmalted corn, rye, and wheat. Distilled as other whiskeys, then aged for 3 or more years. Usually, Canadian whiskeys with different characteristics are blended.
Irish	Malted barley and unmalted barley, corn, oats, rye, and wheat. Product of either the Republic of Ireland or Northern Ireland. The fermented mash, made from barley and other grains, is distilled in 3 stages. (Hence, it it "triple distilled"). Irish whiskeys are aged for at least 4 years. Many are blended.
Rye	Produced from rye and grain and distilled to produce a high content of congeners, then aged in charred barrels.
Scotch	Produced in Scotland from mainly malted barley. The fermented mash is distilled in 2 stages. (Hence, it is "double distilled.") Newly distilled Scotch whiskey is aged from 3 to 4 years in barrels previously used for whiskey or wine, then blended with other whiskeys.

PICTORIAL SUMMARY / 29: Beverages

Most beverages fall into one of the following seven categories: water, carbonated, fruit/vegetable, New Age, aromatic (coffee and tea), dairy, and alcoholic.

WATER

Water contains zero calories (kcal) or vitamins, and its mineral content varies according to region. Water is an essential nutrient required for the very existence of life. Plain water, the simplest of all beverages, from surface or underground sources, is either hard or soft. Other types of water include mineral, deionized, distilled, and sparkling waters.

CARBONATED BEVERAGES

Adding sweetener, flavors, colors, acids, and preservatives to sparkling water creates a limitless number of carbonated drinks, known as soft drinks, sodas, or "pops." Using alternative sweeteners increases the water content from 90 to 98 percent, and decreases the normal 10 percent sugar level to zero.

FRUIT AND VEGETABLE BEVERAGES

Fruit and/or vegetable juices vary widely in their nutrient content. These drinks range from pure juices extracted from fruits and vegetables, to highly diluted, artificially flavored and colored "drinks."

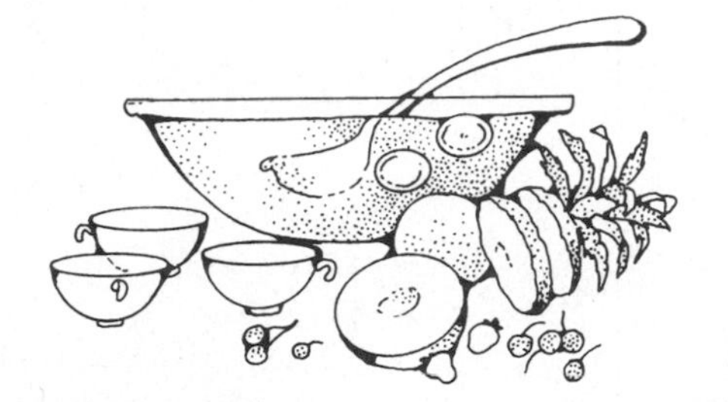

DAIRY BEVERAGES

Milk-based beverages are high in protein, calcium, and B vitamins. Examples of dairy beverages include fluid milk (whole), reduced fat (2 percent), low-fat (1 percent), and fat-free (nonfat) and dried milk. These milks can be incorporated into other beverages such as hot chocolate, milk shakes, and chocolate or other flavored milk drinks.

NEW AGE BEVERAGES

New Age beverages represent all beverages that do not fit into the traditional categorization of drinks, and include beverages identified as: sport or isotonic (low carbohydrate, high electrolytes), herbal (containing herbs), neutraceutical (health benefit), energy (high carbohydrate or caffeine), smart (alleged brain stimulators), and fun (eye-catching).

COFFEE

Coffee is made from beans that are processed by hull removal, roasting, and grinding. The goal of making a good cup of coffee is to extract enough of the compounds contributing to good flavor and aroma, but to limit the extraction of bitter substances.

Factors influencing the quality of coffee:

- Coffee quality
- Coffee freshness
- Water-to-coffee ratio
- Water type
- Water temperature
- Brewing time
- Brewing equipment

Once roasted, coffee beans will stay fresh for two to three weeks, while the freshness of ground coffee lasts only a few days.

TEA

Tea is one of the most popular drinks in the world and can be served hot or iced. The three basic categories of tea—black, green, and oolong—depend on the type of processing, which consists of withering, rolling, "fermenting," and firing. Tea grade is based primarily on leaf size. Tea keeps best in airtight containers placed in a dry atmosphere.

ALCOHOLIC BEVERAGES

Alcoholic beverages fall into three major categories: beer, wine, and spirits (hard liquor).

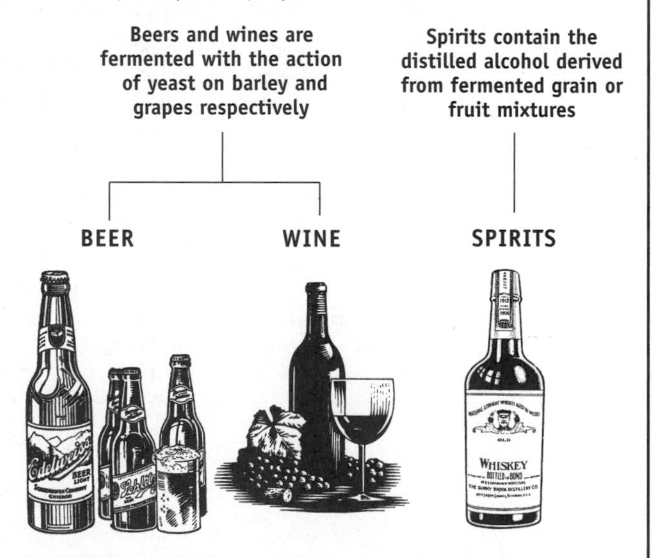

REFERENCES

1. Agarwal R, and H Mukhtar. Cancer chemoprevention by polyphenols in green tea and artichoke. In *Dietary Phytochemicals in Cancer and Prevention Treatment,* American Institute for Cancer Research, 35–50. Plenum, 1996.
2. Baker R. Clouds of citrus juices and juice drinks. *Food Technology* 53(1):64–69, 1999.
3. Banner RJ. A new age for herbal drinks. *Food Processing* 58(7):35–36, 1997.
4. Bellamy G. Drinkable desserts: Hot-selling no alcohol drinks range from bottled beverages to calorie-conscious cocktails. *Restaurant Hospitality* 77(4):112–113, 1993.
5. Berne S, and CD O'Donnell. Filtration systems and enzymes: A tangled web. *Prepared Foods* 165(9):95–96, 1996.
6. Beveridge T. Haze and cloud in apple juices. *Critical Reviews in Food Science and Nutrition* 37(1):75–91, 1997.
7. Beveridge T. Juice extraction from apples and other fruits and vegetables. *Critical Reviews in Food Science and Nutrition* 37(5):449–469, 1997.
8. Borch-Iohnsen B. [Primary hemochromatosis and dietary iron]. *Tidsskrift For Den Norske Laegeforen* 117(24):3506–3507, 1997.
9. Brown D. Consumers dive into fortified drinks. *Journal of the American Dietetic Association* 102(11):1602–1604, 2002.
10. Bubbles fizz while flavor holds appeal. *Food Engineering* 64(12):28, 1992.
11. Byrne M. Sweetners: A matter of taste. *Food Engineering International* 24(1):49–52, 1999.
12. Calbon C, and M Keane. *Juicing for Life.* Avery, 1992.
13. Campbell DL. *The Tea Book.* Pelican, 1995.
14. Carbonated waters. *Consumer Reports* 57(9):569–571, 1992.
15. Charlet K. Hot drinks cool down beverage momentum. *Prepared Foods* 166(5):57–58, 1997.
16. Charlet K. Trickle down theory. *Prepared Foods* 168(4):67–68, 1999.
17. Chung KT, et al. Tannins and human health: A review. *Critical Reviews in Food Science and Nutrition* 38(6):421–464, 1998.
18. Convertino VA, et al. American College of Sport Medicine position stand: Exercise and fluid replacement. *Medicine and Science in Sports and Exercise* 28(1):I–VII, 1996.
19. Couzy F, et al. Calcium bioavailability from a calcium- and sulfate-rich mineral water, compared with milk, in young adult women. *The American Journal of Clinical Nutrition* 62:1239–1244, 1995.
20. Dreosti IE. Bioactive ingredients: Anitoxidants and polyphenols in tea. *Nutrition Reviews* 54(11):S51–S56, 1996.
21. Ensminger AH, et al. *Foods and Nutrition Encyclopedia.* CRC Press, 1994.
22. Genetic engineering paves the way to naturally cafeine-free coffee. *Food Engineering* 63(2):65, 1991.
23. Giese JH. Hitting the spot: Beverages and beverage technology. *Food Technology* 46(7):70–80, 1992.
24. Giese JH. Developments in beverage additives. *Food Technology* 49(9):64–72, 1995.
25. Gil JV, et al. Aroma compounds in wine as influenced by apiculate yeasts. *Journal of Food Science* 61(6):1247–1249, 1996.
26. Hillman H. *Kitchen Science.* Houghton Mifflin, 1989.
27. Hollingsworth P. Functional beverage juggernaut faces tighter regulation. *Food Technology* 54(11):50–54, 2000.
28. Hollingsworth P. Redefining New Age. *Food Technology* 51(8):44–51, 1997.
29. Hollingsworth P. It's tea time. *Food Technology* 56(7):16, 2002.
30. Hollingsworth P. Profits pouring from bottled water. *Food Technology* 56(5):18, 2002.
31. Hoseney RC. An overview of malting and brewing. *Cereal Foods World* 39(9):675–679, 1994.
32. LaBell F. Fresh-squeezed flavorings. *Prepared Foods* 165(12):63, 1997.
33. Lee TA, R Kempthorne, and JK Hardy. Compositional changes in brewed coffee as a function of brewing time. *Journal of Food Science* 57(6):1417–1419, 1992.
34. Levandoski RC. Baby boomers bring big changes in 80's. *Beverage Industry* 81(2):1, 44–46, 1990.
35. MacNeil K. Learning to taste wine with all your senses. *Fine Cooking* 27:20, 1998.
36. McClements DJ. Ultrasonic characterization of foods and drinks: Principles, methods, and applications. *Critical Reviews in Food Science and Nutrition* 37(1):1–46, 1997.
37. McCue N. Beverages fixate on fruit flavors. *Prepared Foods* 165(7):59–60, 1996.
38. McGee H. *On Food and Cooking: The Science and Lore of the Kitchen.* Macmillan, 1997.
39. McMurrough I, et al. Haze formation shelf-life prediction for lager beer. *Food Technology* 53(1):58–62, 1999.
40. Miyagawa C, et al. Protective effect of green tea extract and tea polyphenols against the cytotoxicity of 1, 4-naphthoquinone in isolated rat hepatocytes. *Bioscience Biotechnology Biochemistry* 61(11):1901–1905, 1997.
41. Modified oatrim finds use in healthy beverages. *Food Engineering* 65(1):22, 1993.
42. New generation of beverages will have lower alcohol levels. *Food Engineering* 61(9):60, 1989.
43. Potter NN, and JH Hotchkiss. *Food Science.* Chapman & Hall, 1995.

44. Proliferation of juice-added beverages continues. *Food Engineering* 62(11):31, 1990.
45. Pszczola DE. Drinks for everyone! *Food Technology* 49(9):30, 1995.
46. Pszczola DE. Lookin' good: Improving the appearance of food products. *Food Technology* 51(11):39–44, 1997.
47. Pszczola DE. Sipping into the beverage mainstream. *Food Technology* 53(11):78–88, 1999.
48. Pszczola DE. How ingredients help solve beverage problems. *Food Technology* 55(10):61–73, 2001.
49. Reddy BS, R Hamid, and CV Rao. Effect of oligofructose and inulin on colonic preneoplastic aberrant crypt foci inhibition. *Carcinogenesis* 18(7):1371–1371, 1997.
50. Rediscovering calcium fortification. *Prepared Foods* 162(12):65, 1993.
51. Regular or decaf? Coffee consumption and serum lipoproteins. *Nutrition Reviews* 50(6):175–178, 1992.
52. Rice M. Author claims he found secret recipe for Coke. *Arizona Daily Sun,* April 25, 1993.
53. Sakano T, et al. Improvement of coffee aroma by removal of pungent volatiles using a-type zeolite. *Journal of Food Science* 61(2):473–476, 1996.
54. Semanchek JJ, and DA Golden. Survival of Escherichia coli O157:H7 during fermentation of apple cider. *Journal of Food Protection* 59(12):1256–1259, 1996.
55. Shah N. Functional foods from probiotics and prebiotics. *Food Technology* 55(11):46–53, 2001.
56. Siegal RK. Herbal intoxication: Psychoactive effects from herbal cigarettes, tea, and capsules. *Journal of the American Medical Association* 236:437–476, 1976.
57. Soft drinks and mixes. *Progressive Grocer* 70(7):49, 1991.
58. Spiller GA. The coffee plant and its processing. In *The Methylxanthine Beverages and Foods: Chemistry, Consumption and Health Effects.* Alan Liss, 1984.
59. Van Dusseldorp M, and MB Katan. Headache caused by caffeine withdrawal among moderage coffee drinkers switched from ordinary decaffeinated coffee: A 12-week double-blind trial. *British Medical Journal* 300:1558–1559, 1990.
60. Walker EB, et al. Cranberry concentrate: UTI prophylaxis. *Journal of Family Practice* 45(2):167–168, 1997.

WEBSITES

The Cook's Thesaurus provides detailed information on beer, wines, and liquors:

www.foodsubs.com/alcohol.html

A fairly complete beverage site offered by the University of Nevada at Las Vegas' College of Hotel Administration:

http://www.unlv.edu/Tourism/beverage.html

Appendixes

APPENDIX A

Canada's Food Guide to Healthy Eating

ISBN 0-662-19648-1

APPENDIX B

Approximate Food Measurements. The quantity to purchase for an approximate yield.

Food	*Quantity to Purchase*	*Approximate Yield*
DAIRY		
Cheese		
Cheddar	1 lb	2 C/4 C (grated)
Cottage	1 lb	2 C
Cream	1 lb	2 C
Cream	1 C (1/2 pt)	2 C
EGGS		
Whole	1 lb	1¾ C
Whites (fresh)	8–11	1 C
Yolks (fresh)	12–14	1 C
FATS AND OILS		
Butter/Margarine	1 lb	2 C
Vegetable Oil	1 lb	2⅙ C
Vegetable Shortening	1 lb	2⅓ C
FLOUR		
All-Purpose	1 lb	4 C (sifted)
Cake	1 lb	4½ C (sifted)
Cornmeal	1 lb	3½ C (sifted)
Rye	1 lb	3½–5 C
Whole Wheat	1 lb	3⅓ C (sifted)
FRUIT		
Apples	1 lb/3 med	3 C (sliced)
Bananas	1 lb/3 med	2½ C (sliced)
Berries	1 quart	3½–4 C (sliced)
Coconut	1 lb shredded	5 C
Dates	1 lb whole	2¼ C or 2 C (pitted)
Lemon	1 med	⅓–½ C juice 1½–3 tsp (grated)
Orange	1 med	⅓–½ C juice 1–2 tbs (grated)
Peaches	1 lb	4 C (sliced)
Prunes	1 lb	2⅓ C
Raisins	1 lb	3 C
NUTS		
Almonds	1 lb shelled	3½ C
Pecans	1 lb shelled	4 C
Peanuts	1 lb shelled	3 C
Walnuts	1 lb shelled	4 C
SUGAR/SALT		
Brown	1 lb	2¼–2½ C (firmly packed)
Confectioners	1 lb	4–4½ C (sifted)
Granulated	1 lb	2–2¼ C
Honey	1 lb	1–1¼ C
Salt	1 lb	1½ C

Continued

APPENDIX B (Continued)

Approximate Food Measurements. The quantity to purchase for an approximate yield.

Food	*Quantity to Purchase*	*Approximate Yield*
VEGETABLES		
Beets	1 lb/4 med	2 C
Cabbage	1 lb	4 C (shredded)
Carrots	1 lb/4 med	3 C (diced)
Celery	1 lb/½ bunch	4 C (diced)
Corn	3 ears	1 C (kernels)
Dried Beans	1 lb/2 C	5–6C (cooked)
Green Beans	1 lb	3 C (chopped)
Lettuce	1 lb/med	6 C
Onion	1 med	½ C (diced)
Parsley	1 med bunch	½–1 C (finely chopped)
Potatoes	1 lb/3 med	2½ C (diced)
		3 C (peeled and sliced)
		2 C (mashed)
		2 C (French fries)
Tomatoes	1 med	1 C (chopped)
MISCELLANEOUS		
Bread Crumbs (fresh)	2 slices	1 C
	1 lb loaf	10 C
Chocolate		
Baking	8- ounce pkg	2 C (grated)
Cocoa	1 lb	4 C
Unsweetened	8-ounce pkg	8 1-ounce squares
Coffee	1 lb ground	5 C (about 2½ gallons)
	½ C	10 C
Crackers		
Graham	12	1 C (fine crumbs)
Saltines	18	1 C (coarse crumbs)
	24	1 C (fine crumbs)
Gelatin	1 envelope	1 T gelatin powder
Rice	1 C uncooked	3–4 C cooked

APPENDIX C

Substitution of Ingredients

If Missing	*Measurement*	*Substitute*
DAIRY		
Whole milk	1 C	= ½ C evaporated milk + ½ C water
		= ⅓ C nonfat dry milk + water to make one C + 2 T fat
	1 quart	= 4 oz nonfat dry milk + water to make 1 qt + 1¼ oz fat
		= ½ C heavy cream + ½ C cold water
Sweetened Condensed	1 C	1 C = ¾ C sugar + ⅓ evaporated milk + 2 T butter
Buttermilk/Sour Milk	1 C	= 1 C fresh milk + 1 T fat vinegar or lemon juice (let stand for 5 minutes)
		= 1 C unflavored plain yogurt
Cream		
Half & Half	1 C	= ¾ cup milk + 2 T fat
		= ½ C milk + ½ C light cream
Heavy (Whipping)	1 C	= ¾ C milk + ⅓ C butter or margarine
Sour Cream	1 C	= 1 C yogurt
EGGS		
Whole	one	= 2 egg yolks + 1 T water
		= 2 T dried whole eggs + 2½ T water
Whites, fresh	1 white	= 2 T thawed frozen egg white or 2 tsp dry egg white + 2 T water
Yolks, fresh	1 yolk	= 3½ T thawed frozen egg yolk or 2 T dry egg yolk + 2 tsp water
FATS AND OILS		
Butter/margarine	1 C	= 1 C margarine/butter
		= ⅞ to 1 C hydrogenated fat + ½ tsp salt
		= ⅞ C lard + ½ tsp salt
		= ⅞ C vegetable oil
FLOUR		
All-purpose	1 C sifted	= 1 C unsifted all-purpose flour minus 2 T
		= 1½ C bread flour
		= 1 C rye
		= 1 C + 2 T cake flour
		= 1 C minus 2 T cornmeal
		= 1 C graham flour
		= 1 C minus 2 T rice flour
		= 1 C rolled oats
		= 1 C + 2 T coarsely ground whole wheat or graham flour or 13 T gluten flour
		= 1¼ C rye flour
		= ½ C barley flour
	1 T (as thickener)	= ½ T cornstarch, potato starch, rice starch, arrowroot starch
		= 1 T quick-cooking tapioca, waxy rice flour, waxy corn flour
Self-rising Flour	1 lb	= 4 C of all-purpose flour + 2 T baking powder, 2 t salt
		= 1 C = (1 C of all-purpose flour minus 2 t) + 1½ t baking powder + ½ t salt
Cake Flour	1 C sifted	= ⅞ C sifted all-purpose flour or 1 C minus 2 tablespoons sifted all-purpose flour

Continued

APPENDIX C (Continued)

Substitution of Ingredients

If Missing	*Measurement*	*Substitute*
MISCELLANEOUS		
Allspice	1 T	= ½ t cinnamon + ½ t ground cloves
Baking Powder	1 t	= ¼ t baking soda + ½ t cream of tartar
		= ¼ t baking soda + ½ C buttermilk or sour milk (replaces ½ C of liquid used in recipe)
Broth	1 C	= 1 bouillon C (or 1 envelope powdered broth or 1 t powdered broth) + 1 C boiling water
Catsup	1 C	= 1 C tomato sauce + ½ C sugar + 2 T vinegar
Chili Sauce	1 C	= 1 C tomato sauce + ¼ C brown sugar + 2 T vinegar + ¼ t cinnamon + dash allspice/ground cloves
Chives		Scallion greens
Chocolate		
unsweetened	1 ounce	= 3 T cocoa + 1 T fat
baking	1 square	= 3 T carob powder + 2 T water
semisweet	2 ounces	= 1 ounce unsweetened chocolate + 2 t sugar
Cocoa	3 T	= 1 oz chocolate if recipe reduced by 1 T of fat
		= 3 T carob powder
Cornstarch	1 T	= 2 T all-purpose flour
	1 ounce	= 2 oz all-purpose flour
Garlic	1 medium	= ½ t garlic salt
	clove	= ⅛ t garlic powder
Herbs	1 T (fresh)	= ¼ t dried ground
		= 1 t dried leaf
Lemon Juice	1 T	= ½ T vineagar
Mayonnaise	1 C	= ½ C yogurt + ½ C mayonnaise
		= 1 C sour cream
		= 1 C cottage cheese (pureed)
Pumpkin Pie Spice	1 t	= ½ t cinnamon + ¼ t nutmeg + ⅛ t allspice + ⅛ t cardamon
Tomatoes (canned)	1 C	= ½ C tomato puree + ½ C water
Tomato Juice	1 C	= ½ C tomato puree or sauce + ½ up water
Tomato Purée	2 C	= 1¼ C water + ¾ C tomato paste
Tomato Sauce	2 C	= 1¼ C water + ¾ C tomato paste
SUGAR/SWEETENERS		
Granulated	1 C	= 1⅓ C brown sugar
		= 1½ C Confectioner's sugar
		= 1 C honey minus ¼ to ⅓ liquid in recipe
		= 1¼ to 1½ C corn syrup minus ¼ to ½ liquid in recipe
		= 1⅓ C molasses minus ⅓ C liquid in recipe
Brown	1 C	= ½ C granulated sugar + ½ C liquid brown sugar
		= 1 C granulated sugar + 2 T molasses or dark corn syrup
Confectioners	1 C	= made by grinding 2 C granulated sugar in a processor
Honey	1 C	= 1 ¼ C sugar + ¼ C liquid
Corn Syrup	1 C	= 1 C sugar + ¼ C liquid
Molasses	1 C	= ½ C honey
		1¼ C melted brown sugar

APPENDIX D

Flavorings and Seasonings	
Name	***Uses***
Allspice	Allspice combines the flavors of cloves, cinnamon, and nutmeg. Whole allspice is used for pickling, gravies, broiled fish, and meats. Ground allspice is used for baked goods, fruit preserves, puddings, and relishes.
Almond	Almonds can be used in every dish from soup to dessert. Almond extract is used in cookies, confections, and Chinese cuisine.
Anise	Anise is a popular favorite for a few gourmet dishes such as Oysters Rockefeller. Also, it is used in bakery products, candies (especially licorice candy), certain kinds of cheese, pickles, and many liqueurs and cordials, including anisette and absinthe.
Anise-pepper	It is one of the ingredients in Chinese Five Spices and is commonly used for fish and strongly flavored foods.
Balm	Balm has a pleasant lemon scent and can be chopped and combined with other herbs for use in omelets and salads, and in the production of several liqueurs. Also, balm leaves are used to flavor soups and dressings.
Bay leaves	Bay leaves can be used either fresh or dried. They are one of the ingredients in bouquet garni, and are used in bouillon, marinades, olives, and pickles. They combine well with fish, potatoes, or tomatoes.
Bouquet garni	This is a French term meaning "bundle of sweet herbs." The bouquet garni is used in soups and stews, or any dish in which there is sufficient liquid to absorb the flavors.
Caper	Capers are much used in European cuisine. They are commonly used in making caper sauce, which is usually eaten with boiled lamb. They also go well with fish dishes and with casseroles of chicken and rabbit.
Caraway	The seeds (actually the dried whole fruits) are used in cakes, cheeses, confections, fresh cabbage, meat dishes, rye bread, salads, and sauerkraut. The chopped green leaves can be used in soups and salads. The roots can be cooked and eaten as a vegetable.
Cardamom seed	Freshly ground cardamom has many uses including: breads, cakes, cookies, cheese, curries, custard, liver sausage, meat dishes, pilaus, pork sausage, and punches.
Cassia	The stick cinnamon can be used in dishes to impart flavor, and then removed before serving; for example, some punches are flavored in this manner. Powdered cassia is used in combination with allspice, nutmeg, and cloves for spicing mince-meat, curries, pilaus, meat dishes, desserts and cakes. It is one of the ingredients of the famous Chinese Five Spices.
Cayenne pepper	A little goes a long way, but it is a spice that adds considerable interest to egg dishes, fish, and meat recipes.
Celery salt	This spice is slightly bitter, but it combines well with bouillon, eggs, fish, potato salad, and salad dressing.
Celery seed	Celery seeds have a slightly bitter taste, but they contribute a useful flavoring. They add special interest to many salads and salad dressings.
Chervil, garden	Chervil, which has a mild anise-caraway flavor, is one of the ingredients of Fines herbes, a mixture of chopped fresh herbs extensively used in French recipes. Chervil is used in omelets, soups, salads, sauces, and white wine vinegar. It should not be cooked, but must be added at the last minute; otherwise, it loses its flavor.
Chinese Five Spices	Chinese Five Spices (a blend of anise-pepper, star anise, cassia, cloves, and fennel seed) is an integral part of some of the recipes from the Far East. Also, it can be used to good advantage in flavoring pork dishes.
Chives	Chives are ideal as a garnish because of their delicate onion flavor and bright green color. Chives add interest and flavor to buttered beets, eggs, cottage or cream cheese, potato and other salads, sliced tomatoes, and soups.
Cinnamon	Cinnamon has a more delicate flavor than cassia and is more suitable for sweet dishes, cakes, and cookies.
Citron	Citron peel has a peculiar taste, quite different from other citrus. Is used in the U.S. as candied peel to be added to cakes, cookies, candies, and desserts.
Cloves	Whole cloves are used in many meat dishes, but a little goes a long way. Cloves are stuck into lemon slices for tea, into onions, and into hams for baking; they are also popular for apple cookery and pickle making. In the East, they go into many of the curry dishes. Whole cloves are also included in recipes for spiced wine and some liqueurs. Ground cloves are used in baked goods, borscht (beet soup), chocolate puddings, potato soup, and stews.
Cola	Cola is used in many soft drinks, and for coloring and flavoring some wines.
Coriander (Cilantro)	Coriander leaves are popular in Near, Middle, or Far East recipes, as well as Mexico and South America. The seeds are a principal ingredient of curry. Whole coriander seeds can be used in cakes, cookies, biscuits, gingerbread, green salads, pick-les, and poultry stuffing. Ground seeds are added to many meat and sweet dishes.
Cress Watercress Garden Cress	The cresses are primarily used in salads and sandwiches, but they can be used to flavor soup, cooked greens, or sauces for fish dishes, and to garnish meals.

Continued

APPENDIX D (Continued)

Flavorings and Seasonings

Name	*Uses*
Cumin	Cumin's principal use is in curry powder. It is also used to flavor bread, stuffed eggs, meats, rice dishes, and soups. Commercially, it may be found in cheese, chutney, pickles, meats, and sausage.
Curry (powder)	Curry powder may be added to eggs, chicken, fish, meats, rice, soups, or to a salad made of sweet potatoes and pineapple.
Dill	Dill loses its flavor when cooked, so it should be added at the last minute. Fresh dill leaves can be used for dishes containing chicken, mushrooms, or spinach. The seeds are used in dill pickles and dill vinegar, but they can be added to meat dishes, meat and fish sauces, sauerkraut, salads, and borscht (beet soup).
Fennel	Fennel has an aniselike flavor and is good with many foods: apple pie, candies, fish, liqueurs, pastries, pork, soups, and sweet pickles.
Fenugreek seed	Fenugreek seeds are usually used in Indian curries and chutneys.
Fines herbes	A combination of several herbs such as basil, chervil, chives, marjoram, oregano, parsley, rosemary, sage, tarragon, and thyme. Fines herbes can be used in many dishes such as fish sauces, meat stuffings, omelets, salads, salad dressing, and soups.
Garlic	Garlic blends with a wide range of dishes such as fish, game, meats, and vegetables.
Ginger	Ginger is used in numerous foods including beverages, biscuits, cakes, cookies, fish, gingerbread, ginger beer, ginger wine and cordials, puddings, sauces, and spice mixtures. It is used mostly in sweet preparation in European and North American cooking, but the Orient uses it extensively for chutney, fish, meat, and pickles.
Horseradish	Many cooks limit the use of horseradish to a sauce used on meats, but it can be added to chicken salads, egg dishes, and mayonnaise for use on fish dishes, or tomato combinations.
Leek	The leek is rather like a very mild onion. It is used mostly in soups and chowders. However, the leek may also be used as a bouquet for pork or lamb.
Lemon	Lemon juice can be used on salads instead of vinegar, and it is the predominant favorite for serving with most fishes. Grated lemon rind is added to cakes, cookies, desserts, and sauces, to give an added taste dimension.
Licorice	Licorice is used to flavor candy, chewing gum, and soft drinks. CAUTION: Licorice raises the blood pressure of some people dangerously high, due to retention of sodium.
Lime	Limes impart a unique taste to dishes, which cannot be replaced by lemons. Fish is often marinated in lime juice before cooking.
Mace	Mace can be added to apple dishes, beets, cakes, hot chocolate, coffee cakes, cookies, custards, eggnog, gingerbread, and muffins.
Marjoram	Is related to thyme; hence, they are often used together or to replace each other. It can be added to almost every dish to advantage. It should be added immediately before serving as the flavor is easily lost in cooking. Marjoram is used with egg dishes, lamb, poultry, sausage, soups, stews, and vegetables.
Mint, Peppermint, or Spearmint	Peppermint flavoring is used mostly for candies, cordials, desserts, icings, and liqueurs. Spearmint is the preferred mint for lamb, as well as for iced tea and mint juleps. It can also be used in soups, stews, fish, and meat sauces.
Monosodium Glutamate (MSG)	MSG does not have any flavor of its own, but it intensifies and enhances the flavor in other foods, especially meat and fish.
Mustard Black mustard Brown mustard White or Yellow mustard	Whole mustard seeds add pungency to many foods, including pickles, meats, and salads. Powdered dry mustard is a common kitchen spice. Its sharp, hot flavor develops when the powder is moistened. It is used for roast beef, mustard pickles, sauces, and gravies. Prepared mustard is a mixture of powdered mustard with salt, spices, and lemon juice, with wine or vinegar to preserve the mustard's pungency. It may be used with ham, hamburgers, hot dogs, and sandwich spreads.
Nutmeg	Nutmeg is traditionally used in sweet foods such as cakes, custards, doughnuts, eggnog, pies, and puddings, but it goes very well with meat, sausage, spinach, sweet potatoes, and vegetables.
Onion	Onions are used either as a separate vegetable or as a flavoring for other foods. The leaves of the onion, along, with the bulb, are used in salad.
Oregano	Oregano is used extensively in Italian cooking and can be added to cheese dishes, chili beans, fish, gravies, meats, sauces, sausage, salads, and soups.
Paprika	Paprika is used in many dishes both for its flavor and as a garnish. It can be added to chicken, sweet corn, fish, meats, sausages, tomato catsup, and tomato juice.

Continued

APPENDIX D (Continued)

Flavorings and Seasonings

Name	*Uses*
Parsley	Parsley can be added to fish and fish sauces, meats, sauces, soups, and vegetables. It is commonly used as a decoration for buffet dishes.
Pepper	Pepper loses much in aroma when ground or cooked, so freshly ground pepper should be used whenever possible. Whole peppercorn can be purchased, as well as cracked, and coarsely or finely ground. Except for sweet dishes, pepper can be added to all other dishes.
Poppy seed	Poppy seeds have a pleasant nutlike flavor and aroma and are used primarily in baked goods, the tops of rolls and bread, in cakes and pastries. However, they are also used in confections, fruit salad dressings, and curries.
Rosemary	Rosemary is good with soups, on broiled steaks, or with other meat dishes, sauces, and vegetable. The taste is aromatic, pungent, and slightly bitter.
Saffron	Saffron is used as a flavoring and coloring (yellow) spice in biscuits, confections, boiled fish, fish soup, fancy rolls, and rice, and in some European dishes.
Sage	Sage is available whole, rubbed, or ground. It is used for baked fish, meats, and meat stuffings, sausages, cheeses, and sauces.
Savory	Savory is available whole or ground, and is often combined with other herbs to flavor meats. Also, it can be used in beans, scrambled eggs, peas, salads, sauces, and sausages.
Sesame seed	Sesame seeds develop a beautiful nutty taste when sprinkled on buns, rolls, or cakes, and then baked. They are also used in confections.
Shallot	Shallots can be used in the same way as the onion, although the flavor is much more subtle. Shallots should never be browned, as they turn bitter.
Soy sauce	Soy sauce can be used in a wide array of dishes, especially with beef, chicken, fish, soups, turkey, and vegetable dishes.
Star anise	Star anise has a strong flavor similar to anise, but slightly more bitter and pungent. In Chinese cooking, it is used for duck and pork recipes.
Sweet basil	Basil can be used for green beans, fish, soups, squashes, stews, tomatoes, and vinegar.
Sweet cicely	The plant smells and tastes somewhere between anise and licorice. The tap root can be boiled and used for salads, and the green fruit can be served with salad dressing. Europeans use the leaves in soups and salads. The plant is also used for flavoring desserts and liqueurs.
Tarragon	Tarragon is best known for flavoring vinegar, but it is also used for beef, chicken, eggs, fish, pickles, cookies, salads, and tartar sauce. It has a slightly anise flavor.
Thyme	Thyme is used with fish dishes, meats, poultry, sauces, tomato dishes, and vegetables.
Turmeric	Turmeric and mustard are inseparable partners (it is used to color mustard); and turmeric is superb for almost every meat and egg dish, for pickles, and for curries. It adds yellow color.
Vanilla	Vanilla is almost always used in sweet dishes such as bakery products and desserts.
Wintergreen	Wintergreen is used mainly for candies and lozenges.

APPENDIX E

Garnishes

Fluted Cartwheel Slices:
Use a citrus zester or SNACKER™ citrus peeler. Hold the stem and blossom ends of the unpeeled fruit between your thumb and middle finger. Pull the zester through the peel from end to end, leaving about ¼ to ½ inch between each cut. Cut cartwheel slices of desired thickness. If you don't have a citrus tool, use kitchen shears or a knife to cut notches around the peel of each cartwheel slice.

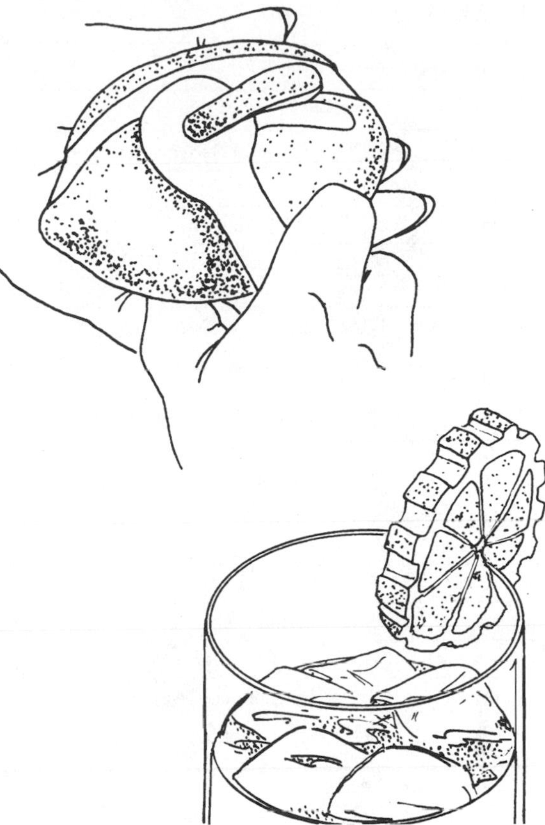

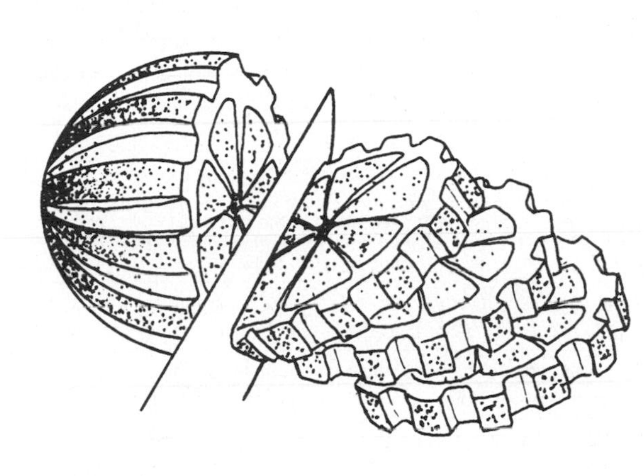

1.

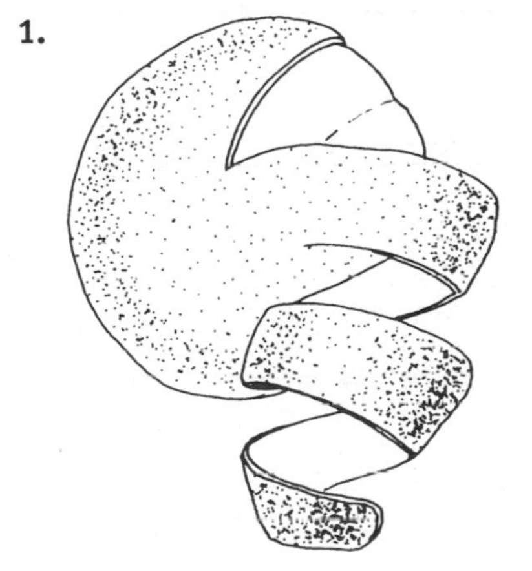

2.

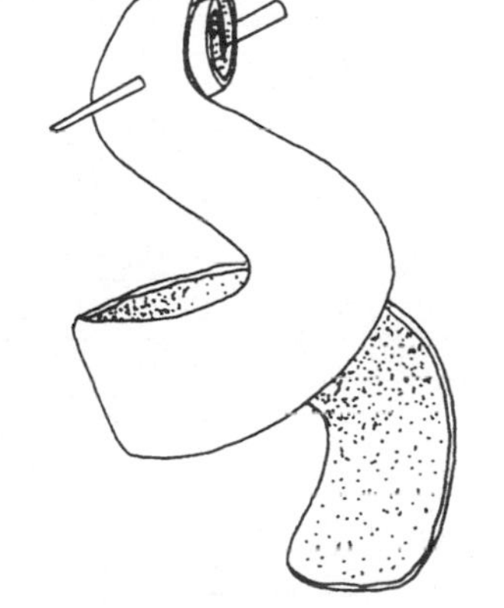

3.

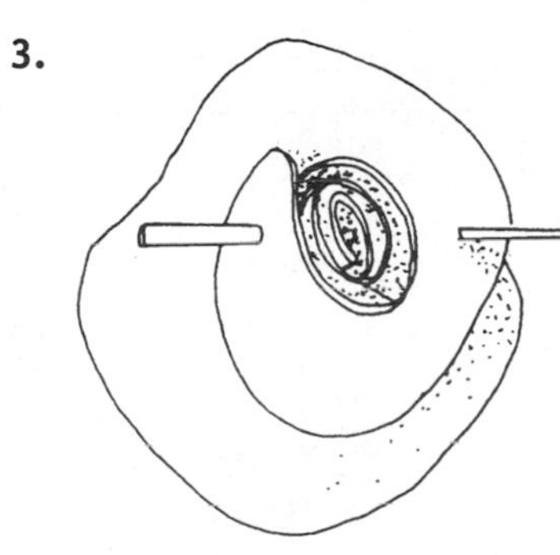

4.

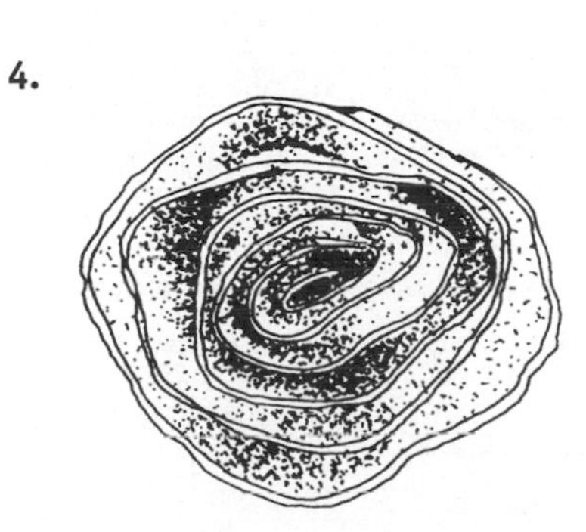

Citrus Roses:
1. Roses can be made from grapefruit, orange, lemon, and tangerine peel. With a vegetable parer or knife, cut a thin continuous spiral from the outer colored layer of the citrus peel, about 1 inch wide. If necessary, place the peel in near-boiling water for 1 to 2 minutes to make it more flexible; cool in cold water for easier handling.
2. To make a rose, wind the peel in reverse, with the colored side in. Starting with the center of the rose, form the peel into a tight "bud"; secure with a toothpick half at the base.
3. Continue winding the peel around the "bud" to form the rose. Secure with toothpick halves at the base, cutting off any excess toothpick.
4. Place the peel in cold water to set the flower. Roses will keep in cold water or a plastic bag in the refrigerator for several days.

APPENDIX F

Common Food Additives

Name	*Function*	*Food Use and Comments*
Acetic acid	pH control; preservative	Acid or vinegar is acetic acid; many food uses.
Adipic acid	pH control	Buffer and neutralizing agent; used in confectionery.
Ammonium alginate	Stabilizer and thickener; texturizer	Extracted from seaweed. Widespread food use.
Annatto	Color	Extracted from seeds of *Bixa orellana*. Butter, cheese, margarine, shortening, and sausage casings; coloring foods in general.
Arabinogalactan	Stabilizer and thickener; texturizer	Extracted from Western larch. Widespread food use; bodying agent in essential oils, nonnutritive sweeteners, flavor bases, nonstandardized dressings and pudding mixes.
Ascorbic acid (vitamin C)	Nutrient; antioxidant; preservative	Widespread use in foods to prevent rancidity, browning; used in meat curing; GRAS additive.
Aspartame	Sweetener, low calorie	Soft drinks, chewing gum, powdered beverages, whipping toppings, puddings, gelatin, tabletop sweetener.
Azodicarbonamide	Flour treating agent	Aging and bleaching ingredient in cereal flour.
Benzoic acid	Preservative	Widespread food use.
Benzoyl peroxide	Flour treating agent	Bleaching agent in flour; may be used in some cheeses.
Beta-apo-8 carotenol	Color	Natural food color. General use not to exceed 30 mg per lb or pt of food.
BHA (butylated hydroxyanisole)	Antioxidant; preservative	Fats, oils, dry yeast, beverages, breakfast cereals, dry mixes, shortening, potato flakes, chewing gum, sausage; often used in combination with BHT.
BHT (butylated hydroxytoluene)	Antioxidant; preservative	Rice, fats, oils, potato granules, breakfast cereals, potato flakes, shortening, chewing gum, sausage; often used in combination with BHA.
Biotin	Nutrient	Rich natural sources are liver, kidney, pancreas, yeast, milk; vitamin supplement.
Calcium alginate	Stabilizer and thickener; texturizer	Extracted from seaweeds. Widespread food use.
Calcium carbonate	Nutrient	Mineral supplement.
Calcium lactate	Preservative	General purpose and/or miscellaneous use.
Calcium phosphate	Leavening agent; sequestrant, nutrient	General purpose and/or miscellaneous use; mineral supplement.
Calcium propionate	Preservative	Bakery products, alone or with sodium propionate; inhibits mold and other microorganisms.
Calcium silicate	Anticaking agent	Used in baking powder, salt, and food; GRAS for use in baking powder and salt.
Canthaxanthin	Color	Widely distributed in nature. Color for foods; more red than carotene.
Caramel	Color	Miscellaneous and color for foods.
Carob bean gum	Stabilizer and thickener	Extracted from bean of carob tree (*Locust bena*). Numerous foods like confections, syrups, cheese spreads, frozen desserts, and salad dressings.
Carrageenan	Emulsifier; stabilizer and thickener	Extracted from seaweed. A variety of foods, primarily those with a water or milk base, especially ice cream.
Cellulose	Emulsifier; stabilizer and thickener	Component of all plants. Inert bulking agent in foods; may be used to reduce caloric content of food.
Citric acid	Preservative; antioxidant: pH control agent; sequestrant	Widely distributed in nature in both plants and animals. Miscellaneous use; used in lard, shortening, sausage, margarine, chili con carne, cured meats, and freeze-dried meats.
Citrus Red No. 2	Color	Coloring skins of oranges.
Cochineal	Color	Derived from the dried female insect, *Coccus cacti*. Provides red color for such foods as meat products and beverages.
Corn endosperm oil	Color	Source of xanthophyll for yellow color. Used in chicken feed to color yolks of eggs and chicken skin.

Continued

APPENDIX F (Continued)

Common Food Additives

Name	*Function*	*Food Use and Comments*
Cornstarch	Anticaking agent; drying agent; formulation aid; processing aid; surface-finishing agent	Digestible polysaccharide used in many foods often in a modified form; example foods include baking powder, baby foods, soups, sauces, pie fillings, imitation jellies, custards, candies, etc.
Corn syrup	Flavoring agent; humectant; nutritive sweetener; preservative	Derived from hydrolysis of cornstarch. Employed in numerous foods, such as baby foods, bakery products, toppings, meat products, beverages, condiments and confections.
Dextrose (glucose)	Flavoring agent; humectant; nutritive sweetener; synergist	Derived from cornstarch. Major users of dextrose are confection, wine, and canning industries; used to flavor meat products; used in production of caramel.
Diglycerides	Emulsifiers	Uses include frozen desserts, lard, shortening, and margarine.
Dioctyl sodium sulfosuccinate	Emulsifier; processing aid; surface active agent	Employed in gelatin dessert, dry beverages, fruit juice drinks, and noncarbonated beverages with cocoa fat; used in production of cane sugar and in canning.
Disodium guanylate	Flavor enhancer	Derived from dried fish or seaweed. Variety of uses.
Disodium inosinate	Flavor adjuvant	Derived from dried fish or seaweed; sodium guanylate a by-product. Variety of uses.
EDTA (ethylenediaminetetraacetic acid)	Antioxidant; sequestrant	Calcium disodium and disodium salt of EDTA employed in a variety of foods including soft drinks, alcoholic beverages, dressings, canned vegetables, margarine, pickles, sandwich spreads, and sausage.
FD&C colors: Blue No. 1, Red No. 40, Yellow No. 5	Color	Coloring foods in general.
Gelatin	Stabilizer and thickener; texturizer	Derived from collagen. Employed in many foods including confectionery, jellies, and ice cream.
Glycerine (glycerol)	Humectant	Miscellaneous and general purpose additive.
Grape skin extract	Color	Colorings for carbonated drinks, beverage bases, ades, and alcoholic beverages.
Guar gum	Stabilizer and thickener; texturizer	Extracted from seeds of the guar plant. Employed in such foods as cheese, salad dressings, ice cream, and soups.
Gum arabic	Stabilizer and thickener; texturizer	Gummy exudate of Acacia plants. Used in a variety of foods.
Gum ghatti	Stabilizer and thickener; texturizer	Gummy exudate of plant growing in India and Ceylon. A variety of food uses.
Hydrogen peroxide	Bleaching agent	Modification of starch, and bleaching tripe; bleaching agent.
Hydrolyzed vegetable (plant) protein	Flavor enhancer	To flavor various meat products.
Invert sugar	Humectant; nutritive sweetener	Primarily used in confectionery and brewing industry.
Iron	Nutrient	Dietary supplements and foods.
Iron-ammonium citrate	Anticaking agent	Used in salt.
Karaya gum	Stabilizer and thickener	Derived from dried extract of *Sterculia urens*. Variety of food uses; a substitute for tragacanth gum.
Lactic acid	Preservative, pH control	Normal product of human metabolism. Numerous uses in foods and beverages; a miscellaneous general purpose additive.
Lecithin (phosphatidylcholine)	Emulsifier; surface active agent	Normal tissue component of the body; naturally occurring in eggs; commercially derived from soybeans. Margarine, chocolate and wide variety of other uses.
Mannitol	Anticaking; nutritive sweetener; stabilizer and thickener; texturizer	Special dietary foods. A sugar alcohol.
Methylparaben	Preservative	Food and beverages.

Continued

APPENDIX F (Continued)

Common Food Additives

Name	*Function*	*Food Use and Comments*
Modified food starch	Drying agent; formulation aid; processing aid; surface finishing agent	Digestible polysaccharide used in many foods and stages of food processing; examples include baking powder, puddings, pie fillings, baby foods, soups, sauces, candies, etc.
Monoglycerides	Emulsifiers	Widely used in foods such as frozen desserts, lard, shortening and margarine.
MSG (monosodium glutamate)	Flavor enhancer	To enhance the flavor of a variety of foods including various meat products.
Papain	Texturizer	Used as a meat tenderizer. Achieves results through enzymatic action.
Paprika	Color; flavoring agent	To provide coloring and/or flavor to foods.
Pectin	Stabilizer and thickener; texturizer	Richest source of pectin is lemon and orange rind; present in cell walls of all plant tissues. Used to prepare jellies and similar foods.
Phosphoric acid	pH control	Used to increase effectiveness of antioxidants in lard and shortening.
Polyphosphates	Nutrient; flavor improver; sequestrant; pH control	Numerous food uses.
Polysorbates	Emulsifiers; surface active agent	Polysorbates designated by numbers such as 60, 65, and 80; variety of food uses including baking mixes, frozen custards, pickles, sherbets, ice creams, and shortening.
Potassium alginate	Stabilizer and thickener; texturizer	Extracted from seaweed. Wide usage.
Potassium bromate	Flour treating agent	Employed in flour, whole wheat flour, fermented malt beverages, and to treat malt.
Potassium iodide	Nutrient	Added to table salt or used in mineral preparations as a source of dietary iodine.
Potassium nitrite	Curing and pickling agent	To fix color in cured products such as meats.
Potassium sorbate	Preservative	Inhibits mold and yeast growth in foods such as wines, sausage casings, and margarine.
Proplonic acid	Preservative	Mold inhibitor in breads and general fungicide; used in manufacture of fruit flavors.
Proply gallate	Antioxidant; preservative	Used in products containing oil or fat; employed in chewing gum; used to retard rancidity in frozen fresh pork sausage.
Propylene glycol	Emulsifier; humectant; stabilizer and thickener; texturizer	Miscellaneous and/or general purpose additive; uses include salad dressings, ice cream, ice milk, custards, and a variety of other foods.
Propylparaben	Preservative	Fungicide; controls mold in sausage casings; GRAS additive.
Saccharin	Nonutritive sweetener	Special dietary foods and a variety of beverages; baked products; tabletop sweeteners.
Saffron	Color; flavoring agent	Derived from plant of western Asia and southern Europe. Used to color sausage casings, margarine, or product branding inks.
Silicon dioxide	Anticaking agent	Used in feed or feed components, beer production, production of special dietary foods, ink dilutent for marking fruits and vegetables.
Sodium acetate	pH control; preservative	Miscellaneous and/or general purpose use; meat preservation.
Sodium alginate	Stabilizer and thickener; texturizer	Extracted from seaweed; widespread food use.
Sodium aluminum sulfate	Leavening agent	Baking powders, confectionery; sugar refining.
Sodium benzoate	Preservative	To retard flavor reversion (i.e., margarine).
Sodium bicarbonate	Leavening agent; pH control	Separation of fatty acids and glyceroil on rendered fats; neutralize excess and clean vegetables in rendered fats, soups, and curing pickles.
Sodium chloride (salt)	Flavor enhancer; formulation acid; preservation	Widespread use of salt in many foods.
Sodium citrate	pH control; curing and pickling agent; sequestrant	Evaporated milk; miscellaneous and/or general purpose food use; accelerate color fixing in baking products.

Continued

APPENDIX F (Continued)

Common Food Additives

Name	*Function*	*Food Use and Comments*
Sodium diacetate	Preservative; sequestrant	An inhibitor of molds and rope-forming bacteria in baking products.
Sodium nitrate (Chile Saltpeter)	Curing and pickling agent; preservative	Used with or without sodium nitrite in smoked, cured fish; cured meat products.
Sodium nitrite	Curing and pickling agent; preservative	May be used with sodium nitrate in smoked, cured fish, cured meat products, and pet foods.
Sodium propionate	Preservative	A fungicide and mold preventative in bakery products, alone or with calcium propionate.
Sorbic acid	Preservative	Fungistatic agent for foods, especially cheeses; other uses include baked goods, beverages, dried fruits, fish, jams, jellies, meats, pickled products, and wines.
Sorbitan monostearate	Emulsifier; stabilizer and thickener	Widespread food usage such as whipped toppings, cakes, cake mixes, confectionery, icings, and shortenings; also many nonfood uses.
Sorbitol	Humectant; nutritive sweetener; stabilizer and thickener, sequestrant	Occurs naturally in berries, cherries, plums, pears, and apples; a sugar alcohol. Examples of use include chewing gum, meat products, icings, dairy products, beverages, and pet foods.
Sucrose (table sugar)	Nutritive sweetener; preservative	Sugar occurs naturally in some fruits and vegetables. The most widely used additive; used in beverages, baked goods, candies, jams and jellies—an endless list including meat products.
Tagetes (Aztec marigold)	Color	Source is flower petals of Aztec marigold. To enhance yellow color of chicken skin and eggs, incorporated in chicken feed.
Tartaric acid	pH control	Occurs free in many fruits, free or combined with calcium, magnesium, or potassium. In the soft drink industry, confectionery products, bakery products, and gelatin desserts.
Titanium dioxide	Color	For coloring foods generally, except standardized foods; used for coloring ingested and applied drugs.
Tocopherols (vitamin E)	Antioxidant; nutrient	To retard rancidity in foods containing fat; used as a supplement.
Tragacanth gum	Stabilizer and thickener; texturizer	Derived from the plant *Astragalus gummifier*.
Turmeric	Color	Derived from rhizome of *Curcuma longa*. Food use in general, except standardized foods; to color sausage casings, margarine or shortening; ink for branding or marking products.
Vanilla	Flavoring agent	Used in various bakery products, confectionery and beverages; natural flavoring extracted from cured, full grown unripe fruit of *Vanilla panifolia*.
Vanillin	Flavoring agent and adjuvant	Widespread confectionery, beverage, and food use; synthetic form of vanilla.
Yellow prussiate of soda	Anticaking agent	Employed in salt.

APPENDIX G

Cheeses

Name	*Origin*	*Consistency*	*Flavor*	*Normal Ripening Period*
American pasteurized process	United States	Semisoft to soft; smooth, plastic body	Mild	Unripened after cheese(s) heated to blend
Asiago, fresh, medium, old	Italy	Semisoft (fresh), medium, or hard (old); tiny gas holes or eyes	Piquant, sharp in aged cheese	60 days minimum for fresh (semisoft), 6 months minimum for medium, 12 months minimum for old (grating)
Bel paese	Italy	Soft; smooth, waxy body	Moderately robust	6–8 weeks
Blue, Bleu	France	Semisoft; visible veins of mold on white cheese; pasty, sometimes crumbly	Piquant, tangy, spicy, peppery	60 days minimum; 3–4 months usually; 9 months for more flavor
Breakfast, Frühstück	Germany	Soft; smooth, waxy body	Strong, aromatic	Little or none (either)
Brick	United States	Semisoft; smooth, open texture; numerous round and irregular-shaped eyes	Mild but pungent and sweet	2–3 months
Brie	France	Soft, thin edible crust, creamy interior	Mild to pungent	4–8 weeks
Caciocavallo	Italy	Hard, firm body; stringy texture	Sharp, similar to provolone	3 months minimum for table use, 12 months or longer for grating
Camembert	France	Soft, almost fluid in consistency; thin edible crust, creamy interior	Mild to pungent	4–5 weeks
Cheddar	England	Hard; smooth, firm body, can be crumbly	Mild to sharp	60 days minumum; 3–6 months usually; 12 or longer for sharp flavor
Colby	United States	Hard but softer and more open in texture than Cheddar	Mild to mellow	1–3 months
Cottage, Dutch, Farmers, Pot	Uncertain	Soft; moist, delicate, large or small curds	Mild, slightly acidic, flavoring may be added	Unripened
Cream	United States	Soft; smooth, buttery	Mild, slightly acid, flavoring may be added	Unripened
Edam	Holland	Semisoft to hard; firm, crumbly body; small eyes	Mild, sometimes salty	2 months or longer
Feta	Greece	Soft, flaky; similar to very dry, high-acid cottage cheese	Salty	4–5 days to 1 month
Gammelost	Norway	Semisoft	Sharp, aromatic	4 weeks or longer
Gjetost	Norway	Hard; buttery	Sweet, caramel	Unripened
Gorgonzola	Italy	Semisoft; less moist than blue	Piquant, spicy, similar to blue	3 months minimum, frequently 6 months to 1 year
Gouda	Holland	Hard, but softer than Cheddar; more open mealy body like Edam, small eyes	Mild, nutlike, similar to Edam	2–6 months

Continued

APPENDIX G (Continued)

Cheeses				
Name	*Origin*	*Consistency*	*Flavor*	*Normal Ripening Period*
Gruyère	Switzerland	Hard, tiny gas holes or eyes	Mild, sweet	3 months minimum
Limburger	Belgium	Soft; smooth, waxy body	Strong, robust, highly aromatic	1–2 months
Monterey Jack	United States	Semisoft (whole milk), hard (lowfat or skim milk); smooth texture with small openings throughout	Mild to mellow	3–6 weeks for table use, 6 months minimum for grating
Mozzarella	Italy	Semisoft; plastic	Mild, delicate	Unripened to 2 months
Muenster	Germany	Semisoft; smooth, waxy body, numerous small mechanical openings	Mild to mellow, between brick and Limburger	2–8 weeks
Neufchatel	France	Soft; smooth, creamy	Mild	3–4 weeks or unripened
Parmesan, Reggiano	Italy	Very hard (grating), granular, hard brittle rind	Sharp, piquant	10 months minimum
Port du Salut, Oka	Trappist Monasteries	Semisoft; smooth, buttery	Mellow or mild to robust, similar to Gouda	6–8 weeks
Primost	Norway	Semisoft	Mild, sweet, caramel	Unripened
Provolone	Italy	Hard, stringy texture; cuts without crumbling, plastic	Bland acid flavor to sharp and piquant, usually smoked	6–14 months
Queso blanco, White cheese	Latin America	Soft, dry and granular if not pressed; hard open or crumbly if pressed	Salty, strong, may be smoked	Eaten within 2 days to 2 months or more; generally unripened if pressed
Ricotta	Italy	Soft, moist and grainy, or dry	Bland but semisweet	Unripened
Romano	Italy	Very hard, granular interior, hard brittle rind	Sharp, piquant if aged	5 months minimum; usually 5–8 months for table cheese; 12 months minimum for grating cheese
Roquefort	France	Semisoft, pasty and sometimes crumbly	Sharp, spicy (pepper), piquant	2 months minimum; usually 2–5 months or longer
Sap Sago	Switzerland	Very hard (grating), granular, frequently dried	Sharp, pungent, flavored with leaves; sweet	5 months minimum
Schloss, Castle cheese	Germany, Northern Austria	Soft; small, ripened	Similar to, but milder than Limburger	Less than 1 month; less intensively than Limburger
Stirred curd, granular	United States	Semisoft to hard	Similar to mild Cheddar	1–3 months
Stilton	England	Semisoft to hard; open flaky texture, more crumbly than blue	Piquant, spicy, but milder than Roquefort	4–6 months or longer
Swiss, Emmentaler	Switzerland	Hard; smooth with large gas holes or eyes	Mild, sweet, nutty	2 months minimum, 2–9 months usually
Washed curd	United States	Semisoft to hard	Similar to mild Cheddar	1–3 months

Appendix H: Basic Chemistry Concepts

This appendix is intended to provide the background in basic chemistry that you need to understand the food concepts presented in this book. Chemistry is the branch of natural science that is concerned with the description and classification of matter, the changes that matter undergoes, and the energy associated with these changes.

Matter: The Properties of Atoms

Every substance has characteristics or properties that distinguish it from all other substances and thus give it a unique identity. These properties are both physical and chemical. The physical properties include such characteristics as color, taste, texture, and odor, as well as the temperatures at which a substance changes its state (from a solid to a liquid or from a liquid to a gas) and the weight of a unit volume (its density). The chemical properties of a substance have to do with how it reacts with other substances or responds to a change in its environment so that new substances with different sets of properties are produced.

A physical change does not change a substance's chemical composition. For example, the three states ice, water, and steam all consist of two hydrogen atoms and one oxygen atom bound together. However, a chemical change occurs if an electric current passes through water. The water disappears and two different substances are formed: hydrogen gas, which is flammable, and oxygen gas, which supports life. Chemical changes are also referred to as chemical reactions.

Substances: Elements and Compounds

Molecules constitute the smallest part of a substance that can exist separately without losing its physical and chemical properties. If a molecule is composed of atoms that are alike, the substance is an element (for example, O_2). If a molecule is composed of two or more different kinds of atoms, the substance is a compound (for example, H_2O).

Just over 100 elements are known, and these are listed in Table K-1. A familiar example is hydrogen, whose molecules are composed only of hydrogen atoms linked together in pairs (H_2). On the other hand, over a million compounds are known. An example is the sugar glucose. Each of its molecules is composed of 6 carbon, 6 oxygen, and 12 hydrogen atoms linked together in a specific arrangement.

The Nature of Atoms

Atoms themselves are made of smaller particles. Within the atomic nucleus are protons (positively charged particles), and surrounding the nucleus are electrons (negatively charged particles). The number of protons (1) in the nucleus of an atom determines the number of electrons (2) around it. The positive charge on a proton is equal to the negative charge on an electron, so the charges cancel each other out and leave the atom neutral to its surroundings.

The nucleus may also include neutrons, subatomic particles that have no charge. Protons and neutrons are of equal mass, and together they give an atom its weight. Electrons bond atoms together to make molecules, and they are involved in chemical reactions.

Each type of atom has a characteristic number of protons in its nucleus. The hydrogen atom (symbol H) is the simplest of all. It possesses a single proton, with a single electron associated with it:

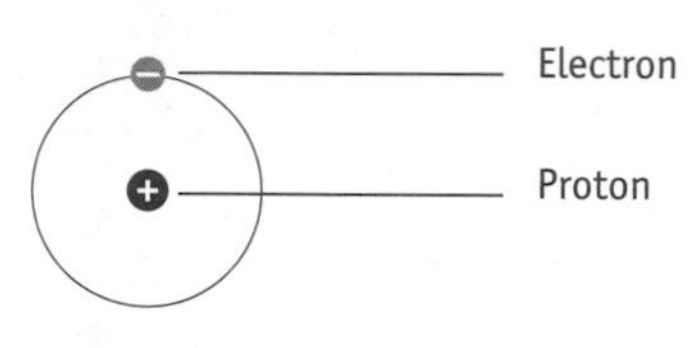

Hydrogen atom (H), atomic number 1.

Just as hydrogen always has one proton, helium always has two, lithium three, and so on. The atomic number of each element is the number of protons in the nucleus of that atom, and this never changes in a chemical reaction; it gives the atom its identity. The atomic numbers for the known elements are listed in Table H-1.

Besides hydrogen, the atoms most common in living things are carbon (C), nitrogen (N), and oxygen (O), whose atomic numbers are 6, 7, and 8, respectively. Their structures are more complicated than that of hydrogen, but each of them possesses the same number of electrons as there are protons in the nucleus. These electrons are found in orbits, or shells :

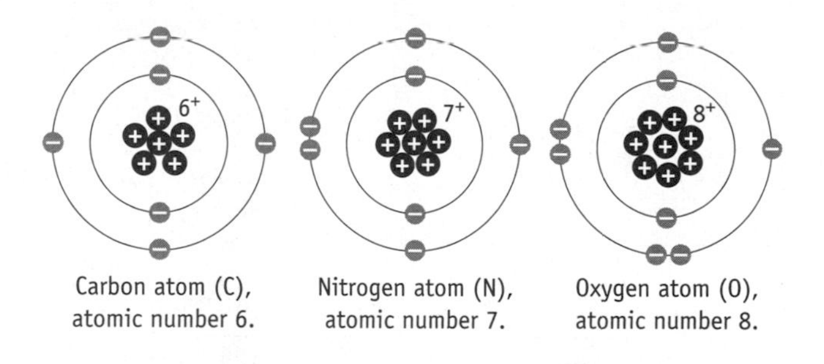

Carbon atom (C), atomic number 6. Nitrogen atom (N), atomic number 7. Oxygen atom (O), atomic number 8.

In these and all diagrams of atoms that follow, only the protons and electrons are shown. The neutrons, which contribute only to atomic weight, not to charge, are omitted.

The most important structural feature of an atom for determining its chemical behavior is the number of electrons in its outermost shell. The first, or innermost, shell is full when it is occupied by two electrons; so an atom with two or more electrons has a filled first shell. When the first shell is full, electrons begin to fill the second shell.

TABLE H-1

Chemical Symbols for the Elements

Number of Protons (Atomic Number)	Element	Number of Electrons in Outer Shell	Number of Protons (Atomic Number)	Element	Number of Electrons in Outer Shell
1	Hydrogen (H)	1	52	Tellurium (Te)	6
2	Helium (He)	2	53	Iodine (I)	7
3	Lithium (Li)	1	54	Xenon (Xe)	8
4	Beryllium (Be)	2	55	Cesium (Cs)	1
5	Boron (B)	3	56	Barium (Ba)	2
6	Carbon (C)	4	57	Lanthanum (La)	2
7	Nitrogen (N)	5	58	Cerium (Ce)	2
8	Oxygen (O)	6	59	Praseodymium (Pr)	2
9	Fluorine (F)	7	60	Neodymium (Nd)	2
10	Neon (Ne)	8	61	Promethium (Pm)	2
11	Sodium (Na)	1	62	Samarium (Sm)	2
12	Magnesium (Mg)	2	63	Europium (Eu)	2
13	Aluminum (Al)	3	64	Gadolinium (Gd)	2
14	Silicon (Si)	4	65	Terbium (Tb)	2
15	Phosphorus (P)	5	66	Dysprosium (Dy)	2
16	Sulfur (S)	6	67	Holmium (Ho)	2
17	Chlorine (Cl)	7	68	Erbium (Er)	2
18	Argon (Ar)	8	69	Thulium (Tm)	2
19	Potassium (K)	1	70	Ytterbium (Yb)	2
20	Calcium (Ca)	2	71	Lutetium (Lu)	2
21	Scandium (Sc)	2	72	Hafnium (Hf)	2
22	Titanium (Ti)	2	73	Tantalum (Ta)	2
23	Vanadium (V)	2	74	Tungsten (W)	2
24	Chromium (Cr)	1	75	Rhenium (Re)	2
25	Manganese (Mn)	2	76	Osmium (Os)	2
26	Iron (Fe)	2	77	Iridium (Ir)	2
27	Cobalt (Co)	2	78	Platinum (Pt)	1
28	Nickel (Ni)	2	79	Gold (Au)	1
29	Copper (Cu)	1	80	Mercury (Hg)	2
30	Zinc (Zn)	2	81	Thallium (Tl)	3
31	Gallium (Ga)	3	82	Lead (Pb)	4
32	Germanium (Ge)	4	83	Bismuth (Bi)	5
33	Arsenic (As)	5	84	Polonium (Po)	6
34	Selenium (Se)	6	85	Astatine (At)	7
35	Bromine (Br)	7	86	Radon (Rn)	8
36	Krypton (Kr)	8	87	Francium (Fr)	1
37	Rubidium (Rb)	1	88	Radium (Ra)	2
38	Strontium (Sr)	2	89	Actinium (Ac)	2
39	Yttrium (Y)	2	90	Thorium (Th)	2
40	Zirconium (Zr)	2	91	Protactinium (Pa)	2
41	Niobium (Nb)	1	92	Uranium (U)	2
42	Molybdenum (Mo)	1	93	Neptunium (Np)	2
43	Technetium (Tc)	1	94	Plutonium (Pu)	2
44	Ruthenium (Ru)	1	95	Americium (Am)	2
45	Rhodium (Rh)	1	96	Curium (Cm)	2
46	Palladium (Pd)	—	97	Berkelium (Bk)	2
47	Silver (Ag)	1	98	Californium (Cf)	2
48	Cadmium (Cd)	2	99	Einsteinium (Es)	2
49	Indium (In)	3	100	Fermium (Fm)	2
50	Tin (Sn)	4	101	Mendelevium (Md)	2
51	Antimony (Sb)	5	102	Nobelium (No)	2

Key:
- Elements found in energy-yielding nutrients, vitamins, and water.
- Major minerals.
- Trace minerals.

The second shell is completely full when it has eight electrons. A substance that has a full outer shell tends not to enter into chemical reactions. Atomic number 10, neon, is a chemically inert substance because its outer shell is complete. Fluorine, atomic number 9, has a great tendency to draw an electron from other substances to complete its outer shell, and thus it is highly reactive. Carbon has a half-full outer shell, which helps explain its great versatility; it can combine with other elements in a variety of ways to form a large number of compounds.

Atoms seek to reach a state of maximum stability or of lowest energy in the same way that a ball will roll down a hill until it reaches the lowest place. An atom achieves a state of maximum stability:

- By gaining or losing electrons to either fill or empty its outer shell.
- By sharing its electrons through bonding together with other atoms and thereby completing its outer shell.

The number of electrons determines how the atom will chemically react with other atoms. Hence the atomic number, not the weight, is what gives an atom its chemical nature.

Chemical Bonding

Atoms often complete their outer shells by sharing electrons with other atoms. In order to complete its outer shell, a carbon atom requires four electrons. A hydrogen atom requires one. Thus, when a carbon atom shares electrons with four hydrogen atoms, each completes its outer shell (as shown in the next column). Electron sharing binds the atoms together and satisfies the conditions of maximum stability for the molecule. The outer shell of each atom is complete, since hydrogen effectively has the required two electrons in its first (outer) shell, and carbon has eight electrons in its second (outer) shell; and the molecule is electrically neutral, with a total of ten protons and ten electrons.

Bonds that involve the sharing of electrons, like the bond between carbon and hydrogen, are the most stable kind of association that atoms can form with one another. They are sometimes called covalent bonds, and the resulting combinations of atoms are called molecules. A single pair of shared electrons forms a single bond. A simplified way to represent a single bond is with a single line. Thus the structure of methane (CH_4) could be represented like this (ignoring the inner-shell electrons, which do not participate in bonding):

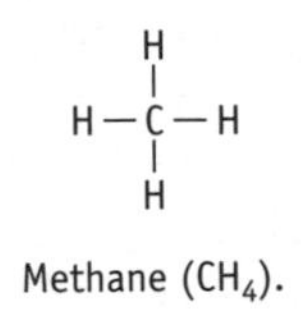

Methane (CH_4).

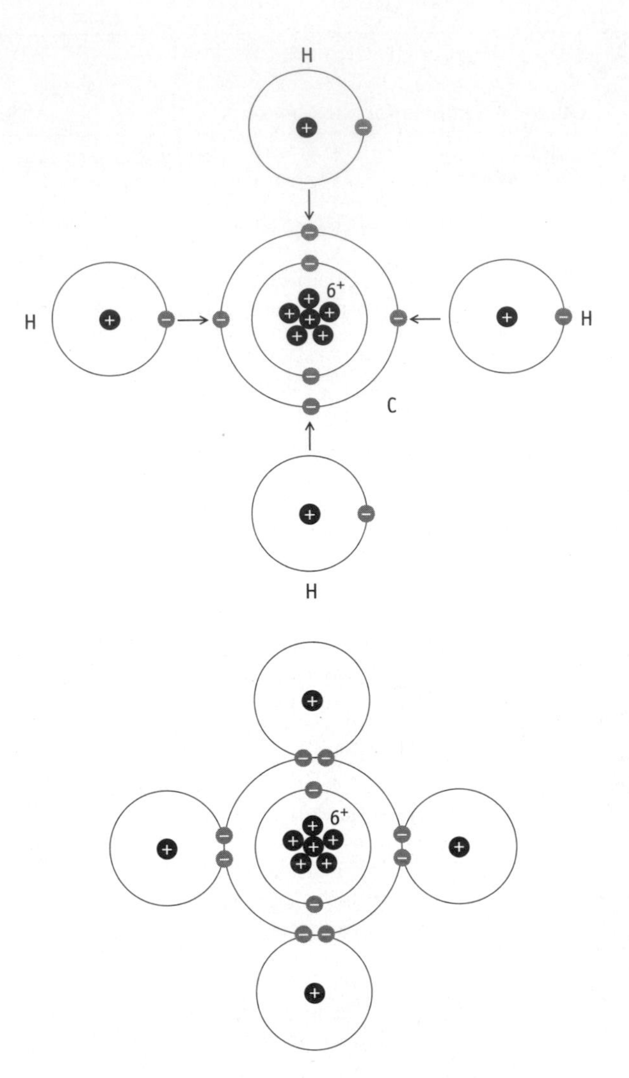

Methane molecule. The chemical formula for methane is CH_4. Note that by sharing electrons, every atom achieves a filled outer shell.

Similarly, one nitrogen atom and three hydrogen atoms can share electrons to form one molecule of ammonia (NH_3) (as shown on page 537).

One oxygen atom may be bonded to two hydrogen atoms to form one molecule of water (H_2O):

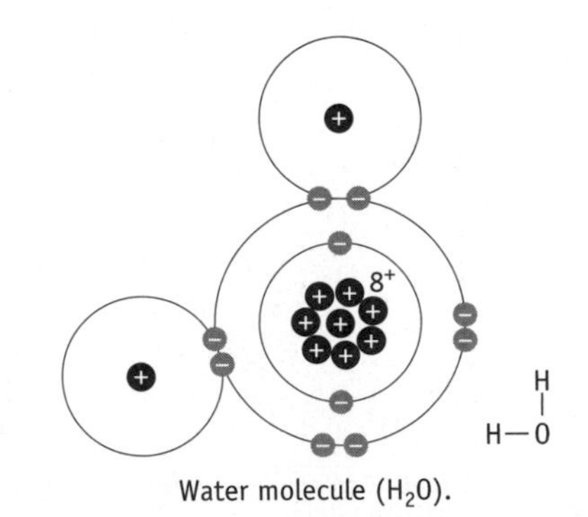

Water molecule (H_2O).

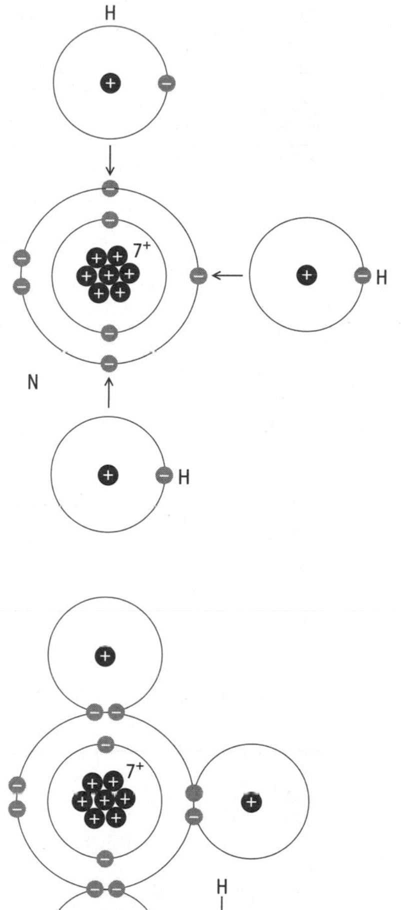

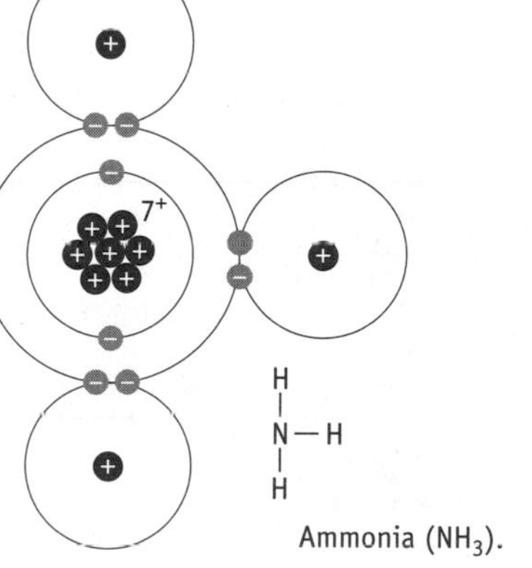

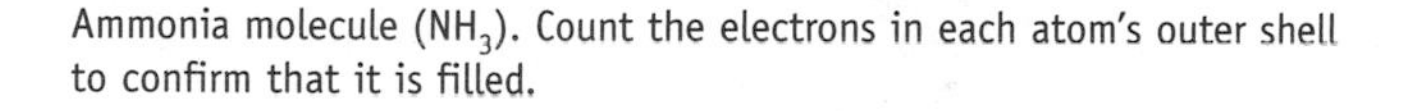
Ammonia molecule (NH_3). Count the electrons in each atom's outer shell to confirm that it is filled.

When two oxygen atoms form a molecule of oxygen, they must share two pairs of electrons. This double bond may be represented as two single lines:

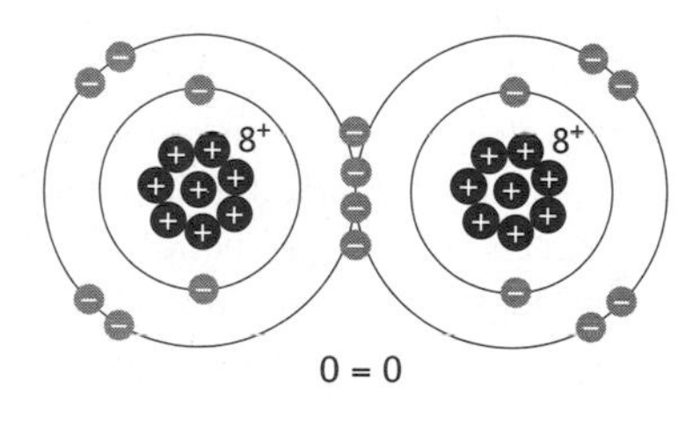

Oxygen molecule (O_2).

Small atoms form the tightest, most stable bonds. H, O, N, and C are the smallest atoms capable of forming one, two, three, and four electron-pair bonds (respectively). This is the basis for the statement that in drawings of compounds containing these atoms, hydrogen must always have one, oxygen two, nitrogen three, and carbon four bonds radiating to other atoms:

H— —O— —N— —C—

The stability of the associations between these small atoms and the versatility with which they can combine make them very common in living things. Interestingly, all cells, whether they come from animals, plants, or bacteria, contain the same elements in very nearly the same proportions. The atomic elements commonly found in living things are shown in Table H-2.

TABLE H-2
Elemental Composition of Living Cells

Element	*Chemical Symbol*	*Composition by Weight (%)*
Oxygen	O	65
Carbon	C	18
Hydrogen	H	10
Nitrogen	N	3
Calcium	Ca	1.5
Phosphorus	P	1.0
Sulfur	S	0.25
Sodium	Na	0.15
Magnesium	Mg	0.05
Total		99.30*

*The remaining 0.70 percent by weight is contributed by the trace elements: copper (Cu), zinc (Zn), selenium (Se), molybdenum (Mo), fluorine (F), chlorine (Cl), iodine (I), manganese (Mn), cobalt (Co), and iron (Fe). Cells may also contain variable traces of some of the following: lithium (Li), strontium (Sr), aluminum (Al), silicon (Si), lead (Pb), vanadium (V), arsenic (As), bromine (Br), and others.

Formation of Ions

An atom such as sodium (Na, atomic number 11) cannot easily fill its outer shell by sharing. Sodium possesses a filled first shell of two electrons and a filled second shell of eight; there is only one electron in its outermost shell:

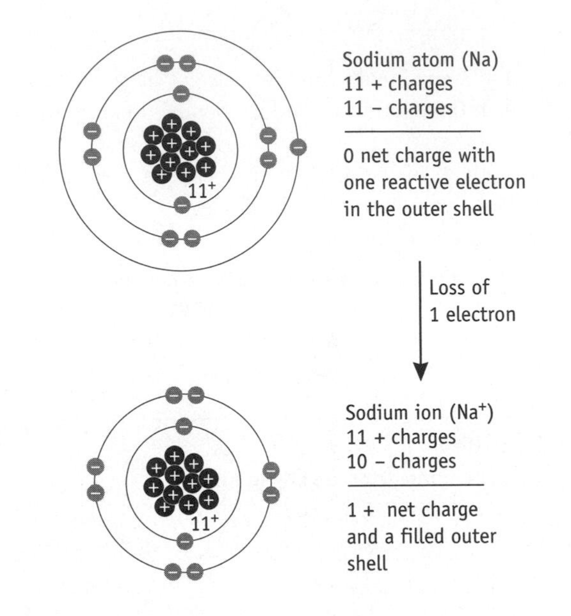

If sodium loses this electron, it satisfies one condition for stability: a filled outer shell (now its second shell counts as the outer shell). However, it is not electrically neutral. It has 11 protons (positive) and only 10 electrons (negative). It therefore has a net positive charge. An atom or molecule that has lost or gained one or more electrons and so is electrically charged is called an ion.

An atom such as chlorine (Cl, atomic number 17), with seven electrons in its outermost shell, can share electrons to fill its outer shell, or it can gain one electron to complete its outer shell and thus give it a negative charge (as shown in next column).

A positively charged ion such as sodium ion (Na^+) is called a cation; a negatively charged ion such as a chloride ion (Cl^-) is called an anion. Cations and anions attract one another to form salts (as shown in next column).

With all its electrons, sodium is a shiny, highly reactive metal; chlorine is the poisonous greenish-yellow gas that was used in World War I. But after sodium and chlorine have transferred electrons, they form the stable white salt familiar to you as table salt, or sodium chloride (Na^+C^{-2}). The dramatic difference illustrates how profoundly the electron arrangement can influence the nature of a substance. The wide distribution of salt in nature attests to the stability of the union between the ions. Each meets the other's needs (a good marriage).

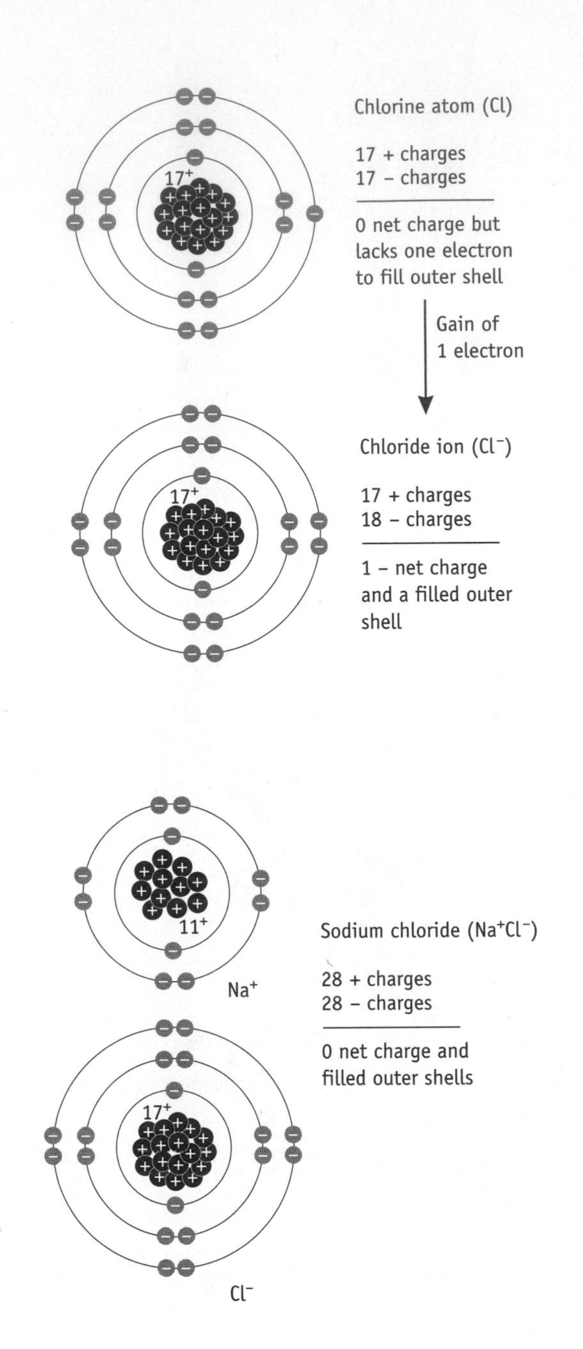

When dry, salt exists as crystals; its ions are stacked very regularly into a lattice, with positive and negative ions alternating in a three-dimensional checkerboard structure. In water, however, the salt quickly dissolves, and its ions separate from one another, forming an electrolyte solution in which they move about freely. Covalently bonded molecules rarely dissociate like this in a water solution. The most common exception is when they behave like acids and release H^+ ions, as discussed in the next section.

An ion can also be a group of atoms bound together in such a way that the group has a net charge and enters into reactions as a single unit. Many such groups are active in the fluids of the body. The bicarbonate ion is composed of five atoms—one H, one C, and three Os—and has a net charge of −1 (HCO_3^-). Another important ion of this type is a phosphate ion with one H, one P, and four O, and a net charge of −2 (HPO_4^{-2}).

Whereas many elements have only one configuration in the outer shell and thus only one way to bond with other elements, some elements have the possibility of varied configurations. Iron is such an element. Under some conditions iron loses two electrons, and under other circumstances it loses three. If iron loses two electrons, it then has a net charge of +2, and we call it ferrous iron (Fe^{++}). If it donates three electrons to another atom, it becomes the +3 ion, or ferric iron (Fe^{+++}).

Ferrous iron (Fe^{++})	Ferric iron (Fe^{+++})
(had 2 outer-shell electrons but has lost them)	(had 3 outer-shell electrons but has lost them)
26 + charges	26 + charges
24 – charges	23 – charges
2 + net charge	3 + net charge

It is important to remember that a positive charge on an ion means that negative charges—electrons—have been lost and not that positive charges have been added to the nucleus.

Water, Acids, and Bases

Water. The water molecule is electrically neutral, having equal numbers of protons and electrons. However, when a hydrogen atom shares its electron with oxygen, that electron will spend most of its time closer to the positively charged oxygen nucleus. This leaves the positive proton (nucleus of the hydrogen atom) exposed on the outer part of the water molecule. We know, too, that the two hydrogens both bond toward the same side of the oxygen. These two facts explain why water molecules are polar: they have regions of more positive and more negative charge.

Polar molecules like water are drawn to one another by the attractive forces between the positive polar areas of one and the negative poles of another. These attractive forces, sometimes known as polar bonds or hydrogen bonds, occur among many molecules and also within the different parts of single large molecules. Although very weak in comparison with covalent bonds, polar bonds may occur in such abundance that they become exceedingly important in determining the structure of such large molecules as proteins and DNA.

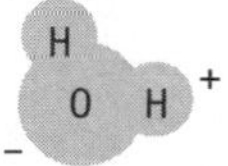

This diagram of the polar water molecule shows displacement of electrons toward the O nucleus; thus the negative region is near the O and the positive regions are near the Hs.

Water molecules have a slight tendency to ionize, separating into positive (H^+) and negative (OH^-) ions. In pure water, a small but constant number of these ions is present, and the number of positive ions exactly equals the number of negative ions.

Acids. An acid is a substance that releases H^+ ions (protons) in a water solution. Hydrochloric acid (HCl) is such a substance because it dissociates in a water solution into H^+ and Cl^- ions. Acetic acid is also an acid because it dissociates in water to acetate ions and free H^+:

```
   H  O              H  O
   |  ||             |  ||
H—C— C—O—H  ⟶  H—C— C—O⁻ + H⁺
   |                 |
   H                 H
```

Acetic acid dissociates into an acetate ion and a hydrogen ion.

The more H^+ ions released, the stronger the acid.

pH. Chemists define degrees of acidity by means of the pH scale, which runs from 0 to 14. The pH expresses the concentration of H^+ ions: a pH of 1 is extremely acidic, 7 is neutral, and 13 is very basic. There is a tenfold difference in the concentration of H^+ ions between points on this scale. A solution with pH 3, for example, has ten times as many H^+ ions as a solution with pH 4. At pH 7, the concentrations of free H^+ and OH^- are exactly the same—1/10,000,000 moles per liter (10^{-7} moles per liter).* At pH 4, the concentration of free H^+ ions is 1/10,000 (10^{-4}) moles per liter. This is a higher concentration of H^+ ions, and the solution is therefore acidic (see figure on next page).

Bases. A base is a substance that can soak up, or combine with, H^+ ions, thus reducing the acidity of a solution. The compound ammonia is such a substance. The ammonia molecule has two electrons that are not shared with any other atom; a hydrogen ion (H^+) is just a naked proton with no shell of electrons at all. The proton readily combines with the ammonia molecule to form an ammonium ion; thus a free proton is withdrawn from the solution and no longer contributes to its acidity. Many compounds containing nitrogen are important bases in living systems. Acids and bases neutralize each other to produce substances that are neither acid nor base.

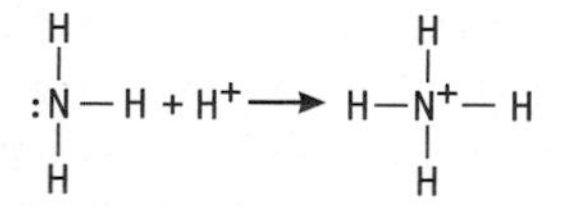

Ammonia captures a hydrogen ion from water. The two dots here represent the two electrons not shared with another atom. These are ordinarily not shown in chemical structure drawings. Compare this with the earlier diagram of an ammonia molecule (p. B-3).

*A mole is a certain number (about 6 x 10^{23}) of molecules. The pH of a solution is defined as the negative logarithm of the hydrogen ion concentration of the solution. Thus, if the concentration is 10^{-2} (moles per liter), the pH is 2; if 10^{-8}, the pH is 8; and so on.

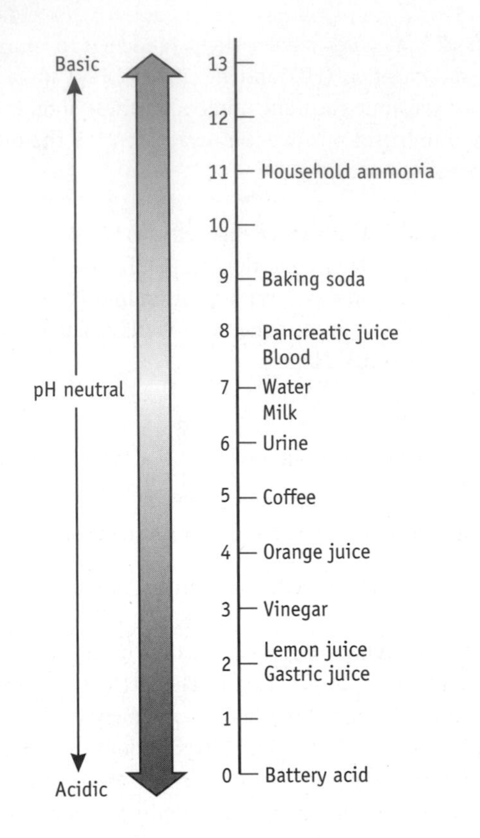

The pH scale.
Note: Each step is ten times as concentrated in base (1/10 as much acid, H^+) as the one below it.

Chemical Reactions

A chemical reaction, or chemical change, results in the breakdown of substances and the formation of new ones. Almost all such reactions involve a change in the bonding of atoms. Old bonds are broken, and new ones are formed. The nuclei of atoms are never involved in chemical reactions—only their outer-shell electrons take part. At the end of a chemical reaction, the number of atoms of each type is always the same as at the beginning. For example, two hydrogen molecules ($2H_2$) can react with one oxygen molecule (O_2) to form two water molecules ($2H_2O$). In this reaction two substances (hydrogen and oxygen) disappear, and a new one (water) is formed, but at the end of the reaction there are still four H atoms and two O atoms, just as there were at the beginning. Because the atoms are now linked in a different way, their characteristics or properties have changed.

In many instances chemical reactions involve not the relinking of molecules but the exchanging of electrons or protons among them. In such reactions the molecule that gains one or more electrons (or loses one or more hydrogen ions) is said to be reduced; the molecule that loses electrons (or gains protons) is oxidized. A hydrogen ion is equivalent to a proton. Oxidation and reduction take place simultaneously because an electron or proton that is lost by one molecule is accepted by another. The addition of an atom of oxygen is also oxidation because oxygen (with six electrons in the outer shell) accepts two electrons in becoming bonded. Oxidation, then, is loss of electrons, gain of protons, or addition of oxygen (with six electrons); reduction is the opposite—gain of electrons, loss of protons, or loss of oxygen. The addition of hydrogen atoms to oxygen to form water can thus be described as the reduction of oxygen or the oxidation of hydrogen.

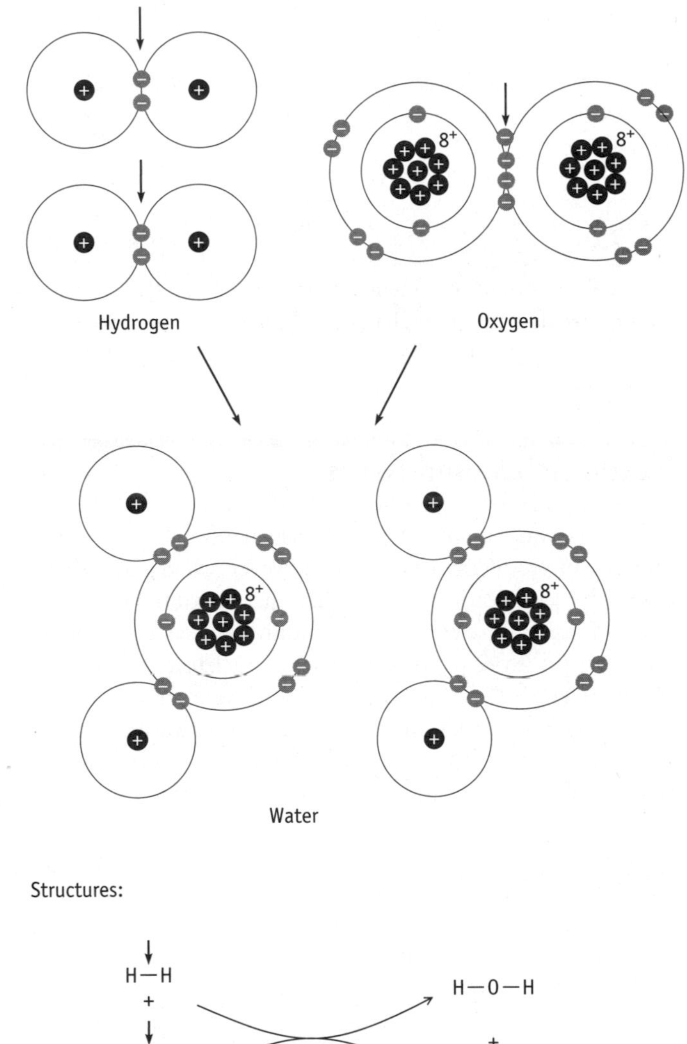

Formulas:

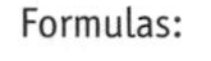

$$2H_2 + O_2 \longrightarrow 2H_2O$$

Hydrogen and oxygen react to form water.

If a reaction results in a net increase in the energy of a compound, it is called an endergonic, or "uphill," reaction (energy, *erg*, is added into, *endo*, the compound). An example is the chief result of photosynthesis, the making of sugar in a plant from carbon dioxide and water using the energy of sunlight. Conversely, the oxidation of sugar to carbon dioxide and water is an exergonic, or "downhill," reaction because the end products have less energy than the starting products. Oftentimes, but not always, reduction reactions are endergonic, resulting in an increase in the energy of the products. Oxidation reactions often, but not always, are exergonic.

Chemical reactions tend to occur spontaneously if the end products are in a lower energy state and therefore are more stable than the reacting compounds. These reactions often give off energy in the form of heat as they occur. The generation of heat by wood burning in a fireplace and the maintenance of human body warmth both depend on energy-yielding chemical reactions. These downhill reactions occur easily, although they may require some activation energy to get them started, just as a ball requires a push to start rolling downhill.

Uphill reactions, in which the products contain more energy than the reacting compounds started with, do not occur until an energy source is provided. An example of such an energy source is the sunlight used in photosynthesis, where carbon dioxide and water (low-energy compounds) are combined to form the sugar glucose (a higher-energy compound). Another example is the use of the energy in glucose to combine two low-energy compounds in the body into the high-energy compound ATP. The energy in ATP may be used to power many other energy-requiring, uphill reactions. Clearly, any of many different molecules can be used as a temporary storage place for energy.

Neither downhill nor uphill reactions occur until something sets them off (activation) or until a path is provided for them to follow. The body uses enzymes as a means of providing paths and controlling chemical reactions. By controlling the availability and the action of its enzymes, the body can "decide" which chemical reactions to prevent and which to promote.

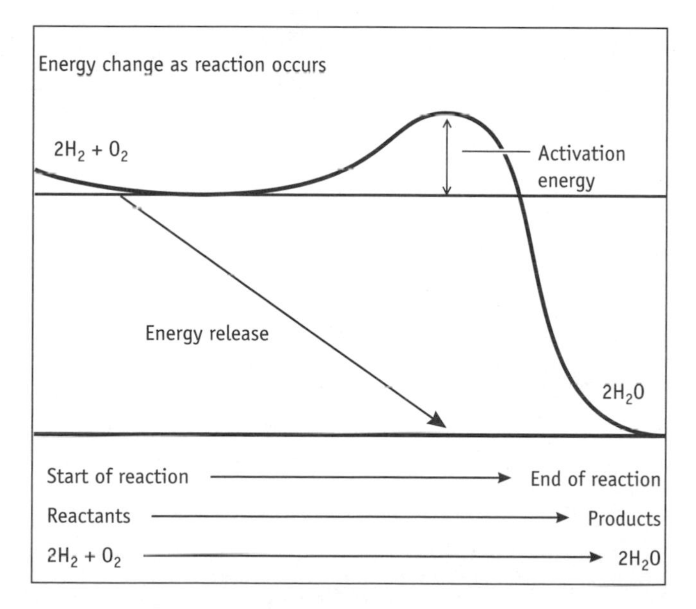

Formation of Free Radicals

Normally, when a chemical reaction takes place, bonds break and re-form with some redistribution of atoms and rearrangement of bonds to form new, stable compounds. Normally, bonds don't split in such a way as to leave a molecule with an odd, unpaired electron. However, weak bonds can split this way, and when they do, free radicals are formed. Free radicals are highly unstable and quickly react with other compounds, forming more free radicals in a chain reaction.

A physical event such as the arrival of an energy-carrying particle of light or other radiation starts the process by breaking a weak bond so that free radicals are formed. A cascade may ensue in which many highly reactive radicals are generated, resulting finally in the disruption of a living structure such as a cell membrane.

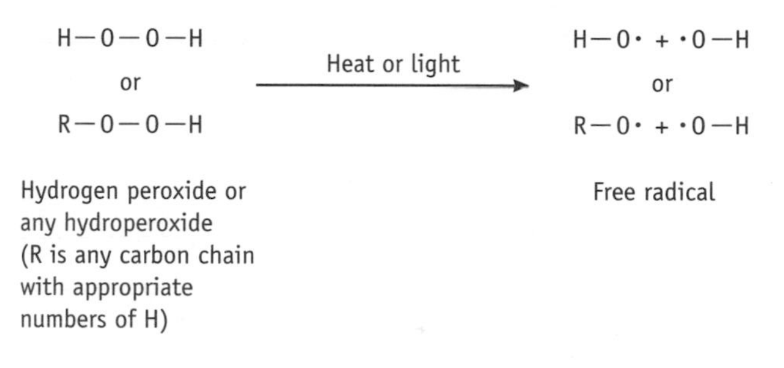

Free radicals are formed. The dots represent single electrons that are available for sharing (the atom needs another electron to fill its outer shell).

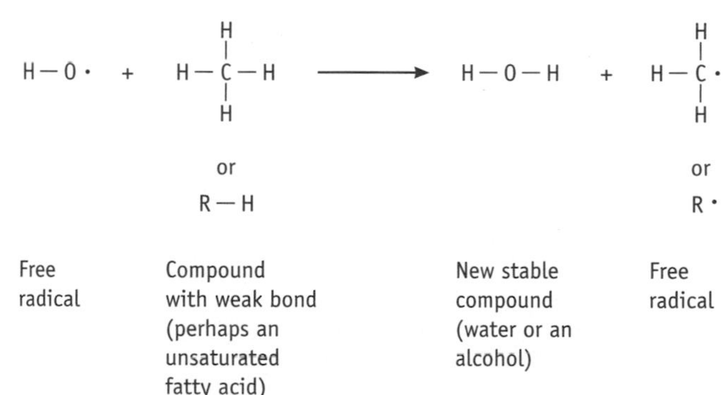

Destruction of biological compounds by free radicals. The free radical attacks a weak bond in a biological compound, disrupting it and forming a new stable molecule and another free radical. This can attack another biological compound, and so on.

Oxidation of some compounds can be induced by air at room temperature in the presence of light. Such reactions are thought to take place through the formation of compounds called peroxides:

Peroxides:

H—O—O—H	Hydrogen peroxide
R—O—O—H	Hydroperoxides (R is any carbon chain with appropriate numbers of H)
R—O—O—R	Peroxide

Some peroxides readily disintegrate into free radicals, initiating chain reactions like those just described.

Free radicals are of special interest in nutrition because the antioxidant properties of vitamins A, C, and E as well as the mineral selenium are thought to protect against the destructive effects of these free radicals. For example, vitamin E on the surface of the lungs reacts with, and is destroyed by, free radicals, thus preventing the radicals from reaching underlying cells and oxidizing the lipids in their membranes.

Glossary

element: a substance composed of atoms that are alike—for example, iron (Fe).

atom: the smallest component of an element that has all of the properties of the element.

compound: a substance composed of two or more different atoms—for example, water (H_2O).

molecule: two or more atoms of the same or different elements joined by chemical bonds. Examples are molecules of the element oxygen, composed of two oxygen atoms (O_2), and molecules of the compound water, composed of two hydrogen atoms and one oxygen atom (H_2O).

matter: anything that takes up space and has mass.

energy: the capacity to do work.

Food	Amount	Weight (g)	Calories	Protein (g)	Fat (g)	Sat. Fat (g)	Cholesterol (g)	Carbohydrate (g)	Fiber (g)	Calcium (mg)	Iron (mg)	Sodium (mg)	Vit A (IU)	Thiamin (Vit B_1) (mg)	Riboflavin (Vit B_2) (mg)	Niacin (mg)	Vit C (mg)	Folate (mcg)
Apples, fresh, w/peel, lrg	1 ea	150	88	0.3	1	0.1	0	23	4.1	10	0.3	0	80	0.03	0.02	0.1	9	4.2
Applesauce, swtnd, w/o salt, cnd	1 cup	255	194	0.5	0	0.1	0	51	3.1	10	0.9	8	28	0.03	0.07	0.5	4	1.53
Apricots, pitted, fresh, whole	3 ea	114	55	1.6	0	0	0	13	2.7	16	0.6	1	2978	0.03	0.05	0.7	11	9.8
Apricots, w/skin, in heavy syrup, cnd, whole	½ cup	120	100	0.6	0	0	0	26	1.9	11	0.4	5*	1476	0.02	0.03	0.5	4	2.04
Asparagus, spears, ckd w/o salt	4 ea	60	14	1.6	0	0	0	3	1	12	0.4	7	323	0.07	0.08	0.6	6	87.6
Avocado, Calif, fresh	½ ea	120	212	2.5	21	3.1	0	8	5.9	13	1.4	14	734	0.13	0.15	2.3	9	78.6
Bagel, plain, 3½" diameter	1 ea	68	187	7.1	1	0.1	0	36	1.6	50	2.4	363	0	0.37	0.21	3.1	0	59.84
Banana, fresh, med	1 ea	140	129	1.4	1	0.3	0	33	3.4	8	0.4	1	113	0.06	0.14	0.8	13	26.74
Bar, granola, hard	1 ea	24	113	2.4	5	0.6	0	15	1.3	15	0.7	71	36	0.06	0.03	0.4	0	5.52
Beans, black, mature, ckd w/o salt	1 cup	172	227	15.2	1	0.2	0	41	15	46	3.6	2	10	0.42	0.1	0.9	0	255.9
Beans, chickpea/garbanzo, mature, ckd	1 cup	164	269	14.5	4	0.4	0	45	12.5	80	4.7	11	44	0.19	0.1	0.9	2	282.0
Beans, frijoles/refried, cnd	½ cup	145	136	8	2	0.7	12	23	7.7	51	2.4	434	0	0.04	0.02	0.5	9	15.95
Beans, green, snap/string, ckd	½ cup	65	23	1.2	0	0	0	5	2.1	30	0.8	2	433	0.05	0.06	0.4	6	21.64
Beans, kidney, red, mature, cnd	1 cup	185	157	9.7	1	0.1	0	29	11.8	44	2.3	631	0	0.19	0.16	0.8	2	93.61
Beans, lima, fordhook, immature, ckd f/fzn w/o salt, drained	½ cup	85	85	5.2	0	0.1	0	16	4.9	19	1.2	45	162	0.06	0.05	0.9	11	18.02
Beans, mung, mature, sprouted, raw	½ cup	52	16	1.6	0	0	0	3	0.9	7	0.5	3	11	0.04	0.06	0.4	7	31.62
Beans, pinto, mature, ckd w/o salt	1 cup	171	234	14	1	0.2	0	44	14.7	82	4.5	3	3	0.32	0.16	0.7	4	294.1
Beef, chuck arm pot roast, brsd, choice, ¼" trim	3 oz	85	296	22.9	22	8.6	84	0	0	8	2.6	50	0	0.06	0.2	2.7	0	7.65
Beef, corned, cnd	3 oz	85	212	23	13	5.3	73	0	0	10	1.8	855	0	0.02	0.12	2.1	0	7.65
Beef, ground, hamburger patty, brld, well done, 16% fat	3 oz	85	225	24.3	13	5.3	84	0	0	8	2.4	70	0	0.06	0.27	5	0	9.35
Beef, ground, hamburger patty, brld, well done, 18% fat	3 oz	85	238	24	15	5.9	86	0	0	10	2.1	76	0	0.05	0.2	5.1	0	9.35
Beef, liver, fried	3 oz	85	184	22.7	7	2.3	410	7	0	9	5.3	90	30689	0.18	3.52	12.3	20	187
Beef, T-bone steak, brld, choice, ¼" trim	3 oz	85	263	19.7	20	7.7	57	0	0	7	2.3	54	0	0.08	0.18	3.4	0	5.95
Beef, top sirloin steak, lean, brld, choice, ¼" trim	3 oz	85	172	25.8	7	2.6	76	0	0	9	2.9	56	0	0.11	0.25	3.6	0	8.5
Beer	12 fl-oz	350	148	1.1	0	0	0	13	0.7	18	0.1	18	0	0.02	0.09	1.6	0	21.6
Beer, light	12 fl-oz	354	99	0.7	0	0	0	5	0	18	0.1	11	0	0.03	0.11	1.4	0	14.51
Beets, cnd, drained, diced	½ cup	30	25	0.7	0	0	0	6	1.4	12	1.5	155	9	0.01	0.03	0.1	3	24.16
Biscuits, homemade	1 ea	35	124	2.5	6	1.5	1	16	0.5	82	1	203	29	0.12	0.11	1	0	21.35
Blueberries, fresh, bilberries	½ cup	73	41	0.5	0	0	0	10	2	4	0.1	4	73	0.04	0.04	0.3	9	4.67
Brandy, 86 proof	1 oz	28	70	0	0	0	0	0	0	0	0	0	0	0	0	0	0	0
Bread, banana, prep f/recipe w/veg shortening	1 pce	50	169	2.2	6	1.5	22	28	0.7	9	0.7	99	46	0.09	0.1	0.7	1	5.5
Bread, cracked wheat	1 pce	25	65	2.2	1	0.2	0	12	1.4	11	0.7	134	0	0.09	0.06	0.9	0	15.25
Bread, French	1 pce	35	96	3.1	1	0.2	0	18	1	26	0.9	213	0	0.18	0.12	1.7	0	33.25
Bread, mixed grain	1 pce	26	65	2.6	1	0.2	0	12	1.7	24	0.9	127	0	0.11	0.09	1.1	0	20.8
Bread, pita pocket, white	1 ea	60	165	5.5	1	0.1	0	33	1.3	52	1.6	322	0	0.36	0.2	2.8	0	57
Bread, pumpernickel	1 pce	32	80	2.8	1	0.1	0	15	2.1	22	0.9	215	0	0.1	0.1	1	0	25.6
Bread, rye	1 pce	25	65	2.1	1	0.2	0	12	1.5	18	0.7	165	2	0.11	0.08	1	0	21.5
Bread, white, f/recipe w/2% milk	1 pce	25	71	2	1	0.3	1	12	0.5	14	0.7	90	20	0.1	0.1	0.9	0	22.75
Bread, whole wheat	1 pce	25	62	2.4	1	0.2	0	12	1.7	18	0.8	132	0	0.09	0.05	1	0	12.5
Broccoli, med stalk, 8" long, ckd w/o add salt	1 ea	140	39	4.2	0	0.1	0	7	4.1	64	1.2	36	1943	0.08	0.16	0.8	104	70
Broccoli, spear, raw, 5" long	1 ea	114	32	3.4	0	0.1	0	6	3.4	55	1	31	1758	0.07	0.14	0.7	106	80.94
Brownie, chocolate, w/walnuts, prep f/rec	1 ea	20	93	1.2	6	1.5	15	10	0.4	11	0.4	69	153	0.03	0.04	0.2	0	5.8
Brussels Sprouts, ckd, drained	½ cup	78	30	2	0	0.1	0	7	2	28	0.9	16	561	0.08	0.06	0.5	48	46.8
Buns, hamburger	1 ea	40	114	3.4	2	0.5	0	20	1.1	56	1.3	224	0	0.19	0.12	1.6	0	38

Food	Amount	Weight (g)	Calories	Protein (g)	Fat (g)	Sat. Fat (g)	Cholesterol (g)	Carbohydrate (g)	Fiber (g)	Calcium (mg)	Iron (mg)	Sodium (mg)	Vit A (IU)	Thiamin (Vit B_1) (mg)	Riboflavin (Vit B_2) (mg)	Niacin (mg)	Vit C (mg)	Folate (mcg)
Buns, hot dog/frankfurter	1 ea	40	114	3.4	2	0.5	0	20	1.1	56	1.3	224	0	0.19	0.12	1.6	0	38
Burger/Patty, vegetarian, Gardenburger, original	1 ea	71	130	8	3	1	11	18	5	84	0	290	50	0.11	0.15	1.1	0	10.08
Burger/Patty, vegetarian, soy	1 ea	71	142	14.9	6	1	0	6	3.3	21	1.5	390	0	0.64	0.43	7.1	0	55.38
Butter, salted	1 Tbs	5	36	0	4	2.5	11	0	0	1	0	41	153	0	0	0	0	0.15
Buttermilk, skim, cultured	1 cup	245	99	8.1	2	1.3	9	12	0	285	0.1	257	81	0.08	0.38	0.1	2	12.25
Cabbage, ckd w/o add salt, drained, shredded	½ cup	85	19	0.9	0	0	0	4	2	26	0.1	7	112	0.05	0.05	0.2	17	17
Cabbage, raw, shredded	½ cup	45	11	0.6	0	0	0	2	1	21	0.3	8	60	0.02	0.02	0.1	14	19.35
Cake, angel food, cmrcl prep	1 pce	60	155	3.5	0	0.1	0	35	0.9	84	0.3	449	0	0.06	0.29	0.5	0	21
Cake, carrot, w/cream cheese icing	1 pce	96	419	4.4	25	4.7	52	45	1.2	24	1.2	236	3310	0.13	0.15	1	1	11.52
Cake, chocolate, w/chocolate icing, 1/8th	1 pce	69	253	2.8	11	3.3	29	38	1.9	30	1.5	230	59	0.02	0.09	0.4	0	11.73
Cake, devils food, marshmallow iced	1 pce	99	408	3.5	21	5.8	52	52	1.2	47	1.3	338	0				0	
Cake, pound, w/butter	1 pce	30	116	1.7	6	3.5	66	15	0.1	10	0.4	119	182	0.04	0.07	0.4	0	12.3
Cake, white, w/chocolate icing	1 pce	71	259	1.8	8	3.7	13	46	0.8	55	0.5	219	166	0.05	0.07	0.3	0	2.11
Calamari/Squid, fried, mixed species	1 cup	150	262	26.9	11	2.8	390	12	0	58	1.5	459	52	0.08	0.69	3.9	6	21
Candy Bar, Almond Joy, fun size	1½ oz	42	196	1.8	11	7.3	2	24	2	26	0.6	61	5	0.01	0.06	0.2	0	
Candy Bar, Mars almond	1 ea	50	234	4.1	12	3.6	8	31	1	84	0.6	85	94	0.02	0.16	0.5	0	9.5
Candy Bar, Milky Way, 2.1 oz bar	1 ea	60	254	2.7	10	4.7	8	43	1	78	0.5	144	65	0.02	0.13	0.2	1	6
Candy Bar, Special Dark sweet chocolate	1 ea	41	226	2	13	8.3	0	25	2	11	1	3	14	0.01	0.03	0.2	0	0.82
Candy, caramels, plain/chocolate	1 oz	28	107	1.3	2	1.8	2	22	0.3	39	0	69	9	0	0.05	0.1	0	1.4
Candy, hard, all flvrs	1 oz	28	110	0	0	0	0	27	0	1	0.1	11	0	0	0	0	0	0
Candy, Kisses, milk chocolate	1 oz	28	144	1.9	9	5.2	6	17	1	53	0.4	23	52	0.02	0.08	0.1	0	2.24
Candy, M & M's peanut chocolate	1 oz	28	144	2.7	7	2.9	3	17	1	28	0.3	13	26	0.03	0.05	1	0	9.8
Candy, M & M's plain chocolate	1 oz	28	138	1.2	6	3.7	4	20	0.7	29	0.3	17	57	0.02	0.06	0.1	0	1.68
Candy, milk chocolate, w/almonds	1 oz	28	147	2.5	10	4.8	5	15	1.7	63	0.5	21	21	0.02	0.12	0.2	0	3.36
Carrots, ckd w/o add salt, drained, slices	½ cup	73	33	0.8	0	0	0	8	2.4	23	0.5	48	17924	0.02	0.04	0.4	2	10.15
Carrots, raw, whole, 7½" long	1 ea	81	35	0.8	0	0	0	8	2.4	22	0.4	28	22784	0.08	0.05	0.8	8	11.34
Catsup/Ketchup	1 Tbs	15	16	0.2	0	0	0	4	0.2	3	0.1	178	152	0.01	0.01	0.2	2	2.25
Cauliflower, ckd, drained	½ cup	63	14	1.2	0	0	0	3	1.7	10	0.2	9	11	0.03	0.03	0.3	28	27.72
Celery, raw, med stalk, 8" long	1 ea	40	6	0.3	0	0	0	1	0.7	16	0.2	35	54	0.02	0.02	0.1	3	11.2
Cereal, 100% Bran, rte, dry	½ cup	33	89	4.1	2	0.3	0	24	9.8	23	4.1	229	0	0.79	0.89	10.5	31	23.43
Cereal, All-Bran, rte, dry	¼ cup	21	55	2.6	1	0.1	0	16	6.8	74	3.1	43	525	0.27	0.29	3.5	10	63
Cereal, Alpha-Bits, rte, dry	1 cup	28	110	2.2	1	0.1	0	24	1.2	8	2.7	178	1235	0.36	0.42	4.9	0	98.84
Cereal, bran flakes, rte, dry	¾ cup	30	96	2.8	1	0.1	0	24	5.3	17	8.1	220	750	0.38	0.43	5	0	99.9
Cereal, Cheerios	1 cup	23	84	2.4	1	0.3	0	18	2	42	6.2	218	958	0.29	0.33	3.8	12	76.59
Cereal, Chex, corn, rte, dry	1 cup	28	105	2	0	0.1	0	24	0.5	94	8.4	270	0	0.35	0	4.7	6	93.24
Cereal, Chex, wheat, rte, dry	1 cup	46	159	4.8	1	0.2	0	37	5.1	92	13.8	412	0	0.34	0.06	4.6	6	92
Cereal, corn flakes, rte, dry	1 cup	25	91	1.6	0	0.1	0	22	0.7	1	7.8	266	625	0.32	0.35	4.2	12	88.25
Cereal, Corn Pops, rte, dry	1 cup	28	107	1	0	0.1	0	26	0.4	2	1.7	111	700	0.36	0.39	4.7	14	98.84
Cereal, Cream of Wheat, quick, ckd w/water	1 cup	244	132	3.7	0	0.1	0	27	1.2	51	10.5	142	0	0.24	0	1.5	0	109.8
Cereal, Crispy Rice, rte, dry	¾ cup	22	87	1.4	0	0	0	19	0.3	4	0.6	161	971	0.41	0.46	5.4	12	108.6
Cereal, Frosted Flakes, rte, dry	1 cup	35	135	1.4	0	0.1	0	32	0.7	1	5.1	226	847	0.42	0.49	5.6	17	105
Cereal, Frosted Mini Wheats, rte, dry	1 cup	55	186	5.2	1	0.2	0	45	5.9	20	15.4	2	0	0.38	0.44	5.4	0	110
Cereal, granola, rte, dry	½ cup	57	257	6	10	1.3	0	38	3.6	43	1.8	92	0	0.18	0.06	0.6	0	8.55
Cereal, Grape Nuts, rte, dry	½ cup	57	205	6.2	1	0.2	0	46	5	19	15.9	348	737	0.37	0.42	4.9	0	98.04
Cereal, Honey Bran, rte, dry	½ cup	30	102	2.6	1	0.2	0	25	3.3	14	4.8	173	1323	0.39	0.45	5.3	16	20.1
Cereal, Life, plain, rte, dry	1 cup	44	167	4.3	2	0.3	0	35	2.8	134	12.3	240	16	0.55	0.62	7.3	0	146.9
Cereal, Mueslix, five grain muesli, rte, dry	1 cup	82	289	6.2	5	0.7	0	63	5.6	67	8.9	107	2488	0.75	0.84	9.8	1	196.8
Cereal, Nutri-Grain, wheat, rte, dry	1 oz	28	101	2.4	0	0.1	0	24	1.8	8	0.8	190	0	0.36	0.42	4.9	15	98.84
Cereal, oatmeal, unsalted, ckd w/water	½ cup	120	74	3.1	1	0.2	0	13	2	10	0.8	1	19	0.13	0.02	0.2	0	4.8

Food	Amount	Weight (g)	Calo-ries	Pro-tein (g)	Fat (g)	Sat. Fat (g)	Choles-terol (g)	Carbo-hydrate (g)	Fiber (g)	Cal-cium (mg)	Iron (mg)	Sodium (mg)	Vit A (IU)	Thia-min (Vit B_1) (mg)	Ribo-flavin (Vit B_2) (mg)	Niacin (mg)	Vit C (mg)	Folate (mcg)
Cereal, raisin bran, rte, dry	1 cup	49	155	3.9	1	0.1	0	38	6.4	22	9	299	623	0.31	0.35	4.2	0	82.81
Cereal, Shredded Wheat, sml biscuits, rte, dry	1 cup	19	68	2.1	0	0.1	0	15	1.9	7	0.8	2	0	0.05	0.05	1	0	9.5
Cereal, Smacks, rte, dry	1 cup	37	141	2.4	1	0.4	0	32	1.3	4	2.5	70	1028	0.52	0.59	6.8	21	136.9
Cereal, Special K, rte, dry	1 cup	21	78	4.3	0	0	0	15	0.7	3	5.9	169	508	0.36	0.4	4.7	10	63
Cereal, Total, wheat, rte, dry	1 cup	33	116	3.3	1	0.2	0	26	2.9	284	19.8	218	1375	1.65	1.87	22.1	66	439.8
Cereal, Wheaties, rte, dry	1 cup	29	106	3.1	1	0.2	0	23	2	53	7.8	215	725	0.36	0.41	4.8	14	96.57
Cheese Puffs/Cheetos	1 oz	28	155	2.1	10	1.8	1	15	0.3	16	0.7	294	74	0.07	0.1	0.9	0	33.6
Cheese Spread, low fat, low sod	1 pce	34	61	8.4	2	1.5	12	1	0	233	0.1	2	92	0.01	0.13	0	0	3.06
Cheese, American, proc, shredded	1 oz	28	105	6.2	9	5.5	26	0	0	172	0.1	401	339	0.01	0.1	0	0	2.18
Cheese, blue	1 oz	28	99	6	8	5.2	21	1	0	148	0.1	391	202	0.01	0.11	0.3	0	10.19
Cheese, cheddar, diced	1 oz	28	113	7	9	5.9	29	0	0	202	0.2	174	297	0.01	0.11	0	0	5.1
Cheese, feta	1 oz	28	74	4	6	4.2	25	1	0	138	0.2	313	125	0.04	0.24	0.3	0	8.96
Cheese, monterey jack, shredded	1 oz	28	105	6.9	8	5.3	25	0	0	209	0.2	150	266	0	0.11	0	0	5.1
Cheese, mozzarella, part skm mlk, low moist, shredded	1 oz	28	78	7.7	5	3	15	1	0	205	0.1	148	197	0.01	0.1	0	0	2.77
Cheese, parmesan, grated	1 Tbs	5	23	2.1	2	1	4	0	0	69	0	93	35	0	0.02	0	0	0.4
Cheese, ricotta, part skm	1 oz	28	39	3.2	2	1.4	9	1	0	76	0.1	35	121	0.01	0.05	0	0	3.67
Cheese, Swiss, shredded	1 oz	28	105	8	8	5	26	1	0	269	0	73	237	0.01	0.1	0	0	1.79
Cheesecake	1 pce	85	273	4.7	19	8.4	47	22	0.4	43	0.5	176	465	0.02	0.16	0.2	0	15.3
Cherries, sweet, fresh	10 ea	75	54	0.9	1	0.2	0	12	1.7	11	0.3	0	160	0.04	0.04	0.3	5	3.15
Chicken, broiler/fryer, breast, rstd	1 ea	98	193	29.2	8	2.1	82	0	0	14	1	70	91	0.06	0.12	12.5	0	3.92
Chicken, broiler/fryer, dark meat, w/o skin, rstd	3 oz	85	174	23.3	8	2.3	79	0	0	13	1.1	79	61	0.06	0.19	5.6	0	6.8
Chicken, broiler/fryer, drumstick, rstd	1 ea	52	112	14.1	6	1.6	47	0	0	6	0.7	47	52	0.04	0.11	3.1	0	4.16
Chicken, broiler/fryer, meat only, w/o skin, rstd	3 oz	85	162	24.6	6	1.7	76	0	0	13	1	73	45	0.06	0.15	7.8	0	5.1
Chips, corn	1 oz	28	151	1.8	9	1.3	0	16	1.4	36	0.4	176	26	0.01	0.04	0.3	0	5.6
Chips, tortilla, chili & lime	18 pce	28	110	2	2	0	0	22	2	60	3.6	200	0				0	
Chips, tortilla, plain	1 oz	28	140	2	7	1.4	0	18	1.8	43	0.4	148	55	0.02	0.05	0.4	0	2.8
Cod, batter fried	3½ oz	100	173	17.4	8	1.6	50	7	0.2	29	0.7	91	30	0.07	0.1	2.3	2	8.71
Cod, stmd/poached	3½ oz	100	102	22.4	1	0.1	46	0	0	9	0.3	80	28	0.02	0.05	2.2	3	6.6
Coffee, brewed	¾ cup	180	4	0.2	0	0	0	1	0	4	0.1	4	0	0	0	0.4	0	0.18
Collards, ckd w/o add salt	½ cup	95	25	2	0	0	0	5	2.7	113	0.4	9	2973	0.04	0.1	0.5	17	88.35
Cone, ice cream, wafer/cake type	1 ea	115	480	9.3	8	1.4	0	91	3.4	29	4.1	164	0	0.29	0.41	5.1	0	117.3
Cookie, chocolate chip, prep w/marg f/rec	2 ea	20	98	1.1	6	1.6	6	12	0.6	8	0.5	72	127	0.04	0.04	0.3	0	6.6
Cookie, chocolate sandwich, creme filled	4 ea	40	189	1.9	8	1.5	0	28	1.3	10	1.6	242	1	0.03	0.07	0.8	0	17.2
Cookie, fig bar	4 ea	56	195	2.1	4	0.6	0	40	2.6	36	1.6	196	18	0.09	0.12	1	0	15.12
Cookie, oatmeal raisin, prep f/rec	2 ea	26	113	1.7	4	0.8	9	18	0.8	26	0.7	140	167	0.06	0.04	0.3	0	7.8
Cookie, peanut butter, prep f/rec	2 ea	24	114	2.2	6	1.1	7	14	0.5	9	0.5	124	144	0.05	0.05	0.8	0	13.2
Cookie, shortbread, cmrcl, plain	4 ea	32	161	2	8	2	6	21	0.6	11	0.9	146	28	0.11	0.11	1.1	0	18.88
Cookie, vanilla, wafer type, 12–17% fat	10 ea	40	176	2	6	1.5	20	29	0.8	19	1	125	11	0.11	0.13	1.2	0	20
Coriander, raw	¼ cup	4	1	0.1	0	0	0	0	0.1	4	0.1	1	111	0	0	0	0	0.41
Corn, yellow, vac pack, cnd	½ cup	83	66	2	0	0.1	0	16	1.7	4	0.3	226	200	0.03	0.06	1	7	40.92
Cornbread, prep f/dry mix	1 ea	60	188	4.3	6	1.6	37	29	1.4	44	1.1	467	123	0.15	0.16	1.2	0	33
Cornmeal, yellow, degermed, enrich, dry	½ cup	120	439	10.2	2	0.3	0	93	8.9	6	5	4	496	0.86	0.49	6	0	224.4
Cottage Cheese, 2% fat	½ cup	113	101	15.5	2	1.4	9	4	0	77	0.2	459	79	0.03	0.21	0.2	0	14.8
Cottage Cheese, creamed, sml curd	½ cup	105	109	13.1	5	3	16	3	0	63	0.1	425	171	0.02	0.17	0.1	0	12.81
Crab, blue, cnd, drained	1 cup	135	134	27.7	2	0.3	120	0	0	136	1.1	450	7	0.11	0.11	1.8	4	57.38
Crackers, cheese	1 ea	10	50	1	3	0.9	1	6	0.2	15	0.5	100	16	0.06	0.04	0.5	0	8
Crackers, graham, plain/honey, 2½ square	2 ea	14	59	1	1	0.2	0	11	0.4	3	0.5	85	0	0.03	0.04	0.6	0	8.4
Crackers, matzoh, plain, svg	1 ea	28	111	2.8	0	0.1	0	23	0.8	4	0.9	1	0	0.11	0.08	1.1	0	32.76

Food	Amount	Weight (g)	Calo-ries	Pro-tein (g)	Fat (g)	Sat. Fat (g)	Choles-terol (g)	Carbo-hydrate (g)	Fiber (g)	Cal-cium (mg)	Iron (mg)	Sodium (mg)	Vit A (IU)	Thia-min (Vit B_1) (mg)	Ribo-flavin (Vit B_2) (mg)	Niacin (mg)	Vit C (mg)	Folate (mcg)
Crackers, rye, wafers	2 ea	14	47	1.3	0	0	0	11	3.2	6	0.8	111	1	0.06	0.04	0.2	0	6.3
Crackers, saltine	1 ea	11	48	1	1	0.3	0	8	0.3	13	0.6	143	0	0.06	0.05	0.6	0	13.64
Crackers, standard, reg, snack type, round	1 ea	3	15	0.2	1	0.1	0	2	0	4	0.1	25	0	0.01	0.01	0.1	0	2.31
Crackers, triscuit	1 ea	5	24	0.5	1	0.2	0	3	0.5		0.2	26						
Crackers, wheat	1 ea	2	9	0.2	0	0.1	0	1	0.1	1	0.1	16	0	0.01	0.01	0.1	0	0.36
Cream Cheese	1 oz	28	98	2.1	10	6.2	31	1	0	22	0.3	83	400	0	0.06	0	0	3.7
Cream, light	1 Tbs	15	29	0.4	3	1.8	10	1	0	14	0	6	95	0	0.02	0	0	0.34
Cream, whipping, heavy	1 Tbs	15	52	0.3	6	3.5	21	0	0	10	0	6	221	0	0.02	0	0	0.56
Croissant, butter	1 ea	57	231	4.7	12	6.6	38	26	1.5	21	1.2	424	424	0.22	0.14	1.2	0	35.34
Cucumber, w/o skin, raw, sliced	½ cup	60	7	0.3	0	0	0	2	0.4	8	0.1	1	44	0.01	0.01	0.1	2	8.4
Dates, fresh, whole	10 ea	83	228	1.6	0	0.2	0	61	6.2	27	1	2	42	0.07	0.08	1.8	0	10.46
Dinner, chicken, cacciatore, w/noodles, low cal, fzn	1 ea	308	311	22.5	10	2.4	59	33	3.4	29	3.2	934	732	0.28	0.4	8	26	32.22
Doughnut, cake	1 ea	47	198	2.3	11	1.7	17	23	0.7	21	0.9	257	27	0.1	0.11	0.9	0	22.09
Doughnut, raised, glazed	1 ea	60	242	3.8	14	3.5	4	27	0.7	26	1.2	205	8	0.22	0.13	1.7	0	25.8
Egg Substitute, Egg Beaters, new	¼ cup	61	30	6	0	0	0	1	0	20	1.1	125	300		0.85		0	32.0
Egg Whites, raw	1 ea	33	16	3.5	0	0	0	0	0	2	0	54	0	0	0.15	0	0	0.99
Egg Yolks, raw, lrg	1 ea	17	61	2.8	5	1.6	218	0	0	23	0.6	7	331	0.03	0.11	0	0	24.82
Eggs, hard ckd/bld, lrg	1 ea	50	78	6.3	5	1.6	212	1	0	25	0.6	62	280	0.03	0.26	0	0	22
Eggs, scrambled, plain, lrg	1 ea	64	106	7.1	8	2.4	225	1	0	45	0.8	179	436	0.03	0.28	0.1	0	19.2
Eggs, whole, fried	1 ea	46	92	6.2	7	1.9	211	1	0	25	0.7	162	394	0.03	0.24	0	0	17.48
Entree, lasagna, w/meat, prep f/rec	1 pce	220	352	20.7	14	7.2	52	36	2.5	243	2.8	351	902	0.21	0.3	3.8	13	17.82
Entree, macaroni & cheese, prep f/rec w/margarine	½ cup	100	215	8.4	11	4.4	21	20	0.6	181	0.9	543	430	0.1	0.2	0.9	0	5.15
Entree, meatloaf, beef	1 pce	111	232	20.2	14	5.6	107	5	0.2	37	2.1	185	148	0.06	0.28	3.3	1	14.28
Entree, quiche, lorraine	1 pce	242	724	20.5	56	25.9	304	34	1	318	2.6	303	1323	0.36	0.67	2.8	1	26.48
Entree, spaghetti, w/meatballs, prep f/rec	1 cup	248	332	18.6	12	3.3	74	39	7.7	124	3.7	1009	1587	0.25	0.3	4	22	9.99
Entree, spaghetti, w/tomato sauce & cheese, prep f/rec	1 cup	250	260	8.8	9	2	8	37	2.5	80	2.2	955	1075	0.25	0.17	2.2	12	8
Figs, dried, unckd	1 ea	21	54	0.6	0	0	0	14	2.5	30	0.5	2	28	0.01	0.02	0.1	0	1.57
Fish Sticks/Portions, heated f/fzn, 4x1x.5	2 ea	56	152	8.8	7	1.8	63	13	0	11	0.4	326	59	0.07	0.1	1.2	0	10.19
Flour, all purpose, white, bleached, enrich	1 cup	125	455	12.9	1	0.2	0	95	3.4	19	5.8	2	0	0.98	0.62	7.4	0	192.5
Flour, whole wheat	1 cup	120	407	16.4	2	0.4	0	87	14.6	41	4.7	6	0	0.54	0.26	7.6	0	52.8
Frankfurter/Hot Dog, beef & pork, 10 pack	1 ea	57	182	6.4	17	6.1	28	1	0	6	0.7	638	0	0.11	0.07	1.5	0	2.28
Frankfurter/Hot Dog, beef, 8 pack	1 ea	57	180	6.8	16	6.9	35	1	0	11	0.8	585	0	0.03	0.06	1.4	0	2.28
Frankfurter/Hot Dog, turkey	1 ea	45	102	6.4	8	2.7	48	1	0	48	0.8	642	0	0.02	0.08	1.9	0	3.6
Frozen Yogurt, vanilla/strawberry, nonfat, sml scoop	4 oz	113	112	5.6	0	0.1	2	22	0	196	0.1	75	7	0.05	0.23	0.1	1	11.99
Fruit Cocktail, in heavy syrup, cnd	1 cup	245	179	1	0	0	0	46	2.5	15	0.7	15	502	0.04	0.05	0.9	5	6.37
Fruit Cocktail, in juice	1 cup	248	114	1.1	0	0	0	29	2.5	20	0.5	10	756	0.03	0.04	1	7	6.2
Fruit Punch, prep f/pwd	1 cup	240	89	0	0	0	0	23	0	38	0.1	34	0	0	0	0	28	0.24
Fudge, chocolate, prep f/rec	1 oz	28	107	0.5	2	1.4	4	22	0.2	12	0.1	17	53	0	0.02	0	0	0.56
Grapefruit, pink, fresh, 3¾" diameter	½ ea	123	37	0.7	0	0	0	9	1.7	14	0.1	0	319	0.04	0.02	0.2	47	15.01
Grapes, tokay/empress/red flame, fresh	10 ea	50	36	0.3	0	0.1	0	9	0.5	6	0.1	1	36	0.05	0.03	0.2	5	1.95
Haddock, fillet, brd, fried	3 oz	85	184	17.1	9	1.9	65	7	0.2	53	1.5	145	69	0.08	0.09	3.7	0	11.6
Halibut, Greenland, fillet, bkd/brld	3 oz	85	203	15.7	15	2.6	50	0	0	3	0.7	88	51	0.06	0.09	1.6	0	0.85
Honey, strained, extracted	1 Tbs	21	64	0.1	0	0	0	17	0	1	0.1	1	0	0	0.01	0	0	0.42
Hot Cocoa/Choc, prep f/rec w/whole milk	1 cup	250	192	9.8	6	3.6	20	29	2	315	1.1	128	515	0.1	0.44	0.4	2	15
Hummus/Hummos, raw	1 cup	246	421	12.1	21	3.1	0	50	12.5	123	3.9	600	62	0.23	0.13	1	19	146.1
Instant Breakfast, prep f/dry mix w/nonfat milk	1 cup	282	216	15.7	1	0.7	9	36	0.2	407	4.8	268	2343	0.4	0.42	5.5	31	118.2

Food	Amount	Weight (g)	Calo-ries	Pro-tein (g)	Fat (g)	Sat. Fat (g)	Choles-terol (g)	Carbo-hydrate (g)	Fiber (g)	Cal-cium (mg)	Iron (mg)	Sodium (mg)	Vit A (IU)	Thia-min (Vit B_1) (mg)	Ribo-flavin (Vit B_2) (mg)	Niacin (mg)	Vit C (mg)	Folate (mcg)
Instant Breakfast, prep f/dry mix w/whole milk	1 cup	281	280	15.4	9	5.4	38	36	0.2	396	4.9	262	2151	0.41	0.47	5.5	31	117.7
Jam/Preserves, pkt	1 ea	14	39	0.1	0	0	0	10	0.2	3	0.1	4	2	0	0	0	1	4.62
Jelly	1 Tbs	18	51	0	0	0	0	13	0.2	1	0	5	3	0	0	0	0	0.18
Juice, apple, unswt'nd, cnd/btld	½ cup	124	58	0.1	0	0	0	14	0.1	9	0.5	4	1	0.03	0.02	0.1	1	0.12
Juice, cranberry cocktail	1 cup	253	144	0	0	0	0	36	0.3	8	0.4	5	10	0.02	0.02	0.1	90	0.51
Juice, grape, unswtnd, btld/cnd	½ cup	127	77	0.7	0	0	0	19	0.1	11	0.3	4	10	0.03	0.05	0.3	0	3.3
Juice, grapefruit, unswtnd, cnd	½ cup	124	47	0.6	0	0	0	11	0.1	9	0.2	1	9	0.05	0.02	0.3	36	12.9
Juice, grapefruit, unswtnd, prep f/fzn conc	1 cup	247	101	1.4	0	0	0	24	0.2	20	0.3	2	22	0.1	0.05	0.5	83	8.89
Juice, lemon, fresh	1 Tbs	15	4	0.1	0	0	0	1	0.1	1	0	0	3	0	0	0	7	1.94
Juice, orange, prep f/fzn	½ cup	125	56	0.9	0	0	0	13	0.2	11	0.1	1	98	0.1	0.02	0.3	49	54.75
Juice, prune, w/o pulp	½ cup	88	60	0.7	0	0	0	14	0.5	2	0.9	4	54				3	
Juice, tomato, w/salt, cnd	1 cup	244	41	1.9	0	0	0	10	1	22	1.4	881	1357	0.11	0.08	1.6	45	48.56
Kale, ckd w/o add salt, drained	½ cup	55	15	1	0	0	0	3	1.1	40	0.5	13	4070	0.03	0.04	0.3	23	7.32
Kiwifruit/Chinese Gooseberries, fresh, med	1 ea	76	46	0.8	0	0	0	11	2.6	20	0.3	4	133	0.02	0.04	0.4	74	28.88
Lamb, leg, whole, lean, rstd, choice, ¼" trim	3 oz	85	162	24.1	7	2.3	76	0	0	7	1.8	58	0	0.09	0.25	5.4	0	19.55
Lamb, loin chop, lean, brld, choice, ¼" trim	3 oz	84	181	25.2	8	2.9	80	0	0	16	1.7	71	0	0.09	0.24	5.8	0	20.16
Lemonade, white, fzn conc	12 oz	340	615	1	1	0.1	0	160	1.4	24	2.4	14	323	0.09	0.33	0.3	60	34
Lentils, sprouts, stir fried	1 cup	124	125	10.9	1	0.1	0	26	4.8	17	3.8	12	51	0.27	0.11	1.5	16	83.08
Lentils, unsalted, ckd	1 cup	200	232	18	1	0.1	0	40	15.8	38	6.7	4	16	0.34	0.15	2.1	3	361.6
Lettuce, butterhead, Boston/bibb, leaf, raw	2 pce	15	2	0.2	0	0	0	0	0.2	5	0	1	146	0.01	0.01	0	1	11
Lettuce, romaine, raw, chpd	1 cup	55	8	0.9	0	0	0	1	0.9	20	0.6	4	1430	0.06	0.06	0.3	13	74.64
Lobster, northern, stmd	1 cup	145	142	29.7	1	0.2	104	2	0	88	0.6	551	126	0.01	0.1	1.6	0	16.1
Lunchmeat Spread, liverwurst, cnd	1 oz	28	87	3.6	7	2.5	33	2	0.5	0	2.3	193	3818				1	
Lunchmeat, bologna, beef & pork	1 pce	28	88	3.3	8	3	15	1	0	3	0.4	285	0	0.05	0.04	0.7	0	1.4
Lunchmeat, bologna, turkey	2 pce	57	113	7.8	9	2.9	56	1	0	48	0.9	500	0	0.03	0.09	2	0	3.99
Lunchmeat, roast beef, deli style, pouch	3 oz	85	96	17.2	3	1.1	41	1	0	5	1.6	860	0				0	
Lunchmeat, turkey breast, rstd, fat free	1 pce	28	24	4.2	0	0.1	9	1	0	3	0.3	334	0				0	
Mayonnaise, imit, low cal	1 Tbs	15	35	0	3	0.5	4	2	0	0	0	75	0	0	0	0	0	0
Mayonnaise, soybean oil, w/salt	1 tsp	5	36	0.1	4	0.6	3	0	0	1	0	28	14	0	0	0	0	0.38
Melon, cantaloupe/musk, med 5" diameter	¼ ea	239	84	2.1	1	0.2	0	20	1.9	26	0.5	22	7705	0.09	0.05	1.4	101	40.63
Melon, honeydew, fresh, wedge, 1/8 melon	1 pce	129	45	0.6	0	0	0	12	0.8	8	0.1	13	52	0.1	0.02	0.8	32	7.74
Milk Shake, chocolate, fast food	10 fl-oz	340	432	11.6	13	7.9	44	70	2.7	384	1.1	330	316	0.2	0.83	0.5	1	11.9
Milk, evaporated, whole, w/add vit A, cnd	½ cup	126	169	8.6	10	5.8	37	13	0	329	0.2	133	500	0.06	0.4	0.2	2	9.95
Milk, low fat, 1%, w/add vit A	1 cup	244	102	8	3	1.6	10	12	0	300	0.1	123	500	0.1	0.41	0.2	2	12.44
Milk, low fat, 2%, chocolate	1 cup	250	179	8	5	3.1	17	26	1.2	284	0.6	150	500	0.09	0.41	0.3	2	12
Milk, low fat, 2%, w/add vit A	1 cup	244	121	8.1	5	2.9	18	12	0	297	0.1	122	500	0.1	0.4	0.2	2	12.44
Milk, nonfat/skim, w/add vit A	1 cup	245	86	8.4	0	0.3	4	12	0	302	0.1	126	500	0.09	0.34	0.2	2	12.74
Milk, whole, 3.3%	1 cup	244	150	8	8	5.1	33	11	0	291	0.1	120	307	0.09	0.4	0.2	2	12.2
Milkshake, strawberry, fast food	10 fl-oz	340	384	11.6	10	5.9	37	64	1.4	384	0.4	282	408	0.15	0.66	0.6	3	10.2
Mixed Vegetables, cnd, drained	1 cup	182	86	4.7	0	0.1	0	17	5.5	49	1.9	271	21198	0.08	0.09	1.1	9	42.95
Muffin, English, plain	1 ea	57	134	4.4	1	0.1	0	26	1.5	99	1.4	264	0	0.25	0.16	2.2	0	46.17
Muffin, English, plain, tstd	1 ea	52	133	4.4	1	0.1	0	26	1.5	98	1.4	262	0	0.2	0.14	2	0	38.48
Muffin, wheat bran, prep f/rec w/whole milk	1 ea	45	130	3.2	6	1.2	16	19	3.2	84	1.9	265	363	0.15	0.2	1.8	4	23.4
Mushrooms, raw, pces/slices	1 cup	35	9	1	0	0	0	1	0.4	2	0.4	1	0	0.03	0.15	1.4	1	4.2
Mustard Greens, ckd w/o add salt, drained	½ cup	70	10	1.6	0	0	0	1	1.4	52	0.5	11	2122	0.03	0.04	0.3	18	51.38
Nuts, almonds, dried, unblanched, whole	¼ cup	36	208	7.7	18	1.4	0	7	4.2	89	1.5	0	4	0.09	0.29	1.4	0	10.44
Nuts, Brazil, dried, shelled, 32 kernels	1 oz	28	184	4	19	4.5	0	4	1.5	49	1	1	0	0.28	0.03	0.5	0	1.12
Nuts, cashews, dry rstd, salted	1 cup	137	786	21	63	12.5	0	45	4.1	62	8.2	877	0	0.27	0.27	1.9	0	94.8
Nuts, coconut, unswtnd, dried	½ cup	65	429	4.5	42	37.2	0	16	10.6	17	2.2	24	0	0.04	0.06	0.4	1	5.85

Food	Amount	Weight (g)	Calories	Protein (g)	Fat (g)	Sat. Fat (g)	Cholesterol (g)	Carbohydrate (g)	Fiber (g)	Calcium (mg)	Iron (mg)	Sodium (mg)	Vit A (IU)	Thiamin (Vit B_1) (mg)	Riboflavin (Vit B_2) (mg)	Niacin (mg)	Vit C (mg)	Folate (mcg)
Nuts, peanuts, oil rstd, unsalted, chpd	1 oz	28	163	7.4	14	1.9	0	5	1.9	25	0.5	2	0	0.07	0.03	4	0	35.2
Nuts, pecans, dried, halves	1 oz	28	193	2.6	20	1.7	0	4	2.7	20	0.7	0	22	0.18	0.04	0.3	0	6.16
Nuts, walnuts, black, dried, chpd	1 oz	28	170	6.8	16	1	0	3	1.4	16	0.9	0	83	0.06	0.03	0.2	1	18.34
Oil, canola	1 cup	218	1927	0	218	15.5	0	0	0	0	0	0	0	0	0	0	0	0
Oil, corn	1 Tbs	15	133	0	15	1.9	0	0	0	0	0	0	0	0	0	0	0	0
Oil, olive	1 Tbs	15	133	0	15	2	0	0	0	0	0.1	0	0	0	0	0	0	0
Oil, peanut	1 cup	216	1909	0	216	36.5	0	0	0	0	0.1	0	0	0	0	0	0	0
Oil, safflower, greater than 70% linoleic	1 Tbs	15	133	0	15	0.9	0	0	0	0	0	0	0	0	0	0	0	0
Oil, soybean	1 tsp	5	44	0	5	0.7	0	0	0	0	0	0	0	0	0	0	0	0
Okra, bindi, ckd w/o add salt f/raw, drained, pods	8 ea	85	27	1.6	0	0	0	6	2.1	54	0.4	4	489	0.11	0.05	0.7	14	38.85
Olives, w/o pits, ripe, lrg, cnd	10 ea	44	51	0.4	5	0.6	0	3	1.4	39	1.5	384	177	0	0	0	0	0
Olives, w/o pits, ripe, sml, cnd	10 ea	32	37	0.3	3	0.5	0	2	1	28	1.1	279	129	0	0	0	0	0
Onions, yellow, ckd w/o add salt, drained, chpd	½ cup	105	46	1.4	0	0	0	11	1.5	23	0.3	3	0	0.04	0.02	0.2	5	15.75
Oranges, fresh, med	1 ea	180	85	1.7	0	0	0	21	4.3	72	0.2	0	369	0.16	0.07	0.5	96	54.54
Oysters, eastern, brd, fried, med	1 ea	45	89	3.9	6	1.4	36	5	0.1	28	3.1	188	136	0.07	0.09	0.7	2	13.95
Oysters, eastern, raw, wild	½ cup	120	82	8.5	3	0.9	64	5	0	54	8	253	120	0.12	0.11	1.7	4	12
Pancake, buckwheat, prep f/incomplete dry mix, 4"	1 ea	27	56	2.1	2	0.5	18	8	0.6	69	0.5	144	63	0.05	0.07	0.4	0	4.59
Pancake, plain, homemade, 4"	1 ea	73	166	4.7	7	1.5	43	21	1.1	160	1.3	320	143	0.15	0.21	1.1	0	27.74
Papaya, fresh, med	½ ea	227	89	1.4	0	0.1	0	22	4.1	54	0.2	7	645	0.06	0.07	0.8	140	86.26
Pasta, egg noodles, enrich, ckd	½ cup	80	106	3.8	1	0.2	26	20	0.9	10	1.3	6	16	0.15	0.07	1.2	0	51.2
Pasta, macaroni noodles, enrich, ckd	½ cup	70	99	3.3	0	0.1	0	20	0.9	5	1	1	0	0.14	0.07	1.2	0	49
Pasta, spaghetti noodles, enrich, salted, ckd	1 cup	140	197	6.7	1	0.1	0	40	2.4	10	2	140	0	0.29	0.14	2.3	0	98
Pasta, spaghetti noodles, whole wheat, ckd	1 cup	125	155	6.7	1	0.1	0	33	5.6	19	1.3	4	0	0.14	0.06	0.9	0	6.25
Pastry, cinnamon danish	1 ea	110	443	7.7	25	6.2	23	49	1.4	78	2.2	408	13	0.33	0.29	3.2	0	68.2
Peaches, fresh, sliced	½ cup	85	37	0.6	0	0	0	9	1.7	4	0.1	0	455	0.01	0.03	0.8	6	2.89
Peaches, in heavy syrup, cnd	½ tsp	96	71	0.4	0	0	0	19	1.2	3	0.3	6	319	0.01	0.02	0.6	3	3.07
Peaches, in juice, cnd, whole	½ tsp	77	34	0.5	0	0	0	9	1	5	0.2	3	293	0.01	0.01	0.4	3	2.62
Peanut Butter, smooth, salted	1 Tbs	32	190	8.1	16	3.3	0	6	1.9	12	0.6	149	0	0.03	0.03	4.3	0	23.68
Pears, bartlett, fresh, med	1 ea	180	106	0.7	1	0	0	27	4.3	20	0.4	0	36	0.04	0.07	0.2	7	13.14
Pears, in heavy syrup, cnd, halves	½ ea	103	76	0.2	0	0	0	20	1.6	5	0.2	5	0	0.01	0.02	0.2	1	1.24
Pears, in juice, cnd, halves	½ ea	77	38	0.3	0	0	0	10	1.2	7	0.2	3	5	0.01	0.01	0.2	1	0.92
Peas, cnd, drained	½ cup	85	59	3.8	0	0.1	0	11	3.5	17	0.8	214	653	0.1	0.07	0.6	8	37.65
Peas, green, ckd f/fzn w/o add salt, drained	½ cup	80	62	4.1	0	0	0	11	4.4	19	1.3	70	534	0.23	0.08	1.2	8	46.88
Peppers, bell, green, sweet, raw, med	1 ea	200	54	1.8	0	0.1	0	13	3.6	18	0.9	4	1264	0.13	0.06	1	179	44
Peppers, bell, red, sweet, raw, sml	1 ea	74	20	0.7	0	0	0	5	1.5	7	0.3	1	4218	0.05	0.02	0.4	141	16.28
Peppers, bell, yellow, sweet, raw, lrg	1 ea	186	50	1.9	0	0.1	0	12	1.7	20	0.9	4	443	0.05	0.05	1.7	341	48.36
Pickles, dill	1 ea	135	24	0.8	0	0.1	0	6	1.6	12	0.7	1731	444	0.02	0.04	0.1	3	1.35
Pickles, sweet, med	1 ea	35	41	0.1	0	0	0	11	0.4	1	0.2	329	44	0	0.01	0.1	0	0.35
Pie, apple, bkd f/fzn, 1/6th of 8"	1 pce	118	280	2.2	13	4.5	0	40	1.9	13	0.5	314	146	0.03	0.03	0.3	4	25.96
Pie, bluberry, prep f/rec, 1/8th of 9"	1 pce	158	387	4.3	19	4.6	0	53	2.2	11	1.9	292	66	0.24	0.21	1.9	1	36.34
Pie, cherry, prep f/rec, 1/8th of 9"	1 pce	118	319	3.3	14	3.5	0	45	1.8	12	2.2	225	483	0.17	0.15	1.5	1	31.86
Pie, chocolate cream, rts, 1/6th of 8"	1 pce	175	532	4.5	34	8.7	9	59	3.5	63	1.9	238	0	0.06	0.19	1.2	0	22.75
Pie, lemon meringue, rts, 1/6th of 8"	1 pce	140	375	2.1	12	2.5	63	66	1.7	78	0.9	204	245	0.09	0.29	0.9	4	18.2
Pie, pecan, rts, 1/6th of 8"	1 pce	138	552	5.5	26	4.9	44	79	4.8	23	1.4	585	242	0.13	0.17	0.3	2	37.26
Pie, pumpkin, rts, 1/6th of 8"	1 pce	114	239	4.4	11	2	23	31	3.1	68	0.9	321	3915	0.06	0.17	0.2	1	22.8
Pineapple, chunks, fresh	½ cup	78	38	0.3	0	0	0	10	0.9	5	0.3	1	18	0.07	0.03	0.3	12	8.27
Pineapple, in heavy syrup, cnd, tidbits	½ cup	128	100	0.4	0	0	0	26	1	18	0.5	1	18	0.12	0.03	0.4	9	5.89
Pineapple, in juice, cnd	½ cup	125	75	0.5	0	0	0	20	1	18	0.3	1	48	0.12	0.02	0.4	12	6
Popcorn, air popped, plain	1 cup	6	23	0.7	0	0	0	5	0.9	1	0.2	0	12	0.01	0.02	0.1	0	1.38

Food	Amount	Weight (g)	Calories	Protein (g)	Fat (g)	Sat. Fat (g)	Cholesterol (g)	Carbohydrate (g)	Fiber (g)	Calcium (mg)	Iron (mg)	Sodium (mg)	Vit A (IU)	Thiamin (Vit B_1) (mg)	Riboflavin (Vit B_2) (mg)	Niacin (mg)	Vit C (mg)	Folate (mcg)
Popcorn, ckd in oil, salted	1 cup	11	55	1	3	0.5	0	6	1.1	1	0.3	97	17	0.01	0.01	0.2	0	1.87
Pork, bacon/cracklings, brld/pan fried/rstd	2 pce	15	86	4.6	7	2.6	13	0	0	2	0.2	239	0	0.1	0.04	1.1	0	0.75
Pork, cured, ham, reg, 11% fat, rstd	3 oz	85	151	19.2	8	2.7	50	0	0	7	1.1	1275	0	0.62	0.28	5.2	0	2.55
Pork, ham, whole, rstd	3 oz	85	232	22.8	15	5.5	80	0	0	12	0.9	51	8	0.54	0.27	3.9	0	8.5
Pork, ribs, spareribs, brsd	3 oz	85	337	24.7	26	9.5	103	0	0	40	1.6	79	8	0.35	0.32	4.7	0	3.4
Potato Chips, plain, salted	10 pce	20	107	1.4	7	2.2	0	11	0.9	5	0.3	119	0	0.03	0.04	0.8	6	9
Potatoes, au gratin, prep w/milk & butter f/dry mix	1 cup	245	228	5.6	10	6.3	37	31	2.2	203	0.8	1076	522	0.05	0.2	2.3	8	16.17
Potatoes, baked, w/flesh & skin, long	1 ea	202	220	4.6	0	0.1	0	51	4.8	20	2.7	16	0	0.22	0.07	3.3	26	22.22
Potatoes, hash browns, prep f/fzn	½ cup	78	170	2.5	9	3.5	0	22	1.6	12	1.2	27	0	0.09	0.02	1.9	5	5.07
Potatoes, mashed, w/whole milk	½ cup	105	31	2	1	0.3	2	18	2.1	27	0.3	318	20	0.09	0.04	1.2	7	8.61
Potatoes, sweet, flesh, bkd in skin, med, peeled	1 ea	146	150	2.5	0	0	0	35	4.4	41	0.7	15	31860	0.11	0.19	0.9	36	33
Pretzels, hard, salted, twisted	1 oz	28	107	2.5	1	0.2	0	22	0.9	10	1.2	480	0	0.13	0.17	1.5	0	47.88
Prunes, dried	5 ea	61	146	1.6	0	0	0	38	4.3	31	1.5	2	1212	0.05	0.1	1.2	2	2.26
Pudding, choc, rte, 5oz can	5 oz	142	189	3.8	6	1	4	32	1.4	128	0.7	183	51	0.04	0.22	0.5	3	4.26
Pudding, tapioca, 5oz can	5 oz	142	169	2.8	5	0.9	1	28	0.1	119	0.3	226	0	0.03	0.14	0.4	1	4.26
Pudding, vanilla, 5oz can	5 oz	142	185	3.3	5	0.8	10	31	0.1	125	0.2	192	30	0.03	0.2	0.4	0	0
Raisins, seedless, unpacked	1 oz	28	84	0.9	0	0	0	22	1.1	14	0.6	3	2	0.04	0.02	0.2	1	0.92
Raspberries, fresh	1 cup	123	60	1.1	1	0	0	14	8.3	27	0.7	0	160	0.04	0.11	1.1	31	31.98
Raspberries, swtnd, fzn	1 cup	250	400	10	15	9.1	12	62	5.5	368	3.1	245	400	0.09	0.53	0.8	1	27.5
Rice, brown, ckd	½ cup	96	107	2.5	1	0.2	0	22	1.7	10	0.4	5	0	0.09	0.02	1.5	0	3.84
Rice, white, reg, ckd	½ cup	103	134	2.8	0	0.1	0	29	0.4	10	1.2	1	0	0.17	0.01	1.5	0	59.74
Rice, wild, ckd	½ cup	100	101	4	0	0	0	21	1.8	3	0.6	3	0	0.05	0.09	1.3	0	26
Rolls, hard, white	1 ea	50	146	4.9	2	0.3	0	26	1.1	48	1.6	272	0	0.24	0.17	2.1	0	47.5
Salad Dressing, blue cheese/roquefort	1 Tbs	15	76	0.7	8	1.5	3	1	0	12	0	164	32	0	0.02	0	0	1.21
Salad Dressing, french	1 Tbs	16	69	0.1	7	1.5	0	3	0	2	0.1	219	208	0	0	0	0	0.67
Salad Dressing, French, low cal	1 Tbs	15	20	0	1	0.1	0	3	0	2	0.1	118	195	0	0	0	0	0
Salad Dressing, Italian	1 Tbs	15	70	0.1	7	1.1	0	2	0	2	0	118	12	0	0	0	0	0.73
Salad Dressing, Italian, diet, 2cal/tsp, cmrcl	1 Tbs	15	16	0	1	0.2	1	1	0	0	0	118	0	0	0	0	0	0
Salad Dressing, ranch	1 Tbs	15	80	0	8	1.2	5	0	0	0	0	105	0				0	
Salad Dressing, thousand island	1 Tbs	15	57	0.1	5	0.9	4	2	0	2	0.1	105	48	0	0	0	0	0.94
Salad Dressing, thousand island, low cal	1 Tbs	15	24	0.1	2	0.2	2	2	0.2	2	0.1	150	48	0	0	0	0	0.84
Salad, chicken, w/celery	½ cup	78	268	10.6	25	3.1	48	1	0.2	16	0.6	201	155	0.03	0.07	3.3	1	8.46
Salad, pasta, garden primavera, prep f/dry	¾ cup	142	280	8	12	2.5	2	34	2	80	1.8	730	200	0.15	0.17	2	1	
Salad, potato	½ cup	125	179	3.4	10	1.8	85	14	1.6	24	0.8	661	261	0.1	0.07	1.1	12	8.38
Salad, tuna	1 cup	205	383	32.9	19	3.2	27	19	0	35	2	824	199	0.06	0.14	13.7	5	16.4
Salami, beef & pork, dry	1 oz	28	117	6.4	10	3.4	22	1	0	2	0.4	521	0	0.17	0.08	1.4	0	0.56
Salmon, pink, w/bone, cnd, not drained	3 oz	85	118	16.8	5	1.3	47	0	0	181	0.7	471	47	0.02	0.16	5.6	0	13.09
Salmon, sockeye, fillet, bkd/brld	3 oz	85	184	23.2	9	1.6	74	0	0	6	0.5	56	178	0.18	0.15	5.7	0	4.25
Salsa, homemade, Mexican sauce	1 Tbs	15	3	0.1	0	0	0	1	0.2	1	0	1	57	0.01	0	0.1	2	1.74
Sandwich, bacon, lettuce & tomato, on soft white	1 ea	130	323	10.8	18	4.7	22	30	1.7	54	2.1	619	271	0.36	0.2	3.4	12	35.5
Sandwich, egg salad, on soft white	1 ea	111	361	9.1	24	4.2	149	29	1.2	67	2.1	499	239	0.25	0.3	1.9	0	35.39
Sandwich, peanut butter & jam, on soft white, unsalted	1 ea	100	348	11.5	15	3.1	2	46	3	60	2.2	290	2	0.27	0.17	5.3	0	40.04
Sandwich, reuben, grilled	1 ea	237	458	27.6	29	9.8	80	25	2.2	286	4.2	1933	453	0.21	0.34	2.8	13	37.79
Sardines, Atlantic, w/bones, cnd in oil, drained	1 oz	28	58	6.9	3	0.4	40	0	0	107	0.8	141	63	0.02	0.06	1.5	0	3.3
Sauce, soy, made f/soy & wheat	1 Tbs	16	9	1.3	0	0	0	1	0.1	3	0.3	871	0	0.01	0.03	0.4	0	2.56
Sauce, teriyaki, rts	1 Tbs	18	15	1.1	0	0	0	3	0	4	0.3	690	0	0.01	0.01	0.2	0	3.6
Sauerkraut, w/liquid, cnd	½ cup	118	22	1.1	0	0	0	5	2.9	35	1.7	780	21	0.02	0.03	0.2	17	27.97

Food	Amount	Weight (g)	Calories	Protein (g)	Fat (g)	Sat. Fat (g)	Cholesterol (g)	Carbohydrate (g)	Fiber (g)	Calcium (mg)	Iron (mg)	Sodium (mg)	Vit A (IU)	Thiamin (Vit B_1) (mg)	Riboflavin (Vit B_2) (mg)	Niacin (mg)	Vit C (mg)	Folate (mcg)
Sausage, pork, smkd, link	1 ea	68	265	15.1	22	7.7	46	1	0	20	0.8	1020	0	0.48	0.17	3.1	1	3.4
Scallops, brd, fried, mixed species, lrg	2 ea	31	67	5.6	3	0.8	19	3	0	13	0.3	144	23	0.01	0.03	0.5	1	11.47
Seaweed, spirulina, dried	1 cup	119	345	68.4	9	3.2	0	28	4.3	143	33.9	1247	678	2.83	4.37	15.3	12	111.8
Shrimp/Prawns, brd, fried, lrg	7 ea	85	206	18.2	10	1.8	150	10	0.3	57	1.1	292	161	0.11	0.12	2.6	1	6.89
Shrimp/Prawns, ckd, lrg	3 oz	85	84	17.8	1	0.2	166	0	0	33	2.6	190	186	0.03	0.03	2.2	2	2.98
Soda, cola	12 fl-oz	369	151	0	0	0	0	38	0	11	0.1	15	0	0	0	0	0	0
Soda, cola/Coke, diet, w/sacc, low sod	12 fl-oz	340	0	0	0	0	0	0	0	14	0.1	54	0	0	0	0	0	0
Soda, ginger ale	12 fl-oz	366	124	0	0	0	0	32	0	11	0.7	26	0	0	0	0	0	0
Soda, lemon lime	12 fl-oz	340	136	0	0	0	0	35	0	7	0.2	37	0	0	0	0.1	0	0
Soda, root beer	12 fl-oz	340	139	0	0	0	0	36	0	17	0.2	44	0	0	0	0	0	0
Sole/Flounder, fillet, bkd/brld	3 oz	85	99	20.5	1	0.3	58	0	0	15	0.3	89	32	0.07	0.1	1.9	0	7.82
Soup, beef bouillon/broth, cnd, prep w/water	1 cup	240	17	2.7	1	0.3	0	0	0	14	0.4	782	0	0	0.05	1.9	0	4.8
Soup, chicken noodle, prep w/water	1 cup	241	75	4	2	0.7	7	9	0.7	17	0.8	1106	711	0.05	0.06	1.4	0	21.69
Soup, clam chowder, Manhattan, prep f/cnd	1 cup	244	112	12.3	2	0	10	11	3.1	41	1.8	725	2297				4	
Soup, clam chowder, New England, prep w/milk	1 cup	248	164	9.5	7	3	22	17	1.5	186	1.5	992	164	0.07	0.24	1	3	9.67
Soup, cream of chicken, prep w/milk	1 cup	248	191	7.5	11	4.6	27	15	0.2	181	0.7	1047	714	0.07	0.26	0.9	1	7.69
Soup, cream of mushroom, prep w/milk	1 cup	245	201	6	13	5.1	20	15	0.5	176	0.6	906	152	0.08	0.28	0.9	2	9.8
Soup, minestrone, prep w/water	1 cup	241	82	4.3	3	0.6	2	11	1	34	0.9	911	2338	0.05	0.04	0.9	1	36.15
Soup, pea, split, w/ham, prep w/water	1 cup	245	184	10	4	1.7	7	27	2.2	22	2.2	975	431	0.14	0.07	1.4	1	2.45
Soup, tomato, prep w/milk	1 cup	248	161	6.1	6	2.9	17	22	2.7	159	1.8	744	848	0.13	0.25	1.5	68	20.83
Soup, tomato, prep w/water	1 cup	245	86	2.1	2	0.4	0	17	0.5	12	1.8	698	691	0.09	0.05	1.4	67	14.7
Soup, vegetable beef, prep w/water	1 cup	245	78	5.6	2	0.9	5	10	0.5	17	1.1	794	1899	0.04	0.05	1	2	10.54
Soup, vegetable, vegetarian, prep w/water	1 cup	250	75	2.2	2	0.3	0	12	0.5	22	1.1	852	3118	0.05	0.05	0.9	2	11
Sour Cream, cultured	1 Tbs	14	30	0.4	3	1.8	6	1	0	16	0	7	111	0	0.02	0	0	1.51
Spinach, ckd w/o add salt, drained	½ cup	103	24	3.1	0	0	0	4	2.5	140	3.7	72	8436	0.1	0.24	0.5	10	150.1
Spinach, raw, chpd	1 cup	55	12	1.6	0	0	0	2	1.5	54	1.5	43	3693	0.04	0.1	0.4	15	106.9
Spinach, w/o add salt, cnd, drained	½ cup	103	24	2.9	1	0.1	0	4	2.5	131	2.4	28	9039	0.02	0.14	0.4	15	100.7
Squash, acorn, ckd	1 cup	245	83	1.6	0	0	0	22	6.4	64	1.4	7	632	0.25	0.02	1.3	16	27.69
Squash, summer, ckd w/o add salt, drained	½ cup	90	18	0.8	0	0.1	0	4	1.3	24	0.3	1	258	0.04	0.04	0.5	5	18.09
Squash, winter, avg, bkd, mashed	½ cup	103	40	0.9	1	0.1	0	9	2.9	14	0.3	1	3664	0.09	0.02	0.7	10	28.84
Strawberries, fresh, whole	1 cup	149	45	0.9	1	0	0	10	3.4	21	0.6	1	40	0.03	0.1	0.3	84	26.37
Strawberries, slices, swtnd, fzn	1 cup	250	240	1.3	0	0	0	65	4.8	28	1.5	8	60	0.04	0.13	1	104	37.25
Stuffing, bread, prep f/dry mix	½ cup	70	125	2.2	6	1.2	0	15	2	22	0.8	380	219	0.1	0.07	1	0	70.7
Sugar, beet/cane, brown, packed	1 tsp	5	19	0	0	0	0	5	0	4	0.1	2	0	0	0	0	0	0.05
Sugar, white, granulated	1 tsp	4	15	0	0	0	0	4	0	0	0	0	0	0	0	0	0	0
Syrup, maple	1 Tbs	20	52	0	0	0	0	13	0	13	0.2	2	0	0	0	0	0	0
Taco Shells	1 ea	10	47	0.7	2	0.3	0	7	0.7	20	0.2	67	33					
Tangerines/Mandarin oranges, fresh, med	1 ea	116	51	0.7	0	0	0	13	2.7	16	0.1	1	1067	0.12	0.03	0.2	36	23.66
Tea, brewed	¼ cup	180	2	0	0	0	0	1	0	0	0	5	0	0	0.03	0	0	9.36
Tempeh	1 cup	166	320	30.8	18	3.7	0	16	9	184	4.5	15	0	0.13	0.59	4.4	0	39.67
Tofu, firm, silken	½ cup	126	78	8.7	3	0.5	0	3	0.1	40	1.3	45	0	0.13	0.05	0.3	0	
Tomatoes, red, ripe, raw, med, whole	1 ea	100	21	0.9	0	0	0	5	1.1	5	0.4	9	623	0.06	0.05	0.6	19	15
Tomatoes, red, ripe, w/o add salt, cnd, in liquid	½ cup	121	23	1.1	0	0	0	5	1.2	36	0.7	179	720	0.05	0.04	0.9	17	9.44
Tortilla/Taco/Tostada Shell, corn	1 ea	148	693	10.7	33	5	0	92	11.1	237	3.7	543	518	0.34	0.08	2	0	8.88
Trout, rainbow, fillet, bkd/brld, wild	3 oz	85	128	19.5	5	1.4	59	0	0	73	0.3	48	42	0.13	0.08	4.9	2	16.15
Tuna, light, cnd in oil, drained	3 oz	85	168	24.8	7	1.3	15	0	0	11	1.2	301	66	0.03	0.1	10.5	0	4.51
Tuna, light, cnd in water, drained	3½ oz	99	115	25.3	1	0.2	30	0	0	11	1.5	335	55	0.03	0.07	13.1	0	3.96
Turkey, average, w/o skin, rstd	3 oz	85	144	24.9	4	1.4	65	0	0	21	1.5	60	0	0.05	0.15	4.6	0	5.95

Food	Amount	Weight (g)	Calo-ries	Pro-tein (g)	Fat (g)	Sat. Fat (g)	Choles-terol (g)	Carbo-hydrate (g)	Fiber (g)	Cal-cium (mg)	Iron (mg)	Sodium (mg)	Vit A (IU)	Thia-min (Vit B_1) (mg)	Ribo-flavin (Vit B_2) (mg)	Niacin (mg)	Vit C (mg)	Folate (mcg)
Turnip Greens, ckd f/fzn, drained	½ cup	73	22	2.4	0	0.1	0	4	2.5	111	1.4	11	5822	0.04	0.05	0.3	16	28.76
Turnips, ckd w/add salt, raw, cubes	½ cup	78	16	0.6	0	0	0	4	1.6	17	0.2	39	0	0.02	0.02	0.2	9	7.18
Veal, loin, brsd	3 oz	85	241	25.7	15	5.7	100	0	0	24	0.9	68	0	0.03	0.26	7.7	0	11.9
Veal, loin, lean, brsd	3 oz	85	192	28.5	8	2.2	106	0	0	27	0.9	71	0	0.04	0.29	8.5	0	12.75
Vinegar, balsamic, 60 grain	1 Tbs	15	21	0	0	0		5	0	2	0.1	3	0	0.08	0.08	0.1	0	
Watermelon, fresh, diced	1 cup	160	51	1	1	0.1	0	11	0.8	13	0.3	3	586	0.13	0.03	0.3	15	3.52
Wheat, bulgur, ckd	1 cup	135	112	4.2	0	0.1	0	25	6.1	14	1.3	7	0	0.08	0.04	1.4	0	24.3
Wheat, flakes, rolled, dry	1 cup	30	97	3.5	0	0	0	21	4.4	18	1	0	0	0.13	0.03	1.4	0	
Wheat, germ, tstd	1 Tbs	6	23	1.7	1	0.1	0	3	0.8	3	0.5	0	0	0.1	0.05	0.3	0	21.12
Whiskey, 90 proof	2 fl-oz	42	110	0	0	0	0	0	0	0	0	0	0	0	0	0	0	0
Wine, cooler	4 oz	113	56	0.1	0	0	0	7	0	6	0.3	9	1	0.01	0.01	0.1	2	1.34
Wine, red	⅛ cup	30	22	0.1	0	0	0	1	0	2	0.1	2	0	0	0.01	0	0	0.6
Wine, Rose	2 fl-oz	59	42	0.1	0	0	0	1	0	5	0.2	3	0	0	0.01	0	0	0.65
Wine, white, med	2 fl-oz	59	40	0.1	0	0	0	0	0	5	0.2	3	0	0	0	0	0	0.12
Yogurt, fruit, low fat, 10g prot/8 oz	1 cup	227	231	9.9	2	1.6	10	43	0	345	0.2	133	104	0.08	0.4	0.2	1	21.11
Yogurt, plain, low fat, 12g prot/8 oz	8 oz	226	143	11.9	4	2.3	14	16	0	413	0.2	159	149	0.1	0.48	0.3	2	25.31

This food composition table has been prepared for West-Wadsworth Publishing Company and is copyrighted by ESHA Research in Salem, Oregon—the developer and publisher of the Food Processor®, Genesis® R&D, and the Computer Chef® nutrition software systems. The major sources for the data are from the USDA, supplemented by more than 1200 additional sources of information. Because the list of references is so extensive, it is not provided here, but is available from the publisher.

Glossary

À la meunière. Fish seasoned, lightly floured, and sautéed in clarified butter or oil and served with a sauce made with butter and parsley.

Acceptable Daily Intake (ADI). The amount of food additive that can be safely ingested daily over a person's lifetime.

Agglomerate. A process in which small particles gather into a mass or ball. In the case of milk, the protein particles regroup into larger, more porous particles.

Aging. Ripening that occurs when carcasses are hung in refrigeration units for longer periods than that required for the reversal of rigor mortis.

Al dente. Meaning "to the tooth" in Italian, it refers to pasta that is tender, yet firm enough to offer some resistance to the teeth.

Albedo. The white, inner rind of citrus fruits, which is rich in pectin and aromatic oils.

Amphoteric. Capable of acting chemically as either acid or base.

Antioxidant. A compound that inhibits oxidation, which can cause deterioration and rancidity.

Arcing. Sparks of electricity generated by microwaves bouncing off metal.

Aromatic compound. A compound that has a chemical configuration of a hexagon.

Artesian water. Water that has surfaced on its own from an aquifer, rather than being pumped.

As purchased (AP). The total amount of food purchased prior to any preparation.

Aspic. A clear gel prepared from stock or fruit or vegetable juices.

Astringency. A sensory phenomenon characterized by a dry, puckery feeling in the mouth.

Atoms. The basic building blocks of matter; individual elements found on the Periodic Table.

ATP. Adenosine triphosphate is a universal energy compound in cells obtained from the metabolism of carbohydrate, fat, or protein. The energy of ATP, which is located in high-energy phosphate bonds, fuels chemical work at the cellular level.

Au gratin. Food prepared with a browned or crusted top. A common technique is to cover the food with a bread crumb/sauce mixture and pass it under a broiler.

Au jus. Served with its own natural juices; a term usually used in reference to roasts.

Bacteria. One-celled microorganisms abundant in the air, soil, water, and/or organic matter (i.e., the bodies of plants and animals).

Baking powder. A chemical leavener consisting of a mixture of baking soda, acid(s), and an inert filler such as cornstarch.

Baking soda. A white chemical leavening powder consisting of sodium bicarbonate.

Barding. Tying thin sheets of fat or bacon over lean meat to keep the meat moist during roasting. The sheets of fat are often removed before serving.

Baste. To add a liquid, such as drippings, melted fat, sauce, fruit juice, or water, to the surface of roasting meat to help prevent drying.

Batter. A flour mixture that contains more water than a dough and whose consistency ranges from pourable to sticky.

Beading. The formation of tiny syrup droplets on the surface of a baked meringue.

Beurre manié. A thickener that is a soft paste made from equal parts of soft butter and flour blended together.

Bile. A digestive juice made by the liver from cholesterol and stored in the gall bladder.

Biological value. The percentage of protein in food that can be utilized by an animal for growth and maintenance. High-quality, complete proteins are considered to have a high biological value.

Bisque. Traditionally, a cream soup made from shellfish. Marketers sometimes label creamed vegetable soups as bisques.

Blanch. To dip a food briefly into boiling water.

Blind bake. To bake an unfilled pie crust.

Bloom. Cottony, fuzzy growth of molds.

Body. The consistency of frozen desserts as measured by their firmness, richness, viscosity, and resistance to melting.

Boiling point. The temperature at which a heated liquid begins to boil and changes to a gas.

Bouillon. A broth made from meat and vegetables and then strained to remove any solid ingredients.

Bouquet garni. A bundle of parsley, thyme, bay leaf, and whole black pepper rolled in a leek and tied together with twine.

Bran. The hard outer covering just under the husk that protects the grain's soft endosperm.

Broth. Stock made from meat or meat/bone combinations and some water with little or no flavoring. Since broths are seldom reduced as stocks, they therefore can contain salt.

Brown stock. The stock resulting from browning bones and/or meat prior to simmering them.

Calorie (kcal). The amount of energy required to raise 1 gram of water 1°C (measured between 14.5° and 15.5°C at normal atmospheric pressure).

Candling. A method of determining egg quality based on observing eggs against a light.

Caramelization. A process in which dry sugar, or sugar solution with most of its water evaporated, is heated until it melts into a clear, viscous liquid and, as heating continues, turns into a smooth, brown mixture.

Carry-over cooking. The phenomenon in which food continues to cook after it has been removed from the heat source as the heat is distributed more evenly from the outer to the inner portion of the food.

Casein. The primary protein (80 percent) found in milk; it can be precipitated (solidified out of solution) with acid or certain enzymes.

Catechins. Flavonoid pigments that are a subgroup of the flavonol pigments.

Chalaza (pl. chalazae). The ropy, twisted strands of albumen that anchor the yolk to the center of the thick egg white.

Chiffon cake. A cake made by combining the characteristics found in both shortened and unshortened cakes.

Clarified butter. Butter whose milk solids and water have been removed and thus will not burn.

Clarify. To make or become clear or pure.

Coagulate. To clot or become semisolid. In milk, denatured proteins often separate from the liquid by coagulation.

Coagulation. The clotting or precipitation of protein in a liquid into a semisolid compound.

Codex Alimentarius Commission. The international organization that develops international food standards, codes of practice, and other guidelines to protect consumers' health.

Collagen. A pearly white, tough, and fibrous protein that provides support to muscle and prevents it from over-stretching. It is the primary protein in connective tissue.

Colloidal dispersion. A solvent containing particles that are too large to go into solution, but not large enough to precipitate out.

Complete protein. A protein, usually from animal sources, that contains all the essential amino acids in sufficient amounts for the body's maintenance and growth.

Compound. A substance whose molecules consist of unlike atoms.

Conduction. The direct transfer of heat from one substance to another that it is contacting.

Congener. Alcohol by-product such as methanol or wood alcohol.

Connective tissue. A protein structure that surrounds living cells, giving them structure and adhesiveness within themselves and to adjacent tissues.

Consistency. Describes a food's firmness or thickness.

Consommé. A richly flavored soup stock that has been clarified and made transparent by the use of egg whites.

Convection. The transfer of heat by moving air or liquid (water/fat) currents through and/or around food.

Court bouillon. Seasoned stock containing white wine and/or vinegar.

Cover. The table setting, including the place mat, flatware, dishes, and glasses.

Creaming. In an emulsion, the collection and rising of the lighter phase, usually oil, to the top of the mixture.

Critical control point (CCP). A point in the HACCP process that must be controlled to ensure the safety of the food.

Cross-contamination. The transfer of bacteria or other microorganisms from one food to another.

Cruciferous. A group of indole-containing vegetables named for their cross-shaped blossoms; they are reported to have a protective effect against cancer in laboratory animals. Examples include broccoli, brussels sprouts, cabbage, cauliflower, kale, mustard greens, rutabaga, kohlrabi, and turnips.

Crumb. (1) The texture of a baked product's interior. (2) The cell structure appearing when a baked product is sliced. Evaluation is based on cell size (called "open" if medium to large, or "closed" if small), cell shape, and cell thickness (thin walls occur in fine crumb, while thick walls predominate in a coarse crumb).

Crumbing. A ceremonious procedure of Russian service in which a waiter, using a napkin or silver crumber, brushes crumbs off the tablecloth into a small container resembling a tiny dust pan.

Crustacean. An invertebrate animal with a segmented body covered by an exoskeleton consisting of a hard upper shell and a soft under shell.

Crystalline candy. Candies formed from sugar solutions yielding many fine, small crystals.

Crystallization. The precipitation of crystals from a solution into a solid, geometric network.

Culture. The ideas, customs, skills, and art of a group of people in a given period of civilization.

Curd. The coagulated or thickened part of milk.

Cure. (1) To preserve food through the use of salt and drying. Sugar, spices, or nitrates may also be added. (2) To expose cheese to controlled temperature and humidity during aging.

Cuticle (bloom). A waxy coating on an eggshell that seals the pores from bacterial contamination and moisture loss.

Cycle menu. A menu that consists of two or more weeks, usually three or four, that "cycles." Cycle menus offer a combination of variety and controlled costs.

Deglaze. Adding liquid to pan drippings and simmering/stirring to dissolve and loosen cooked-on particles sticking to the bottom of the pan.

Degorge. To peel and slice vegetables, sprinkle them with salt, and allow them to stand at room temperature until droplets containing bitter substances form on the surface; the moisture is then removed.

Dehydrate. To remove at least 95 percent of the water from foods by the use of high temperatures.

Delaney Clause. A clause added to the Food, Drug, and Cosmetic Act of 1938 stipulating that "no additive shall be deemed to be safe if it is found to induce cancer when ingested by man or animal."

Denaturation. The irreversible process in which the structure of a protein is disrupted, resulting in partial or complete loss of function.

Density. The concentration of matter measured by the amount of mass per unit volume. Objects with higher densities weigh more for their size.

Dextrinization. The breakdown of starch molecules to smaller, sweeter-tasting dextrin molecules in the presence of dry heat.

Dextrose equivalent (DE). A measurement of dextrose concentration. A DE of 50 means the syrup contains 50 percent dextrose.

Dietary fiber. The undigested portion of carbohydrates remaining in a food sample after exposure to digestive enzymes.

Dietitian (registered dietitian or RD). A health professional who counsels people about their medical nutrition therapy (diabetic, low cholesterol, low sodium, etc.). Registration requirements consist of completing an approved four-year college degree, exam, internship, and ongoing continuing education.

Distillation. A procedure in which pure liquid is obtained from a solution by boiling, condensation, and collection of the condensed liquid in a separate container.

Distilled water. Water that has been purified through distillation to remove minerals, pathogens, and other substances.

Diverticulosis. An intestinal disorder characterized by pockets forming out from the digestive tract, especially the colon.

Dough. A flour mixture that is dry enough to be handled and kneaded.

Drug. A product able to treat, prevent, cure, mitigate, or diagnose a disease or disease symptom.

Drupes. Fruit with seeds encased in a pit. Examples are apricots, cherries, peaches, and plums.

Dry-heat preparation. A method of cooking in which heat is transferred by air, radiation, fat, or metal.

Edible coating. Thin layer of edible material such as natural wax, oil, petroleum-based wax, etc. that serves as a barrier to gas and moisture.

Edible portion (EP). Food in its raw state, minus that which is discarded—bones, fat, skins, and/or seeds.

Electrolyte. An electrically charged ion in a solution.

Emulsifier. A compound that possesses both water-loving (hydrophilic) and water-fearing (hydrophobic) properties so that it disperses in either water or oil.

Emulsion. A liquid dispersed in another liquid with which it is usually immiscible (incapable of being mixed).

Endosperm. The largest portion of the grain, containing all of the grain's starch.

Enriched. Foods that have had certain nutrients, which were lost through processing, added back to levels established by federal standards.

Enrobe. To coat food with melted chocolate that hardens to form a solid casing.

Enzymatic browning. A reaction in which an enzyme acts on a phenolic compound in the presence of oxygen to produce brown-colored products.

Enzyme. A protein that catalyzes (causes) a chemical reaction without itself being altered in the process.

Essential nutrients. Nutrients that the body cannot synthesize at all or in necessary amounts to meet the body's needs.

Essential oil. An oily substance that is volatile (easily vaporized), with 100 times the flavoring power of the material from which it originated.

Eviscerate. To remove the entrails from the body cavity.

Extractives. Flavor compounds consisting of nonprotein, nitrogen substances that are end-products of protein metabolism.

Fermentation. The conversion of carbohydrates to carbon dioxide and alcohol by yeast or bacteria.

Finfish. Fish that have fins and internal skeletons.

Fire point. The temperature at which a heated substance (such as oil) bursts into flames and burns for at least 5 seconds.

Flash point. The temperature at which tiny wisps of fire streak to the surface of a heated substance (such as oil).

Flatware. Eating and serving utensils (e.g., knives, forks, and spoons).

Flavor. The combined sense of taste, odor, and mouthfeel.

Flavor reversion. The breakdown (oxidation) of an essential fatty acid, linolenic acid, found in certain vegetable oils, leading to an undesirable flavor change prior to the start of actual rancidity.

Flavoring. Substance that adds a new flavor to a food.

Flocculation. A partial gel in which only some of the solid particles colloidally dispersed in a liquid have solidified.

Foam. A colloidal dispersion of a gas in a liquid.

Food additive. A substance added intentionally or unintentionally to food that becomes part of the food and affects its character.

Food Code. An FDA publication updated every two years that shows food service organizations how to prevent foodborne illness while preparing food.

Food cost. Often expressed as percentage obtained by dividing the raw food cost by the menu price.

Food infection. An illness resulting from ingestion of food containing large numbers of living bacteria or other microorganisms.

Food intoxication. An illness resulting from ingestion of food containing a toxin.

Foodborne illness. An illness transmitted to humans by food.

Forecast. A predicted amount of food that will be needed for a food service operation within a given time period.

Fortified. Foods that have had nutrients added that were not present in the original food.

Free radical. An unstable molecule that is extremely reactive and that can damage cells.

Freeze-dry. To remove water from food when it is in a frozen state, usually under a vacuum.

Freezer burn. White or grayish patches on frozen food caused by water evaporating into the package's air spaces.

Freezing point. The temperature at which a liquid changes to a solid.

Freshness/quality assurance date. The last day a food product will be of optimum quality.

Fumet. A flavorful fish stock made with white wine.

Functional food. A food or beverage that imparts a physiological benefit that enhances overall health, helps prevent or treat a disease or condition, or improves physical/mental performance.

Gaping. The separation of fish flesh into flakes that occurs as the steak or fillet ages.

Gelatinization. The increase in volume, viscosity, and translucency of starch granules when they are heated in a liquid.

Gene. A unit of genetic information in the chromosome.

Genetic engineering. The alteration of a gene in a bacterium, plant, or animal for the purpose of changing one or more of its characteristics.

Genetically modified organisms (GMOs). Plants, animals, or microorganisms that have had their genes altered through genetic engineering using the application of recombinant deoxyribonucleic acid (rDNA) technology.

Germ. The smallest portion of the grain, and the embryo for a future plant.

Glaze. A flavoring obtained from soup stock that has been concentrated by evaporation until it attains a syrupy consistency with a highly concentrated flavor.

Gluten. The protein portion of wheat flour with the elastic characteristics necessary for the structure of most baked products.

Grade. The voluntary process in which foods are evaluated for yield (a 1 to 5 grading for meats only) and quality (Prime, Choice, AA, A, Fancy, etc.).

Gram. A metric unit of weight. One gram (g) is equal to the weight of 1 cubic centimeter (cc) or milliliter (ml) of water (under a specific temperature and pressure).

GRAS list. A list of compounds that are exempt from the "food additive" definition because they are "generally recognized as safe" based on "a reasonable certainty of no harm from a product under the intended conditions of use."

Gustatory. Relating to the sense of taste.

HACCP. Hazard Analysis and Critical Control Point System, a systematized approach to preventing foodborne illness during the production and preparation of food.

Heat of solidification. The temperature at which a substance converts from a liquid to a solid state.

Heat of vaporization. The amount of heat required to convert a liquid to a gas.

Herb. A plant leaf valued for its flavor or scent.

Hermetically sealed. Foods that have been packaged airtight by a commercial sealing process.

High-conversion corn syrups. Corn syrups with a dextrose equivalent over 58.

Homogenization. A mechanical process that breaks up the fat globules in milk into much smaller globules that do not clump together and are permanently dispersed in a very fine emulsion.

Hops. The dried fruit of the *Humulus lupulus* plant, which grows in the Pacific Northwest of the United States.

Humectant. A substance that attracts water to itself. If added to food, it increases the water-holding capacity of the food and helps to prevent it from drying out by lowering the water activity.

Husk. The rough outer covering protecting the grain.

Hydrogenation. A commercial process in which hydrogen atoms are added to the double bonds in monounsaturated or polyunsaturated fatty acids to make them more saturated.

Hydrolysis. A chemical reaction in which water ("hydro") breaks ("lysis") a chemical bond in another substance, splitting it into two or more new substances.

Hydrophilic. A term describing "water-loving" or water-soluble substances.

Hydrophobic. A term describing "water-fearing" or non-water-soluble substances.

Hygroscopic. The ability to attract and retain moisture.

Imitation milk. A product defined by the FDA as having the appearance, taste, and function of its original counterpart but as being nutritionally inferior.

Incomplete protein. A protein, usually from plant sources, that does not provide all the essential amino acids.

Induction. The transfer of heat energy to a neighboring material without contact.

Intensity. The strength of an odor, taste, or flavor on a scale of 1 to 10 with 10 being the strongest.

Interesterification. A commercial process that rearranges fatty acids on the glycerol molecule in order to produce fat with a smoother consistency.

Interfering agent. A substance added to the sugar syrup to prevent the formation of large crystals, resulting in a candy with a waxy, chewy texture.

Invert sugar. An equal mixture of glucose and fructose, created by hydrolyzing sucrose.

Ionize. To separate a neutral molecule into electrically charged ions.

Irradiation. A food preservation process in which foods are treated with low doses of gamma rays, X-rays, or electrons.

Job description. An organized list of duties used for finding qualified applicants, training, performance appraisal, defining authority and responsibility, and determining salary.

Julienne. To cut food lengthwise into very thin, stick-like shapes.

Kinetic energy. Energy associated with motion.

Knead. To work the dough into an elastic mass by pushing, stretching, and folding it.

Kosher. From Hebrew, food that is "fit, right, proper" to be eaten according to Jewish dietary laws.

Lamination. The arrangement of alternating layers of fat and flour in rolled pastry dough. During baking, the fat melts and leaves empty spaces for steam to lift the layers of flour, resulting in a flaky pastry.

Larding. Inserting strips of bacon, salt pork, or other fat into slits in the meat with a large needle.

Latent heat. The amount of energy in calories (kcal) per gram absorbed or emitted as a substance undergoes a change in state (liquid/solid/gas).

Legumes. Members of the plant family *Leguminosae* that are characterized by growing in pods. Vegetable legumes include beans, peas, and lentils.

Maillard reaction. The reaction between a sugar (typically reducing sugars such as glucose/dextrose, fructose, lactose, or maltose) and a protein (specifically the nitrogen in an amino acid), resulting in the formation of brown complexes.

Marbling. Fat deposited in the muscle that can be seen as little white streaks or drops.

Mealy. A pastry with a grainy or less flaky texture, created by coating all of the flour with fat.

Melting point. The temperature at which a solid changes to a liquid.

Meniscus. The imaginary line read at the bottom of the concave arc at the water's surface.

Methylxanthine. A compound that stimulates the central nervous system.

Microorganism. Plant or animal organism that can only be observed under the microscope—bacteria, mold, yeast, virus, or animal parasite.

Milk solids-not-fat (MSNF). Federal standards identifying the total solids, primarily proteins and lactose, found in milk, minus the fat.

Mineral water. Water from natural springs having a strong taste or odor due to small amounts of salts of calcium, magnesium, and sodium (sodium bicarbonate, sodium carbonate, sodium chloride), and sometimes iron or hydrogen sulfide.

Mirepoix. A collection of highly sautéed, chopped vegetables (a 2:1:1 ratio by weight of onions, celery, and carrots) flavored with spices and herbs (sage, thyme, marjoram, and chopped parsley are the most common).

Modified starch. A starch that has been chemically or physically modified to create unique functional characteristics.

Moist-heat preparation. A method of cooking in which heat is transferred by water, any water-based liquid, or steam.

Mold. A fungus (a plant that lacks chlorophyll) that produces a furry growth on organic matter.

Molecule. A unit composed of one or more types of atoms held together by chemical bonds.

Mollusk. An invertebrate animal with a soft unsegmented body usually enclosed in a shell.

Monograph. A summary sheet (fact sheet) describing a substance in terms of name (common and scientific), chemical constituents, functional uses (medical and common), dosage, side effects, drug interactions, and references.

Mother sauce. A sauce that serves as the springboard from which other sauces are prepared.

Mycotoxin. A toxin produced by a mold.

Myocommata. Large sheets of very thin connective tissue separating the myotomes.

Myotomes. Layers of short fibers in fish muscle.

NOEL. The No-Observed-Effect Level is the level or dose at which an additive is fed to laboratory animals without any negative side effects.

Noncrystalline candy. Candies formed from sugar solutions that did not crystallize.

Nontempered coating. A coating resembling chocolate that is not subject to bloom because it is made with fats other than cocoa butter.

Nuclei. Small aggregates of molecules serving as the starting point of crystal formation.

Nutraceutical. A bioactive compound (nutrients and non-nutrients) that has health benefits.

Nutrients. Food components that nourish the body to provide growth, maintenance, and repair.

Objective tests. Evaluations of food quality that rely on numbers generated by laboratory instruments, which are used to quantify the physical and chemical differences in foods.

Ohmic heating. A food preservation process in which an electrical current is passed through food, generating enough heat to destroy microorganisms.

Olfactory. Relating to the sense of smell.

Omega-3 fatty acids. The polyunsaturated fatty acids eicosapentaenoic (EPA) and docosahexaenoic acid (DHA).

Organizational chart. A descriptive diagram showing the administrative structure of an organization.

Osmosis. The movement of a solvent through a semipermeable membrane to the side with the higher solute concentration, equalizing solute concentration on both sides of the membrane.

Osmotic pressure. The pressure or pull that develops when two solutions of different solute concentration are on either side of a permeable membrane.

Outbreak. Defined by the CDC as the occurrence of two or more cases of a similar illness resulting from the ingestion of a common food.

Oven spring. The quick expansion of dough during the first ten minutes of baking, caused by expanding gases.

Overrun. The volume over and above the volume of the original frozen dessert mix, caused by the incorporation of air during freezing.

P/S ratio. The ratio of polyunsaturated fats to saturated fats. The higher the P/S ratio, the more polyunsaturated fats the food contains.

Parasite. An organism that lives on or within another organism at the host's expense without any useful return.

Parboil. To partially boil, but not fully cook, a food.

Pascalization. A food preservation process utilizing ultrahigh pressures to inhibit the chemical processes of food deterioration.

Pasteurization. A food preservation process that heats liquids to 160°F (71°C) for 15 seconds, or 143°F (62°C) for 30 minutes, in order to kill bacteria, yeasts, and molds.

Patent flour. The finest streams of flour.

Pathogenic. Causing or capable of causing disease.

Peptide bond. The chemical bond between two amino acids.

Percentage yield. The ratio of edible to inedible or wasted food.

pH scale. Measures the degree of acidity or alkalinity of a substance, with 1 the most acidic, 14 the most alkaline, and 7 neutral.

Phenolic. A chemical term to describe an aromatic (circular) ring attached to one or more hydroxyl (-OH) groups.

Plain pastry. Pastry made for producing pie crusts, quiches, and main-dish pies.

Plant stanol esters. Naturally occurring substances in plants that help block absorption of cholesterol from the digestive tract.

Plasticity. The ability of a fat to be shaped or molded.

Polymerization. A process in which free fatty acids link together, especially when overheated, resulting in a gummy, dark residue and an oil that is more viscous and prone to foaming.

Polyphenol. An organic compound with two or more phenols—carbon atoms structured into an aromatic ring with one or more hydroxyl (-OH) groups.

Pomes. Fruit with seeds contained in a central core. Examples are apples and pears.

Prawn. A large crustacean that resembles shrimp but is biologically different. Large shrimp are often called by this name.

Prebiotics. Nondigestible food ingredients (generally fibers such as fructooligosaccharides (FOS) and inulin) that support the growth of probiotics.

Precipitate. To separate or settle out of a solution.

Prime (season). To seal the pores of a pan's metal surface with a layer of heated-on oil.

Prion. An infectious protein particle that does not contain DNA or RNA.

Probiotics. Live microbial food ingredients (i.e., bacteria) that have a beneficial effect on human health.

Processed cheese. A cheese made from blending one or more varieties of cheese, with or without heat, and mixing it with other ingredients.

Product recall. Civil court action to seize or confiscate a product that is defective, unsafe, filthy, or produced under unsanitary conditions.

Proof. (1) To increase the volume of shaped dough through continued fermentation. (2) Alcoholic strength indicated by a number that is twice the percent by volume of alcohol present.

Proof box. A large, specially designed container that maintains optimal temperatures and humidity for the fermentation and rising of dough.

Protein complementation. Two incomplete-protein foods, each of which supplies the amino acids missing in the other, combined to yield a complete protein profile.

Puff pastry. A delicate pastry that puffs up in size during baking due to numerous alternating layers of fat and flour.

Purified water. Water that has undergone deionization, distillation, reverse osmosis, or any other method that removes minerals, chemicals, and flavor.

Quality grades. The USDA standards for beef, veal, lamb, and mutton.

Quick bread. Bread leavened with air, steam, and/or carbon dioxide from baking soda or baking powder.

Radiation. The transfer of heat energy in the form of waves of particles moving outward from their source.

Rancid. The breakdown of the polyunsaturated fatty acids in fats that results in disagreeable odors and flavors.

Reducing sugars. Sugars such as glucose, fructose, maltose, and others that have a reactive aldehyde or ketone group. Sucrose is not a reducing sugar.

Reduction. To simmer or boil a liquid until the volume is reduced through evaporation, leaving a thicker, more concentrated, flavorful mass.

Rennin. An enzyme obtained from the inner lining of a calf's stomach and sold commercially as rennet.

Retail cuts. Smaller cuts of meat obtained from wholesale cuts and sold to the consumer.

Retrogradation. The seepage of water out of an aging gel due to the contraction of the gel (bonds tighten between the amylose molecules). Also known as syneresis or weeping.

Rigor mortis. From the Latin for "stiffness of death," the temporary stiff state following death as muscles contract.

Ripening. The chemical and physical changes that occur during the curing period.

Rope. The sticky, moist texture of breads resulting from contamination by *Bacillus mesentericus* bacteria.

Roux. A thickener made by cooking equal parts of flour and fat.

Saturated solution. A solution holding the maximum amount of dissolved solute at room temperature.

Scalloped. Baked with milk sauce and bread crumbs.

Scampi. A crustacean found in Italy and not generally available in North America. The term is often used incorrectly to describe a popular shrimp dish.

Score. The technique of taking a sharp knife or a special blade called a lame and creating ¼ to ½-inch deep slashes on the risen dough's top surface just prior to baking.

Sear. To brown the surface of meat by brief exposure to high heat.

Searing. Cooking that exposes a meat cut to very high initial temperatures; this is intended to "seal the pores," increase flavor, and enhance color by browning.

Seasoning. Any compound that enhances the flavor already found naturally in a food.

Seed. To create nuclei or starting points from which additional crystals can form.

Shortened cake. A cake made with fat.

Shortening. A fat that tenderizes, or shortens, the texture of baked products, making them softer and easier to chew.

Silence cloth. A piece of fabric placed between the table and the tablecloth to protect the table, quiet the placement of dishes and utensils, and keep the tablecloth from slipping.

Simple syrup. A basic mixture of boiled sugar and water.

Slurry. A thickener made by combining starch and a cool liquid.

Small sauce. A secondary sauce created when a flavor is added to a mother sauce.

Smoke point. The temperature at which fat or oil begins to smoke.

Sol. A colloidal dispersion of a solid dispersed in a liquid.

Solubility. The ability of one substance to blend uniformly with another.

Solute. Solid, liquid, or gas compounds dissolved in another substance.

Solution. A completely homogeneous mixture of a solute (usually a solid) dissolved in a solvent (usually a liquid).

Solvent. A substance, usually a liquid, in which another substance is dissolved.

Specific gravity. The density of a substance compared to another substance (usually water).

Specific heat. The amount of heat required to raise the temperature of 1 gram of a substance 1°C.

Specifications. Descriptive information used in food purchasing that defines the minimum and maximum levels of acceptable quality or quantity (i.e., U.S. grade, weight, size, fresh or frozen).

Spice. A seasoning or flavoring added to food that is derived from the fruit, flowers, bark, seeds, or roots of a plant.

Spore. Encapsulated, dormant form assumed by some microorganisms that is resistant to environmental factors that would normally result in its death.

Spring water. Water that, according to the FDA requirements, flows from its source without being pumped and contains at least 250 parts per million of dissolved solids.

Stabilizer. A compound such as vegetable gum that attracts water and interferes with frozen crystal formation, resulting in a smoother consistency in frozen desserts.

Standardized recipe. A food service recipe that is a set of instructions describing how a particular dish is prepared by a specific establishment. It ensures consistent food quality and quantity, the latter of which provides portion/cost control.

Standards of Fill. Requirements for the amount of raw product that must be put into a container before liquid (brine or syrup) is added.

Standards of Identity. Requirements for the type and amount of ingredients a food should contain in order to be labeled as that food.

Standards of Minimum Quality. Minimum quality requirements for tenderness, color, and freedom from defects in canned fruits and vegetables.

Starter. A culture of microorganisms, usually bacteria and/or yeasts, used in the production of certain foods such as sourdough bread, cheese, and alcoholic beverages.

Sterilization. The elimination of all microorganisms through extended boiling/heating to temperatures much higher than boiling or through the use of certain chemicals.

Stock. The foundational thin liquid of many soups, produced when meat, poultry, seafood, and/or their bones, or vegetables are simmered.

Storage eggs. Eggs that are treated with a light coat of oil or plastic and stored in high humidity at low refrigerator temperatures very close to the egg's freezing point (29° to 32°F/–1.5° to 0°C).

Straight flour. Flour containing all the different types of streams produced during milling.

Stream. A division of milled flour based on particle size.

Streusel topping. A crunchy, flavorful topping that can be strewn over the tops of pies; it is made by combining flour, butter or margarine, brown sugar, and possibly spices (cinnamon) and chopped nuts (pecans, walnuts, or almonds).

Structure/function claims. Statements identifying relationships between nutrients or dietary ingredients and body function. These health claims are held to the standard of "significant scientific agreement."

Subjective tests. Evaluations of food quality based on sensory characteristics and personal preferences as perceived by the five senses.

Sublimation. The process in which a solid changes directly to a vapor without passing through the liquid phase.

Substrate. A substance that is acted upon, such as by an enzyme.

Superglycerinated. A shortening that has had mono- and diglycerides added for increased plasticity.

Supersaturated solution. An unstable solution created when more than the maximum solute is dissolved in solution.

Surfactant. Surface-active agent that reduces a liquid's surface tension to increase its wetting and blending ability.

Surimi. Japanese for "minced meat," a fabricated fish product usually made from Alaskan pollack, a deep-sea whitefish, which is skinned, deboned, minced, washed, strained, and shaped into pieces to resemble crab, shrimp, or scallops.

Suspension. A mixture in which particles too large to go into solution remain suspended in the solvent.

Sweat. The stage of cooking in which food, especially vegetables, becomes soft and translucent.

Tannins. Polymers of various flavonoid compounds, of which some of the larger ones yield reddish and brown pigments.

Temperature danger zone. The temperature range of 40° to 140°F (4° to 60°C), which is ideal for bacterial growth.

Tempering. To heat and cool chocolate to specific temperatures, making it more resistant to melting and resulting in a smooth, glossy, hard finish.

Three-compartment sink. A sink divided into three sections, the first for soaking and washing, the second for rinsing, and the third for sanitizing.

Truss. To tie the legs and wings against the body of the bird to prevent them from overcooking before the breast is done.

Turgor. The rigid firmness of a plant cell resulting from being filled with water.

Ultrahigh-temperature (UHT) milk. Milk that has been pasteurized using very high temperatures, is aseptically sealed, and is capable of being stored unrefrigerated for up to three months.

Ultrapasteurization. A process in which a milk product is heated at or above 280°F (138°C) for at least two seconds.

Unshortened cake. A cake made without added fat.

Variety meats. The liver, sweetbreads (thymus), brain, kidneys, heart, tongue, tripe (stomach lining), and oxtail (tail of cattle).

Vinaigrette. A salad dressing consisting of only oil, vinegar, and seasoning.

Vintage. The year in which a wine was bottled; especially an exceptionally fine wine from a year with a good crop.

Virus. An infectious microorganism consisting of RNA or DNA that reproduces only in living cells.

Viscosity. The resistance of a fluid to flowing freely, caused by the friction of its molecules against a surface.

Vitelline membrane. The membrane surrounding the egg yolk and attached to the chalazae.

Volatile molecules. Molecules capable of evaporating like a gas into the air.

Volume. A measurement of three-dimensional space that is often used to measure liquids.

Water activity (a_w). Measures the amount of available (free) water in foods. Water activity ranges from 0 to the highest value of 1.00, which is pure water.

Weeping (syneresis). The escape of liquid to the bottom of a meringue or the formation of pores filled with liquid.

Well water. Water pumped from an aquifer, an underground source of water.

Whey. The liquid portion of milk, consisting primarily of 93 percent water, lactose, and whey proteins (primarily lactalbumin and lactoglobulin). It is the watery component removed from the curd in cheese manufacture.

White sauce. A mixture of flour, milk, and usually fat.

White stock. The flavored liquid obtained by simmering the bones of beef, veal, chicken, or pork.

Wholesale (primal) cuts. The large cuts of an animal carcass, which are further divided into retail cuts.

Winterizing. A commercial process that removes the fatty acids having a tendency to crystallize and make vegetable oils appear cloudy.

Yeast. A fungus (a plant that lacks chlorophyll) that is able to ferment sugars and that is used for producing food products such as bread and alcohol.

Yeast bread. Bread made with yeast, which produces carbon dioxide gas through the process of fermentation, causing the bread to rise.

Yield grade. The amount of lean meat on the carcass in proportion to fat, bone, and other inedible parts.

Credits and Sources

All nutrient data derived from The Genesis R & D Nutrition Labeling and Formulation Software, ESHA Research, PO Box 13028, Salem, OR 97303.

Color Insert Section
Figures 1, 2, 3, 4, 5, and 6 cuts of meat, Courtesy of the USDA; Figures 1, 4, 5, and 7 (plated food), and Figures 9 (French fries and baked potato) and 17, PhotoDisc; Figures 7 (right side), 8, and 16, Lois Frank Photo; Figures 9 (potato varieties), 10, and 11, Vincent Lee; Figures 12, 13, and 15, Digital Works; Figure 14, Courtesy of Washington State Apple Commission.

Chapter 1
Figures 1-4, 1-5 based on the USDA Food Guide Pyramid; Figure 1-6, U.S. Census Bureau; Figure 1-8, adapted from *FDA Consumer* 29(3) 6–11, 1995; Figure 1-11, Courtesy of Sensory Computer Systems, 800-579-7654; Figure 1-12, the TA.XT2 Texture Analyzer, Texture Technologies Corp., Scarsdale, NY/Stable Micro Systems, Godalming, Surrey, UK; Table 1-3 Courtesy of the USDA.

Chapter 2
Figures 2-10, 2-29, Whitney/Rolfes, *Understanding Nutrition, 8E,* Wadsworth, 1999; Table 2-2, Courtesy of The Carrageenan Company.

Chapter 3
Figures 3-1, 3-3, 3-4, Courtesy of the Centers for Disease Control; Figure 3-2, The Far Side® by Gary Larson © 1992 FarWorks, Inc., All Rights Reserved, used with permission; Figure 3-6, AP/Wide World Photos; Figure 3-15, adapted from Texas Dept. of Health's Food Service Establishment Inspection Report; Table 3-2, adapted from International Commission on Microbiological Specifications for Food (ICMSF) (1986), and Pierson and Corlett, eds., *HACCP Principles and Applications* (New York: Chapman and Hall, 1992); Table 3-9, American Dietetic Association.

Chapter 4
Figures 4-1, 4-3, D.A. Mizer, M. Porter, and B. Sonnier, *Food Preparation for the Professional,* copyright © 1998, reprinted by permission of John Wiley & Sons, Inc.; Figure 4-5, Ashley S. Anderson, *Catering for Large Numbers,* Reed International Books, 1995; Figures 4-7, 4-8, Laura Murray.

Chapter 5
Figures 5-3, 5-17, 5-20, 5-22, 5-24, and 5-26, Laura Murray; Figure 5-3 (digital thermometers) The Baker's Catalogue, 800-827-6836; Figures 5-5, 5-6, Courtesy of The Vulcan-Hart Company; Figure 5-9, The Victory Company; Figure 5-9 (upper right), Courtesy of Kolpak; Figures 5-11, 5-12, Courtesy of the Groen Company; Figure 5-13, Courtesy of Frymaster, A Welbilt Company; Figure 5-15, The Hobart Company; Figure 5-16, Courtesy of KitchenAid®, a registered trademark of Whirlpool, U.S.A; Figure 5-18 (upper left), Courtesy of West Bend; Figure 5-18 (upper right), Courtesy of CecilWare; Figure 5-18 (bottom), Courtesy of Bunn-o-Matic; Figure 5-25, Courtesy of Texas Tech University.

Chapter 6
Figures 6-2, 6-5, 6-6, 6-7, 6-9, and 6-12, Laura Murray; Figure 6-3, Annette Perry/Laura Murray; Figure 6-10, Courtesy of the Avocado Commission; Figure 6-11, Courtesy of Morton Salt, Chicago, IL.; Figure 6-14, PhotoDisc.

Chapter 7
Figure 7-1, SunMaid Raisins; Figure 7-4, Courtesy of Presto®; Figure 7-5, 7-9, Courtesy of Food Technology; Figure 7-6, Courtesy of USDA; Figure 7-8, Courtesy of Cereal Foods World; Figure 7-10, Laura Murray.

Chapter 8
Page 152 logo, Courtesy of the FDA; page 161 logos, Courtesy of the Environmental Protection Agency, the Centers for Disease Control and Prevention, U.S. Dept of Commerce, Federal Trade Commission, and Bureau of Alcohol, Tobacco, and Firearms; Figure 8-3, Laura Murray; page 159 logo and Figures 8-6, 8-8, and 8-9, Courtesy of the USDA; Figure 8-7, Vito Palmisano/©Tony Stone Images.

Chapter 9
Figure 9-1 sugar cane, © Philip Gould, CORBIS; Figure 9-1 sugar beets, © Richard Hamilton Smith/CORBIS; Table 9-5, Courtesy of Food Technology.

Chapter 10
Figure 10-1, reprinted with permission from *Food & Nutrition Encyclopedia* (Ensmiger, ed.), copyright CRC Press, Boca Raton, Florida; Figure 10-7, Courtesy of the USDA; Figure 10-10, Courtesy of the Kelco Company; Table 10-2, adapted from Fat Replacers: Food Ingredients for Healthy Eating, Calorie Control Council, http://www.caloriecontrol.org/fatreprint.html, 2002, and the American Dietetic Association's Position Statement on Fat Replacers, *Journal of the American Dietetic Association* 98(4): 463–468, 1998; Table 10-4, Courtesy of Frosty Acres.

Chapter 11
Figure 11-1, Corbis/Bettmann Archive; Figure 11-2, Courtesy of the American Dairy Products Institute; Figure 11-4, Courtesy of the Milk Board; Table 11-3 and 11-5, Courtesy of the National Dairy Council.

Chapter 12
Figure 12-1, Courtesy of the USDA; Figure 12-3, Courtesy of Dairy Management, Inc.; Figure 12-4, Courtesy of Karoun Dairies, Inc.

Chapter 13
Figure 13-2, Courtesy of the USDA; Figure 13-3, Courtesy of the B. C. Ames Co.; Figures 13-4, 13-5, Laura Murray; Figure 13-9, Courtesy of the California Egg Board; Tables 13-1, 13-2, 13-3, 13-4, and 13-5 based on data from American Egg Board.

Chapter 14
Figures 14-1, 14-2 adapted from C. Starr, *Human Biology,* Wadsworth, 1997; Figure 14-4, © Wolfgang Kaehler/CORBIS; Figures 14-6, 14-10, Courtesy of the USDA; Figures 14-13 and 14-16, Laura Murray; Figure 14-23, PhotoDisc; Figure 14-24, Courtesy of the National Livestock and Meat Board; Tables 14-1, 14-2, 14-3, 14-4, 14-5, 14-7, and 14-8 based on data from USDA.

Chapter 15
Figure 15-2, Laura Murray; Figure 15-3, Sloan Howard; Figure 15-7, Annette Perry/Laura Murray; Figure 15-8, Jim Peterson.

Chapter 16
Figure 16-3, Courtesy of the Dept. of Fisheries, Washington, D.C.; Figures 16-4, 16-6, and 16-7, Laura Murray; Figure 16-8, Lois Frank; Figure 16-9, Mark Ferri; Figures 16-10 and 16-11 from D.A. Mizer, *Food Preparation for the Professional,* John Wiley & Sons, 1998.

Chapter 17
All vegetable icons, Matt Perry/LTS; Figures 17-5, 17-6, 17-7, 17-10, and 17-11, Laura Murray; Figure 17-8, Courtesy of the Mushroom Council.

Chapter 18
All fruit icons, Matt Perry/LTS; Figure 18-4, Anne Thurston; Figures 18-3, 18-5, 18-10, and 18-11, Laura Murray; Figure 18-6, Courtesy of the Watermelon Board; Figure 18-7, Courtesy of the California Olive Industry; Figure 18-8, Courtesy of Dole; Figure 18-10, text adapted from M. Stevens, *Fine Cooking* 25:76, 1998.

Chapter 19
Figure 19-1, Sarah Chester, South River Miso Company, reproduced by permission; Figures 19-2, 19-4, 19-5, 19-6, and 19-8, Laura Murray; Figure 19-3, Digital Works; Figure 19-7, Annette Perry/Laura Murray; Figure 19-9, Courtesy of Food Technology; Table 19-1, Courtesy of Texas Tech University.

Chapter 20
Figures 20-1, 20-4, and 20-5, Courtesy of the Corn Products Company; Figure 20-2 adapted from Whistler and BeMiller, *Carbohydrate Chemistry for Food Scientists* (Eagen Press, 1997); Figure 20-7, Courtesy of CERESTAR USA; Figure 20-8, Courtesy of Food Technology; Figures 20-10 and 20-11, Laura Murray; Figure 20-9, Scott Phillips, courtesy of *Fine Cooking* Magazine #43, February/March 2001, page 78, reproduced by permission; Table 20-1, Courtesy of Cereal Foods World.

Chapter 21
Figures 21-1, 21-4, 21-6, 21-9, 21-12, and 21-13, Laura Murray; Figure 21-3, from F. Sizer and E. Whitney, *Nutrition Concepts and Controversies,* Wadsworth, 1997; Figure 21-5, Lois Frank; Figure 21-7, Scott Phillips, courtesy of *Fine Cooking* Magazine #21, June/July 1997, page 76, reproduced by permission; Figure 21-8, Runk/Shoenberger from Grant Heilman Photography, Inc.; Figure 21-10, Courtesy of the Quaker Oats Company; Figure 21-11, Courtesy of A. Zerega's Sons, Inc.

Chapter 22
Figures 22-1, 22-11, 22-12, 22-15, 22-19, 22-22, 22-23, 22-26, 22-28, and 22-29, Courtesy of C. Schunemann and G. Treu, *Baking: The Art and Science,* Baker Technology, Inc., 1988; Figures 22-3 and 22-5, Courtesy of *Baker's Digest*; Figures 22-6, 22-20, and 22-25, Laura Murray; Figure 22-21, Scott Phillips, *Fine Cooking* Magazine.

Chapter 23
Figures 23-1 and 23-2, Laura Murray.

Chapter 24
Figure 24-1, 24-8, Courtesy of C. Schunemann and G. Treu, *Baking: The Art and Science,* Baker Technology, Inc., 1988; Figures 24-2, 24-4, and 24-9, Laura Murray; Figure 24-3, Scott Phillips, *Fine Cooking* Magazine; Figure 24-5, Annette Perry/Laura Murray; Figure 24-12, Courtesy of King Arthur Flour Baker's Catalogue®.

Chapter 25
Figure 25-1, Courtesy of C. Schunemann and G. Treu, *Baking: The Art and Science,* Baker Technology, Inc., 1988; Figures 25-2, 25-3, 25-4, 25-5, 26-6, and 25-7, Laura Murray; Figure 25-8, © Craig Aurness/CORBIS; Table 25-4, A. J. Dodge, *Fine Cooking* 29:74, 1998.

Chapter 26
Figures 26-1, 26-5, 26-6, 26-14, Laura Murray; Figure 26-2, Courtesy of Cereal Foods World; Figure 26-3, Courtesy of *Baker's Digest,* December 1967; Figure 26-4, Scott Phillips/ *Fine Cooking* Magazine; Figure 26-7, 26-8, 26-9, 26-12, Annette Perry/Laura Murray; Figure 26-11, Courtesy of C. Schunemann and G. Treu, *Baking: The Art and Science,* Baker Technology, Inc., 1988; Figure 26-13, PhotoDisc; Table 26-2, S. Corriher, *Fine Cooking* Magazine, 1996.

Chapter 27
Figure 27-2, from *Food Theory and Applications,* ©1972, John Wiley & Sons; Figures 27-3, 27-4, 27-7, and 27-8, Laura Murray; Figure 27-5, Courtesy of the Chocolate Manufacturer's Association; Table 27-5, R. Chandon, *Dairy-Based Ingredients,* Eagan Press, 1997.

Chapter 28
Figure 28-1, PhotoDisc; Figure 28-5, Laura Murray.

Chapter 29
Figure 29-1, Courtesy of the Perrier Group of America; Figure 29-4, Courtesy of the Tea Board of India; Figures 29-7 and 29-8, Courtesy of The Wine Institute.

Index

WEIGHTS AND MEASURES

Converting Temperature Measurements

To Find Farenheit:
°F = (°C x 1.8) + 32

To Find Celsius:
°C = (°F -32) ÷ by 1.8

Abbreviations for Measurements

Nonmetric			Metric		
Volume/Capacity			*Volume*		
teaspoon	=	t or tsp	milliliter	=	ml
tablespoon	=	T or Tbsp	liter	=	L
fluid ounce	=	fl oz			
cup	=	c or C	*Weight*		
pint	=	pt	microgram	=	ug
quart	=	qt	milligram	=	mg
gallon	=	gal	gram	=	g
pound	=	lb	kilogram	=	kg

Equivalents of Nonmetric and Metric Measurements

Nonmetric		Metric	
		Customary	*Precise*
Volume			
1 teaspoon	=	5 milliliter	(4.9 milliliters)
1 tablespoon	=	15 milliliters	(14.8 milliliters)
1 fluid ounce	=	30 milliliters	(29.57 milliliters)
1 cup	=	240 milliliters	(236.6 milliliters)
1 pint	=	0.50 liter	(0.47 liter)
1 quart	=	0.95 liter	(0.94 liter)
1 gallon	=	3.8 liters	(3.79 liter)
Weight			
1 ounce (dry)	=	30 grams	(28.35 grams)
4 ounces	=	125 grams	(113.40 grams)
8 ounces	=	250 grams	(226.80 grams)
1 pound	=	450 grams	(453.60 grams))
2.2 pounds	=	1 kilogram	(997.92 grams)

Converting Nonmetric & Nonmetric Measurements

When You Know	You Can Find	If You Multiply By
Volume		
teaspoons	milliliters	5
tablespoons	milliliters	15
ounces	milliliters	30
cups	millliliters	237
cups	liters	0.24
pints	liters	0.47
quarts	liters	0.95
gallons	liters	3.8
milliliters	ounces	0.034
milliliters	pints	2.1
liters	quarts	1.06
liters	gallons	0.26
Weight		
ounces	grams	28
pounds	grams	454
pounds	kilograms	0.45
grams	ounces	0.035
kilograms	pounds	2.2

STANDARD CAN SIZES

Can	*Cups*	*Average Net Weight**		*Volume*	
		Non-Metric ounces/lbs	Metric grams	Non-Metric fluid ounces	Metric milliliters
6 oz	¾	6 oz	170	5.75	170
8 oz	1	8 oz	227	8.3	245
No. 1 Picnic	1¼	10½ oz	298	10.5	311
No. 211 Cylinder	1½	12 oz	340	12	355
No. 300	1¾	14 oz	397	13.5	399
No. 303	2	16–17 oz	454–482	15.6	461
No. 2	2½	1 lb 4 oz	567	20	591
No. 2½	3½	1 lb 13 oz	822	28.5	843
No. 3 Cylinder	5¾	3 lb	1360	46	1360
No. 5	6½	3 lb 8 oz	1588	56	1656
No. 10	13	6½–7 lb	2722–2948	103.7	3067

* Net weight/can varies slightly due to food density differences.